Danny Erhardt

BIG BANG
HOAX
EVOLUTION
SCAM
DEMOCRATS
ARE THE RACISTS
NOT THE POLICE
DEMOCRATS
ARE BABY
KILLERS. LIARS.
FALSE ACCUSER.
AND BLACK BABY
EXTERMINATING RACISTS
BLM IS FOUNDED ON LIES
CUZ BLACKS ARE KILLING
BLACKS THE POLICE ARENT
ANTIFA ARE THE
FASCISTS.

RECOMMENDED
WORKBOOK PRESS

ALL THE SECRETS
OF THE UNIVERSE
(All the ones worth knowing)

A Graduate Course on the Truth

WORKBOOK PRESS LLC
187 E Warm Springs Rd,
Suite B285 Las Vegas NV 89119 USA

Website: https://workbookpress.com/
Hotline: 1-888-818-4856
Email: admin@workbookpress.com

Ordering Information:

Quantity sales. Special discounts are available on quantity purchases by corporations, associations, and others. For details, contact the publisher at the address above.

Library of Congress Control Number:

ISBN-13:   978-1-965732-70-0   Paperback Version
              978-1-965732-71-7   Digital Version

REV. DATE: 06/05/2025

ALL ᴛʜᴇ SECRETS
OF ᴛʜᴇ UNIVERSE

(All the ones worth knowing)

A Graduate Course on the Truth

by
Danny Erhardt

CONTENTS

Book II - FIRE

Book III - WATER

Chapter 14:
Fasting and Prayer And Eating at the Table of the Living God   772

The Book of Eternal Life   791

INTRODUCTION

"Evil men do not understand judgment:
but they that seek the Lord understand all things."
(Proverbs 28:5)

The author of those words, written thousands of years ago, would be ridiculed if he spoke them today on any one of our college or university campuses. For the ideological-left elites who control the intellectual discourse there hold the idea of "the Lord" in complete contempt, much less the idea that one could come to an "understanding of all things" merely by "seeking Him." But their scorn notwithstanding, the question remains, was that ancient author right? Did he know what he was talking about? Is it possible to rise above all of the confusion, chaos and senselessness that covers this planet and come to a clear understanding of what is going on, and why, just by seeking the Lord? (And it should be noted that *the Lord* in this context is understood to be the all-powerful, all-knowing and infinitely intelligent Creator of this universe.[1] It is also understood that the Lord and the truth are one and the same, that they are inseparable; for the Lord is the truth. So in seeking Him one is also seeking the truth.)

The truth can be known

So, with all that said, *is* it possible to figure everything out about this life: "this sore travail that God has given to the sons of man to be exercised herewith?" (Ecclesiastes 1:13) Can we know for a certainty the answer to all of life's fundamental questions; those that have puzzled mankind, inspired poets, and tied the minds of philosophers and academics in knots for millennia? Or is anybody's guess just as good as anyone else's? Is anyone's opinion as valid as anyone else's? Can we know what really caused the universe to come into being; why it does exist; how life really originated; why we are here and what is our purpose in being here (*what's the point?*); which ideologies are true and which are false; which religious claims are true and which are false; "WHAT'S GOIN' ON"[2] right now on this panicked planet, and why; and where are we going, if anywhere, after we depart this life? Is it possible to know the actual answers to these and other related questions? Not just to hold an opinion on each one, but to actually uncover their true answers, and thereby come to a clear *understanding of all things*?

Well as strange as it may seem, and despite what the "current wisdom" of this day might assume, the answer is yes. And even though the reader might be skeptical about this possibility at this point, we can assure you that if you follow along through the pages of this book, and if you determine to seek the truth for yourself, with all of your mind and all of your heart and all of your soul, then you

---

1 (And not to jump the gun but as we will see in chapters 6 and 10, and despite what we were all taught, that unlike the scientifically-impossible and imaginary creators of random accidental chance (the theory of evolution) and a real big explosion (big bang theory) true science itself confirms that the Lord is the actual Creator of this Universe as well. Go figure.)
2 (As *screamed from the top of her lungs* in the song "What's up?" by 4 Non Blondes)

will find by the end, and hopefully well before it, that you have come to agree with this conclusion. You will find what millions who have traveled this path before you have already found. That it is possible to leave behind the confusion, fear and anxiety that come from not knowing the truth; from having been taught any number of false assumptions, false beliefs, false religions, false ideologies and unscientific theories. And that by seeking the truth and leaving the lies far behind, it is possible to make sense of all this madness. That it is possible to know what is going on and why. That it *is* possible to understand all things.

"Come unto the truth, all you who labor and are heavy laden, and He will give you rest. Take His yoke upon you, and learn from Him; for He is gentle and humble in heart: and you shall find rest for your souls. For His yoke is easy, and His burden is light." (Matthew 11:28-30, in the third person.)

One of the side benefits of seeking the truth, besides that which comes from any increase in knowledge (it being power if applied correctly), is peace to the mind and rest to the soul. And in this world of unending war, political turmoil, economic collapse, catastrophic tsunamis, horrific tornadoes, violent Islamic insanity, and socialist takeovers here at home, no price can be put on that.

The more I know, the less I understand

There is a difference between knowledge and understanding. (One is knowledge. The other is understanding.) (Shazaam.) There is also an infinite difference between knowing all things and understanding all things. The former is impossible. You would have to have the mind of God. Imagine that you possessed all of the information from every book in every home, school and library across the face of this earth, as well as all of the data stored in the memory and hard drives of every computer on the planet, in addition to every memory and every piece of information in the mind of every human being alive today and everyone who has ever lived, and then multiply that by ten billion to the ten billionth power; and even then you would still only have a mere infinitesimal fraction of the totality of all knowledge.

Understanding all things, fortunately, is a tad easier. For only a relatively limited and also easily accessible volume of knowledge is required to comprehend clearly what's going on in this world, and why it is, and to discover where the truth lies and where it does not. This journey to understanding requires the ability to separate fact from fiction, truth from lies, and that is an innate ability which is also readily available to any and all who will choose to use it. So don't be fooled by the "conventional wisdom" of this world that the answers to all of life's fundamental moral, philosophical and theological questions are not attainable in the absolute; that this is a realm left to the world of opinion and argument, faith and religion, and unscientific theories only; that the sanctity of all the many opinions on these subjects cannot be breached; and that to even *think* otherwise is to be a delusional know-it-all divorced from reality. As it turns out, the exact opposite is true.

Also as it turns out, all the secrets of the universe, at least all the ones worth knowing, aren't all that secret to begin with. Most have already been revealed to us and many of them for quite some time. The problem is that the world has refused to listen. It has not had "ears to hear." People believe what they are taught, and if what they have been taught is not the truth, so often that will keep the truth secret, hidden, and unavailable to them due to their own brainwashing and their own choosing. Therefore the pervading lack of understanding and confusion over the answers to the fundamental questions about this life is more a matter of willful ignorance on mankind's part, than any attempt by the *universe* to keep anything secret.

"For since the creation of the world the invisible qualities of God have been clearly seen, being understood through what has been made, even his eternal power and divine nature; so that men are without excuse." (Romans 1:20)

Follow the evidence

Science over the years has uncovered many secrets of the physical nature of this universe. By making careful observations and by painstakingly collecting facts and evidence, scientists were able to come to accurate conclusions, to reject erroneous assumptions, hypotheses and theories, and to unveil concrete scientific laws that still stand to this day. Laws of motion and of gravity, of gases and electricity, of light and of planetary orbits, and of energy and matter. And from these laws have come every modern invention and convenience that we enjoy, as well as take for granted today; our cars, and TV's, computers and cell phones, air and space travel, the marvels of modern medicine, etc. And all of these extraordinary advancements came about because these early scientists were able to reach accurate conclusions by observing the facts and evidence and by going wherever it led them. As Socrates instructs in Plato's *Republic,* "We must follow the evidence wherever it leads."

In the very same way, we can come to accurate conclusions and unveil correct answers concerning the questions of why we are here, what our purpose is, where we came from, where we are going, how and why this universe came into being, which ideologies are true and which are false, and which religious claims are true and which are false. By looking at the facts and the evidence, and going wherever it leads us, we can accurately answer these questions as well. Just as in science, we are interested in leaving the realm of opinion, conjecture, blind faith, speculation, false theories, false assumptions, false conclusions and false beliefs behind. And in turn we seek to come to conclusions and answers that are demonstrably true. That can be demonstrated to be accurate based on all of the facts and evidence, both historically and scientifically. We'll trust it to the judgment of you, the reader, whether or not this has been accomplished with each of the subjects we will cover. Again, people believe what they have already been taught, so if you disagree with any of the conclusions here just be certain that you are disagreeing based on the preponderance of the facts and evidence, and not because of any pre-existing bias, indoctrination or inaccurate assumptions within your own mind. If you seek the truth, you will find it. Conversely, if one seeks instead to protect the lies and indoctrination and false assumptions that they may now hold dear then, unfortunately, that is exactly what they will receive instead of the truth.

We are going to cover a number of subjects here and, despite the overall length of this tetralogy (4 books in one), we will still be limited in the amount of space we can devote to each one. The facts, evidence and logic that we will bring to bear in each of these areas, however compelling they may be, are obviously not the final word. There is a vast amount more that could also be presented to support these conclusions, and indeed at the end of each chapter there are numerous books and resources listed for this purpose. But it is not our intention nor do we have the space to exhaustively present *all* of the facts and evidence for each of these truths here. All we can do is expose the reader to these conclusions and hopefully enough of the overwhelming evidence that supports them. After that it is up to the individual to decide whether they have found the truth or not. But our attempt here is not to convince the inconvincible, as it is impossible to win an argument with an ignorant man. What we *are* attempting to do is to equip as many as possible with the truth so they are able to withstand the onslaught of lies from the dark side that have controlled our educational system, our entertainment industry and of course our media for some time now. And so they will have the ammunition to stand firm against those teachers, university professors, faux scientists, media charlatans, journalist imposters and late-night clowns who use their power, position and influence to lead the unsuspecting, the naïve, and the misinformed down their chosen path of confusion and lies.

Decision time

There is a war being waged on this planet. It is for possession of the earth. It has been building for six thousand years and it will soon come to its conclusion. (See chapter 9.) It is playing itself out between the nations of men, the "sheep and goat nations," but the real battle has been and is going on behind the scenes, in the spiritual realm, between God and Satan. Now Satan, the god of this world, doesn't stand a chance. God will soon cast him out and retake complete possession of this planet. Indeed the only reason that he has allowed the battle to go on for this long is so another war that is being waged can also play itself out to its conclusion as well. And that war is also occurring behind-the-scenes, in the spiritual realm. It is a battle for the possession of the souls of men, and it has eternal consequences. It is being waged here on the inside, in the realm of the heart and the mind, and it involves an incredible gift that God gave to us; a gift that is both wonderful and terrible, depending upon how it is used; and that is the gift of our own free will. (It's one of the things that makes us in His image and His likeness.) While we are alive on this planet it is decision time. After we die that decision will stay with us for all eternity.

Yet as important as this decision is, most of the people on this planet have never been given the opportunity to look at all of the facts and evidence so they can make a wise and informed choice. Instead most of us have been deceived since birth. For the lies of Satan cover this earth like a blanket—false religions, false ideologies, false scientific theories—and people believe whatever they are taught. And most of us have not been taught the truth. This world is awash in a sea of lies that stem from one known as the Great Deceiver, and the people of this world are drowning in that sea. Satan's lies keep people from coming to the knowledge of the truth, and his lies are many. They are sly. They are pervasive. And their grip is powerful. As such we wanted to leave no stone unturned in exposing them, debunking them, refuting them, and in putting forth a powerful and indisputable case for the truth, so after reading this there could be no doubt or question in anyone's mind as to what the lies are, or to where the truth lies.

This is not a short story. It is a graduate course on the truth. But one that we feel is well worth taking, and passing. It is designed to take one from the lies of Satan (which cover this earth like a blanket) to the knowledge of the truth. It is designed to take one from darkness to the Light. Our goal here therefore is to give the reader the opportunity to do just that. To be able to look at all of the facts and evidence and to be able to do that within the pages of not hundreds of books, but just one work; and from there hopefully to make a wise and informed choice. To that end, good "luck," and may God richly bless you.

In conclusion then, it is our contention that despite what the current *wisdom* of this world might say, <u>the truth can be known</u>. The truth, not based on any man's opinion, but based on all of the facts and all of the evidence both historically and scientifically. A very clear understanding of the answers to the fundamental questions concerning this life *is* readily available to any and all who will seek it. It *is* possible to come to an understanding of all things and to uncover all of the secrets of this universe that are worth uncovering. You just have to seek, and you will find.

BOOK I

AIR

Chapter 1

TRUTH

"Ever learning, yet never able to come to the knowledge of the truth."
(2 Timothy 3:7)

Two thousand years ago a political prisoner standing trial on trumped-up charges made the following statement: "For this reason I came into the world, that I should bear witness to the truth. Everyone who is of the truth hears my voice." (John 18:37) He was then asked, "What is truth?" (John 18:38) But unfortunately for the inquisitor, a Roman governor and judge, it was posed less as a question and more as a sarcastic statement meant to ridicule and demean. That same question is still being asked today but far too often, especially in our once truth-seeking but now left-wing-indoctrinating universities, in much the same way.

Yet it can easily be answered. "The Spirit is truth." (1 John 5:6) And the "word" of that same Spirit also "is truth." (John 17:17) And that same Spirit also is one, fortunately for those who seek the truth, "who desires all men to be saved and to come to the knowledge of the truth." (1 Timothy 2:4) And that if you seek him "he will guide you into all truth." (John 16:13) Now I know that we haven't yet established that *the Spirit* even exists. (We will do that in chapters 3, 4 and 5.) Nor have we yet established that the words of this Spirit, that we have just liberally quoted, also exist as indeed this same Spirit's words, and not as somebody else's; the words of mere man for example. (This we will do in Chapter 6.) But bear with me for now as it is impossible to write a chapter entitled "Truth" without reference to both. Trust that we will tie it all together neatly, and in a bow, before we are all done.

The answer "the Spirit is truth" implies that the truth exists outside of the confines of this physical universe; that truth is transcendent. It was here before the world began. And before man and his lofty intelligence had their beginnings. And if one seeks the truth, it is there where he must find it. To do otherwise, to seek the truth within the confines of this universe of matter alone, is to seek it in vain. It is to look for it, like love, "in all the wrong places." This is the source of one of the major errors in science today, which has turned some fields into more of a primitive religion than science, and one that holds sway over far too many scientists. And that is the religion of *Naturalism*, which believes that there is no *meta*-physical truth or reality—meaning that there is no truth or reality that exists *above*, or outside of, the physical world. So all truth and all reality must be found within and only within the confines of the physical world. Including the origins of this physical world. (This of course is nonsense. Try finding the origins of a car by looking within the car, and only looking within the car... *Uh, excuse me, professor, but you see, there's this place called Detroit...* "SHUT UP, YOU RELIGIOUS NUT! AND KEEP LOOKING INSIDE THE CAR!")

The truth is controversial

"Do not think that I came to bring peace on earth. I came not to bring peace, but a sword. For I have come to set a man against his father, and a daughter

against her mother …and a man's enemies will be those of his own household."
(Matthew 10:34-36) Wow! Those are some fighting words, and not the words one
might expect from One who is called *The Prince of Peace*. Nor from One who
also said this, "Peace I leave with you, my peace I give to you; not as the world
gives." (John 14:27) So how then do we reconcile the apparent contradiction? By
understanding that the peace he gives is a real peace based on truth and one that
will last forever. The peace that the world, and the god of this world, gives is a
false peace based on lies and one that will not last. The division occurs in the battle
between the two contenders for the hearts and minds of men. On the one hand he
came to bring lasting peace, but on the other hand he knew that the truth he spoke
would cause turmoil between those who embraced it and those who were much
more comfortable protecting the lies of the great deceiver already in their mind.
He knew it would not only rend individual hearts and minds asunder, but also
divide one believer in the truth from another who refuses it. (But we will cover
this in more detail in chapters 7 and 8.)

For now, what we want to impart is that the truth is a very controversial thing.
It can tear apart friendships and divide family members. It can separate husbands
and wives, and estrange parents from their children. It is no wonder therefore
that most people don't really want to be bothered with it. They are much more
comfortable sticking with whatever they were taught, and avoiding "making any
waves" or causing any commotion not only within their own minds but among
their family, peers and culture as well. It is better to just go along in order to get
along. And that is also why many will go to great lengths to deny the truth even
when confronted with incontrovertible evidence in favor of it. (*Uh, professor, are
you saying that the car made itself?* "I'm saying you need to stop asking all of your
STUPID questions!" *And just keep looking in the car?* "That's right, IN the car,
and ONLY in the car!" *And forget about Detroit?* "YES! This is "SCIENCE,"
young man, not religion!" *…Uh, professor, is it too late to get my tuition back?*)

The truth is controversial because its adversary, lies, has such a stranglehold on
the peoples, nations, languages, ideologies, religions, faux scientists and cultures
of this world. People believe what they are taught, and the world unfortunately
is in love with the lies that it has been taught. And the crowbar has not yet been
invented that is capable of prying those lies out of the world's mind. They will
defend them against exposure to the truth, even to the death.

Yet this is the first *secret* of our little universe that a person must understand
before he or she will be capable of hearing any others. And that is that the reason
the truth about many different divisive subjects is not known and accepted by the
vast majority of people on this planet is not because it is elusive or so difficult to
come by, but rather it is because the majority of the people on this planet have
a built in, almost insurmountable obstacle to it. They feel way to comfortable
holding on to the lies that they have been taught, and that they have come to
accept, as true. They have a lot of emotion invested in those lies, a lot of fondness
for them and a lot of love, as in most cases do their friends and family, and because
of this they have locked themselves within a prison of their own making. And only
they hold the key.[3]

The controversial nature of the truth in our fallen world is especially evident
with some of the harder truths, like those which expose false religions, or the ones

---

3   (Actually, there is Someone else who holds that key, but that is a discussion for chapter 7)

which exposes idiotic and erroneous scientific theories, or the truth which states that if one calls him or herself a Christian or even a decent person but goes out religiously every two years and votes for the baby-mutilating Democrats then they are either A) a liar, B) a comedian, or C) slightly confused. Yes, there are some hard truths, but if the truth, hard or not, is controversial or offensive or causes one to respond with anger, then that is because they are more interested in protecting lies, and less interested in seeking the truth. We'll assume that this is not the case with anyone interested in reading this book. For those the truth will come as welcome news, the same as for one who learns the truth and all of the knowledge and all of the facts and evidence about a scam-artist like Bernie Madoff, before they lose their life savings. Learning it after you invested with him and lost all of your wealth, is a little too late. And when it comes to the truth about false religions set up by the adversary of truth, and false scientific theories that spit in the face of the Creator of the universe, and voting habits that align one's self with that great child-sacrificer of old, as we shall see there is something at stake far more precious than mere money.

The first decision then, that you have to make if you desire to come to the knowledge of the truth, is whether or not you are willing to seek it with an open and un-imprisoned mind. Or at least that you will have the courage to allow your mind to be challenged by the facts and evidence, even when they conflict with the opinions, assumptions or beliefs that you may now hold dear—with the information that you have been taught and/or indoctrinated with. And that, unfortunately, for far too many is a bridge a little too far. I pray that it is not one for you.

The truth is not "relative"

Also know that the truth is not relative. It is the truth. Relative is relative. (Shazaam again.) Now I know that to anyone with even a shred of common sense it might seem a little silly that we have to waste our time in a book about the truth declaring that the truth is actually the truth, and not something else. (Like mash potatoes for example.) But this is the nature of the insane asylum we live in. The inmates have taken over. (Not only that, they've been given tenure.)

And because this nearly unintelligible and utter stupidity holds sway over much of academia today we unfortunately have to waste a little time here shining a light on it, refuting it (although any reasonably educated eight-year-old could do the same …"Mommy, apple sauce isn't broccoli, right?"), and laying "the truth is relative" to rest alongside a number of other once-popular but now-discredited academic doctrines such as "the earth is flat" and "life arises spontaneously from piles of dirty rags." (Wait a minute. Stop the presses. I think that last one might still be popular, only, hmm, the words have been changed just a little. Let's see… "Life arose spontaneously billions of years ago from piles of dirty rags." No, that's not it. Hmm… "Life arose spontaneously billions of years ago …from an ancient mythological slimy sea." Yeah! *That's* it.)

Unfortunately, this false teaching that the truth is relative has become deeply engrained in the minds of our students-children. It states that no one can really ever assert that something is true because indeed that thing is only *true* to him, and might not be *true* to someone else. Hence the relativity label. The extent to which this idiocy has been ingrained into the minds of so many of our young people can be quickly demonstrated by just asking one of them how they know that what *they believe* is actually true. Instead of backing up their particular beliefs with facts, evidence and logic, many of them will simply say, "It's true because I believe it". (As if their own beliefs were the great creator of truth!?) And conversely they will assert that the same belief is un-true for someone else "because they do not believe it." In this insane asylum everyone is free to create their own reality, within the protected confines of their own mental universe, unhampered by such inconsequential and annoying little details like facts and evidence. In this world of self-delusion their own thoughts and beliefs have been enshrined with supremacy. But in reality of course an individual's own thoughts and beliefs have no bearing whatsoever on what is true and what is untrue. The truth exists irrespective of what one individual or group of individuals might believe or not believe. And the truth itself can only be determined by delving deeply into the area of facts and evidence, and leaving the arena of beliefs far behind.

Calling the truth "relative" is absurd because in fact, by very definition, the truth cannot be relative. If the truth were relative then it wouldn't be the truth. It would be something else—speculation, opinion, conjecture, wishful thinking, theory—but not the truth. Declaring that the truth is relative is also self-refuting, for the very statement "the truth is relative" must then also itself be relative and therefore cannot be true. Indeed the truth is only relative to those intellectually elite but logically challenged individuals[4] who have a problem with it. Calling it relative, and therefore a matter of personal opinion and preference, is a deceitful way of conning the young and the intellectually immature into accepting what is an otherwise obvious scam to emasculate the truth so it can no longer conflict with or undermine ones erroneous and untrue personal ideologies, theologies and belief systems. It is a marvelous scheme that allows the scam artist (usually a left-wing professor, teacher and indoctrinator who were themselves indoctrinated into the scheme) to fill his victims' minds with one fabrication after another, and since "there is no truth," it is impossible to question those lies.

> "I have asked hundreds and hundreds of non-absolutists, from self-styled religionists to independent atheistic materialists, the following question without ever receiving a logical answer: *Mike*: Is their absolute truth?
>
> *Non-Christian*: No. We can't know anything for certain.
>
> *Mike*: Are you certain of that?"[5]

You see, the statement that there is no absolute truth, is in itself an absolute statement of truth. So if it is true, then it is false. Therefore it has to be false. Unless one was to amend their statement with this qualification, "There are no absolute truths, except for the absolute truth that I just stated." But even gullible college students wouldn't fall for that one.

"Education is not necessarily enlightenment. Education, in fact, sometimes casts a dark pall over the truth."[6]

Where did this nonsense, "the truth is relative," come from? Did it spring up from the fertile minds and imaginations of our academic elite out of whole cloth, and then catch on like some new hip fad? No, it evolved over a number of centuries starting back in the 1700's "across the pond" in a period of time in Europe known as *The Age of Enlightenment*,[7] where "a critical questioning of

---

4 (Also known as intellectual idiots- folks who are quite full of themselves, who possess much knowledge but little truth, and even less common sense)

5 (Pgs. 60-61, *There Are Moral Absolutes*, by Michael A. Robinson)

6 (Pg. xi, *The Parthenon Code: Mankind's History in Marble*, by Robert Bowie Johnson)

7 (The more accurate title would be *The Age of En-dark-enment*)

traditional institutions, customs, and morals"[8] began. And there is no doubt that some good did have its roots in this critical questioning: for one, Democracy. But there was also a lot of darkness that descended from it; philosophical darkness. A lot of "isms"—Rationalism, Secularism, Existentialism, Humanism, Positivism, Logical Positivism, Postmodernism, Relativism and Metabolism[9]—all by-products of this Age that have come in and out of popularity in the academic world over the last few centuries. Some have been discredited. Others still hold sway. But it was the philosophy of Relativism that has its roots in the world view of Positivism and Logical Positivism that first advanced the idea that there are no absolutes, that there is no absolute truth. Of course in declaring this, and as we have already pointed out, Positivism breaks its own rule by declaring an absolute truth. It is stating that the only absolute truth is that there are no absolute truths, and of course that statement in itself is an absolute truth. So much for positivism. So much for "there are no absolute truths." So much for "the truth is relative."

But where Positivism and Logical Positivism declared, erroneously, that there are no absolute truths, it is the philosophy of Postmodernism, currently all the rage among our academic intelligentsia, that has taken this stupidity to new levels of absolute gobbledygook. Postmodernists have taken it a step further, to an actual *hatred* of the truth. For Postmodernism declares that truth itself is the "great oppressor," and therefore its followers must subvert all truth.[10] It also demands the tearing down, the "transgression," of all boundaries, moral and otherwise, in order to show that boundaries are *arbitrary*. "Absolute liberation from suppression"[11] is their battle cry. And all they are left with then is the absolute oppression that comes from allowing mindless drivel—bald-faced, nearly incomprehensible and philosophically-disguised lies—to take the place of truth itself within their minds. Lies have a way of enslaving and making fools out of those who embrace them.

Common sense is a term usually reserved for the common man. It is used to describe someone who is able to think clearly and to come to accurate conclusions based on all of the available facts and evidence and using all of the Rules of Logic, even though that person may not have the slightest idea of how to enunciate those same Rules. Conversely, there are many people, many students of philosophy, and many tenured professors in our universities, all highly educated and many quite familiar with the Rules of Logic, who nevertheless lack both common sense and logic...

"Professing themselves to be wise, they became fools." (Romans 1:22)

When man rejects the very existence of truth what is left for him? Only lies. Postmodernist philosophers and the professors who emulate them "profess themselves to be wise" but have only succeeded in muddying their brains with un-intellectual nonsense. Like the emperor marching down the street naked, thinking himself clothed in the finest array, postmodern academia is making a laughing stock of itself in front of the audience. The only ones along the parade route who don't get it are their equally-duped left-wing peers and their hapless, indoctrinated students. They have been deceived by one who we will talk of throughout the book (and especially in chapter 13), the father of lies, the great adversary of the

---

8   ("Age of Enlightenment," from Wikipedia, the free encyclopedia)

9   (Can you find the word that is out of place in that list?)

10   (See the discussion on mans hatred of God in chapter 5)

11   (From lecture "Postmodernism and Society," by Al Mohler, @ ligonierministries.org)

truth who goes forth to deceive the whole world. It is he who hated the Truth long before them, and it is he who inspires foolish men to follow along with him today. Unfortunately, for all their education, they have been bamboozled into "exchanging the truth of God for a lie." (Romans 1:25)

Ultimately men reject the truth, and embrace lies, because they want to be their own god. They want to serve themselves and none other. They want to have the rule over themselves, no matter how illusory that "rule" actually is. And in doing so they unwittingly open themselves up to the rule of someone else. Someone who no one in their right mind would wittingly submit to.

If you reject the truth, then the only thing left to fill the vacuum in your mind is lies. That is the nature of this universe, both the physical and the mental universe. That's just how it works. And that truth-less vacuum within your mind and heart will be filled with the only thing left to fill it. Lies. Reject the Truth which can come only from above, by declaring that it is "relative" or that it "doesn't exist," and you will be filled with the lies that ooze up from below.

> "If everything is relative, then nothing is concrete. Nothing is ultimately real; nothing is ultimately true, because if everything is relative to everything else there is no ultimate reference point. And that is precisely where modern, secular man finds himself in the 20th century; that he lives his life with no ultimate, fixed, stable, absolute reference point that defines his life or the meaning of his existence. ...This means that you have values with no value; you have truths, but no truth; purposes, but no purpose; things that are beautiful, but no beauty. You have no fixed standards by which to measure or to judge values, truth, purpose, beauty. You live in a world of ultimate chaos once you embrace [relativism]"[12]

Also, it is important not to confuse *relativity* with *relativism*. Scientific relativity with regard to motion is accurate:

> "Motion can be considered from more than one reference point. ...If I am moving toward you, we can consider my motion from my perspective or from your perspective, and we can have a different reference point, and there is a sense in which that motion is *relative*. It is relative to a particular reference point."[13]

But *relativism*, saying that "everything is relative" including truth itself, is <u>not</u> accurate.

In addition, there *are* truths that are relative. For example, "broccoli tastes good" might be true to me or to you, but it is not true to many other people. This is an example where truth can be considered relative. Some declarations that are true to me, or true to you, are not true to everyone; but extrapolating from this to say that truth itself is relative, that "everything is relative" or "all truth is relative" is false.

Curiously enough, the academic intellectual's denial of absolute truth does not apply to the scientific realm, where this denial would not be tolerated for it would preclude the very *search* for scientific truth itself. Why look if it doesn't exist? No, the idea of the non-existence of absolute truth only comes into play in the philosophical, theological, ideological and moral realms, for this is the traditional realm of opinions, speculations, beliefs, and ideologies and "the truth is relative" proponents don't want theirs tested. They don't want their opinions, speculations, beliefs, and ideologies held up to the light of truth, so they conveniently deny its existence.

---

12 (From "Renewing Your Mind" by Dr. R. C. Sproul)
13 (From "Renewing Your Mind" by Dr. R. C. Sproul)

What do you love more, the truth, or your own opinions?

"We were talking about the space between us all, and the people who hide themselves behind a wall of illusion, **never glimpsing the truth**, 'til it's far too late, and they pass away. And we were talking about the love that's gone so cold, and the people who gain the world but lose their soul. They don't know. They can't see. Are you one of them?"[14]

Don't be.

Do you want to know the truth? Yes? Then you have won half the battle. The other half lies in a confrontation within your own mind. You must ask yourself, "What do I love more, the truth? or my own opinions?" The answer of course for nearly all of us is, "my own opinions!" Yet this is the obstacle that must be overcome. You must make the decision to love the truth, whatever it may be, more than you love your own opinions. Your own opinions are merely those things inside of your mind that you have no choice but to start out with when you begin the search. It is what you have been taught. It is what you have come to accept as true. It is fine to start out with them—you really have no other choice—just don't be so in love with them that they become an insurmountable obstacle between you and the truth. If you can't let go of your own opinions, but it so happens that your opinions turn out to be true, then you are one of a very rare and lucky few. But if you can't let go of your own opinions, and they turn out to be false, then you're screwed. So why take the chance? The odds are terrible. Just look at all of the facts and evidence. And let them take you wherever they may lead.

"Rather than love, than money, than fame, give me truth."[15]

And keep in mind that the truth, as controversial and as hated as it is for many, is not some dreadful monster that is waiting to disfigure, harm or ruin those who dare to let it in. It is not something to be feared. The lies that fester in far too many people's minds—religious lies, unscientific lies, ideological lies—and that keep them from coming to the knowledge of the truth are the truly ugly and scary things. They are that which truly should be feared, and not the truth. (But if lies are already firmly in place within one's mind, then one can expect that they will put up a fight. They will definitely defend their turf.)

The truth is something that is based on all of the facts and all of the evidence both historically and scientifically. It is as easy as knowing that the earth is round and not flat, that $1 + 1 = 2$, and that your SUV could never ever have evolved accidentally by random chance from the sedan down the street (no matter how many hundreds of millions and billions of years you could fabricate in order to assist with this miraculous transformation). The truth is simple and obvious and un-debatable once you open your mind up to it.

"I'm sick and tired of hearing things
From left-wing-indoctrinated, closed-minded hypocritics.
All I want is the truth. Just gimme some truth."[16]

There are many battles being waged on this earth, but the ultimate war, the primary struggle and the one with the greatest consequence for each individual is the war between truth and lies. We are all pawns in this game; little foot soldiers in this war, whether we think we have enlisted or not. We have all been conscripted by one side or the other; or we are pulled apart by the struggle between the two

---

14   (From song, "Within you, Without you," by The Beatles)

15   (Henry David Thoreau)

16   (John Lennon's "Just give me some truth," updated for today's reality)

within ourselves. The primary combatants in this war are behind the scenes. Truth, and the great enemy of the truth—the father of lies, the great deceiver of mankind—wage their battle in another realm, in the heavenly, but on earth this struggle is played out in the hearts and minds of mankind. Truth is tireless in drawing all those to him who have ears to hear; through "the still, small voice within."[17] But the lies of the great deceiver, that cover this earth like a blanket, seduce men through their own arrogance, and through pride in their own self-righteousness and *lofty* intelligence.

Our purpose and prayer in writing this book is that it might be instrumental in turning one from lies to the knowledge of the truth. That one may come to know the truth, and that the truth will set them free. Free from that inner cloud of darkness created by a swarm of half-truths, un-truths and outright, bald faced lies, that lead a person blindly and unconsciously down a path that bars them from living the "abundant life" that is possible here and now; and that can only end in terrible destruction for them in the life to come.

"The thief [that great liar and deceiver, Satan] does not come except to steal, and to kill, and to destroy. But I am come that they may have life, and that they may have it more abundantly." (John 10:10)

And so, with all that in mind, we return to that one pivotally important question...

What do you love more?
The truth?
Or your own opinions?

What do you love more, the truth, or your own opinions? That is the question.

(To be, or not to be? That is another question, but it has nothing to do with our discussion here.)

Also know that if you choose the truth then be assured that it is not hard to find. In fact, it is guaranteed to you. If you choose to separate yourself from your own opinions long enough to seek it...

"And I say unto you, ask [for Truth], and it shall be given you; seek [the Truth], and you will find it; knock on the door [of Truth], and it will be opened unto you. For every one who asks receives; and he who seeks finds; and to him who knocks it shall be opened." (Luke 11:9-10)

"And you shall seek me, and find me, when you shall search for me with all your heart." (Jeremiah 29:13)

"And you shall know the truth, and the truth shall make you free." (John 8:32)

"The Spirit of truth has come, and he will guide you into all truth." (John 16:13)

"But seek you first the kingdom of God, and his righteousness; and all these things [including the truth] shall be added unto you." (Matthew 6:33)

"If any man lack wisdom, let him ask of God, who gives to all men liberally, and it shall be given him." (James 1:5)

Or as Mr. Twain so aptly put it, "The problem with the world is not that so

---

17   (I Kings 19:12)

many people know so little. The problem with the world is that so many people know so much that just ain't so…"[18]

Amen.

See:
- *There are Moral Absolutes*, by Michael A. Robinson
- "Postmodernism and Society," Al Mohler
- "Renewing Your Mind," Dr. R. C. Sproul

---

18   (Samuel Langhorne Clemens, aka Mark Twain)

Chapter 2

Moral Absolutes

"Woe unto those who call evil good, and good evil; who put darkness for light, and light for darkness!" (Isaiah 5:20)

The truth is not relative, but what about morality? Is there an absolute or *higher* standard of morality upon which we can base our judgments? Or is right and wrong and good and evil merely a matter of personal and/or societies opinion? Of course not. The same folks who have pushed the asinine idea that the truth is relative down our throats have also succeeded in bamboozling much of our society, especially the "elites" of our society, into believing that there is no ultimate code of ethics that they must abide by, and that therefore they are "free" to pretty much do whatever they like, as long as it doesn't hurt anybody. (Of course hurting unborn babies, curiously, is exempt from their moral code.) But as we shall see, rather than being a blueprint for greater freedom, it is a recipe for absolute disaster. For there are unintended consequences for a society that frees itself from the constraints of a higher morality—one that is higher than the thoughts, whims and intents of man.

The undermining of moral absolutes has its roots in academia but this poison has seeped down into every level of our culture. And as we shall shortly see, the results have been destructive and alarming. Yet refuting the moral relativist's position is child's play. For starters, the argument for the existence of moral absolutes is inherently connected to the argument for the existence of God.[19] Indeed the two are inexorably entwined to the extent that the case for the existence of moral absolutes is one of many insurmountable obstacles for those who would deny the existence of an intelligent Creator. And conversely the overwhelming evidence for the necessary existence of a holy, all-powerful, all-knowing, infinitely-intelligent and self-existent Creator is also an insurmountable obstacle for those same people who would deny the existence of moral absolutes. If there is an Intelligent (you know, unlike random chance, One with a brain) and almighty Creator (and there is, see chapter 5) who created us for a purpose (and he did, see chapters 6-8) then there just might be certain things, moral things, right actions and wrong actions, that he might want us to abide by (and there is, see chapters 6-8 again). And it is these required *actions* and *behaviors* that come down from above, which stem from the very *nature* of that Creative Being who is above (and not from any societies or individuals vain opinions) which constitute the basis of that pre-existent and transcendent moral law.

David Berlinski, in *The Devils Delusion: Atheism and Its Scientific Pretensions* (a play on Richard Dawkins intellectually silly yet academically popular *The God Delusion*), reveals the dilemma of facing the question of morals in a godless world by the musings of one famous atheist:

""The problem," the philosopher Simon Blackburn has written, "is one of finding room for ethics, or of placing ethics within the disenchanted, non-ethical order which we inhabit, and of which we are a part." "[20]

Of course, this "non-ethical order" that Mr. Blackburn inhabits is solely within

---

19   (See chapter 5, "God")
20   (Pg. 35, *The Devils Delusion: Atheism and Its Scientific Pretensions*, by David Berlinski)

his own mind. It does not exist out here in the real world. Berlinski continues...

"Blackburn is, of course, convinced that the chief task at hand in facing this question—his chief task, in any case—"is above all to refuse appeal to a supernatural order." It is a strategy that merits admiration for the severity of mind it expresses. It is rather as if an accomplished horseman were to decide that his chief task were to learn to ride without a horse."[21]

Ravi Zacharias, author, radio host, teacher of Christian apologetics and evangelist, posits thus:

"Is there evil in this world? Yes! Well, if there is evil in the world, then there must be good. Yes! If there is good and evil then there must be a moral law, that which tells us what is good and evil. True! And if there is moral law then there has to be a law Giver."[22]

And again, from *The Devils Delusion*:

"We do not believe in any absolute moral truths, my students have always told me, although truths about grading seem a remarkably curious exception.

..."Like so many other positions, moral relativism has been promoted from the back of the college classroom to its podium. "The West," the philosopher Richard Rorty writes, "has cobbled together, in the course of the last two hundred years, a specifically secularist moral tradition—one that regards the free consensus of the citizens of a democratic society, rather than the Divine Will, as the source of moral imperatives." The words the free consensus, although sonorous, come to nothing more than the declaration that just so long as there is rough agreement within society, what its leaders say goes. This was certainly true of Nazi Germany. Many details of the final solution were kept hidden, but the view that the Jews of Europe were a problem requiring solution was so widespread in German society as to have appeared a commonplace. ...The decision physically to kill them all expressed very nicely "the free consensus" of Germany's citizens. Had it not, the final solution could never have taken place. It did not reflect the consensus of citizens in Denmark, Italy, or Bulgaria, and in those countries there was no final solution, there was no mass deportation, and there were no extermination camps, and in all three cases, Nazi officials were left muttering in frustration at the fact that curiously enough these were places where people did not sufficiently grasp the gravity of the Jewish problem. "Curiously enough.

..."If moral imperatives are not commanded by God's will, and if they are not in some sense absolute, then what ought to be is a matter simply of what men and women decide should be. There is no other source of judgment."[23]

Nazi Germany is just one example of the heinous results of man rejecting transcendent Moral Law in favor of his own opinion. More often than not *his own opinion* on what is morally acceptable seems to coincide with pure evils opinion as well.

"These conclusions suggest quite justifiably that in failing to discover the source of value in the world at large, we must in the end retreat to a form of moral relativism, the philosophy of the fraternity house or the faculty dining room—similar environments, after all—whence the familiar declaration that just as there are no absolute truths, there are no moral absolutes.

"Of these positions, no one believes the first, and no one is prepared to live with the second."[24]

---

21 (Pg. 35, *The Devils Delusion: Atheism and Its Scientific Pretensions*, by David Berlinski)
22 (Ravi Zacharias)
23 (Pg. 39-40, *The Devils Delusion*, by David Berlinski)
24 (Pgs. 40-41, *The Devils Delusion*, by David Berlinski)

Least of all those people who suffered unspeakable atrocities from the end result of moral relativism in Nazi Germany, Stalin's Russia and The *People's* Republic of China, and those babies who have suffered equally as horribly under our own form of Planned Parenthood immorality, and its rabid supporters in the media, the ACLU, our educational-indoctrination system and the Democrat Party.

So there is an easy answer to that tired old refrain, issued in the form of a question and endlessly repeated by misinformed teachers and parroted by their equally mis-educated students, "Who's to say what is right and what is wrong!?" God is, stupid! (With all due respect of course.) Of course when you have already rejected the existence of that same God, and in his place have embraced the scientifically-impossible "creators" of evolution (random accidental chance) and the big bang (a big mythical primordial explosion), then you end up searching for morals, like truth, in vain. And more often than not, as we will shortly see, the "morals" (or lack thereof) that you end up with, more closely resemble the contents of a bucket hauled up out of the cesspool of hell. (Or just Google "Silent Scream" and watch the video of a baby being mutilated to death while it is still alive. If you have the stomach for it.)

> "The standard must be based on an unchanging source, or ethics could change. If moral standards can change this would mean: lying and murder on one day are bad. The next day they might be good."[25]

It is easy to refute anyone who does not believe that there are moral absolutes, just ask him if he thinks that it is *wrong* to affirm moral absolutes. If "yes" is his answer then he has just made an absolute statement about morals, and he has refuted himself.

> "The statement that there are no absolutes is an absolute statement. The statement is a contradiction and self-refuting; for it to be true, it would have to be false. There must be absolutes in logic and morality, or we can assert nothing and account for nothing; [and] that is impossible."[26]

We can know that right and wrong either comes down from above—either has its source in that which transcends the intelligence, and even more importantly the *morality* of man—or that it doesn't exist. Either there is higher law or no law. That is because if morality, if what is right and wrong, comes from man and the decisions of men—the decisions of the individual or society as a whole—then it might as well not exist. Because then <u>no</u> objective reliable standard can exist. Morals are then as changeable as the weather, and what is "right" ends up being whatever that society *wants* to be right at any given time. For the ancient Greeks, throwing unwanted babies off of cliffs was "right." For ancient Satan worshipers (worshippers of Baal and Molech) sacrificing their children by burning them alive was right. ("And they built the high places of Baal which are in the Valley of the Son of Hinnom, to cause their sons and their daughters to pass through the fire to Molech." Jeremiah 32:35) For modern unwitting followers of Molech, legislating from the bench the mutilation of a baby to death (and the denying of its most basic right, to live) for the whim of the mother is "right." For the Nazi's in Germany, slaughtering 6 million Jewish men, woman and little children was right. For Nazi Muslim terrorists, blowing up innocent Jewish men, woman and children is right. For racists in the past, hanging blacks was right. For African Tribal leaders, selling

---

25   (Pg. 3, *There are Moral Absolutes*, by Michael A. Robinson)
26   (Pgs. 9, 52, *There are Moral Absolutes*, by Michael A. Robinson)

their fellow countrymen into abject slavery for a few trinkets was right. And on and on. No, it is clear from just a short trip through man's history that either there is a transcendent law of what is right and what is wrong, or there is no law at all.

The same is true of the individual. If the individual decides what is right and wrong, then rape and murder and robbery and assault and lying and cheating[27] and stealing are all just fine, as long as the individual doing it thinks it is "right" for him. No, the only conclusion you can come to is that either morals don't exist, or that they exist independent of the whims and fancies and opinions of any individual or any particular group of individuals. Either anything goes, or there is a universal moral code that transcends the base intellect of humanity at its best, and especially at its worst. It turns out that the latter is the case.

"The law of the Lord is perfect, converting the soul; The testimony of the Lord is sure, making wise the simple; The statutes of the Lord are right, rejoicing the heart; The commandment of the Lord is pure, enlightening the eyes; The fear of the Lord is clean, enduring forever; The judgments of the Lord are true and righteous altogether. More to be desired are they than gold, yes, than much fine gold; Sweeter also than honey and the honeycomb. Moreover by them is your servant warned, and in keeping of them there is great reward." (Psalms 19:7-11)

"Moral relativism *is* the death of right and wrong. It is the necessary component to the destruction of the very idea of morality and values. Think about it. If everything is relative, then there simply can be no judgment. Without judgment, there is no right and wrong. Like God, right and wrong are dead."[28]

In reality then, either right-and-wrong comes down from above, or it comes up from below. Reject the moral law of the Creator, and you are left with the *moral law* of the great deceiver.[29] Above we listed a few examples of barbaric immorality that stemmed from man's *morality*. They show the depths of evil that his morality can so easily descend to. There are many others. Refuse to worship truth and you end up worshipping lies. Refuse the moral council of that which is above and you fall under the council, and the *morals*, of that which is below. Our intellectually elite academics think they have rejected Gods Moral Law so they can "do their own thing." But in reality they have unwittingly opened themselves and their followers up to embracing the morals of demons from hell. (Both of which subjects we will cover later.) They end up doing Satan's bidding. They end up "doing Satan's thing."

Let's take a look at what has happened to our own culture as it has replaced the morals of the Creator with its own…

---

27 (Look at the great increase today in the percentages of high school and college students who not only cheat, but think it is "right" to do so)

28 (Pg. 170, *The Death of Right and Wrong: Exposing the Left's Assault on Our Culture and Values*, by Tammy Bruce)

29 (See chapter 13. The great deceiver, or Satan, the adversary of God on this planet, was the highest of all the angelic beings until he fell, rebelled against the Lord and took a third of the angelic host with him. He is also known as Lucifer, the father of lies, the wicked one, Beelzebub, the serpent, the dragon, the evil one, the prince of darkness, the devil, the prince of devils, ruler of this world, the thief and god of this world.)

America's descent into the *morality* of hell

In the last few decades the "no moral absolutes" - "who's to say what is right and what is wrong" gang have essentially taken over our educational system and entertainment industry, with the support of much of our judiciary and with the uncritical lapdog applause of our mainstream, secular media. The few have succeeded in removing the morals of God, and then forcing their own form of morality down the throats of the many. The majority of Americans are honest and decent folks yet far too many remain uninformed, to the extent that they have sat back and allowed this to happen, and even unwittingly gone out and voted for it. The examples we are going to discuss in this chapter have for the most part been completely ignored by the left-wing, so-called mainstream media (for years it has had next to nothing to do with the mainstream of America and everything to do with the ideology of radical far-left elites). They are news items that most Americans—and most American *voters*—have been purposely kept unaware of. It has been by design. It is called lying by omission, and our left-wing national news media are absolute professionals when it comes to this clever form of deception. (Conversely, when it comes to actually reporting the news in a fair, honest and unbiased manner—which is actually their job—they are clueless.) They are religious in their failure to report, or in their disgraceful under-reporting, of just about anything that would put their progressive, Democrat-Left[30] colleagues in a bad light, or expose the steady, stealth-encroachment of their left-wing ideology throughout our society and the detrimental immorality that accompanies it. In addition to their many outright lies that we will cover throughout the book (and in depth in chapter 13, "Satan"), these lies by omission combine to keep Americans in the dark about what is actually being advanced behind their backs and under their noses, so the Left can get away with in secret what would never fly if exposed to the light of day.

In her eye-opening book, *The Death of Right and Wrong*, Tammy Bruce documents our cultures descent into the toilet of moral relativity at the hands of the "malignant narcissists" that make up the "Left Elite"…

> "While this country's academic intelligentsia work themselves into a frenzy to squelch any kind of speech that challenges the left-wing status quo, they are also busily leading the next generation to embrace a perverse moral relativism that will take us further into a world devoid of right and wrong." [Tammy] exposes the "moral vacuum [that] is engulfing the …Left Elite in this country, from the decision makers in entertainment and academia,

---

30   (The term, "Democrat-Left," which will be used throughout the book, refers to those elements in our society that are hostile to the Judeo-Christian heritage of this country and to traditional Americans and their values. They are the ones looking to "radically transform" America into something that was not envisioned or intended by the Founding Fathers, nor provided for in our Constitution. They include such groups as the ACLU, move-on.org, Americans United for the Separation of Church and State, Planned Parenthood and of course the Democrat Party just to name a few. What we call the Democrat-Left has also been referred to in the past as the "secular-progressives," the "secular-humanists," the "far-left," the "left-elites," the "totalitarian-left," and the "Barack Obombers" (you're right, I just made that up), and they also include America's socialists, communists, neo-Marxists, anti-capitalists and globalists. And, dangerously for America, they have also controlled the dissemination of information in our culture for quite some time through their infiltration and takeover of our media, educational system, entertainment industry and its late-night and Comedy Central clowns.)

to the leaders of the black, feminist, and gay civil rights movements, to politicians and those in charge of the justice system. ...The Left has had to restrict individual freedom of thought and deed in order to destroy the concept of judgment and undermine notions of right and wrong that have been held nearly universally for millennia. This is the result of the wrong people getting control of our culture at a time when we were vulnerable. ... *Tolerance*, once a genuine American Ideal, has become a code word for moral relativism and all its side effects"[31]

Viewer warning: the next segment contains some content that most people still find offensive. It might not be a bad idea to have some bathroom spray handy as Ms. Bruce shines a light on what passes for higher education in the City by the Bay...

The "Art" World

"At the San Francisco Art Institute, the 24-year-old "scholar" Jonathan Yegge, after getting the go-ahead from his instructor, Tony Labat, presented a performance piece for a class in the school's New Genres department. The project involved himself and a student "volunteer." I don't enjoy giving accounts like the one that follows, but it is precisely the reluctance of the main-stream press to give graphic descriptions of disgusting acts that has kept the general public in ignorance of exactly what sort of "art' has been allowed to flourish—often with the support of our tax dollars. Here's how Yegge himself described his "art" to the *San Francisco Weekly*:

"...I engaged in oral sex with him [his student assistant] and he engaged in oral sex with me. I had given him an enema, and I had taken a s**t and stuffed it in his ass. That goes on, he s**ts all over me, I s**t in him. There was a security guard present. There was an instructor from the school present. It was videoed, and the piece was over.

"This depraved trash actually satisfied one of Yegge's course requirements at an institution entrusted with the cultivation and nurturing of our society's next generation of artists."[32]

Yecch! It is almost incomprehensible that something so vile and repugnant could be an acceptable curricular activity in one of our institutions of lower learning. Most folks when presented with this information actually refuse to believe that it is true, since it is so far removed from anything they can comprehend as sanitary or sane, much less "educational." And can you imagine the national media hysteria if a Christian instructor had promoted similar filth? They would have reported it 24/7 and ad infinitum ad nauseum in order to validate their anti-Christian bigotry, but in the real case above they didn't feel the need to mention it. For that would have broken their sacred code of "tolerance," which is a code word for their toleration of anything, including every form of aberrant behavior, except for God and Christianity. Those two they find completely intolerable.

And Yegge's is not an isolated case:

"One of the more famous alumni of the San Francisco Art Institute is Karen Finley. Her "art" involves forcing candied yams into her anus, defecating into a bowl and having another "artist" eat the result, then inviting the audience to lick goo off her naked body. ... [Miss Finley] received funding from the National Endowment for the Arts (NEA) to create this garbage."[33]

---

31   (Pgs. 7, 19, 20, 21, *The Death of Right and Wrong*, by Tammy Bruce)
32   (Pgs. 21-22, *The Death of Right and Wrong: Exposing the Left's Assault on Our Culture and Values*, by Tammy Bruce)
33   (Pg. 22, *The Death of Right and Wrong*, by Tammy Bruce)

And you paid for it! But this is just the end result of slowly ceding control of our culture over the last number of decades to the *who's to say what is right and what is wrong* crowd.

These two examples of academic filth are just a tiny part of the agenda of, as Ms. Bruce describes them, "the malignant narcissists of the left," whose ultimate goal is "legitimizing the depraved." Or as George Will has said, "There is nothing so vulgar left in human experience but that we can find some professor from somewhere to justify it." Again, our national, lame-stream media doesn't feel that we have the need to know that this is the kind of thing that goes on in some of our public universities. The New York Times at least had an excuse; it was too busy printing 50 front page stories of Islamic terrorists getting their picture taken with panties on their heads at Iraq's Abu Ghraib prison. (Good thing for the terrorists they weren't imprisoned at a San Francisco Art Institute.) And that was because our military was the culprit there, and not one of their own leftist universities. When they do find time to report on the disgusting activities of their own such as those listed above, their distorted reportage usually goes something like this, "Right wing Christian prudes tried to stifle the free speech rights of creative college art students, in an attempt to shove their right wing values down their throats!" … No, they just might be a little concerned that these students are shoving something else, of considerably less nutritional value than *right wing* values, down their throats. In a publicly-funded "educational" institution no less, which means that they might also be a little concerned that they are being forced to pay for it.

There was a time in America when these two "artists" and their college instructors would have been committed (maybe there's a reason why many Northern Californians call it "San Fran<u>shit</u>co") so they could be allowed access, not to other impressionable college students and their waste products, but to some serious psychiatric help. But not today! Today the Postmodernists have taken over, and they have done away with the nasty moral constraints of their unenlightened predecessors. *Rage against the suppression of the oppressors* is their liberating battle cry! (As long as one considers eating s**t "liberating.")

> "Irving Kristol defines "postmodern art" as "a politically charged art that is utterly contemptuous of the notion of educating the taste and refining the aesthetic sensibilities of the citizenry. Its goal, instead, is deliberately to outrage those tastes and to trash the very idea of an 'aesthetic sensibility.'""[34]

There are countless other examples of an art world turned upside down by the Postmodernists hatred of truth and beauty, and by their quest to tear down all boundaries and all the restraints of moral decency; and of course to above all trash Christianity in the process. There is the "artist" who photographed a crucifix of Christ in a jar of his urine, called *Piss Christ* and, naturally, funded by the NEA. The left wing media thought it was just another fine example of free speech and a young, aspiring artist at work. But that lie is quickly exposed when you consider what their reaction would be if a young, aspiring artist had made a work entitled *Piss Barack Obama* or *Piss Feminists, Piss Martin Luther King or Piss Mohammed*! We all know it would never be a "free speech" issue then, nor would that art have ever gotten a nickel from the far-left NEA. You see, in the world of the Christian-despising hypocrites in our media and our "artistic" community, it's perfectly fine to trash the real Creator of the universe (as opposed to a big stupid

---

34   (Pg. 26, *Hollywood VS America*, by Michael Medved)

explosion and random accidental chance), the Son of God, who came to earth to die for the sins of mankind so they could be reconciled to him (see chapter 7). And it of course would also be just fine to trash any of the Left's enemies; again *Piss Bush, Piss Sarah Palin, Piss Carl Rove, Piss Jerry Falwell* or *Piss Bill O'Reilly* would also have them reeling with laughter, winking their approval, and defending that "daring and creative" artist's right of "free speech." But, WHOA!, they would go berserk if *Piss Michelle Obama, Piss Islam, Piss Ted Kennedy, Piss NAACP, Piss Bill Maher, Piss Darwin, Piss Letterman, Piss gays, Piss Michael Moore, Piss Ward Churchill, Piss Katie Couric, Piss Stalin* or assorted other *Pisses* had instead been funded by the NEA and presented to the world in a mocking display. But, even more importantly, none of these *Pisses*, hypothetical or otherwise, would ever get out of a morally bankrupt artists loft apartment if it weren't for the destruction of values, the tearing down of decency and morals, and the co-opting of our art world by the malignant, narcissistic, moral relativists of the Left.

Then there is the garbage entitled *The Holy Virgin Mary* by Chris Ofili and put on display in 1999 at the Brooklyn Museum of Art...

> "His depiction of Mary was meant to convey something other than holiness. Pieces of elephant dung were stuck all over the image of Mary, and she was surrounded by photos of vaginas from porn magazines. If elephant feces were smeared on the walls of a mosque, it would be a hate crime. Here, smeared on a Christian icon—voila!—its Modern Art."[35]

And wafting off of nearly every example of this artistic stench we find the offensive, far-left odor of the National Endowment for the Arts...

> "In 1989 the NEA indignantly denied a modest ($10,000) request from the New York Academy of Art to provide young painters with basic skills in representational drawing. In the words of Susan Lubowsky, director of the NEA Visual Arts Programs, "teaching students to draw the human figure is revisionist...and stifles creativity." [Yet] less than a year later, the Endowment paid $70,000 of government funds for a gallery show that featured Shawn Eichman's "Alchemy Cabinet," displaying a jar with the bloody fetal remains from her own abortion. [Another way the Democrat-Left demonstrates their unbearable care, concern and affection for "the children."] The federal arts administrators also found funds to provide grants to performance artists such as John Fleck (whose act includes a sequence in which he publicly urinates on a picture of Christ), and the estimable and outspoken Annie Sprinkle, who masturbates on stage with various sex toys and then invites members of the audience to explore her private parts with a flashlight."[36]

But hey, c'mon, bring the kids. Who's to say what is right and what is wrong?

And finally we have *Corpus Christi*, a play "which features a gay Jesus having sex with his apostles." Christians protested and, of course, our left-wing media rushed to the plays defense:

> ..."a *New York Times* editorial condemned the protesters "bigotry, violence, and contempt for artistic expression" while saying nothing about the anti-Christian bigotry in the play itself.

> ...""In painting and sculpture, the bashing of Christian symbols is so mainstream that it's barely noticed," comments John Leo, one of the nation's most prolific and insightful social critics. [But isn't it interesting that these same cowards...er...artists, never get the inspiration to do the same thing to Muslim symbols? Hmm.] Leo's examples in his *U.S. News & World Report* column include "the Virgin Mary coming out of a vagina, Mary

---

35 (Pg. 39, *The Death of Right and Wrong*, by Tammy Bruce)
36 (Pgs. 26-27, *Hollywood VS America*, by Michael Medved)

encased in a condom, Mary in pink panties with breasts partially exposed, an Annunciation scene with the Archangel Gabriel giving Mary a coat hanger for an abortion, Mary pierced with a phallic pipe." "[37]

There appears to be a trend here and it has something to do with "the Left Elite's popular-culture campaign to marginalize not religion per se, but Christianity specifically. ...The reason is that Christianity represents to the Left the principal threat to their hollow agenda."[38] And that "hollow agenda" consists of replacing age-old moral absolutes, laws of right and wrong that stem from the Jewish-Christian Ten Commandments and other biblical moral codes, with their own liberated ideals.

Tammy Bruce also relates her own compelling personal story in *The Death of Right and Wrong*. She became the head of the Los Angeles chapter of the National Organization for Woman in 1990 and remains "an openly gay, pro-choice, gun owning, pro-death penalty, voted-for-President Bush authentic feminist." But she became alarmed with the direction the movement was heading; that instead of actually trying to advance the cause of women in general, it had morphed over time into a male-hating, far-left, dishonest and unethical organization. Then, as a radio talk show host, she...

..."began to meet and talk with people who impressed me, were thoughtful, *and* were religiously devout. My experience contradicted everything I had been taught. The religious (Christians specifically) were supposedly the enemy, but the people I was now meeting were nothing like the rabid, intolerant species I had been told they were. In fact, unlike my leftist colleagues, they displayed respect for others, tolerance, and thoughtfulness, all the while remaining true to their own beliefs and principles. These are values I feel are inherent in true feminism and yet were absent from the world to which I had devoted myself.

"I was moved, and change, by the experience. It made me realize that perhaps there was more for me to consider when it came to the quality and meaning of life. *Something* had to explain why my Left Elite allies were generally miserable, angry, and paranoid (you could put me on that list at that time as well!), while the "enemy" was secure, comfortable, and a group of generally happy (and determined) people. Funnily enough, at that time, values and virtues didn't even occur to me as the missing pieces of the puzzle. Those ideas, along with religion itself, had become so demonized that I had stopped even thinking about them conceptually.

"My adventure outside the feminist ghetto was initially born of curiosity, but it became much more than that. I learned the truth about people, something I was never afforded as a member of the Left Elite."[39]

Tammy Bruce was the lucky one. Most people indoctrinated into the self-righteous superiority and elitism of the totalitarian-left never escape; like the vast majority of people indoctrinated into and poisoned by the lies of any false religion, unscientific theory or ideology. In the case of the Left-elites, they spend their lives thinking of themselves and their ideology as so much better, smarter, more righteous than, and far superior to the poor, stump-toothed, inbred, traditional Americans down below them in fly-over country, the country in between their world of New York and the Left coast. And they hold the values of those people in the same

---

37　(Pgs. 51-52, *The Death of Right and Wrong: Exposing the Left's Assault on Our Culture and Values*, by Tammy Bruce)
38　(Pgs. 57, 39,*The Death of Right and Wrong*, by Tammy Bruce)
39　(Pg. 36, *The Death of Right and Wrong*, by Tammy Bruce)

contempt as they hold them. This is the way they were seduced into thinking by their indoctrinators in our educational system and by the steady, drip-drip-drip influence of the mockers of traditional America who control our entertainment industry. Yet these same unfortunate, sick-at-heart folks, brainwashed into a way of thinking that holds traditional, common sense values of right and wrong in complete contempt, are defining the values of our popular culture now, and indoctrinating the next generation of artists, and students in general, into accepting as normal what should not even exist in a truly normal society. Obviously the *artists* we have exposed here are not so much avant-garde as they are disturbed. And the point is not to put them down or to disparage, demean and deride them as the Left does its opponents. They are easy targets after all. But the point is that we are now living in a society where the Left-elites who control our media, entertainment industry, educational system, "art" community, late night clowns and much of our government celebrate these disturbed individuals and their professors, because they are both on the same side, holding Christians and the traditional values of regular folks in contempt; and indoctrinating the next generation of Americans to be just like them. And even though most Americans do not agree with them they are still ignorant of their devices, kept in the dark by a complicit media and clueless therefore of the debasement that is happening to their country and to their children right under their noses.

In the past artists strived to capture and interpret the endless beauty, majesty and creativity of our Makers world. They did it in landscapes and city scenes, portraits and in still life. They were motivated by an awe and respect for the wonder of God's creation that shone through in their inspiring works of sculpture and painting. And in doing so they reflected the imagination and workmanship of their Creator God. Now much of art reflects the sheer stupidity of those who have been taught to despise not only God, but all that is good. Rather than striving to emulate, honor and learn from the works of the world's greatest Artist (you know, the one who painted The Painted Desert, carved the Grand Canyon, sculpted that Botticelli who just sashayed past you down the beach the other day, and lit on fire that incredibly beautiful sunset you just witnessed the other evening), many of today's "artists" strive to see who can be the ugliest and the most offensive; for beauty, like truth, is something they have been taught to despise and hold in utter contempt. That is Postmodernisms gift to both them and the nation. Beauty and art were once understood to be synonymous. Not anymore. Not since the ascendancy of moral relativity. Now human feces, elephant dung, trashing the symbols of Christianity and an artist's urine are synonymous with art. And the incomprehensible scribble-scrawl paintings and twisted sculptured wreckage of what is known as "modern art" is all the rage, and has been for some time. Now your two-year-old could produce a priceless masterpiece in less than twenty minutes with some toothpaste, a couple of magic markers, the contents of his diaper and the nearest carpet. As a matter of fact, so could a chimpanzee. Yet on the Upper West Side of New York, financially well-off, left-wing-indoctrinated dupes (Democrat voters all) will pay what amounts to a life's hard-earned wages for the common man, just for the questionable privilege of owning one of these pieces of crap. And I'm not sure, in addition to being financially snookered, if it would add to their solace in knowing that every human being with a brain is laughing at them raucously behind their back. Their emperor is naked, and their artist has no talent, but when they're told they should jump at the chance to admire the invisible clothes, and purchase his non-existent "art," they ask how high, and how much.

Education

"Train up a child in the way he should go, And when he is old he will not depart from it." (Proverbs 22:6)

"Education is a weapon whose effects depend on who holds it in his hands and at whom it is aimed." (That was Joseph Stalin, like his homicidal maniac contemporary, Che Guevara, another hero of the far-left.) And in our country the Left holds the weapon in their hands, having taken over our educational system decades ago, and they have been aiming it at our children ever since. People believe what they are taught, and when they are taught lies, a lifetime of enslavement to those lies is most often the result.

Again we return to Ms. Bruce's book, for a look at what has been allowed to seep into our primary and secondary educational system; after moral absolutes have been effectively locked out...

"The Gay, Lesbian and Straight Education Network [GLSEN] describes itself as "the leading national organization fighting to end anti-gay bias in K-12 schools." This organization, which cloaks itself in a mantra about "tolerance" and "understanding," is implementing programs in public schools that are aimed at nothing less than sexualizing your children.

..."GLSEN claims as its mission the protection of so-called gay, lesbian, and transgendered children from harassment and discrimination. Think about this. Gay and lesbian children? Transgendered children? GLSEN states it is working to make sure all primary- and secondary-school students "accept all of their classmates—regardless of sexual orientation and gender identity/expression." Gender identity/expression in kindergarten?

"I for one, and no doubt you as well, believe we should let children be children without having to "learn" about homophobia, especially when it's used as a pretext to gain access to a child."[40]

In a Youth Only "workshop" in 2000 for ages 14-21 with children as young as 12 attending, cosponsored by GLSEN in Ted Kennedy's state of Massachusetts...

"Dildos were discussed, with tips on how you (a 14-year-old) would know if the dildo was too large or too small. At that point, gay "resources" about similar subjects were offered to the kids. ... there was role-playing too. One female student acted the part of a lesbian attracted to another girl. In addressing the topic "Cum and calories: Spit versus swallow and the health concerns." ...Remember, this conference was supposed to be about "tolerance" and accepting those who are different from you. In fact, it appears to have been nothing more than indoctrinating students into sex—homosexual sex."[41]

Do the words "sick bastards" strike a familiar note? But it gets even worse... Responding to the question "What's fisting?"...

..."a "helpful and enthusiastic" presenter responded by offering the proper hand position for the act, while another described fisting as "an experience of letting somebody into your body that you want to be that close and intimate with.""[42]

---

40 (Pgs. 103-104, *The Death of Right and Wrong*, by Tammy Bruce)
41 (Pgs. 104-105, *The Death of Right and Wrong*, by Tammy Bruce)
42 (Pg. 105, *The Death of Right and Wrong*, by Tammy Bruce)

One can only be thankful the child didn't ask "What's bestiality?" or a demonstration with some barnyard animals and household pets might have ensued. And this is but one example of the toxic refuse that America unknowingly goes out and votes for every time they give Democrats and their Left Elites more and more power. (See chapter 13)

Tammy Bruce continues,

"The death of right and wrong has so permeated our culture that the institutions we rely on to protect children now willingly participate in attempts to destroy them. ...That's another fantastic reason to support school vouchers—to get these parasites out of schools and away from children."[43]

There was a time before in this country when this kind of degenerate, sexually "avant-garde," child-perverting-and-endangering filth would never have been allowed anywhere near our public schools, much less condoned by its administrators and then shoved into the hearts and minds of the most sexually naïve and vulnerable among us, by the most sexually perverted among us. How did this happen? Well, you have to go back a number of decades, to 1962, when a majority of unelected, ACLU, liberal, Constitution-perverting lawyers in black dresses on the Supreme Court legislated from the bench[44] and outlawed prayer in public schools, even though our politicians and officials had prayed publicly in all of our federal, state and local government institutions since our nation's founding; and still do! Go figure. What is OK for our leaders is detrimental and "unconstitutional" for our children. (It *is* however OK for our children, and their teachers, to worship the scientifically-impossible "creators" of random accidental chance and a big explosion. Curiously, *that* religion has been exempt from the mythological, ACLU-created *separation of church and state*.)[45] In 1963 they upped the ante by banning Bible reading in public schools. (But reading and even teaching out of the Koran they have no problem with.)[46] Then, in 1980 another group of ACLU, Constitution-ignoring judges decreed that posting the Ten Commandments in our public schools was now, abracadabra, UN-constitutional. Even though these Commandments *had* been posted (and the Bible had been read and students had prayed) since the founding of this country and with the full approval and expectation of the Founders of this country who actually wrote that Constitution and actually knew what they meant when they wrote it. To say those three things are "un-constitutional" is a bald-faced lie. If those judges had just come out and been honest and declared them un-liberal they would have been telling the truth, but then they wouldn't have gotten away with banning them.

God was out and so were his morals. This opened the door for the great deceiver, and his "values," to slip in. Who's to say what is right and what is wrong became the smoke screen that allowed the disturbing and obviously immoral sexual freedom of the "new morality" to slip in. Nothing was wrong any longer, nothing was indecent, nothing was corrupting to the souls, and health, of our young people. Not according to the new, *enlightened*, progressive keepers of the gates. And all

---

43   (Pgs. 105-107, *The Death of Right and Wrong*, by Tammy Bruce)

44   ("Legislating from the bench," which is expressly *prohibited* by that same Constitution, which they also swore an oath to uphold and defend. Hmm, that wouldn't have anything to do with the word *treason*, would it?)

45   (See chapter 10, and the section "The Separation of Church and State" in chapter 13, "Satan")

46   (See "Islam" section in chapter 8)

hell broke loose. Teenagers having babies, catching incurable venereal diseases, and committing suicide like never before. "Let the games begin!" And now thanks to decades of the ACLU and the Democrat-Left Elite's godless and amoral vision for America we have a public school, left-wing-indoctrination system that is only distinguishable from our prison system in that the kids get to go home each day. Otherwise in so many school districts—especially the poorest—it is equally as chaotic, undisciplined, immoral, disrespectful and nearly as unproductive; with the life-destroying, rotten attitudes and ability to lace a belt of many of the respective inmates also indistinguishable. Congratulations to the ACLU! My, how proud they must be! And now Barack and the other Democrat-Left Elites who have rested control of our government (primarily by lying about both the Bush administration and the War in Iraq and by duping enough gullible and uninformed voters into believing them) have a sure fire plan to remedy the situation! And what is this ingenious policy, this historic *change* that clueless people can believe in, this new ground-breaking and earth-shattering solution coming from the mind of the genius Obama? Why it's the same old, same old! Tax and spend! Tax more and spend even more! Unconstrained liberals throwing vast sums of our money down the drain so their friends at the far-left National Education Association can continue to enrich themselves while blocking any real solutions, and of course funneling vast amounts of money back to the Democrats in massive campaign contributions that come from union-forced dues from all of their members, both Democrats and Republicans, liberals and conservatives. What a sweet deal!

Here's a better idea, and it saves every penny of those hundreds upon hundreds of millions of wasted taxpayer dollars. Let's just go back to the time when the Per Capita Student Expenditure was a fraction of what it is today, and before Left-Wing Indoctrination 101 was the primary course requirement. Back when students actually learned something of value like reading and writing and arithmetic, and how to express themselves in complete sentences. And what was the difference then? Discipline. That was before we allowed the America-hating ACLU[47] and their judges to shove their vision of education down our throats, which included the complete removal of discipline (and of course morals) from our public schools, and the bankrupting through endless, domestic-terrorist lawsuits of any school district that didn't comply. Before the Ten Commandments, that moral and historic document was outlawed. Before God and His Word (see chapter 6) was outlawed. Before prayer was outlawed. Before the moral relativists of the Left shoved their progressive "values" onto all of the rest of us, and onto all of our children. Let's go back to that time, and then turn the tables by suing the ACLU into bankruptcy for the billions of dollars in damages these domestic terrorists have cost decent America.

"The ideological triumph of liberalism among American elites—far from bringing the individual and social enlightenment it promised—has produced unprecedented moral decay. *The principal victims of this decay are the poorest and most vulnerable among us— those most in need of the support of a healthy culture*."[48] [Emphasis mine]

But yet these same Democrat-Left Elite always portray themselves as the

---

47   (See "Take over the criminal justice system" in chapter 13 "Satan")
48   (Robert P. George, Department of Politics, Princeton University.   Inside cover endorsement of *Slouching Towards Gomorrah: Modern Liberalism and American Decline*, by Robert H. Bork)

great champions of the poor and the downtrodden. They "care" so much they can hardly stand it. What a ruse. The Left have turned our classrooms into "living laboratories for left-wing activism"[49] without the least regard for its devastating effect on "the poorest and most vulnerable among us."

And in our universities what can only be described charitably as America-hating liars have been allowed to infiltrate many of our classrooms and poison much of the curriculum. For one example of thousands, the University of Colorado's once-tenured but now-fired imbecile, Ward Churchill, became infamous for calling those innocent folks killed on 9/11 at The World Trade Center "little Eichman's," but he is only the tip of the manure pile. Our colleges and universities, as well as many of our elementary and high schools, are filled with like-minded, far-left zealots who use their positions of power and influence to indoctrinate the minds and poison the souls of their naïve, sequestered children. (Pick up a copy of *Indoctrination U.: The Left's War Against Academic Freedom*, or *The Professors: The 101 Most Dangerous Academics in America*, both by David Horowitz, for a well-documented exposé of the far left's dangerous and damaging infestation of our higher educational system.) These left-wing indoctrinators have slipped into our educational system at all levels, under cover of darkness. A cover afforded them by the same Left's control of our media. They teach African American children that this is a "racist country," that they are "wasting their time" in school "trying to be white" by getting an education, and we wonder why both their dropout and unemployment rate is so high.

"Not everyone who wants to do this country harm is a Radical Islamist armed with a box cutter. Some …use their academic status to infect students with despair, anger, and hopelessness about this country and even their own future. Moral relativism has become a staple of academic life at the hands of a professorate and staff determined to squeeze the last vestiges of decency and morality out of their young charges.

"…the malignant narcissist from the 1960s *became* the establishment in order to attempt to gut it from the inside. And instead of trying to subvert Americans whose characters were already formed, the Left decided to start from the ground up—by parasitically infesting our nation's secondary schools, colleges, and graduate schools to indoctrinate each generation as it came along. They hate themselves and hate this country, and they are determined to not be alone.

"The first step for the Intellectual Elite is to unmake and then remake history itself. Smear the Founding Fathers, cast patriotism as jingoistic, and classify the United States as a genocidal nation bent on terrorism, from its founding to today. On campuses across the country, pride in our country and patriotism are deemed "offensive" and banned. Fewer high schools and colleges are requiring students to take U.S. history courses, and when history is discussed, men like George Washington and Thomas Jefferson are deconstructed and maligned. Someone like Ronald Reagan usually isn't discussed at all, and if he is, it is to complete the vilification. Courses on Western culture and history are attacked as "racist" and often eliminated because the reality of white European men's role in the shaping of world history offends and upsets multiculturalists. Ever since the 1960s, students have been told that everything is relative, judgment is verboten, and if it feels good, do it, while decency and morality are cast as deviant concepts used by the oppressor."[50]

---

49   (Michelle Malkin on Fox News *Hannity*, 9/02/2009)

50   (Pgs. 159,161-162, *The Death of Right and Wrong*, by Tammy Bruce)

The lefts denial of moral absolutes and their embrace of the lie of moral relativism are not orchestrated to get rid of right and wrong. For their disguised but nevertheless obvious answer to their own question *Who's to say what is right and what is wrong* is "We are!" It is done so they can impose their own godless version of "right and wrong" on their unsuspecting and defenseless students; defenseless because without an unchanging higher law of morality they are left without any means of countering that imposition. What are they going to do, bring up The Ten Commandments? They would be laughed out of the building. And then arrested. The very question, "Who's to say what is right and what is wrong?" is meant as an intimidating challenge that removes God from the equation (as they have already removed him from the origins question—replacing him with two scientific impossibilities, random accidental chance and a big explosion ...see chapter 10), and allows the Leftist teacher or professor to step into the vacuum and to start filling it with *his* "truths" (his opinions) and *his* "morals"(or the absence of any thereof). He cannot be challenged because he or she has successfully destroyed any absolute basis from which to judge. Therefore he or she is now the absolute judge. There is no "unchanging source" by which to judge the right and wrong of anything, so the left-wing, godless professor or high school teacher is free to impose his or her doctrine.

And yes, there are of course a great many decent teachers, and a number of decent professors, but they are powerless to stop the onslaught of the Democrat-Left Elites who control the NEA (National Education Association), most of the education bureaucracies across this nation, and the administrations and staffs of most of our colleges and universities. There are countless examples of Christian and conservative teachers in our elementary and high schools being intimidated and silenced by leftist administrators who have no problem with teachers expressing and even pushing their opinions as long as those opinions coincide with the politically-correct, liberal orthodoxy. And it is nearly impossible for a Christian or conservative professor to get tenure from the committed leftists who control most of our colleges and universities, and who wish to stay in control by keeping the doors closed to any ideological miscreants. (See section "Take over the educational system" in chapter 13.)

One of the predictable results of this ideological McCarthyism that has replaced the "free expression of ideas" is a modern day Hitler youth movement that runs amok on many of our college campuses disparaging Christians, despising conservatives, trashing Republicans, mocking anyone with a decent set of morals, and despising anyone who refuses to wallow with them down in the pigpen of their sexually-debauched, baby-mutilating, God-denying, Darwin-loving ideology. (The same clueless, left-wing indoctrinates who are now "occupying" Wall Street.) All in the name of "tolerance" of course. And this is winked at, condoned, if not outright encouraged by the mostly far-left faculty and staff. Students who do not think that mutilating a baby in the womb for the convenience of his or her mother is necessarily such a good or moral thing are harassed and attacked when they attempt to express their views on many of these campuses. Decent, honorable people like George W. Bush are vilified and slandered without evidence because these brainwashed students are taught that conservatives are the enemy, and their opposing views are not even worthy to be heard. (For if those opposing views *were* entertained, their left-wing indoctrination might collapse like a house of cards.) ("See no evil. Hear no evil." ..."See no truth. Hear no truth.") Decent, brave men and women who defend the freedom of this country, including the freedom of these Hitler youth, who come to our college campuses as military recruiters are harassed and harangued because they do not share the children's deranged hatred for everything military, and everything George Bush. Conservative authors and speakers who come to campus are routinely threatened, cursed at and shouted down by these modern-day Nazi thought police. All in the name of their liberal version of "free speech" of course. They are the new book burners of the 21st century. And God forbid anyone who were to stand on a college campus and defend the traditional, morally absolute form of marriage between one man and one woman. They stand a better than 50/50 chance of being beaten to death by the new Hitler youth, again in the name of tolerance. And God forbid also that anyone be so stupid as to forget that affirmative action is good, always has been good and always will be good, even if it is blatant racism in reverse. Moral relativism means it's okay to be absolutely wrong, as long as your secular-progressive professor tells you it's right. And don't make the mistake of letting rationality, free thought or basic logic get in the way of your brainwashing. You would then be committing the absolutely unpardonable sin. For you then might actually start to question their teachings and you then might expose the little naked left-wing professor standing in front of the classroom. It is also interesting to note that these brainwashed, left-wing indoctrinated, completely intolerant, mentally-poisoned and emotionally disturbed youth, who have a green light from most of their respective colleges and universities administration and staff to run rampant, unrestrained and unpunished, also vote Democrat, religiously, whenever they do vote.

Finally, hear the words of Darrell Scott, spoken before the House Judiciary Committee in May of 1999, a month after his daughter, Rachel Scott, was slaughtered along with 12 others in the Columbine High School massacre in Littleton, Colorado.

"Since the dawn of creation there has been both good & evil in the hearts of men and women. We all contain the seeds of kindness or the seeds of violence. The death of my wonderful daughter, Rachel Joy Scott, and the deaths of that heroic teacher, and the other eleven children who died must not be in vain. Their blood cries out for answers.

"The first recorded act of violence was when Cain slew his brother Abel out in the field. The villain was not the club he used. Neither was it the NCA, the National Club Association. The true killer was Cain, and the reason for the murder could only be found in Cain's heart.

"In the days that followed the Columbine tragedy, I was amazed at how quickly fingers began to be pointed at groups such as the NRA. I am not a member of the NRA. I am not a hunter. I do not even own a gun. I am not here to represent or defend the NRA - because I don't believe that they are responsible for my daughter's death. Therefore I do not believe that they need to be defended. If I believed they had anything to do with Rachel's murder I would be their strongest opponent.

"I am here today to declare that Columbine was not just a tragedy -- it was a spiritual event that should be forcing us to look at where the real blame lies! Much of the blame lies here in this room. Much of the blame lies behind the pointing fingers of the accusers themselves. I wrote a poem just four nights ago that expresses my feelings best.

Your laws ignore our deepest needs,

Your words are empty air.

You've stripped away our heritage,

You've outlawed simple prayer.
Now gunshots fill our classrooms,
And precious children die.
You seek for answers everywhere,
And ask the question "Why?"
You regulate restrictive laws,
Through legislative creed.
And yet you fail to understand,
That God is what we need!

"Men and women are three-part beings. We all consist of body, mind, and spirit. When we refuse to acknowledge a third part of our make-up, we create a void that allows evil, prejudice, and hatred to rush in and wreak havoc. Spiritual presences were present within our educational systems for most of our nation's history. Many of our major colleges began as theological seminaries. This is a historical fact. What has happened to us as a nation? We have refused to honor God, and in so doing, we open the doors to hatred and violence. And when something as terrible as Columbine's tragedy occurs -- politicians immediately look for a scapegoat such as the NRA. They immediately seek to pass more restrictive laws that continue to erode away our personal and private liberties. We do not need more restrictive laws. Eric and Dylan would not have been stopped by metal detectors. No amount of gun laws can stop someone who spends months planning this type of massacre. The real villain lies within our own hearts.

"As my son Craig lay under that table in the school library and saw his two friends murdered before his very eyes, he did not hesitate to pray in school. I defy any law or politician to deny him that right! I challenge every young person in America, and around the world, to realize that on April 20, 1999, at Columbine High School prayer was brought back to our schools. Do not let the many prayers offered by those students be in vain. Dare to move into the new millennium with a sacred disregard for legislation that violates your God-given right to communicate with Him. To those of you who would point your finger at the NRA -- I give to you a sincere challenge. Dare to examine your own heart before casting the first stone! My daughter's death will not be in vain! The young people of this country will not allow that to happen!"[51]

Well, unfortunately at this point, the young people of this country and many adults as well, continue to "allow that to happen." Not only that, they have been indoctrinated by their left-wing, progressive teachers and entertainment idols to rush out and vote not only to continue to allow it to happen, but to let it get much, much worse. God help us.

The Courts

"Those who forsake the law praise the wicked, but such as keep the law contend with them. Evil men do not understand justice, but they who seek the Lord understand all things." (Proverbs 28:4-5)

The ACLU-Democrat-Left elite, who for all of their power are nevertheless a distinct minority in this country, have taken over our schools and just about destroyed our educational system in the process. They have also taken over our judicial system and have succeeded in turning that on its head as well. The

---

51   (snopes.com)

absence of moral absolutes to govern the behavior of our primary and secondary school children, and their teachers, has led to an undisciplined, impossible-to-learn hellhole in many areas, especially in the poorest. The extent of damage the advance of moral relativism in our courts has caused is just as bad. Two names in particular come to mind, O. J. Simpson and Robert Blake; a couple of classic case studies that illustrate America's problem with both education *and* injustice. For the teaching of moral relativism throughout our school system over the last few decades has resulted in producing a populace where far too many are incapable of passing judgment on any behavior, no matter how vile. In the above two cases that "education" resulted in two heinous murderers being set free.

But those jurors, as brain dead as they were, didn't operate in a vacuum. They had the help of lawyers who were "educated" that there is no such thing as right and wrong and that therefore the legal system is theirs to use for both the enhancement of their egos and the fattening of their wallets. No matter how many innocent victims get trampled in the process. There is also no such thing as justice in their world, only what they can get away with by twisting and turning the letter of the law until it suits their own purposes. These lawyers also become judges, and of course politicians. Many of today's defense attorneys skipped the course on justice (actually, I don't think it's offered anymore) but passed with flying colors the one on *How to Cheat, Fool, Confuse and Lie to Your Mentally-Challenged Jurors at any cost so your Guilty Client Walks 101*. And there is no better example of this than the case of five year old Samantha Runnion:

> "The Sixth Amendment to the Constitution details the rights of anyone accused of a crime, including the right to the "assistance of counsel for his defense." What does that translate to in a system filled with people who refuse to distinguish between right and wrong? Today, a "vigorous defense" includes lying. It includes deliberately misleading juries. It includes not just defending a client but doing everything possible to get him released—even when the lawyer knows the client is guilty."[52]

In July, 2002, five-year-old Samantha Runnion was kidnapped from her apartment complex in Southern California, raped and murdered and then left in a "provocative" pose by the side of a highway. But the thing that should really make your blood boil is that the sick bastard who did it, Alejandro Avila, should have already been in jail. That is because two years before he had been tried for the sexual molestation of two little 9-year-old girls who lived in the same apartment complex as Samantha. Only he was acquitted because his defense attorney, John Pozza...

> ..."battered away at the girl's credibility and told the jury that Avila's ex-girlfriend had encouraged them to make up their story. Pozza did this although he knew his client had *failed a lie-detector test* about the accusations.
>
> "Despite the girl's dramatic and detailed testimony, John Pozza convinced the jury that they just couldn't be believed. You may be thinking that's his job—to create reasonable doubt. [You may also be an idiot. Or a member of the ACLU.] But at any cost, and when he knew his client was at best not telling the truth, and at worst guilty?
>
> ..."Avila was out on the street again, thanks in part to what Pozza callously touts on his Web site as his track record of "wins." ...[But] Pozza can't dismiss the seriousness of his "win" for Avila. If he had found the thought of a child molester being set free morally repugnant and had withdrawn from the case, or at least provided only the basic assistance

---

52 (Pg. 266, *The Death of Right and Wrong*, by Tammy Bruce)

called for (which does not include telling a jury two young witnesses are essentially *lying*), Samantha Runnion might still be alive today."[53]

But, hey, how can her *life* possibly compare with another win for that equally sick-bastard attorney? The parents of Samantha Runnion should be excused if the next time they hear someone say *who's to say what is right and what is wrong*, they smack them in the face. (In the case of Mr. Pozza, we'd excuse them if they used a baseball bat.)

Then there is the case of the kidnap and murder of Danielle Van Dam by David Westerfield on February 2, 2002. The lead attorney, Steven Feldman, "knew for a fact their client was guilty of this unspeakable crime." That is because "minutes before Danielle's body was found under a bush along a desert highway, Feldman and his colleague Robert Boyce were brokering a deal with prosecutors." They would tell police the location of the seven-year-old girl's body and in turn the prosecutors would not seek the death penalty. "The deal was just minutes away," the *San Diego Union-Tribune* reported, when the poor girl's body was found. However, the fact that they knew for certain that their client was guilty of this heinous murder did not in any way deter them in the trial from seeking experts that were…

…"willing to testify that bugs found under the little girl's body indicated a time line of death that exonerated Westerfield. Danielle was dumped, the defense claimed, at a time when Westerfield was already under police surveillance, making it impossible for him to be the culprit. The indecency of this assertion is absolutely astounding. Here was Feldman, not just countering the prosecution's case, but presenting a scenario to the judge and jury that appears he *knew* was false. It's interesting, isn't it? If a witness lies in the courtroom, it's perjury, a serious crime. When it comes to defense attorneys, however, it seems there are no rules."[54]

"Unfortunately, no agency, no segment of our society is immune from the moral relativism fostered by the Left. The integrity of our justice system depends on the officers of the court—lawyers and judges—being honest, decent men and women. All too often, however, we see moral ambiguity ruling the day—making a mockery of the search for the truth and prosecution of the guilty. Trials are no longer about freeing the innocent, punishing the guilty, and making restitution to the injured. They have devolved into a contest over who will *win*.

…"What disgusts us is a generation of lawyers and judges who are dedicated foot soldiers in the war against our country's principles.

"The majority of lawyers and judges today are first- or second-generation products of the 1960s. As we know, the homegrown Marxists and radical leftists of that period didn't move to Hanoi—they stayed here and became politicians, journalists, and special-interest-group flacks. They also became professors at law schools, and their progeny, indoctrinated with leftism, became today's lawyers and judges.

…"In order for the malignant narcissists to advance their agenda, they must remove morality from the law, or replace it with false morality."[55]

Again, who's to say what is right and what is wrong? God is. Otherwise we are left with ACLU attorney-scum lying to a jury to free heinous child molester-

53   (Pgs. 267-268, *The Death of Right and Wrong*, by Tammy Bruce)
54   (Pgs. 270-271, *The Death of Right and Wrong*, by Tammy Bruce)
55   (Pgs. 264-265, *The Death of Right and Wrong*, by Tammy Bruce)

murderers so they can be free to rape and kill again. Incredibly, when the jury in the above case returned a verdict of guilty and recommended execution for Mr. Westerfield, one of the two lawyers actually turned to their client and said, "I'm so sorry." Sorry for what? What pile of human excrement would be sorry that a child molester and murderer got his just reward, and would never be free again to reap irreparable harm to another victim and their parents, siblings, family and friends? The kind of godless, morally bankrupt pile who was indoctrinated for years in our left-wing education system to believe that the entire universe and all of life is just a complete accident (an "accident" indistinguishable from a miracle), and that therefore there is no such thing as moral absolutes so they are free to do whatever they want. Destroy the morals that come down from above, and you are left with the stinking filth that rises from below.

Today we produce lawyers, steeped in this moral relativity insanity, that leave our universities incapable of understanding the foundation of all law, and therefore incapable of understanding the dispensation of justice. And we are left with an injustice system that is incapable of comprehending evil, or rendering just punishment. These lawyers come out of these colleges believing that there is no transcendent law. But that is so incredibly naïve because we know through the advancements of science that this universe is awash with one scientific law after another. Laws of physics, and of motion, of gravity and of light, laws of chemistry and thermodynamics, of mathematical probability and electromagnetism, etc. And it is through the recognition of the existence of these laws that we set out to discover them in the first place. They are unchangeable and unalterable throughout the universe, and it is through the knowledge of them that we can know what to expect with regard to how this material universe will respond in any given situation under any given set of circumstances. And it is through their discovery that we now enjoy countless technological advancements and modern conveniences—cars, TV, cell phones, electrical appliances, air and space travel, medical and surgical miracles, etc. etc.—that we take for granted but not too long ago were just a fantasy. These laws are built into the very fabric of the universe. (And where did these laws come from if not from a law Giver?)

"God is the author of perfect law, both of morality and of nature."[56]

In science we understand that there are laws of this universe, and we seek to discover them. We don't create them; we don't make them up as we go along. But, conversely, when it comes to the understanding of what is right and wrong, and the laws that govern our moral universe, all of a sudden we are told that "NO!, NO! THOSE LAWS DON'T EXIST!! THERE ARE NO SUCH LAWS THAT GOVERN THE BEHAVIOR OF MAN." So man, the crown of all creation, has no laws to govern him?!@*! And far too many of us are just a little too gullible, or a little too uninformed, and we swallow that line.

Bill O'Reilly in his book Culture Warrior documents the assault on our culture's traditional values of right and wrong by the Secular-Progressives of the left. Of the many examples he gives of this destructive onslaught the most disgusting concerns our old friends at the ACLU. In Massachusetts...

…"that organization volunteered to represent the North American Man-Boy Love Association (NAMBLA) *pro bono* in a civil case after a particularly brutal and (you would think) indefensible crime in Cambridge. ..a ten-year-old boy named Jeffrey Curley was

56   (Pg. 56, *Temple at the Center of Time*, by David Flynn)

raped and murdered by two men, who were caught and convicted. One of the killers had written in his diary that NAMBLA literature, which encourages adult rape of children, gave him some incentive to assault the boy.

..."Outraged as well as heartbroken, the Curly family sought to avenge their young son and prevent any further heinous crimes by filing a $200 million federal lawsuit against NAMBLA. ...And here comes the ACLU, guns blazing, ready to defend free of charge, an organization that promotes the rape of children."[57]

We'll deal in more depth with this anti-American, anti-Christian group that poses as a legitimate legal organization in a later chapter (13 "Satan"), but keep in mind that many attorneys are members of this organization; and that lawyers, both defense and prosecutorial, make up the heart of our judicial system; and that lawyers invariably are the group from which our judges are selected. Yet that is the end result of a generation of lawyers indoctrinated by an educational system steeped in moral relativity, scoured clean of moral absolutes, and running over with the godless values of the Democrat-Left that ooze up from beneath us.

Ann Coulter muses on the Left's schizophrenia when it comes to criminal justice...

"Liberals believe it is important to never, ever punish criminals because—well, I'm not sure why. They produce a constantly scrolling list of reasons: The perpetrator is too young; the perpetrator is too old; the perpetrator has been rehabilitated; the perpetrator will not be rehabilitated in prison; similarly situated perpetrators got a different sentence; the perpetrator wrote a children's book; the perpetrator was making a statement about society; the perpetrator says he didn't do it (and we're too busy writing him mash notes to look at the evidence); the perpetrator was on cold medication when he raped, murdered, and cannibalized a family of four."[58]

When there is no such thing as right and wrong, how can any criminal be wrong, much less guilty? This is the end result of the Democrat-Left's assault on our traditional values, the institutionalizing of their own in the place of those moral absolutes, and the silencing of any debate by shoving *who's to say what is right and what is wrong?* down our throats.

We can also see this clearly in the case of the Boy Scouts of America; where the Democrat-Left in this country would value the advancement of the cause of militant homosexuals over the protection of young boys from possible sexual assault. The Boy Scouts of America, having traditional values (and possessing actual brains), didn't think it was such a good idea to allow homosexuals to be Scout leaders. Not because they hate homosexuals or because they want to discriminate against them, but because they had the extremely profound responsibility of protecting the young boys in their charge. And because they might not have wanted the children under their responsibility to go the same route as all of those Catholic children molested by **homosexual** priests. (The left-wing media licked their collective chops over that story, seeing a golden opportunity to disparage Christianity, but somehow could never put two and two together: that it was homosexuality, and not celibacy, that was the root cause of the problem, math not being a Democrat journalist's strong suit. And neither could they add correctly when it came to their indefensible condemnation of the Boy Scouts.)

57　(Pg. 17, *Culture Warrior*, by Bill O'Reilly)
58　(Pg. 24, *The Church of Liberalism: Godless*, by Ann Coulter)

And why exactly would the Left Elite in America want to force the Boy Scouts to allow homosexuals to be Scout leaders? I'm sure there are many homosexuals that could be trusted in that situation, but as evidenced by the homosexual Catholic priest molestation scandal, there are also many others who obviously cannot. And how exactly are the Boy Scouts supposed to decide who is who? Trial and error? Wait until the results come in? That would be akin to taking a chance on Bill Clinton as Girl Scout Leader. I mean, maybe you can trust him, and maybe you can't (he's a horn dog for goodness sake) (…wasn't he credibly accused of exposing himself to Paula Jones, assaulting Kathleen Willey, and raping Juanita Broderick?), but who would be willing to take that risk? And get sued to New Zealand if they were wrong? Exactly! Well the same is true of the wisdom and common sense demonstrated by the Boy Scouts. But that is not good enough for the social engineering radicals at the United Way, and the America-and-Christian-haters at the ACLU. The former pulled funding and the latter sued. (But then again, who hasn't the ACLU sued?)

And with that in mind think of all those parents who had entrusted the Boy Scouts with their children's well-being. And think of what it would be like for the Boy Scouts to have to come to one of those boy's parents and tell them that their child was molested by one of their *homosexual* scout leaders. Do you think those parents at that point in time would give a damn about how the Boy Scouts were pressured by the politically correct Left Elite in our media, and by the scum at the ACLU, and by the despicable loons who have taken over the United Way, into succumbing and allowing a homosexual to be that Scout leader? Somehow I don't think so. Instead, those parents would be hiring the nearest ACLU lawyer to sue the Boy Scouts of America for every last asset they possessed. Thank God the Boy Scouts did have the guts to stand up against the putrid forces that lined up against them. In fact, when you think about it, maybe that's exactly what the ACLU and their partners-in-crime wanted, the destruction of the Boy Scouts of America. The Boy Scouts, after all, still believe in moral absolutes and, yikes, God! And they have not surrendered those beliefs for the immorality of the Democrat-Left, so they do make a rather inviting target, don't they?

And then there's "Miranda Right's" which were somehow mysteriously found to have lain hidden within our Constitution for two centuries before they were "discovered" in 1966 by some judges with overactive imaginations. What this has resulted in is the throwing out of court of volumes of legitimate evidence that would have otherwise led to the conviction of countless guilty criminals, and their subsequently being put behind bars. Instead decent Americans for four and a half decades now have had their children kidnapped, molested, raped and murdered; their houses broken into and burglarized; their families and friends violated in a thousand different and disgusting ways; their cars hijacked; their wives assaulted, molested, raped and murdered; all because a horde of heinous criminals were allowed to walk scot-free because of one stupid technicality or another that had absolutely nothing whatsoever to do with our Constitutions allowance for them to receive a fair trial. (Ironically, many must enjoy this victimization because they continue to go out and vote for it, religiously, every two years.) It's funny how when morals are dredged up from the sewer beneath us that suddenly we find ourselves bending over backwards for the imaginary rights of heinous criminals, while at the same time spitting in the face and trampling all over the actual rights of all of their many victims. Yes We Can! (No, actually, you can't.)

Space does not allow us in one short section here to do justice to the number of crimes committed within the court rooms of this country, in the name of justice, by the moral relativists of the Democrat-Left who control those court rooms. (In truth an entire 30 volume set to rival the Encyclopedia Britannica could not list them all.) But here are just two more...

"Jonathan Levin, a beloved English teacher at a Bronx high school," was brutally tortured (for his ATM card PIN number) and murdered by two of his students, Corey Arthur and Montoun Hart, who then withdrew a whopping $800, the spoils of their disgusting crime. There was an overwhelming amount of evidence: an 11 page confession in which Hart "gave details of the crime that only someone who had been present would know;" Hart was identified "making a withdrawal from the ATM at the relevant time;" Arthur's "fingerprints were found on the duct tape used to bind Levin to a chair (which Arthur admitted doing); and his girlfriend testified that he had confessed the killing to her." An open and shut case, you would think. Arthur was found guilty but only of the lesser charge of second-degree murder which made him eligible for parole. However, "in the face of overwhelming evidence, Hart was found *not guilty* and *freed.*"[59] Hart got off because...

> ..."The jury said it was the fact that he looked "wasted" in a picture they saw of him after his six-hour interrogation by the police. In a Herculean intellectual epiphany, they determined he must have been drunk or high when he confessed and therefore—*voila!*— his confession didn't count.
>
> "Welcome to a culture where right and wrong have taken such a beating they're no longer recognizable.
>
> ..."Carol Levin, Jon's mother, confessed to a reporter for the *New York Post* that she thought she was going to vomit in the courtroom as Hart, upon hearing "Not guilty," jumped up and shouted, "Ha! Yes! Thank you!" to the jury."[60] [Some small amount of justice might have been rendered if Mrs. Levin had at least been allowed to throw-up on the jury.]

Hart is now free to move about...

> ..."among people many of whom ...probably have ATM cards and remain ignorant of the killer who lives among them, placing them, and their children, at a risk they cannot even fathom.
>
> "The depravity of this story comes not only from Arthur and Hart, but also from a jury that could not, or would not, distinguish right from wrong."[61]

Reminds one of another case involving a certain Orenthal James Simpson where another brain-dead troop of jurors received their 30 pieces of silver... er...thanks for another "Herculean intellectual epiphany." (Which, let's face it, has little to do with either Hercules, intellect or epiphanies and a whole lot to do with a vulgar expression that rhymes with shit-for-brains. With all due respect, of course. ...The head-up-your-ass end result of growing up absurd in a left-wing-indoctrination-education system that effectively removed their ability to actually pass judgment.) (Who are *you* to judge?! Who are *you* to say what is right and what is wrong?!)

---

59   (Pgs. 9-11, *The Death of Right and Wrong*, by Tammy Bruce)
60   (Pg. 11, *The Death of Right and Wrong*, by Tammy Bruce)
61   (Pg. 11, *The Death of Right and Wrong*, by Tammy Bruce)

And then there is the pathetic narrative of former Black Panther Mumia Abu-Jamal who was found guilty of slaying Philadelphia police officer Daniel Faulkner, who died…

…"from five bullet wounds, one of which was to his back. Three witnesses specifically identified Abu-Jamal as the man who fired all the shots at Faulkner and testified that once Faulkner was down, Abu-Jamal stood over him and unloaded more shots directly into his groin and head.

…"With overwhelming evidence against him and because of the special circumstances of killing a police officer on duty, during the penalty phase the jury of ten whites and two blacks deliberated for less than two hours and came back with a sentence of death."[62]

The crime speaks for itself, but the circus that followed was even more revolting. Shortly after this violent, disturbed and dangerous criminal was put on Death Row where he belonged, he became…

…"a *cause célèbre* for the Left, a martyred idealist, if you will.

"The drumbeat to "Free Mumia" began almost immediately after his sentencing. By 1994 it was a favorite slogan for fashionable leftists. With the assistance of international television, the Mumia craze swept the world."[63]

Democrat-Left celebrities rushed to the cause: Ed Asner, Paul Newman, Susan Sarandon and Ossie Davis among them. Those idiots couldn't have cared less about the stone-cold guilt of their hero-killer, nor could they have cared less for the pain and anguish their ill-conceived "cause" was causing Daniel Faulkner's widow. What would be easy for a normal human being to grasp was beyond the ability of their simple minds to comprehend.

"In 2000, the city of Paris, France, in all its anti-American socialistic glory, made Abu-Jamal an honorary citizen… There have been protests supporting Abu-Jamal from Japan to South Africa; "benefit" rock concerts have even been held to raise money for him. …Amnesty international joined the feast, citing "a pattern of events that compromised Abu-Jamal's right to a fair trial."

"Pattern of events? Spare me. The only pattern here was Abu-Jamal pulling the trigger of his .38 five times in order to murder Daniel Faulkner. As for any sign of repentance, after he heard his sentence, Abu-Jamal screamed, "Judge, you have just sentenced yourself to die." With several deputies pulling him out of the chaotic courtroom, his final words were, "You have just convicted yourself, and sentenced yourself to death…" Meet the Left's Ideal Man."[64]

One especially *deep-thinking* moron, the journalist Stuart Taylor…

…"at lease has the guts to weave the obviousness of Abu-Jamal's guilt into his support of him. How does he manage this? As the *New York Times* reported Taylor's artful but morally inane spin, he "speculates that some facts suggest the defendant, found wounded at the death scene with his legally registered gun lying nearby, might indeed have shot the policeman [Ya think? Maybe?!], but in an unplanned confrontation possibly involving elements of provocation and self-defense. He might, in other words, be neither guilty nor innocent."

"Wow! Neither guilty nor innocent! …That's how the liberals would have our world be. No judgment, no conclusions, no reality, no rules, no personal responsibility. No guilt or innocence. The death of right and wrong.

---

62   (Pg. 13, *The Death of Right and Wrong*, by Tammy Bruce)
63   (Pg. 13, *The Death of Right and Wrong*, by Tammy Bruce)
64   (Pg. 14, *The Death of Right and Wrong*, by Tammy Bruce)

"I can't dismiss these liberals as simply confused or stupid. No, I believe the leaders of the Free Mumia campaign, and especially the Black Elite, know Abu-Jamaal is guilty. In fact, that's *their* crime. They know this and they *embrace* it. They not only do not care, they *want* this type of man to be their people's heroes. For blacks, indeed for all of us, this is the ultimate betrayal of our communities."[65]

Tammy Bruce goes on to explain how...

..."During the 2000 Republican and Democratic national conventions, over 3,000 people marched in each convention city in support of Abu-Jamal. Gay-rights and animal-rights activists, feminist, and Hollywood celebrities all poured into the streets of Los Angeles and Philadelphia that year demanding that Abu-Jamal's death sentence be overturned and that he be given a new trial."[66]

But the stench that those folks put off was still outdone by the offensive odor of the students over at Antioch College in Ohio, who actually invited Abu-Jamal to deliver their commencement address. (But in all fairness they had an excuse: Hitler was dead, Pol Pot was unavailable, Stalin didn't return their calls, Jeffrey Dahmer had already committed to speak at Columbia, Mao was sick and Che was too busy hacking up innocent Cubans with a machete while signing t-shirts for American university students and professors with his free hand. So let's give 'em a break.)

Babu's historic address...

..."was recorded over the phone and played for the graduates, faculty, and parents. Besides being overwhelmingly offensive, this represents a much more serious problem. It demonstrates the Left's agenda of infecting young people specifically with a chaotic disregard for life and responsibility. After all, colleges invite people they want their students to *emulate* to deliver the commencement address. Abu-Jamal knows this. Here's part of what he, the Admired One, had to say... that day: "Think of the lives of those people you admire. Show your admiration for them by becoming them."

"Isn't that comforting? The Left is working to create a nation full of Mumia Abu-Jamal's.

..."And so Daniel Faulkner's widow, Maureen, has to face her husband's killer as he writes his Internet column and delivers college commencement speeches, and as he is celebrated on T-shirts and in the media and is compared to Martin Luther King Jr. and Nelson Mandela. She and Danny were newlyweds when he was murdered."[67]

"Free Mumia" is no less revolting than if these cretins were running around with "Free Dahmer" T-shirts (Jeffrey Dahmer was a serial killer and cannibal but unfortunately for him he didn't shoot a cop or eat a conservative, so he didn't rate being sprung by Susan Sarandon and the brain-dead Left); or "Free McVeigh" (Timothy McVeigh, the Oklahoma City bomber, slaughtered 168 people but he wasn't black and, again, he didn't specifically target cops so he also was out of luck).

"At one of the many rallies for Abu-Jamal, Ed Asner made this pronouncement: "We must fight the establishment... This fight is for the nation's soul. Mumia must not die." "[68]

No, Eddie, you pathetic lobotomite, we must fight the excrement in your head and in the poisoned minds of your Left-elite, Democrat friends. That's #1. #2, Mumia <u>must</u> die, for justice's sake. Ask Mrs. Faulkner. And #3, you and your self-

---

65   (Pgs. 14-15, *The Death of Right and Wrong*, by Tammy Bruce)

66   (Pg. 15, *The Death of Right and Wrong*, by Tammy Bruce)

67   (Pgs. 15-16, *The Death of Right and Wrong*, by Tammy Bruce)

68   (Pg. 17, *The Death of Right and Wrong*, by Tammy Bruce)

righteous, utterly mentally incompetent friends are turning "the nation's soul" into vomit. You reveal your character, or total lack thereof, by wasting countless hours trying to free some disgusting killer, while at the same time spitting in the face of a poor, grieving widow. And you and all of your like-minded, liberal friends vote Democrat, religiously, of course.

The Entertainment Industry

"And these words which I command you today shall be in your heart. You shall teach them diligently to your children, and shall talk of them when you sit in your house, when you walk by the way, when you lie down, and when you rise up." (Deuteronomy 6:6-7)

Compare that ancient command to fill one's mind and heart daily with the words of wisdom and instruction, to what both we and our children are being served constantly in America today by the Democrat-Left-controlled entertainment industry... "Be careful little eyes what you see..."

"They [the networks] have not only abandoned my values, they now have sunk to the sewer level, dispensing the foulest of smells that resemble the garbage I take to the curb twice a week."[69]

(But garbage that nevertheless is still critically acclaimed by the Left. Shocker.)

Michael Medved in *Hollywood VS America* speaks of the movie industries "preference for the perverse." In 1990 he went to the screening of what turned out to be a "putrid, pointless, and pretentious piece of filth"—Peter Greenaway's much heralded epic, *The Cook, the Thief, His Wife and Her Lover*. Its female publicist encouraged Medved to take his wife along, "I really think your wife will enjoy it,"[70] she said.

In the opening scene...

..."a group of foppishly dressed thugs tear the clothes off a struggling, terrified victim in order to smear his naked body with excrement."

[Here we go with the s**t again! What exactly is the secret to the Democrat-Left's love affair with excrement?] [...Their minds are full of it?]

..."They force filth into his mouth and rub it in his eyes, then pin him to the ground while the leader of the band proceeds to urinate, gleefully, all over him.

"The "fun" proceeds in much the same spirit, for two all-but-unbearable hours. We see sex in a toilet stall, deep kisses and tender embraces administered to a bloody and mutilated cadaver, a woman whose cheek is pierced with a fork, a shrieking and weeping nine-year-old boy whose navel is hideously carved from his body, a restaurant patron whose face is scalded by a tureen of vomit-colored soup, and an edifying vision of two naked, middle-aged lovers writhing ecstatically in the back of a truck filled with rotting, maggot-infested garbage. The grand finale of the film shows the main character slicing off—and swallowing—a piece of carefully seasoned, elegantly brazed human corpse in the most graphic scene of cannibalism ever portrayed in motion pictures. There is, in short, unrelieved ugliness, horror, and depravity at every turn.

"Naturally, the critics loved it. ... many of [Medved's] critical colleagues enthusiastically applauded this unspeakable film."[71]

---

69   (Columnist Cal Thomas)
70   (Pgs. 19-20, *Hollywood VS America,* by Michael Medved)
71   (Pg. 19, *Hollywood VS America,* by Michael Medved)

Then there is the movie *Cape Fear* which has a scene…

…"in which one of our most distinguished actors (Robert De Niro) bites off the cheek of his victim while he holds her pinned to a bed, then spits the wedge of flesh contemptuously back toward what remains of the young woman's bloody, mutilated face.

And *The Silence of the Lambs* which brings…

…"to life two serial killers: one of whom eats, and the other of whom skins, his victims. [Movies like these routinely receive] innumerable honors as examples of the highest achievements to which today's movie industry can aspire."[72]

And, mercifully, we have our final nominee, *Closetland*, in the category of Best Romantic Comedy:

"The filmmakers apparently intended some sort of searing parable about the universal oppression of women, and therefore [used that lame excuse to show] the male "interrogator" mercilessly and graphically torturing his female victim. He forces her to drink his urine, rips her toenails out with pliers, handcuffs her to a bed, spits a half-chewed clove of garlic into her mouth, administers electric shocks to her genitals, and penetrates her anus with a red-hot metal poker as she howls in agony. Perhaps the most amazing aspect of this "challenging" (and critically acclaimed) drama was the presence of populist filmmaker Ron Howard's name on the credits as co-executive producer. Perhaps he felt a perverse need to demonstrate to the public how far he had traveled from Mayberry, RFD."[73]

When we were children, there was always one seriously disturbed kid in the neighborhood who would take glee in pouring gasoline on a stray cat and setting it ablaze for the unparalleled entertainment of watching it run around and meow to death; or using a frog's legs as a wishbone. What a great laugh he would have! Now these few psychotics have grown up, gotten all together, and taken over an entire industry. And they now have the poisoned souls of a generation of our youth to amuse them in the place of frogs and cats.

Greg Gutfeld, on Fox News' *The Five*, had these thoughts concerning the folks who control our movie industry…

"Oscar voters are 94% white, and 77% male. So, an industry that sees America as racist is as diverse as a David Duke rally. …[and] the median age is 62… Minorities aren't the only outsiders, here's some folks that you never see in the movies: An American soldier who's not a psycho. A Christian not portrayed as a wild-eyed nut. A corporate head who isn't corrupt. An Italian who's not a mobster. A community activist who's really just a protestor living off the government. A journalist who's a lefty propagandist. An academic who's the same. All of these represent reality far more than movies because movies are now defined by the fake edginess of their attitudes, and how cleverly you can dis America. This matters. Hollywood is how America talks to the world. Why put this bunch of coddled geezers in charge of that? Thanks to the relentless drone since the 1960s it's no wonder the world hates us. If America really reflected what's in our movies, wouldn't you? But these films don't reflect us at all. They reflect an America existing in the Viagra-addled minds of Starbucks socialists, who hate our country, [its] values and themselves."[74]

Under the guise of fighting for "freedom of expression," our entertainment industry has justified spewing out its garbage because they can emphasize its…

---

72 (Pg. 24, *Hollywood VS America,* by Michael Medved)
73 (Pg. 28, *Hollywood VS America,* by Michael Medved)
74 (Greg Gutfeld, on Fox News The Five, 02/24/2012)

...“form over [its] content. The prevailing notion is that a piece of work must be judged on some higher standard of excellence, some objective measure of technical brilliance, rather than an evaluation of the attitudes it conveys.

“According to this line of reasoning, a hit song that glorifies gang rape and the genital mutilation of women still deserves praise for its “infectious beat” and “vivid imagery.” Showing a human head exploding onscreen is also considered admirable—so long as the brains are splattered in artful slow-motion, and the special effects are chillingly realistic. By the same token, critics wax rhapsodic about *The Cook, the Thief, His Wife and Her Lover*, because its images of necrophilia, cannibalism, and child abuse are presented with such zest and conviction.”[75]

These are people who embrace evil as good and good as evil. They are like children who never were disciplined, and they possess undisciplined minds (think Charlie Sheen) that are *free* to think whatever they want, without the normal constraints of common sense and decency, and to spew whatever filth they think is “cool,” “hip” and cutting edge enough to gain the approval of the other undisciplined minds that surround them in their fallen industry. Interestingly, this *freedom* from the limitations of traditional morality hasn’t led them into higher ground or greater artistic creativity but has rather sucked them down to the depths of depravity while at the same time allowing them to see themselves as “morally superior” to the rest of us poor idiots who refuse to join in their celebration of evil in all of its forms. There was a time not so long ago in this country when men of their caliber would have qualified for a job in a local strip club’s basement laundry, where they could sneak a few sniffs of the dancers G-strings before tossing them in the machine. Today, thanks to the destruction of a nation’s moral fiber by the relativists, they control the bulk of our entertainment industry.

“In response to [the public’s] anguished expressions of disgust, the defenders of the Hollywood status quo invariably raise the dreaded specter of censorship. They justify the current propensity for repellent material by intoning solemn platitudes about the need to protect “diversity” and “freedom of expression.”

“In one typically sweeping statement, the huge Time-Warner conglomerate recently attempted to disguise an example of its own acutely irresponsible corporate behavior in the cloak of constitutional high-mindedness. “It is vital that we stand by our commitment to the free expression of ideas for all our authors, journalists, recording artists, screenwriters, actors and directors,” the company nobly proclaimed in an official pronouncement of June 11, 1992. Several law-enforcement officials and organizations had bitterly criticized the release of a song called “Cop Killer” in which rap star Ice-T unequivocally encouraged fans to “bust some shots off” and “dust some cops off.” In response, Time Warner insisted that its decision to move ahead to promote and distribute the song “is not a matter of profits, it is a matter of principle... [Sure, and my ass is the third reincarnation of Buddha] We believe this commitment is crucial to a democratic society, where the full range of opinion and thought—whether we agree with it or not—must be able to find an outlet.””[76]

What is *actually* crucial to a democratic society is where morally bankrupt skanks like the corporate heads at Time-Warner and their moronic charge Ice-T are unable to find *any* outlet from which to shower their excrement into the naïve minds of America’s most vulnerable: her children. But as long as our educational system, media and entertainment industry continue to be dominated by the godless

75   (Pg. 24, *Hollywood VS America,* by Michael Medved)
76   (Pg. 31, *Hollywood VS America,* by Michael Medved)

Democrat-Left, we are going to get just that. They have thrown off moral absolutes, have freed themselves from the idea of moral constraints, and have descended into the cesspool of anything goes, no matter how putrid, vile, violent or debased. (Putrid, vile, violent or debased artistic expressions that attack the Democrat-Left and their agenda are, of course, still completely prohibited.)

And finally we have the case of the underhanded treatment of Miss California, Carrie Prejean, in the Miss USA pageant (April 2009). She had the audacity to agree with the majority decision of the voters on Proposition 8 in California where gay "marriage" was rejected. Answering a loaded question from pageant judge Perez Hilton, a radical homosexual, she stated her conviction that marriage should be between one man and one woman. It very likely cost her the crown. Afterwards Hilton called Miss Prejean a "dumb b*tch" because her answer rejected his militant homosexual campaign to despoil the sacred and ancient institution of marriage, under the guise of *equal rights*. After that, on state-run MSNBC (which, according to the Obama administration is not a legitimate news organization, being merely a mouthpiece of the Democrats),[77] he cowardly retracted his apology by adding that what he really was thinking of calling her was the "c-word." (c*nt) Meanwhile, the host of the program, Norah O'Donnell, said "nothing to distance herself from the misogynist attack."[78]

Educating Perez… You see, here are the rules: it's ok to be a disgusting, hate-filled, ignorant bigot as long as you're a homosexual attacking a Christian. Then you can proceed boldly, with impunity and no fear of reprisal from your hypocrite buds who run the Nazi networks. Indeed, just imagine the infernal outrage there would have been if a lesbian beauty contestant for Miss USA had come out and answered that same question by saying that she was in favor of gay marriage, and because of that answer she came in second. And then imagine that it was a Christian pageant judge who asked her the question, who followed up later by calling her a stupid b*tch, and then offered an "apology" by saying that what he really was thinking of was the "f-word." (f*ggot) Our left-wing media and entertainment industry would have gone insane with rage, loudly demanding the removal and banishment of the offender, and continuing 24/7 until their demands were met. But conversely the guilty swine, Perez Hilton, is welcomed with open arms on Democrat-Left networks everywhere.

An e-mailer to the O'Reilly Factor in the midst of the controversy said this: "I look at the way Miss California is being treated and I think: Why isn't this a hate crime?" Answer: Because she's a *Christian*. Besides, hate crime legislation is just a way for the Democrat-Left to criminalize their ideological opponents, silly. That is the hidden intent. If they actually cared about "hate" they would look in their own backyard, inside of their own minds and hearts, and then, just for starters, deal with one Perez Hilton.

And keep in mind that these intolerant, hate-filled bigots are now delirious because their friends, their comrades, their soul-mates have taken over our federal government (2008 election). The morally bankrupt, Secular Progressive leaders in this country that delight in our descent into the morals of hell are the same ones who

---

77   (I know, it was Fox News that he falsely accused of being that, because unlike such Obama-arse-licking lapdogs like Keith Olbermann, Chris Matthews and Ed Schultz on the Democrat networks, Fox refused to carry his water. We just wanted to make sure you were paying attention.)
78   (Michelle Malkin)

for some reason were absolutely mortified over the presidency of George W. Bush (and lied about it constantly through their media and their entertainment industry), and absolutely orgasmic over the election of Barack Obama and the Democrats. And yet millions of uninformed, misinformed and bamboozled Americans, who are directly opposed to most of the values and policies of the Democrat-Left,[79] still nevertheless marched out and voted them into power in 2006 and again in 2008. The disconnect is astounding.

The Music Industry

"O sing unto the Lord a new song; for he has done marvelous things. I will sing to the Lord as long as I live; I will sing praise to my God while I have my being. Praise the Lord. Praise God in his sanctuary. Praise him in his mighty firmament! Praise him for his mighty acts; Praise him according to his excellent greatness! Praise him with the sound of the trumpet; Praise him with the lute and harp! Praise him with the cymbal and dance; Praise him with stringed instruments and flutes! …Let everything that has breath praise the Lord. Praise the Lord!" (Psalms 98:1, 104:33, 150:1-6)

Now contrast that with what you're about to hear in this segment… "Be careful little ears what you hear…" The following is from Robert Bork's[80] book, *Slouching Towards Gomorrah*:

"The music industry, Michael Bywater writes as part of an extended piece of masterful vituperation, "has somehow reduced humanity's greatest achievement— a near universal language of pure transcendence— into a knuckle-dragging sub-pidgin of grunts and snarls, capable of fully expressing only the more pointless forms of violence and the more brutal forms of sex.""[81]

Bork is referring to a particularly delightful form of hip-hop known as "Gangsta rap," which Wikipedia describes as…

…"promoting homophobia, violence, profanity, promiscuity, misogyny, rape, street gangs, drive-by shootings, vandalism, thievery, drug dealing, alcohol abuse, substance abuse and materialism. Some commentators (for example, Spike Lee in his satirical film *Bamboozled*) have criticized it as analogous to black minstrel shows and blackface performance, in which performers – both black and white – were made up to look African American, and acted in a stereotypically uncultured and ignorant manner for the entertainment of audiences."[82]

Like the previous cases of the two s**t-eating artists who merely are representative of a much broader moral-relativity sickness ailing our culture, there is much in the way of disgusting examples of Gangsta rap garbage that we could cite here to make our point, and to offend your sensibilities. But, mercifully, we will use just two. First, though, let us share a few verses of what we like to call the *Quintessential Gangsta Rap Song*…

---

79 (See "Voting Democrat" in section "8- Take over the government" of chapter 13)

80 (You know, the same Robert Bork who Ted Kennedy—the *conscience* of the Senate" and equally eminent lifeguard—unfairly and unconscionably slimed merely because he was neither a rabid abortionist nor was he willing to distort the Constitution, like Teddy, and therefore couldn't be counted on to override the clear intentions of the Framers and the will of the people from the bench. That Robert Bork.)

81 (Pg. 124, *Slouching Towards Gomorrah: Modern Liberalism and American Decline*, by Robert H. Bork)

82 (From *Wikipedia*, the free encyclopedia)

Nigga, nigga, nigga, nigga, bitch bitch ho
Nigga, nigga, nigga, nigga, cop killa ho
Nigga, nigga, nigga, nigga, bust-a-vagina ho
Nigga, nigga, nigga, nigga…

Well, you get the point. Here are our two examples of what happens to music when record producers, their *artists* and their audience are taught that there is no such thing as absolutes of right and wrong, and that therefore they are their own gods and can think, do and "sing" whatever the hell they want:

The first dropping from the rat's bottom comes from a group called Wu-Tang Clan Method Man, one of the more linguistically accomplished of the blackface minstrel artists:

"Yeahhh, torture motherf***er what? Torture nigga what? What? I'll f***in. I'll f***in tie you to a f***in bedpost with you're a** cheeks spread out and s**t. Right? Put a hanger on a f***in stove and let that s**t sit there for like a half hour. Take it off and stick it in you're a** slow like. Tssssssss Yeah, I'll f***in. Yeah I'll f***in lay your nuts on a f***in dresser. Just your nuts layin on a f***in dresser. And bang them s**ts with a spiked fuckin bat. Ooooohhhh. Whassup? BLAOWWW!! I'll f***in. I'll f***in pull your f***in tongue out your f***in mouth and stab the s**t with a rusty screwdriver, BLAOWW!!"

Now these excrement-headed morons obviously lack the mental capacity of a three year old. Orangutan. But the point here is not to disparage apes. It is rather that these mentally rotting crooners get great reviews for their "creativity" and "lyrical prowess" from equally excrement-headed critics of the industry in our Democrat-Left, lame-stream entertainment industry and media. And this is what many kids listen to, thanks to the Left Elite, who find nothing wrong with this at all, because the very concept of right and wrong is rather confusing to them.

And then there is Eminem, whose newest waste product goes like this, "and I'll invite Sarah Palin out to dinner, then nail her." This filth is celebrated with glee by the malignant cancer that is the Left, but if we just substituted the name Michelle Obama—"and I'll invite Michelle Obama out to dinner, then nail her"— in place of Sarah Palin their celebration would immediately turn to a cacophony of self-righteous indignation, the volume of which would trigger world-wide upheaval: avalanches, landslides, earthquakes, tsunamis and irreversible hearing-loss. Rank hypocrisy doesn't even begin to describe them. Putrid hypocrisy still falls way short. (I personally prefer excrement-headed morons but we don't want to wear that one out.)

But that is not the Eminem example we had in mind. That was just to whet your appetite. Here are a few choice verses from his critically acclaimed, *Kill You*, where the Oedipus complex meets a vulgar and unsound mind:

"Shut up slut, you're causin too much chaos. Just bend over and take it like a slut, okay Ma?"

Yes, if you can actual wrap your mind around it, the poor, pathetic "artistically acclaimed" bastard is actually singing about raping his own mother. He continues his lyrical bowel movement…

"Oh, now he's raping his own mother, abusing a whore, snorting coke, and we gave him the Rolling Stone cover?"

Incredibly, in his line of work you get street cred' if you want to f*** your own mother! The fact that he calls her a whore to boot, we're not going to touch…

"You ___ damn right BITCH, and now it's too late. I'm triple platinum and tragedies happen in two states. I invented violence, you vile venomous volatile bitches."

OOOH! <u>Vile</u>, <u>venomous</u>, <u>volatile</u>! What an incredible alliteration! What a *genius*! (Takes the expression "a pompous asshole completely full of himself" to a whole new level.)

"Texas Chainsaw, left his brains all danglin from his neck, while his head barely hangs on. Blood, guts, guns, cuts, Knives, lives, wives, nuns, sluts."

Wow! What a poet! ..."*He's so witty! Oh so witty! So witty n pretty n gay! And he pity's, any girl who isn't him today!*" (What's the Yiddish word for homosexual?[83]) But that's OK, there is justice in the universe, thanks to a little stunt at a recent award show Eminem will now forever be known as the stupid little weenie that let some guy shove his scrotum in his face on National TV; or as Dennis Miller put it, "Some like their M&M's plain. Some like 'em with nuts." I hope the Palins enjoyed a well-deserved laugh.

The lyrics in the above song would actually be hilarious if they were found scrawled on the cell wall of the Hunchback of Notre Dame, or Mumia Abu Jamaal. But these are the putrid ranting's and perverse imaginations of a superstar of the Left Elite. What Bernard Goldberg, in his book *110 People Who are Screwing up America,* calls "moronic, disgusting, inane, ignorant, and soul-deadening, not to mention—with its murderous lyrics that cheapen human life and glamorize perversity—dangerous and destructive;"[84] the stunningly ignorant *New York Times* columnist Frank Rich "offers nothing less than an orgiastic love feast ... 'Any listener with open ears and some affinity for the musical vocabulary of hip-hop can easily become hooked on his [Eminem's] music.'"[85] Sure, Frank, and any art collector with a stuffed up nose and some affinity for the creative expressions of the San Francisco Art Institute can easily become hooked on eating sh*t. In fact, I don't see how an avant-garde intellectual like you, Frank, can't help getting hooked on both. Bon appétit.

""In the end there will be a price to be paid for this," says Herb London, the conservative professor and critic, "the price one always pays for ignoring evil. Some of the best potential minds will be decimated. Culture will be assaulted beyond repair and the nation will be undermined from within.""[86]

There is no intelligent argument in favor of this stench being allowed on our air waves or in our music stores: not "free speech" and not "artistic expression." The only argument is "we don't think there is any such thing as ultimate right and wrong and as far as we're concerned letting these repulsive cretins "express" themselves, and peddling their excrement to children for our profit, is just fine and dandy." Incredibly, *New York Times* columnist Maureen Dowd comes right out and reveals the kind of tasteless attraction that her and some of her fellow female Manhattan liberals feel toward the aforementioned Eminem and his raw sewage:

""A gaggle of my girlfriends are surreptitiously smitten with Eminem. [Ooh! Golly Gee Willakers!] They buy his posters on eBay. They play him on their Walkmen at the gym.... It doesn't feel quite so rebellious to like The Most Evil Rapper Alive.... if your mom is rapping along when he describes how he'd liked to rape and kill his mom." ["Oh, turn him up, daughter! I cain't keep my fingers from betwixt my thighs...."] Dowd's extraordinary irresponsibility in saying that women *enjoy* and essentially fall in love with a man chanting about raping, torturing, and murdering them is repulsive and sick."[87]

---

83   (Yea, what a fagala)
84   (Pgs. 25-26, *110 People Who are Screwing up America*, by Bernard Goldberg)
85   (Pg. 256, *The Death of Right and Wrong*, by Tammy Bruce)
86   (Pg. 28, *110 People Who are Screwing up America*, by Bernard Goldberg)
87   (Pg. 259, *The Death of Right and Wrong*, by Tammy Bruce)

Yes, but what is repulsive and sick when you can earn so much street cred' with your equally brain-dead and sexually disgusting, far-left, Upper West Side, Democrat-voting female pals? This brings to mind an incident in the past when a noted liberal journalist wrote of how her mouth pined in jealousy that *it* wasn't slurping away in the Oval Office instead of Ms. Lewinsky's. (Speaking of which, Talking Points would like to know how much money Ms. Lewinsky has spent on mouth wash over the years.) My how proud their husbands must be! (Where can I get a wife with such character, dignity, sexual discretion, basic moral integrity *and* political charm to boot?[88])

Seriously though, I mean I know we haven't established the existence of Satan yet, but let's pretend now for just a moment that he does. Can you imagine Satan himself, pure evil incarnate, coming up with any lyrics more putrid, more mindless, more moronic than the "torture nigga what" jackass above? And let's say that there is a place called hell, pretend that now for just another moment if you will, can you comprehend any place filthier than what the Left has created here on this earth. I mean what, instead of mutilating 50 million babies to death in four decades, how would hell itself outdo them? 60 million; in three? Instead of having a pile of human excrement sing a song about f***ing and murdering his own "whore" mother, what would hell come up with? One who f***s and murders his ninety year-old grandmother? And here we have New York liberals lusting over this same manure pile and the Left Elite raising him to the level of cultural icon; what could hell do to top that? Have him sleep with barnyard animals live on national TV during primetime and then elect him President? (Actually, with what today's Democrat electorate will march out blindly and vote for, that last one isn't such a stretch...)

Tammy Bruce says this about "the Left's agenda of integrating degenerate and violent rap singers into mainstream culture:"

"This integration has as its goal making the most vile among us into heroes for the next generation, and forcing those who know better into silence lest we be labeled "censors" or enemies of "freedom of expression." Manipulating pop culture is the Left Elite's surefire way of forcing society to bend to their depraved view of themselves and this country. The elevation of rap stars to pop star status is the cultural equivalent of an armed invasion of your home."[89]

It would be unfair not to state here that there are worthy offerings from the world of rap, and that all rap music and lyrics are not disgusting, demeaning, degrading and offensive. (Indeed, far from it, one of my favorite songs is a rap song, "Happy Birthday" by the rapper Piper. ...See chapter 11, Baby Mutilation.) But that we live in a world where any of the filth like the ones quoted above could be published, marketed and sold right out in the mainstream of America with its parents lacking the information or the guts to raise even so much as a whimper in response, is only made possible because of the encroachment into the power centers of our culture that has been made by the malignant, morally-debauched narcissists of the Democrat-Left. As we have stated before, the Secular Progressives, their radical agenda, and their moral relativism are now in control of our media, our educational system and our entertainment industry (and unfortunately now, after 2006 and 2008, also our federal government). And

---

88   (Just go to www.Ivoted4Dumboandneedadate.org)
89   (Pg. 254, *The Death of Right and Wrong*, by Tammy Bruce)

they have used that control to permeate our culture with their progressive kind of morals, the kind which ooze up out of hell. The kind that enjoys listening to a mentally-disturbed, pathetic little bastard boast about raping his own mother. And they have also triumphed in emasculating America's ability to pass judgment on their filth by the institutionalization of their most sacred Commandment: "Who are YOU to say what is right and what is wrong!" (Liberalicus 6:66)

Feminism

"Who can find a virtuous woman? For her worth is far above rubies. The heart of her husband safely trusts in her, so he shall lack nothing of value. She will do him good and not evil all the days of her life." (Proverbs 31:10-12)

Moral relativism opened the door for the rise of the feminist agenda which rapidly increased in its power and influence over the decision makers in our culture during the 1960's and 70's. Whether the "woman's liberation movement" had at one time some noble goals and aspirations, and truly had the best interests of woman and society as a whole in mind is a question we are not dealing with here. But I'm sure that was true of many women involved in it. However, what it has morphed into today has nothing to do with nobility of aspirations, but with the furthering of the far-left agenda at all costs and no matter how many women that agenda steps on, trashes, demeans, imprisons or subjugates in the process. There are hundreds of pertinent cases, but we need look no further than Sarah Palin. When Bill O'Reilly made an innocent and humorous remark concerning the vocal tones of Helen Thomas—an aging, extreme-left, Jew-hating White House reporter—comparing her voice to the Wicked Witch of the West—the Feminists groups came out screaming. But when the left-wing press and entertainment industry trashed Sarah Palin many times over during the 2008 presidential campaign (and continue to do so without let-up to this day); and in far more blatantly dishonest, demeaning and misogynist ways than if they had just poked fun at her voice; and when the aforementioned human-refuse-rapper Eminem sexually abused her verbally in a published song; both of which instances far outweighed any perceived slight to Ms. Thomas and her voice; the silence of the feminist groups was eyebrow-raising, if not expected. And it spoke volumes to the level of their hypocrisy. Like the modern NAALCP,[90] today's feminist movement cares much less about woman and far more about advancing the Democrat-Left's agenda, in addition to "isolating and demonizing men."[91]

Tammy Bruce had this assessment:

"What has been billed as feminism for decades now really hasn't been. It is liberal politics and it masquerades as feminism, saying that it holds the flame for women's rights, but ultimately it really is a disguise to attack anyone who is going to advocate a conservative point of view."[92]

From the outset some of the pronouncements of the feminists were in direct contradiction to the perceived wisdom of the ages, some even flying in the face of common sense. On his *Focus on the Family* radio program, Dr. James Dobson talks about the 1970's, and the release of his best-selling book, *What Wives Wished their Husbands Knew about Women*:

---

90   (*National Association for the Advancement of Left-wing Colored People*, what the NAACP has now become)
91   (Pg. 33, *The Death of Right and Wrong*, by Tammy Bruce)
92   (Tammy Bruce on the O'Reilly Factor, April 8, 2009)

"The women's liberation movement was sweeping through the secular media, and I took on almost every component that was being advanced by the radical feminists at that time. And my book was at one end of the universe and that movement was at the other, and to my delight my book ...sold more than a million copies. ...And it's still available in bookstores today, even though most of the women's liberation agenda has been discredited."

Dr. Dobson goes on to talk about...

..."some of the tenants of the feminist movement ... [of the mid-70s that he took] exception to and was vigorously opposing: ...the most important foolish notion was the idea that males and females are identical except for the ability to bear children. Everybody was saying that at that time, and of course that's been totally discredited now, but it seemed for a while like the whole world was going to buy into that idea. And I opposed it and said so and gave some biological and physiological reasons, and emotional reasons, why I disagreed. Another equally ridiculous belief was that women didn't need men, and they'd be better off without them. Imagine that after 5000 years of humanity! What a revelation. Gloria Steinem wrote that phrase that was quoted in newspapers all across the country, which said "a woman needs a man like a fish needs a bicycle." [Actually, Mrs. Steinem, a woman needs *your asinine declarations* like a fish needs a bicycle.] Can you imagine that? That's what they believed, that's what they taught, and it's what many, many people seemed to accept."[93]

Of course the level of mis-directed anger and ignorance that would allow a supposedly educated and somewhat otherwise intelligent female like Steinem to come up with her "fish on a bicycle" statement has yet to be addressed by her supporters on the Left. Mainly because many of those enlightened individuals think on the same level; as the fish. Dr. Dobson based his values on the absolutes that come from above and even though he was ridiculed and maligned by the Left Elite at the time, it is time that has borne out his ideas as valid and as based on reality, and many of the feminists notions that sprung up from below as truly ridiculous and divorced from the real world.

Yet there is one particular area where without any facts or evidence to back up their progressive claims, radical feminists did succeed in doing incalculable harm over the years to so many of the same women that they purported to be struggling to help. For decades they have been successful in brow beating and deceiving women into outsourcing the raising of their children, or rejecting having them outright, in favor of advancing in a career just like their male counterparts. Under the guise of *equality*. The problem is that many women have found out the hard way that actually becoming a mother, and actually raising their children themselves—as opposed to dropping them off each day to the care of strangers—is exactly what they want to do (because that is how they were hardwired by their Creator), and exactly what gives them the most joy, happiness, satisfaction and fulfillment in life. As opposed to going off to the rat race each day in order to climb the corporate ladder and fight for the almighty dollar like the misinformed radical feminists insisted would. This is because, as most folks with common sense knew all along but which we have now confirmed through scientific studies; men and woman are just hardwired differently. Which doesn't mean one is better than the other, but just means that we are born hardwired to fulfill the different complimentary roles that our Creator (which is not random accidental chance)

---

93   (Dr. James Dobson, *Focus on the Family* radio program)

intended for us. And no amount of feminist indoctrination can overcome that hardwiring. And the reason so many unfortunate women found out "the hard way" is because it was only after a number of decades of buying into this lie and succumbing to the blasting feminist drumbeat of ridicule against the traditional approach to womanhood, that women started to realize they had been deceived, hoodwinked, snookered and bamboozled. But not until countless numbers of them had woken up in their later years, stricken with guilt and overcome with regret, realizing that they had followed a lie, and had foolishly been badgered, cajoled and deceived into wasting their lives; into doing the exact opposite of what would have given them the most happiness, enrichment, satisfaction, delight and fulfillment in life. For all of them, sadly,[94] it was too late. Again, exchanging the moral absolutes that have come down from above, for the "values" and foolish notions that rise up from below, is the culprit.

Militant feminism has been discredited but its ideology still has a free reign over our culture. (Kind of like the theory of evolution.) And it is dominant in our colleges and universities, and throughout our media and our entertainment industry. Bashing men in these venues happens so casually and so often it has come to be expected, but as anyone who has recently taken a shot at women, or feminism itself, knows all too well you had better be prepared to withstand an assault that'd make Genghis Kahn and his Mongol Horde look like a gaggle of poofties, jogging in tights. Just ask Larry Summers, President of Harvard University, who was forced to resign because he had the audacity to do something that has been a near criminal offense in our nation's schools and universities for a number of decades now, he spoke the truth. And worse yet, he spoke the truth about women, or in this case he merely *suggested* a *possible* truth about why women are under-represented at the highest levels of academia in math, science and engineering. He merely suggested that one possible and heretofore overlooked reason (by him and his far-left) for this disparity is a difference in *intrinsic aptitude* between females and their male counterparts. In other words, they're <u>hardwired</u> differently. Woops. You'd of thought he said that "rape was a man's right to choose." It was like someone presented the winner of a high achievement award at a Flat Earth Society banquet with a globe. They went berserk. You see, as we stated earlier, our universities are no longer interested in pursuing, or teaching, the truth. They don't even think it exists. What they <u>are</u> interested in doing is indoctrinating their students in left-wing, progressive, socialist-Marxist ideology. (In case anyone is wondering how one Barack Hussein Obama ended up as a complete, yet highly educated, idiot.) And the doctrines of feminism, no matter how stupid, are at the top of that list.

Here is another example: A number of years ago Katie Couric suggested on national TV that her female guest should castrate her estranged boyfriend, who had slighted her in some way. And not so much as a single Botoxed eyebrow was raised. But can you imagine how apoplectic they would have become if anyone on national TV would have suggested that some ex-boyfriend of Katie Couric's cut

---

94   (And even more sadly is that the Democrat-Left is still able to shove their emotionally-damaging and regrettable ideology down the throats of gullible, naïve and unsuspecting young girls to this day. Because they own our public education/left-wing-indoctrination system, the media, the late-night and Comedy Central clowns, the entertainment industry and countless far-left *woman's studies* departments at universities all across this country. God help our young women)

her breasts off? Or even worse if they had giggled about the idea of an ex-boyfriend slicing off her clitoris? The heavens would have parted and rolled up like a scroll, the earth would have been consumed by fire and Keith Olberman would have torn his garments like the Jewish High Priests of old (after peeing his panties). But it's OK when Couric jokes about male castration, because demonizing men is the accepted pastime of the Left—not only from feminists but supported by the many spineless, neutered males (like Paul Begala, Chris Matthews, Rahm Emmanuel and Jon Stewart) who obediently sniff along behind them sheepishly grasping onto the hems of their skirts, scared to death of disagreeing with, or slighting, them in any way.

And speaking of slicing off clitorises—<u>viewer warning</u>—this next discussion may be a little disturbing: It is a gruesome and highly under-reported fact that right now across this planet there are approximately <u>130 million</u> Muslim women walking around who as a little girl were subjected to an Islamic satanic ritual that actually involves slicing their entire clitoris off. And if that is not brutal, mutilating and vile enough (apparently not), in many cases their parents go a lot further and slice off the little girls minor and major labia as well, leaving behind nothing but the vaginal opening looking like a 2nd anus. This very painful and genital-disease-causing horror is often performed without anesthesia by women who hold the screaming little girls down. It goes on every single day across Africa and the Islamic world, but it is also performed by Muslims in Europe, Indonesia, England and in our own country. We will deal with this further in chapter 8 under the false religion of Islam, but the reason we mention it now is because the silence of the "malignant narcissists" of the feminist Left concerning this horror is revealing. Every year another two million more girls are subjected to this sub-human, life-altering brutality, but our left-wing media and our feminist friends, both the supposed great champions of women's rights, barely give it a mention. You see, blacks from Africa and Muslims are the practitioners of this heinous evil. If whites from Alaska and Christians were the culprits instead then it would be headline news 24/7. Katie Couric would be doing hour-long specials, instead of merely mentioning it in on one of her 60 second *Notebook* segments, and feminists like Kathy Ireland, Bell Hooks and Gloria Steinem would be screaming nonstop until the offenders were arrested and jailed, their children taken away and the practice eradicated. But blacks are a protected minority and Islam a non-Christian religion so let's not make too much of a fuss over it. (And screw the hundreds of millions of genitally-mutilated little girls who get to go through life with a very important part of their womanhood eradicated by mentally disturbed followers of the child-molesting, genocidal-murderer, Mohammed.) (And see chapter 8.)

> "It's mutilation, and every day it's forced on some 6000 little girls. It's female circumcision practiced in Africa, parts of Asia and the Middle East as a rite of passage. It's a savage thing that can cause infection, hemorrhaging, psychological trauma, even death. Worldwide it's been done to about 140 million women. ...It's a ritual steeped in tradition, but this is one tradition that needs to be eradicated. ...Those of us lucky enough to be born in a society without this barbaric ritual should stand up for the girls who are."[95]

Absolutely, Katie, only they're still waiting for you to actually stand up. But then that would require having the guts to expose Islam for the scourge that

---

95   (From the aforementioned "Katie Couric's Notebook: Female Circumcision" CBS News, 02/08/2008)

it actually is, and not the "religion of peace" that you and your Democrat-Left pretend it is. And aren't you really much more concerned about not being labeled by your left-wing friends as someone who is "bigoted against Muslims" (like you falsely accused the majority of Americans who are rightly opposed to the Ground Zero "victory" mosque being shoved down the throats of themselves and the victims of 9-11) than you are with "standing up" for the next generation of clitorally-mutilated little girls? Of course you are.

But it is not just the young female victims of Islamic genital mutilation that the radical feminists and their media don't really care all that much about; just take a look at the harm they and their "values" have done to our own young women. Starting with the *sexual revolution* in the 60's the Left has managed to undo in a few decades what had withstood the test of time for millennium. They have destroyed the sexual morals of the young people of this nation, exchanging the moral absolutes that come down from above with their own values they got from down below (not down *there*, down below their feet). A young woman's sexual purity, her sexual virtue, is one of her greatest gifts. The Democrat-Left elites have succeeded in all but destroying that for far too many of our young women. (To the utter delight of teenage penises everywhere.) They have been in charge of the raising of our children for decades now thanks to their takeover of our educational system and our entertainment industry, and have used that power to turn many of our girls and young woman today into what previous generations, undeterred by a political correctness that protects the cesspool values of the Left, would have called sluts and unpaid whores. That is because the *morals* and the *thoughtful* counsel of the Left doesn't rise much above "if it feels good do it" when it comes to young girls and sex. Their goal appears to be nothing less than the wanton animal fulfillment of their young charges genital urges without regard to any emotional damage, loss of personal dignity, common decency, virtue, health concerns, spreading of infectious and incurable diseases, moral restraint, suicides, unwed pregnancies or of course the slaughter of millions of the products of wanton unrestrained unmarried genital urges: in a word, babies. Their message is like that shoe companies slogan, and it spews out of our entertainment industry and our educational system into the hearts and minds of our youth, rising to the decibel level of a loudspeaker's indignant demand, "JUST DO IT!" "And if you don't JUST DO IT then you are JUST STUPID! You dumb right-wing *Christian*." (That's our educational systems new bad word, *Christian*. F**k you can say all day long and you won't even go to the principal's office—kids do have their *free speech* rights after all—but say words like *Christian* or worse yet, *God*, and you're screwed.) Abstinence, sexual purity, virginity and reputation—all once closely guarded by the vast majority of young women, and for good reason—are also now dirty words, openly mocked by the totalitarian, left-wing clowns who control our culture.

In the past Satan was the unseen tempter when it came to sexual immorality for youth. Now he has the Left Elite in our schools and entertainment industry eagerly assisting him in his cause, and they're not doing it behind the scenes.[96] And as we shall see in chapter 13, his cause is nothing less than the destruction of this nation, by destroying both its morals as well as the family. And he and his Left

---

96 (See the antics of GLSEN, just for one example, under "Education" previously in this chapter)

Elite are well on their way to realizing this dream. (And after the 2006 and 2008 elections, they're delirious.)

That old Liar who entices our youth with sexual immorality promises great satisfaction, fun and happiness, but he delivers none of that; for he is the great deceiver, the hidden evil behind the morals of the Democrat-Left. And all that those who are seduced into following his paths of youthful sexual abandonment receive is dissatisfaction, emptiness, poverty, heartache, incurable genital diseases and even death.

> [For an excellent book on Satan's, and feminisms, Big Lie about the "joys of extra- and pre-marital sex" see *Seduced by Sex: Saved by Love- A Journey out of False Intimacy*, by Jan Kern. I'm surprised they haven't burned it yet.]

And finally we have reached the far limits of contempt and outright hatred for any moral decency whatsoever, and for any pretense for the idea of right and wrong or good and evil, with the case of Andrea Yates, and the feminist-Left's response to it. This unfortunate demonically-possessed individual performed the monstrous deed of drowning her 5 little children in the family's bathtub. Yates confessed that her 7 year old, the last to be slaughtered, asked this heart-rending question, "Mommy, have I done something wrong?" But like the disgusting ACLU and their defense of child-sexual-predator-killers, the Left rushed not to the defense of the 5 little children, screw them, they rushed instead to the defense of their new poster girl, the mentally-deranged Mrs. Yates.

> "Despite (or perhaps because of) its depravity, the Yates case became a perfect gimmick for the Feminists and the Left Elite to use in marketing their special brand of moral relativism to the public... What a coup it would be if the Feminist Elite could persuade the American public to accept the murder of children as something that must be "understood."[97] ...NOW [National Organization of Woman] insisted that Yates murdered her children not of her own volition, but because of the effects of postpartum psychosis. This is a new version of the "Twinkies defense," a term coined in 1979 when a lawyer successfully defended a man accused of a double homicide by saying his client's diet of junk food, primarily Twinkies, was what had caused him to commit murder."[98]

The operative word there being "successfully" defended. And what zoo did that defense attorney go to find those jurors? Twelve Orangutans, for God's sake, hooting and grunting, digging and scratching themselves in a jury box, wouldn't have fallen for that defense. (Makes the OJ jurors look like a troop of veritable Einstein's.)

> "This rallying around postpartum depression, of all things, strongly suggests an anti-childbirth and even anti-child attitude in the modern feminist movement. Certainly not an authentic feminist approach, this is, however, reflective of a movement that desires the complete extirpation of any role for men in women's lives.
>
> "You see, with men come marriage and children (hopefully in that order). And with children many women choose to lead lives not sanctioned by professional feminists. Putting family first—children and husband—is the ultimate blasphemy and must be fought against. That fight includes painting childbirth and the pressures of motherhood as something that drives women mad. Literally.

97　(Of course many would argue that they have already accomplished this with the legalization and proliferation of baby mutilation.) (See chapter 11)

98　(Pg. 62, *The Death of Right and Wrong*, by Tammy Bruce)

...*"Newsweek* columnist Anna Quindlen carried that banner for the morally bankrupt Feminists Elite. In one of her "Last Words" columns, she declared it was the "insidious cult of motherhood" that caused Yates's actions. It is, she wrote, "the hideous sugarcoating of what we are and what we do that leads to false cheer, easy lies and maybe sometimes something much, much worse, almost unimaginable.""[99]

The only *hideous sugarcoating* going on here is the one that covers the indescribable stupidity of any human being, much less a journalist in a position of power, who could somehow twist the casual satanic drowning of five of one's own little children into a cause célèbre for the Left's equally witless assault on men, marriage, motherhood and children. Congratulations, Ms. Quindlen! Because of your success in the far-left, lame-stream print media, you have made it possible for complete imbecile's everywhere to lay claim to some level of social acceptability. ("Hey, I might be a f***ing moron, but look at Quindlen, and she's a journalist!")

"Sally Satel, in an article for the *Wall Street Journal* titled "the Newest Feminist Icon, A Killer Mom," said it best when she wrote: "Andrea Yates is not a symbol of motherhood under duress nor of the embattled state of the American woman. To portray her as such is a cynical move that trivializes a serious mental illness and misinforms women about their risk of committing one of humanity's most unspeakable acts.""[100]

But what a great way for Democrat-Left feminist halfwits to try to scare women away from becoming wives and mothers! Of course this is in many ways just the natural logical progression from abortion *rights*. If it is OK to mutilate a baby to death on one side of the vaginal opening then wasn't it just a matter of time before we "progressed" back to the thinking of the ancient Greeks and other child sacrificers of old where it is OK to throw unwanted babies off of cliffs or burn them alive, and by extension to drown them to death on the other side of that same opening? What's the difference, really? And oh yes, don't forget, *who's to say what is right and what is wrong.*

Baby Mutilation

"They even sacrificed their sons and their daughters to demons, and shed innocent blood, even the blood of their children, whom they sacrificed to the idols of Canaan; and the land was polluted with blood." (Psalms 106:37-38)

Their land is not the only one polluted with innocent blood. Ours is as well, thanks to the success of the malignant narcissists of the Democrat-Left in perverting our own Constitution. We will devote an entire chapter to this abomination later but it wouldn't be right not to mention it in a chapter on moral absolutes, and the lack thereof. But first off, we want to make it clear that we do not condemn, put down or judge any woman or any girl who has had an abortion. We'll talk about this more in chapter 11 but most of them were lied to, deceived and misled by the Left about the reality of what they were doing. Besides, we *all* need to be, and can be, forgiven. But this subject is another great illustration of the moral emptiness of the Democrat-Left and how their self-created *values* actually do ooze up from below. It's not "abortion." No one is coming in to low for a carrier landing and getting waved off, "Abort! Abort!" No, a baby is getting his or her limbs ripped off, his eyes sucked out of their sockets, his head cut off, scissors jammed into the back of her little skull and her brains sucked out, all while they are still alive and

---

99   (Pgs. 63-64, *The Death of Right and Wrong*, by Tammy Bruce)
100   (Pg. 66, *The Death of Right and Wrong*, by Tammy Bruce)

silently scream to their untimely, gruesome and truly torturous death. In what has to be the most disgusting display of near-demonic revelry ever witnessed in America—not since a few sub-human, racist mongrels in the crowd laughed at the heartrending predicament of entire families being auctioned off one by one on the slave blocks, never to see each other again—the Democrats in the Senate during the Clinton administration actually stood up and cheered because they had just barely *won* by one deciding vote the *victory* to continue to jam scissors into the back of the heads of 9 month-old fully-formed babies, and then suck out their brains with a vacuum while the baby is still alive and silently screams to death. (It's called "partial-birth abortion.") To actually cheer that kind of victory is not just value-less, it is sub-human. It is demonic.

Most people are aware of The Ten Commandments; certainly we should assume that most Christians are. But a great many people, including a great many self-described Christians, are not aware of The Eleventh Commandment: THOU SHALL NOT <u>VOTE</u> FOR BABY MUTILATORS! It is actually a part of the Sixth Commandment: THOU SHALL NOT MURDER! But since far too many people can't seem to put two and two together and since they can't seem to understand that if you vote for murder you are supporting murder and therefore you yourself are a murderer and guilty of violating the Sixth Commandment; that it has become necessary to spell it out in The *Eleventh* Commandment:

THOU SHALL NOT <u>VOTE</u> FOR BABY MUTILATORS!
...It's today's handwriting on the wall.

"Blessed are the undefiled in the way, who walk in the law [moral absolutes] of the Lord! Blessed are they who keep his testimonies [moral commandments], who seek him with the whole heart!" (Psalms 119:1-2)

Obamorality and the Audacity of Deceit

"The instant I speak concerning a nation and concerning a kingdom, to pluck up, to pull down, and to destroy it, if that nation against whom I have pronounced, turns from its evil, I will repent of the destruction that I thought to bring upon them." (Jeremiah 18:7-8) (America, you might want to pay a little attention to that verse...)

In this chapter we have documented the moral-relativist Democrat-Left's assault on our values. We have seen what happens when the absolute values that can only originate from a higher source are rejected and are substituted with a version of morality that claims to be "liberating," and that might even appear innocent at first glance, but upon closer inspection looks a whole lot like pure evils version. We have seen what happens when the moral relativists are allowed to slip in under cover of ignorance and take over an entire culture. In this section we are going to take a glimpse at what happens when those same malignant narcissists are foolishly allowed to take over an entire government.

"Honesty is the first casualty when a nation rejects moral absolutes."

In the 2006 Congressional and the 2008 Presidential elections, the Democrats took complete control of our Federal government. And the "malignant narcissists" of the Left Elite were beside themselves with joy. That is because they knew what the deceived voting public obviously did not: that despite the fact that the vast majority of voters do not agree with most of the Left's agenda, they were going to be subjected to it anyway. They were deceived because decades ago the same Left Elite had already taken control of our media, our educational system and our entertainment industry, and through that near absolute control of the dissemination of information in our society they made sure that enough folks remained clueless, uninformed and misinformed, and that they would vote for whom they were told to, the Democrats, no matter what.

In November of 2008 this country elected the first black president and it was truly an historic and inspirational milestone. I too wanted very much to see America accomplish this, but I would have preferred that it not come by way of the election of another far-left Democrat, one committed to the "radical transformation" of this country into something other than it was intended to be. (Obama was, after all, rated the most liberal out of all one hundred United States Senators.) But I did appreciate very much seeing the faces of joy and to feel the warmth in the hearts of millions of our African American brothers and sisters; many of whom had lived a lifetime not believing in their wildest dreams that they would ever see something so unprecedented. And many of whom who had also grown up in the 30s, 40s, 50s and 60s and experienced the degradation of blatant racism firsthand. But even if I could have overlooked his alliance with the radical-far-left, move-on.org, George Soros wing and their ideology that is destroying this country, I was precluded from supporting him because of his intractable position as a rabid, unwavering baby mutilator.[101] Nevertheless, nearly all of America, including most of those who did not vote for him, shared in the pride of the election of the first African-American president of the United States. Many of them also shared in the pride of that historic accomplishment because it was a reflection of the true greatness that is this United States of America.

Conversely though, the election of the wolf in sheep's clothing; the arrogant, angry, extreme-far-left, spoiled, overly-sensitive, used-to-being-coddled, radical Democrat and closet Marxist (although after two years in office I'm not sure how closeted he still is), Barack Hussein Hugo Chavez "ACORN" Obama (his full title), was about the worst thing possible that could have happened to this country at any point in time, and especially at this very vulnerable one. And not just for the people who were compelled to vote for the liberal-lite candidate, John McCain, but for most of the people who voted for Obama as well, but who were also intentionally uninformed about just what they were getting not only with him but with the political Mongol-horde of self-righteous, dangerous elites of the far-left like Nancy Pelosi, Harry Reid, Anthony Weiner and Chuck Schumer who road to increased power along with him—whose values 90% of Americans do not agree with, although half of them still go out and vote for. Go figure. And less than two years into this administration's reign of error, the end results of that vote have been predictable, and alarming.

After watching Obama during the campaign and through the first two years of his administration it has become crystal clear that he is nothing more than another far-left indoctrinate, who as a child was raised as a Muslim (not an American) in Indonesia, mentored in his youth in Hawaii by a communist, indoctrinated into a useful dupe of his Marxist professors at Columbia, and finally discipled by the

---

101   (See chapter 11, "Let every man who names the name of Christ depart from iniquity." 2 Timothy 2:19) And there is nothing more iniquitous than mutilating a baby to death, or voting your support for those who do.)

radical, corrupt, liberal-socialist Democrat thugs that rule Chicago politics. But nevertheless one with charisma and a big, cheesy grin that masks an absence of moral conviction. And that comes in handy when you are wooing an electorate many of whom (in particularly his excitable, youthful voters) can't differentiate between their civic duties, or buying a Honda. And he sure can blather. He can blather with the best of them! And say absolutely nothing whatsoever, but do it authoritatively and with a straight face. And since his ill-informed but excited followers (did Chris Matthews ever get those thrills running up his leg checked out?) can't tell the difference between truth and blather, his never-ending campaign may never lose its support.

There is also no question that he is highly intelligent; but high intelligence coupled with an equally high degree of unfamiliarity with the truth, and a lack of morals, can be a dangerous combination in one in a position of so much power, and with the unwavering loyalty and backing of a state-run, Democrat media. But is Obama dishonest? Karl Rove has an answer:

> "He is a cold, calculating, ambitious politician who feels very comfortable saying one thing on the campaign and doing something entirely different. Whatever was necessary in order to get elected, he was willing to say."[102]

(Translation: Yep, he's a liar.)

He travels around the world trying to impress his far-left friends by taking credit for "stopping torture." But that is a lie. The United States, and the Bush administration, have never tortured anyone. However, he and his Democrat friends torture millions of little innocent babies **to death** every year. He lied about federal funding of abortion in his health-care plan. Every bill produced by his congress so far has it in there. His administration lackeys tried to discredit Fox News by saying they were "not a news organization." That of course, is another lie. (If they had said Fox was not a state-run, Democrat news organization, like NBC, MSNBC, PBS, The New York Times, etc. etc., they would have been telling the truth.) And we could go on, and on. (See chapter 13, Satan) (Right now, May of 2010, he's lying about the Arizona Illegal Immigration Bill.) Indeed, Dick Morris has an entire chapter in his book *Catastrophe* where he "decodes" Barack Obama. He takes what the man says, and then he shows what he really means.[103] But then, this is true of most of the Democrat-Left. The truth has not been one of their closest friends. Lying, doublespeak and blathering, however, has.

And please, no left-wing, Nazi-playbook, Saul Alinsky, false accusations of "racism." This is an accurate assessment of what the man is made of. Besides, I will give my right arm if we can have him replaced with either Dr. Tony Evans, Larry Elder, Supreme Court Justice Clarence Thomas, J.C. Watts, Angela McGlowan, Thomas Sowell, Condoleezza Rice, Les Phillips, Roy Innis, Charles Payne, Alan Keyes, Starr Parker, Alan West, Rev. Jesse Lee Peterson, Michael Steele, Herman Cain or the custodian down at my child's elementary school; African-Americans all, and some even more than 50%. Take your pick. Besides, Obama is not a black-white thing, although the far-left liars will use their Nazi-methodology-of-falsely-accusing-Obama's-disenter's-of-racism and will try to make it that. It's a Left-Right thing. It's a Communism-Capitalism thing.

---

102 (Karl Rove on Fox News *Hannity*, 07/01/2009)
103 (See *How Obama, Congress, and the Special Interests are Transforming... a Slump into a Crash Freedom into Socialism, and a Disaster into a CATASTROPHE ...and How to Fight Back*, by Dick Morris and Eileen McGann)

"Real change actually requires real change. ...What is at stake is the future of this extraordinary experiment in individual human freedom. ...I am not a citizen of the world. I am a citizen of the United States of America. ...No government bureaucrat has the right to take from you the rights that God gave you, and rationing under health care is inevitably limiting your life at the will of a bureaucrat and the manipulation of a politician. ...If we have been endowed by our Creator with certain inalienable rights [and we have] how can a government bureaucrat tell you you don't deserve the best possible medicine, the best possible procedure, the best possible hospital? ...Under the Obama administration, we have fallen back into the utopian fantasies and self-deception of the 1977 Carter administration and the 1993 Clinton administration. ...The great difference between Reagan's rhetorical skills and President Obama's rhetorical skills are that Reagan used his rhetorical skills to shine light on truths and fundamental facts. Obama uses his rhetorical skills to hide from fundamental facts. This president uses rhetoric to hide from reality. I've learned, very sadly, not to pay attention to his words because they are in fact, not good indicators of what he is going to do."[104]

Indeed the meaningless slogan that Obama's presidential campaign was built on was itself built on the assumption that their voters were of a bunch of empty-headed saps, at least when it comes to politics. "Change we can all *believe* in?!" As opposed to what, change you can't believe in?! News flash, change is change, whether you can believe in it or not. It's absolutely meaningless blather, but half of this country soaked it up like a sponge. How about "Policies that we actually want?!" "Policies that will actually help our economy instead of dragging it down to Russia's level?!" That is what you actually vote for. But that slogan would require an actual thought process that could contemplate the differences and the nuances and the ramifications of one *policy* over another. "Change we can all *believe* in?!" Wow! And they called George Bush stupid. Someone should introduce Letterman, Stewart, Maher and the rest of our predictably left-wing entertainment clowns to a mirror, if they're looking for some real fools to mock. Besides, after a couple of years in office it is obvious that the "change we can all believe in" is nothing other than (surprise! surprise!) the same old tax and spend liberalism that this country has been trying to dig its way out of for decades. And now it's back, with a vengeance. And in Obama's case, it is even outright socialism. He is going to "fundamentally transform America" alright. Into Cuba.

It should be clear now to all but the most blind of Democrat supporters, that Obama's campaign and presidency have been nothing other *than* a bumper sticker, with no real substance, no real grasp of the issues, and little clue as to what the hell they're actually doing. Their highly touted levels of heightened intelligence, so much more advanced than the previous "cowboy" and his administration, have turned out to have no more depth than those same bumper stickers. The only thing their vaunted intelligence has led them to do is to continue to ram their ideological agenda down the throats of a country full of now unreceptive citizens with a bad case of buyer's remorse. Because their ideological agenda, it turns out, is crap. It has nothing to do with America. It has everything to do with the socialist, anti-American, far-left beliefs of their university professors, whose cancer has spread from merely wreaking ignorance on naïve students over the years, to now wreaking havoc upon an entire nation. And the policies that flow

---

104   (Newt Gingrich on Fox News *Hannity*, 6/11/09, from newt.org)

from the ideological agenda of their naïve students who now control our federal government, are also crap. They work only in the make believe, fantasy world of far-left professors and their true believing Marxist devotees like Obama. In the real world all their leftist-progressive-socialist government takeover policies can do is destroy freedom, stifle economies and transform representative democracy into tyranny.

> "We are becoming a weak nation. Obama really thinks ...he's going to conquer the world with his soft spoken sweet talk, and ...he's going to bring all the enemies of the world into a little playground where they'll swing each other back and forth."[105]
>
> "We were warned by Hillary Clinton that he had no experience, that he had no qualifications. We were warned by his now vice president Joe Biden, he had no experience. So he was a novice, and now we are getting what we could've expected if we had listened. We have a fellow who's bringing us to chaos [economically] and socialism."[106]

"This decision is by the decree of the watchers, and the verdict by the word of the holy ones: to the intent that the living may know that the Most High rules in the kingdom of men, and gives it to whomever he will, and sets up over it the **lowest** of men." (Daniel 4:17) (And they say the Bible is not relevant for us today…)

- The World Apology Tour To begin his historic presidency Obama went overseas and disparaged our country to our friends and enemies alike; incredibly *apologizing* for America[107] rather than pointing out that we are the greatest country on the face of the earth and that we've done more for good in this world than any country in history; and that we have liberated more people from the shackles of tyranny, totalitarianism, Nazism, Communism and oppression than all other countries combined; through the shed blood of millions of our young men and woman who sacrificed their entire lives for these foreigners he apologized to, many of whom now repay that sacrifice by hating America without a cause!? No, Mr. Obama could not think of any of that. Those facts escaped him. They were never taught to him by his left-wing professors at Columbia U. Who Obama really needs to apologize to is us. America is still waiting.

And then he went to the Islamic countries and lied when he told them that we weren't a Christian country. (Of course he didn't tell the Egyptians or the Saudis that they weren't Islamic countries, which would have been equally dishonest but requiring a few more teabags.)[108] Instead he said we were "a country of citizens with shared values and principles." Well the question is, Mr. President, whose values? Certainly, most Americans don't share your values, although half of them have no clue that they don't because your left-wing, lying media never exposed your extremist, far-left, Maoist, rabid baby-killing and America-condemning mindset. That would not have been advantageous in getting you elected, which

---

105   (Jon Voight speaking at a GOP fundraiser June 8, 2009)
106   (Jon Voight, on *The O'Reilly Factor* June 9, 2009)
107   (He did this in Turkey, an Islamic country increasingly intolerant of Christians, where two years prior to his inane apology three Christians had their throats slit for publishing Bibles. But he failed to mention anything about Turkey's persecuted minority Christian community. Allah Akbar.)
108   (See "Protesting is only for left-wing loons, fool!" coming up in a few pages.)

of course was their main responsibility as opposed to actually reporting the news. But whose values? The values of a baby mutilator so rabid that the first thing he did when he was sworn into office was take our hard-earned tax money and send it to mutilate babies to death in <u>foreign countries</u>! In the middle of the worst recession since the Great Depression, no less. Whose values are those? They're certainly not the values of the Christian Founders of this Christian nation. And I can assure you that no matter how this country is split on this issue, the vast majority of Americans are not so stupid that they would vote to send their hard-earned money overseas to kill innocent babies! So why did <u>you</u>? And what other "values" do you and your far-left friends, and this country, *share*? Not the values that would shove the dangerous practice of homosexuality down the throats of first and second graders. Not the values that have turned our judicial system into the home of far-left social engineering fanatics who have no problem with regularly letting child-molester-killers back out on the street to rape and murder more innocent little children just trying to have some fun playing in their front yard for a while before they get to go back inside to mommy and daddy. (But thanks to you and your friends and your wacko judicial nominees and your "shared values" they don't get to go home anymore.) Certainly we Americans don't share the values of your new Islamic best-friends that you bow to? Their values that have sliced the clitoris off of <u>one hundred and thirty **million**</u> of their women who are alive and walking around today missing a very important part of their anatomy that God gave them, but who a stinking, terrorist-producing, Satan-worshipping, false-religion has robbed them of.[109] That's disgusting. Are those the values you're talking about, Mr. President? No, those are not the values of Americans; those are the values of the far-left. They are cesspool values. Those are your values and the values of your friends in the media that duped America into electing you. They are not the well-established Judeo-Christian values and principles that founded this great country. They are Satan's values, and the values of his left-wing followers who have taken over our schools and our media and our entertainment industry and our universities and now our Federal government. They are the "values" and "morals" that are destroying this country. No, that's what happens when the values from above are rejected, and the values from below are dredged up and covered in the camouflage of "tolerance," and spoken of deceitfully as if they were something to be admired and respected and passed on, instead of something that should be flushed down the toilet. They are the values of your left wing clowns that have been inexorably transforming our country into the proverbial Roman vomit pit. And whereas that might be something that a San Francisco Art Professor may salivate over, it is not something that most Americans find particularly appetizing. Yes, you can waltz around the world with no convictions and a s**t-eating grin and talk authoritatively about "shared values," but the question remains, Obama, whose values? Ours or yours? Americas or Reverend Wrights? Gods or Satan's?

"This president is contemptuous of American values. Watching President Obama apologize last week for America's arrogance - before a French audience that owes its freedom to the sacrifices of Americans - helped convince me that he has a deep-seated antipathy toward American values and traditions."[110]

---

109   (See "Islam's Hatred of the Clitoris," by Jamie Glazov, Front Page Magazine.com, as well as our expose of Islam, both in the upcoming "False Religion" section of chapter 8)
110   (Senator Rick Santorum, in the *Philadelphia Inquirer*, 4/9/2009)

But did anyone with a brain have any doubt about this after knowing that his pal and political mentor was an unrepentant <u>domestic terrorist</u>, Bill Ayers, who said in an interview published coincidentally on the very day of Sept 11, 2001 that he didn't "regret setting bombs," and that "I feel we didn't do enough?" And after Obama spent twenty years soaking up the anti-American lies—"God d**n America!"—spewing out of the mouth of the Liberation-Theology-fool, the "stuck-on-stupid" (his words) *Reverend* Jeremiah Wright? Of course the fact of the matter is that many people with a brain did <u>not</u> know any of this because it was deliberately kept from them by a lying, left-wing, Democrat media.

Karl Rove called it the…

> …"international confession tour. …This president believes that his personal popularity is somehow going to change the world. It didn't in Europe when he went over there for the G20. It didn't in NATO. …It didn't in Turkey. It didn't in Latin America. He is underestimating our enemies, and overestimating his personal popularity and his charm."

We have all heard the story of the Emperor who had no clothes, now we have another story to tell our children and grandchildren about the President who had no sense. And the Democrat cheerleaders who control our media and our entertainment industry mock Sarah Palin?! But all is not lost; at least with Obama we get an extra 7 states for no additional charge.[111]

Charles Krauthammer:

> "This is real amateur hour. It is one thing to go around the world apologizing over and over for a country that is the most benign and beneficent and generous in the history of the world. It is an affront to our dignity, but more than that …it makes no sense at all."[112]

One could argue, Charles, that very little these morons are up to makes any sense at all…

> "This president who bowed at the feet of the Saudi King- well that's the same country that has trained so many of the terrorists that have killed and want to kill Americans. This president who smiles and shakes hands with a ruthless dictator [Hugo Chavez] … who gives [Obama] an anti-American book and he accepts it, …who stands against every democratic freedom that our nation was built on. This president who sits by like a little schoolboy while the one-time Sandinista communist-leader-turned-mainstream-fanatic [Manuel Noriega], lectures about the evils of Imperialist Americans. [For nearly an hour Obama listens to this jerks anti-American rant, yet lacks the guts or the sense to stand up and walk out; or even better, to walk over, smack him upside the head and tell him to shut up!] This president who wants to have tea with a man [Ahmadinejad] who denies the Holocaust ever happened and is hell-bent on turning his nation into the next nuclear power."[113] (And Israel into a parking lot.)

The 9/11 commission report said "they were at war with us, but we were not at war with them." So what is this "change" that we are all supposed to believe in? It is going right back to the naïve mentality of the Left that prevailed before 9/11; that caused three thousand of our country men and woman and children to be slaughtered; and that left a huge hole in New York City where the Twin Towers once stood.

---

111   (During the campaign Obama claimed that he had visited 57 states. The Democrat media covered up for him, saying "He was *tired*." And the late-night left-wing clowns didn't think it was funny. Hussein, after all, is not a stupid Republican. (He's a stupid Democrat.) So it was just an honest mistake. It is interesting to note, however, that in the Muslim religion there are, purely coincidentally, 57 states…) (But he's not a Muslim. He's just tired.)

112   (Charles Krauthammer on *Special Report with Brett Baier*, Monday 07/20/09)

113   (Sean Hannity)

But in keeping with the changing times; and being as how we are now being "radically transformed" into the United *Socialist* States of America; this July 4th we are no longer going to celebrate our Independence, instead we are going to celebrate the Russian Marxist Revolution of 1917 (something the Left doesn't have to apologize for). You can get your hammer and sickle flags down at the local Democrat Party headquarters, or any mainstream media outlet. (Same place actually.)

- School vouchers. Or what's good for Obama and his Democrat friends in Congress is not good for all those poor people who were bamboozled into voting for him... Thurgood Marshall Academy, a charter school in Washington D.C. where poor black and Hispanic children were able to attend thanks to the school voucher program, has a 100% college admission rate. $15,000 dollars per student are spent by the D.C. public schools, yet for half that—only $7,500 per student—a much better education can be had at D.C. private and charter schools through the school voucher program. This program was specifically enacted into law by those "mean-spirited" Republicans and despite the angry protests of Democrats and their teachers union. It was meant to help poor families and struggling single mother households send their children to decent schools at half the cost to the taxpayer of the nearly-insane, discipline-free-zones of the D.C. public schools. So why did the Obama administration and the Democrat Party take advantage of their total control of our government to get rid of this program? Because the teachers unions and the ACLU didn't want it and because despite what the suckers who voted for them were led to think they do not give a damn about the poor black children of those who supported them. It is about their ideology straight up and nothing more. It's about their "values," the ones that ooze up from below:

"The Obama position on abortion makes him someone I could never support. But I look for signs of goodness in the man; some indication that he may be anything other than a cold hearted, calculating, successful politician who has risen on other than his charisma alone. But what I'm about to elaborate on here only re-enforces my belief that Barack Obama is exactly that, a cold hearted calculating politician who would sacrifice and withhold nothing to please his god of political correctness and secular humanism, and also please the others who worship along with him.

"This one issue reveals the deep character flaw of Barack Obama, in his betrayal of the black and Hispanic community (who voted overwhelmingly for him) in something of paramount importance to both those communities, and also the American community as a whole. This betrayal involves the education of minority children in crises in Washington DC, where Barack Obama has now signed a bill which will end the hopes and dreams of Washington DC area children who were attending private schools (including some who were attending a DC area private school where Obama sends his own children) on the federal government's school voucher program. ...Obama has now signed the omnibus spending bill, in which Congressional Democrats put language that will end the school voucher program, and **return 1700 mainly black and Hispanic children to the violent and inferior performing DC public schools**, [emphasis mine] ending a rare federal program which has achieved success.

"I'd like to hear Obama's explanation for this but he has offered none, most likely because there is no defense for his actions. Obama could have vetoed this bill, and told Nancy Pelosi and Harry Reid that he would not stand for a program achieving great success in the education of minority children at risk to be ended. But instead Obama went right along with the program ...and signed a bill sacrificing the future of these children because the powerful teachers unions opposed it, and the Republicans initiated it. The real injustice of this act by Obama and the lackeys in Congress is that they will return these children to the DC public schools, which they refuse to send their own children to. [But why should you be concerned that you are a rank hypocrite when you have a left-wing lying media to run cover, and keep the suckers voting, for you?]

"To anyone reading this, I ask has there ever been a more despicable example of the moral bankruptcy of our so called leadership in Washington, or more compelling and infuriating evidence that under the administration of Barack Obama, we will have business as usual in Washington. Obama states he intends to make the public schools a place where all children can get a decent education, which ends the need for the voucher program, something he failed to accomplish as a Chicago area politician. But I would like to give Barack Obama a chance to prove he's serious about the reformation of public schools, of DC and those throughout America. He can begin by enrolling his two daughters in a Washington DC public school, and urging those members of Congress who wrote this bill to do the same. [Fat chance] It was always the Democratic Party which has claimed to supposedly champion the cause of these children, and instead has betrayed them in a dark deal with the teachers unions, and to political ideology.

"I would like to also give dishonorable mention here to Jesse Jackson, Al Sharpton, the Congressional Black Caucus, and black radio and media who have remained silent as this took place. If this had been anyone other than Barack Obama who signed this bill, and crushed the hopes of these children, they'd be up in arms, screaming from the rooftops. Instead we hear a deafening silence from them all, which allows me to continue to hold them in total contempt, but doesn't help these unfortunate kids."[114]

Amen, brother! But they know that their left-wing, lying, Democrat media (along with the rank black hypocrites mentioned above) will just sweep this under the rug. So they will continue to enjoy, but not *deserve*, the votes of black and Hispanic minorities while at the same time they continue to betray them, to trample-on their hopes and dreams, and to bend them over while their media covers it all up.

- Protesting is only for left-wing loons, fool! The TEA (Taxed Enough Already) parties that took place on April 15, 2009 are very instructive of what happens when the Left Elite take over an entire media. Millions of Americans took to the streets in state capitols and other cities across this nation on "tax day" to oppose the Obama administration and his Democrat Congress taking political advantage of a recession by spending borrowed money, that we don't have, and in record amounts, and that very likely will drive us into bankruptcy. Spending that also sucked nearly a trillion dollars out of the engine of our economy when it could least afford the drain. The Democrats did this under the guise (the lie) of a stimulus bill that was designed to stimulate their socialist agenda, as well as bailing out those Democrat-run cities and states that have already spent themselves into bankruptcy. One of the ways they accomplished this was by working in collusion over the years with their state and local public service and teachers unions to award them with outrageous pay and benefit packages that far exceed what is found in comparable jobs in the private sector; so they could then in turn enjoy the benefit of massive campaign contributions and Democrat-supporting election ads

---

114 (Posted on www.freerepublic.com by Jazzman646, Friday, April 17, 2009)

from those same unions. There were over two thousand of these protests across this country, yet the left wing media either completely ignored them, or vilified, demeaned and falsely portrayed (lied about) the protestors. Broadcasters on CNN, NBC and MSNBC even went so far as to use vulgar references to describe them: referring to the tea party protesters over and over again as "tea-baggers,"[115] a vulgar reference to a homosexual practice. But were these "journalists" and commentators demeaning the protesters or homosexuals? It doesn't matter because it's OK to publicly disparage the sexual practices of homosexuals as long as you are doing it to disparage Obama's critics at the same time. Then you can count on the militant homosexuals and their media, both of whom would normally go ballistic at any perceived slight, in not making so much as a peep.

But can you imagine the uproar if a non-Obama-arse-licking media outlet like Fox News for example had spent an entire day snickering about a Code Pink demonstration by referring to them as "pro-fisters" instead of protesters? "Tee Hee, The Code Pink "pro-fisters" (snort, snort) were at it again today, falsely accusing the Bush Administration of lying about the war in Iraq. Snicker, snicker. The "pro-fisters" appeared angry and agitated." There would be an *Obomic* holocaust, and Fox News would be removed from the air. But not so much as a furrowed brow over the vulgar spectacle from the sophomoric halfwits at CNN, NBC and MSNBC. (Of course now we can add to this list of vulgar, sophomoric halfwits the name of our own president who amused his audience—in May of 2010— by himself referring to tea-party protestors as "tea-baggers." How classy of you, Mr. President.) (…Welcome to the "change really stupid people can believe in," America.)

The New York Times, that "paper of record," whose motto is "all the left-wing spin, distortions and outright lies fit to print," ignored the story—millions of protestors! It was as if it never happened. The Nazi's in Germany didn't have a media propaganda arm like the Democrats do, and they owned the media there, for heaven's sake! (Of course, now that I think about it, the Democrats own the media here as well.) But if ten Code Pink loons showed up at Crawford Ranch in Texas to harass the previous president, this media fell all over one another to get there with more cameras than there were protestors.

"So why is it that the mainstream media finds it so easy to sneer at these protestors, when you've never seen them do it to the bug-infested buffoons protesting environmental ills or the WTO? Well, first: These protests involve people they've never actually met. These are average folks, not professional sign carriers. Most of them work for a living and keep their marching for parades. Also, the media abhors these people because they question their Messiah's mission. These protestors know wealth distribution when they see it and they're calling it out because the media cannot bring themselves to do it for them."[116]

And then there was the arrogance of our madam Speaker:

"Nancy Pelosi was asked about the tea party protests. She denied it's a grass roots movement, saying "We call it *Astroturf*," and that the tea parties are about the wealthiest people in America.

"She's absolutely insane. Did she see pictures and videos from these events? I did

---

115   (Definition of *Tea-bagging*, "Dipping your testicles into the open mouth of another person. Kind of like dipping a *tea bag* in and out of a cup of water." From urban dictionary. com) (And wasn't that professional of those Democrat shills masquerading as journalists at those networks?!)

116   (Greg Gutfeld, Thursday, April 16, 2009, www.dailygut.com)

and I certainly didn't spot a lot of really wealthy looking people. I'm sure there were some, but for the most part everyone I saw seemed to be just your average working class American."[117]

Of course, she knew she was lying about this truly grass roots movement. But she also knew she could get away with it because her media would run cover for her. One protestor, a middle-aged, middle-class woman holding her little baby said Pelosi's face was "revolting." (OOPS! I'm sorry. That was *Pelosi's comments were revolting*! Typo.)

Janeane Garofalo, a perfect example of what happens to a mind when it has been thoroughly poisoned by the lies of the Left, added her 2 cents (going bankrupt in the process) by saying on MSNBC that the tea party protestors were "a bunch of tea-bagging rednecks. [There's that homosexual slur again.] …This is about hating a black man in the White House. This is racism straight up." (Yea, sure. And when your Aunt Sadie asks for the white meat on Thanksgiving, Janeane, that's also "racism straight up.") No, Ms. Garofalo, it is your statement that is "ignorance straight up." It's not about "hating a black man in the White House." It's about hating the fact that there is a left-wing-indoctrinated, America-hating, baby-mutilating, arrogant socialist in the White House who has no respect for America, traditional Americans, the Constitution or the rule of law. Besides, Ms. Garofalo, isn't Barack half-white? So why do you call him a black man, and not a white man? And aren't you, by calling him a "black man" just because he is half-black, a racist yourself? What is black, the polluting influence? Like adding dirt to water, it's now dirty water? Wow!! Doesn't the very fact that you see Obama as "black" just because he is half-black show what a racist you are? …Think about it, after you get your 2 cents back.

- Will the real terrorists please stand up? For years the Democrat-Left has been shouting their lies loud enough and long enough that enough uninformed people have believed them. It's a technique made famous by the Nazis in the 1930s and 40s. But the "regime-change" ones that the Democrat-Left and their media spread about the War in Iraq was a new low. Not long after Iraq's speedy liberation, the Iraqi insurgency began where terrorist elements, many of whom came from outside the country, combined with neighboring Iranian support and Saddam's Sunnis who were displaced and disgruntled after being thrown out of power by the US victory. The Democrats then jumped at the chance to take advantage of this "golden opportunity" for the advancement of their own political fortunes. Instead of helping America attack the Islamic enemy in Iraq they assaulted the war effort and our commander-in-chief here at home, by lying about both. They broadcast everywhere they could the lie that "Bush lied" about the reasons for going in there in the first place even though most of the Democrats themselves originally acknowledged those same reasons (of which WMDs—Weapons of Mass Destruction— was only one). They also supported the Iraq War originally before it became advantageous politically not to. Another ridiculous lie that they shouted was "no blood for oil!" which somehow ignored the fact that not one Iraqi oil well was ever appropriated for our use, and we didn't gain financially from liberating Saddam's oil industry. (But those are facts, which left-wing propagandists are allergic to.) Another lie displayed their historical ignorance: "The Iraq War is an

---

117   (www.lonelyconservative.com, April 16th, 2009)

Illegal War." Of course, and so was the War against Germany because *Japan* and not Germany attacked us at Pearl Harbor. Even Rosie O'Donnell got in on the act, declaring that "radical Christianity is just as threatening as radical Islam,"[118] preferring to wage war against Christians here at home, I guess, rather than on Islamic killers abroad. The liberal mindset in action. But the Democrat-Left's classic lie was "Bush lied and soldiers died!" Whereas the sad reality is that "Democrats lied and our soldiers died." For many of our soldiers actually *were* killed because while they were fighting for our freedom and safety in Iraq, back home the skunks in the Democrat Party and their media were lying to the American people purely for political gain, and emboldening and encouraging our Islamic enemies to continue fighting hard for what was looking more and more like a victory for them, as the war was losing support at home in large part because of the Democrats politically-motivated treachery. The leaders of the terrorists in Iraq were fully aware of what happened here in the 60s and why we lost the War in Vietnam, and because of what the Democrats like Dick Durbin, Jack Murtha, John Kerry and their media were doing, they were seeing the same thing happening again. History was repeating itself because we were again fighting a war on two fronts. (And still are.) And seeing that their deceit was politically profitable for them and that Bush's approval rating was plummeting, our "enemy within" started shouting their lies louder and longer. Dissent is one thing. Theirs was not dissent. It was cold, calculated lying in a time of war, purely for political gain. And it couldn't help but aid and encourage our Islamic enemies in Iraq. There's a name for this, and it rhymes with treason. The Democrats and their media have the blood of many of our soldiers on their hands.[119]

Years ago we had George M. Cohen, America's "Yankee Doodle Dandy," who was awarded the Congressional Gold Medal by FDR for his support of the war effort during World War I, which included writing a number of popular, patriotic, morale-boosting songs like *Over There* and *You're a Grand Old Flag*. But what does America have today? Harry Reid, Democrat Senator and modern-day left-wing equivalent of Benedict Arnold, protected by his Democrat state-run media, whose contribution to the Iraq War effort was declaring on the floor of the United States Senate at a very pivotal and vulnerable moment that "The War in Iraq is lost." And as reward for his support of our enemies this traitor was handed control of the Senate in 2006 by clueless liberal "no-think-um" voters.

Fast forward to 2009, and the Obama administration, apparently believing their own propaganda, has gotten a little confused over who the enemy actually is. In fact, now we don't even *have* a "War on Terror." Instead they've named it the "Overseas Contingency Operation," whatever the hell that is. They bumped their heads. What was World War II, the European Vacation? How can you prosecute a war when your very language betrays the fact that you don't even think it exists? You can't. The same way you can't fight an enemy that you refuse to even identify. For according to these dimwits the Muslim jihadists who hate us and want to see us all dead are no longer terrorists. They are just regular folks who happen to produce "man-caused disasters," like 9/11 or nuking Los Angeles.

---

118   (Rosie O'Donnell on The View, Sept. 15 2006)
119   (See: *Party of Defeat: How Democrats and Radicals Undermined America's War on Terror Before and After 9-11*, by David Horowitz and Ben Johnson; and *Treason: Liberal Treachery from the Cold War to the War on Terrorism*, by Ann Coulter)

So what happened to all of those terrorists then? They had to go somewhere!? Well, they did. They morphed, at least according to Janet Napolitano who as head of the Department of Homeland Security declared in April of 2009 that the real terrorists are law-abiding U.S. citizens who may produce "man-caused disasters" here in the future. (Like voting these clueless bastards out of office for example.) She issued an official report that cautioned U.S. law enforcement to be on the lookout for the following: anti-abortion activists, fundamental Christians, border security protestors, people who attend Bible Prophecy Conferences, conservatives in general, and most incredibly of all, returning <u>veterans</u> from Iraq. All of these decent, law-abiding Americans were defined by her memo as "potential terrorists." And that is because what they all have in common is that none of them voted for the messiah. And to Janet Napolitano and her true-believing kind that is more dangerous than being a homicidal Islamic lunatic.

Colonel Ralph Peters had this take:

"If you look at this [Napolitano] report, there's so much sickness in it. One of the things that really infuriated me is the idea that vets are susceptible to anti-Semitic, anti-Jewish, hate groups. Bill, where is the anti-Semitism today? It's on campus among the Left. It's in the Daily Kos among the Hamas supporters and Hezbollah supporters. This was crazy. But also, as a former intelligence officer I was offended by the pathetic quality of it. In intelligence analysis you do speculation, but it's based upon facts, on things that are corroborated."

O'Reilly responds...

"I'm a taxpayer. I'm paying these peoples salaries and they're putting out this kind of stuff?! ...If we want that kind of whacked out stuff all we have to do is turn on NBC news. [Do] we want our government doing this?"[120]

Jay Sekulow from the ACLJ adds,

"I was skeptical. Now I'm outraged. You should be, too. ...And make no mistake, this unconstitutional report raises serious questions about the leadership and direction of the agency charged with protecting Americans in the ongoing battle against terrorism. Why would the Department of Homeland Security single out groups like pro-life supporters when they should be focusing on identifying and apprehending the real terrorists - like al-Qaeda - groups that have vowed to destroy America?"[121]

Hmm? Maybe it's because they now have as their head someone who is extremely <u>stupid</u>!? (Thought I'd take a wild stab at that one.)

What these memos told our law enforcement is to "be suspicious of anybody who disagrees with the Obama administration," which is right out of the Communist Manifesto—go after anyone who disagrees with the government—and not out of America's tradition of political freedom and robust debate. But then again, he didn't lie to us there. He did say he wanted to "radically transform" America. (What did we think that meant, he was going to make Spanish the official language, and Islam the state religion?) (Hmm, that last one might not be such a stretch.) The next thing we know they're going to bring back water boarding, but not for captured Muslim terrorists, for Christians! (*But wait, Janet, you putz, isn't*

120   (Colonel Ralph Peters on Fox News *O'Reilly Factor*, 04/16/2009)
121   (Jay Sekulow, *American Center for Law and Justice*. Most Americans haven't heard of this group- The ACL<u>J</u>. The left-wing media hasn't told them. They are a group of dedicated, decent, America-<u>loving</u> attorneys who have been fighting the America-hating ACLU for decades to slow them from further shoving their anti-American, anti-Constitutional and anti-Christian agenda down America's throat.)

waterboarding "torture"? "Aw, shucks no, dummy! We just made that up to attack Bush.")

- Pass the Nazi Playbook, please. The Democrat-Left over the years has taken every opportunity to falsely marginalize, vilify, deride and disparage their conservative, Christian or Republican opponents, especially the most popular ones. This also is right out of the Nazi playbook: "shout your lies loud enough and long enough and enough people will believe them." Look at the treatment of Sarah Palin, George Bush, Dick Cheney, Clarence Thomas, Condoleezza Rice, Ronald Regan, Dan Quayle and Newt Gingrich just to name a few. It's interesting how the jokes of the late-night and Comedy Central clowns about Obama are affectionate, whereas their jokes about Christians, conservatives and Republicans are vicious and disparaging, making them out to be downright stupid. But when it comes to their own, the real morons, they'll poke a little good-natured fun perhaps, but they don't deride, disparage or marginalize them or demean their level of intelligence. Because then they would be doing it to themselves.

David Letterman said that he "couldn't think of anything to make fun of Barack Obama because he was *so cool.*" (In Letterman's way of thinking, mutilating another 50 million babies to death is "cool.") But read the following list and imagine what that disingenuous, late-night ass would have said if any, much less all of these had been attributed to George W. Bush. They wouldn't have been so "cool" then…

> Obama "criticized a state law that he admitted he never even read. [He] used a forged document as the basis of the moratorium [on oil drilling in the Gulf of Mexico after the spill] that would render 87,000 American workers unemployed [for no good reason other than to use that crisis as a golden opportunity to throw a bone to his environmental wacko friends]. [He was] the first President to need a TelePrompTer installed to be able to get through a press conference. [He] made a joke at the expense of the 'Special Olympics,' bowed to the King of Saudi Arabia, visited Austria and made reference to the non-existent "Austrian language," [and then] stated that there were 57 states in the United States. [Then Obama flew] all the way to Denmark to make a five minute speech about how the Olympics would benefit him walking out his front door in [Chicago]. [He was] so Spanish illiterate as to refer to "Cinco de Cuatro" in front of the Mexican ambassador when it was the 5th of May (Cinco de Mayo), and continued to flub it when he tried again. [He] misspelled the word "advice" [but the late night clowns and left-wing comics didn't] hammer him for it for years as proof of what a dunce he is like [they did to] Dan Quayle [for his] potatoes comment. [Obama also] burned 9,000 gallons of jet fuel to go plant a single tree on Earth Day. [But he's not an idiot.] [He] Okayed Air Force One flying low over millions of people followed by a jet fighter in downtown Manhattan causing widespread panic. [He] failed to send relief aid to flood victims throughout the Midwest with more people killed or made homeless than in New Orleans."[122] …But he was never accused of being an ignorant racist.

The truth is that if George Bush had done or said ANY of these things, much less all, our Democrat media and our late night clowns would have mocked him to death. For never did any conservative, Christian or Republican produce so much outrageous material. Barack Hussein Obama should have become the laughing

---

122   (From the "Presidential Comparison Quiz," posted on the Hardcore Right Wing News, 8/30/2010)

stock of the entire nation, rather than just those who actually know what's going on, because they don't get their voting orders from Democrat operatives in our lame-stream media, and they long ago saw the late-night and Comedy Central clowns for the one-way streets that they are.

Then there is the case of one Rush Limbaugh, a popular, successful conservative commentator who has been the brunt of the Lefts vilification for decades. They became especially incensed over him because he uses humor as his weapon to ridicule them and their idiotic policies. (And ridicule of course is only allowed when it is used against *their* opponents, not them.) They have found it impossible to argue with his logic, or rebut him on the substance of the issues. They hate him because of the threat he poses to their control over the thinking and the opinions of the electorate. So they must destroy him, at all costs. (In the name of free-speech, of course.)

But with their takeover of the White House, the Left's *Hatred and Vilification of Rush Campaign* has moved from the mainstream media and late-night clowns to an operation running out of 1600 Pennsylvania Ave. And in May of 2009 at the White House Correspondents Dinner they went a step further. They wished him dead. Rush had commented on his radio show that he wanted Obama's liberal policies to fail, but Wanda Sykes, a comedian, distorted that in order to say the following, "Rush Limbaugh: *He hopes the country fails?* I hope his kidneys fail, how about that?" And of course Colonel Klink thought that was hilarious. But, again, as Miss Sykes must have known (unless she is ignorant as well as mean-spirited), Rush never said he wanted his *country* to fail, he said he wanted Obama's left-wing, socialist *policies* to fail. So, Miss Sykes, you are a liar, "how about that?" Also, it is interesting to note that Rush has never wished that his lying, left-wing opponents like Miss Sykes were dead. If he did they would have removed him from the air. But nothing will be done about Miss Sykes' comments, or Obama ignorantly laughing out loud with her. That's because the Left are the champions, not only of lying, and of evil in all of its many forms, but of a double standard that exposes their rank hypocrisy.

During the presidential campaign at one of his rallies John McCain threw someone off stage for using Obama's full name, "Barack Hussein Obama!" (Now there's an unconscionable slur, to actually refer to someone by their own name! God help us!) Conversely Barack Hussein had the decency to return the favor by sitting there and laughing when Sykes wished death on a decent, honest, humorous and *conservative* radio host. Maybe McCain is the most naive person to ever enter the political arena, or maybe he actually thought his spinelessness disguised as civility would earn him brownie points in a knife fight against unprincipled skanks. I'm not sure. But what is certain is that just as the 9/11 report said that Islamic psychos were at war with us but we weren't at war with them, the Democrat-Left and their media declared all-out war on Republicans, conservatives and Christians a long time ago. But nevertheless a number of naïve Republican leaders seem to think they're just in a friendly pie-eating contest down at their churches Sunday social. And they can't figure out why, as they're bent over slurping away on their lemon meringues' like good little boys and girls, the opposition across the aisle are hitting them over the head with the picnic benches. Ouch.

- <u>Skateboarding, snowboarding, wakeboarding, and waterboarding.</u> Incredibly, the Obama administration is kicking around the idea of prosecuting the previous administration for using enhanced interrogation techniques to defend this country; which included waterboarding a few select terrorists. This was something that was done with the full knowledge and approval of Congress, which included the Democrat leadership there as well. Obama has waffled back and forth on this due to the fact that he has no principles (I mean, he's a rabid baby mutilator for heaven's sake) and he just wants to see if he can get away with Nazi show trials right here in the United States. And that is questionable because their media cover is not 100%—they have not shut down Talk radio, Christian radio or the Fox News Channel—yet. The call for these prosecutions is driven by the terrorist-defending ACLU, the lunatics over at moveon.org and the deep pockets of the America-hater George Soros who was very helpful financially in putting Obama in office. Now it's payback time.

All of this Machiavellian intrigue revolves around the Left's absurd attempt to call waterboarding torture, even though it consists merely of pouring water on the subject's cloth-covered face and results in no damage or harm whatsoever to the waterboardee. It is however very uncomfortable, impossible to withstand, and resulted in the divulgence of very useful intelligence information that prevented the deaths of a great many Americans. Yet Obama has refused to use it, preferring instead to give captured terrorists free lawyers, Miranda rights, and milk and cookies in order to get them to talk. The silliness of the Democrat-Left in calling it torture becomes abundantly clear when you compare pouring water on someone's face to actual torture like electric shock to someone's genitals, or ripping off someone's fingernails, or gouging someone's eyes out, or cutting off someone's tongue, or slowly burning someone alive, or of course cutting off someone's head as our no-longer-to-be-called-Islamic-terrorist friends are so fond of doing. And when you also consider the fact that whereas we have only waterboarded a mere handful of terrorists, we have waterboarded thousands of our military personnel, pilots and special forces as a regular part of their training. Yet the Democrat-Left continues to call it torture. Wow! It's so terrible and so torturous that we can't use it on those who would kill millions, if not every one of us if they ever got the opportunity, and we should try to illegally prosecute the Bush administration for using it a few times to keep us safe, but we <u>can</u> use it on thousands of our American military personnel and pilots as a regular part of their training. The only thing tortured here is the Democrat-Lefts thought process. (Or lack thereof.)

Obama then went even further by releasing a number of top-secret CIA memos on those interrogations, which in a time of war could do nothing but aid and assist our Muslim enemies. But that he only released the memos which helped him make his case, revealed his true intentions. The ones that showed how effective those enhanced interrogation techniques were and how many lives they saved from another 911 type attack, *remained* top secret. Then he took the occasion to lecture all of us on how the memos "reflected us losing our moral bearings." What a belly laugh, coming from the people who have no morals, much less "moral bearings!" But they're condemning an administration that actually did have morals, and used them to protect this nation. I would say that we put a pompous, pretentious ass in the White House, but we don't want to offend all of those folks who still blindly support him.

Fox News reporter Jim Angle notes that…

"The odd thing about this ... is that President Obama has decided that waterboarding—which we have done by the way to thousands of our own people in the military; pilots and Special Forces are often trained by being waterboarded—... is too harsh to use on terrorists. On the other hand, days after he took office he approved air strikes on terrorists in their homes in Pakistan, ... presumably with their wives and children so you get a lot of civilians who are killed, you certainly get the terrorists who are killed, so one could argue that waterboarding isn't nearly as bad as being blown up."[123]

Yes, one could argue that, Jim, if one were not arguing with idiots.

At the time of this writing Obama is still flip-flopping. He is getting a lot of pressure from the New York Times, the ACLU, his sugar-daddy George Soros, and those mentally unbalanced politicos who have eluded commitment to the insane asylum by hiding out over at moveon.org, the Huffington Post and the Daily Kos. And it is again important to note that his resistance to this angry mob has nothing to do with his non-existent principles or morals. He's just not sure if it is politically expedient for him at this time. He's not sure he can get away with it. He's got his finger licked and up in the breeze. Keep us posted on how the wind blows, Barack.

Bill Kristol, the editor of the Weekly Standard, had these thoughts:

"Dennis Blair, the current director of national intelligence, in a letter to current employees said we acquired high-value information from the methods that we used. So now we have Obama's own director of national intelligence saying we acquired high-value information from these methods, and yet the president seems to think that they're so out of the question that he's repudiated the use of them and is talking about the possibility of criminal prosecutions of people who wrote in good faith legal analyses which distinguished [those methods] from torture. Eric Holder, Attorney General, said in 2002 the Geneva conventions did not apply to Al Qaeda captives. The idea that we are going back and even raising the possibility of criminal prosecution is so appalling that it renders me almost speechless.

"And then President Obama at the CIA yesterday, in a speech to CIA operatives, trying to reassure them, said "Don't be discouraged by what's happened in the last few weeks. We may have potentially made some mistakes- that's how we learn." I mean, really, we have a president engaging in baby talk [emphasis mine] at a time when there's an ongoing terror threat to this nation."[124]

It would actually be hilarious if it weren't so dangerous. And Letterman still cannot think of anything to poke fun of Obama. But then again, Dave, Kim Jung Il's North Korean comics never made fun of him either.

Dick Morris...

"The people that are trying to find the terrorists are being investigated; the terrorists are getting released from Guantánamo. The enemies of the United States—Hamas, Iran, Russia, Chavez, Castro—merit our friendship; the friends of the United States—Israel, Columbia and Britain—merit our scorn. ... If you want to make time with Obama be an enemy of the United States. ... We have a President whose foreign policy can only be described as anti-American."[125]

Karl Rove...

"What the Obama administration has done in the last several days is very dangerous. What they have essentially said is if we have policy disagreements with our predecessors,

123   (Jim Angle on *The O'Reilly factor*, April 20, 2009)
124   (Bill Kristol on Fox News *The Fox Report with Brett Baer*, April 21, 2009)
125   (Dick Morris on Fox News *The O'Reilly Factor*, April 22, 2009)

what we are going to do is we are going to turn ourselves into the moral equivalent of a Latin American country run by colonels in mirrored sunglasses, and what we're going to do is prosecute systematically the previous administration, or threaten prosecutions, based on policy differences. Is that what we've come to in this country? [Apparently.] That if we have a change of administration from one party to another, we then use the tools of the government to go systematically after the policy disagreements that we've had with the previous administration. Now that may be fine in some little Latin American country run by the latest Junta. It may be the way they do things in Chicago. But that's not the way we do things here in America. I thought this was reprehensible."[126]

Senator Joe Lieberman, himself a Democrat, adds: "This doesn't help the US, but it does help our enemies." And Bernie Goldberg nails the hypocrisy…

"Let's leave aside just for the sake of argument the question of what constitutes torture. And let's also leave aside whether "torture" works or doesn't work. …Barack Obama, Nancy Pelosi and lots and lots of other liberals are against these enhanced interrogation methods 100% of the time. I'm against them, and I suspect most Americans will agree with me, 99% of the time. I'm against them to punish jaywalkers. I'm against them to punish litterbugs. I'm against them to even punish the most vile hate-filled terrorists. I'm against using enhanced methods of interrogation to punish *anybody*. But, if by using these methods it would *prevent*—that's the key word, *prevent*—or foil, a terrorist plot and save thousands of American lives, Sean, the moral position is to *use* these enhanced interrogation methods. That's the *moral* position. Barack Obama talks about "upholding American values." That's why he's against what he calls torture. Please, Mr. President, please tell us what American values would be upheld if you needlessly allowed thousands of Americans to die, needlessly, because you didn't want to use these tough interrogation methods."[127]

Actually, Bernie, I disagree. "Upholding American values" is *not* why the president is "against what he calls torture." He has no real values. Neither does the Left. They pretend like they do, they may even fool themselves into thinking that they do, but they do not. When the rubber meets the road all they have is hypocrisy, impressive levels of irrationality, and the cesspool values that they received from below. (Or, we could always try asking the 50 million dismembered babies what *they* think of their values. But we can't; 'cause they're dead. And they actually *were* tortured. To death.)

Dick Cheney, former Vice President, had this to say:

"When President Obama makes wise decisions, as I believe he has done in some respects on Afghanistan and in reversing his plan to release incendiary photos, he deserves our support. And when he faults or mischaracterizes the national security decisions we made in the Bush years, he deserves an answer. …Our government prevented attacks and saved lives through the terrorist surveillance program, which let us intercept calls and trap contacts between Al Qaeda and persons inside the United States. The program was top secret and for good reason, until the editors of the New York Times got it and put it on the front page."

In World War II there was Tokyo Rose. In the War on Islamic Terror there is the New York Times. Cheney continued…

"The interrogations were used on hardened terrorists after other efforts had failed. They were legal, essential, justified, successful and the right thing to do. The intelligence officers who questioned the terrorists can be proud of their work, and proud of the results,

---

126   (Karl Rove on Fox News *Hannity*, April 21, 2009)
127   (Bernard Goldberg on Fox News *Hannity*, April 27, 2009)

because they prevented the violent death of thousands, perhaps hundreds of thousands, of people. Yet somehow when the soul-searching was done and the veil was lifted on the policies of the Bush administration, the public was given less than half the truth. The released memos were carefully redacted to leave out references to what our government learned through the methods in question. Other memos laying out specific terrorists plots that were averted apparently were not even considered for release. For reasons the administration has yet to explain, they believe the public has a right to know the method of the questions, but not the content of the answers."

Since the administration has yet to explain it, let me give it a shot. They believe the public has a right not to know exactly what they and their media can get away with not telling them, as long as it furthers their plan for totalitarian socialist control. But Mr. Cheney was not through:

"Few matters have inspired so much **contrived indignation** and **phony moralizing** as the interrogation methods applied to a few captured terrorists. I might add that **people who consistently distort the truth in this way are in no position to lecture anyone about values**. [Emphasis mine.] It is a fact that only detainees of the highest intelligence value were ever subjected to enhanced interrogation. You've heard endlessly about waterboarding. It happened to three terrorists. One of them was Khalid Sheikh Mohammed, the mastermind of 9/11, who has also boasted about his beheading of Daniel Pearl. We had a lot of blind spots after the attacks on our country, things we didn't know about Al Qaeda's plans but Khalid Sheikh Mohammed and a few others did know, and with many thousands of innocent lives potentially in the balance, we did not think it made sense to let the terrorists answer questions in their own good time, if they answered them at all. ...

To completely rule out enhanced interrogation in the future is unwise in the extreme. It is **recklessness cloaked in righteousness**, [emphasis mine] and would make the American people less safe."

(Ah, but what does the Democrat-Left care about reckless disregard for the safety of Americans when compared with advancing their incoherent ideology?)

Obama had a different take:

"They [enhanced interrogation techniques] serve as a recruitment tool for terrorists and increase the will of her enemies to fight us, while decreasing the will of others to work with America."

Jim Angle: "Cheney derided that mantra, saying it's a familiar pattern, "It excuses the violent and blames America for the evil that others do. It's another version of that same old refrain from the left, "we brought it on ourselves."""[128]

And then we have Dennis Miller who refuses to treat the Democrat's grave "moral" concerns over waterboarding with the seriousness it deserves:

"Torture is a subjective thing. Some people view this is as torture. Some view it as harsh interrogation. I happen to view it as the first shower some of these slugs in Gitmo have had in around eight years. Waterboarding doesn't break my heart."[129]

A guest on the Fox News Channel challenged Obama about what he would do if his two daughters were kidnapped by Al Qaeda and their heads were about to be cut off. Obama of course has sworn that he wouldn't use waterboarding

128   (On *Special Report with Brett Baer*, May 21st, 2009, Jim Angle reports on Dick Cheney's response to Obama's speech on enhanced interrogation techniques, delivered with the Constitution of the United States as a backdrop. You see, it's OK to pervert, distort and ignore the Constitution as long as you can still use it as a visual aid. See the Democrat-Lefts handbook, *Rules for Radicals*, by Saul Alinsky.)

129   (Dennis Miller on *The O'Reilly Factor*, April 22, 2009)

in any situation because of his "high moral standards." (The same "high moral standards" which include mutilating babies to death in foreign countries with our taxpayer dollars.) The question was, What if we had a terrorist in our custody who knew where Obama's two girls were, and all that our CIA interrogators had to do to get this information was dunk the skunk in a little water; would the president then authorize waterboarding to save his own two daughters? If he would, then he is a liar and a hypocrite. And if he wouldn't then he is not fit to be president of the United States of America, or a father for that matter.[130]

Finally, and this is the *coup de grâce*, the malignant liars on the Left feign hysteria over dunking terrorists in a little water, but they have no problem actually slaughtering living babies in the most demonic and <u>torturous</u> ways imaginable— ripping their limbs off, sucking their eyes out of their sockets, and jamming scissors into the back of their skulls and sucking their living brains out of their skulls with a fricking vacuum! Compare that with pouring water in someone's face and the Left's impressive level of mental incoherence becomes evident. They're literally insane. (See the discussion on what it means to be given over to a reprobate mind in chapter 11, Baby Mutilation.) Waterboarding is torture, but baby mutilation is not? What a joke. What a lie. And now these ACLU members who infest our Injustice Dpt., wholly devoid of any rationality, want to prosecute the decent, moral and ethical people who protected this country for eight years?! But Tiller the baby killer out in Kansas, the one who jams scissors into the backs of 9-month-old babies heads and literally tortures them to death while they scream in horror, they don't want him prosecuted.

"America, America, God shed his grace on thee." Truly, God has shed his grace upon America and blessed it like no other nation in this world's history. But how long will his grace continue to be shed when his kindness is repaid by half of Americans marching out religiously every two years and voting for the baby-mutilating Democrats? The real torturers. Supporting those who reject His morals, His values and even his very existence. And we suffer under the weight of a terrible economy while at the same time we have handed over the reins of power to committed socialists who can do nothing other than continue to destroy our once-free economy.

> There's "a new regime in Washington that stands in stark opposition to the teachings of Scripture (and the will of the voters) with regard to life, marriage, and religious freedom."[131]

It also stands in stark opposition to the *values* of both Scripture and the majority of voters. African Americans in the last election went out in record numbers and percentages to vote for the first Black President even though he actually shares none of their values, except opposing gay marriage, which Barack *says* he opposes out of one side of his mouth but contradicts it with his actions, because every single judicial nominee he submits will be one who will try to legislate gay marriage from the bench. (...And that was written back in 2010,

---

130   (Taken from interview with Bo Dietl, storied New York City Police detective, on Fox News in May of 2009)

131   (From article "The Promise Of Spring," n *Truth & Triumph,* a publication of the Alliance Defense Fund, Volume II, Issue 2, May, 2009, by Alan Sears, President, CEO and General Counsel)

and sure enough, now in March of 2011 Obama has flip-flopped and now states that the Defense of Marriage Act (DOMA) is *unconstitutional*. Which of course is Democrat-speak for saying something is un-liberal. They could care less about the Constitution. And despite the fact that it is his job as the head of the executive branch of the federal government to *enforce* the laws passed by Congress, and not to make dictatorial decrees about their legitimacy; which is the sole purview of the *judicial branch*. But again, that's what the Constitution says, and why should he be bothered with that? He wants to "radically transform" America after all.) ...But those same African Americans couldn't find it in them back in 2006 to vote for Michael Steele, a decent, honest black man who actually does share all of their values, to be the first black Senator in Maryland. They instead went out and voted for a shallow, white, political hack. Why? What was the difference? Well, you see, our far-left media, educational system, entertainment industry and late-night clowns made sure that it was "common knowledge" that if you're black and you voted for Michael Steele, or any Republican for that matter, it was because you were "stupid," and an "Uncle Tom." Or at least not as "smart" and "hip" and "educated" as all of them. And that if you were black and did not vote Democrat then you were also "stupid," and an "Uncle Tom," or at least not as "smart" and "hip" and "educated" as them. And unfortunately, black Americans have bought into this lie, and have come to believe as a monolithic voting bloc in that "common knowledge" which is nothing more than common left-wing lies. When electing a man to the highest office in the land, basic honesty and integrity are not bad things to consider; unless you are told to vote Democrat no matter what, and that you're real stupid if you do otherwise. And I bring this up in this chapter, although we will cover it more extensively in chapters 11 and 13, because this is not so much a political, but a moral issue. And hopefully it is becoming clear that voting for the Democrat-Left is the pinnacle of immorality. But hey, *who's to say what is right and what is wrong*? Barack, pass the scissors please.

[And now in May of 2011 we have a very important postscript to the waterboarding debate. The president just announced yesterday, May 1, that a team of US Navy Seals killed Osama Bin Laden, the world's top terrorist, who has slaughtered tens of thousands of innocents worldwide, and who was the mastermind behind the murder of 3000 Americans on Sept 11, 2001. Praise God, as well as our military! They accomplished this in a daring midnight raid on his compound just outside of Islamabad, the capitol of Pakistan. (It turns out that Osama had been hiding in plain sight of the Pakistani government for years, of course with their full knowledge and protection, which begs the question of why we've been "aiding" them to the tune of 2 billion dollars a year now for the past ten years. So they can use that money to house and protect our greatest enemy? Are we stupid or what?) But the point is that this operation where the United States was finally able to exact justice on this monster was only made possible because we were able to learn of Bin Laden's whereabouts through waterboarding! Congressman Peter King explains...

"We obtained information several years ago; vital information about the courier for Osama. [And we were able to eventually follow that courier to Bin Laden.] We obtained that information through waterboarding. So for those who say that waterboarding doesn't work, who say that it should be stopped and never used again; we got vital information that directly led us to Bin Laden. ...It came from an overseas prison where Khalid Sheik Mohammed was being interrogated, and waterboarding was used."[132]

---

132 (Congressman Peter King on the O'Reilly Factor, 05/02/2011, the day after Navy Seals killed Osama Bin Laden)

So now that president Obama has been successful in killing Osama Bin Laden, the question that remains is, Will he and his Democrats and their left-wing media now change their hypocritical stance on waterboarding, stop lying and calling it "torture," and admit that if it wasn't for the Bush administrations use of it, then he wouldn't be able to stand in front of the nation right now and take credit for killing our #1 enemy in the War on Muslim psychopaths? Will you, Mr. president? Or should we not be holding our breath?]

"It's OK to *kill* the terrorists, but we can't waterboard them." Makes a lot of sense, doesn't it.

- Will the real racists please stand up? Eric Holder's "justice" department has refused to prosecute two Black Panthers who were videotaped intimidating white voters while standing outside of a Philadelphia polling place in paramilitary garb, holding a club and hurling racial epithets on Election Day, November 4, 2008. A clear violation of the Federal Voting Rights law. So why no prosecution? Is it because Holder thought these goons were set to play the part of Officer Krupke in a community theatre production of *West Side Story*, and got lost? No, that's not it. Well then could it be that Eric Holder, and Barack Obama, are a couple of ignorant racists? No? Yes? Well, we're not saying that they are or are not but just ask yourself what George Bush and John Ashcroft would have been labeled if they had refused to prosecute two Ku Klux Klan goons who were videotaped intimidating black voters by standing outside of a Mississippi polling place in white robes, holding a noose and hurling racial epithets? Exactly. They would have been called ignorant racists. And the left-wing media, and the left-wing ladies on the view, and Jesse Jackson, and Alvin Sharpton, and our left-wing late-night clowns would have screamed this accusation until they would have been forced—not to reverse their decision, it would have been far too late for that—but to resign. So, with all that in mind and using the rules of logic or common sense whichever you prefer, why exactly aren't Holder and Obama ignorant racists? Certainly they are according to the very condemnation that would have occurred in the reverse, for what works in one direction certainly works in the other, does it not? The only difference with the real case as opposed to the hypothetical one is that *Democrat* ignorant racists have the advantage of their friends being in control of the dissemination of information in our culture.

Another point is that if Obama, for one, is not an ignorant racist (and again, we're not saying that he is) then how could he have sat under the "preaching" of the ignorant racist Jeremiah Wright for twenty years if he wasn't one himself? What if Bush had sat under the disgusting spewing's of David Duke for twenty years? What would he have been called? Exactly. And in case the reader is not aware of the level of bigotry of Jeremiah Wright, he honored himself and the *Nation of Islam* leader, Louis Farrakhan, in March of 2010 with a "Living Legend" award. This is the same Louis Farrakhan who is only *legendary* for being a disgusting, racist, anti-Semitic manure pile that most decent human beings can barely stand to look at much less listen to; and who is surrounded by some of the more dangerous racist imbeciles ever to walk the face of the earth. (With all due respect of course.) And this is not the first time that Wright has chosen to honor his close racist bosom buddy:

"Barack Obama is a member of Chicago's Trinity United Church of Christ. Its minister, and Obama's spiritual adviser, is the Rev. Jeremiah A. Wright Jr. In 1982, the church launched Trumpet Newsmagazine; Wright's daughters serve as publisher and executive editor. Every year, the magazine makes awards in various categories. Last year, it gave the Dr. Jeremiah A. Wright Jr. Trumpeter Award to a man it said "truly epitomized greatness." That man is Louis Farrakhan.

"Maybe for Wright and some others, Farrakhan "epitomized greatness." For most Americans, though, Farrakhan epitomizes racism, particularly in the form of anti-Semitism. Over the years, he has compiled an awesome record of offensive statements, even denigrating the Holocaust by falsely attributing it to Jewish cooperation with Hitler — "They helped him get the Third Reich on the road." His history is a rancid stew of lies."[133]

And this same ignorant racist, Jeremiah Wright, best friends with the other documented ignorant racist, Farrakhan, was Obama's closest spiritual mentor, friend and pastor for over twenty years (until he had to be thrown off the bus so Barack could be falsely portrayed by his Democrat media as something other than what he actually was). And all three are supported, protected, defended and given cover by the left-wing media, entertainment industry, university professors and late-night clowns. (And if all of them support, protect, defend and give cover to racists, does that make them a bunch of racists as well?) (Just asking.) And they falsely accuse the Tea Party protestors of being racist! Incredible. "If you shout a lie loud enough and long enough, enough foolish, uninformed people will believe you."

- What's next?[134] I don't think it's a stretch to say that we are dangerously close to full blown socialism at this point in America (2009) with our real-life "Three Stooges"—Obama, Pelosi and Reid—now possessing free reign over the Federal Government, and with their indoctrination arm—what was once *America's* media but now is clearly owned by the Left—fully committed to propping them up. The Nazis in Germany couldn't rival the propaganda machine that the Democrat-Left now has in place.

And keep in mind that what we have discussed in the preceding pages is just a partial list, and that much of that was written in only their first 5 months. We could have mentioned much more. For example:

- They are preparing to silence their critics (in the name of free speech of course) on talk radio and on Christian radio by passing the Nazi...er...*Fairness* doctrine.

- They're going to persecute decent, moral doctors and nurses into leaving their chosen profession by forcing them to be involved in the evil of baby mutilation against their conscience and their will. Or else they will lose their job. (Of course the Hippocratic Oath, that all physicians take, states that they will "above all, do no harm." But what are oaths, or the consciences of decent people, to the made-up morals of the Democrat-Left?)

- Cairo, Egypt, June 2009: Every time Hussein calls this country a "torture nation" he's a bald faced liar; unless of course he is referring to the 50 million innocent little babies that he and his left-wing friends have mutilated and tortured to death. Then he would be telling the truth.

---

133   (Richard Cohen of The Washington Post as quoted in article on hotair.com)
134   What's next? Nothing less than the total destruction of America as we all know and love it, but let's not jump ahead, that's a subject for chapter 13)

- Rosa Brooks, the extreme far-left George Soros employee, has been placed inside the Pentagon by this administration. (The next logical step will be to appoint a high-ranking member of Al Qaeda to a Cabinet post.)

- The Democrats in Congress care so much about the energy needs of the American people that they announced a plan to finally allow offshore drilling! Only the jokers failed to mention, and their media failed to report, that they only lifted the ban on drilling <u>outside</u> of the 50 mile limit which accomplishes little because all of the oil is <u>within</u> the 50 mile limit. Oh, those Democrats! They're such kidders! And drilling in the Arctic National Wildlife Refuge (ANWR), a remote, frozen, desolate region on the Arctic Circle loaded with oil, or going after Oil Shale, or Clean Coal, or nuclear energy like the French have are all still off limits as well. But president Obama can bow before the Saudi King, because the Democrat-Left has no problem with the *Saudi's* drilling for oil. (Hey, "No Bowing for Oil!" Now there's a slogan!) The Obama-Left also has no problem sending billions of our dollars over to the Saudi's for their oil, but is vehemently against drilling for oil here even though we have more oil than the Saudi's; and even though we have to pay the Saudi's <u>eight times more</u> for their oil than we would for ours; and even though drilling here would be a great boost to our economy and would create hundreds of thousands of high-paying jobs. Then in turn we are forced to borrow huge sums of money from the Saudi's (money that we just handed them) and the Chinese to try and manage our astronomical debt that the Democrat's wild and ridiculous spending (passed off as an economic *stimulus*) just greatly exacerbated.

And now (June 2010) we have the Gulf Oil Spill which has been gushing for a few months, and the media is trying to assess blame. Of course, George Bush and Cheney are right up there at the top, but what is conspicuously absent from the list is their own complicity and that of their Democrat environmental wacko friends. And that is because the problem with capping this spill and stopping this environmental and economic disaster for the Gulf states is that this well is in deep water; mile deep water. And it is in such very deep water because the environmental wackos influenced our federal government into not allowing drilling in shallow water closer to shore where divers could have gone down in the first few days and capped the leak. They also won't allow drilling for oil in the West, or in ANWAR in Alaska, both places where if an oil leak had occurred it could have been capped in a few hours with little or no environmental damage. And not with the kind of economic damage to the people and workers and families in the Gulf states that we are looking at today. So the left-wing, lying, Democrat media will again protect the real culprits in this disaster (themselves) so their misinformed listeners will continue to go out and vote the way they're told to.

A now-famous slogan of Rahm Emmanuel, Obama's Chicago-thug Chief of Staff, is to never let a crisis go unexploited. He learned it from Saul Alinsky, author of the left's bible for fomenting change titled *Rules for Radicals*. And right in keeping with their game plan, Obama has taken political advantage of this crisis as an opportunity to suspend offshore drilling in the Gulf of Mexico, which in turn has destroyed the jobs and livelihood of a hundred thousand more U.S. workers, in the middle of a terrible economy where we can't afford to lose *any* jobs. But drilling will still continue in the Gulf. How? By our friends the Chinese. They have a contract with the Mexicans. And I'm sure their safety regulations for deep water drilling are some of the best in the business. HA!

- In the works is another push for "comprehensive immigration reform" which is Democrat-speak for giving amnesty to tens of millions of illegal's so they can enjoy their votes (but not deserve them) like they now enjoy (but do not deserve) the African-American and Hispanic vote. To accomplish this they will continue to shout the big lie, "They're just doing the jobs that Americans refuse to do." Which of course is comforting to all of the many black youths in this country who can't find a job, and to all of our unemployed in general in this recession who would do anything to feed their children and to keep from losing their homes, but can't because all of those Mexican illegals are out every day "doing the jobs that Americans refuse to do." What accomplished liars they are! But if you have no real morals, and no problem slaughtering tens of millions of little babies, then lying is as a little thing.

- Obama wants to close Guantanamo Bay because "it is a mess." But that's another bald-faced lie. There is nothing about it that "is a mess." It is actually a God-send of a perfect place to house Muslim psychopaths. Their desire to kill us has been neutered. They cannot harm America; all they can do now is throw their feces at the soldiers guarding them. (Is that taught in the *holy* Koran?) Obama and the far-left also say that Guantanamo is "a recruitment tool for terrorists." Another bald faced left-wing lie, again repeated enough times so enough people will believe it. America itself is a recruitment tool for terrorists; our very existence is what incites them. What should we do, shut down America? (Hmmm. Now that I think about it, the Democrats aren't doing a bad job of that already.)

- "Communism has only killed 100 million people. Let's give it another chance." In the Democrat-Left's ongoing struggle to subvert our American way of life (AKA its "radical transformation") they are paving the way for a socialist state by spending so much money that we will never be able to pay it back. This will lead this country into bankruptcy and then the Democrats and their media can blame it all on Bush. They've certainly had a lot of practice. For the past year (2009) our Lord of the Idiots (see Lord of the Flies) has been ruining this country with his Marxist policies while blaming everything on Bush.

> "His policies are going to prolong this recession because they are not policies designed to end the recession. The recession is [being used as an excuse] to pass his policies. The spending isn't a means to an end. It's the end."[135]

Electing Obama and his fellow socialists to "fix" our economy had about the same chance of success as hiring a baboon to fix your computer.

> "You've heard of voodoo economics. This is doo-doo economics. ...What Obama is doing, intentionally so, is trying to wreak havoc on the private sector. Trying to create as many drones of society as possible, that is people who have to rely on government for their basic food, welfare, medicines. [This is] the most reckless, massive deficit spending in human history."[136]

The end is the socialism that the true believers on the Left have been lusting after for decades. And now they have it in their grasp. Their radical, left-wing professors at Lenin State have been preparing them for this opportunity, and they will not let it pass them by, no matter what the costs; no matter how many of the poor are put out of work by the destruction of our once-free economic engine. It doesn't matter that communism and its cousin, socialism, has never worked

---

135   (Dick Morris on *Hannity* 6/22/09)
136   (Mark Levin)

wherever it has been tried. All it has done is destroy freedom, destroy economies, and saddle its victims with abject poverty and misery. (Look at Greece.) But all that is inconsequential to true believers. They tell themselves that the reason communism and socialism failed everywhere else was because they were not the ones in charge. It is now they who will make all the difference. They are the arrogant genius' who will finally make communism succeed. Just look at what Obama said to Arkansas Democrat Marion Berry to try to alleviate his concern that if Obama didn't stop ramming his reckless socialist policies down the throats of the American voter their Party was in for a bloodbath in the 2010 midterm elections, to rival or even surpass what Clinton experienced in his first midterm in 1994. "Well, the big difference between here and in '94 is, now you've got me," Obama humbly declared. Oh joy. How did we get so lucky? The "big difference" between all the other miserably failed attempts at socialism and outright communism is now we've got him! Oh happy day! That should be comforting to you, America, over the next couple of years as our economy further plunges down his stimulus hole. You get what you pay for, and now "you've got him." Practice your gardening skills. There will always be a need for picking beets down at the local farm collective. Just ask the Russians. Dosvidanya, comrades!

"What they are trying to do is create an America very unlike the America that has existed for centuries. The America that people have been attracted to by the millions from all over the world. The America that many generations of Americans have fought and died for. The thing associated with America, freedom, is precisely what must be destroyed if this is to be turned into a fundamentally different country to suit Obama's vision of the country and of himself."[137]

- The infrastructure of this nation is falling apart. Our bridges, roads, water and gas lines, our sewage pipes, most of our roughly 100,000 private, public and abandoned dams, and our electrical-power grid were only built to last so many years, and time is running out as with each passing year their deterioration gets more and more critical. But the price tag for defusing this ticking time bomb is roughly 2 trillion dollars.[138] So what has the Obama administration done? Well almost half of that 2 trillion he just wasted on his left-wing spendulus bill. The other half of that 2 trillion he wants to waste on a government takeover of our entire health care system, the finest in the world, so they can redistribute more wealth. And as our infrastructure continues to crumble we are literally guaranteed that the money necessary to fix it will never be available.

- They are preparing to pass a "hate Christians" bill disguised as "hate crimes" legislation. It is intended to jail Christian preachers, or anyone else who think that homosexuality is not the wisest choice to make in life; or that clitoris-slicing Islam is not the most intelligent religion one could aspire to. (This is already happening in Western European countries.) But, thank heavens, it *will* protect pedophiles.

"Group dynamics and the decision to sacrifice the individual for the group are the first steps in the march down the road to Thought Crimes (euphemistically termed Hate Crimes) and totalitarianism in its purest form."[139]

---

137   (Thomas Sowell, African-American author, as quoted by Jon Voight on Fox News *Huckabee*, 09/05/2009)
138   (From *The Crumbling of America*, a 2 hour documentary "on the state of the nation's infrastructure from bridges and highways to pipelines carrying water and gas." The History Channel, 09/12/2009)
139   (Pg. 46, *The New Thought Police: Inside the Left's Assault on Free Speech and Free*

Criminalizing politically incorrect beliefs about homosexuality is one of the Democrat-Left's hidden purposes in this legislation. It is a Nazi technique, using the government to coerce people to think like them, and to criminalize and silence those who do not. That is the hidden intent behind hate crimes legislation. Remember, to these clowns the real terrorists aren't even terrorists, Christians are. The terrorist label is now used for their political and ideological opponents.

- Their pals at the ACLU are preparing to outlaw praying in Jesus' name by military chaplains, something they *have* been doing (in this <u>Christian</u> country, Wilbur) for two hundred and thirty some years.

- The Employee Free Choice Act, which allows unions to intimidate workers into signing up by destroying the secret ballot, is being shoved through Congress. It has nothing whatsoever to do with the "free choice" of employees. Fascist's do not give a damn about anyone's "free choice." (Unless they are using it to mutilate babies to death. That kind of freedom of choice fits right in with their value system.)

- There's the "Tax and Stifle Trade" bill (euphemistically called "Cap and Trade"), which is designed to return our energy usage back to that of the Middle Ages, and drastically increase the energy bills of every American at the same time. Al Gore, however, has already been rewarded with over 200 million dollars for his dishonest global warming campaign, and will continue to reap many more millions for his selfless quest to save his financial environment. Besides, the "environmental wacko" movement and global warming scare are nothing more than a convenient vehicle for the Democrat-Left to accomplish their goal of destroying capitalism. Listen to what Van Jones—the self-avowed communist, signer of the "9/11 Truther" letter, and Green Czar appointee of the Obama administration—had to say:

> "Right now we're saying we want to move from suicidal gray capitalism to some kind of eco-capitalism where, you know, at least we're not, you know, fast-tracking the destruction of the whole planet. Will that be enough? No, it won't be enough. We want to go beyond excess and exploitation and oppression altogether, but that's a process."
>
> "Don't stop there. Don't stop there."
>
> "No, we're going to change the whole system."

They want to radically transform *our whole system*, from capitalism to communism of course. The "man-made" global warming scare is just a scam to help them reach their goal.

But it gets even better: another of Obama's far-left appointees, Cass Sunstein, thinks that horses, because they are "more rational and more conversable than an infant baby of one day, one week or one month" should be given the legal right to sue humans. (The ACLU's caseload of financially lucrative lawsuits must be running low.) Ann Coulter responds:

> "This is mainstream liberal thinking as described in *Godless*. [One of Ann's bestsellers.] Peter Singer, professor at Princeton, argues that we should be killing disabled babies but that gorillas should have rights ...perhaps to represent themselves. [Another] guy who had written a piece ...for the L. A. Times or some newspaper compared the slaughter of pigs for food to the holocaust. PETA complained about a donkey being sent in by the PLO to blow up Israeli human beings, because of the donkey. Not because of the Israelis who the donkey was carrying the bomb to try to kill. This is mainstream liberal

Minds, by Tammy Bruce)

thinking because **if you are godless you have no morals**. [Emphasis mine] They are [godless] and they can't formulate an argument against the killing of a baby in the womb but [they can] for a donkey."[140]

- And last but not least there's the Obama-Left's gargantuan attempt at a Government takeover of our entire health-care industry (this was written before they actually did it), "Most of us have health insurance, and we like it. It's not a question of helping those who don't have it; he wants to destroy the healthcare system that most of us like."[141] Or as Mike Huckabee put it, "He's trying to sell a product that no one wants to buy."[142] But what a small price to pay for the level of health care that say, Cubans get to enjoy. Just ask any Crockumentary-making liar like Michael Moore, or go to Canada or England for an urgent medical procedure and wait in line for a number of months until the answer comes to you. Death.

The following is a conversation between Newt Gingrich and Sean Hannity on this subject…

> Newt: "I think it is a very disappointing part of how he is really throwing away the opportunity he had. He was elected on the idea of change you can believe in …on being beyond partisan politics, on trying to bring people together [as long as those people do not include any conservatives, Christians, capitalists or Republicans with intelligent ideas], and then you end up with this petty, partisan blaming."[143]

Gingrich was referring to the president's press conference a few nights previous where he disingenuously tried to blame Republicans for his failure to shove his disastrous, socialized, government-run health care debacle through Congress; when it is clear that Obama's Democrats completely control both houses of Congress and there is absolutely nothing the Republicans can do to stop him. But what are a few bald-faced lies when compared to the noble goal of destroying the finest health care system on the planet in favor of your left-wing professor's barely comprehensible Marxist version? Newt continued…

> "[Obama is] very eloquent, but when you slow down and you look at his eloquence it doesn't hold together and the reason is he can't share candidly with the American people what he's trying to do because his advisors tell him that on almost every issue now the American people are opposed to what he's trying to do. So if he's honest that it's an energy tax ["Cap and Trade"] he's going to lose. If he's honest that it's a huge government-run healthcare program, he's going to lose. He's still very articulate, but he uses the language to actually muddle what people understand.

Translation: he's a liar. He has to be, because Americans are not socialists, and he is. If Hugo Chavez, George Soros, Fidel Castro, Kim Jung Ill, Michael Moore, our Democrat media, our university professors, Keith Olberman, Chris Matthews, moveon.org, and our funny late-night clowns were his only constituents, he could be totally honest. But they are not. The American people are. And they need to be fooled, or to be kept in a willfully ignorant state, in order to keep going out and electing them.

> Hannity replies: "He has broad …sweeping generalizations, platitudes, bumper stickers, slogans, fear mongering, and he's got that down to a science almost, but there are no specifics about what it is and how it's going to actually impact the American people. So it makes me wonder is it just that he's looking for victory. Is he looking to really in his

---

140   (Ann Coulter on Fox News *Hannity*, 9/11/09)
141   (Mark Levin on Fox News *Hannity*, 07/01/2009)
142   (Mike Huckabee on Fox News *Hannity*, 9/03/2009)
143   (Newt Gingrich on Fox News *Hannity*, 07/24/09)

heart and soul change America, alter America because it's not the country that he likes?"[144]

Bingo! It's not Russia. Or Cuba. Or Venezuela. Or Indonesia. And Andrea Tantaros, speaking on *The O'Reilly Factor* about Obama's trip to lecture little school children in the middle of the health-care town hall riots, nailed the president's disconnect…

"We know where we are and we're not happy with it. That's why he has to address a bunch of school children, people who will believe what he's saying because they really were born yesterday. [Emphasis mine.] Look, this is …not about the American economy or the American people anymore. This bill is about Democratic majorities in 2010 and 2012. They're trying to save their own hides."[145]

(That was said in September of 2009, and we all know now how well "saving their own hides" worked out for them in the 2010 elections … We can only pray that the same holds true in 2012.) Miss Tantaros was referring to elementary school children in her "born yesterday" comment. But she could just as well have been referring to all of those uneducated college kids who were duped into bounding out and voting for him last November; tongues hanging out of their mouths like a Jordan lay-up; thrills shooting up their legs and barrels full of hope and change accompanying them into the voting booth; while they carefully checked their brains, and any semblance of common sense, at the door. The end result of 16 years of Democrat-Left indoctrination.

Obama went on to accuse his political opponents of "lying" about his health care catastrophe the other night in his campaign speech to the joint houses of Congress (Sept. 2009). But what his opponents have actually done is to accurately report just what is in his bloated, socialist, idiotic and incomprehensible bill. That's *lying*. Conversely in Obama's speech he said a number of lies that were so obvious they could be documented before he even finished speaking. That's *speaking the truth*, if you're a Democrat. One instance where Obama accused his opponents of lying was over the accusation that his health care monstrosity called for "death panels" where Doctors would council the elderly about their "end-of-life options;" where one option would be saving the government a whole lot of money on expensive medical procedures by talking the elderly person into "ending-their-life-early" for the $good$ of all. He said: "Such a charge would be laughable if it weren't so cynical and irresponsible. It is a lie, plain and simple." But the truth is, those death panels are real and "such a charge by Obama would be laughable if it weren't so cynical and irresponsible. Obama's a liar, plain and simple." (Thank you, Mr. President, we couldn't have said it better ourselves.)

And now (January of 2010) he's still at it; this time by planning to mis-appropriate a congressional technique called *reconciliation* in order to defy the will of the people.

"Obama is a domestic terrorist. He is now terrorizing the nation by threatening to shove his health care monstrosity, which the vast majority of Americans do not want, down their throats anyway. They are very afraid, and justly so, of the disastrous effect it will have on the deficit, the economy and health care itself."[146]

But of course, the problem is the American people just don't "get it." They don't understand that "now we have him."

"For we wrestle not against flesh and blood, but against principalities, against

---

144   (Newt Gingrich and Sean Hannity on *Hannity*, 07/24/09)
145   (Andrea Tantaros, conservative columnist on *The O'Reilly Factor*, 9/03/09)
146   (Oops. Lost the source of this one…)

powers, against the rulers of the darkness of this world, against spiritual wickedness in high places." (Ephesians 6:12) ...and against the forces of darkness that now (in January of 2010) control the White House and both Houses of Congress.

Yvonne Donnelly was a lifelong liberal before 9-11:

"On 9/11 I said "How could this happen?" And [my husband] jumped off the couch and said, "Because of you ...and your liberal friends." And I said, "you're right, and I've been wrong." And I woke up."[147]

Unfortunately, the rest of America didn't wake up along with her. At least not that half of America that went out in 2006 and again in 2008 and elected these liars, these socialists, these baby mutilators. And now we are reaping what we have sown...

"While suturing up a cut on the hand of a 75 year old rancher ...the doctor struck up a conversation with the old man. Eventually the topic got around to Obama, and his being our president. The old rancher said, "Well, ya know, Obama is just a Post Turtle." Not being familiar with the term, the doctor asked, "What's a "Post Turtle"?" The old rancher said, "When you're driving down a country road and you come across a fence post with a turtle balanced on top, that's a post turtle." The old rancher saw the puzzled look on the doctor's face so he continued to explain. "You know he didn't get up there by himself, he doesn't belong up there, he doesn't know what to do while he's up there, he sure as heck ain't goin' anywhere, and you just wonder what kind of dumb ass put him up there in the first place." "[148]

With all due respect of course to Hollywood, our media, our university professors, the late-night clowns and the American voters who did.

"You are of your father the devil, and the lusts of your father you will do. He was a murderer from the beginning, and abode not in the truth, because there is no truth in him. When he speaks a lie, he speaks of his own: for he is a liar, and the father of it." (John 8:44)

- Marxism: The Opiate of the Asses![149] Karl Obama Marx and his merry band of socialists, communists, progressives and Chicago thugs have taken over our federal government. They possess a complete disdain for the United States of America and its Constitution. They are fueled by their arrogant and ignorant leftist notion that they know better than everyone else, and that they and they alone can "fundamentally transform" America into a "much better place for all." But in reality all they will accomplish if they are not stopped is to turn America into the United Soviet Socialist States of America; or a banana republic. Pelosi and Reid and their gang of lapdog Democrats infest the halls of our Congress. They have complete power because of their majorities in both houses (written in 2009 prior to them losing the House of Representatives in November 2010). President Obama bleats incessantly about "the obstructionist Republicans" yet he knows full well that there is nothing they can do to stop him and his anti-American policies. Therefore, he lies. It is his own Democrats in the House and the Senate that stand in his way. They see the reaction of their constituents back home and they are worried about their evaporating chances for re-election. The media, our

---

147   (Yvonne Donnelly on Fox News *Glenn Beck Show*, 09/12/2009)
148   (From email circulating on the net)
149   (Sign at Tea Party Rally, Washington D.C. June, 2010)

entertainment industry, our educational system and the late-night clowns are all firmly in the camp of the Democrat-Left, even though they only represent the views of about 5% of the population. But that's how it works in communist and totalitarian countries. The scum rises to the top. And in this case the barbaric, baby-mutilating few have been handed the power to stomp on the freedom and the rights of the many. And this country is burdened with a misinformed and duped electorate that gets their information from that same lying Democrat media, entertainment industry and their late-night clowns, and even though that media despises the values that most of those voters hold dear. And yet they still march out cluelessly and obediently every two years and vote for them. (If it wasn't for the imminent return of Christ to rid the earth of evil—see Chapter 9, Bible Prophecy—I'd be downright depressed.)

Bill Maher, that left-wing genius, who never met a baby-mutilator he didn't like, says America is stupid because there still remains the possibility, even after years of him and his kind vilifying, disparaging, demeaning and lying about Sarah Palin, that they might still go out and elect her—a decent, honest American who actually loves her country (and coincidentally, does not mutilate babies) (that last one really has him agitated). And he of course is right. America is stupid, but not if they were to elect Sarah Palin. America is stupid if they were instead to continue to go out and vote like him.

But let's be honest, Obama is not the antichrist. In fact, I'm sure Barack Hussein[150] is a real nice guy. You just have to ignore all the left-wing lies; the baby mutilation thing; the attempts to wipe out free-speech; the sabotage of our National security in a time of war; the threat to illegally prosecute those who did keep us safe; the refusal to secure our borders because uneducated, illegal aliens from Mexico are guaranteed Democrat voters even though they will continue to steal jobs from black youths; the screwing of 1700 poor black and Hispanic school children in D.C.; the treating of Islamic war criminals like they were US citizens; the attempted destruction of the greatest health-care system on the planet just so his government can take it over; etc. etc. I'm also sure that Barack is very intelligent. (He certainly is well spoken.) But again you just have to ignore all of the aforementioned things that seem to be "above his pay grade." Also, intelligence in and of itself is not a prevention against becoming one more typical left-wing-indoctrinated campus-lefty; deceived by and indoctrinated with the America-hating, Christian-vilifying, socialism-loving lies of the left.

The following is from Star Parker. She is an African American, conservative, Christian commentator, the founder and president of CURE, the *Coalition on Urban Renewal and Education*, author of *Uncle Sam's Plantation: How Big Government Enslaves America's Poor and What You Can Do About It*, and the black-race-enslaving Democrat-Left's worse nightmare:

"Expanded government has already destroyed the black community. These guys [referring to black race baiters like Charlie Rangel, Jesse Jackson, Cynthia McKinney, Van Jones, etc.] are like leaches. They suck blood from both ends. They've already sucked the life out of black America with education policy and health policy and social security policy and jobs, and now they want the rest of society to buy into this plantation mentality. And

---

150   (It's OK to use his middle name now, because he's already been elected so there's no need to hide it anymore, and besides now it comes in handy overseas with his new Muslim best-friends. McCain, the dunce, won't have to throw anyone offstage anymore.)

what is even more unique is that I am hearing from record numbers of African Americans that are saying "enough of this stuff." "[151]

One can only pray that *is* the case.

"The real betrayal of this presidency; the premise of the campaign last year, which he talked about endlessly and the audience was swooning over this, was he was going to introduce a new politics ...a politics of the people. ...That was what he represented- all that was rubbish last year and now it's all the more so."[152]

"The defining idea of the Obama campaign ...was that he's different from the other candidates. It's not that he had more experience, obviously he didn't. It's not because he knows more about the issues ...or because his ideology is purer. It wasn't that. It was that he would change Washington and he would change business as usual, get rid of the polarization, be bipartisan."[153]

And it was all a lie. Just to get elected. Would it be bad form if we all said on the count of three, "SUCKERS"? Fortunately, though, there is a silver lining. For what the left-wing, lying, Democrat media has gone to great lengths over the last few decades to keep hidden from the American voter—namely the truth about their radical, clueless, socialist, Democrat politicians and their anti-American agenda—the Obama administration and their far-left congress are clearly revealing. That's one gift that America *can* thank them for...

"[Obama] lied. ...He's taken his page right out of Saul Alinski's *Rules for Radicals*.[154] Saul Alinski essentially said, "You've got to sound like you're for the middle class. You've got to act like you're from the middle class and then you destroy the middle class." "[155]

Ultimately, though, Obama is not solely to blame for who he has become. After all, he is just another left-wing indoctrinate. He is a product of our educational system, and of our left-wing universities. It is *we* who bear a part of the blame. We have allowed our educational system, in addition to our media and our entertainment industry, to be taken over by these left-wing ideological wackos, who couldn't solve a problem in the real world if their tenure depended on it. In any educational system the size of ours certainly you're going to have some wacky ideas, unsubstantiated by any facts or evidence, but to have allowed those asinine ideas to end up controlling the curriculum of an entire educational system is a pathetic state of affairs indeed. And we continue, as a clueless nation, to allow our students to waltz through these schools and be brainwashed by them, and then we wonder why we end up with a president that Al Qaeda, or Hugo Chavez, couldn't have done a better job of picking. But this is what you get when you are a country that for years has allowed itself to be intimidated by the question, *who's to say what is right and what is wrong*? That has allowed the moral relativists

---

151   (Star Parker on Fox News *Your World with Neil Cavuto*, 9/03/2009.)
152   (Charles Krauthammer on *Special Report with Brett Baier* 7/01/09)
153   (Fred Barnes on *Special Report with Brett Baier* 7/01/09)
154   (Saul Alinski's famous step-by-step instruction book for the radical Left on how to surreptitiously takeover America. The Clinton's and Obama are some of his most successful followers.)
155   (Mark Levin)

who advance that question to be in charge of the indoctrination of our school children. That has also been duped into accepting the removal of God, and the morals that go with him, from every part of our educational system by a bunch of ACLU, constitution-perverting traitors in our judicial system. It is we as a nation who have ceded control of not only our educational system, but our media, our entertainment industry, our courts and our entire culture to Democrat-Left elites whose minds and "morals" are not too far removed from the depths of a clogged toilet. We are too uninformed and too timid to shout back the answer to their clever, *who's to say* question, <u>GOD IS, STUPID!</u> (see chapter 5) and thus incapable as a nation of pushing back against them, their perverse ideology, and their carefully orchestrated poisoning of the minds of our youth, of whom Barack Obama is only one. And now we have given them their ultimate gift, the power for them to finally, after all these years of plotting and scheming and protesting and lying and waiting, to "radical transform" our country into a socialist utopia. Unfortunately though, it will be transformed into the kind of *third-world* socialist utopia with the kind of *robust* economy where you'll be lucky to go out into the woods each day and eat bushes for a living. Look at Haiti. Yea, YES WE CAN!

The radical Left figured out their strategy a long time ago. It goes something like this: "Hey, let's tell all the people that will listen to us that if they vote Democrat they're smart, and if they vote Republican they're stupid, and it'll work!" And sure enough it has, because again they have cleverly manipulated the spread of information in our country through their iron-fisted control of the media, the educational system, the entertainment industry, and of course the late-night clowns. And their message has gotten across loud and clear and has become "common knowledge" to far too many Democrat-voting disciples who have either failed, or refuse, to educate their selves about what is really going on.

"25 years and my life is still, trying to get up that great big hill of hope, for a destination. I realized quickly when I knew I should that the whole world's made up of this brotherhood of man, for whatever that means. And so I cry sometimes when I'm lying in bed, just to get it all out, what's in my head, and I, I am feeling a little peculiar. And so I wake in the morning and I step outside, and I take a deep breath and I get real high and I scream from the top of my lungs, WHAT'S GOIN' ON?"[156]

Those lyrics are from the group, 4 Non Blondes, but they are indicative of the cry of an entire generation whose heads have been filled with the evolutionary, godless, morally relative mush of the Left that has left them with no foundation, no meaning, no purpose and no real direction in life. They are wanderers across the face of the earth, a lost generation, their heads filled with lies, crying out for direction and meaning, and to make some sense of the madness that surrounds them. They are crying out for the truth, whether they even realize it or not, being as how they have also been taught that it doesn't even exist. And instead of someone to give them what they cry out for, what they desperately need, what do they get? Jon Stewart.

We would ask your forgiveness, however, if our brutal honesty here is found offensive by some whom the morally bankrupt, "malignant narcissists" of our Democrat, lying media, educational system and entertainment industry have been

---

156   (From song, "What's Going On," by 4 Non Blondes)

successful in snookering. People believe what they are taught, and even though we like to think of ourselves as smart and savvy, many times we are not. And buying into the only political message the left-wing media knows how to report— "Democrats good! Christians and conservatives bad!" "Baby mutilators real smart! Christians and conservatives real stupid!" (See Spot Run) (Me Tarzan, you Jane)—is the exact opposite of smart and savvy. It's being made a fool of by rank Nazi propaganda. It's listening to really stupid people tell you how they and their ideologically-brainwashed, Democrat friends are "smart," and how the intelligent thoughtful folks who oppose them are "stupid." And God help you if you are black and you dare to stray from how they insist you must think and vote. They'll do to you what they did to Clarence Thomas and Condoleezza Rice, smearing them even to the point of using disgusting racial stereotypes, including the "N" word. You see, it's OK to be a disgusting vile racist as long as you are a white liberal journalist or cartoonist, and you are attacking conservative blacks. We'll cover this further in the chapter on Satan, the great deceiver, but note for now that he, and his followers, are very adept at "calling evil good and good evil," and smart people *stupid*, and stupid people *smart*.

> "While the people are virtuous they cannot be subdued; but when once they lose their virtue they will be ready to surrender their liberties to the first external or internal invader."[157]

That was from the *prophet* Samuel Adams, one of this country's Founding Fathers. The "death of right and wrong," the destruction of moral absolutes, the surrender of our culture and our government to the *who's to say what is right and what is wrong* crowd has seduced this people into giving up, as a culture, their virtue, and now they are blindly and ignorantly surrendering their liberties to an internal invader, the "malignant narcissists" of the Democrat-Left. And now we have been given their smooth-talking front man, "nihilism with a happy face," who says whatever he needs to say, but then does what he and his far-left elites had in mind all along.

"Now let us hear the conclusion of the whole matter"

"Fear the Lord and depart from evil." (Proverbs 3:7)

Well there you have it, our readers digest version of America's descent into the morals of hell. Tens of thousands of additional examples could have been added if we had thousands of pages for this chapter alone. But we do not. But to anyone who seeks the truth the conclusion is certain, that when you reject moral absolutes that can only come from above you are left with the immorality that rises from below. When you preach the doctrine, *who's to say what is right and what is wrong,* in order to remove moral absolutes, then the answer that comes back is, *For you, stupid, Satan is*. That's not the answer the Democrat-Left is looking for, but it is the reality that the great deceiver has beguiled them into embracing. Ultimately moral relativism is just a smoke screen to cover the real intent, which is simply to call evil good and good evil. (As in, "Woe unto them who call evil good and good evil; who put darkness for light, and light for darkness!" Isaiah 5:20) Rejecting the Creator's moral law of right and wrong is not a liberating thing; it

---

157   (Samuel Adams, one of our Founding Fathers)

is more of a demonic thing. The Left has no firm foundation of moral absolutes to stand on, they have foolishly and categorically rejected that. Therefore "they stand for nothing," and because of that, "they will fall for anything," no matter how stupid, illogical, irrational, immoral, evil or outright insane.[158]

"One principle that today's intellectuals most passionately disseminate is vulgar relativism, "nihilism with a happy face." For them, it is certain that there is no truth, only opinion: my opinion, your opinion. ...To surrender the claims of truth upon humans is to surrender Earth to thugs. It is to make a mockery of those who endured agonies for truth at the hands of torturers. Vulgar relativism is an invisible gas, odorless, deadly, that is now polluting every free society on earth. It is a gas that attacks the central nervous system of moral striving. The most perilous threat to the free society today is, therefore, neither political nor economic. It is the poisonous, corrupting culture of relativism.

...During the next hundred years, the question for those who love liberty is whether we can survive the most insidious and duplicitous attacks from within, from those who undermine the virtues of our people, doing in advance the work of the Father of Lies. "There is no such thing as truth," they teach even the little ones. "Truth is bondage. Believe what seems right to you. There are as many truths as there are individuals. Follow your feelings. Do as you please. Get in touch with yourself. Do what feels comfortable." Those who speak in this way prepare the jails of the twenty-first century. They do the work of tyrants."[159]

"Nothing short of a great Civil War of Values rages today throughout North America. Two sides with vastly differing and incompatible worldviews are locked in a bitter conflict that permeates every level of society. Bloody battles are being fought on a thousand fronts both inside and outside of government."[160]

Nevertheless, our battle here is ultimately not against the Left. They are not the enemy. They are just dupes. Our fight here is against *the lies* of the Democrat-Left, which come from their father Satan, the great deceiver who goes forth to deceive the whole world, who was a liar from the beginning, and the father of lies. Our fight is against him. He is the real enemy. (See chapters 7 and 13)

As we have exposed the seductive trap of moral relativism in this chapter we have singled out a few of its proponents on the Democrat-Left, and they have taken it a little on the chin, and for the most part deservedly so. But nevertheless we want to make it clear that it is not our purpose here to put down or to marginalize, disparage or deride these folks; and despite the fact that they have had a field day unfairly and falsely putting down, marginalizing, disparaging and deriding their opponents for decades. But our main purpose here is not to put down but to expose their lies to the light of truth. And to try to wake up as many as possible in the process. For in reality those on the Democrat-Left are our lost brothers

---

158   (For outright insane, see chapter 10, The Theory of Evolution is a Joke)
159   (Michael Novak at Westminster Abbey on the occasion of receiving the Templeton Prize, May 5th, 1994)
160   (Dr. James C. Dobson, author and host of *Focus on the Family*, a Christian radio program)

and sisters imprisoned by the lies of the god of this world. They are in great need of our prayers, and of whatever seeds of truth we can plant their way. Again, ultimately our battle is not with them but with the lies that, and the liar who, deceives them. Also, it is very important to realize, especially for a right attitude, that "there but for the grace of God go you or I." So it is not a self-righteous thing or a "we're not as stupid and duped as you" thing even though as imperfect human beings it is hard not to come across that way at times. But it is more the thankfulness of those who have been fortunate to have been led from the darkness to the light, from the lies of Satan to the knowledge of the truth, and an attempt to pass on this knowledge without apology so others will have a chance to do the same. When we have fallen short (like suggesting that our president resembles the bumbling prison commandant on the TV series Hogan's Heroes) we can only ask your indulgence and the forgiveness of any folks on the left who might have been offended. Besides, lest anyone think that we feel we are "better than" or "smarter than" or "more righteous than" anyone on the Democrat-Left, 31 years ago, in 1980, I was the <u>stupid, uneducated, left-wing-indoctrinated dupe</u> who ran out and voted for Jimmy Carter, one of the worst presidents in the history of the nation, who now is the most mentally unstable, anti-Semitic ex-president in the history of any nation. So I know from firsthand experience how easy it is to be trapped in the lies of the dark side; and just how important it is to be so fortunate as to have been rescued out of that nightmare. (See chapter 7) Our ultimate goal here is not to condemn or to be self-righteous (although I'm sure we have fallen and will continue to fall short at times), but for the Grace of God to drag as many others out as possible. (Consider this chapter an intervention.) And even if that "many others" is only just one, this project will still have been worth it. Thank you and may God richly bless...

"Behold, I set before you this day a blessing and a curse: A blessing, if you obey the commandments of the Lord your God which I command you this day; And a curse, if you will not obey the commandments of the Lord your God, but turn aside from the way which I command you this day." (Deuteronomy 11:26-28)

See:
- *Right and Wrong: A Case for Moral Absolutes*, Martin R. De Haan II
- *There are Moral Absolutes*, Michael A. Robinson
- *The Devils Delusion: Atheism and Its Scientific Pretensions*, by David Berlinski
- *The Death of Right and Wrong: Exposing the Left's Assault on Our Culture and Values*, by Tammy Bruce
- *Hollywood VS. America*, by Michael Medved
- *Slouching Towards Gomorrah: Modern Liberalism and American Decline*, by Robert H. Bork
- *Indoctrination U.: The Left's War Against Academic Freedom*, by David Horowitz
- *The Professors: The 101 Most Dangerous Academics in America*, by David Horowitz
- *Bamboozled: How Americans are being Exploited by the Lies of the Liberal Agenda*, by Angela McGlowan

- *Temple at the Center of Time*, by David Flynn
- *Culture Warrior*, by Bill O'Reilly
- *The Church of Liberalism: Godless*, by Ann Coulter
- *The ACLU VS America: Exposing the Agenda to Redefine Moral Values*, by Alan Sears and Craig Osten
- *110 People Who are Screwing up America*, by Bernard Goldberg
- *What Wives Wished their Husbands Knew about Women*, by Dr. James Dobson
- *Seduced by Sex: Saved by Love- A Journey out of False Intimacy*, by Jan Kern
- *Catastrophe: How Obama, Congress, and the Special Interests are Transforming... a Slump into a Crash, Freedom into Socialism, and a Disaster into a CATASTROPHE ...and How to Fight Back*, by Dick Morris and Eileen McGann)
- *Party of Defeat: How Democrats and Radicals Undermined America's War on Terror Before and After 9-11*, by David Horowitz and Ben Johnson
- *Treason: Liberal Treachery from the Cold War to the War on Terrorism*, by Ann Coulter
- *The New Thought police: Inside the Left's Assault on Free Speech and Free Minds*, by Tammy Bruce
- *Uncle Sam's Plantation: How Big Government Enslaves America's Poor and What You Can Do About It*, by Star Parker
- *Useful Idiots: How Liberals Got it Wrong in the Cold War and Still Blame America First*, by Mona Charen
- *If Democrats had any Brains, They'd be Republicans*, by Ann Coulter
- *Culture of Corruption: Obama and His Team of Tax Cheats, Crooks, and Cronies*, by Michelle Malkin
- *A Slobbering Love Affair: The True (And Pathetic) Story of the Torrid Romance Between Barack Obama and the Mainstream Media*, by Bernard Goldberg
- *Media Malpractice: How Obama got Elected and Palin was Targeted*, a film by John Ziegler, go to YouTube or howobamagotlected.com
- *Liberty and Tyranny: A Conservative Manifesto*, by Mark R. Levin
- *Real Change: From the World That Fails to the World That Works*, by Newt Gingrich
- *Judicial Tyranny: The New Kings of America?* By Mark I. Sutherland
- *Men in Black: How the Supreme Court is Destroying America*, by Mark R. Levin
- *America Alone: The End of the World as We Know It*, by Mark Steyn
- *Power to the People*, by Laura Ingraham
- *Slander: Liberal Lies about the American Right*, by Ann Coulter
- *Fleeced: How Barack Obama, Media Mockery of Terrorists Threats, Liberals Who Want to Kill Talk Radio, The Do-Nothing Congress, Companies That Help Iran, and Washington Lobbyists for Foreign Governments are Scamming Us... and What to Do About It*, by Dick Morris and Eileen McGann
- *Liberal Fascism: The Secret History of the American Left from Mussolini to the Politics of Meaning*, by Jonah Goldberg
- *The Roots of Obama's Rage*, by Dinesh D'Souza
- *The Amateur: Barack Obama in the White House*, by Edward Klein

CHAPTER 3

THE SPIRITUAL UNIVERSE

"To see a world in a grain of sand
And heaven in a wildflower
To hold infinity in the palm of your hand
And eternity in an hour"[161]

We are going to lighten things up just a little (or get even *heavier*, depending upon your perspective). In the first two chapters we focused upon two foundational lies that have taken root in academia and have spread like a cancer throughout our entire culture. Now we are going to shift gears and look at two foundational building blocks of our everyday reality, the very nature of this universe and the very nature of man. First, in this chapter, science itself will identify the essence of the very substance that comprises this universe, and next, in the following one, our own consciousness will reveal the very essence of who we are. Now we realize that these are discussions that most folks don't have the luxury of spending much time thinking about as they go through their day-to-day lives. Most are far too busy just trying to keep a roof over their heads and food on their families table. (Especially in our new socialist, third-world economy.) But we still consider the truth about both these subjects to be of some foundational importance, especially in coming to an "understanding of all things." Even though gaining this particular knowledge is not nearly as important as that of the three chapters that follow: 5, 6 and 7. But still, it can be very valuable. First, by destroying a number of preconceived notions, assumptions and attitudes that are false, and that many times can be complicit in keeping people from coming to that much more crucial knowledge. Also this information provides a very fundamental, sound and secure foundation, although not in the traditional sense, in which to view one's inner and outer universe. Hopefully you will see what I mean, but if not, if this material seems either too heavy or too irrelevant, then it won't hurt to move on to those more vital *secrets* of this universe covered in the three chapters (5-7) which follow.

The "spiritual" universe? What's up with that? I mean, we all *know* we live in a very <u>physical</u> universe, right? Wrong. Let's see…

But first, know that we are *not* talking about some kind of occultic, one-too-many-hallucinogens, Shirley MacLaine-type view of the universe. We want to look at the actual physical reality that comprises this universe; and in doing this we're not going to rely on any philosophical theories, abstract concepts or mystical leaps of faith. We are going to consider this subject only in terms of what are established scientific facts about the physical nature of this universe. Those facts that have been revealed to us over the past century from nuclear physics and the study of the nature of matter; the study of the basic "building blocks" of the universe. We want to examine the assumptions that humans have held for millennia about the nature of our reality. And we are going to turn them on their head because, as we shall

---

161   (From *Auguries of Innocence*, by William Blake)

see, in this physical universe that we all live in, there is actually nothing *physical* here. Let's take a look...

If we examine any type of physical material—your hand, this book, the desk, the concrete on your driveway, the rocks in your garden, the leaves on the trees, the dirt in the lot across the street—the common denominator is that each is made up ultimately of exactly the same material. Each has the same essential building blocks. And that is, *Atoms*, which are the basic units, or building blocks, of all matter. They are extremely minute in size. One speck of dust is still itself made up of trillions upon trillions of atoms. Yet atoms are themselves made up of even tinier particles. They have a nucleus comprised of protons and neutrons (positively and neutrally charged particles), and electrons (negatively charged particles) which orbit around each nucleus. The number of protons and electrons in each atom determine which element it is; whether hydrogen, oxygen, carbon, potassium, nitrogen, sodium, titanium, iron, copper, cobalt or zinc just to name a few. Every type of matter in the universe is made up of atoms of one element or another, or of a combination thereof.

Now if we take a close-up look at an atom, this infinitesimally small building block from which all matter is comprised, we find as we just mentioned that it is itself made up of much smaller particles: protons, neutrons and electrons, with a great deal of empty space in between them. The atom is therefore not a solid chunk of matter. It's not a hard little marble. It is actually the furthest thing from solid. It turns out that the atom is 99.99999999% empty space. That is because the space within each atom between its nucleus and the orbiting electrons is enormous. It is comparable to the empty space in our solar system, with the sun representing the atomic nucleus and the planets the corresponding electrons which orbit the nucleus. Only on an infinitesimally small scale. To get an idea of the spatial relationships within this tiny atom, imagine that its nucleus was the size of a basketball, which would be millions of times larger than it actually is, and therefore the corresponding electrons would be about the size of a grape seed, or about 1/2000th the size of the nucleus. If the basketball (nucleus) was in your front yard then the orbiting grape seeds (electrons) would be on the other side of town, about 4 miles away. With nothing in between. Just empty space. And that is the major component of each individual atom that makes up everything in the universe... Nothing. Just empty space. 99.99999999% empty space.

"Emptiness here, emptiness there, but the infinite universe stands always before your eyes."[162]

"The diameter of a typical atom, out to where the electrons circulate, is about one ten-billionth of a meter. This can be written as 10^{-10} meter, a length known as one *angstrom*. The thickness of a single page of this book is about one million atoms, or one million angstroms. The atomic nucleus is 100,000 times *smaller* than a single angstrom, written as 10^{-15} meter and known as a length of one *fermi* or *femtometer*. Suppose the nucleus of an atom could be enlarged to the size of a baseball. Then the outer electrons would orbit at a distance of about three miles, or five kilometers. This illustration shows that an atom is mostly empty space."[163]

---

162   (From "The Great Way," by Sengstan (Third Zen Patriarch), translated from the Chinese by Richard Clarke, aka Ram Dass.) (Tell them both I said hi)
163   (Pg. 24, *Thousands... Not Billions: Challenging an Icon of Evolution, Questioning the*

Again if we look at our solar system as a very large atom we can see with just a passing glance at the night sky that, despite the enormous size of the sun and the individual planets, they are dwarfed by the far greater amount of outer space that engulfs them. The stars and their solar systems that comprise the galaxies that comprise the universe, like the tiny atoms that comprise the molecules that comprise all physical matter, are 99.99999999 % empty space. (The macrocosm mirrors the microcosm, and vice versa.)

So take another look at that "solid" table. 99.99999999% empty space. The rocks in your garden. 99.99999999% empty space. The concrete on your sidewalk. This book. The floor underneath your feet. All 99.99999999% empty space. Anything you can think of, from the heaviest metal to the lightest feather. All of it is virtually nothing <u>but</u> empty space.

[Note: Modern quantum mechanics takes this simple orbital model of the atom a little further and explains it in terms of theoretical mathematics and what is called the *wave model*. But for our purposes here (one of which is to avoid a migraine) the traditional model of the atom as a microcosm of the solar system, with electrons whirling around a central nucleus, is still accurate. But either way we are dealing with empty space, and not hard little marbles.]

And we're not finished yet...

Let's take a look at the building blocks within each atom, the protons and neutrons in the nucleus, and the electrons surrounding it, that make up each atom. At least *they* must be solid little chunks of matter, right? The protons and electrons must be tiny hard little marbles. We've got to have something *solid* to stand on here, right?

Wrong. It turns out that these electrons, protons and neutrons, while not being made up of even smaller things themselves, are still not solid definable little chunks of matter. They are the furthest thing themselves from tiny hard little marbles of matter. The best way I know how to explain them in laymen's terms is that they are just energy units. Or *wave particles*, whatever that is. But they are not *solid*, *physical* or *material* in the traditional sense, and according to quantum mechanics, not really in any sense at all.

"Quantum mechanics of the old-fashioned kind assesses the behavior of particles, chiefly by showing that particles are not particles at all but a kind of probabilistic smear."[164]

In scientific terminology each of these subatomic particles has ascribed to them a certain amount of what is called *mass*. But this is not to be confused with mass in the traditional sense of for example the *mass* of a boulder, or a *massive* asteroid out in space. This characteristic of mass has nothing to do with ascribing to those same subatomic particles the characteristic of solidity. The mass of a particle is more a mathematical computation used to determine how that particle will behave, by way of gravity and other forces, with other particles of lesser, equal or greater mass.

"Modern physics says that solid objects are mostly space, that subatomic particles have no definite position..."[165]

So with all that said, what exactly is there that can be said to be material in

---

Age of the Earth, by Dr. Don DeYoung)

164 (Pg. 98, *The Devils Delusion: Atheism and Its Scientific Pretensions*, by David Berlinski)

165 (Pg. 252, *Darwin's Black Box: The Biochemical Challenge to Evolution*, by Michael J. Behe)

this *material* universe? Nothing I'm afraid. There is nothing here. No *thing*. And that is not to say that it is all an *illusion*. It is not. It is definitely real, but it is not physical. It's all just a "vapor." The best way to understand it is to come to the realization that the entire universe is an image in the Mind of God. God spoke the worlds into existence. He created the world from nothing. And nothing is its nature still. Which of course makes perfect common sense… If you build a house out of wood, what is it made of? Wood! If you build a patio out of concrete, what's it made of? Concrete! Well, if you build a universe out of nothing, what do you think it's made of? (Bingo.) And again, this is *not* to say that it isn't real. It is made up of nothing, of "spirit," but you are not hallucinating. It is an infinitely complex image, highly ordered, and interacting with itself according to an incredibly complex and infinitely ingenious set of laws and rules and regulations that define how the image will live and move and have its being, how it will behave with itself in a very highly ordered fashion. Two immediate questions then come to mind, who is projecting the image? And where do the rules, the regulations, the laws and the order come from? And I think of course that the answer is obvious to anyone who still retains even a shred of common sense after "growing up absurd"[166] in our left-wing indoctrination centers. (And see chapter 5)

Again, this is the scientific analysis of the nature of the universe, of the nature of the matter that makes up this universe. And as we mentioned before this is not to be confused with the Hindu concept that the world is Maya or illusion. It may be an image, but I can assure you that the image is very real. As I am sure you can assure me also.

Grab your arm. Pound your fist on the desk. Touch that wall. That is just one part of the image interacting with another part, according to predetermined rules and regulations: the laws of the universe.

But it feels solid! It looks solid! I think it's solid!

But in reality, it's not.

Think of what happens when you go to the movies. There's an image on the screen. It's projected up there. If you turn off the projector the image disappears. That's what the universe is. It's a multi-dimensional projection. Think of it as an astronomically complex image formed in the Mind of God. If God wanted to destroy the universe how long do you think it would take him? Would he have to create some massive, universe-sized fireball that would burn and burn for millions and billions of years until it all finally was vaporized?

Don't kid yourself.

He would just have to blink. …Just for a millionth of a second. And it would all disappear. As if it never was. "All the host of heaven shall be dissolved, and the heavens shall be rolled up like a scroll." (Isaiah 34:4) Like an architect packing up his blueprints in his suitcase and heading home for the day, all God would have to do is just roll up the heavens like a scroll, the entire universe, and poof it would be no more. Thank God He chooses not to do so. God is not only the *Creator* of this universe; he is its *Sustainer* as well. He did not just create it and then walk away. Nor does it stay here blindly and unconsciously of its own accord. He sustains it, and keeps it in place, every hour, every second, every moment, and every instant of every day. (We might want to thank Him for that, occasionally.)

---

166 (*Growing Up Absurd: Problems of Youth in the Organized Society* is a book written by Paul Goodman and published in 1960, but it's not one that we are referencing. We just appropriated the title)

"And he is before all things, and **by him all things are held together**."
(Colossians 1:17)
"The Son is the brightness of God's glory and the express image of his person,
who sustains all things by his powerful word."
(Hebrews 1:3)

Also, most of us have already learned the incredible fact that, according to physicists, if the universe started to contract through gravitational collapse it would end up squeezing down to an object the size of a grain of sand! (This is made possible because all of the space inside of every atom would vanish as all of the electrons and each nucleus in each atom would collapse onto themselves due to the massive gravitational forces!) That's all of the "matter" in this universe—the entire earth and the moon and the sun and all of the planets and billions of stars and planets in this galaxy and billions of galaxies each containing billions of stars and planets—and it could all fit (the entire universe!) into an object the size of a grain of sand! You see, according to science, there is really nothing here in this universe. Just an awesome image in the mind of whomever we might think it is that possessed a mind before the universe began. And that is not a theory. It is a scientific fact.

Interestingly, there is this ancient book that appears to have known this all along...

"For what is your life? It is even a vapor that appears for a little time and then vanishes away." (James 4:14)

"The belief that only the physical world exists cannot even account for itself, since the statement itself is nonphysical. Not only is the statement inadequate, it is self-refuting. If it is true, it is false. Either way, it's not true."[167]

Under the watchful eye of physics, one of the physical sciences, even the very notion of physicality itself in our physical universe, has disappeared. So much for materialism.

"It is impossible for the atheistic scientist to be correct in declaring that nothing exists except the material. For even that declaration is nonmaterial, therefore it is false."[168]

And so much for the material atheist's appropriation of science to suit his purposes and only his purposes, to serve his needs and only his needs, to prop up his lies and to bolster his false ideology. Those days are over, for anyone with knowledge. And as we delve further and further into the nature of this universe on the most basic and astronomically small of levels we are discovering a world far beyond that of what was once thought of as the relative simplicity of electrons, protons and neutrons. Indeed as we seek to continue to discover and make sense of the hidden workings of our micro universe, the mind boggling complexity of this world of subatomic particles, fields and waves makes it clear that in our present level of comprehension we are only barely beginning to scratch the surface of just what is here. (And it all just made itself, accidentally, by random chance, from a big primordial explosion! What absolutely unbelievable dumb luck!) (With the emphasis being on dumb, and not-believable.)

"Popping out of quantum fields, the elementary particles appear as bosons or

167   (Pg. 18, *There are Moral Absolutes*, by Michael A. Robinson)
168   (Pg. 13, *There are Moral Absolutes*, by Michael A. Robinson)

fermions. The fermions are divided into quarks and leptons. Quarks come in six varieties, but they are never seen, confined as they are within hadrons by a force that perversely grows weaker at short distances and stronger at distances that are long. There are six leptons in four varieties. Depending on just how things are counted, matter has as its fundamental constituents twenty-four elementary particles, together with a great many fields, symmetries, strange geometrical spaces, and forces that are disconnected at one level of energy and fused at another, together with at least a dozen different forms of energy, all of them active.

"This is not an ontology that puts one in mind of a longshoreman's view of the material world. ...For the atheist persuaded that materialism offers him a no-nonsense doctrinal affiliation, materialism in this sense comes to the declaration of a barroom drinker that will have whatever *he's* having, no matter *who* he is or *what* he is having. What *he* is having is what he always takes, and that is any concept, mathematical structure, or vagrant idea needed to get on with it. If tomorrow, physicists determine that particle physics requires access to the ubiquity of the body of Christ, that doctrine would at once be declared a physical principle and treated accordingly."[169]

The importance of this truth about the non-physical world we live in and one's ability to grasp it will become more evident as we go on. For one thing, there are those who have difficulty believing that a certain remarkable individual who walked this earth 2,000 years ago could have performed all of those recorded miracles like giving eyesight to the blind, and healing to the leper, and wholeness to the cripple, and life to the dead, and food for five thousand from a few loaves and fishes. But if they understood the nature of this universe and the One who created it, the One who spoke it all into existence and the One who spoke the laws into existence that govern his universe, they would realize that it is no more difficult for him to walk on water or to instantly calm a raging storm than it is for George Lucas to have Yoda levitate a 4-ton x-wing out of a swamp on a distant planet. In fact, there is nothing in the nature of this universe to keep your hand from being able to pass right through this book like it passes through the air. It is only the laws of the universe, the laws that govern the particles of the universe, whatever they actually are, that command it to halt and for the book and your hand to remain separate. There is no real *solidness* stopping it. It is only that every particle and every part of Gods universe is set up to rigorously obey the laws of their motion and their boundaries. However, for the God who created this universe, who controls this image, it is as nothing for him to suspend the laws of these particles and for him to have his way with this image. His image will rigorously obey his laws until he chooses to suspend those laws, which he is free to do at any point in time whether that be for the parting of the Red Sea, or making the sun stand still in the sky for a full day, or instantly rearranging the particles and cells in a man's withered arm to allow it to "stretch" out and be made whole.

"The physical creation obeys God, but we have a tendency, a great tendency, to defy God."[170] And to foolishly deny he even exists. (See chapter 5)

"Oh Great Spirit, whose voice I hear in the wind, whose breath gives life to the world, hear me! I come to you as one of your many children. I am small and weak. I need your

---

169   (Pgs. 53-54, *The Devils Delusion*, by David Berlinski)
170   (Focus on the Family, from "Blueprints for Marriage," by Del Tackett)

strength and wisdom. May I walk in beauty. Make my eyes behold the red and purple sunset. Make my hands respect the things that you have made, and my ears sharp to hear your voice. Make me wise so that I may know the things that you have taught your children—the lessons that you have hidden in every leaf and rock. Make me strong, not to be superior to my brothers, but to be able to fight my greatest enemy: myself. Make me ever ready to come to you with straight eyes, so that when life fades as the faded sunset my spirit will come to you without shame."[171]

It seems that the likes of Christopher Hitchens, Bill Maher, Richard Dawkins and all of their highly educated academic friends who are still holding on desperately to a materialism that doesn't even exist in reality, and despite all of their vaunted intelligence, could still learn a thing or two from the original inhabitants of this land whose insight into the nature of this universe and the Spirit who created, sustains and controls it, far surpasses that of their own.

"Earth is crammed with heaven, and every common bush afire with God, but only he who sees takes off his shoes..."[172]

For now it is enough to be aware that everything is not always as it seems. And the mindless, meaningless and dead physical world that many of us think we live in every day doesn't actually exist in reality. In reality this universe is light years removed from that assumption. It—and the matter that comprises it— is a living, vibrating, astronomically complicated engineering miracle, from the precise workings of the microscopic atoms and molecules, up to the mind blowing wonder and beauty of those far flung spiraling galaxies. It is all a miracle. Or a miraculous image. In the mind of God.

"For in him we live, and move, and have our being"
(Acts 17:28)

Also, this is not to be confused with the Eastern religious view, from Hinduism and Buddhism, that "everything is God" and "God is everything." On the contrary, as we will see in subsequent chapters, God is a Spirit, who is beyond the universe, above the universe, transcends it, and is not limited by it. Yet in his creation, in his image, we dwell.

In reality we are living not in a cold, impersonal and uncaring *physical* world, but in the Mind of God. And we are here at his pleasure. And yet we ignore him the vast majority of the time. And we spit in his face by teaching our children that he doesn't even exist, and by indoctrinating them with the sheer, unscientific insanity that his universe just somehow happened to have made itself. By exploding. Out of nothing. Accidentally. (One afternoon after going for a stroll in the park, as unformed universes are wont to do...)

See:
   - *The Devils Delusion: Atheism and Its Scientific Pretensions*, by David Berlinski
   - *There are Moral Absolutes*, by Michael A. Robinson

---

171   ("Native American Prayer," by John Yellow Lark)
172   (Elizabeth Barrett Browning, as quoted on pg. 248, *The Shack: Where Tragedy Confronts Reality*, by Wm. Paul Young)

CHAPTER 4

MAN

"And the Lord God formed man of the dust of the ground, and breathed into his nostrils the breath of life; and man became a living soul." (Genesis 2:7)

When we look more closely at the very nature of the universe we find that the conventional wisdom falls far short of reality. Now let's take a closer look at the very nature of man. Let's see where all the facts and scientific evidence will take us...

What is man? What are we at the very core of our beings? Just flesh and bone? Or something else entirely? Well, the answer is that our flesh and bones are nothing more than our dwelling place. They make up our "physical" vehicle that we travel through this world in. But that is not who we are. Who we are, here on the inside, is pure consciousness, pure conscious awareness. It turns out that in this universe of images projected onto this screen that we see before our minds eye, we are not even one of the images. Our bodies are a part of the imaging. But we are much more than just that.

"What a piece of work is a man. How noble in reason. How infinite in faculties. In form and moving how express and admirable. In action how like an angel. In apprehension how like a god!"[173]

Let's take a look at the essence of man, this "quintessence of dust," after earth, air, fire, and water, the fifth element of the ancient world. Let's find out just exactly *who* we are...

Smack your arm. Did you *feel* it? Of course. Now let's take a look at what actually happened. Let's take a look at *how* you felt it and at exactly *what,* or *who,* felt it.

Stay with me. (It'll all make sense shortly.) (Hopefully.)

When you smacked your arm, pressure sensitive nerves in your skin were stimulated and "fired" electrical-nerve impulses that traveled instantaneously up your arm along cellular nerve pathways into your spinal column and on up into your brain. As they traveled into the brain these electrical-nerve impulses were immediately translated into information that was instantaneously processed and refined and interpreted and sent to whatever final destination in your cerebral cortex that was already predetermined to be appropriate. But here's the point. What is it up here in your brain that "feels" the final output? What is "it" at the final destination for this information in your brain that *feels* it? That is *conscious* of it? That is *aware* that something (in this case your own hand) smacked its arm? Keep in mind that you are not your arm, or your hand. You are up here in your head. What is it then that is up here in your head that is conscious of your hand and your arm and everything else in here and out there, in your inner and outer universe?

Hmmm...

Let's take sight. Photons of light bombard your eye trillions of times a second, and they cause an enormously complex series of chemical reactions to take place

---

173 (From Shakespeare's *Hamlet*)

in the light-sensing cells at the back of your eyeball, which results in impulses being sent up nerve pathways into your brain where this information is processed, refined and interpreted instantaneously. (All this happens, by the way, in a few *picoseconds*. And a picosecond is the time it takes light to travel the distance of a hairsbreadth. And light travels at 186 thousand miles in one second. So this happens over and over again about a half trillion times every second in every one of the billions of cells in the back of your eye, just so you and I, and every animal, bird, fish, reptile and insect on this planet, can see! But your eye and your brain just formed themselves accidentally over time, by random chance?! Un-believable.) But the question is, what is it up here in your brain that "sees" the final output; that is "viewing" the digital picture that your brain has painted up here in your head? The same is true of hearing and taste and smell. It's all just one nerve hitting another nerve until it ends up in the brain and gets instantaneously processed, refined and interpreted by the extremely high-speed computer that is alive inside of our craniums. But the $100,000 question is this, What is it up here in the brain that is conscious or aware of the final output from your senses? And the answer is just that, consciousness. Consciousness itself. Awareness itself.

You see, the brain itself is not conscious. There is no part of it that is *consciously aware* of itself or of any of the inputs from its senses. Indeed the brain does *respond* to sensory input, and reacts with corresponding output, on a whole host of levels and through a whole complex of inter-related and connected systems. And it does this *sub*consciously, below the level of your conscious awareness. (And it also does this completely *unconsciously*. It does it because it was pre-programmed to do so, many thousands of years ago.) But *consciousness* or *awareness* of itself and what it is doing, it does not have. (Your brain is totally automated and totally unconscious of what it is doing, yet it somehow does it day in and day out for your entire lifetime. The infinite genius that it took to design and manufacture and program it is beyond our ability to comprehend.) (See chapter 10.) Like an extremely high-speed computer the brain is merely a processor of sensory data. A near-infinitely remarkable and fantastic processor of sensory data. But a processor nonetheless. A computer, but not a "viewer." Your desktop or laptop has remarkable capabilities to respond to instructions, to compute extremely complex mathematical problems, to project onto its screen beautiful, intricate, moving images, etc. etc. but no one would suppose that it is consciously aware of what it is doing. Because it is not, nor will it ever be. And neither is your brain.

If you are having trouble with this then it might be because you, like all of the rest of us, identify with the physical part of us—our body, cells, tissues, sensory organs, torso, limbs, hands, eyes, ears, head, etc., including the brain—and therefore we have a hard time separating ourselves, our essence, the conscious awareness that is us, from this purely physical unconscious part that we can see and touch and feel, that we are inseparably attached to (till death), and within which we dwell.[174] We cannot see and feel or touch or taste the pure spirit consciousness

---

174 (But doesn't our consciousness inhabit our entire body? It certainly feels like it. But that is not the case. It's all projected to us up here in our head; everything from sights and sounds and tastes and touch and smells, and including the three-dimensional map of our body and limbs that we *feel* up here in our head, that has all been created by a level of infinite intelligence incapable for us to even comprehend, in order to give us the absolutely real sensation of living out in the world.)

that we are. So it becomes the one thing that we are unaware of. But whether we are conscious of this distinction or not, or whether we care to be bothered about it, nevertheless it exists. Indeed if this distinction did not exist then we would not be conscious or aware of any part of us, including this discussion we are having right now, because there is nothing in the physical universe—the universe of subatomic particles, atoms and molecules, including living cells, tissues, nerve cells and highly sophisticated, living-computer brains—that possess' consciousness or awareness.

Another way of understanding this is through the old dominos analogy. Let's say you set up one of those long convoluted lines of dominos standing on their ends. And then you tap over the first one in line. As it falls it knocks over the next one in line, etc. etc. Until the last domino falls down. Your senses are like those dominos. Take sight for example. Light hits your eye. The first domino falls. It fires cells in the back of your eye. One domino hits another. The nerve impulse travels up into your brain. More dominos falling. When it gets to your brain the nerve impulses shoot along pathways as they are sorted and refined and then a picture is painted instantaneously up here of the world out there. The last domino falls. But who or what is there to "see" or to be aware of the last domino falling? Of the final sensory output in the brain? Of this picture up here of the world out there? Who or what is up here in the brain looking at this picture of the outside world, right now? And the answer (coincidentally) is Y O U. And not the physical brain. The physical brain merely paints the picture inside here, but it is the conscious YOU that sees the picture. You are that pure conscious awareness that inhabits your brain, that is aware of the final outputs of the sensory activity of the brain and of your body, which is merely an extension of your brain. You are that *point* of consciousness inside of your head, that is aware of all of the sights and sounds and touch and smells and tastes, including the feelings and emotions and thoughts, that your computer/brain projects up to you, and creates for you, inside of here. And that YOU has nothing to do with the image—the atoms or electrons or protons or energy units, or cells or organs or brains—of the *physical* world. It is separate from the material world.

In conclusion then it is a scientific fact, or a fact that stems from our cumulative knowledge which stems from our observation of nature, that there is nothing whatsoever in the physical or material world (in this complex universal image), in the world of matter—of electrons and protons and energy—that has consciousness or awareness. It's all just dominos hitting dominos. Only the spirit has consciousness. Only the spirit is aware of the goings on of the universe.

And now we know who we are. Pure consciousness. Pure awareness. A soul. A spirit. (And of course this fact is not new. It has been known and taught since ancient times by a number of religions and their teachers. It is more of a recent phenomenon that those who control academia have put all of their faith, and all of their eggs, in the material basket. A basket that disappeared in the last chapter.) (Somebody shoot off a tweet to Bill Maher.) Even your thoughts are not who you are. That is brain activity. You are aware of your thoughts. You *hear* them inside of here. You can even, to a certain degree, control and direct them. But you are not your thoughts in your brain. (As it turns out, it is not "I think, therefore I am." It is, I am, therefore I am.)[175] Thoughts come and go. But you are consciousness itself. The same is true of your emotions. And your opinions.

---

175   (Or if you want to nitpick, "I am conscious and aware, therefore I am.")

"Then God said, 'Let us make man in our image, in our likeness,' …So God created man in his own image" (Genesis 1:26-27)

The Creator of this universe said that he made man in his own image and likeness. Whether he was talking about our flesh and bones, our appearance, I know not. But that he was referring to the spirit, the consciousness, who we are at the very heart of our beings, where we are consciously aware of our inner and outer universe, there can be no doubt. (He was also referring to the gift of free will, but that's another subject.) It is interesting that science itself, in the study of consciousness and the workings of the mind, confirms what an ancient book said many, many centuries ago.

"That was the true Light, which lights every man that comes into the world." (John 1:9) Our consciousness is a spark of His Consciousness. We are made in His image and likeness. Our consciousness, our spirit, is what "lights" us. And He in turn is the One who lit our consciousness at the beginning of our days.

"You can kiss your family and friends good-bye and put miles between you, but at the same time you carry them with you in your heart, your mind, your stomach, because you do not just live in a world but a world lives in you."[176] (Emphasis mine.)

Also according to the nature of man and his senses, where everything you see and taste and hear and smell and feel is extracted by your senses from the outside world and "shown" to you up here in your head, then in reality the entire universe is inside of your head. It's like what happens when you put on a pair of those cutting-edge, technologically-advanced, virtual-reality goggles and enter into a whole-nother universe. But our minds are many light years ahead of this newfangled yet comparatively "stone-aged" technology. We think we look at the world through the windows of our eyes, but this is completely inaccurate. Our eyes are not like the windows in our home or our car. They are like video cameras that constantly take a movie of the outside world and paint it up here in our heads for us to see. We are "up here" in our brains and what we see is a digital representation of the outside world that our nerve fibers have extracted from the back of our eyes and have "painted" instantaneously up here in our heads for us to *see*. And I'm not saying that it isn't a very accurate representation of what is actually out there in the outside world. I'm sure that it is. And the same is true of hearing, taste, touch and smell. Think of how infinitely powerful and creative and intelligent the Master designer and Maker was who caused the technological wonder and engineering miracle of the human brain and its senses to come into being.[177] He literally took the entire outside universe of sight sound taste touch and smell and gave our brains the astounding power to project it all up here in our heads so we could experience life to the fullest. Indeed, there is no other way (no "direct" way) for us to experience it. We might want to thank Him sometime for the wonderful, extraordinary and fragile gift of life that he has given us. We might want to show Him a little respect…

"In God's hand is the soul of every living thing, and the breath of all mankind."

176   (Frederick Buechner, *Telling the Truth*, as quoted on pg. 209, *The Shack: Where Tragedy Confronts Reality*, by Wm. Paul Young)
177   (And if anyone thinks that random accidental chance was this "Master designer and Maker" then they are in serious need of reading chapter 10, "Evolution is a Joke")

(Job 12:10) And the sight, and the hearing, and the taste, and the touch, and the smell of every living thing as well…

What is the value of this knowledge? Well, for starters it should put all of those materialists who believe, and teach, that nothing exists other than their "physical" world, in a whole new Neanderthalic light. Ultimately they channel this colossal ignorance into the denial of the existence of an intelligent Creator, who necessarily must be a Spirit existing prior to and outside of his creation. But now we know that it is in fact their own "physical" universe that doesn't exist in the material way that they have imagined. And in addition we know now that in denying the existence of God, because he is a Spirit, they deny the very existence of themselves, for they are nothing other than a non-physical spirit, possessing the powerful gift of conscious-awareness as well, and inhabiting a body and a brain. And the only place this powerful gift exists in this "physical" universe is in those living beings where that same Creator-Spirit has placed it. All the rest is just dominoes hitting other dominoes. So much for materialism.

"The Spirit of God has made me, and the breath of the Almighty has given me life." (Job 33:4) Also, being blind to the fact that we are conscious spirits, and not mere physical beings, can also blind us to the fact of the existence of the Spirit who created us. The rejection of the meta-physical universe; that which transcends the physical; that which is above the physical; is another form of self-hatred, of not only denying the existence of God, but of unknowingly denying the existence of one's very self. The idea that "the physical universe is all there is" should be rejected by any truly educated (as opposed to misinformed) conscious being. Indeed to study the physical universe without the knowledge of the metaphysical from which it arose and by which it is sustained, is to be one of the educated ignorant. "Ever learning yet never able to come to the knowledge of the truth." (2 Timothy 3:7) For the two—the physical universe and the meta-physical—the image and that which projects the image—are inseparably entwined, as we, our conscious selves, the soul and spirit, are inseparably entwined with our physical bodies and our physical brains.

"Thus says the Lord, who stretches out the heavens, lays the foundation of the earth, and **forms the spirit of man within him**." (Zechariah 12:1)

One of the arguments in the "Does God exist?" debate is over whether or not there is a Consciousness, a Spirit, behind the physical universe. (What the philosopher and Oxford professor, Richard Swinburne, calls an "omnipresent incorporeal spirit."[178]) And one of the sticking points brought up by the theists to challenge the atheists is "How did consciousness arise from an unconscious universe?" Indeed the question can be answered in two words. It didn't! It could not nor can it arise. The physical universe is unconscious and is absolutely incapable of producing consciousness. Only the spirit has consciousness. You are a conscious spirit, a soul, inhabiting a physical body and a physical brain in a physical universe. (And of course we now understand that the word *physical* refers to the non-physical *image*, and nothing that can be ultimately considered "solid" or material. But for the sake of the ease of communication we will defer to the existing terminology.) And I think it is safe to say that this is scientifically accurate, because it can be shown that it is scientifically impossible for anything in

---

178   (Pg. 71-72, *There is a God: How the world's most notorious atheist changed his mind,* by Antony Flew)

this universe, including the human brain or a computer of any speed or magnitude, to be conscious. Our laptops may be able to respond with increasingly mind boggling speed, but they will never possess mind. An electronic brain, yes. But a conscious mind within that brain, no. Our mind—spirit, soul, consciousness—is yet another proof of the Divine Mind—Spirit, Consciousness—behind both our minds and the physical universe.

"Know thyself"[179]

To be lost in the carnal nature of ourselves, endlessly plodding on the tread mill of life, striving to fulfill this desire or that, or to get this want or that, and to avoid this unwanted thing or that; while at the same time remaining unaware of our true nature, and hiding from any realization of our true selves and in Whose image we were made, is like the undisciplined life, one not particularly worth living. (But you can certainly keep busy at it.)

"Those who live according to the flesh have their minds set on what the flesh desires; but those who live in accordance with the Spirit have their minds set on what the Spirit desires." (Romans 8:5)

It has been said that in each one of us there is this God-shaped hole that we are desperately trying to fill, but with all the wrong things and in all the wrong ways. (Like the Urban Cowboy, we're "looking for Love in all the wrong places.") Who we actually are is a tad more than what most of us have unwittingly come to think and believe about ourselves; which can more or less be summed up as a physical being whose primary purpose and focus in life is *getting what it wants*. Whereas we would be more anatomically correct if we saw ourselves as spiritual beings who have built into their very souls the deep desire to find meaning and purpose in life by focusing on to some degree, and by finding, a relationship with our Creator. With the Spirit who gave us our spirit. The Spirit who made us who we are. The Spirit who bothered to make us in his own image and likeness. Which also segues nicely into the next chapter...

---

179   (Inscription in the Greek Temple of Apollo at Delphi)

CHAPTER 5

GOD

"The heavens declare the glory of God, And the firmament shows his handiwork."
(Psalms 19:1)

Is there a God? Does he actually exist? Or is he, as today's militant atheists insist, just a figment of our "primitive imaginations?" Can this debate over the existence of God be answered once and for all, definitively and beyond any shadow of a doubt?[180] Can we look at all of the facts and evidence, both scientifically and historically, and put this argument to bed as definitively as the flat earth argument has been laid to rest? How can we *know* what the truth, which is not relative, really is?

By seeking it, of course. By looking at all of the facts and evidence concerning this ultimate of all questions, and by accepting the conclusion that they lead us to, the question of the existence of God can be easily answered. Indeed, it is a testament to the extraordinary level of willful ignorance of modern, educated man, and to the intellectually absurd depths to which this "desire-to-be-ignorant-of-the-truth" can cause him to descend, that there is even any debate on this subject. Because it turns out that the evidence, the <u>scientific</u> evidence, clearly and overwhelmingly points to the existence of a somewhat extremely intelligent, pre-existing (existing prior to this created universe), quite powerful (ya think?), and eternal Creator Being.

"When I consider your heavens, the work of your fingers, The moon and the stars, which you have ordained, What is man that you are mindful of him." (Psalms 8:3-4)

I would elaborate on that question just a little… *What is man that you continue to tolerate his ignorance and his arrogance in declaring himself educated and intelligent, while at the same time denying the very One who made him?* Thank God You are so patient and merciful and long-suffering and slow to anger and...

Atheism for idiots

"The fool has said in his heart, 'There is no God.' " (Psalms 53:1)

Let's start with the atheists. The dictionary definition is one who does not believe in a Supreme Being, in God or gods. But even atheists believe in a creator, that this universe came from somewhere, and that life came from somewhere or something. And the concept of God in its most basic form is the concept of a creator, the originator if you will of all of this. And the most basic of questions that are inseparable from the concept of a creator are, *Where did all this come from? How did all this get here?* And even atheists have answers to these. There answers might not be very good, they may even be scientifically absurd, but they are answers nonetheless. So everyone on this planet, even atheists, believes in a creator of one sort or another, whether that creator is an intelligent being, or

---

180 (Psst, over here. The answer is *yes*. Just read these 2 quotes: Psalm 19:1 and Psalm 53:1. Both are on this page)

some raw force of *nature*. And when it comes to this universe and the miracle of life on this planet and who or what created it all, or how it all got here, there are only three possible choices: a big explosion, random accidental chance, or an intelligent Creator.

Let's start with number 1: The first choice is that a big explosion created everything. In faux science this is called the theory of the big bang and, with all due respect to our friend Stephen Hawking, it is scientifically impossible. There is no such thing as an explosion ever creating anything of any value or order. All an explosion can do is to make a big mess—to tear apart, to break down and to destroy—and this universe is about the furthest thing from a big mess you will ever find. Indeed, the universe displays order to an almost unimaginable level. When you learn of all of the laws that govern motion, and gravity, and light, and sound, and that control the behavior of atoms and molecules and their electrons protons and neutrons; the order and engineering complexity of this universe is astronomical and beyond our comprehension; and also far beyond anything that Mr. Big Explosion could ever hope to come up with, accidentally. Also, according to the very laws of science itself, in particular the Second Law of Thermodynamics and the Laws of Mathematical Probability, we know for a scientific fact that it is an absolute impossibility for this universe, this solar system, our Earth Moon system, and life on this planet to have been created by an ancient, cosmic explosion. For the Second Law of Thermodynamics states that nothing can go from a state of disorder to a state of order on its own without being acted upon by an outside force. Hence the extraordinary, mind-boggling order of this universe could never have given rise to itself. Also according to the Laws of Mathematical Probability it is far beyond the possibility of chance for the universe to have come up with the level of order we find within it accidentally.

Beyond all that, and as important as all that is, just use your common sense. Just ask yourself, how many billions and trillions of explosions have there been in just the last few thousand years since the Chinese discovered gunpowder? Has anything of any value or order ever been created by one of these explosions? How about just one simple symmetrical Popsicle stick? Just one. Not a splinter, but just a simple symmetrical Popsicle stick! What have you got? Nothing?! When was the last time a contractor took some dynamite into the woods and said he was going to build his self a house? (The last time Tina Fey hired one of her friends.) No, my friends, despite the God denying high priests of big bang cosmology, and evolution, that push this faux science down our throats in our high schools, universities and in our media, it is total, unscientific nonsense. The theory of the big bang is ludicrous according to the very Laws of science; the same science that is supposed to have given birth to that theory. It is a theory, therefore, that is in direct contradiction to science itself (yet is still taught as a scientific "fact" ...go figure).[181]

Scratch the first choice.

Now for number 2: The second choice for a creator is random chance. This is expressed primarily in the unscientific religious doctrine called the theory of evolution, which has no basis in either science or reality. Supposedly, everything in this universe, all of the astronomical order and complexity, including the unimaginably complex engineering miracle we call life, all supposedly happened

---

181   (And see chapter 10: "Evolution is a Joke")

by itself somehow (miraculously) over time, purely by accident and sheer coincidence, by random chance alone. This is also scientifically impossible. (Besides being so stupid it literally borders on the insane.)[182] Again, according to the very laws of science—*the second Law of Thermodynamics, the Laws of Mathematical Probability, and the Law of Biogenesis*—it is absolutely impossible for random accidental chance to <u>ever</u> create anything of order, or for life to ever come from non-life by itself accidentally. Like explosions, all random chance can do is to randomly wear things down and destroy them—to tear down, to break apart and to destroy. Again, you are left with a colossal mess and not a universe, and definitely not anything that resembles even the "simplest" single-celled living organism (which is actually astronomically complex).

We can scratch the second choice as well.

[Note: Whereas it is true that the majority of folks don't really believe in the theory of evolution, despite it having been constantly shoved down their throats as a "scientific fact" since grade school and on every science channel on TV, it is also true that many others, including all of our "elites," *do* believe in it. So for those readers who might be in this camp let me ask that you sit tight until we get to chapter 10, "The theory of evolution is a joke." For there we will clearly show that—based upon the research and conclusions of countless unbiased and non-evolutionary-indoctrinated scientists, and according to the immutable laws of science itself, in addition to all of the facts and evidence—it is <u>a scientific fact</u> that the theory of evolution <u>is scientifically impossible</u>. Indeed you will find there, if you seek the truth, that this theory is in reality so stupid, so idiotic, so asinine, ridiculous and absurd, and so scientifically impossible that it literally borders on the insane. The question that many might have at this point is, then why doesn't everybody know this? Why isn't this so heavily reported that it has already become common knowledge? And here's why… The theory of evolution is not just scientifically-impossible and idiotic, but more importantly, it is also Satan's genesis. It is *his* origins account, in direct opposition to, and meant to undermine, the true origins account- God's Genesis as found in the first chapter of the Bible. But the kicker is that Satan's genesis, the theory of evolution, is also the foundation of the Democrat-Left (or the secular-humanists, or the secular-progressives, call them what you like). It is what their whole ideology is built upon. Destroy it, and their entire world view comes crumbling down. And since that same Democrat-Left owns the means of disseminating information in this culture—the media, our educational system, the entertainment industry, and of course the science channels—we can understand why the truth about this subject has essentially been blackballed.[183] Why would they allow it to come out? It would destroy their very foundation. And for those who say "wait a minute! Fox News isn't a part of the Democrat-Left media, and they haven't reported it either," we would only reply, you're right. And we can only surmise that they are either ignorant of the truth, or too cowardly to take a stand and confidently report it because it would only open them up to even more ridicule from the elites in the Democrat media, who already hate them, lie about them and falsely disparage them at every opportunity. But you'd have to ask them. …Send them a copy of chapter 10, and see what they do.]

So, what are we left with? Scientifically. Who or what created all this?

---

182 (Again, see chapter 10: "Evolution is a Joke")
183 (And to grasp the level of this censorship, see the documentary, *Expelled: No Intelligence Allowed*, narrated by Ben Stein, as well as the books recommended at the end of chapter 10)

Well, let's review. We know that neither primordial explosions (big bang theory) or accidental random chance (evolutionary theory) could ever create anything of order. And that is a scientific fact. (It is also obvious to anyone with even the proverbial shred of common sense, which is why the vast majority of the "common" folk do not buy into the completely unscientific theory of evolution. That dubious distinction is reserved for the *intellectuals* who have taken over our educational system. The common folk know that something just doesn't add up. They know something with that theory smells of dead fish. And they of course are absolutely right.) (And see chapter 10.)

Where were we? Oh yes, and now let us look at what is behind door number three…

Number 3: And the third choice (the intelligent choice) is that all of this was created by an actual Creator. Shazaam! (A real Creator, you know, like the people who made your car, or your camera. If someone came along and said your car or your camera were made accidentally by random chance, or by a big explosion, you would say they were a fricking moron.) A Creator with the level of extraordinary, more likely infinite, power and intelligence that would be necessary to actually bring into existence something as infinitely complex, highly ordered and mathematically precise as this universe, not to mention being actually capable of imagining, designing and creating all living things on this planet.

The third choice is the only scientifically viable choice. The only choice that is possible according to all of the facts and evidence, and according to all of the Laws of science. But however you choose, whether you choose foolishly and absurdly to believe in random accidental chance or a big explosion, or in some combination of those two. Or you choose wisely and intelligently to believe in the third choice. Either way, you believe in a creator, and so does everyone else on the face of this earth. So much for atheism.

Some atheists worship random chance as their god. And some worship a big explosion as their god. And some worship both. But either way they have a god (creator). Again, so much for atheism.

So, there you have it, everyone believes in some sort of god because everyone believes in a creator. The only question is whether that god is a true God, One who actually has the power, the ability and the intelligence to create a universe of unimaginable complexity and order; or a false god, one who cannot and could not create a damn thing, much less this universe and all the living on this planet.

Interestingly, atheists also have a savior. For their gods, Mr. Explosion and Mr. Random Chance have "saved" them from having to accept the existence of the true God, the real Creator, and therefore from having to live under his laws, and from having to submit to his moral requirements. They are "free." They are "liberated." But their freedom and their liberation is an illusion. In reality, they are imprisoned by the lies of the great deceiver. (Lies that, as we shall see in chapter 7, have a very real and a very dangerous eternal consequence.) In reality atheists are in desperate need of a real savior, One who will save them from lies, not from the truth.

"The great virtue of Darwin's theory, Richard Dawkins has argued, is that it has made it possible to be an intellectually fulfilled atheist."[184] (Albeit an intellectually deluded one.)

---

184   (Pg. 219, *The Devil's Delusion: Atheism and its Scientific Pretensions*, by David Berlinski)

"The great virtue of Darwin's theory," I'm afraid to say Mr. Dawkins, "is that it has made it possible" for foolish, mis-informed, improperly educated academics like you to think that they are intellectually fulfilled, just like that famous emperor who thought he was actually clothed.

"He who sits in the heavens shall laugh; The Lord shall hold them in derision." (Psalms 2:4)

It is also interesting to note that, as it turns out, it is the atheists, and not their more religious counterparts, who are the ones guilty of *blind* faith. Which is faith in an idiotic, unscientific and absurd notion—that the universe and all of life just somehow "made themselves"—that has no basis in reality, and has no evidence to support it. The aforementioned Richard Dawkins, one of the more famous of the militant atheists and author of the book *The God Delusion*, declared that faith in God means "believing not only without evidence, but in the teeth of the evidence." He also has commented on what he sees as "the level of knuckle dragging ignorance" it takes for someone to believe in God and to reject his sacred, false-religious and scientifically-impossible doctrine of evolution. Now we hate to take advantage of an easy opportunity to rub Dawkins nose in his own quotes (being as how we try to suffer fools gently and also know that "there but for the grace of God go you or I") but, in truth, the really remarkable "level of knuckle dragging ignorance" is the level it takes to believe that Mr. Random Chance could actually create anything, accidentally, much less everything. In addition it is science itself which declares that faith in a couple of absurd, unscientific theories means "believing not only without evidence, but in the teeth of the evidence."[185] No, Dick, it is not faith in God that is delusional; it is faith in Mr. Explosion and Mr. Random Chance that truly is. Maybe in your next book you could write about an actual irrational deception. You could call it *The Evolution Delusion*.

> "William James, the father of modern psychology, said, "There's nothing so absurd that if you repeat it often enough, people will believe you." Thus, the ridiculous view that, given enough time, chaos will produce order has taken a firm hold on our culture."[186] (But then again, no one ever accused us of being all that bright.)

And speaking on the subject of "the rise of what the *Wall Street Journal* has called "militant atheism,"" David Berlinski cautions his readers concerning…

> …"the reappearance of that perennial literary character, the village atheist, someone prepared tediously to dispute the finer points of Second Corinthians in time taken from spring planting. A little philosophy, as Francis Bacon observed, "inclineth man's mind to atheism." A *very* little philosophy is often all that is needed. In a recent BBC program entitled *A Brief History of Unbelief,* the host, Jonathan Miller, and his guest, the philosopher Colin McGinn, engaged in a veritable orgy of competitive skepticism, so much so that in the end, the viewer was left wondering whether either man believed sincerely in the existence of the other. Sam Harris's *Letter to a Christian Nation* is in this tradition [Harris is another academically popular village atheist], and if his book is devoid of any intellectual substance whatsoever, it is, at least, brisk, engaging, and short. … If rural atheism is familiar, it is also irrelevant. Religious men and women, having long accommodated the village idiot, have long accommodated the village atheist."[187]

---

185   (And of course see chapter 10, "The Theory of Evolution is a Joke")
186   (From article *The Bible and Evolution*, by Pastor Doug Batchelor, Inside Report Magazine, August 18, 2009)
187   (Pgs. 2-3, *The Devil's Delusion*, by David Berlinski)

Science didn't always find itself in the sad position of being hijacked by atheists. The ones who now control our educational system, and our faux science channels. In fact in the beginning almost all of the great scientists and especially the founders of the fundamental laws of science were very strong believers in the Almighty. Indeed, many were devout Christians. And it was ultimately as a result of their very strong belief in an infinitely intelligent Creator who had ordered this universe that they even set out to try to discover that order in the first place, through experimentation and observation. To those great founding fathers of modern science the very idea that this universe of extreme order, complexity and rigorously unbreakable laws came about by accidental chance would have been insane.[188]

The Design Argument

"God alone was never created, and he exists totally independent from anyone or anything else. "I am" is the one unchanging, eternal God."[189]

If a renowned Egyptologist were to hold a press conference in order to share with the world the momentous news that he had finally uncovered the secret to how the Great Pyramids and the Sphinx at Giza in Egypt were made, you'd be very interested to hear it. But then if he said his theory was that they had just made themselves accidentally by random chance over a number of years, you would say that he was a complete moron. And so would everyone else on the face of this earth.

Yet the design and construction of the Sphinx and the Great Pyramids, as difficult as it may have been, are child's play compared to the design and construction of a solar system, or a spiral galaxy, or this water planet, or our earth-moon system, or compared to the design and creation of light, or of gravity, or of the structure of an atom and the mind boggling forces that control its behavior that we with all of our intelligence and with the greatest of our scientific minds cannot even fully comprehend. Yet those who say that all of that just made itself accidentally by random chance are not called morons. They are called professors and scientists, and are given tenure and science shows on TV.

And if an acclaimed archeologist told you that Machu Picchu, "The Lost City of the Incas," in Peru just made itself accidentally by random chance over many hundreds and thousands of years, you would say he was a complete moron as well. And again, so would everyone else on the face of this earth.

Yet the design and construction of that now ruined city in the clouds, as laborious and ingenious as it may have been, is a joke compared to the design and construction of one microscopic, astronomically-small strand of DNA in a living cell on one-ten-thousandth of a square millimeter on the end of your nose. That DNA strand is about a million times more complicated and complex than any ancient town made of stone and mortar. Not to mention it contains more coded information than a school library, and it is a scientific fact that coded information can never, ever, <u>never</u> happen accidentally on its own by chance. Yet those evolutionists in our public schools who teach this stupidity are not called morons either. Rather they are paid handsomely by us to indoctrinate our children with Satan's genesis, with scientifically-impossible lies.

---

188   (And see the next chapter for a further discussion of those pioneers of modern science)
189   (Pgs. 120-121, *The 100 Most Important Bible Verses*, Thomas Nelson Publishers)

And what if a tenured professor from a local university was introduced to you at a catered dinner party and told you that all of the delicious food spread out on the banquet table before you just made itself by random chance, accidentally, over a very long period of time. You'd think she was joking, and you'd smile. But what if she said she was not joking and that she believed that just by the mere accidental dropping of fruit from trees and fish jumping up out of the sea and pheasants flying into cliffs and carrots and green peppers getting kicked loose by passing herds of wildebeests and lightning striking chickens, that after many years of trial and error the right culinary combinations just accidentally came together to form this wonderful meal. You'd think she was a complete idiot, possibly insane and that her college should consider having her hospitalized for psychiatric evaluation. And so would everyone else on the face of this earth.

Yet as much effort and energy and thought that must go into the design and preparation of a banquet, it is a cakewalk compared to designing and manufacturing just one living cell. A single living cell, the building block of all of life, is literally a billion times more complicated to design, more intricate to prepare, and more mind-boggling in its engineered complexity than the world's most lavish and extravagant state dinner. The dinner couldn't possibly have formed itself no matter how many billions of years we gave the raw materials the time to *evolve* on their own, accidentally. Yet the whole of our endlessly varied and near-infinitely complicated living things and biological systems on this planet did? Because some chemicals in an ancient mythological slimy sea somehow decided to just get together on their own, and design *themselves* somehow miraculously in the process, and form something that has more information and intelligence than your laptop, a DNA molecule; which can't exist outside of a cell, but it did; and then created this cell around it; and then it gets even more ludicrous. If you believe this then you don't necessarily have to be extremely stupid. (But it doesn't hurt.) You just have to be the same as those suckers who were looking for romance by flying planes into tall buildings. You have to be duped. Your brain must be washed with the poison of inane lies; fed to you by individuals—teachers, educators, imams, or faux scientists in your classroom and on your TV—who you were taught to respect and admire and listen to. You have to be indoctrinated. You have to brainwashed.

The lies of Satan cover this earth like a blanket, and people believe what they are taught. And whereas it is not our intention or purpose here to demean, disparage, or make fun of those who have been taught idiotic lies, and therefore have ended up believing them and teaching them to others (we in fact were one of them not too long ago); nevertheless, we should expect a certain degree of consistency in this life. A moron is a moron. And an idiot is an idiot. With all due respect of course to all of our teachers and professors and scientists who have been brainwashed and indoctrinated into believing in and then teaching the idiotic, unscientific lies that a big explosion and random accidental chance are the ingenious creators of both the entire universe and all of life on earth.

"While Ron was speaking at a university on the scientific evidence for creation, a student of physics said to him, "I don't care what you say, I'm still going to believe in evolution!" Ron pulled up his shirtsleeve and showed him his wristwatch. Ron said "You see this watch? I went down to a junkyard and found some rusty, twisted pieces of metal and threw them into a shoebox and began to shake it. I shook it for two weeks, two months, six months, twelve months, and all of a sudden, "bang!" it began to tick off 60 seconds a minute, 60 minutes an hour, 24 hours a day; it tells the day and date all by chance. Amazing!"

"The student laughed and said, "That's impossible!" Ron replied, "You mean to tell me that this watch being created by chance is impossible, yet you tell me my eyes, which see in 3-D and color, or my brain, which has 120 billion cells and 130 trillion electronic/chemical connections, are just a product of chance?"

"We submit to you that it takes far more faith to believe in impersonal chance evolution than it does to believe in a divine Designer, who designed and created this incredible and intricate world we live in."[190]

And then there is this famous anecdote concerning a classic scientist and author which coincidentally has failed to make the required reading list in most of our institutions of higher learning:

"The story is often told of how [Sir Isaac] Newton had a skilled craftsman build him a scale model of our solar system, which was then displayed on a large table in Newton's home. Not only did the excellent workmanship simulate the various sizes of the planets and their relative proximities, but it all was also a working model in which everything precisely rotated and orbited when a crank was turned.

"One day while Newton was in his study, a friend came by who was a great scientist, but who was also an atheist. Examining the model with enthusiastic admiration, he exclaimed: "My! What an exquisite thing this is! Who made it?" Without looking up from his book, Sir Isaac answered, "Nobody."

"Stopping his inspection, the visitor turned and said: "Evidently you misunderstood my question. I asked who made this." Newton, no doubt enjoying the chance to teach his friend a lesson, replied in a serious tone, "Nobody. What you see here just happened to assume the form it now has." "You must think I'm a fool!" retorted the visitor. "Of course somebody made it, and he's a genius. I want to know who he is."

"Laying his book aside, Newton arose and laid a hand on his friend's shoulder, saying: "This thing is but a puny imitation of a much grander system whose laws you know. I am not able to convince you that this mere toy is without a designer and maker. Yet you, as an atheist, profess to believe that the great original from which the design is taken has come into being without either designer or maker!" The atheist was no longer an atheist when he left that day."[191]

We have been discussing some of the factual and logical basis for what is known as the Teleological, or the design argument for the existence of God. Anything that has been designed—your car, your hairdryer, your house, your garden shovel, your face, your brain, and your internal organs—must have a designer. It is called the argument from design. I call it the argument from common sense. And the fact is there exists no independent way of producing the incredible design found not only in the living systems in nature, but also in the design and purpose of the minute particles, laws governing those particles, and the design and order of our own solar system, and our galaxy, and the billions of other galaxies formed across the universe, other than that of the Divine Mind. "Design-like properties are not producible by unguided natural means." And this universe, and this earth, is awash with phenomena—including nearly everything from the tiniest atom and

---

190   (Pg.18, *Fast Facts on False Teachings*, by Ron Carlson and Ed Decker)
191   (From "Who Made It?," Moments with the Book, www.mwtb.org)

the laws that govern it, to living things, to the stars and planets—that possess unfathomably complex, design-like properties. Therefore a Designer must exist "with the *intellectual* properties (knowledge, purpose, understanding, foresight, wisdom, intention) necessary to *design* the things exhibiting the special properties." And those "special properties" can arise only from design, and certainly not from chance, by accident. "*Order* of some significant type is usually the starting point of design arguments. If we are confronted with something which nature unaided by an intelligence truly could not or would not produce (e.g., a DVD player), a design conclusion of some sort is very nearly inescapable."[192]

"The hearing ear and the seeing eye, The Lord has made them both." (Proverbs 20:12) Indeed the very thought that the human eye could have made itself by accident, especially after one is introduced to the astronomical level of chemical complexity necessary for sight to occur, is literally insane. And keep also in mind that the eye is literally tens of thousands of times more complicated, complex and more of an engineering marvel than the most expensive Konica Minolta or Sony video camera on the market. The eye, you see, forms itself automatically and with no conscious direction in the womb, and then repairs and sustains itself for many decades automatically and without conscious direction of any kind, which the designers at Cannon, Sony or Nikon could not reproduce if they worked day and night for the next six thousand years.[193]

> "Look around the world: contemplate the whole and every part of it: you will find it to be nothing but one great machine, subdivided into an infinite number of lesser machines, which again admit of subdivisions to a degree beyond what human senses and faculties can trace and explain. All these various machines, and even their minutest parts, are adjusted to each other with an accuracy, which ravishes into admiration all men who have ever contemplated them. The curious adapting of means to ends, throughout all nature, resembles exactly, though it much exceeds, the production of human contrivance; of human design, thought, wisdom and intelligence. Since therefore the effects resemble each other, we are led to infer by all the rules of analogy that the causes also resemble; and that the Author of Nature is somewhat similar to the mind of man, though possessed of much larger faculties, proportioned to the grandeur of the work which He has executed. By this argument *a posteriori*, [i.e. from experience] and by this argument alone, do we prove at once the existence of a Deity and his similarity to the human mind and intelligence."[194]

Amen! brother Hume. When asked at a symposium "if recent work on the origin of life pointed to the activity of a creative Intelligence" the once world-famous-atheist now-turned-believer, Antony Flew replied:

> "Yes, I now think it does...almost entirely because of the DNA investigations. What I think the DNA material has done is that it has shown, by the almost unbelievable complexity of the arrangements which are needed to produce (life), that intelligence must have been involved in getting these extraordinarily diverse elements to work together. It's the enormous complexity of the number of elements and the enormous subtlety of the ways

---

192   (From "Teleological Arguments for God's Existence," June 10, 2005, Stanford Encyclopedia of Philosophy)
193   (See chapter 10 for an in-depth discussion of the human eye, and the sheer impossibility of it "evolving" on its own, accidentally, by random chance.)
194   (Pg. 53, *Dialogues Concerning Natural Religion*, by David Hume)

they work together. The meeting of these two parts at the right time by chance is simply minute."[195]

The chances are far less than simply minute, they are absolutely non-existent.

"O Lord, how manifold are your works! In wisdom you have made them all. The earth is full of your possessions." (Psalms 104:24)

And speaking of wisdom, it is fascinating that the factually empty assault on the existence of an Intelligent Designer has not come from the ignorant and the uneducated among us, but from the halls of academia where supposedly the most educated and the most intelligent among us reside. Woops.

Further proof

"Before the mountains were brought forth, or ever you had formed the earth and the world, even from everlasting to everlasting, you are God." (Psalms 90:2)

We have called this section "further proof," yet no further proof is necessary. Atheism is finished; at least as a credible intellectual opponent of theism. The only chance it ever had was the theory of evolution but now that has been clearly shown to be scientifically absurd. (Not from our short exposé here, but definitely from the exhaustive case already set forth by true science itself, and presented in summary form for your critical evaluation in chapter 10.) But there is still more information that can shine additional light on this subject and add to our understanding and confidence that the existence of an intelligent Creator is beyond question.

Let's start with the universe. We mentioned that it is scientifically impossible for it to have created itself. We will cover this in more detail in the chapter on evolution, but here are some additional facts…

Cambridge University physicist Brandon Carter PhD calculated "that if gravity had been stronger or weaker by one part in 10^{40}, then life-sustaining stars like the sun could not exist. This would most likely make life impossible."[196] (This would *definitely* make life impossible.) 10^{40} is one with 40 zeroes behind it. A trillion is one with only 12 zeroes behind it. 10^{40} is a trillion times a trillion times a trillion times ten thousand. The chances of winning the lottery are around a million to one. The chances of the force of gravity accidentally being exactly the force needed to render the universe inhabitable is trillions and trillions and billions of times more remote than the chances of winning the lottery. Try to shoot an arrow across the entire known universe with your eyes closed and hit a target in the exact center without being off by even one trillionth of an inch. And then try to do it a billion times in a row. It's impossible. So is the accidental formation of this universe. The force of gravity was specifically created to be exactly the amount needed by a degree of variance of less than one over ten to the fortieth power. And this is only one of a number of examples of the mind-numbing exactness of the laws and constants of this universe. (In addition to the gravitational force mentioned above that holds matter together there is also the three forces—the electromagnetic force and the strong and the weak nuclear forces—which hold the structure of the atom together which are also exactly calibrated so the heavier elements can form and we can have a universe of matter that is able to form stars

---

195 (pg. 75, *There is a God: How the world's most notorious atheist changed his mind*, by Antony Flew)
196 (*International Journal of Systematic and Evolutionary Microbiology*, IJSEM, Collins 1999, 49)

and planets and to support life.) "In light of these and other examples, Collins remarks that "Almost everything about the basic structure of the universe … is balanced on a razor's edge for life to occur.""[197] God- 10^{40}. Atheists- nothing. Game, set, match. Without stars and without orbiting planets made up of heavier elements, there could be no earth and there could be no life.

"God is a mathematician of a very high order and he used advanced mathematics in constructing the universe."[198] What an understatement.

Now let's take a peek inside a living cell, the microscopic building block of all life, and once called by Darwinists, "the simple single cell." With what we know today of the mind boggling, astronomical complexity inside the cell, the absurdity of that statement, though forgivable with the limited amount of knowledge they had at that time, is unrivaled in human history. The actual engineering complexity of *the simple single cell* is greater than that of the entire city of New York. And that is a scientific fact. Yes, it is just that complex. Every second within every single individual cell in your body, a million chemical reactions take place. You have a thousand trillion cells in your body. That's a billion trillion chemical reactions every second just to keep your body alive and functioning. In each cell there are thousands upon thousands of nanobots, micro machines, tiny robots that scurry about all over the cell transporting materials like oxygen and carbon and water from the cell wall to countless places where they are needed, and taking waste products of metabolism like carbon dioxide back to that cell wall for expulsion. They do all of this over and over again millions upon billions upon trillions of times, day in and day out for your entire lifetime. And the miraculous thing is that they do it <u>blindly</u> and <u>unconsciously</u>; everything in the cell is completely automated.

Can you imagine constructing an entire city of New York that could run itself for decades blindly and unconsciously, completely automated, with no human intervention? A car gets in an accident, the tow trucks would have to be able to sense it automatically and drive to the right place and pick it up and then bring it to the garage where it would be repaired by robots working blindly and unconsciously without any human intervention. The same with a water main that breaks. A fire that starts. A hurricane blows the roofs off and the windows out of countless buildings. An elevator or air conditioning or heating system breaks down. The repair robots would have to know exactly where the water leak is, the fire is, the windows are blown out, blindly and unconsciously! Think about it! If we gathered tens of thousands of the world's most brilliant engineers and scientists and allowed them to work night and day for the next ten million years, they could never do it. After a few million years they might get a small scale primitive prototype, consisting of maybe ten square blocks, up and running and it would last maybe half a day until something blew up that the automated city could not handle on its own and the humans would have to rush in and take over. Back to the drawing board. But the Intelligent Creator of this universe accomplished the equivalent of designing and manufacturing an automated city of New York within every cell of your body, and every cell of every living thing on this planet, and it took him just a few days. He spoke it, and matter instantly obeyed his voice. He

197　(*International Journal of Systematic and Evolutionary Microbiology*, IJSEM, Collins 1999, 48)
198　(Paul A. M. Dirac)

breathed it, and it all sprang into being. He wound it all up like a clock, and it's been running itself, unconsciously, ever since.

> "Scientists (even some evolutionists) who understand the amazing complexity inside a living cell know it could never have evolved; it had to be created. But science cannot say who the creator was. It might have been several creators or even "little green men" from Mars. Nevertheless, when one understands the evidence, it is clear that this amazing complexity could not have evolved. It is hard to imagine an unbiased person who understands the evidence reaching any other conclusion. Unfortunately, few educators and scientists have heard this evidence. (Unintended ignorance is excusable. Unwillingness to learn is not. Preventing students from learning is reprehensible.)"[199]

Again, the "level of knuckle dragging ignorance" it takes to suppose that the astronomically complex engineering miracle of a living cell—fraught with highly advanced, completely unconscious and automated nanobot technology that we are literally thousands of years away from even comprehending, much less ever duplicating—could form accidentally by random chance is so far off the charts on the schtupe[200] meter that they have had to employ the Hubble Space Telescope to search for it at the far end of the known universe.

Do I believe God exists? No, I don't believe God exists. I know it. I know it in the same way that I know that China exists, even though I've never been there and I can't see it out of my bedroom window. (I know, Tina Fey, what an unfortunate and useful idiot of the Democrat-Left.) Is there a possibility that China *doesn't* exist? Yes, there is a possibility but what that would entail is that everybody on the face of the earth for thousands of years would have to be in on this scam. Absolutely everybody would have to be in on it except for me, including all of my family members and my children who would have to be secretly taught about the scam before they could blurt out the truth, as little children are apt to do. Every map in every country that was ever made would have to include China in the scam, because even if I traveled to the remotest part of Brazil and a map there did not include China, but some inland sea that actually was there in its place, then the gig would be up. So everybody on the face of the earth would have to be in on the scam, every journalist, every newscaster, every newspaper editor: everyone except for me. So maybe China doesn't exist. Maybe there's just an inland sea there. But the chances of that actually being the case is so far beyond the realm of mathematical probability that it's absolutely impossible. So yes, God exists. I don't believe it; I know it, just like I know China exists. All we have to do is look at all of the facts and evidence. Many other people know it as well. The non-existence of an extremely Intelligent Creator of this universe and of this earth and of the life that teems across it, is scientifically, logically and practically impossible.

Further, further proof

Furthermore, we can know there is a God because the universe had to come from somewhere. Every effect has a cause. Eventually, there has to be a first cause.[201] That is because mathematically there is the impossibility of an infinite

---

199   (Page 408, *In the Beginning: Compelling Evidence for Creation and the Flood*, by Walt Brown, Ph.D.)
200   (stupid)
201   (Even atheists accept this premise, except they believe the first cause was an explosion.

regress of causes. An infinite regress is impossible because infinity itself is logically impossible due to the following argument:

> "If actual infinites that neither increase nor decrease in the number of members they contain were to exist, we would have rather absurd consequences. For example, imagine a library with an actually infinite number of books. Suppose that the library also contains an infinite number of red and an infinite number of black books, so that for every red book there is a black book, and vice versa. It follows that the library contains as many red books as the total books in its collection, and as many red books as red and black books combined. But this is absurd; in reality the subset cannot be equivalent to the entire set. Hence, actual infinites cannot exist in reality."[202]

We also know there is a First Cause, because if there is not, then there would have to be an infinite series of causes going back forever and this is logically and mathematically impossible because if there is an infinite regress of causes than these causes could never have made it to the present moment. (Think about it.) (Without drugs.) "No amount of time can consume an infinite series of events to bring you to the present."[203] Such is the nature of infinity. And time. Therefore there had to be a First Cause. Outside of time. Outside of space. Outside of the concept of outside itself. Self-existent. "Without father, without mother, without genealogy, having neither beginning of days, nor end of life; but made like unto the Son of God" (Hebrews 7:3)

And then there is the following: Could the universe have created itself as big bang cosmologists and evolutionists suppose? Not a chance. It is absolutely impossible and here's why. In order to create itself the universe would have had to exist before it existed, so it would have had to exist (in order to create itself) and not exist (prior to its creation) at the same time, and this is impossible, according to the law of non-contradiction.[204] This law, one of the three classic laws of thought, states that…

> …"two antithetical propositions cannot both be true at the same time and in the same sense. X cannot be non-X. A thing cannot be and not be simultaneously. And nothing that is true can be self-contradictory or inconsistent with any other truth. All logic depends on this simple principle. Rational thought and meaningful discourse demand it. To deny it is to deny all truth in one fell swoop."[205]

We know there is a first cause. And once you eliminate random chance as the first cause for life, and an explosion as the first cause for the universe, there is only one first cause left and that is an infinitely intelligent being, outside of the universe, which has the power to create something from nothing. I know there is a God and that he is a Spirit because the universe itself is not material, the substance of this universe reflects the Spirit that created it. I know there is a God because of my own consciousness and awareness which comes from my mind, my spirit, which is not a part of the physical brain, and that the only Source for my consciousness is the Supreme Consciousness of the Higher Mind. Our consciousness comes from that which *is* Conscious. The "I am that I am." For we were made in his image and likeness. I know there is a God because the laws and the order of this universe

---

Which only punts their scientifically-inaccurate conclusion to "Well then, what *caused* the explosion, stupid?" So either way, their argument is dead.)

202 (from "The Cosmological Argument," The Stanford Encyclopedia of Philosophy)
203 (Ravi Zacharias)
204 (Dr. R. C. Sproul)
205 (From article, "The Law of Contradiction," by Phillip R. Johnson @ spurgeon.org)

demand a lawgiver, and order cannot arise spontaneously from disorder (See the Second Law of Thermodynamics). I know there is a God because his moral law is built into our consciousness, into the soul and the conscience of man, and these laws have no origin in the particles of this universe. They come from that which *is* moral. (And holy, and just, and perfect, and kind, and merciful, and understanding, and compassionate, and long-suffering, and…) I know there is a God because "since the creation of the world his invisible attributes are clearly seen, being understood by the things that are made, even his eternal power and Godhead so that those who deny him are without excuse." (Romans 1:20) His "invisible attributes," like His infinite power and intelligence, are clearly seen, are clearly revealed to us, by His incredible creation, by His universe, by the stars and galaxies, by the wonder, beauty and expanse of the earth, and by the incomprehensibly-complex miracle that is life. And all of this is easily "understood by the things that are made," by us, "so that those who deny Him are without excuse." I don't know how anything could be any clearer, or any simpler, or any more logical, or any more accurate, or more true.

"The heavens declare the glory of God, and the firmament shows his handiwork." (Psalms 19:1) And the skies declare his artistry. The earth demonstrates his creativity and the universe his imagination. The beauty of the earth and of the universe reflect his inherent beauty and grace. As does the beauty of an attractive woman reflect His beauty. ("God, you are so beautiful.") Each waterfall proclaims his majesty. The thunderstorm warns of his awesome power. Volcanoes and their molten lava of his righteous and impending wrath. The mountains testify to his unchanging nature. Our consciences remind us of his justice; the pain we experience of his approaching retribution. And in the very best of human strivings and endeavors we see a reflection of his wisdom and holiness, his faithfulness and love, his mercy and kindness and goodness and graciousness. Truly the heavens do declare the glory of God, and the earth shows his handiwork, so that anyone foolish enough to deny Him will be found in the approaching day of reckoning, which is not all that far off (see chapter 9), to be without any excuse whatsoever. (Time to wake up and come to your senses, Messieurs Dawkins, and Hawkings, and Harris, and Hitchens, and…)

Further, further, further proof

Existence itself cries out for a cause for its existence. It's in the nature of existence. Existence itself not only cries out for a cause, but it calls out for a cause that is itself causeless, that has no cause; that which is outside of existence that caused existence to come into being. In the same way your automobile cries out for a cause, something that caused it to come into existence, which is outside of your automobile. Your automobile is a microcosm of the universe. (Time to look at Ol' Betsy in a whole nother light.) And the microcosm mirrors the macrocosm. Everything in existence including existence itself owes its reality to something other than itself. Except for God, who is responsible for everything in existence, including existence itself, but owes his existence to no one.

Existence presents an impossible problem for the atheist. For starters, we find in all living things the overwhelming presence of purpose. All the intricacies of the cell are there for a purpose. For the survival of the cell, and for the duplication and reproduction of the cell. Forget about how life came about in the first place, which an atheist can't answer. His ludicrous attempt at using accidental random chance just doesn't cut it. But where did life find purpose itself, in a purposeless universe? Purpose is a concept. How would that concept which is not a part of the physical particles of the universe arise on its own, and then plant itself into every aspect and in every corner of every living system? Where does purpose come from in a purposeless universe? This question is unanswerable by someone who denies God. Where do laws come from in the first place, in a universe devoid of the concept of "law?" The laws of the universe demand a Lawgiver. Where does love come from in a universe devoid of feeling or caring?

Where does morality come from in a universe of particles and energy that could care less about right and wrong? Where does the conscience come from?

> "The conscience is not a physical organ: it cannot be seen, have surgery performed on it or be transplanted, and yet it exists. Where does it come from? What is the purpose of it? Who put it in us? Where does this "inner voice" of our conscience come from? ...the conscience [is] God's prosecutor, because it accuses us when we do wrong."[206]

Where does beauty, the very idea of beauty, come from in a universe that has no concept of beauty or ugliness? Where does goodness and kindness come from? Where does logic come from? That man possesses intelligence, along with many other creatures, requires that Intelligence itself must have existed prior to man, and other creatures. Intelligence cannot arise on its own from a universe devoid of thought. These questions are unanswerable by someone who denies God.

And then there is thought and its resulting knowledge. Knowledge itself demands a source of knowing, of knowing anything. A physical universe of matter, light, electrons etc. can know nothing. It can only exist. Knowledge itself demands a knower, and the fact that we know anything at all, in the midst of a barren universe incapable of knowledge, demands a Knower that created us and gave us the ability to know. The universe couldn't. It knows nothing and therefore cannot *bestow* the ability to know on anyone or anything. *Nature*—no matter how much we hear her talked about now (she has replaced God the Creator) on every science channel on TV, and no matter how much knowledge is bestowed upon her by these programs scientists—also knows nothing. The knowledge inherent in the plant and animal and microbial kingdoms—the knowledge of how to survive, and move, and replicate, and migrate, and communicate, and interact—did not come from the unknowing universe, nor from some coroneted, yet imaginary, *Mother Nature*. For both know nothing. The knowledge that we possess and that the rest of life possess' demands a Knower who bestowed on us the ability to know.

> "The existence of God is nowhere defended by Scripture. This fact is taken as being obvious. ...Neither is there any doubt as to His sovereign authority over His creation or what our attitude should be toward Him as Creator. He has the right to set the rules. We have the responsibility *to obey and rejoice in His goodness*, or disobey and suffer His judgment."[207]

In addition to all of this there are also innumerable examples in the physical world of realities that strain credulity if one proposes they happened by coincidence.

---

206   (From "The Conscience: God's Prosecutor," by Norbert Lieth, Midnight Call Magazine, June 2009)

207   (Pg. 80, *The LIE: Evolution*, by Ken Ham)

Here are just two: Who tilted the earth perfectly and spun it at just the right speed so we could enjoy varying seasons? Without which the earth would be uninhabitable. Who made the laws that govern matter in such a way that water is the only element in the entire universe that <u>expands</u> when it freezes? Everything else contracts. Life on earth couldn't exist if it weren't for this incredible anomaly; for if water contracted when it froze the ice particles would <u>sink</u> when they crystallized, rather than rise and melt, and the oceans would long ago have frozen, as would have many rivers. A universal coincidence? Sure, you can think that. If you're a moron.

> "It is philosophically impossible to be an atheist, since to be an atheist you must have infinite knowledge in order to know absolutely that there is no God. But to have infinite knowledge, you would have to be God yourself. It's hard to be God yourself and an atheist at the same time! The Bible says in Psalm 14:1, "The fool says in his heart 'There is no God.'""[208]

(The modern interpretation of that is, "The *idiot* has said in his head, 'There is no God.'") (With all due respect of course to all of those idiots who populate academia and spill over onto our TVs and radios.)

Humanity has been *blessed* with quite a number of atheist philosophers over the centuries who became quite accomplished at performing complicated mental gymnastics in order to try to square their beliefs with reality—logical and linguistic acrobatics that'd put Cirque De Soleil to shame—but they have only come up with feeble, illogical and nearly incomprehensible attempts at answers to all of these questions. After endless attempts at trying, they still have no answers. They only have their stubborn willful ignorance. "In whom the god of this world [that great deceiver] has blinded the minds of those who refuse to believe, lest the [Truth] should shine unto them." (2 Corinthians 4:4) What does it say when so many of the world's most esteemed philosophers can't figure out that God exists, but a high school sophomore with common sense can? You can lose yourself from now until the end of time with classic debates that seek to parlay synthetic distinctions from P-inductive arguments into the existential opposition of any further necessarily contingent propositions and/or obiter dictums, based, a priori, on the C-inductive evidence of the cosmological argument, which categorically states that any hypothetical non-existence or pre-existing epistemological distinctions of a transcendent being would be inversely proportional to the sum of those hitherto unknowable yet semantically deductive pluralities that prove the dialectic supremacy while assuming the self-determination of the mundane; and that furthermore any teleological deductions or logical constructs must be experienced, a posteriori, before the predicate concepts may presume an analytic divergence and combine to falsify the existence, in reality, of actual infinites.[209]

Or, hell, you could go have yourself a beer.

"I form the light, and create darkness: I make peace, and create evil: I the Lord do all these things. I have made the earth, and created man upon it: I, even

---

208   (Pg. 17, *Fast Facts on False Teachings*, by Ron Carlson and Ed Decker)
209   (from The Atheistic Philosophizers Collection: *Introduction to Doubletalk; Doubletalk 101; and The Annotated Doublespeak: Volumes 1 & 2*, by Random Chance Publishers)

my hands, have stretched out the heavens and all their host have I commanded."
(Isaiah 45:7, 12-13)[210]

But what about the very existence of evil?

The universe is unjust; therefore there cannot be a God. And so the argument goes. But the very concept of "justice" could not exist if the Creator of this same universe were not himself perfectly just. A cold, dark, impersonal universe of atoms and molecules has no clue of "justice," nor can the concept arise accidentally by random chance out of the same. The only way we can even conceive of "justice" is that this concept comes down from above. To therefore accuse this same Creator, who exercises perfect justice, of not existing, or of being unjust, because we can perceive of injustice in his universe, is patently absurd; in addition to being a tad presumptuous, and a little offensive in its own self-righteousness. No, the perception of injustice in a world created by a perfectly just and holy God, should spur us, not to reject his existence, but to seek to understand why this perceived injustice exists. And in seeking this we will easily find that it actually is a just universe, and that even though the perfect execution of justice does not always happen in this life, it always happens to everyone in the next one. Perfect justice perfectly executed. (And see chapters 7-13.)

There is evil in the world, therefore there cannot be a God. And so the other argument goes. But the very existence of evil in the world rather than being an argument against the existence of a loving, holy God is actually direct evidence *for* Him. Because the only reason that we can conceive of evil at all as evil is due to the inherent goodness of that same Creator, and the fact that he made us in his own image and likeness, and therefore we can conceptualize both good and evil. (It also has something to do with an old, old story of eating fruit from a certain tree, but that is a discussion for chapter 7.) The universe of matter knows nothing of the concepts of good and evil and therefore cannot bestow that knowledge on anyone. How do we even comprehend the very concept of evil without the existence of ultimate good to compare it to? We cannot. The same as we could have no comprehension of fear without the existence of love.

But why is there such a proliferation of evil in the world? Why would a supposedly loving, holy, all-powerful God allow that to thrive? Answer: because of a powerful gift he also gave us called *free will*, he allows evil temporarily (and only temporarily) to pollute his once perfect world. (See chapters 7, 9 and 13) "And God saw every thing that he had made, and behold it was very good. So the evening and the morning were the sixth day." (Genesis 1:31) In the beginning, when God created the heavens and the earth, there was no evil in the world. Otherwise he could not and would not have stated that "behold it was very good." Evil, and sin, came a little later, with the temptation and fall of man. Indeed, even the most basic understanding of a few rudimentary concepts quickly dispels the *problem* atheist philosophers in particular have had with the existence of evil in the world: 1) that of man's free will; 2) the fall of Lucifer (see chapter 13) one of the most powerful of God's angels, and the fall of a third of the angels along with

---

210   (It is important to realize that atheists are basically calling the Author of that verse a liar. And the only thing that Talking Points can say to them is, "Don't ever die. Whatever you do, don't ever die....")

him, many of whom are already imprisoned in hell[211] but many others roam the earth seeking whom they may possess; and 3) the fall of man. These are things that we will cover in more depth in later chapters, but the fact that an all-powerful God has not *yet* chosen to banish evil is not an argument against his existence, it is only proof that for whatever reason[212] he has not yet banished evil, and destroyed those who practice it.

But what about suffering, death and disease? Why would God make a world with all of this in it, if he really was God? Answer: He didn't. In the beginning God made man perfect and placed him in a perfect world, without sin, death or evil. But he also gave us free will. Man then used his free will to disobey God. He chose evil. He chose sin. And he got everything that comes from sin: suffering, death, disease and evil.

Why does God allow suffering, if he has the power to end it? Answer: *We* allow suffering. (And we could end the vast majority of suffering in the world tomorrow if we weren't so evil, so fallen, so selfish, and so immoral.) He allowed us free will. And he will end suffering soon, for those who side with him. But for those who side with the opposite of God, their payment is coming and it will be in suffering that will never end. Hence the importance of siding with God. (See chapter 7.)

"People try to deny the existence of God by pointing to the injustices in the world. They ask, "Why does God allow pain and suffering? Why hunger and poverty? Why doesn't God just outlaw those things? He is God isn't He?" God isn't callous to our sufferings. In fact He suffered as no human ever suffered on the cross. He paid for the sins of the world in Christ. Our rebellious sin is the root cause of our trouble but people don't want to acknowledge that. There is suffering in this world because there are evil people, and they do evil things. They want power and control. **But just think of the times that you caused suffering to someone. For there to be a lack of suffering in this world, God would have to take us all out**."[213] (Emphasis mine.)

Many people have a problem with God because of the *suffering* question, especially when they have experienced personal suffering of a horrible nature-the death of a child, the untimely loss of a wife or husband to a heart attack, cancer or automobile accident. And their personal pain and frustration over God for allowing it cannot be glossed over nor taken lightly. And hopefully we will provide some comfort in matters such as these as we further uncover the truth in subsequent chapters. (In chapters 7-8 in particular.) But here we are primarily concerned with the facts and evidence for the existence or non-existence of God, and it remains a fact that it is the very existence of suffering and of evil itself that cries out for the existence of an ultimately good Creator.

OK, if God exists, where is he? Is he hiding? No. He's not hiding But the reason he has chosen to remain mostly out of the picture and behind the scenes for the last six thousand years since the fall of man, and except for a short appearance two thousand years ago, is due directly to the fall of man (and woman for any

211   (See: *23 Minutes in Hell*, by Bill Wiese)
212   (And he has some very good ones—see chapter 7—and one of them being... "For God so *loved* the world that he gave his only begotten Son that whosoever..." ...Maybe He's giving mankind as much time as possible to make the right decision before the end of this age collapses all around him...)
213   (From sermon, Sunday, 09/20/2009, by Pastor Ed Regensburg, Faith Community Church, Gambrills, Md.)

sensitive feminists), as well as to the gift of free choice.[214] He stays "behind the scenes" for now because he has given man the free reign to deny and reject him, or to love and embrace him. If he was present in this world like for example water, then no one would be able to deny him. No one here denies the existence of water yet if you were born and raised on the moon you might. (Don't ask me.) He has stated that "without faith it is impossible to please him." He has left us here after our fall with an astronomical amount of evidence so no one has any excuse if they choose to deny him in the face of all of that evidence, but at the same time it takes faith to come to Him. He is not so much hiding as he is waiting for the time of his revelation, when He will clearly reveal himself in all his power and glory to all the world.[215] Before that happens, and it is scheduled to happen shortly, it is decision time for the human race for where, and in whose presence, each person plans to spend eternity.[216]

"Lay not up for yourselves treasures upon earth, where moth and rust corrupts, and where thieves break in and steal; but lay up for yourselves treasures in heaven."[217] (Matthew 6:19-20)

Rapping it up

Many of our young people today are raised and taught by parents who know of the existence of God. But unlike when their own parents were growing up, they live in the midst of a greater culture where the truth now is under vicious attack. And if today's youth are taught the truth at home, and even have that truth reinforced in a Christian or Catholic or traditional Jewish school, but if they do not become rooted and grounded in that truth—in this case the existence of God—then they will be in no position to withstand "all the fiery darts of the wicked one" (Ephesians 6:16) that will be shot at them as they go out into a godless and deceived world. They will be in no position to refute or even withstand all the lies of Satan that will be shoved down their throats by their *esteemed*, yet too often deluded, professors if they go to college, or by their culture and many of their peers if they do not. It is of utmost importance then, not for them to just learn *about* the truth, but to be as rooted and grounded and educated and confident in the truth of the existence of God as they are about the existence of China. Then no intellectual elite, and no professor, and no godless entertainment industry, and no clueless culture, and no uneducated peer, and no late-night clown, and no lie of the father of lies will be able to steal the truth away from them. And that is one of the intents of this chapter, and of this book.

"You shall love the Lord your God with all your heart, with all your soul, with all your mind, and with all your strength." (Mark 12:30)

Hatred of God can manifest itself in a number of ways. Loving the lies of God's adversary is one. Embracing one of the many false religions founded by that same adversary, and refusing to let go of it, is another. Killing 6 million Jews, mutilating

---

214   (Both the fall of man and free choice we will discuss in chapter 7)
215   (See chapter 9, and The Book of Revelation)
216   (And see chapter 7)
217   (And this heaven will be here on earth. In a recreated Garden of Eden. No playing harps, bored to tears, floating on the clouds. Rather, think Hawaii on steroids. And see chapter 9)

50 million babies to death, and slicing the clitoris off of 130 million little Muslim girls are others. And atheism, denial of God's very existence, is just one more in a long line of manifestations of man's hatred of his Maker. Rather than being the result of the more brilliant, or more intelligent, or more logical of what the human mind is capable of achieving, it is the product of our basest of emotions. Just as children in a temper tantrum occasionally will blurt out, "I HATE you, mommy!" so too do many of us go through life harboring deep-seated resentments and anger and outright hatred toward the One who made us; and there are many reasons why. Maybe he didn't make us smart enough, or pretty enough, or handsome enough, or tall enough, or short enough, or popular enough, or charismatic enough, or rich enough, or athletic enough, or strong enough, or quick enough, or fast enough, or healthy enough, or like that other person. Or maybe we didn't get the job we wanted, or the career we wanted, or the girl we wanted, or the husband we wanted, or the children we think we deserved, or the house we wanted, or the promotion we thought we deserved. Or maybe we lost a loved one; perhaps a baby or a child or a son or a daughter or a husband or a wife or a mother or a father. Whatever the reason, many of us, including but certainly not limited to our atheist friends, need to come to terms with our feelings and emotions towards our Creator. See Part 1, "Pure Religion" of chapter 8 for a discussion of prayer and some possible insights into dealing with any deep seated feelings or emotions you might need to come to grips with concerning your Creator. And certainly see the list of recommended books there, as well as seeking out books that may deal with any problems specific to your own situation that you may be dealing with. Talk to Him, you might be surprised at how good he is at listening. And don't be a phony. Be respectful but tell him how you feel about him, no matter how ugly it might be. He's all-knowing, you know, so he already knows perfectly well. He's also very forgiving. Be real with him, and you might be surprised how real He will become to you. For as some of our own poets have said, "in him we live, and move, and have our being; for we are also his offspring." (Acts 17:28) Good "luck," and God bless.

And, hopefully, for some of our atheist friends, if they were to initiate this discussion with their Creator, they might be surprised and delighted to actually be brought to the knowledge of the truth of His existence. ("WOW! I can't believe it! There actually IS a God!") (Shazaam!) This earth is now covered with the lies of Satan like a blanket, but one day soon it will "be filled with the knowledge of the glory of the Lord, like the waters cover the sea." (Habakkuk 2:14) (Kind of takes us back to that water-moon analogy.) (Or not.)

"America has no more stars, now we call them idols."[218]

Everybody has a god. It is in the nature of man and it is unavoidable. In the past man has made the Sun their god, and the Moon. They have worshiped golden images, and statues of wood and of bronze. They have bowed down before images of birds, of beasts and of humans. And today it is no different. We worship money, or sex. We idolize sports stars, TV celebrities, movie idols and rock stars. (Just look at the Elvis and Michael Jackson phenomenon.) (Although we would be a lot better served if we had the sense to idolize our countries real heroes, the men and women who risk life and limb, who serve their country in our military.) And many people today, members of the *me* generation, steeped in a selfishness that

218   (From song, "Lose My Soul" by Tobymac)

few peoples in the past have had the luxury of indulging in, reject God because they want to be their own god. They worship at the altar of their own wants and desires. And too many modern intellectual academics, along with their proselytes, worship at the altar of atheism, where random chance and ancient explosions are enshrined. Denying God's existence, or worshipping any number of false gods, are all seductive and dangerous traps, which is why many years ago the real God, the Creator of this universe, warned mankind about falling into them when he spoke these words:

"I am the Lord your God, You shall have no other gods before me."

(Exodus 20:2-3)

We would be wise to heed them.

We have discussed the existence of God here, not so much in religious or theological terms, not in terms of faith and belief, but on a scientific level, and on the level of logic and common sense. We have looked at all of the facts and evidence and we have gone wherever they have led us. Can science prove the existence of God? Well technically science by definition is the study of the *physical* world, and it consists solely in direct observation and repeatable experiments that lead to testable hypothesis, theories and then laws. It is obvious, by this definition, that you might find some difficulty in direct study or observation of God. (But you can, of course, directly study and observe his *handiwork*.) However, it is a valid scientific question to ask, "What are the origins of this handiwork? Where did it all come from? How did it all come into being?" And the major error of faux scientists across this world is to state that the only valid *scientific* way of discussing origins is to look for that causation within and only within the physical world itself. Hence we see their unfortunate attraction to the scientifically absurd theories of evolution and the big bang. They are looking for origins in all the wrong places. This is as idiotic, as we stated before, as looking for the origins of an automobile only within the automobile itself. No intelligent designer, conceiver or manufacturer outside of the physical limitations of the automobile itself can be considered. Forget Detroit. Forget assembly lines. Forget autoworkers. Only within the physical confines of the automobile itself can you look for its original cause that made it come into existence. Rather insane don't you think? As we have mentioned before, this is known as the religion of *Naturalism* and it has nothing to do with science; even though most teachers and professors of science continue to impose this false religion in a fascist way on students all across this land.

In addition, we have discussed how science can and has eliminated the only other choices for the origin of all this… that of random chance and a big explosion. (Again, see chapter 10 for an exhaustive exposition.) As such I think it is a logical conclusion, supported by all of the facts and evidence and by all of the laws of science, to state that science has proven the existence of an Intelligent Creator (call him, her or it what you will) by eliminating all other scientific possibilities. Therefore the existence of an extremely intelligent and powerful Creator is a scientific fact! ("And you heard it first right here on roller derby.")

"For thus says the high and lofty One who inhabits eternity, whose name is Holy: 'I dwell in the high and holy place, with him who has a contrite and humble spirit'" (Isaiah 57:15)

One day soon, when the world is a little better educated, atheism will be seen in the same light as amoonism or asunism, or flatearthology. *But wait*! I hear. *You're comparing apples and oranges. It's not the same because I can* see *the sun and I can* see *the moon and I can* see *pictures showing me that the earth is round. But I cannot see God.* Ah, but I say that if we were physically blind, we could neither see the sun or the moon either. And God says the reason we cannot *see* (*believe in*) him is because we have chosen not to of our own free will. It is because we have chosen to be blind to him. It has nothing to do with a lack of evidence. It has everything to do with willful ignorance; a direct *denial* of all of the evidence.

"For the invisible things of him (God) from the creation of the world **ARE CLEARLY SEEN**, being understood by the people that he made, even his eternal power and Godhead (Gods mind); so that they (those who choose to deny him) are without excuse." (Romans 1:20)

You see, there actually is a God. And it actually doesn't take a whole lot of faith to believe in him. (Not that there is anything wrong with faith; as long as your faith is in the right thing.) It just takes a little common sense, which that same God happens to have given to us liberally. It's up to each individual to decide whether or not to use it. And the more you decide to use it; and the more you look at the evidence for and the evidence against the existence of an all-powerful, all-knowing, all-wise Creator; the more you will be forced to conclude, scientifically, in the affirmative.

Ideas have consequences. How many of our teenagers have been driven to suicide by the hopelessness of a world with no Creator and thus no purpose? And are being driven to drugs and alcoholism, perverted and aberrant behavior, and sexual promiscuity by the same? All because of a few, a small minority of ill-educated, left-wing indoctrinates in our government, our courts, our schools, our media, our entertainment industry, and faux scientists in our universities, have mandated the teaching of a bald faced, idiotic, asinine and scientifically impossible lie. That the world created itself, and life created itself, on its own, accidentally, by pure chance. And that therefore there is no need for God. They have made these false and idiotic beliefs the state mandated religion of the land. And it is the state mandated religion of our *public* schools. (So much for the "separation of church and state.") They have banned the truth of an Intelligent Creator, and they have perverted the truth of real science. Fundamental lies about fundamental truths have terrible consequences. Just look at Nazi Germany, where Hitler and his henchman—godless, avowed evolutionists—slaughtered millions in order to "improve the species," to help along human evolution. ("Survival of the fittest," you know.)[219] Look at atheist and communist Russia and the hundreds of millions killed; look at imprisoned and suffering Cuba, and miserable North Korea, and communist China where Mao murdered 60 million more; devout atheist governments all. And look at the mutilation and slaughter of the 50 million innocent in our own country. No, America, you've been lied to. The existence of God is a scientific fact. And atheism is an unintelligent and unscientific joke. It has been said that "there is no idea as powerful as one whose time has come." Let's hope the idea of the scientific truth of the existence of an Intelligent Creator, and

---

219   (See "Darwin's Deadly Fruit" in chapter 10)

the truth of the scientifically impossible lies of evolution and the big bang, are a couple of ones whose time has finally come. For the sake of the children.

> "God is in his holy temple, let all the earth keep silent before him."
> (Habakkuk 2:20)

See:

- *The Devil's Delusion: Atheism and its Scientific Pretensions*, by David Berlinski. ("Berlinski eviscerates the supposedly rational and scientific claims of the 'New Atheists' and shows how their anti-religious arguments are in fact *ir*rational, *un*scientific, and very often based on the kind of 'blind faith' and 'wishful thinking' they impute to religious believers.")

- *There Is A God: How the World's Most Notorious Atheist changed his mind*, by Antony Flew. ("Long before the "new atheism" arrived courtesy of best-selling authors Sam Harris, Richard Dawkins, and Christopher Hitchens, the renowned British philosopher Antony Flew was a leading proponent of modern atheist thought. His prolific writings helped shape the atheist agenda for much of the 21st century- until he announced to the world in 2004 that he accepts the existence of God. Now, Flew reveals, for the first time, the scientific discoveries and philosophical arguments that turned him from a staunch atheist to a believer.")

- *In the Beginning Was Information*, by Dr. Werner. ("Information is the very cornerstone of life... The very presence of information reveals a designer.")

- *Darwin's Black Box: The Biochemical Challenge to Evolution*, by Michael J. Behe.

- "Who Made It?" From *Moments with the Book*, www.mwtb.org

- "Teleological Arguments for God's Existence," June 10, 2005, Stanford Encyclopedia of Philosophy

- *Dialogues Concerning Natural Religion, by* David Hume

- "The Cosmological Argument," The Stanford Encyclopedia of Philosophy

- *The LIE: Evolution*, by Ken Ham

- *23 Minutes in Hell*, by Bill Wiese

CHAPTER 6

THE WORD OF GOD

"He was clothed with a robe dipped in blood,
And his name is called The Word of God." (Revelation 19:13)

By way of review, we have found out thus far that much of the "current wisdom" of the day is not wisdom at all, but foolishness. ("For the wisdom of this world is foolishness with God. For it is written, he catches the wise in their own cleverness." 1 Corinthians 3:19) Our once-trusted institutions of higher learning are now failing us miserably in their responsibility of imparting knowledge that is accurate and true. They have gotten it 180 degrees wrong on a number of issues. The truth being *relative*? Wrong! Morality being *relative*? Wrong! The material universe is all there is. Absurd. Atheism is *rational* and *scientific* while belief in God is unscientific and based on blind, primitive religious faith. The exact opposite is true! The theories of evolution and the big explosion are a *scientific fact*? Not only wrong but scientifically-impossible! We see now, that in many regards and on many crucially important subjects, our colleges have descended into godless, left-wing indoctrination camps and many of our professors into brainwashed deceivers uninterested in debate or opposing viewpoints while being openly hostile to the truth in far too many cases. Instead of imparting the truth they disseminate lies. Pray for them...

"And a servant of the Lord must not quarrel but be gentle to all men, able to teach, patient, in humility instructing those who are in opposition, if God perhaps will grant them repentance so that they may acknowledge the truth, and that they may come to their senses and escape the snare of the devil, having been taken captive by him to do his will." (2 Timothy 2:24-26)

God-breathed answers

"Thus speaks the Lord God of Israel, saying, Write all the words that I have spoken unto you in a book." (Jeremiah 30:2)

We have reached the indisputable conclusion that this world actually does have a Creator. And we know that this Creator is not a stupid explosion, nor is he accidental random chance. We know that he has to be someone or something of astronomical intelligence and power that is outside of this universe, that exists prior to this universe, and that is not a part of or a product of the universe itself. He has to be that Transcendent Being who possess' the incomprehensible power and intelligence to have created these worlds in the first place, and to sustain them even now. And we know all of this for a scientific fact.

But this conclusion only leaves us with about a hundred other questions. Like... Who is he? Where did he come from? Is he a he? What is he like? What is the nature of his being? What is God's character? Why the universe? What is the point? Why are we here? Why did he create us? What's up with all this? What does he want from us, if anything? What does he expect of us? And while we're on the subject, God, what happens after we die? Where do we go? Etc. Etc. And it also becomes clear upon a little reflection that there is no way we can answer these questions on our own. We are capable, however, of coming up with a whole lot of *possible* answers, and just a short stroll through some of this world's various religions reveals just how diverse and sometimes bizarre our attempts can be. (See "False religion" section of chapter 8.) But the one thing that man's efforts at answering these questions all have in common is that they are nothing but speculation, guesses, flights of fancy or religiously-held beliefs that are based on opinion only. And that's not what we're looking for. Not in a search for truth. If we just rely on man taking a wild stab at it we're not going to be able to come up with anything whatsoever that we can <u>know</u> to be true.

So how do we do it? How do find the real answers to all of those questions? Well, upon a little more reflection it also becomes clear that the only way to do it would be to actually ask the Creator himself. That is the only hope of getting the real answers because these questions concern that which lies outside of the scope of the *physical* world we live in, the world of matter and space. They concern the <u>meta</u>physical; that which is above the physical; that which is transcendent, because God himself transcends our physical world.

[Agnostics of course would tell us that we are wasting our time, that we cannot find the answers to these questions about God. For where atheists deny the existence of God, an agnostic merely states that if God exists he is unknowable and therefore any discussion about him is futile and a waste of time. And unlike their atheist friends, the agnostics are partially right, for God is unknowable, and can only become known if he chooses to reveal himself to us. And as we will see in this chapter that is exactly what he *has* chosen to do.]

"…man cannot know what God is like, unless God chooses to personally reveal Himself to us. This is what God did, both in His inspired Word, the Bible, and through the incarnation, when God broke into human history and became a man."[220]

So all we have to do is to ask God. Which sounds easy enough but the last time I checked he wasn't giving any interviews. So we are stuck. Or so it would seem, except for the fact that there is this little ancient manuscript that appears to be very unique for a number of reasons, one of which is that it claims <u>over three thousand times</u> within its pages to be written by God. It claims that the Holy Spirit of the Almighty, Eternal and Infinite Creator God was behind the minds, the pens and the writings of every single one of its many and varied authors, inspiring them to pen *exactly* what He had in mind, and *only* what He wanted them to write. In addition to this lofty claim, this "God-breathed" Book also very clearly answers all of the important questions man could have about the true origins and subsequent purpose of this created universe, the nature and character of its Creator, and what is expected of man himself, the crown of God's Creation. And it answers these questions in a manner not only unrivaled but not even approximated by any other book, ancient or religious.

Now there are only two possible choices when confronted with a book that makes such an astounding claim as this- to have been written by God himself.[221] You must conclude either that it was written by some ancient deluded liars, or that it is exactly what it says it is. For if a book claims within its own pages to be written by the Creator of the universe, when in reality it is not, then it would be a most heinous deception. (Which by extension is a condemnation of all other "holy"

220　(Pg. 35, *Fast Facts on False Teachings*, by Ron Carlson and Ed Decker)
221　(Using the minds and the pens of many human writers to be sure, but it claims that the Divine was behind them directing their every word.)

books; those inspired, not by God, but by someone else.) There is no possible middle ground. (God wrote it that way) It is either one or the other. Either it is a terrible deception, or it <u>was</u> written by God. Either it is a lie, or it a treasure of wisdom and instruction written for us, for our edification, for our knowledge, for our warning, and for our eternal blessing, that should not be ignored.

It turns out that the latter is the case. And I know that if you search out the truth about this ancient book, called the "Bible" (perhaps you've heard of it), then you will find as many others have before you, that it has proven itself in the face of absolutely overwhelming evidence to be exactly what it claims to be- the inerrant, inspired communication of the Creator of the universe to man via the written word; in short, the Word of God. This conclusion soon becomes readily apparent to anyone who takes an <u>objective</u> and <u>unemotional</u> look at all of the facts and evidence. The fact that modern academia is lagging woefully behind in coming to terms with this reality is not due to any exhaustive research on its part into the pros and cons, or evidence for and against, this inescapable conclusion; but rather is due to a blind, unsubstantiated, and unreasonable prejudice, known again as *willful* ignorance.

Now I realize that there are many people in this world who because of what they have been taught and have come to believe in, have a strong resistance to the truth, the accuracy and the Divine Authorship of the Word of God. Many just don't want it to be true. But keep in mind that the truth exists irrespective of what any of us want. Wanting the Bible not to be true, or not really to be written by God, has no bearing on whether or not it is. Again, if you seek the truth, you will find it. If you want the truth more than you want to protect the opinions and beliefs that are now in your head, then you will find it. If not, then you don't stand a chance.

Conversely, it is of course also true that the Bible is not true just because many other people <u>want</u> it to be, or believe that it is. Again, the truth exists irrespective of what anyone wants or doesn't want. The truth stands alone and can either be accepted or rejected, but it cannot be altered.

On a personal note, I have found that after weighing all of the facts and evidence and logic (much of which we will discuss in the remainder of this chapter) that can be brought to bear on this subject; it is actually beyond the realm of possibility that the Bible was <u>not</u> written by God. And therefore it is beyond the realm of possibility that the Bible is <u>not</u> true. The Author can be no one else. (In the same way that I think everyone would agree it is beyond the realm of possibility that *On the Origin of Species* was written by anyone other than Charles Darwin. No one disputes that because the great deceiver, the god of this world and the father of lies has no vested interest in calling into question the authorship of that book. When it comes to the Word of God, however, it is an entirely different story.) But I'll leave this for the reader to decide for him or herself, after this chapter has made its case.

Here is what R. A. Torrey, the author of an essay titled, "Ten Reasons Why I Believe the Bible is The Word of God," concluded after his search…

"I was brought up to believe that the Bible was the Word of God. In early life I accepted it as such upon the authority of my parents, and never gave the question any serious thought. But later in life my faith in the Bible was utterly shattered through the influence of the writings of a very celebrated, scholarly and brilliant skeptic. I found myself face to face with the question, Why do you believe the Bible is the Word of God?

"I had no satisfactory answer. I determined to go to the bottom of this question. If satisfactory proof could not be found that the Bible was God's Word, I would give the whole thing up, cost what it might. If satisfactory proof could be found that the Bible was God's Word, I would take my stand upon it, cost what it might. I doubtless had many friends who could have answered the question satisfactorily, but I was unwilling to confide to them the struggle that was going on in my own heart; so I sought help from God and from books, and after much painful study and thought came out of the darkness of skepticism into the broad daylight of faith and certainty that the Bible from beginning to end is God's Word."[222]

"It is written, Man shall not live by bread alone,
But by every word that proceeds out of the mouth of God."
(Matthew 4:4)

Twelve reasons why we can <u>know</u> that the Bible is the Word of God

The following is a list of some of the reasons why both the above author, R. A. Torrey, and countless millions of other intelligent, thoughtful individuals have come to the conclusion that the Bible *is* the Word of God:

1- The unity of the Bible

"The Bible consists of sixty-six books, written by more than thirty different men, extending in the period of its composition over more than fifteen hundred years; written in three different languages, in many different countries, and by men on every plane of social life, from the herdsman and fisherman and cheap politician up to the king upon his throne; written under all sorts of circumstances; yet in all this wonderful conglomeration we find an absolute unity of thought.

"A wonderful thing about it is that this unity does not lie on the surface. On the surface there is oftentimes apparent contradiction, and the unity only comes out after deep and protracted study."[223]

Truly, "back of the human hands that wrought was the Master-mind that thought."[224] The chances of all of these different people, who lived in different times, and had no way to collaborate with one another, coming up with a book of incomparable truth and wisdom and also, incredibly, a **unity** that flows from beginning to end and that builds throughout the book upon each succeeding entry, on their own by chance is beyond the realm of possibility. Just as there were master architects behind the building of the city of Washington DC—the layout didn't happen by accident—so too there had to be a Master Architect behind the scenes directing the construction of this masterpiece. Again, God wrote his Book this way so there could be no doubt as to its super-human origin. It has his signature on it.

"All scripture is given by inspiration of God, and is profitable for doctrine, for reproof, for correction, for instruction in righteousness, that the man of God may be complete, thoroughly equipped for every good work." (2 Timothy 3: 16-17)

---

222 (From "Ten Reasons Why I Believe the Bible is the Word of God," by R. A. Torrey)
223 (From "Ten Reasons Why I Believe the Bible is the Word of God," by R. A. Torrey)
224 (From "Ten Reasons Why I Believe the Bible is the Word of God," by R. A. Torrey)

2- The Bible itself claims to be *inspired* (Written by God using the minds and pen of man, but directing their every letter). "All scripture is given by <u>inspiration of God</u>..." (2 Timothy 3:16) "And so we have the prophetic word confirmed, which you do well to heed as a light that shines in a dark place, until the day dawns and the morning star rises in your hearts; knowing this first, that no prophecy of Scripture is of any private interpretation, for prophecy came not in old time by the will of man: but holy men of God spoke as they were moved by the Holy Spirit." (2 Peter 1:19-21) In fact, as we mentioned before, this book states over three thousand times that God is its author. If the Bible was <u>not</u> written by God then it is a liar. Yet to call a Book a lie that contains more truth than any other book written just doesn't add up, does it?

3- Bible prophecy.

"Remember the former things of old: for I am God, and there is none else; I am God, and there is none like me, Declaring the end from the beginning, and from ancient times the things that are not yet done, saying, My counsel shall stand, and I will do all my pleasure." (Isaiah 46:9-10)

We will devote an entire chapter to this subject (chapter 9) but there our primary focus will be on what the Bible forecasts for our near future and for the end of this age. Here our focus is on the many prophecies that have already been fulfilled and their role in authenticating the Bible as being written by God and not by mere man. But the constraints of time and space do not afford us the opportunity in one tiny section of one chapter to document the overwhelming amount of fulfilled prophetic evidence that alone attests to the Bible as being the work of the Creator. There are far too many. Over 2500 prophecies have been catalogued, 2000 of which have already been fulfilled.

> "Bible prophecy offers the clearest indication that the Bible is the verifiable Word of God. Yet despite the millions of people who attend church and synagogue every week and the existence of countless books on the subject, most people are simply unaware of these prophecies. They feverishly search through the writings of Nostradamus, Edgar Cayce, the Koran, and countless other "sacred texts" in hope of finding a profound truth or revelation concerning the future, only to be disappointed time and again. "Somehow, the most credible source of information on future events, the Bible, escapes notice. Yet the fulfilled prophecies of the Bible are backed by mountains of historical data, archaeological evidence, and the staggering mathematical probability [that discounts] their occurrence merely being the result of chance. No other source of historical knowledge can make the same claims, and for good reason. No other source is the inspired Word of God."[225]

Let's begin by looking at the Messianic Prophecies concerning the once long-awaited Jewish Messiah. Jesus Christ fulfilled all of the over 300 individual prophecies written concerning the Messiahs birth, his life, his death, and his resurrection.[226] The following is a partial list:

- It was foretold that Jesus would be born of a virgin: "Therefore the Lord himself will give you a sign: The virgin will be with child and will give birth to a son, and will call him Immanuel." (Isaiah 7:14) (It was fulfilled in Luke 1:26-35)

- That his birth would take place in the exact town which it did: "But thou, Bethlehem Ephrata, though thou be little among the thousands of Judah, yet out

---

225   (From "Why Study Bible Prophecy?" by Britt Gillette @ www.rapturealert.com)
226   (From "Statistics and Probability," by John R. Funk @ www.rapturealert.com)

of thee shall he come forth unto me that is to be ruler in Israel; whose goings forth have been from of old, from everlasting." (Micah 5:2) (This was fulfilled in Matthew 2:1-7; John 7:42 and Luke 2:47). Even though Mary and Joseph lived in Nazareth and that would invariably have been the place of their child's birth, a decree came forth from the empire to take a census of all the people, and to do so everyone had to return to their ancestral home. (For originally the land in Israel was divided among the Twelve Tribes, so each person's ancestral home was located in a specific part of Israel.) And thus the unrelated desires of the rulers of man were directed to fall perfectly into line with the prophesied will of the Ruler of the Heavens and the Earth!

- That he would come from the tribe of Judah, and be a direct descendant of King David: (Genesis 49:10; Jeremiah 23:5) (Fulfilled in Matthew 1:1-17)

- That a messenger would come before him: (Isaiah 40:1-5, 9) (Fulfilled by John the Baptist in Matthew 3:1-3; 11:10; John 1:23; Luke 1:17)

- He would perform miraculous signs and wonders: "Say to those who are of a fearful heart, Be strong, fear not! Behold, your God will come with vengeance, even God with a recompense; he will come and save you. Then the eyes of the blind shall be opened, and the ears of the deaf shall be unstopped. Then the lame man shall leap like a deer, and the tongue of the dumb sing." (Isaiah 35:4-6) (Fulfilled throughout all 4 Gospels)

- He would enter Jerusalem riding on a donkey: "Rejoice greatly, O Daughter of Zion! Shout, O Daughter of Jerusalem! Behold, your king is coming to you; He is just and having salvation, Gentle and riding on a donkey, A colt, the foal of a donkey." (Zechariah 9:9) (Fulfilled in Luke 35-37; Matthew 21:6-11)

- He would ultimately be rejected by the people, be silent before his accusers, bear our sins, be killed for the transgressions of mankind: (Isaiah 53) (Fulfilled throughout the 4 Gospels; Matthew 27:12)

- He would be betrayed by a friend, for thirty pieces of silver, beaten and spit upon, and buried in the tomb of a rich man: (Psalm 41:9; 55:12-14; Zechariah 11:12-13, Isaiah 50:6, and Isaiah 53:9) (Also fulfilled in 4 Gospels; Matthew 26:49; 26:15; 26:67; 10:4; 26:49-50; 27:57-60; John 13:21)

- The betrayal money would be "cast to the floor of the temple and used to buy the potter's field:" (Zechariah 11:13) (Fulfilled in Matthew 27:5,7)

- He would be "forsaken and deserted by his disciples, accused by false witnesses, wounded and bruised and hated without a cause:" (Zechariah 13:7; Psalms 35:11; Isaiah 53:5; Psalm 69:4) (Fulfilled in Mark 14:50; Matthew 26:59-60; 27:26; John 15:25)

- He would be "mocked, ridiculed and rejected, collapse from weakness, and taunted with specific words:" (Isaiah 53:3; Psalms 109:24-25; 22:6-8) (Fulfilled in Matthew 27:27-31; John 7:5, 48; Luke 23:26; Matthew 27:39-43)

- "Some 400 years before crucifixion was invented, both Israel's King David and the prophet Zechariah described the Messiah's death in words that perfectly depict that mode of execution. Further, they said that the body would be pierced and that none of the bones would be broken, contrary to customary procedure in cases of crucifixion (Psalm 22 and 34:20; Zechariah 12:10). Again, historians and New Testament writers confirm the fulfillment: Jesus of Nazareth died on a Roman cross, and his extraordinarily quick death eliminated the need for the usual breaking of bones. A spear was thrust into his side to verify that he was, indeed, dead."[227] (Fulfilled in 4 Gospels)

---

227   (From "Fulfilled Prophecy: Evidence for the Reliability of the Bible," by Hugh Ross, Ph.D. @ www.reasons.org)

- "People will shake their heads and stare at him:" (Psalms 109:25; 22:17) (Fulfilled in Matthew 27:39; Luke 23:35)

- His "hands and feet will be pierced, he will pray for his persecutors, and his friends and family will stand afar off and watch:" (Psalms 22:16; Isaiah 53:12; Psalms 38:11) (Fulfilled in Luke 23:33; 23:34; 23:49)

- He "will thirst, be given gall and vinegar to drink and will commit himself to God:" (Psalms 69:21; 31:5) (Fulfilled in John 19:28; Matthew 27:34; Luke 23:46)

- His "bones will be left unbroken, his heart will rupture and his side will be pieced:" (Psalms 34:20; 22:14) (Fulfilled in John 19:33,34)

- "Darkness will come over the land at midday:" (Amos 8:9) (Fulfilled in Matthew 27:45)

- He "will die 483 years after the declaration of Artaxerxes to rebuild the temple in 444 BC:" (Daniel 9:24)

- He "will be raised from the dead, ascend to heaven, and be seated at the right hand of God in full majesty and authority:" (Psalms 16:10; 68:18; 110:1) (Fulfilled in Acts 2:31; 1:9; Hebrews 1:3)[228]

All of these specific prophecies, and many more in addition, were fulfilled in the birth, the life, the crucifixion and the resurrection of one person, Jesus Christ. They all came to pass exactly as they were foretold in the Word of God! And these prophecies were written many centuries, some thousands of years, before they were fulfilled. Professor Peter W. Stoner, author of *Science Speaks*, has determined that the probability of just eight of these particular Messianic prophecies being fulfilled in one person is 1 in 10^{17}.

"If we take 1 X 10^{17} silver dollars and lay them on the face of Texas, they'll cover all of the state two feet deep. Now mark one of these silver dollars and stir the whole mass thoroughly, all over the state. Blindfold a man and tell him that he can travel as far as he wishes, but he must pick up one silver dollar and say that this is the right one. What chance would he have of getting the right one? Professor Stoner went on to consider 48 prophecies and says ... "We find the chance that any one man fulfilled all 48 prophecies to be 1 in 10^{157}."

"This is a really large number and it represents an extremely small chance. Let us try to visualize it. The silver dollar, which we have been using, is entirely too large. We must select a smaller object. The electron is about as small an object as we know of. It is so small that it will take 2.5 times 10^{15} of them laid side by side to make a line, single file, one inch long. If we were going to count the electrons in this line one inch long, and counted 250 each minute, and if we counted day and night, it would take us 19,000,000 years to count just the one-inch line of electrons. .. [10^{157}] is approximately **the total number of electrons in all the mass of the known universe.** In other words the probability of Jesus Christ fulfilling 48 prophecies is the same as one person being able to pick out one electron out of the entire mass of our universe.

"Such is the chance of any one man fulfilling any 48 prophecies. Yet Jesus Christ fulfilled not just 48 prophecies, not just 61 prophecies, but **more than 324 individual prophecies** that the Prophets wrote concerning the Messiah."[229] [Emphasis mine]

Mathematicians have determined that any odds greater than 1 in 10^{66} (1 with

---

228   (Source of Messianic prophesies: www.100prophecies.org) (Source of Messianic prophesies and quotes: "Statistics and Probability," by John R. Funk, on www.rapturealert. com)

229   (From "Statistics and Probability," by John R. Funk, on www.rapturealert.com)

66 zeroes behind it) are absolutely impossible. Absolutely nothing with those odds could <u>ever</u> happen by chance even if the universe lasted for an infinite number of years. It is safe to say therefore that the specific fulfillment of the many Messianic prophecies in one person, Jesus Christ, could not have been a chance occurrence, but were indeed both prophesied and caused to be fulfilled by some One who knows the future in all of its intricate and infinite details, before it comes to pass. And that it was that same One who caused these many specific prophecies to be written into the Scriptures. Therefore it is abundantly obvious that these fulfilled prophecies authenticate the Bible as being the Word of God, as having been written by God.

Of the 2000 prophecies already fulfilled, 324 were Messianic prophecies fulfilled by Jesus. That leaves just under 1700 other specific prophecies that were also exactly fulfilled. Here are just a few:

- "The prophet Isaiah foretold that a conqueror named Cyrus would destroy seemingly impregnable Babylon and subdue Egypt along with most of the rest of the known world. This same man, said Isaiah, would decide to let the Jewish exiles in his territory go free without any payment of ransom (Isaiah 44:28; 45:1; and 45:13). Isaiah made this prophecy 150 years before Cyrus was born, 180 years before Cyrus performed any of these feats (and he did, eventually, perform them all), and 80 years before the Jews were taken into exile.

- "Mighty Babylon, 196 miles square, was enclosed not only by a moat, but also by a double wall 330 feet high, each part 90 feet thick. It was said by unanimous popular opinion to be indestructible, yet two Bible prophets declared its doom. These prophets further claimed that the ruins would be avoided by travelers, that the city would never again be inhabited, and that its stones would not even be moved for use as building material (Isaiah 13:17-22 and Jeremiah 51:26, 43). Their description is, in fact, the well-documented history of the famous citadel.

- "The exact location and construction sequence of Jerusalem's nine suburbs was predicted by Jeremiah about 2600 years ago. He referred to the time of this building project as "the last days," that is, the time period of Israel's second rebirth as a nation in the land of Palestine (Jeremiah 31:38-40). This rebirth became history in 1948, and the construction of the nine suburbs has gone forward precisely in the locations and in the sequence predicted.

- "The prophet Moses foretold (with some additions by Jeremiah and Jesus) that the ancient Jewish nation would be conquered twice and that the people would be carried off as slaves each time, first by the Babylonians (for a period of 70 years), and then by a fourth world kingdom (which we know as Rome). The second conqueror, Moses said, would take the Jews captive to Egypt in ships, selling them or giving them away as slaves to all parts of the world. Both of these predictions were fulfilled to the letter, the first in 607 B.C. and the second in 70 A.D. God's spokesmen said, further, that the Jews would remain scattered throughout the entire world for many generations, but without becoming assimilated by the peoples or of other nations, and that the Jews would one day return to the land of Palestine to re-establish for a second time their nation (Deuteronomy 29; Isaiah 11:11-13; Jeremiah 25:11; Hosea 3:4-5 and Luke 21:23-24).

This prophetic statement sweeps across 3500 years of history to its complete fulfillment—in our lifetime.

- "Jeremiah predicted that despite its fertility and despite the accessibility of its water supply, the land of Edom (today a part of Jordan) would become a barren, uninhabited wasteland (Jeremiah 49:15-20; Ezekiel 25:12-14). His description accurately tells the history of that now bleak region.

- "Joshua prophesied that Jericho would be rebuilt by one man. He also said that the man's eldest son would die when the reconstruction began and that his youngest son would die when the work reached completion (Joshua 6:26). About five centuries later this prophecy found its fulfillment in the life and family of a man named Hiel (1 Kings 16:33-34).

- "The day of Elijah's supernatural departure from Earth was predicted unanimously--and accurately, according to the eye-witness account—by a group of fifty prophets (2 Kings 2:3-11).

- "Jahaziel prophesied that King Jehoshaphat and a tiny band of men would defeat an enormous, well-equipped, well-trained army without even having to fight. Just as predicted, the King and his troops stood looking on as their foes were supernaturally destroyed to the last man (2 Chronicles 20).

- "One prophet of God (unnamed, but probably Shemiah) said that a future king of Judah, named Josiah, would take the bones of all the occultic priests (priests of the 'high places') of Israel's King Jeroboam and burn them on Jeroboam's altar (1 Kings 13:2 and 2 Kings 23:15-18). This event occurred approximately 300 years after it was foretold."[230]

Again, the above prophecies are just a small sampling but they make it clear that those words were confidently written by One who knew the future in advance in all of its intricacies and detail.

Then there was the prophesy concerning Tyre, the largest city of Phoenicia which is now called Palestine. This city was located on an island off the Mediterranean coast about a half-mile offshore:

"And it came to pass in the eleventh year, on the first day of the month, that the word of the LORD came to me, saying, Son of man, …thus says the Lord GOD: Behold, I am against you, O Tyre, and will cause many nations to come up against you, as the sea causes its waves to come up. And they shall destroy the walls of Tyre and break down her towers; I will also scrape her dust from her, and make her like the top of a rock. It shall be a place for spreading nets in the midst of the sea, for I have spoken, says the Lord GOD; it shall become plunder for the nations. … Behold, I will bring against Tyre from the north Nebuchadnezzar king of Babylon, king of kings, with horses, with chariots, and with horsemen, and an army with many people." (Ezekiel 26:1-5, 7)

"Three years after this prophecy was given, Nebuchadnezzar came and laid a 13 year siege on that city. [Tyre was a] walled city so all you had to do was …just cut off anything coming into the city and they eventually starved. [It] took him 13 years from 585 to 573 [but] finally the city surrendered because they were all dying. And Nebuchadnezzar broke down the walls and the towers, destroyed the city, did every single thing Ezekiel said he would do and of course he wasn't reading Ezekiel when he did it. He got in the city [but] he didn't find the spoils. … [They] had used their fleet to take all the spoils to an island a half-mile away off the coast and, of course, in chapter 29 Ezekiel said you will gain no plunder, and that's exactly what happened."

"Nebuchadnezzar king of Babylon caused his army to labor strenuously against Tyre; every head was made bald, and every shoulder rubbed raw; yet neither he nor his army received wages from Tyre." (Ezekiel 29:18)

"When he got there they had taken all the valuables off to the island, Nebuchadnezzar had no naval force to go off and get it. The island then became the new city. And it flourished for 250 years out on that island.

230   (From "Fulfilled Prophecy: Evidence for the Reliability of the Bible," by Hugh Ross, Ph.D. on www.reasons.org)

"Only part of the prophecy was fulfilled. The part about Nebuchadnezzar, the part about destroying the walls, smashing it down, slaughtering the people, not getting the spoil, but not all of it was yet complete. The ruins were still on the old sight. The rubble was still there. After 250 years, a 24-year-old guy came by the name of Alexander the Great. He had 33,000 infantry men, he had 15,000 cavalry. He had just defeated the Persians and he was on his way to Egypt. He needed supplies. So he came by the now island city of Tyre and he sent word, "I want you to supply all of my men and all of my horses and all of my army." And they said, "Forget it, buddy, you don't have a navy and we're on an island, we're not going to help you at all."

"He didn't like that. And it wasn't good to get Alexander mad. He didn't have a fleet [either] so he decided he had to get a way to go to that island so he did what Ezekiel, the prophet, said would be done. It said that the place would be scraped bare as rock and all the rubble would be thrown into the sea. Well what conqueror in his right mind would ever do that? Why waste your time once you've conquered the place, picking up everything and throwing it in the ocean? All the stone and all the rest of it. But that's exactly what had to happen. So Alexander did it. He took all the debris and built a 2,000 foot long, 200 foot wide causeway all the way to the island with all the debris. ...It took him seven months, [and then] he went in and murdered 8,000 people, over a period of a few months executed 7,000 more and sold 30,000 into slavery and fulfilled every single detail of the prophecy. And though the city of Jerusalem has been rebuilt 17 times, Tyre has never been rebuilt. And that's exactly what God said: "You will be built no more." And you know what they do? Go there today ...they dry fish nets there, just [as the prophet] said."[231]

Who but God himself could make such an incredibly detailed, specific prophecy and then make certain it came to pass exactly in the manner in which he declared, using two different armies centuries apart to make it happen? When you study even just a little of Bible prophecy the uninformed ignorance of those "educated" elites who mock God's Word becomes embarrassingly evident.

"Knowing this first, that no prophecy of the scripture is of any private interpretation. For the prophecy in old time never had its origin in the will of man: but holy men of God spoke as they were moved by the Holy Spirit." (2 Peter 1:20-21)

And finally there is an incredible prophecy that is on the verge of being fulfilled in our lifetime. It concerns a number of kingdoms to the East of Jerusalem that will march on Israel with an army of two hundred million. "And the number of the army of the horsemen were two hundred thousand thousand: and I heard the number of them." (Revelation 9:16) This will occur shortly in "the last days" at the end of a seven year period of terrible worldwide turmoil, suffering and death known as the Great Tribulation. "And the sixth angel poured out his vial upon the great river Euphrates; and its water was dried up, so that the way of the kings of the east might be prepared. ...For they are spirits of demons, performing signs, which go out to the kings of the earth and of the whole world, to gather them to the battle of that great day of God Almighty. ...And they gathered them together to the place called in the Hebrew tongue, Armageddon." (Revelation 16:12, 14, 16) The astounding thing about this prophecy of a two hundred million man army is that it was made nearly two thousand years ago when the entire population of the known world was far less than that. And for millennium many scholars

---

231 (From "Our God-Breathed Bible," by John MacArthur, Grace To You Ministries @ www. gty.org)

argued that it obviously could <u>not</u> be taken literally; that it must have a figurative meaning only; and they argued this as recently as a few decades ago. But they don't argue that anymore. Today China alone, one of the "kings of the east," with a population of roughly 1.5 billion, has a standing army of nearly that size. So again God's Word the Bible, even in what could be considered at one time as its most outlandish and impossible prophecy, has affirmed itself to be true.

There are also all of the prophecies of *type*, that are found throughout the Old Testament and are too numerous to mention here.

"Everything in the Old Testament- history, institutions, and ceremonies is prophetical. The high priesthood, the ordinary priesthood, the Levites, the prophets, priests and kings are all prophecies. The tabernacle, the brazen alter, the lever, the golden candlestick, the table of showbread, the veil, the alter of incense, the Ark of the Covenant, the very coverings of the tabernacle, are prophecies."[232]

All of these prophecies of type were fulfilled in the life of Jesus, the history of the church, and the doctrines of the New Testament.

The chances of all or even a small part of these prophecies being fulfilled by dumb luck are absolutely nil. Also the tired old lie that these prophecies were written *after* the things occurred is equally as *historically* impossible. (It is as unintelligent as saying today that the Gettysburg Address was written in <u>1963</u>, and *not* a hundred years earlier.) Mathematically, through the laws of statistics and probability, it can be determined <u>scientifically</u> whether a prediction, or a set of predictions, has any chance whatsoever of coming true by accident, by random chance alone. The fulfilled prophecies in the Bible are so far beyond the realm of chance or accidental fulfillment, according to science itself, as to leave no doubt that the book was written by One who knows the future in all of its entirety and in all of its infinite and intricate detail. It leaves no doubt that the book was written by some One who wrote the future. Some One who *created* the future. So science itself, in accordance with its own laws, has placed its stamp of approval on the Bible, and declares that it is exactly what it says it is, the inerrant Word of God!

But what about Nostradamus, Edgar Cayce, and for that matter a whole lot of other ancient seers and modern day psychics? What about all of their prophecies? Why is the Bible so special? Because comparing the cloudy, unspecific forecasting of a Nostradamus or Edgar Cayce to the thousands of crystal clear and specific prophecies of the Bible, is like comparing a mud puddle to the Mediterranean Sea. As far as ancient seers, there is evidence of those even in the Scriptures, but their limited abilities do not compare to, nor detract from, that of the Bible. And modern day psychics are a joke. A ten year old with no knowledge of world affairs could compile a better record by guessing. Also, the prophecies of the Word of God are 100% accurate. Every single one of them has come to pass <u>exactly</u> as they were foretold. And there is not a single one which has been found to be inaccurate. No, the Word of God separates itself from all other historical records of real or imagined prophetic utterances, as the brilliance of the noonday sun distinguishes itself from a lighter held up at a concert. Just look into it further for yourself if need be, and you will see.

---

232   (From "Ten Reasons Why I Believe the Bible is the Word of God," by R. A. Torrey)

"Your word is a lamp unto my feet, and a light unto my path."
(Psalm 119:105)

4- Wisdom and knowledge. "In this book are hidden the infinite and inexhaustible treasures of the wisdom and knowledge of God."[233] "Wisdom with regard to spiritual life, time and eternity is all found in the Word of God. The Bible is the source, and the only source of divine wisdom."[234]

> "It is quite fashionable in some quarters to compare the teachings of the Bible with the teachings of Zoroaster and Buddha and Confucius and Epictetus and Socrates and Marcus Aurelius Antonius and a number of other …authors. The difference between the teachings of the Bible and those of these men is found in three points- First; the Bible has in it nothing but truth, while all the others have truth mixed with error. …Second, the Bible contains all truth. There is not a truth to be found anywhere on moral or spiritual subjects that you cannot find in substance within the covers of [the Bible]. …The third point of superiority is this: The Bible contains more truth than all other books together. Get together from all literature of ancient and modern times all the beautiful thoughts you can; put away all the rubbish; put all these truths that you have culled from the literature of all ages into one book, and as the result, even then you will not have a book that will take the place of this one book."[235]

"The law of your mouth is better unto me than thousands of gold and silver."
(Psalm 119:72)

5- "The Bible has always been hated. For nearly 2000 years every engine of destruction that human science, philosophy, wit, reasoning or brutality could bring to bear against a book has been brought to bear against the Bible to stamp it out of the world, but it has a mightier hold on the world today than ever before."[236] This is because Satan, the god of this world, hates God and he hates his Word, and he educates his earthly followers to mock, ridicule, disparage and demean it. It is funny when you think of how other so called holy books get a pass from the elites of this world. The Book of Mormon, the Koran—you don't find hordes of intellectuals falling all over themselves to see who can write the cleverest or "most damning" criticism of these writings. Why? Because everyone knows they are garbage, so why beat a dead horse? And besides, false holy books have another author, one other than God, and his name is Satan, the great deceiver, "who deceives the whole world." (Revelation 12:9) Why would his followers bother to attack his books? No, it is only God's Word, the Creator's truly Holy Book, that is the target of their vain ridicule and scorn. Yet they have not been able to level one accurate criticism against it. All of their higher critiques have turned out under closer scrutiny to be false. And it is the Bible that has always won out in the end. Let's take a look…

> "I have studied about a hundred alleged errors and contradictions in scripture. Most were easily dismissed and all the others were solved as I dug in a little deeper."[237]

---

233   (From "Ten Reasons Why I Believe the Bible is the Word of God," by R. A. Torrey)
234   (From "The Character of God's Word," a message by John Macarthur)
235   (From "Ten Reasons Why I Believe the Bible is the Word of God," by R. A. Torrey)
236   (From "Ten Reasons Why I Believe the Bible is the Word of God," by R. A. Torrey)
237   (Pg. 136, *There are Moral Absolutes*, by Michael A. Robinson)

There are a few excellent books that answer the critics list of supposed contradictions and errors in the Bible. One is: the *Big Book of Bible Difficulties: Clear and Concise Answers from Genesis to Revelation*, by Thomas Howe and Norman L. Geisler. The other is the *New International Encyclopedia of Bible Difficulties*, by Gleason Leonard Archer. A third, which we will draw upon extensively here, is *The Politically Incorrect Guide to the Bible*, by Robert J. Hutchinson. He does a masterful job of systematically debunking the arguments of today's Bible critics, from "village idiots...er...atheists" like the foul-mouthed (and foul-minded) comedians Penn and Teller—"yet another example of how high you can climb in Hollywood with a high school education"—to such diverse liberal luminaries as Michael Shermer, publisher of *Skeptic* magazine, Robert Reich, of Clinton administration fame, and authors like Richard Dawkins (*The God Delusion*) and Sam Harris (*The End of Faith*). Hutchinson states,

> "It wouldn't be so bad if these attacks on the Bible represented something genuinely new- something witty and entertaining on the level of, say, a Nietzsche or a Swineburne- but instead they are merely repetitions of allegations made for about 1,800 years. ...What the village atheists of today fail to realize (in their monumental arrogance and ignorance of history) is that every seeming contradiction found in the Bible- every discrepancy, every apparent historical or scientific error, grammatical mistake, or puzzling fact or comment- has been noted and argued about and debated literally thousands of times, for thousands of years, by the greatest minds in history. Nothing- literally nothing- they say is new."[238]

-- *The real danger to world peace*. Page by page and chapter after chapter Mr. Hutchinson debunks criticism after empty accusation using historical facts and statistical data. On the "village atheists" claim that religion, and of course Biblical religion in particular, is "the greatest danger to world peace," Hutchinson points out that "as horrible as the examples of religious intolerance may be" (such as The Spanish Inquisition and religious "witch hunts"—which I would argue were the result of *false* Christian religion and not true biblical faith), yet...

> ..."according to research conducted by the political scientist Rudolph Rummel at the University of Hawaii, the officially atheist states of the Communist bloc committed more acts of genocide than any societies in history. [Totaling about 170 million]... and these numbers don't even include the people killed in the *wars* initiated by these officially anti-Christian states. ...For more than three hundred years , from the French *philosophes* to Marx, Lenin, and the "death of God" theologies of the 1960's, we have been assured that, freed from the superstitions and imbecilities of organized religion, rational secularists could make the world into a utopia. The results have invariably been horrific- from the Terror of the French Revolution to the terrors of Nazism and Communism. ...Rummel's conclusion is as shocking as it is inescapable: War wasn't the most deadly evil to afflict humanity in the twentieth century. Government was! And not just any government, but atheist government."
>
> "[Yet these popular, Bible-attacking] secular fundamentalists talk about the "intolerance" and "violence" of biblical religion. ...A century of firsthand, bloody experience with "rational" atheism has proven that it is atheism, not the Bible or religious belief, that is the greatest danger to world peace."[239]

Of course we made this same point in chapter 2 "Moral Absolutes," that a society which rejects the moral absolutes of their Creator, and instead *frees*

---

238 (Pgs. 2,8,51, *The Politically Incorrect Guide to the Bible*, by Robert J. Hutchinson)
239 (Pgs. 13,14,15,17, *The Politically Incorrect Guide to the Bible*, by Robert J. Hutchinson)

themselves to do as they please and to make up their own morals, invariably comes up with the morals of hell. And it is the progressive moral-relativists of the Democrat-Left today, kin to those moral-relativist, genocidal monsters who slaughtered 170 million in the twentieth century, who constitute the real "danger to world peace." (And anyone who disagrees should ask the 50 million babies they mutilated to death while they were still alive in order to protect the "rights" of their mother, while ignoring the baby's far more sacred right- to live.) Along with their friends the Islamic terrorists, whom they can't even bring themselves to call terrorists, because in the mind of the Democrat-Left, and in the mind of their father Satan, Islamic terrorists are not the enemy, Christians and their Bible are.

-- *Dr. Laura*. Mr. Hutchinson also sets the record straight about a now well-known assault on a popular radio host. We're referring to…

…"the infamous "Open Letter to Dr. Laura" …One of the wittiest attempts to undermine the authority of the Bible in recent years- and which was, perhaps inadvertently, a fairly vicious attack on the very foundation of Judaism."

This "Open Letter" was an internet posting that got a lot of traction with many on the Democrat-Left because they saw it as another "powerful criticism" of the Bible and as a further marginalization of all of those who hold it in reverence. But it clearly distorted the truth in order to lampoon one tiny section of arcane laws found in the Old Testament. The point was to discredit Dr. Laura Schlessinger, popular author and radio talk show host, because she had become successful in speaking out about the Bible's strong negative stance concerning the sexual preference of homosexuals. The more militant among them found this, and her free speech right to do so, intolerable. (In the name of "tolerance" of course.) So both had to be demeaned, vilified, marginalized, and destroyed. (Free Speech is another right the Democrat-Left supposedly champions yet systematically tramples whenever it is advantageous to do so.) This in turn spawned a segment on the Democrat-propaganda-TV-serial-disguised-as-a-drama, *The West Wing* (aka *The Left Wing*) where "the fictional (and liberal) President Bartlet (played by actor Martin Sheen) ridicules a conservative character named Dr. Jenna Jacobs, modeled on Dr. Laura Schlessinger." Sheen (Bartlet) quizzes her about selling his youngest daughter into slavery, putting to death his chief of staff for working on the Sabbath, and burning his mother for wearing garments made from two different threads.[240]

"For the deep thinkers in Hollywood, this little exchange represents a fatal, unanswerable blow to the Bible and all it represents (even though it deliberately misrepresents what the Bible actually says regarding these laws). …In typical Hollywood fashion, however, the writer …deliberately misrepresents what the Torah actually says … just to take a cheap shot at Dr. Laura and strike a blow for homosexuality. So what if he had to misrepresent the texts in the process and malign the Jewish people and religion to do it?"

[And as far as the grievous penalties the show misstated in order to make fun of these arcane laws]… …"The TV producers and writer just made that part up out of whole cloth- to make the biblical laws seem more harsh than they actually are."[241]

Mr. Hutchinson then poses these questions:

"But what about the *substance* of this assault? Does the Bible *really* teach that President Bartlet should "burn his mother in a small family gathering" for wearing garments made from two different threads? Or that he should stone his brother for planting

---

240 (Pgs. 95, 97, *The Politically Incorrect Guide to the Bible*, by Robert J. Hutchinson)
241 (Pgs. 99,100, *The Politically Incorrect Guide to the Bible*, by Robert J. Hutchinson)

different crops side by side? Or sell his daughter into slavery? And, if so, how could the most brilliant minds in history—from Augustine, Aquinas, and Maimonides to John Locke and Sir William Blackstone—insist that the biblical law is the foundation upon which the great innovations of Western law and civilization, including the recognition of basic human rights, was based?"[242]

Hmm, could it be because the Democrat-Left wants to marginalize the Bible as well as its foundation of our entire judicial system so they can then be free to make up their own laws based on their own morality that oozes up from below? They have to disparage and discredit the Bible and its laws and its morality so they can be free to mutilate babies at will; to falsely prosecute their political rivals for actually attempting to defend this country; so they can be free as judges to take an oath to uphold and defend the Constitution of the United States that they know they are going to break, as they proceed to brazenly pervert, distort and ignore its clear demands at every turn ("Thou shalt not lie"); to allow heinous child molesters to go free so they can kidnap rape and murder more of our little boys and girls (it's "Societies fault" after all, not the perverts); to defile God's sacred institution of marriage, etc. etc. The Left can't actually dispute the Bible on its substance or merits, so they do what they do best. They lie. They mock, disparage, marginalize and belittle it just like they do their political opponents. Remember the only political news that they and their lame-stream media know how to report is "Democrats <u>good</u>! Republicans <u>bad</u>!" "Baby mutilators <u>good</u>! Christians <u>very bad</u>!" And now we can add "Bible stupid! We smart!" to their See Spot Run, Me Tarzan You Jane, intellectual level.

Hutchinson then proceeds further into the subject of Old Testament law where he reveals some pertinent historical facts concerning the era in which these biblical commands in the Torah were written that provide context and a greater understanding.

> "On the one hand, some of the penalties (death!) given for specific acts (bestiality, incest, homosexuality, striking one's parents) seem harsh by contemporary standards-although in most cases there is little evidence that such penalties were ever actually enforced. "In practice...these [physical] punishments were almost never invoked, and existed mainly as a deterrent and to indicate the seriousness of the sins for which they were prescribed," said the late biblical scholar Rabbi Aryeh Kaplan. "The rules of evidence and other safeguards that the Torah provides to protect the accused made it all but impossible to actually invoke these penalties.""[243]

It is also instructive that Jesus in the New Testament stopped a crowd dead in their tracks—who were hell bent on stoning a woman to death because she was "caught in the very act of adultery"—by answering them with "he among you who is without sin, let him cast the first stone." (John 8:4, 7) And in this we see into the true heart of God- mercy, forgiveness, kindness, patience, and long-suffering; the same God who *wrote* that Old Testament and it's sometimes *hard-for-us-today-to-understand*[244] archaic laws. And by contrast we can also see into the juvenile minds of those who have rested control of our entertainment industry, who in their ignorance become giddy over their cleverness in mocking the Bible, and, by extension, its Author as well.

---

242   (Pg. 99, *The Politically Incorrect Guide to the Bible*, by Robert J. Hutchinson)
243   (Pgs. 103-104, *The Politically Incorrect Guide to the Bible*, by Robert J. Hutchinson)
244   (Especially when our understanding is clouded by our own self-righteousness and arrogance)

"Be not deceived; God is not mocked: for whatsoever a man sows, that shall he also reap." (Galatians 6:7)

Hutchinson goes on to point out a number of things that the Torah does forbid, like…

..." Committing murder… Doing nothing when someone's life is in danger… Committing adultery… Homosexual sex… Incest… Bestiality… Castration of man or beast… Forcing a newlywed man to join the army or perform government service… Ill-treating widows and orphans… Cheating people in business… Committing robbery… Drinking blood… Bodily mutilation… Selling your daughter into prostitution."[245]

In addition to expressly forbidding ritual child sacrifice, a practice that was all too commonplace in many cultures of the ancient world. In fact, all of those above things that the Torah forbade were practiced near-universally by all of the tribes, societies, cultures and countries that the nation of Israel was surrounded by in biblical times when the Old Testament, and its laws, were written. I'd like to have Mr. Sheen explain why he, and the author of that *Open Letter to Dr. Laura*, thought themselves intelligent in disparaging the Bible and its laws, as opposed to reporting on just how cutting edge and progressive those laws actually were when compared to everything else in their day. In complete sentences. Also, speaking of ritual child sacrifice which the Bible expressly forbade, it seems the barbarity of *that* particular act is still a little above the pay grade of Mr. Sheen and the other baby-killing geniuses on the Democrat-Left.

"Modern atheist crusaders, such as Richard Dawkins and Christopher Hitchens, are largely ignorant of the influence of biblical law on the development of both Western morality generally and the law specifically."

…"The biblical law mocked in the "Open Letter to Dr. Laura" and in episodes of *The West Wing*—however harsh some of its penalties may seem today—represented a radical break from the general barbarism and social indifference that was typical of the ancient world.

"It introduced unprecedented concepts of individual morality, equality before the law, the preciousness of human life, and the illegitimacy of vicarious or brutal punishments that stood in stark contrast to the legal codes of the time.

…"If parts of the biblical law seem harsh to us, we are only able to recognize that harshness because we live in a society that has been shaped by 2,000 years of reflection upon, and experience with, the principles of justice and mercy found in that very same law."[246]

So much for the "Letter to Dr. Laura/West Wing" argument against the truthfulness, reliability and credibility of God's Word.

-- *Science*. Then there is the following criticism that "enemies of the Bible—from Voltaire and David Hume to Penn and Teller and Richard Dawkins" have leveled that…

…"for millennia, Christianity in general, and the Roman Catholic Church in particular, attempted to keep mankind locked in a dark prison of superstition and irrational dogma—and it was only when mankind threw off the shackles of revealed religion during the Renaissance and Enlightenment that modern science was able to develop.

…"This has been the conventional view of atheist intellectuals throughout most of the twentieth century."[247]

Yet it is total hogwash. The truth is that…

---

245   (Pgs. 104, 105, *The Politically Incorrect Guide to the Bible*, by Robert J. Hutchinson)
246   (Pgs. 107, 116, *The Politically Incorrect Guide to the Bible*, by Robert J. Hutchinson)
247   (Pg. 137, *The Politically Incorrect Guide to the Bible*, by Robert J. Hutchinson)

..."biblical religion was not the enemy of science but rather the intellectual matrix that made it possible in the first place. ...Only in Christian Europe, among millions of pious, churchgoing believers, did mankind begin the systematic study of nature—quantified by precise measurement and experimentation—which led to a whole new way of understanding. ...most of the true *giants* of empirical science—the people who founded entire scientific disciplines or who made landmark scientific discoveries—were primarily devout Christians who believed that their scientific studies, far from being in conflict with their religious faith, ultimately were dependent upon it."[248]

"Here is just a sampling:

"Nicolaus Copernicus (1473-1543), pioneer of modern astronomy, ...often referred to God in his works: "When a man is occupied with things which he sees established in the finest order and directed by divine management, will not the unremitting contemplation of them and a certain familiarity with them stimulate him ...to admiration for the Maker of everything, in whom are all happiness and every good?"

..."Johannes Kepler (1571-1630), astronomer and physicist, "...now I see how God is, by my endeavors, also glorified in astronomy, for "the heavens declare the glory of God.""

"Sir Isaac Newton (1642-1727), founder of modern physics and devout Christian: "This most beautiful system of the sun, planets, and comets could only proceed from the counsel and dominion of an intelligent and powerful Being. This Being governs all things, not as the soul of the world, but as Lord of all; and on account of his dominion he is wont to be called Lord God, or Universal Ruler."

"Sir Francis Bacon (1561-1627), philosopher and devout Anglican: "It is true, that a little philosophy inclineth man's mind to atheism."

"Rene Descartes (1596-1650), mathematician, scientist: "I am convinced that those who examine carefully my arguments about God's existence will find that ...they are clearer than any geometrical demonstrations."

"Robert Boyle (1627-1691), the founder of modern chemistry and devout Christian.

"Michael Faraday (1791-1867), inventor of the electrical generator and the transformer, "A Christian finds his guide in the Word of God."

"Matthew Maury (1806-1873), the father of modern oceanography followed the Psalms expression "paths of the sea" and found warm and cold continental currents

"James Prescott Joule (1818-1898), a Bible-believing Christian, authored the first law of thermodynamics: "It is evident that an acquaintance with natural laws means no less than an acquaintance with the mind of God therein expressed."

"James Clerk Maxwell (1831-18790, physicist credited with pioneering statistical thermodynamics, field equations of electricity, magnetism and light; devout Christian: "Almighty God, Who has created man in Thine own image, and made him a living soul that he might seek after thee, and have dominion over Thy creatures, teach us to study the works of Thy hand, that we may subdue the earth to our use, and strengthen the reason for Thy service."

"Lord William Kelvin (1824-1907), physicist, inventor of absolute temperature scale in his name; committed Christian who diligently studied the Bible: "Overwhelmingly strong proofs of intelligent and benevolent design lie around us...**the atheistic idea is so non-sensical that I cannot put it into words**." [emphasis mine]

"Werner Karl Heisenberg (1901-1976), Nobel Prize winner in physics for the creation of quantum mechanics and the Uncertainty Principle and a devout Lutheran.

---

248   (Pgs. 139, 148, *The Politically Incorrect Guide to the Bible*, by Robert J. Hutchinson)

"Werner Von Braun (1912-1977), <u>first director of NASA, pioneer of space exploration</u>: "Scientific concepts exist only in the minds of men. Behind these concepts lies the reality which is being revealed to us, but only by the grace of God."

"Albert Einstein (1875-1955) "I want to know how God created this world. ...I want to know His thoughts, the rest are details." "[249]

Keep in mind that the men listed above are literally the founders of science as we know it, and that without them and their discoveries of the laws of this universe those atheists who deny the integral and essential part played by Christians and their Bible in the development of the laws of science and thus the modern world would still be riding on their donkeys. The dishonesty of the people who control our media, entertainment and educational system in how they suppress the truth about the Bible is staggering.

"On the one hand, we have scientific (let's be charitable) amateurs—from Nietzsche and Ingersoll to Christopher Hitchens and Sam Harris—insisting that science and biblical religion are fundamentally incompatible.

"On the other hand, we have the greatest minds in the history of science, the people who actually made most of the discoveries that created modern science to begin with ... who insist that, not only is religion not at odds with science, but biblical religion is *what made science possible in the first place*.

"Whom should we believe?

"Should we believe the attorney Clarence Darrow, who said, "I don't believe in God because I don't believe in Mother Goose" or should we believe Albert Einstein."[250]

Other areas where Mr. Hutchinson debunks those who dismiss the Bible as God's Word, and exposes their re-writing of history are:
-- *Slavery*.

"Today, secular liberals continue to lie to themselves that it was the atheist "free thinkers" of the Enlightenment who created the abolitionist movement, but better-informed historians know the truth: That while a few Enlightenment writers like Rousseau and Thomas Paine wrote abolitionist pamphlets, the people who actually risked life and limb to *end* slavery were almost all, without exception, devout Christians. They included Dr. Beilby Porteus (the Anglican bishop of London), classicist and biblical scholar Granville Sharp, deacon Thomas Clarkson, and Tory member of Parliament William Wilberforce, Irish political leader Daniel O'Connell, and the U.S. Baptist journalist William Lloyd Garrison."[251]

"The truth is that the savage cruelty of slavery has existed on a massive scale all over the world for most of human history—and *still exists today* in parts of the Islamic world and Asia—and yet it was first officially banned, by force of law, only in Christian Europe. **No culture on earth questioned the morality of slavery until Christians did the questioning**. [Emphasis mine.]

"The golden age of ancient Greece and Rome, celebrated by the "enlightened" pagans of the eighteenth century, was built almost entirely by slave labor: By some estimates, fully one-third of Roman society was made up of slaves who could be killed at will by Roman householders.

---

249   (Pgs. 148-152, *The Politically Incorrect Guide to the Bible*, by Robert J. Hutchinson)
250   (Pg. 155, *The Politically Incorrect Guide to the Bible*, by Robert J. Hutchinson)
251   (Pg. 159, *The Politically Incorrect Guide to the Bible*, by Robert J. Hutchinson)

"Socrates and Plato could sit around in symposia, drinking wine and discussing the essence of justice, largely because their civilization was maintained by an unimaginable number of slaves.

...[In Attica] there were nineteen slaves supporting every philosophizing Greek citizen. ...[The slaves in Sparta] outnumbered citizens ...by seven to one.

"Far from condemning such cruelty, many of the anti-Christian writers of the Enlightenment, such as Edward Gibbon (c 1737-1794) and David Hume (c. 1711-1776), justified it as a regrettable but necessary price to be paid for civilization.

"Slavery was, Gibbon said, "almost justified by the great law of self-preservation." According to the atheist hero David Hume, "the negroes ... [are] naturally inferior to the whites"—a sentiment that the Enlightenment's great moral philosopher, Immanuel Kant, cited approvingly.

""The Negroes of Africa have by nature no feeling that rises above the trifling," Kant declared, in his 1764 essay, *Observations on the Feeling of the Beautiful and Sublime*. "Mr. Hume challenges anyone to cite a single example in which a Negro has shown talents, and asserts that among the hundreds of thousands of blacks who are transported elsewhere from their countries...not a single one was ever found who presented anything great in art or science or any other praiseworthy quality...So fundamental is the difference between these two races of man, and it appears to be as great in regard to mental capacities as in color.""[252]

So much for the *Enlightenment*. Maybe we should rename that period, reminisced so affectionately by the Democrat-Left, as *The Other Dark Ages*.

"Ironically enough, modern atheist critics who would indict the Bible as supporting or tolerating slavery actually make common cause with Southern racists who tried to do the same thing—and were easily and roundly refuted by nineteenth century biblical scholars and theologians."[253]

So it was the Bible and Christianity, and not those *caring* and *compassionate* atheists of old, that precipitated the end to slavery everywhere in our world today, except in non-Christian, atheist-communist and Muslim countries where it is still unfortunately practiced.

-- *Human rights*. "We hold these truths to be self-evident, that all men are created equal, that they are endowed by their Creator with certain unalienable Rights"[254] The very concepts that certain truths are self-evident, and that certain rights come from God and not the state and therefore cannot be taken away by government, are "derived not from secular philosophy, but from biblical religion." (See Romans 1:19-20, 2:14, Isaiah 10:1-2, Jeremiah 5:26, 28-29) Indeed, the founding fathers were all "steeped, from childhood, in the stories and values and ideas of the Bible."[255]

"The ancient world did not have any true conception of human rights. ... It was not until after the legalization of Christianity by Constantine in 313 AD that Christian thinkers developed a theory of human rights separate from and above the power of the state.

"The early Christian philosophers, strongly influenced by the teaching of Jesus and the theocentric ideas inherited from Judaism, believed in what is now called natural law. In essence, natural law is simply the idea that there is an objective moral order, established by God and grounded in an essential humanity, which stands above mere human law and against which mere human law must be judged.

---

252 (Pgs. 157-159, *The Politically Incorrect Guide to the Bible*, by Robert J. Hutchinson)
253 (Pg. 160, *The Politically Incorrect Guide to the Bible*, by Robert J. Hutchinson)
254 (From *The Declaration of Independence*)
255 (Pgs. 180-181, *The Politically Incorrect Guide to the Bible*, by Robert J. Hutchinson)

...."As the Christian church began to think through the *political* implications of the belief that every human being is a child of God, created in his image and likeness and possessing an eternal destiny, canon lawyers began to insist that there were God-given *rights* that could not be justly abrogated by government officials. As these ideas developed, in both Catholic and Protestant countries, the Christian tradition of natural rights eventually developed into a theory of self-government that directly influenced the founders of the American republic.

"Those modern philosophers who rejected the truths of biblical Christianity—such as David Hume, Thomas Hobbes, Rousseau, and Karl Marx—not surprisingly also rejected the biblical, Christian theory of human rights."[256]

So it was the Bible, and not the philosophical ancestors of today's progressive Democrat-Left, that gave birth to our modern concept of human rights.

-- *"Egalitarianism in early Christianity."* "Now all who believed were together, and had all things in common. And great grace was upon them all. Nor was there anyone among them who lacked; for all who were possessors of lands or houses sold them, and brought the proceeds of the things that were sold, and laid them at the apostles' feet; and they distributed to each as anyone had need." (Acts 2:44, 4:33-35)

-- *Feminism*:

"You'd never know it from the shrill talk of "patriarchy" and sexism among today's aging feminists, but Christianity was undoubtedly the most pro-female religion in history. ...That Christianity accorded women greater social status, based on its conception of basic human rights, can be seen in the peculiar early Christian rejection of abortion, infanticide, and divorce.

"The Christian rejection of divorce tended to give women greater social status, not less, as some modern feminists allege. That's because it conferred upon woman a fundamental economic and social security that both Jewish and pagan women lacked. ...Another way in which human equality and the concept of fundamental human rights were reinforced was through the Christian ban against infanticide and abortion. ...The fact that all human infants—including even female infants—possessed, in Christian eyes, a fundamental right to life, naturally reinforced in the Christian community the belief that human beings did possess certain rights that no one could legitimately abrogate."[257]

Then there is the following passage that has had feminists all in an uproar for years:

"Submitting yourselves one to another in the fear of God. Wives, submit yourselves unto your own husbands, as unto the Lord. For the husband is the head of the wife, even as Christ is the head of the church: and he is the saviour of the body. Therefore as the church is subject unto Christ, so let the wives be to their own husbands in every thing. Husbands, love your wives, even as Christ also loved the church, and gave himself for it." (Ephesians 5:21-25)

Feminists and many liberal women mock and deride that command to submit to their husband's authority as head of the home, in particularly concerning the making of final decisions. (But note that the same passage commands the husband to love their wives as Christ loved, and died for, the church, which means that the husband should consider prayerfully those decisions and to keep her wants and needs above his own.) But unfortunately in so doing, feminists are really

---

256   (Pgs. 192-193, 199, *The Politically Incorrect Guide to the Bible*, by Robert J. Hutchinson)
257   (Pgs. 191-192, *The Politically Incorrect Guide to the Bible*, by Robert J. Hutchinson)

mocking themselves, and displaying their own ignorance. For after even the most cursory examination of marriage it becomes obvious that it is not, nor can it be, a democracy. (God, conversely, figured that out a long time ago.) Democracy is actually impossible in a marriage because there are only two people involved, and in decision making if the two disagree, where is the tiebreaking vote? It doesn't exist. So God in his infinite wisdom, as opposed to our mocking ignorance, set up the man to have the ultimate authority in the marriage. If a woman disobeys His command, and refuses to obey her husband, then she chooses to disrespect him, to demean him in front of his children and peers, to lessen his God-ordained position in the home as the head of the family as Christ is the head of the church, and usurps his place of authority by in essence demanding to be the head herself. Because if the wife refuses her husband's authority, then whose authority is she going to turn to? Her own, of course. In refusing to respect and to submit to her husband's divinely ordained authority, she unwittingly (or perhaps wittingly) demands that the husband submit to her. And in doing so those women are calling God, who created both man and woman and ordained marriage itself, stupid; and themselves smart. What a hoot! (And unfortunately it is not just feminists, many Christian woman are equally at fault.) (However it is also true that the wisest of women know that whereas the man is the head of the home, <u>she is the neck</u>, and "the neck can turn the head wherever it wants...")[258] They have reversed the natural God-made order of things by putting themselves at the head of the home, and instead have submitted to the unnatural Beelzebub-made disorder of things. And we wonder why our marriages don't work, and why our children turn out many times to be an unruly mess.

So again we see that even in the most contentious and hardest to comprehend rules of God's Word, those that are refused with the strongest emotions and derided with the most passion, it is God and his Word that turn out to be right, and wise, and looking out for our own best interests; and it is those who mock it that turn out to be the truly foolish ones, poking a stick in the spokes of their own wheels.

--*Civil law*. In contrast to "modern, rather cynical theories of jurisprudence ...that basically hold that law is whatever the state says it is," and that "Law and morality have nothing to do with one another, ...[conversely] the classic Judeo-Christian view ...has always been that governmental elites must answer to a higher law than mere human legislation." And therefore, "Governments that fail to respect the "unalienable rights" endowed by God are tyrannies and, therefore, illegitimate."[259]

The opposite of this Christian, biblical view is an aberration known as *legal realism* or *legal positivism* (reminds one of another aberration that we exposed in chapter 2, *logical positivism*), and it...

> ..."was a highly influential theory of jurisprudence throughout the first half of the twentieth century. But the horrors of World War II and Communist and Nazi totalitarianism made many law professors rethink whether it is a good idea to teach the doctrine that what is legal is whatever the state says is legal."

Conversely...

> "The notion of a Divine law above mere human law was expressed clearly by Thomas Aquinas in his *Summa Theologica*, ratified by John Calvin in his *Insitutes*, and summarized

---

258 (As quoted in the movie, *My Big Fat Greek Wedding!*)
259 (Pgs. 194-195, *The Politically Incorrect Guide to the Bible*, by Robert J. Hutchinson)

succinctly by Sir William Blackstone in his *Commentaries on the Laws of England* (1765), one of the chief sources used by Jefferson (and all the colonists) in crafting the new American government.

"According to Blackstone, civil law is given, not to *create* rights, but to protect already *pre-existing* natural rights;

..."Once again, these ideas stem, not from atheistic philosophers, but from Christian theologians reflecting upon the truths found in the Bible."

Mr. Hutchinson goes on to explain that...

"This great tradition of classical natural right—which extended from the biblical prophets and the teaching of Christ through the medieval scholastics and Protestant divines up to the U.S. Declaration of Independence—was challenged directly by what is sometimes called "political atheism."

"Modern people, indoctrinated as they often are, tend to think of the Renaissance as a great "rebirth" of humanism and enlightenment, but in terms of human rights and of the rights of women it was actually a profound step backwards.

"This is because the Renaissance was a return to the ideas of Roman and Greek civilization and a rejection of medieval Christianity—yet it was precisely medieval Christianity which championed what we today call human rights."[260]

And this would be common knowledge among our youth if it weren't for the fact that historical revisionists, who have little interest in disseminating the truth if it conflicts with their left-wing ideology, have taken over our educational system at every level. (And see section 4 "Destroy the historical foundations of America" in chapter 13)

-- "*Liberal democracy and political freedom.*" Both stem from biblical religion, and not despite it, as again the Left likes to falsely assert...

"The historical evidence clearly shows that the enemy of political liberty is not Christianity or Judaism, as false liberals like Robert Reich and Sam Harris like to claim, but the dangerous idea that there is no Higher Authority to which politicians must be accountable and that, therefore, the ends always justify the means.

"Societies that have actually done what atheist polemicists now advocate and deliberately rejected "bourgeois" Christianity in favor of a more "rational," more "modern," more "scientific" approach to politics—such as revolutionary France, Communist Russia, and fascist Germany—ended up, not in the utopia of their dreams, but in a bloody tyranny.

..."In fact, it was the explicit *rejection* of Christian natural law theory—first by people like Machiavelli and Thomas Hobbes and then, in the nineteenth and twentieth century, by the early theoreticians of socialism, communism, and fascism—that led to the horrors of the twentieth century totalitarianism."

..."In conclusion, therefore, we can say that the enemies of Christianity and Judaism have it exactly backwards: Far from being a threat to liberal democracy and political freedom, biblical religion is, in fact, the intellectual matrix out of which both arose."[261]

I have only quoted Mr. Hutchinson's summaries and conclusions on this subject as space won't allow for more. For the facts and historical evidence that expose the dishonesty of the academic Left in denying the Bible as the true source of liberal democracy and political freedom, read chapter 12, "Put Not Your Trust in Princes," in *The Politically Incorrect Guide to the Bible*.

-- *The Jesus Seminar.* The real historical Jesus is actually, according to all

---

260 (Pgs. 194-196, *The Politically Incorrect Guide to the Bible*, by Robert J. Hutchinson)
261 (Pgs. 202,210,216, *The Politically Incorrect Guide to the Bible*, by Robert J. Hutchinson)

of the facts and evidence, exactly who the gospels say he is, and not who modern liberal critics, like those from the Jesus Seminar, want to remake Him, in their own image, by distorting and rewriting the clear, unambiguous declarations of history. Again we find the Bible wholly accurate and the liberal revisionists either misinformed or less than honest, or both.

"This modern quest for the "real" or the "historical" Jesus ...usually operates out of one overriding, often unquestioned assumption: that the "real" Jesus was something quite other than what his followers, and the New Testament, said that he was."[262]

Of course this is a false assumption, and it leads its followers on a wild goose chase fraught with ill-conceived flights of fancy that end in idiotic yet self-serving conclusions. But then again, they appear to have no intention of coming to accurate conclusions, nor of finding the "real" or the "historical" Jesus. Their intention is only to *deny*, under the guise of academic scholarship, the real, historical Jesus.

"The Jesus Seminar represents the lunatic fringe of the radical Left. ...Their *scholarship* is pseudo-scholarship that is pure drivel.

"There's less reason today than at any time in church history to be skeptical about the veracity and integrity of the biblical witness. ...No book has been subjected to the most rigorous scrutiny, the most vicious attacks. ...If we look clearly at the historical record and at the claims, no document from antiquity has been manifested more frequently to have authenticity than the scriptures, particularly the New Testament."[263]

In conclusion then...

"Reverent students have long believed, as I do, that the preservation of the Word of God down through the centuries is miraculous and supernatural. How Satan hates the Bible! The conspiracies of Romish priests, burning multiplied thousands of copies, the putting to death of those that harbored the Bible and insisted on reading it- these could not do away with the Word of God."[264]

Yes, "the Bible has always been hated," and the bible-haters are still going strong today. They are not threatening to kill those who own one now (at least not yet...give 'em a few more years as we slide further into the hell of the end times and all bets are off) but their friends in communist and Islamic countries have been doing that for quite some time. But here in the West they are doing everything else in their power to marginalize, demean and lie about God's Word so very few will want to own one, much less open it up and absorb its timeless and all-important message (see chapter 7).

The lies of Satan cover this earth like a blanket and people believe what they are taught. And the lie that the Bible is something other than what it says it is— that it's something to be made fun of; something that deserves to be mocked, scorned and ridiculed by the clueless, uneducated buffoons of the Democrat-Left such as Robert Reich, Penn and Teller, Sam Harris, Richard Dawkins, Christopher Hitchens and Martin Sheen; that it is something that cannot be trusted; something that isn't absolutely true; something that doesn't contain all wisdom; something that wasn't written by God for the benefit of all mankind—is one of the devils biggest whoppers. And a whopper that only continues to exist, in the face of overwhelming evidence to the contrary, because of the eager complicity of our

---

262　(Pg. 222, *The Politically Incorrect Guide to the Bible*, by Robert J. Hutchinson)
263　(From "The Authenticity of Scripture," by Dr. R. C. Sproul, radio broadcast 12/27/2010)
264　(Pg. 265, *Prayer: Asking & Receiving* by John R. Rice)

Democrat-Left media, educational system and entertainment industry; America's "disseminators of information." The proud disseminators of countless lies. And poor unfortunate, misinformed folks the world over have swallowed this lie, hook, line and sinker. And, as we will see in the next chapter, they have swallowed it at their own eternal peril.

As we quoted earlier, "For nearly 2000 years every engine of destruction that human science, philosophy, wit, reasoning or brutality could bring to bear against a book has been brought to bear against the Bible to stamp it out of the world. [Yet] it has a mightier hold on the world today than ever before."[265] And that is because it is a book that stands apart from all others, and one that is covered in divine protection. It is also a book that is truly holy; one that is separate from all the rest.

"Heaven and earth shall pass away, but my words shall not pass away."
(Matthew 24:35)

6- Archeological evidence, coupled with other secular histories of the same era, supports the historical information recorded in the Bible. Many times in the last few centuries critics of the Bible have leveled charge after empty accusation that it is inaccurate with regard to this historical statement, or that this ruler never existed, or that this city is fictional, etc. etc. But time and time again, with each new archeological dig and discovery, and with each new historical document that is unearthed, it is shown that yes indeed, this historical statement is accurate, and this ruler did exist, and this city was real, just like the Word of God says. (We are still waiting, of course, for the academic Left to issue the appropriate statements retracting each of their false accusations. And, billions of years from now, long after the sun burns off its last ounce of nuclear fuel, we'll still be waiting.)

"There are twenty five thousand archaeological references in the Bible that have been confirmed in modern times."[266]

"Time and again, archaeology has verified the existence of places, people, and events we once knew solely from biblical records. These discoveries ...bolster the credibility of the biblical text as containing real historical records."[267]

Here is just a small sampling of archaeological discoveries (again sourcing from *The Politically Incorrect Guide to the Bible*, by Robert J. Hutchinson) that have embarrassed Bible critics, overturning their claims of biblical fiction with regard to certain nations, rulers, cities or events:

-- *The Merneptah Stela.*

"In the nineteenth century, as in our own, some scholars questioned the reliability of biblical narratives. Then, as now, some even suggested that the Israelites never really existed at all as a separate people—that is, until a seven-foot slab of black granite ... discovered in a temple in Thebes, Egypt, in 1896."[268]

It was dated to 1209/1208 BC, and it spoke of the Israelites and of Israel being laid to waste! So much for Israel "never existing..."

-- *To the critics claim* that Solomon and even King David never existed...

---

265 (From "Ten Reasons Why I Believe the Bible is the Word of God," by R. A. Torrey)
266 (Bob Cornuke, Bible Archaeology Search and Exploration (BASE) Institute)
267 (Pg. 20, *The Politically Incorrect Guide to the Bible*, by Robert J. Hutchinson)
268 (Pg. 21, *The Politically Incorrect Guide to the Bible*, by Robert J. Hutchinson)

...ʺon July 21, 1993, archaeologists working at Tel Dan in northern Israel ...uncovered a basalt stone, written in Old Aramaic, that explicitly mentions the house of David. Pottery fragments ...date it to the end of the ninth or the beginning of the eighth century BC. ... The Stela refers to events recorded in the O d Testament Book of II Kings.ʺ[269]

So much for King David ʺnever existing...ʺ

-- Then there is the *Moabite Stone*.

ʺCompared with other ancient historical works—such as Thucydides' *History of the Peloponnesian War* or Caesar's *Gallic Wars*—the Bible has a tremendous amount of archaeological and historical support. [The] Moabite Stone [was] discovered in 1868 [and it contains] the personal testimony of a Moabite king that mentions numerous places in the Bible.ʺ[270]

So much for the Bible's geographical references ʺnot existing...ʺ

-- *The Samaritan Ostraca* silenced some modern scholars who had called into question the very existence of Israel's Twelve Tribes. It was...

...ʺdiscovered in 1910, under the direction of Harvard archaeologist G. A. Reisner, in Samaria—sixty three potsherds bearing inscriptions in Old Hebrew script written in ink, called ostraca. Among the oldest samples of ancient Hebrew writing ...dated to around 784-783 BC... the ostraca are commercial records of shipments of wine and oil, but they have one characteristic of historical significance: Thirty of them name the clan or district name of seven of the ten sons of Manasseh mentioned in Joshua 17:2-3.ʺ[271]

So much for the Twelve Tribes of Israel ʺnever existing...ʺ

-- *The Pontius Pilate Inscription*.

ʺWhen contemporary skeptics complain that there is ʺno archaeological proofʺ that a certain biblical figure existed—such as Moses—what they often don't admit is that there is little or no archaeological evidence for many figures of ancient history that no one seriously doubts existed. This ʺargument from silenceʺ has been refuted so many times you would think that modern skeptics would stop using it. For example, until 1962 there was no archaeological proof that Pontius Pilate ever existed—and skeptics made much of the fact. But in that year, an Italian archaeologist working at Caesarea Maritima, on the coast of Israel south of Haifa—the center of government for the Roman administration in the time of Christ—found the long-sought proof. It came in the form of an inscription that mentioned ... ʺTiberius [the Roman emperor of the period]/Pontius Pilate/Prefect of Judea.ʺʺ[272]

So much for... are we sensing a pattern here?

-- *The Ebla Tablets*.

ʺIn 1964, Italian archaeologists from the University of Rome began excavating a palace found at Tell Mardikh, in northern Syria. ...they found a virtual library of 15,000 well-preserved cuneiform tablets dating from around 2300 BC. Not only do these amazing tablets, written in Sumerian and Akkadian reveal laws, customs, and events surprisingly in harmony with the account in Genesis, they also mention explicitly the five undiscovered ʺcities of the plainʺ mentioned in Genesis 14:8 that modern skeptics insist never existed [emphasis mine]—Sodom, Gomorrah, Admah, Zeboiim, and Zoar.ʺ[273]

-- *Hezekiah's Tunnel*.

ʺMost biblical scholars have long believed the 1,750-foot-long tunnel, rediscovered

---

269 (Pg. 21-22, *The Politically Incorrect Guide to the Bible*, by Robert J. Hutchinson)

270 (Pg. 22, *The Politically Incorrect Guide to the Bible*, by Robert J. Hutchinson)

271 (Pg. 24, *The Politically Incorrect Guide to the Bible*, by Robert J. Hutchinson)

272 (Pgs. 25-26, *The Politically Incorrect Guide to the Bible*, by Robert J. Hutchinson)

273 (Pg. 26, *The Politically Incorrect Guide to the Bible*, by Robert J. Hutchinson)

in 1838, did indeed date back to the time of Hezekiah; a minority insisted the passage was built centuries later. ...But once again, archaeology has proven the textual skeptics to be wrong. In 2003, Israeli and British scientists tested organic material within the plaster lining of the tunnel and dated it back to around 700 BC- just as the Bible says."[274]

-- *The Nuzi Tablets.*

"A common tactic that critics use when attacking the historicity of the biblical narratives is to point out cultural anomalies—customs that (allegedly) didn't exist in the period in question. For example, skeptics point to a number of customs in the Pentateuch that, they claim, do not correspond to what we "know" about the Middle East in the second millennium. But excavations in 1925 at Nuzi, in northern Iraq, discovered 4,000 tablets written in Akkadian cuneiform script that dated back to 2300 BC. The Nuzi tablets describe customs that parallel those described in Genesis. ...Once again, archaeological discoveries have disproven what was once "known." "[275]

And we could go on, and on. Time and time again the Bibles critics have been proven to be wrong, and the Bible right. So much so that you would have thought by now they would have given up, or at least would have been driven to the fringes of academic scholarship due to an absence of all credibility. But no! That's not how it works. Bible critics don't have to admit when they're wrong, because their continuous errors are rarely if ever reported by the Christian- and Bible-bashing folks who control the dissemination of information in our culture. All of their false claims and false attacks are immediately treated as gospel, and these lies are spread far and wide throughout academia and our media. But time and again when they are proven to be wrong, it never merits the same amount of coverage (if any at all).

"I rejoice at your word, as one who finds great spoil."
(Psalm 119:162)

7- God is the infinite mathematician as well as being the very Source of mathematics itself, and it would be natural therefore to suspect that His Word might reflect some of those abilities. There are a number of eye-opening books on this subject (as well as chapter 12 "The Mathematical Signature of God in the Words of Scripture" in Grant R. Jeffrey's book *The Signature of God: Astonishing Biblical Discoveries*). One is *Bible Mathematics: Keys to Scripture Numerics* by Ed F. Vallowe. Another is *Bible Code Bombshell: Compelling Scientific Evidence that God Authored the Bible*, by Edwin Sherman who "describes numerous examples of encoded phrases and sentences that are both lengthy and relevant to the text where they were found." Sherman started, however, as a skeptic who set out to prove that the encoded messages found in the biblical text were nothing but chance occurrences, the same as you could find, for example, in the Jerusalem phone book with a high speed computer...

...."But he was intrigued enough to develop his own software and begin analyzing the Masoretic text of the Old Testament. "I was very skeptical about the whole thing," he says. "I started a project to try and show the whole thing was bogus." Instead, he says he found many examples of messages that went beyond simple words and phrases – and they often were contextually similar to the Biblical passage in which they were found. "Finding

---

274   (Pgs. 26-27, *The Politically Incorrect Guide to the Bible*, by Robert J. Hutchinson)
275   (Pg. 27, *The Politically Incorrect Guide to the Bible*, by Robert J. Hutchinson)

dozens of lengthy encoded messages on the same topic in one short section of text is about as likely as winning a one-in-a-million jackpot ten times in a row. Basically, it cannot happen by chance."

"The process of searching for encoded messages involves analysis of the biblical Hebrew text in digital form. The Scriptures are encoded by removing all spaces between words and creating long strings of letters. According to Sherman, vowels are inserted in the strings of letters – the Hebrew alphabet lacks vowels – following standard rules based on the sequence of consonants. Software then analyzes the strings in search of patterns based on equidistant letter sequences. "The shorter an expression, the easier it is to find," Sherman notes in Southern Oregon News, "but when we find longer statements that have a connection to the actual biblical passage, and we find those longer statements with frequency, it leads us to believe that those statements were purposely implanted in the Bible by God." Some of the most compelling evidence of a mathematical pattern in the Hebrew text comes from the 53rd chapter of Isaiah, a passage most Bible scholars see as messianic and which Christians have traditionally seen as prophecy about Jesus.

"Sherman developed a baseline using non-encoded Hebrew texts as his standard of comparison for determining whether the number of messages he found in a Biblical passage were statistically significant. Isaiah 53 proved to be a rich cluster of hidden messages, containing 42 encoded statements relating to Jesus' death, resurrection and ascension, far more than his baseline predicted. As evidence, Sherman points to statements such as "Gushing from above, my mighty name arose upon Jesus, and the clouds rejoiced," "dreadful day for Mary," "In his name as he commanded, Jesus is the way," "Resurrection of Jesus, he is risen indeed," and others that echo Isaiah's prophecy. It is the coherence between the hidden messages and the Hebrew text from which they are drawn that excites Sherman. ...The messages plucked from the text are more like divine fingerprints.

""The Bible itself claims to be written by God, and when the subject of the encoded messages ties in so closely with the subject of the literal text, it has to make you take notice," Sherman told the paper. "I just want ... to capture the curiosity of skeptics and cause them to consider the possibility that the Bible is not written by men, but by God, and should therefore be taken very seriously.""[276]

And then there is another remarkable work titled *Temple at the Center of Time*, by David Flynn. Drawing on the discoveries of Isaac Newton and other ancient Jewish and Christian writings, the author reveals a striking relationship between the dates of significant events in this world's history, and the connection in space and time between the Temple Mount in Jerusalem and both ancient and modern cities that were at the center of these events. It is difficult to read this book and to walk away without being in awe of the infinitely powerful, supremely intelligent, omniscient being who was clearly the Author of not only the Scriptures themselves, but of the geography of this world, its nations, and the exact locations of their cities. Nothing was left to chance... "And he has made from one blood all nations of men for to dwell on all the face of the earth, and has determined their pre-appointed times, and the bounds of their habitation." (Acts 17:26)

"I understand more than the ancients, because I keep your precepts, and meditate on your testimonies all the day." (Psalm 119:100)

8- The Bible also speaks with absolute moral authority. A right that comes from it having been written by the One who is the source of truth and righteousness, from the One who wrote all of the laws of the universe, both scientific and moral. It doesn't give us ten suggestions, but The Ten Commandments. And its laws and codes of justice are the backbone of modern jurisprudence. The Bible is the only source this world has of moral absolutes and that clearly reveals its Author.

"For the word of God is living and powerful, and sharper than any two-edged sword, piercing even to the dividing asunder of soul and spirit, and of joints and marrow, and is a discerner of the thoughts and intents of the heart." (Hebrews 4:12)

9- Science. The Bible also speaks authoritatively and accurately on scientific subjects. It is not just a Spiritual book...

"The same God who ordered science wrote this book. And whenever it talks about scientific things or matters in the physical world, it speaks with accuracy."[277]

Long before man began to uncover a little of the true nature of this universe and began to understand the scientific laws that govern it and that give it order, the Word of God had already spoken with perfect understanding and knowledge on much of these matters. Here are a few examples:

-- *How old?* The Bible says that the universe was created about six thousand years ago. Faux scientists say that the universe was created billions of years ago by a big explosion and then formed itself, all that we survey today, miraculously, and completely accidentally, by random chance alone. The Biblical account is scientifically accurate. The explosion and random chance account is scientifically absurd. The current, widely-accepted notion that the universe is billions of years old is nothing more than a deliberate fabrication, and a distortion and covering up of the evidence, in order to provide cover for the completely asinine yet academically popular theory of evolution. All the scientific evidence—the magnetic fields of the earth and the planets; Carbon-14 being present on a worldwide scale in rocks "dated" at *hundreds of millions* and even *billions* of years old even though any rock older than 100,000 years should contain no carbon-14 whatsoever because it would have already decayed into nitrogen; the depth of the dust found on the surface of the moon; the recession rate of the moon; known sedimentation rates of rivers as compared to the amount of sedimentation present in our oceans and river deltas; the elemental makeup of the spectrum of light traveling from distant galaxies; and yes even the evidence from the speed of light itself—all agree with the Biblical age. And this list is just a tiny fraction of the body of scientific evidence that goes far beyond just calling seriously into question these preposterously old yet well "established" ages taught in our schools and universities. It flat out disproves them, as you will find documented in chapter 10, "The Theory of Evolution is a Joke."

-- *Noah.* The Bible says that there was a catastrophic worldwide flood that occurred about 4,500 years ago. Faux scientists say that all of the features found on the earth today, including the Grand Canyon, can be explained by the slow processes of upheaval and erosion continuing slowly over millions of years. Again, the Biblical account is scientifically accurate. The evidence for a worldwide flood is overwhelming and incontrovertible, yet it is blackballed by the evolution

---

277   (From "The Character of God's Word," by John Macarthur, Grace To You, www.gty.org)

zealots who possess a stranglehold over our godless media and our Bible-mocking universities. There are countless trillions of recently deposited seashells on the tops of the highest mountains. Coal and oil formations demand the rapid and very deep burial of an extraordinary volume of living matter. The fossil record which covers this earth could only have been caused by the rapid worldwide burial of untold trillions of living things because fossils can <u>only</u> form under conditions of rapid and <u>deep</u> burial—a dead fish lying on an ocean floor, or a dead elephant lying on the ground have no chance whatsoever of being fossilized—which is why we find no fossils forming today. It is nothing other than feeble-minded to presume that the Grand Canyon, including all of its incredible <u>side</u> canyons, was formed by a river slowly eroding a *raised plateau* over any period of time. All the scientific, geological, erosion rate and sedimentary evidence call for the Canyons rapid formation by the catastrophic runoff of a massive amount of water over a very short period of time. On a worldwide scale sedimentary layers thousands of feet deep, with absolutely no evidence of time passage or erosion between each successive layer (their transition surfaces are perfectly smooth and flat), and extending in many places uniformly for hundreds of thousands of square miles cry out for only one possible scientific explanation: that of a catastrophic worldwide flood! And we could go on. (As we do in chapter 10) But once again the Bible proves itself scientifically and historically accurate.

People believe what they are taught. Satan is acutely aware of this. He is also acutely aware that if you shout a lie loud enough and long enough, enough uninformed people will believe you. And that's why the bald-faced, scientifically impossible lie of these ridiculous ages for the earth and the universe that are used to prop up the equally bald-faced, scientifically impossible lies of evolution and the real big explosion, have become common knowledge. Again, as Mark Twain said "The problem with the world is not that so many people know so little. The problem with the world is that so many people know so much that just ain't so." And these ridiculously old ages for the earth and the universe- they just ain't so. (And see chapter 10)

-- *The conservation of mass and energy.* Then there is the discovery of the First Law of Thermodynamics, the law of the conservation of mass and energy, which states that matter and energy cannot be created or destroyed, but can only change from one form to another. And yet the Bible knew this thousands of years before modern man discovered this fundamental principle: "You, even you, are Lord alone; You have made heaven, the heaven of heavens, with all their host, the earth and everything on it, the seas and all that is in them, **and You preserve them all**." (Nehemiah 9:6) And this: "Is there any thing of which it may be said, '<u>See, this is new</u>?' <u>No, it has already been from of old</u>." (Ecclesiastes 1:10)

-- *The water cycle.* Water evaporates from the seas, condenses into clouds, moves over the land, drops down as rain and snow, and then returns to the seas by streams and rivers, but this cycle was not understood until the seventeenth century. But the Bible understood the cycle of evaporation, condensation and precipitation thousands of years ago: "All the rivers run into the sea, yet the sea is not full; unto the place from which the rivers come, there they return again." (Ecclesiastes 1:7) And where is "the place from which the rivers come?" ... "For He draws up drops of water, which distill as rain from the mist, which the clouds drop down and pour abundantly on man." (Job 36:27-28) That is the entire water cycle described thousands of years before modern man figured it out.[278]

---

278   (From "The Character of God's Word," by John Macarthur, Grace To You, www.gty.org)

-- *The lifeblood.* The Bible says "For the life of the flesh is in the blood." (Leviticus 17:11) Yet not more than a hundred years ago when someone got sick they bled them! They took out their blood! Now we know better. Now we *give* sick people blood. Because we finally figured out what God knew and wrote in His Word thousands of years ago.[279]

-- *Circumcision.* "Then Abraham circumcised his son Isaac when he was eight days old, as God had commanded him." (Genesis 21:4) Why did God choose the eighth day after birth for circumcision? Did the Author of the Bible know something that man did not figure out until thousands of years later? ...

> "S.I. McMillen, a medical doctor, wrote a little book called "None of These Diseases" that's really a classic. He showed how following God's principles corrects and prevents some of our most devastating diseases. And one of the things that he points out is that an important blood-clotting agent, prothombrin, is at its highest natural level in the body on the 8th day after birth. No wonder God commanded circumcision to be performed on the 8th day. Abraham didn't know anything about prothombrin. He didn't have to."[280]

God knew.

-- *Starry, starry night.* The Bible says that the stars are too numerous to be counted: "As the host of heaven cannot be numbered, neither the sand of the sea measured." (Jeremiah 33:22) Yet long after the Word of God had made that declaration, early astronomers thought they knew better. Hippocrates "said there were 1,022 stars. Ptolemy said, 'No, there are 1,056.' And Kepler said, 'You're both wrong, there are 1,055.' Why didn't they read Jeremiah? ...Now we know there are 100 billion stars in a typical galaxy and there are billions of galaxies."[281]

-- *De old Klug.* "Old ideas about the earth and about the solar system were very strange. Most people thought that the earth was a CD, a flat circular disk. ...One ancient eastern book says it's on seven layers of sugar, honey and butter. Fortunately we're not on the gooey side. The Koran says it's on the back of elephants that produce earthquakes when they shake. ... [But] Job, the oldest book in the Bible, says God hangs the earth on nothing." "He stretches out the north over empty space, and hangs the earth upon nothing." (Job 26:7) Job uses the word Klug, which means a <u>sphere,</u> to describe the earth. Also, "the oldest book in the Bible says He turns the earth like the clay to the seal, [which simply means] it rotates on an axis."[282]

-- *The air.* "Job said God imputed weight to the wind. It wasn't until about the seventeenth century that anybody understood that there was weight to the wind."[283] "For he looks to the ends of the earth, and sees under the whole heaven, to establish a weight for the wind, and apportion the waters by measure." (Job 28:24-25)

-- *In the Beginning.* And then there is this incredible discovery...

> "There was a man ...named Herbert Spencer ...a [much] heralded scientist because he had made a unique discovery. He had come up with the classification of everything in existence. He said all that exists, all that is knowable "falls into one of these five categories... time, force, action, space and matter." And Herbert did very well. But Herbert

---

279   (From "The Character of God's Word," by John Macarthur, Grace To You, www.gty.org)
280   (From sermon, December 28, 2008, Pastor Ed Regensburg, Faith Community Church, Gambrills, Md.)
281   (From "The Character of God's Word," by John Macarthur, Grace To You, www.gty.org)
282   (From "The Character of God's Word," by John Macarthur, Grace To You, www.gty.org)
283   (From "The Character of God's Word," by John Macarthur, Grace To You, www.gty.org)

was a little late in coming to something that the very first verse of the Bible makes clear. Genesis 1:1, "In the beginning (that's time) God (that's force) created (that's action) the heavens (that's space) and the earth (that's matter)." Every classification of the knowable is placed in the first verse of the Bible. The genius of the mind of God is beyond anything that man could conceive."[284]

Amen! The Bible is a scientific book. And it is the oldest, and the only completely accurate one ever written. Science itself was created by the Bibles Author. And science itself is built upon the foundation of the *laws* of science (and not stupid theories). And these laws of science came of course from a law *Giver*. (Where else did they come from, Big Bird?) The same person who wrote the laws of science into the fabric of this universe is the same person who wrote the Bible. Therefore the Bible is a scientific book. Yet how it is mocked by the lame-brains and faux scientists in our colleges, universities and media today! God have mercy on them.

"Concerning your testimonies,
I have known of old that you have founded them for ever."
(Psalm 119:152)

10- The Bible is inerrant. "The Bible in its original autograph, that is to say in the very document written by the inspired writer, was without error." It was exactly what God intended to be written, down to the every word. But then the argument has been put forth that certainly over the many years in the past, the document must have changed, and errors must have crept in as it was copied and recopied and copied again and again. And that is a valid question. Or at least it might have been up until the discovery of the Dead Sea Scrolls. These scrolls dated to "450 years before Christ, B.C., and they verified with exact accuracy the current manuscripts that are being used for the translations of the Bible. And we saw that not only were the originals inerrant but the superintending work of the Holy Spirit as well as the careful diligence of the scribes had preserved that Scripture through all those centuries. In its origins it is inerrant..."[285]

"For ever, O Lord, your word is settled in heaven."
(Psalm 119:89)

11- Miracles. The Bible is full of recorded miracles—thousands of miracles with eyewitnesses—from the parting of the Red Sea to the ascension of Jesus into heaven, and countless signs and wonders in between. There was the water from the rock, and the manna from heaven experienced by hundreds of thousands of Israelites for years in the desert; the collapse of the impregnable walls of Jericho, and the historical record of the sun standing still in the sky for a day; calling down fire from heaven to consume a crowd of Satan worshippers; raising of the dead and the healing of lepers. And then came Jesus who turned water into wine, healed all manner of sickness and deformity and disease, cleansed lepers, raised the dead, made the lame whole, walked on water, calmed the raging winds and

---

284 (From "The Character of God's Word," by John Macarthur, Grace To You, www.gty.org)
285 (From "The Character of God's Word," by John Macarthur, Grace To You, www.gty.org)

seas with a few words from his mouth, fed thousands with a few loaves and fishes, and finally himself was resurrected from the dead after three days and three nights in the tomb. All of these miracles were done in the sight of many witnesses and were recorded as a part of the historical record. Now either they happened or they didn't. If they didn't then the Bible is lying. But it is virtually impossible for thousands of bald faced lies to be recorded as history and passed down with perfect accuracy for generation after generation. No one would waste their time on such a monumental task, generation after generation, for a known historical lie. If this were the case then every false religion on the planet would have done the same in order to lend some semblance of credibility and divine authority to their bogus claims. Mohammed did not perform a single miracle and it is recorded as such because it is virtually impossible to pass down lies through the historical record. (And that's because history is His story.) Just look at how so many left-wing zealots in our universities have been trying for years to rewrite the history of this nation to conform to their America-despising beliefs. But they have only been successful in duping many of their students and uneducated faculty. The historical record is still intact and has already recorded them as the lying revisionists that they are.

"For this reason we also thank God without ceasing, because when you received the word of God which you heard from us, you welcomed it not as the word of men, but as it actually is, the word of God." (1 Thessalonians 2:13)

12- Greek Mythology: In a ground-breaking book, *The Parthenon Code: Mankind's History in Marble*, the author, Robert Bowie Johnson, Jr. unlocks the true identity of the "mythological" gods and goddesses of ancient Greece. The evidence clearly shows that the Greek's temple sculptures, frieze's, vase paintings and writings do not deal with myth, but rather tell the story of their ancestors, and indeed the ancestors of the entire human race: and in doing so they also present to the world "a startling testament to the validity of Scripture."[286] In the ancient Greek's temple sculptures and frieze's, vase paintings and writings we clearly see the depictions of Adam and Eve and their fall from the Garden of Eden; the serpent who "enlightened" them with the fruit from the tree; Cain (who slew Abel), Seth, Noah and the great worldwide flood, and Nimrod all represented. The Greeks tell the story from a different perspective than the Bible, though. "Greek myth and Greek art are inseparable. Greek art depicts the myth: Greek myth explains the art. And Greek myth, it turns out, is the story of mankind told from the Greek's religious perspective."[287] You see, the ancient Greeks were turned 180 degrees from the Bible in their view of man's earliest history. In fact, they saw it from the serpent's perspective. They believed that...

> ..."the serpent did not delude Adam and Eve in the ancient garden, but rather, enlightened them. ...To the Greeks, the serpent freed mankind from bondage to an oppressive God, and was therefore a savior and illuminator of our race. ...There is no Creator-God in the Greek religious system. The ancient Greek religious system is about getting away from the God of Genesis, and exalting man as the measure of all things. You

---

286   (James R. Coram, Co-editor, Unsearchable Riches Magazine)
287   (Pg. xi, *The Parthenon Code: Mankind's History in Marble*, by Robert Bowie Johnson, Jr.)

may think to yourself that the Greeks are exalting gods, not man; but haven't you ever wondered why the Greek gods looked exactly like humans? The answer is the obvious one: for the most part, the gods represented the Greeks' (and our) human ancestors. ...You have no doubt heard of the supposedly great philosopher, Socrates. In Plato's *Euthydemus*, he referred to Zeus, Athena, and Apollo as his "gods" and as his "lords and ancestors." ... Greek religion was thus a sophisticated form of ancestor worship."[288]

It also was a sophisticated form of Satan worship, the worship of the serpent.

The famous Parthenon itself and the entire array of ancient Greek art and sculpture is profound extra-biblical evidence for the validity of Scripture. "The Greeks ...directly connected Zeus [their name for Adam] and Hera [their name for Eve] to a paradise, a serpent, and a fruit tree." The Greeks called this paradise the "Garden of the Hesperides [which] is, with little doubt, the Garden of Genesis. ...Greek myth/art and the Book of Genesis tell basically the same story, because that's what really happened in our past."[289]

On the Parthenon's west pediment, "Poseidon, the power of the sea, is depicted as having struck the ground with his trident, causing the earth to quake and split. When he did that, water gushed forth. This matches the account of the Deluge in Genesis. Water came first from below, then from above."[290] Greek "mythology" (read: *history*) is teeming with corroborating references to the world-wide flood of Noah's day depicted in the Bible and yet dismissed out-of-hand by those who predominate the dissemination of information in our educational/left-wing-indoctrination system.

The ancient Greeks also confirm the Biblical story of the Tower of Babel and the division of languages in their recognition of Hermes, who is the Bibles "Cush, the son of Ham, and the grandson of Noah." Hermes (Cush) led humanity in "the building of the infamous tower" of Babel. The Greeks "understood him to be ...the "divider of the speeches of men." According to Plato, Hermes invented language and speech."[291] This corresponds exactly with the Biblical account of the building of the Tower of Babel and God's decision to confuse their languages and scatter them abroad over the earth.

"What the Greeks meant to be an unparalleled, intricately chiseled monument to the glory of mankind turns out to be a detailed history of mankind's delusion, and a clear-cut validation of the truth of the Word of God."[292]

Thank you, Greece.

"Blessed are those who hear the word of God and keep it!"
(Luke 11:28)

---

288 (Pgs. 7, 10-12, *The Parthenon Code: Mankind's History in Marble*, by Robert Bowie Johnson, Jr.)

289 (Pgs. 13, 15, 36, *The Parthenon Code: Mankind's History in Marble*, by Robert Bowie Johnson, Jr.)

290 (Pg. 52, *The Parthenon Code: Mankind s History in Marble*, by Robert Bowie Johnson, Jr.)

291 (Pg. 221, *The Parthenon Code: Mankind's History in Marble*, by Robert Bowie Johnson, Jr.)

292 (Pg. 258, *The Parthenon Code: Mankind's History in Marble*, by Robert Bowie Johnson, Jr.)

Conclusion

"For this reason also we thank God without ceasing, because, when you received the word of God which you heard from us, **you received it not as the word of men, but as it is in truth, the word of God**, which effectively works also in you who believe." (1 Thessalonians 2:13)

And with those twelve reasons, we will rest our case. Much more could be added, but we will leave it up to the reader to come to his own conclusion. At least now he or she can do so based not on the bald-faced lies of Satan and his unwitting accomplices, but on the facts and evidence, which the god of this world has determined to keep hidden from the vast majority. Lies about the Bible he has made sure are on the front page of every left-wing magazine and newspaper, and taught as *gospel* in every college and university in the land; but the truth about it he and his followers have made sure is kept "secret."

Hear what Pastor John Macarthur, who we have drawn from heavily in this chapter, has himself concluded:

"Why do I believe the Bible? I believe the Bible because of its scientific accuracy. I believe the Bible because of its prophetic accuracy. I believe the Bible because of its historic accuracy. Archaeologists have found all kinds of verifications of the absolute accuracy of the Scriptures. I believe the Bible because of its experiential accuracy. ...I've experienced its power in my life. But more than anything, I think, dominating all those reasons why I believe God wrote this book is because of the one singular theme of this book, the Lord Jesus Christ. Forty plus authors writing over fifteen hundred to eighteen hundred years all over the place in all different circumstances writing on their own could never have maintained the continuity of glory and majesty and honor that is given to the Lord Jesus Christ from Genesis to Revelation. This is a book written by one author about one person. Jesus said, "Search the Scriptures for in them you're going to find out about Me." Every book points someway to Christ, from Genesis all the way to Revelation He is the theme. He Himself on the road to Emmaus in Luke 24 opened the Bible and began to speak of things concerning Himself out of the Scripture, the law, the prophets and the writings. You go into the law and He is the one that the law points to. He is the one who will fulfill all law. He is the perfect sacrifice who will bear all sin. He is the one of whom the prophets spoke. He is the one who is to be praised in all the holy writings. He is the one who is wisdom personified. He is the theme of the gospels. His life, death and resurrection are the subject of the epistles and He's the returning King of Revelation.

"The dominating theme of Jesus Christ speaks of divine authorship. And I'll tell you why; because human beings could never conceive of such a person. It's beyond human capability. Look at the gods of the nations, look at the gods of pagan religions. They're like men. There's no explanation for Jesus Christ other than He is who He said He was, He is who God said He is and that He is God in human flesh. There's no explanation for the Bible other than that God wrote it. Nothing else explains its scientific accuracy. Nothing else explains its prophetic accuracy. Nothing else explains its historic accuracy. Nothing else explains its spiritual and penetrating accuracy and its ability to transform lives. Nothing else explains its miracles which are from front to back in this book and verified by eyewitnesses. And nothing else can explain Christ. This monumental, inexplicable, inconceivable personality, who could not be the figment of any human imagination, let alone the collective imaginations of over forty different writers all over the place. This is God's Word."[293]

---

293 (From *The character of God's Word*, by John Macarthur, Grace To You, www.gty.org)

Again, as we said at the beginning of this chapter, the reason the Bible is not widely accepted as true by the intellectual elites who control the means of communication and information dissemination in our world is not due to any real intellect on their part, but rather it is due to a blind and unreasoning prejudice. They are willfully ignorant. The god of this world hates the truth, and his followers blindly follow him to their unfortunate end. Many others in this world have been born and bred and indoctrinated into the lies of a false religion that despises Christianity, Judaism, and the Jewish-Christian Holy book, and therefore are cut off from ever even considering the truth of the Bible. And many others are just plain duped by the lies of their Democrat-Left professors, or are swayed and intimidated by the mockery of ignoramus' like Penn and his halfwit sidekick Teller, leveled against God's Word and against those who, unlike them, have come to the knowledge of the truth.

This is one of the greatest *secrets of the universe* there is today. For it is hidden from the vast majority of the people in this world. Not because this information about this book, or the book itself, is secret. But because the vast majority of people today have been lied to and have been duped into believing that this ancient manuscript has been discredited, that it's "just another book," that it's been "proven to be historically inaccurate," or if you are a Muslim, that it is "the book of the infidels." I pray that the truth that this ancient book actually is the Word of God is no longer a secret to you. If that is the case than you can consider yourself uniquely blessed among the people of this world. And also can consider yourself in a unique position of responsibility to pass this secret along to as many of your family, friends, neighbors and coworkers as you can in a prayerful, kind and considerate way.

One of my sons, while serving in Iraq with the United States Army infantry, sent me a camouflaged, pocket-sized copy of the *New Testament, Psalms and Proverbs* distributed by The Gideons International. The following is written in the front as way of an introduction…

"The Bible contains the mind of God, the state of man, the way of salvation, the doom of sinners, and the happiness of believers. Its doctrines are holy, its precepts are binding, its histories are true, and its decisions are immutable. Read it to be wise, believe it to be safe, and practice it to be holy. It contains light to direct you, food to support you, and comfort to cheer you.

"It is the traveler's map, the pilgrim's staff, the pilot's compass, the soldier's sword, and the Christian's charter. Here Paradise is restored, Heaven opened, and the gates of hell disclosed.

"Christ is its grand subject, our good the design, and the glory of God its end.

"It should fill the memory, rule the heart, and guide the feet. Read it slowly, frequently, and prayerfully. It is a mine of wealth, a paradise of glory, and a river of pleasure. It is given to you in life, will be opened at the judgment, and be remembered forever. It involves the highest responsibility, will reward the greatest labor, and will condemn all who trifle with its sacred contents."[294]

---

294 (Introduction to *New Testament, Psalms & Proverbs*, by The Gideons International)

The Character of God

"God's glory refers both to God's unfathomable beauty and to his incomparable expression of power."[295]

Because of the existence of this ancient Holy book written by God Himself, mankind has easy access to the answers to all of those questions that we were stuck with at the beginning of this chapter; after we were able to determine that there is a God, and that his existence is a clearly delineated scientific fact. The Bible discloses what God is like. His nature and his character are revealed within its pages. He is the Living God, a Spirit, Perfect and Holy. He is not an impersonal, petulant or uncaring god like the gods of false religions ("gods" that are really no different than man, which kind of reveals that it was men who thought them up). He is far above man's sins and shortcomings but he is close enough to us to feel our pain and our suffering and to know our needs. And He knows just how much we actually need him, even though this fact escapes far too many of us, far too much of the time. He is also very patient, slow to anger, not wanting that any should perish without him. He is a personal God. He sees, he hears, he listens, and he answers prayer. He is infinite, eternal, unchanging, all powerful, all knowing, omnipresent, and possesses all wisdom. He is the Heavenly Father, eternal Creator, compassionate, faithful, invisible, beyond comprehension, long-suffering, patient, kind, merciful, slow to anger but capable of irreversible, eternal judgment and wrath for those who die in their sins after rejecting His free gift of eternal life in paradise restored here on this earth. (See chapters 7 and 9) He is also perfectly just, and righteous. He is self-existent and sovereign and his wisdom and knowledge are unsearchable. He *is* love, and he is the truth. He is beyond time and space; they are his mere creations. He is the Beginning and the End. He is in no way limited by his created universe. Yet he resides within the hearts of men (and women); in those who open the door when he knocks.[296] (And He is knocking all of the time.) (Like right now.)

And he has stated clearly within the Bibles pages what our purpose is in being here, and what he expects of us. He created us to glorify him, not because he is vain, but because He knows that is the highest possible expression of our existence, and our love. And he wants us to live to the fullest. We also know from the Books pages that he loves us and cares for us even more than we love our own children. He loves us so much he would die for us. Just like any of us would die for our own children if we had to in order to save one of them.

He knows everything. There is nothing that you can possibly conceive of that he is not aware of. And his knowledge is not partial or limited in any way. It is complete. He knows everything there is to know about any problem you have or any relationship you are involved in. When he acts, and when he answers prayer, he acts and answers based upon perfect and complete knowledge, for he is all-knowing.

And He is all-powerful. He has the power of course to create and sustain this universe, and all the living here on this planet. But beyond that, nothing has ever happened, nothing is happening now, and nothing will ever occur that does not perfectly coincide with His Sovereign Will. Which even includes all of those evil,

---

295   (Pg. 141, *The 100 Most Important Bible Verses*, Thomas Nelson Publishers)
296   (See chapter 3, "The Character of God," from *Fast Facts on False Teachings* by Ron Carlson and Ed Decker)

wicked and sinful things that He temporarily tolerates until the gift of free will that He bestowed upon us has played itself out to its fruition. Until the harvest is complete. Until all of those whom he has chosen, and who have chosen Him, have been brought safely in…

"Another parable he put forth unto them, saying, The kingdom of heaven is like a man who sowed good seed in his field. But while men slept, his enemy came and sowed weeds among the wheat, and went his way. But when the grain had sprouted up, and brought forth a crop, then appeared the weeds also. So the servants of the owner came and said unto him, Sir, did you not sow good seed in your field? From where then did the weeds come? He said unto them, An enemy has done this. The servants said unto him, Would you have us go and gather them up? But he said, No; lest while you gather up the weeds, you root up also the wheat with them. Let both grow together until the harvest: and in the time of harvest I will say to the reapers, Gather you together first the weeds, and bind them in bundles to burn them: but gather the wheat into my barn."[297] (Matthew 13:24-30)

We know what is right and what is wrong because he has written his law on our conscience and in our hearts, and also clearly spelled it out throughout his Word. He gave us the Ten Commandments, and a new commandment that we love one another even as he loves us. He has commanded us to "fear not," to "cast our burdens upon him for he cares for us" and to "love our neighbor as our self." In reading and studying his Word you learn more about him, you understand his Character, and you draw closer in relationship to the One who created you, the One who loved you enough to die for you, and the One who himself desires that you learn to live in intimate fellowship with Him.

The Word of the Lord

"For all flesh is as grass, and all the glory of man as the flower of grass. The grass withers, and its flower falls away; but the word of the Lord endures for ever." (1 Peter 1:24-25)

Another source for determining whether the Bible is what it says it is, is the Bible itself. Just read it. But make sure that you determine to read it because you are prayerfully seeking the truth with all of your heart, and all of your mind, and all of your strength, and all of your soul; because you love the truth more than you love your own opinions. Then you can be assured that He will answer your prayer, because God answers prayer. And a great place to start is with the Gospel of John in the New Testament. Followed by the first few chapters of Genesis. Then the other three gospels, and then the Psalms. And then continue through the rest of His Word. And as you're doing so just ask yourself, are these the Words of God? Or are they just the words of men? Is God speaking to you? Do you hear His voice? …

"My sheep hear my voice, and I know them, and they follow me." (John 10:27)

"If you hear God's word, you belong to him. If you don't hear his word, you

---

297   (The "enemy" is Satan, of course. And if you have any trouble understanding the rest of that parable of the sower and the seed, it will become clear after reading the next chapter.)

don't belong to him. The Bible is the determining factor. The Bible is infallible, inerrant, complete, authoritative, sufficient, effective and determinative."[298]

Here then are a few selections from his Word…

"Hear, O heavens, and give ear, O earth: for the Lord has spoken." (Isaiah 1:2)

"For the word of God is living and powerful, sharper than any double-edged sword, it penetrates even to the dividing of soul and spirit, joints and marrow, it judges the thoughts and attitudes of the heart." (Hebrews 4:12)

"Blessed is the man who walks not in the counsel of the ungodly, nor stands in the way of sinners, nor sits in the seat of the scornful. But his delight is in the law of the Lord, and in his law does he meditate day and night. He shall be like a tree planted by the rivers of water, that brings forth its fruit in its season, whose leaf also shall not wither; and whatsoever he does shall prosper." (Psalms 1:1-3) Blessed is the man who delights in the Words of the Lord. The converse of that is, cursed is the man, or woman, who mocks the Words of the Lord. And our media, our educational system and our entertainment industry do seem to fall into that latter category. But why would anyone let themselves become so deceived that they would reject the wisdom of this book, much less hate it or mock it? I mean, it's such a cool book. It contains the secret of eternal life, for one, which no other book has. The secret to spending an eternity in paradise here on this soon-to-be-re-formed and renewed earth with the God of Creation as your best friend. And we are going to reveal this ultimate secret in the following chapter. Now what could be more awesome than that?

"Be not deceived; God is not mocked: for whatsoever a man sows, that shall he also reap." (Galatians 6:7)

"This book of the law shall not depart from your mouth, but you shall meditate therein day and night, that you may observe to do according to all that is written therein. For then you shall make your way prosperous, and then you shall have good success." (Joshua 1:8)

"The law of the Lord is perfect, converting the soul; the testimony of the Lord is sure, making wise the simple. The statutes of the Lord are right, rejoicing the heart; the commandment of the Lord is pure, enlightening the eyes. The fear of the Lord is clean, enduring for ever; the judgments of the Lord are true and righteous altogether. More to be desired are they than gold, yes, than much fine gold; sweeter also than honey and the honeycomb. Moreover by them is your servant warned: and in keeping of them there is great reward." (Psalms 19:7-11)

"I will worship toward your holy temple, and praise your name for your loving kindness and for your truth; **for you have magnified your word above all your**

---

298   (From "The character of God's Word," by John Macarthur, Grace To You, www.gty.org)

name." (Psalms 138:2) It's probably safe to say that he considers his word very important. And certainly not something to be made fun of by the ignorant, the misled and the misinformed.

The longest chapter in the Bible, Psalm 119, is devoted entirely to praising, appreciating and rejoicing over the invaluable gift of God's Word. His *Word* is also referred to in these verses as his <u>testimonies</u>, his <u>law</u>, his <u>statutes</u>, his <u>commandments</u> and his <u>precepts</u>. Here are some verses from Psalm 119:

"Blessed are those who keep his <u>testimonies</u>, who seek him with the whole heart." (2) "How can a young man cleanse his way? By taking heed according to your <u>word</u>." (9) "Your <u>word</u> I have hid in my heart, that I might not sin against you." (11) "Open my eyes, that I may behold wondrous things out of your <u>law</u>." (18) "Remove from me the way of lying: and grant me your <u>law</u> graciously." (29) "Give me understanding, and I shall keep your <u>law</u>; indeed, I shall observe it with my whole heart." (34) "Let your mercies come also unto me, O Lord, even your salvation, according to your <u>word</u>." (41)

"The proud have had me in great derision, yet I do not turn aside from your <u>law</u>." (51) "Your <u>statutes</u> have been my songs in the house of my pilgrimage." (54) "I entreated your favor with my whole heart; be merciful to me according to your <u>word</u>." (58) "I thought about my ways, and turned my feet to your <u>testimonies</u>." (59) "The earth, O Lord, is full of your mercy: teach me your <u>statutes</u>." (64) "It is good for me that I have been afflicted; that I might learn your <u>statutes</u>." (71) "The <u>law</u> of your mouth is better unto me than thousands of gold and silver." (72) "Your hands have made me and fashioned me; give me understanding, that I may learn your <u>commandments</u>." (73) "Let your tender mercies come to me, that I may live: for your <u>word</u> is my delight" (77) "Your <u>word</u>, O Lord, is eternal; it is settled in heaven." (89) "Oh, how I love your <u>law</u>! It is my meditation all the day." (97) "I have more understanding than all my teachers, for your <u>testimonies</u> are my meditation." (99) "I understand more than the ancients, because I keep your <u>precepts</u>, and meditate on your testimonies all the day." (100) "How sweet are your <u>words</u> to my taste, yes, sweeter than honey to my mouth!" (103) "Your <u>word</u> is a lamp unto my feet, and a light unto my path." (105) "You are my hiding place and my shield; I hope in your <u>word</u>." (114) "Depart from me, you evildoers, for I will keep the <u>commandments</u> of my God!" (115) "Your <u>testimonies</u> are wonderful; therefore my soul keeps them." (129) "Rivers of waters run down my eyes, because men do not keep your <u>law</u>." (136) "Your righteousness is an everlasting righteousness, and your <u>law</u> is truth." (142) "You are near, O Lord, and all your <u>commandments</u> are truth." (151) "Concerning your <u>testimonies</u>, I have known of old that you have founded them for ever." (152) "I rejoice at your <u>word</u>, as one who finds great spoil." (162) "Let my cry come near before You, O Lord; give me understanding according to your <u>word</u>." (169) "I have longed for your salvation, O Lord; and your <u>law</u> is my delight." (174)

"Then **the word of the Lord** came to me, saying..." (Ezekiel 33:23) This phrase, *the word of the Lord*, quoted here from the book of Ezekiel, is also found in nearly every book of the Bible. It is repeated hundreds and hundreds of times from Genesis and Exodus, Numbers and Deuteronomy, Joshua and Samuel, in the book of Kings and of Chronicles, in the Psalms, and in the prophets from Isaiah and

Jeremiah to Daniel and Jonah and Zechariah, and throughout the writings of the New Testament. But yet Satan, the father of lies and the great deceiver, who goes forth to deceive the whole world, has the world believing that the Bible is lying and he is telling the truth! That it is not the word of the Lord. And we believe the worlds all-time greatest liar, instead of believing he who can never lie. Well, let's face it, we've never been accused of being all that bright.

Listen to these verses and see the paramount importance God gives to **His Word**…

"In the beginning was **the Word**, and **the Word** was with God, and **the Word** was God." (John 1:1) "Now the parable is this: The seed is **the word** of God." (Luke 8:11)

"And **the Word** became flesh and dwelt among us, and we beheld his glory, the glory as of the only begotten of the Father, full of grace and truth." (John 1:14) "And many more believed because of [Jesus'] **own word**." (John 4:41) "Most assuredly, I say unto you, he who hears **my word** and believes in him who sent me has everlasting life, and shall not come into condemnation, but has passed from death into life." (John 5:24) "But you do not have **his word** abiding in you, because you do not believe him who he sent." (John 5:38) "Why do you not understand my speech? Because you are not able to listen to **my word**." (John 8:43) "Most assuredly, I say to you, if anyone keeps **my word** he shall never see death." (John 8:51) "Jesus answered and said unto them, If a man love me he will keep **my word**, and my Father will love him, and we will come unto him and will make our abode with him." (John 14:23) "**Your word** is truth." (John 17:17) "Then **the word** of God spread" (Acts 6:7) "Therefore they who were scattered abroad went everywhere preaching **the word**." (Acts 8:4) "Now the apostles and brethren who were in Judea heard that the Gentiles had also received **the word** of God." (Acts 11:1) "**The word** of God grew and multiplied." (Acts 12:24) "This man called for Barnabas and Saul and sought to hear **the word** of God." (Acts 13:7) "On the next Sabbath almost the whole city came together to hear **the word** of God." (Acts 13:44) "And **the word** of the Lord was being spread throughout all the region." (Acts 13:49) "So **the word** of the Lord grew mightily and prevailed." (Acts 19:20) "So then faith comes by hearing, and hearing by **the word** of God." (Romans 10:17) "And take the helmet of salvation, and the sword of the Spirit, which is **the word** of God:" (Ephesians 6:17) "He was clothed with a robe dipped in blood, and his name is called **The Word of God**." (Revelation 19:13-14)

Again… "**The word of God is quick**, and **powerful**, and <u>**sharper than any two-edged sword**</u>, piercing even to the dividing asunder of soul and spirit, and of the joints and marrow, and is a discerner of the thoughts and intents of the heart." (Hebrews 4:12)

The Bible <u>is</u> God's Word. But it was written for those "who have ears to hear." And it was also purposely written in such a way to allow men to deny it if that is their choice. In other words, it allows for man's free will. The evidence is overwhelming of course, and undeniable, that it was written by God, but it is still "subtle" enough, if you will, to allow us to choose to deny and reject and mock and vilify it. Otherwise man would not be free to exercise his own free will. Keep in mind that God has the power to appear out in space before the world in the form of a man the size of Jupiter, standing there on nothing and crying out with a thunderous voice that would shake the earth to its foundations and every man to his core, "I WROTE THE BIBLE! IT IS MY WORD! READ IT! STUDY IT! LET IT BE A GUIDE UNTO YOU!" But He chooses not to. Why? Again, it is because of the incredible, yet double edged gift (depending upon how it's used) of free will that he bestowed upon us. It is what makes us human. If he did appear as shown above, then we would not be <u>free</u> to love him and follow him. We really would have no choice. (Unless you could actually find someone so dense that they would still deny Him even after such an incarnation. Hmm... Bill Maher?) So he remains "hidden" and leaves the decision up to us. But though "hid" he also has left us with more than ample proof of both Himself and his Word. So we are now free to accept or reject Him. And we are free to accept or reject his Word. And, of eternal importance, we are free to accept or reject the priceless gift of his only begotten Son...

See:
- *Fast Facts on False Teachings*, by Ron Carlson and Ed Decker
- "Ten Reasons Why I Believe the Bible is the Word of God," by R. A. Torrey
- "Why Study Bible Prophecy?" by Britt Gillette, from www.rapturealert.com
- "Statistics and Probability," by John R. Funk, from www.rapturealert.com
- "Fulfilled Prophecy: Evidence for the Reliability of the Bible," by Hugh Ross, Ph.D. on www.reasons.org
- www.100prophecies.org
- "Our God-Breathed Bible," by John MacArthur, *Grace To You Ministries*, on www.gty.org
- "You can trust the Bible," a pamphlet by John MacArthur
- *Science Speaks: Scientific Proof of the Accuracy of Prophecy and the Bible*, by Peter W. Stoner
- "The Character of God's Word," a message by John Macarthur
- *There are Moral Absolutes*, by Michael A. Robinson
- *The Big Book of Bible Difficulties: Clear and Concise Answers from Genesis to Revelation*, by Thomas Howe and Norman L. Geisler
- *New International Encyclopedia of Bible Difficulties*, by Gleason Leonard Archer
- *The Politically Incorrect Guide to the Bible*, by Robert J. Hutchinson
- *Temple at the Center of Time*, by David Flynn
- *The Parthenon Code: Mankind's History in Marble*, by Robert Bowie Johnson, Jr.
- *New Testament, Psalms & Proverbs*, by The Gideons International
- *The Signature of God: Astonishing Biblical Discoveries*, by Grant R. Jeffrey
- *The Bible*, by I Am that I Am, the Alpha and Omega, the Beginning and the End, the First and the Last, the Eternal Creator, the Lord of lords, the King of kings, the Everlasting Father, the Prince of Peace, the Holy One, the Most High, the Savior of the world and the Only True God.

CHAPTER 7

JESUS CHRIST

AND

GODS FREE GIFT OF ETERNAL LIFE

"In the beginning was the Word, and the Word was with God, and **the Word was God**. The same was in the beginning with God. All things were made by him; and without him was not any thing made that was made. In him was life; and the life was the light of men. And the light shone in the darkness; and the darkness comprehended it not." (John 1:1-5)

Jesus *is* that Word. He is also the One who spoke both the worlds and His Word (the Bible) into existence. He is the Alpha and Omega, the Beginning and the End, the First and the Last. He is Almighty God, the Author of Life, Holy and True, the Great I Am. He is the Christ, the Anointed One, the Judge of the living and the dead. He is Lord of Lords and King of Kings, the Root and the Offspring of David, and the Bright and Morning Star. He is the Light of the world, and the Way, the Truth and the Life. He is the Bread of Life. He is the Water of Life. He who comes to him shall never hunger; and he who believes on him shall never thirst. He is the Mediator between God and man, the Great High Priest, and the Image of the Invisible God. He is the Door; by him if any man enter in, he shall be saved. He is the Good Shepherd, who lays down his life for the sheep. He is the Lamb of God who takes away the sins of the world. He is the only begotten Son of God, yet if you have seen him, you have seen the Father. He who hears his Word, and believes on the One who sent him, shall have everlasting life, and shall not come into condemnation, but is passed from death unto life. But if you believe not that Jesus is God, then you will die in your sins. For he is the only Savior of this world. No man comes to the Father but through him. He is the Messiah, the Author of Salvation. His name is Salvation in the Old Testament, Yeshua in the Jewish Tenach, and he is promised throughout it. He is the Resurrection and the Life; he who believes in him, though he were dead, yet shall he live. For he and the Father are One and the same.

A Warning

"For what is a man profited, if he shall gain the whole world, and lose his own soul? Or what shall a man give in exchange for his soul?" (Matthew 16:26)

Of all the valuable information God has shared with us in his word, and there are volumes of knowledge unveiled therein, the most important is that of His plan of salvation. In truth it is the entire theme of the Book. Imagine it is a dark and stormy night and you are standing on the side of the road just a few hundred feet from a rusty old bridge that spans a rivers deadly rapids, now even more swollen after days of monsoon rains. Then abruptly the screeching sounds of steel being mangled and twisted pierce the night and you watch amazed as the road collapses into the raging torrent below. A few moments later, before your mind has time to process what you've just witnessed, you see headlights approaching. So you do what any decent person would do. You try to warn the occupants. You wave frantically and shout "the bridge is out!" But to no avail. They cannot hear. So they fly on by, wondering what this nut is doing out in the rain, jumping up and down on the side of the road. But a few seconds later it hits them, and you hear the squeal of wheels and the muffled screams of that terrified family as their car flies off the road and hauls them down to their watery grave. Staring in shock and disbelief, you are overcome with the guilt of being unable to do anything to save them. But then you are jolted back to reality as you hear the sound of another car fast approaching…

The Word of God is like that person on the side of that road. It's shouting out a warning to "the people, nations and languages" of this world. But few there be who are listening. Few there are who have ears to hear. "For many are called, but few are chosen." We rush on through our busy lives, never realizing the fate that awaits us after we die, and never coming to the knowledge of the simple way of escape. We are all of us driving down this well-traveled road. It's called life. And up ahead there is a bridge that is out. It's called death. But for those who are unprepared to face it, it is not just the end of life, it is the beginning of something more horrible than sleep's worst nightmare. The Bible calls it hell or Sheol, the Lake of Fire, a place of unending suffering and torment, of excruciating pain that never ends. A place of wailing and gnashing of teeth, with no hope of escape. Imagine being thrown into a flow of molten lava, but without the hope of death to relieve your agony. Because you are already dead, but suffering forever in actual fire, in hell.

If you are someone who is skeptical about this subject of hell for whatever reason, argument or education that you may have, then I can only suggest that you suspend disbelief long enough to finish this chapter (and the section on hell coming up in a few pages), and long enough to digest the reasons, arguments, facts, logic and evidence from the Light side. There is also a book we would suggest to help you make an informed decision. It is written by someone who claims to have experienced this nightmare firsthand. (You know, like all of those accounts from heart attack and accident victims who have been revived, and come back telling of "going towards the light" in some instances, yet in others of ending up in a place of unspeakable torment and suffering.) It is called *23 Minutes in Hell* and it is written by Bill Wiese, who was chosen by the Lord and sent there through no fault of his own and for that period of time so he could come back and be a witness to "those who do have ears to hear." If anyone is having difficulty with this subject then it might be profitable to take a little time to read his account; as well as our discussion a few pages hence.

But "*how*" I hear someone saying "*could a loving, caring God send anyone to such a horrible eternal fate? There's no way hell can be real. It's just something religious nuts use to try and scare people into believing like them.*" And the answer is, God doesn't send us to hell. We send ourselves there by our own free will, by our own free choice. He has done everything possible to stop us from ending up there, everything within the bounds of allowing for our free will. He himself came down to earth in the form of a man and let us humiliate him, curse him the King of Heaven, falsely accuse him, and then crucify him on a cross where he suffered and died in order to pay the penalty *himself* for our sin. No, he doesn't send us there. It's our choice. And we choose it ourselves by slapping away his outstretched nail-pierced hand, by refusing his offered assistance, by laughing in his face, by foolishly being duped one way or another by the god of this world into rejecting his free gift of eternal life, of rejecting the wonderful alternative- an eternity in a restored paradise here on this planet, dwelling in his presence forever.

"I dreamed I searched heaven for you
In vain I searched heaven for you
Friend, won't you prepare
To meet me up there
Lest I should search heaven for you."[299]

The Scarlet Thread of Redemption

"Jesus said unto her, I am the resurrection, and the life: he who believes in me, though he were dead, yet shall he live: And whosoever lives and believes in me shall never die. Do you believe this?" (John 11:25-26)

The Bible unveils the way that God has provided for us to escape from that approaching eternal horror. And Jesus is that Way. "I am the way, the truth, and the life. No man comes to the Father but by me." (John 14:6) "Neither is there salvation in any other: for there is none other name under heaven given among men, whereby we must be saved." (Acts 4:12) An eternity of excruciating and unbearable pain was never intended for man. It was prepared for the devil and his angels. They have no way out. But man does if he will just stop long enough to look and to listen, and to allow God and his Word to make their eternal case.

Let's start at the beginning...

There is a central message in the Bible which is repeated and expanded upon from beginning to end, from Genesis to Revelation, and it tells of the sinful nature of man and his eternally lost condition on the one hand, and the good news of salvation from this predicament that God has provided on the other. This theme, which has been called "the scarlet thread of redemption," begins in Genesis with the creation of man from the dust of the earth, and the creation of woman from man himself. (And this origin, despite the problem today's feminists have with it, does not make woman less than man, but in a sense, even more valuable; for whereas man was formed from inanimate matter, woman was made from life itself.)[300] They began their lives in the Garden of Eden,[301] a place which God prepared for them on this water planet, surrounded by all the living, and it was a paradise, far beyond anything that exists on earth today. (Think Hawaii on steroids.) They had perfect health. Everything they had need of, and anything they could have desired for their happiness, joy, fun, entertainment, satisfaction, peace and fulfillment, was freely provided. They lived in the presence of their Creator. They spoke with him. They walked with him in the cool of the day. Speaking of being "in love," they dwelled in the presence of Love himself, like a whale dwells in the waters of the sea. God was not hidden or separated from them because sin had not yet entered into the world. Pain, suffering, sorrow and even death, as yet, did not exist.

---

299   (From song by Kitty Wells)
300   (That's called, "Throwing the feminists a bone")
301   (Called the *Garden of the Hesperides* in ancient Greek *Mythology*)

[We should take a moment here to acknowledge the fact that many of our intellectual giants in academia and the media hold what they see as "the mythological fairy tale" of an Adam and Eve in complete contempt. And one reason is because they have bought into the faux science of evolution and the fabricated millions and billions of years used in a futile attempt to prop it up. They know "for a *fact*" that the first man and woman somehow created themselves sloooowly over many hundreds of millions of years from some tall chimpanzees. Even though if you asked them if the universe could ever create the pencil on their desk, starting from scratch, and even if given an infinite amount of time to perform this comparatively simple task, they will say "of course not!" ...Thus displaying their difficulty when it comes to the simple task of putting two and two together.

Also, in addition to being sorely uninformed about the proven scientific accuracy of God's Word (see chapters 6 and 10), they are also unaware (because their media and universities have failed to inform them) of the recent scientific studies on the mutation rate of the human cells mitochondrial DNA, which comes only from the mother, and which now places the original ancestral female of all mankind at only 6 to 7 thousand years ago, and *not* 200 thousand years ago as was previously thought! (See "Mitochondrial Eve" section of chapter 10.) So when the Bible talks about the creation of man as occurring about six thousand years ago, you can trust it as being very accurate in and of itself but also, as we shall see in chapter 10, "The Theory of Evolution is a Joke," it is also very accurate based on all of the scientific facts and evidence now at our disposal. Check it out for yourself. (And leave those mythological, evolutionary fairy tales behind.)]

But when God created man he bestowed upon him that very wonderful and powerful gift we have already spoken of; which was, again, one of the gifts that made us in His own image and likeness; the gift of free will. The freedom to choose. For what pleasure would God have gained from creating a being that was nothing more than a preprogrammed robot? A yes man? (An Eric Holder.) None at all. And we were created for God's pleasure, and to be a reflection of his glory. We were created with the freedom to glorify God and to serve him; to worship and love him, and to please him. Or to choose to do otherwise. So that gift came with the awesome responsibility to choose wisely. And this responsibility remains to this day.

God cautioned Adam and Eve that if they chose to disobey him and to turn from his commandments and to go their own way, the consequences would be severe. They would be driven out of paradise. Death would enter into the world, and with it terrible sorrow and suffering and hardship. And finally, of greatest import, because of their stain of sin they would also be banished from his Presence for all eternity.

"And the Lord God commanded the man, saying, Of every tree of the garden you may freely eat: But of the tree of the knowledge of good and evil, you shall not eat of it: for in the day that you eat of it, you shall surely die." (Genesis 2:16-17)

And this death was not just physical death, it was spiritual death- eternal separation from God and from the paradise he had created and intended for man. God is not dead; nor is he hiding from us today. It is our sin that has separated us from Him, and from his Presence.

Hell

"He who believes on the Son has everlasting life: and he who believes not the Son shall not see life, but the wrath of God abides on him." (John 3:36)

The place of this eternal separation is of course called hell, the Lake of Fire, or the bottomless pit, where "the fire is never quenched and the worm never dies." It is described throughout the Bible as a place of excruciating pain (imagine the unbearable agony for someone who is burned alive) and incomprehensible horror and torment in the company of demons- fallen angels now transformed into evil beings possessed of the blackest hatred for God and all that is good. I wish I could say that hell's suffering will be a little less than that or that hell will be a little more bearable, but if I did I'd be lying. Either the Bible is God's Word or it is not. (And it is.) Either the Bible is telling us the truth or it is not. (And it is.) And here is a small sampling of the truth about hell, of what God's Word says about this place of eternal suffering and torment:

"…but he will burn up the chaff [unrepentant man] with unquenchable fire." (Matthew 3:12)

"…for wide is the gate, and broad is the way, that leads to destruction, and there are many who go in there." (Matthew 7:13)

"…it is better for you to enter into life maimed, than having two hands to go into hell, into the fire that never shall be quenched." (Mark 9:43)

"…the rich man also died, and was buried; And in hell he lift up his eyes, being in torments." (Luke 16:22-23)

"The same shall drink of the wine of the wrath of God, which is poured out without mixture into the cup of his indignation; and he shall be tormented with fire and brimstone [sulfur]… And the smoke of their torment ascends up for ever and ever: and they have no rest day nor night." (Revelation 14:10-11)

"These both were cast alive into a lake of fire burning with brimstone." (Revelation 19:20)

"And the devil who deceived them was cast into the lake of fire and brimstone, where the beast and the false prophet are, and shall be tormented day and night for ever and ever." (Revelation 20:10)

"And whosoever was not found written in the book of life was cast into the lake of fire." (Revelation 20:15)

"But the fearful, and unbelieving, and the abominable, and murderers, and whoremongers, and sorcerers, and idolaters, and all liars, shall have their part in the lake which burns with fire and brimstone." (Revelation 21:8)

The Bible could not be more clear concerning this place of eternal torment. Yet the intellectual-elites of this world see themselves as just way too brilliant to believe in some "religious fairy tale" like hell. But keep in mind that these are the same people who actually believe that the truth is relative, that morals are relative, that God doesn't exist, and that the entire universe, the earth, and all of life were all actually created by Mr. Big Explosion and Mr. Random Chance, purely by accident. They are idiots. (With all due respect, of course.) So if anyone has been intimidated by their mockery, all I can do is encourage you to seek the truth (obviously), to keep an open mind, and to consider the source. (They are idiots.)[302] But beyond the questionable intelligence of those who mock something because they are improperly educated as to its reality, or are afraid to face it because of what it has to say about their future if they continue on their present course; there

---

302 (Also, if the reader feels the need, go ahead and skip ahead to chapter 10, and then come back here after dispelling any doubts as to the scientific impossibility of Satan's genesis.)

is still the necessity of determining, for a fact, whether or not this hell thing is real. Whether it is true or not. And to that end we would first say, and by way of repetition, that the Word of God has conclusively proven itself to be exactly what it says it is. (The Word of the actual Creator.) And since God cannot lie (unlike men or the god of this world, Lucifer), and since the Bible speaks continually about hell as a very real place (see above quotes just for starters), and as one of the main reasons the Book was written in the first place, as a warning to man of a destination to seriously try and avoid, then I think we can pretty much take it to the bank that hell is real, from that source alone.

Secondly, there is also the testimony of those who have been there, temporarily, and who have come back to warn the rest of us. See aforementioned book, *23 Minutes in Hell* by Bill Wiese, along with the testimony and books of others who have been there, done that. Also see the real account (it's history, not a story) of Lazarus and the rich man in Luke 16:20-25. And last but certainly not least see the testimony of Jesus Christ himself throughout the New Testament. Jesus spoke more often of hell than he did of heaven. For example, there is this verse from the gospel of Matthew: "So shall it be at the end of the age: the angels shall come forth, and separate the wicked from among the just, and cast them into the furnace of fire. There shall be wailing and gnashing of teeth." (Matthew 13:49-50) Now you have to ask yourself, who is the liar here? Jesus? Or those who deny the existence of hell?

Also, the very concept of salvation entails the reality of something to be saved *from*. If you are standing on a busy street corner, and the woman next to you begins to step off the curb and you grab her by the arm and pull her back just before the fatal impact of a speeding city bus, then you would have saved her from a horrible and untimely demise. Again, she would have been saved from some*thing*. In the case of eternity, the thing that we need to be saved from is far more horrible and excruciatingly painful than stepping in the path of a speeding bus. Who could bear even the thought of falling into a pit of molten lava, much less the agony of being cast there forever? What could be so horrible that it would cause the Sovereign God of Heaven and Earth, the Creator of the universe, He who spoke the worlds into existence, to come down to earth in the form of one of us, to allow himself to be humiliated, vilified, mocked and crucified, in order to save us from it? Hell, a real literal lake saturated with the unbearable stench of burning flesh and burning sulfur.

And thirdly, it might be helpful for the more scientific and analytical mind to consider the very nature of this universe itself, and that is, it is based on duality. Nothing something; high low; hot cold; space matter; up down; here there; good bad; light dark; big small; God devil; happiness sorrow; us them; subject object; me you; pleasure pain; heaven hell; joy and suffering. Like our electronic computers that work off the binary code—off/on, zero/one—and then go on to create entire mathematical, textual and visual electronic worlds from that simple foundation, the universe was created from a similar mathematical origin. (And so were our brains.) And without getting into a long drawn out dorm room discussion that we wouldn't have the slightest idea how to moderate, suffice it to say for our purposes here that in this life we get a taste of heaven and a taste of hell. Sometimes something happens to someone that is so horrific—like being burned horribly in a fire centuries ago before the advent of morphine or other pain killers and living day to day with constant unrelenting suffering—that we can get a glimpse of the horror of hell. And many times we experience incredibly pleasurable and joy-filled occasions such as falling in love with the right individual, or having a baby, or interacting with our growing little children on a spectacular spring day, that we can get just a glimpse of what heaven will be like. But here we can never fully experience the extremes of either. Until we die. Then we inherit either one or the other. And one thing that God's Word says about heaven is this, "But as it is written, eye has not seen, nor ear heard, neither has it entered into the heart of man, the things which God has prepared for those who love him." (1Corinthians 2:9) In other words, we can't even begin to comprehend or imagine how incredible heaven—the Garden of Eden restored here on this earth—is going to be. And with that in mind just imagine how equally horrible hell is going to be for those who unwisely choose to go there instead of heaven.

Another stumbling block to coming to the knowledge of the reality of hell for some is that they have acquired an unrealistic opinion of their own morality, compassion and sense of justice. This can be evidenced by various arrogant and uninformed declarations such as: "There is no such thing as hell because I believe that God, if he exists, is a loving God and wouldn't send anybody to such a horrible place." Of course, if the author of that declaration were God, he or she would have done so much better. They would have created a so much better world than what God came up with. They would have just let everyone do whatever they wanted to and whenever they wanted to do it with no consequences no matter how much pain they inflicted on others and evil they brought into the world, and with no One to answer to. (Kind of like our public school system, now that the ACLU has destroyed it.) Their idea for a universe would be so much wiser than the actual Creator's plans and purposes, and of course they would be so much more *loving* and *compassionate* than God. What a hoot! What an arrogant delusion! But these are some of the very thoughts that steer people away from any chance of embracing their only hope of salvation provided by a truly loving and compassionate and all-knowing and infinitely intelligent and merciful and just Creator.

Also, when pondering the concept of perfect justice, consider this: Nero, emperor of Rome during the first century, was fond of taking Christian men, woman and children and dipping them in wax, tying them to posts and then setting them on fire at night to light his garden. Now what does perfect justice demand of Nero? Where do you think he has been passing the time for the last nineteen hundred and fifty years? In a garden? With 72 virgins? Or in a Lake of Fire? How about Hitler? And Stalin? And Pol Pot, Mao and Che? How about Mohammed and his suicidal murdering followers? How about Tiller the baby killer? And Teddy Kennedy, that great champion of unfettered baby mutilation? What does perfect justice demand of those? No, let us not be so arrogant, so foolish, so pompous and so uninformed as to raise our infinitesimal notions of love, compassion, mercy and justice over those of the God of all Creation, the perfect Judge, who coincidentally is the *Source* of all love, compassion, mercy and justice.

And finally then, with all that said, it is our conclusion—based on all of the facts and all of the evidence and all that we can know to be true in this life, and based on the scientific fact that there is a God and that he has revealed himself and his plan for our eternal destiny in a Book that has also been conclusively determined beyond a shadow of doubt to be one that He did write—that in the next life there exists the absolute extremes of both an unimaginably wonderful heaven and a diabolically horrible and excruciatingly painful hell. The world would be wise to open its eyes sometime to the reality of this. Before it dies…

The Scarlet Thread (continued)

"Jesus answered and said unto her, Whosoever drinks of this water shall thirst again:

But whosoever drinks of the water that I shall give him shall never thirst; but the water that I shall give him shall be in him a well of water springing up into everlasting life." (John 4:13-14)

Where were we? Oh yes… God warned Adam and Eve[303] of the terrible consequences of sinning against him. And I would imagine that for some period they dwelled in that Paradise before they fell, but the Bible does not reveal what that length of time was. (We're guessing six years.) What it does say is that a time came when their willingness to obey God was put to the test. That old serpent, the devil, the great deceiver "who goes forth to deceive the whole world," approached Eve when she was alone in the garden:

"And the serpent …said to the woman, Did God really say, You shall not eat of every tree of the garden? And the woman said to the serpent, We may eat of the fruit of the trees of the garden: But of the fruit of the tree which is in the middle of the garden, God has said, You shall not eat of it, neither shall you touch it, or you will die. And the serpent said unto the woman, You shall **not** surely die:" (Genesis 3:1-4)

Woe! "And the serpent said unto the woman [are you ready for this] you shall **NOT** surely die." How incredible is that! He called God a liar! Then Satan continued to deceive the woman even further… "For God knows that in the day you eat of it your eyes will be opened and you will be like God, knowing good and evil. And when the woman saw that the tree was good for food, and that it was pleasant to the eyes, and a tree desirable to make one wise, she took of its fruit and ate." Woe again! He calls God a liar, and then makes up a cock and bull story about *why* God would lie to them, supposedly to cheat them out of an even "better deal" than what they had been given, that of being just like God. And, unfortunately for her and all of her descendants, she bought it. She believed him! She took the word of Satan over God! (As far too many folks are still doing today.) And Adam was no better than Eve… "she also gave to her husband with her; and he did eat. And then the eyes of both of them were opened, and they knew that they were naked." (Genesis 3: 5-7)

> [And what of the prophecy "for in the *day* that you eat of it, you shall surely die?" Well, in that *day* they *did* surely die, not just spiritually but physically as well. For a day unto the Lord is as a thousand years, and a thousand years as one day: "But, beloved, be not ignorant of this one thing, that one day is with the Lord as a thousand years, and a thousand years as one day." (2 Peter 3:8) Adam was 930 years old when he died, and no one before the flood, when the lifespan of man was far greater than it is today, lived to 1000 years of age. Many came close, but all fell short. So, both spiritually and physically, the prophesy of God was fulfilled. For in that *day* (24 hours) they died spiritually. And in that *day* (thousand years) they died physically.]

---

303   (Known as Zeus and Hera in ancient Greek "Mythology")

In these early passages in Genesis we see the original lie coming from the mouth of the father of lies... "You shall **not** surely die." And less we be too hard on Adam and Eve for falling for it, keep in mind that this lie is still being broadcast, and fallen for today: "It's OK. Go ahead! Do what you want! Do your own thing! Don't listen to those *Bible thumpers*! They'll just bring you down!" (and here's the killer...) "C'mon, you can't tell me with a straight face that a loving God would send anyone to hell!!!" (He doesn't, Mel, people send themselves there by rejecting his free gift of eternal life.) "Besides, *there's no such thing as hell* anyway." ...And blah, blah, blah.

And more incredibly still, even though we have a whole lot more information available to us today than the first man and woman did, and even though we now have six thousand years of recorded history that clearly show the catastrophic results— the evil, the death, the destruction, the disease, the barbarity, the oppression, the heartache, the suffering and the pain—that came into the world because our first ancestors believed the bald-faced lies of Satan; incredibly, mankind for the most part still buys his lies today! For, again, the lies of Beelzebub blanket this earth, and people believe what they are taught. So let's not be too hard on great-great-great...great grand mom and granddad. They might have been dumb, but we're even dumber.

After their temptation by Satan and their fall into sin, Adam and Eve "heard the voice of the Lord God walking in the garden in the cool of the day," and for the first time they knew fear... "And Adam and his wife hid themselves from the presence of the Lord God among the trees of the garden." ...Then God said to Adam, "Because you have heeded the voice of your wife, and have eaten from the tree of which I commanded you, saying, you shall not eat of it: cursed is the ground for your sake; in toil you shall eat of it all the days of your life; thorns also and thistles shall it bring forth to you, and you shall eat the herb of the field. In the sweat of your brow shall you eat bread, till you return to the ground; for out of it were you taken; For dust you are, and to dust you shall return." (Genesis 3:17-19)

"Cursed is the ground" because of their sin. "In toil ...in the sweat of your face..." We all know how tough life is, and that is why. In one hour man went from paradise to a constant struggle to stay alive and clothed and fed and sheltered. And keep in mind that modern man, especially in America, really has no concept of just how tough life has been for the overwhelming majority of those who have lived in the last six thousand years. For even the most impoverished among us in this country today would still be considered rich, and as having a very easy life, compared to the brutally hard existence that most of the people on this earth in the past have struggled with; and that most around this globe still struggle with to this day.[304] So, truly it can be said that on this earth the *ground* has indeed been *cursed*.[305]

Because of their sin Adam and Eve were cast out of paradise and forced to live in a fallen, cursed world. They could no longer live in God's presence because they had the stain of sin upon them. And this stain of sin that separated them from

---

304 ("Be thankful for the good things that you got. For the good things that you got are for many just a dream, so be thankful for the good things that you got.")
305 (In addition, all of the material universe, all of matter, was also cursed for this was when the second Law of Thermodynamics was brought into existence)

God would be passed down to each and every generation. This is the Doctrine of Original Sin and despite numerous attacks on its validity, in particularly by 17th and 18th century philosophers from the so-called Ages of Reason and Enlightenment, it still stands not only as well-supported by the Bible itself but also by the exercise of pure reason. As Jonathan Edwards eloquently pointed out in…

> …"his masterful treatise …in the 18th century [where] he argued for the doctrine of original sin not only exegetically which he did do, showing the Biblical case for original sin, but he also has a lengthy section in his treatise on *Original Sin Proved by Natural Reason.* And again to simplify this, what Edwards was saying was if the Bible didn't tell us anything about a historical fall, or anything about the doctrine of Original Sin, pure unvarnished reason would have to construct a theory of Original Sin to account for the universality of corruption and that all people everywhere struggle with sin. If we're all born neutral and not fallen one would expect not to find this kind of universality of human corruption."[306]

The reply to this by some enlightenment philosophers is that "we are all born free of original sin and born actually in a virtuous state," but that it is *civilization, society* or *government* that corrupts man. Of course the naivety of this argument is demonstrated by the following question, "How did civilization, [society or government] get corrupted in the first place? Because civilization or society is simply made up of individuals."[307]

So I think it is safe to move forward and to accept as logically and historically accurate what God's Word says in Romans, that: "Wherefore, just as by one man [Adam] sin entered into the world, and death through sin; and so death passed upon all men, inasmuch as all have sinned" (Romans 5:12) We are all born in the original sin of our original parents.[308]

Adam and Eve were not only cast out of the Garden of Eden but they also became subject to a much more terrible fate, that of being separated from God for all eternity. But fortunately for them and for all of the rest of us, on the very day of their fall, God provided a means of escape, a way of salvation from this terrible fate. God prophesied to Satan on that day… "Then the Lord God said unto the serpent, Because you have done this, you shall be cursed above all animals, and above every beast of the field; upon your belly you shall crawl, and dust you shall eat all the days of your life: And I will put enmity between you and the woman, and between your offspring and hers; **he will strike your head**, and you shall bruise his heel." (Genesis 3:14-15) This is called the *Protoevangelium*, or the first prophecy of many that point to the work of salvation to be accomplished through Jesus Christ. It is the first of many Old Testament prophecies about the Messiah, the Savior of mankind, all of which were precisely fulfilled in the birth, life, death and resurrection of Jesus Christ. (See previous chapter.) It is also the beginning of the *scarlet thread of redemption* that weaves its way throughout the Bible and builds with each successive revelation. The Protoevangelium prophesied the solution that God would implement to redeem fallen man, and provide a way for him to enter back into eternal fellowship with him in the Garden (paradise, heaven) as was originally planned. God foretold that the seed of the woman, her

---

306   (From "The Enlightenment (Pt. 2)," by Dr. R. C. Sproul, Renewing Your Mind series)
307   (From "The Enlightenment (Pt. 2)," by Dr. R. C. Sproul, Renewing Your Mind series)
308   (And no, that does not mean that babies or innocent young children go to hell if they die. They do not)

offspring, Jesus Christ, would strike Satan with a mortal wound to the head, and that Satan would strike Christ with a lesser wound, both of which happened at Calvary. There Satan bruised Jesus' heel by having him put to death, but Christ crushed Satan's head and his goal for all of mankind's eternal destruction by dying on the cross for our sins, substituting himself for us and our condemnation, paying the penalty for our sin and the sin of our father Adam, reconciling the world to his Heavenly Father, and finally triumphing over the death of the cross and over death itself by rising from the dead three days later.

[An important note: An opening for critics of the Bible has been presented because of the apparent discrepancy regarding the number of days between Jesus' crucifixion and his resurrection three days later. And that is because it is generally accepted throughout Christianity that Friday was the day of the Lord's crucifixion and death. Hence the world wide celebration of "Good Friday." But this is contradictory because the Bible clearly says that Jesus was in the "belly of the earth" for three days and three nights. ("For as Jonah was three days and three nights in the belly of the whale, so shall the Son of man be three days and three nights in the heart of the earth." Matthew 12:40) And we know that Jesus rose on Sunday, just two days from Friday. Is the Bible wrong, and therefore its critics right that it is not accurate? No, it is Christian tradition and not the Bible that has gotten it wrong. And here is why: Jesus was arrested Passover night, and crucified the next day. And we know that Passover was Thursday. There is no argument about that. But the historical mistake occurs in thinking that the last supper or Passover supper occurred on Thursday evening. And that is not true. Because if we understand the Jewish day, of when it began and when it ended, and indeed if we understand when a day actually began and ended as ordained by God from the very first week of Creation, we can solve this apparent dilemma. The Jewish day began and ended at sundown or sunset, as indeed did each day in the Creation week—"And the **evening** and the morning were the first day" Genesis1:5, 8, 13, 19, 23, 31—and not at midnight or at dawn as has been our custom for some time. Therefore it could not be the case that Passover supper occurred on Thursday evening as is generally thought, because then the last supper would not have taken place on Passover. And that is because for the Jews both Passover and Thursday ended at sundown on Thursday evening. Therefore the last supper, as did all Passover suppers, took place the night before the day, in this case on Wednesday evening, because that was the beginning of Passover. This is also a part of the historical record for the celebration of Passover in Biblical times, or for the start of the Sabbath, the seventh day of rest. Both commenced from sundown. Therefore all of the events attributed to Thursday night and to "Good Friday" actually occurred on Wednesday night and on "Good Thursday" (if we may call it that). This is not only historically accurate but also coincides with Scripture which we know to be true, that says Jesus was in the "belly of the earth" for three days and three nights.

Another reason for the historical error of thinking Christ was crucified on Friday is that the Scripture clearly says (referring to the day after Jesus' crucifixion) that "because **the next day was the Sabbath day.**" (Mark 15:42) And since the weekly Sabbath day was Saturday, the seventh day of the week, it was assumed that the next day referred to in the aforementioned Scripture was Saturday, and that the previous day, the day of the Lord's crucifixion, therefore had to be Friday. But this is also in error. Because besides the weekly Sabbath days that of course occurred every Saturday throughout the year, there were in addition a number of **special** Sabbath days that occurred on other days of the week during certain Jewish feasts. And during this feast of Passover, *Friday* **was** a **special** Sabbath day. So in fact, the **next day** spoken of in Scripture was not a Saturday or ordinary Sabbath day, but it was <u>Friday</u>, the *special* Sabbath day referred to! See John 19:31 NIV referring to Jesus' death: "Now it was the day of Preparation, and the next day was to be a **special** Sabbath. Because the Jews did not want the bodies left on the crosses during the Sabbath." So, it is a historical fact that the Last Supper took place on Passover evening or <u>Wednesday</u> night; that Jesus was crucified and died the following day Thursday; that the "next day" Friday was a *special* Sabbath day; that he was in the grave for three days and three nights (Thursday and night, Friday and night, Saturday and night) as he himself had prophesied, and that he rose from the dead *three* days later on Sunday morning.

And the score is still, God: a whole lot, Bible critics: nothing.]

The Bad News

By one man, Adam, sin entered into the world, and death because of sin, and eternal death as well. And this curse of eternal death in hell passed upon all of mankind, because all of us have sinned. Everyone is a sinner, as well as Eve and Adam. ("For all have sinned, and come short of the glory of God." Romans 3:23)

But here's the good news, "For God so loved the world, that he gave his only begotten Son, that whosoever believes in him should not perish, but have everlasting life. For God sent not his Son into the world to condemn the world; but that the world through him might be saved. He that believes on him is not condemned: but he that believes not is condemned already, because he has not believed in the name of the only begotten Son of God. And this is the condemnation, that light is come into the world, and men loved darkness rather than light, because their deeds were evil." (John 3:16-19)

Jesus reversed the curse:

"Adam came into the world by supernatural means; he was not born like you and me. Jesus also came into the world by supernatural means. Adam's "father" was God, for He created him. Jesus' Father is God. Adam entered this world without guilt. Jesus came into this world without sin and he also left it without sin. Adam lived in closest fellowship with God before the fall of man. Jesus also lived in closest fellowship with the Father. Adam lived in a perfect world before the fall of man. Jesus also lived in a perfect world before His incarnation. He was equal with God (compare Philippians 2:6). We lost our innocence through Adam. We lose our guilt through Jesus. Sin entered the world in a garden, the Garden of Eden, where man decided for himself, against the will of God. The Son of God began to overcome sin in a garden, the Garden of Gethsemane, where he decided to do the will of God, "not my will, but thy will, be done.""[309]

The Nation of Israel

"The woman said unto him, I know that Messiah will come, who is called Christ: when he is come, he will tell us all things. Jesus said unto her, I who speak unto you am He." (John 4:25-26)

But we have jumped ahead. God's scarlet thread of redemption wound its way through a number of other events from the time of the fall of man and before Jesus' triumph on the cross four thousand years later. After that fateful day in the Garden of Eden, the next major step in God's salvation plan was to establish a nation of people that would be separated from all other nations, a unique people, through

---

309   (From "The Miracle of the Seven Statements of Jesus on the Cross," by Norbert Lieth, Midnight Call Magazine online, www.midnightcall.com)

whom God would unfold his will. Thus he called a man, Abraham, to leave the country of his birth and to journey to a foreign land. "The Lord said unto Abram, Go forth from the land of your kinsfolk, and from your father's house, unto a land that I will show you: And I will make of you a great nation, and I will bless you, I will make your name great, so that you will be a blessing. And I will bless those who bless you, and curse those who curse you: and through you shall all families of the earth be blessed." (Genesis 12:1-3)

["And I will bless those who bless you"—this is one of the reasons why the United States temporarily continues as the most prosperous nation on earth (and despite mutilating 50 million babies to death), because of our defense of the tiny nation of Israel which is surrounded by an overwhelming horde of Mohammedans sworn to their annihilation. "And I will curse those who curse you"—conversely this is why the Arab nations are among the most backward, poor and miserable people on earth even though they sit atop a gold mine of oil, and why they are imprisoned by a horrendous, evil religion that slices the clitoris off of hundreds of millions of their own little girls. …Prophecy does have a way of being fulfilled.]

"And through you shall all families of the earth be blessed." God prophesied to Abraham that through his offspring, the future nation of Israel, the Messiah would come, the Savior of the world, who would pour out His blessing of salvation to all who through faith would receive it. The Hebrew Scriptures, the Christian Old Testament, contains hundreds of specific prophecies about this Messiah, so that when the time came for him to appear, there could be no confusion as to who he was. There could be no mistaking his identity. Jesus Christ, in his birth, his life, his death and his resurrection, fulfilled all of these prophecies. He is the promised Messiah not only for the Jewish people and the nation of Israel but for all mankind.

The nation of Israel was instructed to offer, on a regular basis, unblemished lambs on the alter as a sacrifice and atonement for their sins. This of course was prophetic, for Jesus, "the lamb of God who takes away the sins of the world," was the sinless sacrifice, the unblemished lamb that once and for all would pay the penalty for the sins of mankind. God so loved the world, and therefore so wanted to provide a way for man to escape hell and live with him forever in paradise here on this earth in the renewed Garden of the Hesperides, that he came here to die in our place. He was our substitution. For as a Holy God and as one who cannot lie, he could not just overlook our sin nor could he just forget our transgression. The penalty had to be paid. The justice of an absolutely Just God demanded it. So in essence, by coming to earth and living a sinless life, and by allowing himself to be nailed to a cross and crucified for our sins, God himself as the perfect unblemished sacrifice, paid the penalty for us. For Jesus Christ and God the Father are one and the same.

"And all this is from God, who has reconciled us to himself through Jesus Christ, and has given to us the ministry of reconciliation; namely, that God was in Christ, reconciling the world unto himself, no longer counting their trespasses against them" (2 Corinthians 5:18,19)

"For even the Son of man came not to be ministered unto, but to minister, and to give his life a ransom for many." (Mark 10:45)

The exact place of the Messiah's birth was prophesied. "But thou, O Bethlehem

Ephrata, though you are little among the thousands of Judah, yet out of you shall he come forth unto me that is to be ruler in Israel; whose goings forth have been from of old, from everlasting." (Micah 5:2) And was of course fulfilled in Jesus Christ

The unique manner of the Messiah's birth was also prophesied. "Therefore the Lord himself shall give you a sign; Behold, a virgin shall conceive, and bear a son, and shall call his name Immanuel." (Isaiah 7:14) And again was fulfilled in Jesus Christ. Immanuel means *God among us*. This prophecy was made 700 years before the virgin birth of Jesus in Bethlehem.

[Also, keep in mind that without this virgin birth, if Jesus had been conceived by the seed of a man, then he too would have been born with original sin, and he would have been a *blemished* lamb, and one unsuitable to be sacrificed for our sins. But it was the Holy Spirit (or God Himself) and no man who caused Mary to be with child. So the virgin birth of Jesus, besides being a well-documented historical fact,[310] is also absolutely essential to the entire Christian faith. For without it we are all still dead in our sins. Someone might want to explain that to all of those liberal, so-called Christians who don't feel it is all that important to believe in the miracles of God's Word, including the crucially important one for our salvation, that of Jesus' virgin birth. Wake up! My uneducated faux-Christian brothers and sisters!]

Scripture also prophesied the exact year of his death: "From the going forth of the command to restore and build Jerusalem [which occurred in 445 B.C.] until Messiah the Prince, there shall be seven weeks [a *week* in prophecy means a seven year period, so this would be 49 years and take us to 396 B.C.] and sixty-two weeks [another 434 years would take us to 32 A.D. and the crucifixion of Christ!]... And after the sixty-two weeks Messiah shall be cut off, but not for himself." (Daniel 9:25-26) "But not for himself" because he died ("was cut off") for the sins of mankind and not because of any transgression he himself was guilty of. Again we see the amazing foreknowledge of Biblical prophecy.

Jesus was born to die for the sins of mankind. We celebrate his birth on Christmas day and we give one another gifts as a memorial to the greatest gift of all, the gift of eternal life that Jesus gave to us, to all who will gladly receive it. He shed his blood for us, for the remission of our sins. "And almost all things are by the law purged with blood; and <u>without shedding of blood there is no forgiveness</u>." (Hebrews 9:22) "For this is my blood of the new covenant, which is shed for many for the remission of sins." (Matthew 26:28) The miracles of Jesus were shocking and unbelievable for those who witnessed them, and they displayed the power, majesty and mercy of one who was truly "Immanuel," God among us. The eternal words that he spoke, from the mouth of One who was "wiser than Solomon" (what an understatement), were powerful and sharper than any double-edged sword. But without Jesus' death on that cross and his resurrection from the dead three days later, his entire life would have been of little use to any of us, for we would still be trapped in our eternal predicament, dead in our sins, and under the penalty of everlasting punishment.

The Good News

"Verily, verily, I say unto you, he who hears my word, and believes on him

---

310   (remember it is nearly impossible to pass down a lie throughout history, or Buddha, Joseph Smith and Mohammed would all have walked on water)

who sent me, has everlasting life, and shall not come into condemnation; but is passed from death unto life." (John 5:24)

But the good news is that Jesus *did* suffer and die on that cross 2000 years ago. And three days later he *did* rise from the dead. And it would be good to point out here that the resurrection of Jesus Christ, along with his life and death, far from being some "ancient religious fable" or fabrication, is actually one of the best supported events in all of history:

> "The evidence for the life, the death, and the resurrection of Christ is better authenticated than most of the facts of ancient history."[311]

Nevertheless, the lies concerning the resurrection of Jesus began the day the stone was rolled away from the tomb. The Roman soldiers were deathly afraid because to lose someone they were guarding, even a dead body, meant their execution. So they went to the chief priests and told them what had happened. Then "when [those priests] had assembled with the elders and had taken counsel, they gave a large sum of money to the soldiers, saying, "Tell them [their military superiors], 'His disciples came by night, and stole him away while we slept.'"" (Matthew 28:12-13) This story of course was ludicrous because the disciples would have been slaughtered in that idiotic and vain attempt, so they never would have attempted such a daring and impossible mission to begin with. And for what? To snatch what for all of them at that point was nothing more than a dead body?! They were all running scared after scattering from the Garden of Gethsemane three and a half days before. With all due respect to all of them, the "dead body" of Jesus was the last thing on their minds. And besides, Roman soldiers would never have been "sleeping" because again if that massive stone was rolled away from the mouth of the tomb and the body disappeared they would have paid for it with their life. (What did the disciples do, sprinkle pixie dust to make all of the soldiers fall into a catatonic sleep, so deep that they couldn't hear a massive boulder being rolled aside by a bunch of hysterically fearful disciples?) But the lies of Satan don't have to make any sense, they just have to be shouted loud enough and long enough so enough uninformed people will believe them. Then, in addition to the hefty bribe, the soldier's minds were put at ease concerning the risk of the commander of their legion executing them for "falling asleep" and allowing their prisoner to "escape." If that were to happen, the priests said "we will persuade him, and secure you. So they took the money, and did as they were taught: and this saying is commonly reported among the Jews until this day." (Matthew 28:14-15) Never mind that it was all a bunch of crap. Satan will say absolutely anything, no matter how asinine and downright stupid on the face of it, as long as he can get away with it and he can keep people from coming to the knowledge and the embrace of the saving grace of Jesus Christ as their Messiah, Lord and Savior.

Jesus Christ, his virgin birth, his life, his many miraculous signs and wonders, his death, his burial, his resurrection from the dead three days later, and his fulfillment of all of the many specific prophecies in the Old Testament of the Bible: these are all proven historical facts, just as Julius Caesar's life and death and his accomplishments are proven historical facts. If you can believe that Julius Caesar was exactly who history said he was: the emperor of Rome and not a pizza delivery boy from Naples, then you can believe that Jesus Christ was exactly who history said *he* was as well. They are both a part of the historical record; not myth,

---

311   (E. M. Blaiklock, Professor of Classics Auckland University)

not wishful thinking, not religious beliefs. They are both historical facts. The only difference between the two is that Satan and his academic dupes have no reason to try and lie about the life of Caesar. Caesar is of no concern to the god of this world. Jesus, however, is; which is why the great deceiver seduces "educated" men, in positions of great influence, into assaulting Jesus' reality by questioning his place in history. (Which leads us back to the folks at *The Jesus Seminar*. They have no interest whatsoever in finding the "real, historical Jesus." For he's already been found, by anyone with ears to hear.)

> "There exists no other document from the ancient world, witnessed by so excellent a set of textual and historical testimonies... Skepticism regarding the historical credentials of Christianity is based upon an **irrational bias**."[312] (Emphasis mine)

Or consider this from Professor Thomas Arnold, author of the *History of Rome* and appointed to the chair of modern history at Oxford:

> "I have been used for many years to study the histories of other times, and to examine and weigh the evidence of those who have written about them, and I know of no one fact in the history of mankind which is proved by better and fuller evidence of every sort, *to the understanding of a fair inquirer* [emphasis mine], than the great sign which God hath given us that Christ died and rose again from the dead."[313]

And then there are the words of the Lord Himself, who prophesied to his disciples what his end would be, for he knew the reason he came to this earth: "Behold, we go up to Jerusalem, and the Son of man shall be betrayed unto the chief priests and unto the scribes, and they shall condemn him to death, and shall deliver him to the Gentiles to mock, and to scourge, and to crucify. And the third day He shall rise again."[314] (Matthew 20:18-19) "And he began to teach them, that the Son of man must suffer many things, and be rejected of the elders, and of the chief priests, and scribes, and be killed, and after three days rise again." (Mark 8:31) "You know that after two days is the feast of the Passover, and the Son of man is to be betrayed to be crucified." (Matthew 26:2) "The Son of man must suffer many things, and be rejected of the elders and chief priests and scribes, and be slain, and be raised the third day." (Luke 9:22) "And as Moses lifted up the serpent in the wilderness, even so must the Son of man be lifted up:" (John 3:14) "For as Jonas was three days and three nights in the belly of the whale; so shall the Son of man be three days and three nights in the heart of the earth." (Matthew 12:40) Those who refuse to accept Jesus' resurrection as the historical fact that it is are basically calling Him a liar, rather than themselves. Personally, I wouldn't recommend that, but everyone does have free will.

"And the angel answered and said unto the women, why seek you the living among the dead? He is not here, but is risen." (Matthew 28:5, 6)

"What must I do to be saved?"

"Neither is there salvation in any other: for there is none other name under heaven given among men, whereby we must be saved." (Acts 4:12)

You see, people don't refuse to believe in Jesus because of a lack of evidence.

---

312 (Clark Pinnock, McMaster University)

313 (Professor Thomas Arnold, author of the *History of Rome* and appointed to the chair of modern history at Oxford) (Is there an echo?)

314 (An excellent book on the subject of the historical accuracy of the resurrection of Christ is, *The Resurrection of Jesus: A New Historiographical Approach*, by Michael Licona)

Quite the contrary. God would never have given us so great a salvation without also giving us ample proof, both historical and otherwise, to support the bona-fides of that great gift. No, people refuse to believe in Jesus because the great adversary of Jesus—"the great dragon …that old serpent, called the Devil, and Satan, who deceives the whole world" (Revelation 12:9)—has duped them into rejecting the greatest Christmas gift of all time. And some are not happy at just rejecting it themselves. They want to make certain that no one else receives the gift as well. One of the more famous of those, Ted Turner, has publicly stated that "Christians are losers." But he of course is deceived. Christians are not losers. True Christians, that is. They may be many things. Some are financially poor. Others are timid and meek. Many are not highly educated. Those in Muslim, atheist-communist, and some Western European countries are terribly persecuted. All are sinners. But true Christians are the only winners that there are on this planet. Eternal winners in life's grand lottery. Recipients of Jesus' free ticket to eternal life. (As in… ADMIT ONE TO PARADISE.) Mr. Turner, however, is a loser, no matter how many billions of dollars he may give away to the corrupt, Jew-hating United Nations in a vain attempt to sooth his conscience. (Or *buy* his way into the afterlife. He might be trying to hedge his bets.) Because Ted will soon be weeping and gnashing his teeth in unending torment and unbearable pain that he will nonetheless have to bear for all eternity, unless through the grace of God he can repent and turn to Jesus and accept that free gift of eternal life and become another happy "loser" like the rest of us followers of Christ. We be a-prayin fur ya, Theodore.

The salvation God has provided for us is a free gift, because there is nothing we could ever hope to do to earn it if it wasn't free. "For God so loved the world, that he **gave** his only begotten Son, that whosoever believes in him should not perish, but have everlasting life." (John 3:16) Gave! He gave his only begotten son. It's a gift. It only has to be received through ones belief and faith in the Giver of the gift. If you don't believe that a gift exists, how can you possibly receive it? Hence the importance of believing.

"Getting saved is trusting in Jesus, that he is who he said he was, and that he will do what he said he would do (and that is give you eternal life) if you put your trust and faith in him." If you trust in his mercy and his grace. Mercy is not getting what you do deserve. (Hell) And grace is getting what you do not deserve. (Heaven)

Of all the gifts that God has given us—life, love, imagination, children, companionship, free will, sports, intelligence, sex, 3D sight, friendship, intimacy, creativity, hearing—the greatest of all is the gift of eternal life. We accept all of his other gifts without question. Why would anyone be so stupid as to reject the most important gift of all? And what did any one of us do to deserve the gift of life? Not a thing. And what did any one of us do to deserve all the other gifts? Nothing either. And in the same way we can do nothing to *deserve* the gift of eternal life, nor could we ever hope to do anything to *earn* it. All anyone can do, if they have any wisdom, is to humbly accept it.

This is not just *one* of the greatest secrets of the universe. This is it! This is the big enchilada! This is undoubtedly the greatest secret of the universe to the great number of people living in this world today. And again, it is not because this incredibly vital information is actually *secret*. It has been shouted from the mountaintops for 2,000 years. But the fact that this information is actually true (and not just some *religious fable* or some "infidel's religion") is still a *secret*, sadly, to the vast majority of people living today. "But if our gospel be hid, it is hid to them who are lost: In whom the god of this world [your basic Satan] has blinded the minds of them who believe not, lest the light of the glorious gospel of Christ, who is the image of God, should shine unto them." (2 Corinthians 4:3-4) They have been duped by one variation or another of Satan's original lie to Eve in the garden... "You shall not surely die." (What is not recorded in the Bible is what Satan snickered underneath his breath right afterwards... "suckers")

Now if you, the reader, are hearing about this information here for the first time, or perhaps in a way that hasn't had the same effect before, then we implore you not to go forward in this book, or in life any longer without making the most important decision that you ever could. And that is to decide right now to be saved; to accept Christ and his free gift of eternal life. To come from darkness to the Light. To come from the seductive lies of the serpent to the knowledge of the Truth. To escape the fate of the Devil and those many poor unfortunate souls who are being drawn by him into that same horrible eternal fate.

And "what must I do to be saved?"

"And the keeper of the prison... fell down trembling before Paul and Silas ... and said, Sirs, what must I do to be saved? And they said, Believe on the Lord Jesus Christ, and you shall be saved." (Acts 16:30-31) It's that simple. All you have to do is believe in the One who is "the author and finisher of our faith, who for the joy that was set before him endured the cross, despising the shame, and sat down at the right hand of the throne of God." (Hebrews 12:2)

"Wait a minute! There is no way eternal life is that easy to attain? I'm sure there must be something else! Don't I have to say a thousand Hail Mary's? How about crawling to Mecca on my knees? How about meditating in the lotus position for years in a Tibetan cave? How about donating all of my money to the poor? How about trying to be really good for the rest of my life?"

No, you don't. Because there is nothing that you can **do** in order to get to heaven! Nothing! If you could do something to earn eternal life than God would not have sent his Son here to pay the price for your sins? Why bother? Not if you could "pay the price" yourself. No, the only thing that you can do is to have faith that he did it for you. That he paid the price. And with that faith in him, with that trust in his Word, you gladly accept that gift and then go forward in life following him and striving do his will and seeking how you can best serve him. You can never *repay* God for his unspeakable gift but if you *accept* it then you definitely can strive to live your life in a way that is pleasing to him who has freely given you so much, at such a cost.[315]

It is that simple. Remember you did nothing to deserve the gift of life that God freely gave you. What then could you possibly do to *earn* the gift of eternal life? Nothing.

> "We must remember that man originally fell into sin by believing Satan's words rather than the Word of God. Now man can be reconciled to God by believing the words that Jesus Christ died for his sin."[316]

---

315 (And nor does this mean that someone who gets saved is free to live like the devil. For if one becomes a *follower* of Christ through faith, he or she becomes imbued with the desire to be more and more like Christ, and less and less to follow the old sin nature and "live like the devil." It just loses its deceitful attractiveness.)

316 (Pg. 120, *Mark of the Beast*, by Arno Froese)

"Not by works of righteousness which we have done, but according to his mercy he saved us, by the washing of regeneration, and renewing of the Holy Spirit." (Titus 3:5)

This is one of the great errors of Roman Catholicism, and other denominations of Christianity as well. They think they are going to <u>work</u> their way into heaven. They're going to get there by being "good enough," or by "doing the best they can." By avoiding all of the "really bad" sins; the "mortal" ones. By going to church every Sunday. They're basically hoping that their good works will "outweigh" the bad. Well, not to be flip, but lots of luck, because the Bible clearly states that "all of our <u>righteousness</u> is as filthy rags." (Isaiah 64:6) Not "all of our <u>sins</u> are as filthy rags" (which they of course are) but "all of our <u>righteousness</u>!" So how could we possibly be good enough? What are we going to do, bring all of our good works, our righteousness, our <u>filthy rags</u> to set before God at the judgment seat to try to buy or earn our way into heaven? While still covered in our sin? No, our only hope is to put off our own pretentious righteousness, and to put on the real righteousness of Christ. There is nothing wrong with your own righteousness, with your own good works, unless you try to use it as a ticket into heaven, while rejecting the free gift of Christ's righteousness.

This is the very warning given to us in the parable of the wedding garment:

"And Jesus answered and spoke to them again by parables, and said, The kingdom of heaven is like a certain king who arranged a marriage for his son, and sent forth his servants to call those who were invited to the wedding: and they were not willing to come. [That refers to the rejection of Christ by the Jews in general.] Again, he sent forth other servants, saying, Tell those who are invited, Behold, I have prepared my dinner: my oxen and my fatted calf are killed, and all things are ready. Come to the wedding. But they made light of it and went their ways, one to his own farm, another to his business. And the rest seized his servants, and treated them spitefully, and killed them. [A reference to the Jews killing their own prophets over the centuries.] But when the king heard about it, he was furious. And he sent forth his armies, and destroyed those murderers, and burned up their city. [This occurred some 37 years later in 70 AD when Jerusalem was leveled by the Romans.] Then he said to his servants, The wedding is ready, but those who were invited were not worthy. Therefore go into the highways, and as many as you shall find, invite to the wedding. [A reference to the early Jewish Christians, Apostles and disciples spreading the gospel to the Gentiles or non-Jews.] So those servants went out into the highways and gathered together all whom they found, both bad and good. And the wedding hall was filled with guests. But when the king came in to see the guests, he saw <u>a man there who did not have on a wedding garment.</u> So he said to him, **Friend, how did you come in here without a wedding garment**? <u>And he was speechless. Then the king said to the servants, Bind him hand and foot, take him away, and cast him into outer darkness; there shall be weeping and gnashing of teeth. For many are called, but few are chosen.</u>" (Matthew 22:1-14)

The wedding feast is the gathering of the saints in heaven to the marriage supper of the Lamb, when the bride of Christ, the church, which consists of those who have accepted him as their Savior and who have therefore followed him in this life, will be presented to the Bridegroom Jesus to dwell with him for all eternity in paradise restored here on this earth. All those who are saved will be clothed in the righteousness of Christ, they will be washed in the blood of the Lamb, and that will be their wedding garment, the garment that they attained in this life by embracing Jesus and his free gift of eternal life. The man who didn't have on a wedding garment was clothed in his own righteousness. He was deceived in this life by a false Christian religion that taught him to look to his own righteousness and good works and to avoid really bad sins, as a way to get to heaven. And therefore he neglected the only way there is, and that is faith in Jesus alone, and not faith in your own good works, or faith in another false Christian religion no matter how popular it may be with the peoples of this earth. Jesus said, "I am the way." There is no other way to come to the Father but by Him.

"Good works are evidence of a growing faith in God and demonstrate the power of God's grace to transform a life. Good works bring honor to God, blessings to others, and joy to the one who performs them. But they have no power to open the doors of heaven."[317]

"For by grace are you saved through faith; and that **not of yourselves**: it is the **gift of God:** Not of **works**, lest any man should boast." (Ephesians 2:8-9)

"And as Moses lifted up the serpent in the wilderness, even so must the Son of man be lifted up: That whosoever believes in him should not perish, but have eternal life. **For god so loved the world, that he GAVE his only begotten Son, that whosoever believes in him should not perish, but have everlasting life.** For God sent not his Son into the world to condemn the world; but that the world through him might be saved. He that believes on him is not condemned: but he that believes not is condemned already, because he has not believed in the name of the only begotten Son of God. And this is the condemnation, that light is come into the world, and men loved darkness rather than light, because their deeds were evil. For every one that does evil hates the light, neither comes to the light, lest his deeds should be reproved." (John 3:14-20)

"He that believes on the Son has everlasting life: and he that believes not the Son shall not see life; but the wrath of God abides on him." (John 3:36)

It is that simple. If you come to him seeking his salvation, you will receive it. "For every one who asks receives; and he who seeks finds; and to him who knocks it shall be opened." (Matthew 7:8) "Seek the Lord while he may be found, call upon him while he is near. Let the wicked forsake his way, and the unrighteous man his thoughts: and let him return unto the Lord, and he will have mercy upon him; and to our God, for he will abundantly pardon." (Isaiah 55:6-7) "All that the Father gives to me shall come to me; and he who comes to me I will in no wise cast out." (John 6:37) "But as many as received him, to them gave he power to become the sons of God, even to them who believe on his name:" (John 1:12)

"God was in Christ reconciling the world to himself, not imputing their trespasses to them; and has committed to us the word of reconciliation. For he made him who knew no sin to be sin for us, that we might become the righteousness of God in him." (2 Corinthians 5:19, 21) He suffered and died on the cross to pay the penalty for your sins. Then he arose again from the dead three days later to crush Satan ("Christ shall bruise his head") and to crush the evil design that the Devil had for the entire human race to perish with him in eternal damnation. Jesus established victory over death, and forged the Way of Salvation for the world whom he so loves. Won't you accept his gift of eternal life and reconciliation with the Father? What other choice is there? Only eternal damnation in the Lake of Fire, a place of everlasting horror, reserved for the devil and his angels and for those who foolishly reject the simple way of escape through Jesus Christ.

317　(Pg. 61, *The 100 Most Important Bible Verses*, by Thomas Nelson Publishers)

Salvation 101

In order to be saved you must first realize that you have a <u>need</u> for salvation. For who can believe in a Savior who they don't even think they need? Realize that you are a sinner as is everyone else. "For all have sinned, and come short of the glory of God" (Romans 3:23) "As it is written, There is none righteous, no, not one:" (Romans 3:10) "But we are all as an unclean thing, **and all our righteousness's are as filthy rags**; and we all do fade as a leaf; and our iniquities, like the wind, have taken us away." (Isaiah 64:6) "Wherefore, as by one man [Adam] sin entered into the world, and death by sin; and so death passed upon all men, for that all have sinned." (Romans 5:12)

Second, you must understand that there is a grave penalty for your sinful condition; that there is something that you are in dire need of being saved *from*-eternal separation from God and from everything that is good. "For the wages of sin is death; but the gift of God is eternal life through Jesus Christ our Lord." (Romans 6:23) "And as it is appointed unto men once to die, but after this the judgment." (Hebrews 9:27) "And it came to pass, that the rich man also died, and was buried; **And in hell he lift up his eyes, being in torments**." (Luke 16:22-23) We need help. We need someone to save us from this horrible eternal fate. Who can do this? Only God has that kind of power.

Third, you must believe that Jesus came down from heaven and paid the penalty for your sin on the cross. "Who himself bore our sins in his own body on the tree, that we, having died to sins, might live for righteousness: by whose stripes[318] you were healed." (1 Peter 2:24) "But God demonstrates his own love toward us, in that while we were still sinners, Christ died for us." (Romans 5:8) God himself provides the help we need.

And finally, you must <u>accept</u> the gift of salvation that God has provided for you through His Son, the Jewish Messiah Jesus Christ. "Believe on the Lord Jesus Christ, and you shall be saved." (Acts 16:31) "That if you shall confess with your mouth the Lord Jesus, and shall believe in your heart that God has raised him from the dead, you shall be saved. For with the heart man believes unto righteousness; and with the mouth confession is made unto salvation. For the Scripture says, Whosoever believes on him shall not be ashamed. For there is no difference between the Jew and the Greek: for the same Lord over all is rich unto all who call upon him. **For whosoever shall call upon the name of the Lord shall be saved**." (Romans 10:9-13)

<u>"For whosoever shall call upon the name of the Lord shall be saved."</u>

Just <u>accept</u> the gift. Embrace it now while you can. Don't wait until you feel you are "good enough" to receive it. You will <u>never</u> be good enough and thank God you don't have to be. (Remember that God's word says that "while we were still sinners, Christ died for us." Romans 5:8) And don't wait until you feel you are intellectually satisfied that you *should* accept the gift. The prince of darkness is very clever at keeping people just intellectually confused enough so they will *never* make the decision to follow Christ. If the Holy Spirit is drawing you then the

---

318   [Jesus was "scourged," a common practice of the Romans where the back was repeatedly whipped with leather strips that had nails and razor-sharp pieces of metal embedded in the ends. These "stripes" were the lines where his flesh was literally ripped off his back.]

time is now. You could die in your sleep tonight, and your eternal fate is hanging by a thread. You think you are standing on firm earth, on terra firma, but that is an illusion. All matter is nothing but a vapor, an image in the Mind of God, and all that stands between you and a lake of molten lava is the thinnest of veils, with only your untimely death waiting to erase that flimsy barrier. And your decision to put off getting saved till some mythological future time when you might be a little better disposed or a little more comfortable is a no decision. It is saying no to Jesus, and no to God, and yes to the great deceiver. It is nothing more than the father of lies inside of your mind drawing you away from your own salvation. You've been his for all this time and he is fighting to keep you his. And a no decision to accept Jesus as your Lord and Savior is a yes decision to let Satan be your Lord and Master. Either way, you *will* have a Lord. Choose wisely then.

"For what does it profit a man, if he gains the whole world, and loses his own soul? Or what shall a man give in exchange for his soul?" (Matthew 16:26) "Yet indeed I also count all things but loss for the excellence of the knowledge of Christ Jesus my Lord, for whom I have suffered the loss of all things, and do count them but dung, that I may win Christ." (Philippians 3:8) "Behold, **now** is the accepted time; behold, **now** is the day of salvation." (2 Corinthians 6:2) "Thanks be unto God for his unspeakable gift." (2 Corinthians 9:15) "My sheep hear my voice, and I know them, and they follow me: And I give to them eternal life, and they shall never perish" (John 10:27-28)

If you hear his voice now, if you feel the Holy Spirit's calling to accept Christ as your Lord and Savior, then just take a minute now, put the book down, and pray to him. He will hear your prayer. Ask him to forgive you for your sins. And ask him to take you to heaven when it is your time to leave this life. Tell him you believe in Him. Thank him for paying your penalty on that cross 2000 years ago. And tell him you have faith in Him and that you believe in his death, burial, and His resurrection from the dead three days later as he himself prophesied. Then rejoice. For your name is now written down in heaven, in the Lambs Book of Life from the foundation of the world!

"Amazing love! How can it be, that You, my God, should die for me? ...I'm forgiven, because you were forsaken. I'm accepted, you were condemned. I'm alive and well, you're spirit is within me, Because you died and rose again. Amazing love! How can it be, that You, my King, should die for me? Amazing love, I know it's true, and it's my joy to honor you. In all I do, I honor you."[319]

It truly is amazing, and nearly incomprehensible, when you think about it, that God would die for us. But then again, it also makes perfect sense, because that is the nature of true love. What parent wouldn't sacrifice their life for the life of one of their little children?

The assurance of eternal life

"And this is the record, that God has given to us eternal life, and this life is in his Son. He that has the Son has life; and he that has not the Son of God has not life. These things have I written unto you that believe on the name of the Son of God; that you may **KNOW** that you have eternal life, and that you may believe on the name of the Son of God." (1 John 5:11-13)

And don't be led into doubts by those who falsely preach that you must

---

319   (From Hymn, "And can it be that I should gain" by Charles Wesley.)

experience some wild emotional or *spiritual* sensation when you accept Christ as your Savior. Or that you must start speaking in some unknown tongue. For your salvation is a step of faith. It's a *decision* and not an "experience." You may very well have an overwhelming and overpowering emotional and/or spiritual experience, some people do, but again salvation is not an experience, it is an act of will; by the grace of God. Besides, it is far more important what you do after you get saved—that you have a serious desire to learn about your Savior through his Word, to depart from evil and iniquity of every kind (like voting for baby mutilators), and to follow him and his will for your life—than what you felt or didn't feel at the time of your prayer of salvation. Just know that God cannot and does not lie. He said he would save you if you asked him to. And if you did, then he did. And that's that. Your assurance in your own salvation and in the utter truthfulness of God's Word will grow as you surround yourself with other believers and as you grow in the reading and study of his Word. Your *faith* will become *knowing*. Your initial step of *trusting* that this salvation/Jesus thing is actually real and you want to have a part in it will be followed by the actual *knowledge* that it is true, as you continue to look at the evidence. God is not interested in *blind* faith. Blind faith is what causes people to believe lies.[320] God has backed up the truth with overwhelming evidence. It is there for the taking, for anyone who exercises their free will and chooses to accept it.

"Man, carnal man, has an insatiable desire to be saved by his own works, or his own feelings or emotions or experiences, instead of by simple faith. The idea of God's unmerited grace is alien and foreign and distasteful to the carnal mind. Most of the world wants to be saved by its good works. And if good Christians are forced, by the Bible, to abandon the idea of salvation by works, then they like to substitute salvation by feeling, by earnestness, by certain experiences and emotions. There is no special way a person has to feel in order to be saved. There is no special experience one has to go through, emotionally, in order to be saved. One who believes in Christ, in the sense of committal, dependence, trust, is already born again. ...God's Word tells me that if I trusted in Him, I am saved, forgiven. And that is the sure evidence."[321]

...And not some feeling or emotional experience, as wonderful as that may be to those whom it has been given at the time of their belief in Christ.

Be assured also that, once you have called upon the name of the Lord and accepted his free gift of eternal life, you are stuck with it. Don't be fooled into thinking that you can lose it. There are a number of faux Christian churches out there that preach that foolishness; because if you can lose your gift of eternal life then it definitely wasn't *eternal*. For the life of me I can't figure out exactly what it might have been, but if you had it at one time, and then you lost it by some subsequent action, then it wasn't eternal life. I guess it was temporary life. And Jesus did not waste his time, effort and energy to come down here and die on a cross so you could receive a temporary reprieve from eternal death; one that you could easily lose if you mess up. Of what value would that have been? Because we

---

320   (Blind faith is what causes followers of the Devil to send themselves straight to an eternity in the Lake of Fire because through incredible heartless selfishness they sought 72 virgins by flying planes into tall buildings and slaughtering thousands of innocent, and by strapping bombs on themselves and blowing up innocent Israeli and Iraqi men, woman and children. Pray for those souls who are still alive and deluded into believing this madness, so they might avoid the same horrible and irreversible mistake.)

321   (Pg. 197, *Prayer: Asking and Receiving*, by John R. Rice)

are all going to mess up every day. We are all flawed sinners, both before we get saved, and afterwards. Saved Christians are accurately called *saints* and children of God because they have been adopted by Him and he has become our Father, but we are never called sinless. When Jesus died on that cross for your sins he did so for all of your sins, past, present and future. And no, this fact is not a license to go out and sin freely because if you get saved and have a desire to go out and sin wantonly then you can be assured that your prayer and your salvation were not sincere. God's Word clearly says that you cannot work your way to heaven, but it also clearly states that after you make your profession of faith in Jesus Christ you will be known by your works. Good works and a path that leads to more personal righteousness are the *evidence* of your salvation, of your faith in Christ, because "faith without works is dead."[322]

"Let your light so shine before men, that they may see your good works, and glorify your Father which is in heaven." (Matthew 5:16)

"For we are his workmanship, created in Christ Jesus unto good works, which God has before ordained that we should walk in them." (Ephesians 2:10)

"They profess that they know God; but in works they deny him, being abominable, and disobedient, and unto every good work reprobate." (Titus 1:16)

"And let us consider one another to provoke unto love and to good works." (Hebrews 10:24)

"Yes, a man may say, you have faith, and I have works: show me your faith without your works, and I will show you my faith by my works." (James 2:18)

"But will you know, O vain man, that faith without works is dead?" (James 2:20)

And then there is this: "If we deliberately keep on sinning after we have received the knowledge of the truth, no sacrifice for sins is left, but only a fearful expectation of judgment and of raging fire that will consume the enemies of God." (Hebrews 10:27-28) In other words, if you have no desire to follow him after you are saved, then you did not get saved. It was not a conversion of the heart. It was just empty words. He was knocking at the door but you didn't let him in. "Behold, I stand at the door, and knock: if any man hear my voice, and open the door, I will come in to him, and will dine with him, and he with me." (Revelation 3:20-21) And if you do have a desire to follow him, then you will also have a desire to please him, by growing in his knowledge and grace, and by struggling to avoid sin, sinful lifestyles, iniquity itself and a sinful life. Thus leading to a life of faith with works.

But if you do fall into sin, or in your Christian walk are struggling with sin, then remember, "My little children, these things I write unto you, that you may not sin. And if anyone sins, we have an advocate with the Father, Jesus Christ the righteous." (1 John 2:1)

Jesus himself in his Word clearly speaks of the comfort and the assurance we can have in our salvation; the eternal security that we can have and the joy that we can share in this life with other believers in Him who can trust that he will never let us go...

"My sheep hear my voice, and I know them, and they follow me: And I give unto them **eternal** life; and they shall **never** perish, neither shall any man pluck

---

322 (Conversely, refusing to depart from iniquity, like voting religiously every 2 years for the baby-mutilators, is *evidence* that one is *not* saved.)

them out of my hand. My Father, who gave them to me, is greater than all; and no man is able to pluck them out of my Father's hand. I and my Father are one." (John 10:27-30)

"For I am persuaded, that neither death, nor life, nor angels, nor principalities, nor powers, nor things present, nor things to come, nor height, nor depth, nor any other creature, shall be able to **separate us from the love of God**, which is in Christ Jesus our Lord." (Romans 8:38-39)

"And this is the record, that God has given to us **eternal** life, and this life is in his Son. He that has the Son has life; and he that has not the Son of God has not life. These things have I written unto you that believe on the name of the Son of God; that you may **KNOW that you have eternal** life, and that you may believe on the name of the Son of God." (1 John 5:11-13)

Follow me

"And when he had called the people unto him with his disciples also, he said to them, whosoever will come after me, let him deny himself, and take up his cross, and follow me." (Mark 8:34)

Once you have believed in Jesus Christ as your Lord and Savior the next step is to obey his commandment to "follow him." And the first step in following him is to be baptized...

"Then Peter said unto them, repent, and be baptized every one of you in the name of Jesus Christ for the remission of sins, and you shall receive the gift of the Holy Ghost." (Acts 2:38)

"Therefore we are buried with him by baptism into death: that just as Christ was raised up from the dead by the glory of the Father, even so we also should walk in newness of life." (Romans 6:4)

"All authority has been given to me in heaven and on earth. Go therefore and make disciples of all the nations, **baptizing them in the name of the Father and of the Son and of the Holy Spirit**, teaching them to observe all things that I have commanded you; and lo, I am with you always, *even* to the end of the age. Amen." (Matthew 28:18-20)

All of the early Christians were baptized as soon as they made the decision to put their faith in Christ. This is an important first step in your Christian walk. Baptism is symbolic of the death, burial and resurrection of Jesus Christ. Going down into the water is symbolic of his death on the cross. Under the water you are identifying with his burial. Staying under for three minutes—just kidding, for three seconds—represents Christ being in the grave for three days and three nights. Re-emerging from the water you are identifying with Christ's resurrection from the dead. This is also symbolic in that, when you put your faith in Christ, you were raised from once being dead in your sins to the newness of eternal life, and to being able to access the abundant, spirit-filled life of knowing, following and serving Jesus as Lord and Savior and Guide in this life. Just as you were born the first time out of water from your mother's womb, so too coming up out of the water is symbolic of the second birth; when you accepted Christ as Savior; when you believed on the name of the Only Begotten Son of God; when you were born of the Spirit...

"Verily, verily, I say unto you, Except a man be born again, he cannot see the kingdom of God. ...Except a man be born of water and of the Spirit, he cannot enter into the kingdom of God. That which is born of the flesh is flesh; and that which is born of the Spirit is spirit." (John 3:3, 5-6)

With your first birth you were born into the family of man. The second birth you are born into the family of God, the Creator of the entire universe, who spoke all the worlds into existence and all the living into being by the breath of his mouth. And you can now cry "Abba Father" (Romans 8:15 ...not the rock group) for He is now your Father and you are his son or daughter; just as He is Jesus' Father. And Jesus is not only your Lord and Savior, but he is now your brother as well. However, it is important to point out that you do not get saved by being baptized, either as an adult or as an infant. But it is a very important first step in obeying his command to "follow Him" after you have become a Christian.

Fellowship with other believers in a true local church is also a very important step for a new believer in Christ. But beware that "all that glitters is not gold" and there are as many false Christian churches out there that preach false doctrine, as there are true ones. We will try to give you enough basic information here to discern between the two. (One recent example is the church of Obama's mentor, the America-hating theatrical clown, Reverend Wright. "Black Liberation Theology" is not a Christian religion anyway. It comes from another place. ...See part III "False Religion" in the next chapter.)

"But there were false prophets also among the people, even as there shall be false teachers among you, who secretly shall bring in damnable heresies, even denying the Lord that bought them, and bring upon themselves swift destruction." (2 Peter 2:1)

Prayer is the most important tool you can utilize in finding a true body of believers in your area. Ask God to help you find a good local church that is faithful to his Word, and conversely to protect you from all of the false Christian churches and cults that are out there preying upon the unsuspecting. (And see the following chapter.)

One way of determining whether a potential church is doctrinally sound, true to the Word of God and therefore worth your time is to read their "statement of faith." Ask the Pastor. (If they don't have a statement of faith, tell him to get one. No, just kidding, ask him to write down a list of what they believe.) Here is a sample that you can use for comparison; it is from Bay Country Church in Cambridge, Md.:

Bay Country Church Statement of Faith

The doctrinal position of Bay Country Church is summarized in our eleven-article Statement of Faith. We Believe:

1. The scriptures both Old and New Testaments, to be the inspired Word of God, without error in the original writings, the complete revelation of His will for the salvation of men and the divine and final authority for Christian faith and life.

2. In one God, Creator of all things, infinitely perfect and eternally existing in three persons: Father, Son and Holy Spirit.

3. That Jesus Christ is true God and true man, having been conceived of the Holy Spirit and born of the virgin Mary. He died on the cross, a sacrifice for our sins according to the Scriptures. Further, He arose bodily from the dead, ascended into heaven, where, at the right hand of the Majesty on High, He is now High Priest and Advocate.

4. That the ministry of the Holy Spirit is to glorify the Lord Jesus Christ and, during this age, to convict men, regenerate the believer sinner, and indwell, guide, instruct and empower the believer for godly living and sacrifice.

5. That man was created in the image of God but fell into sin and is, therefore, lost and only through the regeneration by the Holy Spirit can salvation and spiritual life be obtained.

6. That the shed blood of Jesus Christ and His Resurrection provide the only ground for justification and salvation for all who believe, and only such as receive Jesus Christ are born of the Holy Spirit and, thus become children of God.

7. That water baptism and the Lord's Supper are ordinances to be observed by the church during the present age. They are, however, not to be regarded as means of salvation.

8. That the true church is composed of all such persons who through saving faith in Jesus Christ have been regenerated by the Holy Spirit and are united together in the body of Christ of which He is the Head.

9. That Jesus Christ is the Lord and Head of the church and that every local church has the right, under Christ, to decide and govern its own affairs.

10. In the personal and pre-millennial and imminent coming of our Lord Jesus Christ and that this "Blessed Hope" has a vital bearing on the personal life and service of the believer.

11. In the bodily resurrection of the dead; of the believer to everlasting blessedness and joy with the Lord; of the unbeliever to judgment and everlasting conscious punishment.

"And I will give you pastors according to my heart, who shall feed you with knowledge and understanding." (Jeremiah 3:15)

I would also make sure that the Pastor, staff and all of the elders believe in all of the miracles of God's Word, both Old and New Testament. This is an easy way to smoke out a liberal congregation; one that is Christian in name only. And one that is referred to in these verses, "For such are false apostles, deceitful workers, transforming themselves into apostles of Christ. And no wonder! For Satan himself transforms himself into an angel of light. Therefore it is no great thing if his ministers also transform themselves into ministers of righteousness, whose end will be according to their works." (2 Corinthians 11:13-15) Find out what he and his church leaders think of the first few chapters of Genesis. Are they an historic, and scientifically accurate, depiction of the origin of the universe and the origin of life on earth? What does he think of the theory of evolution? Make sure his answer is that it is a scientifically-impossible joke, Satan's genesis, perpetrated by the father of lies; and one that spits in the face of the Creator of the universe. (Or something that approximates that.)

There are in addition, two other subjects that do not separate us as believers, but I think still might be good to discuss. (Before asking these questions, though, you may have to do a little homework.) Ask the Pastor where his church stands on "Replacement Theology."[323] "This view teaches that the Church is the replacement for Israel and that the many promises made to Israel in the Bible are fulfilled in the Christian Church, not in biblical, literal, Israel."[324] It interprets allegorically (as opposed to literally) the promises of God surrounding the Jews and the nation

---

323   (See "The Roots of Replacement Theology," article by William L. Krewson in Israel My Glory Magazine, May/June 2007, and "Upholding the Truth" by Richard D. Emmons and "Replacement Theology: The Black Sheep of Christendom (Part 1)" by James A. Showers, both articles in Israel My Glory Magazine, March/April 2007. Also see "Replacement Theology" section in Chapter 13)

324   (From article "Replacement Theology" by Alan Torres @ www.biblicist.org)

of Israel and transfers them to the church. (How this can still be done after 1948 is beyond me.) (See chapter 9, and also the "Replacement Theology" section of "Part I – Destroy Israel" in Chapter 13.) The other question is whether or not they believe in the premillennial rapture of the church. (See chapter 9) As you study these two topics ("to show yourself approved") you will find they are related.

And finally, and most importantly, find out what the Pastor thinks about those who call themselves *Christian* and yet go out and vote for the baby mutilating Democrats. Does he understand that "Thou shalt not vote for baby mutilators" is an unspoken part of the Sixth Commandment "Thou shalt not murder." (If he doesn't understand what "baby mutilation" is, or that the Democrat Party is the Party of baby mutilation, that it is their Unholy Grail, then run screaming from the place.)[325]

Respectfully yet boldly ask the Pastor all of these things. With boldness because it is the proper spiritual mentoring of your soul, and not his vanity, that is your primary concern here. (It should be his as well.) Besides, if he does anything other than enthusiastically welcome this type of inquiry then flee from him and his "church," kicking the dust off your feet as you go. ("And whosoever will not receive you, when you depart from there, shake off the very dust from your feet for a testimony against them." Luke 9:5) Because a true preacher of God's Word would not only be excited if a new believer was interested in his church, but would also be impressed and pleased if he or she were also asking all of these very important questions. Conversely, a false preacher might get angry or uncomfortable, be condescending or argumentative, attempt to change you to his more superior and "tolerant" way of thinking, or all of the above.

One should also be wary of a doctrinally sound church, but one that is also overly judgmental and legalistic, emphasizing the letter of the Law over justice, mercy, long-suffering and love. Signs might be phrases that are used such as "SMO's" which is a derogatory term for "Sunday Morning Only" Christians; ones who are just not as *spiritual* as the "really committed" members who go 2, 3, or even 4 times a week. Be careful if this kind of self-righteousness, and condemnation of others, is present.

Of course there are no perfect churches just as there are no perfect people, but when choosing which mate you will spend the rest of your life with it is a good idea not to choose a liar, a deceiver, a false witness, an emotional abuser, a woman of low morals, a bum who refuses to actually get a job and work for a living, a pathological controller or a hyper-critical person. Similar discretion should be used in choosing a church family as well.

It is also of necessity that every new Christian (and old ones as well) read and study God's Word. "As newborn babes, desire the sincere

---

325   (And for a concise yet comprehensive study of this subject see: "How to Know a Counterfeit Minister of Religion," by Dr. James A. McBean D. Th. MCC. Revival Tract & Book Publishing Ministries @ netfirms.com)

milk of the word, that you may grow thereby:" (1 Peter 2:2) "But he answered and said, It is written, man shall not live by bread alone, but by every word that proceeds out of the mouth of God." (Matthew 4:4) As we mentioned in the previous chapter, start by reading the Gospel of John in the New Testament. Then read Genesis. Then read all four Gospels. And then continue through the Old and New Testaments. "Study to show yourself approved unto God, a workman that need not to be ashamed, rightly dividing the word of truth." (2 Timothy 2:15) Knowing God entails learning about him, his ways, his character, what he desires of you; and his Word is what he has provided for doing just that. The Bible is God's way of talking directly to you. And, conversely, prayer is your way of talking to God. Get in the habit early of doing both on a daily basis.

And then there's prayer. Pray like your life depended on it! Your spiritual life does. "Pray without ceasing." (I Thessalonians 5:17) Give thanks always. And ask for whatever you need; for your physical and spiritual needs. Ask for the indwelling and guidance of the Holy Spirit, to lead you into all truth. "Be anxious for nothing, but in everything by prayer and supplication, with thanksgiving, let your requests be made known to God. And the peace of God, which surpasses all understanding, will guard your hearts and minds through Christ Jesus." (Philippians 4:6-7) And James 4:2 says, "You have not because you ask not." So ask; for whatever your needs or shortcomings are. Pray without ceasing. No matter how many times you fall, get back up. Cast every burden upon him. Trust in him. And he will not fail you. He will answer your prayer. "Most assuredly, I say to you, **whatever you ask the Father in my name he will give you.** Until now you have asked nothing in my name. Ask, and you will receive, that your joy may be full." (John 16:23-24) Pray in Jesus' name, because he is our High Priest, the mediator between God and man. (Which is why there is also no need for any human priests, and why the Catholic priesthood runs contrary to the will of God and to his Word.) Also it is through him and by being clothed in his righteousness, that you may boldly approach the throne of grace; that you may boldly approach the Father with your needs and your hearts desires. "Cast all your care upon him, for he cares for you." (I Peter 5:7) Get in the habit of immediately casting on Him anything and everything that might arise on a daily basis that upsets you, or causes you to fear, or to be anxious, or to worry. You'll save yourself a lot of grief.

You are a child of God now. You are a new creature in Christ. You have been born a second time by the Spirit, adopted into the family of God. You have also been bought with a terrible price. Remember that he is very near and you will be less apt to do things that would grieve him… "Amazing love, how can it be, that You, my King, should die for me? Amazing love, I know it's true, and it's my joy to honor You, in all I do, to honor You." God is now your Father. You are his adopted son (or daughter), just as Jesus is his Son. And Jesus is now both your Lord and brother.

"The Cost of Discipleship"

"Then spoke Jesus again unto them, saying, I am the light of the world: he that follows me shall not walk in darkness, but shall have the light of life." (John 8:12)

God's gift of eternal life is free, but nevertheless Jesus is not a "get-out-of-hell-free" card, one that you pick up at some time in your life, put in your wallet, and go on about your merry way. Salvation is as easy as accepting Jesus and his free gift of eternal life, but that is not to say that it is done without a cost. Jesus paid the ultimate price for our salvation. Are we then in turn to pay nothing after accepting his gift? There are many who have made a profession of faith in Christ at some time in their life, possibly as a young child, but then have shown absolutely no evidence of that "conversion" as they go through life. Are they then truly saved? Or have they been misled? Only Jesus knows what is going on in their heart and who are his own but there are many verses where he warns of the kind of Christianity that has no evidence and is unwilling to pay any price. We would be wise to heed his warning and to make sure that we are indeed followers of Christ, those who are "working out their own salvation with fear and trembling." (Philippians 2:12) Again not trying to "work your way into heaven" but making sure that we are truly saved and striving to do the best we can, through his grace, to follow him no matter what the cost. "With fear and trembling" because we know the horrible end for those who have made an easy profession of faith in Christ but were never truly converted; who might even go to church and have a lot of the outward signs of being a Christian but who are still covered in their own righteousness and not His; and who therefore are just as lost as any real religious Muslim, Buddhist, Mormon or Jehovah's Witness.

"The gospel is not a doctrine of the tongue, but of life. It cannot be grasped by reason and memory only, but it is fully understood when it possesses the whole soul and penetrates to the inner recesses of the heart."[326]

The following excerpts are from "The Cost of Discipleship," a radio address by John MacArthur of Grace to You ministries. We would recommend going to their website @ gty.org and reading the entire 10 page sermon, or you can download and listen to it as well:

"A pastor from behind the Iron Curtain said to me, one time, "There's no easy believism in our churches. There's no shallow professions of faith. Nobody is taking Jesus who isn't willing to lay their life down because that's the price in many, many cases. The cost of naming Christ," he said, "is so high that we don't have false conversions. If they aren't willing to pay the price," he said, "they don't want to be associated with Jesus Christ in any way at all.""[327]

Here are a few verses that show us the seriousness with which Jesus held, not only believing in Him, but also in wholeheartedly *following* Him...

"Then said Jesus to those Jews which believed on him, If you continue in my word, then are you my disciples indeed." (John 8:32)

"And he said to them all, If any man will come after me, let him deny himself, and take up his cross daily, and follow me." (Luke 9:23)

"Herein is my Father glorified, that you bear much fruit; so shall you be my disciples." (John 15:8) And the fruits of the spirit are "love, joy, peace, patience, kindness, goodness, faithfulness, gentleness and self-control" (Galatians 5:22), as well as being as instrumental as is individually possible in leading others to Christ.

---

326 (John Calvin.) (Old man Calvin's son)
327 (From "The Cost of Discipleship" by John MacArthur, Grace to You ministries)

"Whosoever therefore shall confess me before men, him will I confess also before my Father who is in heaven. But whosoever shall deny me before men, him will I also deny before my Father who is in heaven. Think not that I am come to send peace on earth: I came not to send peace, but a sword. For I have come to set a man at variance against his father, and the daughter against her mother, and the daughter-in-law against her mother-in-law. And a man's foes shall be those of his own household. He who loves father or mother more than me is not worthy of me: and he who loves son or daughter more than me is not worthy of me. And he who does not take his cross and follow after me, is not worthy of me. He who finds his life shall lose it: and he who loses his life for my sake shall find it." (Matthew 10:32-39)

"And whosoever does not bear his cross, and come after me, cannot be my disciple. For which of you, intending to build a tower, sits not down first, and counts the cost, whether he has enough to finish it?" (Luke 14:26-28)

"John Stott wrote in his helpful little book Basic Christianity, "The Christian landscape is strewn with the wreckage of derelict half-built towers. The ruins of those who began to build and were unable to finish. For thousands of people still ignore Christ's warning and undertake to follow Him without first pausing to reflect on the cost of doing so. The result is the great scandal of Christendom today, so called nominal Christianity. In countries to which Christian civilization has spread, large numbers of people have covered themselves with a decent but thin veneer of Christianity. They have allowed themselves to become somewhat involved, enough to be respectable but not enough to be uncomfortable. Their religion is a great soft cushion. It protects them from the hard unpleasantness of life while changing its place and shape to suit their convenience. No wonder the cynics speak of hypocrites in the church and dismiss religion as escapism.""[328]

How do you know if you are a "nominal" Christian or not? Just be certain that you are not covered in your own righteousness, and that you have put on a wedding garment by truly accepting Jesus as your Savior through faith and by the Grace of God, and therefore that you are clothed in the righteousness of Christ. If that is the case then you will want to truly follow Him, whatever the cost.

"Again, the kingdom of heaven is like unto treasure hid in a field; which when a man has found, he hides, and for the joy thereof goes and sells all that he has, and buys that field. Again, the kingdom of heaven is like unto a merchant man, seeking goodly pearls: Who, when he had found one pearl of great price, went and sold all that he had, and bought it." (Matthew 13:44-46)

"Now we know those parables. A man found a treasure, sold all he had and bought it. A man found a pearl, sold all he had and bought it. Now that says that when a man comes across the saving gospel of Jesus Christ, he gives up all he has. He turns his back on all of life to embrace Christ."[329]

…"I believe that every Christian is a disciple; every Christian is a follower of Christ. Some of us are following more faithfully than others, but every true believer has committed himself or herself to follow Jesus Christ. I do believe that as we have seen you can be a follower of Jesus and not be a real Christian. You could follow along without having a changed heart and say, "Lord, Lord," and He would say, "I never knew you." Like John 6:66, "Many of His disciples walked no more with Him." Like those who wanted to follow Jesus in Matthew chapter 8, but they certainly were not willing to make whatever

---

328   (From "The Cost of Discipleship" by John MacArthur, Grace to You ministries)
329   (From "The Cost of Discipleship" by John MacArthur, Grace to You ministries)

commitment had to be made. They were the ones of whom we read earlier who said, "Let me go do this and let me go do that, and let me go do the other thing." And He said, "You're not worthy to be My disciple."

"So, there are some "disciples" who aren't real, but there are no believers who aren't disciples. It simply means that we have entered into a relationship with Jesus Christ in which we follow Him. We don't follow perfectly and please, we don't follow out of our own will and our own flesh, we follow because God in His sovereign grace transformed us into followers.

…"You see, salvation is not an experiment. Salvation is a life-long commitment. Salvation is not "try Jesus," see if He works. Salvation is a life-long transformation. Those who would tell us that a person can become a Christian without becoming a disciple do a great disservice to Scripture and they do a great disservice to people who then live under the illusion that they can be saved without following Christ in obedience. They can be saved without giving up all they are and have and ever hope to be unconditionally to Christ. That's tragic."[330]

And to that end I pray that anyone who has accepted Christ as their savior either while reading this chapter or at some time earlier in their life, that their conversion was wholehearted and sincere, and the kind that will lead them, through his grace, to follow him in true discipleship, not perfectly and not without stumbling, but to "deny himself while taking up their cross and following Him daily," to "give up everything for the Pearl of great price," to "sell all that they have for the treasure hidden in a field" and to "lose their life for Jesus' sake." Amen.

The Bread of Life

"Man shall not live by bread alone, but by every word that proceeds out of the mouth of God." (Matthew 4:4)

The entire Bible is food for the Christian, whether newly born again into the kingdom of God, or one who has been under the adoption of God as Father for many years. Just as any child who lacks proper nutrition will see his growth stunted; and any man or animal who does not eat what is necessary to survive and thrive will physically wither away; so too will a Christian *die on the vine* without regularly ingesting the Word of God into his mind and heart. With that in mind, here are some selected verses from God's Word, some spiritual cuisine if you will, for your soul's dining pleasure…

"And Jesus said unto them, I am the bread of life: he who comes to me shall never hunger; and he who believes on me shall never thirst." (John 6:35)

"Whosoever therefore shall confess me before men, him will I confess also before my Father who is in heaven." (Matthew 10:32) Now that you are saved, don't be silent, don't be ashamed of the truth, and don't try to avoid ridicule by those who are lost in lies. Just study to show yourself approved so you can confidently answer their ridicule with the truth. (Indeed, that is the point of this *College Course on the Truth*.) And you will be equipped to help turn some, through the power of the Holy Spirit, from the deceit of Satan to the knowledge of the Truth of Jesus Christ as Lord and Savior.

---

330   (From "The Cost of Discipleship" by John MacArthur, Grace to You ministries)

"For what is a man profited, if he shall gain the whole world, and lose his own soul? or what shall a man give in exchange for his soul?" (Matthew 16:26)

"For many are called, but few are chosen." (Matthew 22:14) (as in, "two roads diverged in a wood…")

"And the angel answered and said unto the women, Why seek you the living among the dead? He is not here, but is risen" (Matthew 28:5, 6)

"When Jesus heard it, he said unto them, They who are whole have no need of the physician, but they who are sick: I came not to call the righteous, but sinners to repentance." (Mark 2:17) "I came not to call the righteous to repentance." What did Jesus mean by that? Here is Dr. John Gerstner's answer:

> "So often the thing that stands between God and the sinner is the sinners "virtue." He has no righteousness in reality, but he thinks his righteousness is real. Because he will not give up his trust in his own goodness and acknowledge his sin and trust in Christ, these form an impenetrable barrier between the sinner and the savior. We have nothing to contribute to our salvation, my friends, except one thing: our sin. That is our total contribution. Our faith and our repentance are the work of God's grace in our hearts. Our contribution is simply the sin for which Jesus Christ died."[331]

Don't make the mistake of being one of the *righteous*, which is in reality one of the *false* righteous. Many false Christians suffer under this illusion, and the devil is happy to keep them in that state of mind until they die. Rather, put on the righteousness of Christ, which is righteousness indeed.

"Those eighteen, upon whom the tower in Siloam fell, and slew them, do you think that they were sinners above all the men who dwelt in Jerusalem? No, I tell you, but <u>unless you repent, you shall all likewise perish</u>." (Luke 13:4-5)

"There was a certain rich man, which was clothed in purple and fine linen, and fared sumptuously every day: And there was a certain beggar named Lazarus, which was laid at his gate, full of sores, And desiring to be fed with the crumbs which fell from the rich man's table: moreover the dogs came and licked his sores. And it came to pass, that the beggar died, and was carried by the angels into Abraham's bosom: the rich man also died, and was buried; **And in hell he lift up his eyes, being in torments**, and seeing Abraham afar off, and Lazarus in his bosom. And he cried and said, Father Abraham, have mercy on me, and send Lazarus, **that he may dip the tip of his finger in water, and cool my tongue; for I am tormented in this flame**. But Abraham said, Son, remember that you in your lifetime received many good things, and likewise Lazarus evil things: but now he is comforted, and you art tormented. And beside all this, between us and you there is a great gulf fixed: so that they which would pass from here to you cannot; neither can they pass to us, that would come from there. Then he said, I pray you therefore, father, that you would send him to my father's house: For I have five brethren; that he may testify unto them, less they also come into this place of torment. Abraham said unto him, They have Moses and the prophets; let them hear them. And he said, Nay, father Abraham: but if one went unto them

331   (Dr. John Gerstner 1914-1996, Author, teacher, professor, theologian)

from the dead, they will repent. And he said unto him, If they hear not Moses and the prophets, **neither will they be persuaded, <u>though one rose from the dead</u>.**" (Luke 16:19-31) Another prophecy, for Jesus of course did rise from the dead, and how many are still not persuaded?

"And he said unto Jesus, Lord, remember me when you come into your kingdom.
And Jesus said unto him, Verily I say unto you, Today you shall be with me in paradise." (Luke 23:42, 43) What did the thief on the cross do to work his way into heaven? What did he do to *earn* his way in? Nothing. He simply asked.

"And the Word [Jesus Christ] was made flesh, and dwelt among us, and we beheld his glory, the glory as of the only begotten of the Father, full of grace and truth." (John 1:14)

"Jesus answered and said unto him. Verily, verily, I say unto you, Except a man be born again, he cannot see the kingdom of God. Nicodemus said unto him, How can a man be born when he is old? can he enter the second time into his mother's womb, and be born? Jesus answered, Verily, verily, I say unto you, Except a man be born of water and of the Spirit, he cannot enter into the kingdom of God. That which is born of the flesh is flesh; and that which is born of the Spirit is spirit." (John 3:3-6) Our first birth is "of water" and we are born into the family of man with a human as our father. If we embrace the second birth then we are born again "of the Spirit" and we are adopted as sons and daughters into the family of God, with God as our Father, and we as his adopted children through faith in the atoning blood of Jesus Christ. "I am a child of God. God is my Father; heaven is my home; every day is one day nearer. My Savior is my brother; every Christian is my brother too."[332]

"He who believes on the Son has everlasting life: and he who believes not the Son shall not see life, but the wrath of God abides on him." (John 3:36)

"Jesus answered and said unto her, Whosoever drinks of this water shall thirst again:
But whosoever drinks of the water that I shall give him shall never thirst; but the water that I shall give him shall be in him a well of water springing up into everlasting life." (John 4:13-14)

"The woman said unto him, I know that Messiah will come, who is called Christ: when he is come, he will tell us all things. Jesus said unto her, **I who speak unto you am He.**" (John 4:25-26)

"Verily, verily, I say unto you, he who hears my word, and believes on him who sent me, has everlasting life, and shall not come into condemnation; but is passed from death unto life." (John 5:24)

"Most assuredly, I say to you, he who believes in me has everlasting life. I

---

332   (Pg. 228, *Knowing God*, by J. I. Packer)

am the bread of life. …I am the living bread which came down from heaven. If anyone eats of this bread, he will live forever." (John 6:47-48, 51)

"Then spoke Jesus again unto them, saying, **I am the light of the world**: he that follows me shall not walk in darkness, but shall have the light of life." (John 8:12)

"Your father Abraham rejoiced to see my day: and he saw it, and was glad. Then said the Jews unto him, you are not yet fifty years old, and have you seen Abraham? Jesus said unto them, Verily, verily, I say unto you, **Before Abraham was, I am**." (John 8:56-58) "Before Abraham was, **I AM**." **Wow!** Talk about getting your mind blown! They thought they were sharp enough to spar with the Creator of the universe. With infinite intelligence himself! Well, nobody ever accused us of being all that smart. It is kind of funny when you think about it. Yet how many are attempting to do the same thing today? (Richard Dawkins? Paging Mr. Dawkins?)

"I am the door: by me if any man enter in, he shall be saved, and shall go in and out, and find pasture. The thief [that would be your basic Satan again] comes only to steal, and to kill, and to destroy: I am come that they may have life, and that they may have it more abundantly. I am the good shepherd: **the good shepherd gives his life for the sheep**." (John 10:9-11)

"Jesus said unto her, I am the resurrection, and the life: he who believes in me, though he were dead, yet shall he live: And whosoever lives and believes in me shall never die. **Do you believe this**?" (John 11:25-26) Do *you* believe this? At this point I surely hope so…

"I came forth from the Father, and am come into the world: again, I leave the world, and go to the Father." (John 16:28)

"And now, O Father, glorify me with your own self with the glory which I had with you before the world was." (John 17:5)

"I went down into the countries underneath the earth, to the peoples of the past. But you lifted my life from the pit. Lord! My God!" (Jonah 2:6)[333]

"I do not pray for these alone, but also for those who will believe in me through their word; that they all may be one, as you, Father, are in me, and I in you; that they also may be one in us, that the world may believe that you sent me. And the glory which you gave me I have given them, that they may be one just as we are one: I in them, and you in me; that they may be made perfect in one, and that the world may know that you have sent me, and have loved them as you have loved me. Father, I desire that they also whom you gave me may be with me where I am, that they may behold my glory which you have given me; for you loved me before the foundation of the world." (John 17:20-24)

---

333 (From Jerusalem Bible translation, spoken by the Apostle John in Franco Zeffirelli's film *Jesus of Nazareth* in the scene where Lazarus is raised from the dead)

"Jesus answered, My kingdom is not of this world: if my kingdom were of this world, then would my servants fight, that I should not be delivered to the Jews: but now is my kingdom not from here." (John 18:36)

"Then said Pilate unto him, Do you not speak to me? Do you not know that I have power to crucify you, and have power to release you? Jesus answered, You would have no power over me if it had not been given to you from above." (John 19:10-11)

"But Thomas, one of the twelve, called the Twin, was not with them when Jesus came. The other disciples therefore said to him, We have seen the Lord. But he said to them, Unless I see in his hands the print of the nails, and put my finger into the print of the nails, and thrust my hand into his side, I will not believe. And after eight days his disciples were again inside, and Thomas with them. Jesus came, the doors being shut, and stood in the midst, and said, Peace be unto you. Then said he to Thomas, Reach your finger here, and look at my hands; and reach your hand here, and thrust it into my side. Do not be faithless, but believe. And Thomas answered and said to him, My Lord and my God. Jesus said unto him, Thomas, because you have seen me, you have believed. **Blessed are those who have not seen, and yet have believed**. [That would be us.] And truly Jesus did many other signs in the presence of his disciples, which are not written in this book; but these are written, that you may believe that Jesus is the Christ, the Son of God; and that believing you might have life through his name." (John 20:24-31)

"**Neither is there salvation in any other: for there is none other name under heaven given among men, whereby we must be saved**." (Acts 4:12) ...Mohammed? ...Buddha? ....Pope Johann XXXIII? ...Allah? ...Moroni? Not a chance. (Those who know the truth must speak the truth in love, but for heaven's sake, let's not keep our mouths shut because we "don't want to offend anyone." Besides, there is nothing in this universe more offensive than travelling through this life as one of the deceived and then waking up in hell.)

"For in it the righteousness of God is revealed from faith to faith; as it is written, 'The just shall live by faith'." (Romans 1:17) Living a life pleasing to God is only possible through faith: for "without faith it is impossible to please him;" and the attaining of eternal life comes only through faith as well. But so too, "faith without works is dead."

"Therefore no one will be declared righteous in his sight by observing the law; rather, through the law we become conscious of sin. But now a righteousness from God, apart from the law, has been made known, to which the Law and the Prophets testify. This righteousness from God comes through faith in Jesus Christ unto all and upon all who believe. ...For what does the Scripture say? 'Abraham believed God, and it was counted unto him for righteousness.' Now when a man works, his wages are not credited to him as a gift, but as an obligation. However, to the man who does not work but trusts and believes on him who justifies the ungodly, his faith is counted as righteousness." (Romans 3:20-22, 4:3-5) In these verses from Paul's letter to the Romans we see further clarification of the relationship between saving faith and works under the Law. This was just as much a hurdle to grasp back then as it is for many now, and also was a golden opportunity for false teachers and preachers to come in and lessen the effectiveness and perfect truth of the gospel of Jesus Christ. But it can be broken down to its essence and summed up in these two verses, "the just shall live by faith" (which we would add that the just shall live *eternally*, or shall *gain eternal life*, by faith as well.), and "faith without works is dead."

"God will credit righteousness to us who believe in him who raised Jesus our Lord from the dead. He was delivered over to death because of our sins, and was raised to life for our justification." (Romans 4:24-25)

"But God demonstrates his own love toward us, in that while we were yet sinners, Christ died for us. Since we have now been **justified by his blood**, how much more shall we be saved from God's wrath through him." (Romans 5:8-9) This concept of being "washed in the blood of the Lamb" has been prone to ridicule by some as being a little "barbaric." But the Bible says in Hebrews 9:22 that "according to the law almost all things are purified with blood, and without shedding of blood there is no forgiveness [of sins]." And this was also prophesied right at the fall when God killed an animal for the now-naked Adam and Eve to cover themselves. "Unto Adam also and to his wife did the Lord God make coats of skins, and clothed them." (Genesis 3:21) It is a little arrogant, and of course ignorant, of puny man to challenge the means that a perfect, infinitely intelligent and loving God would use to provide for our salvation. Being washed in the blood of the Lamb is to be clothed in the righteousness of Christ, without which no one can stand before God, or spend an eternity with him in Paradise. Jesus is "the Lamb of God who takes away the sins of the world." And along these lines the following analogy is very instructive...

"When Ron was lecturing in New Zealand and Australia, the sheep ranchers told him what often happens in a large flock of sheep. When the mother ewes are giving birth to lambs, there will often be a mother that dies while giving birth to a live lamb. But somewhere else in the flock a mother ewe gives birth to a dead lamb. The sheep ranchers bring the orphan lamb to the mother who lost her baby, in order for the orphan to nurse and feed. But the mother ewe can smell that it is not her baby, and she will always kick it away and not allow it to suckle. But the sheep ranchers have discovered that they can take the blood of the stillborn lamb and smear it as a covering over the fleece of the orphan lamb. Then when they bring that lamb to the mother who lost her baby, she will smell the blood, sense that it is her lamb, and allow it to nurse and feed.

"It's the same way with God. God is holy and will not look upon our sin. But when the blood of Jesus Christ covers us and cleanses us and forgives us, the holy God looks down upon us and does not see our sinful nature. Instead, he sees the blood of Jesus Christ that covers us. So he accepts us as his own. It is the blood of Jesus Christ that covers us and cleanses us and reconciles us to a relationship with God."[334]

Indeed, this biological instinct that the Creator built into lambs is also a prophecy of what is required for a restored relationship with Him, through the shed blood of the Lamb of God, Jesus.

"Without the precious blood of Jesus Christ there is no forgiveness, no redemption, no cleansing, no justification and no peace. Without the precious blood of Jesus, there is no

---

334   (Pg. 51, *Fast Facts on False Teachings*, by Ron Carlson and Ed Decker)

sanctification, and without sanctification no way into the presence of God. And without the precious blood of Jesus there is no victory and no glory."[335]

"What can wash away my sins? Nothing but the blood of Jesus. What can make me whole again? Nothing but the blood of Jesus. Oh! Precious is the flow That makes me white as snow. No other fount' I know, Nothing but the blood of Jesus."[336]

"Therefore, just as through one man [Adam] sin entered the world, and death through sin, and thus death spread to all men." (Romans 5:12) And through one man, Jesus, salvation entered the world, and through his death, life spread to all those who believe.

"For if by the trespass of one man [Adam] death reigned through that one, how much more will those who receive the abundance of God's grace and his gift of righteousness, reign in life through the one, Jesus Christ." (Romans 5:17)

"We shall all stand before the judgment seat of Christ. For it is written [in Isaiah 45:23], As I live, says the Lord, every knee shall bow to me, and every tongue shall confess to God." (Romans 14:10-11) "Every knee shall bow and every tongue shall confess" that Jesus Christ is Lord. And this includes both the saved and the unsaved. There will be no one at that final judgment who will accuse the Lord of being unjust for sending them to an eternity in hell. Keep this in mind when thinking of who will be spending an eternity in paradise and who will not, that God is an absolutely perfect and wholly righteous Judge. He makes no mistakes. "But what about all of those scattered around the world, like Aborigines, Eskimos, Africans and South American Indians, who died centuries in the past without ever having the opportunity of hearing the gospel due to their geographical location?" Answer: they will be judged exactly the same as all of the Old Testament saints who are now in heaven, for they also did not have the opportunity to hear the gospel due to their historical location, but they were thought worthy nonetheless. They had faith in God according to the level of information that they had at their disposal, and it was accounted unto them for righteousness. "Abraham believed God, and it was imputed unto him for righteousness." (James 2:23) And then there is this passage from Romans; "For when the Gentiles, who have not the law, do by nature the things contained in the law, these, although having not the law, are a law unto themselves, who show the work of the law written in their hearts, their conscience also bearing witness, and their thoughts the mean while accusing or else excusing one another; In the day when God shall judge the secrets of men by Jesus Christ according to my gospel." (Romans 2:14-16) And this should not be used as an excuse for us to sit at home and not go and witness to the lost, whether your neighbor across the street or someone in a country across the globe. But this information should be used as a cause to further glorify God, who is a God of perfect justice, judgment and righteousness; and as a way to refute the foolish who would accuse him of being otherwise.

---

335   (From "The Blood of Jesus," by Thomas Lieth. Midnight Call Magazine, January 2009)
336   (By Robert Lowry (1826-1899), author of many popular hymns)

"Christ died for our sins according to the Scriptures." (1 Corinthians 15:3)

"For the preaching of the cross is to those who perish foolishness; but unto us who are saved it is the power of God. For it is written, I will destroy the wisdom of the wise, and will bring to nothing the understanding of the prudent." (1 Corinthians 1:18-19)

"But as it is written, Eye has not seen, nor ear heard, neither has it entered into the heart of man, the things which God has prepared for those who love him." (1 Corinthians 2:9)

"We implore you on Christ's behalf, **be reconciled to God**." (2 Corinthians 5:20) The Sovereign Almighty God, the Creator of heaven and earth, is imploring the world to be reconciled to him; or to be cast into a Lake of Burning Sulfur for all eternity with no hope of escape. Wow! What an easy choice! I mean, this is not a tough choice! World, are you listening? Sadly...

"Christ has redeemed us from the curse of the law, being made a curse for us: for it is written, Cursed is every one who hangs on a tree:" (Galatians 3:13) "The Creator of the universe, who owes us nothing, became a curse for us, so that we may be free from the curse of sin."[337]

"For by grace you have been saved through faith, and that not of yourselves; it is the gift of God, not of works, lest anyone should boast." (Ephesians 2:8-9) We are saved by the **grace** of God. It is something we don't deserve. Even our faith in him is his gift to us, so Jesus gets all the glory; as it should be.

"Yes doubtless, and I count all things but loss for the excellency of the knowledge of Christ Jesus my Lord: for whom I have suffered the loss of all things, and do count them but dung, that I may win Christ" (Philippians 3:8) Paul is saying that everything you can acquire in this life—money, possessions, houses, lands, cars, boats, gold, jewels, TV's, laptops, electronics etc. etc.—is nothing other than stinky doggy-doo compared to "the excellency of the knowledge of Christ Jesus my Lord;" when compared to being saved.

"He has rescued us from the power of darkness, and transferred us into the kingdom of his dear Son." (Colossians 1:13-14)

"For there is one God, and one Mediator between God and man, the man Christ Jesus; who gave himself a ransom for all, to be testified in due time." (1 Timothy 2:5-6) Jesus is our High Priest. We, as true Christians, can go directly to the Father through him. We have no need for any other, but pastors and preachers are there for our instruction in the Word and guidance in better following Him.

"God, who at many times and in various ways spoke in time past to the fathers by the prophets, has in these last days spoken to us by his Son, whom he has appointed heir of all things, and through whom he made the universe. The Son is

---

337   (Pastor Ed Regensburg, Faith Community Church, Gambrills, Md.)

the brightness of Gods glory and the exact image of his being, upholding all things by his powerful word. When he had by himself purged our sins, he sat down at the right hand of the Majesty in heaven." (Hebrews 1:1-3)

"For we also had the gospel preached unto us, just as they did; but the message they heard was of no value to them, because those who heard did not combine it with faith." (Hebrews 4:2) Ouch.

"Therefore, since we have a great High Priest who has gone through the heavens, Jesus the Son of God, let us hold firmly to the faith we profess. For we do not have a High Priest who is unable to sympathize with our weaknesses, but we have one who has been tempted in every way just as we are, yet was without sin. Let us therefore come boldly unto the throne of grace, so that we may obtain mercy, and find grace to help us in our time of need." (Hebrews 4:14-16)

"For this Melchisedek, king of Salem, priest of the most high God, who met Abraham returning from the slaughter of the kings, and blessed him; To whom also Abraham gave a tenth part of all; first being by interpretation King of righteousness, and after that also King of Salem, which is, King of peace; Without father, without mother, without genealogy, having neither beginning of days, nor end of life; but made like unto the Son of God; abides as a priest continually. Now consider how great this man was, unto whom even the patriarch Abraham gave the tenth of the spoils." (Hebrews 7:1-4) This is an example of a Theophany, or a preincarnate appearance of Jesus Christ.

"It is appointed unto men once to die, but after this the judgment:" (Hebrews 9:27)

"But without faith it is impossible to please him, for he who comes to God must believe that he is, and that he is a rewarder of those who diligently seek him." (Hebrews 11:6)

"Blessed be the God and Father of our Lord Jesus Christ, who according to his abundant mercy has given us a new birth to a living hope through the resurrection of Jesus Christ from the dead." (1 Peter 1:3) In a hopeless world there is only one place to turn for real hope.[338] To the one who can give us hope in this life both for the strength and wisdom and help to live it triumphantly, and for the eternal joy that will be ours in the life to come.

"But you are a chosen generation, a royal priesthood, a holy nation, a peculiar people; that you should show forth the praises of him who has called you out of darkness into his marvelous light." (1 Peter 2:9)

"If we say that we have no sin, we deceive ourselves, and the truth is not in

---

338   (And how sad it is that so many deceived people who call themselves "Christians" voted for a baby mutilator like "The man from Hope" (Clinton) and then again for another one, the author of "The Audacity of Hope"(Obama), rather than having the good sense to turn from them and their evil, to the One who alone offers real hope. How unfortunate to be so deceived)

us." (1 John 1:8-9) Here John is addressing those Christians who fell under a false doctrine that taught that because they were saved, they were without sin in this life. "But when Jesus comes [the Second Coming], the sin question will be settled once for all. *Now* we have *forgiveness* of sins. *Then* we will have complete eradication of sin. [Now] we have the first-fruits of salvation, but when Jesus comes, we get the complete salvation."[339] Christians will still struggle with sin (which is not the same as *iniquity*- see chapter 11) until either they die, or they are raptured at the trumpet of the Lord. (See chapter 9.)

"God is love, and he who abides in love abides in God, and God in him." (1 John 4:16) "Behold, I stand at the door, and knock: if any man hear my voice, and open the door, I will come in to him, and will dine with him, and he with me." (Revelation 3:20) "And let him who thirsts come and let him take the water of life freely." (Revelation 22:17)

"And Abraham took the wood of the burnt offering, and laid it upon Isaac his son; and he took the fire in his hand, and a knife; and they went both of them together. And Isaac spoke unto Abraham his father, and said, My father: and he said, Here am I, my son. And he said, Behold the fire and the wood: but where is the lamb for a burnt offering? And Abraham said, My son, God will provide himself a lamb for a burnt offering: so they went both of them together." (Genesis 22:6-8) Another of the Old Testament prophecies of Jesus the Messiah. Abraham's son was spared, but God did not spare his only Son, for our sakes.

"The Lord bless you, and keep you: The Lord make his face shine upon you, and be gracious unto you: The Lord lift up his countenance upon you, and give you peace." (Numbers 6:24-26)

"The Lord is nigh unto all those who call upon him, to all who call upon him in truth. He will fulfill the desire of those who fear him: He also will hear their cry, and will save them." (Psalm 145:18-19)

"Come now, and let us reason together, says the Lord: though your sins be as scarlet, they shall be as white as snow; though they be red like crimson, they shall be as wool." (Isaiah 1:18)

"Therefore the Lord himself shall give you a sign; Behold, a virgin shall conceive, and bear a Son, and shall call his name Immanuel." (Isaiah 7:14) Again, Immanuel means *God among us*.

"For unto us a Child is born, unto us a Son is given: and the government shall be upon his shoulder: and his name shall be called Wonderful, Counselor, The mighty God, The everlasting Father, The Prince of Peace. Of the increase of his government and peace there shall be no end, upon the throne of David, and upon his kingdom," (Isaiah 9:6-7)

"Behold, my servant shall deal prudently, he shall be exalted and extolled

---

339   (Pgs. 107-108, *Prayer: Asking And Receiving*, by John R. Rice)

and be very high. [God's servant, which is Jesus, is bestowed upon with qualities normally reserved for the Most High, because He and the Father are one and the same.] As many were astonished at you; so his visage was so marred more than any man and his form more than the sons of men: [That refers of course to the brutal bodily harm done to Jesus during his scourging and crucifixion.] So shall He sprinkle many nations, [with his own blood, for the salvation of those who hear his voice.] Kings shall shut their mouths at him: [Referring to his triumphant Second Coming which will shortly come to pass.] for what had not been told them they shall see, and what they had not heard they shall consider." (Isaiah 52:13-15) These verses are just one of many Old Testament prophecies that were fulfilled in Jesus.

"Who has believed our report? and to whom is the arm of the Lord revealed? For he shall grow up before him as a tender plant, and as a root out of a dry ground: he has no form nor comeliness; and when we shall see him, there is no beauty that we should desire him. He is despised and rejected of men; a man of sorrows, and acquainted with grief: and we hid as it were our faces from him; He was despised, and we esteemed him not. Surely he has born our grief's, and carried our sorrows: yet we did esteem him stricken, smitten of God, and afflicted. But he was wounded for our transgressions, He was bruised for our iniquities: the chastisement of our peace was upon him; and with his stripes we are healed. All we like sheep have gone astray; we have turned every one to his own way; and the Lord has laid upon him the iniquity of us all. He was oppressed, and he was afflicted, yet he opened not his mouth: he is brought as a lamb to the slaughter, and as a sheep before her shearers is dumb, so he opened not his mouth. He was taken from prison and from judgment: and who shall declare his generation? for he was cut off out of the land of the living: for the transgression of my people was he stricken. And he made his grave with the wicked, and with the rich in his death; because he had done no violence, neither was any deceit in his mouth. Yet it pleased the Lord to bruise him; he has put him to grief: when you shall make his soul an offering for sin, he shall see his seed, he shall prolong his days, and the pleasure of the Lord shall prosper in his hand. He shall see of the travail of his soul, and shall be satisfied: by his knowledge shall my righteous Servant justify many; for he shall bear their iniquities. Therefore will I divide him a portion with the great, and he shall divide the spoil with the strong; because he has poured out his soul unto death: and he was numbered with the transgressors; and he bore the sin of many, and made intercession for the transgressors." (Isaiah 53) Probably the most widely quoted of the Old Testament prophecies concerning the Jewish Messiah, Jesus Christ.

"I saw in the night visions, and, behold, one like the Son of man came with the clouds of heaven, and came to the Ancient of days, and they brought him near before him. And there was given him dominion, and glory, and a kingdom, that all people, nations, and languages, should serve him: his dominion is an everlasting dominion, which shall not pass away, and his kingdom that which shall not be destroyed." (Daniel 7: 13-14)

Decisions have consequences in this life. Some may have unintended and even very drastic consequences, like a young driver who is in a real hurry and in a flash decides to run a red light and **BOOM** he smashes into a car and kills a mother and her two little children, and then ends up going to prison for thirty years and loses most of his life. Not what he had in mind, but nevertheless the unintended consequences of a very bad decision. But the decision you make in this life for whether you will reject or embrace Jesus Christ and his incomparable gift of eternal life has the most terrible or wonderful consequences. Intended consequences if you choose wisely. Unintended if you choose otherwise. No one chooses hell consciously. They are suckered, duped by Satan the great deceiver, the father of lies, into rejecting heaven by rejecting Christ. And like the driver who didn't intend to go to prison but must suffer the consequences, they too must suffer forever the consequences of their extremely poor and misinformed decision.

There is a God, and we can know this for a scientific fact. The Bible was written by Him for our information, our instruction and most importantly for the knowledge of salvation which it imparts; this too we can know to be true based on all of the facts and evidence. And finally Jesus Christ is God. He and the Father are one and the same. And he came to this earth for one over-arching purpose, to be a sacrifice for the sins of mankind. "Behold! The Lamb of God who takes away the sin of the world!" (John 1:29) This is the greatest secret of the universe. And again, not because the universe has tried to keep it hidden. It has been shouted from the highest hills and the deepest valleys, preached from a thousand pulpits and passed down from generation to generation, from father to son and from mother to daughter, from grandmother to grandson, from one friend to another and from one traveler to the next, and from countless missionaries all across the face of the entire earth, for two thousand years, in spite of terrible and unending persecution, to anyone who has had ears to hear. No, it is secret because the people, nations and languages of this earth have been hoodwinked into ignoring it. The educated Left elite are deceived into thinking it is a "religious fable" and flattered by the father of lies that they could never be so stupid as to fall for something like that. The Jewish people are deceived by their own Rabbis and indoctrinated that Jesus is not their Messiah, when of course he obviously is. The Muslims are lied to by their own Mullahs and taught to actually hate the truth; and to hate and kill those who would try to witness to them about this greatest of all truths; a truth that could save them from an eternity of indescribable pain and suffering, burning alive in the actual fires of hell with their false prophet Mohammed and his demonic entity Allah screaming along with them, and of course powerless to help. ("Mohammed! Where are my damn virgins?!") And Catholics are fooled into thinking that they can work their way into heaven by "doing the best they can," by "avoiding those really bad *mortal* sins," by making sure their good works outweigh the bad, and by putting their trust, faith and allegiance in "Holy Mother Church;" and thus they make the cross of Christ of no effect. And every day salt-of-the-earth type of people are just too busy struggling to get by in this life, and just too skeptical that there even is a God who would love them so much and care for them so much as to come down here and die on a cross for them personally ("I mean, if he loves us so much, why did he makes us so dang poor?"); so many just don't want to be bothered with it at all. Satan whispers in their ear that it's all just a bunch of hooey and they confidently pass it off as something "that's just for all those holy rollers." And so it remains a secret, to the majority of people, nations, religions and cultures across this earth.

By far the most important thing that you can attain in this life is the gift of eternal life. The most crucial thing that you can accomplish is coming to the knowledge of the truth of Jesus Christ as your Lord and Savior. In fact it is literally millions upon millions of times more important than anything else. Indeed nothing that you can attain or accomplish in this life will have any lasting meaning or value whatsoever if you fail in this one regard.

This chapter is the focal point of this book, as Jesus is the focal point of the entire Bible, and as he being the Creator and Sustainer of the entire universe is the focal point of all things. Grasping the truth of this and believing in his power to save you from an eternity in hell, and taking advantage of it by appropriating through faith his gift of eternal life, is also the key to "Understanding All Things." For what good is all the truth and understanding in the world if this most important of all truths is not understood?

Indeed it is extremely difficult to come to the knowledge of the truth in many other areas if this one most important area of all is neglected. For then the father of lies will still have the rule over you, and your mind will still be operating under the cloud of his deceit. This is why many times people have found that it is only after they have accepted Christ as Savior that, for example, the lie of evolution, Satan's genesis, suddenly becomes so painfully obvious as the balderdash that it is, that it is almost impossible to comprehend that just a short while before they actually believed it was true. It is as if a great dark fog of deceit and blindness had suddenly been lifted from their eyes. But before that moment of accepting Christ the lie still had a stranglehold on their thought process. The same is true of people who are trapped in various false religions. It is only after surrendering to Christ as Lord and Master does the truth again become immediately and painfully obvious. Up until that point, however, no amount of talking or explaining, no matter how many irrefutable facts and evidence were brought to bear, seemed to make a dent in the impregnable fortress inside of their mind where their false religion was enshrined and protected. Not until Christ, the Truth Himself, was embraced as Lord and Savior did all of their inner walls and defenses come crashing down like a house of cards, and the lies of Satan that blanket this earth and hitherto covered their mind, were exorcised once and for all. Again, Christ was the key to their being freed from the lies that had previously imprisoned their mind and their thinking and had kept them shackled and enslaved to the will of the father of lies. And that is because Jesus is the truth, and without him it is impossible to free your mind from the powerful prison of Lucifer's lies. And this is especially important to keep in mind before we delve into Part III of the following chapter.

"If the Son therefore shall make you free, you shall be free indeed." (John 8:36)

See:

- *The Case for Christianity*, by C. S. Lewis

- *The Case for Christ: A Journalist's Personal Investigation of the Evidence for Jesus*, by Lee Strobel with Jane Vogel
- *Why I Believe: In The Bible, God, Creation, Heaven, Hell, Moral Absolutes, Christ, The Resurrection, Christianity, The Second Birth, The Holy Spirit, The Return of Christ*, by D. James Kennedy
- *Skeptics Who Demanded a Verdict*, by Josh McDowell
- *Christianity: Hoax or History?* by Josh McDowell

- *Prayer: Asking and Receiving*, by John R. Rice
- *How I Know Christ Rose from the Dead*, by D. James Kennedy, Coral Ridge Ministries
- *Biblical Mathematics*, by Ed F. Vallowe
- *More Than A Carpenter*, by Josh McDowell
- *23 Minutes in Hell*, by Bill Wiese
- *90 Minutes in Heaven: A True Story of Death and Life*, by Don Piper and Cecil Murphey
- *Knowing God*, by J. I. Packer
- *The Resurrection of Jesus: A New Historiographical Approach*, by Michael Licona
- *The Star of Bethlehem DVD: Unlock the Mystery of the World's Most Famous Star*, by Presenter Frederick A. Larson. What sign in the heavens did the Magi, the wise men from the east, see that led them to Judea and the birth of the Messiah? Was it Jupiter aligned with Venus, the brightest star ever seen in the night sky? Incredibly, even two thousand years later, astronomy is still able to confirm the accuracy of the Biblical account.

Book II

FIRE

CHAPTER 8

TRUE RELIGION

(SEE PREVIOUS CHAPTER.)

Part I- Pure Religion
"Pure religion, undefiled before God and the Father is this, to visit orphans and widows in their affliction, and to keep oneself unsoiled by the world." (James 1:27)

A religion is a system of beliefs that puts forth answers, whether they are accurate or not, to those questions that we had at the beginning of chapter 6 concerning the Creator; his nature and character; how he brought the universe and life on earth into being; and what was his purpose in doing so. And there is only one system of beliefs that answers those questions factually and accurately. There is only one set of beliefs that is true. All the rest are false. There is only one that is based on all of the facts and all of the evidence both historically and scientifically. And it is the claims of only one religion that would hold up in a court of law under the intense scrutiny of a rational and unbiased inquiry. The claims of all the others, unfortunately, would fall far short.

This subject is clearly quite controversial, even to the point where the truth itself is found to be downright offensive to a great many of the peoples, nations, languages, cultures and religions of this earth. (Of course if someone finds the truth offensive then the problem is not with the truth, but with that individual and his set of false beliefs.) The lies of the great deceiver cover the minds of the people of this earth like a dense fog, and people believe what they are taught, and the vast majority have of course bought into the false religious systems that they have been born and bred into. And the natural reaction of those people is to strongly defend them, no matter how indefensible they may be.[340] As we stated earlier, that is why Jesus said the following: "Think not that I am come to bring peace on earth. I came not to bring peace, but a sword." (Matthew 10:34) He knew the anger and the strife that would occur between those people who would choose to hold onto the lies that they were taught, and those who would embrace the truth that he taught. The two, lies and truth, are completely incompatible, as God and the devil are as well. But those of us who are blessed with the priceless gift of eternal life are commanded to share that gospel with our lost brothers and sisters on this earth no matter what the cost or sacrifice. "I have become all things to all people, that by all means I might save some." (1 Corinthians 9:22) We are to try to be as inoffensive as possible, to speak the truth and to speak it in love, but ultimately, offensive or not, we are commanded to set our light upon a hill and not hide it under a basket. "You are

---

340   (I mean, not to single out our unfortunate Muslim friends, but how incredible is it that someone could actually be brainwashed into believing that they will be rewarded by God with 72 virgins for slaughtering innocent men, women and children? How their Mullahs keep a straight face while shoving this particularly disgusting religious lie down the throats of their young charges is beyond me.)

the light of the world. A city that is set on a hill cannot be hidden. Neither do men light a lamp and put it under a basket, but on a lampstand, and it gives light unto all who are in the house." (Matthew 5:14-15)

There is also, in addition to the claims, doctrines, practices and beliefs of any particular religion, one paramount question of crucial importance that concerns the next life, and that is: What must I do to get to heaven? What must I do to attain eternal life? What must I do to be saved? And again there is only one religion that reveals the true answer to this ultimate and all important question. There is only one way to get to heaven, and only one way to enter into a personal relationship of fellowship, eternally, with the God of Creation. Jesus is that way. "I am the way, the truth, and the life: no man comes unto the Father, but by me." (John 14:6) All the other ways are leading to someplace else and to an eternity with someone else.

> "Religion is man trying to get to God. but Christianity is God reaching down to man. God took the initiative and sent Jesus Christ into the world to take our sin that separated us from God, and to nail that sin onto a cross. He then covered that sin with His blood as the payment to restore us back into full relationship with Himself."[341]

"Jesus said unto them, If God were your Father, you would love me: for I proceeded forth and came from God; neither came I of myself, but he sent me. Why do you not understand my speech? even because you cannot hear my word. You are of your father the devil, and the desires of your father you will do. He was a murderer from the beginning, and does not abide in the truth, because there is no truth in him. When he speaks a lie, he speaks from his own depths: for he is a liar, and the father of it. And because I tell you the truth, you believe me not." (John 8:42-45)

Remember what was said at the beginning of this work about the all-importance of loving the truth more than your own opinions. The previous verses from John 8 state the reason why. **"Why do you not understand my speech? Even because you cannot hear my word. ...And because I tell you the truth, you believe me not**." The religious leaders that Jesus was speaking to not only did not love the truth more than their own opinions, beliefs and wicked purposes; they hated the truth, which is why they could not hear his word, and why they hated him. They were too much in love with their own opinions of what the Messiah should be- someone to save them not from their sins, but from the Romans. They were too much in love with their own power as well; and were incensed that Jesus, being so popular with the people but rejecting the hypocrisy of elites like them, might siphon it away. He was clearly not "one of them," so he had to go. But when one gives themselves over to the search for truth, it is the lies in one's head and the evil intents in one's heart that must go, and certainly not Jesus.

"God changes the heart. Religion is just something you rub on the outside."[342] That's why Jesus said "You must be born again." (John 3:7) As well as this, "Woe unto you, scribes and Pharisees, hypocrites! for you are like unto whited sepulchers, which indeed appear beautiful outward, but are within full of dead men's bones,

---

341   (Pg. 30, *Fast Facts on False Teachings* by Ron Carlson and Ed Decker)
342   (Dr. G. Vernon McGhee)

and of all uncleanness." (Matthew 23:27) The scribes and Pharisees had the rub-on form of religion, and had rubbed their own righteousness all over themselves, for all the world to see. But their holiness was a sham. Inside they were the furthest thing from holy. The same is true of so many devoutly religious people today, from so many of the world's false religions. They're trying real hard to be good and holy and righteous, but as long as they refuse to surrender to the righteousness of Christ, which is righteousness indeed, and as long as they steadfastly hold onto the false religious lies that they were taught, they are worshipping God in vain.

The Body of Christ

In Greek the word for church is *ecclesia*, meaning *the called out*. The true church of Jesus Christ on this earth is made up of all those people who have been "called out" of the world to follow him. All those who have accepted his gift of eternal life, and who follow him and his commands as they go forward and struggle to live their lives through his grace in a way that is pleasing to him and that brings glory to him, and in a way that will help lead others to a saving knowledge of Him. All of these people make up what is called in God's word, the *body of Christ,* with Jesus as the *head* and all believers as *members* of this body. Christ lives within the members of his body and exercises his Lordship over them, to the extent that they allow him by following him and surrendering to his will. The members of this body are the physical temples where he also resides on this earth; in our hearts and minds and souls. ("Know you not that you are the temple of God, and that the Spirit of God dwells in you? If anyone defiles the temple of God, him shall God destroy; for the temple of God is holy, which temple you are." 1 Corinthians 3:16-17) The *body of Christ* refers to the physical aspect of his people that you can see and hear and touch and relate to as opposed to his Spirit which you cannot see or hear or touch with your physical senses. Also, as the many different parts of the human body—the nose, the foot, the eye, the heart, etc.—all vary in their usage and their purposes but each one is indispensable; so too the body of Christ on this earth is made up of many different people with varying talents and abilities, but each is an indispensable part of the whole.

"For as the body is one and has many members, but all the members of that one body, being many, are one body, so also is Christ. For by one Spirit we were all baptized into one body ...and have all been made to drink into one Spirit. For in fact the body is not one member but many. ...But now God has set the members, each one of them, in the body just as he pleased. ...there are many members, yet one body. And the eye cannot say to the hand, 'I have no need of you'; nor again the head to the feet, 'I have no need of you.' No, much rather, those members of the body which seem to be weaker are necessary. ...But God composed the body ...that there should be no schism in the body, but that the members should have the same care for one another. And if one member suffers, all the members suffer with it. ...Now you are the body of Christ, and members individually." (1 Corinthians 12:12-14, 18, 20-22, 24-27)

To become a member of this body of Christ, to be a true Christian, you must be born again. You must be saved. That is the only requirement. Becoming a member of one organization or another, or joining some church or another, has nothing to do with it. And don't misunderstand, there are some excellent churches out there that are full of saved people and it is important to be involved in one of them once you have accepted Christ. (Or while you are being led in that direction by the calling of the Holy Spirit.) But don't confuse any one church or any one Christian organization with the church of Jesus Christ on this earth. For the latter is made up of all saved people no matter what denomination they happen to subscribe to or to what church they happen to go.

"And he is the head of the body, the church: who is the beginning, the firstborn from the dead; that in all things he might have the preeminence." (Colossians 1:18)

Conversely, there are a number of false Christian churches spread across this earth that have nothing to do with the church of Jesus Christ, and whose members, unfortunately, are not members of His body of believers; as much as they may know about and believe about Him; being covered in their own righteousness and not the righteousness of Christ; having failed to put on a wedding garment... "Friend [He's speaking to one who thought they were a follower], how did you come in here without a wedding garment? And he was speechless. Then the king said to the servants, Bind him hand and foot, and cast him out into the outer darkness [hell]; there shall be the weeping and the gnashing of teeth. For many are called, but few chosen." (Matthew 22:12-14)

Witnessing

"And he said unto them, Follow me, and I will make you fishers of men." (Matthew 4:19)

What if you were on a camping trip and you saw a child pick up a can of gasoline and start walking over to a campfire, with the obvious intent of pouring it on for a bigger flame? What would you do? Wouldn't you jump up and rush over to stop him, screaming for him to get away from the fire, and/or alerting anyone who was closer that he was about to burn himself horribly?! Of course you would. If you're a human being. Well, that's what the Word of God is doing. That's why it was written. It's shouting at all of mankind to wake up; to throw down the *can* they're holding onto in their mind and in their heart that is filled with the combustible false-religious lies of Satan; to turn from their path that is heading toward an eternal fire, and a far worse fate than that of the little boy, as excruciatingly painful as it is to be burnt alive by gasoline. And that is also what we are called to do, we who have been given the unspeakable gift of eternal life through faith in the death and resurrection of our Lord Jesus Christ; to warn this world and the people in it of where they are heading, and in so doing to turn some of them, those who have "ears to hear," from that terrible eternal doom.

"For whosoever calls upon the name of the Lord shall be saved. How then shall they call on him in whom they have not believed? And how shall they believe in him of whom they have not heard? And how shall they hear without a preacher? And how shall they preach unless they are sent? As it is written: How beautiful are the feet of those who preach the gospel of peace, who bring glad tidings of good things!" (Romans 10:12-15)

Once you have "joined the true religion" by believing in Jesus Christ and following him ("And why do you call me, Lord, Lord, and do not the things which I say?" Luke 6:46), you are then commanded to share this good news with anyone who you would normally share great news with, your friends and family and neighbors and co-workers for example. We are not to "be ashamed of the gospel of Jesus Christ." But many Christians shrink from the thought of sharing the good news ("gospel") because they feel they just couldn't do a decent job of explaining it; or that they would be unable to answer an unsaved friends questions, doubts or arguments; and all they would get for their trouble is to end up being ridiculed and looking foolish. (Of course, Jesus Himself suffered the ultimate in ridicule and humiliation—being scourged with a leather whip that had imbedded in it chunks of jagged metal that ripped the skin off his back; having a crown of nail-like thorns mercilessly driven into his scalp and covering his head, face and neck in blood; forced to drag his own heavy cross to Calvary on shoulders already laid bare from the scourging; being nailed to the cross naked with jagged metal spikes driven through his feet and wrists and then left there to suffer and die, the Lord of all creation, exposed and mocked before his own created beings—in order to pay the penalty for our sins. A Christian should count it all joy if he or she is called upon to share in the mocking and humiliation of their Savior... "Blessed are you, when men shall revile you, and persecute you, and shall say all manner of evil against you falsely, for my sake. Rejoice, and be exceeding glad: for great is your reward in heaven: for so persecuted they the prophets who were before you." Matthew 5:11-12)

But beyond that, it is also expected of every follower to grow as a Christian, in your faith, in your knowledge of the truth, and in your knowledge of Jesus Christ; so the world can see the truth of the gospel reflected in you, in your character and in your life; and so when called upon to testify to the truth you can also do so confidently, in an informed manner. Indeed that is one of the goals of this book, to equip Christians with the knowledge of the whole truth so they can expose the misplaced ridicule of the militant atheist, the unscientific foolishness of the confused evolutionist and anyone else who opposes the gospel of Christ, and the truth, on either substantive or intellectual grounds.

And yet in the meantime, it is possible for even the newest of Christians, even while they're just starting down the road of growing in faith and in the knowledge of the truth, to be an effective witness for Jesus Christ; through the leading of the Holy Spirit. It is a worn out expression but nevertheless still true that "people don't care how much you know until they know how much you care." Our motivation to share the good news that we have found should come from a deep compassion and concern for the eternal plight of others, and not from any sense of superiority, arrogance, or self-righteousness. And leaning less on our own devices, but surrendering to the indwelling power of the Holy Spirit, is one of the keys to effective witnessing that focuses on the needs of others. Go to God "always with all prayer and supplication in the Spirit... be anxious for nothing, but in everything by prayer and supplication, with thanksgiving, let your requests be made known to God." (Ephesians 6:18; Philippians 4:6) Let sharing the gospel be a time also for silent prayer inside of your mind and heart, asking the Holy Spirit to lead you in what you should say and do, and asking the Holy Spirit to do His work and accomplish His will.

However, if you do find it difficult to personally witness to your friends, one of the best techniques is to just keep inviting them to church until they break down and come. I say that because it is extremely rare when someone will come to church the first time they are invited. This is normal as people are naturally apprehensive in situations like this. They may need to get to know you a little better, or see some positive changes in your own life, before they trust you enough to visit your church. Be persistent but respectful, make sure they know that you care about them and their needs and you're not just inviting them for your own glory. Of course, it goes without saying that you first have to find a really good, Bible-believing, evangelical, soul-winning church that doesn't vote for baby mutilators and openly displays the love of Jesus Christ to all who come.

Sharing the good news is <u>not</u> the *righteous* leading the *unrighteous* to Christ. For we are <u>all</u> unrighteous, we "have all sinned, and come short of the glory of God," "we have all like sheep gone astray." (Romans 3:23, Isaiah 53:6) It is just one well-fed beggar telling another starving beggar where they can get some bread; in this case, the Bread of Life.

Always have a Bible handy—in your home, in your car, in your desk at work or on your laptop—so when the opportunity presents itself you can open it and show someone how accessible eternal life is. Remember it is the Word of God that "is living and powerful, and sharper than any two-edged sword, piercing even to the dividing asunder of soul and spirit, and of the joints and marrow, and is a discerner of the thoughts and intents of the heart." (Hebrews 4:12) Let it do its work. Also, you don't need to be a biblical scholar, or to have an in-depth knowledge of the entire Old and New Testaments (but it certainly doesn't hurt). Indeed, even a brand new Christian can be a pretty effective witness armed with the excitement of their "first love" (see Revelation 2:4) and with four simple verses that are found in the book of Romans: Romans 3:23, Romans 6:23, Romans 5:8, and Romans 10:9-13:

"**For all have sinned** and come short of the glory of God." (Romans 3:23)

"**For the wages of sin is death**, but the gift of God is eternal life in Christ Jesus our Lord." (Romans 6:23)

"But God demonstrates his own love toward us, in that while we were still sinners, **Christ died for us**." (Romans 5:8)

"**That if you shall confess with your mouth the Lord Jesus, and shall believe in your heart that God has raised him from the dead, you shall be saved**. For with the heart man believes unto righteousness; and with the mouth confession is made unto salvation.

For whosoever shall call upon the name of the Lord shall be saved." (Romans 10:9-10, 13)

Those verses contain the essence of the salvation message. They provide a simple and effective road map in leading someone to Christ. 1) We are all sinners. 2) The penalty for our sin is eternal separation from a Holy God in a place of severe suffering. 3) God made a way of escape for us by sending his own Son to pay our penalty. 4) This incredible gift is of no use unless it is accepted, embraced and appropriated by believing on the Lord and Savior Jesus Christ. Other verses that mirror and further explain this simple gospel message are John 3:16, John 14:6, Matthew 7:8, Isaiah 55:6, John 1:12, 2 Corinthians 5:19, 21, Romans 3:10, Isaiah 64:6, Romans 5:12, Hebrews 9:27, Luke 16:22, 23 and Acts 16:31, just to name a few. And keep in mind that we don't *get* anyone saved, the Holy Spirit does; Jesus does. All we are required to be is a witness, one who will get our own self out of the way, and let Jesus and the Holy Spirit work through us to do their work, and their will.

We should also keep in mind that at the Final Judgment, when we all stand before the holy and Almighty God, it would be far better to be accused by someone, just before they are cast into the horror of an eternity in hell, that you were a "holy roller," than for it to be said, "you know, I knew him, and I believe he was a Christian, but he (or she) was no different than anyone else I knew." Ouch. It's a matter of what we prioritize in our mind. And when we think about the reality of what is to come then it should serve as some motivation to be a real Christian and a real witness to those around us, as opposed to practicing one form or another of a much more popular brand known as *stealth* Christianity.

"Let me always be focused on You, Oh Lord, and on the mission: spreading the gospel, the main purpose for every Christian in this life."

God calls us to a life of godliness after we accept Christ as Savior and Lord, as a sign of our reverence for his sacrifice on the cross for us, as a way for us to honor his unspeakable gift, and as a means to draw others to faith in him. And that leads us to the parable of the friend who knocks on his Neighbors door late at night seeking Bread...

"And he said to them, Which of you shall have a friend, and go to him at midnight and say to him, 'Friend, lend me three loaves; for a friend of mine has come to me on his journey, and I have nothing to set before him;' and he will answer from within and say, 'Do not trouble me; the door is now shut, and my children are with me in bed; I cannot rise and give to you.' I say to you, though he will not rise and give to him because he is his friend, yet because of his persistence he will rise and give him as many as he needs." (Luke 11:5-8)

This parable is about every Christian. For each of us has these two friends:

"One Friend is God, who has plenty of bread for sinners. ["I am the bread of life. He who comes to me shall never hunger, and he who believes in me shall never thirst." (John 6:35)] The other friend is the sinner himself. Every Christian in the world stands between God and the lost world. We are the channel that God has chosen to carry bread to sinners."[343]

But the interesting thing here is that Jesus did not answer right away the prayer of the believer for Bread for his unsaved friend. He only answers the prayer after much pleading and insistence.

"Isn't it strange for God to say, "Trouble Me not; don't bother Me. I don't have any time nor disposition now to give you bread for sinners"? And yet that is what the Scripture seems to teach! What can Jesus mean by these words?

"I think I can help you to see it by this illustration. ... will the mother immediately turn over an expensive sewing machine to the little girl the first time she wants to use it to sew? Or what foolish parent would turn an expensive, high powered automobile over to the little fellow the first time he wants to drive it? And so, I tell you it is foolish to expect the infinite God to place the dynamite of Heaven in the careless hands of a Christian who does not know the travail of soul that Christ had, and has not entered into the burden for sinners, and the sense of shame over sin. How can God give to the light-hearted Christian who casually asks to be filled with the Holy Spirit, asks to be made a great soul-winner— how can God, I say, give them the power to defeat Satan, to transform lives, yea, put in his hands all the infinite power that raised Jesus Christ from the dead?

"Certainly God means for us to get this lesson: there are some blessings that a Christian will never have without pleading, importunate [persistent] waiting on God! Those who want soul-winning power, want to be able to carry the bread of life to sinners, must learn the secret of *praying through*! ...Here is encouragement, then, for every child of God who has longed to win souls but never has. God has sent someone your way for you to win to Christ. A friend in his journey has come to you and you have had nothing to set before him, and when you asked God, He seemed to say "No." He seemed to disregard your prayer. You never received the power you requested and you never bore fruit as you longed to do.

---

343   (Pg. 80, *Prayer: Asking & Receiving*, by John R. Rice)

But you can! God …meant for you to keep on asking and praying and for you to refuse to be denied. How eagerly God waits for his people to pray. …How gladly He will give bread to those who wait before Him with importunity."[344]

"That I may know him, and the power of his resurrection, and **the fellowship of his sufferings**, being made conformable unto his death;" (Philippians 3:10)

"Christ would teach us to enter into His experience in the garden of Gethsemane, yea, into the very agony of the cross itself. Christ died to save sinners. For this purpose He left Heaven and came to earth, and was despised and abused and betrayed and deserted and spit upon and scourged and mocked and stripped and nailed to the cross and slain. Christ paid the price to get the bread of life to sinners. He Himself is that Bread. None of us can ever take His place. None of us can atone for man's sin. …But still it is true that He has granted that Christians who will may learn to enter into the intercessory work of Christ, to bear His burdens, suffer His travail, and so carry the bread of life from God to the sinner."[345]

Paul sought to know Christ… "and the fellowship of His sufferings." We should seek the same. Indeed Jesus' agony was so great that even the thought of what he was about to endure for the sake of our sins, when he was praying in the Garden of Gethsemane the night before, was almost too much to bear, so much so that he actually sweated blood. Yet far too many Christians, as they juggle through their busy lives, show very little concern and give even less thought for the lost. We need to go to the Lord daily and ask him to teach us to suffer for the lost, to bear a burden for them, to see and feel their eternal plight, and then instead of being comfortable and self-involved we might be more effective in reaching them. "Dear Lord, I don't suffer or hurt for the lost like I know that I should, like I want to and like I know that you want me to, but I'm trusting in you to change me." Be persistent! And then trust Him to lead you and grow you into the fellowship of his sufferings. And subsequently give God the glory when he allows you to reach the lost in a way that before might have been unavailable.

"There is no winner like a soul winner."[346]

Walking

"And Enoch walked with God." (Genesis 5:24)

"God was Enoch's first thought when he awoke in the morning, and his last thought when he went to sleep at night. He worked with God and he lived with God within his family. When he met friends or neighbors he let God accompany him. His conversations were coloured by his faith in God, and when he failed he clung to God. Everyone knew where they were with Enoch. It was his life's task to walk with God. He did not do this now and again, but continually. He lived with God in the broad daylight and in the darkest night. Whether he was alone or in the company of others made no difference. God was always the center of Enoch's life."[347]

Walking with the Lord should be the daily focus of every Christian. "As you have therefore received Christ Jesus the Lord, so walk in him…" (Colossians 2:6) Yet let's face it, "running with the devil" often times seems to come a whole lot easier. For we still struggle with the sin nature, and many things that we found so

---

344   (Pg. 87-88, 90 *Prayer: Asking & Receiving*, by John R. Rice)
345   (Pg. 82, *Prayer: Asking & Receiving*, by John R. Rice)
346   (Pastor John A. Krach Jr. Church on the Rock, Millersville, Md.)
347   (From article, "The Rapture of Enoch," by Norbert Lieth, Midnight Call Magazine May 2008)

attractive in the past before we were saved, now even though we know they are wrong and we struggle against falling back into them, still have a hold and a pull over us. But it might be comforting to know that no less of a "spiritual giant" than Paul (he wrote most of the New Testament) struggled with the same problem...

"For I know that in me (that is, in my flesh,) dwells no good thing: for to will is present with me; but how to perform that which is good I find not. For the good that I want to do, I do not: but the evil which I do not want to do, that I do. Now if I do that which I would not, it is no more I that do it, but it is sin dwelling in me that does it. I find then a law, that when I would do good, evil is present with me. For I delight in the law of God in my inward being, but I see another law in my members, warring against the law of my mind, and bringing me into captivity to the law of sin which is in my members. O wretched man that I am! who shall deliver me from this body of death? I thank God through Jesus Christ our Lord! So then with the mind I myself serve the law of God; but with the flesh the law of sin." (Romans 7:18-25)

Here then are some thoughts and verses that might be helpful in waging this internal war, and in coming to a closer walk with Him...

"For this reason I bow my knees before the Father of our Lord Jesus Christ, from whom every family in heaven and earth receives its name, that he would grant you, according to the riches of his glory, to be strengthened with power through his Spirit in the inner man, that Christ may dwell in your hearts through faith; so that you, being rooted and grounded in love, may be able to comprehend with all the saints what is the breadth, and length, and depth, and height; And to **know** the love of Christ, which surpasses all knowledge, that you might be filled with all the fullness of God. Now unto him who is able to do exceeding abundantly above all that we ask or think, according to the power that works within us, unto him be glory in the church through Christ Jesus throughout all generations, world without end. Amen." (Ephesians 3:14-21)

That's some powerful stuff! Along the lines of "Your word have I hid in my heart, that I might not sin against you." (Psalms 119:11)

And equally as powerful,

"Finally, my brethren, be strong in the Lord and in the power of his might. Put on the whole armor of God, that you may be able to stand against the wiles of the devil. For we wrestle not against flesh and blood, but against principalities, against powers, against the rulers of the darkness of this age, against spiritual wickedness in high places. Wherefore take unto you the whole armor of God, that you may be able to withstand in the evil day, and having done all, to stand. Stand therefore, having your loins girded about with **truth**, and having on the breastplate of **righteousness**; And your feet shod with the preparation of **the gospel of peace**; Above all, taking the shield of **faith** with which you shall be able to quench all the fiery darts of the wicked one. And take the helmet of **salvation**, and the sword of the **Spirit**, which is **the word of God**: Praying always with all prayer and supplication in the Spirit, and watching thereunto with all perseverance and supplication for all saints." (Ephesians 6:10-18)

It is interesting to note that all of the battle equipment in that passage is *defensive* except for the last, the "sword of the Spirit, which is the word of God;" the only *offensive* weapon in the group. Too many times we as Christians go out into the world unprotected, failing to "put on the whole armor of God," and ill-prepared to do battle, failing to equip ourselves with the truth, having neglected to study the word of God. ("Study to show yourself approved unto God, a workman that need not be ashamed, rightly dividing the word of truth." 2 Timothy 2:15) Forgetting to grow in the righteousness of Christ which we freely received, yet at great cost to him. Unfamiliar with the power of the simple gospel message so we can pass it along to others. Neglecting to grow in the faith that first led us to Christ. And never realizing the incredible power of divine protection and communication that we are heir to, joint heirs with Christ no less, because of our salvation, and adoption. And these failures lead to an unsatisfying and unfulfilled Christian life; one constantly buffeted by and undefended against the attacks of Satan. However, if we begin to take the admonition of Ephesians seriously and follow its recommendations, then we will find ourselves entering into the abundant life that Jesus promised. "I am come that [you] may have life, and that [you] may have it more abundantly." (John 10:10)

And finally, when we do attempt to do battle with the lies of Satan by witnessing to a friend or co-worker, many times we forget to use the one offensive weapon we have at our disposal, the Word of God, and just how effective it can be. Let's revisit the following verse, "For the word of God is quick and powerful, and sharper than any two-edged sword, piercing even to the dividing asunder of soul and spirit, and of the joints and marrow, and is a discerner of the thoughts and intents of the heart." (Hebrews 4:12) Again, we should not rely on our own wit and wisdom, not even so much on our own words, rather open up the Bible and share some passages and let God speak directly to your friends, family or coworkers, relying on his Word to change hearts and minds, and relying on his wisdom to guide us also in what we add to it. (For many people, including many who consider themselves Christians, perhaps because their parents were, it might be the first time they have ever seen the Bibles salvation verses, or read for themselves its clear salvation message.) "But when they deliver you up, take no thought about how or what you should speak. ...For the Holy Spirit shall teach you in the same hour what you should say." (Matthew 10:19 and Luke 12:12) And let God's promises fulfill themselves, "So shall my word be that goes forth out of my mouth; it shall not return unto me void, but it shall accomplish that which I please, and it shall prosper in the thing whereto I sent it." (Isaiah 55:11) And certainly one of the things "whereto he sends it," the same place that he sends us, is to the lost. Witnessing to others and walking with the Lord are intricately tied to one another.

"You have made us for yourself, and our hearts are restless, until we find our rest in you."[348]

Every Christian should be striving daily to serve the Lord who bought us with a terrible price, to walk with him, to be like him; by submitting to his will for our life; by "dying to self" where our selfish desires are in conflict with what he desires for our life. Paul pressed on "toward the goal for the prize of the upward call of God in Christ Jesus." (Philippians 3:14) And that goal was "that I may know him and the power of his resurrection, and the fellowship of his sufferings, being conformed to his death." (Philippians 3:10) And in this quest we have nothing less than the Almighty on our side, for it is God's desire to give us the desires of our heart, especially when the primary desire of our heart is for Him.

"Delight yourself also in the Lord, and he shall give you the desires of your

---

348 (From the *Confessions*, by Saint Augustine)

heart." (Psalms 37:4) "But seek first the kingdom of God and his righteousness, and all these things shall be added unto you." (Matthew 6:33) "Father, let me seek you, and let me find you. Let me search for you with all my heart." (Jeremiah 29:13)

"To delight in God is to find your deepest pleasure, your highest ecstasy, and your richest fulfillment in life through your relationship with him. ...As your relationship deepens, what is dear to God's heart becomes dear to your heart. This aligns your prayers with God's will. As your prayers are answered, you discover that your deepest desires are fulfilled."[349]

Just as an earthly father wants what is best for his children, even though what is best for them many times is not the same as what they want at any given moment, so too our Heavenly Father having perfect knowledge of what is best for us, is in a position far better than us to direct our paths. "My Father, show me how to pray in Thy will, I want to ask what will please Thee. I want to have what You want me to have."[350] "You have not, because you ask not." (James 4:2)

"Love focuses on what it desires. When you desire God, you spend your time and energy getting to know him better and doing what pleases him."[351]

But then, not every need for every person is meant to be immediately fulfilled. Sometimes that need or that problem or that crisis is placed there as a wake-up call because there is something in that person's life which is keeping him or her from a closer relationship with Jesus. Therefore it is important to "pray through" to get your requests answered. To keep on praying until the prayer is answered.[352] And along the way Jesus will open your mind to certain things in your life and in your heart that you need to address before your request is answered. That might be fear, or worry, or unforgiveness, or not trusting in him, or selfishness, or some unrepented sin. For unsaved people what needs to be addressed is accepting Christ as Lord and Savior, but so often they will pray for one pressing need or another but then give up if they do not get immediate results, and sometimes give up on belief in God entirely. They never come to the point where the Holy Spirit can convict them of their primary need, which in fulfilling would open the door to the fulfillment of other needs as well. "Seek you first the kingdom of God and his righteousness, and all these things shall be added unto you." (Matthew 6:33) And, "you shall seek me, and find me, when you shall search for me with all your heart." (Jeremiah 29:13) For those of us who are saved the point is *not* to give up, or we will miss out on the much greater blessing that Jesus has in mind, in addition to getting our original request fulfilled; as long as the original request was not contrary to the will of God... "If you shall ask any thing in my name, I will do it." (John 14:14)

"As the deer pants for the water brooks, so pants my soul after you, O God. My soul thirsts for God, for the living God." (Psalms 42:1-2)

"I urge you therefore, brethren, in view of God's mercy, to offer your bodies as

---

349   (Pg. 101, *The 100 Most Important Bible Verses*, by Thomas Nelson Publishers)

350   (Pg.59, *Prayer: Asking & Receiving*, by John R. Rice)

351   (Pg. 86, *The 100 Most Important Bible Verses*, by Thomas Nelson Publishers)

352   (Conversely it is not necessary to "pray through" to be saved. God will not withhold his salvation for even an instant to anyone who sincerely comes to him seeking eternal life. But for some of our other legitimate needs and requests sometimes he does withhold the answer until a higher purpose is met. ...See Chapter XIV, "Praying Through," of *Prayer: Asking and Receiving*, by John R. Rice.)

living sacrifices, holy and pleasing to God, which is your act of worship. And do not be conformed to the pattern of this world, but be transformed by the renewing of your mind. That you may be able to prove what is the good and acceptable and perfect will of God. For I say, through the grace given to me, to every one of you, do not think of yourself more highly than you ought, but think of yourself with sober judgment, in accordance with the measure of faith that God has given you. For as each of us has one body with many members, and these members do not all have the same function, so in Christ we who are many form one body, and each member belongs to all the others. We have different gifts according to the grace given us. If a man's gift is prophesying, let him use it in proportion to his faith. If it is serving, let him serve. If it is teaching, let him teach. If it is encouraging, let him encourage. If it is contributing to the needs of others, let him give generously. If it is leadership, let them govern diligently. If it is showing mercy, let them do it cheerfully. Love must be sincere. Hate what is evil; cling to what is good. Be devoted to one another in brotherly love. Honor one another above yourselves. Never be lacking in zeal, but keep your spiritual fervor, serving the Lord. Rejoice in hope, be patient in affliction, continuing diligently in prayer. Share with God's people who are in need. Practice hospitality. Bless those who persecute you; bless and do not curse. Rejoice with those who rejoice, mourn with those who mourn. Live in harmony with one another. Do not be proud, but be willing to associate with people of low position. Do not be conceited in your own opinion. Repay no one evil for evil. Be careful to do what is right in the sight of all men. If it is possible, as far as it depends on you, live peaceably with all men. Beloved, do not take revenge, but leave room for God's wrath, for it is written, Vengeance is mine, I will repay, says the Lord. Therefore If your enemy is hungry, feed him; if he is thirsty, give him something to drink. In doing this, you will heap coals of fire on his head [i.e. make him burn with shame]. Do not be overcome by evil, but overcome evil with good." (Romans 12)

> "Righteousness is more than just doing the right thing. It is being the right person, which begins by being in a right relationship with God. ...As you allow God to become more involved in your life, his righteousness begins to take hold in your own character. ... Being righteous is fulfilling God's expectations for you. It is living up to who he created you to be. It's impossible to achieve this through self-effort; rather, you achieve this through God's Spirit and your will working together in harmony."[353]

Unrepentant sin in one's life is a hindrance to an effective prayer life and an obstacle to a close relationship with Jesus. We will struggle with sin as long as we remain in these bodies, and until we receive our glorified, sinless bodies when we meet the Lord in the air prior to his return. (See discussion of the Rapture in the next chapter, "Bible Prophecy.") But it is crucial to confess ones sins every day, so we can remain in fellowship with Jesus. This doesn't mean that if you don't confess your sins you will "lose your salvation." It means that you will lose your ability to live the abundant life. "The thief does not come except to steal, and to kill, and to destroy. I have come that they may **have life**, and that they **may have it more abundantly**." (John 10:10) Daily confession of sin is also the point of the incident when Jesus washed the Apostle's feet. (See John 13:5-10) Peter at

---

353   (Pgs. 178-179, *The 100 Most Important Bible Verses*, Thomas Nelson Publishers)

first protested and then after being rebuked for refusing, he asked the Lord to wash his hands and his head as well. But Jesus explained that "He who is bathed needs only to wash his feet" (John 13:10) which means that they were already clean, they were saved, they had already been "washed in the blood of the Lamb" (see Revelation 7:14), but they *did* need to wash their feet because it is ones feet that comes in contact with the soil of this world. This alludes to one's daily sins because one is still bodily in this world and still comes in contact with it and still has the sin nature which will cause him or her to fall, whether in thought, emotion or deed.

"Therefore if any man be in Christ, he is a new creature: old things are passed away; behold all things are become new." (2 Corinthians 5:17)

But that discussion is still different from having unrepentant sin in your life that you are holding onto and refuse to turn from, such as continued adultery, or lesbianism, or homosexuality, or lust, or sloth (i.e. the lazy refusal to go out and find a job and support your family instead of lying around watching TV and making excuses), or gluttony, or drunkenness, or supporting murder disguised as a woman's right to choose, or constant worry, or the abuse of drugs, just to name a few. Allowing any of these sins to take hold in your life and to not struggle against them, and to not daily pray for release from them, and to not daily cry out in repentance from them is a major obstruction to living the abundant life, to having your prayers heard at all, and to a closer walk with Jesus.

"Therefore submit to God. **Resist the devil and he will flee from you**. Draw nigh to God and he will draw near to you." (James 4:7-8) And that's the problem, most of us, including most of us Christians, never even put up a fight. Yet that is a clear promise from God- that if you resist Satan, he will flee from you. Put him to the test. Whatever your major obstacle in life, or sin problem is—whether it's temper, or cigarettes, or sexual sins, or alcohol, or cursing, or drugs, or worry, or whatever—take your primary sin that you seek to overcome, let's say it is temper, and make your primary goal in life to wait like a cat ready to pounce the next time your temper rears its ugly head. Don't be asleep and let it hit and run its course until eventually you realize, "Hey, wasn't I supposed to resist that?!" No, your goal in life should be to look forward to the next time that something comes along to make your temper flare up, so you can put God and his promise to the test. So you can resist the devil and see whether or not he will flee from you. I assure you he will, if you resist. The same with cigarettes. The same with alcohol. The same with anything. Put him to the test. See if that promise is true. It is. Because God cannot lie. It might take you a number of battles, it might take numerous times of resisting Satan, but if you are faithful, you will find release.

"If we say that we have no sin, we deceive ourselves, and the truth is not in us. If we confess our sins, he is faithful and just to forgive us our sins and to cleanse us from all unrighteousness. If we say that we have not sinned, we make him a liar, and his word is not in us." (1 John 1:8-10)

"No temptation has overtaken you except such as is common to man; but God is faithful, who will not allow you to be tempted beyond what you are able, but with the temptation will also make the way of escape, that you may be able to bear it." (1 Corinthians 10:13) And what is that "way of escape?" "Resist the devil and he will flee from you…"

But just as in our previous discussion of the parable of the man seeking bread for a late night guest, when we saw that God does not immediately answer a

Christians prayer for soul winning power, for the Bread of life, but only after much persistence, so too in the battle to overcome addictions or bad habits we find many times that our prayers are not immediately answered. God wants to see if we want him more than we want to drink, or smoke, or fulfill lusts, or eat, or do drugs. "You shall have no other gods beside me." Which is why he said "And you shall seek me, and you shall find me, when you shall search for me with all your heart." How badly do we want Him? More than eating religiously three or four times a day? More than your secretary? More than a margarita on the rocks with salt? More than _____? (Fill in the blank.) God is waiting for each one of us Christians at our own time and at our own speed, but he is also anxious for many of us to start the journey, to start the internal war, to seek him like the deer searches for that mountain stream to keep from dying of thirst. "As the deer pants for the water brooks, So pants my soul for you, O God." (Psalms 42:1) Once you enter into the fray, once you begin praying for growth in Christ, then there will begin a battle inside of you between the Christian, the follower of Jesus, and the pagan, the flesh, the one who was used to following self, sin and Satan.

Also, if impatience overcomes you because one or more of your prayers have not yet been answered, it might be helpful to realize that our time is not Gods time. For "one day is with the Lord as a thousand years, and a thousand years as one day." (2 Peter 3:8) So let's do the math. If you've been praying for something, let's say release from anger or a bad temper, for a year and you haven't quite completely licked it—it still sneaks up on you and erupts every once and a while when you least expect it—in Gods time you've only been praying for about 1 & 1/2 minutes. And if he answers your prayer after two years then he will have responded in three minutes! Wow! That would have to be considered an immediate response! The point is we just need a little more patience. And we need to be persistent, and faithful, never wavering, trusting in Him, and if we do waver then we need to get right back on track after we recover from the waver.

"I can do all things through Christ who strengthens me." (Philippians 4:13)

"Have mercy upon me, O God, according to your loving kindness; according to the multitude of your tender mercies blot out my transgressions. Wash me thoroughly from my iniquity, and cleanse me from my sin. For I acknowledge my transgressions, and my sin is ever before me. Against you, you only, have I sinned, and done this evil in your sight: that you may be justified when you speak, and blameless when you judge. Behold, I was shaped in iniquity; and **in sin did my mother conceive me**.[354] Behold, you desire truth in the inward parts: and in the hidden part you will make me to know wisdom. Purge me with hyssop, and I shall be clean: wash me, and I shall be whiter than snow." (Psalms 51:1-7)

"Blessed is the man who walks not in the counsel of the ungodly, nor stands in the path of sinners, nor sits in the seat of the scornful. But his delight is in the law of the Lord, and in his law he meditates day and night. And he shall be like a tree planted by the rivers of water, that brings forth its fruit in its season; his leaf also shall not wither; and whatsoever he does shall prosper." (Psalms 1:1-3)

"Most assuredly, I say to you, he who believes in me, the works that I do he will do also; and greater works than these he will do, because I go to my Father. And whatever you ask in my name, that I will do, that the Father may be glorified in

---

354   (One more verse where the Bible reaffirms the doctrine of original sin, that we all are under the sin of Adam from our earliest beginnings.)

the Son. If you ask anything in my name, I will do it." (John 14:12-14) When you want only *what* your Heavenly Father wants, and *when* he wants it to happen, then all of your prayers will be answered, and exactly when you want them to be. In the meantime, keep praying.

Walking with the Lord also requires the humble attitude of a servant:
"Out of anyone who ever lived on this earth, Jesus had more reason to demand that others put him on a pedestal than anyone else. He deserved to be honored and served. Instead, out of an attitude of humility, he chose to be the servant of all. ... A selfish, self-centered attitude will eventually lead to self-destruction. A humble, God-centered attitude will lead to deeper personal peace and a greater positive impact on the world around you."[355]

"You know what kind of life it is that Christ calls you, as his disciple, to live. His own example and teaching in the Gospels (to look no further in the book of God than this) make it abundantly clear. You are called to go through this world as a pilgrim, a mere temporary resident, travelling light, and willing, as Christ directs, to do what the rich young ruler refused to do: give up material wealth and the security it provides and live in a way that involves you in poverty and loss of possessions. Having your treasure in heaven, you are not to budget for treasure on earth, nor for a high standard of living—you may well be required to forego both. You are called to follow Christ, carrying your cross."[356]
"Lord, give me a humble spirit, and a servant's heart."

"Then Job arose, and rent his mantle [tore his robe], and shaved his head, and fell down upon the ground, and worshipped, and said, Naked came I out of my mother's womb, and naked shall I return thither [to the womb of our Mother Earth]: the Lord gave, and the Lord has taken away; blessed be the name of the Lord. In all this Job sinned not, nor charged God foolishly." (Job 1:20-22) If you want to walk with God, you have to be prepared for whatever comes down the pike. You hope for the best but are prepared for the worst. If the worst comes, you might not be thrilled with it, but you don't stop trusting God. Like Job, you deal with it as best you can without whining or complaining, knowing that God is sovereign and that *He* was not surprised by this occurrence. And you come to realize that everything is a blessing in disguise, only sometimes it's not disguised. When good things come it is easy to see them as a blessing- they're not disguised, it's only the bad things that don't seem to be blessings, while they're happening. But so many times after they are over, we can see the blessing in them, or that resulted from it. If you learn to be thankful for all the good things, the obvious blessings, then it will be easier to handle the trials, the disguised blessings, prayerfully and in faith, always trusting in Him, when they come.

And certainly we are not saying that the terrible things that occasionally are visited upon some of our lives—like the death of a young child or the accidental death of a husband or wife or other loved one—are "disguised blessings." No, they are horrible occurrences. But even then we can be assured that God had a purpose in allowing it to happen. And along those lines there is no one who suffered like Job, who had to deal with the sudden loss of all of his property, his oxen and sheep and camels, the slaughter of all of his many servants, as well as the untimely death

---

355   (Pg. 81, *The 100 Most Important Bible Verses*, by Thomas Nelson Publishers)
356   (Pg. 268, *Knowing God*, by J. L. Packer)

of all of his sons and daughters. But in response he said this to his emotionally distraught wife (for she had lost all of her children as well) and her angry demand that he give up his faith in God, after he was also smitten with painful boils all over his body...

"Then said his wife unto him, Do you still retain your integrity? curse God, and die. But he said unto her, you speak as one of the foolish women speak. What? shall we receive good at the hand of God, and shall we not accept adversity? In all this Job did not sin with his lips." (Job 2:9-10)

Besides, we were never promised a rose garden, we were promised this, "These things I have spoken unto you, that in me you may have peace. In the world you will have tribulation; but be of good cheer, I have overcome the world." (John 16:33)

Therefore... "Rejoice in the Lord always and again I say, Rejoice!" (Philippians 4:4) Now our situation in life will not always be something that we can rejoice in. We might have problems, or tragedies, or setbacks or heartaches. So we can't always rejoice in our circumstances, finances, situation, etc., but we can always rejoice in the Lord, and His sovereignty, and His salvation, and the unimaginable wonder of the life to come.

> "Knowing God encourages a state of mind that is peaceful and secure. Yet this kind of knowledge isn't a treasure trove you acquire the moment you invite God into your life. It is something that is revealed one gem at a time as you dig deeper into the Bible. Focus on Jesus. Get to know him intimately, with your whole being."[357]

"Be still, and know that I am God:
I will be exalted among the heathen,
I will be exalted in the earth."
(Psalm 46:10)

Trusting

And then we are *commanded* to trust in God. It's not a suggestion, although it is treated as such. It is an actual command and to disobey any command of God is to sin against Him. And even though we will all fall short in this regard it is still important to treat this command with the respect it deserves, instead of the passing glance it usually gets.

"Trust in the Lord with all your heart; and lean not on your own understanding. In all your ways acknowledge him, and he shall direct your paths. Be not wise in your own eyes: fear the Lord and depart from evil. It shall be health to your flesh, and strength to your bones. Honor the Lord with your possessions, and with the first fruits of all your increase: So shall your barns be filled with plenty, and your vats will overflow with new wine. My son, despise not the chastening of the Lord, neither be weary of his correction: For whom the Lord loves he corrects; even as a father the son in whom he delights." (Proverbs 3:5-12)

Hold that thought, of those verses. We'll refer back to it in a moment. First let's look at the promises of this Psalm...

"He who dwells in the secret place of the Most High shall abide under the shadow of the Almighty. I will say of the Lord, he is my refuge and my fortress:

---

357 (Pg. 105, *The 100 Most Important Bible Verses*, by Thomas Nelson Publishers)

my God; in him will I trust. Surely he shall deliver you from the snare of the fowler, and from the noisome pestilence. He shall cover you with his feathers, and under his wings shall you trust: his truth shall be your shield and buckler. You shall not be afraid for the terror by night; nor for the arrow that flies by day; nor for the pestilence that walks in darkness; nor for the destruction that wastes at noonday. A thousand may fall at your side, and ten thousand at your right hand; but it shall not come near you. Only with your eyes shall you behold and see the reward of the wicked. Because you have made the Lord, which is my refuge, even the Most High, your habitation, no evil shall befall you, neither shall any plague come near your dwelling. For he shall give his angels charge over you, to keep you in all your ways. They shall bear you up in their hands, lest you dash your foot against a stone. You shall tread upon the lion and adder: the young lion and the dragon shall you trample under feet. Because he has set his love upon me, therefore will I deliver him: I will set him on high, because he has known my name. He shall call upon me, and I will answer him: I will be with him in trouble; I will deliver him, and honor him. With long life will I satisfy him, and show him my salvation." (Psalm 91)

There are many incredible, seemingly *unattainable* blessings that are promised in that Psalm; a life without fear, angelic protection against harm and calamity, and long life. Now we know that many great men of God including most of the Apostles and many early church saints did not escape harm or calamities, nor did many lead a long life (of course one could also argue that eternal life is rather long…). Are these promises therefore for our comfort alone? Or can they still be taken literally? I would say both, depending upon the individual and each person's circumstance, but nevertheless all of them are conditional upon the completion of one fundamental requirement. And it is found in the first verse: "He who dwells in the secret place of the Most High." But how do we *dwell* in the secret place of the Most High? And where is that *secret place*?

Let's answer the second question first. The secret place of the Most High is inside of you; in your heart and in your mind and in your soul. "Behold, the kingdom of God is within you" (Luke 17:21) If you have sought it; if you have chosen to let Him in… "Behold, I stand at the door and knock: if any man hear my voice, and open the door, I will come in to him, and will dine with him, and he with me." (Revelation 3:20) "God is love, and he who abides in love abides in God, and God in him." (1 John 4:16) "If a man loves me, he will keep my words: and my Father will love him, and we will come unto him, and make our abode with him." (John 14:23) "That Christ may dwell in your hearts through faith." (Ephesians 3:17) "And to know the love of Christ, which surpasses all knowledge, that you might be filled with all the fullness of God." (Ephesians 3:19)

And what is the secret to *dwelling* in this secret place? Trusting in God, and praying without ceasing. (Look back now to Proverbs 3:5-12 quoted above.) Trusting in him every moment of every day; to meet our every need; to guide us along our/his paths; to overcome sin; to be with us in trouble; to comfort us through affliction. The exact opposite of trusting in God is worry. ("Faith is the opposite of worry and worry is the opposite of faith.") Worry is a constant, nagging churning inside of your mind of negative fear-filled thoughts concerning all of the bad things that could possibly happen in the near or far future. Sometimes it is because of some impending negative situation like for example the possible loss of one's home due to the loss of their job; or the possible loss of one's life due to the diagnosis of cancer. Also, there are many other unfortunate people who have gotten into the bad habit of worrying constantly no matter what their circumstances may be. But worry is the direct opposite of faith and trust. And remember "without faith it is impossible to please God." (Hebrews 11:6) The undisciplined mind will tend to go on and on like the proverbial drunken monkey on your back, or in the back of your mind, churning and incessantly worrying over the endless negative possibilities that may occur for any given difficult situation. Prayer, and "casting your burdens upon him for he cares for you" (1 Peter 5:7), is the antidote to the mentally and emotionally debilitating sins of fear, worry and anxiety.

"Why pray when you can worry?"[358]

Worry, if unchecked, also has the power to become a self-fulfilling prophecy. Your subconscious mind is a very powerful, goal-oriented computer, and if you concentrate on the negative it will seek that result because that is the *goal* you are unwittingly putting before your mind's eye. This is because we are made in the image and likeness of God. And there are four things that this refers to. The first is obviously that we are a spirit that has consciousness. Just as God is a conscious Spirit. The essence of God's being has nothing to do with the physical world and nor does the essence of who we are have anything to do with it. The physical world, including our physical bodies, is just what we dwell inside of and *drive* through this world, like a pilot in an aero plane or a driver in a car. Secondly, we are eternal. Our spirit is eternal; it lives forever, just like God is eternal. This is why rejecting his free gift of eternal life is so absolutely horrible, because what you choose in this life will stay with you forever, either in heaven or in hell. The third is free will which we have discussed previously. And the fourth thing that makes us in His image and likeness is the creative power of our mind. Just as our Father and Creator is imbued with the power to speak an entire universe into existence and to speak all of the living into being, so too do we have a tiny bit of that power to create. For example, in the material world we can see how man has changed and shaped and improved it, and designed and created everything from buildings that touch the sky, to dams that tame the power of mighty rivers, to great walls that can be seen from outer space, to machines that fly, to computers and electronic devices, to pyramids that rival the ability of modern man to replicate. And then in the world of the mind and heart, man has the power to bring kindness and goodness and generosity and comfort and love and peace and joy to the lives of the family and people around him, or to create a hell hole of physical and verbal abuse, of violence and stupidity and unimaginable cruelty, of lying and cheating and stealing, and of ignorantly refusing to support the physical needs of one's wife and children by being a lazy worthless individual who refuses to go out and find a job and actually work for a living. We can see how in this regard man can wield as much power to create or destroy as he can in the physical world of construction, design and engineering.

So we have powerful minds as a gift from our Creator. And we need to be careful what we place in front of our mind on a regular basis. If it is focusing on and praying for your goals and needs and dreams, then your subconscious mind will help you achieve them. If it is constantly focusing on all of the bad things that you can conceive and worry about and place in front of your mind's eye, then you can unwittingly create exactly what you do not want to happen. We have all heard

---

358   (Dr. Walter Lewis Wilson, M.D. 1881-1969, author, preacher, Bible teacher)

of the power of positive thinking in the business world, and in the power of writing down your goals and reading and praying over them every day. Again, the mind is a very powerful goal-oriented computer on the subconscious level and it will respond to the input that you give it.[359] "Garbage in, garbage out" as the computer expression goes. So keep this in mind as one motivation to strongly resist Satan's trap of worry and anxiety and negative thinking, to avoid ruining your own world. Instead "cast all your cares [and burdens and troubles and financial difficulties and addictions and sins and family problems] on him because he cares so much for you." And then try focusing on all the good things that could happen instead. So, indeed, "why pray when you can worry?"

"Therefore humble yourselves under the mighty hand of God, that he may exalt you in due time, casting all your care upon him, for he cares for you." (1 Peter 5:6-7)

[A warning, though, is in order here: this power of the goal-oriented nature of the subconscious mind; and of speaking and thinking your future into existence, can also be perverted for the enrichment of charlatan's as it has in the "prosperity gospel" movement. Which tells their congregants that they can "name it and claim it" to get any and every financial blessing that they can conceive of; as long as they donate every last cent in the meantime to the Mercedes-driving preacher at the center of this gospel distortion. And if they don't "get rich quick" like their happy preacher, then they are told there must be something wrong with them and with their "claiming." Of course the only thing wrong with anyone suckered into this movement is that by following their own greed they bought into this lie in the first place. It has little to do with the true gospel of Jesus and its demands on the sacrifice and hardship and **poverty** and relentless persecution of the vast majority of those who have "named the name of Christ" over the centuries and in many countries today (see *Persecuted Church* later in this Chapter). It has more to do with the greedy financial enrichment of those who preach this deception. And for a more thorough exposé see Jim Bakker's book, *I was wrong*. Bakker was one of the more notorious of prosperity gospel preachers (remember Tammy Faye?) but he has now renounced it.]

"Worry is a horrible sin. Worry is the opposite of faith and trust. John Wesley said, "I would no more fret than to curse and swear." And God's cure for worry is to pray about *everything*."[360]

And to "pray without ceasing."

"Worry, feeling uneasy or troubled, seems to plague multitudes of people in our world today. It's human nature to be concerned about the bad situations in our world and in our personal lives, but if we're not careful, the devil will cause us to worry beyond what's reasonable. Worry is like a rocking chair—it's always in motion but it never gets you anywhere. So why do we struggle with it? And what good does it do? Worry is the opposite of faith, and it steals our peace, physically wears us out, and can even make us sick. When we worry, we torment ourselves—we're doing the devil's job for him! Worry is caused by not trusting God to take care of the various situations in our lives. Too often we trust our

---

359   (See *Psycho-Cybernetics, A New Way to Get More Living out of Life*, written back in the dark ages of 1960 by a plastic surgeon, Maxwell Maltz, but it still has relevance today. He was a little ahead of his time)

360   (Pg.131, *Prayer: Asking and Receiving*, by John R. Rice)

own abilities, believing that we can figure out how to take care of our own problems. Yet sometimes, after all our worry and effort to go it alone, we come up short, unable to bring about suitable solutions."[361]

"But my God shall supply all your need according to his riches in glory by Christ Jesus." (Philippians 4:19)

Here is a powerful daily prayer by author and speaker, Joyce Meyer:

"God I am nothing without you. You're my joy. You're my peace. You're my righteousness. I will never be anything in life if you don't help me. I can't change myself. I can't make myself be the way I'm supposed to be, **but I'm trusting in you.** [Emphasis mine] Now God I'm just going to go about my day and do the best that I can, but I know that even doing my best it's still going to be a mess. But I thank you God for your grace and your mercy."

... "And [as you pray this prayer each day] you'll just find that you're not doing as much wrong as you used to do, you'll find yourself walking more in the fruit of the Spirit, and you'll find yourself being kinder, and you'll find that you're getting closer to God, and all you can say is 'well thank you, Jesus! Thank you, Lord, because it's you, and not me. Amen'"[362]

And in that way God gets all the glory, as He should. Also, as you strive and pray each day for spiritual growth, to be filled with the fruit of the Spirit ("love, joy, peace, patience, kindness, goodness, faithfulness, gentleness, self-control." Galatians 5:22-23), to be filled with the Holy Spirit, to overcome sin, to draw closer to God, you can <u>know</u> that this is also the will of the Father for you and your life. So keep this in mind and it can add power to both your prayers and your growth. "Father I know you want me to overcome ____, to change ____, to love ____, to forgive ____, to stop ____, so I am trusting in You."

"And why is it that we can have faith in God for eternal life, but when it comes to day to day things, which are tiny in comparison, we struggle with faith in him, with trusting in him?"[363]

"As your character grows stronger, so does your hope. Experiencing firsthand how God can use difficult circumstances in a positive way solidifies your hope for the future as it strengthens your trust in him. This character-building process hones the resulting hope by persevering through difficulties. This is what allows you to find genuine joy, even in the middle of suffering."[364]

"Trust in the Lord with all your heart; and lean not on your own understanding. In all your ways acknowledge him, and he shall direct your paths." (Proverbs 3:5-6)

"For I have learned whatever circumstances I am in, to be content. I know

---

361　(From "The Cause and Cure for Worry,' by Joyce Meyer, Joyce Meyer Ministries)
362　(Joyce Meyer, Joyce Meyer Ministries) (And yes, I am aware that she is a part of the aforementioned "Prosperity Gospel" movement. Pray for her ...and the hundred million dollars she rakes in every year.)
363　(Pastor Ed Regensburg, Faith Community Church, Gambrills, Md.)
364　(Pg.133, *The 100 Most Important Bible Verses*, Thomas Nelson Publishers)

what it is like to be in need, and I know what it is to have plenty. I have learned the secret of being content in any and every situation, whether well fed or hungry, whether living in plenty or in want." (Philippians 4:11-13) *Stress*: when you start to realize that God is in control, that He is absolutely sovereign, it takes a lot of the pressure off of you.

"Good Morning! This is God. I will be handling all of your problems today. I will not need your help. So, relax and have a great day!"

"Do not worry about anything, be anxious for nothing, but pray and ask God for everything you need, with thanksgiving, let your requests be made known to God. And the peace of God, which surpasses all understanding, will keep your hearts and minds in Christ Jesus." (Philippians 4:6-7)

> "There is an easy, effective way to eliminate worry- pray. ...Peace and worry cannot coexist. ...At the first sign of worry, practice the life-changing principle found in Philippians 4:6-7. Tell God what you're anxious about. Thank him for who he is, for what he has done, and for the peace he provides."[365]

Overcoming the pain of worry and changing that habit to one of trusting in the Lord takes work. Every time that worry comes up you have to counter it with prayer, and with positive thoughts. "God is all powerful. I have faith in you, Oh Lord. I trust in You. I cast this burden on you, Oh Lord. I have faith in you to work this situation out. My life, my family, my job, my home are all in your hands. I have faith in you for my financial future."

Of course, it also goes without saying that thoughtful consideration of a problem or an obstacle is not worry. Thinking about what to say or prayerfully formulating in your mind what actions to take to resolve a situation or overcome an obstacle is productive; and is an expected part of "casting your burdens on the Lord." Just make sure it is not accompanied by an overture of negative emotions like fear and anxiety, or worried thoughts that focus only on the worst that could happen.

Fear not

Jesus also *commands* us to fear not.

"Fear not, for I am with you; be not dismayed, for I am your God. I will strengthen you, yes, I will help you, I will uphold you with my right hand of righteousness." (Isaiah 41:10) "And fear not them which kill the body, but are not able to kill the soul: but rather fear him which is able to destroy both soul and body in hell." (Matthew 10:28) "Fear not therefore, you are of more value than many sparrows." (Matthew 10:31)

Many times in his word he commands us to *fear not*. It's not a suggestion. It is an actual command. And the secret to not fearing is trusting in the Lord, and casting our burdens upon him, and praying without ceasing, and…

As a practical application, let's say that you are afraid that you are going to lose your job, or you have already lost your job and you are afraid of losing your house, or anything else you may be afraid of happening. The first thing to do is to make sure that you are trusting in God and have surrendered your life to his will. This is an ongoing process so it doesn't matter at what point along the way your trust and your surrender has come, just as long as you are in the game. Then you must realize that if you lose your job (or your house or your health or whatever)

---

365   (Pg. 56-57, *The 100 Most Important Bible Verses*, Thomas Nelson Publishers)

it is Gods will, for he is absolutely sovereign and nothing occurs that does not fit in with His perfect will. So what is there actually to fear? At that point all you can do is to trust in God. Maybe this is a blessing in disguise and he has something in mind for you and your family even better down the road. Or maybe you are being tested and tried like Job. Then it is just best to do everything you can to help yourself (prayerfully look for another job, or ways to keep your home, or maybe fasting and praying for your health (see chapter 14), etc.) but to "cast your burden on him" knowing that he knows the future and exactly what is going to happen to you, good or bad, and that he cares for you and he will never forget you or forsake you, and also that an eternity in paradise, the restored Garden of Eden here on this earth, is waiting for you. So, suck it up! Be strong! Have faith! ("Without faith it is impossible to please him" ..he who died on a cross for our sins. So please him! And what better place to have faith than now when your future may be in doubt and your shelter, security, or your very life may be hanging by a thread?) See this as a golden opportunity to please him with your faith and trust and your struggle against the fear inside of you, as opposed to an opportunity to sink into the will of Satan which is to worry and fret and fear. Fear not! As for you and your house, you will serve the Lord ("but as for me and my house, we will serve the Lord." Joshua 24:15), even in trials and tribulations and great uncertainty. See it as a golden opportunity to please him by having faith in Him. We see through a glass darkly but he sees all of our futures in all of their intricate detail. Perhaps if we could see what He does then we wouldn't be so worried or afraid…

"Yesterday is history, tomorrow is a mystery, but today is a gift. That is why it is called the present."[366]

Wisdom

Pray for wisdom, for "if any of you lack wisdom, let him ask of God, who gives to all men liberally and without reproach, and it will be given to him. But let him ask in faith, without wavering." (James 1:5-6)

"Happy is the man who finds wisdom, and the man who gains understanding. For her merchandise is better than the merchandise of silver, and the gain thereof than fine gold. She is more precious than rubies, and all the things you can desire are not to be compared unto her. Length of days is in her right hand, and in her left hand riches and honor. Her ways are ways of pleasantness, and all her paths are peace. She is a tree of life to those who lay hold upon her, and happy is every one who retains her. The Lord by wisdom has founded the earth; by understanding he has established the heavens. By his knowledge the depths are broken up, and the clouds drop down the dew." (Proverbs 3:13-20)

"I, wisdom, dwell with prudence, and find out knowledge of witty inventions. …I love those who love me, and those who seek me diligently will find me. Riches and honor are with me, yes, durable riches and righteousness. My fruit is better than gold, yes, than fine gold, and my revenue than choice silver. I lead in the way of righteousness, in the midst of the paths of justice. …The Lord possessed me at the beginning of his way, before his works of old. I was established from everlasting, from the beginning, before there ever was an earth. When there were no depths, I was brought forth; when there were no fountains abounding with water. Before the mountains were settled, before the hills, I was brought

---

366   (Oogway, great turtle master, quoting an ancient saying in the movie, *Kung Fu Panda*)

forth; While as yet he had not made the earth or the fields, or the primeval dust of the world. When he prepared the heavens, I was there; when he drew a circle on the face of the deep [this refers to the initial creation of matter in Genesis 1:1, see "In The Beginning" section of Chapter 10]; When he established the clouds above, when he strengthened the fountains of the deep [this refers to the initial creation of the waters under the earth's crust that were released 1500 years later to cause the great flood. See "A World Wide Catastrophic Flood" section of Chapter 10]; When he assigned to the sea its limit, that the waters should not transgress his command, when he appointed the foundations of the earth: Then I was beside him as one brought up with him; **and I was daily his delight, rejoicing always before him**; Rejoicing in the habitable part of his earth, and my delight was with the sons of men. Now therefore listen unto me, my children, for blessed are those who keep my ways. Blessed is the man who listens to me, watching daily at my gates, waiting at the posts of my doors. For whoever finds me finds life, and shall obtain favor from the Lord. But he who sins against me wrongs his own soul; all those who hate me love death." (Proverbs 8:12, 17-20, 22-36)

Wow! "I, wisdom, was daily God's delight." With all that in mind I think wisdom is something not to be overlooked. "And with all your getting, get wisdom, get understanding." (Proverbs 4:7)

"From where then does wisdom come? And where is the place of understanding? Seeing it is hidden from the eyes of all the living, and concealed from the birds of the air. Destruction and death say, 'We have heard the fame thereof with our ears.' God understands its way, and he knows its place. For he looks to the ends of the earth, and sees under the whole heaven, to establish a weight for the wind, and apportion the waters by measure. When he made a law for the rain, and a path for the lightning of the thunder. Then he saw wisdom and declared it; he prepared it, indeed, he searched it out. And unto man he said, '**Behold, the fear of the Lord, that is wisdom, and to depart from evil is understanding**.' " (Job 28:20-28)

Life is a struggle between good and evil, and for the Christian this struggle is an ongoing one within himself. As Paul said, the good which he wanted to do, he did not, and the things which he hated, that he did. (Romans 7:15) A Christian is keenly aware of this struggle, whether it is against hatred and anger, selfishness, arrogance or lust, or even worry, fear and lack of faith. It is good to keep in mind that for the most part we are powerless over this struggle against the evil within, and in order to make any lasting inroads it is necessary to turn this fight over to the Lord: to trust in him. Wisdom then can be said to be in relying completely, and for everything, on the One who alone is wise.

"Casting down imaginations, and every high thing that exalts itself against the knowledge of God, and bringing into captivity every thought to the obedience of Christ." (2 Corinthians 10:5)

"For my thoughts are not your thoughts, neither are your ways my ways, says the Lord. For as the heavens are higher than the earth, so are my ways higher than your ways, and my thoughts than your thoughts." (Isaiah 55:8-9) Teach me your ways, Oh Lord, teach me Your ways.

Love

"And be you kind to one another, tenderhearted, forgiving one another, just as God in Christ forgave you." (Ephesians 4:32)

We are *commanded* to love. Again, it is not a suggestion. "A new commandment I give unto you, That you love one another; as I have loved you, that you also love one another." (John 13:34 (KJV) "Then one of them, who was a lawyer [figures], asked him a question, tempting him saying, 'Master, which is the greatest <u>commandment</u> in the law?' Jesus said to him, '<u>You shall love the</u> Lord <u>your God with all your heart, and with all your soul, and with all your mind</u>. This is the first and great commandment. And the second is like unto it: '<u>You shall love your neighbor as yourself</u>. On these two <u>commandments</u> hang all the law and the Prophets." (Matthew 22:35-40) "For this is the love of God, that we keep his commandments." (1 John 5:3) "God is love, and he who abides in love abides in God, and God in him. ... There is no fear in love; but perfect love casts out fear, because fear involves torment. He who fears has not been made perfect in love. We love him because he first loved us. If a man says, 'I love God,' and hates his brother, he is a liar; for he who does not love his brother whom he has seen, how can he love God whom he has not seen?" (1 John 4:16-20) "My little children, let us not love in word, neither in tongue; but in deed and in truth." (1 John 3:18) "By this we know love, because he laid down his life for us. And we also ought to lay down our lives for the brethren." (1 John 3:16) "Behold, what manner of love the Father has bestowed upon us, that we should be called the sons [and daughters] of God." (1 John 3:1)

"Though I speak with the tongues of men and of angels, but have not love, I have become as sounding brass or a clanging cymbal. And though I have the gift of prophecy, and understand all mysteries and all knowledge, and though I have all faith, so that I could remove mountains, but have not love, I am nothing. And though I bestow all my goods to feed the poor, and though I give my body to be burned, but have not love, it profits me nothing. Love suffers long and is kind; love does not envy; love does not parade itself, is not puffed up; does not behave rudely, does not seek its own, is not provoked, thinks no evil; does not rejoice in iniquity, but rejoices in the truth; bears all things, believes all things, hopes all things, endures all things. Love never fails. But whether there are prophecies, they will fail; whether there are tongues, they will cease; whether there is knowledge, it will vanish away. For we know in part and we prophesy in part. But when that which is perfect has come, then that which is in part will be done away. When I was a child, I spoke as a child, I understood as a child, I thought as a child; but when I became a man, I put away childish things. For now we see in a mirror, dimly, but then face to face. Now I know in part, but then I shall know just as I also am known. And now abide faith, hope, love, these three; but the greatest of these *is* love." (1 Corinthians 13)

"But as it is written: 'Eye has not seen, nor ear heard, nor has it entered into the heart of man the things which God has prepared for those who love him.'" (1 Corinthians 2:9)

"If you love me, keep my commandments." (John 14:15)

"This is my commandment, That you love one another, as I have loved you. Greater love has no man than this, that a man lay down his life for his friends. You are my friends, if you do whatsoever I command you." (John 15:12-14)

We say we love God but do we? Do we love his will? He is all-powerful and everything that is happening to us right now is according to his will. Are you about to lose your job, your house or your business to this socialist, Democrat, Obam-economy? God could have you go out right now in your backyard with a shovel and dig a 3 foot hole and uncover a perfect uncut diamond the size of a grapefruit. *That* would certainly solve your financial problems. Why doesn't He? There must be an extremely good reason. Do we love his will? That's a tough one, because in this fallen world his will can be hard at times. But we can't love God without loving all that He is because, unlike our spouses for example, he actually is perfect.

Praying without ceasing

"Rejoice always. Pray without ceasing. In every thing give thanks; for this is the will of God in Christ Jesus for you." (1 Thessalonians 5:16-18)

"Pray without ceasing." Pray for anything and everything you want.

"Every Christian should take every desire to God in prayer. It is a sin to want anything that you cannot honestly pray for, and you should ask God to remove the desire, if it is wrong. And if the desire itself is not wrong, then you ought to ask God to fulfill it."[367]

God's Word is full of verses encouraging us to "take it to the Lord in prayer"... "Delight yourself also in the Lord, and he shall give you the desires of your heart." (Psalms 37:4) "Ask, and it shall be given you; seek, and you shall find; knock, and it shall be opened unto you: For every one that asks receives; and he that seeks finds; and to him who knocks it shall be opened." (Matthew 7:7-8) "Whatsoever things you desire, when you pray, believe that you receive them, and you shall have them." (Mark 11:24) "If you ask anything in my name, I will do it." (John 14:14) "If you abide in me, and my words abide in you, you shall ask what you will, and it shall be done unto you." (John 15:7) "Be anxious for nothing, but in everything by prayer and supplication, with thanksgiving, let your requests be made known to God." (Philippians 4:6) "You have not, because you ask not." (James 4:2) "Call unto me, and I will answer you, and show you great and mighty things, which you know not." (Jeremiah 33:3)

"Prayer is not a way to draw God's attention to your needs. It is a way to draw your attention to how much you need God."[368]

Praying is one thing, but it's the <u>without ceasing</u> part that would appear to be a little much to ask. (Along the lines of "Be you therefore perfect, even as your Father in heaven is perfect." Matthew 5:48) How does one pray while ordering a pizza, for example? You could get the wrong toppings and *then* what would you do? But upon further inspection the task becomes a little more attainable. For we are not called to endlessly repeat some formulaic prayer like "Hail-Mary-full-of-grace-the-Lord-is-with-thee-blessed-art-thou-among-woman-and-blessed-is-the-fruit-of-thy-womb-Jesus-Holy-Mary-mother-of-God-pray-for-us-sinners-now-and-at-the-hour-of-our-death-amen.[369] The Bible says that the vain repetitions of the heathens are of no use. "But when you pray, do not use vain repetitions as the heathen do: for they think that they shall be heard for their many words."[370] (Matthew 6:7) God is not deaf, and he doesn't want us to try to get his attention by endless and mindless repetition. But God does command us to pray without ceasing, and to disobey that command is sin. So how then do we accomplish this?

---

367 (Pgs. 127-128, *Prayer: Asking and Receiving*, by John R. Rice)
368 (Pg. 13, *The 100 Most Important Bible Verses*, by Thomas Nelson Publishers)
369 (Wow! After 50 years of complete neglect it still springs forth effortlessly from the deep recesses of my brain, having been indelibly etched on my electrical circuitry through years of endless repetition. God bless those nuns.)
370 (The NIV says "do not keep on babbling like pagans.")

The clue is in the entirety of the verse…

"Rejoice always. Pray without ceasing. In every thing give thanks; for this is the will of God in Christ Jesus for you." (1 Thessalonians 5:16-18)

> "How does one pray continually? We cannot always be on our knees. With the daily demands on our busy lives, we are fortunate to kneel in prayer even a few minutes each day. However, the context of this passage gives us a clue. This passage focuses on heart attitude. "Rejoice always" is an attitude of joyfulness. Giving thanks in everything also requires a mental attitude of thankfulness. How do we rejoice and give thanks? Through prayer! Therefore, effective prayer is a proper heart attitude: a mental outlook of joyful thanksgiving. It expresses itself throughout the day with silent prayers of vital communication with the LORD."[371]

God intends for each of us to have the same kind of intimate contact and communion with him like the great ones of the Bible- Moses, Daniel, Elijah, Stephen, Enoch, David, Peter, Paul and Mary. (Not the folk group.) For those great men (and women) of God are no more "special" than you or I or anyone else who has believed on the name of the Lord Jesus Christ and who has passed from death to life and who can now cry, "Abba, Father." Indeed, the Bible also clearly records the sins and the shortcomings and the weaknesses of many of those great men so we can see that they were just like us, and therefore the relationship that they had we too can have. If we seek to be mindful of the Lord throughout the day,[372] seeking him with our verbal prayers and requests, praises and Bible verses, thoughts that are focused on Him, and a heart attitude of joy and thanksgiving.

Here are some various ideas for praying without ceasing, to keep it from becoming repetitive, monotonous or boring. They are: Praising, Confessing, Thanking, Asking, Focusing, Being, and The Word of God.

- Praising For this, let's go to the Psalms. They are a great source not only for praises, but for thanksgiving as well. "Praise God from who all blessings flow." "O Lord my God, I will give thanks unto you for ever. …In God we boast all the day long, and praise your name for ever. …Let the peoples praise you, O God; Let all the peoples praise you. …I will praise the name of God with a song, and will magnify him with thanksgiving. …I will praise you, O Lord my God, with all my heart: and I will glorify your name for evermore. …I will sing unto the Lord as long as I live: I will sing praise to my God while I have my being. …You are my God, and I will praise you: you are my God, I will exalt you. …Praise the Lord! Praise God in his sanctuary: praise him in the firmament of his power. …Let everything that has breath praise the Lord. Praise the Lord!" (Psalms 30:12, 44:8, 67:3, 69:30, 86:12, 104:33, 118:28, 150:1, 6) "O Lord, our Lord, how excellent is your name in all the earth!" (Psalms 8:9) "Let the word of Christ dwell in you richly in all wisdom; teaching and admonishing one another in psalms and hymns and spiritual songs, singing with grace in your hearts to the Lord." (Colossians 3:16) "Rejoice in the Lord always, and again I say, Rejoice!" (Philippians 4:4) The variety of things available at any moment to praise Jesus for is endless.

It is also a good suggestion that in the asking part of our prayer life we start praising as well. For example, when God told the Israelites that they were going

---

371   (From article "What does it mean to pray without ceasing?" @ allaboutprayer.org)

372   (Praying without ceasing in many ways is just remembering Him and his presence as we go through our hectic day)

to take Jericho, He said, "See, I <u>have</u> delivered Jericho into your hand." (Joshua 6:2) He said it as if it had already happened, <u>before</u> it happened. As we pray then, knowing that what we are praying for is in his will, we should also <u>praise</u> Him for it as if he had already delivered it into our hand. "Therefore I say unto you, whatsoever things you desire, when you pray, <u>believe that you receive them</u>, and you shall have them." (Mark 11:24) "I thank you and praise you for answering my prayers, Oh Lord."

- <u>Confessing</u> "If we say that we have no sin, we deceive ourselves, and the truth is not in us. If we confess our sins, he is faithful and just to forgive us our sins, and to cleanse us from all unrighteousness." (1 John 1:8-9) I can't pray without ceasing, Lord, there's no way, but I'm trusting in you. Father, forgive me for my hatred of the liars on the Left. Let me see them as you do. Let me have compassion for the lost. Forgive my sin, Jesus, and fill me with your love and your understanding. God, forgive me for failing to be a better father to my children (husband to my wife) (wife to husband) (mother to child) (boss to my employees). Father God Jesus, forgive that other driver…er…me for the dangerous, selfish way he…er…I drive. Help me to overcome this sin; I cast this burden on you, Oh God. Jesus, forgive me for my lust (for my addiction to alcohol) (my addiction to cigarettes) (my addiction to drugs or gambling or to food) (my habit of worrying); forgive me, Jesus, for putting these gods before you. Forgive me for trusting in uncertain riches and not in you. ("I am the Lord your God, who brought you out of the land of Egypt, out of the house of bondage. You shall have no other gods before me." (Exodus 20:2-3) Forgive me, Father God Jesus, for my sins, "cleanse me from all unrighteousness," "wash me and I shall be whiter than snow." (Psalm 51:7) Father, if there is anything in my life or in my heart, any unconfessed sin, that is keeping me from getting my prayers answered and my needs met and from having a closer walk with you, please show it to me now so I can confess it and get it under the blood.

- <u>Thanking</u> "Giving thanks always for all things unto God and the Father in the name of our Lord Jesus Christ." (Ephesians 5:20) Thank you, Lord, for this day. Thank you, Jesus, for this breath you have given me. Thank you, Father God, for the light; there are so many people on this earth who are blind and yet you have given me the stone-cold, mind-boggling, flat-out miracle of 3 dimensional, video-camera sight. (See chapter 10) Thank you, Jesus, for dying on that cross for me, and for taking me from darkness to the Light. There are so many spiritually blind, lost souls on this earth, and yet you have chosen to give me the Light of Life. Thank you, Lord, that the godless, baby-mutilating socialists have not completely destroyed our country. Yet. (Come quickly Lord Jesus.) Thank you for all of your blessings.

Be thankful in your own words, of course, for your own things. Thank him for your job, for your family, for your health, for your children, for your parents, for your home, the country you live in, the food you eat, the sky you see, the birds you hear… Again, the array of things available at any moment to be grateful to God for is also endless.

"<u>Rejoice</u> always. <u>Pray</u> without ceasing. In every thing <u>give thanks</u>; for this is the will of God in Christ Jesus concerning you." (1 Thessalonians 5:16-18)

"This cycle ["<u>Rejoice</u> …<u>pray</u> …<u>give thanks</u>"] enables you to be thankful in every situation. That doesn't mean you're thankful *for* every situation. God doesn't ask you to thank him if you lose your job—or your child. Yet he provides reasons to be thankful even in the midst of tough times. Responding to those reasons deepens your joy, leading back to the never-ending circle of rejoicing, prayer, and thanks."[373]

- Asking (God Answers Prayer!!!)[374] And you can take that to the bank. "Be anxious for nothing; but in every thing by prayer and supplication with thanksgiving let your requests be made known unto God." (Philippians 4:6) Ask him for anything and everything you can think of that you and your family and your friends need; from physical needs to spiritual needs. Lord, change my desires and turn them to you. "Let the words of my mouth, and the meditation of my heart, [and the things that I do] be acceptable in thy sight, O Lord, my strength, and my redeemer." (Psalms 19:14) Remove my temper, Jesus. Father, supply my friend with a job to meet his family's needs. God, help my son (or daughter) to overcome this (or that) temptation as they go off to our godless, left-wing, false-religious and atheistic public schools. Jesus, I need three hundred and eighty seven dollars and thirty five cents to fix the transmission in my car, and I don't know where the money is going to come from; but I cast this burden upon you, oh Lord. I am trusting in you to fulfill my need. Jesus, show me the way to overcome my addiction to alcohol (or cigarettes) (or drugs) (or women) (or golf) (or all five): I cast this burden upon you, Oh Lord. Father, let me be a witness for the love and saving grace of Jesus to my son, my daughter, my friend, my co-worker, my husband, wife, father or mother. Bring them from darkness into the Light of truth. Save them, oh Lord, from an eternity in hell!

- Focusing Or you could just hold onto this simple yet powerful prayer for a while… "Heavenly Father Holy Spirit Lord Jesus Eternal Creator God, I want to be pleasing to you, in all that I think, and do." Repeating this prayer every time you find yourself getting off track during your busy and stress-filled day will bring you back into focus, and will direct you toward your prayers ultimate goal. It is also a good prayer to recite when you find yourself falling into temptation (i.e. cigarettes, drugs, alcohol abuse or extra-curricular sex either in thought or deed). And the "and think" part is very helpful in cleaning up your mind when it attempts to go astray.

Another good one is "Jesus, I love you, I worship you, I adore you, I need you. I thank you for dying on that cross for me. I praise you, I trust you, I have faith in you. I cast my burdens upon you. I want to follow you and serve you and glorify your name."

- Being "Be still, and know that I am God. I will be exalted among the nations, I will be exalted in the earth!" (Psalms 46:10) Praying without ceasing also includes quiet meditation on the presence of the Almighty. Just be silent for a while, and know that he is God. ("Turn off your mind, relax and float down stream. This is not dying. This is not dying. Lay down all thought, surrender to the Lord. This is believing. This is believing.")[375] Find a quiet place in your house or your yard or

---

373   (Pg. 191, *The 100 Most Important Bible verses*, Thomas Nelson Publishers)
374   (In case anyone has any doubts)
375   ("Tomorrow never knows" by the Beatles, with slight improvements)

in your neighbor's yard (you might want to get permission first) where you can go and sit and quietly contemplate on the joy and the love and the infinite Being that is behind all of this creation. Just behind the veil. (But don't get so carried away that you travel off to Tibet and join a Buddhist monastery.)

"Everywhere I go I see You."[376]

This life of course is the veil. Like the veil that hung in the Holy of Holies that separated the Jewish priests from the Shekinah glory of God, this life is the veil that separates us from the glory of God. What we see, what we hear, what we feel and sense and smell and taste—it's all the veil. There's nothing wrong with the veil. It's a gift of God and it's incredible. But what we ultimately seek and ultimately crave, is what is behind the veil. "God, my Father, I am the prodigal son, and I'd like to come back home."

- The Word of God Meditate on verses from God's Word. Repeat them in your mind, quietly, and pray for the Lord to open up your understanding to a full meaning of the text. (Scripture memorization is of course very useful for this, as many times you may be driving or in some situation where reading directly from the pages of the Bible is not possible.) This can be done with just one verse or with a series of verses related to a particular topic. For example:

-- *Love.* "If a man loves me, he will keep my word: and my Father will love him, and we will come unto him, and make our abode with him." (John 14:23) "God is love, and he who abides in love abides in God, and God in him." (I John 4:16) "Greater love has no man than this, that a man lay down his life for his friends." (John 15:13) "And you shall love the Lord your God with all your heart, and with all your soul, and with all your mind, and with all your strength. This is the first commandment. And the second is like unto it, You shall love your neighbor as yourself." (Matthew 22:37-39) "For God so loved the world that he gave his only begotten Son, that whosoever believes in him should not perish, but have everlasting life." (John 3:16)

-- *Sin.* "Have mercy upon me, O God, according to your loving kindness: according unto the multitude of your tender mercies blot out my transgressions. Wash me thoroughly from my iniquity, and cleanse me from my sin. For I acknowledge my transgressions: and my sin is ever before me. Against you and you only have I sinned, and done this evil in your sight: that you may be justified when you speak, and be clear when you judge. Behold, I was formed in iniquity; and in sin did my mother conceive me. Behold, you desire truth in the inward parts: and in the hidden part you shall make me to know wisdom. Purge me with hyssop, and I shall be clean: wash me, and I shall be whiter than snow." (Psalms 51:1-7) "Your word have I hid in my heart, that I might not sin against you." (Psalms 119:11) "Therefore will I divide him a portion with the great, and he shall divide the spoil with the strong; because he has poured out his soul unto death: and he was numbered with the transgressors; and he bore the sin of many, and made intercession for the transgressors." (Isaiah 53:12) "The next day John saw Jesus coming unto him, and said, Behold the Lamb of God, who takes away the sin of the world." (John 1:29) "Let not sin therefore reign in your mortal body, that you should obey it in the lusts thereof." (Romans 6:12) "And if any man sin, we have an advocate with the Father, Jesus Christ the righteous." (1 John 2:1)

---

376 (From song, "I See You" by Michael W. Smith)

"Submit yourselves therefore to God. Resist the devil, and he will flee from you." (James 4:7) "For if we deliberately keep on sinning after we have received the knowledge of the truth, there no longer remains a sacrifice for sins, but a certain fearful expectation of judgment, and of raging fire that will consume the enemies of God. ...Of how much worse punishment, do you suppose, will he be thought worthy who has trampled the Son of God underfoot, counted the blood of the covenant by which he was sanctified a common thing, and insulted the Spirit of grace?" (Hebrews 10:26-27, 30)

-- *Forgiveness*. "Then Peter came to him, and said, Lord, how often shall my brother sin against me, and I forgive him? Up to seven times? Jesus said unto him, I say not unto you, until seven times: but, until seventy times seven." (Matthew 18:21-22) "He who is without sin among you, let him cast the first stone." (John 8:7) "God has exalted him with his right hand to be a Prince and a Savior, to give repentance to Israel, and forgiveness of sins." (Acts 5:31) "In whom we have redemption through his blood, the forgiveness of sins, according to the riches of his grace." (Ephesians 1:7) "There is therefore now no condemnation to those who are in Christ Jesus, who walk not according to the flesh, but according to the Spirit." (Romans 8:1) "The mercy of the Lord is from everlasting to everlasting upon them that fear him, and his righteousness unto children's children." (Psalms 103:17) "And according to the law almost all things are purified with blood, and without the shedding of blood there is no forgiveness." (Hebrews 9:22) "If we say that we have fellowship with him, and walk in darkness, we lie, and do not the truth: But if we walk in the light, as he is in the light, we have fellowship one with another, and the blood of Jesus Christ his Son cleanses us from all sin. If we say that we have no sin, we deceive ourselves, and the truth is not in us. If we confess our sins, he is faithful and just to forgive us our sins, and to cleanse us from all unrighteousness." (1 John 1:6-9) "Who is a God like unto you, who pardons iniquity, and passes over the transgression of the remnant of his heritage? He retains not his anger for ever, because he delights in mercy." (Micah 7:18)

-- *The Character of God*. "There is none holy as the Lord: for there is none beside you: neither is there any rock like our God." (1 Samuel 2:2) "Holy, holy, holy, is the Lord of hosts: the whole earth is full of his glory." (Isaiah 6:3) His is a perfect nature: "The Lord is righteous in all his ways and holy in all his works." (Psalms 145:17) "This then is the message which we have heard from him, and declare unto you, that God is light, and in him is no darkness at all." (1 John 1:5) "Can a mortal be more just than God? Can a man be more pure than his maker?" (Job 4:17) "Great is our Lord, and of great power: his understanding is infinite." (Psalms 147:5) "For my thoughts are not your thoughts, neither are your ways my ways, says the Lord. For as the heavens are higher than the earth, so are my ways higher than your ways, and my thoughts than your thoughts." (Isaiah 55:8-9) "'Can anyone hide himself in secret places that I shall not see him?' says the Lord. 'Do I not fill heaven and earth?' says the Lord." (Jeremiah 23:24) "Every good gift and every perfect gift is from above, and comes down from the Father of lights, with whom there is no variation, neither shadow of turning." (James 1:17) "And God said unto Moses, I AM THAT I AM." (Exodus 3:14)

-- *Salvation*. "Neither is there salvation in any other: for there is none other name under heaven given among men, whereby we must be saved." (Acts 4:12) "For God sent not his Son into the world to condemn the world; but that the world through him might be saved. He who believes on him is not condemned: but he who believes not is condemned already, because he has not believed in the name of the only begotten Son of God. And this is the condemnation, that light is come into the world, and men loved darkness rather than light, because their deeds were evil." (John 3:17-19) "The Lord is my strength and song, and he is become my salvation." (Exodus 15:2) "The Lord is my light and my salvation; whom shall I fear?" (Psalms 27:1) "And I heard a loud voice saying in heaven, Now is come salvation, and strength, and the kingdom of our God, and the power of his Christ." (Revelation 12:10-11)

Of course the list of verses and themes to meditate on is nearly inexhaustible. There's wisdom and rejoicing, faith and the glory of God, promises and blessings; enough to keep one busy praying and meditating on God's Word for quite some time.

Praying without ceasing can also be an effective way of overcoming worry and anger. Whatever thoughts of worry, anxiety or anger that your mind is constantly churning away with because of some situation or some person, you can crowd these out just by saying for example: "Lord, fill me with your love. Lord, fill me with your understanding. Jesus, fill me with your peace. Lord, control my thoughts. Father God Jesus, I can't overcome my worry but I'm trusting in you. Father, I can't stop these thoughts and feelings of anger towards this situation or this individual but I'm trusting in you. Lord, I can't overcome the fear inside of me but I'm trusting in you to wipe it away and replace it with faith in you. Jesus, I cast this burden on you." If you keep praying that, then all of a sudden you are praying without ceasing instead of worrying without ceasing, or hating without ceasing, or wasting your life churning away with a bunch of negative, destructive thoughts. Praying without ceasing is not that hard when you realize that most people think without ceasing anyway. They are in the habit of doing it and do not think about it. Change your habit. It's just a matter of changing your thoughts from garbage, to gold. (The alchemist's secret.) "Therefore if any man be in Christ, he is a new creature: old things are passed away; behold, all things have become new." (2 Corinthians 5:17)

An actual meaningful relationship with the infinite and all powerful Creator of the universe is available to all those who have accepted the gift of eternal life through Jesus Christ his Son, and who desire to go further into the joy of knowing God by obeying his command to "pray without ceasing." It is there for the taking. Until eventually, in accordance with another of Jesus' commands to "follow him," one's life itself becomes a prayer, just as his was. A prayer of rejoicing always, and praising, and giving thanks, and worshipping, and asking, and confessing, and trusting at all times, and casting every burden on Him, and of course of never forgetting who is always there walking, and sitting, and working, and sleeping alongside of us.

"Father, let me seek you, and let me find you. Let me search for you with all my heart."

He who sits on the throne

It is also helpful to be mindful when you pray of just who you are praying to. A holy God, yes, and one who should be held in the utmost reference and awe, but those are words and sometimes it is helpful to be able to visualize just Who we are approaching at the throne of grace. And along those lines we have found a couple of chapters in the Book of Revelation to be profitable…

"I was in the Spirit on the Lord's day, and heard behind me a great voice, as of a trumpet, saying, I am Alpha and Omega, the first and the last… And I turned to see the voice that spoke with me. And being turned, I saw seven golden candlesticks; and in the midst of the seven candlesticks one like unto the Son of man, clothed with a garment down to the feet, and girded about the chest with a golden band. His head and hair were white like wool, as white as snow; and his eyes were as a flame of fire; his feet like unto fine brass, as if they burned in a furnace; and his voice as the sound of many waters. And he had in his right hand seven stars: and out of his mouth went a sharp two-edged sword: and his countenance was as the sun shining in its strength. And when I saw him, I fell at his feet as dead. But he laid his right hand upon me, saying unto me, Fear not; I am the first and the last. I am he who lives, and was dead; and, behold, I am alive for evermore. Amen. And I have the keys of hell and of death." (Revelation 1:10-18)

"After this I looked, and, behold, a door was open in heaven: and the first voice which I heard was as it were a trumpet talking with me; which said, Come up here, and I will show you things which must take place hereafter. And immediately I was in the spirit: and, behold, a throne was set in heaven, and One sat on the throne. And he who sat there was to look upon like jasper and a sardine stone: and there was a rainbow round about the throne, in sight like unto an emerald. And round about the throne were four and twenty thrones: and upon the thrones I saw four and twenty elders sitting, clothed in white robes; and they had on their heads crowns of gold. And out of the throne proceeded lightning and thundering and voices: and there were seven lamps of fire burning before the throne, which are the seven Spirits of God. And before the throne there was a sea of glass like unto crystal: and in the midst of the throne, and round about the throne, were four living creatures full of eyes in front and behind. And the first beast was like a lion, and the second beast like a calf, and the third beast had a face as a man, and the fourth beast was like a flying eagle. And the four living creatures had each of them six wings about him; and they were full of eyes around and within: and they rest not day or night, saying, Holy, holy, holy, Lord God Almighty, who was, and is, and is to come. And when those beasts give glory and honor and thanks to Him who sits on the throne, who lives forever and ever, then the four and twenty elders fall down before Him who sits on the throne, and worship Him who lives forever and ever, and cast their crowns before the throne, saying, You are worthy, O Lord, to receive glory and honor and power: for you have created all things, and for your pleasure they are and were created." (Revelation 4:1-11)

Wow! If that doesn't get you in the right frame of mind to pray then nothing will. Let your prayers ascend up before His throne like sweet-smelling incense, knowing they are heard, and will be answered in his perfect time.

"Holiness will express itself with greatest integrity when our primary motive in life is the passionate desire to please God with our whole hearts."[377]

Commands and Promises

"I have come that they might have life, and that they may have it more abundantly." (John 10:10)

---

[377] (Pg. 61, *The Power of Prayer and Fasting: God's Gateway to Spiritual Breakthroughs*, by Ronnie W. Floyd)

The Word of God is full of commandments; some that we readily recognize, like the Ten Commandments; and some that many times we do not recognize as commands, such as: "Fear not, be anxious for nothing, resist the devil, pray without ceasing, and cast your burdens upon the Lord." But these are all commands as well. And obeying them is one of the keys to entering into the abundant life. Jesus came to give us life and to give us life more abundantly. Conversely if one disobeys these commands—number one that is sin, to disobey God—and number two that means that individual is going to miss out on the fruits of the Spirit—Love, Joy, Peace, Patience, Kindness, Goodness, Faithfulness, Gentleness and Self-Control—that lead to the abundant life.

These commands are also followed by promises, if the command is obeyed. "Draw near to God," the command, "and he will draw near to you," the promise. (James 4:8) "Draw near" is not a suggestion. It's a command. "Resist the devil," the command, "and he will flee from you," the promise. (James 4:7-8) Commands followed by promises.

Most of us think of sin as murder, adultery, stealing, homosexuality, lust, envy, etc., and those of course are sins. But most of us do *not* think of disobeying the commands of God that lead to the abundant life as being sin. But they are. They are just as much sin. They might not be as heinous as some of those other ones like murder or voting for murderers, but they're sin nonetheless because sin is disobeying God. And this disobedience again will keep one from experiencing the fruits of the Spirit and the abundant life that Jesus promised.

Commands and promises…

"*Delight yourself also in the* LORD, and he shall give you the desires of your heart." (Psalms 37:4)

"*Submit yourselves therefore to God. Resist the devil* and he will flee from you. *Draw near to God* and he will draw near to you. Cleanse your hands, you sinners; and purify your hearts, you double-minded. Lament and mourn and weep: let your laughter be turned to mourning and your joy to heaviness. *Humble yourselves in the sight of the Lord*, and he shall lift you up." (James 4:7-10)

"*Trust in the Lord with all your heart; and lean not on your own understanding. In all your ways acknowledge him*, and he shall direct your paths." (Proverbs 3:5-6)

"*Do not worry about anything, be anxious for nothing, but pray and ask God for everything you need, with thanksgiving, let your requests be made known to God*. And the peace of God, which surpasses all understanding, will keep your hearts and minds in Christ Jesus." (Philippians 4:6-7)

"*Rejoice in the Lord always: and again I say, Rejoice!*" (Philippians 4:4)

"*Fear not*, for I am with you; be not dismayed, for I am your God. I will strengthen you, yes, I will help you, I will uphold you with my right hand of righteousness." (Isaiah 41:10)

"*You shall love the* LORD *your God with all your heart, and with all your soul, and with all your mind.*" (Matthew 22:37-38)

"*Pray without ceasing. In every thing give thanks*; for this is the will of God in Christ Jesus for you." (1 Thessalonians 5:17-18)

"*Cast your burden on the* LORD, and he shall sustain you; he shall never permit the righteous to be moved." (Psalms 55:22)

"If you *abide in me*, and *my words abide in you*, you shall ask what you will, and it shall be done unto you." (John 15:7)

Hallowed be thy name

"Jesus was God incarnate. ...He was God who took on humanity. As God He was the second person of the trinity. As Jesus of Nazareth He was a human being like you and me. In the Gospels Jesus is called "The Son of Man" eighty four times. It was His favorite title to describe Himself. Jesus walked this earth as a human being. He had a nature like ours except for one difference. He was sinless. He came to die as the perfect sacrifice for our sins, but He came for something else too. He came to show us what it would be like for a human being to be totally dependent on the Father in heaven."[378]

And because of that he prayed often and many times at great length. It was apparent that he was in constant communion with the Father. His disciples of course were aware of this and also that his prayers were much more intimate and meaningful than their more formalized ones that they had learned at home and at the local synagogue. They wanted to pray like Jesus so eventually they came to him and asked, "Lord, teach us to pray." (Luke 11:1) So Jesus answered their request: "When you pray, do not use vain repetitions as the heathen do. For they think that they shall be heard for their many words. Therefore do not be like them. For your Father knows what things you have need of before you ask him. After this manner, therefore, pray: 'Our Father who are in heaven, **Hallowed be thy name**. Your kingdom come. Your will be done, on earth as it is in heaven. Give us this day our daily bread. And forgive us our sins as we forgive others. And lead us not into temptation; but deliver us from the evil one." (Luke 11:2, Matthew 6:7-13)

It is interesting to note that the first thing Jesus told his disciples to pray for, before asking for any of their needs to be met, was that the name of God would be made sacred among them and all across the earth.

"How do you make prayer meaningful? We don't start off asking for ourselves. We're to ask for something for God: "Hallowed be Your name." Jesus who is God incarnate instructs us to pray first of all for Him, ...That His Name Would Be Honored. When we pray that the name of God will be hallowed, we're praying that the name of God will receive the glory and the praise, the holy, reverent attention of all people everywhere. We're praying that God's name would be honored above all. This is prayer for the glory of God - that God would be glorified by all people. Now think about that. Why is the glory of God the subject of the first request of the prayer that Jesus taught us to pray? It's because the glory of God has to be our primary concern, the glory of God has to be our central interest. What are we here for? Why did God create us in the first place?"[379]

"For of Him and through Him and to Him are all things, to whom be glory forever." (Romans 11:36)

"For by Him all things were created that are in heaven and that are on earth, visible and invisible, whether thrones or dominions or principalities or powers. All things were created through Him and for Him." (Colossians 1:6)

"Therefore, whether you eat or drink, or whatever you do, do all to the glory of God." (I Corinthians 10:31)

"We were created by God and for God. The reason life can seem so empty at times is because we try to live it for us. So many times we make life about us instead of about God and the result is we're not fulfilling our created purpose. We're here to honor the name

---

378 (From sermon by Pastor Ed Regensburg @ Faith Community Church, Gambrills, Md. 10/25/09)

379 (Pastor Ed Regensburg, Faith Community Church, Gambrills, Md. 10/25/09)

of God. That's the reason we were created in the first place. God created man in His own image and God wants man to be a part of His greatest work. God wants nothing more than to receive the glory and the praise that He deserves from His creation, but He wants to receive that glory and praise as a result of the prayers of His people.

"When we come to God in prayer, we generally come with a bag full of concerns and worries and needs and they are all legitimate and understandable. But now think about how Matthew 6:9 turns this all upside down. God tells us that the first thing we're to pray for is that God may be glorified. The first thing on the list isn't us, it's God. We read that and we're astounded. We think, wait a minute, doesn't God understand? He doesn't need our prayers. We don't have to learn how to pray like that. We need this thing and we need that to happen and we need this person we love to get better. That's what we're all caught up in. And then suddenly we see that when the glory of God becomes our first concern everything else falls into place."[380]

"But seek you first the kingdom of God, and his righteousness; and all these things shall be added unto you." (Matthew 6:33)

"The agenda of bringing glory to God should be first in our prayer life. ...The reason that so many of us are frightened and concerned and nervous and tense is that we usually put the glory of God last in our thinking. When we don't prioritize God, we turn our focus in on ourselves. We become dissatisfied and unhappy and unfulfilled. Some of us may succeed in getting all kinds of possessions for ourselves, but in the end life is like a sack of sawdust. Once we begin to pray and work for the hallowing of God's name, we begin to see everything in a radically different light. We become free. We are freed from the hardest taskmaster there is, and that's ourselves. It is so simple. Jesus said, when you pray, begin by asking God "Hallowed be Your name." Start by asking that you and your family and your brothers and sisters in Christ will learn to know Him and honor Him. Guess who's asking for prayer. God is! And if you learn to pray for Him first, your prayers will be effective and your life will be meaningful."[381]

"Let the earth be filled with the knowledge of the glory of the Lord, as the waters cover the sea." (Habakkuk 2:14)

What is Gods will for my life?

(In a word, to glorify Him.) (Well, three words.)

"Whether therefore you eat, or drink, or whatsoever you do, do it all to the glory of God." (1 Corinthians 10:31)

One of the most asked question to pastors, preachers and teachers of the Word by their students and young people is this, *How do I know the will of God for my life?* They are usually referring more so to the major decisions in life such as where to go to school, what job to take and who to marry. And there are of course many commentaries, articles and books on this subject and we would recommend that any young, or older, reader who is struggling with this question should seek that material out. Our brief comment here is that if you want to know the will of God for your life then you should get as close to him as you possibly can by making certain that you have accepted his gift of eternal life, and the adoption through his Son Jesus Christ where you can cry "Abba, Father!" And that you are daily with him in prayer, seeking the indwelling direction of his Holy Spirit as your guide in sharing this Great News with as many friends, family and associates as

---

380 (Pastor Ed Regensburg, Faith Community Church, Gambrills, Md. 10/25/09)
381 (Pastor Ed Regensburg, Faith Community Church, Gambrills, Md. 10/25/09)

possible. And that you are seeking to walk with him closer every day, by "drawing nigh to God so he will draw near to you." That you are working with the Lord to overcome sin in your life by "resisting the devil so he will flee from you." That you are "trusting in him so he shall direct your paths," (which would appear in and of itself to be the Bible's short answer to the question of how to know His will for your life) which means you will not be anxious or worried over the major, or minor, decisions of life. That you are "delighting yourself also in the LORD, so he can give you the desires of your heart." That you are obeying his command to "fear not" and are constantly seeking to rely on him in difficulties by "casting your cares upon him for he cares for you." That in the trials, problems and obstacles in your life when things are not going so well, you are using that situation as a golden opportunity to have faith in him because "without faith it is impossible to please him." That you are praying for wisdom and to be filled with his love. That you are praying for anything and everything that you need, and "with all prayer and supplication" you are letting "your requests be made known unto him." And finally that you are seeking, through the use of the ancient art of alchemy, to turn your thinking and worrying and fantasizing without ceasing into *praying without ceasing* so "the peace of God, which surpasses all understanding, will keep your hearts and minds in Christ Jesus." I think if we do all of those things then the will of God for our lives cannot help but make itself known.

"Delight yourself also in the Lord; and he shall give you the desires of your heart." (Psalms 37:4) Also know that the desires of your heart, as long as they are good and righteous and not sinful ones, have been put there by God. Following them, while relying on His guidance and direction to fulfill them, is another simple way to "know the will of God for your life." Or at least a pretty good way to get started in the right direction. And to that end, coming to know his will, and receiving all the truly *good* things that you may desire in this life—the right school, the right spouse, the right career, etc.—may God make His face to shine upon you, and richly bless you.

Prayer and Fasting

"And he said unto them, This kind can come forth by nothing, but by prayer and fasting." (Mark 9:29)

Why don't we get our prayers answered like the followers of Christ we read of in the New Testament? Why can't we raise the dead and make the crippled whole, according to His will? There seem to be credible reports of them doing it over in China, where the church is horribly persecuted; reports of the dead being raised and the sick being miraculously cured. I guess they haven't heard that you're not supposed to take all of the Bible literally; that those gifts were "only for the Apostles and disciples, and only in their day." Or maybe it's because that faith such as this "comes forth by nothing, but by prayer and fasting."

"Jesus said unto them, ...verily I say unto you, if you have faith as a grain of mustard seed, you shall say to this mountain, Be you removed and be cast into the sea, and it shall move; and nothing shall be impossible for you. However, this kind does not go out but by prayer and fasting." (Matthew 17:20-21, Mark 11:23)

Jesus also said that when the bridegroom was with them they would feast, but when the bridegroom was taken away they would fast. "And he said unto them, Can you make the friends of the bridegroom fast while the bridegroom is with them? But the days will come when the bridegroom shall be taken away from them, and then they shall fast in those days." (Luke 5:34-35) But we're still feasting, at least in America.

"And I set my face unto the Lord God, to seek by prayer and supplications, with fasting, and sackcloth, and ashes." (Daniel 9:3)

Just a thought. (And see chapter 14.)

Meaning

"As you act on what you understand about Jesus, you'll mature more fully into who you were created to be. By reading the Bible, spending time with God in prayer, and doing what you feel God wants you to do, you will come alive in the deepest, most authentic sense of the word. The true meaning of life will become evident **in** you."[382] (Emphasis mine)

What is the meaning in life? What is its purpose and its significance? Countless philosophers and lay people alike have taken a stab at that one over the years, but it should hopefully be obvious after reading this far that God has already answered it within the pages of His book. And He has answered it so well and so completely that the question itself no longer has any meaning. Because if someone still feels the need to ask it, then they are asking the wrong question. The right question would be, "Where is the meaning in *my* life?" And the answer to that invariably is that the meaning in their life is either hidden behind a wall of atheism- denying the existence of He who alone can give any true meaning, joy, purpose, happiness and fulfillment in life. Or false religion, "in vain do you worship me, teaching for doctrine the commandments of men," where one has neglected their own salvation and the establishing of a *meaningful* relationship with Jesus/God/their Creator. Or behind a wall of selfishness; something that we all are guilty of to one degree or another but when it gets out of control, especially after years of nurturing, it can block any true meaning, or joy, or purpose in one's life. Or a wall of callous disregard for the needs of others. Or obsessive self-pity. Or a bad, intensely-negative attitude toward life in general which, like out-of-control and obsessive selfishness, can destroy any chance of a meaningful life for those who fall under that spell. There is also the very popular trap of looking for meaning in all the wrong places...

"I the Preacher was king over Israel in Jerusalem. And I set my heart to seek and search out by wisdom concerning all things that are done under heaven; this sore travail [burdensome assignment] that God has given to the sons of man by which they may be exercised. I have seen all the works that are done under the sun; and behold, all is vanity and vexation of spirit [grasping for the wind]. That which is crooked cannot be made straight: and that which is wanting cannot be numbered." (Ecclesiastes 1:12-15)

King Solomon (the *Preacher* in the above verse) was the richest man who ever lived. He had a thousand wives and concubines. Silver was so abundant in his kingdom that they made musket balls out of it.[383] There was nothing material that he lacked. But he came to the conclusion that if you seek for your meaning and purpose in life in the material things that you may acquire, no matter how extensive and extravagant those possessions may be, or in the goals that you may

---

382   (Pg. 221, *The 100 Most Important Bible Verses*, Thomas Nelson Publishers)
383   (You're right, gunpowder hadn't even been invented then. It was a trick question)

accomplish, or in any pleasure, activity, craft or pursuit.... then you will search in vain. Why? Because the meaning and purpose of life, and happiness itself, is not out there in the material world. It's in here. In the spirit realm. Behind the veil. "The kingdom of heaven lies within." This does not mean that all of the gifts of the material world that God has richly bestowed upon us, including families and friends and wives and husbands and children and grandchildren, are to be looked down upon, or scorned or rejected. It just means to look to the Giver of the gift for your meaning and purpose and happiness, and not to the gift itself alone, as wonderful as it may be. (Which is also a way of pleasing him by recognizing his importance in your life, rather than failing to thank Him for his many gifts or outright ignoring him as so many lost individuals foolishly do.) Besides, the Giver is so much more wonderful than his gifts that it is beyond the ability of man to conceive.

> "To find meaning in life, one has to look to the true and living God. He is the God of the living and he gives life purpose and meaning. Life is a sacred gift to be unwrapped every day with the joy of a child's birthday."[384]

As one begins to establish a meaningful relationship with their Creator and Lord they will find meaning and purpose in life. They will begin to fill the "God-shaped hole that everyone has inside of them," an emptiness that only Jesus can fill.

> "When [God] made us, his purpose was that we should love and honor him, praising him for the wonderfully ordered complexity and variety of his world, using it according to his will, and so enjoying both it and him. And though we have fallen, God has not abandoned his first purpose."[385]

Another way to answer the question is to say that this unfathomable, incredible, mind-boggling miracle that we call life on this glorious, elegant and nurturing water planet floating in the middle of this astonishing, infinite universe of incomprehensible beauty and order is so saturated with meaning that it is like a large bath towel that falls into your pool. You drag it up out of the water and hang it over the railing, soaked, dripping. Life is soaked, saturated and dripping with meaning like that towel is soaked with water.[386] And this meaning is still evident despite the fact that we live in a fallen, evil, difficult, problem-ridden, persecuting, unjust and many times incomprehensibly stupid world. (Just listen to Joy Behar.) (With all due respect of course.) After the Lord returns, shortly, and after He "shall wipe away all tears from their eyes; and there shall be no more death, neither sorrow, nor crying, neither shall there be any more pain: for the former things are passed away" (Revelation 21:4), this world will again drip with meaning, just like that soaked towel, but for all peoples and at all times. ("Let the earth be saturated with meaning and soaked with the knowledge of the glory of the Lord, like the waters cover the sea." Habakkuk 2:14, slight return) Which certainly is something to look forward to and to be eternally grateful for. (And see the next chapter.)

Conversely, it is also a pathetic testimony to the extent that people in this country, and especially our youth, befuddled after twelve years in our left-wing indoctrination camps, have been so poisoned by the godless, amoral, unscientific

---

384 (Does anyone know where this quote came from? I seem to have misplaced its source.)
385 (Pg. 92, *Knowing God*, by J. L. Packer)
386 (Life is saturated with Jesus, with the Creator Himself, but we are hopelessly blind to that fact.) (Too busy driving like a madman to the next stop on our list...)

and bald-faced lies of the Democrat-Left that so many have been made blind to this meaning. (Just look at all of those clueless, angry young people inhabiting the "Occupy Wall Street" movement. They would be far better off if instead they spent some time removing the lies of the great deceiver that are occupying their minds.) (Pray for them.)

The Persecuted Church

"I am crucified with Christ: nevertheless I live; yet not I, but Christ lives in me: and the life which I now live in the flesh I live by the faith of the Son of God, who loved me, and gave himself for me." (Galatians 2:20)

The true Church of Jesus Christ, which is made up of all true Christians, has been horribly persecuted for the past two thousand years. They have suffered death, imprisonment, torture, rape, having their children stolen from them by the state and/or the "church," humiliation, public degradation and the confiscation of their homes and earthly belongings. And this continues to this day in Muslim and Communist countries even though far too many Christians in this country unfortunately are oblivious to, and seemingly unconcerned with, their fate. The left-wing media hasn't bothered to inform them. That would not further their Christian bashing agenda here in this country. And sadly, many of these uninformed Christians go out religiously every two years and vote for that same lying Democrat-Left media. Ouch.

"Indeed, all who desire to live a godly life in Christ Jesus will be persecuted." (2 Timothy 3:12)

The persecution of followers of Christ started immediately after his resurrection. The established Jewish leadership tried to eradicate their fellow Jews who became believers in "The Way" (nearly all of the early Christians were Jews); and then it was taken up mercilessly by the Romans; and then relentlessly by the Roman Catholic Church; and it is still going strong today in atheist Communist countries like North Korea, China and Cuba, and in the "religion of peace" Muslim countries like Iran, Pakistan, Egypt, Sudan, Saudi Arabia and Indonesia. Here is an account of what is going on—unreported by our lame-stream, moron media—behind the bamboo curtain:

A New Wave of Terror

"The current persecution of the underground church has spread throughout all the provinces of China and beyond to North Korea and Vietnam. While the Dragon's[387] persecution has been steady since the communist takeover in 1949, there have been several "tidal waves of terror" during this time. One such "new phase" of persecution is underway now. Fearing another Tiananmen Square, the hardliners have ordered the police to renew their sadistic attacks on innocent Christians and the house churches.

"...Our brothers and sisters in China are ...suffering. ...The communist police beat to death one of the young preachers (a sister) who works ...in Hunan. In another area, Northeast China, they took five into the station and poured scalding water over them until their skin began to peel and beat them until their front teeth came out.

"They then mocked and challenged them to praise the Lord as they had been doing

---

387 (The Chinese worship *The Dragon*, but most are unaware that the Dragon is Satan, as he is clearly identified in Scripture... In the Book of Revelation alone, there are 11 verses which refer to Satan as "the dragon.")

in church when they were arrested. The Chinese Christians began to sing beautiful songs of praise. These wicked men were so ashamed of what they had done when they saw the presence of the Lord in these young preachers that they released them all. Incidents like this occur daily all over China."[388]

That story, with many variations in brutality, cruelty, method of torture and outcome, is being repeated many times over throughout the Communist and Islamic world today. Compare their suffering with the comfortable and care-less condition of much of the Christian community in our country and the disconnect is striking. We have little idea of what it means to be a follower of Christ, the sacrifice that it entails and has entailed for the vast majority of those who call and have called upon the name of the Lord today and over the centuries.

"In the early church, no Jew or Gentile would publicly say that Jesus was God unless he or she truly believed it. ...It was commonplace for people who chose to follow Jesus to suffer relationally, financially, and physically for what they believed. They often lost their property, their position, and sometimes even their lives."[389]

"Confirming the souls of the disciples, and exhorting them to continue in the faith, and that we must through much tribulation enter into the kingdom of God." (Acts 14:22)

How many of us who call ourselves Christians in this country today would deny Christ in a moment if we were ever faced with what those five Christians suffered at the hands of the Chinese police? We need to pray that if this kind of persecution and torture ever comes to this country we will have the strength to stand like them. "Wherefore take unto you the whole armor of God that you may be able to withstand in the evil day, and having done all, to stand. Stand therefore" (Ephesians 6:13-14) "I can do all things through Christ who strengthens me." (Philippians 4:1) "For unto you it has been granted on behalf of Christ, not only to believe on him, but also to suffer for his sake." (Philippians 1:29)

"Christians are under fire around the world: in Sudan, where Muslim slave masters have abducted, abused, and tortured Christian slaves ... in India, where anti-Christian persecution is widespread ... and now – in some of the most brutal instances of anti-Christian violence ever – in Pakistan. Christian homes have been looted and torched, churches desecrated and burned. Christian families have been forced to flee their hometowns. There have been shootings and bombings ... and the list goes on. ...just weeks ago, seven Christians were burned to death in the town of Gojra. ...Christians are being singled out and murdered for their faith because of pro-Islamic "blasphemy laws" that encourage Muslim-on-Christian violence."[390]

The travesty is that none of this is reported by our left-wing, Democrat media. You can be sure, though, that if instead of Muslims it was American soldiers doing the killing and burning people alive, then it would be front page New York Times, 24-7 MSNBC, Katie Couric's month-long special, and every Hollywood intellect-retardant's newest crusade.

"Yet if anyone suffer as a Christian, let him not be ashamed; but let him glorify God because of this." (1 Peter 4:16)

"A mob of about 150 people led by Hindu extremists stormed the funeral of a 50-year-old Christian, pulled the coffin apart, and desecrated a cross relatives of the deceased were

---

388   (Pgs. 61-62, *The Gospel Under Fire*, by Wm. Thomas Bray with Nora Lam)

389   (Pgs. 207, 209, *The 100 Most Important Bible Verses*, Thomas Nelson Publishers)

390   (From Oct. 2009 "Religious Freedom Update," Jay Sekulow, ACLJ-American Center for Law and Justice)

carrying. "They threw the body into a tractor and dumped it outside saying his burial would have contaminated Indian soil and his body should be buried in Rome or the United States," said Compass Direct.

..."Much of the world scorns Christians and operates on the rationale that slaughtering them is a legitimate undertaking that will bring little or no punishment to the murderers either in their own countries or internationally.

"Most disturbing is the lamentable lack of outcry from the church and some Christian leaders who should be in the forefront of raising awareness about the situation and explaining what it portends for the body of Christ."[391]

"A Sudanese Christian boy has his knees and feet nailed to a board and he is left to die. When rescued he says he forgives the man who did this because Jesus was also nailed and forgave him. A Vietnamese pastor is sentenced to two years in prison. When he is offered an early release, he declines stating that he has a group of new Believers in the prison he has to disciple. A Colombian missionary is kidnapped and told she only has two hours to live. She tells her captors that if she only has two hours to live, she wants to spend it telling them about Jesus. The persecution of Christians around the world is a tragic reality. Our brothers and sisters are beaten and tortured simply for their faith in Jesus Christ. And some pay with the ultimate price. However, in the midst of this persecution is some of the most courageous stories of faith you will ever read."[392]

And in the Muslim Middle East, the lands of that "religion of peace," there is no less than a serious ethnic cleansing going on against Christian and Jews without so much as a peep out of our Democrat-Left, Christian-hating media here. Here is a list of some acts of Muslim genocide against Christians around the world, compiled by Jihad Watch:

"1. In 2002, [in Indonesia] 10,000 Christian men, women and children were murdered by Muslim attackers after a radical cleric declared them 'belligerent infidels' who deserved no mercy. The international community looked on as 10,000 defenseless Christians were slaughtered.

2. Iraqi Shia and Sunni have used the war to persecute Christians. Nearly half of Iraq's 700,000 Christians have fled for their lives.

3. Muslims in southern Sudan have murdered an estimated 2 million Christians and displaced another 5 million.

4. In Nigeria, Muslim mobs have torched churches, placed Sharia law on Christians, even whipped Christian female college students deemed improperly dressed.

5. In major cities in Europe and England, including Paris and London, non-Muslim women are routinely shouted down, pushed, spat on, or worse for not wearing head scarves. Authorities have done nothing.

6. In England, the highest justice in the land ruled that Sharia law could be used in Muslim communities instead of that nation's law.

7. Following their Friday evening prayers, Egyptian Muslim mobs routinely take to the streets to attack Coptic Christians and burn their homes and businesses, many times in front of the Egyptian authorities.

8. Since its takeover of the Gaza in 2007, Hamas has systematically burned Christians' churches and homes, and killed and brutalized Christians who have tried to stay there.

---

391   (From article, "They Cry in Silence: A Look at the Persecution of Christians Around the World," by Elwood McQuaid, *Israel My Glory* magazine, July/August 2010)

392   (From the American Family Association @afa.net)

9. Pakistani radical Muslims, enraged by the Knighthood of author Salman Rushdie and armed with guns and sticks, attacked Christians worshiping in a Salvation Army church.

10. In 2007, the State Department's Annual Report on International Religious Freedom listed virtually every member of the Organization of Islamic Countries for their abuse of Christians and their ethnic and religious cleansing."[393]

Yet as terrible as the suffering that persecuted Christians are enduring in this life and have had to endure, it still cannot be compared with the everlasting suffering and torment in the life to come that a soul cast into hell will have to bear. "And fear not those who kill the body, but are not able to kill the soul. But rather fear him who is able to destroy both soul and body in hell." (Matthew 10:28) As horrible as the persecutions, murders, tortures and beheadings are in Darfur, Iran, Pakistan, Palestine, Egypt, Saudi Arabia, Indonesia, North Korea, China or Cuba, it cannot begin to compare with what awaits the perpetrators if they do not come to repentance; if they die trapped in Satan's lies. Jesus said, "Love your enemies." The love that a Christian shows to his or her tormentors might be the only chance they will have of experiencing the love of Christ who died on a cross for their sins, and through that of being brought from darkness into the light.

"If we suffer, we shall also reign with him: if we deny him, he also will deny us." (2 Timothy 2:12)

And finally, all Christians can take comfort in these words...

"If the world hates you, know that it hated me first. If you were of the world, the world would love its own: but because you are not of the world, but I have chosen you out of the world, therefore the world hates you. Remember the word that I said unto you, The servant is not greater than his lord. If they have persecuted me, they will also persecute you; if they have kept my teaching, they will keep yours also. But all these things will they do unto you for my name's sake, because they know not the One who sent me. If I had not come and spoken unto them, then they would not have sinned: but now they have no cloak for their sin. He who hates me hates my Father also. If I had not done among them the works which no other man did, they would not have sinned: but now have they both seen and hated both me and my Father. But this came to pass, that the word might be fulfilled that is written in their law, They hated me without a cause." (John 15:18-25)

For further reading on the persecution of Christians today by the atheist/ communists and the Muslims see: *The Voice of the Martyrs* @ www.persecution. com; *The Gospel Under Fire*, by Wm. Thomas Bray with Nora Lam; *Nora Lam Chinese Ministries International*; *Tortured for Christ*, by Richard Wurmbrand; and *Unshaken: A Story of Faith and Hope*, by ACLJ Films.

* * *

See:
- *Fast Facts on False Teachings*, by Ron Carlson and Ed Decker
- *Prayer: Asking & Receiving*, by John R. Rice
- *The 100 Most Important Bible Verses*, by Thomas Nelson Publishers
- *Knowing God*, by J. L. Packer
- *Nave's Topical Bible*, by Orville J. Nave

---

393   (@ jihadwatch.org)

- *The Holiness of God*, by R. C. Sproul
- *My Utmost for His Highest*, by Oswald Chambers
- *God Came Near*, by Max Lucado
- *Signs of Life: Back to the Basics of Authentic Christianity*, by Dr. David Jeremiah
- *Angels: Who They Are and How They Help...What the Bible Reveals*, by Dr. David Jeremiah
- *Psycho-Cybernetics, A New Way to Get More Living Out of Life*, by Maxwell Maltz
- *I Was Wrong: The Untold Story of the Shocking Journey from PTL Power to Prison and Beyond*, by Jim Bakker
- *Salvation Crystal Clear*, by Dr. Curtis Hutson
- *Salvation is more than Being Saved*, by Dr. Jack Hyles
- *Great Preaching On Soul Winning*, compiled by Curtis Hutson
- *To Seek and To Save: Winning and Building Committed Followers of Jesus Christ*, by Paul Chappell
- *The Power of Prayer and Fasting: God's Gateway to Spiritual Breakthroughs*, by Ronnie W. Floyd
- *The Voice of the Martyrs* @ www.persecution.com
- *The Gospel Under Fire*, by Wm. Thomas Bray with Nora Lam
- *Nora Lam Chinese Ministries International*
- *Tortured for Christ*, by Richard Wurmbrand
- *Unshaken: A Story of Faith and Hope*, by ACLJ Films

Part II- The Jewish Religion
"Can a maid forget her ornaments, or a bride her attire?
Yet my people have forgotten me days without number." (Jeremiah 2:32)

The Jewish religion is the one true religion. Wait a minute! How can that be? I thought the Christian religion was the one true religion!? Well, they both are. You see, the Jewish religion doesn't stop at the Old Testament, what is traditionally called the Jewish Scriptures. It also includes the New Testament as well, the Christian New Testament. And that is because Jesus, the Christ of Christianity, who obviously was a Jew, is also the Jewish Messiah, and he wrote not only the New Testament but the Old Testament as well. And despite the fact that the vast majority of Jews the world over today might reject the idea that Jesus is their Messiah, He still is! For God set up the nation of Israel, the people of Abraham, a unique people, the people through whom he would send his Son, the Messiah, the Christ, the Savior of the world. They do not control his Jewish religion; He does. They do not control who their Messiah is; He does.

"For you are a holy people to the Lord your God; the Lord your God has chosen you to be a people for himself, a special treasure above all the peoples on the face of the earth. The Lord did not set his love on you nor choose you because you were more in number than any other people, for you were the least of all peoples; but because the Lord loves you, and because he would keep the oath which he swore to your fathers, the Lord has brought you out with a mighty hand, and redeemed you from the house of bondage, from the hand of Pharaoh king of Egypt. Therefore know that the Lord your God, he is God, the faithful God who keeps covenant and mercy for a thousand generations with those who love him and keep his commandments; and he repays those who hate him to their face, to destroy them." (Deuteronomy 7:6-10)

God is the origin of the Jewish people and the Jewish religion and it can no more be high-jacked by people who deny their own Messiah, then the Christian religion can be taken over by people who are Christians in name only, who are not actually saved, and who follow the great deceiver, and his child-sacrificing slaughter of the innocent, rather than following Jesus as Lord. In fact, the Jewish people should know that it was many of these same false Christians, and not true followers of Christ, who were guilty of the widespread hatred and persecution of the Jews by "Christians" over the centuries.

"O Jerusalem, Jerusalem, who kills the prophets and stones those who are sent unto her! How often would I have gathered your children together, as a hen gathers her brood under her wings, and you would not! Behold, your house is left unto you desolate: and verily I say unto you, You shall not see me until the time comes when you shall cry, Blessed is he who comes in the name of the Lord." (Luke 13:34-35)

One of the reasons that many of the Jewish leaders rejected Christ (besides the main reason of "my sheep hear my voice" and they just weren't God's sheep; they were wolves in sheep's clothing, leading their people astray) was that they were looking for a Messiah who would crush their Roman oppressors and be a great King and ruler in Israel and across the known world. And as a result of this they figured that they, being in a position of leadership in the nation of Israel, could expect to reap great benefits if they were alive when the Messiah finally arrived: increased political power, vast wealth and social position. This Jesus offered none of this and therefore they would have none of him and his new teaching. But if they had searched the Scriptures, and their hearts, they would have seen that Christ first had to come as a humble servant who was to give his life for the sins of many. (See Isaiah 53) When he comes again (the Second Coming of Christ), which is going to happen very shortly (see the next chapter), then he will come as the conquering King and all-powerful ruler to destroy the enemies of God, of Israel and of the cross. And *then* his true followers will rule and reign with Him for a thousand years.

The following Scripture is instructive of the blindness of the Jewish leaders of Jesus' day, and of their incredible refusal to recognize him as their Savior…

"[Jesus said] 'I know that you are Abraham's descendants, but you seek to kill me, because my word has no place in you. I speak what I have seen with my Father, and you do what you have seen with your father.' They answered and said to him, 'Abraham is our father.' Jesus said unto them, 'If you were Abraham's children, you would do the works of Abraham. But now you seek to kill me, a man who has told you the truth which I heard from God. Abraham did not do this. You do the deeds of your father.' Then they said to him, 'We were not born of fornication; we have one Father, God.' Jesus said unto them, 'If God were your Father, you would love me, for I proceeded forth and came from God; nor have I come of myself, but he sent me. Why do you not understand my speech? Because you are not able to listen to my word. You are of your father the devil, and the desires of your father you want to do. …Verily, verily, I say unto you, if anyone keeps my word he shall never see death.' Then the Jews said to him, 'Now we know that you have a demon! Abraham is dead, and the prophets; and you say, "If anyone keeps my word he shall never taste death." Are you greater than our father Abraham, who is dead? And the prophets who are dead. [In a word, yes.] Whom do you make yourself out to be?' Jesus answered, 'If I honor myself, my honor is nothing. It is my Father who honors me, of whom you say that he is your God. Yet you have not known him, but I know him. And if I say, "I do not know him," I shall be a liar like you; but I do know him and keep his word. Your father Abraham rejoiced to see my day, and he saw *it* and was glad.' Then the Jews said to him, 'You are not yet fifty years old, and have you seen Abraham?' Jesus said unto them, 'Verily, verily, I say unto you, before Abraham was, I AM.' Then they took up stones to throw at him; but Jesus hid himself and went out of the temple, going through the midst of them, and so passed by." (John 8: 37-44, 51-59)

They took up stones to throw at him because they knew that in saying "I AM," Jesus was identifying himself as the same eternal Jehovah God whose answer to Moses at the burning bush was, "I AM that I AM." What they did not know however, in their blind rage, was that Jesus was not blaspheming when he said "I AM," but was speaking the truth. The other remarkable thing is how the scripture casually mentions that Jesus "hid himself and went out of the temple, going through the midst of them" as if this was as easy as eating a burger and some fries. Try disappearing from right in the middle of a crowd of false-religious zealots, spurred on by the devil himself, who were hell-bent on stoning you to death right then and there! (Like a street full of enraged Muslims in Islamabad, Pakistan who had just had the words of your American T-shirt translated for them, "Mohammed's mother wore combat boots.") An impossible task, except for the Lord of all Creation. He, after all, created the entire universe out of nothing.

"The woman said unto him, I know that Messiah will come, who is called Christ: when he is come, he will tell us all things. Jesus said unto her, **I who speak unto you am He**." (John 4:25-26)

And not to be controversial or offensive here, or too blunt, but those Jews who deny Jesus should understand that they are calling their own Messiah a liar. How unfortunate. How very sad. The lies of the evil one truly have a hold on this earth like a strait jacket and people clench onto whatever they have been taught, no matter how self-degrading those lies may actually be.

"I know the blasphemy of those who say they are Jews, and are not, but are of the synagogue of Satan." (Revelation 2:9)

If you are a Jew who is struggling with the idea that Jesus is the Messiah, take to heart this thought from Moran, a Jewish believer in Jesus:

> "I want to challenge each one of the people, not to believe what I say, but simply just to say one thing, "God, if you are there, please reveal it to me." If it's not true, nothing will happen. If it is true your life will change."[394]

"For Moses truly said unto the fathers, 'The Lord your God will raise up for you a Prophet like me from your brethren. Him you shall hear in all things, whatever he says to you. And it shall come to pass that every soul who will not hear that Prophet shall be utterly destroyed from among the people.'" (Acts 3:22-23) Wake up! Israel.

---

394   (from *Forbidden Peace*, a DVD from Jews for Jesus)

One question that unsaved Jews often raise is that if Jesus is the Messiah why isn't his name mentioned in the Hebrew Scriptures (the Old Testament). And the answer is, it is! His name is Yeshua, which means salvation, and it is specifically mentioned twenty nine times in the Tenach, the Old Testament scriptures.

"Every time the Old Testament uses the word SALVATION (especially with the Hebrew suffix meaning "my," "thy," or "his"), with very few exceptions (when the word is impersonal), it is the very identical and absolutely same word YESHUA (Jesus) used in Matthew 1:21."[395]

Also as we mentioned before, as far as the persecution of Jews over the centuries by so-called Christians is concerned, the fact is that it was *false* Christians and *false* Christian organizations that were responsible. No one who "names the name of Christ" would ever involve themselves in the persecution of others, or other religions, and they would flee from any religious organization involved in the same. True Christians over the centuries were the ones being persecuted, not the ones persecuting others. One case in point is the United States of America, which is the first example of a country founded by Bible-believing Christians. The religious freedom that our founders established in this country has resulted in great tolerance of all religions. And it is not by accident that this same country, mainly because of the influence of Bible-believing Christians, protects and defends the tiny, vulnerable, heavily slandered and persecuted nation of Israel. (Of course, as the Democrat-Left takes even greater control over the White House and Congress—November, 2008—all bets are off.)

"And I will make of you a great nation, and I will bless you, and make your name great; and you shall be a blessing. And I will bless those who bless you, and I will curse him who curses you; and in you all the families of the earth shall be blessed." (Genesis 12:2-3)

The nation of Israel in general, and the city of Jerusalem in particular, will continue to be the focal point of the worldwide conflict unfolding in the next few years (and perhaps decades) up until the culmination point of the Lord returning to earth to rescue his chosen people from their imminent slaughter. It is for this reason that Satan, and therefore his followers, hate the Jews and the nation of Israel with such a religious fervor. For if Satan can destroy the nation of Israel before the Lord returns then Scripture can not be fulfilled. In Isaiah it was foretold that Messiah would return to earth and establish his earthly kingdom from Jerusalem, and from the line of David: "Of the increase of his government and peace there will be no end, upon the throne of David and over his kingdom, to order it and establish it with judgment and justice from that time forward, even forever." (Isaiah 9:7) And if prophecy is not fulfilled then Satan could prevent God from finishing his plan of salvation, and therefore escape his eternal judgment (see the Book of Revelation, and the next chapter). There has to be a nation of Israel for Christ to return to. And this is why we find such an irrational, diabolical hatred for the Jews and for the nation of Israel, not only from Muslim Arabs in the Middle East, and a great many "liberals" in this country, but from most of the rest of the nations of this world as well. Satan is the god of this world, and they follow after him and his lusts.

---

395 (From "YESHUA IN THE TENACH: The name JESUS in the Old Testament," by Arthur E. Glass, The Messianic Hebrew-Christian Fellowship Inc., at messiahpa.org)

"He who hates me hates my Father also." (John 25:23) Ouch.

Jesus is the Jewish Messiah, not only for the Jews but for all the peoples of the world who will accept him and receive his gift. There are many Jews who have accepted Jesus as their Messiah, and the day will come soon when all of Israel will finally recognize that it *was* Jesus who was their long awaited, and now too long rejected, Lord and King. As Zechariah prophesied: "And I will pour upon the house of David, and upon the inhabitants of Jerusalem, the spirit of grace and of supplications: and they shall look upon me whom they have pierced [Jesus' hands and feet and side were pierced on the cross], and they shall mourn for him, as one mourns for his only son, and shall be in bitterness for him, as one who is in bitterness for his firstborn." (Zechariah 12:10) "And one shall say unto him, 'What are these wounds in your hands?' Then he shall answer, 'Those with which I was wounded in the house of my friends.'" (Zechariah 13:6) Both of these prophecies of Jesus the Messiah are found of course in the Jewish scriptures.

The Jewish people need to heed the words of their own Messiah: "Search the scriptures; for in them you think you have eternal life: and they are that which testifies of me. But you are unwilling to come to me, that you might have life. I receive not honor from men. But I know you, that you have not the love of God in you. I have come in my Father's name, and you receive me not; if another shall come in his own name, him you will receive. [That is a prophesy of the soon-to-come world-wide embrace of the antichrist.] How can you believe, who receive honor from one another, and seek not the honor that comes from God only? Do not think that I will accuse you to the Father: there is one who accuses you, even Moses, in whom you trust. For had you believed Moses, you would have believed me: for he wrote of me. But if you believe not his writings, how shall you believe my words?" (John 5:39-47)

"Seeing then that we have such hope, we use great boldness of speech. And not as Moses, who put a veil over his face that the children of Israel could not look steadfastly at the end of that which was passing away. But their minds were blinded, for until this day the same veil remains unlifted in the reading of the Old Testament, because the veil is taken away in Christ. But even to this day, when Moses is read, a veil lies on their heart. Nevertheless when one turns to the Lord, the veil is taken away." (2 Corinthians 3:12-17) Turn to Jesus! Oh house of Israel…

"Say unto them, As I live, says the Lord God, I have no pleasure in the death of the wicked, but that the wicked turn from his way and live. Turn you, turn you from your evil ways! For why should you die, O house of Israel?" (Ezekiel 33:11)

Jesus was, and is, a Jew. He worshiped in a synagogue. The Book he read from, the Bible, is a Jewish book. Gods Son was, and is, a Jew. The first Christians were all Jews. That's something that seems to have escaped the minds of many Christians of the false variety today. The Lord of all Christians was the child of a Jewish virgin. And the Messiah of the Jewish people is the Savior of Christendom. *Therefore, the Jewish religion and the Christian religion are one and the same.* They are inseparably entwined. It's really that simple. And that truth remains despite any protestation from faux Christians deceived by liberal ideology and any Jews who still reject Jesus as their Messiah despite all of the overwhelming evidence.

"Behold, I will make those of the synagogue of Satan, who say they are Jews though they are not, but are liars. Behold, I will make them to come and worship before your feet, and to know that I have loved you." (Revelation 3:9)

If you are in a Jewish congregation or synagogue that does not embrace Jesus as the Messiah, then get out. You are not in the Jewish religion. You are in a perversion of the Jewish religion, one that denies their own Messiah. If you are in a Christian congregation or church that does not love, embrace and totally support the Jewish people and the Nation of Israel, get out. You are not practicing the Christian religion. You are in a false Christian church. Because the Jewish religion and the Christian religion are one and the same. And the Jews are still God's Chosen People, and their nation of Israel, with its sacred city of Jerusalem and with its Temple mount that holds the place of bedrock upon which the Holy of Holies stood and will stand once again inside the soon to be rebuilt Third or Millennial Temple, is also sacred in the Lords sight.

The Messiah came to Israel two thousand years ago. And he is preparing to return to Jerusalem very soon. Then all the world will know that...

The Jewish religion and the Christian religion are one and the same.

Just as Jesus and his Father are one and the same.

"Behold, the virgin shall be with child, and bear a Son, and they shall call his name Immanuel, which is translated 'God with us.'" (Matthew 1:23)

> O come O come, Emmanuel
> And ransom captive Israel
> That mourns in lonely exile here
> Until the Son of God appear
> Rejoice! Rejoice!
> Emmanuel has come to thee O Israel

Let's end this Part II on *The Jewish Religion* with an incredible prophecy concerning Israel's future, as spoken by the prophet Jeremiah...

"Behold, the days are coming, says the Lord, when I will make a new covenant with the house of Israel, and with the house of Judah: Not according to the covenant that I made with their fathers in the day that I took them by the hand to bring them out of the land of Egypt; my covenant that they broke, though I was a husband unto them, says the Lord. But this shall be the covenant that I will make with the house of Israel; After those days, says the Lord, I will put my law in their inward parts, and write it on their hearts; and I will be their God, and they shall be my people. And they shall no more teach every man his neighbor, and every man his brother, saying, Know the Lord: for they shall all know me, from the least of them unto the greatest of them, says the Lord. For I will forgive their iniquity, and I will remember their sin no more. Thus says the Lord, who gives the sun for a light by day, and the ordinances of the moon and of the stars for a light by night, who disturbs the sea when its waves roar; The Lord of hosts is his name. If those ordinances depart from before me, says the Lord, then the seed of Israel also shall cease from being a nation before me for ever. Thus says the Lord; If heaven above can be measured, and the foundations of the earth searched out beneath, I will also cast off all the seed of Israel for all that they have done, says the Lord." (Jeremiah 31:31-37)

Sounds like we're stuck with the nation of Israel, forever. Someone might want to tell the Muslims. (Break it to them gently.)

See:
- *Forbidden Peace*, a DVD from Jews for Jesus
- Jews for Jesus @ www.jews-for-jesus.org
- "Questions and Answers," a booklet from Jews for Jesus
- *The Tenach*, *The Jewish Scriptures*, *The Old Testament*. Study them and you will learn of Jesus, the Messiah, Yeshua which means Salvation, for he is prophesied and spoken of on nearly every page…
- Isaiah 53

Part III- False Religion
"And many false prophets shall arise, and shall deceive many."
(Matthew 24:11)

Why include a section on false religion in a chapter titled "True Religion?" Because it is only from being filled with the light of true religion that we are able to recognize the darkness of false religion. It is only by coming to the knowledge of the truth that we are able to recognize the false religious lies of Satan that, again, envelop the globe. People believe what they are taught, and most of us on this earth were not blessed to have been taught the truth. Most of us were taught religious lies that our parents were taught, and their parents before them, and the curse of that sin is cast down to the thirtieth and fortieth generation. And it is only by seeking the truth later in life that we can be set free from the chains of Satan's many false religions. ("And you shall know the truth, and the truth shall set you free." John 8:32) And in this section we are going to unapologetically shine the light of truth on the major false religions that curse this planet. We will try to speak the truth in love, and to be as inoffensive as possible, although that is a difficult task because it is the truth itself about this subject that many people trapped in and loyal to these religions will find offensive. Nevertheless, our purpose here is not to "offend" but rather that every reader would be equipped with enough knowledge to defend themselves, their minds, their hearts, their souls and that of their friends and their loved ones against the <u>blatant, eternally damning lies</u> of these many false religions.

Every religion on this planet except for true Christianity (which is totally separate from false Christianity) is a system whereby man tries to get to heaven or to be right with God on his own; by being "good enough;" by doing the "best he can;" by his own merits; and in some of the most mentally-disturbed and vile of cases, by blowing up the innocent. True Christianity is a step of faith whereby man is able to attain heaven and righteousness with God through absolutely <u>nothing</u> he can do himself and through <u>no</u> merit of his own. In fact, quite simply, heaven is absolutely impossible to attain through any effort of man. Therefore, all religions on this planet which preach thus (which includes all religions except true Christianity) are false. This includes Islam, Buddhism, Hinduism, Mormonism, Jehovah's False Witnesses, Roman Catholicism and other forms of false Christianity, and the "mainstream" of the Jewish religion who's Rabbis and congregants remain in Messiah denial; in addition to many other lesser religions and bizarre cults that also plague some of the people of this earth. If you have been raised and indoctrinated into any one of these religions I can only encourage you to leave it, and to embrace the truth in its place. What else is there? Only lies, and a horrible, unspeakable, evil eternity that they will bring one to.

"And you shall seek me, and find me, when you shall search for me with <u>all your heart</u>." (Jeremiah 29:13) We must ask ourselves, What do we love more, the truth, or that which we have been taught, and have come to believe in, as *true*?

Keep in mind that Satan hates the truth and he hates salvation because his descent from his once-lofty perch has taken him to a place of pure evil within his heart and mind. That is his reward for rebelling against God and taking a third of the entire angelic host with him. And unlike men while they live, Satan has no hope of being saved. And being pure evil he is "hell bent" on dragging as many as he can down with him to his eternal resting place, the Lake of Fire. In his attempt to "be like the Most High"[396] he has fallen in the opposite direction. And it is through his inspiration and encouragement that all of these false religions have been set up. With the intent of keeping men confused; of keeping them lost in darkness and imprisoned in strongly held false beliefs that are nevertheless lies—loving them instead of the truth—and keeping these chains shackled around them until they die in their sins. The wicked one does everything within his power to keep men from hearing the truth and being set free; to keep them from accepting Gods free gift of eternal life. And Satan's success is closely tied to all of those multitudes of deceived people who assist him in his endeavor with such a religious fervor. The Jewish-Christian-and-truth-hating fervor of the followers of Mohammed is legendary, and deadly to all, whether Muslim or "infidel," who dare turn from, or dispute, their false beliefs. The Christian-and-truth-hating religious fervor of the Democrat-Left is becoming legendary. They passionately strive to carry out the will of their father the devil, whether that be mutilating millions of babies to death under the absurd guise of women's *rights*; or teaching the bald-faced, scientifically-impossible lie of evolution—Satan's genesis—as *fact* and spitting in the face of the real Creator of the universe in the process; or assisting the Islamists in spreading their lies about the nation of Israel so Satan can fulfill his goal of their annihilation. (See chapters 9, 10, 11 and 13) These are some examples of those who don't just love what they have been taught far more than the truth, but who are so far gone that they hate the truth, and the true religion Christianity, with a demonic rage, reflecting the emotions of their father and hidden counselor, the devil.

It is also important to know that the true religion—which is true Christianity (as distinct from false Christianity)—is not defined by any particular religious organization. God saves individuals, not any particular group. Saved individuals can and have gotten together and formed groups and organizations, but unfortunately so have many <u>un</u>saved individuals gotten together and formed groups and organizations under the banner of Christianity. Hence it is very

---

396   (See Isaiah 14:12-20, or chapter 13, for a description of the fall of "Lucifer, son of the morning.")

important for one who seeks the truth to become knowledgeable of which groups are which, so they can know who to fellowship with and what to avoid, and so they also can know who to witness to. Because of this there are a number of false Christian religious organizations in this world where very few of their followers are actually saved. And there are others where some of their followers have a saving relationship with Jesus Christ. And still other churches where most of the congregation has truly accepted Christ as Savior. It is up to the individual to truly follow Jesus, and not a matter of what particular denomination he or she happens to belong to. Christ died for everyone, and God is calling everyone to accept the salvation he has provided, but only those who actually turn from doing it their way ("But in vain they worship me, teaching for doctrines the commandments of men." Matthew 15:9) and choose to accept Gods way, are chosen. ("Many are called, but few are chosen." Matthew 22:14) And along those lines Jesus himself made it perfectly clear that there would be those who would profess Christianity and even faith in him, but that their Christianity would be found to be of the false variety; a form that might sound great to those with "itching ears"[397] but one that confuses and perverts the simple gospel of Jesus to serve their own traditions and their own ideology.[398] These false preachers and their followers want to do it their way, but that is unacceptable to the Lord whom they profess to follow: **"Why do you call me 'Lord, Lord,' and do not do the things that I say?"** (Luke 6:46) "Not everyone who says to me, Lord, Lord, shall enter the kingdom of heaven, but he who does the will of my Father in heaven. Many will say to me in that day, 'Lord, Lord, have we not prophesied in your name, cast out demons in your name, and done many wonders in your name?' And then I will say unto them, Depart from me, you workers of iniquity, I never knew you." (Matthew 7:21-23)

There are many decent, moral and God-fearing Muslims. As there are many decent, moral and God-fearing Catholics. As there are many decent, moral and God-fearing Mormons, and Jehovah Witnesses, etc. But they are trapped in religions that do not trace their origins from the Creator of the universe. They trace their origins from someone else. For the Creator of the universe and the truth are one and the same. And he has no part in lies or false doctrine thought up by man, and inspired by the adversary of God. And keep in mind that if we could get to heaven by being *decent, moral and "God-fearing"* then there would have been no purpose for Jesus to come to earth and pay the penalty for our sins. Also, accepting Jesus as your Savior requires that you leave false religion behind; that you "follow him." For you cannot "serve two masters." "No servant can serve two masters; for either he will hate the one and love the other, or else he will be loyal to the one and despise the other. You cannot serve God and mammon." (Luke 16:13) Neither can you follow Jesus while clinging to any of Beelzebub's false religions.

"But if our gospel is hidden, it is hidden to them who are lost: Whose minds the god of this world has blinded, so they cannot see the light of the glorious gospel of Christ." (2 Corinthians 4:3-4)

---

397 ("For the time will come when they will not endure sound doctrine, but after their own lusts, and because they have itching ears, they shall heap up for themselves teachers; and they shall turn their ears away from the truth, and shall be turned aside unto fables." 2 Timothy 4:3-4)

398 (Obama's Reverend Wright, who preaches a perversion of the gospel known as *Liberation Theology*, is a prime example of this- where a twisted ideology has been substituted for the true gospel of Christ. Liberation theology has done a great job of *liberating* a large number of people from any hope of salvation.)

Satan has many religions. God has only one. Or, as George Owen put it, "The world has many religions; it has but one gospel."

False Christianity

"O full of all subtlety and all mischief, you child of the devil, you enemy of all righteousness, will you not cease to pervert the right ways of the Lord?" (Acts 13:10)

He is talking right now today to all of those false Christian ministers and preachers who cover this land, who have infiltrated Christianity, and who are leading their congregations to hell. They, and the false Christians who follow them, can be recognized by how they are "getting to heaven." Which is by their own righteousness, by their own good works, by being good enough, by making sure their good outweighs the bad. But their only hope of heaven is through the righteousness of Christ. And many black preachers, unfortunately, are more shills for the Democrat party than preachers of the gospel of Christ, openly encouraging and even expecting that all of their congregation will continue to go out and vote for the baby-mutilating Democrats, religiously, every two years. Despite the fact that one of the very fundamentals of Christ's word, and of Christianity, is "Thou shall not murder!"

"That we should no longer be children, tossed to and fro, and carried about with every wind of doctrine, by the sleight of men, and cunning craftiness, whereby they lie in wait to deceive." (Ephesians 4:14)

"We have all heard the gospel presented as God's triumphant answer to human problems-problems of our relation with ourselves and our fellow humans and our environment. Well, there is no doubt that the gospel does bring us solutions to these problems, but it does so by first solving a deeper problem—the deepest of all human problems, the problem of man's relation with his Maker. And unless we make it plain that the solution of these former problems depends on the settling of this latter one, we are misrepresenting the message and becoming false witnesses of God—for a half-truth presented as if it were the whole truth becomes something of a falsehood by that very fact. No reader of the New Testament can miss the fact that it knows all about our human problems—fear, moral cowardice, illness of body and mind, loneliness, insecurity, hopelessness, despair, cruelty, abuse of power and the rest—but equally no reader of the New Testament can miss the fact that it resolves all these problems, one way or another, into the fundamental problem of sin against God."[399]

The gospel message is so simple. A child can understand it. So why do we have so much division and argument going on within the many denominations of Christianity worldwide? Jesus said that men would know us by the love we would have for one another. Why then all the division and separation? The answer lies in the extent that false doctrine, false gospels, false preachers, faux Christians and false Christian organizations have inserted themselves into the picture over the centuries.

"Beloved, believe not every spirit, but test the spirits, whether they are of God; because many false prophets have gone out into the world." (1 John 4:1)

To understand the extent that this has occurred, it is best to go back to the beginning. During the first 300 years of Christianity there was terrible persecution

---

399   (Pg. 189-190, *Knowing God*, by J. I. Packer)

throughout the Roman Empire. First by the Jews who rejected this "new religion" within Judaism that claimed that Jesus of Nazareth was the long awaited Messiah. Then when Christianity started to take a foothold throughout the Roman Empire by its conversion of a great many non-Jews or Gentiles as well, the Romans themselves took up this persecution, because this new sect of Judaism was becoming a direct threat to their established order. Followers of Christ refused to bow to the emperor and his claim of divinity, and this new, rapidly spreading faith threatened their financially lucrative religious system that worshipped many gods. In both instances the power behind this early cruelty was Satan, whose goal was to stamp out the message of the Gospel of salvation by killing all of its messengers. The church, however, grew and prospered despite relentless persecution, torture and murder. But at the outset of the early church the Apostles clearly prophesied that one day Satan and his followers would pretend to stop their persecutions and instead present themselves as ministers of light, as "Christians."

Roman Catholicism

"For such are false apostles, deceitful workers, transforming themselves into the apostles of Christ. And no marvel; for Satan himself is transformed into an angel of light. Therefore it is no great thing if his ministers also be transformed as the ministers of righteousness; whose end shall be according to their works." (2 Corinthians 11:13-15)

This prophecy began its fulfillment soon after the words were spoken, as many false preachers plagued the early church with their distortions and perversions of the true gospel, and this continues up to our present time. The Apostles, inspired by the Holy Spirit, pronounced the eternal sentence on these wolves in sheep's clothing, "But even if we, or an angel from heaven, preach any other gospel to you than what we have preached to you, let him be **eternally condemned**. As we have said before, so now I say again, if anyone preaches any other gospel to you than what you have received, let him be **eternally condemned**." (Galatians 1:8-9) And with the advent of the Roman Catholic Church in the 3rd century AD the prophecy of 2 Corinthians began its universal fulfillment. For this church held sway over most of the world for twelve hundred years, and used its power to mercilessly persecute the true church of Jesus Christ in an attempt to destroy it. Those who were actual *followers* of Christ and therefore refused to follow the false Church of Rome and its gospel distortions, doctrinal fabrications and the usurping of the title of the Church of Jesus Christ on this earth, were forced into hiding over those dark centuries.[400]

The Roman Catholic Church can trace its beginnings to 312 AD when Emperor Constantine, right before a crucial battle, had a vision of a large cross, which was a well-known symbol of Christianity even then. He took this as a providential sign from the gods and placed this cross on his army's banners and his soldier's shields as they went into battle the next day against the rebel army of Maxentious which outnumbered them two to one. Constantine's army overcame these odds and won a decisive victory, and he credited this to the power of the cross. This in

---

400   (The Catholic Church doesn't have the desire today, nor the power, to openly persecute followers of Christ like it did in its past. That dubious honor is now held by Muslim countries like Saudi Arabia, Sudan, Egypt, Iran and Pakistan, and Communist nations such as China and North Korea.)

turn led to the Edict of Milan in 313 which officially ended the Roman persecution of Christians. The historical record, however, indicates that Constantine never became a true Christian himself, but that he was far more interested in using the widespread advance of the Christian faith for his own power and for the security of his empire. He saw the futility of his predecessor's barbaric attempts to stamp it out, so instead he absorbed this new powerful and unstoppable Christianity and made it the state-sponsored religion of the Empire. Unfortunately what resulted was a pagan-ized version of Christianity, because many of Rome's already established heathen and idolatrous practices were not forsaken but instead were incorporated into this new *Christian* religion, and the pure Gospel message of Jesus was muddied down and distorted. Constantine did not embrace Christ or Christianity.   He merely absorbed it's names—Jesus Christ, Joseph, Mary, the heavenly Father, the Holy Spirit, the Apostles, etc.—and it's events, history, and some of its doctrines into the already established Roman pagan religious practices.

"So under the leadership of Emperor Constantine there comes a truce, a courtship and a proposal of marriage. The Roman Empire through its emperor seeks a marriage with Christianity. Give us your Spiritual power and we will give you of our temporal power."[401]

This Roman, "Christian" state religion exists to this day as the worldwide Roman Catholic Church. It claims to be "descended from Peter" but it is not. Peter died centuries before the birth of the Roman Catholic religion. It is descended from Constantine and other successive pagan, non-Christian emperors. And besides, neither the Roman Catholic Church *nor* the true church of Jesus Christ was ever descended from Peter. Jesus never built his church upon the "tiny rock" of Peter. That would have been an absurd and flimsy foundation. He built his church upon himself, the "massive rock" of Matthew 16:18. In fact, it is a gross misinterpretation of this verse that the Catholic Church has used to claim that the church of Jesus Christ on this earth is built somehow on a meager man, Peter, and that their Popes are the heirs to the authority of this man. But an accurate reading of Matthew 16:18 makes the truth clear. "And I say also unto you that you are Peter [*petros*, a piece of rock], but upon This Rock [*Petra*, a mass of rock- Jesus Christ] I will build my church, and the gates of Hell shall not prevail against it." (Matthew 16:18) In addition, it is Jesus Christ who is spoken of throughout the Old and New Testament as the Chief Cornerstone, the solid foundation of bedrock, upon which his church is built. "Therefore thus says the Lord God: 'Behold, I lay in Zion a stone for a foundation, a tried stone, a precious cornerstone, a sure foundation." (Isaiah 28:16) "Jesus Christ himself being the chief corner stone, in whom the whole building [the church] fitly framed together, grows into a holy temple in the Lord." (Ephesians 2:20-21) Truly Jesus is the Rock that his church is built upon. Indeed, to say that Jesus' church is built upon a sinful man, Peter, is, with all due respect to Peter, laughable. And I'm sure Peter would agree, as he would give all the glory and honor and praise, as well as the rightful ownership and Foundation of the church, to Jesus Christ.

"The papal system today is venerated as "Christian"... It is portrayed in movies, in government and in everyday language as Christian. ... [But] to worship the Son of God, and Mary, the chosen vessel of God, with paganism and paganistic ritual, was the most clever and deceptive system of philosophy that Satan could hope for."[402]

---

401   (Pg. 16, "The Trail of Blood," by J. M. Carroll)

402   (Pg. 64, *The Unveiling: A Journey Through the Book of Revelation*, by Keith Harris)

The Roman Catholic Church also claims to be *the one true church*. But this also obviously is false. We have already seen in Part I that the true church of Jesus Christ on this earth is made up of all those who have accepted Jesus' gift of eternal life. (See *The Body of Christ* section in Part I.) And it is certainly *not* made up of all the members of a Roman religious organization that has actually stood in opposition to the true gospel of Jesus Christ and its followers over the centuries. So what is the Roman Catholic Church then? In truth I tell you that it is the organization that the Word of God identifies as the "great whore" in the Book of Revelation. Let's see…

"Then one of the seven angels who had the seven bowls came and talked with me, saying to me, 'Come, I will show unto you the judgment of the <u>great whore</u> that sits upon many waters, with whom the kings of the earth have committed fornication, and the inhabitants of the earth have been made drunk with the wine of her fornication.' [*Fornication* means false doctrine.] So he carried me away in the spirit into the wilderness. And I saw a woman sitting upon a scarlet colored beast, full of names of blasphemy, having seven heads and ten horns. And the woman was arrayed in purple and scarlet color, and adorned with gold and precious stones and pearls, having a golden cup in her hand full of abominations and the filthiness of her fornication. [A description of the vast worldly wealth of the Roman Church, and the filthy, hell-enlarging result of her false doctrines- the "doctrine of demons."] And upon her forehead a name was written, MYSTERY, BABYLON THE GREAT, THE MOTHER OF HARLOTS AND ABOMINATIONS OF THE EARTH. And I saw the woman, drunk with the blood of the saints and with the blood of the martyrs of Jesus. [The Roman Catholic Church in the past 1700 years has martyred millions of true Christians who refused to bow to her earthly authority or follow her false doctrine.][403] And when I saw her, I marveled with great amazement. But the angel said unto me, 'Why did you marvel? <u>I will tell you the mystery of the woman and of the beast that carries her</u>, which has the seven heads and the ten horns. <u>The beast that you saw</u> [which is Apollyon, the angel of the bottomless pit] was, and is not, and shall ascend out of the bottomless pit, and go into perdition. And those who dwell on the earth will marvel, whose names are not written in the Book of Life from the foundation of the world, when they behold the beast that was, and is not, and yet is. **And here is the mind which has wisdom. <u>The seven heads are seven mountains on which the woman sits</u>.**" (Revelation 17:1-9)

---

403   (See *Foxe's Book of Martyrs* by John Foxe; and "The Trail of Blood," by J. M. Carroll. And why did the Catholic Church become drunk with the blood of the saints? Why did she martyr so many true Christians over the centuries? Because true Christians refused to partake in the perversion of the Roman Catholic Church. They instead continued to preach the true gospel, which stated that the only way to be saved—the only way to escape the flames of an eternity in hell—was to accept Christ as Savior and to depart from false religions and false gospels. And by declaring the true gospel they were saying that all of those priests and bishops and leaders of the Catholic Church were on their way to hell; and that they were not the righteous, sanctimonious followers of Christ that they pretended to be. And this incensed them. It drove them to fits of rage, and throughout the countries of Europe and the Mediterranean over the centuries they slaughtered these true Christians. They burned them alive. They (the real heretics) falsely accused them of being heretics. They also kidnapped and enslaved their children, and they raped their wives. Yes, the Catholic Church is truly drunk with the blood of the saints. And, unfortunately, this <u>truth</u> about the Catholic Church will still offend those who have no interest in the truth today.)

There is only one place on earth that this could be and that is Rome. Rome has been known from antiquity as the city that sits on seven hills. The Roman Catholic Church can be none other than the great whore described in the book of Revelation. She fulfills every single description and requirement. The reason Jesus calls her, this "Holy Mother Church," the great whore is because she has perverted and distorted the pure Gospel message; because for most of her 1700 year history she was persecuting and slaughtering true Christians by the millions; and because she has been carrying out the designs of her father Satan by leading untold millions to an eternity in hell by teaching them to trust in the Catholic Church and their own good works, rather than trusting in the shed blood of Jesus and his righteousness alone. And this last point is not a minor quibble. Just ask any Catholic if they are 100 percent sure that when he or she dies they are going to heaven, and the answers you will get are very revealing... "I'm doing the best I can," "I attend Mass regularly," "I haven't committed any mortal sins" or "I'm a good Catholic." What you will not hear is "yes, I am 100 percent sure I will spend eternity with Jesus because I have trusted in his shed blood and in his righteousness alone and I am in no way relying whatsoever on my own righteousness or in any church membership." And that makes all the difference in the world.

"We are of God. He who knows God hears us; he who is not of God does not hear us. By this we know the spirit of truth and the spirit of error." (1 John 4:6)

Constantine turned the tables 180 degrees. He stopped fighting Christianity and instead joined it. But he joined it in the same way that Judas, the betrayer, became one of Christ's disciples. It was outward appearance only. This Roman emperor absorbed the outward symbols and most of the doctrines of Christianity, while at the same time perverting the very substance of the religion. He turned faith in Christ alone, and the determination to follow him and his teachings no matter what, into faith in the state sanctioned church and a determination to follow it and to be loyal to it, although unbeknownst to its practitioners doing so would be in direct contradiction and in opposition to the teachings of Jesus Christ.

> "This was not a revolution of change, but a resolution to incorporate; not a conversion to Christianity but an adaptation of paganism in order to appear Christian. The rituals and observances remained basically the same as those of paganism; only the names and images were changed to appear Christian."[404]

Truly, "Satan himself *was* transformed into an angel of light," and "his ministers also were transformed as the ministers of righteousness." The con was complete.

> "A church had been born that would become the most renowned religious machine and political influence of all time. The true church would be pushed into the background and eventually into hiding through the severest of persecutions."[405]

Then when the Roman Empire fell...

> ..."(approx. AD 450) the power had begun to shift from the emperors to the popes. Thus, in all actuality Rome never fell, but only shifted its power. In AD 1200, the papacy had reached its pinnacle of power. Innocent III (ruled 1198-1216) declared, "As Vicar of Christ, all power is given to me in heaven and in earth." "[406]

The "Vicar" was, of course, a tad delusional, not to mention megalomaniacal. These popes exercised great authority because "they claimed the power to

---

404   (Pg. 63, *The Unveiling: A Journey Through the Book of Revelation*, by Keith Harris)
405   (Pg. 62, *The Unveiling: A Journey Through the Book of Revelation*, by Keith Harris)
406   (Pg. 62, *The Unveiling*, by Keith Harris)

excommunicate and condemn to Hell" those who dared to disobey them. Of course that claim is laughable, only God has that power and he never bequeathed it to a false religion or its popes, yet fear of that same lie causes many to cling to this false religion today. The Catholic Church uses the fear of hell (which is a legitimate fear) *not* to bring people to a saving, personal relationship with Jesus Christ, but to keep them Catholic. So the god of this world, the power behind the Roman Catholic Church, uses the fear of hell in this instance to suck people down to it. No one ever said that he wasn't terribly clever; and diabolical.

If you are trapped in this false Christian religious organization, if you were *born and raised Catholic* as I was, all I can tell you is to get out! Do not pass go, do not collect 200 dollars. Just get out! And take as many of your friends and relatives with you as have ears to hear. Bring them into the light of the glorious Gospel of our Lord and Savior Jesus Christ. Far too many people are seduced by this false religion. That is because it appears righteous, it appears godly on the outside, but on the inside, at its core, it is "full of dead men's bones, and all corruption." (Matthew 23:27) It has "a form of godliness, but denying the power thereof: from such turn away." (2 Timothy 3:5) The book of Revelation talks of "the inhabitants of the earth have been made drunk with the wine of her fornication." (Revelation 17:2) They are intoxicated with the wealth and pageantry and earthly glory and apparent righteousness of Rome. But it is a false righteousness, and her doctrine is false. It is the doctrine of demons, leading multitudes to hell.

People are also seduced because of all of the earthly good the Catholic church has done like helping the poor, and doing good works, and standing for freedom in Poland and other oppressed nations, and standing up for the unborn and for right morals in this country and this world. And the Catholic church should be commended for all of the good that it has done and is doing. But it should also be condemned for all of the terrible things it has done to the true church of Jesus Christ over the centuries, and exposed for the distortion of the gospel that it has taught for 1700 years and continues to propagate today. Doing good on this earth is truly laudable, but leading people to an eternity in hell is truly wicked. Just ask Jesus what he thinks about it; about those who make the cross of Christ of no effect.

Also many people are trapped because all of their friends and family are in, and their social structure, indeed in many instances their very lives, revolves around the local church, so the very thought that it is a false Christian religion would be abhorrent to them. They are thus powerfully separated from the knowledge of the truth. Satan has erected a great wall between them and the truth, and there are very few who have any desire to scale it. "For many are called, but few are chosen." (Matthew 22:14) Which is why Jesus said this…

"Think not that I am come to send peace on earth: I came not to send peace, but a sword. For I am come to set a man at variance against his father, and the daughter against her mother, and the daughter in law against her mother in law. And a man's foes shall be those of his own household. He who loves father or mother more than me is not worthy of me: and he who loves son or daughter more than me is not worthy of me. And he who does not take his cross, and follow after me, is not worthy of me. He who finds his life shall lose it: and he who loses his life for my sake shall find it." (Matthew 10:34-39)

"False religions have done more to doom people's souls than anything else on Earth. While promising life, they give death. They offer salvation while hoarding souls for damnation."[407]

"For true and righteous *are* his judgments: for he has judged the great whore, which did corrupt the earth with her fornication, and has avenged the blood of his servants at her hand." (Revelation 19:2)

And no, this is not "Catholic bashing." This is the historical record. And it is also clearly a part of the biblical record. If anyone finds this exposé *offensive*, and what we are writing here is the truth (and it is), then that individual is offended by the truth, and that is a terrible predicament indeed. For if one is offended by the truth, what is left? Only lies, and the evil one behind those lies. Besides, as I mentioned, I was "born and raised Catholic," and most of my extended family is still Catholic. So I know firsthand what a terribly persuasive, eternal trap it is to be indoctrinated from a youth in <u>any</u> false religion—whether that is Catholicism (as it was in my and my families case), or Islam, or Messiah-denying Judaism, or Jehovah's False Witnesses, or Mormonism, or Algore-Gaya-Mother-earth-environmentalism, or whatever. "For I, the Lord your God, am a jealous God, visiting the iniquity of the fathers onto the children to the third and fourth generations of those who hate me." (Exodus 20:5) People believe what they are taught, and unfortunately if they are taught lies or subtle distortions they will believe that as well; neither will they see anything wrong with it at all. And the father of lies will stop at nothing to keep them blinded and trapped. Thus the lies and the eternal loss are perpetuated from one generation to the next, affecting countless millions of hopeless souls... until someone comes along in the line and breaks the chain. And thankfully this is happening the world over, not only in Catholicism, but Muslim converts have come to Christ despite the fear of being slaughtered for their faith; and for years Jewish men and woman have been led to faith in Jesus, the Jewish Messiah, despite the specter of their own families utterly disowning them, even going so far as to perform a mock funeral to "bury the dead" relative; *dead* because they accepted the truth of Jeshua, the Jewish Messiah.

"Beware lest anyone cheat you through philosophy and empty deceit, according to the tradition of men, according to the basic principles of the world, and not according to Christ." (Colossians 2:8) "Cheat you" out of what? Out of eternal life.

There are many Catholics today who are truly wonderful people who are certainly doing "the best they can." Just as there are many Muslims and Jews and Mormons and Jehovah's False Witness' who also are "wonderful people doing the best they can." But that is <u>not</u> a formula for attaining eternal life. Only in choosing between Catholicism and Jesus, or Islam and Jesus, or the-Judaism-that-still-denies-their-own-Messiah and Jesus, can eternal life be gained. I would encourage anyone trapped in any false belief system to choose Jesus, and "to let the dead bury the dead." Again, you must love the truth more than you love the false religion that you were born into and that has been bred into your mind and heart, or you don't stand a chance. You have to choose between the Catholic Church and Jesus because it is not possible to have **faith** in the Catholic Church to take you to heaven when you die, and at the same time to have **faith** in Jesus Christ

---

407   (Pg. 347, *The Unveiling: A Journey Through the Book of Revelation*, by Keith Harris)

to do the same. You can't have faith in both. You can't hedge your bets. If you have found faith in Jesus Christ then you don't need false religion. "No man can serve two masters: for either he will hate the one, and love the other; or else he will hold to the one, and despise the other. You cannot serve God and *the devil's false religions at the same time*." (Matthew 6:24) (I know, it's supposed to be *mammon*) As we said previously, Jesus pointed this choice out very clearly in two pertinent parables: The Hidden Treasure, and The Pearl of Great Price:

"Again, the kingdom of heaven is like treasure hidden in a field, which when a man has found it, he hides it; and for joy over it he goes and **sells all that he has** and buys that field."

"Again, the kingdom of heaven is like a merchant seeking beautiful pearls, who, when he had found one pearl of great price, went and **sold all that he had** and bought it." (Matthew 13:44-46)

Notice the repetition of the point that they "sold all that they had." They gave up everything for the pearl of great price, which is Jesus and the eternal life he offers. That doesn't square with someone professing faith in Jesus Christ and at the same time holding on to Catholicism, or any false religion. It just doesn't work. If you can't be "sold out" for Jesus, at least when it comes to divesting yourself from a false religious system set up by the adversary of Christ, then you are lost. If that is the case then you haven't found the *pearl of great price*. You might know all about it, as most Catholics do, but you haven't taken possession of it. You're still holding on to lies. And if you continue in that vein until you perish, then you are in grave danger of hearing Jesus speak the following words to you, "Depart from me, you worker of iniquity... for I never knew you."

"Many will say to me in that day, 'Lord, Lord, have we not prophesied in your name? and in your name have cast out devils? and in your name done many wonderful works?' And then will I profess unto them, 'I never knew you: depart from me, you who practice iniquity.'" (Matthew 7:22-23)

In these verses Jesus is not talking to Muslims, or atheists, or Jews who reject their Messiah, for these people would never call Him "Lord." They might exclaim something like "Oh my God, how could I have been so foolish to be so deceived!" but they will *not* be mistaking him for their Lord! No, it is definitely Christians, or rather people who *thought* they were Christians because they were "good Catholics," or "good Methodists," or "good Episcopalians," or because they clapped their hands and shouted "JESUS! JESUS!" every Sunday morning while their false preacher blathered a bunch of non-Christian lies and utter nonsense from the liberation theology handbook to a congregation the vast majority of whom went out every two years, religiously, and voted for Lucifer and his modern-day practice of child sacrifice.[408] These people are those who think they are Christians but are not, because they have never put their faith and trust in Jesus; nor have they made him their Lord. "Not everyone who says to me, 'Lord, Lord,' shall enter the kingdom of heaven, but he who does the will of my Father in heaven." (Matthew 7:21) "But why do you call me 'Lord, Lord,' and do not do the things which I say?" (Luke 6:46) "Then he will also say to those on the left hand, 'Depart from me, you cursed, into the everlasting fire prepared for the devil and his angels.'" (Matthew 25:41)

But what about *Judge not, and you shall not be judged*? Well, this verse has more to do with not standing in self-righteous condemnation of others then it does with compassionately issuing a warning, and giving one an opportunity to escape

---

408   (AKA "baby-mutilation." AKA "abortion")

a terrible eternal fate. It also has nothing to do with a clever attempt by the lost to stifle the exercise of one's God-given ability to discern between truth and lies, to distinguish between a true and a false Gospel, to differentiate a true preacher of the Word from a distorter of the truth, and to ascertain that certain people are in all probability not the *believers* that someone or some false Christian organization has deceived them into thinking they are. And equipped with this discernment, and differentiation, and ascertainment, and proper judgment one can then proceed to witness (see "Witnessing" section earlier in this chapter) to those who are so deceived, and can then possibly reap the joy and the privilege of drawing someone from the lies of Satan to the knowledge of the truth; which will never happen if the old *Judge not and you shall not be judged* crowd has their way. In truth, the verse that more aptly applies in this case is "When I [Jesus/God] say to the wicked, 'You shall surely die,' and **you give him no warning, nor speak to warn the wicked from his wicked way**, to save his life, that same wicked man shall die in his iniquity; **but his blood I will require at your hand**. Yet, if you warn the wicked, and he does not turn from his wickedness, nor from his wicked way, he shall die in his iniquity; but you have delivered your soul." (Ezekiel 3:19)

So, yes, we are not to stand in righteous judgment or holier-than-thou condemnation of others, whether they be Catholics or anyone else we may come in contact with who are lost. However, we are to humbly and compassionately share the truth with the lost as we are able, in order to warn them (as we are commanded to do in the above verse from Ezekiel) about the dire consequences of continuing down a path of wickedness which includes holding on to any of Satan's false religions.

> "All too many people are willing to hear the words of peace and faith from Jesus but are reticent to hear the Jesus who upbraided those teachers who would lead their followers into darkness (Matthew 15:14)."[409]

"Let them alone: they are blind leaders of the blind. And if the blind lead the blind, both shall fall into the ditch." (Matthew 15:14)

Another very powerful parable is that of the man without a wedding garment, which we looked at in the previous chapter but is worth revisiting here:

"And Jesus answered and spoke unto them again by parables and said: 'The kingdom of heaven is like a certain king who had arranged a marriage for his son, and sent out his servants to call those who were invited to the wedding; and they were not willing to come. Again, he sent out other servants, saying, "Tell those who are invited, 'See, I have prepared my dinner; my oxen and fatted cattle are killed, and all things are ready. Come to the wedding.'" But they made light of it and went their ways, one to his own farm, another to his business. And the rest seized his servants, treated them spitefully, and killed them. But when the king heard about it, he was furious. And he sent out his armies, destroyed those murderers, and burned up their city. Then he said to his servants, "The wedding is ready, but those who were invited were not worthy. Therefore go into the highways, and as many as you find, invite to the wedding." So those servants went out into the highways and gathered together all whom they found, both bad and good. And the wedding hall was filled with guests. **But when the king came in to see the guests, he saw a man there who did not have on a wedding garment. So he said to him, "Friend, how did you come in here without a wedding garment?" And**

---

409 (Pg. 10, *Fast Facts on False Teachings*, by Ron Carlson and Ed Decker)

he was speechless. Then the king said to the servants, "<u>Bind him hand and foot, take him away, and cast him into outer darkness; there will be weeping and gnashing of teeth</u>." For many are called, but few are chosen.'" (Matthew 22:1-14)

The marriage supper of the lamb will occur in heaven when the bridegroom, Jesus, is united for eternity with his bride, the true church. In this parable the invited guests are the Jews, many of whom two thousand years ago rejected Jesus as Messiah and refused to come to the wedding. The verse concerning the anger of the King and the burning of the city is a prophecy of the destruction of Jerusalem, which occurred a few decades later in 70 AD. The folks gathered from the highways are the Gentiles, the non-Jews, who are invited to accept Christ as Lord and Savior. But the really sad part of the parable is the man who was found at the wedding without a wedding garment. He was someone who thought he was saved—see how the Lord calls him "friend"—but he was not. He tried to enter heaven clothed in his own righteousness, but he had never put on the righteousness of Christ. He had never been washed in the blood of the Lamb, who was "slain from the foundation of the world" (Revelation 13:8). He had never put on a wedding garment. He thought he was a "Christian" but he had never accepted Jesus' free gift of eternal life. He hadn't sold everything he had for the pearl of great price. He hadn't forsaken all pride and trust in his *own* righteousness.

"Brethren, my heart's desire and prayer to God for Israel [and Catholics] is, that they may be saved. For I bear them witness that they have a zeal for God, but not according to knowledge. **For they being ignorant of God's righteousness, and going about to establish their own righteousness, have not submitted themselves unto the righteousness of God**." (Romans 10:1-3)

There are so many people today who call themselves Christians but who have never put on a wedding garment. They are "going about [in this life] to establish their own righteousness," but they have not "submitted themselves unto the righteousness of God." They have not been "washed in the blood" of Jesus Christ... "And without the shedding of blood there is no remission [of sins]." (Hebrews 9:22) They are clothed instead in their own righteousness, and not the righteousness of Christ. As the above verse (Romans 10:1-3) says, they even "have a zeal for God," but it is not based on the truth. In other words, Catholics may be very zealous for God, but just like all other false Christians—Mormons, Jehovah's False Witnesses, Liberation Theologists, etc.—and all other people trapped in any non-Christian false religion as well—Muslims, Buddhists, etc.—they are determined to continue to worship God according to their way, according to how they were taught, and <u>not</u> according to His requirements. "But in vain they do worship me, teaching for doctrines the commandments of men." (Matthew 15:9) Satan has them all eternally trapped. Unless someone comes along and shares the truth with them; with those among them who "have ears to hear." Otherwise their zeal will be in vain, it will not save them from their sins, and they will die in their iniquity. Ouch!

"Yet indeed I also count all things loss for the excellence of the knowledge of Christ Jesus my Lord, for whom I have suffered the loss of all things, and do count them but dung, that I may gain Christ and be found in him, **not having my own righteousness**, which is from the law, but that **which is through faith in Christ, the righteousness which is from God by faith**." (Philippians 3:8-9)

"Salvation didn't come to you by your confirmation, by your baptism, your church attendance, your church membership. It didn't come to you by giving money. It doesn't come to you by communion, keeping the Ten Commandments, living by the Sermon on the Mount, giving to charity, believing in God, being a good neighbor, living a respectable life, none of those things. In fact, hell will be loaded with people who did all of those. Salvation is through faith, from sin, by love, into life, with purpose, through faith."[410]

Finally, in his letter to the Galatians, Paul said this, "Know that no man is justified by the works of the law, but by faith in Jesus Christ. So we also have trusted in Christ Jesus, that we might be justified by faith in Christ and not by the observation of the law; for by the works of the law no one shall be justified." (Galatians 2:16) He then chided them for leaving the truth and simplicity of faith in Christ to that of trusting in their own works, in their own righteousness: "Are you so foolish? Having begun in the Spirit, are you now trying to attain salvation by human effort?" (Galatians 3:3) The same letter could be written today and sent to Roman Catholics for that same descent into error is still being practiced. Salvation is *unto* good works, it should *result* in good works. The result of putting your faith in Jesus *is* good works. Indeed "faith without works is dead," so works are a good indicator of one's faith. Just don't put the cart before the horse.

"But I fear, lest by any means, as the serpent beguiled Eve through his craftiness, so your minds may be corrupted from the simplicity that is in Christ." (2 Corinthians 11:3)

Salvation doesn't come from knowing about Jesus, and that is the error of so many people who call themselves Christian, or Catholic. They know all about Him but that's not how you are saved. In fact, you're not saved even by believing that Jesus is who he said he was, for so do the demons, and they tremble... "You believe that there is one God. You do well. The devils also believe, and tremble!" (James 2:19) You are saved by believing in Jesus to the extent that you have placed your trust and your faith in Him, and in him alone. For example, I know all about Julius Caesar, in fact I believe he is exactly who history says he was, the emperor of Rome, but I don't believe in Julius Caesar. I don't trust in him for anything.

Here is how the theologian Dr. R. C. Sproul puts it, from his discourse, "Faith Alone:"

> "Rome believes that justification is by faith, it is by grace, and it is by Christ. What Rome doesn't believe is that justification is by faith alone, or by grace alone, or by Christ alone. But rather it combines other elements, for example in the Roman Catholic view it is faith plus works that gives us justification. It is grace plus merit... It is Christ plus me and my inherent righteousness that gives me justification. That's the formula in a nutshell. [Emphasis mine]

[And that is Roman Catholicism's fatal error in a nutshell as well.]

> "But we are justified by faith alone in this sense, that faith is the instrument by which we embrace Christ. We put our trust in Him and in Him alone as the grounds for our salvation. And when we put our trust and our faith in Christ, the moment we trust in Christ, God in a legal action transfers or imputes or counts or reckons the righteousness of Jesus to our account. So that at the end of my life I stand before the judgment seat of God and I am clothed, not in my own *inherent* righteousness, but I am covered by the righteousness of Christ. And God declares me just in the beloved, in Christ.

[Dr. Sproul goes on to say that if you take away this doctrine of justification by faith alone, like the Roman Catholic Church does, then "you take away the gospel..."]

---

410   (From "Coming Alive in Christ," by John MacArthur & Grace To You Ministries)

...That's why this doctrine of the imputation of the righteousness of Christ to the helpless believer is at the very heart and soul of the gospel itself. If you reject that then you are rejecting the gospel. [And you, being the Roman Catholic religion, are sending people to hell.] Nothing less than the gospel of Christ is at stake."[411]

"For by grace are you saved through faith; and that not of yourselves: it is the gift of God: Not of works, lest any man should boast. For we are his workmanship, created in Christ Jesus unto good works, which God has before ordained that we should walk in them." (Ephesians 2:8-10)

"I do not set aside the grace of God; for if righteousness comes through the law, then Christ died in vain." (Galatians 2:21)

"This know also, that in the last days perilous times shall come. For men shall be lovers of their own selves, covetous, boasters, proud, blasphemers, disobedient to parents, unthankful, unholy, without natural affection, trucebreakers, false accusers, incontinent, fierce, despisers of those that are good, traitors, heady, high minded, lovers of pleasures more than lovers of God. Having a form of godliness, but denying the power thereof: from such turn away. For of this sort are they which creep into houses, and lead captive silly women laden with sins, led away with many lusts. Ever learning, and never able to come to the knowledge of the truth. Now as Jannes and Jambres withstood Moses, so do these also resist the truth: men of corrupt minds, reprobate concerning the faith." (2 Timothy 3:1-8)

And then, in addition to all of that, there is this...

"But in vain they do worship me, teaching for doctrines the commandments of men." (Matthew 15:9)

Besides their major error of teaching a false gospel, the Catholic Church is also guilty of disseminating a number of lesser doctrines that are also false and in direct conflict with God's Word. Here are some of their inventions and fallacies, and the dates when they were decreed. Again, not to beat a dead horse, but to leave no stone unturned in identifying as truly false a religious organization that has trapped millions of poor unfortunate souls over the centuries, and continues to do so today:

- Praying for the dead- 300 AD. After you're dead, it is too late for prayers, but it is a great way for a corrupt church to rake in the cash.

- The veneration of angels and of dead saints, and the use of images- 375AD. Yet the Bible clearly says, in the Ten Commandments no less, that "You shall have no other gods before me." And neither shall you "make for yourself a carved image, or any likeness of anything that is in heaven above, or that is in the earth beneath, or that is in the water under the earth." (Exodus 20:3-4) What could be a more obvious disregard of the clear commands of God? ..."But in vain do they worship me...

- Worshipping Mary- 431 AD. "And the devil said unto Him, 'All this power will I give you, and their glory: for this has been delivered unto me, and to whomsoever I wish I give it. If you therefore will worship me, all shall be yours.' And Jesus answered and said unto him, 'Get thee behind me, Satan: for it is written, "You shall worship the Lord your God, and him only shall you serve."'" (Luke 4:6-8)

- The creation of Purgatory- 593 AD. There of course is no such thing as this

411   (From "Faith Alone," by Dr. R. C. Sproul, Renewing Your Mind)

'half-way house' between heaven and hell "where we go after death to be cleansed and purified [by fire] of our venial (minor) sins before we can be allowed into heaven."[412] (If anything, *this* life is purgatory.) But it is a great way to get living relatives to pay for prayers and indulgences to hoist their dead loved ones up into heaven.

- Worshipping the cross, images and relics- 786 AD. See "The veneration of angels and of dead saints" above.

- The celibacy of the priesthood- 1079 AD. "Therefore a man shall leave his father and his mother, and shall cleave unto his wife: and they shall be one flesh." (Genesis 2:24) Think of the millions of unfulfilled lives over the centuries since that unscriptural doctrine was decreed, and all for a lie. (But then again, there's always the homosexual angle...)

- Monotonous praying of the rosary- 1090 AD. "And when you pray, do not use vain repetitions as the heathen do. For they think they will be heard because of their many words." (Matthew 6:7)

- The Inquisition- 1184 AD. "Blessed are you, when men shall revile you, and persecute you, and shall say all manner of evil against you falsely, for my sake. Rejoice, and be exceeding glad: for great is your reward in heaven: for so persecuted they the prophets who were before you." (Matthew 5:11-12) But woe unto them and their eternal destination who do the persecuting.

- The sale of indulgences- 1190 AD. "Step right up! Just light a candle and toss in a few coins and you can take away one year from your loved ones suffering in Purgatory! We also have a great deal on Dr. Barnum's miracle potion, the Energy-Boost! Step right up!" In addition, many wealthy Catholics over the years have tried to buy themselves into heaven by donating vast portions of their estate to the Catholic Church upon their death. (Another way the "great whore" has enriched herself. See Revelation 17:4... "And the woman was arrayed in purple and scarlet color, and adorned with gold and precious stones and pearls") This is an attempt to make sure that their good works outweigh their bad, as if there's a scale at the doors of heaven. (Yet the Bible clearly says that... "All our righteousness are as filthy rags." Isaiah 64:6) Unfortunately they find out the hard way that it is only by having on a wedding garment, by being clothed in Jesus' righteousness, that they will be admitted. Something that the Roman Catholic Church failed to tell them, and now it is too late.

- Confessing your sins out loud to the priest- 1215 AD.[413] I could be wrong but I believe the second oldest known profession is blackmail. Furthermore, the very idea of a priesthood that you must confess your sins to is false and has nothing to do with true Christianity. Jesus Christ is our high priest, he is our mediator between God and man. He is the one through whom we pray directly to the Father. The Father will hear our prayers because he is the mediator, and not any man. "For there is one God, and one mediator between God and men, the man Christ Jesus; Who gave himself a ransom for all, to be testified in due time." (1 Timothy 2:5-6)

All of these things are examples of worshiping God in vain by following the commandments of men as if they were the doctrines of God. They are all the doctrines of men, and yet Catholics follow these religiously, so in vain do they worship God.

---

412   (From "Doctrine of Purgatory Extinguished by Grace," at contenderministries.org)

413   (List of fallacies from "Inventions by the Catholic Church," at juststopandthink.com)

The Roman Catholic Church, and their pomp, and their circumstance, and the Vatican in all of its earthly glory, and the Pope and all of his worldly regalia, and all of their wealth, and all of their property, and their air of self-righteousness, and despite all of their good works, and despite being a force for freedom in Poland and elsewhere, and despite being a force for moral truth, has absolutely nothing whatsoever to do with the church of Jesus Christ on this earth. In fact, for the great majority of their history they were in direct opposition to the church of Jesus Christ and horribly persecuted it, Jesus' true followers.[414] Please don't be seduced by the great deceiver into putting your eggs in the wrong basket.

For too many people, the Catholic Church is their god. It is sacred to them, they are married to it from their earliest youth, they are attached to it at the hip like Siamese Twins, and they will hold on fast to it no matter what the truth, or Jesus or God (all the same) says. The same is true of many Muslims, Mormons, Jehovah Witnesses and Evolutionists. They have made their religion their god, and therefore Lucifer has them trapped eternally.

Again, this is not "Catholic bashing" nor is it "putting anyone down." For there but for the grace of God go I. But it is a very clear warning. And it is not me who is issuing the warning. It is the Word of God. It is Jesus himself, whom they say that they serve.

"For I know this, that after my departure savage wolves shall enter in among you, not sparing the flock. Also from among yourselves shall men arise, speaking perverse things, to draw away the disciples after themselves. Therefore watch, and remember that for three years I did not cease to warn everyone night and day with tears." (Acts 20:29-31)

In the Book of Revelation Jesus reveals the coming judgment of the Roman Catholic Church... "For true and righteous are his judgments: for **he has judged the great whore, who corrupted the earth with her adulteries**, and **has avenged the blood of his servants shed by her**. And again they shouted, 'Alleluia. **Her smoke rises up for ever and ever**.'" (Revelation 19:2-3)

And he also pleads with her congregants to flee from her... "And I heard another voice from heaven, saying, Come out of her, my people, that you be not partakers of her sins, and that you receive not of her plagues. For her sins have reached unto heaven, and God has remembered her iniquities." (Revelation 18:4-5)

And I plead with my own Catholic friends and family members to heed his warning, before death arrives, and the time for heeding is long past...

For further reading on the truth about the Roman Catholic Church see: *Foxe's Book of Martyrs*, by John Foxe; "The Trail of Blood," by J. M. Carroll; *Preparing Catholics for Eternity*, by Mike Gendron; and *A Primer on Roman Catholicism*, by John H. Gerstner.

There are a number of other prominent churches and church organizations that operate under the heading of "Christian," but in truth are not...

Jehovah's *False* Witnesses

"Beloved, while I was very diligent to write to you concerning our common

---

414   (Again, see: *Foxe's Book of Martyrs*, by John Foxe, and "The Trail of Blood," by J. M. Carroll)

salvation, I found it necessary to write to you exhorting you to contend earnestly for the faith which was once delivered for all the saints. **For certain men have crept in unnoticed**, who long ago were marked out for this condemnation, ungodly men, **who turn the grace of our God into lewdness and <u>deny</u> the only Lord God, and our Lord Jesus Christ**. But I want to remind you, though you once knew this, that the Lord, having saved the people out of the land of Egypt, **afterward destroyed those who did not believe**. And the angels who did not keep their proper domain, but left their own abode, **he has reserved in everlasting chains under darkness for the judgment of the great day;** as Sodom and Gomorrah, and the cities around them in a similar manner to these, having given themselves over to sexual immorality and gone after strange flesh, are set forth as an example, **suffering the vengeance of eternal fire**. Likewise also these dreamers defile the flesh, reject authority, and speak evil of dignitaries. …they speak evil of whatever they do not know; and whatever they know naturally, like brute beasts, in these things they corrupt themselves. Woe unto them! …These are spots in your love feasts, while they feast with you without fear, serving only themselves. They are clouds without water, carried about by the winds; late autumn trees without fruit, twice dead, pulled up by the roots; raging waves of the sea, foaming up their own shame; wandering stars **for whom is reserved the blackness of darkness forever**. Now Enoch, the seventh from Adam, prophesied about these men also, saying, 'Behold, the Lord comes with ten thousands of his saints, to execute judgment on all, to convict all who are ungodly among them of all their ungodly deeds which they have committed in an ungodly way, and of all the harsh things which ungodly sinners have spoken against him.' **These are grumblers, complainers, walking according to their own lusts; and they mouth great swelling words, flattering people to gain advantage**. But you, beloved, remember the words which were spoken before by the apostles of our Lord Jesus Christ: how they told you that **there would be mockers in the last time** who would walk according to their own ungodly lusts. These are sensual persons, who cause divisions, not having the Spirit." (Jude 1:3-8, 10-19)

I start this section with that Scripture because, although it refers to any false preacher of the word, Jude could have written it today specifically for the *Jehovah's False Witnesses*, and mailed it to them. A century ago they "crept into" Christianity unnoticed and "turned the grace of our God into lewdness by denying the only Lord God, and our Lord Christ." They "are spots in your love feasts …clouds without water, carried about by the winds …raging waves of the sea, foaming up their own shame; wandering stars for whom is reserved the blackness of darkness forever. …They mouth great swelling words, flattering people to gain advantage" (while at the same teaching unchristian lies). Such is the contempt and derision that the Almighty holds towards those false preachers who would pervert his religion and lead millions to an eternity in hell.

"Who is a liar but he who denies that Jesus is the Christ? He is antichrist, who denies the Father and the Son. Whosoever denies the Son, the same has not the Father: (but) he who acknowledges the Son has the Father also." (1 John 2:22-23)

The Jehovah's False Witnesses are exactly that, what the above scripture says, they are "liars who deny that Jesus is the Christ." They mock the truth of the gospel of God and his Christ. They deny the deity of Christ, the trinity, that Jesus rose bodily from the dead, and the Holy Spirit. In the place of these eternal Christian truths they teach this man-made rubbish:

...“that Jesus was only a man when on earth, not “the Word become flesh,” ...that heaven's doors are open to only 144,000 people, ...that salvation is found only through the organization,

... [and] that salvation can be maintained only by energetic works for the Organization until the end when one may then merit eternal life on a paradise earth.”[415]

Sound familiar? It's just another tired, same-old-same-old *salvation by works* false doctrine.

“There are five important facts to remember about the Jehovah's Witnesses and the Watchtower Organization.

1. They have accepted the Organization as the prophet of God.

2. They have accepted the Organization as God's sole channel for His truth.

3. They believe that to reject the Organization is to reject God.

4. They believe that only the Organization can interpret the Bible, as individuals they are unable to do so.

5. They believe that the *Watchtower* magazine contains God's truth, directed by Him, through the Organization.”[416]

They, of course, are deluded. They have been seduced by the prince of darkness into putting their faith and trust in an *organization* that runs contrary to the word of God. They have broken the very first commandment, “You shall have no other gods [and no deceitful organizations] before me.” They have neglected to seek the truth... “And you shall seek me, and find me, when ye shall search for me with all your heart.” (Jeremiah 29:13)... and instead are following some of the great deceiver's more obvious lies. The lies of Satan hover over this earth like a stench, and people believe what they have been born, or roped, into.

“But I fear, lest by any means, as the serpent deceived Eve through his trickery, so your minds may be corrupted from the simplicity that is in Christ. For if he who comes preaches another Jesus whom we have not preached, or if you receive another spirit which you have not received, or another gospel which you have not accepted--you may well put up with it!” (2 Corinthians 11:3-4)

The Jehovah's Witnesses cannot substantiate their religion or back up their beliefs with any facts or evidence—it is a historic fact that their founder just made it all up out of whole cloth—so it is astounding that they cannot recognize its deception, and unfortunate that they remain trapped under its spell. But that is another demonstration of the power of Satan, and the grip that his false-religious lies usually have. And under this spell they continue to doggedly proliferate this deception, by selling it internationally to the unsuspecting and the uninformed, when indeed it has nothing whatsoever to do with actual Christianity. And like so many suckers in the past who forked over their hard earned money to fast-talking snake-oil salesmen for a pint of their energy-boosting miracle-cures, there are millions of the uninformed today who are still buying into it.

“The Jehovah's Witnesses have mapped out the entire United States so that every residence will be contacted at least once or twice a year by a team of door-to-door workers. They claimed recently that in one year over 3.6 million members spent over 835 million hours of door-to-door witnessing for the Watchtower. Out of that sheer grueling persistence they have been able to harvest many people who have not been grounded in God's Word and were easily led astray by this counterfeit religion.”[417]

---

415   (Pg. 122, *Fast Facts on False Teachings*, by Ron Carlson and Ed Decker)

416   (Pg. 121, *Fast Facts on False Teachings*, by Ron Carlson and Ed Decker)

417   (Pg. 117, *Fast Facts on False Teachings*, by Ron Carlson and Ed Decker)

"Woe unto you, scribes and Pharisees, hypocrites! [Woe unto you, Jehovah's False Witnesses, deceivers!] For you travel sea and land to win one convert, and when he is made, you make him twofold more the child of hell than yourselves." (Matthew 23:15)

And they are coming to a front door near you... When they show up I would be careful in any attempt to engage them in debate, as unfortunately "it is impossible to win an argument with an ignorant man, or woman." And they have been strongly brainwashed into their organizations particular form of spiritual ignorance. But if the Holy Spirit leads you to try to get through to them with the actual witness of Christ, then here are a few suggestions...

First of all, you cannot "offer him any hope of heaven, since Jehovah's Witnesses teach and believe that only 144,000 people are going to make it to heaven, and the odds are that the visitor at your door isn't one of them." This scam masquerading as a "revelation" was unveiled in 1917 in order to increase membership, but as their ranks swelled past that number in 1935 they had a problem; which they conveniently resolved by another "revelation" that said the first 144,000 were going to heaven, and everyone after them "would stay here on earth and live in a new paradise."[418] Whew! That was some quick thinking! And we would say that the idiots bought it except Talking Points is trying to be as inoffensive as possible. But yes, they bought it.

In response to their claim that Jesus is not God, have them read this verse: "Therefore take heed to yourselves and to all the flock, over which the Holy Spirit has made you overseers, to shepherd the church of <u>God</u>, which <u>he</u> has purchased with <u>his own blood</u>." (Acts 20:28) Then ask them, who purchased the church "with his own blood?" That could only be Jesus. So ask them to reread the verse a couple of times themselves so the Holy Spirit himself can show them that Jesus is God—"<u>God</u> purchased the church with his own blood!"—if they have ears to hear. If they are not interested then shake the dust off your shoes ..."for a testimony against them. Verily I say unto you, it shall be more tolerable for Sodom and Gomorrah in the Day of Judgment, than for [those people]." (Mark 6:11) You can also share with them the following verse concerning the deity of Christ: "I and my Father are one." (John 10:30) Then ask them who is the one not telling the truth; Jesus? Or their Organization? And why would it do that if it was "the sole channel for God's truth?" Challenge them in as kind a way as you can. Put them on the spot. It might be the only time they've heard the truth, and it could be their only chance of coming from darkness to the Light. For the vast majority of doors they knock on, unfortunately, are opened by people who, even if completely uninterested in their sales pitch, are still ignorant of the truth and completely powerless to challenge them and their Christ-denying, Bible-distorting inventions. And unlike many of the parents of brainwashed Moonie or Hare Krishna children, you can't kidnap them, take them to a hotel room and deprogram them. So pray for them and try to plant as many biblical seeds as you can, and let God's word do its work.

You can also have them turn to these verses in the Book of Revelations: 1:8, 21:5-7, and 22:13, which all clearly establish that "the Alpha and the Omega, <u>the First and the Last</u>" is another title for their very own <u>Jehovah God</u>. They have no choice but to agree with this as they are also taught to believe the Bible. (Only

---

418   (Pgs. 125-126, *Fast Facts on False Teachings*, by Ron Carlson and Ed Decker)

they have been constantly drilled with their leaders "superior" yet perverted interpretation of it.) After those verses have them read Revelation 1:17-18: "And when I saw him, I fell at his feet as dead. But he laid his right hand upon me, saying unto me, Fear not; <u>I am the First and the Last</u>. I am he who lives, and was dead, and behold, I am alive for evermore." Then you can ask them, "When did the Alpha and the Omega, the First and the Last, die? When did Jehovah God die?" And the answer of course is that *Jesus* died and is now alive, and therefore clearly Jesus *is* their very own <u>Jehovah God</u>, "the First and the Last" from Revelation 22:13![419] But don't expect them to jump up and down in amazement and joy as they embrace this new found truth. False religions are deeply ingrained and they die very hard. People learn to love their religion, no matter how obviously false and absurd, to the extent that they have no real interest in discovering whether it is actually true or not. They are also heavily swayed by their companionship, fellowship, love and friendship with many other of the deceived, and the thought of losing all of that love and friendship, and breaking that fellowship by learning that their religion is false, is a little too much to bear. And therefore the evil one is able to keep them tightly bound in his powerful grip. The power of the Almighty, however, is infinitely greater, but God has given each of us free choice, and it is only if one freely chooses the truth over the lies they have been taught, that they will be set free. Pray for them. There is not a more hollow or more loathsome way to spend one's life than being seduced into thinking you're working for God when in reality you're wasting your time, energy and talents working for the dark side of the force; "travelling sea and land to win another convert, and when he is made, you make him twofold more the child of hell than yourself." Congratulations. And their final reward for all of that vain effort is even worse.

"I am Alpha and Omega, the Beginning and the End, says the Lord, who is, and who was, and who is to come, the Almighty." (Revelation 1:8)

Keith Harris, author of *The Unveiling*, concludes:

"Jesus is the Almighty. This is very plain and easily understood although many have not grasped this profound truth. Jesus is God. Many think of Jesus as less than God. This is oftentimes due to the title given Jesus as the Son of God. Note that in the gospels Jesus is also the son of David, the son of Abraham, and the son of man. Why? Because as the son of David, Jesus has a right to the throne of David (Lu. 1:32). As the son of Abraham, Jesus is entitled to the land of Israel, which includes the royal grant to Abraham. He is the son of man revealing His title to the Earth and the world. He is the Son of God, indicating that He is the heir of all things. Jesus is the second person of the three-fold nature of Almighty God. He Is God.

"Note that Jesus thought it not robbery to be equal with God (Phil. 2:6). Why? Because He is Almighty God. In Isaiah 43:11, Jehovah God says, "I, even I, am the Lord; and beside me there is no saviour." Yet the angel of the Lord proclaims, "For unto you is born this day in the city of David a Saviour, which is Christ the Lord" (Luke 2:11). Why? Because the Saviour, the Lord, Jesus, is God! Matthew says, "and they shall call his name "Emmanuel," which means, "God with us." Wow!

"John tells us in the first few verses of his gospel that Jesus is God. In the beginning was the Word, and the Word was with God, and the Word was God." The Greek for "Word," here, is "Logos," meaning "something said, utterance." John goes on to say that the "Word became flesh and dwelt among us." That's Jesus, God in the flesh! Note what the apostle

---

419   (See pgs. 129-131, *Fast Facts on False Teachings*, by Ron Carlson and Ed Decker)

Paul says concerning the deity of Jesus: "Looking for that blessed hope, and the glorious appearing of the great God and our Saviour Jesus Christ; Who gave himself for us, that he might redeem us from all iniquity, and purify unto himself a peculiar people, zealous of good works. Titus 2:13-14

"Paul says that Jesus is the great God and Saviour who gave "himself." Please don't let those who claim to be Jehovah's Witnesses tell you that Jesus is a lesser God. Jesus is God, the Almighty. Praise God, that's a revelation!"[420]

Jehovah False Witnesses are also taught to mock the idea of a triune God, three persons in one God. They have been taught to think of themselves and their fellow witnesses as far superior than all of those Christians who do not think like them. (Of course no one wants to think of themselves as the dupe of Satan that they actually are.) But yet in Zechariah we find this prophetic verse, "And I will pour upon the house of David, and upon the inhabitants of Jerusalem, the spirit of grace and of supplications: and they shall look upon **me** whom they have pierced, and they shall mourn for **him**, as one mourns for his only son, and shall be in bitterness for him, as one who is in bitterness for his firstborn." (Zechariah 12:10) It is interesting to note in this verse how the writer of the Bible, God, refers to Jesus as himself (**me**) and also as another person (**him**), as his Son. This is just one of the many indications in God's Word of his triune nature. Father-Son-Spirit; three ...but only One. In the New Testament of course we find countless references to this three-in-one concept, as Jesus spoke abundantly about his relationship with the Father, that he was the Son of man and the Son of God, and that he and the Father were One and the same; in addition to abundant references to the third person of the Trinity, the Holy Spirit.

"I said therefore unto you, that you shall die in your sins: for **if you believe not that I am he, you shall die in your sins**." (John 8:24) Jehovah's False Witnesses believe not that Jesus is God, and if they do not repent and turn from the lies of the great deceiver, then they will die in their sins. Warn them if you can.

For further reading on the truth about Jehovah's False-Witnesses see: *How to answer a Jehovah's Witness: How to Successfully Take the Initiative When They come to Your Door*, by Robert A. Morey; and Chapter 8 "Jehovah's Witnesses" from *Fast Facts on False Teachings*, by Ron Carlson & Ed Decker.

Mormonism or the Church of Joseph Smith[421] of Latter-day Deceived

"But there were false prophets also among the people, even as there shall be false teachers among you, who secretly shall bring in damnable heresies, even denying the Lord who bought them, and bring upon themselves swift destruction. And many shall follow their pernicious ways; because of whom the way of truth will be blasphemed. By covetousness they will exploit you with deceptive words." (2 Peter 2:1-3)

It is a sad reality that millions of people have been raised and taught to believe in this false religion, and for most Mormons this has happened due to their birth and not through any choice of their own. (Although as adults they, along with everyone else, are wholly responsible for what they *continue* to believe. For God has given each of us a conscience and a truth meter inside of us and we can choose

---

420   (Pgs. 20-21, *The Unveiling: a Journey through the Book of Revelation*, by Keith Harris)
421   (For it has nothing to do with Jesus Christ.)

to either listen to it or ignore it.) And certainly our heart goes out to these many people thus trapped, and to the many people raised and thus trapped in all the other false religions of this world as well. And our prayers also go out that they might be rescued from it, and brought into the light of the saving grace of our Lord Jesus Christ. That is our intent, and not to insult or demean them in any way. However, sometimes something is just so downright stupid[422] that it is almost impossible to treat it with anything other than a measure of richly deserved ridicule and scorn. The religion that is, and not the people brainwashed into it.

> "Most Mormons are victims of a deception as clever as anything thrown at the world since the days of Adam. But, tragically, even though they may be victims, they do great harm to the true cause of Christ."[423]

And that "true cause" is to lead people to heaven, and not to hell...

Mormons are all the generational victims of a known con man turned spiritual huckster by the name of Joseph Smith, the founder of the "Church of Latter-Day Saints." He lived in the first half of the 1800's and engaged in a scam known as "money digging," and used it to defraud his victims of their cash...

> ""Money-digging" (also sometimes called "glass-looking") was a con or a fraud that was practiced in the Northeastern US at that time. The con man would have a "magic stone" which he would place in his hat, and then pull his hat over his face, excluding all light. The stone would then supposedly shine and the money-digger could locate hidden treasure. People would pay the money-digger to tell them where to dig, but there was another part of the scam. When you got "close" to the treasure, the money-digger would usually tell you that the treasure had moved. The whole thing was like looking into a crystal ball or doing palm-reading."[424]

Smith would later graduate from separating folks from their money, to separating them from eternal life as well. The following is a description, from one of Joseph Smith's early followers, of how this money-digging con man "dug up" the Mormon *holy* book:

> "David Whitmer, another of the three witnesses, wrote: "I will now give you a description of the manner in which the Book of Mormon was translated. Joseph Smith

---

422 (Let me take a moment here to address our use of certain words such as "stupid" and "dumb" and "ignorant" and (our favorite) "idiot" that we will use occasionally throughout the book to refer to certain people, as far as some of their beliefs, ideology or voting habits are concerned. Our intention is not to offend, but to be brutally and completely honest... The truth doesn't need to be sugar-coated. And along those lines it is interesting to note that the Lord himself calls all of us stupid, many times throughout His Word, because He refers to us as sheep hundreds of times... "All we like sheep have gone astray." (Isaiah 53:6) "And he shall set the sheep on his right hand." (Matt 25:33) "My sheep hear my voice." (John 10:27) "Feed my sheep." (John 21:17) And the fact is that sheep are internationally recognized as the dumbest animal on the planet. For example, if one sheep starts walking around in a circle then all the other sheep will follow it, thinking it must be going *somewhere*. And around and around they go. They're dumb. And our own Creator says that we are *all* dumb as sheep. So if we occasionally refer to certain folks as stupid or idiots or ignorant or imbeciles, know that we know that we are all stupid. And that in realizing this fact it can go a long way toward freeing someone from the sort of pride that blinds them from seeing just how stupid some of their beliefs, ideologies or voting habits actually are, as some of ours once were as well. Thank you and may God richly bless.)

423 (Pg. 166, *Fast Facts on False Teachings*, by Ron Carlson & Ed Decker)

424 (From "Rejecting the Mormon Claim," Part 4, Joseph Smith and "Money-Digging," @ bibletopics.com)

would put the seer stone into a hat, and put his face in the hat, drawing it closely around his face to exclude the light; and in the darkness the spiritual light would shine. A piece of something resembling parchment would appear, and on that appeared the writing. One character at a time would appear, and under it was the interpretation in English. Brother Joseph would read off the English to Oliver Cowdery, who was his principle scribe, and when it was written down and repeated to Brother Joseph to see if it was correct, then it would disappear, and another character with the interpretation would appear.'"[425]

Abracadabra and bibbidee bobidee boo and just like that the world has another "holy" book; and Satan has another false Christian religion. What Joseph Smith and his *scribes* came up with is a bunch of words, phrases, paragraphs and chapters that—although they cleverly mirror (for credibility's sake) much of the old English language of the Bible itself—have less spiritual insight than The Book of Elephant Jokes. And I'm not trying to be smart or rude by saying that. The Book of Mormon actually does have less spiritual insight than The Book of Elephant Jokes. (Read both and compare for yourself.) Get a bunch of college kids together for a weekend, get them rip roaring drunk, give them a copy of the Bible as a guide, and have them make up as much "prophecy" as they care to spew, using a bunch of *beholds* and *unto's* and *came to pass'* to sound real ancient and biblical; write it all down, and you will have come up with a spiritual work comparable to the Book of Mormon. Indeed, you probably will have surpassed it.

> "All men have heard of the Mormon Bible, but few except the "elect" have seen it, or, at least, taken the trouble to read it. I brought away a copy from Salt Lake. The book is a curiosity to me, it is such a pretentious affair, and yet so "slow," so sleepy; such an insipid mess of inspiration. It is chloroform in print. If Joseph Smith composed this book, the act was a miracle — keeping awake while he did it was, at any rate. If he, according to tradition, merely translated it from certain ancient and mysteriously-engraved plates of copper, which he declares he found under a stone in an out-of-the-way locality, the work of translating was equally a miracle, for the same reason."[426]

Mormons are taught not to trust the Bible, but instead to trust their Book of Mormon; a book which can't pass the laugh test. (People, again, believe whatever they are taught.) They also run a TV commercial every year around Christmas that says, "There's another testament of Jesus Christ, and it's called *The Book of Mormon.*" And they encourage you to call for a free copy. I always do… "Yes, can you send me a copy of the "other" testament of Jesus Christ, you know, the one written by Satan." There's usually some silence, followed by a polite response. But at least a seed has been planted, and even sometimes a short but interesting discussion will ensue. Pray for the day I can witness further and perhaps lead someone from the idiotic lies of Satan, to the glory of the Lord. What a day of rejoicing that would be.

"For I testify unto every man that hears the words of the prophecy of this book: If any man shall add unto these things, God shall add unto him the plagues that are written in this book. And if any man shall take away from the words of the book of this prophecy, God shall take away his part from the Book of Life, from

---

425   (From "Rejecting the Mormon Claim," Part 4, Joseph Smith and "Money-Digging," @ bibletopics.com)
426   (Pgs. 58-59, *Roughing It* (On the Book of Mormon), by Samuel Langhorne Clemens, aka Mark Twain)

the holy city, and from the things which are written in this book." (Revelation 22:18-19)

As far as their equally incomprehensible theology, space does not permit a detailed exposé of it here. I would suggest reading Chapter 11 "Mormonism: The Church of Jesus Christ of Latter-Day Saints" from Ron Carlson and Ed Decker's book, *Fast Facts on False Teachings*. Here are a few excerpts:

"The major heresy of Mormonism is summed up in its central theological axiom, the doctrine of the law of eternal progression. To believe in and teach this doctrine is to be so separated from Christian orthodoxy that the unrepentant adherent is consigned to a Christless eternity. It is stated as follows:

As man is, God once was, And as God is, man may become.

Roll that through your mind a time or two: "*As man is, God once was, And as God is, man may become.*"

This all starts with the LDS teaching that there are a great number of planets scattered throughout the vastness of outer space which are ruled by countless exalted men-gods who once were human like us.

This may all sound like "Battlestar Galactica" to the average person, but upon this axiom is based the entire theology of Mormonism: from the temple rituals for the living to those for their dead; from the teachings that families are forever, to the pressure on parents to send their youth to the mission fields across the world.

The Mormon people are committed to a controlled program that maps out their entire lives as they seek their own exultation and godhood,[427] their own planet to rule and reign over."[428]

(Only to find to their utter disappointment that the really choice planets have already been taken. Mr. Applewhite and 39 other castrated members of the Heaven's Gate cult beat them to it by catching the red eye on the comet Hale-Bopp.)

Authors Carlson and Decker then go on to quote the script from their movie *The God Makers* which summarized Mormon doctrine:

"Mormonism teaches that trillions of planets scattered throughout the cosmos are ruled by countless gods who once were humans like us. They say that long ago on one of these planets, to an unidentified god and one of his goddess wives, a spirit child named Elohim was conceived. This spirit child was later born to human parents who gave him a physical body.

…"Mormons believe that Elohim is their heavenly Father and that he lives with his many wives on a planet near a mysterious star called Kolob. [Which is in a little less mysterious galaxy called Bulls**t.] Here the god of Mormonism and his wives, through endless celestial sex, produced billions of spirit children. [They also produced a two-million-year back-order of Viagra.] [Kolobians have the government option.]

…"Early Mormon prophets taught that Elohim and one of his goddess wives came to Earth as Adam and Eve to start the human race. Thousands of years later, Elohim in human form once again journeyed to Earth from the star base Kolob [on the good ship Lollipop], this time to have physical relations with the Virgin Mary in order to provide Jesus with a physical body.

…"By maintaining a rigid code of financial and moral requirements, and through performing secret temple rituals for themselves and the dead, the Latter-day Saints hope

---

427   ("You shall have no other gods before me." Exodus 20:3)
428   (Pg. 167, *Fast Facts on False Teachings*, by Ron Carlson & Ed Decker)

to prove their worthiness and thus become gods. The Mormons teach that everyone must stand at the final judgment before Joseph Smith... [Sure, and right next to him will be Bernie Madoff, hawking some sweetheart deals in Kolobian time-shares.]

"Those Mormons who are sealed in the eternal marriage ceremony in LDS temples expect to become polygamous gods or their goddess wives in the Celestial Kingdom, rule over other planets, and spawn new families throughout eternity. The Mormons thank God for Joseph Smith, who claimed that he had done more for us than any other man, including Jesus Christ. The Mormons claim that he died as a martyr, shedding his blood for us so that we too may become gods. [Actually, he was shot to death by an angry mob while in jail awaiting trial. ...Their treasure had been moved one time too many.]

"Shocking? Incomprehensible? Maybe to you and to us, but *this is the core of Mormon theology*. It binds its believers away from the real Jesus, the real gospel, and the real spirit of truth as surely as though they were locked away in chains of metal."[429]

People believe whatever they are taught, no matter how idiotic, asinine or absurd, which is why it is so crucially important to be taught the truth, and if not then to seek it out for oneself before this life is over. For then it is too late. One will have already made their eternal choice. That's Joseph Smith's real gift to his followers, he gave them the way of eternal death. And it wouldn't hurt to add here how some of the early Mormons in Utah dressed up as Indians and ambushed wagon trains full of hopeful pioneers, slaughtering the men folk and taking the attractive woman and girls alive, for celestial "marital" purposes of course. And how today a substantial number of this con man's religion practices a form of polygamy that in truth is nothing more than the raising of little girls like cattle for the sexual use of older men. With their moms kept in check by having their minds brainwashed from early childhood with the fear of "spending an eternity in hell" if they try to escape from this perverted madness. The only way, of course, that anyone can spend an eternity in hell is if they reject the gift of the Lord Jesus Christ, and *not* if they reject the lies of one more bizarre, satanically-inspired, false religion.

But in fairness, the Mormon child pervert sects, however disgusting, are a minority. The majority of Mormons are for the most part wonderful people who preach wonderful family values and do subscribe to many of the Bible's basic moral teachings. They are also just unfortunately trapped in another false Christian religion. One not founded by Jesus, but instead inspired at its inception by the father of lies and then brought to fruition by one of his more accomplished con-men. And one that has no interest whatsoever in leading its people into a saving relationship with Jesus Christ. Pray for their escape, and pray for those ex-Mormons who have come to know Christ that they may be a powerful witness for the freeing of many others caught in that eternal trap.

"You hypocrites, well did Isaiah prophesy of you, saying, 'This people draws nigh unto me with their mouth, and honors me with their lips; but their heart is far from me. But in vain do they worship me, teaching as doctrines the commandments of men.'" (Matthew 15:7-9) They teach the "commandments made up by men" as if they were the doctrines of God. And they lead unsuspecting and uninformed millions down to an eternity in hell. Somebody might want to tell Mitt Romney, and Glenn Beck.

---

429   (Pgs. 168-179, *Fast Facts on False Teachings*, by Ron Carlson & Ed Decker)

Liberal Christianity

"But there were false prophets also among the people, even as there shall be false teachers among you, who secretly shall bring in destructive heresies, even denying the Lord who bought them, and bring upon themselves swift destruction. And many shall follow their pernicious ways, because of whom the way of truth shall be evil spoken of. And through covetousness they shall exploit you with deceptive words. ...The Lord knows how to deliver the godly out of temptations and to reserve the unjust unto the day of judgment to be punished. ...They are presumptuous, self-willed. ...They are spots and blemishes, sporting themselves with their own deceptions while they feast with you, having eyes full of adultery and that cannot cease from sin, beguiling unstable souls. They have a heart trained in covetous practices, and are cursed children. ...These are wells without water, clouds that are carried by a tempest, for whom the blackness of darkness is reserved forever." (2 Peter 2:1-3, 9-10, 13-14, 17)

Liberal Christianity is "a part of the unbelief that denies God, denies the inspiration of the Bible, denies the virgin birth, the bodily resurrection and the atonement of our Saviour, and denies the need for and the possibility of an actual regeneration, the new birth."[430] It is a gross perversion of the Christian faith, and it is a distortion of the gospel of Jesus Christ. But it is running rampant throughout this country, and the world. The disciples of liberal Christianity have taken over many Christian church organizations, many pulpits, many Christian radio stations, and many Bible colleges as well where they are free to pervert the minds and dilute the faith of many susceptible youths looking for a lifetime of service in the ministry; who enroll there having been taught the basics of the Christian faith but having not been rooted and grounded in the truth so they are unequipped to withstand the onslaught of Satan's, and their liberal "Christian" professors, distortions and denials of the basic Christian faith. So they come out of there, instead, rooted and grounded in lies. And the beat goes on...

Liberal Christians embrace Satan's genesis while denying the truth of the Bible's Genesis account of creation, unfortunately uneducated as to the scientific accuracy of it. (See chapter 10) Instead of believing the truth given to them from the mouth of the Creator who, coincidentally, was actually *there* at the time of creation, they embrace the scientifically impossible lies of Darwinism taught to them by faux scientists who were not. Many even go so far as to celebrate once a year what they call "Darwin Sunday" where they spend the day apologizing to the dead (and smokin') Darwin because faux Christianity took so long to believe him. (Real Christianity never has, and never will.) They swallow the great deceiver's ridiculous fairy tale and reject the very foundations of their own religion, the bible's Genesis. ("Satan Sunday" is coming next where they will throw off all pretense whatsoever and apologize to the father of lies for ever believing in *any* part of Christianity whatsoever.)

Liberal Christianity also denies the miracles of the Bible, being ignorant of the fact that even the most extraordinary miracles of God's Word like parting the Red Sea or making the sun stand still in the sky for an entire day are less difficult for the Creator to perform than it is for them to scratch their rear ends. They deny that the Bible is the Word of God. They deny the virgin birth, without which there would be no acceptable sacrifice for the sins of mankind, preaching instead the

---

430   (Pg. 249, *Prayer: Asking & Receiving*, by John R. Rice)

blasphemy that Jesus was a blemished Lamb. They deny the miracles of Jesus which authenticated and identified him as the Christ, the Messiah, and the Son of the Living God. They even deny his resurrection without which there actually is no such thing as Christianity. (Which of course was their father Satan's intention by conjuring up liberal Christianity in the first place.)

"And if Christ is not risen, then our preaching is in vain, and your faith also is in vain." (1 Corinthians 15:14)

And what is just as incriminating, the great majority of liberal Christians go out and vote for the baby mutilating Democrats, religiously, every two years. And most are proud of it. Indeed that seems to be the only thing that they *do* passionately believe in; killing babies. Not the Bible, not Genesis, not Jesus, not his virgin birth or his resurrection; just killing babies. All they retain is the *title* of Christian church. But they are the church of their father Satan. Flee from them, people, like you would scramble out from below a dam of raw sewage that has just burst.

"Therefore as the fire devours the stubble, and the flame consumes the chaff, so their root shall be as rottenness, and their blossom shall ascend like dust: because they have cast away the law of the Lord of hosts, and despised the word of the Holy One of Israel." (Isaiah 5:24) Liberal Christianity truly "despises the word of the Holy One of Israel." They deny the Bible's historical accuracy, its truthfulness, and that He even wrote it.

The liberal church is more interested in spreading its "social" gospel, a doctrine that takes precedence over the true gospel. It has also become known as the *Emerging Church*, which is accurate as it has emerged from the constraints of the truth of Christ into the liberation of the lies of Christ's adversary. (If we can call that *emerging*.)

> "This movement reinvents Christianity. It takes your eyes off the cross and has you focus on experience. Scripture is no longer the authority. There are no absolutes, even in the Bible. ...Hell, sin, and repentance are downplayed so that no one is offended. ... In a nutshell, social action trumps eternal issues; and subjective feelings are preferred over absolute truth. Experience trumps reason."[431]

Liberal Christianity can best be described as *Churchianity*. A self-righteous going through the motions of religiosity, holiness and caring so they can feel superior to everyone else while at the same time not offending anyone; anyone but Jesus that is.

> "Biblical creation, the rapture of the Church, the millennium of peace, the restoration of Israel, eternal salvation or eternal condemnation are considered vague terms or called false doctrines."[432]

One of the few actually unassailable doctrines that liberal Churchianity does have is that it doesn't matter whether or not you believe in certain traditional "controversial" and "divisive" Christian doctrines like the virgin birth, the miracles of Jesus, the resurrection, etc.; all that matters is that you have **faith**.[433] But the

---

431   (From article, "Will the Emergent Church Submerge Yours?" by Jan Markell, Israel My Glory Magazine, March/April 2009)

432   (From article, "Israel, A Stumbling Block for Churchianity," by Reinhold Fedorolf, News From Israel Magazine, April 2009)

433   (This is the Rodney King version of Christianity—"Can't we all just get along?" ..."So what if I'm driving drunk and wasted on drugs, recklessly careening about endangering the lives of many of my fellow citizens, in addition to stubbornly resisting arrest." Or the

question is, Faith in <u>what</u>? Doesn't matter! Not to liberal Churchianity. You can have your *faith* in anything whatsoever, just so long as you have it. As if faith were an end in and of itself. Faith such as that is not a virtue. It's a delusion. For faith is nothing but an empty word unless you have faith in *something* worthy of that faith. Something that is true. Faith standing alone with no deserving object to place that faith in—whether some truly worthy individual like the real Jesus of Scripture or those doctrines from that same Scripture that came forth from the Creator—is ludicrous. It's like going to a ball game without any teams on the field and telling your confused little boy, "Son, it doesn't matter whether any teams show up to play, we can still have "watch." We can watch the clouds. We can watch the bases. We can watch the grass grow. Look, we're watching! See, my son, how special you and I are, we're better than those other less-progressive folks, we don't need teams, for we have *watch*."

Also, faith in some jolly-old Santa Claus conjured up in the minds and imaginations of Chinos (Christians in name only) to replace the real God, one who demands nothing in the way of obedience or adherence to His biblical truths, is faith in nothing at all. God's Word talks of people "with itching ears" who have no interest in taking up their cross and following Jesus but instead are drawn, like flies to road apples, to pure gobbledygook. "Preach the word. Be ready in season and out of season. Convince, rebuke, exhort with all longsuffering and **doctrine**. For the time will come when **they will not endure sound doctrine**, but after their own lusts, because they have itching ears, they shall heap up for themselves teachers; and they shall turn away their ears from the truth, and be turned aside to fables." (2 Timothy 4:2-4)

(But, hey, at least they have plenty of *watch*.)

"Dear friends, do not believe every spirit, but test the spirits to see whether they are from God, because many false prophets have gone out into the world. This is how you can recognize the Spirit of God: Every spirit that acknowledges that Jesus Christ has come in the flesh is from God, but every spirit that does not acknowledge Jesus is not from God. This is the spirit of the antichrist, which you have heard is coming and even now is already in the world." (1 John 4:1-3)

> "God's Spirit sets you apart and becomes your personal teacher and trainer. He helps you understand Scripture, see things more clearly from God's perspective, and discern between God's truth and others' religious-sounding lies."[434]

As we mentioned before, liberal Churchianity either outright denies the miracles of God's Word or says that it is of no concern whether you believe in them or not, as long as you don't think that your belief is better than anyone else's. (The old "truth is relative" slop served-up out of another dirty bowl.)

> "Every opponent of Christianity opposes it on the point of miracles. Evolutionists do not believe in a direct, immediate, supernatural, miraculous, creation. They do not believe that God made something out of nothing, that He did it instantly and not by a process.

"all-inclusive" vacation version of the same—"Let's not actually believe anything that might exclude or make someone feel uncomfortable." And don't anyone confuse them by quoting Matthew 10:34-36: "Think not that I am come to bring peace on earth. I came not to bring peace but a sword. For I have come to set a man at variance against his father, and the daughter against her mother, and the daughter in law against her mother in law. And a man's foes will be those of his own household.")

434   (Pg. 183, *The 100 Most Important Bible Verses*, Thomas Nelson Publishers)

They do not believe that God made man directly out of dust. Infidels do not believe that the Bible was directly, miraculously inspired of God. They may think good men wrote it; they do not think that God miraculously gave the book, making it entirely different in kind from other books and infallibly correct. Critics particularly deny the recorded miracles of the Bible, such as the flood, the miracles of Jonah and the whale, and of the sun standing still for about a day in its relation to the earth in Joshua's time, for example. Those who do not believe that Christ is the Son of God, of course, mean that they do not believe in the miracles recorded about Him. They believe that there was a man named Jesus Who lived, that He was a good man, a great teacher, a fine example. But they do not believe He was God incarnate in human form, that He was born of a virgin, without a human father, that He actually miraculously rose from the dead and ascended bodily into Heaven."

…"You may have heathen religions without miracles. You may have the evolutionary theory without miracles. You may have modernism without miracles. You may have atheism without miracles. BUT YOU CANNOT HAVE CHRISTIANITY WITHOUT MIRACLES! …Christianity stands or falls on its miracles. Christianity is a miracle religion."[435]

"If anyone teaches false doctrines and does not consent to the sound instruction of our Lord Jesus Christ, and to godly teaching, he is proud, knowing nothing, but is obsessed with disputes and arguments over words, from which come envy, strife, reviling, evil suspicions, and useless arguments between men of corrupt minds, who have become destitute of the truth, and who suppose that godliness is a means to financial gain [see "Prosperity Gospel"]. From such withdraw yourself." (1 Timothy 6:3-5)

Another way to recognize a false Christian or liberal church is by how they treat the nation of Israel. False Christians hate Israel and love its enemies just as their father Satan does. Here is a headline from the "World Focus" (news) section of Midnight Call Magazine, October, 2009:

"The United Church of Canada has been accused of anti-Semitism by the Canadian Jewish Congress. The National Council of Canada's largest Protestant denomination is due to debate proposals calling for the boycott of Israeli institutions."

This report is about a Canadian faux-Christian organization, but the same could have been written about many liberal Christian organizations throughout America and Europe. The article states that United Church's proposal…

…"call[s] for a "comprehensive boycott of Israeli academic and cultural institutions at the national and international levels" and refer to the recent assault on Gaza as a "visible reminder of the ongoing Israeli regime of exclusion, violence and dehumanization directed against Palestinians." They also state that Israel was "built mainly on land ethnically cleansed of its Palestinian owners." "[436]

Their proposal is full of nothing but bald-faced lies and false accusations that have been shouted "loud enough and long enough" by the Muslim-Arabs and their complicit, Western, leftwing media that these false-Christian simpletons have come to believe them. We will shine the light of truth on these Jew-hating, Israel-exterminating Muslim lies for the caustic bilge that they are in chapter 13 "Satan," so if you have previously been snookered by this propaganda just be patient. The above article goes on to accurately conclude that…

---

435   (Pg. 254-255, *Prayer: Asking & Receiving*, by John R. Rice)
436   (From article, "United Church Boycotts Israel," Midnight Call Magazine, October 2009)

"'The whole purpose of this material is to vilify Israel and to present it in a crude caricature as the "new apartheid" state, allegedly based on a state-sanctioned policy of racial superiority. ...The purpose of this hateful invective is to deny the legitimacy of Israel because it is a Jewish state.'"[437]

False Christians are allied with their Muslim brother's hate-drenched obsession to annihilate the nation of Israel and exterminate its Jewish occupants. Satan will settle for nothing less. *Peace* is just a game the Muslims play while they doggedly pursue this, their ultimate goal. And deluded, liberal faux-Christians play the part of their "useful idiots" to Oscar-winning perfection.

"Among Christians, there are also people like Judas Iscariot, good counterfeits. They may attend church over a long period, but they were never converted to Jesus Christ. Very few people notice this, but the heavenly Expert sees through everything that is faked."[438]

See: *Wide Is the Gate: The Emerging New Christianity*, a DVD by Video Journalist, Caryl Matrisciana

Black Liberation Theology (Aka Black Enslavement Theology)
"If the world hates you, know that it hated me first." (John 25:18)

"If God is not for us and against white people, then he is a murderer, and we had better kill him. [I'd hate to be in that lunatic's shoes 6 seconds after he expires...] ...Black theology will accept only the love of God which participates in the destruction of the white enemy."[439]

That idiotic, racist bilge came from the founder of this movement. Black Enslavement Theology is an obvious perversion of the gospel of Christ, one that "liberates" its followers from the truth. It's another hell vacuum, and one that our illustrious president sat under the influence of for 20 years. (Sure, he's a "Christian.") Personally I think it should be renamed, "Blame it All on Whitey Theology,"[440] but that might be a tad too honest.

This distortion of Christianity was made famous by Obama's preacher and mentor of twenty years, the "stuck-on-stupid" Jeremiah Wright. It is nothing more than false Christians following a racist, hating-whitey, Marxist political religion under the guise of Christianity, who interestingly all march out and vote for Lucifer and his child-sacrifice, religiously every two years. (And also for their own genocide; but that's a discussion for chapter 11 "Baby mutilation.") They call it the "social gospel" but it is really a socia<u>list</u> gospel, having little to do with following Jesus Christ. Why they don't just throw off all pretenses and worship Karl Marx instead is beyond me. But that would not help in suckering as many uninformed, weak-minded individuals with itching ears into their fathers trap. (With all due respect to president Obama.) I would suggest Stalin, Lenin and Trotsky as their holy trinity, and The Communist Manifesto their bible. But brutal honesty is never one of the hallmarks of the great deceiver's religions. Pray for them. Helping the poor is a worthy Christian undertaking. Advancing Marxism-socialism while

---

437   (From article, "United Church Boycotts Israel," Midnight Call Magazine, October 2009)
438   (From article "The Calling of the Twelve Disciples—Judas Iscariot," by Marcel Malgo, Midnight Call Magazine, October 2009)
439   (Black theology founder, and racist, James H. Cone. As quoted in "Black Liberation Theology," a handbook from The Abe Lincoln Foundation, J.A. Parker, President)
440   (See *Hating Whitey: and Other Progressive Causes*, by David Horowitz)

stealing from the rich to give to the poor (income redistribution) and perverting the true gospel of Jesus Christ is just another ticket to an eternity in hell.

Prosperity Gospel: The "Gospel of Greed"

We warned of this in Part 1. Here we would like to share their theme song:

"Oh Lord, won't you buy me a Mercedes Benz?
My friends all drive Porsches, I must make amends.
Worked hard all my lifetime, no help from my friends,
So Lord, won't you buy me a Mercedes Benz?
Oh Lord, won't you buy me a color TV?
Dialing For Dollars is trying to find me.\
I'll wait for delivery each day until three,
So Lord, won't you buy me a color TV?
Oh Lord, won't you buy me a night on the town?
I'm counting on you, Lord, please don't let me down.
Prove that you love me and buy the next round,
Oh Lord, won't you buy me a night on the town ?"[441]

The greasy, greedy charlatans at the top of this perversion of the gospel of Christ, "preachers" like Kenneth Copeland, Paula White, Benny Hinn, Creflo Dollar (previously known as Jackass Scratch-off), just to name a few, will soon receive their eternal reward. They live lavishly by sucking the hard-earned moneys, sometimes even the life savings, from naïve folks who actually think they are going to be blessed physically, financially and spiritually by donating to these clowns. Their followers are another fulfillment of the following prophecy, "For the time will come when they will not endure sound doctrine; but after their own lusts shall they heap to themselves teachers, having itching ears." (2 Timothy 4:3) "Itching ears" that need to be scratched by the lies of the great deceiver. Wake up, suckers! You'd be better off investing with Bernie Madoff. God's not for sale. And His blessings can't be bought. He demands that we worship Him in spirit and in truth. Which is slightly different than worshipping money in ignorance and greed. Pray for them. (And see "The Persecuted Church" section for insight in what is really demanded of followers of Christ.)

Speaking in Tongues

"If any man speaks in an unknown tongue, two, or at the most three, should speak, one at a time, and someone should interpret. But if there be no interpreter, let him keep silent in the church, and let him speak to himself, and to God." (1 Corinthians 14:27-28)

Without interpretation speaking in tongues in church or in public is nothing more than babbling, and little more than a show.[442]

Christian Cults

"The greatest curse that God can possibly send upon the people in this world is to give

---

441   ("Mercedes Benz" by Janis Joplin)
442   (See *Speaking with Tongues*, by John R. Rice)

them up to blind, unregenerate, carnal, lukewarm and unskilled guides. And yet in all ages we find that there have been many wolves in sheep's clothing. …As it was formerly, so it is now. There are many that corrupt the word of God and deal deceitfully."[443]

There are, in addition to the more well-known groups previously exposed, many other false Christian churches. There's the aforementioned "Black liberation theology" and the speaking in tongues movement. We have the prosperity gospel group that also erroneously teaches that a Christian can lose their salvation. (See eternal security discussion in previous chapter.) There's also the famous cults of Jim Jones and David Koresh that the media has already exposed, but not until most of their members were slaughtered. And there's the ecumenical movement that has been popular since the 60s, which is all about love-love-love but at the same time preaches a unity free from the constraints of truth or sound doctrine, and therefore will tolerate every manner of false doctrine and damnable heresy; lies that have their roots in the hatred of Satan and not in the love of Christ. It is related to the charismatic movement, another branch of the "feel good" gospel whose disciples might be better served by drinking beer and smoking pot then pretending to actually be following Jesus. (But that's probably a little harsh.) For the latter comes with some very heavy demands. It also comes with a peace that passes all understanding, and a real love that is deeper than whooped-up praising and dancing which can't be sustained for the long run because it has no foundation in Spirit and in truth. (And not that there's anything wrong with whooped-up praising and dancing—see "And David danced before the Lord with all his might." 2 Samuel 6:14—as long as it is coupled with an equally excitable passion for sound doctrine and the truth.) And finally, there is even a group out there now that is rapidly spreading, called *Holy Laughter*, which is appropriate because Christian cults really are a hoot.

"The cults that are now thriving in the Western World are feeding on the pathetic spiritual naiveté of people who have become interested in Christianity, or who have even been converted to Christ through Christian evangelism. The victims of the cults are not hardened atheists who cannot be converted to anything; rather they are gullible people who are willing to believe anything that is spoken with a straight face in the name of God, Christ, or the Spirit. Sincere but sentimental Christians who are **ignorant of Biblical doctrine** [emphasis mine] are sitting ducks for the clever cultist who is pushing for a piece of the action in today's religious scene."[444]

Which is why it is so important *not* to be "ignorant of Biblical doctrine," but rather to heed the biblical admonition to "study to show yourself approved... rightly dividing (understanding) the Word of God," so as not to be led astray by spiritual con men hawking variations of the gospel of Christ that seem exciting on the surface but underneath are full of "dead men's bones and all corruption." (Matthew 23:27)

"Beloved, believe not every spirit, but try the spirits whether they are of God: because many false prophets are gone out into the world." (John 4:1)

Read and study the Bible until you are familiar with what *it* says, so you will be able to quickly recognize when someone is perverting it. "Study to show yourself approved unto God, a workman that does not need to be ashamed, and

---

443   (From *The Method of Grace*, by George Whitefield)
444   (From article in *Midnight Call Magazine* 11/97)

can rightly divide the word of truth." (Timothy 2:15) "If any man will do his will, he shall know of the doctrine, whether it be of God." (John 7:17) Again, if you seek the truth, God will lead you into it. If you seek to do the will of God, if you seek the truth, his promise is that he will lead you into all truth, and he will keep you from falling under the spell of Satan's lies. But you have to seek the truth, you have to desire it, you have to want it with all your heart, more than your next breath of air.

"No longer should we be children, tossed to and fro and carried about with every wind of doctrine, by the trickery of men, and in cunning craftiness, whereby they lie in wait to deceive." (Ephesians 4:14)

Freemasonry

"Beware of the scribes, who desire to walk around in long robes, love greetings in the markets, the best seats in the synagogues, and the chief places at feasts; that devour widows' houses, and to make a big show pray long prayers. The same shall receive greater damnation." (Luke 20:46-47)

Many Christians are members of the Masonic Lodge, but they are unaware that they are involved in Baal worship, an ancient form of devil worship. In fact, a *sacred* word of Freemasons is "Abaddon" which is taught to initiates who attain the 17th degree of the Scottish Rite. It is also found in Revelation: "And they had as king over them the angel of the bottomless pit, whose name in Hebrew is Abaddon, but in Greek he has the name Apollyon." (Revelation 9:11)[445]

The Shriners, an association within the Masons, who do give much of their time and money to charitable works, nevertheless reveal the evil, demonic roots of their organization in the very colorful caps called red fezzes that they proudly wear in their parades and circuses:

> "The fez itself is an example of the double meaning behind most of Freemasonry's facade. Worn by every Shriner and even carried to the grave with pompous dignity, the history of the fez is both barbaric and anti-Christian. In the early eighth century, Muslim hordes overran the Moroccan city of Fez, shouting, *"There is no god but Allah, and Muhammad is his prophet."* There they butchered the Christian community. These men, women, and children were slain because of their faith in Christ, all in the name of Allah, the same demon god to whom every Shriner must bow in worship, with hands tied behind his back, proclaiming him the god of his fathers in the Shrine initiation, at the Altar of Obligation."

> ..."During the butchering of the people of Fez, the streets literally ran red with the blood of the martyred Christians. The Muslim murderers dipped their caps in the blood of their victims as a testimony to Allah. These bloodstained caps eventually were called *fezzes* and became a badge of honor for those who killed a Christian. The Shriners wear that same red fez today, with the Islamic sword and crescent encrusted with jewels on the front."[446]

Most people who join the Freemasons do so primarily for the social and business contacts afforded by the organization. They move up through the ranks over the years, some becoming Shriners, and have no idea of the demonic nature of what they're involved in.

---

445   (See Pgs. 87-88, *Fast Facts on False Teaching*, by Ron Carlson & Ed Decker)
446   (Pg. 74, *Fast Facts on False Teaching*, by Ron Carlson & Ed Decker)

For more information exposing Freemasonry see chapter 5 "Freemasonry and the Masonic Lodge," in *Fast Facts on False Teaching*, by Ron Carlson & Ed Decker.

ExChristian.net

"First off, you must understand that scoffers will come in the last days, mocking [the truth] and following their own evil desires." (2 Peter 3:3)

ExChristian.net is a website run by self-described "ex-Christians" although in truth they are ex-*faux*-Christians, those who were Christians in name only; who were never converted in the heart and never came to the knowledge of the truth. (Being born to Christian parents, being a member of a Christian organization, or attending church services regularly or occasionally, does not make you a Christian. Just as driving by a golf course, having dinner in the clubhouse and walking around the links doesn't make you a golfer.) They are an example of those who the seed of the Word of God fell on "stony places" and "among thorns." (See the parable of the sower of the seed in Matthew 13.) There is no such thing as an "ex-Christian," for once a true Christian you're always a Christian, for Jesus will let no one pluck a Christian out of his hand… "And no man is able to pluck them out of my Father's hand." (John 10:29) (See "eternal security" discussion in chapter 7.)

The following quote is their "mission statement" if you will…

> "God had to kill himself to appease himself so that he wouldn't have to roast us (his beloved creation) alive for all eternity. He loves us more than we can comprehend but, if we don't love him back, He will send us to HELL to suffer forever and ever. That really is AMAZING GRACE!"

Actually it is, but why confuse them? The truth is you don't have to "love him back" (that is a natural response from one who realizes what an unfathomable gift he or she has been given, and "what is the height and depth and length and breadth" of the love of God who sacrificed his only begotten Son to make that gift possible), you just have to accept his free gift of eternal life, to put off your own false righteousness and clothe yourself in his perfect righteousness, which these ex-faux-Christians never did, so they were prime targets for the great deceiver to steal away the little bit of truth that they did have.

"They went out from us, but they were not one of us; for if they had been of us, they would no doubt have continued with us: but they went out, that they might be made manifest that they were never one of us." (1 John 2:19) Like Judas Iscariot who spent much time in the very presence of the Lord himself, but who never was a true follower or believer, these ex-faux-Christians were never one of us.

"Being born in a Christian home doesn't make you a Christian any more than being born in a bakery makes you a bagel."[447]

Their mission statement is also a perfect example of how the wisdom of God is foolishness to man, and vice versa. "For the wisdom of this world is foolishness with God." (1 Corinthians 3:19) They are also the fulfillment of the above verse from 2nd Peter about how scoffers and mockers "will come in the last days." But, even so, Jesus still battles Satan for the souls of even those who mock Him and his priceless gift. "The Lord is not slack concerning his promise, as some men count

---

447   (Moishe Rosen, founder of Jews for Jesus, from article "A Jew for Jesus: Moishe Rosen put Christian belief on the map for American Jews," by Edward E. Plowman, *World* Magazine, June 19, 2010)

slackness; but is patient with us, <u>not wanting that any should perish, but that all should come to repentance</u>." (2 Peter 3:9) But nevertheless if these bagels…er… ex-faux-Christians do not wake up they are in for the rudest of awakenings when they die… "But the day of the Lord will come as a thief in the night." (2 Peter 3:10) We should pray for their conversion, and for the launch of their new website called *ex-exfauxchristians.net*.

Drugs

"My peace I give unto you; not as the world gives do I give unto you. Let not your heart be troubled, neither let it be afraid." (John 14:27)

I don't know if you can classify drug abuse as a false religion but I couldn't find any other place for this... (But it definitely *is* in many ways the worshipping of a false god.) Abuse of drugs and alcohol is another form of looking for love in all the wrong places; and happiness and peace and escape from ones problems in the same areas. There is no shortcut to peace on earth. It doesn't come through popping a pill, taking a trip, ingesting an herb, smoking a 'shroom, or drowning your sorrows in a bottle. (Although you certainly can give it a good shot.)

"Enter in by the narrow gate; for wide is the gate and broad is the way that leads to destruction, and there are many who go in by it. <u>But narrow is the gate and difficult is the way which leads to eternal life</u> [and happiness, and peace, and joy, and the abundant life], and few there are who find it." (Matthew 7:13-14)

"For many are called, but few are chosen." (Matthew 22:14)

"And when it was demanded of him by the Pharisees, when the kingdom of God should come, he answered them and said, the kingdom of God comes not by observation: Neither shall they say, Lo here! or, lo there! for, behold, <u>the kingdom of God is within you</u>." (Luke 17:20-21)

Drug and alcohol addiction can be overcome by trusting in the Lord to take it away, by "casting your burden upon him for he cares for you" and by "resisting the devil and he will flee from you." But again, just as in the parable of the man seeking bread for a late night guest when we saw that God answers a Christian's prayer for soul winning power after much persistence, so too in the battle to overcome terrible addictions (like the one for food) we find many times that our prayers are not immediately answered. God wants to see if we want him more than we want to drink alcohol, or smoke weed, or do drugs. ("You shall have no other gods beside me.") Which is why he also said "And you shall seek me, and <u>you shall find me</u>, when you shall search for me with <u>all your heart</u>." A lot of us (yours truly included) are still working on the "with all our heart" part. How badly do we want him? More than a fifth of vodka at ten o'clock in the morning?[448] More than smoking weed? More than getting "high?" More than the demon of crack or heroin? God is waiting for each one of us at our own time and at our own speed, but he is also anxious for many of us to start the journey; to wage the internal war; to seek him like the deer who is dying of thirst after being chased for miles through the forest by, and finally alluding, a pack of ravenous wolves, who now searches for that life-giving mountain stream of cold thirst-quenching water. "As the deer pants for the water brooks, so my soul pants for you, O God." (Psalms 42:1)

Let us leave you with these sobering thoughts from Dr. David Jeremiah:

"The second prominent sin of the tribulation will be drug-related occultic activities.

---

448   (The Lord Jesus is faithful and true. May he exorcise those demons, Schu)

... The word sorceries in Revelation 9:21 comes from the Greek word *pharmacea* which means pharmacy, and refers to the practice of the occult with the use of drugs. In biblical times drug use lead to astrology, witchcraft and demon possession. Now the Bible says that in the tribulation, along with rampant murders, there will be untold addiction to drugs. Mind-changing drugs, demon-producing drugs—are we on the path? In our day drug addiction has swelled into a flood across the nation. And with it has come an unprecedented tide of witchcraft and demonism. They always go together. Demons are no strangers to drugs. When you get on drugs, and you start messing with your mind you open yourself up to the occult, to witchcraft, to demonology. And where the vacuum of your mind is left because of mind-burning drugs, the demons will come in and take possession. The devil and drugs go together. Kids, if you are messing with that stuff, don't do it. You are not only playing with your mind, you're not only playing with your own body which is the temple of the Holy Spirit, but you are opening yourself up to demonic activity over which you may not ever be able to have control."[449]

The kingdom of heaven lies within but there are no shortcuts to getting there. There are however a number of seductive, deceptive paths that ultimately lead to somewhere else, whether that place is addiction, or death, or hell, or the destruction of one's life. Drugs are one of them.

Evolution, Satan's genesis

"They changed the truth of God into a lie, and worshipped and served the creature more than the Creator, who is blessed for ever. Amen." (Romans 1:25)

Yes, this "scientific" theory can only be classified as a false religion. It no longer has anything to do with science for we know now that it is in direct opposition to the Laws of science and to all of the facts and evidence. It is an intellectually idiotic, absurd, asinine, ridiculous and scientifically impossible theory that has been exalted to the realm of the sacred and the unquestionable; just like the bogus claims of any other false religion. And the killer is that it is <u>religiously</u> taught as doctrine in our supposedly *secular, non-religious,* public schools. It would be quite a hoot if it were not so damaging to the students. But we'll leave this discussion for later because this is one false religion for which we've devoted an entire chapter.

Scientology

"As a dog returns to his vomit, so a fool returns to his folly." (Proverbs 26:11)

You would have a better shot making sense out of the mutterings of a lobotomy victim than in deciphering any real meaning from this pork rind. L. Ron Hubbard, however, has succeeded in one important thing: giving new life to a couple of old adages, "There's a sucker born every minute," and "A fool and his money are easily parted." (With all due condolences to Mr. Cruise.)

Liberals, Progressives, the Left, and the Democrat Party

(All the same thing.)

"Then he shall say also unto them on the left hand, Depart from me, you cursed, into everlasting fire, prepared for the devil and his angels." (Matthew 25:41)

I think it is appropriate to catalog the ideology of these folks as a false religion. Indeed it is in many ways a religion of pure evil. Oh, they fool themselves into

---

449   (From "Hell on Earth Part 4," by Turning Point with Dr. David Jeremiah)

how *righteous* they are, how *superior* they are to everybody else, and how *caring* they are; how much they *care*, for example, about the poor (as long as they are stealing other people's money to do the caring, while at the same time trapping the poor into dependence on the state, fatherless homes and abject poverty as well). How much they *care* about the rights of the mother (but it is not the mother's right to live that is being taken away, it is the babies). But if one looks carefully at what they do, and not just what they say, then their true nature becomes evident. They are of course the party of baby mutilation. It is their Unholy Grail, and the heart and soul of their politics (and the heart and soul of their father Satan). There is nothing more iniquitous on this earth for them to embrace, but they have tried. They embrace cop killers like Mumia Abu-Jamal. They embrace child killers like Andrea Yates who had a bad hair day when she drowned her 5 children in the family tub; brought on of course by the *oppression* of marriage and children. They are the Party of child perverts, who their left-wing judges release to kidnap, rape and murder again. They are the Party of the ACLU who defended pro bono the child-pervert, rapist killers of the North American Man Boy Rape Association. And we are to overlook all of this, as they are somehow able to do, in light of "how much they care." What a crock!

Back in the days of President Kennedy the Democrat Party was one of our great political parties. Now he would find it unrecognizable. Karl Marx, Joseph Stalin and Mao Tse-Tung, however, would feel right at home. (Lucifer certainly does.) John Kennedy believed firmly in a strong national defense, not the Neville Chamberlain "peace in our time" appeasement mentality and moral equivalency of the Obama administration and modern liberals in general. He also spoke passionately of the need for tax cuts to spur economic growth, and was an ardent supporter of the freedom granted by our capitalist system. The outright socialist, closet-communist Democrats of today would make him sick to his stomach. The Democrat-Left likes to reminisce about the good ole' days of Camelot, and to invoke Kennedy's memory as if they have anything to do with him, or him with them. How pathetic. The truth is that if he were alive today they would marginalize, disparage, demean, vilify and spit in his face just like they do to all of those conservatives, Christians and Republicans who believe in the same things today that JFK did back then. Kennedy famously said, "Ask not what your country can do for you, but what you can do for your country." The Democrat Party of Obama, Pelosi and Reid today says, "Demand what your country can do for you, and steal what you can from your fellow countrymen without working for it."

Ann Coulter nails the orthodoxy of the Left in her book, *Godless: The Church of Liberalism*:

"Though Liberalism rejects the idea of God and reviles people of faith, it bears all the attributes of a religion. ...Coulter throws open the doors of the Church of Liberalism, showing us its sacraments (abortion), its holy writ (*Roe v. Wade*), its martyrs (from Soviet spy Alger Hiss to cop-killer Mumia Abu-Jamal), its clergy (public school teachers), its churches (government schools, where prayer is prohibited but condoms are free), its doctrine of infallibility (as manifest in the "absolute moral authority" of spokesmen from Cindy Sheehan to Max Cleland), and its cosmology (in which mankind is an inconsequential accident)."[450]

"Liberal doctrines are less scientifically provable than the story of Noah's ark, but their belief system is taught as fact in government schools, while the Biblical belief system is

---

450   (From front flap, *Godless: The Church of Liberalism*, by Ann Coulter)

banned from government schools by law. As a matter of faith, liberals believe: Darwinism is a fact, people are born gay, child-molesters can be rehabilitated, recycling is a virtue, and chastity is not. If people are born gay, why hasn't Darwinism weeded out people who don't reproduce? (For that, we need a theory of survival of the most fabulous.) And if gays can't change, why do liberals think child-molesters can? Pedophilia is a sexual preference. If they're born that way, instead of rehabilitation, how about keeping them locked up? Why must children be taught that recycling is the only answer? Why aren't we teaching children "safe littering"?

"We aren't allowed to ask. Believers in the liberal faith might turn violent—much like the practitioners of Islam, the Religion of Peace, who ransacked Danish embassies worldwide because a Danish newspaper published cartoons of Mohammed. This is something else that can't be taught in government schools: Muslims' predilection for violence. On the first anniversary of the 9/11 attack, the National Education Association's instruction materials exhorted teachers, "Do not suggest that any group is responsible" for the attack of 9/11."[451]

Again, people believe what they are taught, and liberalism is as seductive a trap as any other false religion. And it will keep its followers imprisoned in those lies until they die in their sins and are subject to a rude awakening in the Lake of Fire. Pray for them to be led from the darkness of a self-righteousness that deludes them into seeing themselves as better than everybody else while at the same time supporting and embracing every form of evil, to the Light of true righteousness that can only be found in Christ.

The Unification Church

"And Jesus answered and said unto them, 'Take heed that no man deceives you. For many shall come in my name, saying, 'I am Christ;' and shall deceive many.'" Matthew 24:4-5)

The Charlatan...er...Reverend Sun Myung Moon is another in a long line of false Christ's. Yet the tactics that his church and his brainwashed leaders use to keep their poor conscripts in line are particularly diabolical. To give you an idea, here are some excerpts from the diary of a former Moonie, quoted in *Fast Facts on False Teachings*:

""Like thousands of Moonies across the country, I worked all day every day, selling carnations, to raise money for the movement. Up at 4:00 A.M., rattling through the streets with other teens in a seatless van by 5:00, heading for shopping centers or business districts. Breakfast was Chinese rice balls or cereal and candy served in the van, with milk spilling all over.

"Team Captains whip us into evangelical frenzy with songs, Bible verses, prayers and chants until we shout out, in determination, the amount we will personally raise that day. No one shouts out an amount less than 100 dollars. ...On the streets till the money is made no matter how long it takes; rarely back before midnight or 1:00 A.M. ...Sometimes we are so tired we don't even eat. ...We collapse and sleep until the next day begins before dawn.

"All emotion, everything is handled by the Center Director. If I sing too loud, he tells me how to sing. If I want to eat or sit with different people, he says "no." If I feel like crying, he says don't cry for yourself. We have no newspapers, no TV, no radio and no

---

451   (Pg. 2, *Godless: The Church of Liberalism*, by Ann Coulter)

talk of the outside world. Only later, after I had left the cult, did I see what was happening here. ...The lack of sleep, the poor food, the ceaseless noise and commotion, the isolation, the chanted prayers and songs weaken my resolution and make me desperately afraid of trying to break free. I never wake without wondering what I am doing here, but [soon] I am thinking only of how to sell enough flowers to meet my quota. ...Like the others, I lie. The money goes for drug rehabilitation programs, I tell some people, or for Christian youth projects. Sun Myung Moon ...calls it heavenly deceit. ...Some days I raise over $240.00, but never less than one hundred. ...My legs ache from pounding the pavement. ... I am sick with fever. I can't get up. Other Moonies sprinkle holy salt around my bed to drive away bad spirits. A doctor? Forget it.

"I know of one Moon girl who developed an eye ailment. Other members were brought to see her writhing in pain, as an example of someone who had been possessed. By the time she was finally allowed to get medical treatment, she was partially blind. ... I am given conditions or punishments that free me from Satan's influence: cold showers, longer and longer ones, reading Moon's doctrine over and over, praying all night long in repentance, ...Oh, God, please help me. I am so afraid."

[The authors respond...] "Most people do not fully understand such intense brainwashing, but the tactics that Sun Myung Moon uses are very similar to those which some of the prisoners of war underwent in North Korea during the Korean War."[452]

And here is an excerpt from an online article by John Gorenfeld:

"Years ago, Moon was widely considered a dangerous madman, the next Jim Jones. He inspired TV specials with names like "Escape From The Moonies." His cult separated college students from their families, persuaded them to take to the streets by the hundreds to sell flowers and underwrite Moon's mansions and yacht. So completely did they surrender to Moon that he even assigned them spouses at fabulous stadium weddings."[453]

The sad thing is that this demented anti-Christ has now become main stream because of the vast wealth he has accumulated off the backs of all of those poor suckers, those "college students separated from their families," who got trapped in his web. The only difference between this hell-bound devil and David Koresh is that he has bought his way into respectability, so Janet Reno won't be smoking him and his followers out any time soon. Money talks. It could be argued that Moon has caused an infinite amount more harm to this country and her youth than a nut like Koresh; but since this cashew owns the *Washington Times* and vast amounts of real estate and businesses worldwide, and donates significant sums to Republican candidates, our government will continue to wink at him and look the other way. Screw all of the children and all of their families whose lives have been, and are being, ruined.

Hare Krishna

(See previous segment. Same scam, different charlatan.)

The New Age Religion

"For many deceivers have entered into the world, who do not confess that Jesus Christ has come in the flesh. This is a deceiver and an antichrist." (2 John 1:7)

As we have stated in a previous chapter, the first lie that emanated from the

---

452   (Pgs. 155-157, *Fast Facts on False Teaching*, by Ron Carlson & Ed Decker)
453   (From "King of America" @ Gorenfeld.net)

mouth of the father of lies was that <u>there was no such thing as eternal death in a place of severe suffering called hell</u>, and this was quickly followed by the second lie which stated that <u>we could be like God</u> if we just entered into the *secret occult knowledge* that God was unfairly keeping from us. All we had to do was disobey God and obey the serpent, and believe that God the Creator was a liar and that he, Satan, was telling the truth. (And whereas Chimpanzees wouldn't have fallen for that one, we did!) "And the woman said unto the serpent, 'We may eat of the fruit of the trees of the garden; but of the fruit of the tree which is in the middle of the garden, God has said, "You shall not eat of it, neither shall you touch it, or you will die."' And the serpent said unto the woman, [Lie #1] "**You will not surely die**. For God does know that in the day that you eat thereof, your eyes will be opened, and [Lie #2] **you will be like God**, knowing good and evil [*secret occult knowledge*]." (Genesis 3:2-5) And these same stupid, tired old lies are being recycled today by a modern conglomeration of "progressive" spiritual and religious beliefs known as the New Age Movement. It has its roots in "Witchcraft and shamanism, Astrology, Hinduism and Yoga." It teaches that God is "just a cosmic force;" that Jesus is "a man who evolved into an Ascended Master (god-like being) through occult and metaphysical disciplines;" that Christ is "an impersonal 'force' that rested on the man Jesus and made him special, but which has also rested on others, and can even rest on us;" that the Bible is "at best a work of cabalistic secrets that can only be understood by "Masters." At worst a stupid book of Jewish legends." It also rejects the only way of salvation offered through the shed blood of Jesus Christ and his atoning sacrifice on the cross, and instead offers a salvation "by **works** of occult discipline"(emphasis mine ...*working your way to heaven*, now there's a novel concept), and by the "law of karma" and "belief in reincarnation."[454]

Originally popularized by Shirley MacLaine in her book, *Out on a Limb* (which she has yet to come down from), this religion has more recently been espoused and promoted by Oprah Winfrey. On her daily talk show she feeds her unsuspecting viewers a smorgasbord of hell-enlarging fabrications. She preaches...

"- Jesus Christ is no more God than you are. And yes, you *are* God.

- God is everything: people, rocks, blades of grass- it's all God.

- Jesus was just trying to teach us that we are all God- you are "I Am."

- There are many paths to God besides Jesus.

- As a divine being, you create your own reality and morality- including your own gender. A man can call himself a woman- and he is! [Another man can call Oprah Winfrey a horses ass, and she is!] [That was too easy.]

- Being "born again" means discovering your own "God consciousness." "[455]

And blah, blah, blah. There's absolutely nothing new there, just more of Satan in the Garden. And uneducated folks keep swallowing it. Mrs. Winfrey has also been affectionately hawking Eckhart Tolle's book, *The New Earth*, which is nothing more than a repackaging of eastern pantheist religion...

"Of course, we have heard this before: There are many ways to heaven, Jesus only embodied the God-consciousness that showed us how God wanted us to live, Jesus is not the only way to God, Christianity is a restrictive ideology that actually keeps us from God, God is bigger than religion, all religions should be sampled, there is no authority except for yourself and you can choose from different religions to piece together what is best for you, you listen within for your own definitions of spirituality, etc., etc. Every day, Oprah

---

454   (From pg. 184, *Fast Facts On False teaching*, by Ron Carlson & Ed Decker)
455   (Source not found.)

is speaking this to millions and millions of people. In reality, this book is really just a restating of Buddhism (we must escape desire to enter enlightenment). Eckhart just says that we have to bypass our "ego" to tap into our true self where true peace and spiritual enlightenment is attainable. Again, straight Buddhism and Oprah thinks it's wonderful and has made it compatible with her new form of "Christianity." If you combine Oprah's philosophy with that of the prosperity gospel that also tells people that life is all about us and that we can gain God's blessings through our gifts, our thoughts on spirituality in America are devolving into a mish-mash of heresy and paganism."[456]

"This is the dawning of the Age of Aquarius," so the sixties song goes, but for those who refuse the truth and settle instead for sweet-sounding lies, their part will be in the Lake of Fire and boiling sulfur, and not in the promised millennium of peace and prosperity where the lion *will* lie down with the lamb and swords *will* be beaten into plowshares. Certainly their intentions are good and their desire for peace and love is understandable.[457] But they are following the great deceiver and his distortions and rejecting the Prince of Peace and the only Source of Love in the process. You can't choose your own path by which to follow, serve and worship God, no matter how creative, imaginative, and full of candles and crystals it may be. "Because narrow is the gate and difficult is the way which leads to life, and few there are who find it." (Matthew 7:14) God wants us to follow him and to believe his truths and doctrines, and not the serpent and Eckhart Tolle and Oprah Winfrey and their gobbledygook. "But the hour is coming, and now is come, when the true worshippers shall worship the Father in spirit and in truth; for the Father seeks such to worship him. God is a Spirit: and they that worship him must worship him in spirit and in truth." (John 4:23-24)

"I am astonished that you are so quickly deserting the one who called you into the grace of Christ and are turning to a another gospel, which is really no gospel at all. Evidently there are some who are throwing you into confusion, and want to pervert the gospel of Christ." (Galatians 1:6-7) And who is "throwing people into confusion?" The envelope, please? And the winner is... Oprah Winfrey! Let's give her a big round of applause.

And on a personal note, at one point a number of decades ago the author deeply swallowed the New Age lie on his way to finding the truth. I would suggest that the reader, and any poor, unfortunate Oprah viewer, avoid that detour.

"I, even I, am the LORD; and beside me there is no savior." (Isaiah 43:11)

For further reading on the truth about The New Age Movement see: *Don't Drink the Kool-Aid: Oprah, Obama [and the] Occult*, by Carrington Steele; and Chapter 12 "The New Age Movement" in *Fast Facts On False teaching*, by Ron Carlson & Ed Decker.

Reincarnation

The Hindu concept of reincarnation—the belief in the return of the soul after death to live again in another person or animal—is an unassailable doctrine among New Age disciples. It states that we are trapped in an almost never ending cycle

---

456   (@ downshoredrift.com, April 05, 2008 in Spiritual News)
457   (Except it is very revealing that the one thing that Oprah and everyone else in this wildly diverse movement do seem to agree on is voting for the baby mutilating Democrats, religiously, every two years. It's their sacrament. Hmm. Like father like daughter? The apple doesn't fall too far from the tree?)

of coming back to this life after our death again and again and again, as another human, or as a cow or a dog or a pig or a rat or a bush, until *finally* we gain release and move on to some mythical state of Nirvana which no one has yet to actually attain. (But don't let that put a damper on your enthusiasm.) The state of confusion, however, *has* been reached. Conversely the Bible, which has proven itself to be actually *true*, says clearly that "it is appointed unto men once to die, but after this the judgment:" (Hebrews 9:27)

Yet those who cling to the idea of reincarnation will offer as proof a number of cases of people (before the internet) who can recite exact and confirmable details of places where they say they have lived before, details that could only have been known by someone who had actually been there, when it is clear that they have not. So isn't this "proof" positive of reincarnation?! (No, it is proof positive of demonic possession...) There are a number of plausible explanations for this phenomenon put forward by those who have investigated the subject, but the best answer I've found is in Keith Harris' study on the Book of Revelation, *The Unveiling*, on pg. 184. There he discusses the fate of fallen angels, "disembodied spirits," who are not yet cast into hell, but roam the earth "ever searching for new bodies to inhabit." There are numerous examples in the Gospels of Jesus where he cast out these demons that had possessed certain individuals and caused them a great deal of suffering. Mr. Harris states...

> "This is why so many who tend toward the occult, through psychics, mediums, familiar spirits, and so on, believe in reincarnation. Upon opening themselves up to the occult, that is, the world of "familiar" spirits, they become possessed by these immortal beings who have lived in previous bodies. The re-embodied spirit, having the thoughts and memories of previous victims, conveys those thoughts and memories to the new victim. The new victim is convinced that he or she has lived before in a previous life. This is a very deceptive tactic of Satan."[458]

The authors of *Fast Facts On False Teaching* reveal how this concept of reincarnation was first introduced to America:

> "In India [reincarnation] is known as transmigration. Hinduism teaches that based on the law of karma, your good and bad deeds will determine how you will come back in your next life. If you live a bad life and do not do the things required in Hinduism and Buddhism to renounce this world of illusion, you may come back as a lower form. The possibility of returning as a cow or rat have made both animals sacred in India. You don't kill a cow in India, since it may be somebody's reincarnated uncle or aunt. That's also why you don't kill a rat. The United Nations now estimates there are over three times more rats in India than the human population. These rats eat nearly one-fourth of the total grain crop!

> "The idea of transmigration was introduced in 1891 at the World's Fair on Religions in Chicago by a man named Swami Vivikananda. ...[But] he discovered that Americans were not very excited about the idea of coming back as a rat, a frog, or a snail. So the concept was changed to reincarnation, which says that you can only come back as another human being. This was much more palatable for Western consumption."[459]

Of course the Swami first consulted with all of the proper celestial authorities before coming up with such a monumental change to one of India's most sacred and time-honored doctrines. Don't think for a minute that he just made it all

---

458   (Pg. 184, *The Unveiling: A Journey Through the Book of Revelation*, by Keith Harris)
459   (Pg. 97-98, *Fast Facts On False Teaching*, by Ron Carlson & Ed Decker)

up out of whole cloth to fool gullible Westerners. No way! Unfortunately, the celestial authorities have yet to make this change known in India where people are still starving to death while bargain beef wanders their streets and rats eat their breakfast. Maybe Oprah can do some kind of chant or sprinkle some flower pedals...

For further information see: *The Reincarnation Sensation*, by Norman L. Geisler and J. Yutaka Amano.

Buddhism and Hinduism

"Speak not in the ears of a fool: for he will despise the wisdom of your words." (Proverbs 23:9)

False religions come with terrible consequences. Eternal death is of course the most horrific; for those who are born and bred into one, and do not find escape. Which is why God in his word has made it clear a number of times that... "The Lord is longsuffering, and of great mercy, forgiving iniquity and transgression, but by no means clearing the guilty, visiting the iniquity of the fathers upon the children unto the third and fourth generation." (Numbers 14:18-19) So the eternal curse is passed down from one generation to the next and entire family trees, groups of families, and cultures are affected. It becomes the gift that never stops giving.

But other false religions, as if to add insult to injury, bestow some terrible consequences in this life as well. If Jesus came to give life and to give it more abundantly then Satan has come to destroy lives. Islam mutilates the genitals of hundreds of millions of its little girls and robs them of their God-given birthright to sexual pleasure, fulfillment, intimacy and release with their husbands. While Buddhism and Hinduism, with their concept of an impersonal god, one who embodies the universe itself ("all is one and all is god") have successfully destroyed human value and any real meaning in life for hundreds of millions of its adherents. Under their ideology, human suffering is not something to be relieved but rather is a result of each individual's attachment to the "world of illusion." *Bliss* and ultimate release from suffering can only be attained by extricating oneself, through meditation and concentration, from the world of *Maya*, of illusion. Of course the only world of illusion that Buddhists and Hindus inhabit is the one inside their head, clouded by false beliefs about the actual incredible world of reality that God created...

"The problem with an impersonal universe is that it destroys personality and all the special characteristics that make us human. First, the highest expression of human emotions and personality is love and compassion. An impersonal universe never loved or cared about anyone. The meaning of love as a personal commitment of an individual's will is lost. There is no basis for love or human compassion in an impersonal universe. This is why in Asia, Hindu and Buddhist cultures do not build hospitals or schools; there is no basis for human value."[460]

"In Thailand ...along the Cambodian border over 300,000 refugees were caught in a no-man's-land. ...Here, in this Buddhist country of Thailand, with Buddhist refugees

---

460   (Pg.96, *Fast Facts On False teaching*, by Ron Carlson & Ed Decker)

coming from Cambodia and Laos, there were no Buddhists taking care of their Buddhist brothers. There were also no Hindus or Muslims taking care of those people. The only people there, taking care of these 300,000 people, were Christians from Christian mission organizations and Christian relief organizations.

"[A man who headed a large part of the relief effort explained why] ... "Have you ever seen what Buddhism does to a nation or a people? Buddha taught that each man is an island unto himself. Buddha said, '*If someone is suffering, that is his karma.*' [Which, roughly translated, means "f**k 'em."] You are not to interfere with another person's karma because he is purging himself through suffering and reincarnation!"

"[He continued] "The only people that have a reason to be here today taking care of these 300,000 refugees are Christians. It is only in Christianity that people have a basis for human value, that people are important enough to educate and to care for. For Christians, these people are of ultimate value, created in the image of God, so valuable that Jesus Christ died for each and every one of them. You find *that* value in no other religion, in no other philosophy, but in Jesus Christ!""[461]

The following is from the laws of the Hindus...

"Do not interfere with the karmaic suffering of the destitute, the poor, the afflicted, and the diseased. Their suffering is for their own good... Throw the widow on the funeral pyre of the deceased husband for this is pleasing to Brahma (Hindu laws)."[462] (And don't forget the marshmallows.)

"The Hindus consider the Ganges river to be sacred and are commanded to bathe, drink, and swim in its polluted waters. An *Associated Press* release, in the spring of 2002, recounted how the faithful Hindu immerses himself into the Ganges. The result, of this polluted bath, is many of them get skin, intestinal, and stomach ailments from the "holy river." A river that is filled with rotting corpses, ashes of the dead, sewage, and other wastes. Many devout Hindus die from the religious rituals of the Ganges that are theologically imposed on the faithful. All religions are not basically the same."[463]

No, they certainly are not. But the lies of Satan flood this earth like a river of sewage, and people believe what they're taught. And Hindus and Buddhists are cursed by the lies that have been indoctrinated into their minds just assuredly as final-stage lung cancer patients, who suffer horribly before they die, have been cursed by the cigarette smoke that for years they foolishly sucked into their lungs. And this indoctrination will remain unless someone is sent to give them the opportunity to remove their curse, and to come from darkness into the Light.

"How beautiful upon the mountains are the feet of those who bring good news, who proclaim peace, who bring glad tidings of good things, who proclaim salvation." (Isaiah 52:7)

"For whosoever shall call upon the name of the Lord shall be saved. How then shall they call on him in whom they have not believed? And how shall they believe in him of whom they have not heard? And how shall they hear without a preacher? And how shall they preach, except they be sent? As it is written, 'How beautiful are the feet of those who preach the gospel of peace, who bring glad tidings of good things!" (Romans 10:13-15)

"To open their eyes, and to turn them from darkness to light, and from the

---

461   (Pg.28-29, *Fast Facts On False teaching*, by Ron Carlson & Ed Decker)
462   (Pg. 5, *There are Moral Absolutes*, by Michael A. Robinson)
463   (Prajnan Bhattacharya, "Hindu's Risk Health for Holiness" (Las Vegas, NV; Review Journal, May 14, 2002, Associated Press), 7B. From Pg. 24, *There are Moral Absolutes*, by Michael A. Robinson)

power of Satan unto God, that they may receive forgiveness of sins, and inheritance among them who are sanctified by faith that is in me." (Acts 26:18)

And for an excellent first-hand account that unmasks another one of India's "holy" men see: *Lord of the Air: Tales of a Modern Antichrist*, by Tal Brooke.

Islam

(We've saved the best for last.)

Before we get started let me point out here that what we are going to attack in this section is the *religion* of Islam, and not all of those hundreds upon hundreds of millions of poor, unfortunate folks who have been born into and trapped in it. Indeed, most Muslims are just ordinary folks like you or me who want to live their lives in peace, raise their children securely, and find some amount of happiness and fulfillment. And most of them have also never been taught, nor had the opportunity to learn, the truth about their religion...

Islam (with all due respect of course) is a manure pile. And that's being kind. It is actually the most vile, disgusting horrific excuse for a religion that has ever tormented a semi-intelligent species.[464] And if you feel these words are harsh or offensive then you need to reserve judgment until you've finished this section and learn all of the seedy details that themselves condemn this religion of Mohammed. The only thing that could surpass its degree of evil would be if you could find another religion somewhere where the followers actually eat their young. The Praying Mantis religion or something. (Or one where the followers actually mutilate their own babies to death?) (Hmm...)

Islam is... "The religion of the false prophet Mohammed which encourages the worship of the demonic entity, Allah."[465] That's it in a nutshell, and all you really need to know, but being as how our left-wing, lying media and education system has sugar-coated this satanic religion and concealed the truth about it [466] from the uninformed masses, it is important that we spend a little time here setting the record straight...

The *word* Islam translates as peace but the *religion* Islam translates as brutality, violence, bloody conquest, Jihad, mass-murder, forced conversions, female genital mutilation, unrivaled ignorance, double-dealing, rape, the commandment to lie and deceive, unprovoked war, the sword, and the abject subjugation of women in every culture, language, nation or other religion that has had the extreme misfortune of getting in the path of its demonically-inspired, ever-expanding hordes. It was begun in Saudi Arabia by a sexually incontinent lecher[467] by the name of Mohammed who had about as much to do with the living God as O. J.

---

464   (But with that said it is also important to point out that *all* false religions are vile and disgusting in the fact that they all originate from the father of lies and they are all his very clever hell vacuums, sucking untold millions of unsuspecting folks down to an eternity in the Lake of Fire. Just ask all of those poor souls writhing in sheer agony in the fires of hell what they think of their particular religion now, and whether they are still as loyal to it.)

465   (From article in The Christian Edition of the Jerusalem Post)

466   (In the same way they concealed the truth about their left-wing, lying, capitalist-hating and oil-adverse Marxist friend, Hussein ACORN Obama.)

467   (Just for starters, Mohammed <u>consummated</u> his marriage to his third wife, Aisha, when she was just 9 years old. Wrap your mind around that if you care to. And that is just one of this perverts numerous historically documented sexual crimes. Prophet? Absolutely. Prophet of hell.)

Simpson had with marriage counseling. (Barney Frank could come up with a more credible religion, for goodness sakes.) His early followers saw it as a great way to conquer rival tribes and towns, to plunder their riches, slaughter their males both young and old, and take their wives and daughters as sexual booty. Yea boy! Now there's a Gestapo outfit that every disgusting bastard on the block would butt-up in line for! And they were the early followers of Mohammed. (And sadly today, many fools on our prison blocks are still butting up in that line.)

Today this perverse excuse for a religion has subjugated much of the world by the sword and holds it in its satanic grip. But if present day Muslims had any idea of the truth about their Islam, and the fact that many of them are descended from those distant conquered and subjugated woman and little girls, enslaved and raped by evil bastards who are now lighting up hell as their reward, they might question their misplaced loyalties. For it is very likely that many Muslims today are the distant descendant of a woman whose husband and children or father were slaughtered before her eyes, as she was being dragged off as the next sex slave for some disgusting pig. The decision today is whether to claim as their descendant the disgusting pig—in which case they should remain a proud, loyal and obedient follower of Islam (and that ancient disgusting pig)—or the kidnapped and brutalized woman or little girl, in which case they might want to rethink their affiliation with the congregation of Mohammed. But the lies of the ruler of this world cover Islam like a pompous Anthony Weiner lecture debases congress,[468] and its followers believe exactly what they were taught.

Islam is absolutely intolerant of the truth. The most volatile of its followers become rather emotionally disturbed when it comes to anything that threatens keeping their people, and especially their woman, in the dark. Our good friend Bill Maher had this eloquent take on it:

> "There is one religion in the world that *kills* you when you disagree with them. And they say, 'Look, we're a religion of peace, and if you disagree, we'll f***ing cut your head off.' And nobody calls them on it, or very few people call them on that."[469]

Well, let's spend the rest of this section doing just that...

"'He who hates me hates my Father also. If I had not done among them the works which no other man did, they would not have sinned: but now they have both seen and hated both me and my Father. But this came to pass, that the word might be fulfilled that is written in their law, They hated me without a cause." (John 15:23-25)

- Beelzebub's bible The Koran is the holy book of the Muslim religion and yet it makes less sense than an Al Franken book or a Michael Moore Crockumentary, and that is quite a feat. Here are just a few examples:

- Muslims are told not to raise their eyes toward heaven when praying. If they do they are told that their eyes will be snatched out of their sockets, so instead they lift up their buttocks. (And what that says about Allah, we won't speculate.)

- Orthodox Islam also has rules on proper bathroom hygiene that prohibits the use of toilet paper. (No, we're not going to touch that one either.)

---

468   (And that was written before his sexting scandal)
469   (Bill Maher)

- One of the commandments of Mohammed is to "follow camels, and drink their urine for health." (Good thing for all of those Muslims that he wasn't from San Francisco...)

- In Islamic societies a man can divorce his wife simply by saying "I divorce you." Of course this can happen so easily, especially in a fit of anger that it can also lead to regret. In that case they can get remarried, but this can only happen three times. The fourth time they are absolutely forbidden to remarry until the woman goes out and marries another and copulates with him. This inane law of Mohammed, followed religiously by devout Muslims, has "given rise to the phenomenon of "temporary husbands." After a husband has divorced his wife in a fit of pique [for the third time], these men will "marry" the hapless divorcee for one night [of copulation] in order to allow her to return to her husband and family."[470] It would be truly hilarious if it wasn't actually true.

- Another commandment from this "religion of peace" is this, "Slay the idolaters, wherever you find them, take them captives and besiege them and lie in wait for them and ambush them. (Sura 9, Koran)"[471] (*Idolater* being roughly translated as: "he who refuses to slice off little girl's clitorises.")

- Mohammed enticed his followers into expanding their slaughters with the allure of woman and booty from the neighboring tribes, villages and cities. But when asked what one would gain if they died in the battle, he came up with this famous dung-cranium epiphany: in that case they would immediately go to paradise and be given 72 virgins for their unending pleasure. (Think of the deceived departed's surprise and horror when they found themselves being cast into a Lake of Burning Sulfur. Satan's lies, when swallowed, and if not regurgitated, will make one eternally sick.)

Satan counterfeits everything. God wrote the Bible, so Satan wrote the Koran; and the Book of Mormon for that matter. Pity all those who are trapped in such sludge due to the unfortunate circumstance of their birth, and pray that some brave soul might give them the opportunity to repent and turn away from that sludge before they die.

"Do not fret yourself because of evildoers, neither be envious of the wicked; For there shall be no reward for the evil man. The candle of the wicked shall be put out." (Proverbs 24:19-20)

- The Crime of *getting* raped

But the utter stupidity of many of Mohammed's decrees pale in comparison to the sheer ignorance of his religions treatment of the fairer sex, or as they say in Islam, the genitally mutilated sex. If Bobby Brown and O. J. Simpson got together and started their own religion they could not top the level of physical and mental abuse, and outright degradation, of women accomplished by this singularly disgusting religion. (Not to say that they wouldn't give it a damn good try, but they couldn't possibly top it.)

Men are free to physically abuse, rape and even murder women in Muslim societies with virtual impunity. "Honor killings" are expected to be, and are, performed by relatives, usually *loving* brothers, fathers or husbands, if their wives or daughters, sisters or little nieces dress improperly, date or have sex with an

---

470   (Pg. 72, *The Politically Incorrect Guide to Islam*, by Robert Spencer)
471   (Pg. 5, *There are Moral Absolutes*, by Michael A. Robinson)

"infidel" or become "westernized" in any way. Mothers, daughters and sisters are killed by family members or imprisoned by the state if they are the victims of rape. (Yes, you did hear that correctly.) It doesn't matter that they were RAPED! The concept of this heinous crime seems to have escaped the pay grade of the leaders of this religion. Indeed, it is decreed in the inane...er...holy Koran that "four male witnesses are required" to corroborate a woman's, or even a little girl's, claim of rape, and that "they must have seen the act itself."[472] Of course everyone knows that rape mostly occurs in the median strip of a busy highway during rush hour, or in front of the stands at your daughters soccer game, so there are always ample numbers of male witnesses on hand to fulfill this catatonically stupid demand of the "religion of peace."

> "It is almost impossible to prove rape in lands that follow the dictates of Sharia. Men can commit rape with impunity: As long as they deny the charge and there are no witnesses, they will get off scot-free, because the victim's testimony is inadmissible. Even worse, if a woman accuses a man of rape, she may end up incriminating herself. If the required male witnesses can't be found, the victim's charge of rape becomes an admission of adultery. That accounts for the grim fact that as many as 75 percent of the imprisoned women in Pakistan are, in fact, behind bars for the crime of being a victim of rape. Several high-profile cases in Nigeria recently have also revolved around rape accusations being turned around by Islamic authorities into charges of fornication, resulting in death sentences that were modified only after international pressure."[473]

It would be easy here to call these Muslims complete morons but you have to keep going back to the fact that people believe whatever they are taught no matter how idiotic or asinine it may be (and we Americans are certainly not exempt from that rule: see chapter 10 on the theory of evolution, just for starters), and that there but for the grace of God go you or I. Remember the words of William James, the father of modern psychology, quoted earlier, "There's nothing so absurd that if you repeat it often enough, people will believe you." Especially when they are growing up.

- Old Burka And we still haven't gotten to the dress code...

> "One of the most demeaning practices of Islam is its barbaric treatment of women. Women are considered property in the fundamental sects of Islam. They are not allowed to have ownership of any kind of property. ...They are dressed from head to toe in clothes that cover all but the eyes, and often even these are covered by a veil.

> "It is interesting that what an illiterate nomadic tribesman wore in the desert in seventh-century Arabia is still mandated as the dress code for Muslim women today! It is a clear denial of civil rights to women and is reflective of the Islamic Arabian culture and its low view of women."[474]

> "In European cities such as London and Amsterdam, some non-Muslim women won't go outside without a veil for fear of Muslim men shouting threats and insults like "infidel whore" or worse. Remember seeing this on CBS news? Of course you don't."[475]

---

472 (Pg. 74, *The Politically Incorrect Guide to Islam*, by Robert Spencer)
473 (Pg. 76, *The Politically Incorrect Guide to Islam*, by Robert Spencer)
474 (Pg. 106, *Fast Facts On False teaching*, by Ron Carlson & Ed Decker)
475 (As reported by *National Review*)

"A graphic example of the oppression that Islamic dress regulations for women engender came in March 2002 in Mecca, when fifteen girls were killed in a fire at their school. Saudi Arabia's religious police, the *muttawa*, wouldn't let the girls out of the building. Since only women were in the school, the girls had shed their all-concealing garments. The *muttawa* [translation: *shit4brains*] preferred the girls' death to transgression of Islamic law—to the extent that they actually battled police and firemen who were trying to open the school's doors."[476]

Humans treat their dogs better. And keep in mind that woman are forced to wear these potato bags in some of the most extremes of heat found on the planet. (Perhaps we can gain some insight into why Muslims are flooding into the colder climes of Europe and England from the sun baked lands of North Africa and the Middle East, besides their religions aim to subdue the world for Allah through copulation.) But all this falls short in comparison to the crème de la crème of female brutalization in the Islamic world, which we touched on briefly in chapter 2... welcome to Mohammed's world of Female Genital Mutilation. ("Step right up, little girls, does Mohammed have a surprise for you!") ...

- Lose weight, without dieting

"Islam breeds a culture that practices female genital mutilation. The young women and little girls are forced to have their clitoris removed by knives and razor blades. Many die a horrible death from this ghastly genital removal. In Egypt, 6000 girls a day have their genital organs cut away and it is a common practice in Muslim Africa, Indonesia and the Middle East (**130 million** women have survived this barbaric procedure)."[477] [emphasis mine]

"Approximately 75 percent of Muslim women suffer female circumcision in a most barbaric, painful ritual **designed to make them obedient and docile**."[478]

Wrap your mind around that one! That means that approximately 3 out of 4 Muslim women are walking around today without their clitoris because it was sliced off to the bone when they were a happy little girl! Another gift that the well-done Mohammed left to the world.

Jimmy Carters friend, "Gad el-Haq, head of one of the Muslim world's leading religious institutions, ruled that female circumcision was an Islamic duty."[479] (Their little girls would only be so lucky if circumcising el-Haq (above the shoulders) was "an Islamic duty" instead.)

This is more than enough to make one *beyond* sick to their stomach. (It's called *Islam-sick-to-the-stomach*.) What these demonically-possessed followers of Mohammad do to their own children is a horror. But the real question here is why the feminist Left that controls our lame-stream media hasn't made this their loudest outcry, or their hottest topic of discussion. Why have they all but covered up this international horror? (Oh, darn, that's right. I keep forgetting, Islam is not Christianity. They have no reason to expose the barbaric practices of our Muslim friends, that would take away face time from reporting on Christians who *spank*

476   (Pg.68, *The Politically Incorrect Guide to Islam*, by Robert Spencer)
477   (Pg. 66-67, *There are Moral Absolutes*, by Michael A. Robinson)
478   (Pg. 106, *Fast Facts On False teaching*, by Ron Carlson & Ed Decker)
479   (As reported @ David Horowitz's *The Terrorism Awareness Project*)

their children.)

And why hasn't the NOW gang, the other liberal feminist confabs, and those international left-wing groups disguised as "human rights" organizations shouted the demand on every single one of their press releases and policy statements that this horror be ended immediately; that we save millions upon millions of these little girls from suffering the same macabre fate, and that these insane Muslims be stopped? Where is the moral outrage from all of the Bush-haters on the Democrat-Left? They certainly got all worked up over 3 Islamic lunatics who got their faces soaked. They even demanded *prosecutions* over it, but hardly a stir over the terrible fate of hundreds of millions of innocent little girls whose only crime was being born into the most putrid excuse for a religion since dinosaurs roamed the earth. Oh, they'll mention it here and there. They'll give it the same attention that they might give to cockfighting in Cuba (and much less attention than they gave to Michael Vick's dog-fighting in America), but their overall hands-off approach to this horror is instructive as to the agenda that takes real precedence; which is marginalizing, vilifying and disparaging Christianity. Vigorously attacking the horrendous, clitoral-mutilating sins of the religion of Mohammad would take away from their primary narrative. Besides, Islam and the Left are two sides of the same satanic coin, so why would the evil, baby-mutilating Democrat-Left expose the true nature of the evil, clitoral-mutilating Islamists? Think about it, before you think that is an unfair indictment. (Both religions have the same father, the same progenitor, after all.)

Here is an excerpt from an excellent article on the subject by Jamie Glazov, titled, "Female Genital Mutilation on British Turf:"

"It may come as no surprise to the knowledgeable political and cultural observer that the poor victims of these crimes are not from Christian or Jewish families, nor from Hindu or Buddhist ones. They are to be found predominantly in Muslim households. And being Muslim is a status that gives the victims, and all future victims, the unfortunate distinction of being part of a group that society can't help, because the lib-Left has made sure that the Muslim culture can never be criticized and, therefore, that its sufferers can never be protected or saved.

"Fact: female circumcision is illegal in Britain. But this doesn't mean that British law enforcement is doing anything about this crime that Muslim communities are perpetrating against their little girls.

"The reality: five hundred girls' genitals are mutilated every year in Britain. Not one arrest. Not one incarceration.

"You think protecting little girls' genitals is more important nowadays than protecting oneself from the charge of being Islamophobic? Think again. Islamic women haters, therefore, are reigning free in Britain. Enraged at even the thought of female sexuality, the self-appointed guardians of Islamic purity make sure to obliterate the clitorises of little girls before the girls begin to get the concept of their own human agency and the magic of love. In a fascistic effort to deny women even the possibility of personal happiness, individuality and sexual satisfaction, these mutilators start cutting girls at the age of seven or eight—before their menstrual periods begin—so that their sexuality will be amputated forever.

"Despite its gruesome terror, this crime is widely practiced throughout the world, for it is a crucial ingredient of Islamic gender apartheid and is known as female genital mutilation (FGM). Its ideological premise has been carefully constructed: a girl's genital area is dirty and unacceptable. How much is amputated varies among cultures. In Egypt only the clitoris is amputated; in countries like Sudan the woman-haters are not so kind. In a savagery called infibulation, the girl's external genital organs are completely removed: the clitoris, the two major outer lips (*labia majora*) and the two minor inner lips (*labia minora*). In Sudan, the term used for this is *tahur*—which means "cleansing" or "purification."

"More than 130 million women living today have been subjected to this horrifying practice, and more than two million girls are assaulted by it each year. That is more than five thousand girls every day. Many girls lose their lives during FGM, which is often done with broken glass. Most victims suffer from chronic infection and pain for the rest of their lives. The mutilation robs women of their ability to enjoy the fullness of their sexuality and, therefore, the fullness of their lives. Approximately 75 percent of women cannot achieve orgasm without clitoral stimulation; thus, the possibility of sexual satisfaction has been obliterated for millions of women in the Muslim world.

"The Muslim communities who practice FGM will not easily abandon their barbarity. The Egyptian government, for example, banned FGM in 1996, but an Egyptian court overturned the ban in July 1997. The problem is that the clitoris mutilators point to traditional teachings [of the devil] that sanction FGM. Islamic tradition, for instance, records the Prophet [of hell] Muhammad emphasizing that circumcising girls is "a preservation of honor for women." A legal manual of the Shafi'i school of Islamic jurisprudence, *'Umdat al-Salik*, which is endorsed by Al-Azhar University of Cairo -- the oldest and most prestigious university in the Islamic world -- states that circumcision is *obligatory* for both boys and girls.

"Underlying this brutality is the obvious belief that the sexual mutilation of women will help keep the structure of Islamic gender apartheid in place. Keeping FGM legitimized and institutionalized is one of the most effective means to keep women subjugated and caged. The assumption is that amputating the clitoris will kill the woman's sexual desire and thereby reduce the chances that she will ever toy with the notion of self-determination.

"Thanks to the Left's policy of multiculturalism, where no value can be said to be worse or better than any other (except, of course, if American society and culture is the subject of discussion), FGM is now being widely practiced on *Western* territory. A study estimates that 66,000 women living in England and Wales have suffered FGM, most of them before emigrating from their home country. More than 7,000 girls in Britain alone, meanwhile, are at a high risk of being victims of the crime. At present, we know that more than 500 girls in Britain are being mutilated every year.

"This horror show demands a certain question: Where are the leftist feminists in the West crying out in opposition to this crime against their sisters being carried out on their own shores? In what pages, in what demonstrations, are they denouncing the theology that serves as an inspiration to this crime and calling it to account? Where are the Women's Studies Departments on Western campuses demanding that Islam be confronted on these grotesque elements of misogyny? Where are the columnists of the *Nation Magazine*, the supposed leader of the humanitarian Left, repudiating this practice and the texts on which is based?

"The Left's cries of indignation are not to be heard because admitting the inferiority of an adversarial culture might very well legitimize Western civilization, a recognition that no leftist can allow if he hopes to retain his identity and social belonging. That's why the leftist forces in our society do their best to excuse FGM with the tired old mantra: *it's not only Muslims that do it* [a number of non-Muslim African tribes practice it as well] – as if inaction to save human beings from evil is somehow justified because a sin might exist somewhere else."[480]

---

480   (From article: "Female Genital Mutilation on British Turf," by Jamie

Again, it's enough to make one *Islam-sick-to-the-stomach*. The Democrat-Left with their control of the dissemination of information in our country through their dominance of our media, educational system, entertainment industry and the late-night and Comedy-Central clowns has the power to fully expose this, to shame Muslims and could be instrumental in putting an end globally to this atrocity. But that would be asking far too much of them, for shedding a bad light on their Muslim brothers would not help their broader narrative of vilifying Israel and disparaging Christianity. And unfortunately the right's record on this subject is not much more compelling. For example, the other night (10-28-09) Bill O'Reilly—who should be applauded for being a welcome break in an overwhelmingly left-wing, lying, *slobberingly*[481] Democrat media—on his nightly *O'Reilly Factor* program on The Fox News Channel said that we shouldn't "offend" the religion of Islam, as it is a "legitimate" religion. Well, Bill, no one wants to "offend" anyone if at all possible, but what about actually reporting the truth no matter whom it might "offend?" Isn't that your job, big guy? So why have you never reported the truth about the Muslim propensity to hack the genitals off of hundreds of millions of their little girls? Too controversial? Too *dangerous*? (Look at what happened to that Danish cartoonist.) Maybe because it's not the safest way to insure the continuance of those $high$ ratings? I thought you were "looking out for all of us," O'Reilly, but I guess "all of us" does not include millions of poor, clitoral-mutilated, sexually-stunted little Muslim girls. No, William, *you* might feel comfortable being a coward "hiding under his desk," calling Islam a "legitimate" religion. I call it a butcher factory. And refusing to call it what it actually is just helps to continue to keep trapped all of those hundreds of millions of poor, unfortunate souls, and their desperate little girls, who were cursed to be born into it. Congratulations. (Maybe you'll get another Emmy.)

- Child rape

And then we progress from female genital mutilation to their godly practice of child sexual perversion…

"Fatwa rationalizes pedophilia

"…A recent fatwa issued by a top Saudi cleric regarding child brides and sex with female children [makes a] statement …regarding how old girls should be before marriage. Dr. Salih bin Fawzan asserts that '…the only criterion is that they are capable of being placed beneath and bearing the weight of men.' [Sweet Jesus, what a f***ing disgusting pig, and a leader of this religion of piece.]

"For those who claim this can't be happening in America, we have seen reports of Muslim men immigrating to Europe, the UK, and to the United States, and identifying their child brides as relatives, such as daughters, in order to avoid scrutiny from Western laws that prohibit pedophilia."[482]

And this can come as no surprise for, as we mentioned earlier, their prophet, the incontinent pig Mohammed, began the daily rape of his third wife when she was only 9 years old. Hey Bill, let me ask you a question, Is child rape, like clitoris-hacking, another "legitimate" part of this "legitimate" religion, that we

---

Glazov, FrontPageMagazine.com, Friday, March 20, 2009)
481   (See Bernie Goldberg's book, *A Slobbering Love Affair: The True (And Pathetic) Story of the Torrid Romance Between Barack Obama and the Mainstream Media*)
482   (From *Act for America* email, 08/08/2011)

want to be so careful "not to offend?" Come out from under your desk long enough to answer that one...

- Karma And then there's this poignant lament from Spain:

"I walked down the street in Barcelona, and suddenly discovered a terrible truth – Europe died in Auschwitz. We killed six million Jews and replaced them with 20 million Muslims. In Auschwitz we burned a culture, thought, creativity, talent. We destroyed the chosen people, truly chosen, because they produced great and wonderful people who changed the world. The contribution of this people is felt in all areas of life: science, art, international trade, and above all, as the conscience of the world. These are the people we burned.

"And under the pretense of tolerance, and because we wanted to prove to ourselves that we were cured of the disease of racism, we opened our gates to 20 million Muslims, who brought us stupidity and ignorance, religious extremism and lack of tolerance, crime and poverty due to an unwillingness to work and support their families with pride.

"They have blown up our trains and turned our beautiful Spanish cities into the third world, drowning in filth and crime. Shut up in the apartments they receive free from the government, they plan the murder and destruction of their naïve hosts. And thus, in our misery, we have exchanged culture for fanatical hatred, creative skill for destructive skill, intelligence for backwardness and superstition.

"We have exchanged the pursuit of peace of the Jews of Europe and their talent for hoping for a better future for their children, their determined clinging to life because life is holy, for those who pursue death, for people consumed by the desire for death for themselves and others, for our children and theirs.

"What a terrible mistake was made by miserable Europe."[483] (Yes. It's called Karma.)

Geert Wilders, author of *Marked for Death: Islam's War Against the West and Me*, had these thoughts on Fox News *Stossel*...

Islam is "a dangerous ideology rather than a religion. ...It has to be compared to other ideologies like communism or fascism. ...You are not free to leave Islam. If somebody wants to leave Islam when he becomes apostate, you are eligible for death. You should be killed according to the Koran. This is not the case with Christianity or Judaism. This was the case with communism and fascism. I'm proud to say that our values that are based on Christianity are far better, even superior, than the barbaric Islamic values. You're not a racist if you say that."[484]

No, you're just speaking the truth, which can also get you killed, by the followers of the "religion of peace."

Bernie Goldberg describes the events of 9/11 in his book *Bias* as the day...

..."when a band of religious lunatics declared war on the United States of America to punish us for not wanting to dwell in the fourteenth century, where they currently reside, and, of course, to show the world that their intense hatred of Israel—*and of Israel's friends*—knows no bounds."[485]

---

483 ("All European Life Died In Auschwitz," by Sebastian Vilar Rodrigez)

484 (Geert Wilders, author of *Marked for Death: Islam's War Against the West and Me*, on Fox News *Stossel*, 05/19/2012)

485 (Pg. 202, *Bias: A CBS Insider Exposes How the Media Distort the News*, by Bernard

His reference to the fourteenth century is an accurate assessment of Islam, which is in reality not so much a religion but a culture, and a very backward one at that...

"In his excellent book *Islamic Invasion* (Harvest House Publishers, 1992), Dr. Robert Morey says that what Mohammed did was to raise the seventh-century culture in which he was born to the status of divine law. In fact, Islam is the deification of seventh-century Arabian culture. ...Islam imposes its seventh-century Arabian culture in its political expression, in its family affairs, in its dietary laws, in its clothing, in its religious rites, and in its language. Muslims are religiously compelled to impose seventh-century Arab culture on the rest of the cultures of the world.

"Muhammad took the political laws which governed seventh-century Arabian tribes and literally made them the laws of Allah, their God. In such tribes, the sheik or chief of the nomadic tribes had absolute authority. There was no concept of civil or personal rights in seventh-century Arabia. This is why Islamic countries are inevitably ruled by dictators or strong men who rule as despots. There are 21 Arab nations today, and not one of them is a democracy. Democracy cannot flourish in Islam. [Common sense cannot flourish in Islam.]

"The more Islamic fundamentalism gains dominance, the more a nation is plunged back into the dark ages of seventh-century Arabia. Iran is a good example of this. The despots today of Libya, Iran, Iraq, Syria, the Sudan, and Yemen are merely examples of such Arabian tyranny grafted into modern times.

"Because there was no concept of personal freedom or civil rights in the tribal life of seventh-century Arabia, Islamic law today does not recognize freedom of speech, freedom of religion, freedom of assembly, or freedom of the press. This is why non-Muslims (such as Christians) are routinely denied the most basic of human rights and are often physically attacked and jailed."[486]

This is also why so many Islamic countries are showcases for how nations saturated with oil riches can remain illiterate backwaters, religiously perpetrating a modern day dark age for their citizens that makes 13th century equatorial Africa look like Atlantis. No, my hate-filled, brainwashed Muslim friends, it is not the Jews or their nation of Israel that is the cause of all of your terrible problems and poverty. It is your pathetic Islam. Get a clue, and dump it.

- The Crusades As a related subject to the scourge that is Islam I think it is important to point out here how the Democrat-Left that controls our media, our entertainment industry and of course our educational system has distorted the historical record concerning the Crusades. This has been done in order to conceal the disturbing truth about Islam and its quest to dominate the world by violent conquest, and of course to further lie about and disparage Christianity. (Surprise, surprise.) Let us now set the record straight...

The first Big Lie in the Left's rewriting of this time in history is that "the Crusades were an unprovoked attack by Europe against the Islamic world...

..."Wrong. The conquest of Jerusalem in 638 [by the Muslims] stood at the beginning of centuries of Muslim aggression, and Christians in the Holy Land faced an escalating spiral of persecution. A few examples: Early in the eighth century, sixty Christian pilgrims from Amorium were crucified [they were caught with cartoons of Muhammad drinking his camel's urine.]; around the same time, the Muslim governor of Caesarea seized a group of pilgrims from Iconium and had them all executed as spies; ...and Muslims demanded money from pilgrims, threatening to ransack the Church of the Resurrection if they didn't pay.

Goldberg)

486   (Pgs. 103-104, *Fast Facts On False teaching*, by Ron Carlson & Ed Decker)

"...Brutal subordination and violence became the rules of the day for Christians in the Holy Land. ...Conversions to Christianity were dealt with particularly harshly. In 789, Muslims beheaded a monk who had converted from Islam and plundered the Bethlehem monastery of Saint Theodosius, killing many more monks. Other monasteries in the region suffered the same fate. Early in the ninth century, the persecutions grew so severe that large numbers of Christians fled to Constantinople and other Christian cities. More persecutions in 923 saw additional churches destroyed, and in 937, Muslims went on a Palm Sunday rampage in Jerusalem, plundering and destroying the Church of Calvary and the Church of the Resurrection. [Practicing their "religion of peace" in the traditional, time-honored way.]

"In reaction to this persecution of Christians, the Byzantines moved from a defensive policy toward the Muslims to the offensive position of trying to recapture some of their lost territories. In the 960's, General Nicephcrus Phocas ...carried out a series of successful campaigns against the Muslims, recapturing Crete, Cilicia, Cyprus, and even parts of Syria. In 969, he recaptured the ancient Christian city of Antioch."

However, the Muslims declared jihad and fought back hard. "In 1004, the sixth Fatimid caliph ...turned violently against the faith of his Christian mother and uncles (two of whom were patriarchs), ordering the destruction of churches, the burning of crosses, and the seizure of church property. He moved against the Jews with similar ferocity. Over the next ten years, thirty thousand churches were destroyed. ...In 1009, al-Hakim gave his most spectacular anti-Christian order He commanded that the Church of the Holy Sepulcher in Jerusalem ...the traditional site of Christ's burial ...be destroyed. ...Al-Hakim commanded that the tomb within be cut down to the very bedrock. [Imagine how these fascist bastards would react if their Dome of the Rock was bulldozed off of the Temple Mount?] [On second thought, that's not such a bad idea. Israel, why don't you rub the doggies noses in their own doo-doo?] He ordered Christians to wear heavy crosses around their necks (and for Jews, heavy blocks in the shape of a calf). [But that's OK. He's had a heavy block of burning sulfur hanging around his neck for the past ten centuries. Perfect justice in God's perfect world.]

...."The Christian empire of Byzantium, which before Islam's war of conquest had ruled over a vast expanse including southern Italy, North Africa, the Middle East, and Arabia, was reduced to little more than Greece. ... [So they] appealed for help. And that is how the First Crusade came about: It was a response to the Byzantine Emperor's call for help."[487]

So the truth is that:

- "The Crusades were *not* acts of unprovoked aggression by Europe against the Islamic world, but were a delayed response to centuries of Muslim aggression, which grew fiercer than ever in the eleventh century.

- "These were wars for the recapture of Christian lands and the defense of Christians, *not* religious imperialism.

- "The Crusades were not called in order to convert Muslims or anyone else to Christianity by force."[488]

"And now you know the rest of the story..." that the history-rewriting liars in our Democrat-Left media and educational system would not tell you if their lives depended on it. They'd rather shout their Muslim-supporting, Christian-hating lies loud enough and long enough so enough of our uninformed adults and

---

487   (Pgs. 122-125, *The Politically Incorrect Guide to Islam (and the Crusades)*, by Robert Spencer)

488   (Pg. 121, *The Politically Incorrect Guide to Islam (and the Crusades)*, by Robert Spencer)

young students will become indoctrinated with them. Which leads us into our next segment...

For further reading on the truth about the Crusades see: *The New Concise History of the Crusades*, by Thomas F. Madden, and "Part II: The Crusades," from *The Politically Incorrect Guide to Islam (and the Crusades)*, by Robert Spencer.

- The educated ignorant Knowing all that we have just reported here about the "religion of peace," one would think that our educational system might find it of some importance to actually educate our children about the horrors of this religion. One would be wrong. In fact, they are doing the exact opposite. Like our lame-stream, lying media, they are concealing the truth about this clitoral-mutilating religion, and instead are endeavoring to get our school children to embrace it by indoctrinating them, in many of our public schools and universities, with a completely false and sanitized version.

"Islamists, or those who believe that Islam is a political and religious system that must dominate all others, are focusing less on the military and more on the ideological. It turns out that Western liberal democracies can be subverted without firing a shot.

"Nowhere is this more evident than in the educational realm. Islamists have taken what's come to be known as the "soft jihad" into America's classrooms and children in K-12 are the first casualties. Whether it is textbooks, curriculum, classroom exercises, film screenings, speakers or teacher training, public education in America is under assault.

"Capitalizing on the post-9/11 demand for Arabic instruction, some public, charter and voucher-funded private schools are inappropriately using taxpayer dollars to implement a religious curriculum. They are also bringing in outside speakers with Islamist ties or sympathies. As a result, ...children [are] receiving a biased education. ...Consider the following cases:

"- Last month, students at Friendswood Junior High in Houston were required to attend an "Islamic Awareness" presentation during class time allotted for physical education. The presentation involved two representatives from the Council on American-Islamic Relations, an organization with a record of Islamist statements and terrorism convictions. According to students, they were taught that "there is one God, his name is Allah" and that "Adam, Noah and Jesus are prophets." Students were also taught about the Five Pillars of Islam and how to pray five times a day and wear Islamic religious garb...

"- Earlier this year at Lake Brantley High School in Seminole County, Fla., speakers from the Academy for Learning Islam gave a presentation to students about "cultural diversity" that extended to a detailed discussion of the Quran and Islam. The school neither screened the ALI speakers nor notified parents...

"- As reported by the Cabinet Press, a school project last year at Amherst Middle School transformed "the quaint colonial town of Amherst, N.H., into a Saudi Arabian Bedouin tent community." Male and female students were segregated, with the girls hosting "hijab and veil stations" and handing out the oppressive head-to-toe black garment known as the *abaya* to female guests. Meanwhile, the boys hosted food and Arabic dancing stations because, as explained in the article, "the traditions of Saudi Arabia at this time prevent women from participating in these public roles." An "Islamic religion station" offered up a prayer rug, verses from the Quran, prayer items and a compass pointed towards Mecca. The fact that female subjugation was presented as a benign cultural practice and Islamic religious rituals were promoted with public funds is cause for concern...

"Equally problematic are the textbooks used in American public schools to teach

Islam or Islamic history. Organizations such as Southern California's Council on Islamic Education and Arabic World and Islamic Resources are tasked with screening and editing these textbooks for public school districts, but questions have been raised about the groups' scholarship and ideological agenda. The American Textbook Council, an organization that reviews history and social studies textbooks used in American schools, and its director, Gilbert T. Sewall, have produced a series of articles and reports on Islam textbooks and the findings are damning. They include textbooks that are factually inaccurate, misrepresent and in some cases, glorify Islam, or are hostile to other religions...

"Such are the complaints about "History Alive! The Medieval World and Beyond," a textbook published by the Teachers' Curriculum Institute. Similarly, a San Luis Obispo mother filed an official complaint several years ago with her son's school authorities over the use of Houghton Mifflin's middle school text, "Across the Centuries," which has been widely criticized for whitewashing Islamic history and glorifying Islam...

"The forces in opposition [to the widespread objections of this Islamification of America's *public* schools] are powerful and plenty. They include public education bureaucrats and teachers mired in naiveté and political correctness, biased textbook publishers, politicized professors[489] and other experts tasked with helping states approve textbooks, and at the top of the heap, billions of dollars in Saudi funding. These funds are pouring into the coffers of various organs that design K-12 curricula. The resultant material, not coincidentally, turns out to be inaccurate, biased and, considering the Wahhabist strain of Islam promulgated by Saudi Arabia, dangerous. And again, taxpayer dollars are involved. National Review Online contributing editor Stanley Kurtz explains:

"The United States government gives money - and a federal seal of approval - to a university Middle East Studies center. That center offers a government-approved K-12 Middle East studies curriculum to America's teachers. But in fact, that curriculum has been bought and paid for by the Saudis, who may even have trained the personnel who operate the university's outreach program. Meanwhile, the American government is asleep at the wheel - paying scant attention to how its federally mandated public outreach programs actually work. So without ever realizing it, America's taxpayers end up subsidizing - and providing official federal approval for - K-12 educational materials on the Middle East that have been created under Saudi auspices. Game, set, match: Saudis."

"...While groups such as People for the American Way, Americans United for Separation of Church and State, and the ACLU express outrage at any semblance of Christianity in America's public schools, very little clamor has met the emergence of Islam in the same arena."[490]

"Very little clamor" because these organizations don't have anything against false or even dangerously evil religions that hack the genitals off of hundreds of millions of little girls. They're only opposed to the true religion, Christianity. They hide behind a mask of overriding "concern" for the Constitution, but in truth they will jump at any chance to pervert and distort it in order to further their father's ideological agenda. Their tired old lie, *the separation of church and state*, is just a red herring. All they are really interested in doing is separating *Christianity* from not only the state, but from every aspect of American life. Besides, as we mentioned before, these groups—People for the American Way, Americans United for Separation of Church and State, and the ACLU—and other Left-wing groups

---

489 (Useful idiots all of the Islamists)
490 (From investigativeproject.org, "Islam in America's public schools: Education or indoctrination?" by Cinnamon Stillwell, *SFGate (San Francisco Chronicle online)* June 11, 2008)

like them are on the same side as the Islamists (see chapter 13 "Satan"), so why would they make any trouble for them? (Besides risking their heads.)

- Stealth Jihad The following is from a publication by Coral Ridge Ministries titled "Radical Islam on the march:"

"Their Target: You and Your World

"The average everyday Muslim may not be a danger. But radical leaders of Islam are advancing the cause of global domination. As Dr. Peter Hammond has written, Islam is not simply a "religion" or a "cult," but a "system of life"—with religious, legal, political, economic, social, and military components. When there are enough Muslims in an area to begin agitating for religious privileges, the process of "Islamicization" begins. The process is gradual. Eventually a culture begins to regard tolerance of Islamic religious practices as politically correct; then the rest of Islam's components begin creeping in.

"Using oil money Islamic nations have already gained undue influence in impoverished regions like Africa. This phenomenon makes President Obama's approach to Islam— reflected in his tragic Muslim-friendly speech to Islamic elites in Cairo—particularly dangerous. He said it's important to create "a modern international community that is respectful," where "there are not tensions." *The tension between Islam's commitment to destroying 'infidels' and the Christian view of grace and love is apparently inappropriate.*

""We in the USA," the President said on another occasion in Turkey, "do not consider ourselves a Christian nation!"

"As author David Limbaugh has pointed out, the U.S. is truly and uniquely a Christian nation. "The concept of unalienable rights inheres in the Judeo-Christian precept that an all-loving God created man in his own image," he writes, "thus entitling him to dignity, freedom and rights that cannot be divested by the state." The bloodthirsty agenda of radical Islam could not be more opposite from American values and Judeo-Christian ethics!

"Islamicization is creeping into every crevice of modern life. A California-based non-profit organization dedicated to "teaching about Islam" in U.S. schools is headed by various extremists including one who reportedly delivered a speech entitled "Jihad is the Only Way." The group claims to have made some 750 classroom visits in a single year, not to mention encounters in churches, senior centers, businesses, and forums for policemen and healthcare workers. They've apparently created a curriculum about Islam for grades 7-12.

"The problem with America's President embracing Islam is that a "peaceful world community" is not Islam's ultimate goal. Its ultimate goal is the entire world subjected to sharia law—the strict extremist authoritarian system under which millions are already suffering. The American worldview, based on Judeo-Christian values of peaceful living, cannot co-exist with the Islamic worldview, based on the agenda of "convert or kill." The Palestinian example demonstrates this conflict.[491]

The "problem with America's President" is that he is either too naïve or too incredibly uneducated to know this or, even more insidiously, he doesn't care. It is as stupid and dangerous as if we had let Nazi Germany come over to our country in the middle of World War II and indoctrinate our young people with their Aryan Master Race ideology of exterminating Jews and dominating the world. That's not an earthquake you just felt, that's our entire World War II generation turning over in their graves; and puking their guts out.

Coral Ridge Ministries continues...

---

491   (From Coral Ridge Ministries Media, Inc. 2009)

"America saw Islamic extremism on display on 9/11. Radical Islam's goals have not changed since then. Author Walid Shoebat, in the book *Why We Want to Kill You*, writes: "Islamic terrorism wants to subjugate the entire world to Muslim rule, in which the non-Muslims, called Dhimmi, are relegated to second-class, non-citizens, lacking any legal or human rights, living nearly as slaves for the benefit of Muslim survival... Religious conditioning taught to Muslim masses, by using allusions of misery, historic manipulation, and illusion of the virtues of a distant past by continual reflection of glory days long ago, in order to convert masses into angry, pride-filled, remorseless killers and seekers of salvation by death. The goal is to intimidate non-Muslims by fear and threats, in order to re-establish a utopian theocratic world in which Islam and Muslims are dominant and all non-Muslims are subservient."

"Author Raymond Ibrahim, in *The Al Qaeda Reader*, writes: "The matter is summed up for every person alive: either submit, or live under... Islam, or die." "[492] [I think we'll opt for the fourth alternative, Mr. Muslim Jihadi man. If it's all the same to you. We'll kill you first...]

Conversely, real Americans still cry. "Give me liberty or give me death."[493] But how many real Americans are still left after years of the Democrat-Left's dissemination of lies, misinformation and cowardice is the question that remains.

And the beat goes on...

"In the United States, Muslims have demanded – and received – special religious preferences in public schools, city government buildings, even city-owned municipal airports. Minnesota public schools, for example, have installed foot baths in student bathrooms so Muslims can pray during school."[494] [But this should come as no surprise from the same folks who actually elected, with the help of thousands of ACORN's fraudulent votes, a mentally disturbed, humorless ex-comic to the United States Senate.]

"The University of Michigan recently spent $25,000 to install foot baths for Muslim students in campus rest rooms. [But conversely, University administrators *have* steadfastly resisted, at least for now, their demands for vending machines that dispense knives, razor blades and sharp rocks. Protests are expected. (And keep the little girls indoors.)]

"At the University of Pennsylvania, the Muslim Student Association receives $50,000 a year in student funds – conversely, campus Democrats and Republicans do not receive a single dollar."[495] Another shocker.

- Taqiyya "You shall not lie" is one of the Ten Commandments found in God's word. In Islam they have a slightly different one, "You shall lie..."

"Those familiar with sharia law understand the role "taqiyya," (deception) plays in the advance of Islamic conquest and subjugation."[496]

And what is that role? Muslims are taught that bald-faced lying and total deception are required of them in advancing their religion across this globe, and in duping any gullible useful idiots that they can into aiding them in their quest. This includes of course our Democrat-Left media, educational system and entertainment

---

492 (From Coral Ridge Ministries Media, Inc. 2009)
493 (Patrick Henry, hero of the American revolution and one of the Founding Fathers of our country, March 23, 1775)
494 (As reported by founder Robert Spencer of *Jihad Watch*, a project of the David Horowitz Freedom Center)
495 (From a *Collegiate Network* mailing, T. Kenneth Cribb, President)
496 (From an *Act! For America* online report, 01/07/2011, by Brigitte Gabriel)

industry. For example, Katie Couric and many of her uninformed liberal associates in the media and entertainment industry see the vast opposition to a despicable Muslim "victory mosque" (see section 10 "Weaken America from within" in chapter 13) at Ground Zero in New York City by the majority of Americans (those who, unlike her and her Democrat friends, actually *are* informed) as evidence of "Muslim bigotry." She of course, is wrong. But this is just another one of countless examples in our media and our educational system where uninformed, useful liberals play right into the hands of the Muslims and their religions mandated practice of deception.

> "The famous circus promoter P.T. Barnum once said "There's a sucker born every minute." How true this is when it comes to gullible politicians, journalists, and academics who persist in believing without question everything shoveled at them by Muslims who practice taqiyya. We believe that ultimately we will prevail in defeating the threat of radical Islam because the truth about sharia law, jihad, etc., is impossible to defend. Taqiyya will only work for so long. Americans are beginning to wise up, and eventually even many of those in political leadership shackled by political correctness will "see the light" when they get enough pressure from the grassroots."[497]

One can only hope. (Couric, are you paying attention?)

In April of 2011 the Assembly of Muslim Jurists, not of Saudi Arabia but of America issued a fatwa that, incredibly, opposes democracy, which is in keeping with their goal to shove sharia law down the West's throat.[498] And these people pretend to be Americans!?#@! When will liberal America wake-up and smell the camel urine?

And yet as Muslims try to force their religion and their culture down our throats, this is what they do in countries that they have already occupied. The following is just a small sampling of reports that expose the *religion of peace* for what it really is. Thousands more could have been offered:

> "- 40-year-old Muslim cleric Jaffar Umar Thalib proclaimed Indonesian Christians "belligerent infidels" and called on radical Muslims to kill them until their last drop of blood was spent. By declaring Christians "belligerent infidels," Thalib was not only inciting his followers, but telling them they were justified in the eyes of their God [Beelzebub], as well. The international community looked on as 10,000 defenseless Christians were slaughtered. [Katie Couric covered this story by hosting a two-hour special on the evils of Christian parents who home school.]

> "- Christian clergy in Iraq are increasingly singled out and murdered. The Assyrian Orthodox Priest at St. Paul and Mark church in Baghdad was murdered in a drive-by shooting in front of his home this past April. [There is no truth to the rumor, however, that the killers completed their driver's ed training in south L.A..] Just weeks prior, the Archbishop of the Chaldean Catholic Church in Mosul was kidnapped and later murdered along with three of his parishioners.

> "- Every Friday, following afternoon prayers, Egyptian Muslims take to the streets of Cairo, Alexandria and other cities and brutalize that nation's Coptic Christians. Scores of Coptic's have been killed, many more seriously injured, and their churches, businesses and homes vandalized- all while Egyptian authorities do nothing to stop it. [Now where was

---

497 (From Brigitte Gabriel's *Act! for America* online @ actforamerica.org)

498 (From Brigitte Gabriel's *Act! for America* online @ actforamerica.org, April 4, 2011)

it that President Hussein ACORN Obama went and delivered his "whitewashing Islam" speech? Oh yes, Cairo, Egypt...]

..."These are just a handful of the thousands of attacks on Christians. And since this genocide is drawing no response from world leaders, the radicals are growing bolder and more brazen with each new day."[499]

Will the real Islam please stand up..

"Islam isn't in America to be equal with any other faith, but to become dominant. Koran should be the highest authority in America, and Islam the only accepted religion on earth." [Well, that s certainly what the wicked one wants.] This was said by "Omar Ahmad, the chairman of the board of CAIR, the Council on American-Islamic Relations. CAIR's spokesmen appear regularly in the media complaining about the treatment of Muslims and giving us the message that "Muslims are part of the fabric of this great country and are working to build a better America." [Yea, and Democrats are the Party of hope and change...] But when speaking to a Muslim audience, as Omar Ahmad was, the message is different."[500]

Compare their disingenuous whining with the slaughter of countless Christians in the nations they have already taken over and you're apt to come down with another bad case of *Islam-sick-to-your-stomach*.

"When 17 Canadian Islamic terrorists were arrested last May, suspected of plotting to bomb the government buildings and to behead the prime minister, Canadian officials bent over backwards to avoid describing these criminals as Muslims. And so did the U.S. media. ...It would be offensive, you see, to associate Muslims with terrorism! [It would also be the truth, but that in and of itself would offend the sensibilities of our left-wing media.]

..."California middle school students study Islam for several weeks as part of world history. Their textbook devotes 55 pages to Islam – far more than any other religion. It presents Islam in a wholly positive way and presents Muslim religious beliefs as factual and Mohammed as a hero. [A child-molesting, genocidal *hero*, but a hero nonetheless to the Left's elites. ...Heroism, like beauty, being in the eye of the beholder.] By contrast, Judaism and Christianity are given little space, and nothing positive is said about Christianity.

"Furthermore, the schools use handouts supplied by Saudi Arabia to give teachers ideas for activities. One teacher asked students to fast during daylight hours for one, two or three days as Muslims do during ...Ramadan. At other schools, students were taught the history of Islam by adopting Muslim names, dressing as Muslims, learning the Five Pillars of Islam, and memorizing an Islamic prayer that extols the greatness of Allah. And at one school, the front grounds of the school featured a banner that read "There is one God, Allah, and [Bingo was his name, sir?] Mohammed is his prophet." Can you imagine what the ACLU would do if a teacher taught Christianity this way?"[501]

(Can you imagine what the ACLU would do if it came up with a single rational thought? The shock to their brains would cause them mass hysteria and uncontrollable epileptic fits.)

If we want to understand the power that is behind the relentless advance of this religion of demons, just follow the money...

---

499 (As reported by founder Robert Spencer of *Jihad Watch*, a project of the David Horowitz Freedom Center)

500 (From a *Freedom Center* circular. David Horowitz, President and Founder)

501 (From a *Freedom Center* circular. David Horowitz, President and Founder)

"Saudi money has funded (and continues to fund) the creation of whole departments on Middle Eastern Studies at many of our most influential universities. Billionaire Saudi Prince al-Wahid bin Talal gave grants of $20 million each to universities like Harvard, Yale, Georgetown and Columbia. These schools are not merely havens of pro-Arab, anti-Israel, anti-U.S. hatred. These schools educate an overwhelming number of our foreign diplomats – men and woman who determine our Middle Eastern policies!

"Saudi money has influenced the content of curricula and *public school textbooks* which now magnify and glorify the history of Islam, and distort and diminish the histories of Christianity and Judaism.

"In California, "Muslim Understanding Week" is an approved activity in school districts across the state. [Yet if they really wanted to "Understand Muslims" they could do it in less than an hour; just have the students read this exposé, and the rest of the week maybe they could learn some simple math, and how to actually read and write. Just a thought.] Students memorize verses from the Koran, ...dress in traditional Muslim garb, bring prayer rugs to class, pray Muslim prayers, *drink camel urine and slice off a couple clitoris'* [yea, I made up those last two], and learn about Mohammed's journey to *hell*. [I'm sorry, that should read *Mecca*. Journey to Mecca. It was a typo.] Can you imagine children spending a week memorizing the Lord's Prayer and learning about Jesus' Sermon on the Mount? [Sure, I can. It's called education, a concept completely foreign to today's left-wing, Islamo-indoctrinating educational system.] Textbooks used in the Islamic Saudi Academies in Northern Virginia teach that, among other things, it's alright to kill adulterers and Muslims who *escape* [that should have read *convert*] to other religions. [On a completely unrelated topic, following this disclosure it has been reported that in the future President Clinton will avoid Northern Virginia entirely.]

"Islamic holy war – jihad – apologists like the Council on Islamic American Relations (CAIR) and subversive and violent groups like al Qaeda and the Muslim Brotherhood are often the recipients of funding from the Saudi's. Working in concert with such groups, Arab immigrants routinely ask for special privileges – special footbaths in public schools, special prayer rooms at schools or jobs, and even days off for certain Muslim holy days – and readily claim victim status if such request cannot be met. ["Waah! Waah! Waah!" ... Those damn Muslim crybabies. They're everywhere.]

"Experts believe the Saudis have funded the building of nearly 80% of all the new mosques built in America since 2001. Undercover reconnaissance has found that more than 75% of the mosques and Islamic schools across the country preach jihad warfare and Islamic supremacy. This type of radical teaching is called "Wahhabism" – it's the same virulent strain of the Muslim faith preached by Osama Bin Laden. And, as a CIA report noted, Wahhabism is preached in nearly every Saudi mosque nearly every Friday evening."[502]

As previously mentioned, that is like letting the Nazi's, right in the middle of WWII, come over here en masse and foment for German victory. Are we a nation of complete jackasses or what? Or maybe one whose media, educational system, entertainment industry, judicial system and government have all been hijacked by horse's asses, those of the Democrat-Left...

"Experts estimate that Saudi Arabia has spent nearly a billion dollars on pro-jihad mosques, schools, and free books in the United States during the last 3 decades."[503]

---

502   (From another *Freedom Center* circular. David Horowitz, President and Founder)
503   (From a *David Horowitz Freedom Center* circular)

Note that the pertinent and disturbing phrase there is "<u>pro-jihad</u> mosques, schools!" In the middle of a war <u>against</u> those jihadists no less!

And then there is this from a communiqué from the *Intercollegiate Studies Institute*:

"America's university system is being hijacked. Revisionists liberals and apologists ...are attempting to disconnect our students from the values of the West. [They] are willingly doing to America what Islamic terrorists cannot accomplish with bombs and bullets. They are attempting to destroy America from the inside-out. <u>And they are being helped directly by the Saudi Government</u>. **Saudis, who want their propaganda validated by major university programs, are joining forces with apologist administrators to indoctrinate a generation of American leaders with an incomplete and biased version of Islam.**

"Consider these examples at Harvard and Georgetown (recipients of $40 MILLION in Saudi funds):

- "Harvard's Islamic Chaplain claimed there is "great wisdom" in the "preserved position" of killing Islamic apostates. Can you imagine the commotion and ensuing backlash Harvard would receive if the Christian Chaplain called for the death of those who converted from Christianity?

- "In response to the requirement for Muslim woman to remove their "hijab" or head cover to enter secure areas in Maryland court houses, Georgetown Professor John Voll claimed that was the same as requiring a woman to "take off her brassiere right there in the inspection section.""[504]

Upon closer investigation, though, it was discovered that Mr. Voll suffers from a rare mental disorder that renders its victims incapable of distinguishing between a Muslim woman's breast and her face. (Known as "ACLulosis Demingitis Leftonella disease." It is highly contagious.) (Stay away from Jon Stewart.) Mr. Voll has since been promoted to Reich's chancellor of the university.

And from another communiqué from *ISI*, this one presented by Senator Rick Santorum:

"The Enemy Within The Ivory Tower

"- The University of California, Berkeley Center for Middle East Studies is reported to "run programs funded by groups and individuals the U.S. State Department links to terrorism."

"- A newsletter of the University of Arkansas' Middle East Studies Center (established with support from King Fahd) contained translations by Arabic language students of a poem praising "martyrdom and death."

"These outrages are now the norm as blame-America academics disconnect young Americans from the Judeo-Christian values and institutions that sustain a free and amazingly generous America.

"While they teach a "tidy" view of radical Islam and Sharia law, they vilify America's traditional culture and heritage through wildly popular books such as Howard Zinn's Marxist inspired *A People's History of the United States*.

"Consider these latest outrages:

"Following Sharia law, Harvard has a gym that now bans male students during set regular hours so that Muslim coeds can be segregated from males...

"Arab-American student Oubai Shahbandar was impeached from his student senate seat at Arizona State University after he introduced a resolution to display an American flag in a campus dining hall. His colleagues complained that the resolution might result

---

504   (From an *Intercollegiate Studies Institute, Inc.* mailer, T. Kenneth Cribb Jr. President)

in "offending international students." [They were going to cut his head off but, after much heated debate and a rousing tongue-yelping chorus of "Allah Akbar's," decided on impeachment instead.]

"An administrator told several students at Central Michigan University to remove patriotic symbols—a flag, an eagle, and such—from their dorm room. The administrator described these symbols of our country as "offensive."

"And I could go on and on."[505]

On a recent documentary on Fox News, Tucker Carlson had this to say:

"Islam is taking over in [our] children's schools. ...Out of 307 million Americans, 1.8 million identify themselves as Muslim Americans. That's less than 1% of the population. And yet for the past 2 decades Muslims have had considerable influence over textbook content."

Gilbert Sewall, author of the study *Islam in the classroom*, added:

"In almost all textbooks terrorism is not identified as Islamic. ...Makes it hard to understand what their motives were, in a cowardly way, of being deferential to the Council on Islamic Education, which Houghton-Mifflin [textbook publisher] uses as a consultant. The editors removed Jihad and Sharia from their high school books."

And Dennis Ybarra, co-author of *The Trouble with Textbooks*, reports:

"It's not a level playing field at all in American public schools with the three religions. Muslim beliefs are taught as historical facts whereas Christianity and Judaism are qualified... When they talk about Christianity, *Jesus is believed to be the Messiah*. The tone of skepticism on, *Moses claimed to have received the Ten Commandments*. He could've made the whole thing up. But *the Koran is the revelation received from God by Mohammed*."[506]

So with the aid of Saudi billions, the disgusting, clitoral-hacking false religion of the child-molesting, genocidal monster Mohammad is put forth into the minds of our young people, in our *public* schools no less, as revealed truth from God instead of the filth from Satan that it is; and the true religion based on all the facts and all the evidence, the Jewish-Christian religion, is painted with incredulity, all to the delight of the Democrat-Left elitists who control our educational system and our media. One possibly inconspicuous absence from this section on Islam, though it becomes glaring when pointed out, is the fact that of all the factual and disturbing information found here, virtually none of it came from the reporting of the "mainstream" media. And that is because the so-called "mainstream" media does not consider any of this news. You see, *news* use to actually be news, but now to them news is just a code word for *that which furthers their left-wing, godless, immoral, Christian-hating, baby-mutilating ideology*. And that ideology fits hand in glove with the fascist, demonically-inspired, Christian-and-Jew-hating, clitoral-mutilating, satanic ideology of their Islamic brothers. The Left in this country and the Muslims around this world are just two sides of the same satanic coin. Indeed it not only conveniently fits hand in glove but, as we will establish with ample proof and facts and evidence in the chapter on Satan, the Democrat-Left elites in this country and the Islamic Wahhabists waging jihad (both by violence and by stealth) across this world are truly one and the same. Both are working for the

---

505   (From the *Intercollegiate Studies Institute*, and the Honorable Rick Santorum)
506   (From Fox News Reporting, "Do You Know What Textbooks Your Children are Reading?" 09/04/2009)

same end, destroying America, in order to remake her in one of two of Satan's visions, the Islamic one or the far-left one. Either way it works out just fine for him. Which saddles America with having to fight this Third World War on two fronts. The enemy is among us, some hiding, plotting, making bombs, and waiting to strike, and some quietly working for them in our schools, our media, and now (in 2009)—thanks to the misled, uninformed and thoroughly duped American voter—even in our Federal Government. The same godless, value-less, immoral Left elite that we exposed in Chapter 2, are opening up the floodgates so the world's most hideous false religion—one that has plagued the Middle East, North Africa and the Far East for centuries under its satanic, dictatorial, woman-subjugating, genital-mutilating and imbecilic grip—is now free to hawk its hell-enlarging snake oil to America's unsuspecting, naïve and gullible youth in our colleges, universities, elementary and high schools. The truly sad thing is that many of these Muslims across this world, and many of our students, actually might "have ears to hear" the truth if ever given the chance. But they cannot hear. The truth is being religiously kept from them by Islamic fanatics on the one hand and left-wing fascists on the other. Instead of being given the chance of hearing the Truth of the love of Jesus Christ, and the free gift of eternal life that he offers, our youth are being fed the lies and distortions and inanities of the far Left elites on the one hand, and of Mohammed on the other. And we'll use that as another excuse for the apparent disgust we feel towards an electorate that continues to march out religiously every two years and vote for these liars, these distorters of the truth, these Democrat-Left elites who proudly (in their supreme ignorance) condemn so many of our uninformed youth to not only a life, but an eternity, without Christ. (But at least we can be thankful that our girls still have their clitorises.) (For now.) (Allah Akbar!)

"As I live, says the Lord GOD, I have no pleasure in the death of the wicked, but that the wicked turn from his way and live. Turn, turn from your evil ways! For why will you die?" (Ezekiel 33:11)

It is right to hate Islam, but not Muslims. It is right to hold false religion in complete contempt, but not those who are trapped in it. These poor unfortunate people are stuck in the world's most vile, disgusting religion. Even if they are one of the lucky ones who didn't get their genitals mutilated in this life they are still being guided into hell for all of eternity in the life to come. And nobody has the guts, or very few people have the guts, to help them get out of it. Bill O'Reilly doesn't have the guts. He's hiding under his desk. He thinks it is "a legitimate religion." But he is desperately wrong, because Muslims are in desperate need of hearing the words of truth, not lies that tell them how legitimate their manure pile is. They are in desperate need of hearing of the incomparable love that the real God has for them. A love so great He came to earth in the form of one of us in order to offer himself as the perfect sacrifice for our sins on that cross at Calvary. Muslims need to know that "neither is there salvation in any other, for there is no other name under heaven given among men whereby we must be saved." (Acts 4:12) And their leaders who would rather kill their people than allow them to hear the words of truth need also to know that, despite all of their self-righteous, "godly" pretensions, "he who is of God hears God's words; therefore the reason you do not hear is because you are not of God. You belong to your father the devil, and the desires of your father you want to do. He was a murderer from the beginning, and did not hold to the truth, because there is no truth in him. When he speaks a lie, he speaks his native language, for he is a liar and the father of it." (John 8:47, 43-44)

Our Muslim brothers and sisters need to wash themselves clean of this putrid, clitoral-mutilating religion of Beelzebub. They need to wash their minds clean with the truth of the Bible, the real Holy Book, and they need to wash their hearts clean with the love of Jesus Christ. The time is ripe for Muslims to be led from the darkness of the kingdom of the father of lies, to the light of the glorious gospel of Jesus Christ. What else is there but an eternity of sheer agony in hell?

"And anyone not found written in the Book of Life was cast into the lake of fire." (Revelation 20:15)

(And, just as a side note, isn't it true that our illustrious president was born Muslim, somewhere, and then raised Muslim—<u>not</u> American—in Indonesia? And did he ever renounce the clitoral-mutilating religion of his youth? No? Hmm…)

For further reading on the truth about Islam see:
- *They Must Be Stopped: Why We Must Defeat Radical Islam and How We Can Do It*, by Brigitte Gabriel
- *A God Who Hates: The Courageous Woman Who Inflamed the Muslim World Speaks Out Against the Evils of Islam*, by Wafa Sultan
- *Unveiling Islam*, by Ergun and Emir Caner
- *The Truth About Islam*, by Anees Zaka & Diane Coleman
- *Because They Hate: A Survivor of Islamic Terror Warns America*, by Brigitte Gabriel
- *Leaving Islam: Apostates Speak Out*, by Ibn Warraq
- *The Life and Religion of Mohammed*, by Reverend J. L. Menezes
- "Jihad in America! Militant Islam: The Religious War Against America," by Martin Mawyer
- *Answering Islam: The Crescent in Light of the Cross*, by Abdul Saleeb and Norman L. Geisler
- "What Americans Need To Know About Jihad," by Robert Spencer
- *The Politically Incorrect Guide to Islam (and The Crusades)*, also by Robert Spencer
- *Unholy Alliance: Radical Islam and the American Left*, by David Horowitz
- *The Grand Jihad: How Islam and the Left Sabotage America*, by Andrew C McCarthy
- *They Must Be Stopped: Why We Must Defeat Radical Islam and How We Can Do It*, by Brigitte Gabriel
- *A God Who Hates: The Courageous Woman Who Inflamed the Muslim World Speaks Out Against the Evils of Islam*, by Wafa Sultan
- *Militant Islam Reaches America*, by Daniel Pipes
- *The Terrorist Next Door: How the Government is Deceiving You About the Islamist Threat*, by Erick Stakelbeck
- *Marked for Death: Islam's War Against the West and Me*, by Geert Wilders
- Or go online to David Horowitz's *The Terrorism Awareness Project*; *Jihad Watch*, a project of the David Horowitz *Freedom Center*; The *Intercollegiate Studies Institute*; or contact *Coral Ridge Ministries Media, Inc.* Conversely, if you would prefer to be indoctrinated into the <u>lies</u> about Islam, contact The *Council on American-Islamic Relations (CAIR)*; show up at virtually any American university or public school; or try listening to our "mainstream" media, if you have the

stomach for it.

* * *

In conclusion

"This then is the message we have heard from him, and declare unto you, that God is light, and in him there is no darkness at all. If we say that we have fellowship with him, and yet walk in darkness, we lie, and do not have the truth: But if we walk in the light, as he is in the light, we have fellowship one with another, and the blood of Jesus Christ his Son cleanses us from all sin." (1 John 1:5-7)

Those trapped in false religions across this globe think that they are worshipping God, but in fact they are walking in darkness.

"In vain they do worship me, teaching for doctrines the commandments of men." (Matthew 15:9)

Their righteousness, their worship, their religiosity, their good works and their sacrifices are all in vain, unless they hear the word of truth and turn from Satan's falsehoods to the wonderful gift of eternal life through Jesus Christ, the Son of the living God.

"Let us hear the conclusion of the whole matter: Fear God, keep his commandments, and depart from evil [which includes false religion], for this is the whole duty of man." (Ecclesiastes 12:13 and Proverbs 3:7)

False religions are the devils curse upon the peoples of this world. False religions are the most disgusting things on this planet. They are hell vacuums, sucking their unsuspecting adherents down to an eternity in the Lake of Fire. But they are protected from exposure because they are "sacred." And anyone who dares to expose them is said to be "offensive." But they are not sacred. They are demonic. And all those brave souls and missionaries across this globe who risk their lives and the safety of their families on a daily basis to expose them are not offensive. They are courageous. Let all religions be held up to the light of truth, so the false ones can be exposed for the garbage that they are. In Jesus' name, I pray.

See:
- "The Trail of Blood," by J. M. Carroll
    - *The Unveiling: A Journey Through the Book of Revelation*, by Keith Harris
    - *Foxe's Book of Martyrs* by John Foxe
    - *Coming Alive in Christ*, by John MacArthur & Grace To You Ministries
    - *Faith Alone*, by Dr. R. C. Sproul, Renewing Your Mind
    - *The Spirit of Truth and the Spirit of Error*, by Keith L. Brooks
    - *Preparing Catholics for Eternity*, by Mike Gendron
    - *The Christian Research Institute* @www.equip.org
    - "Christ and the Pope Contrasted," by Wendell Holmes Rone
    - "List of Catholic Heresies and Human Traditions- Adopted and Perpetrated by The Roman Catholic Church in The Course of 1600 Years," Compiled by Rev. Stephen L. Testa (www.jesus-is-savior.com)
    - *How to answer a Jehovah's Witness: How to Successfully Take the Initiative When They come to Your Door*, by Robert A. Morey

- *The Method of Grace*, by George Whitefield
- *Godless: The Church of Liberalism*, by Ann Coulter
- Wide Is the Gate: The Emerging New Christianity, a DVD by Video Journalist, Caryl Matrisciana
- *Speaking with Tongues*, by John R. Rice
- *The Truth about Tongues and the Charismatic Movement*, by Hugh F. Pyle
- *Don't Drink the Kool-Aid: Oprah, Obama [and the] Occult*, by Carrington Steele
- *The Reincarnation Sensation*, by Norman L. Geisler and J. Yutaka Amano
- *Lord of the Air: Tales of a Modern Antichrist*, by Tal Brooke
- *There are Moral Absolutes*, by Michael A. Robinson
- *The Politically Incorrect Guide to Islam (and the Crusades)*, by Robert Spencer
- *Bias: A CBS Insider Exposes How the Media Distort the News*, by Bernard Goldberg
- *The Truth About Islam*, by Anees Zaka & Diane Coleman

CHAPTER 9

BIBLE PROPHECY

("It's the end of the world as we know it ...and I feel fine.")

"Remember the former things of old: for I am God, and there is none else; I am God, and there is none like me, Declaring the end from the beginning, and from ancient times the things that are not yet done, saying, my counsel shall stand, and I will do all my pleasure." (Isaiah 46:9-10)

God knows exactly what is going to happen in the future, down to the minutest detail. Obviously. Otherwise he wouldn't be God. (He would be Nostradamus.) He has known from before the beginning of time. For us, foreseeing the future is an amazing ability completely beyond our reach. But for God it is no more difficult than it would be for one of us to *foretell* the ending of a movie we've already seen. For His sight is not limited like ours to the present and the past. His gaze also peers into the future, as far as His eye can see; which is as far as time exists. In fact, today he could tell us, not just future events, but absolutely everything that will happen tomorrow down to even the minutest of details. Every leaf that will fall and the exact moment down to the millionth of a second when each will leave the branch, and when each will touch the ground. The exact movement and location in space at any moment in time tomorrow of every one of the trillions times trillions of grains of sand on the shores and ocean beds and deserts and sand boxes and kitchen floors of this world. The undulation of every wave in all the lakes and rivers and streams and oceans and swimming pools and bath tubs, and their exact size, amplitude, speed, salinity and pollutant levels at any given moment in time tomorrow. The shape and form and changing nature of every cloud at any moment tomorrow across the entire globe. And you can extrapolate that to include everything that will occur tomorrow in this entire infinite universe, and on every other day until the end of time. It's what is known as all-knowledge. It's one of the attributes of the Creator.

With that in mind I hope it is evident what an entirely puny exercise it is for that same Creator to know exactly what will happen tomorrow in the affairs of men. And the day after tomorrow. And the day after that. And on and on. In fact from the very beginning of his creation God knew exactly what was going to happen on every single day in the affairs of man forever. Because he wrote the book. (And that is *not* an endorsement for *hyper-Calvinism* which, "simply stated, is a doctrine that emphasizes divine sovereignty to the exclusion of human responsibility."[507] In other words, just because he knows what each one of us will choose to do, it is still our choice.) All of life—present, past and future—is not only His story, because he wrote it, but to God even the future is history, because to him it has already happened. Six thousand years into the past and a thousand years into the future from our teensy weensy little perspective are for God happening at the same time, or have already happened. (Depending upon how you want to look at it.) Time is not an absolute like truth or love. Time is a creation, like the universe itself. God

---

507   (From article, "A Primer on Hyper-Calvinism," by Phillip R. Johnson @ spurgeon.org)

created both space and time, and he exists outside of the confines of both.[508] And he is not limited or contained by either. Therefore, foretelling the future in an exact and absolutely accurate manner is no more difficult or strenuous for the Lord, than it would be for you to tell us what you had for lunch yesterday.

"The Lord of hosts has sworn, saying, surely as I have thought, so shall it come to pass; and as I have purposed, so shall it stand:" (Isaiah 14:24)

I point all of this out because before we can even look at the subject of Bible prophecy, we must first be able to grasp the concept that the future <u>can</u> be <u>known</u>. At least by the Creator. And then it becomes an easier thing to grasp that the future can also be known by man, if it is the will of that same Creator to reveal it to us. And it turns out that it is his will to have done just that. His book is full of many hundreds, even thousands, of prophecies. Many of which have already come to pass exactly as they were foretold. And a number of others that are still yet to come which have a direct bearing on this "time of the end" in which we live. These remaining prophecies clearly outline a series of major world-wide events, centered on the conflict over Jerusalem and the nation of Israel, which must shortly come to pass.

> "Some of you may think, as I have heard others say, that *prophecy* has no personal bearing on what's happening today. Well if ever there was a day when that was not any longer possible for us to say, we're living in that day right now, because the prophetic word has everything to do with what is happening to us today.
>
> "Prophecy is telling the future before it happens. In the prophecies of the Bible there are two ways that we are blessed as we study it. First of all, to know ahead of time what's going to happen. That's one thing. But secondly, to be able to see the prophecies that were made way back in the beginning and to follow them through and to see how they were literally fulfilled. Historically we can trace many of the prophecies, in fact, all of the issues of the first coming of Christ to this earth were prophesied before He came. And we're able now, looking back in history, to see that they were literally fulfilled. But the blessing of that goes even beyond tracing the fulfillment of prophecy through history. It is the blessing of knowing that the God Who said back there that it would happen here, and it happened, is the same God Who is speaking here and telling us this is going to happen out there. If God spoke truthfully in the beginning and it was fulfilled, you can be sure that what God is telling us of the future will also be fulfilled. Bible prophecy always comes true... the standard of prophecy in the Bible is it is 100% fulfilled."[509]

"Know this first, that no prophecy of Scripture is of any private interpretation. For the prophecy came not in old time by the will of man: but holy men of God spoke as they were moved by the Holy Spirit." (2 Peter 1:20,21)

Compared to any other time in history, global events are unfolding now at a frenzied pace. Calamitous and earth shattering events in the affairs of men that in times past would have filled an entire century, now take place every decade, and even every few years. At this writing in the Spring of 2010 we just had the gulf oil spill, preceded by world-wide economic collapse, and the horrific earthquake in

---

508　("Time has come today. Hey! And I'm thinking about the subway...")

509　(From, *What in the World is Going On: 10 Prophetic Clues You cannot afford to Ignore, Chapter 1, The Israel Connection*, by Dr. David Jeremiah)

Haiti followed by the terrible earthquake in Chile, and 5 years ago there was the catastrophic Indonesian tsunami, and before that we had 9/11. And, unfortunately, the Bible says that terrible events like these are only going to continue to increase in frequency until the Lord returns and the earth is cleansed. There is no epic movie adventure yet imagined that can possibly compare with the shock and awe of just watching the news unfold in the next number of years. (And this is not to be callous or unsympathetic towards all of those of us who will be adversely affected, even killed.) Indeed, to those who don't understand the meaning and the purpose behind these events, they will appear random, senseless and impossible to fathom. They will also be a source of great fear. (And now, in March of 2011, we have the record-breaking Japanese earthquake with its catastrophic tsunami that in a matter of minutes wiped hundreds of square miles—entire towns, farmlands and cities—from off the face of the earth.) (And then just a month or so later, at the end of April, we have the record-breaking tornado outbreak in the Southeast that wiped entire swaths of houses and towns from off the face of the earth.) In fact the Bible says that in the years soon to come the distress will be so great that men will drop dead from fear...

"And there shall be signs in the sun, and in the moon, and in the stars; and upon the earth distress of nations, with perplexity; the sea and the waves roaring [tsunamis]; Men's hearts failing them for fear, and for looking after those things which are coming on the earth: for the powers of heaven shall be shaken. And then shall they see the Son of man coming in a cloud with power and great glory." (Luke 21:25-27)

But for those who have accepted Jesus' free gift of eternal life, for those who know where they are headed, and in particularly for those true Christians who have also taken advantage of the foreknowledge that God's Word freely offers, what is happening now, and the events soon to unfold, will not only make perfect sense, but they will herald the coming of a time of righteousness and sanity and prosperity and peace and love beyond anything man has had the pleasure to experience since Adam and Eve were cast out of the garden.

"Blessed is he who reads, and they who hear the words of this prophecy, and keep those things which are written therein, **for the time is at hand**" (Revelation 1:3)

We tend to think of the Bible's Old Testament and its historical clashes between good and evil, between God and the devil, between God's chosen people and Satan's wicked nations, as history. We tend to think of all of it as having happened a long, long time ago (in a continent far, far away), and as having no real meaning or relevance to our lives, or to what is going on now. But the truth of the matter is that we have never really left the Old Testament as far as that ancient struggle is concerned. It's still going on. It has never stopped. We're actually still living in "biblical times." We're just not, as a culture, paying very close attention. (Too busy sucking those lies from Beelzebub's teat.) And according to Bible prophecy the culmination of this ancient war will happen very soon. And the focal point of this age old struggle is still the tiny nation of Israel and God's Holy City of Jerusalem, just as it was thousands of years ago.

The End of the Age

In the beginning when God created all things he worked for six days, and then he rested on the seventh day. "For in six days the Lord made heaven and earth, the sea, and all that in them is, and rested the seventh day: wherefore the Lord blessed the Sabbath day, and hallowed it." (Exodus 20:11) This is not because he needed six days to do the creating or because he got tired afterwards, but merely because that was the exact time frame that he chose. (It has something to do with mathematical perfection.) And there is a prophetic correlation between the time-line he chose for the creation of the world and the time-line he chose for human history. "But, beloved, be not ignorant of this one thing, that one day is with the Lord as a thousand years, and a thousand years as one day." (2 Peter 3:8) The early church, the body of believers in the first few hundred years following the resurrection and ascension into heaven of the Lord Jesus, understood that after six thousand years (six *days* of work) had elapsed from the beginning[510] of the creation of the universe that the end of the world as we know it—the end of this age—would occur. And that following this the world would enter into a thousand year period of peace (the *day* of rest) brought about by the triumphant return of Jesus Christ to the earth; his *Second* Coming. This time though, he isn't coming as a Lamb to the slaughter but as an all-powerful King to retake possession of the earth, to reassert his sovereign ownership and rule over this planet, and to cast aside the rule of Satan, the god of this world, who has been exerting his power and has been in partial control since the fall of man. Adam and Eve were given dominion over all the earth, but they relinquished that dominion to Satan when they sinned against God by obeying the lies of the great deceiver over the truth of Almighty God.

"And God said, Let us make man in our image, according to our likeness: and let them have dominion over the fish of the sea, and over the fowl of the air, and over the cattle, and over all the earth, ...and over every living thing that moves upon the earth." (Genesis 1:26, 28) (When we fell, that dominion was appropriated by the evil one who then became the god of this world.)

From the time of the Creation of the world to the time of Jesus of Nazareth 4,000 years had passed. From the time of Jesus to today 2,000 more years have come and gone. That's 6,000 years, and that alone can be taken as a prime indicator that the end of this age is very near. But there are a number of other indicators that are also significant...

The Fig Tree

Throughout the Bible, many symbols are used to refer to different things. They include plants and elements, animals and weapons, numbers and colors, food and drink, clothing and jewels, shelters and parts of the body. Here are just a few examples:

"Water - represents truth." "But whosoever drinks of the water that I shall give him shall never thirst. But the water that I shall give him will become in him a fountain of water springing up into everlasting life." (John 4:14)

"Sea - represents restless masses of society, usually unruly and lawless." "But the wicked *are* like the troubled sea, when it cannot rest, whose waters cast up mire and dirt." (Isaiah 57:20)

..."Wheat - represents the children of the Truth. In the parable of the wheat and the tares [Matthew 13:24], Jesus said the good seed [the wheat] are the children of the

---

510   (as in "In the beginning")

Kingdom, the product of Truth which the sower had sowed.

"Tares - represent children of error. Tares stand tall and proud, while wheat bow humbly. Tares try to pose as true Christians, but are imitations. Tares are the product of false teachings.

"Barn - represents the place where the ripe wheat is taken, away from the field and the tares; ultimately, heaven itself. The wheat is gathered from worldly surroundings to be with other wheat. True Christians gather together to study God's Word. In the final sense, the wheat, true Christians, are garnered to the heavenly barn. [Which will be the restored Garden of Eden covering the entire earth.]

"Reapers - represent God's servants. Reapers are spirit begotten Christians who are gathering the wheat [other faithful Christians] into His barn [separate from worldly interests, and finally into heaven itself]." "Then he said unto his disciples, The harvest truly *is* plentiful, but the laborers *are* few. Pray you therefore the Lord of the harvest, that he will send forth laborers into his harvest." (Matthew 9:37-38)

…"Rain - represents blessings." "The LORD shall open unto you his good treasure, the heavens to give the rain to your land in its season, and to bless all the work of your hand." (Deuteronomy 28:12) (See also, Matthew 5:45, Isaiah 55:10, Deuteronomy 32:2, Leviticus 26:3-4 and Exodus 16:4)

…"Hand - represents power." "Therefore humble yourselves under the mighty hand of God, that he may exalt you in due time." (1 Peter 5:6)

…""7" represents heavenly perfection. Seventh day is God's commanded Sabbath. There were seven creative days… There are seven colors in the rainbow."

…"Wine - represents doctrine." "And no man puts new wine into old wineskins; or else the new wine will burst the wineskins and be spilled, and the wineskins will be ruined." (Luke 5:37)

"Bread - represents life-sustaining, Jesus or Truth." "And Jesus said unto them, I am the bread of life. He who comes to me shall never hunger, and he who believes in me shall never thirst." (John 6:35) "Therefore let us keep the feast, not with old leaven, neither with the leaven of malice and wickedness, but with the unleavened *bread* of sincerity and truth." (1 Corinthians 5:8)

"Famine - represents lack of spiritual nourishment." "Behold, the days are coming, says the Lord God, that I will send a famine on the land, not a famine of bread, nor a thirst for water, but of hearing the words of the Lord." (Amos 8:11) [That famine has already hit.]

"White - represents purity." "Come now, and let us reason together, says the LORD, Though your sins be as scarlet, they shall be white as snow." (Isaiah 1:18) "Purge me with hyssop, and I shall be clean: wash me, and I shall be whiter than snow." (Psalms 51:7)

…"Silver - represents the truth." "The words of the LORD *are* pure words, *like* silver tried in a furnace of earth, purified seven times." (Psalms 12:6)

"…Linen - represents righteousness." "And to her it was granted that she should be arrayed in fine linen, clean and white, for the fine linen is the righteousness of the saints." (Revelation 19:8)

"Crown - represents immortality." "Be thou faithful until death, and I will give you the crown of life." (Revelation 2:10) "And when the chief Shepherd shall appear, you shall receive a crown of glory that does not fade away." (1 Peter 5:4) "Everyone who competes for the prize undergoes diligent training. Now they *do it* to obtain a corruptible crown; but we an incorruptible." (1 Corinthians 9:25)

"Robe - represents righteousness." "He has clothed me with the garments of salvation, he has covered me with the robe of righteousness…" (Isaiah 61:10)

"Sword - represents the Word of God." "…the sword of the Spirit, which is the word of God." (Ephesians 6:17) "For the word of God *is* quick, and powerful, and sharper than any two-edged sword..." (Hebrews 4:12)

...."Lamps - represent the Holy Scriptures." "Your word *is* a lamp unto my feet, and a light unto my path." (Psalms 119:105)

...."Lamb - represents Jesus and the Church." "And looking upon Jesus as he walked, he said Behold the Lamb of God!" (John 1:36) "I am the good shepherd, and I know my *sheep*, and am known by my own." (John 10:14)

"Serpent - represents Satan." "Then the serpent said to the woman, You will not surely die." (Genesis 3:4) "And he laid hold on the dragon, that serpent of old, which is the Devil, and Satan, and bound him a thousand years..." (Revelation 20:2)[511]

And then there is the *fig tree*. God uses it in his Word as a symbol for the nation of Israel. "[Jesus] also spoke this parable; A certain man had a fig tree planted in his vineyard, and he came seeking fruit on it, and found none. Then he said to the keeper of his vineyard, Behold, these three years I have come seeking fruit on this fig tree, and have found none. Cut it down; for why does it encumber the ground?" (Luke 13:6-7) For three and a half years of his earthly ministry Jesus performed countless miracles and spoke eternal words of truth and salvation to the people and to the leaders of Israel, but as a nation they refused to accept him as their Messiah. The fig tree refused to bear its fruit.

And long before the time of Christ a prophecy was made that the nation of Israel would be reborn in 1948. This was foretold even down to the very day.[512] And on May 14th, 1948 the state of Israel declared its independence...

"Now learn this parable from the fig tree; When its branch is yet tender, and puts forth leaves, you know that summer is nigh. So you also, when you shall see all these things, know that it is near, even at the doors. Verily I say unto you, this generation shall by no means pass away, till all these things be fulfilled." (Matthew 24:32-34) This is a prophecy of the nation of Israel coming back to life, which it did miraculously in 1948. (But notice that where the fig trees branches put forth leaves- the nation is reborn, it still does not bear fruit- it doesn't yet recognize Jesus as their Messiah. That won't happen as a nation until the end of the Great Tribulation.)

Jesus said in the above quote from Matthew that within one generation of the rebirth of the nation of Israel that the end times would come to pass. A generation is anywhere from 40 to 70 years. We are that generation.

[However, although the Jewish nation was re-born in 1948, they did not re-possess Jerusalem until the six-day war in 1967. And Jerusalem is the heart and soul of the Jewish nation. It is the center of worship with its Temple Mount and its place opposite the Eastern gate where the bedrock is visible and over which the Holy of Holies of the first two temples stood, and over which the Holy of Holies of the third or millennial temple must be built. So the prophetic timeline probably started from that year.]

---

511   (This partial list of symbols and their accompanying bible quotes were taken from "Footnote on Biblical Symbols" @ revelation-today.com. Copyright 2001 John Class. For the full list go to their website.)

512   (See Ezekiel 4:3-6, and for the interpretation see the third chapter, titled "Ezekiel's Vision of the Rebirth of Israel in 1948," of Grant R. Jeffrey's book, *Armageddon: Appointment with Destiny*)

This rebirth of the nation of Israel from a people that had been scattered across the face of the earth for nearly 2,000 years is unprecedented in human history. "Who has heard such a thing? Who has seen such things? Shall the earth be made to give birth in one day? Or shall a nation be born at once?" (Isaiah 66:8)

"Nowhere else in history were a people who were scattered, disassembled, removed from their nation, assimilated into multitudes of other nations on the face of the earth, were ever gathered back together once again, in their original land, and established as a recognized nation in the world."[513]

And in the book of Ezekiel we find this prophecy…

"The hand of the Lord came upon me and brought me out in the spirit of the Lord, and set me down in the midst of the valley; and it was full of bones. Then he caused me to pass by them all around: and, behold, there were very many in the open valley; and indeed they were very dry. And he said unto me, Son of man, can these bones live? So I answered, O Lord God, you know. Again he said to me, Prophesy to these bones, and say to them, O dry bones, hear the word of the Lord! Thus says the Lord God to these bones; Surely I will cause breath to enter into you, and you shall live. And I will lay sinews upon you and bring flesh upon you, and cover you with skin, and put breath in you; and you shall live. Then you shall know that I am the Lord. So I prophesied as I was commanded: and as I prophesied, there was a noise, and suddenly a shaking, and the bones came together, bone to bone. Indeed, as I looked, the sinews and the flesh came upon them, and the skin covered them over: but there was no breath in them. Then he said to me, Prophesy unto the wind, prophesy, son of man, and say to the wind, Thus says the Lord God; Come from the four winds, O breath, and breathe upon these slain, that they may live. So I prophesied as he commanded me, and the breath came into them, and they lived, and stood up upon their feet, an exceeding great army. Then he said to me, Son of man, these bones are the whole house of Israel. They indeed say, Our bones are dry, and our hope is lost, and we ourselves are cut off! Therefore prophesy and say to them, Thus says the Lord God: Behold, O my people, I will open your graves and cause you to come up out of your graves, and bring you into the land of Israel." (Ezekiel 37:1-12)

Never before in this world's history had a people and a nation that were exiled and persecuted and scattered to the four corners of the earth for thousands of years, been reborn in the land of their inheritance. There are many historical examples of nations of people that once were distinct and thriving, and yet now little or no trace of them can be found. The rebirth of the nation of Israel in 1948 can truly be spoken of as a miracle in man's history. Yet it occurred exactly as the Bible foretold.

The Rapture

But there is another unprecedented event, also prophesied by God's Word that is still to happen in the near future. And that is the rapture of the true Church of Jesus Christ. When all those who have trusted in him, all those who are *known* by him, will be caught up in the air to meet the Lord.

"For this we say to you by the word of the Lord, that we who are alive and remain until the coming of the Lord will by no means precede those who are

---

513   (Nick Triveri, Biblical teaching and Application, on The Final Prophecies DVD, an Ingenuity Films Production)

asleep. For the Lord himself will descend from heaven with a shout, with the voice of an archangel, and with the trumpet of God. And the dead in Christ will rise first. <u>Then we who are alive and remain shall be caught up together with them in the clouds to meet the Lord in the air</u>, and so shall we ever be with the Lord." (1 Thessalonians 4:15–17)

"Then two men will be in the field; the one shall be taken, and the other left. Two women shall be grinding at the mill; the one shall be taken and the other left." (Matthew 24:40-41)

"Behold, I tell you a mystery; We shall not all sleep, but we shall all be changed, in a moment, in the twinkling of an eye, at the last trumpet. For the trumpet will sound, and the dead shall be raised incorruptible, and we shall be changed. For this corruptible must put on incorruption, and this mortal must put on immortality. So when this corruptible has put on incorruption, and this mortal has put on immortality, then shall be brought to pass the saying that is written, 'Death is swallowed up in victory.'" (1 Corinthians 15:51-54)

> "The Rapture is a glorious event which God has promised to the Church. The promise is that someday very soon, at the blowing of a trumpet and the shout of an archangel, Jesus will appear in the sky and take up His Church, living and dead, to Heaven. The term, Rapture, comes from a Latin word that means to catch up, to snatch away, or to take out. It is a Biblical word that comes right out of the Latin Vulgate translation of the Bible. The word is found in 1 Thessalonians 4:17. ...The concept of the Rapture was not revealed to the Old Testament prophets because it is a promise to the New Testament Church and not to the saints of God who lived before the establishment of the Church."[514]

Yet the idea of being raptured or caught up bodily from this earth into heaven was not unknown in the Old Testament. In Genesis we find the example of Enoch... "And Enoch walked with God; and he was not, for God took him." (Genesis 5:24) And then later Elijah was taken up into heaven in a whirlwind... "And so it was, when they had crossed over, that Elijah said to Elisha, 'Ask! What may I do for you, before I am taken away from you?' Elisha said, 'Please let a double portion of your spirit be upon me.' So he said, 'You have asked a hard thing. Nevertheless, if you see me when I am taken from you, it shall be so for you; but if not, it shall not be so.' Then it happened, as they continued on and talked, that suddenly a chariot of fire appeared with horses of fire, and separated the two of them; and Elijah went up by a whirlwind into heaven. And Elisha saw it, and he cried out, 'My father, my father, the chariot of Israel and its horsemen!' So he saw him no more." (2 Kings 2:9-12)

There are also these two verses from the New Testament:

"In my Father's house are many mansions; if it were not so, I would have told you. I go to prepare a place for you. And if I go and prepare a place for you, <u>I will come again and receive you to myself</u>; that where I am, there you may be also." (John 14:2–3)

"For our citizenship is in heaven, <u>from which we also eagerly wait for the Savior</u>, the Lord Jesus Christ, <u>who will transform our lowly body that it may be conformed to his glorious body</u>, according to the working by which he is able even to subdue all things to himself." (Philippians 3:20-21)

And in the last book of the Bible we find a prophetic allusion to the rapture:

"After these things I looked, and behold, a door was opened in heaven.

---

514   (From "The Rapture of The Church," by David Regan, on prophecyfellowship.org)

And the first voice which I heard was like a trumpet speaking with me, saying, 'Come up here, and I will show you things which must take place hereafter.' And immediately I was in the spirit...in heaven." (Revelation 4:1,2)

"John's ascent into heaven is a type (forerunner, example) of the "calling out" of the Church (born-again believers). John heard the sound of the trumpet and was immediately taken (v.2) Likewise the first sound we will hear will be the trump of God, and we will immediately be taken (I Cor. 15:52; I Thess. 4:16)."[515]

There is also this from Patrick Heron, author of the book, *Return of the AntiChrist and the New World Order*, who states that the correct translation of 2 Thessalonians 2:3 is not a "falling away" but a "departure... "Let no man deceive you by any means: for that day shall not come, except there come *a departure* first, and that man of sin be revealed, the son of perdition." (2 Thessalonians 2:3)[516] And clearly that "departure" is referring to the disappearance of the elect from this earth.

When will the rapture occur? "But of that day and hour **no one knows**, not the angels of heaven, **but my Father only**. But as the days of Noah were, so also will the coming of the Son of Man be. For as in the days before the flood, they were eating and drinking, marrying and giving in marriage, until the day that Noah entered the ark, and they did not know until the flood came and took them all away; so also will the coming of the Son of Man be. Then two men will be in the field; the one shall be taken, and the other left. Two women shall be grinding at the mill; the one shall be taken and the other left. Watch therefore: for you do not know what hour your Lord is coming." (Matthew 24:36-42)

"The most controversial aspect of the Rapture is its timing. Some place it at the end of the Tribulation, making it one and the same event as the Second Coming. Others place it in the middle of the Tribulation. Still others believe that it will occur at the beginning of the Tribulation. "The reason for these differing viewpoints is that the exact time of the Rapture is not precisely revealed in scripture. It is only inferred. There is, therefore, room for honest differences of opinion, and lines of fellowship should certainly not be drawn over differences regarding this point, even though it is an important point."[517]

No one knows the day or the hour when the rapture will occur but I think the best interpretation of the prophecies have it occurring before the final tribulation period, the last seven year period of this present fallen world, which culminates with the Second Coming of the Lord. For when the Rapture does occur and the church is removed, the restraining presence and power of the Holy Spirit working through that church, will also be removed. (The Holy Spirit of course will still be here on earth as God is ever present everywhere, and He will still be bringing people to Christ even during the Great Tribulation.) At that point Satan and his followers will be free to truly run rampant and wreak havoc upon the earth and its inhabitants for a short, seven-year period of time.

There are a number of verses that bolster the pre-tribulation position for the rapture. "Jesus who delivers us from the wrath to come." (1 Thessalonians 1:10) "The wrath to come" being interpreted as the seven year tribulation period at the end of this age, and not hell. Hell is not *coming* in the future, it has been there for

---

515   (Pg. 108, *The Unveiling: A Journey Through the Book of Revelation*, by Keith Harris)
516   (Pgs. 156-158, *Return of the AntiChrist and the New World Order*, by Patrick Heron)
517   (From *The Rapture of The Church*, by David Regan, on prophecyfellowship.org)

quite some time. And this: "Because you have kept my command to persevere, I also will keep you from the hour of trial, which shall come upon all the world, to test those who dwell upon the earth." (Revelation 3:10) Clearly Jesus is promising to rapture his believers *before* the great tribulation, "the hour of trial." And this as well: "Therefore be you also ready: for in such an hour as you think not the Son of man will come." (Matthew 24:44) "The only time frame I can think of when we believers would not be expecting Jesus to return would have to be before the tribulation."[518] And that is because if the rapture does not occur before the Great Tribulation and we as Christians find ourselves in that clearly delineated period of time, we would then know for certain that it must happen during it or at the end of it, and none of us would be thinking not that the Son of man is coming.

And then there is this: "Ephraim the Syrian said, in 373 AD, 'For all the saints and Elect of God are gathered, **prior to the tribulation that is to come**, and are taken to the Lord **lest they see the confusion that is to overwhelm the world** because of our sins.'"[519] (Emphasis mine) So, hey, if the pre-trib rapture was good enough for old Ephraim the Syrian, then it's good enough for me. Seriously though, this is a subject where true Christians can hold varying opinions. (Unlike the subjects of the deity of Christ, his virgin birth, his countless miracles, his Resurrection from the dead and his ascension into heaven, where other "varying opinions" which deny these can be held, but only by faux Christians deceived by the father of lies.)

What would be the worldwide effect, though, of such a disappearance of so many people? Well, certainly in many families, small towns and small businesses the loss would be profound, and answers would be sought. And I'm sure that many who for whatever reason refused Christ in the past would be a tad more receptive to the gospel message, only there would be no one left to tell them. But there would be books and tapes and salvation pamphlets and movies left behind that people could easily seek out and find Jesus' words of comfort and truth during this disturbing time. (Perhaps you are one in the near future who was left behind, and are reading these words now!?) (No way!) (Yes, way.) Also, there will be many who have heard the gospel message but sat on the fence without responding to Christ who will then make the decision and be available to lead others. In addition there will be those Christians who are not saved, who are clothed in their own righteousness and not the righteousness of Christ, many of whom go out and vote for the baby-mutilating Democrats religiously every two years, who hopefully would be woken up to the fact that while we are gone, they are still here, and this would then prompt them to make the decision to put off faith in themselves and their own good works, to accept Jesus as their Lord and Savior, to put on his righteousness and to finally wake up and depart from iniquity, and to therefore become true Christians who could also lead others to the Lord. (Dr. Anderson? Paging Dr. Anderson.)

However, on the other hand, no one would be missing from our elite media, or from the educational staff of our colleges and universities. (Except for the one or two actual Christians they've let slip through the cracks. Ideological miscreants

---

518   (From article "Defending the Pre-Trib Rapture," by Todd Strandburg @ raptureready. com)
519   (From article "Defending the Pre-Trib Rapture," by Todd Strandburg @ raptureready. com)

are not tolerated in the ranks of totalitarian-left McCarthyites. Just like Jews were not tolerated in the ranks of Hitler's Youth.) So these people, who control the dissemination of information in our society, would just spin it any way they wanted; any way that best suited their mind-set which abhors true Christianity to begin with. They might just attribute the whole thing to UFOs and their alien pilots, taking a page out of Richard Dawkins book. This militant evolutionist and village atheist answered the question of where life came from in the first place (which is absolutely unanswerable by Darwinists) by saying that he thought it was planted here by aliens travelling from another galaxy.[520] Unfortunately Mr. Dawkins failed to comprehend that he wasn't actually answering the question; he was just punting it to another galaxy… "OK then, Richard, where did life come from in the first place on THAT galaxy?"

With that in mind, our media might spin the disappearance something like this: "Alien beings, the source of life on earth according to our scientists, have been watching over our planet for some time now (as many of our government officials already know). Last night they came and beamed up millions of fundamental Christians in order to take them back to laboratories on their home planet, possibly one Kolob that Mormons are familiar with, to be reprogrammed. As we all know these Christians were the cause of all of our problems. [This they will steal from the Nazi's and their Big Jewish Lie.] They were destroying our environment by denying global warming, and they mostly voted for those mean-spirited, obstructionist Republicans. They also kept harassing our Muslim friends, and they actually spanked their children while trying to shove their morality on the rest of us. Our Alien creators knew this better than anyone and could no longer afford to watch their precious and costly experiment of life on earth being destroyed. And now, Praise Algore-Gaya-Mother-Earth![521] we are finally free to achieve peace in our time and to live in harmony." (As long as one thinks it's harmonious to cut off little girl's genitals and torture little babies to death.)[522]

For further reading see: *The Great Mystery of The Rapture*, by Arno Froese.

The battle of Gog and Magog

But before the final seven year tribulation period that will end this age, and which we will cover shortly, another major event will happen, and that is the battle of Gog and Magog. This will be a multinational assault against Israel involving Russia and the Muslim nations (no surprise there). We find the prophecy in Ezekiel:

"Now the word of the Lord came to me, saying, Son of man, set your face against Gog, the land of Magog, the chief prince of Rosh, Meshech and Tubal, and prophesy against him, And say, Thus says the Lord God; Behold I am against you, O Gog, the prince of Rosh, Meshech and Tubal. I will turn you around, put hooks into your jaws, and lead you out, with all your army, horses and horsemen,

---

520   (See Ben Stein's interview with Dawkins at the end of the movie *Expelled: No Intelligence Allowed.* We should "suffer fools gently" but Dawkins' *answer* is something you would expect to hear from an uneducated imbecile, or a six-year old, and not from a well-read and highly educated author. But then blind adherence to an asinine false religion, in this case evolution, has a way of making a fool out of you every time.)

521   (Al Gore will by then be considered a god)

522   (And for a further discussion see "UFO" section later in this chapter)

all splendidly clothed, a great company with bucklers and shields, all of them handling swords. Persia, Ethiopia, and Libya are with them, all of them with shield and helmet: Gomer and all its troops; the house of Togarmah from the far north and all its troops- many people are with you. ...After many days you will be visited. In the latter years you will come into the land of those brought back from the sword and gathered from many people on the mountains of Israel, which had long been desolate; they were brought out of the nations, and now all of them dwell safely. You will ascend, coming like a storm, covering the land like a cloud, you and all your troops and many peoples with you. ...And you will come up against my people Israel like a cloud, to cover the land. It will be in the latter days that I will bring you against my land, so that the nations may know me, when I am sanctified in you, O Gog, before their eyes." (Ezekiel 38:1-6,8-9,16)

> "Ezekiel 38 through 39 tells of a future invasion of Israel by a vast coalition of nations that surround it. As we read the headlines in the newspapers of today, and witness the conflict in the Middle East, it's not hard to imagine that this invasion prophesied over 2600 years ago, could be fulfilled in our lifetime. Ezekiel 36-37 predicts a gathering of the Jews to the nation of Israel, which will be followed by this massive invasion. For 19 centuries the Jewish people were scattered throughout the world, and until May 14, 1948 there was no nation of Israel to invade. With the nation of Israel now a reality, the stage seems set for the war that will usher in the tribulation and the rise of the Antichrist; a war that will end with the destruction of Israel's enemies by God Himself, and lead to the signing of a peace treaty with the Antichrist."[523]

But all of these parties, Gog (the leader of Magog), Magog (Southern Russia and Central Asia), Rosh (Russia), Meshech and Tubal (Turkey), Persia (Iran), Ethiopia, Libya, Gomer (in Turkey), Togarmah (Georgia and Armenia) and the other Arab-Muslim nations involved in this intended genocide will be destroyed by the intervention of the Lord himself:

"Therefore, son of man, prophesy against Gog, and say, Thus says the Lord God: Behold, I am against you, O Gog, the chief prince of Rosh, Meshech and Tubal: and I will turn you around and lead you on, bringing you up from the far north, and will bring you against the mountains of Israel. Then I will knock the bow out of your left hand, and cause the arrows to fall out of your right hand. You shall fall upon the mountains of Israel, you and all your troops and the people who are with you: I will give you to the birds of prey of every sort and to the beasts of the field to be devoured. You shall fall upon the open field: for I have spoken it, says the Lord God. And I will <u>send fire</u> on Magog and on those who live in security in the coastlands. Then they shall know that I am the Lord. And it shall come to pass in that day, that I will give Gog a burial place there in Israel, the valley of those who pass by east of the sea: and it shall stop the noses of the travelers: because there they will bury Gog and all his multitude. Therefore they will call it The valley of Hamon Gog. **For seven months the house of Israel will be burying them, that they may cleanse the land. Indeed all the people of the land will be burying them**; and they will gain renown for it on the day that I am glorified, says the Lord God." (Ezekiel 39:1-6,11-13)

---

523   (From article "The Coming War of Gog and Magog, An Islamic Invasion?" By Jennifer Rast, @ contenderministries.org)

For seven months the people of Israel will be burying the dead! The Bible does not say specifically how this defeat will be brought about, whether miraculously by God directly, or possibly by God using the United States (Israel's main defender in a very hostile world), but however it occurs what a stunning judgment will the Lord bring upon the enemies of his nation Israel.

But the battle of Gog and Magog, as cataclysmic as it will be, is still only a precursor of far more terrible things to come…

The Rise and Fall of Satan's Fourth Reich

"We live in a surveillance society. There's no question that detailed profiles are being created and stored about each and every one of us. There's no question that this is only going to increase with every passing day, and with every new technological advancement. There's no question that with every new threat, new crisis, and consolidation of world resources, knowledge and currencies, our world moves closer to a one world governmental system. This new system will not be a friend of Christianity or Judaism. Basic rights like the freedom of speech, the freedom of religion and the freedom to own land will disappear. Already today there are very powerful and influential people who have dedicated themselves to seeing this comes to pass. I have no doubt that it will, and soon."[524]

Big Brother is watching you, and it's only going to get worse as the world plunges further toward global government…

"The entire globe is moving in the direction of totalitarianism. Our world is literally becoming a prison grid."[525]

And that is because the government's ability to keep an eye on and to track you is exploding exponentially. Surveillance cameras are everywhere; many have already been linked to computers with sophisticated facial recognition software. National ID cards are coming here soon. They are already being distributed in China, with Mexico and India soon to follow, and they have GPS (Global Positioning System) locaters with radio frequency identification (RFID) devices inside them so the government can know where you are at any given moment, and can track your every move. The same technology will soon be put in every cell phone, and car. Eventually they will put those identity cards and tracking devices into your right hand or forehead. Will this then be the prophesied "Mark of the Beast"??! Only time will tell…

A worldwide government and religion is coming upon us very soon. All around we see the push toward this globalization, which will entail the unification of the earth's political, religious, economic and financial systems. Most of the nations of the world are either already on board with this or are heading in that direction. The sovereignty of the United States, the world's most powerful nation, is one of the sticking points in reaching this goal, and that is why she must be "radically transformed" into something other than what the founders, and our Constitution, had in mind. But even so, this one world government, when it soon comes, will not have to be forced upon the citizens of the world. For the wars, the political and social unrest, the economic and financial turmoil and the religious extremism/terrorism happening all across the globe will also soon lead the people of this planet, including most of the people in this country, to demand the one

---

524   (Grant R. Jeffrey, on *Shadow Government DVD*)
525   (From "32 Signs That Big Brother Technology Is Growing" @ endoftheamericandream. com)

world, unified system of both government and religion that has been prophesied in Scripture for thousands of years. The world will think that this will bring them peace and prosperity. But they will get neither. They will instead get the antichrist, the false prophet, and a terrible seven-year period of unparalleled death and destruction that will culminate with the return of Christ and the end of this age...

"And for this reason God shall send them strong delusion, that they should believe a lie." (2 Thessalonians 2:11) In this case, the antichrist will be that lie...

> "Satan is the master of the counterfeit. And in the last days, he will possess such power over Earth that most people will not realize they have placed their faith in a fake. In an attempt to keep the earth in its corrupted state for eternity and thus avoid his own eternal judgment, Satan will establish a visible, political form of his kingdom by using a counterfeit Christ (messiah). Then he will use the political control to seek the elimination of Israel.
> "Israel is critical to Satan's end times strategy because if he could eliminate Israel, then Messiah could not return to earth to rule on the throne of King David, as promised in Scripture (Isa. 9:7). Thus Satan would prevail over God and prevent the Almighty from completing His plan of redemption for humanity. Of course, he will not prevail. But he will deceive millions before he plunges to his defeat."[526]

Satan is the father of lies. He is also the god of the people of this world. It is true that Satan will deceive the world when the Antichrist appears on the scene, but the world is already deceived by him. The world rejects the truth of God and embraces the lies of Satan already. And indeed the people of this world are already a part of the kingdom of Satan for we all have sinned, and "he who sins is of the devil" (1John 3:8), until through the washing and regeneration by the shed blood of the Lord we are rescued from the kingdom of Satan and adopted into the family and kingdom of God. Again, the lies of the wicked one cover this earth—in the form of false religions, baby-killing ideologies, and scientifically-absurd theories about our origins that spit in the face of our Creator—and most people are heavily under their influence. Thus it will be an easy thing for him to deceive them even further in the near future when he entices them into embracing his false messiah, the Antichrist. Indeed they will flock to him and fawn all over him just like the Germans embraced Hitler, and like the *not-so-well-informed* (or intentionally *mis*informed) voters in this country fawned all over Obama and ran out and elected him and his party of traitors, socialists, American-soldier-killers, baby-mutilators, pathological spenders and professional liars (like Chucky Schumer, whose back-stabbing, obscenely partisan and deceitful rants regularly desecrate the halls of Congress, but who nevertheless gets reelected religiously). The world in fact is primed for the embrace of Lucifer's man. They have rejected Jesus Christ as their Savior, and like their rejection of moral absolutes which has left them with the morals of hell, they will be left with the "savior" from hell as well.

> "Satan will begin by introducing his counterfeit messiah, the Antichrist, who will come "with all power, signs, and lying wonders" (2 Th. 2:9) The Rapture of the church and

---

526   (From article, "The Coming Counterfeit," by James A. Showers, Israel My Glory Magazine, July/August 2007, a publication of The Friends of Israel)

ensuing removal of the restraining work of the Holy Spirit will pave the Antichrist's way (vv. 1-12) Without the restraining work of the Holy Spirit, the world will fall away from God into worldwide apostasy (v. 3). Satan will seize this opportunity and move quickly to establish a visible, political form of his kingdom and seek to make it permanent."[527]

"Now we beseech you, brethren, concerning the coming of our Lord Jesus Christ and our gathering together unto him, we ask you not to be soon shaken in mind or troubled, either by spirit or by word or by letter, as if from us, as though the day of Christ had come. Let no man deceive you by any means: for that day shall not come unless there comes a falling away first, and the man of sin is revealed, the son of perdition. Who opposes and exalts himself above all that is called God, or that is worshiped, so that he sits as God in the temple of God [profaning the Third Temple, soon to be rebuilt on the temple Mount in Jerusalem], showing himself that he is God. Do you not remember that when I was still with you I told you these things? And now you know what is restraining, that he may be revealed in his own time. For the mystery of iniquity is already at work; only he who now restrains [the Holy Spirit] will do so until he is taken out of the way. And then shall that Wicked one be revealed, whom the Lord shall consume with the breath of his mouth, and shall destroy with the brightness of his coming. Even the lawless one [Antichrist] whose coming is according to the working of Satan with all power and signs and lying wonders, and with all unrighteous deception among those who perish, because they received not the love of the truth, that they might be saved. And for this reason God shall send them strong delusion, that they should believe the lie; That they all might be damned who believed not the truth, but had pleasure in unrighteousness." (2 Thessalonians 2:1-12)

"According to the prophet Daniel, this kingdom will be a revived form of the Roman Empire (Dan. 2:41-44; 7:7). It will be a federation of 10 regions or divisions of nations that will rule the world (Rev. 13:7). ["And authority was given him over every tribe, and tongue, and nation."] The Antichrist will rise to power and seize control of the entire revived Roman Empire (Dan. 7:24; Rev. 13:1). And as Satan's human agent, he will be empowered to perform amazing wonders "with all unrighteous deception among those who perish" (2Th. 2:10).

"In other words, the Antichrist will use unbridled satanic power to deceive the world into believing he is the Promised One.

"Initially, Satan will align the revived Roman Empire with the apostate religion of the Western world to gain worldwide support for the Antichrist's leadership. However, the alliance will be temporary until Satan's counterfeit Christ has consolidated his power. Eventually, the Antichrist will break the alliance and replace the apostate religion with the worship of himself (Rev. 17:16).

"These events will occur over a seven-year period known as the Tribulation. At the midpoint, Antichrist will be killed, and it will appear this mighty man of peace was cut down before his work was finished (13:3). ["And I saw one of his heads as if it were wounded to death; and his deadly wound was healed. And all the world marveled and followed after the beast."] With the pseudo messiah now dead, Satan's scheme to stymie God's plan of redemption would seem to be in jeopardy.

---

527   (From article, "The Coming Counterfeit," by James A. Showers, Israel My Glory Magazine, July/August 2007, a publication of The Friends of Israel)

"At the same time, Satan will suffer another loss: a defeat in his war with Michael the archangel and the angels of heaven. As a result, Satan and his demons will be cast out of heaven permanently and confined to Earth (12:7-12)."[528]

"And there was war in heaven: Michael and his angels fought against the dragon; and the dragon fought and his angels, And they did not prevail; neither was their place found any more in heaven. And the great dragon was cast out, that serpent of old, called the Devil and Satan, who deceives the whole world: he was cast to the earth, and his angels were cast out with him. Then I heard a loud voice saying in heaven, Now is come salvation, and strength, and the kingdom of our God, and the power of his Christ: for the accuser of our brethren, who accused them before our God day and night, has been cast down. And they overcame him by the blood of the Lamb, and by the word of their testimony; and they loved not their lives unto the death. Therefore rejoice, O heavens, and you who dwell in them. Woe to the inhabitants of the earth and the sea! For the devil has come down to you, having great wrath, because he knows that he has but a short time." (Revelation 12:7-12)

"Knowing his time is short before Messiah Jesus will return, the Devil will increase his wrath on Earth, using all means possible to prevent Jesus from coming (v. 12).

"First he will heal the Antichrist's deadly wound (13:3-4). The counterfeit messiah will then have died, "risen," and come a second time."[529]

It is interesting to note that Satan counterfeit's the death and resurrection of Christ with his own version involving the Antichrist. It is also interesting to note how many people in this life have been duped into thinking that the death and resurrection of Jesus Christ is a fable, yet obviously Satan knows otherwise, and in his "seeking to be like the Most High," he tries to duplicate God's unique triumph.

"Moving swiftly to consolidate his political power, he will kill three of the 10 kings in the revived Roman Empire (Dan. 7:24). The remaining seven kings will pledge allegiance and submit to the Antichrist. To complete his coup, he will break his alliance with the Western apostate religion and eliminate it (Rev. 17:16). He then will have control to rule the world.

"Next, through Antichrist, Satan will establish a religion to exalt and glorify himself, aided by the False Prophet who will act as a counterfeit Holy Spirit (13:11-15)."[530]

As Satan counterfeits the death and resurrection of Jesus, he also counterfeits the Holy Trinity of God with his own un-holy version: himself, the antichrist and the false prophet. (Another point here is that if the trinity were not real, as certain false Christians insist, why then would Satan want to counterfeit it?)

---

528   (From article, "The Coming Counterfeit," by James A. Showers, Israel My Glory Magazine, July/August 2007, a publication of The Friends of Israel)
529   (From article, "The Coming Counterfeit," by James A. Showers, Israel My Glory Magazine, July/August 2007, a publication of The Friends of Israel)
530   (From article, "The Coming Counterfeit," by James A. Showers, Israel My Glory Magazine, July/August 2007, a publication of The Friends of Israel)

"Then I beheld another beast coming up out of the earth, and he had two horns like a lamb, and he spoke like a dragon. And he exercises all the authority of the first beast before him, and causes the earth and those who dwell therein to worship the first beast, whose deadly wound was healed. And he performs great wonders, so that he even makes fire come down from heaven on the earth in the sight of men. And he deceives those who dwell on the earth by those miracles which he was granted power to do in the sight of the beast; saying to those who dwell on the earth, that they should make an image to the beast that was wounded by the sword and lived." (Revelation 13:11-14)

The people of this world, having rejected God and in turn having embraced the evil-god, Satan, and his Antichrist, will get just what they wanted. God gives them over to their lust for the father of lies and allows Satan to even work miracles for his deceived worshippers.

> "Empowered by Satan, the False Prophet will perform signs, wonders, and miracles to deceive the world and force humanity to worship the Antichrist. It is he who will require everyone to take the infamous mark of the Beast:

"He was granted power to give breath to the image of the beast, that the image of the beast should both speak and cause as many as would not worship the image of the beast to be killed. He causes all, both small and great, rich and poor, free and slave, to receive a mark on their right hand or on their foreheads, and that no one may buy or sell except one who has the mark or the name of the beast, or the number of his name. Here is wisdom. Let him who has understanding calculate the number of the beast, for it is the number of a man: His number is 666." (Rev 13:15-18)

> "With his political and religious power consolidated, Satan will now concentrate on eliminating Israel."[531]

The beginning of the Great Tribulation, the final seven year period of this age, with all of its horrible plagues, mass deaths, world-wide military conflicts and terrible judgments, will start with the Antichrist, the leader of the revived Roman Empire, making a treaty with the nation of Israel. (All indications from prophecy are that this revived Roman Empire will not just be limited to the European Union as many have suggested, but will be a worldwide political and military system based on the Roman-European model of government, and headed by the Antichrist.)[532] He breaks this pact after 3 & 1/2 years and sets himself up as God in the rebuilt temple of Solomon in Jerusalem,[533] and demands to be worshipped

---

531   (From article, "The Coming Counterfeit," by James A. Showers, Israel My Glory Magazine, July/August 2007, a publication of The Friends of Israel)

532   (See pgs. 109-121, *119 Most Frequently Asked Questions About Prophecy*, by Arno Froese)

533   (Besides the prophesied Battle of Gog and Magog and the rapture, which we conclude will both happen before the tribulation, another sign to look for is the building of the Third or Millennial Temple in Jerusalem, <u>on</u> the Temple Mount, over the bedrock of the Holy of Holies, opposite the Eastern Gate. Even though the Muslims have been given control over the Temple Mount and their mosques and shrines to Allah/Beelzebub pollute this holy ground, the preparations have been and are being made by devout Jews, and the Jewish Temple will be rebuilt there, very soon. Possibly right after the battle of Gog and Magog. Keep your eye out.)

by the people of the world. Many will follow this Antichrist who will be a charismatic, charming, even outwardly righteous, yet utterly godless individual. "The plight of the world is to reject Christ and follow such an individual."[534] Jesus warned the world of this in the gospel of John, "I have come in my Father's name, and you do not receive me; if another shall come in his own name, him you will receive." (John 5:43) Antichrist will put on the trappings of religiosity and righteousness but will embrace all that is evil. (Similar to a certain un-named charismatic, baby-mutilating, big-eared, authoritative-sounding yet incompetent leader in this country today. The antichrist, however, will not be incompetent.) Those with wisdom will reject him even though the consequences will be brutal. The pressure to surrender to the Antichrist will be extreme. Those who hold out will not be allowed to buy or sell, or they will simply be killed. "And he caused that as many as would not worship the image of the beast should be killed. ...And that no man might buy or sell, save he that had the mark, or the name of the beast, or the number of his name." (Revelation 13:15-17) God stresses the importance of being able to recognize this impostor by revealing the numerical equivalent of his name, 666. "Here is wisdom. Let him who has understanding calculate the number of the beast, for it is the number of a man; and his number is six hundred threescore and six." (Revelation 13:18) There are many different interpretations of the meaning of this verse, of exactly how the number 666 will be associated with this man, and due to the limitations of space here we will have to leave this search up to the reader. Personally I think that the exact way the Antichrist is associated with 666 will be revealed when he is revealed, and at that time those with wisdom will know.

Whereas those without wisdom will not. Which brings us to a question that often arises in any discussion of biblical prophecy, and that is, After all the publicizing of the mark of the beast and the number of his name, 666, how could anyone be so foolish as to accept the mark and worship the beast when that time comes. And the answer is that a great many more people than have heard about this information over the last few decades have not heard it, or have closed their ears and just blown it off. Besides, the Democrat-Left has long ago blown off the Bible and its prophecies. They are much more impressed with Nostradamus or the human-sacrificers, the Mayans; anyone but God. So most people ignore whatever the Bible has to say. The following verse prophesies this type of denial:

"Knowing this first, that there shall come in the last days scoffers, walking after their own lusts, and saying, Where is the promise of His coming? For since the fathers fell asleep, all things continue as they were from the beginning of the creation." (2 Peter 3:3-4)

In other words, there are those, and they are many, who say that from the beginning of time until now everything has remained pretty much the same, and that despite many "Chicken Little" predictions that "the sky is going to fall," and that the world as we know it is going to end, everything will still continue to go on pretty much as it ever has. These folks of course are ignorant of the geologic reality and scientific fact of the worldwide flood (see chapter 10) for one thing, and of the 100% accuracy of biblical prophecy on the other. But that is the information that they have chosen of their own free will to not allow into their head.

Also, those who have bought into the lie that this "material" world is the

---

534   (Pg. 143, *The Unveiling: A Journey Through the Book of Revelation,* by Keith Harris)

ultimate reality and that this life is "all that there is" will be easy marks for the mark of the beast, along with the mockers spoken of above. In addition, those *religious* folks who are not rooted and grounded in the truth will find giving up the ability to buy and sell, and therefore to *live*, in this life to be a price too high to pay for their barely believed promise of a much better life in the next. Having faith in what you can see and feel, and eat and drink, is a whole lot easier than having faith in that which you cannot, especially for the carnal mind. That is why Christ said that "without faith it is impossible to please Him." By exercising faith in him daily, and thereby growing in the hope in what cannot presently be seen, will increasingly give more power to the spiritual over the carnal mind; will move one to trust in Him and his promises as they experience his faithfulness in answering their prayers and in fulfilling his promises to them. But for the mockers, the materialists and the Christians-in-name-only the decision between the mark of the beast and starving to death out of doors will be an easy one to make, despite its horrific eternal consequences.

The Antichrist will finally succeed where all others have failed, from Alexander The Great, to Caesar and to Hitler, in ruling over a truly world-wide empire. The dream of so many tyrants, dictators and despots will finally come true but thankfully for only a short while. And the groundwork for this has been going on for some time now. The United States is the main obstacle to this goal, which is why people like George Soros and the Democrat-far-left in general seem to have such a passion for our "fundamental transformation," which as we will see in chapter 13, translates into our "fundamental destruction." Their goal is not the preeminence of our country. That preeminence is an obstacle to their goal. They zealously pursue their dream of a utopia here on earth, which they can only visualize as coming from the rule of a One World Government. With them in charge of course. Or so they think. In reality they are all unwittingly doing the bidding of their father the devil where a one-world government <u>will</u> soon be set up with <u>him</u>, and his antichrist, in charge. Besides, evil men trying to create their own world-wide utopia will never work. But that's what the globalists are trying to do, and in doing so they are paving the way for the Antichrist; who consequently they will embrace with joy and excitement, just like they did with their "anointed one," Barack Hussein Obama.[535]

The second beast spoken of in the Book of Revelation, called the false prophet, will be the head of the one world religious system. He will use his power and his religious authority to influence vast multitudes to fall in line with the wishes of the Antichrist, and ultimately to worship him as he demands. It is also pretty clear that the identity of this second beast will be a future Pope of Rome, and that he may very well be duped into believing, along with many other unsaved Christians, that the Antichrist is actually the returned Christ. Which is why they will have no problem worshipping him when he demands it. The false prophet will probably be a sincere man, thinking he is furthering the cause of peace and goodness and of helping the poor. But he will be deceived, and consumed, like everyone else

---

535 (It is also fairly clear that the antichrist will also be the Shia Muslim's long awaited "Madhi" or "12th Imam" who they herald as their "Islamic messiah." - From "Prophecy News Watch" online, 11/29/2011)

who follows Satan… "who goes about like a roaring lion, seeking whom he may devour." (1 Peter 5:8)

After the Antichrist breaks the terms of his peace accord with Israel, at the halfway point of the seven year tribulation period, and sits himself down on the throne of the rebuilt temple in Jerusalem, all hell will break loose on the earth. Massive persecution, plagues, natural disasters and earth-shattering catastrophes will follow for the final 3 & 1/2 years of the Antichrist's reign. And then the nations of the world will be gathered together at the plain of Megiddo in northern Israel for the battle of Armageddon.

> "[Satan] actually will put his strategy into motion three and one-half years earlier with the signing of a covenant between Antichrist and Israel (Dan. 9:27). This covenant will appear to initiate a period of unprecedented security in modern Israel; it also seems to open the door to rebuilding the Temple and resuming the Old Testament sacrifices. It will promise the peace and security Israel has long sought, as Antichrist promises to protect Israel from its enemies.
>
> "However, the covenant will turn out to be a deceitful trap set by Satan. As a result of the agreement, Israel will let down its guard and put its trust in Satan's man. During the first three and one-half years of the covenant, Israel's confidence in the Antichrist will appear to be well founded. When the kings of the North and South arrive to make war with Israel, the Antichrist will come to Israel's defense with his armies and defeat them soundly (11:40-45). Subsequently, he will defend Israel at the battle of Gog and Magog, although God will intervene to destroy the king's armies before the Antichrist arrives (Ezek. 38-38)."[536]

According to the Bible, the Antichrist makes a covenant with Israel for one *week* and then breaks it half way through the *week*. But in this and other instances the Bible is using the term "week" prophetically to refer to a *seven year period*. Also, when the Antichrist's deadly wound is healed, Satan himself enters into the man and takes total control from that point forward.

The book of Revelation chronicles a series of judgments that are unleashed on mankind against those who have the Mark of the Beast in their right hand or in their forehead; against those who have chosen to follow the devil and worship his false messiah. "And he (the Antichrist) causes all, both small and great, rich and poor, free and slave, to receive a mark in their right hand, or in their foreheads." (Revelation 13:16) But these earthly judgments are only a foreshadowing of a far worse eternal fate for those who are so deceived: "And the third angel followed them, saying with a loud voice, If any man worship the beast and his image, and receive his mark in his forehead, or in his hand, the same shall drink of the wine of the wrath of God, which is poured out without mixture into the cup of his indignation; and **he shall be tormented with fire and brimstone in the presence of the holy angels, and in the presence of the Lamb: And the smoke of their torment will ascend up for ever and ever: and they have no rest day nor night, who worship the beast and his image, and whosoever receives the mark of his name.**" (Revelation 14:9-11) Again, the consequences for being seduced by Satan, his Antichrist and the false prophet, for following their lies, and for rejecting the truth of Jesus Christ the Creator of this universe, will be an eternal horror beyond

---

536   (From article, "The Coming Counterfeit," by James A. Showers, Israel My Glory Magazine, July/August 2007, a publication of The Friends of Israel)

our ability to imagine or for anyone to bear. But bear it, nevertheless, they will…

"Because I have called, and you refused; I have stretched out my hand, and no one regarded; Because you have disdained all my counsel, and would have none of my rebuke: I also will laugh at your calamity; I will mock when your terror comes; When your fear comes like desolation, and your destruction comes as a whirlwind; when distress and anguish comes upon you. Then shall they call upon me, but I will not answer; they shall seek me diligently, but they shall not find me. For they hated knowledge, and did not choose the fear of the Lord: They would have none of my counsel, and they despised my rebukes." (Proverbs 1:24-30) Jesus will not answer them, because they will already be in hell.

The earthly judgments that are poured out on a disobedient and unrepentant mankind and that mark the tribulation period, begin with the opening of the seven seals, as described in the Book of Revelation. The first four seals release the famous four horsemen of the Apocalypse: a white horse, a red horse, a black horse and a pale horse with their riders coming forth when each of their seals is opened. They go forth, one to conquer; one to take peace from the earth; one to bring famine and disease; and one to kill: "And I looked, and behold a pale horse: and his name that sat on him was Death, and Hell followed with him. And power was given unto them over the fourth part of the earth, to kill with sword, and with hunger, and with death, and with the beasts of the earth." (Revelation 6:8) One fourth of the people of the earth will die under the trampling of these four horsemen.

"And when he had opened the fifth seal, I saw under the altar the souls of them who were slain for the word of God, and for the testimony that they held. And they cried with a loud voice, saying, How long, O Lord, holy and true, before you judge and avenge our blood on those who dwell on the earth?" (Revelation 6:9-10) Well, that question is about to be answered, in spades…

And then the sixth seal was opened: "And I beheld when he had opened the sixth seal, and, lo, there was a great earthquake; and the sun became black as sackcloth of hair, and the moon became like blood. And the stars of heaven fell to the earth, as a fig tree drops its late figs when it is shaken by a mighty wind. And the sky receded as a scroll when it is rolled up, and every mountain and island were moved out of its place. And the kings of the earth, the great men, the rich men, the chief captains, the mighty men, every slave and every free man, hid themselves in the caves and in the rocks of the mountains, and said to the mountains and rocks, 'Fall on us and hide us from the face of him who sits on the throne and from the wrath of the Lamb! For the great day of his wrath has come; and who is able to stand?'" (Revelation 6:12-17)

These catastrophic judgments from God are brought about because of the unrepentant wickedness of man. The seals are then followed by the seven trumpet judgments: "The first angel sounded [his trumpet], and there followed hail and fire mingled with blood, and they were thrown to the earth. And a third of the trees were burned up, and all green grass was burned up. Then the second angel sounded, and something like a great mountain burning with fire was thrown into the sea, and a third of the sea became blood. And a third of the living creatures in the sea died, and a third of the ships were destroyed. And the third angel sounded: And a great star fell from heaven, burning like a torch, and it fell on a third of the rivers and on the springs of water. The name of the star is Wormwood. A third part of the waters became wormwood, and many men died from the water, because it was made bitter. Then the fourth angel sounded, and a third of the sun was struck, a third of the moon, and a third of the stars, so that a third of them were darkened. A third of the day did not shine, and likewise the night." (Revelation 8:7-12)

Talk about ecological disasters! It is poignant how aggressive and zealous many are becoming over *saving the planet*, which of course at face value is a noble and worthy undertaking, but many of the far-left people at the forefront of this cause are the first to ridicule the notion of getting saved themselves, through faith in Jesus Christ who created and sustains our Mother Earth. Nor are they the least bit concerned about saving millions of unborn babies from being mutilated to death, so their ignorance, hypocrisy and evil is apparent. Man was given dominion over the earth and over all the creatures of the earth by God, to tend it and to nurture it as it nurtured him, but he relinquished that dominion to Satan, the god of this world, when he followed him and rejected the Word of the Lord. Trying to "save" the earth while rejecting God is a fool's errand. And we can see in the previous four trumpet judgments what the end result is going to be for those left after the Rapture, and what will happen to the environment of the earth because of their continued blasphemy and iniquity.

"Because the creature itself also shall be delivered from the bondage of corruption into the glorious liberty of the children of God. For we know that the whole creation groans and travails in pain together until now." (Romans 8:21-22)

In truth, *Mother Earth*[537] "groans and travails" and convulses under the contamination of sinful man. Her skin crawls because of him and his wickedness. She was not created for this type of evil inhabitant to arrogantly, ignorantly and unthankfully stake his dwelling upon her, and scurry about on her pursuing his selfish and godless desires. She was created in perfection and her purpose was to nurture and sustain all that God had made, and all that he had made was "very good." ("And God saw every thing that he had made, and behold, it was very good." Genesis 1:31) That was before the fall of man and his casting out of the Garden.[538] And now, after six thousand years of groaning under this itching, noisome rash, she will finally say "ENOUGH" of Algore and his wicked, lying, Christian-hating, clitoral-slicing and baby-slaughtering kind. Now she is rejoicing that the Lord has given her the go ahead to free herself from the curse of unrepentant man. Today's earthquakes, tornados, hurricanes, floods, droughts, tsunamis, plagues, famine, disease and volcanic eruptions are just a mild taste of what is to come as she prepares to finally "vomit out" this cancerous skin disease. (I know, that sounds a little harsh, but it is not us saying it...)

---

537   ("Our Mother, the Holy Spirit of God, envelops us in her womb [within this spirit-ual universe], and we are connected to her by the umbilical cord of breath." ...One question that we are not going to answer in this book—mainly because we cannot at this point be sure of the answer—is that of the relationship, if any, between our Mother Earth and the Holy Spirit of God; between the universe of matter, this image in the mind of God, and the Second Person of the trinity (is it really the Third?). The macrocosm mirrors the microcosm. Human father, mother, and child. The Heavenly Father, the Holy Spirit, the only begotten Son. And it is interesting to note that in biblical marriage counseling Christians are taught that the husband is a type of Christ, and the wife is a type of the Holy Spirit, the Comforter. If God is our Father, and Jesus is our Lord and brother, then Who is our Mother? Something to think about.)

538   (People often ask that if Adam and Eve were cast out of the Garden of Eden, then where is it? Well, whatever its actual physical position on the earth prior to the Flood is a moot point since the entire earth was covered in thousands of feet of mud, sediment, vegetation, dead bodies and debris. But the answer for where it is *going* to be when the Lord returns is: right here. The restored Garden of Eden, that initial perfection that "was very good," that heaven here on earth, will be the entire earth herself.)

"For the land is defiled: therefore I visit the iniquity thereof upon it, and the land itself <u>vomits out</u> her inhabitants." (Leviticus 18:25)

Therefore, even more terrible judgments are going to come…

"And I looked, and I heard an angel flying through the midst of heaven, saying with a loud voice, 'Woe, woe, woe, to the inhabitants of the earth, because of the remaining blasts of the trumpet of the three angels who are about to sound!'" (Revelation 8:13)

When the fifth angel sounds his trumpet, a star falls from heaven to earth. "To him was given the key of the bottomless pit. And he opened the bottomless pit; and smoke arose out of the pit like the smoke of a great furnace. Then out of the smoke locusts came upon the earth. And they were given authority to torment man for five months. Their torment was like the torment of a scorpion when it strikes a man. In those days men will seek death and will not find it; they will desire to die, and death will flee from them." (Revelation 9:1-3,4-6)

Patrick Heron, in two of his books, *Return of the AntiChrist and the New World Order* and *The Nephilim and the Pyramid of the Apocalypse*, argues that this "star" that "falls from heaven to earth" is a fallen angel, and that these locusts from the bottomless pit are in reality other fallen angels who were cast out of heaven because of their rebellion with Lucifer, were cast down to earth before the flood, and slept with the beautiful women of their choosing…

"And it came to pass, when men began to multiply on the face of the earth, and daughters were born unto them, that the sons of God saw the daughters of men that they were fair; and they took them wives of all which they chose. …There were giants in the earth in those days; and also after that, when the sons of God came in unto the daughters of men, and they bare children to them, the same became mighty men which were of old, men of renown." (Genesis 6:1-2, 4)

When angels, who are spiritual beings, appear in physical form on earth, whether good or fallen angels, the Bible always refers to them as men; because that is what they are. But unlike mere mortal men, they are far more powerful both physically and mentally. It is natural to assume therefore that in sleeping with women, taking to them "wives of all which they chose," that their offspring would be "super human" or the giants referred to in the above verses. The author also proposes that Apollyon, the lead angel who arises out of the bottomless pit and is king over the "locusts" or other fallen angels, is none other than the antichrist… "And they had a king over them, which is the angel of the bottomless pit, whose name in the Hebrew tongue is Abaddon, but in the Greek tongue hath his name Apollyon." (Revelation 9:11) This would also make sense in that fallen angels, as the antichrist and the false prophet, would be more apt to be able to work "signs and miracles" to further deceive unregenerate men, than mere humans could. In addition, if these fallen angels and their progeny were the "mighty men which were of old, men of renown," then we can also see how the antichrist would be revered and idolized when he returns from the bottomless pit and becomes this world's savior (think of Michael Jackson, Arnold Schwarzenegger, Einstein and Elvis all wrapped up in one).[539] At any rate, the reader can decide for him or herself by checking out Mr. Heron's books.

---

539   (See chapters 6 and 7 in *Return of the AntiChrist and the New World Order*, by Patrick Heron)

Then the sixth angel sounds, and a third of mankind is killed, because of the two hundred million man army that marches across the dried up river Euphrates, and because of the fire, smoke and brimstone that comes from their weapons:

"And the sixth angel sounded, and I heard a voice from the four horns of the golden altar which is before God, saying to the sixth angel who had the trumpet, 'Release the four angels who are bound in the great river Euphrates. And the four angels, who had been prepared for the hour and day and month and year, were released to kill a third of mankind. And the number of the army of the horsemen were two hundred thousand thousand [200 million]: and I heard the number of them. And thus I saw the horses in the vision, and those who sat on them, having breastplates of fiery red, hyacinth blue, and sulfur yellow; and the heads of the horses were like the heads of lions; and out of their mouths issued fire and smoke and brimstone. By these three plagues a third part of mankind was killed: by the fire and by the smoke and by the brimstone, which issued out of their mouths." (Revelation 9:13-18)

But incredibly those who are left still do not repent of their sins... "And the rest of mankind, who were not killed by these plagues, did not repent of the works of their hands, that they should not worship demons. Neither did they repent of their murders [like torturing little babies to death; a practice that will still be going strong ...YES WE CAN!] or their sorceries or their sexual immorality or their thefts." (Revelation 9:20-21)

So he sends them their final judgments, the seven bowls:

"Then I heard a loud voice from the temple saying to the seven angels, 'Go and pour out the bowls of the wrath of God upon the earth.' So the first went and poured out his bowl upon the earth, and a foul and loathsome sore came upon the men who had the mark of the beast and those who worshipped his image. Then the second angel poured out his bowl on the sea; and it became as the blood of a dead man;[540] and every living creature in the sea died. Then the third angel poured out his bowl on the rivers and springs of water, and they became blood. And I heard the angel of the waters say, 'You are righteous, O Lord, The One who is and who was and who is to be, because you have judged these things. For they have shed the blood of saints and prophets, and you have given them blood to drink. For it is their just due.' Then the fourth angel poured out his bowl upon the sun, and power was given to him to scorch men with fire. And men were scorched with great heat [finally, the Democrat-Left will get their global warming, but not in the way they had envisioned], and they blasphemed the name of God who has power over these plagues; and they did not repent and give him glory. Then the fifth angel poured out his bowl on the throne of the beast; and his kingdom became full of darkness; and they gnawed their tongues because of the pain. They blasphemed the God of heaven because of their pains and their sores, and did not repent of their deeds." (Revelation 16:1-6, 8-11)

---

540 (It is June 2010, and we are in the middle of the catastrophic oil spill in the Gulf of Mexico. But what is interesting to note for our discussion in this chapter, and in particularly for the verse just quoted, "the second angel poured out his bowl on the sea; and it became as the blood of a dead man," is that as you look at the pictures of the Gulf from planes and helicopters, the oil spreading over the waters looks exactly like blood. Are we seeing one way where biblical prophecy of the seas being turned to blood could be fulfilled in the future, possibly on a more global scale as massive undersea earthquakes, and not oil drilling mistakes, release vast quantities of oil that have been trapped there since the flood?)

Men will curse God, and will blame everyone and everything else for the horror that has descended upon their world. They will blame "man-made" global warming, capitalism, Republicans, George Bush, Christians (even though they will more than likely be gone), Israel, conservatives, the Jews in general, and even God. The only ones they will not blame are themselves.

"Then the seventh angel poured out his bowl into the air; and a loud voice came out of the temple of heaven, from the throne, saying, 'It is done!' And there were noises and thundering and lightnings; and there was a great earthquake, such a mighty and great earthquake as had not occurred since men were on the earth. And the cities of the nations fell. And great Babylon came in remembrance before God, to give her the cup of the wine of the fierceness of his wrath. And every island fled away, and the mountains were not found. And great hail from heaven fell upon men, each hailstone about the weight of a talent. Men blasphemed God because of the plague of the hail; for that plague was exceedingly great." (Revelation 16:17-21)

This is why this time is called the Great Tribulation. Thank God that the church, the body of believers in Christ, will be rescued prior to this terrible period of suffering and death for those left on the planet. (Unless of course you're post-trib, in which case you will be allowed to stay for all of the festivities.)☺

> "After his "resurrection" and consolidation of power, Antichrist will violate his covenant with Israel, putting a sudden end to the sacrifices and offerings in the Temple (Dan. 9:27). Then he will desecrate the Temple by placing his throne there, declaring himself to be God and demanding everyone worship him (Dan. 7:8, 20, 25; 11:36-37; 2Th. 2:3-4; Rev. 13:4-8, 11-17).
>
> "Those who refuse to worship him will be killed. Antichrist also will erect an image of himself in the Temple, which Christ referred to as the " 'abomination of desolation,' spoken of by Daniel the prophet" (Mt. 24:15). Antichrist will defame the sacred place where Israel worships; but that is only the beginning.
>
> "He will unleash a period of unparalleled persecution on Israel, referred to as the "time of Jacob's trouble" (Jer. 30:4-7; Mt. 24:21-28). The Bible warns Jewish people to flee into the wilderness. Antichrist's purpose will be to secure Satan's political kingdom on Earth by annihilating Israel and cutting it off from being a nation so that the name of Israel will no longer be remembered (Ps. 83:4). Without Israel, the Messiah could not return to rule on David's throne (Isa. 9:6-7) and Satan would avoid his eternal judgment.
>
> "Antichrist will carry out his anti-Israel campaign for three and one-half years, [He will carry it out "for a time and times and half a time" (Daniel 7:25). A time is one year, times- plural of time- is two years, and a half a time is ½ year, hence the three and one half years.] ...reaching a crescendo with the Battle of Armageddon, which is actually a military campaign that will last for several weeks or months (Dan. 11:40-45; Joel 3:9-17; Zech. 14:1-3; Rev. 16:14-16). Antichrist will assemble his armies from the nations of the world to complete the elimination of Israel."[541]

"For behold, in those days and at that time, when I shall bring back the captives of Judah and Jerusalem, I will also gather all nations and will bring them down to the Valley of Jehoshaphat; And I will enter into judgment with them there

541   (From article, "The Coming Counterfeit," by James A. Showers, Israel My Glory Magazine, July/August 2007, a publication of The Friends of Israel)

concerning my inheritance and my people Israel, whom they have scattered among the nations, and divided up my land. ...Let the nations be wakened, and come up to the Valley of Jehoshaphat: for there will I sit to judge all the surrounding nations. ... Multitudes, multitudes in the valley of decision: for the day of the Lord is near in the valley of decision." (Joel 3:1-2, 12, 14)

The Bible tells us that a route will be opened up for the armies of China and of Asia to descend on the Middle East:

"And the sixth angel poured out his bowl on the great river Euphrates; and its water was dried up, so that the way of the kings from the east might be prepared. And I saw three unclean spirits like frogs coming out of the mouth of the dragon, and out of the mouth of the beast, and out of the mouth of the false prophet. For they are the spirits of demons, performing signs, which go out to the kings of the earth and of the whole world, to gather them to the battle of that great day of God Almighty. ...And they gathered them together to the place called in the Hebrew tongue, Armageddon." (Revelation 16:12-14,16)

Next the size of the army is foretold:

"And the sixth angel sounded, and I heard a voice from the four horns of the golden altar which is before God, saying to the sixth angel who had the trumpet, Loose the four angels which are bound in the great river Euphrates. And the four angels were loosed, which were prepared for an hour, and a day, and a month, and a year, for to slay the third part of men. And the number of the army of the horsemen were **two hundred thousand thousand**: and I heard the number of them." (Revelation 9:13-16)

That's 200 million! The People's Republic of China along with the other "kings of the east" could field an army of that size today. And keep in mind that this prophecy was written two thousand years ago when the population of the entire world was less than that! (The population of the *known* world was a fraction of it.) This is also why for very many years people insisted that this prophecy could not be taken literally. But God, of course, two thousand years ago when he caused this prophecy to be written, was fully aware of the size of present day China, and of its army. But the armies from the other three corners of the earth will be summoned also: Russia, Africa, Europe and one would assume the United States.

"Behold, the day of the Lord is coming, and your spoil will be divided in your midst. For I will gather all the nations against Jerusalem to battle; and the city shall be taken, and the houses rifled, and the women ravished. Half of the city shall go into captivity, but the remnant of the people shall not be cut off from the city. Then the Lord will go forth, and fight against those nations, as he fights in the day of battle. And in that day his feet shall stand upon the mount of Olives, which faces Jerusalem on the east. And the mount of Olives shall be split in two, from east to west, making a very large valley. Half of the mountain shall move toward the north, and half of it toward the south." (Zechariah 14:1-4)

This final war has the potential of destroying mankind and even all living things, from nuclear holocaust, but the Lord intervenes before it can go that far. "And unless those days were shortened, no flesh would be saved: but for the elect's sake those days will be shortened." (Matthew 24:22)

> "When there seems to be no hope, those Jewish people still living will turn to God with all their hearts and call on Him to deliver them (Hos. 6:1-3) During this time millions of

Gentiles will perish. And only one-third of Israel will survive the Antichrist's persecution (Zech. 13:8); but God will again have mercy on Israel (1:16).

"Answering Israel's cry for salvation, Messiah will return to the Earth in the same way He left: in the clouds (Mt. 24:30; Acts 1:9-11). The great Son of David will slay the Antichrist and destroy the armies of the world that sought to annihilate the Jewish people (Joel 3:12-13; Hab. 3:13; Zech. 14:12-15; 2Th. 2:8; Rev. 14:19-20). The prayer of Psalm 122:6 seeking "the peace of Jerusalem" will finally be realized."[542]

"And I saw heaven opened, and behold a white horse. And he who sat on him was called Faithful and True, and in righteousness he judges and makes war. His eyes were like a flame of fire, and on his head were many crowns. He had a name written that no man knew, but he himself. He was clothed with a robe dipped in blood, and his name is called The Word of God. And the armies which were in heaven followed him on white horses, clothed in fine linen, white and clean. And out of his mouth goes a sharp sword, that with it he should smite the nations. And he himself will rule them with a rod of iron. He himself treads the winepress of the fierceness and wrath of Almighty God. And he has on his robe and on his thigh a name written, KING OF KINGS, AND LORD OF LORDS." (Revelation 19:11-16)

This is the final event to occur before the millennium begins. It is the end of this first part of his story, the first six thousand years. It is called the Second Coming of the Lord Jesus Christ. The first time he came to Earth he was born in a humble manger. No throne. No earthly crown. Destined to die a humiliating and painful death for the sins of mankind at the hands of the Romans, after being betrayed by his own Jewish people. But when he returns to Earth it is going to be a slightly different story. He is coming with all of his power, unrestrained, and with legions of Angels. (A legion is five to ten thousand strong, and one Angel alone could easily dispose of hundreds of men.)

"Then shall the Lord go forth, and fight against those nations, as when he fought in the day of battle. And his feet shall stand in that day upon the Mount of Olives, which faces Jerusalem on the east, and the Mount of Olives shall be split in two, from east to west, forming a very great valley; and half of the mountain shall remove toward the north, and half of it toward the south. ... And this shall be the plague with which the Lord will smite all the people who fought against Jerusalem: Their flesh shall dissolve away while they stand upon their feet, and their eyes shall consume away in their sockets, and their tongues shall dissolve away in their mouths." (Zechariah 14:3-4, 12)

This last verse is a clear image of the effects of a nuclear explosion, prophesied 2,000 years ago, before even gunpowder was discovered, and before any of our modern weaponry had even been conceived. (Hello, Nostradamus.)

As we mentioned previously, these two separate appearances of Jesus the Messiah are one cause for confusion among unsaved Jews. Many Rabbis tell their flocks that Jesus can't be the Messiah because Messiah will come as a great and powerful king to conquer Israel's enemies, and not as one of the common folk who ended up being humiliated and crucified as a common "criminal." But they are

---

542   (From article, "The Coming Counterfeit," by James A. Showers, Israel My Glory Magazine, July/August 2007, a publication of The Friends of Israel)

wrong of course. Jesus comes as both. First he came as a servant to shed his blood, to be brought "as a Lamb to the slaughter," to die, and to rise again on the third day for the salvation of those who believe *in* him (and not just *about* him). Then He comes for the second time as King of Kings and Lord of Lords to rescue his chosen people Israel and to defeat Satan and his earthly armies that have gathered at the valley of Megiddo, and to reclaim sovereign rule over his planet.

"I saw in the night visions, and, behold, one like the Son of man came with the clouds of heaven, and came to the Ancient of days, and they brought him near before him. And there was given him dominion, and glory, and a kingdom, that all people, nations, and languages should serve him: his dominion is an everlasting dominion, which shall not pass away, and his kingdom that which shall not be destroyed." (Daniel 7:13-14)

> "Jesus is often called the "son of man" referring to His entitlement to the earth and the world. He is also called the "son of David," which reveals His right to the throne. He is called the "son of Abraham," revealing His title to the land of Israel and all the royal grant to Abraham. He is called the "son of God" as He is heir to all things. These attributes reveal His inheritance as well as ours (Rom. 8:17)."[543]

The Lord will descend from heaven with his Raptured church. Upon this final victory he will set up his millennial kingdom where he will rule over the nations from Jerusalem. At this time all of those Jews who remain alive after this terrible final seven year period will recognize and accept Jesus as their Messiah. And the veil that has been covering their eyes as a people for the last 2000 years will be lifted.

"And I will pour upon the house of David, and upon the inhabitants of Jerusalem, the Spirit of grace and supplication; and they shall look upon me whom they have pierced, and they shall mourn for him, as one mourns for his only son, and shall grieve for him, as one who grieves for his firstborn." (Zechariah 12:10)

And what of the fate of our un-holy trinity: Satan, his Antichrist and the false prophet?

"Then I saw an angel come down from heaven, having the key to the bottomless pit and a great chain in his hand. And he laid hold of the dragon, that old serpent, which is the Devil, and Satan, and bound him for a thousand years. And cast him into the bottomless pit, and shut him up, and set a seal on him, that he should deceive the nations no more, till the thousand years should be fulfilled. ... Then the beast [the Antichrist] was captured, and with him the false prophet who wrought miracles before him, with which he deceived those who had received the mark of the beast, and those who worshipped his image. These two were cast alive into the lake of fire burning with brimstone. ...And the smoke of their torment will ascend up for ever and ever: and they have no rest day nor night." (Revelation 20:1-3, 19:20, 14:11)

Heaven on earth

As horrible as it is to comprehend the fate of those three and their followers, it is equally as difficult now to appreciate how incredible our future will be. In the midst of the struggles and hardships and trials and pain and tragic accidents and untimely deaths and loss of loved ones and suffering and financial turmoil

---

543   (Pg. 128, *The Unveiling: A Journey Through the Book of Revelation*, by Keith Harris)

and constant obstacles to peace and happiness and joy in this fallen world that we are so accustomed to, it is hard to imagine the perfection, the peace of mind, the happiness and fulfillment that will be ours to enjoy forever in the life to come; the eternal life on this earth that awaits those whose names are written down in the Lambs Book of Life. "As it is written, no eye has seen, nor any ear heard, neither has any mind conceived the things which God has prepared for those who love him." (1 Corinthians 2:9) We can't adequately envision it but what we can do is let the last few chapters of the final book of the Bible *reveal* a glimpse of just what "he has prepared" for us. Which is paradise here on earth. Which is the earth as it was meant to be…

"And I saw thrones, and they who sat upon them, and judgment was given unto them. Then I saw the souls of those who had been beheaded for their witness of Jesus, and for the word of God, who had not worshiped the beast, nor his image, and neither had received his mark upon their foreheads, or in their hands. And they lived and reigned with Christ for a thousand years. But the rest of the dead lived not again until the thousand years were finished. This is the first resurrection. Blessed and holy is he who has part in the first resurrection. Over such the second death has no power, but they shall be priests of God and of Christ, and shall reign with him a thousand years." (Revelation 20:4-6)

In those verses we see the promise of God to all those who are called by his name. A thousand years of living and reigning with Jesus on this earth. But our life in paradise will not end even after those one thousand years, the seventh "day" of rest. We will go on to live with him in a new heaven and a new earth, and in a city called the New Jerusalem which will come down to earth from out of heaven.

"And I saw a new heaven and a new earth, for the first heaven and the first earth had passed away; and there was no more sea.[544] Then I, John, saw the holy city, New Jerusalem, coming down from God out of heaven, prepared as a bride adorned for her husband." (Revelation 21:1-2)

> "This new city, new Jerusalem, is the place that Jesus has presently gone away to prepare (Jn. 14:1-7). It is the Father's "*oikia*," His residence, His abode, His house of many mansions. …We've got mansions, many mansions, and they are in a brand new city. The materials used for this new city are not mere wood, brick or mortar, but they are the best of the best. It is a very real and literal place. We call it Heaven. God calls it new Jerusalem. It's the city that Abraham searched for, whose Builder and Maker is God (Heb. 11:8-10)."[545]

"And I heard a mighty voice from heaven saying, Behold, the tabernacle of God is with men, and he will dwell with them, and they shall be his people, and God himself shall be with them, and be their God." (Revelation 21:3)

> "Note that the people of Earth do not go to where God dwells, that is, to the third Heaven into which Paul and John were caught up. It is God who comes to dwell with man"[546]

"And God shall wipe away all tears from their eyes; and there shall be no more death, neither sorrow, nor crying. Neither shall there be any more pain, for the former things have passed away." (Revelation 21:4)

---

544 ("The absence of the seas does not mean that there will be no more waters at all, for waters will once again flow in abundance on the new Earth." Pg. 424, *The Unveiling*, by Keith Harris)

545 (Pg. 424, *The Unveiling: a Journey through the Book of Revelation*, by Keith Harris)

546 (Pg. 425, *The Unveiling: a Journey through the Book of Revelation*, by Keith Harris)

"No more death." As it was in the beginning so shall it be in the end. Unlike the idiotic, bald-faced, scientifically-impossible lie that is the theory of evolution (see the next chapter) that says there was death from the very beginning and it was death that was the *great creator* by causing the appearance of ever more successful types of organisms that <u>somehow</u> <u>miraculously</u> altered themselves by random chance, accidentally, into increasingly more complicated and near-infinitely-complex forms (which of course is an absolute scientific impossibility) …unlike that <u>un</u>scientific fairy tale, God said that he created a perfect world from the beginning and that there was *no* death, until the sin of man and subsequent curse on the universe and all of life.

And there shall be no tears, "neither sorrow, nor pain." What a life! With nothing to fear, nothing that can come along and ruin ones happiness or destroy one's life. No accidents or injuries or sickness or untimely deaths; no homes and lives being devastated by fire, or flood, or divorce, or tornados, or hurricanes, or tsunamis or volcanic eruptions. Peace will not only have come to earth in the form of an absence of communist, socialist, dictatorial government or mindless Muslim oppression, and an absence of violent conflicts and all out wars between nations, but peace will have come to every man, woman and child in the form of a quiet and completely secure mind and heart. Think about that. It is almost impossible to comprehend because of the innumerable dangers and evils and thefts and breaking-and-entering's and rapes and murders and car accidents and plane crashes and hunger and molestations and kidnappings and disappearances and job loss and loss of income and not being able to pay ones bills and fear of homelessness that lie in wait at every turn for the people of this world today, and for as far into the past as anyone can remember. (To my knowledge there are no stories, passed down from one generation to another, of Adam and Eve's life in the Garden before the fall. If there were, then we might have at least some distant collective memory to help envision the next life. But we do not.)

"The new Heaven and the new Earth will far exceed anything this world has to offer. The waters will be pure and crystal clear. The nights won't freeze… Everything will be perfect, for God himself will be there and our home will be permanent. I truly look forward to the things that await. Don't miss it! I know you'll love it, too.

…"John [the writer of the Book of Revelation] was allowed to give us only small portions of information… We can only imagine the grandeur of the city through our spiritually dimmed eyes. However, the information we do have helps us to readily see that the new city and the new Earth will be a tremendous and glorious place."[547]

That is one of the values of learning about God's plan for your future if you are a true Christian. No matter how difficult and laden with suffering your situation in this life may be, it is only temporary. Certainly you should seek to lighten your burden and those of the people around you (whether they are saved or not) in any way you can. But either way know that this life is but an instant compared to eternity, and know what is in store for you and yours in the coming world without end. Heaven is not sitting around in the clouds "fee fie fiddle-di-i-o, strumming on the old banjo," or on the old golden harp, being bored to tears. (No offense to harpists.) It is living on this phenomenal water planet, fraught with such awe inspiring beauty and glory and majesty, which God has specifically formed for us to enjoy forever. Knowing what is the come—"the earnest of our inheritance until

---

547   (Pg. 455-456, *The Unveiling: a Journey through the Book of Revelation*, by Keith Harris)

the redemption of the purchased possession" (Ephesians 1:14)—we can have a certain amount of peace even in these tumultuous economic times exacerbated by left-wing indoctrinated, Marxist revolutionaries who have taken over our Federal Government, whose leader campaigned as a "bipartisan moderate" and whose Democrat media covered up for him. But I digress. Jesus "purchased" this earth, indeed he spoke it into existence, and he purchased *us* with his own blood. And he is shortly coming back to claim what is rightly his. The 100% accurate, absolutely reliable prophetic revelations of God's Word concerning our near future *can* give you great peace of mind and comfort in the midst of the kind of economic turmoil we are experiencing today, and in the midst of the further environmental, financial, political, economic, social and military chaos that will soon follow. And there is no price that you can put on peace of mind. "And the peace of God which passes all understanding, shall guard your hearts and minds through Christ Jesus." (Philippians 4:7) "Peace I leave with you, My peace I give unto you; not as the world gives do I give unto you. Let not your heart be troubled, neither let it be afraid." (John 14:27)

"Then he who sat upon the throne said, Behold, I make all things new. And he said to me, Write, for these words are true and faithful. And he said unto me, It is done! I am Alpha and Omega, the Beginning and the End. I will give unto him who thirsts of the fountain of the water of life freely. He who overcomes shall inherit all things, and I will be his God and he shall be my son. But the cowardly, and unbelieving, and the abominable, and murderers, and the sexually immoral, and sorcerers, and idolaters, and all Democrats [oops, that should read all liars] shall have their part in the lake which burns with fire and brimstone [melted sulfur], which is the second death." (Revelation 21:5-8)

"What is the second death? We see in the passages above that the lake of fire and brimstone "is" the second death. ...those who die without accepting God's plan of escape will go immediately to Hades, the Old Testament "*Sheol*." ...this place of torment is within the Earth. ...[They] are held in Hades until the Great White Throne Judgment ([Rev.] 20-12). At that time, they (the "dead" without Christ) will be resurrected ...[and] they will be cast into the lake of fire, the everlasting Hell. It is the second death.

"You may have heard the saying, "Born once, die twice. Born twice, die once." That is, if you have only had a natural birth, you will die twice: the physical death followed by the second death. However, if you have been naturally born, then at some point in your life accept Jesus as Lord and Saviour, thus becoming "born-again" by the Spirit of God (born twice), you die only once, physically."[548]

"And there came unto me one of the seven angels who had the seven bowls filled with the seven last plagues... And he carried me away in the Spirit to a great and high mountain, and showed me that great city, the holy Jerusalem, descending out of heaven from God. Having the glory of God; and her light was like a most precious stone, even like a jasper stone, clear as crystal. ...The city is laid out as a square, and its length is as great as its breadth. And he measured the city ...twelve thousand furlongs [fifteen hundred miles!]. Its length, breadth, and height are equal. And then he measured the wall: one hundred and forty-four cubits, according to the measure of a man... And the construction of its wall was of jasper; **and the city was pure gold, like unto clear glass**." (Revelation 21:9-11,16-18)

---

548   (Pgs. 427-428, *The Unveiling: a Journey through the Book of Revelation*, by Keith Harris)

What an astronomical structure! The New Jerusalem, God's holy city, will be one thousand five hundred miles square! That's nearly the size of the moon. Of course the new earth will also have to be many times larger than its present size in order to support a city of that size without wobbling on its axis. If you took a drive around this city you would have to travel <u>six thousand miles</u>! That would be like driving from New York to the coast of California and back again! The new Jerusalem would cover an area fifty percent of the size of the United States. (That's one half for those in Congress.) And the *shape* of this "mega-structure" would appear to be like a mountain peak or a pyramid, rather than a cube:

> ..."The main reason for this assumption is that the wall is only 216 feet thick (144 cubits). A relatively thin wall with a cubical shape would seem to be completely out of proportion. Remember that the walls extends 1500 miles in length and in width.
>
> "The wall, like the entire city, is crystal clear and is constructed without the use of human hands. It is the city whose builder and maker is God (Heb. 11:10)."[549]

There is also the evidence from the Great Pyramid in Egypt of what the New Jerusalem will look like. In *The Nephilim and the Pyramid of the Apocalypse* Peter Heron postulates, according to all of the facts and evidence, that the Great Pyramid was not built by ancient Egyptian slaves—indeed that idea is preposterous as modern man himself with all of his advanced space-age technology couldn't replicate it—but by those same fallen angels who were cast out of heaven with Lucifer and came down to earth and slept with the daughters of men. These angels, or Nephilim, lived in heaven before they were evicted and experienced the New Jerusalem there. When cast down to earth they built a replica of it here. And again, the reader can decide for him or herself the accuracy of this hypothesis by picking up a copy of Mr. Herons thought provoking book.[550]

"And he showed me a pure river of water of life, clear as crystal, proceeding out of the throne of God and of the Lamb. In the middle of its streets, and on either side of the rivers, was there the tree of life, which bore twelve kinds of fruit, each tree yielding her fruit every month; and the leaves of the tree were for the healing of the nations. And there shall be no more curse, but the throne of God and of the Lamb shall be in it; and his servants shall serve him. And they shall see his face; and his name shall be on their foreheads. And there shall be no night there; and they need no lamp, neither light of the sun; for the Lord God gives them light. And they shall reign for ever and ever." (Revelation 22:1-5)

God said to Adam in the Garden after he and his wife had sinned against him, "Cursed is the ground for your sake." (Genesis 3:17) And this curse was not just on the topsoil, for "ground" here meant all of the material universe; so the curse was on all matter, and on all of the Laws of matter, and on all of the living. This curse was discovered in the 19th century when Sir William Thomson (Lord Kelvin) uncovered the universal scientific principle of *entropy* known as *The Second Law of Thermodynamics*, which states that everything in the universe is running down, decaying, going from order to disorder. (And <u>not</u> in the other direction, my Darwinist friends.) He came to this discovery through the inspiration of this verse in the Bible "All of them shall wax old like a garment." (Psalm 102:26)[551] But

---

549   (Pgs. 432-433, *The Unveiling: a Journey through the Book of Revelation*, by Keith Harris)

550   (See *The Nephilim and the Pyramid of the Apocalypse*, by Peter Heron)

551   (From article, "The Second Law of Thermodynamics," by Dr. Richard Paley @ objectiveministries.org)

now we see in the above verse from Revelation (22:3) that "there shall be no more curse." Jesus has reversed the curse that was brought upon the entire universe and the entire human race by the sin of Adam, and after this earth is cleansed, then the new heaven and the new earth will be brought about where there will be no more entropy and no more Second Law of Thermodynamics.

It is hard for us to imagine the awe, the wonder and absolute amazement of standing on the new earth and watching this city, fifteen hundred miles square, descending down from out of heaven, shining with marble and gold and all manner of precious stones, gleaming in the sun, tens of thousands of waterfalls coming from the river of life cascading down everywhere. Trees of every kind with innumerable forests and fields and all variety of foliage and flowers on thousands upon thousands of levels. Millions of infinity pools spilling off of as many marble terraces. The Hanging Gardens of Babylon, one of the Seven Wonders of the Ancient World, would look like a small potted plant next to it.[552] Imagine the feelings, and knowing that you are going to live there forever! Think of how incredible it was when you first saw Minas Tirith, that astonishing, gleaming-white city carved out of the side of a mountain in Tolkien's *The Lord of the Rings* movie. Minas Tirith, as incredible as it is, would be just a shed next to God's Holy Jerusalem. And the enormous trees of the Redwood and Sequoia National forests will be like fence posts compared to the forests of the new heaven and the new earth. The home of the elves in the trees at Lothlórien in the *Lord of the Rings* movie, and the tree-house home of the Na'vi in the movie *Avatar*, as awesome as they were, can give us just a glimpse of what the imagination and creativity of the Lord will produce, and give to the inhabitants of the new earth to build upon, in the future. There will be no boredom in heaven either, for every day will be as exciting and interesting as the one before, no matter how many thousands of years we will have already spent there. Every day will be the dawning of a new day, for "Behold, He makes all things new." (Revelation 21:5) ...Truly "eye has not seen, neither has any ear heard, nor has it entered into the imagination of man, what God has in store for those who love Him" and accept His free gift of eternal life.

We must keep in mind also that what we see on this earth is an aberration. It is not the norm. God's perfect will is done throughout this universe. It is only here on this earth where he has given man free will, and where six thousand years ago we chose evil, that *our* will is being done, and for the most part in opposition to his. But He is still sovereign because it is his choice to give us our own free will, and to allow us to choose to reject his will and act in accordance with the will of the dragon, the god of this world. But that is soon going to come to an end. Come quickly, Lord Jesus... "He who testifies to these things says, 'Surely I come quickly.' Amen. Even so, come, Lord Jesus." (Revelation 22:20) Not my will, Lord, but thy will be done... "And he was withdrawn from them about a stone's cast, and knelt down, and prayed, saying, Father, if you are willing, remove this cup from me. Nevertheless not my will, but yours, be done." (Luke 22:41-42) "After this manner therefore pray: Our Father who are in heaven, hallowed be your name. Your kingdom come. Your will be done in earth, as it is in heaven."

---

552   (We speculate that the new earth will be roughly a hundred times larger than the present earth, again, in order to accommodate this massive city without wobbling in its rotation. (But then again, what do we know?) Think of all the countless thousands of years it would take just to explore all of the wonders on a water planet of that size!)

(Matthew 6:9-10) …Even so, come quickly, Lord Jesus…

Paradise is coming back to this planet, and it's not coming by way of a charismatic, radical-left-indoctrinated leader like Barack Obama and his merry band of Marxist, baby-mutilating, Islam-protecting, capitalist-hating clowns, striving to create their own "Utopia" here on earth. (*Utopia*, as in *Venezuela*.) Nor is it coming by way of the George Soros' of the world, and their soon-to-be-revealed Antichrist's One World Government which will suddenly be upon us. No, it is coming by way of He who spoke all the worlds into existence and all the living into being. And anyone who refuses to know him and be known by him, is going to miss out on this eternal Garden of Eden II. They will instead be going to another place, a place of unending suffering, and gnashing of teeth. "There shall be weeping and gnashing of teeth." (Luke 13:28) "Where their maggots do not die, and the fire is never quenched." (Mark 9:48)

For the last hundred years, ever since World War I, man has been crying out for world peace, but there will be no world peace for man, not until the Prince of Peace returns to reclaim what is rightly his. When all "the earth shall be filled with the knowledge of the glory of the Lord, as the waters cover the sea." (Habakkuk 2:14) When that happens we will live in the knowledge of Him. We will be immersed in His love like the sun floods the countryside on a summer day. His glory will cover the earth like a monsoon washing over India. And in that state of consciousness there will be no room for sin. There will be no more liars, or murderers. There will be no more baby-mutilator voters for there will be no more baby-mutilators to vote for. There will be no more clitoral mutilators either, for there will be no more false religions to delude their followers into performing vile acts on their little girls in the name of a demonic god. There will be no more suicide bombers, or rapists. No more thieves or con-men. And no more death.

The Book of Revelation is a treasure chest of information concerning this time of the end that we find ourselves in. It should be coveted as far more valuable than the Hope diamond or the Crown jewels. For what price can be put on knowing what is going to happen in our near future and being able to understand why, especially in these turbulent times? Many Christians complain about "how hard it is to understand," but it is not. The Book of Revelation is what separates the men from the boys. Those who are serious about the entirety of God's Word, including those parts that disclose what God has in mind for our near future, no matter how terrible that vision might be, and those who are not. Many Christians shrink from Revelation for that very reason, because it is too fearful for them to contemplate (forgetting His admonition to "fear not!"), but in doing so they miss the incredible vision of what is coming *after* the Great Tribulation, an incomprehensible paradise that lasts for all eternity. It would be comparable to a woman who closes herself off from the joy of children and motherhood because she is too afraid of the pain of childbirth.

Revelation is a wellspring of foreknowledge, from the Mind of God to the mind of man, that is there for the taking. It just needs to be opened up. And one of the best books that we have found to do so, one that we have referenced heavily in this chapter, is *The Unveiling, a Journey through the Book of Revelation*, by Keith Harris. We recommend it highly.

Addendums

The Lost Tribes of Israel

"And I will gather the remnant of my flock out of all countries where I have driven them, and will bring them again to their pasture; and they shall be fruitful and increase in number." (Jeremiah 23:3)

We have mentioned the miraculous ingathering of the Jews who were scattered abroad after the Roman destruction of Jerusalem in 70 AD. These were the "descendants of those who were taken to Babylon under King Nebuchadnezzar [in 598 BC] and then returned during the reign of Cyrus"[553] 70 years later, which included the tribes of Judah, Levi and Benjamin. They made up the Southern Kingdom of Israel which earlier was divided from the northern tribes during the reign of King Solomon's son, Rehoboam. Yet equally as remarkable has been the gathering together of the preserved remnants of the other ten tribes of Israel— Reuben, Gad, Asher, Simeon, Dan, Naphtali, Issachar, Zebulun, Ephraim and Manasseh— from the Northern Kingdom who were conquered and carried off into captivity by the Assyrians from 740 to 722 BC. They have been discovered in these modern times and after all of these many centuries, from India to China and from Spain to Africa. And they still preserve many of their ancient Old Testament heritage.

In the Bnei Menashe, the "Sons of Manasseh," of Northeastern India we find a people who use "the Hebrew name of God as it appears in the Bible," and speak of "Mount Sinai, Mount Moriah, and Mount Zion." They also practice "biblical circumcision using a stone," and observe "other Old Testament ordinances such as those regarding skin disease. In addition to having songs, handed down from one generation to the next, such as this one commemorating the Exodus from Egypt:

"We must keep the Passover festival
Because we crossed the Red Sea cn dry land
At night we crossed with a fire
And by day with a cloud
Enemies pursued us with chariots
And the sea swallowed them up
And used them as food for the fish
And when we were thirsty
We received water from the rock" [554]

There are also lost tribes found in Ethiopia, India, Spain, China and Afghanistan. "If you visit Israel ... you will see Asian Jews ... dark-skinned Ethiopian Jews ... Hispanic Jews ... fair-skinned Nordic Jews ... Jews from India ... Jews from the four corners of the earth!"[555] The significance of this lies in showing us one

---

553   (Pg. 7, "A Rabbi Looks at the Lost Tribes of Israel," by Jonathan Bernis, Jewish Voice Today magazine: Proclaiming Messiah to the World, July/August 2010)

554   (Pgs. 8-9, "The Bnei Menashe of Northeastern India," by Sarah Weiner, Jewish Voice Today magazine: Proclaiming Messiah to the World, July/August 2010)

555   (Pg. 5, "About this Issue," by Sarah Weiner, Jewish Voice Today magazine: Proclaiming Messiah to the World, July/August 2010)

more example of God's ability to fulfill His promises in Scripture by doing the miraculous. He not only is gathering all of these scattered lost tribes back to Israel today, but He allowed them to keep from being completely assimilated, and from disappearing, into the cultures and nations they were exiled to over a period of 2700 years. Absolutely remarkable. The nation of Israel is truly God's Chosen People.

UFO's

What *are* these Unidentified Flying Objects that have been seen for decades by millions of eyewitnesses all over the globe? They are very probably demons, fallen angels, in their flying machines. But before we discuss the thinking behind this answer, let's explore whether some other, more popular explanations could be valid:

Could they be aliens from another galaxy? Probably not. Darwinists have come up with this explanation, believing as they do that life can arise spontaneously from lightning-struck-mud; so why not from "mud" elsewhere in the universe?! They of course are willfully ignorant of the law of Biogenesis that states, after centuries of countless scientific observations, that life comes only from life. Back to the drawing board, UFO-Darwinists. "There is simply no evidence at this point of any extraterrestrial life anywhere in the universe."[556]

Are they a hoax? "The problem with this view is that there are too many reliable eyewitness accounts of UFO sightings by people trained to know what is in the sky, including Air Force pilots and astronauts."[557] They are in all probability not a hoax.

Could they just be a result of various natural phenomenon? "Perhaps people have seen a satellite go over, or a weather balloon, or marsh gas, or a flock of birds, or reflected light from some other object."[558] However, this explanation doesn't fly as the source of *all* sightings because there are simply far too many that cannot be explained away by "natural phenomenon."

How about a top secret government military program? Unfortunately, "we have no technology anywhere in the world that can fly at the speed or angles of UFOs."[559] Scratch that explanation as well.

So what *are* UFOs? And who really is behind the wheel?

"Many researchers who study UFOs are now saying that the only possible explanation remaining for Unidentified Flying Objects is that they are of some ultradimensional nature, a spirit realm outside our three-dimensional, naturalistic world. ...What we believe ...and we emphasize that this is only an educated opinion ...is that we are seeing the world being prepared for the coming of the Antichrist on a UFO. In other words, we believe UFOs are of the demonic realm.

..."Our world today is not looking for a spiritual messiah; it is looking for a technological savior. If some UFO landed on planet earth and out walked a Christlike, benevolent creature, an E.T. who claimed to be a higher intelligence, who could solve all our economic and environmental and technological problems, the world would flock to him. [Look at how easily the wool was pulled over the eyes of those who flocked to

---

556   (Pg. 249, *Fast Facts on False Teachings*, by Ron Carson and Ed Decker)
557   (Pg. 249, *Fast Facts on False Teachings*, by Ron Carson and Ed Decker)
558   (Pg. 249, *Fast Facts on False Teachings*, by Ron Carson and Ed Decker)
559   (Pg. 250, *Fast Facts on False Teachings*, by Ron Carson and Ed Decker)

elect—and now, 2012, re-elect—Obama.]

"And for this reason God shall send them strong delusion, that they should believe a lie." (2 Thessalonians 2:11)

"In virtually all "fourth encounters," where people claimed to have been taken on UFOs, they all relate the same experience: The creatures told them they could save themselves. They don't need God because they are gods themselves. They don't need Jesus as their Savior because they can save themselves through cyclic rebirth. It's the same old doctrine of demons and deceitful spirits."[560]

"Now the Spirit expressly says that in latter times some will depart from the faith, giving heed to deceiving spirits and doctrines of demons." (1 Timothy 4:1)

It's the same old lie of Satan in the garden, "You shall be as God." Satan, after all, is the prince of the power of the air, and the ruler of the darkness of this age. "Wherein in time past you walked according to the course of this world, according to the prince of the power of the air, the spirit who now works in the children of disobedience." (Ephesians 2:2)

"For we wrestle not against flesh and blood, but against principalities, against powers, against the rulers of the darkness of this world, against spiritual wickedness in high places." (Ephesians 6:12)

Remember earlier how we saw that God will allow Satan and his false prophet (who quite possibly are the fallen angel, Apollyon, and his fallen angel sidekick) to delude the people of the earth with miracles and signs and wonders. "And he performs great wonders, so that he even makes fire come down from heaven on the earth in the sight of men. [UFOs firing their laser weapons?] And he deceives those who dwell on the earth by those miracles which he was granted power to do…" (Revelation 13:13-14) What greater *miracle* could there be for Beelzebub to perform than to have the skies alive with his demon-operated flying saucers and he himself striding out of one of them to rule the earth?

* * *

"But if we hope for what we do not see, then we do with patience eagerly wait for it." (Romans 8:25)

Very soon, after a horrific but thankfully brief seven-year period at the end of which her labor will have been completed, our "Mother Earth" will give birth to a glorious new age for this planet. One in which "the earth shall be filled with the knowledge of the glory of the Lord, as the waters cover the sea." (Habakkuk 2:14) Something to look forward to for those who choose to partake in it; for those who choose to believe in the truth of God, rather than the lies that they were taught.

"For the grace of God that brings salvation has appeared to all men, Teaching us that, denying ungodliness and worldly lusts, we should live soberly, righteously, and godly, in this present world; Looking for that blessed hope, and the glorious appearing of the great God and our Saviour Jesus Christ." (Titus 2:11-13)

See:
- *What in the World is Going On: 10 Prophetic Clues You cannot afford to Ignore*, by Dr. David Jeremiah
- *The Great Mystery of the Rapture*, by Arno Froese
- *The Unveiling: a Journey through the Book of Revelation*, by Keith Harris

---

560   (Pg. 250, *Fast Facts on False Teachings*, by Ron Carson and Ed Decker)

- *How Democracy Will Elect The Antichrist*, by Arno Froese
- *119 Most frequently Asked Questions About Prophecy*, by Arno Froese
- *Preparing For the Mark of The Beast*, by Arno Froese
- *Escape the Coming Night*, by Dr. David Jeremiah with C.C. Carlson
- *Armageddon: Appointment with Destiny*, by Grant R. Jeffrey
- *Prince of Darkness: Antichrist and the New World Order*, by Grant R. Jeffrey
- *Apocalypse: The Coming Judgment of the Nations*, by Grant R. Jeffrey
- *The Next World War: What Prophecy Reveals About Extreme Islam and the West*, by Grant R. Jeffrey
- *Messiah: War in the Middle East & the Road to Armageddon*, by Grant R. Jeffrey
- *Heaven: The Last Frontier*, by Grant R. Jeffrey
- *Final Warning: Economic Collapse and the Coming World Government*, by Grant R. Jeffrey
- *On Prophecy: Man's Fascination with the Future*, by J. Vernon McGee
- *Last Things*, by Evangelist Ron Comfort
- *The Edge of Time: The Final Countdown has Begun*, by Peter & Patti Lalonde
- *Beyond Iraq, The Next Move: Ancient Prophecy and Modern Day Conspiracy Collide*, by Michael D. Evans
- *The End Times are Here Now*, by Charles Halff
- *China, The Last Superpower: The Dragon's Hunger for World Conquest*, by Joseph Lam with William Bray
- *What the Bible Says About The End Times: Prophecies You Can't Ignore*, by Robert Taylor
- *Hidden Signs in the Olivet Discourse,* by Norbert Lieth
- *Planet Earth 2000*, by Hal Lindsay
- *The Final Prophecies* DVD, an Ingenuity Films Production
- "Hell on Earth Parts 1-4," from Turning Point with Dr. David Jeremiah
- *The Real Meaning of the ZODIAC*, by D. James Kennedy, Ph.D.

Are the stars themselves a prophecy given to us by our Creator from ancient times, from the very beginning, that foretold the plan of salvation that God was about to unfold over the ages? Do the names of the constellations and their accompanying minor signs have a far greater meaning than anything modern astrology is aware of? In this fascinating book, Dr. Kennedy opens up the meaning of the ancient zodiac, the real zodiac, before it was distorted over the years and turned into the daily *predictor* for pagans that it is today. Dr. Kennedy goes back to the very beginning of Genesis and starts with this quote, "Then God said, 'Let there be lights in the firmament of the heavens to divide the day from the night; and let them be <u>for signs</u> and seasons, and for days and years.'" (Genesis 1:14) Everyone understands what "seasons, days and years" refers to. But what of "signs?" "A sign is something which proclaims a message." And the message in the zodiac is nothing less than the birth, death, and resurrection of Jesus Christ- "The Gospel in the stars."

"The Bible tells us that God called all of the stars "the host of heaven"- that he numbered them, ordered them, and set them in the firmament to be signs. Their original meanings have been corrupted into something which is... counterfeit, something which has given birth to what is known as modern astrology, and which the Bible repeatedly condemns and warns Christians against.

..."Where did these signs come from? ...Scholars [have] discovered that their antiquity cannot be ascertained; no matter how far back we go, they are always still there!

…"It is interesting that the signs bear little or no resemblance to the stars in the heavens. If a hundred people were to go cut and name the constellations, they would not come up with the pictures that have been given to them and that have been held by peoples all over the world.

"According to Arabic tradition, the signs came from Seth and Enoch. This tradition is interesting since it links these signs to the grandson of Adam and says that Enoch and his father Seth (both men of faith) were the founders of this ancient understanding of the heavens."[561]

Dr. Kennedy's book is a fascinating and informative read for anyone who seeks the truth behind the ancient Zodiac.

- *Return of the Antichrist: And the New World Order*, by Patrick Heron
- *The Nephilim and the Pyramid of the Apocalypse*, by Patrick Heron
- *Shadow Government DVD*, by Grant R. Jeffrey
- *NIV Prophecy Marked Reference Study Bible*, by Grant R. Jeffrey
- *Psalm 83: The Missing Prophecy Revealed – How Israel Becomes the Next Mideast Superpower!* By Bill Salus … is there a pending war in the Middle East with far-reaching consequences that actually precedes the war of Gog and Magog? According to Bill Salus and his interpretation of the prophecy in the 83rd Psalm, and in other old Testament books, there is! According to the author in the very near future the nations of Egypt, Saudi Arabia, Iraq, Iran, Syria and Jordan, along with Hamas, Hezbollah and the Palestinians launch an attack to annihilate Israel. But they themselves are soundly defeated by the Israel Defense Forces and controlled, along with their wealth, by Israel herself. Stay tuned for further updates…

---

561   (Pgs. 8-9,12-13, *The Real Meaning of the ZODIAC*, by D. James Kennedy, Ph.D.)

Book III

WATER

CHAPTER 10

THE THEORY OF EVOLUTION IS A JOKE

"In the beginning was the Word, and the Word was with God, and the Word was God. The same was in the beginning with God. **All things were made by him; and without him nothing was made that has been made.**" (John 1:1-3)

Without Him, **nothing** that you see in existence today was made. Nothing! Not by random accidental chance. Not by random accidental mutations. Not by a big explosion and random accidental collisions of matter in space. Nothing! It is not The Word of God that is *anti-science*—the Big Lie that the totalitarian-left that controls the dissemination of information in our culture has been shouting for so long—it is the theory of evolution that is anti-science. In fact, as we will see in this chapter, it is a scientific **fact** that the theory of evolution is absolutely scientifically impossible according to the laws of science and all of the facts and evidence. (And remember, as Plato has instructed, we are following the facts and the evidence wherever they may lead.) We will also see that it is so scientifically-impossible, so idiotic, so asinine, absurd and ridiculous, and such a flat-out, bald-faced lie, that it literally borders on the insane. (With all due respect to those who, like myself, were taught it and believed it.) The theory of evolution is nothing more than a false religion. It is "a fairy tale masquerading as science."[562] It is Satan's genesis, and it has everything to do with his plan to keep people from coming to the knowledge of the truth of Jesus Christ as their Lord and Savior by undermining the credibility of God's word by attacking the real origins account, the Bibles Genesis, and replacing it with his own idiotic lie.[563] And even though we have begun this chapter with reference to the real Creator, our conclusions actually have nothing to do with religion or theology. Religion and theology *are* indeed very important subjects that are indispensable to the discussion of the *real* origin of the universe, of life on earth, and the astronomical complexity and diversity of species. But neither of these subjects are necessary for exposing the outright un-scientific hoax—the world's greatest hoax—that is the theory of evolution. What the indisputable and overwhelming facts and evidence and information that we are going to present in this chapter have already clearly established beyond any reasonable doubt is that based on the immutable laws of science, and on all of the scientific facts and evidence, according to science itself and the scientific method alone, the theory of evolution is a joke.

And in addition, the same holds true for its companion theory, that of the big bang, which seeks to explain in purely natural terms the origin and "evolution" of the universe and our solar system from some mythical massive primordial explosion, out of which the entire universe just formed itself somehow, and which is equally as un-scientific and absurd. We will expose these twin hoaxes along with setting the record straight on a number of other related subjects such as the fossil record, Dinosaurs, anthropology, the worldwide flood, the speed of light and the ridiculous yet widely accepted notion that the universe and our solar

---

562   (from a Creation Studies Institute newsletter, Sept 2013)
563   (Evolution = Evil delusion)

system are billions of years old. These incomprehensible ages did not become our scientific and educational establishments preferred doctrine as a result of the unbiased search for scientific truth, but rather were the product of the desperate need that evolution had for virtually limitless periods of time in which to work its "miracle." In this case *necessity* truly was *the mother of invention*.

Macro vs. micro evolution

"As you know not what is the way of the spirit, nor how the bones do grow in the womb of her that is with child: even so you know not **the works of God, who made everything**." (Ecclesiastes 11:5)

First it is important to understand the difference between these two terms, macro and micro evolution. Charles Darwin in his book, *On the Origin of Species by Means of Natural Selection, or the Preservation of Favoured Races in the Struggle for Life*, first published 150 years ago in November of 1859, attempted to show that the mind boggling diversity of life on earth happened by purely natural means. That by slow minor adaptations to their environment, each living species had slowly accumulated more advantageous characteristics and therefore had survived, whereas those competitors who had *not* evolved or somehow self-created those or other beneficial changes had become extinct. How *exactly* this could happen scientifically, the turning of one species into another entirely new, different and more complex organism by accident, was never explained by Darwin nor by anyone ever since; except to say with the devotion of pure, blind faith that "chance mutations over vast periods of time" did the trick. Nor did Darwin come up with an even remotely plausible scientific explanation as to *how* life could have accidentally formed itself originally, and nor has anyone since; except to say with the zeal of unwavering blind faith that somehow "lightning striking an ancient mythological atmosphere, or sea" did the trick. But the fundamental error of Darwin that led him to his erroneous conclusions in the first place was that he mistook *micro*evolution (limited change within one *kind* of organism), which is common to all of life, as being evidence for *macro*evolution (the massive change of one *kind* of organism into an entirely new and different one), which has never been observed, nor, as we shall see, has any credible evidence ever been found in the fossil record, or anywhere else on earth, to support it.

"Micro-evolution [is defined as] small adaptations within a population of organisms which allow a certain trait to be expressed to a greater or lesser degree than before; variation within a given category. This is regularly observed to occur within living populations.

"Macro-evolution [is defined as] large hypothetical changes which are thought to occur in an individual or in a population of organisms that produce an entirely new category or novel trait. These changes have never been observed to occur within living populations."[564]

...Or within dead populations (the fossil record) for that matter.

"Often textbooks and newspapers cloud the truth by confusing terms. While sometimes referred to simply as "change," evolution ...implies *enormous* changes in organisms. People supposedly came from ape-like ancestors, which came from a rat-like animal, which came from a fish, which came from an animal with no skeleton, which came from a single-celled organism. The proper term for this large-scale change is *macro*evolution. This macroevolution has... never been observed.

---

564   (Pg. 8, *The Young Earth: The Real History of the Earth – Past, Present, and Future*, by John Morris, Ph.D.)

[And keep in mind that the very *definition* of science is "the study of the physical and natural world and phenomena, especially by using systematic **observation**."[565] (Emphasis mine.) Macro-evolution has never been observed, therefore by the very *definition* of science, evolution is *not* science. It is a belief. And, as we shall see, like the Muslim suicide-bomber's faith that he'll get his 72 virgins, it is an idiotic belief that is not based in reality.]

"Frequently, what is observed scientifically is *micro*evolution. Microevolution refers to small changes within an animal type. No new features are produced, but varying traits are expressed. This is also called adaptation or variation. *All* the proposed examples of observed change are *micro*evolutionary changes.

"Scientifically proven examples of microevolution are: certain microbes' acquired resistance to antibiotics, pesticide-resistant insects, a color pattern shift in the peppered moth in England, and taller people today than those who lived long ago. Do these changes tell us anything about how people evolved from fish? Obviously not!"[566]

So microevolution, small changes within a kind, happen all the time. But this fact has absolutely nothing to do with, nor does it support in any way whatsoever, the scientifically-impossible theory that one organism can develop into another, higher, more advanced organism through accidental changes in its DNA. Macroevolution, as we shall prove throughout this chapter, is absolutely impossible. Never has occurred, never will. Back to the drawing board, Mr. Dawkins. We have big doggies and little doggies, short haired and long haired, Cocker Spaniels and Pit Bulls, Great Danes and Dalmatians. And we have tall humans and short humans, smart and dumb, black and white, Chinese and Indian, Eskimo and Pygmy. That's micro evolution; variation within a *kind*. ("And God said, Let the earth bring forth the living creature after its kind, cattle, and creeping thing, and beast of the earth each according to its kind: and it was so. And God made the beast of the earth after its kind, and cattle after their kind, and every thing that creeps upon the earth after its kind: and God saw that it was good." Genesis 1:24-25) But we don't have any flying humans. And there aren't any talking doggies. And that's because macroevolution doesn't exist in the real world, and Darwin's theory of evolution, with all due respect to misinformed scientists everywhere, is impossible.

But what about mutations? Don't these cause major changes in an organism's DNA and couldn't they eventually change one living thing into an entirely new one? No, not a chance...

"One possible source of significant change is that of mutations – the sudden, random, altering of a particular gene, which could then be passed on to offspring. Again, true science must ask, "What has been observed?" Although thousands of mutations have been observed, and thousands more have been induced, **never once has a mutation benefited an organism** or produced any *new* trait or functioning organ. **No mutation which could contribute to macroevolution is known to science**. Most mutations are neutral, many are harmful, and some are fatal. Therefore, it appears that random chance in the genetic code cannot improve it."[567] (Emphasis mine.)

"No mutation which could contribute to macroevolution is known to science."

565   (Encarta Dictionary)
566   (Pg. 6, *Dinosaurs, the Lost World, and You*, by John D. Morris, Ph.D.)
567   (Pgs. 6-7, *Dinosaurs, the Lost World, and You*, by John D. Morris, Ph.D.)

Beyond that, according to the laws of mathematical probability, it is abundantly clear, as we shall see later in this chapter, that "random chance in the genetic code" has no chance whatsoever of improving it, or of changing one viable organism into another. It is simply mathematically impossible for this to happen accidentally by random chance. Most of the systems in living organisms are irreducibly complex,[568] including most of the complex functions within the cell itself, and as such they could never have been produced by "small, incremental changes" (random mutations) within the genetic code. (See later section, "Irreducible Complexity.")

So, when we speak of *evolution* in this chapter, and in this book, we are referring to <u>macro</u> evolution, which we shall show is scientifically impossible; and not to <u>micro</u> evolution (small alterations, adaptations, variations and changes within a species or *kind*) which occur frequently but have never and could never lead to macroevolution.

Life is not the transforming miracle that evolutionists think it is. Like a much, much slower, organic version of the changing autobot robots in the movie *Transformers*, evolutionists suppose that life can just morph from one distinct life form into another, seemingly effortlessly, if given enough time. They make this gross misjudgment by observing the changes within a biological *kind* (read: species) that are pre-programmed into each organisms DNA, and surmising that therefore these organisms can miraculously transform themselves into whatever type of organism they might wish, again over vast periods of time. But, as we shall see, this is absolutely scientifically impossible. There is near infinite variety in living forms, and each of these living forms have pre-programmed into their DNA the ability for a fair degree of variation in response to their environments. (And this is because their Maker was not an idiot. He is a genius to dwarf all ingenuity, and He created his living things to be able to adapt and survive.) But there is great limitation within each organisms DNA as to how far that adaptive change is allowed to go. And it is <u>not</u> written in there to allow it to *transform* into another different *kind* of organism. Nor, as we just reported and as we will show in greater depth, can accidental mutations ever cause these supposed miraculous transformations. Again, life is <u>not</u> the transforming miracle that evolutionists suppose, and erroneously teach, that it is.

"And they shall turn away their ears from the truth, and shall be turned unto fables." (2 Timothy 4:4)

Part I - The Greatest Hoax on Earth

"And for this reason God shall send them a strong delusion, that they shall believe the lie." (2 Thessalonians 2:11)

"The theory of evolution is a fact!"[569] We have heard this stated so many times and so often by the high priests of this faux-science (a false religion disguised as science) in our educational system, and by their followers in our media and

---

568   (This is a term coined by Dr. Michael J. Behe, author of *Darwin's Black Box: The Biochemical Challenge to Evolution* and *The Edge of Evolution: The Search for the Limits of Darwinism*. We will discuss its fatal ramifications for the theory of evolution later in this chapter, in the section also titled, "Irreducible Complexity,".)

569   (Sure, and so is Santa Claus and his 9 flying reindeer.)

entertainment industry, that it has become ingrained in the psyche of far too many. But in reality the exact opposite is actually true.

> "The idea of evolution has come to be so firmly entrenched in our educational system that most people assume it is true. Scientific facts are [falsely and deceptively] placed within this interpretive scheme. End of discussion! Remember and repeat. Never mind the fact that no one has ever seen evolution take place, neither have the fossils documented evolutionary trends in the past, scientific law refutes the whole idea of evolution, and evolution is contrary to logic. Many people intuitively suspect evolution is not true, but still "believe" it anyway, because it is all they've been taught. "All educated people believe in evolution," they're told. Only ignorant, bigoted Christian fundamentalists still deny it."[570]

Yet the truth is that it is a total lack of proper education which bamboozles people into believing in evolution in the first place, and the real ignorance and bigotry lies in those evolutionists who refuse to even entertain the thought that there might be another side to the fairy tale...er...story. ("See no evil, hear no evil.") But let's start at the beginning. Let's take a look at all of the "evidence" that has been taught for decades as *proof* for the theory of evolution. Keep in mind, as we will clearly show, that <u>real</u>, <u>scientific</u> evidence is nonexistent.

Fabrications, exaggerations, hoaxes and lies...
... the "evidence" for evolution.
- <u>Miller-Urey's 1952 "amino acid in a bottle" experiment</u> is still reported in classrooms to this day as evidence that the essential building-blocks-of-life just formed themselves accidentally from lightning striking inorganic chemicals in earth's "early" atmosphere. But that is absurd. First of all, there are a number of problems with their experiment that render it completely useless for those who believe in the accidental origin of life. One is that the gases used in their experiments were, with what geochemists know today, nothing like the gases that we could expect in a hypothetical "early" atmosphere of earth. (Again you have to assume an astronomically large age for the earth which flies in the face of all of the actual scientific <u>evidence</u> that is supportive of a far younger earth and solar system. See "The Young Earth" section later in this chapter.) "The concentrations of methane and ammonia were carefully selected [by the experimenters] to <u>insure</u> the production of organic molecules." (Emphasis mine.) But in the real world, "a methane-ammonia reducing atmosphere would be fatal to life forms." Also, Miller and Urey used mild sparks to "simulate" lightning but that "is unrealistic." Because "Actual lightning would have destroyed any organics which may have been present."[571] Back to the drawing board.

But all of that, as damning as it is, is irrelevant because it is a fact that even if the test tube conditions of Miller-Urey's experiment were exactly the same as in earth's supposed early atmosphere and included real lightning, it is absolutely impossible for the next step in the accidental formation of life to ever have occurred. And that is for those Miller-Urey amino acids (the simplest and most primitive of organic molecules) to just accidentally form themselves into the complex protein chains essential for cellular life. To believe in that sheer impossibility is to take a

---

570   (Pg. 6, *The Young Earth: The Real History of the Earth – Past, Present, and Future*, by John Morris, Ph.D.)
571   (Pg. 114, *The Collapse of Evolution*, by Scott M. Huse)

leap of blind faith comparable to finding a substance spewed from a volcano that resembles asphalt and then concluding that the entire interstate highway system made itself accidentally by random chance. It would be like growing rock crystals in a jar and then teaching your students that this was proof that all of the precisely cut gems and jewelry on the face of the earth just made themselves accidentally by random chance. That of course would be idiotic. But in those two cases no one has a vested interest in denying the existence of man as creator of highways and jewelry. In the case of life, evolutionists not only have a vested interest in denying the existence of God as Creator of life, their entire careers depend upon it. Unfortunately for them, though, instead of being a strong evidence for the spontaneous formation of the building blocks of life and therefore the beginning steps of evolution, these early experiments of "life in a bottle," so highly touted then and still incredibly taught to this day, are evidence for the exact opposite. If the smartest scientific minds cannot recreate under completely controlled conditions in a laboratory the accidental formation of the necessary organic compounds for a living cell, then how in the hell could the earth have done so unconsciously?

"The probability of getting hit by lightning is about 1 in 600,000 (fortunately). The probability of winning a lottery grand prize with a single ticket is about 1 in 5.2 million (unfortunately). …the chance formation of even the simplest replicating protein molecule is 1 in 10^{450}."[572]

That is ten with 450 zeroes behind it! And remember that it has been scientifically established that anything with odds greater than 10^{66} could <u>never</u> happen, even if the universe lasted for an infinite number of years.

"Thus, we find that it is mathematically impossible for even the most elementary form of life to have arisen by mere chance. Life is no accident. It is not even something which brilliant scientists can synthesize. The bewildering complexity of even the most basic organic molecules completely rules out the chance of life originating apart from super-intelligent design and planning."[573]

To think that the essential building blocks of life, amino acids, could have somehow formed themselves accidentally and *then* somehow could have come together completely unconsciously to link up and create a single protein molecule (much less the very large number that are necessary for the life of a cell), is beyond preposterous. And this is not speculation, theory or conjecture. It is established scientific fact. So much for the fable that life just formed itself accidentally because a couple of scientists in the 1950's made an amino acid in a bottle.

But for those who prefer a more thorough analysis of this, we would recommend reading pgs. 113-115 in Scott M. Huse's *The Collapse of Evolution*; and the article "The Miller-Urey experiment," by J. H. John Peet, BSc, MSc, PhD, CChem, FRSC, online @ truthinscience.org.

- Recapitulation theory Another phony proof of evolution that we were all taught comes from the fudged drawings of a German zoology professor, Ernst Haeckel. He was an avid promoter of Darwinism and in 1866 advanced a theory that claimed the developing human embryo passes through all of its "previous evolutionary stages"—e.g. a worm, a fish with alleged gill slits, a frog and then

---

572   (Pgs. 65,68, *The Collapse of Evolution*, by Scott M. Huse)
573   (Pg. 69, *The Collapse of Evolution*, by Scott M. Huse)

reptile[574]—and "showed" this progression in a series of drawings. It is called recapitulation theory which states that "ontogeny recapitulates phylogeny," or an organism's embryonic development (ontogeny) mirrors its supposed evolutionary path (phylogeny). But it has long been known that Haeckel made up his diagrams, and the actual pictures we now have of human embryo development show his bald-faced distortions for what they are. His theory was and is a lie, but incredibly that still hasn't kept our public schools from continuing to use it to indoctrinate our children in the greater lie. (Would it be fair to ask, What the hail is going on here, in what is supposedly a "science" curriculum?!#$)

"Eventually, Ernest Haeckel admitted this fraud, but the deplorable aspect is that this theory is still taught in many universities, schools, and colleges throughout the world."[575]

"Evolutionists have taught for over a century that as an embryo develops, it passes through stages that mimic an evolutionary sequence. In other words, in a few weeks an unborn human repeats stages that supposedly took millions of years for mankind. A well-known example of this ridiculous teaching is that embryos of mammals have "gill slits," because mammals supposedly evolved from fish. (Yes, that's faulty logic.) Embryonic tissues that resemble "gill slits" have nothing to do with breathing; they are neither gills nor slits. Instead, those embryonic tissues develop into parts of the face, bones of the middle ear, and endocrine glands.

"Embryologists no longer consider the superficial similarities between a few embryos and the adult forms of simpler animals as evidence for evolution. Ernest Haeckel, by deliberately falsifying his drawings, originated and popularized this incorrect but widespread belief. Many modern textbooks continue to spread this false idea as evidence for evolution."[576]

"Valid science must have integrity, dependability, reliability, and be trustworthy. How can you come to true conclusions when experimental data is falsified?"[577]

Incredibly, evolutionists even tried to raise this hoax to the status of a *law* of science by calling it *The Biogenetic Law*.[578] However, this "law"—which merely attempts, fraudulently, to elevate the status of Haeckel's bogus theory—is scientifically absurd:

"The field of molecular genetics establishes the impossibility of the Biogenetic "Law." DNA is very specific and uniquely programmed for each type of organism. It simply does not recreate passing developmental stages of other organisms. It only produces after its own kind.

"Almost all scientists now reject the Biogenetic "Law." Only naïve or poorly informed evolutionists still cite this concept in defense of their theory for it has no valid scientific foundation whatsoever."[579]

---

574   (Pg. 113, *The Collapse of Evolution*, by Scott M. Huse)
575   (Pg. 105, *The Collapse of Evolution*, by Scott M. Huse)
576   (Pg. 11, *In The Beginning: Compelling Evidence for Creation and the Flood*, by Walt Brown, Ph.D.)
577   (Pg. 14, *Evidence for Creation: Intelligent Answers for Open Minds*, by Tom DeRosa)
578   (Which is not the same as the Law of Biogenesis, an actual law of science, "attributed to Louis Pasteur, [that] states that life arises from pre-existing life, not from nonliving material." From Wikipedia, the free encyclopedia.)
579   (Pg. 113, *The Collapse of Evolution*, by Scott M. Huse)

Scratch one more "evidence" for evolution.

- England's peppered moth But what about the case of England's famous peppered moths? Aren't they "proof" of evolution, as we were all taught?

> "Peppered moths have always existed in light, intermediate, and dark-colored varieties. Before the advance of the industrial revolution, the tree trunks were light-colored and the light-colored moths were camouflaged; whereas, the dark-colored moths were easily spotted and eaten by birds. Consequently, the dark colored moths constituted a very minor proportion of the total population.

> "As the industrial revolution progressed, however, and pollution increased, the tree trunks became darker and within 45 years the situation was reversed. In the Manchester vicinity, for example, 95% of the moths were of the dark-colored variety.

> "But is this really evolution? Certainly not! This process did not produce anything new. It did not result in increased complexity and organization. The dark-colored moths always existed. ...absolutely no evolutionary change occurred in these moths... Despite these obvious facts, many textbooks and encyclopedias continue to cite the peppered moth as an example of evolutionary development."[580]

England's peppered moths are an example of *micro*evolution, change within a kind, and not of Darwinian evolution. Through natural selection the fittest moths did survive; first the light-colored, and then as the trees darkened, the dark-colored. But they did <u>not</u> evolve into a new organism with any new, different or more complex organs; because that is impossible.

- "Vestigial" organs In the past evolutionists have tried to use the case of organs, like the thyroid, pituitary glands and the appendix, whose function scientists had yet to uncover, as evidence for their theory:

> "Vestigial organs are those structures which are presumed by evolutionists to be the useless remains of an organ which was once fully developed and operational in ancestral types. Such structures have long been cited as evidence for evolution since they are assumed to represent former evolutionary changes. ... [However] advances in our understanding of physiology have shown that supposed vestigial organs are actually quite useful and even essential. ...Today all organs formally classified as vestigial are known to have some function during the life of the organism. "[581]

- The Galapagos Islands But how about Darwin's famous Galapagos finches? Aren't they indisputable proof for the "fact" of evolution? No, they have just been taught that way. They are just one of thousands of examples we see across this planet of *variation within a species*, or <u>micro</u>evolution. Again, all organisms have pre-programmed within their DNA the ability to change and alter and adapt to the changing circumstances of their environments. This is because their Creator had foresight. But they do <u>not</u> have the ability to change into another species slowly over time by many different accidental mutations in their genetic code. That is scientifically impossible.

> "Christians look to Jerusalem; Muslims look to Mecca; but evolutionists look to the Galapagos as the spiritual center of their scientific faith."[582]

---

580 (Pg. 108, *The Collapse of Evolution*, by Scott M. Huse)
581 (Pg. 107, *The Collapse of Evolution*, by Scott M. Huse)
582 (Doug Phillips, executive producer of *The Mysterious Islands* DVD. -See list of

When Darwin visited the Galapagos Islands in the 1800's his observations "helped convince him that all living things evolved from a common ancestor." But science has come a long way in the last 150 years, and Darwin's fanciful but scientifically unsound conclusion has been thoroughly debunked. We now know that the "simple single cell" is an astronomically complex engineering marvel that could never, ever have happened accidentally by random chance. And we now know that the "endless forms most beautiful and most wonderful" that Darwin erroneously concluded "have been, and are being, evolved,"[583] are in fact displaying the built-in ability for great variation and adaptation that their Designer/Creator wrote into their DNA blueprints. It turns out that the Galapagos Islands are *not* the hallowed "laboratory of evolution" that devout Darwinists have long-claimed, but rather are a stunning "showcase for creation."[584]

"The random chaos that is the fundamental idea of evolution is altogether at odds with the rationality and design that we observe in creation. The mechanisms proposed for evolution do not square with the tenets of true science. Imagination, exaggeration, and in some cases, pure deception, are the hallmarks of the evolutionary movement.

"Evolution has been advertised and marketed to the public as the only credible and intellectual explanation of the origins of the universe and all life. Many have accepted it without question because it has been dressed up with false scientific integrity. There are those, usually at the universities, that have zealously dogmatized the world with the canons of evolutionary thinking. They have become the priests in their own religion and work with an evangelistic energy to spread the faith of evolution. Yet [it is all a] lie."[585]

- Other *evidence* There are a number of other things that evolutionists have cited as "proof" for their unscientific theory, and that, even though they have all been discredited, they still teach. Space precludes our listing all of them here but you can read chapter 6, "Commonly Cited "Proofs" of Evolution," from Scott Huse's exposé, *The Collapse of Evolution.* We'll let his words wrap up this section:

"To summarize, the commonly cited *proofs* of evolution—the fossil horse series, vestigial organs, the peppered moth, the duck-billed platypus, Archaeopteryx, the Biogenic "Law," the Miller-Urey experiment, and comparative anatomy—are either mere unfounded assumptions or known erroneous concepts. Since evolution has not been evidenced in the fossil record and is not observed today, evolutionists are completely deprived of any empirical evidence with which to support their dubious hypothesis."[586]

And speaking of "the fossil record," that just happens to be our next subject...

The fossil record

"Before the mountains were brought forth, or ever you had formed the earth and the world, even from everlasting to everlasting, you are God." (Psalms 90:2)

But what about the fossil record? Isn't this, finally, strong proof of the theory of evolution? Not hardly. We were all taught that this was such a powerful

---

recommended books, and DVD's, at end of chapter)
583 (From Charles Darwin's *On the Origin of the Species*)
584 (From "Impact," November 2009, a circular from Coral Ridge Ministries)
585 (Pgs. 69-70, *Evidence for Creation: Intelligent Answers for Open Minds*, by Tom DeRosa)
586 (Pg. 117, *The Collapse of Evolution*, by Scott M. Huse)

evidence, that proved the theory of evolution was a *fact*. But it is not. The problem is that after literally millions of evolutionary scientists and their devotees have eagerly, expectantly and exhaustively searched for trillions of hours and for well over a century and all over the globe, they have not found a single missing link in the fossil record. And that's because <u>there aren't any</u>. If evolution was real than for every single species we see there would have to be millions of transitional forms that led to each one of those species. For every bird that supposedly successfully developed or "evolved" an actual functioning set of wings on its own, out of thin air, accidentally,[587] there would have to be millions of fossils of semi-birds with really bad, unusable wings that died off in the blind and unconscious struggle for survival, and in this case the *search* for flight. (How an organism searches or "struggles" for something they don't even know exists is beyond me.) The same would have to be true of every species. But there are none whatsoever. All the fossils that we find are fully functioning, fully developed organisms. Even Charles Darwin, the author of the theory, stated that the fossil record would be the final authority as to the validity of his proposal. But the scientific fact of the total lack of evidence in the fossil record is conveniently ignored, so Satan's genesis can still be shoved into the minds of our unsuspecting school children.

> "The sudden appearance of advanced and diverse life forms that lack evolutionary ancestors, the permanence of kinds throughout time, and the complete absence of transitional forms in the fossil record testify vigorously for Biblical creationism. ...The evidence supplied by the fossil record substantiates the Biblical account of creation and the Flood while at the same time it soundly refutes evolution and uniformitarianism."[588]

> "All species appear fully developed, not partly developed. They show design. There are no examples of half developed feathers, eyes, skin, tubes (arteries, veins, intestines, etc.), or any of the vital organs (dozens in human alone). Tubes that are not 100% complete are a liability; so are partially developed organs and some body parts. For example, if a leg of a reptile were to evolve into a wing of a bird, it would become a bad leg long before it became a good wing."[589]

Besides, common sense itself dictates that virtually <u>any</u> complex organ could never be mutated from another previous, less-complex organ because along the mythological way from <u>A</u> (less-complex organ) to <u>Z</u> (more-complex organ), every letter (every *step*) in between would be a lousy alteration of <u>A</u> and an incomplete version of <u>Z</u> and therefore due to the survival of the fittest these organisms would be killed or die almost immediately. So therefore, even if it were possible for a random mutation to even get the transition from a lizard leg to a bird wing started (which it is not) it would merely result in a crippled leg and the lizard would be eaten as it tried to slunk away from its predator. And this holds true with any supposed advancement on Darwin's mythological tree of death (yes it is death and not life). Whether it's fish to snakes, or deer to giraffe, or elephant to whale,

---

587   (Think for a moment for how many centuries, even millenniums, man with all of his intelligence and design capabilities strived to learn how to fly but failed miserably at every attempt until finally the Wright brothers were successful. And then try to convince yourself that birds and insects and lizards could have each stumbled upon flight accidentally. It is truly idiotic.)

588   (Pg. 54, *The Collapse of Evolution*, by Scott M. Huse)

589   (Pg. 7, *In the Beginning: Compelling Evidence for Creation and the Flood*, by Walt Brown, Ph.D.)

or reptile to bird, or bush to beaver. End of evolution. The only possible way for a random mutation to turn a lizard into a bird is to do it all at once in a single generation, and this would be indistinguishable from a miracle. And the theory of evolution is not based on miracles. (Creation is.) It is based on small incremental accidental mutations that slowly advance the organism forward into another better, more highly developed, and better adapted organism, which is logically, factually and scientifically impossible. So much so that it surpasses the ridiculous and the irrational and borders on the insane. Again, as we mentioned in the beginning of this book, we live in an insane asylum, and the inmates have taken over. Literally. And I don't say that arrogantly or judgmentally because I was taught it and I believed it along with everyone else I went to school with. The lies of Satan cover our educational/left-wing-indoctrination system like a muddy layer of sediment, and we all believed the idiocy that we were fed.

"At the most fundamental level, a big gap exists between forms of life whose cells have nuclei (eukaryotes, such as plants, animals, and fungi) and those that don't (prokaryotes such as bacteria and blue-green algae). Fossil links are also missing between large groupings of plants, between single-celled forms of life and invertebrates (animals without backbones), among insects, between invertebrates and vertebrates (animals with backbones), between fish and amphibians, between amphibians and reptiles, between reptiles and mammals, between reptiles and birds, between primates and other mammals, and between apes and other primates. In fact, *chains* are missing, not *links*. The fossil record has been studied so thoroughly that it is safe to conclude that these gaps are real; they will never be filled."[590]

It is a flat-out deception to teach the school children of America that the fossil record is evidence for evolution and for vast periods of time.

"The "evolutionary tree" has no trunk. In what evolutionists call the earliest part of the fossil record (generally the lowest sedimentary layers of Cambrian rock), life appears suddenly, full-blown, complex, diversified, and dispersed—worldwide. Evolution predicts that minor variations should slowly accumulate, eventually becoming major categories of organisms. Instead, the opposite is found. Almost all of today's plant and animal phyla—including flowering plants, vascular plants, and vertebrates—appear at the base of the fossil record. **In fact, many more phyla are found in the Cambrian then exist today**. [Emphasis mine.] [Oops, back to the drawing board. Next theory: devolution.] Complex species, such as fish, worms, corals, trilobites, jellyfish, sponges, mollusks, and brachiopods appear suddenly, with no sign anywhere on earth of gradual development from simpler forms. Insects, a class comprising four-fifths of all known animal species (living and extinct), have no known evolutionary ancestors. The fossil record does not support evolution.

..."Frequently, fossils are not vertically sequenced in the assumed evolutionary order. For example, in Uzbekistan, 86 consecutive hoofprints of horses were found in rocks dating back to the dinosaurs. Hoofprints of some other animal are alongside 1,000 dinosaur footprints in Virginia. A leading authority on the Grand Canyon published photographs of horselike footprints visible in rocks that, according to the theory of evolution, predate hoofed animals by more than 100 million years. [Oops!] Dinosaur and humanlike footprints were found together in Turkmenistan and Arizona. Sometimes, land animals, flying animals, and marine animals are fossilized side-by-side in the same rock. Dinosaur, whale, elephant,

---

590   (Pgs. 11-12, *In The Beginning: Compelling Evidence for Creation and the Flood*, by Walt Brown, Ph.D.)

horse, and other fossils, plus crude human tools, have reportedly been found in phosphate beds in South Carolina. Coal beds contain round, black lumps called *coal balls*, some of which contain flowering plants that allegedly evolved 100 million years after the coal bed was formed. In the Grand Canyon, in Venezuela, in Kashmir, and in Guyana, spores of ferns and pollen from flowering plants are found in Cambrian rocks—rock supposedly deposited before flowering plants evolved. Pollen has also been found in Precambrian rocks deposited before life allegedly evolved. ["Houston, we have a problem."]

"Petrified trees in Arizona's Petrified Forest National Park contain fossilized nests of bees and cocoons of wasps. The petrified forests are reputedly 220 million years old, while bees (and flowering plants, which bees require) supposedly evolved almost 100 million years later. Pollinating insects and fossil flies, with long, well-developed tubes for sucking nectar from flowers, are dated 25 million years before flowers are assumed to have evolved. Most evolutionists and textbooks systematically ignore discoveries which conflict with the evolutionary time scale."[591]

Wait a minute! Don't pass by that last statement too quickly. "Most evolutionists and textbooks **systematically ignore** discoveries which conflict with the evolutionary time scale." Wow! Is that *science*, to systematically ignore evidence, to ignore observations? No, actually, that is what is known as lying by omission, and not science. To "systematically ignore discoveries" is to be ignorant, at least the last time I checked. And yet it happens all the time, every day, all across this great land, in the name of *science*. In the name of *evolutionary* science.

- Living Fossils And then there are "living fossils" …

"Did you know that flamingos, sandpipers, penguins, cormorants, parrots, owls and many other creatures living today, including numerous types of mammals, reptiles, amphibians and arthropods are found in supposedly 65-plus-million-year-old rock layers, when dinosaurs and other "pre-historic" beasts once roamed the earth? So, why don't we ever see them displayed in our museums or depicted together in books and textbooks?

Even though many fossils of these kinds of extent (or living) fauna have been found in allegedly "ancient" rock layers, this information is purposely censored from public view because it doesn't fit the evolutionist's millions-of-years' timeline."[592]

Ignoring the evidence in order to protect ones coveted yet false theory; what does that have to do with science?

- The "Geologic Column" The Darwinists tell us that all across this world there is a "geologic column" of different sedimentary layers or strata that show us the orderly march of time, and of evolution, over hundreds of millions and even billions of years. They call this *uniformitarianism*. Yet nothing could be further from the truth. All the evidence in the sedimentary strata across this world cries out for their cataclysmic and sudden deposition through the process of *liquefaction* (water saturated with earthen material that were deposited under enormous pressure), and the subsequent *rapid* burial of an entire planets life forms, both animal and plant, from both land and sea. (See "A World Wide Catastrophic Flood" coming up shortly.)

---

591 (Pg. 12, *In the Beginning: Compelling Evidence for Creation and the Flood*, by Walt Brown, Ph.D.)

592 (From "Living Fossils" article in a Creation Studies Institute circular, June 2011, by Tom DeRosa, Executive Director and Founder)

"**The entire geologic column was founded and built on the assumption that organic evolution was a fact**. The fact that modern historical geology is based on the assumption of evolutionary biology is a blatant case of circular reasoning. The only basis for placing rock formations in chronological order is their fossil content, especially index fossils. The only justification for assigning fossils to specific time periods in that chronology is the assumed evolutionary progression of life. In turn, the only basis for biological evolution is the fossil record so constructed. In other words, the assumption of evolution is used to arrange the sequence of fossils, then the resultant sequence is advanced as proof of evolution. Consequently, **the primary evidence for evolution is the assumption of evolution!** Since the arrangement of fossils is completely arbitrary (i.e., based on the assumption of evolution) the geologic column cannot be used to demonstrate evolution or a vast geologic age.

"It is [also] important to realize that **nowhere in the world does the geologic column actually occur**. It exists only in the minds of evolutionary geologists."[593]

"The primary evidence for evolution is the <u>assumption</u> of evolution!" Incredible, isn't it? And they get away with calling it science.

"Practically nowhere on earth can one find the so-called "*geologic column*." Most "geologic periods" are missing at most continental locations. Only 15-20% of the Earth's land surface has even one-third of these periods in the correct order. Even within the Grand Canyon, 150 million years of this imaginary column are missing. Using the assumed geologic column to date fossils and rocks is fallacious."[594] (…and stupid, and unscientific, and ignorant, and bigoted)

"The fossil record presents yet another serious problem for evolutionary uniformitarianism: large-scale fossilization is not occurring anywhere in the world today! When a fish dies, it does not sink to the bottom and become a fossil. Instead, it either decomposes or is destroyed by scavengers. Similarly, there is hardly a trace left of the millions of buffalo carcasses, which were slaughtered all over the plains only 2 generations ago.

"In sharp contrast to the virtual lack of fossilization transpiring today, there is an incredible amount of fossilization that occurred sometime in the past. The billions and billions of fossils we find preserved in the fossil record simply could not have been formed by processes observable in the world today as uniformitarian geologists have assumed and taught. Such preservation is very abnormal, the exception, and not the rule. Our global fossil record, therefore, attests to a strictly atypical world-wide, cataclysmic, hydraulic event. The fossil record of the geologic column is not a history of evolving life forms, but rather it is a great memorial to the sudden mass extermination of life from another age—the annihilation of the antediluvian world, around **2,346** B.C., by the Genesis flood.

"Evolutionists insist that the progression of life forms found in the fossil record from simple to complex prove the evolutionary progression of life. However, it should be pointed out the same general progression would be expected from the hydraulic sorting action of a worldwide cataclysmic flood."[595]

What would also be expected, and is also found, is countless anomalies in this fossil record where fossils of animals or advanced life forms are found in

---

593   (Pgs. 14-15, *The Collapse of Evolution*, by Scott M. Huse)
594   (Pg. 67, *In the Beginning: Compelling Evidence for Creation and the Flood*, by Walt Brown, Ph.D.)
595   (Pg. 47, *The Collapse of Evolution*, by Scott M. Huse)

sedimentary strata hundreds of millions of years before they supposedly appeared/ evolved. Who, or what, put them there? In addition to countless other examples of *polystratic*[596] fossilized trees, and even animals, going up through "millions of years" of sediment. Who put them there if it took millions of years for these sediments to be laid down? (It has to be Bush's fault.)

"Geologists also have discovered polystrate animal fossils. Probably the most famous is the fossilized skeleton of a whale discovered in 1976 near Lompoc, California. The whale is covered in diatomaceous earth. Diatoms are microscopic algae. As diatoms die, their skeletons form deposits—a process that evolutionists say is extremely slow. But the whale (which is more than 75 feet long) is standing almost on its tail at an angle, and is completely covered by the diatomaceous earth. There is simply no way a whale could have stood upright for millions of years while diatoms covered it, because it would have decayed or been eaten by scavengers. It is clear from this extraordinary evidence that the long ages attached to the geologic column simply are not correct."[597]

"Fossils all over the world show evidence of rapid burial. Many fossils, such as fossilized jellyfish, show by the details of their soft, fleshy portions that they were buried rapidly, before they could decay. (Normally, dead animals and plants quickly decomposed.) The presence of fossilized remains of many other adult animals, buried in mass graves and lying in twisted and contorted positions, suggests violent and rapid burial over large areas. These observations, together with the occurrence of compressed fossils and fossils that cut across two or more layers of sedimentary rock, are strong evidence that the sediments encasing these fossils were deposited rapidly—not over hundreds of millions of years. Furthermore, almost all sediments that form today's rocks were sorted by water. The worldwide fossil record is, therefore, evidence of rapid death and burial of animal and plant life by a worldwide, catastrophic flood. The fossil record is not evidence of slow change."[598]

The fossil record is full of animals and fish that died quickly and were buried rapidly. There is even a fossil of a fish *in the process* of swallowing another fish. Half of it is still sticking out of its mouth for goodness sake! How quickly and how deeply did they have to be buried for that to happen? There are countless examples of fossils of animals and fish with their backs bowed downward because something very heavy came upon them suddenly, and buried them instantly. Also, if these sediments, and these fossils, were laid down over hundreds of millions of years then "the fossil record should show continuous and gradual changes from the bottom to the top layers." Yet there are vast gaps and many discontinuities throughout. The fossil record was supposed to be the Darwinists best friend, but truly it is his worst nightmare.[599]

"Typically, fossils are found in fossil graveyards where the animals have been dismembered and the remains broken and scattered about. The bones are mixed with

---

596   (A *polystratic*, "many layers," fossil is a fossil that extends through many layers of strata, and therefore through "many eons of geologic time," which of course is impossible; but yet there they are, all over the world.)

597   (Pg. 55, *Dinosaurs Unleashed: The True Story about Dinosaurs and Humans*, by Kyle Butt and Eric Lyons)

598   (Pg. 11, *In the Beginning: Compelling Evidence for Creation and the Flood*, by Walt Brown, Ph.D.)

599   (Pgs. 10-11, *In the Beginning: Compelling Evidence for Creation and the Flood*, by Walt Brown, Ph.D.)

bones of other individual animals, making reconstruction extremely difficult. Animal and plant remains from vastly different environments and habitats are often mixed together. For instance, the dominant fossil at Dinosaur National Monument Park is a fossil clam. Many times, the deposits in which the fossils are found cover great areas on the earth's surface – at times hundreds of thousands of square miles. The event which deposited these sediments was not only of large geographic extent, but involved intense energy levels. Truly, what could have been larger and more powerful than the worldwide flood? [Hmm... evolutionists attempts to deny it?]

"For any creature to be fossilized in a flood, it must be buried rapidly. Otherwise, it will decompose or be eaten within a fairly short period of time. Obviously, these vast fossil graveyards were collections of creatures who died in the global flood of Noah's day."[600]

"Fossils rarely form today, because dead plants and animals decay before they are buried in enough sediments to preserve their shapes. We certainly do not observe fossils forming in layered strata that can be traced over thousands of square miles. ... [Only a worldwide deluge can explain] why animals and plants were trapped and buried in sediments that were quickly cemented to form the fossil record and why fossils of sea life are found on every major mountain range."[601]

It is also instructive that before the spread of the error of Darwinism, scientists all across this globe and for centuries <u>observed</u> the features of the earth and concluded that they were the result of a worldwide deluge. They could think of <u>nothing</u> else to explain them that made any sense whatsoever. And that's <u>not</u> because they were stupid. It's because they were not brainwashed and deceived by having it drilled into them that "the theory of evolution is a fact!" and "if you don't agree with us then you're stupid!" ... like our professors today are. They did not *see* everything in the world through the distorted lenses of evolutionary glasses. They were not blind.

"Let them alone. They are blind leaders of the blind. And if the blind lead the blind, both shall fall into a ditch." (Matthew 15:14)

And for a further discussion on the actual cause of the world's fossils and sedimentary strata, see the section "A World Wide Cataclysmic Flood" later in this chapter...

Natural Selection

"You have made heaven, the heaven of heavens, with all their host, the earth and everything therein, the seas and all that is therein, and you preserve them all." (Nehemiah 9:6)

But what about survival of the fittest? Doesn't the struggle of all living things to survive cause evolution to occur? No, not hardly. It can cause only <u>micro</u>evolution to occur, where the ability to adapt to certain changing environmental pressures is already pre-programmed into the organisms DNA. And if that specific adaptive ability is not already preprogrammed then nothing, and no known mechanism, can cause it to miraculously appear within that organisms DNA. There is no question that in the wild the "strong do survive" and the weak perish, but this fact has nothing whatsoever to do with the idea that these *strong* can also develop,

---

600   (Pgs. 34-35, *Dinosaurs, the Lost World, & You*, by John D. Morris, Ph.D.
601   (Pg. 116, *In the Beginning: Compelling Evidence for Creation and the Flood*, by Walt Brown, Ph.D.)

accidentally by random chance, more sophisticated and more highly advanced appendages or organs even to the point of morphing by accident into entirely new creatures that would allow them to be even *stronger*. For anything like that to ever happen, accidentally, within an organisms DNA is completely beyond the realm of possibility, or mathematical probability. It would be no different than if an old computer were able to change itself from Windows 95 to Windows 7 without a download or an upgrade to its software and hardware. Impossible. Again, were it to occur in either a computer or in life's DNA, it would be indistinguishable from a miracle.

"Acquired characteristics—characteristics gained after birth cannot be inherited. For example, large muscles acquired by a man in a weight-lifting program cannot be inherited by his child. Nor did giraffes get long necks because their ancestors stretched to reach high leaves.

…"However, stressful environments for some animals and plants cause their offspring to express various defenses. New genetic traits are not created; instead, the environment can switch on **genetic machinery** *already present*. [Emphasis mine] The marvel is that optimal genetic machinery already exists to handle some contingencies, not that time, the environment, or "a need" can produce the machinery."[602]

"Natural selection cannot produce *new* genes; it *selects* only among pre-existing characteristics. As the word "selection" implies, variations are reduced, not increased.

…"The variations Darwin observed among finches on different Galapagos islands is another example of natural selection producing micro- (**not** macro-) evolution. While natural selection sometimes explains the survival of the fittest, it does not explain the *origin* of the fittest. Today, some people think that because natural selection occurs, evolution must be correct. Actually, natural selection *prevents* major evolutionary changes."[603]

"Mutations are the only known means by which new genetic material becomes available for evolution. Rarely, if ever, is a mutation beneficial to an organism in its natural environment. Almost all observable mutations are harmful; some are meaningless; many are lethal. No known mutation has ever produced a form of life having greater complexity and viability than its ancestors.

…"A century of fruit fly experiments, and involving 3,000 consecutive generations, gives absolutely no basis for believing that any natural or artificial process can cause an increase in complexity and viability. No clear genetic improvement has ever been observed in any form of life, despite the many unnatural efforts to increase mutation rates."[604]

In addition, even if macroevolution were feasible (which it is not) and genetic mutations could somehow prove positive and then slowly change one organism into another over time with incremental alterations (which it also cannot), evolution would still be absolutely impossible for purely practical reasons. For (as we discussed previously, but let us reiterate) long before an organism changed *completely* into a more advanced and stronger form, it would have to try to survive as a transitional form which would have much less of an ability to survive

---

602   (Pgs. 5-6, *In the Beginning: Compelling Evidence for Creation and the Flood*, by Walt Brown, Ph.D.)
603   (Pgs. 6-7, *In the Beginning: Compelling Evidence for Creation and the Flood*, by Walt Brown, Ph.D.)
604   (Pg. 7, *In the Beginning: Compelling Evidence for Creation and the Flood*, by Walt Brown, Ph.D.)

than either the organism it was "struggling" away from, or the organism it was supposedly somehow blindly and unconsciously "struggling" towards. Let's look at the "classic" evolutionary example that reptiles, in particularly Archaeopteryx, evolved into birds...

"How could a walking, ground-dwelling – or even tree-dwelling – reptile become a bird that could fly? [Why the same way that your pick-up truck could slowly, accidentally, become a Piper Cub; in the Alice-in-Wonderland fantasy world of the human mind.] For the front limbs to be effective wings, multiple mutations would have to result in progressively longer fingers or arms with a membrane between each finger. Long before the walking or climbing creature would have good wings, it would have inadequate legs and would be at a tremendous disadvantage against predators. In fact, it is hard to imagine how a creature that could not yet fly, but which possessed long fingers or arms, could even survive, let alone have an advantage in the struggle for existence. [Unless Darwinists were to come up with a variation of their theory, called "survival of the silliest." In this scenario predators would be laughing too hard to catch these prey, and the deformed transitions would therefore survive to "struggle" on...] Natural selection would weed out these mutant misfits.

[And that is why evolutionists, after scouring the globe for a hundred and fifty years, have not found one single transitional life form in the fossil remains—even though there should be trillions of them—not only because none ever existed, but because they could never have survived even if they had.]

"Furthermore the breathing apparatuses of birds and reptiles are entirely different. In birds, air flows in one air pipe and out another, while in reptiles, the air goes in and out through the same opening, as in mammals. Since breathing is essential for survival, and cannot be halted for even a few minutes, how could such a major transition occur by mutation and natural selection? [Answer: it could not.] A similar problem exists with the totally different design of the heart. Clearly birds did not evolve from dinosaurs or other reptiles."[605]

So, any way you slice it, and as kind and as bend-over-backwards accommodating as you might care to be, evolution still comes out smelling of dead fish. It's a joke.

The father and hero of modern evolutionists, Darwin, viewed "all beings, not as special creations, but as lineal descendants of some very few beings." And whereas that's a nice view, it has no foundation in science, or in reality...

"The complete title of his book, *The Origin of Species by Means of Natural Selection or the Preservation of Favored Races in the Struggle for Life*, illustrates the power that Darwin attributed to the concept of natural selection. His hope, centered on limited observations, was that the struggle for life would give us an upward movement of change in which the best of the best would succeed. ... [However] Darwin knew nothing of genetics and DNA in his day. He was forced to make suppositions about mechanisms at a much more simplistic level, but our understanding of genetics today actually invalidates those assumptions.

"The bedrock of Darwin's theory, "natural selection," is shown ...to have no creative power whatsoever. Natural selection only has the ability to eliminate populations with traits less suited to survival, or to cause traits that were already designed into the DNA to become more or less dominant, but it cannot create new characteristics (such as new body parts). Darwin's belief in the ability of organisms to acquire new characteristics through

---

605   (Pg. 22, *Dinosaurs, the Lost World, & You*, by John D. Morris, Ph.D.)

environmental influence, effort, exercise, use or disuse is also shown to be without merit. We know today that none of these factors has any influence upon the fixed coding of the DNA that is passed on to future offspring."[606]

Which is why a very wise individual said a very long time ago, "The fool has said in his heart, There is no God." (Psalms 14:1) Which translates into today's language as, "The improperly educated evolutionist has foolishly determined in his mind, There is no intelligent Creator." (Sciences 15:2)

On his visit to the Galapagos Islands in 1835, Charles Darwin observed...

..."changes and diversification. [But then] Darwin takes a great leap of faith: "From so simple a beginning endless forms most beautiful and most wonderful have been and are being evolved." Darwin concludes that small changes within a kind, something we can observe, are proof of something which has never been observed, evolution, or the transformation of one kind of animal into another."[607]

However, we know a lot more now than the folks in Darwin's day, and it is clear from all the facts and evidence that we now have at our disposal that the natural selection of certain dominant or recessive traits within an organism, depending on the demands of its environment, is not proof of evolution. But it is proof that a Supreme intelligence was behind all of creation, and that the amazing power of *change, adaptation and variation within each species or biblical "kind"* is also proof that a Supremely intelligent Creator was there pre-programming that adaptability into every species DNA. Just like the operating system in your computer is pre-programmed to handle a whole host of contingencies, and viruses, and errors.

Anthropology

"Then God said, Let us make man in our image, according to our likeness." (Genesis 1:26)

But what of anthropology? What of all of those charts that we see everywhere we go—in our schools, our museums, our universities and on our science channels—that show an ape slowly morphing into a man? And what of all those "ape-men" discoveries over the years, those "transitional" forms in between apes and men that have "proven" that man descended from apes? Answer: "Sorry, Wally World's closed." As far as the first question, the charts are fabrications. They are not based on science or the scientific evidence, but on the fertile imaginations and evolutionary biases of those who desperately want the unscientific idiocy of Darwinism to be true; and will tolerate no alternative. They have staked their entire lives and their reputations on it, and what we see in their "ape-men" progression charts is their unscientific fantasies.

As far as the second question, there has not been found one single, solitary transitional form between apes and men. Every single *heralded discovery* over the years that claimed a fossil of a primitive and transitional ape-man was found, has turned out to be a gross exaggeration; either an outright hoax, or a chimpanzee, or an ape, or a pig. But our evolutionist friends who control the dissemination of information in our Democrat-Left media, educational system and entertainment

---

606   (From, Creation Studies Institute circular, October 2009, by Tom DeRosa)
607   (From, *The Mysterious Islands: A Surprising Journey to Darwin's Eden*, An Erwin Brothers Film)

industry have not thought it necessary to pass along this information to the public, or to their students. Like the truth about the Bible or the truth about far-left, Muslim-born-and-bred-in-Indonesia, Democrat candidates for the presidency, this is all the news that is <u>not</u> deemed fit to print, or to share in any way with the unwashed masses. After all, they might come to the "wrong" conclusion... like evolution is a scientifically-impossible joke.

Let's take a peek...

"All flesh is not the same flesh, but there is one kind of flesh of men, another flesh of beasts, another of fishes, and another of birds." (1 Corinthians 15:39)

- Piltdown "man" In 1912 a few fragments of teeth, bones, and primitive implements were allegedly found in a gravel pit in Piltdown, Sussex, England.

"The remains were acclaimed by anthropologists to be about 500,000 years old. A flood of literature followed in response to this discovery with Piltdown Man being hailed in the museums and textbooks as the most wonderful of finds. Over 500 doctoral dissertations were performed on Piltdown Man."[608]

"[However] it is now universally acknowledged that Piltdown "man" was a hoax, yet Piltdown "man" was in textbooks for more than 40 years."[609]

Reader's Digest, in October, 1956, published an article, *The Great Piltdown Hoax* in which it reported the findings that the jaw-bone was that of an ape that had died 50 years before, and that the teeth had been filed down and biochromate of potash had been put on them in order to hide their identity.

"And so, Piltdown Man was built upon a deception which completely fooled all the "experts" who promoted him with the utmost confidence. ...The person responsible for placing the fake fossils in the pit at Piltdown [had] authored several philosophical books in which he attempted to harmonize evolution and Christianity. Exasperated by the lack of convincing evidence for Darwin's theory, [he] was apparently motivated into *assisting* the theory of evolution by fabricating the needed missing link.

"It should be noted that Piltdown man was viewed in stately museums and studied in major textbooks for several generations."[610]

How incredible is all of that? "<u>Over 500 doctoral dissertations</u> were performed on" a flat-out hoax!? "Piltdown man was viewed in stately museums and studied in major textbooks for several generations." Yet they were all viewing and studying a bald-faced lie! Darwinists were overcome with pride over their *great find*. It was such a wonderful *proof* of their hallowed theory, yet not one of them thought to actually <u>check it out</u> to see if it was valid, as actual scientists are supposed to do. No, they were too busy calling anyone who disagreed with them "stupid," "<u>un</u>scientific" and (worse yet) "religious." How pathetic. And how utterly opposed to the cause of true science.

One *transition* between apes and men down, just a few more to go...

- Rama Lama Ding Dong And then there's our old friend *Ramapithecus*, made famous by the Leakey's archeological work in Kenya, and splashed all over our

608   (Pg. 100, *The Collapse of Evolution*, by Scott M. Huse)
609   (Pg. 12, *In the Beginning: Compelling Evidence for Creation and the Flood*, by Walt Brown, Ph.D.)
610   (Pgs. 100-101, *The Collapse of Evolution*, by Scott M. Huse)

TVs and newspapers for years as the missing link; and it was splashed confidently and without question by the evolutionary true believers who saturate our media. But, oops, it was nothing more than an ape:

"Before 1977, evidence for *Ramapithecus* was a mere handful of teeth and jaw fragments. We now know these fragments were pieced together incorrectly by Louis Leakey and others into a form resembling part of the human jaw. ... Some textbooks still claim that *Ramapithecus* is man's ancestor, an intermediate between man and some apelike ancestor. This mistaken belief resulted from piecing together, in 1932, fragments of upper teeth and bones into... two large pieces. ...This was done so the shape of the jaw resembled the parabolic arch of man... In 1977, a complete lower jaw of *Ramapithecus* was found. The true shape of the jaw was not parabolic, but rather U-shaped, distinctive of apes. ... *Ramapithecus* was just an ape."[611]

However, the problem here is that this is the truth. And Biology textbooks on evolution are exempt from the normal requirements of educational material, like printing the truth for example. It is a religion after all.

"Fossils of supposedly ape-like man are speculative, incomplete, falsified and imagined."[612]

- Nebraska "man" Next the Cornhuskers have an entry:

"Nebraska man was discovered in 1922 by Harold Cook in the Pliocene deposits of Nebraska. A tremendous amount of literature was built around this supposed missing link which allegedly lived 1 million years ago. [But, shocker, it was all a lie.]

"The evidence for Nebraska Man was used by evolutionists in the famous Scopes evolution trial in Dayton, Tennessee in 1925. William Jennings Bryan was confronted with a battery of "great scientific experts" who stunned him with the "facts" of Nebraska man. Mr. Bryant had no retort except to say that he thought the evidence was too scanty and to plead for more time. Naturally, the "experts" scoffed and made a mockery out of him. After all, who was he to question the world's greatest scientific authorities? [Like the "great scientific authorities" who wrote those 500 plus doctoral dissertations on the Piltdown hoax.]

"But what exactly was the scientific proof for Nebraska Man? The answer is a tooth. That's right; he found one tooth! The top scientists of the world examined this tooth and appraised it as proof positive of a prehistoric race in America. What a classic case of excessive imagination!

[What a classic case of a bunch of pompous, ignorant horse's asses hurling false accusations at a great American patriot, and slandering his character and reputation in the process. (Kind of reminds you of the congressional Democrats and their media.)]

"Years after the Scopes trial, the entire skeleton of the animal from which the initial tooth came was found. As it turns out, the tooth upon which Nebraska man was constructed belonged to an extinct species of pig. **The "authorities" who ridiculed Mr. Bryan for his supposed ignorance, created an entire race of humanity out of the tooth of a pig!**"[613]

But, again, that's the truth. Which won't in any way hinder our public school system from continuing to indoctrinate our children (mostly out of ignorance I am sure, and hopefully not out of any intentional desire to deceive) in the falsehood

---

611 (Pg. 13, *In the Beginning: Compelling Evidence for Creation and the Flood*, by Walt Brown, Ph.D.)
612 (Pg. 61, *Evidence for Creation: Intelligent Answers for Open Minds*, by Tom DeRosa)
613 (Pgs. 97-98, *The Collapse of Evolution*, by Scott M. Huse)

that the Scopes trial was a "victory" for evolution; rather than teaching them the truth that it was an historical showcase for how those who are obsessed with a scientifically-impossible theory can make complete fools out of themselves by combining just one pigs tooth with their over-heated imagination, and coming up with an entirely new and non-existent species in the process.

- A cup of Java Another false alarm was *Pithecanthropus erectus* (not to be confused with an erectile dysfunction drug of a similar name) or Java Ape-Man. In fact, it was the discoverer, Eugene Dubois himself who forty years later...

> ..."conceded that it was not a man, but was similar to a large gibbon (an ape). In citing evidence to support this new conclusion, Dubois admitted that he had withheld parts of four other thigh bones of apes found in the same area."[614]

> "The dubious nature of Java Ape-Man (and human evolution as well) is either conveniently ignored or concealed behind the mask of "scientific fact."

> "One final note regarding Java Ape-Man. Another *Pithecanthropus* was found in Java in 1926. Typically, this discovery was also billed as a prodigious breakthrough, the missing link for sure. It turned out to be the knee-bone of an extinct elephant."[615]

Again, how pathetic. It is obvious that these people were never out there for the purposes of scientific observation. They were searching for what they wanted to see. And no matter what they found, they were going to force it to fit into their pre-conceived notion of evolution, even if that meant presenting the knee-bone of an elephant as *evidence* for man's descent from apes.

> "Preliminary analysis has revealed that chimp and human DNA are much more different than claimed by the evolutionary spin-doctors. Furthermore, each is rapidly accumulating mutations today. These changes in the genomes cause damage to the genome and result in birth defects, not evolution. In fact, the genomes seem to be deteriorating so rapidly that eventual extinction of both is only a matter of time."[616]

- Sum Ting Wong But what of Peking man or *Homo erectus?* Is this finally the missing link? (Or just a San Franciscan with good posture?)

> "Many experts consider the skulls of Peking "man" to be the remains of apes that were systematically decapitated and exploited for food by true man. Its classification, *Homo erectus*, is considered by most experts to be a category that should never have been created."

And here's another dud...

> "The first confirmed limb bones of *Homo habilis* [not to be confused with *Homo fabulous*, a close cousin of *Homo erectus*] [Oh boy, could be a lawsuit there.] were discovered in 1986. They show that this animal clearly had apelike proportions and should never have been classified as manlike (*Homo*)."[617]

614　(Pg. 13, *In the Beginning: Compelling Evidence for Creation and the Flood*, by Walt Brown, Ph.D.)

615　(Pg. 100, *The Collapse of Evolution*, by Scott M. Huse)

616　(Pg. 76, *The Young Earth: The Real History of the Earth – Past, Present, and Future*, by John Morris, Ph.D.)

617　(Pg. 13, *In the Beginning: Compelling Evidence for Creation and the Flood*, by Walt

"Your hands have made me and fashioned me" (Psalms 119:73)

- Southwest Colorado Man was another fantastic discovery that added to the *reams of evidence* (bupkis in reality) of apes morphing into men, but it was also nothing but a tooth, which turned out to be that of a horse.

> "How resourceful and imaginative scientific "experts" can be at times. Give them one tooth, not necessarily human, and they can create an entire race of prehistoric humanity."[618]

> "Bones of modern-looking humans have been found deep in undisturbed rocks that, according to evolution, were formed long before man began to evolve. Examples include the Calaveras skull, the Castenedolo skeletons, Reck's skeleton [not a distant relative of Red Skelton, another great American patriot] and others. Remains such as the Swanscombe skull, the Steinheim fossil, and the Vertesszollos fossil presents similar problems. Evolutionists almost always ignore these remains."[619]

"See no evil, hear no evil, speak..." "Acknowledge no truth, publish no truth, teach no truth..." The mission statement of evolution instruction.

- Neanderthal Man was once hailed as another "intermediary link between man and apes," but is now "classified as *Homo Sapiens*, completely human."[620]

> "For about 100 years the world was led to believe that Neanderthal man was stooped and apelike. This false idea was based upon some Neanderthals with bone disease such as arthritis and rickets. ... Neanderthal man, Heidelberg man, and Cro-Magnon man are now considered completely human. Artist's drawings of "ape-men," especially their fleshy portions, are often quite imaginative and are not supported by the evidence."[621]

Another one bites the dust. Scratch one more *transition* between apes and men. But don't give up yet! There is still hope. Anthropologists are now very excited about the possibility of studying a missing link that, incredibly, is still alive! ("Bill Maher, could you please report to the Smithsonian.")

"Your hands have made me... You have clothed me with skin and flesh, and have knit me together with bones and sinews." (Job 10:8, 11)

- I Love Lucy And finally there is Lucy. Not Ricky Ricardo's wife, but an *australopithecine* trumpeted by National Geographic Magazine (which has never met an evolutionary hoax that it did *not* embrace) in a December 1976 article as another missing link, and as having "walked on two legs." However, an unbiased reading of her bone fragments now suggests that Lucy was a chimpanzee. In addition...

> ..."there is evidence that people walked upright before the time of Lucy. This would include the Kanapoi hominid and Castenedolo Man. Obviously, if people walked upright

---

Brown, Ph.D.)

618 (Pg. 98, *The Collapse of Evolution*, by Scott M. Huse)

619 (Pg. 14, *In the Beginning: Compelling Evidence for Creation and the Flood*, by Walt Brown, Ph.D.)

620 (Pgs. 101-102, *The Collapse of Evolution*, by Scott M. Huse)

621 (Pg. 13, *In the Beginning: Compelling Evidence for Creation and the Flood*, by Walt Brown, Ph.D.)

before the time of Lucy, then she must be disqualified as an evolutionary ancestor."[622]

No! Don't be silly. Haven't you gotten it yet? <u>Only the truth can be disqualified</u>. Get with the program.

"The australopithecines, made famous by Louis and Mary Leakey, are quite distinct from humans. Several detailed computer studies of australopithecines have shown that their bodily proportions were not intermediate between those of man and living apes. Another study, which examined their inner ear bones, used to maintain balance, showed a striking similarity to those of chimpanzees and gorillas, but great differences from those of humans. Likewise, their pattern of dental development corresponds to chimpanzees, not humans. ...[Lucy] likely swung from the trees and was similar to pygmy chimpanzees."[623]

The same is true of all the other so-called missing links between apes and men. They are frauds. Indeed, finding a few bone fragments in the dirt and extrapolating that as proof for the idea that apes could somehow turn into men, is no less absurd than if you found a rusty bolt and a bent license plate in a glacier in Alaska and hypothesized that Detroit does not exist and that mid-size sedans somehow changed on their own, accidentally by random chance, over many, many years, into SUV's.

"And now let us here the conclusion of the whole matter." The theory of evolution is a joke. And a bad one at that. It is absurd. There are no transitional forms from apes to man, because man did not come from apes. Man proceeded forth from the command of the Almighty. As did all the living. He spoke them all into being with the breath of his mouth. And that *is* a scientific fact.

'The Spirit of God has made me, and the breath of the Almighty has given me life." (Job 33:4)

[However, and not to contradict everything we just said, but there have been some recent developments that have shed new light on this entire subject. And in the interest of fairness and honesty, we would be remiss if we did not report these facts. A new species of human-like creatures <u>has</u> been discovered. And they are called *Pithe-yeswecanus Leftus-Pelosius*, or Democrat Man. The last sighting of these primitive, ape-like, pre-humans was in Washington D.C. where they were observed destroying the U.S economy, apologizing for the greatest country on the face of the earth, undermining our National Defense and attempting to eviscerate the world's greatest Health Care system. Maybe evolution isn't so crazy after all? Stay tuned for further updates...]

So, in conclusion, there is no scientific evidence that supports the theory of evolution, because evolution doesn't exist. It's built on lies, fabrications and hoaxes. It truly is a joke. Now let's take a look at all of the real evidence. (Hey! Is this fun or what?)

Part II – "Follow the evidence wherever it leads"

"In the beginning God created the heaven and the earth... the seas, and all that in them is... And God said, Let the waters bring forth abundantly the moving

---

622   (Pgs. 102-103, *The Collapse of Evolution*, by Scott M. Huse)

623   (Pgs. 13-14, *In the Beginning: Compelling Evidence for Creation and the Flood*, by Walt Brown, Ph.D.)

creature that has life, and fowl that may fly above the earth in the open firmament of heaven. And God created great whales, and every living creature that moves, which the waters brought forth abundantly, after their kind, and every winged fowl after his kind: and God saw that it was good. …And God said, Let the earth bring forth the living creature after his kind, cattle, and creeping thing, and beast of the earth after his kind: and it was so. And God saw that it was good." (Genesis 1: 1, Exodus 20:11, Genesis 1:20-21, 24-25)

We have seen how Darwinists over the last century have taken physical evidence and, with the help of their willing accomplices in our anti-Christian, anti-Bible, anti-Creation media, educational system and entertainment industry, have distorted, lied about, fabricated and repressed it in order to further their religion-of-the-absurd. Now it's time to see what science, and not pseudo-science, has to say about all of the evidence. Let's see what happens when you look at the world without evolution-colored glasses. It all suddenly becomes very clear…

The Laws of science
"For God, who commanded the light to shine out of darkness, has shined in our hearts." (2 Corinthians 4:6)

The laws of science by themselves render the theory of evolution impossible. We have mentioned them before: the Law of Biogenesis, the Second Law of Thermodynamics and the Laws of Mathematical Probability. There are others that could be included, like the First Law of Thermodynamics, but those three will be sufficient. And remember that laws of science are so well-established that they never have been, and cannot be, contradicted. Otherwise they would not be laws. Theories, conversely, are literally a dime a dozen. They come and go as scientific observation either confirms or refutes them, but under any other circumstance except for when it comes to evolution, no theory would ever be proposed that contradicts well-established laws of science. Because the person proposing that theory would be laughed out of the scientific community. Evolution, however, gets a special exemption. You see, evolution has to be true. Its devotees have far too much invested in it and their professional humiliation cannot be allowed; and they are protected from widespread exposure because of their control over our media, educational system and entertainment industry. So, "Damn the truth. Full speed ahead! Evolution is a *fact*, kids, and don't ever dare to question this decree! Ignore that little man behind the curtain, and never forget, you're stupid if you think otherwise!"

- The Law of Biogenesis[624] is an established law of science. It states that life comes only from life. The theory of evolution stands in direct contradiction to this law. It states that life arose spontaneously on its own by random accidental chance. Yet, in addition, according to the Laws of Mathematical Probability, that is absolutely impossible. Again, nothing with odds greater than 1 in 10^{66} (ten with sixty six zeroes behind it, which is one million times a trillion times a trillion times a trillion times a trillion times a trillion) can ever happen even if the universe were to last for an infinite number of years. And as we shall see in the pages that follow,

---

624   (Again, not to be confused with the so-called "Biogenic Law" which is a complete falsehood, as we saw in an earlier section, and no "Law" of science.)

the odds of even a tiny fractional part of a living cell making itself accidentally are far beyond 1 in 10^{66}.

"Spontaneous generation (the emergence of life from nonliving matter) has never been observed. [And never will be. For the same reason that Santa Claus will never be observed. But that doesn't stop little children from still believing in him.] All observations have shown that life comes only from life. This has been observed so consistently it is called *the law of biogenesis*. The theory of evolution conflicts with this scientific law when claiming that life came from nonliving matter through natural processes."[625]

- The Second Law It is also scientifically impossible for life to have arisen on its own according to the Second Law of Thermodynamics which states that this universe and everything in it is running down, deteriorating, becoming less ordered. Life of course is order and complexity to the highest magnitude and for it to have arisen on its own accidentally would be in direct violation of this well established law of science.

"The observed laws of science provide a huge barrier to evolution. [He's being kind-they destroy it.] A basic law of science, known as the "second law of thermodynamics," has been observed and verified in every field of science. This law reveals a general trend toward deterioration in all of nature. Everything moves in a downward spiral toward a less ordered state. Stars burn out. The moon's orbit is decaying. Cars wear out. Wood rots. Living plants wither and fade. Animals die. People get sick and grow old. No exception to this rule of decline has ever been observed."[626]

*But wait! What about rock crystals? Aren't they an example of order arising spontaneously? And doesn't **that** overturn the Second Law of Thermodynamics?!* No it does not…

"A crystal growing out of a solution… only follows well-known laws of molecular attraction and bonding and merely conforms to a repetitious pattern. There is no comparison to the much more complex patterns found in living molecules. No known law exists to order the growth of life systems."[627]

The same is true of snowflakes. They, like crystals, follow the laws of molecular attraction. But there are no laws of science or nature that could cause your camera or your computer to form itself, nor your eye or your brain. The Second Law of Thermodynamics destroys the theory of evolution all by itself.

   - Statistics and Probability Also, both the Laws of Mathematical Probability—as we will see in the upcoming section "What are the odds?"—and the Second Law of Thermodynamics render the idea of an organism accidentally turning into another life form by unconsciously altering its DNA as equally scientifically impossible.

"Many molecules necessary for life, such as DNA, RNA, and proteins, are so incredibly complex that claims they evolved are questionable.

[That author is also being kind, the claims are laughable. Nay, they are beyond laughable. They have reached the critical tipping point where you could actually *die* laughing. So be careful.]

---

625   (Pg. 5, *In the Beginning: Compelling Evidence for Creation and the Flood*, by Walt Brown, Ph.D.)
626   (Pg. 7, *Dinosaurs, the Lost World, and You*, by John D. Morris, Ph.D.)
627   (Pg. 8, *Dinosaurs, the Lost World, and You*, by John D. Morris, Ph.D.)

...."There is no reason to believe that mutations or any natural process could ever produce any new organs—especially those as complex as the eye, the ear, or the brain. For example, an adult human brain contains over 10^{14} (a hundred thousand billion) electrical connections, more than all the soldered electrical connections in the world. The human heart, a ten-ounce pump that will operate without maintenance or lubrication for about 75 years, is another engineering marvel."[628]

That's what the <u>laws</u> of science say. And next we will present a treasure trove of facts and evidence that add to their incontrovertible decree...

"Just the facts, ma'am. Just the facts."[629]

There are one hundred trillion living cells in your body. Touch your skin with the tip of your finger and you are touching millions of cells just on the surface of your skin.

"If all the DNA in [just] one of your cells were uncoiled, connected and stretched out, it would be about seven feet long. It would be so thin its details could not be seen, even under an electron microscope. If all this very densely coded information from *one cell* of one person were written in books, it would fill a library of about 4000 books. If all the DNA *in your body* were placed end-to-end, it would stretch from here to the moon more than 500,000 times!

[The average distance from the earth to the moon is 240,000 miles, so all the DNA in your body would *stretch out* to 120 <u>billion</u> miles! And we were made by "random chance, accidentally." What a colossal joke! Truly we are "fearfully and wonderfully made." (Psalm 139:14)]

..."In book form, that information [from *all* of your cells] would fill the Grand Canyon almost 100 times. [Yet] If one set of DNA (one cell's worth) from every person who ever lived [approximately 50 billion people] ...—enough to define the physical characteristics of all those people in microscopic detail— ...were placed in a pile, [it] would weigh less than an aspirin! Understanding DNA is just one small reason for believing that you are *"fearfully and wonderfully made."* "[630]

"Understanding DNA is just one small reason for" knowing that it is a scientific fact that the theory of evolution is scientifically absurd. Don't skip over that last paragraph too quickly. Read it slowly and make sure the information there—staggering facts that are almost impossible for us to comprehend—sinks in deeply and pushes out once and for all any possibility of thinking that any of it could ever have evolved. The micro-miniaturization of this information, crammed into a seven-foot-strand of a molecular computer that is so microscopically thin that just the DNA in your body would stretch from here to the moon a half million times! If we humans assembled all of the engineers on the planet and took all of the cables of all of the suspension bridges in the world and tried to slice them thin enough to go from here to the moon even <u>once</u> we would find it nearly impossible. And all of those cables would probably fill a football stadium a number of times over. But just the DNA in just one human body, stretched out end to end, would go from here to the moon 500,000 times. Who did that? Who had the limitless power

---

628   (Pg. 7, *In the Beginning: Compelling Evidence for Creation and the Flood*, by Walt Brown, Ph.D.)
629   (A catch phrase made famous by Sergeant Joe Friday in the classic TV series *Dragnet.*)
630   (Pgs. 3, 75, *In the Beginning: Compelling Evidence for Creation and the Flood*, by Walt Brown, Ph.D.)

and the infinite intelligence necessary to do that? Keep in mind that we think it is incredible when a South Korean artist can carve Michelangelo's *Madonna and Child* on a rice kernel, and rightly so. But this is like taking a million IBM mainframe computers from NASA, each filled to capacity with data, and reducing them to the *tip* of a pin. And it all happened by random chance?!#@#!? Accidentally!?#@%#@?! Don't get me wrong. The obvious intention of this book is to educate people as to the truth, and not to put people down, or to demean, disparage, ridicule and mock the other side as they have so delighted in doing to us over the years. But, c'mon! Let's set the record straight. Anyone at room temperature looking at this information can't help but laugh out loud at the theory of evolution. I was not exaggerating when I said that it literally borders on the insane. And we are just getting started...

"Cells require a remarkable array of sophisticated signal processing behaviors that rivals or surpasses that of modern computers."[631]

Question: Would any parent allow their children to be indoctrinated by some irrational human being who taught that his laptop just formed itself accidentally by random chance, and said that his students were idiots if they believed in Microsoft?

..."the arrangement of multiple genetic elements into sophisticated logic circuits similar to those of computers is also well beyond the edge of Darwinian evolution."[632]

The DNA molecule, the information and command center of the cell, contains more coded information than a school library. And it is a scientific fact that coded information can never, ever, ever arise on its own by dumb luck, by random chance. Coded information requires a *Coder*. Now I've met Random Chance and I can tell you, he's no Coder. He cannot see, he cannot speak, he cannot think, he cannot design, he cannot create, and he definitely cannot code. In fact, he is indistinguishable from an idiot. And "those who make him into something that he is not are like unto him; and so is everyone who has faith in him." (Psalms 115:5-8) The worshipping of idols of carved stone and wood in biblical times has given way to the bowing down to the more *intellectually palatable* idols of today: Mr. Random Chance and Mr. Explosion. Otherwise known as the great god "Chaos." They are the ones who are now worshipped in place of the Creator God, who clearly warned us, "You shall have no other gods before me." Exodus 20:3)

"*A code is a set of rules for converting information from one useful form to another.* Examples include Morse code and Braille. Code makers must simultaneously understand at least two ways of representing information and then establish the rules for converting from one to the other and back again. [And the idea that random chance could accomplish this is literally insane.]

"The genetic material that controls the physical processes of life is coded information. Also coded are complex and completely different functions: the transmission, translation, correction, and duplication systems, without which the genetic material would be useless, and life would cease. It seems most reasonable that *the genetic code* and the accompanying transmission, translation, correction, and duplication systems were produced simultaneously in each living organism by an extremely high intelligence.

---

631   ("Bhattacharyya, R. P., Remenyi, A., Yeh, B. J., and Lim, W. A. 2006. Domains, motifs, and scaffolds: the role of modular interactions in the evolution and wiring of cell signaling circuits. *Annu. Rev. Biochem.* 75:665-80. As quoted on pg. 270 of, *The Edge of Evolution: The Search for the Limits of Darwinism,* by Dr. Michael J. Behe)

632   (Pg. 275, *The Edge of Evolution: The Search for the Limits of Darwinism,* by Dr., Michael J. Behe)

[Indeed, it is "most reasonable" because there is no other way in the known universe that it could be produced. Unless you know of another way that your laptop, your fridge, your toaster, your house and your car could all be produced, accidentally…]

"Likewise, no natural process has ever been observed to produce a program. *A program is a planned sequence of steps to accomplish some goal.* Computer programs are common examples. Because programs require foresight, they are not produced by chance or natural processes. The information stored in the genetic material of all life is a complex program. Therefore, it appears that an unfathomable intelligence created these genetic programs. [Ya think?]

"Life contains matter, energy, and **information**. All isolated systems, including living organisms, have specific, but perishable, amounts of information. No isolated system has ever been shown to increase its information content significantly. Nor do natural processes increase information; they destroy it. Only outside intelligence can significantly increase the information content of an otherwise isolated system. All scientific observations are consistent with this generalization, which has three corollaries:

- Macroevolution cannot occur.

- Outside intelligence was involved in the creation of the universe and all forms of life.

- Life could not result from a "big bang."

…"As explained above, only intelligence creates codes, programs, and information (CP&I). Each involves senders and receivers. Senders and receivers can be people, animals, plants, organs, cells, or certain molecules. (The DNA molecule is a prolific sender.) The CP&I in a message must be understandable *and* beneficial to both sender *and* receiver; otherwise, the effort expended in transmitting and receiving messages (written, chemical, electrical, magnetic, visual, and auditory) will be wasted.

"Consider the astronomical number of links (message channels) that exist between potential senders and receivers: from the cellular level to complete organisms, from bananas to bacteria to babies, and across all of time since life began. All must have compatible understandings (CP&I) and equipment (matter and energy). Designing compatibilities of this magnitude requires one or more *super*intelligences. Furthermore, these superintelligence(s) must completely understand how matter and energy behave over time. In other words, the superintelligence(s) must have made, or at least mastered, the laws of chemistry and physics wherever senders and receivers are found. The simplest, most parsimonious way to *integrate all of life* is for there to be only one superintelligence.

"Also, the sending and receiving equipment, including its energy sources, must be in place and functional before communication begins. But the preexisting equipment provides no benefit until useful messages begin arriving. Therefore, intelligent foresight (planning) is mandatory—something nature cannot do."[633]

Isn't knowledge a wonderful thing? And with that in mind we'd like to offer a group discount to Richard Dawkins, Christopher Hitchens, Bill Maher and Sam Harris if the four of them sign up together for the next class on *Remedial Re-education 101: Basics in Biology*. Shucks, we'll even throw in a second one for free. It's called, *If you feel compelled to be a completely ignorant, arrogant, abusive, intolerant, militant and thoroughly misinformed atheist it might help to actually know what the hell you are talking about first, 201.* David Berlinski had this observation…

---

633   (Pg. 9, *In the Beginning: Compelling Evidence for Creation and the Flood,* by Walt Brown, Ph.D.)

"When asked what he was in awe of, Christopher Hitchens responded that his definition of an educated person is that you have some idea of how ignorant you are. This seems very much as if Hitchens were in awe of his own ignorance, in which case he surely has found an object worthy of his veneration."[634]

But let us get back to the reasons for being in awe of the Creator of this universe, rather than in our own ability to deny Him and to hide that denial behind an idiotic and unscientific theory...

"The notion that not only the biopolymers but the operating programme of a living cell could be arrived at by chance in a primordial organic soup here on Earth is evidently nonsense of a high order."[635]

But "nonsense of a high order" that is nevertheless taught, as fact, with a straight face, every day to the public school children all across America. And we wonder why our country is going to hell in a handbasket, and why our voting-age youth have no more sense than to rush out and vote for the left-wing jackasses that they've been told to vote for.

"...consider the information in the DNA molecule. No undirected process produces intelligent information that can be read and understood. Its coded information surpasses our own ability to understand it. Today's molecular biologists cannot write such a code or even devise a way to do the things a cell can do. Certainly a level of intelligence beyond our own was behind the writing of the DNA [and] the functions of a cell."[636]

Mutations never result in an increase of information in the DNA code; they always result in a decrease. Mostly they result in horrible defects and distortions and cause death. Therefore, from this <u>observation</u> alone (and remember that science is based on systematic observation) the evolutionists belief that one organism (i.e. an ape) can change into another (i.e. a man) by random accidental genetic mutation over whatever period of time they would care to create, is completely false. *Everyone* knows that it is scientifically impossible for your computer, your camera, your tape recorder, your automobile, the wiring in your house, your smoke detector, our Star Wars Defense System, your refrigerator or your cars fuel system to have arisen accidentally over time. If anyone said otherwise they would be called a moron. Well it is even more scientifically impossible (according to mathematical probabilities) for your brain, your eyes, your ears, your skeletal system and muscles, your nerves, your nose, your immune system, your sweat glands or your digestive system to have arisen accidentally over time. And if anyone says otherwise then by extension they would have to be a m___n as well, wouldn't they? (Or at least someone who was duped into believing the nonsense that they were taught.)

"In the human body, DNA "programs" all characteristics such as hair, skin, eyes, and height. DNA determines the arrangement for 206 bones, 600 muscles, 10,000 auditory nerve fibers, 2 million optic nerve fibers, 100 billion nerve cells, 400 billion feet of blood vessels and capillaries, and so on. Further, the capacity of DNA to store information vastly

634 (Pg. 208, *The Devil's Delusion: Atheism and its Scientific Pretensions*, by David Berlinski)
635 (British astronomer, Sir Fred Hoyle)
636 (Pg. 123, *The Young Earth: The Real History of the Earth – Past, Present, and Future*, by John Morris, Ph.D.)

exceeds that of modern technology. The information needed to specify the design of all the species of organisms which have ever lived could be held in a teaspoon and there would still be room left for all the information in every book ever written... Such extraordinary sophistication can only reflect intelligent design."[637]

Did you catch that? 400 <u>billion</u> feet of blood vessels and capillaries in your body! That's 75 million, 757 thousand, 575 **miles** of blood vessels and capillaries in one person's body! That's incomprehensible! And how absolutely amazing of random chance to have made all of those blood vessels and tiny microscopic capillaries just the right size, and to have known <u>exactly</u> where to put all of them, in precisely the right locations and circuits all throughout your entire body, and to hook all of them up, accidentally, so they made a complete circuit back to your heart, and didn't end up just squirting out of your ears and your elbow instead!? Just by trial and error. Accidentally. 76 million miles worth. Man, random chance is unbelievable! (Literally) Let's all sing songs of worship and praise to random chance, shall we? Christopher? Richard? Bill? Are you all ready? In G-minor... "Praise Random Chance from whom all blessings flow! Praise Random Chance all creatures here below!"

"Furthermore, computer scientists have demonstrated conclusively that information does not and cannot arise spontaneously. Information results only from the expenditure of energy (to arrange the letters and words) and under the all-important direction of intelligence. Therefore, since DNA is information, the only logical and reasonable conclusion that can be drawn is that DNA was formed by intelligence.

"The intricately ordered structure of the DNA molecule is truly an engineering wonder, almost beyond comprehension. This amazing biochemical system could never have arisen apart from Divine creation. Without question, DNA remains one of the greatest testimonies of special creation.

..."The bewildering complexity, diversity, beauty, order, and astonishing perfection of life forms thoroughly defy evolutionary explanation. Fascinating interrelationships such as mimicry, symbiosis, parasitism, and so on, among organisms clearly indicate purposeful, intelligent design. The widespread existence of amazingly sophisticated migratory and other instincts do not lend themselves to an evolutionary elucidation. The numerous instances of convergence and double convergence preclude evolutionary rationale. The incomprehensible complexity of the DNA molecule can only be explained in terms of special creation. And most importantly, the theory of organic evolution lacks a mechanism! Thus, the consistent and overwhelming witness of biology is in support of Biblical creationism and soundly against mythological evolution."[638]

That is, of course, if one has the unfettered and un-brainwashed use of his or her mind. Remember the lies of Satan cover this earth like a blanket, and people believe what they are taught. (And did I mention that we live in an insane asylum, and the inmates have taken over?)

DNA obviously is necessary for life. Without it a cell cannot exist. Just as a human being cannot exist without a brain. (Well, that could be debatable. Where's my ear scope? Mr. Dawkins, could you come over here and tilt your head for a second?) **Yet DNA cannot function without the presence within the cell**

---

637   (Pg. 94, *The Collapse of Evolution*, by Scott M. Huse)
638   (Pg. 95, *The Collapse of Evolution*, by Scott M. Huse)

of at least 75 pre-existing proteins. *Yet proteins are ONLY produced by DNA*!! And neither can exist without the cell wall encasing them and surrounding them with all of the cellular chemicals and enzymes and buffers and tens of thousands of nanobot transportation robots, etc. etc. So even if a DNA molecule formed itself somehow by random chance, which by the way is impossible—about the same chances as the contents of The Library of Congress forming themselves by random chance—but even if it did it wouldn't do any good because it would immediately fall apart into slop. The cell wall is a miracle of engineering. It is far more technologically advanced than even our most secret and highly protected military installations or sophisticated anti-missile defense systems (and unlike those comparatively crude mechanisms, the cell wall operates completely UN-consciously). But in order for evolution to be viable not only would the DNA molecule have to, *POOF*, form itself miraculously by accident, but 75 separate proteins, and the cell wall, and all the other multitudinous contents of the cell would have to also form themselves miraculously by accident, at exactly the same time and at exactly the same place! This, of course, "would be indistinguishable from a miracle!" But evolution is not in the business of miracles. Only God, the Creator, is. Evolution is in the non-existent, imaginary business of small, slow, incremental, mythological changes over vast spans of time, which also don't exist.

"It's evolution that doesn't exist, stupid, not God."

> "The complexity of the simplest known type of cell is so great that it is impossible to accept that such an object could have been thrown together suddenly by some freakish, vastly improbable, event. **Such an occurrence would be indistinguishable from a miracle.**"

> "Alongside the level of ingenuity and complexity exhibited by the molecular machinery of life, even our most advanced artifacts appear clumsy."[639]

"Alongside the level of ingenuity and complexity exhibited by the molecular machinery of life, even our most advanced artifacts appear clumsy." Yet no one would ever dare be so stupid as to suggest that our clumsy artifacts could just make themselves, if we waited around long enough.

Let's look even further inside the engineering miracle of a living cell, the building block of all of life…

Again, there are a hundred trillion cells in the human body. Every second there are a million chemical reactions within just one of these many cells. That's a hundred million **trillion** chemical reactions every second just to keep your body alive and functioning. If you started right now to write down each of those chemical reactions on paper, and you did not stop day or night until you were done, it would take you well over a trillion years to do so. And that's just the chemical reactions that occur within one second within one human body.

> "Although serendipity certainly plays its part in nature, advancing sheer chance as an explanation for profoundly functional features of life strikes me as akin to abandoning reason altogether."[640]

Again, not only according to common sense but according to the *Second Law*

---

639   (Pg. 264, *Evolution: A Theory in Crisis*, by Michael Denton)
640   (Pg. 165, *The Edge of Evolution: The Search for the Limits of Darwinism*, by Dr., Michael J. Behe)

of Thermodynamics and the <u>*Laws* of Mathematical Probability</u> sheer chance can <u>never</u> create anything of order. All random chance can do is to tear down, to break apart and to destroy, and you are left with a colossal mess and not anything that resembles even the simplest living thing.

Also, as we have stated in a previous chapter but it's worth repeating here, the actual engineering complexity of what was once called the "simple single cell" out of primitive scientific ignorance, is greater than that of the entire city of New York. And unlike the joke that is evolution, that actually <u>is</u> a scientific fact. In each cell there are thousands upon thousands of nanobots, micro machines, tiny robots that scurry about all over the cell transporting materials like oxygen and carbon and water from the cell wall to countless places where they are needed, and taking waste products of metabolism like carbon dioxide back to the cell wall for expulsion. They do all of this over and over again millions upon billions upon trillions of times, day in and day out for your entire lifetime. And the miraculous thing is that they do it <u>blindly</u> and <u>unconsciously</u>. Everything in the cell is completely automated.

> "Cells are robots. Or rather, because they are so small, "nanobots." They work by unconscious, automatic mechanisms. To perform the routine tasks of their microscopic lives, cellular nanobots need sophisticated molecular machinery that works without conscious guidance."[641]

By way of repetition, can you imagine constructing an entire city of New York that could run itself for decades blindly and unconsciously, completely automated, with no human intervention? A car gets in an accident, the tow trucks would have to be able to sense it automatically and drive to the right place and pick it up and then bring it to the garage where it would be repaired by robots working blindly and unconsciously without any human intervention. The same with a water main that breaks; a fire that starts; a hurricane that blows the roofs off and the windows out of countless buildings; an elevator or air conditioning or heating system that breaks down. The repair robots would have to know exactly where the water leak is, the fire is, the windows are blown out, blindly and unconsciously! Think about it! If we gathered all of the world's smartest engineers and scientists and allowed them to work day and night for the next ten million years, they could never do it. After a few million years they might get a small scale primitive prototype, consisting of maybe ten square blocks, up and running and it might last for maybe half a day until something blew up that the automated city could not handle on its own and the humans would have to rush in and take over. Back to the drawing board. But the Intelligent Creator of this universe accomplished the equivalent of designing and manufacturing an automated city of New York within every cell of your body, and every cell of every living thing on this planet, and it took him just a few days. He spoke it, and all matter instantly obeyed his voice. He breathed it, and it all sprang into being. He wound it all up like a clock, and it's been running itself, unconsciously, ever since.

> "Most people don't think of cells as robots, probably because cells are made of organic materials rather than metal. But cells truly are self-replicating nanoscale robots. Self-replicating, because of course they reproduce themselves. Nanoscale, because most cells are quite tiny and all can manipulate single molecules. Robots, because their activities are

641 (Pg. 19, *The Edge of Evolution: The Search for the Limits of Darwinism*, by Dr., Michael J. Behe)

carried out unconsciously and automatical:y by precision machinery that follows ordinary physical laws."[642]

"One crucial way in which machinery in the nanobot differs from machinery in our everyday experience is that cellular machines have to *assemble themselves*. There is no conscious agent walking around in the cell putting pieces of machinery together, as there might be in a factory making, say, flashlights. Needless to say, the requirement for self-assembly enormously complicates the task of building a functional nanomachine."[643]

"Needless to say" we are talking about a level of astronomical complexity and supreme design ingenuity that absolutely dwarfs anything man has ever conceived. Again, the "level of knuck:e dragging ignorance" (Mr. Dawkins, I am eternally in your debt) it takes to suppose that the astronomically complex, engineering miracle of a living cell (fraught with highly advanced, completely unconscious and automated nanobot technology that we are literally thousands of years away from even fully comprehending, much less ever duplicating) could form itself accidentally by random chance is literally insane. And so is the theory of evolution.

[It is spring, and I'm sitting outside looking at all of the life splashed across my backyard. And I'm thinking about the near-infinite number of chemical reactions that are occurring constantly right before my eyes in every cell of every blade of grass and weed and flower and shrub and tree, and all of the nearly as infinite number of microscopic little robots that are unconsciously scurrying around each cell, and all of this is happening just so we can experience life. Life is happening microscopically and miraculously in its infinite and rigidly ordered astronomical complexity in order to bring us, life. And yet it all happened by random, accidental chance?!#%&*! What an ignorant slap in the face to an incredible and unfathomable and marvelous Creator who has blessed us with this perpetual miracle that he has showered across the face of this earth. It's a wonder he doesn't just stomp on us.

The wonder of life is far beyond our ability to even comprehend it, much less fully appreciate it. It is so true that "you don't know what you got till it's gone." Pray for all of those who are mis-educated into the false-religious lies of the great deceiver, who stand to lose this miracle of life for all eternity, never again to have the joy of experiencing it.]

We will leave the incredibly-complex, molecular-sized world of the living cell now and look at the evidence from other perspectives. Again, "just the facts, ma'am, just the facts"...

I bet you never thought that even altruism could refute evolution. Neither did I...

"Humans and many animals will endanger or even sacrifice their lives to save another—sometimes the life of another species. Natural selection, which evolutionists

---

642   (Pg. 242, *The Edge of Evolution: The Search for the Limits of Darwinism*, by Dr., Michael J. Behe)
643   (Pg. 255, *The Edge of Evolution: The Search for the Limits of Darwinism*, by Dr., Michael J. Behe)

say select individual characteristics, should rapidly eliminate altruistic (self-sacrificing) "individuals." How could such risky, costly behavior ever be inherited? Its possession tends to prevent the altruistic "individual" from passing on his genes for altruism? If evolution were correct, selfish behavior should have completely eliminated unselfish behavior. Furthermore, cheating and aggression should have "weeded out" cooperation. Altruism contradicts evolution."[644]

(Life contradicts evolution, but let's not pile on.)

And then there is language...

"Children as young as seven months can understand and learn grammatical rules. Furthermore, studies of 36 documented cases of children raised without human contact (feral children) show that language is learned only from other humans; humans do not automatically speak. So, the first humans must have been endowed with a language ability.

..."If language evolved, the earliest languages should be the simplest. But language studies show that the more ancient the language (for example: Latin, 200 B.C.; Greek, 800 B.C.; Linear B., 1200 B.C.; and Vedic Sanskrit, 1500 B.C.), the more complex it is with respect to syntax, case, gender, mood, voice, tense, verb form, and inflection. The best evidence shows that languages devolve; that is, they become simpler instead of more complex. Most linguists reject the idea that simple languages evolve into complex languages.

..."If humans evolved, then so did language. All available evidence indicates that language did not evolve, so humans probably did not evolve either."[645]

Probably.

"Speech is uniquely human. Humans have both a "prewired" brain capable of learning and conveying abstract ideas, and the physical anatomy (mouth, throat, tongue, larynx, etc.) to produce a wide range of sounds. Only a few animals can approximate some human sounds. Because the human larynx is low in the neck, a long air column lies above the vocal cords. This helps make vowel sounds. Apes cannot make clear vowel sounds, because they lack this long air column. The back of the human tongue, extending deep into the neck, modulates the air flow to produce consonant sounds. Apes have flat, horizontal tongues, incapable of making consonant sounds.

"Even if an Ape could evolve all the physical equipment for speech [which he of course could not], that equipment would be useless without a "prewired" brain for learning language skills, especially grammar and vocabulary."[646]

But don't worry, this little annoying fact hasn't deterred evolutionists in the least. In fact, as we speak, they are working feverishly on a new theory of how human speech somehow *evolved* from the grunts of apes, and they're calling it *kreehayshun evolution*. It's where evolution skips all of those little annoying, incremental steps that take so long and are so scientifically-impossible, and just *evolves* any organ, any organism and any ability it wants, instantly! Shazaam! But of course no higher power or outside intelligence is involved in any way. Don't be silly.

---

644   (Pg. 8, *In the Beginning: Compelling Evidence for Creation and the Flood*, by Walt Brown, Ph.D.)
645   (Pg. 8, *In the Beginning: Compelling Evidence for Creation and the Flood*, by Walt Brown, Ph.D.)
646   (Pgs. 8-9, *In the Beginning: Compelling Evidence for Creation and the Flood*, by Walt Brown, Ph.D.)

And then there is this vaudevillian sleight-of-hand disguised as another brilliant Darwinian *explanation*...

"When the same complex capability is found in related organisms, but not in their alleged evolutionary ancestors, evolutionists say that a common need caused identical complexities to evolve. They call this *convergent evolution*. [It's also called *horseshit personified*, but hey! let's try to keep the vulgarity to a minimum, shall we?]

"For example, wings and flight occur in some birds, insects, and mammals (bats). Pterosaurs, an extinct reptile, also had wings and could fly. These capabilities have not been found in any of their alleged common ancestors. Other examples of convergent evolution are the three tiny bones in the ears of mammals: the stapes, incus, and malleus. Their complex arrangement and precise fit give mammals the unique ability to hear a wide range of sounds. Evolutionists say that those bones evolved from bones in a reptiles jaw. If so, the process must have occurred at least twice—but left no known transitional fossils. How did the transitional organisms between reptiles and mammals hear during those millions of years? [Uh, hearing implants?] Without the ability to hear, survival—and reptile-to-mammal evolution—would cease.

"Concluding that a miracle—or any extremely unlikely event—happened once requires strong evidence or faith; claiming that a similar "miracle" happened repeatedly requires either incredible blind faith [called *willful ignorance*] or a cause common to each event, such as a common designer."[647]

Common ancestry is an evolutionary lie. Common design is a scientific truth. Of course, convergent evolution is downright hilarious. How they can even keep a straight face at the lectern while speaking it into their microphones is beyond me. By the way, in case you're not sure, it's OK to laugh out loud. Not at my lame attempts at humor, but at evolution. It really *is* a joke.

But here are even more problems for Darwinists:

"Many single-celled forms of life exist, but no known forms of animal life have 2, 3, 4, or 5 cells. Known forms of life with 6-20 cells are parasites, so they must have a complex animal as a host to provide such functions as respiration and digestion. If macroevolution happened, one should find many transitional forms of life with 2-20 cells--filling the gap between one-celled and many-celled organisms."[648]

No. Not necessarily. Not if one understands *hippety-hop evolution*. Closely related to *kreehayshun* evolution and similar to *horsh...er...convergent* evolution, this variation to standard Darwinism calls for the convenient leaping over of those potential 2- to 20-celled organisms not found either alive or in the fossil record. You see, when life first evolved it didn't have a calculator (*that* was the problem), so it is possible that it just missed a couple of simple numbers, like 2, 3 and 4 for example; numbers that *we* all take for granted.

Oh, but it gets even better! Much, much better...

"The sea slug lives along the sea coast within the tidal zone where it feeds primarily on sea anemones. Sea anemones are not exactly the most inviting of dinners as they are equipped with thousands of small stinging cells on their tentacles which explode at

---

647   (Pgs. 9-10, *In the Beginning: Compelling Evidence for Creation and the Flood*, by Walt Brown, Ph.D.)
648   (Pg. 10, *In the Beginning: Compelling Evidence for Creation and the Flood*, by Walt Brown, Ph.D.)

the slightest touch, plunging poisoned harpoons into intruders. The speared intruder is paralyzed and drawn into the anemone's stomach to be digested.

"Although this is an impressive defense system, the remarkable sea slug is able to eat sea anemones without being stung, exploding the stinging cells, or digesting them. One of the most fascinating mysteries in nature is what the sea slug does with the poor anemone's stinging cells. The undigested stinging cells are swept along through ciliated tubes which are connected to the stomach and end in pouches. The stinging cells are arranged and stored in these pouches to be used for the sea slug's defense! And so, whenever the sea slug is attacked, it defends itself using the stinging cells which the ill-fated anemone manufactured for its own protection.

"The highly complicated series of modifications that would have had to occur to produce this incredible relationship completely defies evolutionary explanation. First of all, in order to prevent the stinging cells from exploding, the sea slug would have to evolve some sort of chemical means to temporary neutralize them. The sea slug would also have to evolve a new digestive system, which would digest the tissues of the anemone but not the stinging cells. The sea slug would also have to cleverly evolve the sophisticated ciliated tubes and pouches as well as a highly complex mechanism for arranging, storing, and maintaining the stinging cells. Finally, and contrary to evolutionary expectations, the anemone would have to endorse the sea slugs plans by refraining from evolving countermeasures.

"Obviously, there is no satisfactory evolutionary explanation for the existence of such extraordinary adaptive design. The only reasonable solution to this fascinating relationship is offered by Biblical creationism. These organisms were specifically created and carefully designed by their Creator to fit into their respective ecological niches."[649]

"O the depth of the riches both of the wisdom and knowledge of God! How unsearchable are his judgments and his ways past finding out!" (Romans 11:33)

"The Arctic Tern, a bird of average size, navigates across oceans... with the skill normally associated with navigational equipment in modern intercontinental aircraft. A round-trip for the tern might be 22,000 miles. The tern's "electronics" are highly miniaturized, extremely reliable, maintenance free, and easily reproduced. Furthermore, this remarkable bird needs no training. If the equipment in [an intercontinental aircraft] could not have evolved, how could the terns more amazing equipment have evolved?"[650]

Really. How, Mr. Maher? You arrogant, uneducated ass. (With all due respect of course.) Please tell us how. As you and your imbecilic friends laugh at all of us who are just not *sharp* enough, like you, to believe in the scientifically-impossible and clearly idiotic fairy tale called evolution; and as you deride us for being stupid, ignorant, religious and *unscientific*, please explain to us how. (Here's a crowbar in case you'd like to pry your head out of your ass first.)

And the beat goes on. (We're not even warmed up yet.) The avalanche of scientific facts that buries Darwinism for good, continues to thunder down the mountain of life...

As we quoted earlier "an adult human brain contains over 10^{14} (a hundred

---

649   (Pgs. 74-75, *The Collapse of Evolution*, by Scott M. Huse)
650   (Pg. 21, *In the Beginning: Compelling Evidence for Creation and the Flood*, by Walt Brown, Ph.D.)

408

thousand billion) electrical connections, more than all the soldered electrical connections in the world." [651] Yet the human brain just formed itself by random accidental chance mutations over many years? Could all of the electrical appliances on this planet have made themselves if we just waited around long enough? Could all of the electrical wiring in the city of New York just happen by random chance if we waited long enough? Even trillions of years? Of course not. You would have to be <u>insane</u> to actually think that, wouldn't you? (I'm sorry, you'll have to excuse me for a minute. I can see there are some people in white coats over at Sam Harris' front door. I'd better go over and see what all the fuss is about...)

OK, I'm back...

"The electric eel can pack an electrical punch greater than the shock that you would get from sticking your finger into an electrical socket. The electricity from this amazing creature arises from a group of highly compacted nerve endings found all along its body. Each one of these nerve endings has a small electric voltage that, when added together, can be a shocking experience! A full-grown eel can produce enough electricity—600 volts—to stun a horse. Eels use this electrical current as self-defense, and a way to stun their prey. Yet, they live in water, don't shock themselves, and can recharge without an extension cord.

[And they were created by Mr. Accidental Mutation! Wow! Wasn't that extraordinarily brilliant of him?! How did he get so smart?]

"Lightning bugs (also called fireflies) are little beetles that carry their own "flashlights." These insects have a special chemical called luciferin, which they store in their abdomens. When this chemical comes into contact with oxygen from the air they breathe, fireflies give off a bright flash of light. The whole process is called bioluminescence. According to scientists, this light helps fireflies find their mates. The male flashes his light first, and when the female sees it, she flashes her light. This tells the male where she is. What an incredible way to find a mate!

"Perhaps one of the most amazing animals on Earth today is a little insect known as the bombardier beetle. This bug packs a powerful, explosive defense mechanism. Inside its body, it has tiny glands that secrete two chemicals into a "storage tank" (known as a "collecting vesicle"). Those chemicals are hydrogen peroxide and hydroquinones. If an enemy attacks the beetle, it empties these two chemicals into a special "explosion chamber." Then it adds special enzymes to the mixture. As a result, the chemicals form a solution that reaches the boiling point of water (100°Celsius or 212°Fahrenheit) in just a couple of seconds. The beetle can then take aim with two small gun-like projections on the rear of its body, and fire the boiling mixture into the face of its attacker. What a remarkable creature!"[652]

People believe what they are taught. No matter how idiotic, asinine, absurd or ridiculous it may be (don't forget the "72 virgins" thing). And it should be becoming painfully obvious, just from the last three examples alone, that the theory of evolution falls into each of those four categories.

And then there is the woodpecker:

"A woodpecker's chisel-tipped bill hammers wood at the rate of 16 times a second,

---

651　(Pg. 7, *In the Beginning: Compelling Evidence for Creation and the Flood*, by Walt Brown, Ph.D.)

652　(Pgs. 62-63, *Dinosaurs Unleashed: The True Story about Dinosaurs and Humans*, by Kyle Butt and Eric Lyons)

or nearly 1,000 pecking blows per minute—a "rate of fire" doubly fast as a submachine gun—with an impact velocity of 1,300 mph.

"While drilling, the woodpecker's head travels at more than twice the speed of a discharged bullet. At this speed (over 1900 feet/second), any slight whiplash rotation of the head during drilling would tear away the bird's brain upon impact. To prevent this, superbly coordinated neck muscles keep the head in perfect alignment with the beak during impact. The head and beak thus drive straight back and forth with no side movement at all. Shock-impact is further minimized by special muscles in the head which pull the woodpecker's braincase away from its beak every time it strikes a blow.

"Unlike most birds, which have their bill fused directly to the bones of the cranium, the woodpecker's bone-reinforced skull is physically separated from its beak by a remarkable sponge-like cartilage (recognized by scientists as being better than any shock absorber manufactured by man). This padding is essential to the bird's survival when one considers that the suddenness with which the woodpecker's head is brought to a halt during each peck results in a stress equivalent of 1,000 times the force of gravity The head thus snaps back with an impact of deceleration more than 250 times that of the G-force gyrations experienced by the astronauts during a launch-pad liftoff!"[653]

How did all of that slowly evolve by small incremental steps? What, as the beak of the bird slowly got stronger, the brain slowly detached from it and slowly grew that shock-absorber cartilage that we cannot duplicate, and slowly grew those "special muscles in the head" and those "superbly coordinated neck muscles," and this process slowly repeated itself thousands of times over and over, in millions and millions of years' worth of birdie generations until we finally accidentally ended up with this absolute miracle of engineering and design that our greatest scientists and most ingenious minds could not duplicate in a thousand years? Incrementally. Slowly. Accidentally. Hmm.

There are so many more examples of biological design that defy any evolutionary explanation; indeed they are too numerous to list here. How does evolution explain those creatures? How did they *evolve* those incredibly complex and mind-boggling features that we humans can't even fully comprehend much less duplicate? How could accidental mutation and random chance design and create those astonishing wonders of nature? Answers: 1) They can't explain them. All they can do is lie and distort and hide and deceive and fabricate and then refuse to ever talk about it or debate it. 2) They didn't *evolve* them. God designed and created them. 3) They couldn't. Accidental mutation and random chance can't design or create a damn thing. Like the idols of wood, stone and statues of gold worshipped by ignorant people of old; *gods* that couldn't talk or think or hear or act, much less answer a single prayer or request; so too the *gods* of accidental mutation and random chance worshipped by faux scientists, ignorant professors and their unfortunately duped students today, couldn't design and create a single, stupid lead pencil much less the wonders of God's nature.

And then there is metamorphosis. To attempt to explain that with Darwinism is like trying to explain advanced calculus with the farts of buffalos:

"Most insects (87%) undergo complete metamorphosis. It begins when a larva (such as a caterpillar) builds a cocoon around itself. Then its body inside disintegrates into a

---

653   (From *The Wonderful Woodpecker: Jehovah's Jaw-Jarring Jackhammer*, by David V. Bassett, M.S. posted on astronomy.net)

thick, pulplike liquid. Days, weeks, or months later, the adult insect emerges—one that is dramatically different ...amazingly capable, and often beautiful, such as a butterfly. Food, habitat, and behavior of the larva also differ drastically from those of the adult.

...."The millions of changes inside the thick liquid never produce something survivable or advantageous in the outside world until the adult completely forms. How did the genetic material for both larva and adult develop? Which came first, larva or adult? What mutations could transform a crawling larva into a flying monarch butterfly that can accurately navigate 3,000 miles using a brain the size of a pin head? Indeed, why should a larva evolve in the first place, because it cannot reproduce?

"Charles Darwin wrote, *If it can be demonstrated that any complex organ existed which could not possibly have been formed by numerous successive, slight modifications, my theory would absolutely break down.* [See ya, evolution.] Based on metamorphosis alone, evolution "breaks down."

"Obviously, the vast amount of information that directs every stage of a larva's and adult's development, including metamorphosis, must reside in its genetic material at the beginning. This fits only creation."[654]

And then there are numerous examples of symbiotic relationships between living things, each dependent upon the other for their survival, that are still waiting for an intelligent and scientifically viable evolutionary explanation. The silence has been deafening for decades now. (Mr. Hitchens, we're still awaiting with bated breath. Christopher, where are you? Come out, come out, wherever you are.)

"Different forms of life are completely dependent upon each other. At the broadest level, the animal kingdom depends on the oxygen produced by the plant kingdom. Plants, in turn, depend on the carbon dioxide produced by the animal kingdom.

"More local and specific examples include fig trees and the fig gall wasp, the yucca plant and the yucca moth, many parasites and their hosts, and pollen-bearing plants and the honeybee. Even members of the honeybee family, consisting of the queen, workers, and drones, are interdependent. If one member of each interdependent group evolved first (such as the plant before the animal, or one member of the honeybee family before the others), it could not have survived. Because all members of the group obviously have survived, they must have come into existence at essentially the same time. In other words, creation."[655]

And then there is the whole idea of sex. Evolutionary explanations are incomprehensible.

"If sexual reproduction in plants, animals, and humans is a result of evolutionary sequences, an unbelievable series of chance events must have occurred at each stage.

a. The amazingly complex, radically different, yet complementary reproductive systems of the male and female must have **completely** and **independently** evolved at each stage at about the **same time** and **place**. [Of course, that is a scientific impossibility.] Just a slight incompleteness in only one of the two would make both reproductive systems useless, and the organism will become extinct.

b. The physical, chemical, and emotional systems of the male and female would also

---

654 (Pgs. 17-18, *In the Beginning: Compelling Evidence for Creation and the Flood,* by Walt Brown, Ph.D.)
655 (Pg. 18, *In the Beginning,* by Walt Brown, Ph.D.)

need to be compatible.

    c. The millions of complex products of a male reproductive system (pollen or sperm) must have an affinity for and a mechanical, chemical, and electrical compatibility with the eggs of the female reproductive system.

    d. The many intricate processes occurring at the molecular level inside the fertilized egg would have to work with fantastic precision—processes that scientists can describe only in a general sense.

    e. The environment of this fertilized egg, from conception through adulthood and until it also reproduced with another sexually capable adult (who also "accidentally" evolved), would have to be tightly controlled.

    f. This remarkable string of "accidents" must have been repeated for millions of species.

    Finally, to produce the first life form would be one miracle. But for natural processes to produce life that could reproduce itself would be a miracle on top of a miracle."[656]

And it is! It's called Creation...

"For in six days the Lord made heaven and earth, the sea, and all that in them is, and rested the seventh day: wherefore the Lord blessed the Sabbath day, and hallowed it." (Exodus 20:11)

Next we have the immune system:

    "How could immune systems of animals and plants have evolved? Each immune system can recognize invading bacteria, viruses, and toxins. Each system can quickly mobilize the best defenders to search out and destroy these invaders. [And they do that automatically and without conscious intervention, thousands of trillions of times, over and over again, every minute of every day.] Each system has a memory and learns from every attack.

    "If the many instructions that direct an animal's or plant's immune system had not been pre-programmed in the organism's genetic system when it first appeared on earth, the first of thousands of potential infections would have killed the organism. This would have nullified any rare genetic improvements that might have accumulated. In other words, the large amount of genetic information governing the immune system could not have accumulated in a slow, evolutionary sense. Obviously, for each organism to have survived, all this information must have been there from the beginning. Again, creation."[657]

"It's impossible. Ask a baby not to cry. It's just impossible. Evolution is a lie. It's so impossible."[658] (Sorry, for some strange reason I had the irresistible urge to break out in song.)

How many times have you found yourself watching a science channel on your TV and you hear something like this example from the National Geographic Channel, "The Great White Shark glides through the sea with minimum effort and perfect grace. A shape of ideal hydrodynamic **design**." [emphasis mine] And it

---

656   (Pg. 19, *In the Beginning: Compelling Evidence for Creation and the Flood*, by Walt Brown, Ph.D.)

657   (Pg. 19, *In the Beginning*, by Walt Brown, Ph.D.)

658   (Sung to the tune of *It's Impossible*, song and lyrics by Perry Como)

is such an obvious oxymoron for those evolutionists to <u>ever</u> be using that word. I've heard that word "design" used hundreds of times for the snakes in the deserts and the sheep on the mountains, for the gophers in the prairie and the foxes and polar bears in the snow-covered tundra of Alaska, and I find myself shouting at the TV, "So, who #*&#*@% *designed* them then, you knuckleheads?!" Anyone with a brain and some common sense knows that design requires a **designer**! And that Great White Shark is more beautifully and precisely designed for its purpose than a flipping Lamborghini. But they just don't get it, because they love the lie more than they love their own Creator (and Designer). So He gives them over to a reprobate and illogical mind. One that is "ever learning, yet never able to come to the knowledge of the truth." And one that has been imprisoned by the asinine, idiotic, scientifically-impossible, bald-faced lie of evolution, Satan's genesis.

"Most complex phenomena known to science are found in living systems—including those involving electrical, acoustical, mechanical, chemical, and optical phenomena. Detailed studies of various animals also have revealed certain physical equipment and capabilities that the world's best designers, using the most sophisticated technologies, cannot duplicate. [And that, of course, is because the One who designed these living systems is infinitely more intelligent than any human designer.] Examples of these designs include molecular-size motors in most living organisms; advanced technologies in cells; miniature and reliable sonar systems of dolphins, porpoises, and whales; frequency-modulated "radar" and discrimination systems of bats; efficient aerodynamic capabilities of hummingbirds; control systems, internal ballistics, and the combustion chambers of bombardier beetles; precise and redundant navigational systems of many birds, fish, and insects; and especially the self-repair capabilities of almost all forms of life. No component of these complex systems could have evolved without placing the organism at a selected disadvantage until the component's evolution was complete. All evidence points to intelligent design.

[But our left-wing, lying media, entertainment industry and educational system will continue to shout the fiction that those who believe in intelligent design are *ignorant, unscientific fundamentalists*, and those who believe in evolution are the "rational" "scientific" ones. What a hoot! What a lie!] [Kind of like their "Democrats smart! Conservatives stupid!" myth, but we'll get into that a little further in chapter 13.]

"Many bacteria, such as *Salmonella, Escherichia coli,* and some *Streptococci,* propel themselves with miniature motors at up to 15 body-lengths per second, equivalent to a car traveling 150 miles per hour—in a liquid These extremely efficient, reversible motors rotate at up to 100,000 revolutions per minute. Each shaft rotates a bundle of whiplike flagella that acts as a propeller. The motors, having rotors and stators, are similar in many respects to electrical motors. [If someone actually believed that your refrigerator motor just evolved accidentally by random chance, you would call them a moron.] However, their electrical charges come from a flow of protons, not electrons. The bacteria can stop, start, and change speed, direction, and even the "propeller's" shape. They also have intricate sensors, switches, control mechanisms, and a short-term memory. All this is highly miniaturized. **Eight million** of these bacterial motors would fit inside the circular cross section of a human hair.

[And yet rather than allowing our students to revel in the awesome, mind-boggling imagination and creative power of their Maker, they shovel this bald-faced, scientifically-absurd lie down their throats: the great deceiver's creation account. And their uninformed parents just sit back and let it happen.[659] How sad. How very sad.]

"Evolutionary theory teaches that bacteria were one of the first forms of life to evolve, and, therefore, they are simple. While bacteria are small, they are not simple. They can even communicate among themselves using chemicals.

"Some plants have motors that are one-fifth the size of bacterial motors. Increasing worldwide interest in nanotechnology is showing that living things are remarkably designed—beyond anything Darwin could have imagined."[660]

Darwin has an excuse. He could not possibly have imagined it. Today we have no excuse. Only willful ignorance.

"The hearing ear and the seeing eye, The Lord has made them both." (Proverbs 20:12)

How about a movie camera? Could the universe make a movie camera, by random chance, starting from rocks and dirt, using small incremental accidental steps? Of course not. You'd have to be a moron to think so. Well how about the human eye? Could the universe make a human eye, by random chance, starting from a mixture of chemicals and water, using small incremental accidental steps? Of course not. You'd have to believe the ridiculous lies that you were taught in order to think so. And keep in mind that the human eye is literally hundreds of thousands of times more complicated than a movie camera. First of all, the human eye forms itself in the womb blindly and unconsciously from just the DNA blueprints contained inside of each cell. It performs this incredible task with no conscious intervention, all completely automated, millions of tiny little molecular machines running around forming each specific type of cell and groups of cells and putting countless millions of them in precisely the right spot, building your eye in the womb, using the coded information from a tiny strand of DNA. And then your eye will go on to operate itself 24/7 without any conscious intervention for the rest of your life. It is a completely automated system that operates constantly, repairs itself, and captures and interprets trillions of pixels of visual information every picosecond, transmitting it all reliably and accurately to your brain; all accomplished since the beginning of life on earth over and over again, trillions upon trillions of times, in every organism with sight capabilities; again, completely automated with no conscious intervention. Konica Minolta, Cannon or Nikon could not reproduce the human, or any animals, eye or even a bulkier facsimile if they worked day and night for the next six million years. But according to evolution, random chance did!?!?#$

Also, consider this... What evolved first? The eye or the eye lid? Why evolve an eye lid without an eye? (As if *random chance* could even formulate that question, much less actually make the decision.) And if the eye evolves first then it dries up and shrivels without a lid? They would have to evolve together. At exactly the same time. In the same organism. And don't forget tear ducts. Also at exactly the same time. In the same organism. And that's exactly how it did happen. It's called creation, stupid.

---

659   (And then rush out and vote Democrat. Yes, I just had to throw that in)
660   (Pg. 19, *In the Beginning: Compelling Evidence for Creation and the Flood*, by Walt Brown, Ph.D.)

"Evolutionists are hard-pressed to explain the step-by-step step chance evolution of the eye which is characterized by a staggering complexity. Furnished with automatic aiming, automatic focusing, and automatic aperture adjustment, the human eye can function from almost complete darkness to bright sunlight, see an object the diameter of a fine hair, and make about 100,000 separate motions in an average day, faithfully affording us a continuous series of color stereoscopic pictures. All of this is performed usually without complaint, and then while we sleep, it carries on its own maintenance work.

"The human eye is so complex and sophisticated that scientists still do not fully understand how it functions. Considering the absolutely amazing, highly sophisticated synchronization of complex structures and mechanisms which work together to produce human vision, it is difficult to understand how evolutionists can honestly believe that the eye came about through the step-by-step, trial and error evolutionary process. This is especially true where we realize that the eye would be useless unless fully developed. It either functions as an integrated whole or not at all. Thus, the piecemeal evolution of the eye is completely outlandish and unreasonable."[661]

If anyone said that his car or his computer or his camera or his tape recorder had just evolved by random accidental chance everyone on earth would say that he was literally insane. Yet here's the kicker. The electronic and engineering equipment in your muscles, your brain, your eye and your ears are far more complicated technological wonders then these man-made artifacts. Indeed, compared to your brain, your ears and your eyes, all of our "man-made artifacts are primitive and clumsy."

"The cells macromolecular machines contain dozens of or even hundreds of components. But unlike man made machines, which are built on assembly lines, these molecular machines assemble spontaneously from their protein and nucleic-acid components. It is as though cars could be manufactured by merely tumbling their parts onto the factory floor."[662]

How about the latest robots from Japan? They are called Dancing Japanese Robots and you can watch them on YouTube. They not only walk and dance, but can even negotiate stairs. Could the universe make these robots by random chance, starting from minerals and dirt, using small incremental accidental steps? Of course not. You'd have to be a moron to think so. Well how about a human being, that can not only make itself in the womb from just one microscopic little cell, from the information contained inside of that cell on a minute strand of DNA, unconsciously, and repair itself, completely automated, with no outside conscious intervention to help it, but it can run and walk and talk and jump and sit and negotiate a thousand obstacles hundreds of times more complicated than just a simple set of stairs or a couple of basic dance moves? Could the universe make a human being by random chance, starting from water and chemicals, using small incremental accidental steps? Of course not. You'd have to believe the idiotic lies that you were taught in order to think so.

---

661   (Pgs. 71-73, *The Collapse of Evolution*, by Scott M. Huse)
662   ("Woodson, S. A. 2005. Biophysics: assembly line inspection. *Nature* 438:566-67." As quoted on pg. 126, *The Edge of Evolution: The Search for the Limits of Darwinism*, by Dr. Michael J. Behe)

What are the odds?

"Understand, you stupid among the people: and you fools, when will you be wise? He who planted the ear, shall he not hear? He who formed the eye, shall he not see?" (Psalms 94:8-9)

Previously we have discussed the Law of Mathematical Probability and its role in disproving evolutionary theory. Here we will delve a little deeper into this fascinating subject.

If you flip a quarter, the odds of it landing on heads are fifty/fifty, or a probability of 1 in 2. If you flip that same coin a trillion times, the odds of it landing *on its edge* every single time are zero. If the universe lasted for an infinite number of years, and you flipped that quarter over and over trying to get it to land on its edge a trillion straight times (I know, once would be hard enough) it would never happen. After a million years, you might even get up to 2 in a row! Those are the same odds that life, not all of life, but just one single cell, arose by random accidental chance. (Actually, your odds at the coin toss are a lot better.) If you took 1,000 jigsaw puzzles—each one having 10,000 pieces and of extreme difficulty and taking many days for a team of humans to assemble—and you took them up in a jet and dumped all of the pieces of all of those puzzles at 30,000 ft. into the jet stream over California, what are the odds of all of them landing in the same place, and assembling themselves perfectly, right next to each other, and in alphabetical order, one puzzle after another, in the parking lot of the Ravens football stadium in Baltimore, Md.? Impossible, of course, but that has a far better chance of happening accidentally by random chance than your eye forming itself in the same way. In fact, the odds that those same Ravens will go undefeated each year and win the next 20 Super Bowls in a row by a combined score of 500 to nothing without incurring a single offensive or defensive penalty in those 380 games, are still a trillion times better odds than just one protein forming by random accidental chance outside of a living cell in some mythological, lightning-struck, primordial mud-pond.

You see, it is a scientific fact that macro evolutionary theory, the whole thing and just about any single part of it, is absolutely scientifically impossible according to the mathematical laws of probability.

As we have previously pointed out, it has been determined that statistical odds that exceed one out of 10^{66} can never ever happen, even if the universe lasted for an infinite number of years. The probability of one protein molecule forming by random chance is one in 10^{175}. And that's just one protein...

> "The statistical improbability for the next step, the formation of a single cell from all these improbable proteins, is beyond comprehension. ...It is statistically impossible for life to originate from nonlife."[663]

> "The chemical evolution of life... is ridiculously improbable.
>
> "No evolutionary theory has been able to explain why earth's atmosphere has so much oxygen. Too many substances should have absorbed oxygen on an evolving earth. Besides, if the early earth had oxygen in its atmosphere, compounds (called *amino acids*) needed for life to evolve would have been destroyed by oxidation. But if there had been no oxygen, there would have been no ozone (a form of oxygen) in the upper atmosphere.

---

663   (Pgs. 53-55, *Evidence for Creation: Intelligent Answers for Open Minds*, by Tom DeRosa)

Without ozone to shield the earth, the sun's ultraviolet radiation would quickly destroy life. The only known way for both ozone and life to be here is for both to come into existence simultaneously—in other words, by creation.

"Clays and various rocks absorb nitrogen. Had millions of years passed before life evolved, the sediments that preceded life should be filled with nitrogen. Searches have never found such sediments. Basic chemistry does not support the evolution of life.

"Living matter is composed largely of proteins, which are long chains of amino acids. ...amino acids cannot link together if oxygen is present. That is, proteins could not have evolved from chance chemical reactions if the atmosphere contained oxygen. However, the chemistry of the earth's rocks, both on land and below ancient seas, shows that the earth had oxygen before the earliest fossils formed.

..."To form proteins, amino acids must also be highly concentrated in an extremely pure liquid. However, the early oceans or ponds would have been far from pure and would have diluted amino acids, so the required collisions between amino acids would rarely occur. Besides, amino acids do not naturally link up to form proteins. Instead, proteins tend to break down into amino acids. Furthermore, the proposed energy sources for forming proteins (earth's heat, electrical discharges, or solar radiation) destroy the protein products thousands of times faster than they could have formed. The many attempts to show how life might have arisen on earth have instead shown (a) the futility of that effort, (b) the immense complexity of even the simplest life, and (c) the need for a vast intelligence to precede life.

"If, despite virtually impossible odds, proteins arose by chance processes, there is not the remotest reason to believe they could ever form a membrane-encased, self-reproducing, self-repairing, metabolizing, living cell.

"There is no evidence that any stable states exist between the assumed formation of proteins and the formation of the first living cells. No scientist has ever demonstrated that this fantastic jump in complexity could have happened—even if the entire universe had been filled with proteins.

"Living cells contain thousands of different chemicals, some acidic, others basic. Many chemicals would react with others were it not for an intricate system of chemical barriers and buffers. If living things evolved, these barriers and buffers must also have evolved—but at just the right time to prevent harmful chemical reactions. How could such precise, seemingly coordinated, virtually miraculous events have happened for each of millions of species?

[Hmmm. *Kreehayshun* evolution? Whadaya think?]

"All living organisms are maintained by thousands of chemical pathways, each involving a long series of complex chemical reactions. For example, the clotting of blood, which involves 20-30 steps, is absolutely vital to healing a wound. However, clotting could be fatal if it happened inside the body. Omitting one of the many steps, inserting an unwanted step, or altering the timing of a step would probably cause death. If one thing goes wrong, all the earlier marvelous steps that worked flawlessly were in vain. Evidently, these complex pathways were created as an intricate, highly integrated system. ['Hello!]

"The genetic information in the DNA of each human cell is roughly equivalent to a library of 4,000 books. Even if matter and life (perhaps a bacterium) somehow arose, the probability that mutations and natural selections produced this vast amount of information is essentially zero. ...To produce just the enzymes in one organism would require more than $10^{40,000}$ trials. (To begin to understand how large $10^{40,000}$ is, realize that the visible universe has fewer than 10^{80} atoms in it.)

"DNA cannot function without at least 75 preexisting proteins, but proteins are produced only at the direction of DNA. Because each needs the other, a satisfactory explanation for the origin of one must also explain the origin of the other. The components of these manufacturing systems must have come into existence simultaneously.

...."When a cell divides, its DNA is copied, sometimes with errors. Each animal and plant has machinery that identifies and corrects most errors; if it did not, the organism would deteriorate and become extinct. If evolution happened, which evolved first, DNA or its repair mechanism? Each requires the other."[664]

Ooh! That was a nasty blow! Evolution is reeling and down for the count. Creation has been ordered to a neutral corner. Folks, we're going to have to make the call. This fight is over.

A snowman can't even make itself. An intelligent being is required in order for it to come into existence. Someone has to roll the snow, and stack it, and stick the carrot in its face. It has to be **designed**, and then **manufactured**. No one would try to contradict this. And yet all a snowman is, is a couple of balls of snow, some buttons, a carrot and a hat. If you took trillions of video cameras and then set them up at the bottom of every slope of every snow covered mountain and hill in the world, and then waited for a snowball to roll down the hill and stop. And before it melted another ball would have to roll down and land right on top of the first without busting up either itself or the bottom snowball. Then this process would have to be repeated a third time for the head of our self-created snowman. And we're not finished. We would still need a tornado somewhere to rip a carrot out of the ground and send it through the atmosphere and fly it right into the middle of the top snowball for a nose, without damaging the three balls in the process. And then another high wind would have to snatch up a couple of acorns and send them randomly flying through the sky until they accidentally landed above the carrot in the place of two eyes. Simple enough? Trillions of times simpler than even a single celled organism. Yet with all of its simplicity, you could set up all of those cameras and you could wait an infinite number of years and you would never, ever, ever get a snowman. Never. Unless you walked around in front of one of the cameras and made the damn thing yourself. The theory of evolution is absolutely idiotic.

"To claim that life evolved is to demand a miracle. The simplest conceivable form of single-celled life should have at least 600 different protein molecules. The mathematical probability that even one typical protein could form by chance arrangements of amino acid sequences is essentially zero—far less than one in 10^{450}. To appreciate the magnitude of 10^{450}, realize that the visible universe is about 10^{28} inches in diameter."

"From another perspective, suppose we packed the entire visible universe with a simple form of life, such as bacteria. Next, suppose we broke all their chemical bonds, mixed all their atoms, then let them form new links. If this were repeated a billion times a second for 20 billion years under the most favorable temperature and pressure conditions throughout the visible universe, would even one bacterium of any type reemerge? [Uh, no?] The chances are much less than one in $10^{99,999,999,873}$. Your chances of randomly drawing one preselected atom out of a universe packed with atoms are about one chance in 10^{112}—much better."[665]

---

664 (Pgs. 14-16, *In the Beginning: Compelling Evidence for Creation and the Flood*, by Walt Brown, Ph.D.)

665 (Pg. 17, *In the Beginning*, by Walt Brown, Ph.D.)

Let's state the odds thing one more way. The universe has a better chance to create a car accidentally from scratch over millions of years than even one DNA molecule. It has a better chance of creating an entire small town accidentally by random chance over millions or billions of years than one single-celled amoeba. And keep in mind that these are scientific facts. Not theory, conjecture, speculation or wishful irrational thinking.

But what about the atheist's stubborn old argument in favor of evolution that—despite the apparent astronomical technical complexity of living cells, organs and tissues—given enough time, anything is possible? Of course, this is hogwash, for according to the mathematical laws of probability even a stupid pencil is beyond the scope of possibility for random chance even if given an *infinite* number of years…

> "A few years ago a curious fellow decided to test the old saw that, given enough typewriters and enough time, an army of monkeys would eventually produce the works of Shakespeare. A computer with keyboard was placed in a cage containing six macaques in a British zoo and left for four weeks. The result? "The macaques—Elmo, Gum, Heather, Holly, Mistletoe and Rowan—produced just five pages of text between them, primarily filled with the letter S. There were greater signs of creativity towards the end, with the letters A, J, L and M making fleeting appearances." The five pages have been published under the ironic title "Notes toward the Complete Works of Shakespeare." "[666]

If it were not for their complete control of the media, the educational system and our entertainment industry, Darwinists might have already realized that they've become the laughing stock of the entire universe, and their theory would have already been exposed to mankind for the sheer lunacy that it is. And maybe then, just maybe, we could stop shoving this insanity down the throats of our naïve and unsuspecting school children in our colleges, universities, elementary and high schools. It is an insane asylum. And the inmates are in charge.

Irreducible Complexity

"For you formed my inward parts; you have covered me in my mother's womb. I will praise thee; **for I am fearfully and wonderfully made**: marvelous are thy works; and that my soul knows right well." (Psalm:139:13-14)

Irreducible Complexity is a term coined by Lehigh University biochemist Michael Behe, Ph.D. He studies the workings of the living cell on a molecular level, and is the author of two books on the subject that clearly show that the machinery inside of cells and inside of living organisms are so mind-boggling in their complexity that they are far beyond what could ever be formed accidentally or created by random mutations.

> "Darwin knew that his theory of gradual evolution by natural selection carried a heavy burden: "If it could be demonstrated that any complex organ existed which could not possibly have been formed by numerous, successive, slight modifications, my theory would absolutely break down."[667]

> "It is safe to say that most of the scientific skepticism about Darwinism in the past century has centered on this requirement. …What type of biological system could not be formed by "numerous, successive, slight modifications"?

---

666   (Pg. 104-105, *The Edge of Evolution: The Search for the Limits of Darwinism*, by Dr., Michael J. Behe)
667   (Darwin, C. (1872) *Origin of Species*, 6th ed. (1988) New York University Press, New York, p. 154.)

"Well, for starters, a system that is irreducibly complex. By *irreducibly complex* I mean a single system composed of several well-matched, interacting parts that contribute to the basic function, wherein the removal of any one of the parts causes the system to effectively cease functioning. An irreducibly complex system cannot be produced directly (that is, by continuously improving the initial function, which continues to work by the same mechanism) by slight, successive modifications of a precursor system because any precursor to an irreducibly complex system that is missing a part is by definition nonfunctional."[668]

Dr. Behe uses the example of a Rube Goldberg contraption, which is a "deliberately overengineered apparatus that performs a very simple task in a very complex fashion, usually including a chain reaction."[669] Most of us have seen one of Goldberg's drawings that involve numerous successive and sometimes silly steps to put toothpaste on his brush or wipe his mouth with a napkin. The point is though, that if you were to remove just one of those many successive steps then the entire machine would be useless. The same is true with so many living systems. However, unlike a Rube Goldberg machine, an *irreducibly complex* cellular or biological system is one that is deliberately and *exquisitely* engineered to perform an extremely vital task in the most effective fashion possible, usually including many successive and highly complex steps that also can travel down multiple pathways, depending upon any number of various situations or contingencies it may encounter, in order to fulfill its function and accomplish its preprogrammed goal. But like the Goldberg contraption, if you remove just one of those steps the entire cellular or biological task fails and the organism dies. So, because of that, there is absolutely no conceivable process of "numerous, successive, slight modifications" that could arrive at these complex and vital cellular machines or biological systems. Either the entire machine is made at once as a fully functioning, completed system, or it is worthless, and even detrimental, to the organism. In other words, these systems could never, ever, ever have *evolved*, and therefore by their existence alone, Darwinism "has absolutely broken down."

"Modern science has learned that, ultimately, life is a molecular phenomenon: All organisms are made of molecules that act as the nuts and bolts, gears and pulleys of biological systems. Certainly there are complex biological features (such as the circulation of blood) that emerge at higher levels, but the gritty details of life are the province of biomolecules. Therefore the science of biochemistry, which studies those molecules, has as its mission the exploration of the very foundation of life.

…"It was once expected that the basis of life would be exceedingly simple. That expectation has been smashed. Vision, motion, and other biological functions have proven to be no less sophisticated than television cameras and automobiles.

…"Life is based on *machines*--machines made of molecules! Molecular machines haul cargo from one place in the cell to another along "highways" made of other molecules, while still others act as cables, ropes, and pulleys to hold the cell in shape. Machines turn cellular switches on and off, sometimes killing the cell or causing it to grow. Solar-powered machines capture the energy of photons and store it in chemicals. Electrical machines allow current to flow through nerves. Manufacturing machines build other molecular machines, as well as themselves. Cells swim using machines, copy themselves with machinery, ingest

---

668   (Pg. 39, *Darwin's Black Box: The Biochemical Challenge to Evolution*, by Michael J. Behe)
669   ("Rube Goldberg machine," from Wikipedia, the free encyclopedia)

food with machinery. In short, highly sophisticated molecular machines control every cellular process. Thus the details of life are finely calibrated, and the machinery of life enormously complex."[670]

Let's take a look at the "enormously complex machinery" of the human eye:

"When light first strikes the retina a photon interacts with a molecule called 11-*cis*-retinal, which rearranges within picoseconds to *trans*-retinal. (A picosecond is about the time it takes light to travel the breadth of a single human hair.) The change in the shape of the retinal molecule forces a change in the shape of the protein, rhodopsin, to which the retinal is tightly bound. The protein's metamorphosis alters its behavior. Now called metarhodopsin II, the protein sticks to another protein, called transducin. Before bumping into metarhodopsin II, transducin had tightly bound a small molecule called GDP. But when transducin interacts with metarhodopsin II, the GDP falls off, and a molecule called GTP binds the transducin. (GTP is closely related to, but critically different from, GDP.)

"GTP-transducin-metarhodopsin II now binds to a protein called phosphodiesterase, located in the inner membrane of the cell. When attached to metarhodopsin II and its entourage, the phosphodiesterase acquires the chemical ability to "cut" a molecule called cGMP (a chemical relative of both GDP and GTP). Initially, there are a lot of cGMP molecules in the cell, but the phosphodiesterase lowers its concentration, just as a pulled plug lowers the water level in a bathtub.

"Another membrane protein that binds cGMP is called an ion channel. It acts as a gateway that regulates the number of sodium ions in the cell. Normally the ion channel allows sodium ions to flow into the cell, while a separate protein actively pumps them out again. The dual action of the ion channel and pump keeps the level of sodium ions in the cell within a narrow range. When the amount of cGMP is reduced because of cleavage by the phosphodiesterase, the ion channel closes, causing the cellular concentration of positively charged sodium ions to be reduced. This causes an imbalance of charge across the cell membrane that, finally, causes a current to be transmitted down the optic nerve to the brain. The result, when interpreted by the brain, is vision.

"If the reactions mentioned above were the only ones that operated in the cell, the supply of 11-*cis*-retinal, cGMP, and sodium ions would quickly be depleted. Something has to turn off the proteins that were turned on and restore the cell to its original state. Several mechanisms do this. First, in the dark the ion channel (in addition to sodium ions) also lets calcium ions into the cell. The calcium is pumped back out by different proteins so that a constant calcium concentration is maintained. When cGMP levels fall, shutting down the ion channel, calcium ion concentration decreases, too. The phosphodiesterase enzyme, which destroys cGMP, slows down at lower calcium concentration. Second, a protein called guanylate cyclase begins to resynthesize cGMP when calcium levels start to fall. Third, while all of this is going on, metarhodopsin II is chemically modified by an enzyme called rhodopsin kinase. The modified rhodopsin then binds to a protein known as arrestin, which prevents the rhodopsin from activating more transducin. So the cell contains mechanisms to limit the amplified signal started by a single photon.

"*Trans*-retinal eventually falls off of rhodopsin and must be reconverted to 11-*cis*-retinal and again bound by rhodopsin to get back to the starting point for another visual cycle. To accomplish this, *trans*-retinal is first chemically modified by an enzyme to *trans*-retinol--a form containing two more hydrogen atoms. A second enzyme then converts the

---

670 (Pgs. X, 4-5, *Darwin's Black Box: The Biochemical Challenge to Evolution,* by Michael J. Behe)

molecule to 11-*cis*-retinol. Finally, a third enzyme removes the previously added hydrogen atoms to form 11-*cis*-retinal, a cycle is complete."[671]

And here's the kicker, "The above explanation is just a sketchy overview of the biochemistry of vision."[672]

Don't be upset if you couldn't quite understand the complexity of all of that. Indeed, that's the point! If you, an intelligent human being with an extremely complex computer-brain find it difficult to even comprehend the complexity of just these few initial steps in the process that leads to sight within your head, how in the hell could random accidental chance ever have comprehended, or "stumbled" upon, it to begin with? If anyone at this point still thinks we were going a tad overboard when we said that the theory of evolution was so stupid, idiotic and ridiculous that it literally borders on the insane, then hopefully they have changed their opinion right now. For, from just those last few paragraphs alone, it is abundantly clear that the theory of evolution is not only scientifically-impossible, but it is a scientific fact that it is insane! (Richard Dawkins, are you listening? This is not a laughing matter and I am not poking fun at or mocking you now. I am telling you that you need professional help. I would suggest giving Jesus, the Great Physician, a call. He specializes in eternal diseases of the mind.)

And in case anyone might have glossed over the beginning of Dr. Behe's description, let me emphasize that he did say that all of that happens within a few **picoseconds**. And "a picosecond is about the time it takes light to travel the breadth of a single human hair." Hmm? That seems pretty darn fast. Let's see, light travels at 186,000 miles per second, and if we multiply that by 5280 ft. per mile, times 12 inches per foot, times 250 hairbreadths in an inch we get 2,946,240,000,000! That's two trillion, nine hundred and forty six billion, two hundred and forty million astronomically-complicated-sequenced chemical reactions every single second that occurs in the back of your eye in just one cell! And that is just one of a number of chemical and electrical sequences that continually occur to allow sight to happen. Multiply that by the billions of retinal cells in the back of your eyes, and the billions of human eyes and the trillions upon trillions of other animal's eyes across this planet, and this occurs every single second of every single day. The idea that this all happened by random chance is absolutely insane. It is beyond absurd, it far surpasses mere stupidity and it eclipses the idiotic and the asinine. Yea, it's insane.

Dr. Behe goes on to explain the equally enormously complex biochemical-mechanical machinery of the cilium, of blood clotting, of animal development and of intracellular transport. They rival the intricacy of sight in the human eye.

> "The function of the cilium is to be a motorized paddle. In order to achieve this function microtubules, nexin linkers, and motor proteins all have to be ordered in a precise fashion. They have to recognize each other intimately, and interact exactly. The function is not present if any of the components is missing. Furthermore, many more factors besides those listed are required to make the system useful for a living cell: the cilium has to be positioned in the right place, oriented correctly, and turned on or off according to the needs of the cell."[673]

---

671   (Pgs. 18-21, *Darwin's Black Box: The Biochemical Challenge to Evolution*, by Michael J. Behe)

672   (Pg. 22, *Darwin's Black Box*, by Michael J. Behe)

673   (Pg. 204, *Darwin's Black Box: The Biochemical Challenge to Evolution*, by Michael J. Behe)

"The structural elegance of systems such as the cilium, the functional sophistication of the pathways that construct them, and the total lack of serious Darwinian explanations all point insistently to the same conclusion: They are far past the edge of evolution. Such coherent, complex, cellular systems did not arise by random mutation and natural selection, any more than the Hoover Dam was built by the random accumulation of twigs, leaves, and mud."[674]

Yes but the difference is that the Hoover Dam does not have an entire false religion disguised as science built up around it that insists that Dams just *build themselves* by accident, driven of course by the unconscious struggle to create lakes and hydroelectric power.

The mechanism of blood clotting is so complicated. There are so many steps to get from a cut to a clot. There are so many options for starting the clot and stopping it at the right time. There are so many different controls and failsafe mechanisms and redundancies built into the system so a clot does not clog the artery or the veins or the capillaries and kill the organism it is supposed to heal. There are so many different pathways from which the blood clotting mechanism can go down or change course depending on each unique circumstance. To suppose that this mechanism could have made itself accidentally by random chance is absurd. It would be akin to teaching that the entire mechanized assembly line at Ford or General Motors just formed themself, somehow, accidentally, over millions of years. No design. No intelligent foresight, direction, manufacturing of all of the many parts, and then assembly of the line into a functioning machine. Just dumb luck.

"The function of the blood-clotting system is as a strong, but transient barrier. The components of the system are ordered to that end. Fibrinogen, plasminogen, thrombin, protein C, Christmas factor, and the other components of the pathway together do something that none of the components can do alone. When vitamin K is unavailable or anti-hemophilic factor is missing, the system crashes just as surely as a Rube Goldberg machine fails if a component is missing. The components cut each other in precise places, align with each other in exact ways. They act to form an elegant structure that accomplishes a specific task.

"The function of the intracellular transport system is to carry cargo from one place to another. To do this packages must be labeled, destinations recognized, and vehicles equipped. Mechanisms must be in place to leave one enclosed area of the cell and enter a different enclosed area. The failure of the system leaves a deficit of critical supplies here, a choking surplus there. Enzymes that are useful in a confined area wreak havoc in another area.

..."The designing that is currently going on in biochemistry laboratories throughout the world—the activity that is required to plan a new plasminogen that can be cleaved by thrombin, or a cow that gives growth hormone in its milk, or a bacteria that secretes human insulin—is analogous to the designing that preceded the blood-clotting system. The laboratory work of graduate students piecing together bits of genes in a deliberate effort to make something new is analogous to the work that was done to cause the first cilium."[675]

---

674   (Pg. 102, *The Edge of Evolution: The Search for the Limits of Darwinism*, by Dr., Michael J. Behe)
675   (Pgs. 204-205, *Darwin's Black Box: The Biochemical Challenge to Evolution*, by Michael J. Behe)

Except the design work that caused the first cilium was brought forth effortlessly by an Infinite Intelligence beyond our ability to imagine, and beyond the capability of Darwinists to demean. An Infinite Intelligence who holds in derision those who would deny Him and His Creative Power and His Divine Authorship over all of creation. "He who sits in the heavens shall laugh. The Lord shall hold them in derision." (Psalms 2:4)

"For every house is built by someone, but he who built all things is God." (Hebrews 3:4) We all understand blueprints; every construction project has them. An architect's blueprints specify everything, down to the last detail of a home addition or a skyscraper, or of any other relatively simple or highly complex construction project. Could those blueprints have ever just happened by random chance? Of course not, you'd have to be an idiot to say so. Well, every living cell has a set of construction blueprints as well. The blueprints for animal bodies are inside the coded information in each animals DNA. And they are far more complicated and intricate and more ingenuously designed than any set of man-made construction plans. Could these blueprints have ever just happened by random chance? Of course not, you'd have to be a... you'd have to be taught an idiotic lie to think that as well.

"The inadequacy of Darwinism to account for the intricacies of animal development has not been lessened by recent discoveries; it has been greatly exacerbated."[676]

"The bottom line is that, while great progress has been made towards understanding how animals are made, and has revealed unexpected, stunning complexity, no progress at all has been made in understanding how that complexity could evolve by unintelligent processes."[677]

That's because, in a word, it can't. And this same utter incomprehensible complexity is true not only of the eye, and the blood clotting system, and the motor apparatus of cilium, and the making of an adult animal from a tiny cell, and the cargo delivery systems of the cell, but also of the immune systems of plants and animals, and the flagellum motors of bacteria, and the information coding and decoding in the DNA, and on and on and on. They are all irreducibly complex, Rube Goldberg machines to the umpteenth power. No way any single one of them, much less *all* of them, could ever have *evolved* by random accidental unintelligent chance mutations.

And if not evolution, then the only other choice is intelligent design:

"Here, then, is the argument for design in a nutshell: (1) We infer design whenever parts appear arranged to accomplish the function. (2) The strength of the inference is quantitative and depends on the evidence; the more parts, and the more intricate and sophisticated the function, the stronger is our conclusion of design. With enough evidence, our confidence in design can approach certitude. If while crossing a heath we stumble across a watch (let alone a chronometer), no one would doubt ...that the watch was designed; we would be as certain about that as about anything in nature. (3) Aspects of life overpower us with the appearance of design. (4) Since we have no other convincing explanation for that strong appearance of design, Darwinian pretensions notwithstanding, then we are rationally

---

676 (Pg. 192, *The Edge of Evolution: The Search for the Limits of Darwinism*, by Dr., Michael J. Behe)
677 (Pg. 183, *The Edge of Evolution*, by Dr., Michael J. Behe)

justified in concluding that parts of life were indeed purposely designed by an intelligent agent.

"A crucial, often-overlooked point is that the overwhelming appearance of design strongly affects the burden of proof: in the presence of manifest design, the onus of proof is on the one who denies the plain evidence of his eyes. For example, a person who conjectured that the statues on Easter Island or the images on Mount Rushmore were actually the result of unintelligent forces would bear the substantial burden of proof the claim demanded. In those examples, the positive evidence for design would be there for all to see in the purposeful arrangement of parts to produce the images. Any putative evidence for the claim that the images were actually the result of unintelligent processes (perhaps erosion shaped by some vague, hypothesized chaotic forces) would have to clearly show that the postulated unintelligent processes could indeed do the job. In the absence of such a clear demonstration, any person would be rationally justified to prefer the design explanation.

"I think these factors account to a large degree for why, much to the consternation of Darwinian biologists, the bulk of the public rejects unintelligent processes as sufficient explanations for life. People perceive the strong appearance of design in life, are unimpressed with Darwinian arguments and examples, and will reach their own conclusions, thank you very much. Without strong, convincing evidence to show that Darwin can do the trick, the public is quite rational to embrace design. (Of course other factors besides the quality of the evidence, such as social pressure, can affect a person's judgment. In the scientific and academic communities as a whole there is strong social pressure to dismiss design explanations for life out of hand. The social situation is quite different for the general public.)"[678]

But what do evolutionists have to say about this frontal assault on their doctrine? What is the passionate defense that they have presented for all the world to hear and to debate. It doesn't exist:

"Molecular evolution is not based on scientific authority. There is no publication in the scientific literature—in prestigious journals, specialty journals, or books—that describes how molecular evolution of any real, complex, biochemical system either did occur or even might have occurred. There are assertions that such evolution occurred, but absolutely none are supported by pertinent experiments or calculations. Since no one knows molecular evolution by direct experience, and since there is no authority on which to base claims of knowledge, it can truly be said that—like the contention that the Eagles will win the Super Bowl this year—the assertion of Darwinian molecular evolution is merely bluster.

""Publish or perish" is a proverb that academicians take seriously. If you do not publish your work for the rest of the community to evaluate, then you have no business in academia (and if you don't already have tenure, you will be banished). But the saying can be applied to theories as well. If the theory claims to be able to explain some phenomenon but does not generate even an attempt at an explanation, then it should be banished. Despite comparing sequences and mathematical modeling, molecular evolution has never addressed the question of how complex structures came to be. In effect, the theory of Darwinian molecular evolution has not published, and so it should perish."[679]

---

678   (Pgs. 265-266, 309, *Darwin's Black Box: The Biochemical Challenge to Evolution*, by Michael J. Behe)

679   (Pgs. 185-186, *Darwin's Black Box: The Biochemical Challenge to Evolution*, by Michael J. Behe)

"The impotence of Darwinian theory in accounting for the molecular basis of life is evident not only from the analyses in this book, but also from the complete absence in the professional scientific literature of any detailed models by which complex biochemical systems could have been produced... In the face of the enormous complexity that modern biochemistry has uncovered in the cell, the scientific community is paralyzed. No one at Harvard University, no one at the National Institutes of Health, no member of the National Academy of Sciences, no Nobel Prize winner—no one at all can give a detailed account of how the cilium, or vision, or blood clotting, or any complex biochemical process might have developed in a Darwinian fashion. But here we are. Plants and animals are here. The complex systems are here. All these things got here somehow: if not in a Darwinian fashion, then how?"[680]

Hmm, I'm going to take a wild stab in the dark at that one but, uh, anyone ever heard of Genesis? The Bible? Mr. Hitchens, ever heard of God??

"If the great majority of cellular protein-protein interactions are beyond the edge of evolution [beyond the ability of random chance to *ever* produce], it is reasonable to view the entire cell itself as a nonrandom, integrated whole—like a well-planned factory, as National Academy of Sciences president Bruce Alberts suggested. ...Nonrandomness isn't a rare property of just a handful of extra-complex features of the cell. Rather **it encompasses the cellular foundation of life as a whole**."[681] (Emphasis mine.)

What that means is that if it is a scientific <u>fact</u> that random chance could never ever have produced the complexity of life's cellular systems (and it is), then evolution is impossible and the only other plausible scientific theory that remains for the origin and diversity of life is that it was deliberately and purposefully <u>designed</u> and then <u>manufactured</u> by an intelligence and a power far beyond our ability to comprehend. All we can do is to stand back and look on in absolute awe and wonder. An awe and a wonder and a respect that is being stolen from our school children every day across this nation by the shoveling down their throats of the bald-faced, idiotic, asinine, scientifically-impossible, poisonous, noxious, satanically-inspired, rancid lie that is the theory of evolution! Congratulations, America! My, how proud you must be... "My boy Charlie is an honor student at Leftos Indoctrinato Junior High, and he got straight A's in evolutionary biology." And if you had just a tad more sense and a little more actual education then you might have been a little more proud if he had told his teacher she was wrong, that she was teaching a lie as if it were true, and had gotten straight F's instead. (Fascists are highly intolerant of dissent.)

Well there you have it. I think that should be more than sufficient for anyone but the saddest of religious fanatics. For that is what in truth the theory of evolution is. It is a religion and not science. And a false and absurd religion at that. And keep in mind that we have only scratched the surface of the tip of the iceberg in our discussion of the scientific evidence here. (See the list of recommended books at the end of this chapter, many of which we have quoted from.) For there is in truth a crushing avalanche of scientific facts and evidence that obliterates the *theory* of evolution. And there are no scientific facts or evidence that can be used to credibly

---

680   (Pg. 187, *Darwin's Black Box: The Biochemical Challenge to Evolution*, by Michael J. Behe)
681   (Pg. 146, *The Edge of Evolution: The Search for the Limits of Darwinism*, by Dr., Michael J. Behe)

support this theory. It is an unscientific and blind religious faith in impossible happenstance, in order to afford its proselytes the dubious distinction of denying their own Creator.

"But ask now the beasts, and they shall teach you; and the fowls of the air, and they shall tell you. Or speak to the earth, and it shall teach you: and the fishes of the sea shall declare unto you. Who knows not in all of these that the hand of the Lord has wrought this?" (Job 12:7-9) If the beasts of the field and the fishes of the sea can figure out who wrought all of this, you'd of thought it would have at least by now been within the grasp of the intellectual elites in our universities and in our media. You'd of thought wrong.

The Big Bang, All Blown Up

"God, who at many times and in various ways spoke in time past unto the fathers by the prophets, has in these last days spoken unto us by his Son, whom he has appointed heir of all things, through whom also He [and not some stupid explosion] **made** the universe. Who being the brightness of his glory and the exact image of his being, sustaining all things by his powerful word, when he had by himself purged our sins, sat down on the right hand of the Majesty on high." (Hebrews 1:1-3)

Well, I think poor evolution has had about all it can take. So let's give it some time to die peacefully as we shift our attention now to its companion theory and the other half of Satan's genesis, the big bang. This theory states that the entire universe was created by an enormous explosion billions of years ago and that in the proceeding eons of time everything you see today in the night sky just made itself; just *evolved* somehow accidentally and miraculously by random chance. Of course this is almost as idiotic as the notion that life just made itself, and equally as scientifically-impossible, as we shall see…

Again we start with the Laws of science:

"The Second Law of Thermodynamics… tends to bring a system to disorder. …When one observes the universe, the second law is apparent everywhere. The sun is wearing down slowly; stars are burning out and even exploding. … Big Bang theory contradicts the Second Law because it requires particles to organize and cohere on a cosmic scale. There is no scientific evidence for this claim. …What we see in the universe is directly opposite to the expectation of evolutionary cosmologists. We observe a decaying universe whose order of complexity is in decline. Evolution cosmology directly defies this great law of science."[682]

Evolution cosmology also defies all of the facts and evidence, and common sense as well…

Faux science is enamored with the theory of the big bang, popularized by the British scientist, Stephen Hawking. Many people have heard of his bestseller, *A Brief History of Time*, "a book that was widely considered fascinating by those who did not read it, and incomprehensible by those who did."[683] It is "incomprehensible" because it champions a theory that is clearly scientifically impossible. Again (as we stated before in chapter 5), there is no such thing as

---

682 (Pg. 30, *Evidence for Creation: Intelligent Answers for Open Minds*, by Tom DeRosa)
683 (Pg. 98, *The Devil's Delusion: Atheism and its Scientific Pretensions*, by David Berlinski)

an explosion ever creating anything of any value or order. All an explosion can do is to make a big mess—to tear apart, to break down and to destroy—and this universe is about the furthest thing from a big mess you will ever find. Indeed, the universe displays order to an almost unimaginable level. When you learn of all of the laws that govern motion, and gravity, and light, and sound, and that control the behavior of atoms and molecules and their electrons, protons and neutrons; the order and engineering complexity of this universe is *astronomical* and beyond our comprehension. Also, according to the very laws of science itself, in particular the Second Law of Thermodynamics and the Laws of Mathematical Probability, we know for a scientific fact that it is an absolute impossibility for this universe, this solar system, our Earth-Moon system, and life on this planet to have been created by an ancient, cosmic explosion.

Again (as we reported in chapter 5) Cambridge University physicist Brandon Carter PhD calculated "that if gravity had been stronger or weaker by one part in 10^{40}, then life-sustaining stars like the sun could not exist. This would most likely make life impossible."[684] (This would *definitely* make life impossible.) 10^{40} is one with 40 zeroes behind it. A trillion is one with only 12 zeroes behind it. 10^{40} is a trillion times a trillion times a trillion times ten thousand. The chances of winning the lottery are around a million to one. The chances of the force of gravity accidentally being <u>exactly</u> the force needed to render the universe inhabitable is trillions and trillions and billions of times more remote than the chances of you winning the lottery. Try to shoot an arrow across the entire known universe with your eyes closed and hit a target in the <u>exact</u> center without being off by even one trillionth of an inch. And then try to do it a billion times in a row. It's impossible. So is the accidental formation of this universe. The force of gravity was specifically created to be <u>exactly</u> the amount needed by a degree of variance of less than one over 10^{40}. And this is only one of a number of examples of the mind-numbing exactness of the laws and constants of this universe. (In addition to the gravitational force mentioned above that holds matter together there is also the three forces—the electromagnetic force and the strong and the weak nuclear forces—that hold the structure of the atom together which are also <u>exactly</u> calibrated so the heavier elements can form and we can have a universe of matter that is able to form stars and planets and to support life.) "In light of these and other examples, Collins remarks that 'Almost everything about the basic structure of the universe ... is balanced on a razor's edge for life to occur.'"[685] God, 10^{40}. Big bang theory, zip. Game, set, match. Without stars and without orbiting planets, there could be no earth and there could be no life.

"God is a mathematician of a very high order and he used advanced mathematics in constructing the universe."[686] Again, what an understatement.

"There is no empirical evidence to support the star formation theory proposed by evolutionary cosmologists. [It is no more than wishful thinking, using one's fertile imagination to try and make sense of a scientifically-impossible theory.] ... The explanation offered is that as cooling occurs, particles slow down and clump together. The problem

684   (*International Journal of Systematic and Evolutionary Microbiology*, IJSEM, Collins 1999, 49)
685   (IJSEM, Collins 1999, 48)
686   (Paul A. M. Dirac)

is, however, that the celestial objects are moving at relatively high speeds away from each other. No star or galaxy has ever been seen to form in space from star gas. As Harvard astrophysicist, Abraham Loeb stated, "The truth is that we don't understand star formation at a fundamental level." "[687]

The truth is they never will "understand star formation at a fundamental level" until they understand that someone of infinite intelligence and power *formed* the darn things to begin with.

"The heavens declare the glory of God; and the firmament shows his handiwork." (Psalms 19:1)

There is no compelling or even credible scientific evidence for, nor is there even an intelligent idea of, how the universe could have just formed itself. Again, both our solar system and the universe are highly ordered. And order cannot come out of disorder by an explosion, or accidentally, or by random chance, or by any combination thereof. Thus speaks The Second Law of Thermodynamics, The Laws of Mathematical Probability, logic, common sense, and as we shall see, all of the facts and evidence as well...

"The evidence that the earth-sun system was designed by God far outweighs any possibility that it all just happened to come together by mere chance. Here are a few features of the earth-sun system which appear to be specifically and very carefully designed for the unique purpose of supporting life:

"1. The earth is positioned at just the right distance from the sun so that we receive exactly the proper amount of heat to support life. The other planets of our solar system are either too close (too hot) to the sun or else too far (too cold) to sustain life.

"2. Any appreciable change in the rate of rotation of the earth would make life impossible. For example, if the earth were to rotate 1/10th its present rate, all plant life would either be burned to a crisp during the day or frozen at night.

"3. Temperature variations are kept within reasonable limits due to the nearly circular orbit of the earth around the Sun.

"4. Temperature extremes are further moderated by the water vapor and carbon dioxide in the atmosphere which produce a greenhouse effect.

"5. The moon revolves around the earth at a distance of about 240,000 miles causing harmless [and quite essential] tides on the earth. If the moon was located 1/5 of this distance away, the continents would be completely submerged twice a day!

"6. The thickness of the earth's crust and the depths of the oceans appear to be carefully designed. Increases in thickness or depth of only a few feet would so drastically alter the absorption of free oxygen and carbon dioxide that plant and animal life could not exist.

"7. The earth's axis is tilted 23&1/2 degrees from the perpendicular to the plane of its orbit. This tilting, combined with the earth's revolution around the sun, causes our seasons, which are absolutely essential for the raising of food supplies.

"8. The earth's atmosphere (ozone layer) serves as a protective shield from lethal solar ultraviolet radiation, which would otherwise destroy all life.

"9. The earth's atmosphere also serves to protect the earth from approximately 20 million meteors that enter it each day at speeds of about 30 miles per second! Without this crucial protection the danger to life would be immense.

"10. The earth is the perfect physical size and mass to support life, affording a careful balance between gravitational forces (essential for holding water and an atmosphere) and atmospheric pressure.

---

687   (Pg. 30, *Evidence for Creation: Intelligent Answers for Open Minds*, by Tom DeRosa)

"11. The two primary constituents of the earth's atmosphere are "Nitrogen (78%) and oxygen (20%)." This delicate and critical ratio is essential to all life forms.

"12. The earth's magnetic field provides important protection from harmful cosmic radiation.

"13. The earth is uniquely blessed with a bountiful supply of water that is the key substance of life due to its remarkable and essential physical properties.

"Many other examples of this type could be cited that would also support the idea that the earth was created and carefully designed for a purpose. Such numerous perfect and complex combinations of interrelated conditions and factors essential to delicate life forms unequivocally point to an intelligent purposeful design. To believe that such an intricately planned and carefully balanced life support system is the result of mere chance is absolutely senseless."[688]

But don't tell Mr. Evolutionist that. It might hurt his feelings. And protecting his feelings is so much more important than stopping him from continuing to lie to our school children about the existence of an intelligent Creator called God. After all, not *that* many of our kids commit suicide because of his lie. And not all *that* many of them end up hooked on drugs for the rest of their lives or fall into debased sexual immorality because of the purposelessness that this untruth has instilled in them. So go ahead, America, continue blissfully in your stupidity and ignorance and just let this lie be taught unchallenged and unabated. Oh, gosh! I'm sorry, America, now I've gone and hurt your feelings.

"The evolutionary "Big Bang" theory proposes that dust particles in space coalesced to form galaxies, stars, planets and other celestial objects, and that 4.5 billions of years ago, the beginnings of our sun (a "proto-sun") formed out of a large gas cloud. [All of which is physically impossible.] The leftover gas from the cloud flattened into a disc [miraculously] [why the gas chose the disc over a cube, a pyramid, a pretzel or a Botticelli, remains a mystery], and then little particles coalesced, got bigger and bigger and developed into planet-like clumps that eventually became the planets of our solar system. Unfortunately for Big Bang proponents, this "common origin" theory of our solar system directly conflicts with known facts about the planets, including their vastly different compositions, rotations, distances between them, densities, magnetic fields, etc. The situation is summarized by physicist Lambert Dolphin, who states, "Theories for the origin of our solar system have come and gone, yet even today no satisfactory model exists that both explains all the facts and is consistent with the known laws of physics." In fact, physical laws inform us that far more than gravity is required to form a star, planet or any other heavenly body out of particles in a giant gas cloud. [You also need pixie dust.] Simple physics tells us that (kinetic) energy tends to disperse particles in a gas cloud, or cause them to move away from each other, not bring them together. In other words, gas clouds naturally expand rather than condense. It is actually preposterous—and defies the laws of science—to believe that without an overwhelming outside force, gas can contract to the point where nuclear fusion can take place, as in our sun."[689]

"Many undisputed observations contradict current theories on how the solar system evolved."[690] And remember, in science as in life, observations trump

---

688 (Pgs. 55-58, *The Collapse of Evolution*, by Scott M. Huse)
689 (From Creation Studies Institute circular, March 2011, Tom DeRosa, Executive Director and Founder)
690 (Pg. 25, *In the Beginning: Compelling Evidence for Creation and the Flood*, by Walt

theories all day long. If a theory is not backed up by our observations, then that theory must go. According to the theory of the big explosion and the subsequent "evolution" of the stars and galaxies of this universe, and of our own solar system, we should expect that certain things would be true. But they are not...

"All planets should spin in the same direction, but Venus, Uranus, and Pluto rotate backwards.

...Each of the almost 200 known moons in the solar system should orbit its planet in the same direction, but more than 30 have backward orbits. Furthermore, Jupiter, Saturn, Uranus, and Neptune have moons orbiting in both directions. ...The orbit of each of these moons should lie very near the equatorial plane of the planet it orbits, but many, including the Earth's moon, are in highly inclined orbits. ...The orbital planes of the planets should lie in the equatorial plane of the Sun. Instead, the orbital planes of the planets typically deviate from the Sun's equatorial plane by 7 degrees, a significant amount. ...The Sun should have about 700 times more angular momentum than all the planets combined. Instead, the planets have 50 times more angular momentum than the Sun."[691]

There is no known way for our earth to have acquired all of its water by natural, evolutionary means. Every explanation that the evolutionary cosmologists try to come up with turns out to be absurd. Comets couldn't have "brought" it here, neither could meteorites, neither could any other fantasized water transport system. It had to be placed here in the beginning, plain and simple.

"If the Earth and solar system evolved from a swirling cloud of dust and gas, almost no water would reside near Earth's present orbit. Any water (liquid or ice) that close to the Sun would vaporize and be blown by solar wind to the outer reaches of the solar system, as we see happening with water vapor in the tails of comets."[692]

Textbooks teach our children that the early earth was molten and bombarded by meteorites. But that is false as well:

"Had Earth ever been molten, dense, nonreactive chemical elements such as gold would have sunk to Earth's core. Gold is 70% denser than lead, yet it is found at the Earth's surface."[693]

Also, radioactive dating of certain zircon minerals, that had to have formed on a cold earth, contradicts a molten "early" Earth.

"Evolutionists claim that the solar system condensed out of a vast cloud of swirling dust about 4,600,000,000 years ago." This however is scientifically and astronomically impossible:

"Contrary to popular opinion, planets should not form from just the mutual gravitational attraction of particles orbiting the Sun. Orbiting particles are much more likely to be scattered or expelled by their gravitational attraction than they are to be permanently pulled together. Experiments have shown that colliding particles almost always fragment rather than stick together. (Similar difficulties exist in trying to form a moon from particles orbiting the planet.)

Brown, Ph.D.)

691   (Pg. 25, *In the Beginning: Compelling Evidence for Creation and the Flood*, by Walt Brown, Ph.D.)

692   (Pg. 26, *In the Beginning*, by Walt Brown, Ph.D.)

693   (Pg. 26, *In the Beginning: Compelling Evidence for Creation and the Flood*, by Walt Brown, Ph.D.)

"Despite these problems, let us assume that pebble-size to moon-size particles somehow evolved. "Growing a planet" by many small collisions will produce an almost *nonspinning* planet, because spins imparted by impacts will be largely self-canceling."

"The growth of a large, gaseous planet (such as Jupiter, Saturn, Uranus, or Neptune) far from the central star is especially difficult for evolutionist astronomers to explain...

..."Based on demonstrable science, gaseous planets and the rest of the solar system did not evolve."[694]

"Thou burning sun with golden beam, Thou silver moon with softer gleam, Oh praise Him! Oh praise Him!" (Francis of Assisi, circa 1225)

And then there's Saturn...

"Planetary rings... have nothing to do with a planet's origin. ...Supposedly after planets formed from a swirling dust cloud, rings remained... [But that is just not so.] Rings form when material is expelled from a moon by a volcano, a geyser, or the impact of a comet or meteorite. Debris that escapes a moon because of its weak gravity and a giant planets gigantic gravity then orbits that planet as a ring. If these rings were not periodically replenished, they would be disbursed in less than 10,000 years. Because a planet's gravity pulls escaped particles away from its moons, particles orbiting the planet could never form moons--as evolutionists assert."[695]

And the score at halftime is: Planetary rings, 10,000. Big bang, nothing. Let's see what strategy big bang has up his sleeve for the second half. Ah! He and coach Hawking have gone to the chalkboard, and are drawing feverishly...

"Evolutionary theories for the origin of the Moon are highly speculative and completely inadequate. The Moon could not have spun off from Earth, because its orbital plane is too highly inclined. Nor could it have formed from the same material as Earth, because the relative abundances of its elements are too dissimilar from those of Earth. The Moon's nearly circular orbit is also strong evidence that it was never torn from nor captured by Earth. ...[Nor was the Moon torn off by] a Mars-size impactor.

..."These explanations have many other problems. Understanding them caused one expert to joke, "The best explanation is observational error--the Moon does not exist." Similar difficulties exist for evolutionary explanations of the other (almost 200) known moons in the solar system.

"But the Moon does exist. If it was not pulled or splashed from Earth, was not built up from smaller particles near its present orbit, and was not captured from outside its present orbit, only one hypothesis remains: the Moon was created in its present orbit."[696]

"He. Could. Go. All. The. Way." Yes! It's another touchdown. In the "Origin of the Moon" game, Creation has scored dozens of touchdowns, extra points and field goals, and evolution has yet to get a first down. But that's OK! Do not let your hearts be troubled! They will still lie to our school children and tell them evolution has won. Ooh, I cain't wait 'till the next time I can go out and vote Democrat! How about you? (The theory of evolution is *the* foundation of the godless, baby-mutilating Democrat-Left, as we will see in a later section of this chapter.)

694 (Pgs. 27-28, *In the Beginning*, by Walt Brown, Ph.D.)
695 (Pg. 27, *In the Beginning: Compelling Evidence for Creation and the Flood*, by Walt Brown, Ph.D.)
696 (Pgs. 27-28, *In the Beginning*, by Walt Brown, Ph.D.)

"The Big Bang theory [is] now known to be seriously flawed...

"The redshift of starlight is usually interpreted as a Doppler effect; that is, stars and galaxies are moving away from Earth, stretching out (or reddening) the wavelengths of light they admit. Space itself supposedly expands—so the total potential energy of stars, galaxies, and other matter increases today with no corresponding loss of energy elsewhere. Thus, the big bang violates the law of conservation of energy [or The *First* Law of Thermodynamics], probably the most important of all physical laws.

…"Many objects with high redshifts seem connected, or associated, with objects having low redshifts."[697]

This, and a number of other observational evidences, is **in**consistent with the Doppler effect and with the big bang theory in general. Evolutionists also teach that the *cosmic microwave background* (CMB) *radiation* is evidence of the big bang theory. That is false…

"Because the CMB is so uniform, many thought it came from evenly spread matter soon after a big bang. But such uniformly distributed matter would hardly gravitate in any direction; even after tens of billions years, galaxies and much larger structures would not evolve. In other words, the big bang did not produce the CMB.

"Contrary to what is commonly thought, the big bang theory does not explain the amount of helium in the universe; the theory was adjusted to fit the amount of helium. Ironically, the lack of helium in certain types of stars (B type stars) and the presence of beryllium and boron in "older" stars contradicts the big bang theory.

"A big bang would produce only hydrogen, helium, and lithium, so the first generation of stars to somehow form after a big bang should consist only of those elements. Some of these stars should still exist, but despite extensive searches, none has been found.

…"If the big bang occurred, we should not see massive galaxies at such great distances, but such galaxies are seen. …A big bang should not produce highly concentrated or rotating bodies. Galaxies are examples of both. Nor should a big bang produce tightly clustered galaxies. Also, a large volume of the universe should not be—but evidently is—moving sideways, almost perpendicular to the direction of apparent expansion."[698]

The evidence just keeps pouring in, but as fast as it does, big bang cosmologists just keep ignoring it, or dismissing it out of hand. (How they get away with doing that to the observable evidence, while still calling themselves *scientists*, we've yet to figure out.) (Let's ask Jon Stewart. He's so smart. Maybe he knows.) "Hear no evil. See no evil. Hear no evidence. See no facts. Monkey see, monkey do. Evolution is a fact. Evolution is a fact. Evolution is a fact…" Remember, people believe what they are taught and, *if you shout a lie loud enough and long enough* and academically blackball anyone who dissents,[699] your lie will stick.

"Evolutionists claim that stars form from swirling clouds of dust and gas. For this to happen, vast amounts of energy, angular momentum, and residual magnetism must be

---

697   (Pg. 30, *In the Beginning: Compelling Evidence for Creation and the Flood*, by Walt Brown, Ph.D.)

698   (Pg. 30, *In the Beginning: Compelling Evidence for Creation and the Flood*, by Walt Brown, Ph.D.)

699   (See *Expelled: **No Intelligence Allowed***, a documentary narrated by Ben Stein that exposes the fascism of today's evolutionists who control academia, and who not only don't tolerate dissent, but are openly hostile to it and punish anyone foolish enough to cross their doctrinal beliefs. Of course, the fact that doing so is the antithesis of scientific inquiry, doesn't appear to phase them in the least.)

removed from each cloud. This is not observed today, and astronomers and physicists have been unable to explain, in an experientially verifiable way, how it all could happen.

..."If stars evolve, star births should about equal star deaths. ...We have seen hundreds of stars die, but we have never seen a star born."

Also, stars could not have "evolved in globular clusters, where up to a million stars occupy a relatively small volume of space. ...Wind and radiation pressure from the first star in the cluster to evolve would have blown away most of the gas needed to form the other stars in the cluster. In other words, if stars evolved, we should not see globular clusters, yet our galaxy has about 200 globular clusters. To pack so many stars that tightly together requires that they all came into existence at about the same time.

"Poor logic is involved in arguing for stellar evolution, which is assumed in estimating the ages of stars. These ages are then used to establish a framework for stellar evolution. This is circular reasoning.

"In summary, there is no evidence that stars evolve, there is much evidence that stars did not evolve, and there are no experimentally verifiable explanations for how they could evolve and seemingly defy the laws of physics."[700]

So the big bang theory is not only absurd according to common sense and our everyday experience, it is scientifically impossible according to the very laws of science itself—the first and second laws of thermodynamics and the mathematical laws of probability that tell us that explosions can't create anything of order. So an explosion could not and did not create the universe. But the big bang theory is also false according to all of our *observations* of the universe itself. And the efforts of evolutionary cosmologists to ignore this evidence, and their feeble attempts to try to *fix* the many problems these observations have created for them (only a few of which we have listed here) have been tedious and at times absurd. It is also professionally incriminating as to their close-mindedness at even entertaining the thought that there might exist another theory of the origin of the universe that is in total agreement with both the laws of science and with all of the observational data: Creation. But entertaining anything other than their accepted dogma is beyond the scope of religious zealots (in this case, devotees to the religion of *Naturalism*)[701] so they nevertheless doggedly persist in their vain and useless attempts at supporting the insupportable. The big bang theory is a big bust. Why then is it still taught as *fact* to our unsuspecting and gullible school children in our colleges, universities, elementary and high schools?

"The hard work of many scientists across many scientific disciplines in the past century unexpectedly demonstrated that both the universe at large and the earth in particular were designed for life. The heavens and earth—and life itself—alike are fine-tuned."[702]

"And to make all men see what the fellowship of the mystery is, which from the beginning of the world has been hid in God, who created all things through Jesus Christ." (Ephesians 3:9)

- A Bounded Universe Another somewhat complicated but nevertheless elucidating fact about this subject of cosmology that has been completely ignored

---

700   (Pg. 32, *In the Beginning*, by Walt Brown, Ph.D.)
701   (We mentioned this religion earlier in the beginning of chapter 1 "Truth," and will discuss it further at the end of this chapter)
702   (Pg. 218-219, *The Edge of Evolution: The Search for the Limits of Darwinism*, by Dr., Michael J. Behe)

by our disseminators of information is that a fundamental mathematical error entered into the equations of evolutionary cosmologists from the outset and thus falsified their conclusions. Conclusions that led them to the idea of that big, *primordial explosion*. This error stems from their erroneous assumption that we live in what has been called an *unbounded* universe.

"Most non-experts (in fact, most scientists not trained in cosmology) are unaware that the universe assumed by the Big Bang theorist has no boundaries, no edge and no center.

…"Why do Big-Bang cosmologist use as their starting point the assumption (which seems quite contrary to common sense) that the universe has no boundary? Is there some good scientific reason, or is it perhaps demanded or even suggested by well-established, experimentally-backed theory, like general relativity?

The answer is no. It is an *arbitrary assumption*, called the "cosmological principle," or more recently, the "Copernican principle." This assumes that (whether the universe is finite… or infinite) there is no edge and no center. On a large enough scale matter is evenly distributed around us."[703]

And the reason they have assumed that is because what we *do* observe from our vantage point here on earth is that all of the matter in the universe (stars and galaxies) does appear to be evenly distributed all around us. Therefore, rather than entertaining the thought that the even distribution of matter around us belies the fact that our earth is in a special and unique position at the center of God's universe (which of course it is), big bang cosmologists instead jumped to the more secularly acceptable assumption (God being such a hard pill for them to swallow) that there is no edge and no center. Because if there *were* an edge and a center then…

…"why don't we see more matter and more galaxies on one side of us than the other? This would be easy to explain if we were in a special place close to the center. … [However] Such a "special arrangement" is exceedingly improbable on a chance basis. It therefore strongly smacks of purpose, and [here's the kicker] is **thus unpalatable to most theorists today, who prefer to believe in a universe ruled by randomness**. [Emphasis mine] [Which is far better than a universe ruled by God because, you see, you don't have to answer to randomness.] …Of course, this idea of randomness is the essence of Darwinism… So it is simply assumed that there *is* no center, and no boundary. In this assumption, every part of the universe will appear to have matter evenly distributed around it as well.

…"It may not be unfair to suggest another possible reason for the near-universal acceptance of this assumption. To allow the possibility of anything "outside" the universe (perhaps God?) makes it harder to hold the position that the universe is "all there is" (the popular position of philosophical materialism)."[704] [And a popular position that was destroyed earlier in this book. See Chapter 3 "The Spiritual Universe."]

So it is easy to see what has happened here. Because our cosmological scientists are in love with Mr. Random Chance, and worship him as the creator of the universe, and of course of life here on earth, they refuse to acknowledge that this universe was created with a purpose and that the creation of life, and especially man, is at the forefront of this purpose. And that, therefore, of course this extraordinarily unique water planet is at the center of this universe! And therefore of course we would see the same amount of stars and galaxies, and matter, all around us.

---

703   (Pgs. 14, 18-19, *Starlight and Time: Solving the Puzzle of Distant Starlight in a Young Universe*, by D. Russell Humphreys, Ph.D.)

704   (Pgs. 19, 87, *Starlight and Time*, by D. Russell Humphreys, Ph.D.)

Conversely, if one starts out with these correct assumptions then, mathematically, we get an entirely different cosmology. And one that also neatly coincides, rather than blatantly conflicts, with all of the observed data:

"What if we begin our calculations with the opposite assumption, equally scientifically valid, namely that matter in the universe *has* a center and an edge (is bounded)? This makes more common sense and is also Scripturally far more appropriate. When we feed in this, plus the same observations, into general relativity, quite a different cosmology falls out."[705]

And that cosmology is one where God "stretched out the heavens like a curtain." (Isaiah 40:22) It never *exploded*. He created matter and then formed the stars and the galaxies and stretched them out to their present positions. This is one of the conclusions that are reached when the assumption of a *bounded* universe with earth at its approximate *center* is plugged into the general theory of relativity. It is also a historical fact of creation according to God's Word which we have already determined to be scientifically, and historically, accurate. (See chapter 6, "The Word of God") Besides the verse in Isaiah quoted above, there are numerous others that reveal this. Here are a few:

"Who alone spreads out the heavens, and treads upon the waves of the sea." (Job 9:8)

"Who covers yourself with light as with a garment, Who stretches out the heavens like a curtain." (Psalms 104:2)

"He has made the earth by his power, he has established the world by his wisdom, and has stretched out the heavens at his discretion." (Jeremiah 10:12)

"Thus says the LORD, who stretches out the heavens, and lays the foundation of the earth, and forms the spirit of man within him." (Zechariah 12:1)

"Indeed My hand has laid the foundation of the earth, and My right hand has stretched out the heavens." (Isaiah 48:13)

"I have made the earth, and created man upon it: I, even my hands, have stretched out the heavens, and all their host have I commanded." (Isaiah 45:12)

"Thus says God the LORD, he who created the heavens and stretched them out, he who spread forth the earth and that which comes from it, he who gives breath to the people upon it, and spirit to those who walk therein." (Isaiah 42:5)

This "stretching out" or expansion also caused the redshifts of distant stars and galaxies.

"A... misconception is that the red shifts of the galaxies are Doppler shifts, i.e. caused by the velocity of the galaxies away from us at the time the light starts its journey toward us. But [even] one undergraduate textbook... and many graduate textbooks... make it clear that the red shifts are an *expansion* effect. As space is stretched out, the lengths of all electromagnetic waves passing through the space are similarly stretched out. Consequently, the *speed* of recession doesn't matter, only the *amount* of expansion that takes place as the light travels to us, whether the expansion is fast or slow."[706]

It is also the reason, along with gravitational time dilation, why we can see the light from those distant galaxies (some as far as 12 billion light years away) even though our earth and our universe are very young, less than 10,000 years old. (See upcoming sections, "The Young Earth" and "Starlight and Time.")

---

705 (Pg. 21, *Starlight and Time: Solving the Puzzle of Distant Starlight in a Young Universe*, by D. Russell Humphreys, Ph.D.)

706 (Pg. 98, *Starlight and Time: Solving the Puzzle of Distant Starlight in a Young Universe*, by D. Russell Humphreys, Ph.D.)

So, if you feed the <u>ideological assumption </u>of an **un**bounded universe into Einstein's theory of general relativity then mathematically you will come out with the absurd theory of the big bang—that an explosion made the universe—which as we have seen is also scientifically impossible and completely unsatisfactory as a creator of all that we see now in this highly ordered and complex universe, solar system, earth-moon system, and living planet.

Conversely, the mathematical result of feeding into the general theory of relativity the accurate assumption that the universe is bounded, has a center and an edge, and that the earth is approximately at this center, is that "the cosmos has been expanded" (and *not* exploded) and that the universe is very young.[707] And, coincidentally, that is exactly what the Bible says.

Keep in mind also that the theory of the big bang was not hypothesized in a vacuum, or in the pure unbiased search for scientific truth. It was formulated under the weight of the extreme bias of 1) assuming evolution was true, and 2) assuming that the universe had to have formed itself as well and therefore *had to be* billions of years old in order to have had the time to do that. We know now that the first assumption is absolutely false, and the second assumption has already been dealt a fatal blow and will soon expire as we continue through the remainder of this chapter.

The Young Earth

"He stretches out the north over the empty place, and hangs the earth upon nothing."(Job 26:7) (...And He did it about 6 thousand years ago, and not hundreds of millions or billions.)

We have already introduced a number of facts and evidences throughout this chapter that contradict the established conclusion of the main-stream of the scientific community (at least that part of the scientific community that maintains near-totalitarian and fascist control of our media, our entertainment industry and our educational system) that the earth and this universe are "billions of years old." But in this section we will introduce enough further evidence that will establish that the age of the earth, the solar system and the universe is less than ten thousand years old. Again, the evil one's origins account poisons our educational system, and those who pass through it have become contaminated with the Darwinist drivel they were told. Indeed, the unscientific but evolutionary compatible lie that the universe is billions of years old has been so universally shoveled into the minds of America and from every conceivable source in order to firmly establish it as another so-called scientific fact, that it is of great importance to provide as much information here as possible so we can shovel it all back out. So here we go...

> "Before the nineteenth century, the vast majority of scientists interpreted earth history in terms of Biblical creationism and catastrophism (Genesis Flood), and consequently, believed in a relatively short time scale. However, the more recent acceptance of a principle known as *uniformitarianism* has successfully promoted the idea of an ancient earth. Uniformitarianism is the belief that the origin and development of all things can be explained exclusively in terms of the same natural laws and processes operating today. According to this dogma, nature can be satisfactorily explained according to natural causes and therefore, "the present is the key to the past." This concept was introduced by James

---

707   (Pgs. 79-80, 99-100, *Starlight and Time: Solving the Puzzle of Distant Starlight in a Young Universe*, by D. Russell Humphreys, Ph D.)

Hutton, popularized by Sir Charles Lyell, and greatly influenced the thoughts and works of Charles Darwin. Uniformitarianism has been the backbone of modern historical geology and is responsible for the current widespread assumption that the earth is billions of years old. Consequently, the earth has *aged* from just a few thousand years to about 5 billion years in a little more than a century!"[708]

"Most Scientific Dating Techniques Indicate That the Earth, Solar System, and Universe Are Young. For the last 150 years, the age of the Earth, as assumed by evolutionists, has been doubling at roughly a rate of once every 15 years. In fact, since 1900 this age has multiplied by a factor of 100!"[709]

That is because their theory collapses without hundreds of millions and billions of years for evolution to work its *miracle*. (But even that is moot because we now know that evolution collapses no matter how many eons you give it.)

"Our media and textbooks have implied for over a century that these almost unimaginable ages are correct. Rarely do people examine the shaky assumptions and growing body of contrary evidence. Therefore, most people today almost instinctively believe that the Earth and universe are billions of years old. Sometimes, these people are disturbed, at least initially, when they see the evidence."[710]

"And I will send them strong delusion, that they should believe a lie." The Creator is basically saying, *you want lies, you got lies. You love them more than me, then help yourself. Go back for seconds. You want to ignore the evidence and believe a fairy tale, then go ahead. I won't stand in your way. I gave you free will, and you are free to choose the bald-faced lies of Satan over my scientific truth.* (Where does scientific truth come from anyway? Scientists? Guess again.) You get what you ask for. Which is why it is so important at some time in your life, no matter what you were taught, to ask for the truth. So you can receive.

Let's look at some facts:

"Humanlike footprints, supposedly 150--600 million years old, have been found in rock formations in Utah, Kentucky, Missouri, and ...Pennsylvania."[711]

Of course these supposed ages for the rock predate man by hundreds of millions of years, according to evolutionary time lines. Obviously something is very wrong there.

"The strength of the earth's magnetic field has been [continuously weakening]. [It has been] decaying exponentially at a rate corresponding to a half-life of 1400 years. ... If we extrapolate back as far as 10,000 years, we find that the earth would have had a magnetic field as strong as that of a magnetic star! This is, of course, highly improbable, if not impossible. Thus, based on the present decay rate of the earth's magnetic field, 10,000 years appears to be an upper limit for the age of the earth. ...**any objections to this conclusion must be based on rejection of the same uniformitarian assumption which evolutionists utilize to derive a great age for the earth.**"[712]

---

708   (Pg. 7, *The Collapse of Evolution*, by Scott M. Huse)

709   (Pg. 37, *In the Beginning: Compelling Evidence for Creation and the Flood*, by Walt Brown, Ph.D.)

710   (Pg. 37, *In the Beginning*, by Walt Brown, Ph.D.)

711   (Pg. 35, *In the Beginning*, by Walt Brown, Ph.D.)

712   (Pg. 21, *The Collapse of Evolution*, by Scott M. Huse)

"Scientists have known for some time now that cosmic dust particles enter the earth's atmosphere from space at an essentially constant rate. Eventually these dust particles settle down to the earth's surface. Hans Peterson has made accurate measurements of this influx, and has determined that the earth receives about 14 million tons per year. Now, if it is true that the earth is around 5 billion years old, as evolutionists insist, there should be a layer of meteoric dust that is about 182 feet thick all over the world! No such dust layer exists anywhere, of course. Even on the moon where it should be at least as thick, little sign of it was found by the astronauts (about an eighth of an inch). The fear that the astronauts would sink into such a layer of meteoric dust when they landed on the moon proved to be completely unwarranted."[713]

Astronauts, 182. Billions of years old, nothing.

"The Mississippi River delta offers additional evidence to support the concept of a relatively young earth. Approximately 300 million cubic yards of sediment are deposited into the Gulf of Mexico by the Mississippi River each year. By carefully studying the volume and rate of accumulation of the Mississippi River delta and then dividing the weight of the sediments deposited annually into the total weight of the delta, it can be determined that the age of the delta is about 4,000 years old."[714]

Indeed, if the earth was just <u>one</u> million years old (much less the *hundreds* of millions and *billions* claimed) then <u>there would be no Gulf of Mexico</u>! It would be a sand lot, and not a sea. The same can be said of all the other river deltas across this world. In addition, the known rate per year that river sedimentation enters the oceans and the total amount of accumulated ocean sediment also vastly contradicts these ridiculously old evolutionary ages for our planet.[715]

Are the oceans themselves hundreds of millions and even billions of years old? Not a chance...

"Rivers carry dissolved elements such as copper, gold, lead, mercury, nickel, silicon, sodium, tin, and uranium into the oceans at very rapid rates when compared with the small quantities of these elements already in the oceans. ...far fewer than a million years' worth of metals are dissolved in the oceans. There is no known means by which large amounts of these elements can come out a solution. Therefore, the oceans must be much younger than a million years"[716]

And then there are all of those natural gas and oil deposits across the globe which we know had to be deposited rapidly by the cataclysmic burial of a planets worth of vegetation, animal and marine life. (See "A World Wide Catastrophic Flood" coming up shortly) But evolutionists have insurmountable problems explaining *their* age for these deposits:

"Petroleum and natural gas are contained at high pressures in underground reservoirs by relatively impermeable cap rock. In many cases, the pressures are extremely high. Calculations based upon the measured permeability of the cap rock reveal that the oil and gas pressures could not be maintained for much longer than 10,000 years in many instances. Thus, the generalization that such fossil-fuel deposits have been confined for

---

713   (Pgs. 22-23, *The Collapse of Evolution*, by Scott M. Huse)
714   (Pgs. 23-24, *The Collapse of Evolution*, by Scott M. Huse)
715   (Pg. 37, *In the Beginning*, by Walt Brown, Ph.D.)
716   (Pg. 38, *In the Beginning: Compelling Evidence for Creation and the Flood*, by Walt Brown, Ph.D.)

millions of years, having not leaked out through their cap rock, becomes preposterous.

"Furthermore, recent experiments have demonstrated conclusively that the conversion of marine and vegetable matter into oil and gas can be achieved in a surprisingly short time. For example, plant-derived material has been converted into a good grade of petroleum in as little as 20 minutes under the proper temperature and pressure conditions. Wood and other cellulosic material have also been converted into coal or coal-like substances in just a few hours. These experiments prove that the formation of coal, oil, and gas did not necessarily require millions of years to form as uniformitarian geologists have assumed and taught.

"Creationists believe that the great coal deposits of the world are the transported and metamorphosed remains of the extensive vegetation of the antediluvian world. This catastrophic interpretation is further supported by the presence of polystrate fossils in coal beds which indicate rapid formation. Also, the type of plants involved and the texture of these deposits testify of turbulent waters, not a stagnant swamp.

"Evolutionists propose that coal was formed millions of years before man evolved. However, human skeletons and artifacts, such as intricately structured gold chains, have been found in coal deposits. [Oops, we have another problem, Houston.] In Genesis 4 we learn that metal working was already highly developed; Tubalcain was an instructor of every artificer in brass and iron. In Genesis 7 and 8, the Deluge buried the antediluvian civilizations in the sedimentary layers of the earth's crust.

…"The rotation of the earth is gradually slowing due to the gravitational drag forces of the sun, moon, and other factors. If the earth is billions of years old, as uniformitarian geologists insist, and it has been slowing down uniformly, then its present rotation should be zero! Furthermore, if we extrapolate backwards for several billion years, the centrifugal force would have been so great that the continents would have been sent to the equatorial regions and the overall shape of the earth would have been a flat pancake."[717]

Today's early bird special: eggs, hash browns, and two blueberry earthcakes.

"The present rate of recession of the moon is known and clearly indicates a young age for the earth-moon system. Calculations based on the known recession speed of the moon and presumed evolutionary age of 4 to 5 billion years require that the moon should be much farther away from earth than it is. Obviously, the earth-moon system is not as old as evolutionary scientists have assumed. The vast time span essential for the presumed evolution of life forms is apparently mythological and nonexistent."[718] (Apparently.)

Also, the amount of atmospheric helium is consistent with an age of the earth of less than 10,000 years. Helium is produced by radioactive decay within rocks and if the earth were even one million years old there would be far more helium in our atmosphere.[719] The size of the world's human population is another indicator of a relatively young earth.

And then there is the conclusive testimony of zircon crystals:

"Lead diffuses (or leaks) from zircon crystals at known rates that increase with temperature. Because these crystals are found at different depths in the Earth, those at greater depths and temperatures should have less lead. If the Earth's crust is just a fraction

---

717   (Pgs. 24-25, *The Collapse of Evolution*, by Scott M. Huse)
718   (Pgs. 25-26, *The Collapse of Evolution*, by Scott M. Huse)
719   (Pg. 37, *In the Beginning*, by Walt Brown, Ph.D.)

of the age claimed by evolutionists, measurable differences in the lead content of zircons should exist in the top 4,000 meters. Instead, no measurable difference is found.

"Similar conclusions are reached based on the helium content in the same zircon crystals. Because helium escapes so rapidly and so much helium is still in zircons, they (and the Earth's crust) must be less than 10,000 years old. Furthermore, the radioactive decay that produced all that helium must have happened quite rapidly, because the helium is trapped in *young* zircons."[720]

"Based on the measured helium retention, statistical analysis gives an estimated age for the zircons of 6000 +/- 2000 years. This age agrees with literal biblical history ["Well in that case, it can't be true. Throw that science out! And didn't we tell you to burn that book?!"] and is about 250,000 times shorter than the conventional age of 1.5 billion years for the zircons. The conclusion is that helium diffusion data strongly supports the young-earth view of history."[721]

In addition there are those gosh darn fission tracks and radiohalos. (At what point do they stop ignoring all of the evidence, or stop calling themselves scientists? What is the breaking point? Is there one?)

"The abundance of fission tracks and radiohalos …in crystalline solids… provides evidence for a recent creation. This follows because the host rocks have not experienced serious heating since the track and halo formation. Just hundreds of degrees are sufficient to erase the crystal defects, yet they remain. It is difficult to imagine the rock formations remaining cool over vast ages of time with accompanying episodes of volcanic and tectonic activity. In the young-earth view, the radiohalos and tracks remain relatively recent and freshly made."[722]

"Abnormally high oil, gas, and water pressures exist within relatively permeable rock. If these fluids had been trapped more than 10,000 to 100,000 years ago, leakage would have dropped these pressures far below what they are today. This oil, gas, and water must have been trapped suddenly and recently."[723]

When you take the rate that volcanoes eject material into the atmosphere each year, and the fact that only "25% of Earth's sediments are of volcanic origin" then the age of the Earth becomes only a tiny fraction of the 4.5 billion that evolutionists insist.[724]

"The continents are eroding at a rate that would level them in much less than 25 million years. However, evolutionists believe that fossils of animals and plants at high elevations have somehow avoided this erosion for more than 300 million years. Something is wrong."[725]

---

720 (Pg. 37, *In the Beginning: Compelling Evidence for Creation and the Flood*, by Walt Brown, Ph.D.)

721 (Pg. 76, *Thousands… Not Billions: Challenging an Icon of Evolution, Questioning the Age of the Earth*, by Dr. Don DeYoung)

722 (Pg. 106, *Thousands… Not Billions: Challenging an Icon of Evolution, Questioning the Age of the Earth*, by Dr. Don DeYoung)

723 (Pg. 37, *In the Beginning*, by Walt Brown, Ph.D.)

724 (Pg. 37, *In the Beginning*, by Walt Brown, Ph.D.)

725 (Pg. 37, *In the Beginning*, by Walt Brown, Ph.D.)

Something is terribly wrong. ("Quick, we need a new story that explains away those high altitude fossils "resistance to erosion." Think fast.") ("See no truth. Hear no evidence.")

In sedimentary layers there are *conformities* and *unconformities*. An unconformity between layers is where you have signs of water and wind erosion, animal burrowing, and tree rooting's that show a passage of some period of time. Layers that are parallel, with a perfectly straight laser-line separating them and with no sign of unconformities or the passage of time, are called *conformities*. They "imply continuous, relatively rapid deposition."

> "Frequently, two adjacent and parallel sedimentary layers contain such different index fossils that evolutionists conclude they were deposited hundreds of millions of years apart. However, because the adjacent layers are conformable, they must have been deposited without interruption or erosion. ... Often, in sequences showing no sign of disturbance, the layer considered older by evolutionists is on top! ... Evolutionary dating rules are self-contradictory."[726]

"OK! Who's the wise guy? Who put the older sedimentary layer on top of the younger one? You've had your laugh, now put them back where they're supposed to be, and then report to the principles office."

The sodium levels in the oceans, the erosion rates of the continents and "the fact that convincing soil layers, or even soil materials, are seldom found in the geologic record" also demand a much younger earth than what Darwinists claim.[727] A massive buildup of soil layers and soil materials due to so many endless years of plant and tree activity should have occurred in between each successive sedimentary layer if indeed they are hundreds of millions of years old with millions of years passing between each successive layer. But there is none. No soil layers. No soil materials. No sign of wind or water erosion. Because absolutely no time passed between the deposition of each layer. They were all put down, one on top of the other, during the great global flood that occurred 4,500 years ago. (Again, see "A World Wide Catastrophic Flood.")

> "Many people have the mistaken impression that geology has proved that the earth is billions of years old. As we have seen, nothing could be further from the truth! ...If the Bible is really true, then the geologic evidence must support it—and indeed it does! The evidence not only supports the Bible, but a great deal of geologic evidence is quite incompatible with an old-earth scenario."[728]

> "Estimated old ages for the Earth are frequently based on "clocks" that today are ticking at extremely slow rates. For example, coral growth rates were thought to have *always* been very slow, implying that some coral reefs must be hundreds of thousands of years old. More accurate measurements of these rates under favorable growth conditions now show that no known coral formation need be older than 3,400 years. A similar comment can be made for growth rates of stalactites and stalagmites in caves. ... [In one

---

726   (Pgs. 36-37, *In the Beginning*, by Walt Brown, Ph.D.)
727   (See pgs. 89-91, 92-93, 99-100, *The Young Earth: The Real History of the Earth – Past, Present, and Future*, by John Morris, Ph.D.)
728   (Pgs. 115-118, *The Young Earth: The Real History of the Earth – Past, Present, and Future*, by John Morris, Ph.D.)

recent instance] water from an underground spring was channeled to [a] spot on a river bank for only one year. In that time, limestone built up around sticks lying on the bank."[729]

...Accumulating "millions of years" worth of deposits in just one year. Oops. Better take another look at the ages of all those caves again...

In addition to a young earth, the evidence for a relatively young universe and solar system is overwhelming and incontrovertible. The very presence of comets, which should have had their masses "boiled off" to nothing by the sun in less than 10,000 years cry out for a very young solar system. And star clusters which contain...

...."hundreds or thousands of stars moving, as one author has put it, "like a swarm of bees." ...in some star clusters, the stars are moving so fast that they could not have held together for millions or billions of years. Thus, the presence of star clusters ...indicates that the age of the universe is numbered in the thousands of years."[730]

And you're wondering why no one in the media has ever had the sense to report any of this. So am I. (Maybe we should ask Bill O'Reilly.)

"Some stars are so large and bright that they radiate energy anywhere from 100,000 to 1 million times as fast as our own sun! These stars could not have contained enough hydrogen to run the atomic fusion energy production process at such rates for millions or billions of years because their initial mass would have been absolutely implausible. Therefore, these stars must be ...only thousands of years old."

...."It is important to remember that just because most people believe in something does not necessarily make it true. The argument from a majority opinion is not an impressive one. In fact, it is completely irrelevant since scientific truth is never determined by taking a vote. Majorities can be, and often have been, completely wrong."[731]

(Just look at the 2006 and 2008 elections...)

The evidence from meteorites is also on the side of a relatively young earth and solar system...

"Experts have expressed surprise that meteorites are almost always found in young sediments, very near Earth's surface. Even meteoric particles in ocean sediments are concentrated in the topmost layers. If Earth's sediments, which average about a mile in thickness on the continents, were deposited over hundreds of millions of years, as evolutionists believe, we would expect to find many deeply buried iron meteorites. Because this is not the case, the sediments were probably deposited rapidly, followed by "geologically recent" meteorite impacts.

...."Similar observations can be made concerning ancient rock slides. Rock slides are frequently found on Earth's surface, but are generally absent from supposedly old rock."[732]

Also, the present temperature of the earth, the recession rate of the moon, the "amount of heat ...flowing out of the Moon from just below its surface, [although] the Moon's interior is relatively cold," and the lack of hundreds of millions or billions of years' worth of meteoric material on the moon all point to a very young earth-moon system. There is also the absence of any signs of "flattening out"

---

729   (Pg. 34, *In the Beginning: Compelling Evidence for Creation and the Flood*, by Walt Brown, Ph.D.)
730   (Pgs. 29-30, *The Collapse of Evolution*, by Scott M. Huse)
731   (Pgs. 30-31, *The Collapse of Evolution*, by Scott M. Huse)
732   (Pg. 38, *In the Beginning*, by Walt Brown, Ph.D.)

(called "creep") of the large, steep-walled craters on either the Moon, Venus or Mercury (which should have happened if they were formed over 4 billion years ago as evolutionists say). All of these are further strong scientific evidence of youth and not great age.[733]

"As comets pass near the Sun, some of their mass vaporizes, producing a long tail and other debris. Comets also fragment frequently or crash into the Sun or planets. Typical comets should disintegrate after several hundred orbits. For many comets this is less than 10,000 years. There is no evidence for a distant shell of cometary material surrounding the solar system, and there is no known way to add comets to the solar system at rates that even remotely balance their destruction. Actually, the gravity of planets tends to expel comets from the solar system rather than capture them. So, comets and the solar system appear to be less than 10,000 years old."[734]

"Jupiter, Saturn, and Neptune each radiate away more than twice the heat energy they receive from the Sun. Uranus and Venus also radiate too much heat. Calculations show that it is very unlikely that this energy comes from nuclear fusion, radioactive decay, gravitational contraction, or phase changes within those planets. This suggests that these planets have not existed long enough to cool off."

"The Sun's radiation applies an outward force on particles orbiting the Sun. Particles less than about one 100,000th of a centimeter in diameter should have been "blown out" of the solar system if it were billions of years old. Yet these particles are still orbiting the Sun."[735]

And concerning those particles which are larger than one 100,000th of a centimeter, there is "a large disk shaped cloud that orbits the Sun between... Venus and the asteroid belt. [It consists] of dust particles larger than about one 100,000th of a centimeter in diameter." However, the well-established *Poynting-Robertson effect* states that "forces acting on these particles should spiral most of them into the sun in less than 10,000 years. ...Known forces and sources of replenishment cannot maintain this cloud, so the solar system is probably less than 10,000 years old."[736] Probably.

"Huge quantities of microscopic dust particles also have been discovered around some stars. Yet, according to the theory of stellar evolution, those stars are many millions of years old, so that dust should have been removed by stellar wind and the Poynting-Robertson effect. Until some process is discovered that continually resupplies vast amounts of dust, one should consider whether the "millions of years" are imaginary."[737]

Uh, we've already considered it. Trust me, they're imaginary.

"In galaxies similar to our Milky Way Galaxy, a star will explode violently every 26 years or so. These explosions, called *supernovas*, produce gas and dust that expand outward thousands of miles per second. With radio telescopes, these remnants in our galaxy should be visible for a million years. However, only about 7,000 years' worth of supernova debris are seen. So, the Milky Way looks young."[738]

733   (Pg. 39, *In the Beginning*, by Walt Brown, Ph.D.)

734   (Pgs. 39-40, *In the Beginning*, by Walt Brown, Ph.D.)

735   (Pg. 40, *In the Beginning*, by Walt Brown, Ph.D.)

736   (Pg. 40, *In the Beginning: Compelling Evidence for Creation and the Flood*, by Walt Brown, Ph.D.)

737   (Pg. 40, *In the Beginning*, by Walt Brown, Ph.D.)

738   (Pg. 40, *In the Beginning*, by Walt Brown, Ph.D.)

The Milky Way looks <u>real</u> young. And the lies they have concocted to support the absurdity that is Satan's genesis, that they are indoctrinating our children with, are looking real disgusting.

In addition there are: galaxies that are connected yet "have vastly different redshifts;" the existence of so many spiral galaxies whose internal dynamics cannot allow them to maintain their highly esthetic shape for even "a small fraction of the universe's assumed evolutionary age;" and the existence of "many stars in spiral galaxies and gas clouds that surrounds some galaxies [that] have such high relative velocities …they should have broken their "gravitational bonds" long ago if they were billions of years old." These are also strong astronomical evidence for a very young universe.[739]

"These observations have led some to conclude, not that the universe is young [shocker!], but that unseen, undetected mass is holding these stars and galaxies together. For this to work, the hidden mass, sometimes called *dark matter*, must be 10 - 100 times greater than all visible mass, and the hidden mass must be in the right places. However, many experiments have shown that the needed "missing mass" does not exist. Some researchers are still searching, because the alternative is a young universe."[740] (O. J. is still searching too. He couldn't find his wife's killer out on the golf course, so now he's looking in prison.) (Are there mirrors in prison?)

If these folks were actual scientists, they might come up with a different theory to explain these observations, one that doesn't include a tooth fairy called "missing mass." But as high priests of a false religion, they want to protect their sacred doctrine at all costs, even at the expense at their own scientific credibility. In every other scientific discipline, all of those which have no relevance to evolution or the Bible, when observations can't be fit into existing theories it is the theory that goes. Instead evolutions high priests will conjure up any explanation no matter how laughable ("Here, missing mass. Here, boy. Tweet, tweet, tweet. Come out, come out, wherever you are.") rather than even entertaining the idea that their astronomically inflated timeframes don't exist.

- Mitochondrial Eve In 1987 three scientists did a study of mutations within the DNA of the mitochondria, which are energy-producing units in the cell and that come only from the mothers DNA. They studied 147 people from all over the world and determined that all of them had the same ancestral mother. There mistake however was that they used the *assumption of evolution* in determining the mutation rate of the mtDNA (mitochondrial DNA) and thus how long ago the first woman lived…

"**When** did mitochondrial Eve live? To answer this, one must know how frequently mutations occur in mtDNA. Initial estimates were based on the following *faulty* reasoning: "Humans and chimpanzees had a common ancestor about 5 million years ago. Because the mtDNA in humans and chimpanzees differ in 1000 places, one mutation occurs every 10,000 years." …These estimated rates, based on evolution, led to the mistaken belief that mitochondrial Eve lived 100,000-200,000 years ago. [But even] this surprised evolutionists who believe that our common ancestor was an apelike creature that lived 3 1/2 million years ago."

---

739   (Pg. 41, *In the Beginning*, by Walt Brown, Ph.D.)
740   (Pg. 41, *In the Beginning*, by Walt Brown, Ph.D.)

[So even when using their faulty circular reasoning to determine the mutation rate of mtDNA, they still came up with an age for the first woman that was 30 times less than what they needed for their theory. But it gets even worse...]

"A greater surprise, even disbelief, occurred in 1997, when it was announced that mutations in mtDNA occur 20 times faster than had been estimated. Without assuming that humans and chimpanzees had a common ancestor 5 million years ago... mutation rates can now be determined directly by comparing the mtDNA of many mother-child pairs. Using the new, more accurate rate, *mitochondrial Eve lived only about 6,500 years ago.*"[741]

Bingo! So science confirms the time line of roughly 6,000 years ago that the Word of God gives through its genealogical record for the first woman, Eve, and one would assume therefore for the first man, Adam. Isn't science a wonderful thing?

- Radioactive dating techniques But what about all of those scientific dating methods like carbon-14 and the radioisotopes of rocks; aren't they *conclusive evidence* for the earth being hundreds of millions and billions of years old? No, not hardly.

First, let's look at carbon-14:

"*For some years there has been a growing realization that carbon-14 atoms are found where they are not expected.* With a half-life of 5730 years, C-14 should no longer exist within "ancient" fossils, carbonate rocks, or coal. Yet small quantities of C-14 are indeed found in such samples on a worldwide scale. ...Measurable levels of C-14 are found in every case for both coal and diamond samples. This evidence supports a limited age for the earth. There is a widely held misconception that carbon-14 dating is in direct conflict with creation and a young-earth view. Instead, however the carbon-14 findings strongly support a recent supernatural creation."[742]

Whoops! There should be no carbon-14 whatsoever in all of the rocks and sediments and deposits that are supposedly more than a million years old, much less hundreds of millions of years old! But there they all are!

"To understand the significance of this carbon-14 finding, consider a comparison. Suppose an archaeologist investigates an Egyptian mummy. The outer covering is carefully removed to reveal the ancient, undisturbed interior. As the last wrapping is removed an amazing discovery is made. Inside the mummy is a wind-up clock which is still ticking! Perhaps the mummy is not as old as the archaeologist initially thought. The discovery of carbon-14 in "ancient" samples is just as startling to the conventional radio isotope dating community.

[Must have been real startling, because they've already ignored it. For health reasons of course.]

..."The presence of C-14 in "very old" fossils, rocks, coal, and diamond samples is clearly in major conflict with the long age time scale... The conclusion is that the pervasive presence of C-14 is strong evidence for a young earth."[743]

Radioisotope dating, that which has been used for decades to "prove" that the earth is billions of years old, has recently been proven to be completely unreliable

---

741 (Pgs. 319-320, *In the Beginning: Compelling Evidence for Creation and the Flood*, by Walt Brown, Ph.D.)
742 (Pg. 175-176, *Thousands... Not Billions: Challenging an Icon of Evolution, Questioning the Age of the Earth*, by Dr. Don DeYoung)
743 (Pg. 49-50, 56, 58, *Thousands... Not Billions: Challenging an Icon of Evolution, Questioning the Age of the Earth*, by Dr. Don DeYoung)

due to a number of false assumptions.[744] Here is just one glaring discrepancy, the vast difference between the *known* ages of certain rocks and their *radioisotope age*:

"We can gather samples, for example, from recent volcanic eruptions and date them. If the dating process is accurate, then the date derived should be almost equivalent to zero, or too young to be measured. In the scientific literature, research results have often been reported where rocks of known ages have been dated. In almost every case, the age of these recent lavas has come back from the lab in terms of excessively high ages, not essentially zero, as one would predict.

"Let me give you a few examples. It is known that Sunset Crater in northern Arizona was formed by a series of recent volcanic eruptions. ...Tree-ring dating accurately dates the eruption to about A.D. 1065, and it is instructive to compare this historical date with the radioisotope date obtained.

"Two of the lava flows were dated by the potassium-argon method. Much to everyone's surprise, the lava flows gave the "ages" of 210,000 and 230,000 years!

..."Consider another example. A coal mine in Queensland, Australia, required a vertical ventilation shaft to provide air to the miners. On the way down the drill encountered a basalt layer, and underneath the basalt they found pieces of unfossilized wood. Multiple carbon dating studies of the wood fragments yielded an "age" of 30,000 to 45,000 years, while the basalt, using the potassium-argon method, dated 39-58 million years!

"[Again, rocks were dated] from a volcano in ...New Zealand, Mt. Ngauruhoe. It has often erupted in recent decades, and it was these recently formed rocks that were gathered and dated by multiple methods. While K-Ar model ages dated from 270,000 years to 3.5 million, the Rb-Sr isochron dated over 133 million years, the Sm-Nd method nearly 200 million years and the lead-lead ratio method indicated a date of **3.9 billion years**! [Emphasis mine.] All this from rocks less than 60 years old. Do radioisotope-dating results warrant our trust?"[745] (Absolutely! If we're complete idiots.)

Wow! 3.9 billion years old! From rocks less than a hundred years old. (And they call Sarah Palin stupid!) And there are countless other examples of volcanoes that have erupted recently and yet yield astronomically inflated ages.

"Diamonds are a crystalline form of carbon, thought to have formed in earth's earliest days under extreme conditions. Because of their supposed great age, inorganic origin, and complete impermeability, no one had ever even suggested they might contain C14. Until [recently] ...diamonds from several varied sources were obtained and tested. Once again, each specimen contained measurable C-14 and dated just thousands of years old."[746]

But please don't tell anyone! We don't want to expose that little man behind the curtain. And then there's this little known fact...

"The igneous rocks on the rim of the Grand Canyon date "older" than the igneous rocks at the bottom, according to radioisotope dating."[747]

---

744   (For a complete discussion you can see Chapter 5, "Radioisotope Dating," from Dr. John Morris' book, *The Young Earth: The Real History of the Earth – Past, Present, and Future*.)

745   (Pgs. 51-52, *The Young Earth: The Real History of the Earth – Past, Present, and Future*, by John Morris, Ph.D.)

746   (Pg. 67, *The Young Earth*, by John Morris, Ph.D.)

747   (Pg. 78, *The Young Earth: The Real History of the Earth – Past, Present, and Future*, by John Morris, Ph.D.)

Oops, now how did that fly get in the ointment? (But Darwinists have an answer. They got away with convergent evolution so now, emboldened by that success, they've proposed reverse sedimentation, where the most recent sedimentary layers actually fall *below* the older ones. Shazaam!)

In addition, the very assumptions that radioisotope dating are based upon are themselves questionable…

"The public has been greatly misled concerning the consistency and trustworthiness of radiometric dating techniques (such as the potassium-argon method, the rubidium-strontium method, and the uranium-thorium-lead method)…

"A major assumption underlying all radioactive dating techniques is that decay rates, which have been essentially constant over the past 100 years, have also been constant over the past 4,600,000,000 years. This is a huge and critical assumption that few have questioned. [And since Darwinists depend on that assumption being true, why would they question it?] Several lines of evidence show that radioactive decay rates were once much faster than they are today."[748]

And then there is DNA "the immortal"!

"When an animal or plant dies, its DNA begins decomposing. Before 1990, almost no one believed that DNA could last 10,000 years. This limit was based on measuring DNA disintegration rates in *well-preserved* specimens of known age such as Egyptian mummies. DNA has now been reported in supposedly 17-million-year-old magnolia leaves and 11-to-425-million-year-old salt crystals. …DNA fragments are also said to be in alleged 80-million-year-old dinosaur bones buried in a coal bed and in the scales of a 200-million-year-old fossilized fish. DNA is frequently reported in insects and plants encased in amber, both assumed to be 25-120 million years old.

[These discoveries alone should totally discredit these asinine ages, but religious fanatics will fight against *heresy* (in this case *the truth*) even to the death! And in this case the death of the scientific method.]

"These discoveries have forced evolutionists to re-examine the 10,000 year limit [of DNA]. [Shocker! Curiously, they never stopped to re-examine their fraudulently old time scales.] They now claim that DNA can be preserved longer if conditions are dryer, colder, and freer of oxygen, bacteria, and background radiation. However, measured disintegration rates of DNA, under these more ideal conditions, do not support this claim."[749] ("Hear no truth. See no evidence.")

Live bacteria spores have been found in bees preserved in amber that is dated to 25 to 40 million years old. They have also been found in rocks dated 250 to 650 million years old. And "proteins, soft tissues, and blood compounds [have been found] preserved in dinosaur bones." None of which could last for even a tiny fraction of the supposed ages of the material in which they were found. Human artifacts such as "a thimble, an iron pot, an iron instrument, an 8-karat gold chain, three throwing spears, and a metallic vessel inlaid with silver" have been discovered in coal deposits that are, according to the "established" dates, *300 to 350 million years old!* In addition "nails, a screw, a strange coin, a tiny ceramic doll" and other artifacts were found inside of deeply buried rock, again *dated* to hundreds of millions of years old. Yet according to the evolutionists' time scales,

---

748   (Pg. 34, *In the Beginning: Compelling Evidence for Creation and the Flood*, by Walt Brown, Ph.D.)
749   (Pgs. 35-36, *In the Beginning*, by Walt Brown, Ph.D.)

man is less than <u>one</u> million years old! (And what do they say when presented with such obvious and incontrovertible contradictions? Nothing. They just ignore them.) Obviously, these dates are absurd. This amber, these rocks, those dinosaurs, that coal and this earth are all very young, at the maximum about 10,000 years old. The Bible again is found to be scientifically accurate, and the dates claimed by the uniformitarian evolutionists are found to be astronomically exaggerated to fit their preconceived and intransigent worldview.[750]

"For in six days the Lord made the heavens and the earth, the sea, and all that in them is." (Exodus 20:11)

As we can see the evidence is staggering. It is overwhelming. (And we have only presented a tiny fraction of it here.) The earth is very young. Why this isn't common knowledge is due to the fact that we live in an insane asylum, and the inmates have truly taken over. They own our schools, our entertainment industry, our late-night clowns and of course our media, and their unscientific lies cover all of it like a shroud, and people believe what they have been taught.

Again, as we stated in chapter 6, "The Word of God," the Bible says that the universe was <u>created</u> about six thousand years ago. Faux science says that the universe sprung up billions of years ago from a big explosion and then formed itself, all that we survey today, accidentally by random chance. The Biblical account is scientifically accurate. The explosion and random chance account is scientifically absurd. And the unscientific notion that the universe is billions of years old did not arise from unbiased observations or a dispassionate compilation of the facts and evidence in order to ascertain their meaning. Instead it came from a biased search that had at the outset what the "find" was going to be, *astronomically* old ages that could be used to "validate" the ridiculous theory of evolution. And to keep selling their find today it is necessary to distort the evidence, ignore the observations, cover up the facts, and to deliberately overlook the truth. What any of this has to do with actual science, you'd have to ask them.

Starlight and Time

"Bless the Lord, O my soul! O Lord my God, you are very great; you art clothed with honor and majesty. Who covers yourself with light as with a garment. Who stretches out the heavens like a curtain. Who …makes the clouds his chariot, who walks upon the wings of the wind." (Psalms 104:1-3)

Both scientifically and biblically we know that the universe is less than 10,000 years old. But a major question still remains, *How did the light from all of those distant galaxies, some as far as twelve billion light-years away, get here so quickly?* And it's a very good question. Light travels at 186,000 miles per second, which is 5,865,696,000,000 miles every year. A distance of 12 billion light years is 70,388,352,000,000,000,000,000 miles away. (Or roughly the size of our National Debt after the Democrats get through with their hopeless change.) That is also the distance light must travel, going extreeeemely quickly, in order to arrive here 12 billion years later from that distant galaxy 12 billion light-years away. However, if the entire universe is only around 6 <u>thousand</u> years old, which it of course is, then how did that light get here already? It should still be 11 billion, 994 thousand

---

750   (Pg. 36, *In the Beginning*, by Walt Brown, Ph.D.)

miles away, or just slightly farther than Tina Fey can see out of her bathroom window.

The answer lies in exactly <u>how</u> God made the universe. It turns out, as we mentioned before in the section on "The Big Bang," that he "stretched out the heavens like a curtain" (Isaiah 40:22) He created matter and then formed the stars and the galaxies and then stretched them out to their present positions. And the beams of light from those stars and galaxies were stretched out as well. So the light never had to "get here" to earth. It was here from the beginning. This "stretching out" or expansion also caused the redshifts of distant stars and galaxies. The following was quoted earlier in "The Big Bang" section but it bears repeating here:

> "A... misconception is that the red shifts of the galaxies are Doppler shifts, i.e. caused by the velocity of the galaxies away from us at the time the light starts its journey toward us. But one undergraduate textbook... and many graduate textbooks... make it clear that the red shifts are an *expansion* effect. As space is stretched out, the lengths of all electromagnetic waves passing through the space are similarly stretched out. Consequently, the *speed* of recession doesn't matter, only the *amount* of expansion that takes place as the light travels to us, whether the expansion is fast or slow."[751]

"He has made the earth by his power, he has established the world by his wisdom, and has stretched out the heavens by his discretion." (Jeremiah 10:12)

This "stretching out" or expansion is the cause of both the redshifts of distant stars and galaxies, and the reason, along with gravitational time dilation,[752] why we can still see the light from those distant galaxies (some as far as 12 billion light years away) even though the earth and our universe are very young, less than 10,000 years old.

But what about the Dinosaurs?

"Behold now the behemoth, which I made along with you; he eats grass like an ox. See now, his strength is in his hips, and his power is in his stomach muscles. He moves his tail like a cedar; the sinews of his thighs are tightly wrapped together. His bones are as strong beams of brass, his ribs are like bars of iron." (Job 40:15-18)

Evolutionists tell us that Dinosaurs roamed the earth hundreds of millions of years ago. They say they *know* this because their fossils are found in the "geologic column" during this time period, and that they have been extinct for about the same amount of time. But all that of course is a lie. We know now that the sedimentary layers that make up the Darwinists imaginary geologic column are not hundreds of millions or billions of years old, but only thousands of years. We also know that their geologic column is a fabrication; it does not exist in reality, nor in the earth's sediments; nor in real science; it only exists in their own minds and imaginations.

The scientific evidence actually points to these Dinosaur creatures existing alongside of "modern" man just in the recent past. We mentioned in a previous segment examples of fossilized human footprints being found next to Dinosaur prints, but it bears revisiting...

---

751 (Pg. 98, *Starlight and Time: Solving the Puzzle of Distant Starlight in a Young Universe*, by D. Russell Humphreys, Ph.D.)
752 (For an explanation of "gravitational time dilation" see pgs. 11-13 of *Starlight and Time: Solving the Puzzle of Distant Starlight in a Young Universe*, by D. Russell Humphreys, Ph.D. We would explain it here but it gives us a headache)

"Scientists in the former Soviet Union have reported a layer of rock containing more than 2,000 dinosaur footprints alongside tracks resembling human footprints. Obviously, both types of footprints were made in mud or sand that later hardened into rock. If some are human footprints, then man and dinosaurs lived at the same time. Similar discoveries have been made in Arizona. Were it not for the theory of evolution, few would doubt that these were human footprints. [But hey, who are you going to believe, evolution or your lying eyes?]

"Soft tissue has now been recovered from several dinosaurs: three tyrannosaurs (*T. Rex*) and one hadrosaur. It is ridiculous to believe that soft tissue can be preserved for more than 60 million years, but it could be preserved for 5000 years."[753]

It is also ridiculous to believe that a genital-mutilating, genocidal child molester is the prophet of God, but that is the nature of false religions: people believe what they are taught. And if Dinosaurs are supposed to be 60 million years old, well then that's that. (I once met a Rastafarian who said he was 75 million years old, and he had the face to prove it.)

There is also… "the existence of numerous contemporaneous human and dinosaur prints found in Mexico, New Mexico, Arizona, Missouri, Kentucky, Illinois, and other US localities. These tracks are widely distributed and are usually only exposed by flood erosion or bulldozers. They have been carefully studied and verified by reliable paleontologists and cannot be dismissed as frauds. Furthermore, there are places in Arizona and Rhodesia were dinosaur pictographs have been found drawn on cave or canyon walls by man. The obvious implication is that man once lived contemporaneously with dinosaurs."[754]

But if Dinosaurs lived alongside of man on this earth in the recent past shouldn't we expect to find stories and legends and historical writings about them? Yes, we should by gosh by golly. And we do! Thousands of them! They're called Dragons! (Bingo!) You see, these "ancient" Dinosaurs and their *mythological* counterpart, Dragons, are one and the same.

"From countries all over the world, we read and hear stories of ancient dragons. Some flew, while others lived in caves or murky swamps. Some of the dragons protected emperors and cities, but most of the dragons ravaged the countryside killing animals and people.

"In far-eastern countries such as China, dragons often are found in the ancient writings. Some of them were even used to pull the chariots of Chinese rulers. Also, many of the ancient Chinese people used "dragon bones" for special medicines and potions. While visiting the continent in the 1200s, the Italian explorer Marco Polo said that he saw long reptiles called Lindworms that easily ran as fast as a horse!

"In the British Isles, **hundreds** of dragon stories have come down to the present day. [Emphasis mine] One legend told of an animal with a crested head, teeth like a saw, and a long tail. Also, in 1449 in England, it is reported that two huge reptiles were seen fighting on the banks of the river Stour.

…"Countries all over the world have stories of dragon slayers."[755]

Really? Could it be?! Could "Dragon" and "Dinosaur" actually be two words for the same creatures?! Of course! What else could they be? It's only because of the widespread acceptance of the idiotic lie of evolution and these hundreds of

---

753 (Pg. 350, *In the Beginning: Compelling Evidence for Creation and the Flood*, by Walt Brown, Ph.D.)

754 (Pgs. 16-17, *The Collapse of Evolution*, by Scott M. Huse)

755 (Pgs. 64-65, 75, *Dinosaurs Unleashed: The True Story about Dinosaurs and Humans*, by Kyle Butt and Eric Lyons)

millions and billions of years/imaginary time-spans, that Darwinists have been able to pull off the scam of calling Dinosaurs ancient, and Dragons a "myth." It is truly a wonder that they have been able to keep so many people in the dark for so long. The look of wonder and enlightenment (like a bright light was suddenly switched on inside of their head) on a person's face when they first learn that Dinosaurs and Dragons are one and the same is absolutely priceless. It's like, "Yea! Wow! Why didn't I think of that before? It's so obvious now." ("I could've had a V8!") We didn't think of it because we have all been lied to constantly in our schools, in our media and in our entertainment industry, and we believe what we are taught, and we become lost in the powerful blinding grip of a paradigm that, as completely false and ridiculous as it is in reality, can still blind us from seeing the obvious truth. (Just ask the Swiss why they sold the battery-operated watch patent for a song and a dance, and gave away what became an astronomically lucrative market to the Japanese.)

"The ancient legends of dragons – huge reptilian beasts with long necks and plated spines – depict beasts very similar to our modern reconstruction of dinosaurs. Mariners have told of dragons in the oceans even up until recent times. Almost every culture around the globe has legends of dragons, and these legends show striking similarities.

"Some dragon stories recorded in human history involve people who actually did exist. Alexander the Great encountered dinosaur-like beats, and so did King Beowulf of the British Isles. The ancient historian Berosus wrote of such beasts. Clearly, something that fits our modern ideas of dinosaurs existed in human history… **the legends of dragons are actually the faded memories of human encounters with dinosaurs.** [Emphasis mine]

…"Dragons are reputed to have lived up through the Middle Ages when brave knights would slay such beasts. **The scientific listings of animals of that day included dragons as rare, but still living creatures**… In South America, Africa, China, Australia, Europe, North America, and elsewhere, man engraved on rock walls the images of dinosaur-like beasts."[756] [Emphasis mine]

"Consider the many dragon legends. Most ancient cultures have stories or artwork of dragons that strongly resemble dinosaurs. *The World Book Encyclopedia* states that:

The dragons of legend are strangely like actual creatures that have lived in the past. They are much like the great reptiles [dinosaurs] *which inhabited the earth long before man is supposed to have appeared on earth. Dragons were generally evil and destructive. Every country had them in its mythology.*

The simplest and most obvious explanation for so many common descriptions of dragons around the world is that man once knew the dinosaurs."[757]

And because these Drago-saurs did not go extinct in some imaginary age 200 million years ago, but only disappeared in the last few thousand years, there is a very good possibility, and some evidence as well, that a few may still be alive today.

"For the past three centuries, reports have come from the Congo in western Africa that dinosaurs exist in remote swamps. Eyewitness stories are often from educated people who can quickly describe dinosaurs. Two expeditions to the Congo, led by biologist Dr.

---

756 (Pgs. 25, 33, *Dinosaurs, the Lost World, & You*, by John D. Morris, Ph.D.)
757 (Pg. 351, *In the Beginning: Compelling Evidence for Creation and the Flood*, by Walt Brown, Ph.D.)

Roy Mackal of the University of Chicago, never saw dinosaurs, but interviewed many of these witnesses and concluded that their reports were about dinosaurs and were apparently true."[758]

And then there is the most important witness of all, and the last one Mr. Evolutionist would ever think of entertaining, and that is the Word of God. We know that the Bible's witness is true. So what does it to say about Dinosaurs and Dragons? Well, in the Book of Job we find some fascinating verses…

"Behold now the behemoth, which I made along with you; he eats grass like an ox. See now, his strength is in his hips, and his power is in his stomach muscles. **He moves his tail like a cedar**; the sinews of his thighs are tightly wrapped together. His bones are as strong beams of brass, his ribs are like bars of iron." (Job 40:15-18)

Job goes on to describe how he lived in swamps and was unconcerned with a rampaging river. "Indeed the river may rage, yet he is not disturbed; he is confident, though the Jordan should gush against his mouth." (Job 40:23) Sounds like a Brontosaurus[759] to me. This behemoth could not have been an elephant or a hippopotamus because they both "have tails like ropes… [Yet] any animal with a tale as huge and strong as a cedar tree is probably a dinosaur."[760] Probably.

But what of *fire-breathing* dragons? Legends abound from across the globe of a Tyrannosaurus Rex type of creature who could shoot fire from his mouth and nostrils. (Of course, after putting together the bombardier beetle this was just an afterthought for the Creator. He had some leftover explosive material lying around and didn't want it to go to waste…) Could these "legends" be actual true accounts of real dinosaurs that plagued man for centuries? Let's look at the next chapter of Job…

"Can you draw out Leviathan with a hook, or snare his tongue with a line which you let down? Can you fill his skin with harpoons? Or his head with fishing spears? None is so fierce that he would dare stir him up. Who can open the doors of his face, with his terrible teeth all around? His rows of scales are his pride, shut up tightly as with a seal. One is so near another that no air can come between them. They are joined one to another; they stick together and cannot be parted. **His sneezings flash forth light… Out of his mouth go burning lights, and sparks of fire shoot out. Out of his nostrils goes smoke, as out of a boiling pot or caldron. His breath kindles coals, and a flame goes out of his mouth**. When he raises himself up, the mighty are afraid; because of his crashing about they are beside themselves. Though the sword reaches him, it cannot avail; nor does the spear, dart, or javelin. He regards iron as straw, and brass as rotten wood. The arrow cannot make him flee; sling stones are turned into stubble by him. Darts are regarded as straw; he laughs at the threat of spears. His undersides are like sharp stones." (Job 41:1, 7, 10, 14-21, 25-30)

If the writer of Job wasn't describing dinosaurs that coexisted with him and his contemporaries then he must gotten a bootlegged copy of *Jurassic Park*.

---

758   (Pgs. 350-351, *In the Beginning: Compelling Evidence for Creation and the Flood*, by Walt Brown, Ph.D.)

759   (See "The Flintstones." …It's now called a *Brachiosaurus*)

760   (Pg. 350, *In the Beginning: Compelling Evidence for Creation and the Flood*, by Walt Brown, Ph.D.)

"Leviathan in Job, chapter 41, is described as a marine creature (v. 1, 7) that spent at least some of his time on land (v. 26-30). He was incredibly fierce (v. 10), with large teeth (v. 14) and scales (v. 15-17). This beast was so vicious that he was even compared to Satan himself (Isa. 27:1). Most remarkable is his ability to breath fire (v. 18-21). Perhaps this gives us a clue as to the fire-breathing dragon legends of the past.

"Today, we may wonder how an animal could have breathed fire, but actually, a fire-breathing dinosaur is not impossible. The chemistry necessary to achieve this feat is not nearly as complicated as that possessed by other animals today, such as the firefly, the electric eel, and others. Most dinosaurs were big plant eaters. The digestion of plant material produces methane gas, which can burn. A methane belch only needs a spark to ignite it.

"Several known chemicals which burst into flame in the presence of oxygen would be able to provide the necessary spark if combined with methane. Producing these chemicals in living things would not be out of the question. Perhaps some dinosaurs had glands which could secrete these chemicals, which when combined with oxygen and methane, would produce a powerful flame thrower! ...the biblical account makes sense out of the legends of "fire-breathing dragons." "[761]

The truth about Dinosaurs is they are Dragons. And they lived and breathed and walked the earth with man since the dawn of creation six thousand (and not hundreds of millions of) years ago. They didn't *evolve* from any other organism, nor did any other organisms *evolve* from them. The Bible is absolutely accurate, and the theory of evolution is a joke. (Somebody might want to tell The New York Times.) (And Fox News.)

A World Wide Catastrophic Flood

"In the six hundredth year of Noah's life, in the second month, the seventeenth day of the month, on that day were all the fountains of the great deep broken up, and the windows of heaven were opened. And the rain was upon the earth forty days and forty nights." (Genesis 7:11-12)

The overwhelming evidence from the Earth Sciences indicate that a worldwide flood did occur in the recent past, just as the Bible records. And this evidence clearly shows, among many other things, that...

...."coal, oil, and methane did not form over hundreds of millions of years; they formed in months. Fossils and layered strata did not form over a billion years; they formed in months. The Grand Canyon did not form in millions of years; it formed in weeks. Major mountain ranges did not form over hundreds of millions of years; each formed in hours."[762]

Before we get to all of the facts, though, there is the historical evidence from the many so-called "myths" that all tell of a catastrophic deluge and can be found in nearly every culture:

"A gigantic flood may be the most common of all legends—ever. Almost every ancient culture had legends telling of a traumatic flood in which only a few humans survived in a large boat. This cannot be said for other types of catastrophes, such as earthquakes, fires,

---

761   (Pgs. 36-37, *Dinosaurs, the Lost World, & You*, by John D. Morris, Ph.D.)
762   (Pg. 103, *In the Beginning: Compelling Evidence for Creation and the Flood*, by Walt Brown, Ph.D.)

volcanic eruptions, disease, famines, or drought. More than 230 flood legends contain many common elements, suggesting they have a common historical source that left a vivid impression on survivors of that catastrophe."[763]

"Anthropologists have noted remarkable similarities between the historical folklore of nearly all cultures. Hundreds of widely dispersed people groups have a similar legend of a flood, sent by God because of man's wickedness, but survived by a favored righteous family, who built a huge boat for survival which eventually landed on a high mountain. Their common themes speak of a common ancestor that alone survived the Flood and passed the story of it on to their descendants.

"Less well-known are the common creation legends, again held by groups in all corners of the globe. Typically, the legend tells of a great golden age in which food was abundant, life spans were long, and the language was the same. This wonderful situation was lost due to disobedience and punishment, eventually leading to a great watery cataclysm. A smaller but not insignificant number of legends tells of a God-induced dispersion of tribes, followed by migration and reestablishment of civilization. The similarities with Genesis are obvious, and fit the thesis that all people alive today descended from Noah, and remember their legendary histories."[764]

God's word is true. Every part of it. And the common legends of a world-wide flood from every corner of humanity bear witness to it.

The Bible talks of "all the fountains of the great deep" being broken up in one day. (See Genesis 7:11 above.) Those words are the key to understanding exactly what happened to cause this great, worldwide flood. When God created the earth there were vast reservoirs of water under the earth's crust. This is borne out by ample physical evidence all across the globe that came from the aftermath of Noah's Flood, but let's also see what the Bible says:

"And God made the firmament, and divided the waters which were under the firmament from the waters which were above the firmament; and it was so. And God called the firmament Heaven. And the evening and the morning were the second day." (Genesis 1:7-8)

An accurate interpretation of the word *firmament* (firm) in this verse is not *sky*, but earth, or more accurately, the earth's *crust*. Also in this verse when God calls the firmament *Heaven* he was not referring to the sky or to outer space or to the place where He dwells, he was referring to the paradise here on earth that he had created originally before the fall of man and the subsequent curse.[765] So there is both ample evidence from the earth and from the Bible that there were these vast subterranean caverns of water, these waters of the "great deep," held under the earth's crust at enormous pressure, that, when released, produced a catastrophic worldwide flood.

"New evidence shows that the earth has experienced a devastating, worldwide flood, whose waters violently burst forth from under the earth's crust. Standard "textbook" explanations for many of the earth's major features are scientifically flawed. We can now

763 (Pg. 41, *In the Beginning*, by Walt Brown, Ph.D.)
764 (Pg. 73, *The Young Earth: The Real History of the Earth – Past, Present, and Future*, by John Morris, Ph.D.)
765 (For an in-depth discussion of this see pgs. 365-370, *In the Beginning: Compelling Evidence for Creation and the Flood*, by Walt Brown, Ph.D.)

explain, using well-understood phenomenon, how this cataclysmic event rapidly formed so many features."[766]

These features—which include the Grand Canyon, the mid-oceanic ridges, continental drift, continental shelves and slopes, the ocean trenches and the ring of fire, earthquakes, magnetic variations on the ocean floor, submarine canyons, coal and oil deposits, methane hydrates, the Ice Age, frozen mammoths, major mountain ranges, overthrusts, volcanoes and lava, geothermal heat, strata and layered fossils, limestone, metamorphic rock, plateaus, the Moho, salt domes, jigsaw fit of the continents, comets, asteroids, and meteorites—are all…

…"a consequence of a sudden, unrepeatable event—a global flood, whose waters erupted from interconnected, worldwide subterranean chambers with an energy release exceeding the explosion of 300 trillion hydrogen bombs."[767]

Dr. Walt Brown explains in his book, *In the Beginning: Compelling Evidence for Creation and the Flood*, how this water was released along a rapidly-spreading, north-south crack in the earth's crust above the line of what is now the Mid-Atlantic Ridge, and then spread globally along the Mid-Oceanic Ridge, which "wraps around the Earth and is the world's longest mountain range—46,000 miles."[768] This water was under so much pressure, 62,000 psi, that it quickly eroded the sides of this ever-widening split in the earth's crust, and shot tremendous volumes of water and earthen debris into the atmosphere and out into outer space.

"When the flood began, the pressure in the jetting SCW [supercritical water][769] dropped in seconds from at least 62,000 psi (4,270 bars) to almost zero. The energy released was huge. Because the 46,000-mile-long fountains continued this release for several weeks, one should not think of it as a single explosion. Instead, the jetting water was a powerful, earth-size engine that launched considerable mass from earth."[770]

"As the crack raced around the earth, the 10-mile-thick crust opened like a rip in a tightly stretched cloth. Pressure in the subterranean chamber directly beneath the rupture suddenly dropped to nearly atmospheric pressure. This caused supercritical water to explode with great violence out of the 10-mile-deep slit that wrapped around the earth like the seam of a baseball.

"All along this globe-circling rupture, whose path approximates today's Mid-Oceanic Ridge, a fountain of water jetted supersonically into *and far above* the atmosphere. Much of the water fragmented into an "ocean" of droplets that fell as rain great distances away. This produced torrential rains such as the earth has never experienced--before or after."[771]

This breaking up of "all the fountains of the great deep" was also the origin of comets, asteroids and meteoroids, as much water along with mass quantities of

---

766 (Pg. 105, *In the Beginning*, by Walt Brown, Ph.D.)
767 (Pg. 105, *In the Beginning: Compelling Evidence for Creation and the Flood*, by Walt Brown, Ph.D.)
768 (Pg. 105, *In the Beginning*, by Walt Brown, Ph.D.)
769 (As the pressure on water increases, so does the temperature at which it will boil. When water reaches pressures above 3,200 psi (called the *critical point*) it becomes *supercritical* and cannot boil. See pg. 118, *In the Beginning*, by Walt Brown, Ph.D.)
770 (Pg. 119, *In the Beginning*, by Walt Brown, Ph.D.)
771 (Pgs. 120-121, *In the Beginning*, by Walt Brown, Ph.D.)

rock debris from the rapidly eroding sides of the great "globe-circling rupture," had enough velocity to escape earth's orbit.

"The most powerful jetting water and rock debris escaped the earth's gravity and became the solar system's comets, asteroids, and meteoroids."[772]

Again, in the beginning God created a perfect world, without all of this solar debris orbiting the sun and threatening to hit earth with a large enough chunk to wipe out all of life. That's not what God had in mind when he finished Creation and "saw every thing that he had made, and, behold, it was very good. And the evening and the morning were the sixth day." (Genesis 1:31) The very bad came a little later...

"When the fountains of the great deep erupted, rocks were crushed, eroded, and sometimes reduced to clay. Mixed with that debris was carbonate-rich, salty, subterranean water... Organic compounds, including methane and ethane, are found in comets, because that water contained pulverized vegetation from preflood forests (as well as bacteria and other traces of life) from within hundreds of miles of the globe-encircling rupture."[773]

These "strange bodies, sometimes called "the mavericks of the solar system," have several remarkable similarities with planet earth. They contain considerable water. (About 38% of the mass of comet Temple 1 was frozen water.) Water is rare in the universe, but both common and concentrated on earth--often called "the water planet." Most of the remaining mass of a comet is dust, primarily the crystalline mineral olivine. Solid material that formed in space would not be crystalline. Olivine may be the most abundant of the almost 4,000 known minerals in the earth's crust and mantle. Asteroids and meteorites are similar in many ways to earth rocks. Surprisingly, a few meteorites contain salt crystals, liquid water, and living bacteria! Some asteroids have a chemical substance (kerogen) found in plants."[774]

Wow! Finally we have solid evidence that life must have *evolved* elsewhere in the universe?! No, we have evidence for a cataclysmic worldwide flood caused by the breaking up of "all the fountains of the great deep" and the shooting out into space of a volume of earthen debris from those fountains, exactly like God's Word describes.

It also caused torrential rains; but some of the water reached outer space where the temperature is minus 270°F, or absolute zero, before falling back to earth and instantly freezing all of those mammoths and other animals we find, who have remained frozen in some of the coldest climes like Siberia and Alaska.[775]

---

772   (Pg. 121, *In the Beginning*, by Walt Brown, Ph.D.)

773   (Pgs. 278-279, *In the Beginning, Compelling Evidence for Creation and the Flood*, by Walt Brown, Ph.D.)

774   (Pg. 116, *In the Beginning*, by Walt Brown, Ph.D.)

775   (And no, that super frozen water would not have been heated up upon re-entry into the earth's atmosphere like space capsules or the Space Shuttle. These vehicles must orbit the earth at extremely high velocities (apprcx. 18,000 mph) so their centrifugal force will counteract the pull of earth's gravity, and therefore allowing them to stay in orbit. When they re-enter the earth's atmosphere at those velocities, the friction causes extreme heat. Not so for water frozen to absolute zero that *falls* back to earth at a comparatively low speed and much less friction.)

"Some jetting water rose above the atmosphere, where it froze and then fell on various regions of the earth as huge masses of extremely cold, muddy "hail." That hail buried, suffocated, and froze many animals, including some mammoths."[776]

"Fleshy remains of about 50 elephant-like animals called *mammoths*, and a few rhinoceroses, have been found frozen and buried in Siberia and Alaska. One mammoth still had identifiable food in its mouth and digestive tract. To reproduce this result, one would have to *suddenly* push a well-fed elephant (dead or alive) into a very large freezer that had somehow been precooled to -*150°F*. Anything less severe would result in the animal's internal heat and stomach acids destroying the food. If the animal remained alive for more than a few minutes, one would not expect to find food in its mouth."[777]

There is only one plausible explanation for these instantly frozen creatures and that is the one given above. All others are pathetic attempts to explain away the obvious. Our academics have been taught to laugh at the idea of a worldwide flood (but in reality the laughs on them) so they just close their ears and their minds to all of the overwhelming evidence, write off all of those who believe otherwise as "stupid religious nuts" and continue to proudly parade around our colleges and universities buck naked. And unfortunately their condition is still hidden from the great majority of their students.

The waters rushing out from under the crust quickly eroded the sides of the ten-mile-high walls and spilled out across the earth, and this muddy deluge formed the sedimentary layers that cover the earth today, in the process sweeping up mass quantities of animal and plant life, an entire planets worth, and then rapidly burying and fossilizing them deep within those sediments. This is what we see today preserved in the "worldwide fossil record." (And if you'd like to get an idea of what this muddy flood spilling out across the earth looked like, we have just been given a glimpse in the recent Japanese tsunami, in March of 2011. Just look at all of those videos of the surging black waters that roared in from the ocean, filled with homes and barns and debris that marched across many miles of farm lands and towns and cities devouring everything in its path, and you can see a small-scale replica of what occurred 4,500 years ago, on a worldwide scale!)

"Each side of the rupture was basically a 10-mile-high cliff. Compressive, vibrating loads greatly exceeded the rock's crushing strength in the bottom half of the cliff face, so the bottom half of the cliff continuously crumbled, collapsed, and spilled out into the jetting fountains. That removed support for the top half of the cliff, so it also fragmented and fell into the pulverizing supersonic flow. Consequently, the 46,000-mile-long rupture rapidly grew to an average width of about 800 miles all around the earth.

"About 35% of the eroded sediments were from the basalt of the chamber floor. Sediments swept up in the escaping flood waters gave the water a thick, muddy consistency. These sediments settled out over the earth's surface in days, trapping and burying many plants and animals, beginning the process of forming the world's fossils."[778]

As the escaping water opened an ever-widening crevasse, the point was eventually reached where the opening became wide enough to substantially reduce the pressure on the earth's mantle directly below it. This caused the mantle

---

776   (Pg. 121, *In the Beginning: Compelling Evidence for Creation and the Flood*, by Walt Brown, Ph.D.)

777   (Pg. 110, *In the Beginning*, by Walt Brown, Ph.D.)

778   (Pg. 121, *In the Beginning*, by Walt Brown, Ph.D.)

there to buckle up which created the Mid-Oceanic Ridges. All other attempts at explaining the origin of these ridges in order not to contradict the evolutionary, old-earth model have proven somewhat fanciful, as well as failing to jibe with the existing scientific evidence. (Which is what scientific explanations are supposed to do, aren't they? As opposed to, say, supporting a particular group's strongly held religious beliefs.)

"Material within the earth is compressed by overlying rock. Rock's slight elasticity gives it springlike characteristics. The deeper the rock, the more weight above, so the more tightly compressed the "spring"—all the way down to the center of the earth.

"The rupture path continuously widened during the flood phase... Eventually, the width was so great, and so much of the surface weight had been removed, that the compressed rock beneath the exposed floor of the subterranean chamber sprung upward."[779]

The currently accepted plate tectonics notion that "supposedly, material deep inside the earth is rising toward the crest of the entire Mid-Oceanic Ridge... [and then] moves laterally away from the ridge"[780] unfortunately defies the laws of physics and also is a poor explanation for the observed phenomenon; one that seeks to deny the more simple and much more obvious one.

As the mantle buckled up along the Mid-Atlantic Ridge, gravity caused the now-separated continents to slide away on a cushion of the remaining "waters of the great deep."

"As the Mid-Atlantic Ridge began to rise, creating slopes on either side, the granite **hydroplates** [continents or *tectonic plates* riding on water, hence *hydro*] started to slide downhill. This removed even more weight from what was to become the floor of the Atlantic Ocean. As weight was removed, the floor rose faster and the slopes increased, so the hydroplates accelerated, removing even more weight, etc. The entire Atlantic floor rapidly rose almost 10 miles. [It has, of course, since subsided.]

"As the first segment of the Mid-Atlantic Ridge began to rise, it helped lift adjacent portions of the chamber floor just enough for them to become unstable and spring upward. This process continued all along the rupture path, forming the Mid-Oceanic Ridge. Also formed were fracture zones and the ridge's strange offsets at fracture zones. ... For a day or so, the sliding hydroplates were almost perfectly lubricated by water still escaping from beneath them."[781]

Also, these "fracture zones and strange offsets" all along the Mid-Oceanic Ridges cannot be sensibly or even intelligently explained by uniformitarian geologists.

As these continents travelled farther and farther apart the remaining water in the subterranean chambers spilled out, eventually covering the entire earth. "And the waters prevailed exceedingly upon the earth; and all the high hills that were under the whole heaven were covered." (Genesis 7:19) Then, as the lubricating water under these sliding continents was exhausted, they ground to a "screeching" halt, buckling up all of the great mountain ranges of today:

"Eventually, the hydroplates ran into resistance of two types. The first happened as the water lubricant beneath each sliding plate was depleted. The second occurred when a plate collided with something. As each massive hydroplate decelerated, it experienced a gigantic

---

779   (Pg. 122, *In the Beginning*, by Walt Brown, Ph.D.)
780   (Pg. 108, *In the Beginning*, by Walt Brown, Ph.D.)
781   (Pg. 124, *In the Beginning: Compelling Evidence for Creation and the Flood*, by Walt Brown, Ph.D.)

compression event--buckling, crushing, and thickening each plate. ... crashing hydroplates at the end of the continental-drift phase crushed and thickened each hydroplate for many minutes. Mountains were quickly squeezed up.

...“Naturally, the long axis of each buckled mountain was generally perpendicular to its hydroplates motion--that is, parallel to the portion of the Mid-Oceanic Ridge from which it slid. So, the Rocky Mountains, Appalachians, and Andes have a north-south orientation.

...“Friction at the base of skidding hydroplates and below sinking mountains generated immense heat, enough to melt rock, produce huge volumes of magma, and begin earth’s volcanic activity. ... Sometimes magma escaped to the earth’s surface, producing volcanic activity and “floods” of lava outpourings, called *flood basalts*, as seen on the Pacific floor and the Columbia and Deccan plateaus.

...“As the new postflood continents rose out of the flood waters, water drained into newly opened ocean basins.”[782]

And now we know the rest of the story, the true one that we were never taught in school. This is what actually happened to form the features of the earth that we see today. This is a scientific explanation, as opposed to the fairy tale pawned off as science that is still being taught to our school children today; not because of any preponderance of facts or evidence, but merely because it fits into the framework of a false, unscientific religion that long ago usurped the place of science, truth and reason. (Hats off to the father of lies for his accomplishment in getting his genesis account to its place of preeminence in our culture.)

Also, attempting to explain the features and dynamics of the earth’s continents, volcanoes and earthquakes with *plate tectonics theory* is fraught with problems and scientific impossibilities. Earth scientists were right initially in advancing the idea of continental drift, but by refusing to acknowledge the earth’s present features as the aftermath of a cataclysmic flood, they have been spinning their wheels ever since. And their attempts at explaining all that we see now, as a result of this refusal, have proven implausible. For one, the idea of *subduction*, of a huge continental plate slowly sliding underneath another plate over, what else, “hundreds of millions of years,” is scientifically impossible according to the laws of physics. It would be akin to you trying to shove your flattened stiffened hand slowly through your skull. Even if you could stand the pain you wouldn’t get subduction, you would get destruction.

“Pressure inside the earth increases with depth. So, if one tried to depress a plate 30 miles or more below another plate, tremendous upward pressure from below would quickly prevent that much depression. Consequently, subduction—necessary for plate tectonics— could not begin, even if the plate were colder and, therefore, denser.”[783]

Now let’s leave the study of *how* that global flood occurred and look further into the worldwide compelling evidence *for* that same deluge. Let’s begin with the sedimentary layers that cover the earth:

“Sedimentary rocks are distinguished by sharply defined layers called **strata**. Fossils almost always lie within such layers. Fossils and strata, seen globally, have many unusual

---

782 (Pg. 125, *In the Beginning: Compelling Evidence for Creation and the Flood*, by Walt Brown, Ph.D.)
783 (Pg. 159, *In the Beginning*, by Walt Brown, Ph.D.)

characteristics. A little-known and poorly-understood phenomenon called **liquefaction** explains these characteristics. It also explains why we do not see fossils and strata forming on a large scale today.

…"*Liquefaction--associated with quicksand, earthquakes, and wave action--played a major role in rapidly sorting sediments, plants, and animals during the flood. Indeed, the worldwide presence of sorted fossils and sedimentary layers shows that a gigantic global flood occurred. Massive liquefaction also left other diagnostic features such as cross-bedded sandstone, plumes, mounds, and fossilized footprints.*"[784]

"Sedimentary rocks need water for cementing the tiny particles of sediment together. Where did the water come from? Evolutionists resolve this dilemma by waving the magic wand of "time." They insist that the sediments are washed into rivers and lakes and oceans and then fall to the bottom and pile up in layers. After many years, the sediments harden in layers, with those on the bottom hardening first. The problem with this theory is that there are beds of sedimentary rock everywhere on top of the earth's crust. Most of the large mountain chains are made of sedimentary rocks. How did all those sediments collect and become cemented together to make towering mountains? The logical answer is a worldwide Flood."[785]

The problem also is that these sedimentary layers not only cover the entire earth and are a mile thick in many places, but they also cover thousands of square miles where each successive layer is dropped onto the one below it with absolutely no sign of water or wind erosion, chemical deterioration, plant roots or animal burrowing, or the passage of any time whatsoever. Look at the walls of the Grand Canyon and you will see that these lines in between each layer, going for mile after mile, could have been drawn with a laser! Plant rooting, animal burrowing and wind and rain erosion over hundreds of millions of years would produce tremendous signs of the passage of time between each successive layer. (And therefore, for one, the lines on the side of the Grand Canyon would be zigzagged and jagged and discontinuous in places.) But there is none whatsoever. Where is the evidence of *any* passage of time, Mr. Uniformitarian evolutionist, much less "hundreds of millions" of years?! So much for the Darwinists obviously erroneous and unsupportable claim that these sedimentary layers all across this earth were laid down over hundreds of millions of years.

In the Grand Canyon… "the lack of evidence for time passing between the layers is compounded by the immensity of the supposed time gaps between layers."[786]

"The earth's sedimentary layers are typically parallel to adjacent layers. Such uniform layers are seen, for example, in the Grand Canyon and in road cuts in mountainous terrain. Had these parallel layers been deposited slowly over thousands of years, erosion would have cut many channels in the topmost layers. Their later burial by other sediments would produce nonparallel patterns. Because parallel layers are the general rule, and the earth's

---

784 (Pg. 169, *In the Beginning: Compelling Evidence for Creation and the Flood*, by Walt Brown, Ph.D. For a more in-depth study of liquefaction and the sedimentary layers that cover this earth, see pages 169-181, *Liquefaction: The Origin of Strata and Layered Fossils* in Dr. Brown's book.)

785 (Pgs. 35-36, *Evidence for Creation: Intelligent Answers for Open Minds*, by Tom DeRosa)

786 (Pg. 102, *The Young Earth: The Real History of the Earth – Past, Present, and Future*, by John Morris, Ph.D.)

surface erodes rapidly, one can conclude that almost all sedimentary layers were deposited rapidly relative to the local erosion rate—not over long periods of time."[787]

The sedimentary rock layers that cover this earth are incontrovertible evidence for a global flood, and for nothing else. There is no other intelligent interpretation of the data.

"Rock formation can be explained by means of two processes, compaction and cementation, both of which could have occurred within one year after the waters of the Flood began to subside.

"Global floodwater would have collected a huge amount of sediment. When the flood water came to rest, the sediment would have begun to fall out. At the beginning… there would have been a great deal of pressure on the bottom layers. This would cause compaction.

…"The second step is cementation. The glue that hardens the rock is generated by the warm water and minerals produced by the effects of the Flood. It does not take a long time for the rock to harden into layers. …Remember, making sedimentary rock is like preparing cement."[788]

And cement can become as hard as rock in a few hours.

The Grand Canyon[789] is one of the world's most striking testimonies to the worldwide flood. The idea that a measly river carved this canyon and all of its side canyons through a *raised plateau* by defying gravity is a testament both to man's stupidity and to the level of willful ignorance that he can stoop to if so driven by tightly-held, false-religious beliefs: "There is no God. There was no worldwide biblical Flood. Evolution is true. And that is the end of the discussion." And science, common sense, and all the evidence be damned.

Also, all across this world there are countless examples of these layers of sedimentary rock being turned, twisted, compacted, accordioned (folded back and forth upon them self like the folds of that musical instrument) and tilted. The Grand Canyon has many such places. Only a worldwide catastrophic aqueous event could have produced this. And this evidence is absolutely damning for uniformitarian geology, which cannot *see* the signs of this catastrophe written all across the earth. There are innumerable instances around the world where we can see that a number of these sedimentary layers have together been squeezed, bent, folded, curved and compacted into wave formations (as well as accordioned) which could only have occurred when these sediments were still *in a putty-like state*. Yet supposedly, according to evolutionists and uniformitarian geologists, these layers were laid down over hundreds of millions of years, with millions of years separating the beginning of the deposition of each successive layer on top of the one beneath. Which means that, according to them, all of the layers would have long-since been hardened into solid rock, not putty. And this *putty-like state* can only occur for a very short period of time, measurable in days or weeks, immediately after the sediments were deposited and before they had a chance to completely harden. The sedimentary layers therefore had to have been laid down very rapidly one upon the other up to thousands of feet thick and covering hundreds of thousands

---

787   (Pg. 11, *In the Beginning: Compelling Evidence for Creation and the Flood*, by Walt Brown, Ph.D.)

788   (Pg. 36, *Evidence for Creation: Intelligent Answers for Open Minds*, by Tom DeRosa)

789   (For an excellent in-depth study on what really made the Grand Canyon and all of the features associated with it, see pages 183-219, "The Origin of the Grand Canyon" in Dr. Brown's book, *In the Beginning*.)

of square miles and then very soon thereafter, before they had time to harden and cement themselves, were deformed and bent, folded and curved, squeezed into wave and accordion formations. And of course this is exactly what did occur as the sediments were rapidly laid down by the muddy flood waters, separated into distinct layers by the process of liquefaction, began the process of cementing, and then as the sliding continents came to a halt they were compressed, bent, folded and buckled up all over the earth. The explanation of hundreds of millions of years for the deposition of these layers is absurd, and was only thought up to support the theory of evolution by aiding it with preposterous spans of time where "anything is possible."

"The Grand Canyon is far from unique. There are many, many other places where **rocks have deformed while in a soft, unconsolidated condition**. [Emphasis mine.] The Rocky Mountains are full of such occurrences. The Appalachian Mountains are as well. One such occurrence might be passed off as an anomaly, but the world is full of examples of soft sediment deformation, just as it should be if the earth is young and the Flood really is responsible for most of the world's geologic features."[790]

Hardened rock is brittle, and it doesn't squeeze, bend, fold, curve or compact into wave formations. It breaks! It cracks! It busts up! If these sedimentary layers were sifted down over hundreds of millions of years as Darwinists claim, then how could all of those layers still have been in a putty-like state, millions upon millions of years later, in order to have been able to be squeezed and bent and folded and curved and compacted into wave formations before they hardened and became brittle? Answer: of course they were not sifted down over hundreds of millions of years. They were laid down rapidly by a global flood.

Before the time of Charles Lyell[791] scientists used their common sense and saw the features of the earth's surface—massive and deep sedimentation, the Grand Canyon, the white cliffs of Dover, seashells on the tops of every mountain range, etc.—and logically and naturally concluded that they were the result of the Great Flood. But that was before logic, rationality and science took a back seat to the religion of Darwinism, when God and His Bible had to go, and anything attributed to God or His Word (like a world-wide, biblical flood) had to go as well.

And the evidence is overwhelming. . .

"There are caves, fissures, and mass burial sites throughout the world that are literally packed with masses of fossils; often times the fossils of these various animals come from widely separated and differing climactic zones, only to be thrown together in disorderly masses. Such phenomenon can only be satisfactorily explained in terms of a worldwide aqueous cataclysm.

...."Further startling evidence of the fact that a great and sudden cataclysm once struck

---

790   (Pgs. 112-113, *The Young Earth: The Real History of the Earth – Past, Present, and Future*, by John Morris, Ph.D.)

791   (Lyell, an atheist determined to undermine the Bible, was a geologist in the 1800's who championed the idea of uniformitarianism (slow, uniform changes in the earth's features over great periods of time), and rejected the biblical concept of catastrophism as well as the obvious effects on this planet of a global flood. The idea of uniformitarianism also fit in neatly with Charles Darwin's new theory of evolution (indeed Darwin had a copy of Lyell's book on his famous Beagle voyage) which soon become all the rage with the European intellectuals of the day, and which demanded the outrageous lengths of time—provided conveniently by Lyell and his unscientific interpretation of geologic features—to give Darwin's theory a better shot at believability.)

the earth is found in the millions of mammoths and other large animals that were killed instantly in the north polar regions (northern Siberia and Alaska). Many of these have been found preserved whole and undamaged (except for being dead, of course) with flesh and hair intact, and in some cases, either kneeling or standing upright with food on their tongues. The eyes and red blood cells were found to be extremely well preserved, and the separation of water in the cells was only partial, which speaks of extremely sudden and sustained freezing conditions.

"All uniformitarian explanations fail dismally when attempting to interpret this phenomenon."[792]

Of course, we know for a scientific fact that it is impossible for all of these cellular tissues to have been preserved for hundreds of millions of years. But this fact is ignored so the Big Lie can remain, and can continue to be taught.

"Evolutionists have been baffled by the fact that strong, well-established groups of animals such as dinosaurs and trilobites suddenly disappeared from the fossil record. ... Creationists simply attribute their misfortune either directly or indirectly to the Genesis Flood. ...The collapse of the vapor canopy ["and the windows of heaven were opened" Genesis 7:11)] resulted in a post-Flood climate which was dramatically different from the pre-Flood climate. Instead of continuing to enjoy the stable, warm, and mild climate which they were accustomed to, these animals found themselves thrust into a relatively unstable and more hostile environment characterized by cooler temperatures, severe storms, and bitter winter conditions. ...Cold-blooded animals, such as dinosaurs, whose body temperature is regulated by the temperature of the environment were probably especially affected in an adverse manner by the collapse of the vapor canopy and the subsequent change in climate.

"Petrified logs ...are not being formed anywhere in the "uniformitarian world" today. They do, however, occur by the thousands in the fossil record having their fibers and cell structure perfectly preserved by replacement of silica.

"All of the evidence concerning these logs points to a sudden and catastrophic event. None ...are found to be standing. All have had their branches stripped off and many appear to have had their bark still intact, indicating rapid burial before rotting could occur. It is apparent that the original forests were uprooted by some sort of hydraulic cataclysm of enormous power, which also transported and deposited them in their present locations, where they became petrified (under unique conditions). ...the only reasonable and plausible explanation for such a phenomenon is the Genesis Flood.

"Polystratic trees are fossil trees which extend through several layers of strata, often 20 feet or more in length. There is no doubt but that this type of fossil was formed relatively quickly; otherwise it would have decomposed while waiting for strata to slowly accumulate around it. In some cases, these trees bridge a presumed evolutionary time span of millions of years. Obviously, the more reasonable interpretation calls for the simultaneous transportation and deposition of the trees and their surrounding sediments. Such logical conclusions cast serious doubts upon the common uniformitarian assumption that these sedimentary strata were laid down gradually over millions of years."[793]

Science is ruled by observation. That is the very method that gives rise to science itself. How can anyone in his or her right mind look at all of these polystratic trees, trees that stand upright through "millions" of years of sediment and are found all over the globe, as well as animals, including a whale in California

---

792   (Pgs. 47-48, *The Collapse of Evolution*, by Scott M. Huse)
793   (Pgs. 51-52, *The Collapse of Evolution*, by Scott M. Huse)

standing on its tail, and conclude anything other than those trees and those animals and that whale were placed there when the sediments were, and therefore both were deposited rapidly, and therefore the ridiculous ages for these sediments that the lie of evolution demands are patently absurd? Indeed, the mental gymnastics one has to perform in order to conclude otherwise are truly amazing. So they just ignore it. Unfortunately, people do believe what they are taught. No matter how lame. And they keep on teaching it to generation after generation. And the curse goes on and on...

"Ephemeral markings such as ripple marks, rain imprints, worm trails, and animal tracks are found in great abundance in the fossil record. This special type of transient fossil is originally formed as an evanescent marking on the surface of a recently deposited layer of sediment. These structures are very perishable and easily destroyed by normal weather conditions or by erosion and sedimentation. The formation and preservation of ephemeral markings is not observed today under normal uniformitarian conditions. Their preservation depends entirely upon abnormally rapid and complete burial associated with chemical lithification processes. The fact that these structures are found in great abundance throughout the world in the fossil record testifies to a sudden and worldwide cataclysmic sedimentary event—the Genesis Flood.

"There are numerous cases of preservation of actual soft parts (tissues) in the fossil record. This is true even in what is presumed to be the most ancient of strata. Such fossils are commonly discovered together in large masses...

"Such remarkable deposits could not have formed by normal, slow, uniformitarian, *in situ* processes. Atypical transportation of the organisms and rapid burial by sediments are clearly indicated. The uniformitarian assumption that such deposits of soft tissues have remained untouched by decay or erosion for millions of years seems absolutely ludicrous. The only reasonable alternative for such a phenomenon is a worldwide hydraulic cataclysm of relatively recent occurrence which would have quickly destroyed, transported, deposited, and lithified these organisms in the engulfing sediments. The Genesis Flood stands alone as the only sensible explanation for the observed facts."[794]

And in summary...

"The evidences of fossilization such as woolly mammoths, petrified logs, polystratic trees, ephemeral markings, soft parts, fossil graveyards, and so on, all designate the workings of a sudden and powerful worldwide aqueous cataclysm of unmatched proportions—the Genesis Flood! ...The evidence supplied by the fossil record substantiates the Biblical account of creation and the Flood while at the same time it soundly refutes evolution and uniformitarianism."[795]

And this would all be common knowledge in our universities, colleges, elementary and high schools and throughout our country, if it weren't for the fact that the lies of the evil one cover this earth like a thick sediment, and people believe what they are taught.

"You who laid the foundations of the earth, So that it should not be moved forever, you covered it with the deep as with a garment; The waters stood above the mountains. At your rebuke they fled; At the voice of your thunder they hastened away. They went up over the mountains; They went down into the valleys, To the place which you founded for them. You have set a boundary that they may not pass over, That they may not return to cover the earth." (Psalms 104:5-9)

---

794   (Pgs. 52-54, *The Collapse of Evolution*, by Scott M. Huse)
795   (Pg. 54, *The Collapse of Evolution*, by Scott M. Huse)

Again, as we stated in chapter 6, "The Word of God" …The Bible says that there was a catastrophic worldwide flood that occurred about 4,500 years ago. Faux scientists say that all of the features found on the earth today, including the Grand Canyon, can be explained by the slow processes of upheaval and erosion continuing slowly over millions of years. Again, the Biblical account is scientifically accurate. The evidence for a worldwide flood is overwhelming and incontrovertible; yet it is blackballed by the evolution indoctrinates who possess a stranglehold over our media and our educational system. There are trillions of recently deposited seashells on the tops of the highest mountains. Coal and oil formations demand the rapid and very deep burial of an extraordinary volume of living matter. The fossil record which covers this earth could only have been caused by the rapid worldwide burial of trillions upon trillions of living things (fossils can <u>only</u> form under conditions of rapid and <u>deep</u> burial- a dead fish lying on an ocean floor, or a dead elephant lying on the ground have no chance whatsoever of being fossilized). It is nothing more than an unrealistic and unscientific fantasy to think that the Grand Canyon and all of her side canyons were formed by a river slowly eroding a *raised plateau* over any period of time. All the scientific and erosion rate and sedimentary evidence call for the Canyons rapid formation by the catastrophic runoff of a massive amount of water over a very short period of time. On a worldwide scale sedimentary layers thousands of feet deep, with absolutely no evidence of time passage or erosion between each successive layer (their transition surfaces are perfectly smooth and flat), and extending in many places uniformly for hundreds of thousands of square miles cry out for only one possible scientific explanation, that of a catastrophic worldwide flood! And we could go on. But once again the Bible proves itself scientifically and historically accurate.

"Then the Lord saw that the wickedness of man was great in the earth, and that every intent of the thoughts of his heart was only evil continually. And it sorrowed the Lord that he had made man on the earth, and it grieved him at his heart. And the Lord said, I will destroy man whom I have created from the face of the earth." (Genesis 6:5-7)

Approximately 4,500 years ago God allowed the waters under the crust of the earth, the "fountains of the great deep," to erupt and flood the entire planet. He did this because of the wickedness of man. But how can anyone think that the evil of this present generation is any less than theirs? And in this country God has shed his grace on us in ways never even dreamt of by those people thousands of years ago, and yet how do we repay his kindness and mercy and abundance? For starters, we shove the bald-faced, scientifically-absurd, Beelzebub-inspired lies of evolution and the big explosion down the throats of our school children; and then we top that off with mutilating millions of his babies to death each year, and going out and foolishly voting for the same. If we look across this earth with open eyes we can see the aftermath of his last judgment. The next one will not be by water, but it will be no less severe. God's approaching final judgment of wicked, unrepentant man will be by fire. (See chapter 9.)

But as noted in chapter 7, he has made a way of escape for us. Indeed the history of Noah's ark and the worldwide deluge are themselves a prophecy of Christ and His work of redemption. For as the ark saved righteous Noah and his family from destruction in the flood, so Jesus saves all who place their trust and faith in Him and His righteousness, from the horror of eternal separation from

God and all that is good in a place of everlasting torment. Our school children are in desperate need of hearing this truth. Who has the guts to tell them? O'Reilly? Fox news? Anybody there?

Human "Races"

Question: *OK then, the earth is young just like the Bible and all of the scientific evidence says, and all 6 billion of us alive today are descended from Noah and his wife and children who walked off of the ark roughly 4,500 years ago. But what about all of the races of man? Eskimos, and Caucasians, and Negroes, and Australian Aborigines, and Chinese, and Japanese, and Polynesians, and American Indians, and India Indians; tell us where all of these races came from if they didn't have millions of years to "evolve?"*

Answer: We don't have to tell you. God's Word already has. It's in Genesis 11; in the story of the Tower of Babel...

"Now the whole earth was of one language and of one speech. And it came to pass, as they journeyed from the east, that they found a plain in the land of Shinar, and they dwelt there. And they said to one another, 'Come, let us make bricks and bake them thoroughly.' And they had brick for stone, and they had asphalt for mortar. And they said, 'Come, let us build ourselves a city, and a tower whose top may reach unto heaven; let us make a name for ourselves, lest we be scattered abroad over the face of the whole earth.' But the Lord came down to see the city and the tower which the children of men had built. And the Lord said, 'Behold, the people are one and they all have one language, and this is what they begin to do; and now nothing that they propose to do will be restrained from them. Come, let Us go down and there confuse their language, that they may not understand one another's speech.' So the Lord scattered them abroad from there over the face of all the earth, and they stopped building the city. Therefore its name is called Babel, because the Lord confused the language of all the earth; and from there the Lord did scatter them abroad over the face of all the earth." (Genesis 11:1-9)

[A side note, look at "Come, let Us go down..." That is another reference to the triune nature of the One God.]

This "confusion of the languages" was also the beginning of the differentiation of the distinct "racial" types that we see today across this globe. The reason "racial" is in quotes is because there really is only one race, the human race, and all the differences we see are merely the result of "variations within a kind" that we discussed earlier. And whether these variations in skin color, body size and body type, and distinct facial features were brought about by the Lord at the same time that he confused the languages, or whether they developed over successive generations after each language separated each group from one another, is a question that remains. But the fact that that is where the separation of "racial" types occurred, or began to occur, is fairly obvious.

For one, variations in skin color are something that is brought about merely by varying within the epidermal cells the amount of the skin-coloring pigment *melanin*...

"We all have the same coloring pigment in our skin—melanin. This is a dark-brownish pigment that is produced in different amounts in special cells in our skin. If we had *none* (as do people called albinos, who inherit a mutation-caused defect, and cannot produce melanin), then we would have a very white or pink skin coloring. If we produced a little melanin, we would be European white. If our skin produced a great deal of melanin, we would be a very dark black. And in between, of course, are all shades of brown. There are no other significant skin pigments.

...."This situation is true not only for skin color. Generally, whatever feature we may look at, no people group has anything that is essentially different from that possessed by any other. For example, the Asian, or almond, eye differs from a typical Caucasian eye in having more fat around them. Both Asian and Caucasian eyes have fat—the latter simply have less.

"What does melanin do? It protects the skin against damage by ultraviolet light from the sun. If you have too little melanin in a very sunny environment, you will easily suffer sunburn and skin cancer. If you have a great deal of melanin, and you live in a country where there is little sunshine, it will be harder for you to get enough vitamin D (which needs sunshine for its production in your body). You may then suffer from vitamin D deficiency, which could cause a bone disorder such as rickets."[796]

So the need for vitamin D alone could have caused the divergence, by the natural selection of those humans better suited for their sunny or cold environments, of the various skin colors we find in the different "races" such as European Caucasian and African Negro. For these continents obviously have entirely different climates, average temperatures and levels of sunshine. Or God could have allowed for those different needs by altering skin color and other features when he confused the languages, knowing in advance where these separate language-groups would end up in their migrations after He "scattered them abroad over the face of all the earth." Either way though, we can dispose of the false anthropological notion that because there are obvious differences between racial groups and sub-groups, there had to be vast periods of time for these differences to be brought about. That is a false notion that only survives because of a rejection of the accuracy of God's Word concerning this matter, and an acceptance of ridiculous time periods for life on earth in order to support the lie of evolution.

Part III - The Religion of Naturalism
"For they exchanged the truth of God for a lie, and worshipped and served the creature more than the Creator..." (Romans 1:25)

It should be obvious now to any reader who seeks the truth that the theory of evolution is not only a joke, and a lie, but it is also not science. It is of course a religion. No mere theory this scientifically-impossible, this bereft of any credible evidence whatsoever and indeed this irrational and idiotic could have survived within the confines of science for even five minutes much less have thrived for over 100 years. No, there has to be something going on here much more powerful than just a false theory. There has to be a deep religious fervor involved in order to drive such a colossal lie as this into a position of total dominance in the classrooms and the cultures of an entire world. It is no different than the reason why the satanic, genital-mutilating religion of Islam has been driven into a position of total dominance throughout the Muslim world. Blind, irrational, intolerant religious fervor; supported also by the fear of academic ridicule and exile on the one hand or death on the other for anyone who would dare to challenge either religions' beliefs.

---

796   (From article, *Where did the human races come from?* @ christiananswers.net)

"Evolution dismisses intelligence and assumes random chance to be the mechanism responsible for material reality—a physical presupposition that is not drawn from empirical evidence. Evolution, therefore, is an irrational belief. It assumes randomness and chance, not design, to be the governing principle of reality, a view, again, without hard empirical data to support it."[797]

"A view, again," that is also logically and scientifically absurd.

Satan is the god of this world, and the father of lies. He hates the gospel of Jesus Christ and will do anything in his power to keep as many people as possible from hearing it and believing in it. One of his greatest achievements has been in using science, indeed *usurping* it, to serve his diabolical goal. The gospel of Jesus Christ is intimately tied to the integrity of the Bible. So Satan, the great deceiver, has used every means he can to lie about the Bible, to vilify it, to falsely demean and disparage it, in order to discredit it in the eyes of a gullible and uneducated world. And the integrity of the Bible is intimately tied to the truth of the Book of Genesis. Indeed Genesis is the *very foundation* of God's Word. Look at the very first verse in Genesis, "In the beginning God created the heaven and the earth." (Genesis 1:1) Satan took aim at this verse with the theory of evolution (as well as with that of the big explosion) and his dupes have been continuing the assault ever since. Destroy the foundation of God's Word through the lies of faux science and you have destroyed the integrity of the Bible in the eyes of this foolish and deceived world. And you have also cut off the same world from their only hope of salvation, the glorious gospel of Jesus Christ. This is what Satan has achieved through the widespread acceptance of his religion of Naturalism and its theories of the absurd.

Two worldviews are in conflict here. The creationist worldview is based on God's Word, has faith in a Creator-Intelligence outside the universe, believes in Absolutes in morality and truth, and that man is accountable to God. The evolutionist worldview is based on man's word, has faith in unknown internal processes within the universe, believes in relativism in morality and truth, and that we are accountable to ourselves only.[798] The creationist worldview is supported by all of the facts and evidence, the laws of science, and common sense. The evolutionist worldview is scientifically absurd and built on a foundation of lies.

"In recent decades, a grave change has taken place that limits the parameters in which scientific study is allowed. [It is known as "scientific fascism."] The change has not so much happened as it has been foisted upon us. Previously, science was defined as "the search for truth," but now it is nearly always equated with naturalism, the search for a naturalistic answer to all questions, even those ultimate questions of the long-ago past that defy normal explanations. The very possibility of supernatural involvement is denied, excluded by definition."[799]

We have already discussed this error in the first chapter, "Truth," but since that was a while ago it is worth repeating here: The biblical statement, "the Spirit is

---

797   (Pgs. 16-17, *Evidence for Creation: Intelligent Answers for Open Minds*, by Tom DeRosa)
798   (Pg. 14, *Evidence for Creation: Intelligent Answers for Open Minds*, by Tom DeRosa)
799   (Pg. 6, *The Young Earth*, by John Morris, Ph.D.)

truth" (1John 5:6), implies that truth exists outside of the confines of this physical universe; that truth is transcendent. It was here before the world began; and before man and his lofty intelligence had their beginnings. And if one seeks the truth, it is there where he must find it. To do otherwise—to seek the truth within the confines of this universe of matter alone—is to seek it in vain. It is to look for it, like love, "in all the wrong places." This is the source of one of the major errors in science today, which has turned certain disciplines within it into more of a primitive *religion*; and one that holds sway over far too many scientists. And that is the religion of *Naturalism*, which declares that there is no *meta*-physical truth or reality—meaning that there is no truth or reality that exists *above* or—outside of the physical world. So all truth and all reality must be found within and only within the confines of the physical world. Including the *origins* of this physical world.

"Many evolutionists believe in evolution simply because that is the only concept they have ever been taught. Their mentors, from high school on up, have drilled into them the false notion that only ignorant fundamentalists—flat-earthers—believe in creation, so young evolutionists reject creation thinking without investigation.

[And even though the exact opposite of this accusation is true: only ignorant, uneducated evolutionists put their blind faith in the nonexistent creative power of accidental random chance.]

"They have never heard a credible case for creation, and so they perpetrate the lie that evolution is the only legitimate view. ["For I the Lord your God am a jealous God, visiting the iniquity of the fathers upon the children unto the third and fourth generation of those who hate me" (Exodus 20:5) …and of those who teach that His Genesis is a lie.] This fallacy is furthered by the redefinition of science as *naturalism*, which denies the possibility of creation."[800]

"The existence of God is nowhere defended by Scripture. This fact is taken as being obvious. …Neither is there any doubt as to His sovereign authority over His creation or what our attitude should be toward Him as Creator. He has the right to set the rules. We have the responsibility *to obey and rejoice in His goodness*, or disobey and suffer His judgment.

…"Those who oppose the Creator are opposing the One who is the absolute authority—the One who sets the rules and *keeps them*.

"In the Book of Judges it is stated: "In those days there was no king in Israel; but every man did that which was right in his own eyes" (Judg. 17:6). People today are little different. They want evolution taught as fact and the belief in creation banished because they, too, want to be a law unto themselves. They want to maintain the rebellious nature they have inherited from Adam, and they will *not* accept the authority of the One who, as Creator and law-giver, has the right to tell them exactly what to do. This really is what the creation/ evolution conflict is all about."[801]

And this is also the reason why such an obviously ridiculous theory that has no basis in reality would still be held onto with such a religious fervor. They have made the alternative, accountability to the Creator, unthinkable. They will stop

---

800   (Pg. 6, *The Young Earth: The Real History of the Earth – Past, Present, and Future*, by John Morris, Ph.D.)
801   (Pgs. 80-81, *The LIE: Evolution*, by Ken Ham)

at nothing to remain their own "rulers." No lie, and no distortion, fabrication or ignoring of the evidence is too big. Nothing is more important than their self-delusion of mastery and control over their own lives and destinies. Both truth and science have become as worthless things, or even worse, as obstacles that get in their way. Unlike the real Creator, their god random chance demands nothing but also gives nothing in return for their blind faith except for an eternity of absolute suffering and horror.

And like any good proselyte of any false religion, evolutionary fascists respond with closed-mindedness, intolerance and even rage to any disagreement with, or opposition to, their beliefs, no matter how rational or based on all of the facts and evidence that opposition may be. In his book, *The Devils Delusion*, David Berlinski describes one particularly fanatical zealot, Hector Avalos, "a professor of religious studies at Iowa State University, and an avowed atheist ...[who is] much occupied in denouncing theories of intelligent design..."

> "He is a member in good standing of the worldwide fraternity of academics who are professionally occupied in sniffing the underwear of their colleagues for signs of ideological deviance."[802]

Still other academics, which have little inclination for the sniffing, are still resigned to accept their situation quietly, meekly and yet safely, as demonstrated by this comment from a colleague of Berlinski's... ""Darwin?" a Nobel laureate in biology once remarked to me over his bifocals. "That's just the party line.""[803] And Ben Stein, in the movie *Expelled: No Intelligence Allowed*, had a field day with these underwear-sniffing, modern-day McCarthyites and their relentless witch-hunt of those scientists and professors who dare stray from the evolutionary party line. ("Sieg Heil!") Of course the reason it's the party line, and you must stick to it or else, is that, unlike scientists, religious fanatics have no interest in entertaining conflicting views or evidence. They only seek to defend their false doctrines from any exposure, and at all cost.

> "Teachers, professors, and educators are gagged and told not to utter the words **God, Creator, Creation** and even **Intelligent Design**. When such words are spoken, or mere inferences are posed about alternatives to evolution in a classroom, sirens and alarms go off, alerting "gravediggers" like the ACLU, the National Academy of Sciences, the American Association for the Advancement of Science, and many others. Armed with their legal clubs and shovels, they rush to silence such utterances, hoping to bury them forever. If this sounds like something out of a police state, you're right. Unfortunately, this is happening in America." [In *Nazi* America.]

> "The goal of the self-appointed censors mentioned above is to turn the U.S. into a secular state, where belief in God is outlawed from the public square and denied all government support. One of their tactics is to suppress all truth about the existence of a Creator in connection with science in our nation's schools. God is to be completely censored out and replaced with naturalistic evolution. These "elites" use the fallacious cry of "separation of church and state"-- which cannot be found anywhere in our Constitution, but is actually part of the former Russian Constitution (1936)-- used to silence all critics."[804]

Dosvidanya, Comrades. The only difference between the totalitarian

---

802   (Pg. 52, *The Devil's Delusion: Atheism and its Scientific Pretensions*, by David Berlinski)

803   (Pg. 192, *The Devil's Delusion: Atheism and its Scientific Pretensions*, by David Berlinski)

804   (Tom De Rosa, *Creations Studies Institute*, in a March 2009 Newsletter)

Democrat-Left in America and the leaders of outright communist countries like North Korea, Cuba and Vermont, is that here their power is still limited by our Constitution. But at the rate they are shredding it, it is only a matter of time before they succeed in getting that obstacle out of their way as well.

But there is hope. Because just as Genesis is the foundation of God's Word, from which we can know the truth of Christianity and the saving grace of Jesus Christ; so also is Satan's genesis, the theory of evolution, the foundation of his godless Democrat-Left and all of their America-hating lies and baby-mutilating iniquity. Destroy this theory and you destroy the foundation of the Left-Elites and their godless, immoral agenda for this country. There are so many conservative and Christian commentators and authors and pundits and leaders and politicians who are fully engaged in the battle against the Left, its immorality and iniquity, and its destructive Marxist ideology, policies and agenda, but many of those are clueless when it comes to the most effective weapon that they have in this war. And that is because as much as they appear to know about the truth on so many important subjects like proper education; our national defense; the role of the judiciary; insane left-wing judges and their incredibly stupid rulings; the war against Islamic lunatics; our precious freedoms and our economic prosperity; and as much as they expose many of the clueless policies of the Left who have no idea how to intelligently steer the course of this country;[805] they still apparently remain ignorant and uneducated on the very fundamental lie that props up the entire ideology and backs up the entire agenda of the secular-progressive Democrat-Left Elites. And if they do not become educated enough to expose and overturn this most foundational of the Left's lies, then they fight this war in vain. They blindly flail away, beating nothing but the air, trying to expose the left and educate the next generation of voters; and then they watch as another wave of reinforcements— Democrat-Left, evolution-indoctrinated youth—rise up against them. They nobly try to shovel the sewage out of our national basement, while at the same time are uneducated enough to shut off the damn valve, evolution, that is letting it all gush in. And that is because they too believe what they were taught; and unfortunately, like their adversaries on the Left, they too have been taught the scientifically-impossible, idiotic, ridiculous, asinine and bald-faced lie that is the theory of evolution. Wake up, my brothers, before the stinking tide overflows our basement, and the war is lost.

In Chapter 2, "Moral Absolutes," we saw how the godless Democrat-Left Elites who have taken over our lame-stream media, our educational system, our courts and our entertainment industry are destroying the decency and the moral underpinnings of our culture. And in chapter 13, Satan, we will see how they are destroying America herself. But here we have exposed the *foundation* of the totalitarian-left's godless philosophy. And just as Satan has sought to destroy Christianity and the gospel of Christ by undermining the Bibles scientifically accurate Genesis account of the origin of the universe and all of life; so too can we turn the tables on him (if we have a brain); so too can we destroy the godless ideology and the socialist, America-destroying agenda of the Democrat-Left by exposing the lie that is the theory of evolution. Destroy evolution, and you will defeat the Left. It is the quicksand foundation on which they stand. If you want to "Take Back America" this is what you must do. Are you listening, Christians? Conservatives? Tea Partiers? O'Reilly? Beck? Hello? Anybody out there?

---

805   (See chapters 2 and 13)

In New Guinea, in…

    … "one cannibal tribe …men would race into a village, grab a man by the hair, pull him back, tense his abdominal muscles, use a bamboo knife to slit open his abdomen, pull out his intestines, cut up his fingers, and while he was still alive, eat him until he died. People hear that and say, "Oh, what primitive savages!" They are not "primitive" savages; their ancestor was a man called Noah. … Noah had the knowledge of God and could build ships. His ancestors could make musical instruments and they practiced agriculture. What happened to those New Guinea natives is that, somewhere in history (as Rom. 1 tells us), they rejected the knowledge of God and His laws. And God turned them over to foolish, perverse, and degenerate things.

    "However, this same degeneracy (this same rejection of God's laws) can be seen in so-called civilized nations that cut people up alive all year long (one and a half million of them in the United States each year), and it is legalized. This is what abortion is—cutting up people alive and sucking out the bits and pieces. The so-called "primitive tribes" had ancestors who once knew the true God and His laws. As they rejected the true God of creation, their culture degenerated in every area. The more our so-called "civilized nations" reject the God of creation, the more they will degenerate to a "primitive culture.""[806]

And this is the same thing that is happening in our nation; the Democrat-Left and their father's theory of evolution are "progressing" us into a primitive culture that is controlled by intellectual savages. They too have "rejected the knowledge of God and His laws." They deny Him as Creator, and instead worship someone they can more easily relate to, the clueless imbecile, Random Accidental Chance. (With all due respect, of course.)

Darwin's Deadly Fruit

    "For a good tree does not bring forth corrupt fruit, neither does a corrupt tree bring forth good fruit. For every tree is known by its own fruit. For men do not gather figs from thorns, nor do they gather grapes from a bramble bush." (Luke 6:43-44)

    The fruit of Darwin's godless and unscientific theory has poisoned this world in the last century in ways that he could never have envisioned, and certainly never intended. But lies have consequences. The hidden demonic nature and evil intent of the theory of evolution can not only be seen in its attack on the truth of Genesis, Christianity, the gospel of Jesus Christ and science itself; and on its destruction of the moral basis of our society by helping to create and then empower the liberal, progressive and godless Democrat-Left; but it is also made evident in a number of cases of mass genocide in the twentieth century that it bares direct responsibility for.

    Most people do not know that the full title of Darwin's book ends with this prophetic little phrase, "the preservation of favored races in the struggle for life." And, oh my, what a toll that *struggle* has exacted on the **un**-favored human races in the last hundred years.

    "Darwin's theory is an idea that wicked men have used as a rationale for mass murder; it is also an instrument of spiritual death for millions who embrace its lie."[807]

---

806   (Pg. 84, *The LIE: Evolution*, by Ken Ham)
807   (Pg. 190, *Evolution's Fatal Fruit: How Darwin's Tree of Life Brought Death to Millions*, by Tom DeRosa)

"The two most notorious and blood-soaked political movements of the twentieth century, Nazism and Communism, both rejected God and were animated by the idea of evolution.

"It was Darwin's theory—carried to its logical conclusion—that led to the death of some 11 million people at the hands of German Nazis. Darwin's notebook jottings became fodder for Hitler's guns. The German leader was a devout evolutionist. [And not even remotely a "Christian" like the historical revisionists on the Democrat-Left like to claim.] He was determined to create a super race by eliminating so-called inferior races. Hitler wrote in *Mein Kampf*: "The stronger must dominate and not mate with the weaker, which would signify the sacrifice of its own higher nature."

"Hitler tried to speed up evolution—to help it along. "The German Fuhrer, as I have consistently maintained, is an evolutionist," British evolutionist Sir Arthur Keith wrote in the 1940s. "He has consciously sought to make the practice of Germany conform to the theory of evolution." And millions suffered and died in unspeakable manners because of it.

"Likewise, Karl Marx, the founder of communism, found in evolution exactly what he needed: a pseudo-scientific foundation for his godless worldview. So too did Vladimir Lenin and communist henchman Josef Stalin.

"Combined, these communist leaders and others killed more people than all those killed in all religious wars. The "rough estimate" from the *Black book of Communism* is that the "total approaches 100 million people killed." Stalin, Mao, Pol Pot, and all the rest are the greatest mass murderers of all time—all complements of evolution."[808]

"There are those, usually at the universities, that have zealously dogmatized the world with the canons of evolutionary thinking. They have become the priests in their own religion and work with an evangelistic energy to spread the faith of evolution. **Yet this lie has had terrible practical consequences—among them abortion, Nazism, communism and different forms of socialism.**"[809] [Emphasis mine.]

...Hmm. Like the *different form of socialism* being shoved down the throats of Americans today (2009) by the Obama administration and his Democrat congress.

"In the limitation of this living space lies the compulsion for the struggle for survival, and the struggle for survival, in turn, contains the precondition for evolution." (Adolf Hitler)

And then we have that visionary hero of so many American university professors, Karl Marx, who wrote this in a letter to Ferdinand Lassalle:

"Darwin's book is very important, and serves me as a basis in natural science for the class struggle."[810]

Karl's *class struggle* (a variation of which our hapless president Obama is trying to orchestrate today—April 2012—for his reelection purposes of course) ultimately resulted in the slaughter of a hundred million human beings.

In addition to being the ideological inspiration for some of the world's greatest

---

808 (From *Impact*, November 2009, a circular from Coral Ridge Ministries)
809 (Pgs. 69-70, *Evidence for Creation: Intelligent Answers for Open Minds*, by Tom DeRosa)
810 (Quote from *The Mysterious Islands: A Surprising Journey to Darwin's Eden*, a film by the Erwin Brothers)

genocidal maniacs, Darwin's theory also has on its resume being the rationale behind modern-day racism:

> "Arguments for racism may have been common before 1859, but they increased by orders of magnitude following the acceptance of evolutionary theory."[811] So says leading evolutionist, Stephen Jay Gould.

In his book, *The Descent of Man*, Darwin himself had this to say, "at some future period, not very distant as measured by centuries, the civilized races of man will almost certainly exterminate, and replace, the savage races throughout the world." Those "savage races" he was referring to included, among others, the aborigines and the blacks. What the racist Adolf Hitler visited upon the Jews is common knowledge to everyone except for a number of half-witted Muslims (like Iran's Ahmadinejad) and other holocaust deniers. But what is not common knowledge, nor is even whispered in our educational system or by our left-wing media, is that it was the theory of evolution that gave him the "scientific" basis, as well as the "moral" imperative, to carry out his horrific crime against humanity.

And then there is the founder of Planned Parenthood, Margaret Sanger, another devout Darwinist who we will deal with in more depth in the following chapter. Like Hitler, she was committed to the idea of improving the human race by helping evolution along, in her case by eliminating the Negro. In a December 19th, 1939 letter to Dr. Clarence Gamble she said this, "We don't want the word to get out that we want to exterminate the Negro population."[812] And keep in mind that she definitely *wanted* to exterminate the Negro population; she just didn't want the word about it to get out. (And where are most Planned Parenthood clinics located in this country today? You guessed it, poor black neighborhoods.) (And who do 90% of blacks go out and vote for religiously every two years? You guessed it, Planned Parenthood and the Democrat Party, and therefore the genocide of their own race. Ouch! Wake up, my brothers and sisters! And see the next chapter.)

PETA

Another of the unintended consequences of the acceptance of Darwinism throughout our society—specifically the belief that man is a mere animal, no different than any other animal or creature along the "evolutionary" ladder—is that it opened the sanatorium doors and let out a uniquely-modern lunatic fringe. Evolution serves...

> ..."as the philosophical foundation for the animal rights movement. After all, if man is simply another evolving animal, what gives him rights superior to bugs, and birds, and bats? This is the point frequently made by the leaders of People for the Ethical Treatment of Animals [PETA], perhaps the most visible and outspoken element of the animal rights community."[813]

PETA president, Ingrid Newkirk, actually thinks that the holocaust should be equated with the horror of eating fried chicken...

---

811 (From *Ontology and Phylogeny*, 1977, by leading evolutionist Stephen Jay Gould. As quoted in *The Mysterious Islands: A Surprising Journey to Darwin's Eden*, an Erwin Brothers film)

812 (Quoted from *The Mysterious Islands: A Surprising Journey to Darwin's Eden*, an Erwin Brothers Film)

813 (From *The Mysterious Islands* video featurette, "Galapagos Whaling Controversy: A Christian Perspective," by Doug Phillips)

"Six million Jews died in concentration camps, but six billion broiler chickens will die this year in slaughterhouses."[814]

(Lucky for Frank Purdue the Nuremburg trials are over.)

And that humans and rats are one and the same...

"There is no rational basis for saying that a human being has special rights. A rat is a pig is a dog is a boy. They're all mammals."[815]

Ms. Newkirk is a rat is a pig is a dog is a flipping idiot. But that is the result of being brainwashed in the evolutionary lies of the wicked one. God help her.

And then there is this gem, from another animal rights lunatic...er... advocate...

"The life of an ant and the life of my child should be granted equal consideration."[816]

And with that said, social services should remove his child immediately. (Give him an ant farm instead.)

Conversely, this is what God says, the actual Creator of life...

"And God said, Let us make man in our image, according to our likeness: and let them have dominion over the fish of the sea, and over the birds of the air, and over the cattle, and over all the earth, and over every creature that moves upon the earth. So God created man in his own image, in the image of God he created him; male and female he created them. And God blessed them, and God said unto them, Be fruitful, and multiply, and replenish the earth, and subdue it: and <u>have dominion over</u> the fish of the sea, and over the birds of the air, and over every living thing that moves upon the earth." (Genesis 1:26-28)

"What is man that you are mindful of him, and the son of man, that you care for him? For you have made him a little lower than the angels, and have crowned him with glory and honor. You have made him to <u>have dominion over</u> the works of your hands; you have put all things under his feet: all sheep and oxen, and the beasts of the field, the birds of the air, and the fish of the sea, and whatsoever passes through the paths of the seas." (Psalms 8:4-8)

"From these two passages we learn two important points: first, that man alone was made in the image of God. And second, that man was given dominion over all the creatures. ...[Having dominion] doesn't mean that man can abuse or exploit creation as he pleases. Nor should we demonstrate cruelty to animals. ... 'The righteous man cares for the life of the beast.' (Proverbs 12:10) ...We ought to be wise stewards of the rich resources we have been given, and with thankful hearts use them for the good of mankind.

"We should also recognize that the notion that it's criminal and cruel to harvest animals because they are big, or cute, or fluffy, or beautiful is a neo-Darwinian philosophy born out of self-loathing for humanity. ...It's not accidental that many of the most vocal advocates for animal rights are also defenders of the rights of women and physicians to kill human babies within the womb... Save the whales. Kill the babies. It makes perfect sense in a Darwinian world. And it's this type of perverse perspective that inspires leaders of this movement to compare the genocide of Jews by Adolph Hitler to the so-called sin of eating fried chicken."[817]

---

814   (Ingrid Newkirk, President and Founder, PETA)
815   (Ingrid Newkirk, President and Founder, PETA, People for the Early Treatment of Asininity)
816   (Michael Fox, Vice President, The Inane Society of the United States. Oops, <u>Humane</u> Society, that should have read "Humane." Another typo.)
817   (From *The Mysterious Islands* video featurette, "Galapagos Whaling Controversy: A Christian Perspective," by Doug Phillips)

PETA and our animal "rights" friends have passionately and self-righteously immersed themselves and their lives in a crusade that is all built upon a lie; the idiotic fantasy of evolution. What a waste of a lifetime. They need to immerse themselves in the truth, and the real righteousness that can only come through faith in Jesus Christ, life's real Creator. Pray for them. And pray that the bald-faced, scientifically-impossible lie that enslaves their minds and their hearts will soon be exposed nationally and internationally for what it is.

"Where were you, stupid, when I laid the foundations of the earth? Tell me, if you have understanding. …On what are its foundations fastened? Or who laid its cornerstone; when the morning stars sang together, and all the sons of God shouted for joy? …When I made the clouds its garment, and thick darkness its swaddling band" (Job 38:4, 6-7, 9)

(I know, "stupid" isn't really in there. I just threw it in for comic relief.)

"Theistic" Evolution

Evolution's bastard child. We need to mention it here, briefly, because sadly many millions of uninformed Christians (usually of the faux variety) have been bamboozled into accepting this farcical amalgamation that attempts to blend belief in God and His Word with the "scientific fact" (read: lie) of evolution. It's a fool's bargain, and one that is completely unnecessary for any believer because the only thing the lie of evolution needs to be amalgamated with is the manure pile. Evolution poses no challenge to belief in God, but since so many uninformed Christians remain ignorant of this fact, they have "comforted" themselves by believing in its theistic version. *Theistic evolution* goes something like this, "Well, we *know* that evolution is *science* and therefore it is a *fact*, but we also believe in God, so we believe God helped evolution along the way. It was God who was behind all of the evolving over all of the hundreds of millions of years."

As such, theistic evolution is possibly more stupid than the actual theory itself. It is a pathetic attempt to reconcile the God of Creation with the lie of evolution. It is insulting, and it only exists because so many Christians, and other believers in God, have never been educated as to the scientific impossibility of the bald-faced lie that macroevolution is. God is not an idiot. (Man is another story entirely.) God is all-powerful, and all-intelligent. The idea that he had to muddle around for millions of years trying to change one completely viable and incredibly engineered organism into another, rather than just making them all at once, is patently absurd. I mean, as so-called Christians, who are we going to believe? Brainwashed Darwinists, faux scientists all, who weren't even there, who haven't the slightest idea what they are talking about, or God who actually was there at Creation? (Take your time.) Indeed, theistic evolution was merely created so those Christians who had bought into the scientifically absurd fiction that is evolution, could try to square it with the Word of God. It doesn't work. Either Satan is right (which of course he isn't) or the Bible is absolutely true, and exactly what it says it is thousands of times within its pages and that would be, written by God! And God is not a liar. Evolutionists, however, are. And so too are "theistic" evolutionists. And both are also calling God a liar. Wake up! Learn the truth. If you're going to call yourself a Christian, stop aiding and abetting the enemy. (Something that many on the far-left who call themselves Americans might also learn.)

"All too many Christians have chosen not to see or become involved in the battle around us, in effect surrendering to the enemy, and in so doing, abandoning all those who come under the influence of the enemy.

...."The evidence is especially on our side in real science. Creation far excels evolution as a scientific model. Evolution survives only by suppression of alternatives. The tactics of evolutionists include ridicule, personal attacks, bureaucratic policies, and court rulings. Mostly, evolution survives because so few people have ever been allowed to hear a credible case for creation. All that most people know is what they have been taught."[818]

Either Genesis is scientifically true or it is not. If it is not then the Bible is a lie and Jesus quoted from a lie... and his death on the cross to pay the price for our sins, his resurrection from the dead three days later, and the salvation he offers are all highly suspect. And that is just the way the evil one wants it. As we have previously stated, the theory of evolution is Satan's genesis. It is his origins account. It is his lie. And he is quite content in having foolish, uneducated "Christians" believing in his lie instead of God's accurate account, and to then try and make his lie "compatible" with the Bible by coming up with the utter stupidity known as theistic evolution. God have mercy on us idiots.

Part IV - In the Beginning
"These are the generations of the heavens and of the earth when they were created, in the day that the Lord God made the earth and the heavens." (Genesis 2:4)

The first chapter of the first book of the Bible records the sequence of events that culminated in God's finished work of creation: this universe of stars and galaxies, this solar system with our water-planet earth, and all of life upon it. But not only is it obvious that a great amount of specifics and details have been left out, have been left unsaid in this first chapter of Genesis, but there has also been a great deal of confusion over the meaning of what actually has been said. For example, what exactly do phrases like "the earth was without form and void," and "darkness was upon the face of the deep," and "the Spirit of God was hovering over the face of the waters," actually mean? What was "the deep," and what was "the face of the waters" that "the Spirit of God was hovering over?" Is there any possible way for us to accurately answer these questions, and to come to a clear understanding of what literally happened in those first few days?

I think there is, and that the answer may have come by way of a book by D. Russell Humphreys, a Ph.D. in physics who for years "worked for Sandia National Laboratories (New Mexico) in nuclear physics, geophysics, pulsed-power research, and theoretical atomic and nuclear physics."[819] It is titled, *Starlight and Time: Solving the Puzzle of Distant Starlight in a Young Universe*, and his unveiling of the mysteries of Genesis, using his extensive knowledge of physics and astronomy, can be found in chapter 2, "Creation Week: A Possible Scenario."

818   (Pg. 121, *The Young Earth: The Real History of the Earth – Past, Present, and Future*, by John Morris, Ph.D.)
819   (From back cover of *Starlight and Time: Solving the Puzzle of Distant Starlight in a Young Universe*, by D. Russell Humphreys, Ph.D.)

Dr. Humphreys begins his analysis on Day One (Genesis 1:1-5):

"In the beginning God created the heaven and the earth. And the earth was without form, and void; and darkness was upon the face of the deep. And the Spirit of God was hovering over the face of the waters." (Genesis 1:1-2)

According to Dr. Humphreys calculations, God began by creating both 3 dimensional space ("the heavens") a billion light years in diameter, and a sphere of liquid water which contained all of the matter of the universe ("the earth") over 2 light-years in diameter at the center of this "large 3-D space." It is fitting that the Creator should begin this material world with water, as water is life. "And he said unto me, It is done. I am Alpha and Omega, the Beginning and the End. I will give unto him who thirsts of the fountain of the water of life freely." (Revelation 21:6)

> "The ball [of matter-water] is greater than two light-years in diameter, large enough to contain all the mass of the universe... Two light-years is surprisingly small compared to the later size of the universe, but it is still huge (about 12 trillion miles or 20 trillion kilometers) compared to us, being more than a thousand times greater than the diameter of our solar system. Imagine floating on the face of the deep and gazing down into its unimaginable depths! ...The earth at this point is merely a formless, undefined region of water at the center of the deep, empty of inhabitant or feature. The deep is rotating slowly and there is no visible light at its surface."[820]

What other explanation fits with this description in Genesis of the future, physical earth being "without form, and void," which would have been at that point just an infinitesimal spec in the center of this immense ball of water-matter? And what other explanation fits with the phrase "the deep" other than a massive sphere of water? "The Spirit of God was hovering over the face of the waters" because that was all there was, water. And then there is this in the New Testament, "By the word of God the heavens existed long ago and **the earth was formed out of water and by water**." (2 Peter 3:5) That kind of nails it right there. Dr. Humphreys continues:

> "In the early 1980's, I based a theory about the origin of planetary magnetic fields on the possibility that the earth and other bodies in the solar system were originally created as pure water. The theory has been remarkably successful, even to the point of correctly predicting the Voyager spaceprobe's measurements of the magnetic fields of the planets Uranus and Neptune.

[It's been almost 30 years since then so the New York Times must be getting real close to a decision about whether this is a part of "all the news that's fit to print." It's a big decision after all and they don't want to feel rushed. Not like they did when they immediately printed the lie concerning a McCain affair that never happened, in order to aid in the election of their candidate, Hussein ACORN Obama.]

> "The theory could not work with the present elements composing the solar system bodies, but only with water as the original material. Thus it seems that transformation of water on day one (the modern word is "nucleosynthesis") is a distinct biblical and scientific possibility.

---

820   (Pgs. 32-33, *Starlight and Time: Solving the Puzzle of Distant Starlight in a Young Universe*, by D. Russell Humphreys, Ph.D.)

..."If the "nucleosynthesis" scenario above is correct, then it means that at the instant of creation, the earth was merely a small region of water at the center of a much larger ball of water, the deep. That region had no distinguishing marks and was empty of any other kind of matter. This, I suggest, is the meaning of the much-discussed phrase, "formless and void" [in Genesis 1:1-2]."[821]

"Because the enormous mass of the whole universe is contained in a ball of (relatively) small size, the gravitational force on the deep is very strong, more than a million trillion "g"s. This force compresses the deep very rapidly toward the center, making it extremely hot and dense. The heat rips apart the water molecules, atoms, even the nuclei into elementary particles."[822]

"And God said, Let there be light: and there was light." (Genesis 1:3)

"Thermonuclear fusion reactions begin, forming heavier nuclei from lighter ones and liberating huge amounts of energy. As a consequence, an intense light illuminates the interior, breaking through to the surface and ending the darkness there."[823]

"And God saw the light, that it was good: and God divided the light from the darkness." (Genesis 1:4)

"As the compression continues, gravity becomes so strong that light can no longer reach the surface, re-darkening it. Psalm 104:2, "Covering Thyself with light as with a cloak," in context appears to refer to Day One. This suggests to me that at this point the Spirit of God, "moving [or hovering] over the surface of the waters" (Genesis 1:2), becomes a light source, in the same way as He will again become a light source at a future time (Revelation 21:23, 22:5). This would give the deep a bright side and a dark side, thus dividing light from darkness and inscribing "a circle on the face of the waters, at the boundary of light and darkness." (Job 26:10)."[824]

"And God called the light Day, and the darkness he called Night. And the evening and the morning were the first day." (Genesis 1:5)

"The deep speeds up its rotation as the compression continues, as a whirling ice skater speeds up when she brings her arms inward. We can imagine a reference point on the surface rotating around to the dark side and continuing further around to the bright side again, marking off evening and morning. Rough calculations show that all of the events from the beginning to this point had to take place in a very short time... To calculate the time exactly would go beyond the frontiers of modern relativity, but I suspect that... [it was] about 24 hours from the instant of Creation to the end of Day One."[825]

I would have to agree, mainly because it is abundantly clear from linguistic studies of the Bible that during Creation week God was talking about *literal 24-hour days*, and not some vast eons of time as proposed by theistic evo-delusionists.

Dr. Humphreys continues his analyses of Genesis on Day Two (Genesis 1:6-8):

"And God said, Let there be an expanse [firmament] in the midst of the waters, and let it divide the waters from the waters. And God made the expanse, and divided the waters which were below the expanse from the waters which were

---

821   (Pg. 73, *Starlight and Time,* by D. Russell Humphreys, Ph.D.)
822   (Pg. 33, *Starlight and Time: Solving the Puzzle of Distant Starlight in a Young Universe,* by D. Russell Humphreys, Ph.D.)
823   (Pg. 33, *Starlight and Time,* by D. Russell Humphreys, Ph.D.)
824   (Pgs. 33-34, *Starlight and Time: Solving the Puzzle of Distant Starlight in a Young Universe,* by D. Russell Humphreys, Ph.D.)
825   (Pg. 34, *Starlight and Time,* by D. Russell Humphreys, Ph.D.)

above the expanse: and it was so." (Genesis 1:6-7)

"By direct intervention... God began stretching out space, causing the ball of matter to expand rapidly. [Who covers *thyself* with light as *with* a garment: **Who stretches out the heavens like a curtain**" (Psalms 104:2)] ...He marks off a large volume, the "expanse" ("firmament" in the KJV) within the deep, wherein material is allowed to pull apart into fragments and clusters as it expands.

..."Normal physical processes cause cooling to proceed as rapidly as the expansion. Heat waves are stretched out to much longer wavelength as a relativistic consequence of the stretching of space. Eventually the stretched-out waves will become the cosmic microwave background radiation.

"Matter beneath the expanse expands until the surface reaches ordinary or present temperatures, becoming liquid water underneath an atmosphere. God collects various heavier atoms beneath the surface (formed from fusion reactions as mentioned earlier) and constructs minerals of them, laying "the foundations of the earth" (Job 38:4), i.e., its core and mantle. Gravity at the surface drops to normal or present values. Out in the expanse, matter is drawn apart, leaving irregular clusters of hydrogen, helium, and other atoms formed by the nuclear processes of the first day."[826]

"And God called the expanse Heaven. And the evening and the morning were the second day." (Genesis 1:8)

..."These heavens are interstellar space. Since the sun has not yet been created, the Spirit of God continues to be the light source close to the rotating waters below, giving them a light and dark side. The expansion started at the beginning of this day will continue until at least the end of the fourth day."[827]

Day Three...

"And God said, Let the waters under the heaven be gathered together into one place, and let the dry land appear: and it was so." (Genesis 1:9)

"Rapid radioactive decay occurs... possibly as a consequence of the rapid stretching out of space. The resulting heating forms the earth's crust and makes it buoyant relative to the mantle rock below it, causing the crust to rise above the waters, thus gathering the waters into ocean basins. I hypothesize that rapid volume cooling of molten rock deep within the earth also occurs, again as a result of their rapid expansion of space... solidifying the rock. ...There are no stars yet, only clusters of hydrogen, helium, and other atoms left behind in the expanse by the rapid expansion."[828]

Day Four...

"And God said, Let there be lights in the firmament of the heaven to divide the day from the night; and let them be for signs, and for seasons, and for days, and years: And let them be for lights in the firmament of the heaven to give light upon the earth: and it was so. And God made two great lights; the greater light to rule the day, and the lesser light to rule the night: he made the stars also. And God set them in the firmament of the heaven to give light upon the earth, And to rule over the day and over the night, and to divide the light from the darkness: and God saw that it was good. And the evening and the morning were the fourth day." (Genesis 1:14-19)

"During this ordinary day [the fourth day] as measured on earth, billions of years' worth of physical processes take place in the distant cosmos. [This is due both to the

---

826   (Pgs. 34-36, *Starlight and Time*, by D. Russell Humphreys, Ph.D.)
827   (Pg. 36, *Starlight and Time*, by D. Russell Humphreys, Ph.D.)
828   (Pgs. 36-37, *Starlight and Time*, by D. Russell Humphreys, Ph.D.)

expansion of space and matter, and gravitational time dilation.] In particular, gravity has time to make distinct clusters of hydrogen and helium atoms more compact. Early on the fourth morning, God coalesces the clusters of atoms into stars and thermonuclear fusion ignites in them. The newly-formed stars find themselves grouped together in galaxies and clusters of galaxies. As the fourth day proceeds on earth, the more distant stars age billions of years, while their light also has the same billions of years to travel to the earth. While the light is on its way, space continues to expand, relativistically stretching out the light waves ...and shifting the wavelengths toward the red side of the spectrum. Stars which are now farthest away have the greatest red shift, because the waves have been stretched the most. This progressive redshift is exactly what is observed."[829]

We will skip to Day Six...

"And God saw every thing that he had made, and behold, it was very good. And there was evening and there was morning, the sixth day." (Genesis 1:31)

"God stops the expansion... before the evening of the sixth day. Therefore, Adam and Eve, gazing up for the first time into the new night sky, can now see the Milky Way, the Andromeda galaxy, and all the other splendors in the heavens that declare the glory of God."[830]

Wow! "Lord of all creation, of water earth and sky. The heavens are your Tabernacle. Glory to the Lord on high. God of wonders beyond our galaxy. You are holy, holy. The universe declares your Majesty. You are holy, holy. Lord of Heaven and Earth."[831]

He truly is a "God of Wonders!" The "Lord of all creation, of water earth and sky." How awe inspiring it is to catch a glimpse of how it *really* all began! Not from some mindless, stupid, accidental explosion, ill-conceived by those who ignorantly reject God and any consideration of His creative work. Who "professing themselves to be wise, became fools. ...Who changed the truth of God into a lie, and worshipped and served the creature more than the Creator." (Romans 1:22, 25) An explosion that couldn't create a damn thing, much less this universe of incredible beauty, exact physical laws and precisely-tuned forces, all spoken into existence by the holy "Lord of Heaven and Earth" who possesses an intelligence that stands in infinite contrast to those *scientists* and *intellectual elites* who refuse to acknowledge him as their Creator. And what could be more perfect then for He who *is* the Water of Life to have begun his universe by creating water? "My people have committed two evils; they have forsaken me the fountain of living waters, and hewn them out cisterns, broken cisterns that can hold no water." (Jeremiah 2:13) We as a nation and a people have forsaken our own Creator, and worship random chance, a "broken cistern," in his stead.

And then he stretched the whole thing out, and changed the water into all of the known and unknown elements, and made stars, countless numbers of them and formed them into billions of galaxies. The universe truly declares His majesty! This "God of wonders beyond our galaxy," beyond all galaxies, beyond all time and space. And at the center of it all He made this tiny solar system with this

---

829   (Pgs. 37-38, *Starlight and Time: Solving the Puzzle of Distant Starlight in a Young Universe*, by D. Russell Humphreys, Ph.D.)

830   (Pg. 38, *Starlight and Time*, by D. Russell Humphreys, Ph.D.)

831   (From song *God of Wonders*, by Third Day. ...Check out the video on YouTube)

little water-planet-earth and orbiting-moon system uniquely designed to sustain and nurture life in the midst of the cold dark expanse of outer space; a single planet made up of an infinitesimal amount of the universes matter, yet of more significance than all the rest of the universe combined; where he would breathe all the living into being and crown this planet teeming with all variety and splendor of life with man, made in His image and His likeness. "Early in the morning, I will celebrate the Light,"[832] because he made us from the light of His own Conscious Spirit, to serve Him and glorify Him, and to have a part in letting "the earth be filled with the knowledge of the Glory of the Lord, as the waters cover the sea." And behold He saw that it *was* very good.

"Thus says the Lord, who stretches out the heavens, lays the foundation of the earth, and forms the spirit of man within him..." (Zechariah 12:1)

How certain can we be, though, that Dr. Humphreys unveiling of Creation Week is completely accurate? Roughly 85.67 %. (And I did that without a calculator.) Seriously though, this is just one scientist, a brilliant one to be sure, and one unfettered by the shackles of an absurd, idiotic theory that leads its followers on a wild goose chase looking for origins in all the wrong places. And this is that one scientist's interpretation of Genesis. It is his version of the specifics of creation week not spelled out in God's Word, using the known laws of the universe and those parts of the general theory of relativity that appear to be accurate. It is not "written in stone." But God's version *is* written in Stone. It is written in stone by the Chief Cornerstone, and it accurately describes what happened in the beginning. But at the same time much is obviously left unsaid, and because of that much has also been left wrapped up in mystery for thousands of years. A mystery that has defied an accurate and scientifically sound interpretation, perhaps until now. Whether Dr. Humphreys has absolutely accurately unveiled that mystery, I can't say. You decide. (I would think, though, that even at the worst, he has come very close.) But until some non-deluded, non-faux scientist comes up with a better and more accurate scientific interpretation, it works for me. Dr. Hawking, are you up to the challenge? Can you leave the scientific absurdity of big bang cosmology alone long enough to search out the truth? The world of true science awaits your decision; with bated breath.

"And, you, O Lord, in the beginning have laid the foundations of the earth; and the heavens are the work of your hands." (Hebrews 1:10)

Conclusion

"God, who at various times and in sundry manners spoke in time past unto the fathers by the prophets, has in these last days spoken unto us by his Son, whom he has appointed heir of all things, by whom also he made the worlds. Who being the brightness of his glory, and the express image of his person, and upholding all things by the word of his power, when he had by himself purged our sins, sat down on the right hand of the Majesty on high." (Hebrews 1:1-3)

Science is based upon observation. No one has ever observed macro evolution, nor will they. But there is someone who observed Creation. He was the world's first Scientist, and science can know that His witness is true. It would be wise for

---

832   (From song *God of Wonders*, by Third Day)

America to wake up to this reality.

The theory of evolution's preeminence in our educational system has everything to do with the Left's fascist control over that system as we have already stated, but the *roots* of evolution's total preeminence can be traced back to the height of the cold war, and to the earliest years of the "space race:"

"The USSR's ability to place a satellite in orbit raised a cry of alarm about the "Space Gap" between superpowers—how had we let American resolve become so flabby? One result was a federal imperative to improve high-school science education—which led, not incidentally, to evolution instruction being mandated across the board."[833]

Well, we made it to the moon, with no thanks to the "across the board mandating" of the teaching of a scientifically-impossible lie, and the Russians still haven't. But before we start overly congratulating ourselves, the fact remains that we are too scientifically ignorant and uneducated to mandate teaching our school children the truth about how the moon *got* there in the first place. The *space race* is over, but the question that still remains is, *Why are we so stupid as to continue to allow bald-faced, scientifically-impossible lies to be taught, under the guise of science, and as if they were fact, to our naïve and unsuspecting school children in our colleges, universities, elementary and high schools?*

And the answer of course is that the lies of Satan cover this earth like a noxious primordial gas cloud. (And did we mention before that people believe what they are taught?) It is so easy for us here in America to look at all of those poor, unfortunate Muslim jihadists who actually thought they were going to go to an eternal paradise replete with 72 virgins for their unending pleasure and use, by flying a plane into a tall building and killing thousands of innocent men, women and children. And we cry out, "HOW COULD ANYBODY BE SO STUPID?" Yet we should be shouting that into a mirror, because the main reason that many of us don't believe that peculiarly Islamic idiocy is because we weren't taught it. But the bald-faced, just-as-idiotic, evolutionary lies we have been taught, those we do believe. (And for the record, I am not pointing the finger because I believed them too, before I was brought from the lies of the devil to the knowledge of the truth. ..."He rescued me from the power of darkness, and transferred me into the kingdom of his dear Son." Colossians 1:13-14) (Hallelujah!)

When we look at the stunning miracle that is life—the computing marvel that is the human brain with its hundred trillion electrical connections; the sixty thousand miles of blood vessels that course through our bodies; the astonishing imagination that designed the metamorphosis of insects and the resulting breathtaking beauty of many of its end results such as the butterfly; the creativity and unfathomable intelligence that could place an entire library of coded information in a seven-foot-long strand of DNA and somehow coil it up and put it into a living cell a billionth of a cubic inch in size; the mind-boggling, nanobot, engineering intricacy of the living cell that functions and reproduces for years without any conscious intervention—it is all so absolutely overwhelming. It is almost beyond belief that it all even exists. Yet He merely spoke, and it all came into being. By the breath of his mouth. "By the word of the Lord were the heavens made, and all the host of them by the breath of his mouth" (Psalm 33: 6) And yet the sad thing is that instead of glorifying and honoring and praising and thanking the God of Wonders, the Lord of all Creation, for bestowing this incomprehensible miracle gift upon us,

---

833   (From *One Large Misstep*, by Janie B. Cheaney, article in World Magazine, Aug. 2009)

we spit in his face by giving all the credit to an idiot, random accidental chance. Again, may God have mercy on us morons.

"You are worthy, O Lord, to receive glory and honor and power; for you have created all things, and for your pleasure they exist and were created." (Revelation 4:11)

Let us give glory to the God of Creation, the author of the Bible, who inspired Moses to record in the Book of Genesis the scientifically accurate history of the creation of the world. Who, unlike our friend Beelzebub who has inspired a competing genesis account, is not a liar. Who spoke all the universe into existence during a six-day period six thousand years ago; and who breathed all the living into being during the same week; and who then "rested the seventh day, wherefore that same Creator Lord blessed the Sabbath day, and hallowed it." (Exodus 20:11) And that, unlike evolution, is a scientific fact!

"Lift up your eyes on high, and behold who has created these things" (Isaiah 40:26)

"He was in the world, and the world was made by him, and the world knew him not." (John 1:10) Because the world did not care to know Him, nor to be bothered with His truth. For if it did care to know him and the truth, both would be very easy for them to find. As we said earlier, the universe is not the one keeping certain very important things secret from us. We are. We just refuse to hear the truth, so certain things remain secret, like the scientific fact that evolution really is a joke.

"So I say unto you, ask, and it shall be given you; seek, and you shall find; knock, and it shall be opened unto you." (Luke 11:9)

"And you shall seek me, and find me, when you shall search for me with all your heart."(Jeremiah 29:13)

"And you shall love the Lord your God with all your heart, and with all your soul, and with all your mind, and with all your strength." (Mark 12:30) And the "Lord your God" and the Truth are one and the same… "I am the truth." (John 14:6)

"Fear God and give glory to him, for the hour of his judgment has come; and worship him who made heaven and earth, the sea and the fountains of water." (Revelation 14:7)

"But sanctify the Lord God in your hearts: and be ready always to give an answer to every man who asks you a reason for the hope that is in you with meekness and fear." (1 Peter 3:15)

Now we know the truth about the evolution-creation debate. There is none. All that remains is a battle between the myths of the great deceiver and the truth of God our Creator. And in many ways that truth is extremely privileged information, and you and I are extraordinarily blessed that it has been revealed to us. It is another incredible "secret of the universe;" and *secret* only because it has been systematically and deliberately kept hidden from the majority of people today by the dragon's sly influence over the means of disseminating information in our culture. There are countless multitudes of people who have not received this blessing. And you and I both will be required to answer for what has been freely given us. ("For unto whomsoever much is given, much shall be required." Luke 12:48) Will we hide our light under a bushel, or will we shine it on a hill? (See Matthew 5:15). The light of the truth of the glorious gospel of Jesus Christ has been hid from this world in many ways because of the darkness bestowed upon it by Satan's genesis. It is therefore incumbent upon anyone who has been blessed with the knowledge of this truth to pass it along to as many as possible. To this end, whether you are young or old, teacher or student, in a position of great power and influence or of humble means, may God give you the courage and the boldness to do so.

"I can do all things through Christ who strengthens me." (Philippians 4:13)

"For by him were all things created that are in heaven and that are on earth, visible and invisible, whether thrones or dominions or principalities or powers. All things were created by him and for him." (Colossians 1:16)

Amen.

See:

-*The Young Earth: The Real History of the Earth - Past, Present, and Future*, by John Morris

- *Dinosaurs, The Lost World, & You*, by John D. Morris, Ph.D.

-*The Collapse of Evolution*, by Scott M. Huse

-*In The Beginning: Compelling Evidence for Creation and the Flood*, by Walt Brown, Ph.D.

-*Evidence for Creation: Intelligent Answers for Open Minds*, by Tom DeRosa

- *The Mysterious Islands: A Surprising Journey to Darwin's Eden*, An Erwin Brothers Film

-*Dinosaurs Unleashed: The True Story about Dinosaurs and Humans*, by Kyle Butt and Eric Lyons

- *The Edge of Evolution: The Search for the Limits of Darwinism*, by Dr. Michael J. Behe

- *The Devil's Delusion: Atheism and its Scientific Pretensions*, by David Berlinski

- *Evolution: A Theory in Crisis*, by Michael Denton

- *Darwin's Black Box: The biochemical Challenge to Evolution*, by Michael J. Behe

- *Expelled: No Intelligence Allowed*, a documentary movie narrated by Ben Stein

-*Starlight and time: Solving the Puzzle of Distant Starlight in a Young Universe*, by D. Russell Humphreys, Ph.D.

- *The Greatest Hoax on Earth? Refuting Dawkins on Evolution*, by Dr. Jonathan Sarfati

-*Thousands... Not Billions: Challenging an Icon of Evolution- Questioning the Age of the Earth*, by Dr. Don DeYoung

- *The Lie: Evolution*, by Ken Ham

-*Evolution's Fatal Fruit: How Darwin's Tree of Life Brought Death to Millions*, by Tom DeRosa

-*Taking Back Astronomy: The Heavens Declare Creation*, by Dr. Jason Lisle

-*Grand Canyon: a different view*, written and compiled by canyon river guide, Tom Vail

- *Ice Age Civilizations*, by James I. Nienhuis

- *The Genesis Factor: Myths and Realities*, by Ron J. Bigalke Jr. Compilation Editor

- *Bones of Contention: A Creationist Assessment of Human Fossils*, by Marvin

L. Lubenow

- *Darwin on trial*, by Phillip E. Johnson (Also see: *Darwin Found Guilty, Darwin Sentenced, Darwin Executed* and, the final one in the series, *Darwin Buried in Bill Maher's Back Yard*, by I. M. Sighents, Ph.D.)

- *The Privileged Planet: How Our Place in the Cosmos is Designed for Discovery*, by Guillermo Gonzalez and Jay W. Richards

- *In the Beginning was Information: A Scientist Explains the Incredible Design in Nature*, by Dr. Werner Gitt

- *The Cells Design: How Chemistry Reveals the Creator's Artistry*, by Fazale Rana, Ph.D.

CHAPTER 11

BABY MUTILATION

"Then the word of the LORD came unto me, saying, Before I formed you in the womb I knew you; and before you came forth and were born I sanctified you." (Jeremiah 1:4-5)

Before we get into this subject we want to make it very clear from the outset that the point of this chapter is <u>not</u> to in any way condemn, demean, judge, put down, or disparage any girl or woman who has suffered through having an abortion. They have already been beaten up enough just by what they have gone through, and they don't need any further condemnation. They need the love of Christ which surpasses all knowledge, can heal any wound and forgive any sin. (Paul murdered Christians before his conversion, for goodness sake.) Indeed we are all sinners, some of us saved by grace, and repentance and forgiveness (in that order) is open to all. In addition, the only time we will use the term *abortion* is when talking about those who have suffered through one, out of empathy for what they have experienced. For indeed, the very term "abortion" is the lie of the enemy, Satan, and it is meant to whitewash and "put a happy face" on the horror that is actually occurring. No one is flying a jet in too low for a carrier landing: "Abort! Abort!" A baby is being mutilated to death while it is still alive. His arms are being ripped out of his shoulders and her eyes are being sucked out of their sockets; scissors are being jammed into the back of their skulls and their brains are being sucked out of their head while they silently scream to death. Also, many of those who have had an abortion are indeed victims themselves (along with their destroyed babies) of the lies constantly spewed by the fascist Democrat-Left that controls our educational system, media and entertainment industry: "It's your *right!*" "Why mess up your life?" "It's just a simple procedure." "It's your *choice.*" "It's not a baby anyhow." "It's just a lump of tissue." "It's only a *fish embryo* right now."

[Remember in the last chapter that we spoke of the fraud perpetrated by Ernest Haeckel; how he fabricated drawings to *show* that as an embryo develops in the womb it passes through previous *evolutionary stages*. Well, Planned Parenthood found a "silver lining" in that hoax...

"Some abortion clinics in America have taken women aside to explain to them that what is being aborted is just an embryo in the fish stage of evolution, and that the embryo must not be thought of as human. These women are being fed outright lies."[834]]

These girls should have been told the truth so they could have at least made a truly informed decision. Instead they were lied to, deceived, misled and purposely kept in the dark as to the horrible reality of what baby-mutilation actually is: mutilating a live baby to death. (Coincidentally.) When women are given the opportunity to hear the truth; by seeing sonograms of their infant; by hearing the truth about how much of an actual baby that "lump of tissue" really is; by seeing the real disgusting horror-show that mutilating him or her to death actually is by being shown videos of the crime or by going to websites like durarealidad.com

---

834   (Pg. 105, *The LIE: Evolution*, by Ken Ham)

("hard truth"); oftentimes they decide against going through with it. Which is why when they go into a Planned "Parent" hood clinic they are purposely kept in the dark: too many greenbacks at stake. What we do condemn, judge and denounce therefore, is the Prince of Darkness and his followers in high places in our media, our educational system and our entertainment industry, who perpetrate these lies and the resulting horror, as well as politically and financially benefiting from them, because so many uninformed idiots are duped into continuing to go out and vote for them. (With all due respect, of course.)

Miracles...

"As you know not what is the way of the spirit, nor how the bones do grow in the womb of her who is with child: even so you know not the works of God who creates all." (Ecclesiastes 11:5)

...and abominations

"These six things does the Lord hate, yes, seven are an <u>abomination</u> unto him: A proud look, a lying tongue, and <u>hands that shed innocent blood</u>, a heart that devises wicked imaginations, feet that are swift in running to evil, a false witness who speaks lies, and he who sows discord among brethren." (Proverbs 6:16-19)

> "A pre-born life is more than a clump of cells or a by-product of conception. It's a newly conceived human being whose life is sacred and inviolable at every moment. From conception until its natural end. It's a living growing person whose DNA is infused with potential, with talent, with love, with life. Whose tiny heart begins to beat by day 25. Who practices inhaling and exhaling by day 90. Whose delicate frame can kick, twist, flail, grasp, squint, frown, grimace, suck a thumb, by day 105. Every human life, whatever the stage or condition, is endowed by our Creator with certain unalienable rights and dignity. We cannot diminish the value of one category of human life -the unborn- without diminishing the value of all human life. A pre-born life is an impression of God's own image and likeness, a sign of His Presence, a reflection of His Glory. And, thus, should be entitled to life, liberty and the pursuit of happiness."[835]

What greater miracle is there on the face of this earth than a baby's birth? No event is more joyous and life-affirming for the infant's parents. No gift is more sacred, fragile, protected and precious to them and their family and close friends. And I believe this miracle is more precious to Jesus, the Lord of all Creation, than any of His more extra-ordinary ones; including walking on water, or multiplying the loaves and fishes, or healing the blind, or raising the dead, or parting the Red Sea, or making the sun stand still in the sky for one full day, or even speaking all the worlds into existence in the very beginning. A baby is truly sacred, beyond the formation of any other kind of living thing. For we alone are made in His image and likeness and, in a way, he is creating a little spark of Himself. And once that baby is born into the world, he or she is so precious, and so protected, and so coveted, and rightly so, that it is hard to even comprehend what kind of disgusting monster would come along and rip that babies arms off, or suck his eyes out of his sockets, or jam scissors into the back of her skull while he or she is still alive. ... <u>After</u> it is born.

---

835 (Words from short film, *Sanctity of Life*, by Beamer Films. Google it and watch the video. It's very powerful.)

Of course *before* it is born, nowadays, in our *enlightened, progressive,* totalitarian-liberal,[836] Democrat-Left society, is another story entirely. Disgusting monsters who would rip these precious babies from limb to limb, torturing and mutilating them to death while they are still alive, are now legally sanctioned by the state, well-compensated for their butcheries, and supported by tens of millions of deceived individuals—people who have not come to the knowledge of the truth—with their votes. What an incredible difference a few inches can make, from inside the vaginal opening to just outside of it. Contrast the sacredness of the miracle of the birth of a baby, and the precious, invaluable treasure that he or she is upon arrival into this world; with the demonic mutilation of that same baby while he or she is still alive, but also still inside of his mother's womb. It is almost impossible to comprehend how humans could be so blind, and so numbed, as to choose to be so willfully ignorant of the connection. (Until one considers the effects of sin, and the deceptive power of the great deceiver, and his lies that cover this earth like the dirt on a tiny coffin.) Imagine for a moment if you will the heart-rending screams of agony of a mother whose baby or little child has just died in an untimely and unexpected way or who has just disappeared from their front yard, and contrast that with the soul-less decision of a nation of people to allow, indeed to encourage ("Go ahead! It's your *right!*"), the mutilation to death of those same precious, priceless babies in the womb. Two overlapping parallel universes, polar opposites, yet both somehow inhabited simultaneously by roughly half the people in this country. Quite an accomplishment.

> "The most significant thing about abortion legislation in biblical law is that there is none. It was so unthinkable that an Israelite woman should desire an abortion that there was no need to mention this offense in the criminal code."[837]

Yet today we must suffer through the delusion of an entire generation of feminists, beaming over their *accomplishment* in obtaining "reproductive rights." "You've come a long way, baby!" ("A long way" travelling down AC/DC's "Highway to Hell.") …Yeah, you go girl. And the blood of fifty million babies cries out from the ground beneath your feet over just how far you've actually gone. Sixty years ago if you asked this country if they thought baby mutilation should be legal, only about 2 percent would have answered yes. Now roughly half this country thinks so. My, how *progressive* we now are and how *enlightened* we have become. (Read: *deceived* and *brainwashed*.)

Again, the few times in this book when we have said that we live in an insane asylum, and that the inmates have taken over, I'm sure many thought it was a stretch. But when you remove all of the brainwashing and lies that we were all served, growing up in our left-wing, lying, state-run schools and under the spell of our Democrat-Left media and entertainment industry; and when you take a clear, undistorted look at this subject, as the Creator Himself views it; then the stunning lack of rationality of an entire generation, duped by the god of this world, yet still able to somehow see themselves as so terribly clever (and even more incredibly, so "morally superior"), becomes poignantly obvious.

---

836   (Credit William O'Reilly for that term, unveiled on his O'Reilly Factor program, 03/21/2011.) (Give me a call, Bill, if you want to discuss royalties)

837   (Meredith Kline)

Reprobate minds

"And even as they did not like to retain God in their knowledge, God gave them over to a reprobate mind, to do those things which ought not be done [like mutilating babies to death, for example]." (Romans 1:28)

Satan is the god of this world, literally. It is he whom so many unwittingly worship, for it is his lies that they embrace. They have turned their minds and their hearts and their backs on the truth of God, so God gives them what they want. He gives them over to a reprobate mind that will allow them to safely wallow in the lies that they love, by *protecting* them from any truth that might creep in and expose those myths.

"Now the Spirit expressly speaks, that in the latter times some shall depart from the faith, giving heed to seducing spirits, and doctrines of demons. Speaking lies in hypocrisy; having their conscience seared with a hot iron." (1 Timothy 4:1-2)

When your conscience is "seared with a hot iron" it will no longer bother you, and you will be free to do that "which ought not be done," like mutilating babies to death, and voting for the same. You will be free to follow the "doctrines of demons," like an African-American pastor the other day—responding angrily to a comment I had just made on a Christian radio call-in show—who actually said, incredibly, that "abortion is a woman's right to choose." WOW! From a "Christian" pastor, no less! And we're not trying to pick on African-Americans. This country is full of white and Hispanic and other pastors who, sadly, say the same. They have given heed to doctrines of demons: "mutilating a baby to death is a woman's right to choose;" their consciences have been seared; and God has given them over to a reprobate mind.

There are varying degrees of being lost in a reprobate mind. Sometimes it is mild, hopefully like the case of the—I'm sure sincere, yet deceived—*Christian* pastors mentioned above. Many of them, as well as many of their parishioners, are just waiting for someone to come along and bring them from darkness into the light, and into the knowledge of the truth. Yet we're not sending many missionaries to America. But there are others whose hearts are so hardened that they will never change. They are the true believers. The "vessels of wrath fitted for destruction" spoken of in Romans 9:22.[838] They already have one foot in the Lake of Fire. They have swallowed Satan's program hook, line and sinker; are fully committed to it; and will let no one and no thing sway them from their appointed task, which is remaining faithful inside of their minds to the evil one's inventions that they think are true. Among them are many leaders of the feminist movement, the media and the Democrat Party, totally devoted to baby-mutilation at all costs. And many committed Darwinists whose hatred of the very idea of The God of Creation knows no bounds. And deluded Islamic Mullahs, poisoning the minds of another generation of hell-bound suicide killers. Like Pharaoh in Egypt thousands of years ago, many of these folk's hearts are hardened to the point of no return.[839]

---

838 (See "Vessels of Wrath" section in chapter 13 for a further discussion of this.)

839 (However, in saying that we must keep in mind that it *is* true that only the Lord knows who these are, so it is our task to witness to any and all that we can, but at the same time knowing that we are instructed not to cast pearls before swine and realizing that there are those who, sadly, will never repent, and will never come to the knowledge of the truth, and will never embrace Jesus as their Lord and Savior. They are just mired way too deep in the pen, and in the lies of their father Satan.)

And then we have the far end of the spectrum, that of pure evil, where the reprobate mind is corrupt, degenerate, debased, depraved, foul, immoral, lewd, unprincipled, vile, detestable and abominable. A perfect example is the Nazi's in Germany who snuffed the life out of six million Jews; whose "doctors" performed diabolical operations and fatal tests, even <u>dissection,</u> on Jewish men, women and children <u>while they were still alive.</u> (Of course PETA would just say that they got what they deserved, because they used cosmetic products that were tested on animals.) And yet those Nazis thought of themselves as intelligent and even *righteous* in doing so. (They are suffering their just recompense at this very moment in that Lake of Fire and boiling sulfur.) Other reprobate minds are merely lost in an illogic and irrationality and absence of common sense that allows them, for example, to support an evil like baby mutilation. They haven't "thought it through," no matter how much they might think that they've *thought* about it. This is how millions upon millions of people who consider themselves *compassionate* and *intelligent* and *righteous*, including millions of false Christians, can go out and vote for the horror of mutilating a baby to death. It makes no sense whatsoever; and it is decidedly stupid both spiritually and politically. ("Stupid politically" because morally reprehensible baby mutilators also tend to be morally reprehensible liars and cheats and distorters and deceivers and crooks and fakes and hypocrites and false-accusers and economy-destroying-bankrupters and socialists on other issues as well.) There are no valid, rational, logical or sane reasons. You have to be delusional, which is another aspect of the reprobate mind.

The reprobate mind also has the ability to rationalize anything. In the case of those who support killing babies, they don't see themselves as evil, heinous or even immoral. They see themselves as "caring" and "compassionate." They are the "righteous" ones. They care about the <u>mother,</u> you see. They care about the rights of the mother, which includes the mothers "right" over her own body. They see themselves as the great "protector" of women's rights, but somewhere along the line they overlooked protecting the rights of a little innocent baby that's getting tortured and mutilated to death while it's still alive, by them and their votes. They also care so much about the quality of the rest of the mother's life. But the mother doesn't get killed; the baby does, so the illogic in pretending to care about one life by destroying another is lost on them. It can't get through the wall of self-righteous irrationality that allows them to see themselves, like the Nazis in Germany did, as being morally *superior* to everyone else. In their delusion they are blind to just how much they do <u>not</u> care about the life of the baby. If the mother decides not to abort then again she does not die, her life is changed for sure, but she is still alive. Her life might be a little tougher or a little less free than if she had killed her baby, but where does it say that we get to murder people in cold blood in order to make our lives a little easier? Is that in the Bible? (No, you'd have to look for that in the satanic bible.) And if she decides to give the baby up for adoption then that might be a life-altering and heart-rending decision as well; and one that would have emotional consequences for the rest of her life. (But again the baby would as a result actually have a life to live.) But so does abortion have life-long emotional consequences despite the constantly shouted lies to the contrary that we are forced to hear from the rabid baby-mutilating far-left because they own the media. And where does it say that women have the right to slaughter babies in order to trade one form of emotional trauma (giving their baby up for adoption) for another (killing the poor baby)? I bet if you asked the baby he would choose the former. No, Democrat supporters don't see themselves as wicked vile baby killers. They see themselves as the righteous ones standing up for the mother. (Just like the Nazis didn't see themselves as wicked, vile Jew-killers, but as the righteous improvers of the human race, standing up for evolution. (...Darwin's "Survival of the fittest," you know. And the Jews just weren't "fit" according to the Nazis.) That is how Satan, using the irrational thought process of a reprobate mind, deceives people into embracing total wickedness, under the guise of being righteous and caring. It would be a hoot if there weren't millions of babies getting murdered as a result, and if the ones trapped in this deception weren't headed for the eternity of the damned. Wake up! America!

The reprobate mind can also become indistinguishable from that of an idiot. A case in point would be Eric Holder, our addle-brained Attorney General, who attempted to explain his *reasons*, in front of a Congressional committee, for breaking 235 years of precedence and holding the trials of 5 Islamic war criminals in our civilian courts. He truly sounded like a child, or an adult idiot, take your pick. There was no logic or common sense behind his words, but he still could not see clear to alter his position because he is blinded by his belief that he is far superior to everyone else; which makes the fact that he is in reality an irrational jackass unthinkable to him. He is also bound to do the bidding of his masters, Obama and the radical ACLU. (And all of them are bound to do the bidding of their master, Lucifer. ...See chapter 13.) He has embraced the lies of the Far Left: "Bush lied!" "America is killing millions of innocents in Iraq." "The terrorists are just freedom fighters." "America, and not the Muslims, is to blame." "Waterboarding is "torture."" (The illogic and irrationality in that is evidenced by the fact that for them it's OK to kill terrorists with drones, or by shooting them in the face, like Osama Bin Dyin' had the pleasure to experience, but pouring water in their nose is <u>not</u> OK.) "We must *protect* the [non-existent] Constitutional rights of terrorists because we are a nation of laws and blah, blah, blather." And he has separated himself from the truth. He is another victim of that self-righteous, anti-American dementia common to the far-left known as *ACLU disease*. (The kind of mental disease that can also allow one to actually say with a straight face that dipping the faces of three Islamic lunatics in water is "torture," but mutilating 50 million babies to death while they are still alive is not.)

The reprobate mind can go on for hours speaking in complete sentences but saying nothing that makes any logical sense whatsoever. Listen to any totalitarian-Democrat on TV explaining the "reasons" behind their incomprehensible health care plan; their outrageous spending that has bankrupted the states where voters have foolishly put them in charge, and now has bankrupted our federal government; their cap and trade idiocy meant to "solve" their man-made crisis of global warming; trying war criminals in our civilian courts; or why our country should continue to refuse to defend its own borders so millions of potential uneducated Democrat voters can stream in. (Just listen to Pelosi, Reid, Weiner and Schumer, during the budget debates of 2011, blather about why we can't cut any program whatsoever, as they strive to continue spending us even further into their socialist hell.) They will go on and on, but they will make no sense whatsoever, except to another reprobate mind. Then it will make perfect sense. They will say nothing that even remotely approaches the truth; lies and distortions being the reprobate minds best friend. Obama is a perfect example of this. He can go on for hours and sound so authoritative and impressive to those who are deceived like him, but much of the time he makes no sense whatsoever in the real world. Again, it's all just blather.

People who have been given over to a reprobate mind are not aware of it. They think they are smart, even extraordinarily intelligent, and it is everybody else who is stupid. (Just listen to Bill Maher and Jon Stewart. They are very proud of their own ignorance.) Many patriotic citizens today are beside themselves trying to figure out how anybody could be so stupid as to destroy this country like these Democrat Marxist buffoons who have taken over our federal government (in 2009) seemed determined to do. They look at these people and see that in some ways they are literally insane. Why, oh why would they be doing these things? They are perplexed because they do not see them in terms of the reprobate mind, and of the depths of irrationality that God will allow that mind to descend to. If one chooses the lies and the evil that come from the great deceiver, and rejects the truth and the morals that come from our holy Creator, then that Creator will give that individual what they have chosen. Including the level of irrationality necessary for them to remain secure in their choice.

Many people see the Democrat-Left as "basically decent" people who "just have a differing opinion," and I can't disagree with that on the one hand. But on the other hand what exactly is "basically decent" about having a hand in mutilating 50 million babies to death? And counting...? The same thing that was "basically decent" about having a hand in slaughtering 6 million Jews. Nothing. Wake up, America! Most people think that, hey, these Democrat-Left broadcasters, entertainers, professors, comics and politicians are mostly nice people. And who could disagree. Of course they are. But so were the Nazis, as long as you weren't Jewish. And so is the Democrat-Left, as long as you're not an unborn baby. (Or a conservative, or a true Christian, or Sarah Palin, or Rush Limbaugh, or Bill O'Reilly, or anyone who opposes their godless agenda...)

"They even sacrificed their sons and their daughters to demons, and shed innocent blood, even the blood of their children, whom they sacrificed to the idols of Canaan; and the land was polluted with blood." (Psalms 106:37-38) Canaan is not the only land "polluted with innocent blood." Ours is as well: the blood of millions upon millions of the most innocent. And just as God said that Abel's blood cried out from the ground to him ("And God said, What have you done? The voice of your brother's blood cries out to me from the ground." Genesis 4:10) so too does the innocent blood of many millions of mutilated babies cry out to Him from the ground beneath our feet.[840] God definitely has "shed his grace on thee," America. We have been more blessed than any country in the history of the world, and this is how we repay him? Wow! And I'm sure millions of people who have lost their jobs in this terrible economy (January, 2010) are praying to God for help, but half of them will march out in November of this year and vote again for the baby mutilating Democrats. That's like a Palestinian praying to God to lift him and his family out of the poverty and misery of the economic hell-hole created by his Jew-hating Muslim leaders, while at the same time he marches out and votes for a Jew-hating and Jew-killing terrorist organization, Hamas, to run his country. What's wrong with both of those pictures?

---

840   (Contributed by Tim Getz, Evangelist and homeless outreach, Baltimore, Md.)

"It's baby mutilation, stupid!"

Remember "It's the economy, stupid!" that famous, oh-so-clever slogan that our Democrat friends used to dupe the "no-think-ums" into electing them in 1992? Well, I think ours is a tad more relevant. Socialists, after all, know next to nothing about stimulating an economy. (Destroying it, however, they are good at.) For if all people ever hear is the phony word "abortion" and they are never exposed to the gruesome reality of that vile practice, then they will continue to allow it, and will not rise up in the numbers necessary to put an end to it. This is why you will never hear the words "baby mutilation" spoken on National TV, or see them written in our newspapers, even though that is a completely accurate description of what occurs.[841] Instead we hear *abortion*. But our Democrat media or education system would never think of calling date-rape *satisfaction*, even though for one of the parties that is what occurs. "Today a Columbia University student was accused of *satisfaction* on his date last night with a freshman coed." No, of course not, because they don't want to whitewash date-rape. Nor would they ever call what Bernie Madoff did *appropriation*. "Madoff was indicted today on 3,500 counts of alleged *appropriation* after it was discovered he spent billions of his client's hard-earned money on his wife's face." (Oops. That was a little harsh.) (…Unless you're one of their victims.) No way, because their father, the great deceiver, has no interest in whitewashing grand larceny, embezzlement and investment fraud. He has, however, always had a penchant for child sacrifice, going back to biblical times, and he knows that our modern day, *enlightened*, Democrat progressives wouldn't go for burning their children alive as an offering to Moloch (another name for the Prince of Darkness) like their ancestors did thousands of years ago. Instead he dupes them into presiding over the slaughter of millions under the guise of "a woman's right to choose," while at the same time whitewashing their murder with the sanitized term "abortion."

It is estimated that the Nazi holocaust in Germany killed 20 million people. But the modern day Nazi genocide in this country has mutilated 50 million innocent babies to death. And, after this last election (November 2008), they are thrilled at the idea of doing it to another 50 million more. YES WE CAN! I remember talking to a twenty-something-year-old during another good walk spoiled on the golf course a number of years ago and the subject of baby-mutilation came up. His immediate response was, "Oh, we had that debate back in High School. It's definitely a woman's right to choose!" Definitely. (Wouldn't you love to have been a fly on the wall in that classroom listening to the *impartial* moderator of that debate?) I told him that he needed to go back to debate class and he looked at me funny. But here's the truth, kids. <u>Demons</u> mutilate babies. <u>Angels</u> protect and defend them. And this world is full of both angels and demons. For now. And depending upon the choices you make in this life, you will spend eternity with either one or the other. And that is also an accurate assessment of the "two sides" of this *debate*. You can either be possessed of a reprobate mind and be on the side of demons; or you can align yourself with the truth, with sanity, with all of the heavenly hosts of angels, and be on the same side as the Creator of this universe.

---

841 (This is also why you will never see pictures or video on National TV of the revolting procedure that results in a baby's life being snuffed out. But they have no problem showing the bodies of dead babies killed collaterally in a war zone by our military, or in an earthquake, or in a tsunami. Go to durarealidad.com for a graphic video of the results of this demonic ritual the Democrat-Left refuses to show.)

(Take your time, kids. This is another tough decision.)

We have already used the following quote in the previous chapter under *The Religion of Naturalism* section, but it is worth repeating here:

> In New Guinea, in "one cannibal tribe …men would race into a village, grab a man by the hair, pull him back, tense his abdominal muscles, use a bamboo knife to slit open his abdomen, pull out his intestines, cut up his fingers, and while he was still alive, eat him until he died. People hear that and say, "Oh, what primitive savages!" They are not "primitive" savages; their ancestor was a man called Noah. … Noah had the knowledge of God and could build ships. His ancestors could make musical instruments and they practiced agriculture. What happened to those New Guinea natives is that, somewhere in history (as Rom. 1 tells us), they rejected the knowledge of God and His laws. And God turned them over to foolish, perverse, and degenerate things. [And a reprobate mind.]
>
> "However, this same degeneracy (this same rejection of God's laws) can be seen in so-called civilized nations that cut people up alive all year long (one and a half million of them in the United States each year), and it is legalized. This is what abortion is—cutting up people alive and sucking out the bits and pieces. The so-called "primitive tribes" had ancestors who once knew the true God and His laws. As they rejected the true God of creation, their culture degenerated in every area."[842]

As our culture now too is "degenerating in every area," and yet is beaming with pride, and ignorance, over it. ("O-BA-<u>MA</u>! O-BA-<u>MA</u>!") Again, the truth is that…

<u>Demons mutilate babies. Angels protect and defend them.</u>

You've been lied to, kids. It's time to go back to debate class.

A woman's right to *what?*

You have to be borderline insane to think that a woman actually has the *right* to mutilate her baby to death. (See *reprobate mind* discussion above.) She might have the legal permission of a godless and immoral state, and the greedy assistance of demon-possessed "doctors,"[843] but <u>she has no right</u>. For human rights come only from God. Our Founding Fathers certainly knew this when they immortalized the following words into our nations Declaration of Independence:

> "We hold these truths to be self-evident, that all men are created equal, that **they are endowed by their Creator with certain unalienable Rights**, that among these are Life, Liberty, and the pursuit of Happiness."

It is the Creator, and not the state, who bestows human rights. That's why they are *unalienable*; you can't be alienated from them; they can't be taken away. If they came from the state and not from an unchangeable Creator, they would be temporary, at the mercy of the whims of those in power, and meaningless. Many

---

842   (Pg. 84, *The LIE: Evolution*, by Ken Ham)

843   (Isn't it funny how the half-wits in Hollywood are always putting down the terrible greed of corporations and of Wall Street? But they never say anything about their own greed, their own astronomical salaries, or how their friendly neighborhood Planned Parenthood "doctors" get rich by slaughtering babies. The hypocrisy is stunning, but it completely escapes them. May I suggest a title for their next "cutting edge" movie: *Hollywood: Where the Reprobate Mind Feels Right at Home*)

of those who now enjoy the fruits of, and the unparalleled freedom provided by, those unalienable human rights recognized by our Founders and written into our Constitution, are completely ignorant of this fact. Indeed, if our Founding Fathers were alive today and you told them that 7 unelected, left-wing, ACLU lawyers in black dresses discovered a "right" in their Constitution that they wrote, for a woman to mutilate a baby to death in her womb, they would vomit.

The godless, morally-perverse Left goes on and on about the fabricated *right* of a woman to slaughter her own baby, as long as this murder is performed before it is born; but they are either ignorant of, or willfully ignore, the most basic human right of all—and one that *does* comes from the Creator—and that is <u>the right of a baby to be born</u>. The right of a baby <u>to live</u>. The baby mutilator's never talk about that right. But their reprobate minds can lead them even further into the depths of irrationality. During this month, December of 2009, with regard to the debate over whether to include funds for killing babies in the Democrats attempt to destroy our health-care system (which the liar...er...president, Obama said would never be in there), the Left is now saying, if you can actually believe this, and I quote, that "abortion is a *god*-given right." And, honestly, they are absolutely right when *they* say that. Because their god is Satan. He is the god of baby-mutilators and baby-mutilator voters. And baby mutilation is indeed a Satan-given right, as is rape and murder and assault. But it is not, never was, nor ever will be a <u>God</u>-given right. That would be insane.

"If you shout a lie loud enough and long enough, enough people will believe you." And Satan has come up with some real whoppers over the years. "The Jews are the cause of all of our problems." That was the Nazi's favorite 70 years ago. "The theory of evolution is a fact." That's faux scientist's favorite. And then there's this, "Abortion is a woman's right to choose." The Left's favorite demonic lie. But not only does a woman <u>not</u> have the right to mutilate a baby to death in her womb according to the Creator of the universe, she also does <u>not</u> have that right even according to our own Constitution. Back in 1973 seven unelected, left-wing, ACLU lawyers in black dresses, in the *landmark* Roe V. Wade decision, perverted the Constitution of the United States of America in order to fabricate a "right" which miraculously lay hidden and undiscovered inside that Constitution for nearly 200 years until those seven bald-faced-liars' came along. (Again, the very writers of the Constitution themselves were somehow not even aware of this previously undiscovered *right*.)

Judge Robert Bork:

> "Without any basis whatsoever in the Constitution they found that somehow there was a right to an abortion. Roe V Wade really surprised me. I just couldn't believe that they would make up something like that. You can read that—what is it something about 58 pages—and there's not a line of legal reasoning in it."[844]

And that is because they weren't interested with legal reasoning, or the Constitution. They felt they were empowered by their progressive ideology to "improve" this country, to remake it in their own totalitarian-liberal image, by circumventing the clear provisions of the Constitution and then legislating directly from the bench. They, you see, were (and still are today) so much wiser than both the Founding Fathers and the "poor, stupid rubes" whose votes stood

---

844   (From Fox News Special, "The Right, All Along: The Rise, Fall & Future of Conservatism Part 2")

in the way of their progressive agenda. And even though those seven lawyers took an oath to uphold and defend the Constitution of the United States, they in effect pulled down their pants and defecated on it. They should have been put in prison, and tried for treason. But instead America sat back and allowed them to pass a law, to *legislate*, from the bench, which is expressly forbidden by the separation of powers of the three branches of government enumerated by that same Constitution. This includes the Legislative Branch (Congress which has the power to pass laws), the Executive Branch (the President which has the power to enforce those laws), and the Judicial Branch (the Supreme Court which has the power to destroy this country if constitution-perverters keep getting appointed to it by Democrats). Subsequently, as a result of their corruption, a woman may have the legal protection from prosecution for the crime of murder, given to her by only seven, Constitution-perverting lawyers, but she has no moral right, and no Constitutional right, to mutilate her baby to death. So much for a "woman's *right* to choose."

Choices

Now for the "to choose" part.

"Behold, I set before you this day a blessing and a curse: A blessing, if you obey the commandments of the Lord your God, which I command you this day; And a curse, if you will not obey the commandments of the Lord your God, but turn aside out of the way which I command you this day, to go after other gods, which you have not known. ... I call heaven and earth to be witnesses this day against you, that I have set before you life and death, blessing and cursing; therefore choose life, that both you and your descendants may live." (Deuteronomy 11:26-28, 30:19)

Unfortunately, we have not obeyed the commandments of the Lord our God, especially the one that says , "Thou shall not murder!" so we are cursed with a curse. (And we wonder why our economy is going to hell in a handbasket.) We also have turned aside from the way which he has commanded, "to go after other gods;" specifically Satan, the great god of this world himself, and child-sacrificer of old; but in addition the gods of selfishness, personal convenience, and sexual licentiousness freed from the constraints of its normal consequences (birthed babies). And therefore we have chosen death, and the eternal death in a Lake of melted sulfur that comes along with that choice. You might say, with regards to those verses in Deuteronomy, that we have scored the hat trick. Congratulations, America.

When did we get so smart that we learned to call murder a *choice*? I must have been out of the country during that collective epiphany. Mutilating a living baby to death is a woman's *choice*?!#@ Wow! That's incredible! Well how about racism, should we now rename those hangings of a number of innocent black men by the Ku Klux Klan, a *bigot's choice*? How about robbery, can we call that a *criminal's choice*? And rape, is that a *man's choice*? (Well yes, actually it is, but only when the Democrat-Left confers that choice upon a certain lecherous ex-president, who will remain nameless (liBl ntoiCnl), and his lip-biting unwanted forced fornication on a woman whose name rhymes with Juanita Broderick.) How about the Left's good ole friends the Nazis, was it their *right to choose* to burn millions of Jews in ovens? How about kidnapping, raping and murdering little children, can we call that a *pervert's choice*? Or (and I know the left is going to love this one) how about beating your wife, can we call that a *husband's choice*. Of course not! The genius' on the left would never stand for any of that. (Except for Christian and conservative bashing, that is not only the Democrat-Left's choice, it has become their national sport.) But when it comes to the gruesome murder of innocent babies, the left disguises that heinous crime as a woman's "choice." How terribly clever. My, they must be so proud of themselves. The devil and his angels certainly are. They're giving each of them a standing ovation right now, as they die and are welcomed into eternity. (If you listen closely you can hear the cackles and the applause, and the throat-scarring screams, right below your feet.)

"Baby mutilation might be a *woman's* choice, but it is never a baby's choice."

It is no more a woman's choice (or her "right") to mutilate a baby in her womb than it would be her choice (or her "right") to grab a machete and start whacking the limbs off her new born; or off her five year old for that matter. (Of course, as the case of Andrea Yates has shown, the Left will excuse you as long as you choose the drowning method. The machete being just a little too messy for them right now.) (Give 'em a few more years...) It is no more a woman's choice to mutilate a baby in her womb than it would be her choice if she was a Nazi, or a hate-filled Muslim, to slaughter Jews in ovens; or her choice if she was an ignorant racist to hang blacks; or her choice if she was an ancient and unenlightened Greek to throw her baby off of a cliff. And speaking of the Nazis, we would never have called what they perpetrated a "choice." *It was the Nazi's right to choose to abort the lives of a number of Jews*. NO! We called it what it was: disgusting, vile, genocide, a holocaust! So let's stop using the terms of the enemy. Why don't we, as true Christians, start having the non-politically correct guts to call it what it is, murder, and stop enabling the liars on the Democrat-Left in our media, our entertainment industry, our educational system, our late-night clowns and our government to call it what it is not.

One reprobate's *fetus* is another man's baby

How many times have we heard the reprobates proclaim, "But it's just a *fetus*." "It's not a baby." "It's just a lump of tissue." "It's not *human*."

Really? What makes *it* human then? The vaginal opening? A baby is just an inhuman, fetus, lump of tissue until it emerges into the world from out of the vaginal opening? And then Wallah! Shazaam! Abra Ka Dabra! and Bibbidi Bobbidi Boo! It's human!! Look, it's a <u>BABY</u>! (What a surprise! It could have been a fish.) Now, but a few seconds earlier it was just an inhuman, fetus, lump of tissue and you could have sliced it and diced it and served it over some steamed rice as a side dish. (And if you think that's flip and callous go to durarealidad.com and see what they actually *do* to babies- slicing and dicing is an understatement.) No, the vagina may be many things to many people (for an exhaustive compilation of its uses, see *The Vagina Monologues*) but it is <u>not</u> the author of humanity. God is. He makes us human, and not the vaginal opening. Remember what we said earlier about the reprobate mind: it is irrational, illogical, debased, lacking in common sense and literally insane. The Left's *argument* over when a baby suddenly becomes an actual baby is just one more in a long line of clear illustrations of this.

But let's see what the King of Kings says. In addition to the verse at the beginning of this chapter that clearly states that a human is a human at the moment of conception—"Then the word of the Lord came unto me, saying, Before I formed you in the womb I knew you; and before you came forth and were born I sanctified you." (Jeremiah 1:4-5)—the Author of life also inspired the Psalmist to declare this: "Behold, I was shaped in iniquity; and in sin did my mother conceive me." (Psalm 51:5) God is clearly saying that the psalmist <u>was</u> the psalmist (he was "me") even at the moment of conception inside of his mother's womb. And then there is this, "For you formed my inward parts: you covered me in my mother's womb." (Psalms 139:13) A baby receives his or her soul in the womb, at conception, and then the body begins its 9 month journey to cover the soul. Conception is the point where one is made in the image and likeness of God. And the completeness of our soul is not limited by the size or development of our physical body. Remember from chapter 4, "Man," that we are not physical beings. We are a spirit, which is pure consciousness, pure awareness: "You are that pure conscious awareness that inhabits your brain; that is aware of the final outputs of the sensory activity of the brain and of your body which is merely an extension of your brain. And that Y O U has nothing to do with the image—the atoms or electrons or protons or energy units, or cells or organs or brains—of the *physical* world. It is separate from the material world."[845] And that pure being, inside of these physical bodies, is a gift of God. And only Satan would desire to snuff it out at its very beginning. Only Satan would destroy it, and those who are duped by him.

The past

"Moreover he ...burned his children in the fire, according to the abominations of the heathens whom the Lord had cast out before the children of Israel." (2 Chronicles 28:3)

"Moreover they ...mutilated their children in the womb, according to the abominations of the Democrats whom the Lord will cast into the Lake of Fire at His Second Coming." (2 Washingtons 23:8) (From the second letter to the Washingtonians)

Satan is the great counterfeit. He mirrors everything God does, but it's a mirror image, it's backwards, the exact opposite. God sacrificed his only Son on the cross for man's sins. The devil dupes men into sacrificing their children to him. In the olden days it was burning them alive to him. Today it is mutilating them to death to him, disguised as the god of convenience and "choice."

History clearly records that this demonic practice of killing children has been with us for many years. We were vile then, and we're *progressive* now. (Hello, Hillary.) As much as things change, the more they stay the same. Except today we like to fool ourselves about how intelligent and enlightened we are. Unlike for example those ignorant pagans who would burn their children alive as a sacrifice to the gods. Or those barbaric Greeks of old who would throw unwanted, handicapped or disabled babies off of cliffs. (What was it that they used to call it? Oh yes, a Greek's "right to choose.") We know how stupid and ignorant they were, right? Not like us. We're much smarter. We would never throw an unwanted baby off of a cliff, would we? Of course not. We're not that barbaric! No, we just cut them up in the womb, before they are born. We're much more progressive. What was it that Someone once said about a speck and a beam? ... "And why

---

845   (See Chapter 4, "Man")

do you behold the speck that is in the Greek's eye, but do not consider the beam that is in your own eye?" (Matthew 7:3) Don't stare, but the *intellectually-elite, progressive, enlightened* Democrat-Left has a 2-by-4 stuck in their collective eye.

"Many ancient pagan societies believed that parents possessed an unqualified right to kill their own children for any or no reason. "The Law of the Twelve Tables," a Roman legislation circa 450 BC, actually *required* a father to put to death any deformed child (*Cito necatus insignis ad deformitatem puer esto*). (Modern moral philosophers, like Joseph Fletcher and Princeton University's Peter Singer, advocate the same thing.) [Surprise, surprise.]

"The killing of female children was so widespread that, just as in Asia today, the ancient world had a large abortion- or infanticide-caused imbalance in the sexes. ...A letter written in 51 BC from a pagan husband in Egypt to his wife, revealed the casual way pagans viewed killing infants, particularly young girls:

" "Know that I am still in Alexandria... I ask and beg you to take good care of our baby son, and as soon as I receive payment I shall send it up to you. If you are delivered [before I come home], if it is a boy keep it, if a girl, discard it." [Get that man's number! There's another highly qualified prospect for Planned Parenthood of China.]

"Ritual child sacrifice was also widely practiced.

"According to Plutarch, for example, the Carthaginians "offered up their own children, and those who had no children would buy little ones from poor people and cut their throats as if they were so many lambs or young birds; meanwhile, the mother stood by without a tear or moan.""[846]

That is what happens to a human being when God gives them over to the depths of a reprobate mind. They become unrecognizable as human. And many mothers, and "doctors," and abortion clinic staffers all across this country—set up by the lies of the Satan-worshiping, Democrat-Left—walk out of baby-mutilation clinics every day "without a tear or moan." Enlighten me, Barack, when you finally get to a high enough pay grade, what exactly is the difference?

History also records the slaughter of the innocents after Jesus' birth in Bethlehem: "And then King B. Hussein sent out a decree throughout all the land that millions of the <u>unborn</u> children should be put to death..." Whoops! Wait a minute! That's the wrong quote, dog gone it. That's not King Herod! That's the Democrat Party. Let's get the right quote… "Then Herod, when he saw that he was mocked of the wise men, was exceeding wroth [read: ticked off], and sent forth, and slew all the children that were in Bethlehem, and in all the coasts thereof, from two years old and under, according to the time which he had diligently enquired of the wise men." (Matthew 2:16) It is interesting to note that at the first coming of Jesus as a baby in Bethlehem, the government as ruled over by King Herod ordered the slaughter of all of the innocent children up to 2 years old in the entire metropolitan area. And now, two thousand years later, as we stand on the verge of the second coming of the Lord Jesus, our government and indeed most of the governments of this entire world have been slaughtering innocent unborn babies to death in numbers that dwarf anything King Herod could have conceived. "The thing that has been, it is that which shall be; and that which is done is that which

---

846   (Pgs. 81-82, *The Politically Incorrect Guide to the Bible*, by Robert J. Hutchinson)

shall be done: and there is no new thing under the sun." (Ecclesiastes 1:9)[847]

The present

"You shall not worship the Lord your God in their way, for every abomination that the Lord hates, they have done in worshipping their gods. For even their sons and their daughters they have burned in the fire as sacrifices to their gods." (Deuteronomy 12:31)

Tonight, April 16th 2007, is the date of the infamous Virginia Tech massacre where 32 people were gunned down by a psychotic who had no business being free among the general population much less on a college campus—but for political correctness run amok (he was a minority psychotic after all). And we're watching Fox News' *On The Record* and Greta asks this question, "How could anyone be so cruel?" ("Why, oh why?") Well, the answer, Greta, lies in these words from the film *Sanctity of Life* quoted earlier in this chapter: "We cannot diminish the value of one category of human life -the unborn- without diminishing the value of all human life."[848] I, and millions of others in this country, Greta, have been asking your question for years: "How could any political party be so cruel as to mutilate 50 million babies to death?" And we're still waiting for an answer. And how could anyone be so senseless and so cruel and so absolutely stupid as to casually go out and vote for these baby mutilators every 2 years, religiously? We're still waiting for an answer to that one as well. Thirty two innocent college students and professors were slaughtered for no reason! Why? Juxtapose that with: fifty million innocent little precious babies had their limbs ripped off, and their eyes sucked out of their sockets, and scissors jammed into the back of their skulls, all while they were still alive, for no reason. Why? So before we all sit around self-indulgently as a nation ringing our hands and crying out "Why-oh-why?" and "How could anyone be so cruel?" and "What has happened to the basic respect for life?" we might want to put a sock in it just long enough to walk into the bathroom and take a peek in the mirror. For there is your answer, America. You f***ing voted for it, if anyone possesses the rudimentary intelligence necessary to make the connection. (The same could be said of Columbine, but Darrell Scott already nailed that one. See chapter 2.)

"Be not deceived; God is not mocked: for whatsoever a nation sows, that shall it also reap." (Galatians 6:7)

America has no problem understanding the heinous nature of the Virginia Tech massacre, but they do seem to have a problem comprehending the horror that happens every day in a baby-snuffing clinic. And that is because unlike the doggie that poops on the carpet, this country has never had its nose rubbed in the filth that they keep voting for. The Democrat-Left has systematically kept them from viewing it on their TV's. Instead they are treated constantly to far-left imbeciles mocking, disparaging, demeaning and marginalizing any prominent Christian or conservative who is not an ignorant baby mutilator like them. Sarah Palin is *so stupid*. Sure. She's "only run a state …a town …and a commercial fishing operation." Conversely, Barack Obama is a genius, even though the only thing he'd ever run was his mouth.[849] That is what passes for logic in the reprobotic minds of left-wing luminaries such as Chris Matthews, Keith Olberman, Paul

---

847   (From Tim Getz, Evangelist and homeless outreach, Baltimore, Md.)

848   (From short film, *Sanctity of Life*, by Beamer Films.)

849   (Mark Steyn on Fox News *Hannity*, 2/10/10)

Begala, Bill Maher, Jon Stewart and David Letterman. Them ridiculing Sarah Palin is like that s**t-eating artist from San Francisco mocking a vegetarian.)

The number of baby-killings that occurs on a regular basis in this country alone is staggering, 3000 mutilations a day, 1.2 million per year, but the nature of what actually occurs is abhorrent. (Go to www.durarealidad.com and watch the video.) At the end of World War II America was not kept from seeing uncensored news reels of the newly liberated Nazi concentration camps and the ditches piled full of emaciated bodies, but Americans have been kept for decades from seeing the videos of dead babies, ripped apart, bloodied, beheaded, mutilated and discarded like so much unwanted trash. (Go to www.durarealidad.com and watch the video.) A holocaust covered up by so much unwanted left-wing trash in our media, our educational system and our entertainment industry.

And here is the crowning achievement in their legacy. What is commonly known as Partial Birth Baby Mutilation involves taking a 9 month old baby about to be delivered, turning it around in the birth canal so they can let the rest of the baby be born except for its head. (Begala has something really *special* planned for that.) While making sure the head stays inside of the mother's vagina (this is for purely *legal* reasons because if they let the entire baby be born and did what they are about to do then they would go to jail for murder... my, what a difference a few inches can make) then "thrill-up-the-leg" Matthews takes scissors and jams them into the back of the baby's head at the base of his or her skull and pries the scissors apart to make a convenient hole for the monsters next step. Then Olberman, aided by his hilarious sidekick, Stewart, gets a vacuum tube out and shoves it up into the skull so they can suck out the babies still living brains. And as that 9 month old, fully-formed little baby is letting out her dying, muffled and horrified screams, Letterman sits behind his desk and yuks it up with his audience of uneducated, Democrat-loving cretins. And incredibly, half of this country goes out religiously every two years and votes for these same disgusting monsters.

Let me ask you a question, America,[850] what kind of a "human being" would support letting someone take scissors and shove them into the back of the neck of a nine month old precious little baby, just waiting to be born and to start its journey through life? Just waiting for the first time he or she walks, and sees the look of excitement and delight on the faces of their mommy and daddy. Just waiting for the first time he plays catch with his dad, and the first time she rides a bike as her mommy looks on anxiously with pride. What kind of a reprobate, deluded, uneducated animal would casually go out and vote for the perpetrators of this horror? I mean, what's next, skinning babies alive? Would you wake up then, America, if the demons started an abortion procedure that involved skinning the little babies alive? Well what's the difference? What could be more evil and disgusting and inhuman than sucking a baby's brains out of its skull while it is still alive? If we did this to one Islamic lunatic, the New York Times, and Katie Couric and the Democrat-Left (including those same 5 baby-mutilating clowns mentioned above) would be screaming so loud and for so long that every man, woman and child in North America would wind up totally and irreversibly deaf. And what if we withheld ourselves from the scissors-jamming and brains-sucking and instead just tore the limbs off and the eyes out and hacked the heads off our psychotic Mohammedan friends? We'd get the same deafening result. Yet these

---

850   (I learned this technique from Bernie Mac, may he rest in peace)

same screamers sit back and laugh while they support, nay demand that babies continue to be subjected to this vile, inhuman horror. Excuse me for no longer finding Tina Fey and the rest of her kind all that funny. And for holding them in complete and utter contempt.

"Were they ashamed when they had committed abomination? No! They were not at all ashamed, neither did they blush. Therefore they shall fall among those who fall; at the time I punish them they shall be cast down, says the Lord." (Jeremiah 6:15)

Baby mutilation is the stinking, rotting piece of demon-flesh that the entire Democrat-Left is choking on. Their voters need to stick their fingers down their throats and vomit it out. (Or go to hell, whichever they prefer...)

> "However common it may be, abortion is a direct attack on what God intended to be a life-giving place, thus turning the womb into a killing field."[851]

For a woman to turn her own *womb into a killing field* is almost beyond comprehension, but that is the deceptive power of the great deceiver, the father of lies and the god of this world. On the one hand, decent human beings are willing to die for their children. On the other hand, twisted and deceived human beings kill their children for their own convenience, lifestyle and financial considerations. The Democrat-Left therefore has reached the absolute limit of indecency, yet how they righteously revile their opponents. It is such a hoot to see them in action. The clown Tina Fey demeaning the intelligence of Sarah Palin because she is not a disturbed baby-mutilator like her. And not just cracking a few jokes but purposefully disparaging her in ways that Fey and her left-wing kind would never think of doing to one of their baby-mutilating Democrat friends. They become apoplectic at the thought of a decent person gaining political power. Palin just doesn't rise to their high standard of "intelligence." The kind of intelligence that supports sucking the brains out of the skulls of infants while they are still alive and then weeps over the oratorical genius of their president as he "radically transforms this nation" into a pile of socialist dog shit. Our TVs are littered with like-minded Neanderthals such as Bill Maher, Michael Moore, Bill Moyers, David Letterman, Chris Matthews and Keith Olberman. Everybody is a cretin but them. The Democrat-Left is the height of indecency, uneducated ignorance, and moral depravity, killing children rather than dying for them, yet they parade themselves across our radio and TV waves as the pinnacles of righteousness and intelligence, the tolerant defenders of all that is good and right. It would be mildly amusing if it weren't for the fact that they are complicit in the cold blooded execution of so many millions of innocent babies. "YES WE CAN!" ...Yea, you sure can, you f***ing idiots...

We can only imagine what the Creator of this universe thinks of all this, looking down on it from above. When you put yourself in his position, you create this incredible universe, and then you create this water-planet that has the unique ability to sustain life—a one in a million to the millionth power shot in an otherwise cold, dark and lifeless expanse of space. You create life on it, and then you crown this achievement with the creation of intelligent life, made in your own image and likeness, for the express purpose of mirroring your beauty and grace, your wisdom and your holiness, and glorifying You. And then this is what it stoops to?!

---

851   (From *Creation Studies Institute* circular, April 2009, by Tom DeRosa)

Ripping living babies from limb to limb in the womb. Jamming scissors into the back of the necks of nine month old babies. Wouldn't you just want to vomit it all out of your mouth?[852] But what do we hear every time there is a terrible natural disaster—a catastrophic Indonesian or Japanese tsunami, or Haitian earthquake, or devastating tornados that hit the Southwest like a nuclear bomb, or a hurricane that floods an entire city in Louisiana, or an inundating flood of the Mississippi that wipes out many entire counties, towns and cities, or fires that destroy many thousands of square miles of forests and homes—what do we hear from most of these nauseatingly-self-righteous, deluded Democrat journalists who control our TV news? "These people need our prayers." And of course they do, but this request of "prayers to God" is coming from the same people who just went out in the last election and voted for the mutilation to death of millions more of that same God's creation; and who made sure with their obviously slanted reportage that they influenced as many other stupid people as possible to go out and do the same. Their ignorance, and their disconnect, is beyond comprehension. ...Unless you understand the nature of the reprobate mind.

"But woe unto you, scribes and Pharisees, hypocrites! For you shut up the kingdom of heaven against men; you neither go in yourselves; neither do you suffer those to enter who are trying to." (Matthew 23:13)

Woe unto you, Democrats and progressives, hypocrites! For you shut up the truth of God against men; you neither learn it yourselves; nor do you allow access to it for those who would. (All The Secrets 11:1-2)

"Woe unto you, scribes and Pharisees, hypocrites! For you compass sea and land to make one proselyte, and when he is made, you make him twofold more the child of hell than yourselves. Woe unto you, you blind guides." (Matthew 23:15-16)

Woe unto you, left-wing teachers and professors! For you indoctrinate your students to make another baby-killing convert, and when you do, you make him twofold more the child of hell than yourselves. Woe unto you, you blind guides. (All The Secrets 11:3-5)

"Woe unto you, scribes and Pharisees, hypocrites! For you pay tithe of mint and dill and cummin, but have neglected the weightier matters of the law: justice, mercy, and faith. These you ought to have done, without leaving the former undone. You blind guides, who strain out a gnat, and swallow a camel." (Matthew 23: 23-24)

Woe unto you, baby-mutilator voters, hypocrites! For you think yourselves compassionate, as caring for the poor, and as the great protectors of the "rights" of women, but you have neglected the more important commands of God: justice for the unborn, mercy for the unborn, and Thou shalt not murder My innocent babies! These you ought to have done, without leaving the former undone. You blind fools, who care so much about the whales, the Polar bear and the baby seals, and then slaughter by the millions the far greater and more precious miracle. (All The Secrets 11:6-8)

"Woe unto you, scribes, Pharisees [and Democrats], hypocrites! For you are like unto whitewashed tombs, which indeed appear beautiful on the outside, but inside are full of dead men's bones and all corruption. Even so you also outwardly appear righteous unto men, but within you are full of hypocrisy and evil. You

---

852   (See Revelation 3:16)

serpents, you brood of vipers, how can you escape the damnation of hell?" (Matthew 23:27-28, 33)

Woe unto you, Barack, Michelle, Bill, Hillary, James, Kathy, Christiane, Nancy and Barney, hypocrites! For you are like whited sepulchers, which indeed appear beautiful and well-spoken on the outside, but inside are full of dead men's bones and all corruption. Even so you also appear righteous unto the fools who support you, but in reality you are full of hypocrisy, murder, deceit and all evil. You serpents, you brood of vipers, how can you escape the damnation of hell? (All The Secrets 11:9-12)

How can they escape the damnation of hell indeed? Unless they repent, and change, and turn 180 degrees from the kingdom of Satan, and ask God and America to forgive them for their transgressions, and spend the rest of their lives working for Jesus Christ, for the truth, for innocent babies and for the Kingdom of Heaven. Pray for them. None of us can possibly comprehend the terror, the anguish, the excruciating and endless pain, and the everlasting horror that is in store for them otherwise.

The future

"For they have sown the wind, and they shall reap the whirlwind." (Hosea 8:7)

Too many folks in this country are oblivious to the true satanic nature that underlies the entire feminist-progressive-Democrat-totalitarian-left-elite baby-mutilating movement, and what that movement is ultimately capable of. (See chapter 13, Satan.) And this is especially true of the extreme Left that has taken control over the Democrat Party, and our government (prior to the Republicans gaining back control of one-third of it, the House of Representatives, in November, 2010). The George Soros, moveon.org, Daily Kos, extreme radical socialist Marxist, Christian-hating, Constitution-despising, America-hating, wacko-university-professor, New York Times wing of the Democrat Party. We like to think of what happened in Nazi Germany as some kind of aberration, something that happened "over there" and could never happen here (unless of course those "eeevil" Republicans get back in power). But we are mistaken. Evil is evil. If you are capable of exterminating 6 million Jews, then you are capable of mutilating 50 million babies to death. And if you can mutilate 50 million babies to death, then you can exterminate 6 million Jews. Nazism, as the embodiment of pure evil, is alive and well. It reared its disgusting head in Germany 70 years ago. (And remember it was accompanied by that government's embrace of the satanic, unscientific lie of evolution.) And it has been rearing its ugly head in this country for decades. Both movements have Satan as their head. Up until now in this country it has been limited to just mutilating babies to death. But that is no guarantee of the safety of everyone else. Because, if they can mutilate 50 million babies to death, they can exterminate 6 million Jews, or they can imprison, torture and murder 6 million Christians (just like their totalitarian socialist-communist soul mates in China and North Korea have been doing for decades now) (see the Mao White House Christmas tree ornament reference in the next paragraph), or 6 million conservatives who refuse to swear allegiance to them in the future. When people finally wake up it is usually too late. Just ask the Germans.

"If the world hates you, you know that it hated me before it hated you. If you were of the world, the world would love its own: but because you are not of the world, but I have chosen you out of the world, therefore the world hates you."

(John 15:18-19)

In helping to make this historic connection, and in understanding what the far-left is ultimately capable of, one need only go so far as to look at their heroes. On college campuses everywhere young and old imbecile's alike walk around sporting their Che Guevara t-shirts. They also deify him in their films, yet this mass murderer was responsible for the deaths of tens of thousands of Cubans in his role as "the Castro regime's chief executioner."[853] They admire the philosophy of Mao Tse Tung who said "power comes from the barrel of a gun," and they hang his picture-ornament on the White House Christmas tree, yet Mao slaughtered "some 70 million Chinese, along with countless Tibetans, Mongolians, Manchus, Koreans, Hmong, Uyghurs, and other nationalities ...during his long and brutal reign."[854] (Wow! George Bush was an "idiot," but their heightened intelligence allows them to honor someone like Mao on our very own White House Christmas tree?!) And then there is Comrade Stalin, or "Joey" as he is fondly recalled by our academic intelligentsia who share in his dream of a "socialist utopia." The best estimates now show that *Joey*—a real fun guy who once quipped "A single death is a tragedy; a million deaths is a statistic"—was responsible for the genocide of over 60 million human beings. (They despise Ronald Reagan, but they share in the political ideology of a mass murderer?!) Che Guevara, Mao Tse Tung, Joseph Stalin; these three are some of the greatest mass murderers in history, far surpassing the butchery of Hitler and his Third Reich, and not coincidentally heroes all of the Democrat-Left.

> [And, their heroes also include rapists (I believe one William Blythe Clinton was credibly accused of raping one Juanita Broderick—even the FBI, who interviewed her, said her testimony was trustworthy) and racist eugenicists (see the information on Margaret Sanger coming up shortly in "Planned Parenthood's Stealth Genocide of Blacks."]

If you are having trouble making this connection let me say that our point here is not to equate any particular person who may vote Democrat with the leaders of the Third Reich. But it is the *movement itself* that can be equated. Remember that there are 130 million Muslim women walking around without their genitals, and these same baby mutilators who control the means of disseminating information in our country—the media, the entertainment industry, the educational system and now even the federal government—cannot find enough compassion for these once-little girls who were thus brutalized to even inform the American people, and the rest of this clueless world, about it. Instead they let the despicable practitioners of this religion continue to hack away at the genitals of 2 million more little girls every year (while that same "see-no-evil" Left hacks away at millions of unborn babies), without exposure or condemnation for their crimes and for their false-religious insanity. If a Christian parent spanked one of their little girls to the point where it left a small red mark, Christiane Amanpour (CNN anchor and chief international correspondent) would do an hour-long special on it calling for their children to be taken away and for the parents to be prosecuted. But not a word about the heinous crimes of her Muslims friends. Why? For the same reason why they would allow a "happy-faced" version of the Muslim religion to run rampant through the curriculum of so many of our public schools (see section on "Islam" in chapter 8)—in the middle of a war against the dangerous jihadists

---

853   (From "A Guide to the Political Left" @ *discoverthenetworks.org*)
854   (From "A Guide to the Political Left" @ *discoverthenetworks.org*)

of that same Muslim religion no less—while going ballistic if Christianity is even whispered there; and why they can't even identify Islamic killers as terrorists, or even mention to our school children that they were the ones responsible for 9/11 and so many other despicable attacks. But conversely, it's a natural reaction for them and their Janet Napolitano-run, Homeland Security Dpt. to call all decent, patriotic and non-left-wing-indoctrinated Americans "terrorists." Why? Because the Left in this country and the Islamics across this world are <u>one and the same</u>. They are soul-mates. Just like the Jewish religion and the Christian religion are one and the same because they have the same Founder, God; the baby-mutilating Left, the Jew-hating and clitoral-mutilating Muslims and the Jew-killing Nazis have the same founder, Lucifer. You might want to make the connection, America, and rethink your voting habits, before the damage becomes irreversible.

The Nazi's hated the Jews with a demonic hatred. The Democrat Left holds the rights of the unborn, and the intelligence of those who defend them, in a demonic contempt. They disguise themselves as the Party that "cares so much for the poor and the downtrodden" that they can barely stand it. And millions have bought into this lie. But the Democrats are in fact the party of pure hatred.[855] You don't mutilate millions of babies to death because of your boundless love. They hate the unborn, as they hate those who defend them. They hate this country, as they hate the Christian religion which this country was founded on. And they hate God, as they love both their father the devil, and his lies. And with this satanic hatred as the driving force underlying their ideology and ultimately their actions, like the Nazi's, and the Islamists, there is no limit to the evil they are capable of in the future depending on the amount of power they are foolishly allowed to accumulate. God help us if the idiots continue to go out and vote for them.

"Then the second angel poured out his bowl on the sea; and it became as the blood of a dead man; and every living creature in the sea died. Then the third angel poured out his bowl on the rivers and springs of water, and they became blood. And I heard the angel of the waters say, 'You are righteous, O Lord, The One who is and who was and who is to be, because you have judged these things. For they have shed the blood of saints and prophets [<u>and the blood of tens of millions of innocent babies</u>], and you have given them blood to drink. For it is their just due.'" (Revelation 16: 3-6)

The justice and retribution of a Holy God is coming down the pike in the very near future, for all of those who refuse to turn from iniquity.

Just a simple question

"Let everyone who names the name of Christ depart from iniquity." (2 Timothy 2:19)

Before we ask this "simple question" we must make it clear that nothing we say here is to be taken as an endorsement for the Republican Party, as it is also clear that in far too many ways they are the Party of cowardice disguised as "thoughtful moderation," because they lack the intestinal fortitude or the intelligence or both to really stand up for what is truly the most fundamental moral issue of our day. It is akin to the issue of slavery in the time of Lincoln. Also, what is being addressed in this section is ultimately a spiritual and a moral, as opposed to a political, issue.

My question therefore is this: "How can anyone call themselves a *Christian*

---

855   (See *United in Hate: The Left's Romance with Tyranny and Terror*, by Jamie Glazov)

when they go out and vote for baby mutilators?" And let there be no confusion, the Democrat Party is the Party of baby mutilation. It is their un-holy Grail. During the Clinton administration, when the Democrats still controlled Congress, they stood up and cheered when they barely won the *victory*, by one vote, that allowed them to continue to jam scissors into the backs of the skulls of nine-month old little babies, seconds before their birth, and to suck their brains out while they were still alive; and they applauded themselves on their *stirring accomplishment*. How many demons were in the room, unseen and applauding along with them, cackling in delight over their ignorance as to their eternal fate? And they haven't gotten any better since then. Early in 2009, as soon as Obama was sworn in the very first thing he did was to issue an executive order to take our tax dollars—in the middle of a terrible recession where federal tax dollars are scarce and they are already spending us into bankruptcy—and send them to mutilate babies in foreign countries!?#@ Why? Because in truth mutilating babies to death is more sacred to the Democrat Party than the Dome of the Rock is to the Muslims. It is as important to them to keep baby-mutilation going strong, as continuing to kill Jews was to the Nazi's, or wiping Israel off the map is to the Muslims. The Democrat-Left lies to the American people with the aid of their complicit, left-wing media about their phony outrage over the Bush administrations supposed "torture" of terrorists, which in fact consisted of pouring water over the faces of three, count 'em, three Islamic lunatics. That's *torture* according to these psychotics, but what they do to millions of babies every year is *not*?!#@

I ask again, "How can anyone call themselves a *Christian* or even a decent person (as opposed to, say, a disgusting Nazi) when they go out and vote for these baby mutilators?" Just a simple question for "those who have ears to hear."

The Nazi's in Germany slaughtered 6 million Jewish men woman and children, and the thought of them makes us sick. But the Democrat-Left in America have mutilated 50 million babies to death, and yet half of this country, and over 90 percent of African Americans, and over 70 percent of Hispanics, and over 50 percent of Catholics, and incredibly over 40 percent of so-called evangelicals go out and vote for them religiously, every two years?!#* Even though the most recent polls say that 83 percent of Americans call themselves *Christians*! What is wrong with this picture? Is it just me, or are a whole lot of people fooling themselves as to whom they serve? In the *Lord of the Rings* movie, the evil wizard Saruman asks, "Whom do you serve?" And his loyal half-human, half-goblin Uraki growls out his answer, "Saruman!" My question to these loyal baby mutilator voters is, "Whom do you serve?" And the answer is, "Satan!" Certainly not Christ, the author of life and God of the living and Creator of every precious little fully human baby upon conception in the womb.

"They profess that they know God; but **in works they deny him**, being abominable, and disobedient, and unto every good work reprobate." (Titus 1:16)

The Democrats latest meaningless little witty slogan is "YES WE CAN!" but my question is "Yes you can what?" ...Mutilate another 50 million babies to death, while at the same time destroying this once-Christian nation and turning it into another godless, socialist *paradise*? (You know, like Cuba, Russia, Venezuela, North Korea, or California.) And this is what millions of *Christians* went out and voted for? My goodness, we might have been better served if millions of avowed *Satanists* had gone out and voted instead.

Now there are those who might argue that the Democrats and the Left are not the ones responsible for all of those mutilated babies; that those were the decision and the responsibility of the women who made that choice. But that is a lie. What would those same people say if the Democrats made it legal to enslave African Americans again? (Which they might try to do if blacks ever start voting Republican.) Would they say that they are not responsible, only the individual slave owners are? Of course not. For no one would let them off the hook for a second on that one, but they are so quick to grasp at any straw that might allow them to soothe their conscience, and wash their hands, of their responsibility in the baby mutilation issue.

We have stated this before but it is worth repeating, since obviously far too many *Christians* seem to be ignorant of this fact,[856] Demons mutilate babies, Angels protect and defend them. Indeed I would reiterate here, as we first mentioned in chapter 2, that there is an implied Commandment in the Ten Commandments. One that perhaps for far too many has not been spelled out clearly enough in our *progressive* times. So let's spell it out here. We'll call it the Eleventh Commandment and it is this: **THOU SHALT NOT VOTE FOR BABY MUTILATORS**! Why? Because of the sixth Commandment, **THOU SHALT NOT MURDER**![857] Like the Nazi's who worked at the death camps as guards, administrators and processors and who tried to escape indictment for *real* war crimes[858] but were found to be just as guilty of murder and genocide as the leaders at the top; so too those who vote for these baby mutilating murderers, and therefore grant them the power to continue their crimes against God, humanity and countless little babies, are just as guilty of murder of the most innocent before the Ultimate Judge as those who jam in the scissors or rip off the limbs.

"He who is not with me is against me." (Matthew 12:30)

Just before the 2008 election I had the occasion to walk down the halls of a Christian school attached to a Christian church, and all over the walls were posters, and colorful flyers stuck on bulletin boards, declaring the love of Jesus and verses from the Bible and encouragement for the young students to "follow" Him. Yet out in the parking lot nearly every other car had an Obama sticker on the back, even the parking places reserved for pastors and staff. What is wrong with that picture? Indeed what is wrong with *Christian* America? Maybe it has never had the truth spelled out clearly enough for them. So let's give it another try…

"You shall love the Lord your God with all your heart, and with all your soul, and with all your mind, and with all your strength." (Matthew 22:37) And you shall NOT vote for baby mutilators!

"You shall love your neighbor as yourself." (Leviticus 19:18) And don't vote for baby mutilators!

"And be you kind one to another." (Ephesians 4:32) And don't vote for baby mutilators!

"Follow me." (Mark 2:14) And don't vote for baby mutilators!

"You shall not murder." (Exodus 20:13) And voting for baby mutilators is what? That's right, it's murder!

---

856   (See II Corinthians 2:11: "Lest Satan should take advantage of us: for we are not ignorant of his devices.")

857   (As the bumper sticker I saw the other day pointed out: "ABORTION? What part of "Thou Shalt Not Murder" don't you understand?")

858   (As opposed to the phony "war crimes" that the lying Democrat-Left has accused the Bush administration of.)

That should about cover it. Hopefully.

Millions of babies are getting mutilated to death every year in this country by the votes of so-called Christians, just like millions of Jews in Germany 75 years ago were slaughtered and then burnt in ovens because of the support of so-called Christians. (They even sung louder in their churches on Sunday as the trains rolled on by, so they couldn't hear the cries of the Jews jammed inside like cattle. Where exactly do you think those particular "Christians" are now spending eternity?) Could a true follower of Christ have voted for the Nazis in Germany knowing they were slaughtering millions of Jews? No way, the same as a true Christian could not kidnap, molest rape and then murder a young child either. And could a real (as opposed to a pretend) Christian vote for Hamas, or Hezbollah, or the Muslim Brotherhood knowing that they are drooling over wiping out Israel and exterminating all of the Jews in it? Of course not. And keep in mind that the "good Christians" in Nazi Germany didn't lift a finger to help the Jews there either. The "good Christians" here not only refuse to lift their fingers to help the plight of millions of slaughtered babies, but they actually go a step further; they lift their finger to the voting lever, to help the murderers. "The more things change, the more they stay the same."

("Oh no, the holocaust could never happen here in America! Not in *this* country!" Actually it not only could, it has. And it's being perpetrated with the proud support of "Christians." How absolutely revolting.)

"Not everyone who says to me, "Lord, Lord," shall enter the kingdom of heaven, but he who does the will of my Father in heaven. Many will say to me in that day, 'Lord, Lord, have we not prophesied in your name, cast out demons in your name, and done many wonders in your name [*and gone to church in your name, and taught at a Christian school in your name, and donated our moneys in your name, and cried "Jesus! Jesus!" in your name, and been a Pastor or a church leader in your name, and done all the little things that Christians are supposed to do in your name*]?' And then I will declare unto them, 'I never knew you; depart from me, you workers of iniquity!' " (Matthew 7:21-23)

There are two points that are significant in this verse as far as our discussion here is concerned. The first is that there is nothing more iniquitous than the horror of mutilating a baby to death while it is still alive, and no more callous worker of iniquity than those who support these killers with their votes. No one is coercing them, or tempting them to do this evil. They are not like those who are struggling with the sins of alcohol, drug or crack addiction, overeating or lust after the flesh, and the terrible tormenting temptation to fall that goes along with it. They are also not like the young unmarried girl who may be freaked out over suddenly and unexpectedly being in the family way, and who is full of fear and anxiety over the financial, social, parental, peer and lifelong ramifications of what has happened. No, they vote this way proudly of their own volition, thinking themselves *smart*, *savvy* and *educated*, yet obviously wholly ignorant of both the earthly horror they are helping to sow and the eternal spiritual consequences they are preparing to reap.

The second point is that Jesus is not speaking to unbelievers in this verse. He is clearly speaking to those who would call him "Lord, Lord" which could only be those who think they are one of his, who were fooled into thinking that they were Christians but who obviously were not saved. (Like those Germans who just sang louder.) Because they were never converted, they never put on a wedding garment;[859] they were never clothed in the righteousness of Christ and instead lived this life covered in their own righteousness not knowing that his Word says that "all of our righteousness are as filthy [menstrual] rags." (Isaiah 64:6) Indeed the very point of this has not so much to do with politics but that perchance one who reads this and is in need of hearing these words might take an inner look and make the decision to come from darkness into the light before they hear those very alarming words from Matthew 7:23 spoken to them on the other side of the grave. Incidentally, of course, and not to be thought of as a small thing, if enough of the millions of *Christians* who go out and religiously vote for the baby mutilators every two years were to repent and be converted, a side benefit is that we might be able to save the next 50 million babies from being mutilated to death. That is if anyone actually cares enough to be "the salt of the earth" that we are supposed to be, and speak this very hard truth ("dura realidad"), in love, to all of the many millions of *Christians* who sorely need to hear it. Like many of our friends, family members, co-workers and fellow church goers, for example. (Again, where are the missionaries to *Christian* America?)

(The next paragraph was already included in chapter 8 in the section on "Roman Catholicism," but since it is equally relevant here it bears repeating...)

But what about *Judge not, and you shall not be judged*? Doesn't that verse tell us that we should not be saying all of this in this section?! Not really. What that verse tells us is not to stand in self-righteous condemnation of others, which is also not what we are advocating here. But it also is not saying that we should stifle the exercise of our God-given ability to discern between truth and lies, to distinguish between a true and a false Gospel, to differentiate a true preacher of the Word from a distorter of the truth, and to ascertain that certain people are in all probability not the *believers* that someone or some false Christian church or organization has deceived them into thinking they are. And equipped with this discernment, and differentiation, and ascertainment, and proper judgment we can then proceed to witness to those who are so deceived, and can then possibly reap the joy and have the privilege of drawing someone from lies to the knowledge of the truth; which will never happen if the old *Judge not and you shall not be judged* crowd has their way. In truth, the verse that more aptly applies in this case is "When I [Jesus/God] say to the wicked, 'You shall surely die,' and **you give him no warning, nor speak to warn the wicked from his wicked way**, to save his life, that same wicked man shall die in his iniquity; **but his blood I will require at your hand**. Yet, if you warn the wicked, and he does not turn from his wickedness, nor from his wicked way, he shall die in his iniquity; but you have delivered your soul." (Ezekiel 3:19)[860] That is what God commands, yet we seem to be far more concerned with "not offending anyone."

Indeed, in many ways, "not offending anyone" has become the *New Christianity*...

"Oh, Mary, by the way, are you inoffensive?"

Oh, yes, I'm inoffensive. I wouldn't offend anyone.

"OK, you're a Christian then. How about you, Bob, are you inoffensive?"

---

859   (See The Parable of the Wedding Feast in Matthew 22: 1-14, or our discussions concerning it in two previous chapters, 7 and 8)

860   (It's also interesting to note how God, who incidentally has memorized the entire Bible and not just eight words, says the exact opposite of the world's clever but confused admonishers.)

Oh yes, I'm inoffensive.
"And how about you, Suzie? Are you offensive?"
Oh no, I never say anything that is offensive to anyone.
"Okay, then you're a good Christian too."[861]

Now contrast that with what Paul said to the Athenians on Mars hill, or what Stephen said to the Jews before they stoned him to death, or what Paul said to the Corinthians:

Paul's letter to the Corinthians is very relevant to this subject. He scolded the church at Corinth because they had tolerated open sin in their midst without confronting the offender and giving him the opportunity to repent. (See 1 Corinthians 5) Then in his second letter to them he rejoices that they had heeded his words and because of that the guilty one had repented. (See 2 Corinthians 2: 1-11) It's a shame that we don't have Paul with us today as we could sure use a similar letter to the *Americans*. For the sin that he spoke of in Corinth can't compare with the iniquity that we have sown in this country. Yet we as a church do not confront, in love, the guilty parties in order to give them the opportunity to repent and come to the knowledge of Jesus Christ as Lord and Savior. "And why do you call me, 'Lord, Lord,' and yet don't do the things that I say?" (Luke 6:46) ("The things that he says" like "You shall not murder" for example.)

The late George Tiller, known as "Tiller the Baby Killer," became famous for performing the late term *Partial Birth Abortion*, a procedure as we mentioned before that jams scissors into the backs of the skulls of nine-month-old babies and then sucks their brains out while they are still in the birth canal, just moments before they are born and under the legal protection of the state. Women would come to his clinic in Kansas from all over the country to have him perform this barbaric and revolting crime, and he became quite wealthy in the process. But the real killer is that this monster went to a Lutheran church in Kansas every Sunday where he was an usher, and where his wife sang in the choir! And the people in that church, supposedly followers of Jesus Christ, just sat there and allowed such "a worker of iniquity" to be in their midst without confronting him, in love, about his heinous business, and giving him the opportunity to either repent or leave their fellowship. As Paul instructed the church at Corinth to do in his letter of admonition to them, "I have written unto you not to keep company, if any man who is called a brother be a fornicator, or covetous, or an idolater, or a reviler, or a drunkard, or an extortioner [or a baby-killer]; **with such a person do not even eat**." (1 Corinthians 5:11) If Tiller the Baby Killer and his church group are "Christians" for goodness sake then Hitler and Himmler and Goering and Goebbels and every member of the Nazi Party were all "Christians" too. Hell, let's all be "Christians." *Hey, Jeffrey Dahmer, want to be a "Christian?"* ...*No problem. What? Do you have to stop killing and eating young men? Of course not! Don't be silly. You can join the same church as Idi Amin. They have a great pot-luck dinner.* ...(What a disgusting joke liberal "Christians" have made out of their false version of Christianity.)

"No man can serve two masters: for either he will hate the one, and love the

---

861 (Yet our concern with "not offending" might be better placed in the opposite direction, for... "It would be better for him if a millstone [approx. 3,000 lbs.] were hung around his neck and he were cast into the sea, than he should **offend one of these little ones**." (Luke 17:2) And what could be more offensive to those little ones than voting to give power to those who mutilate them to death?? ...Take your time Democrat voters, that's another tough one.)

other; or else he will hold to the one, and despise the other. You cannot serve God , *and support Satan the child sacrificer at the same time.*" (Matthew 6:24)

Recently I had a discussion on this subject with the principal of a local Christian church-school, where a certain percentage of teachers, staff members and church members were open about their support of Obama and the Democrat religion. Here is an excerpt from her written response to my concerns about a "Christian" church-school refusing to confront, in love, this situation:

> "The LCMS [Lutheran Church Missouri Synod] believes that abortion is contrary to God's Word and "is not a moral option …except as a tragically unavoidable byproduct of medical procedures necessary to prevent the death of another human being, viz., the mother" (1979 Res. 3-02A).
>
> …"We respond to abortion with grace and forgiveness when there is repentance. We have Lutherans for Life group that promotes Pro-Life policy. It is preached against in our pulpit. **We do not however, tell people how to vote in a political election**." (Emphasis mine.)

In response I said that I didn't see where hiding behind "politics" was an excuse for refusing to do their Christian duty as outlined in First and Second Corinthians; which is to confront, firmly and in love, the so-called Christians in their church and on their staff in order that they might be given the opportunity to be educated and informed, to hopefully then to repent of the disgusting evil of voting for baby-mutilation, and to become true Christians who have "departed from iniquity." I also said that since they didn't "tell people how to vote in a political election" (as if that level of "polite restraint" was some righteous, commendable act in and of itself) then perhaps they might at least explain to the people who don't seem to get it, the connection between "THOU SHALL NOT MURDER!" and "THOU SHALL NOT VOTE FOR MURDER EITHER!" (The eleventh commandment.) I'm still waiting for an answer…

"God is love, and he who abides in love, abides in God, and God in him." (1John 4:16) God is also truth, and he who abides in truth, abides in God, and God in him.

The two go together. You can't have one without the other. You can't pretend to embrace the love of Christ while at the same time denying the truth. (And if you do then you are a Chino, a Christian in name only.) Especially a truth as fundamental and basic and inarguable as supporting the Nazi's in Germany 70 years ago, especially knowing that they were exterminating the Jews, was to be a genocidal murderer, and no "follower of Christ." Or another truth just as fundamental and basic and inarguable that to vote for baby mutilators is to be accomplice to murder, and a deceived "follower" of Christ. To "abide in love" you must depart from evil, as the verse that began this segment clearly states: "Let everyone who names the name of Christ depart from iniquity." (2 Timothy 2:19)

"And hereby we know that we have come to know him, if we obey his commandments. He who says, 'I know him,' but does not keep his commandments, is a liar, and the truth is not in him." (1 John 2:3-4) And what commandment could be more essential and more basic and more fundamental than *Thou shall not murder*!? As Christians we are all sinners but we are not all child molesters. We are all sinners but we are not all unrepentant murderers. Sin is sin, but iniquity is iniquity. If you say you know him and yet you vote for iniquity, and you refuse therefore to "depart from iniquity," I'm real sorry, but you do not know him, and he doesn't know you… "'I tell you I do not know you, from whence you are.

Depart from me [into the everlasting fires of hell], all you workers of iniquity." (Luke 13:27)

Repentance from evil must come before salvation can enter in. John the Baptist, repentance, came before Jesus Christ, Salvation. John the Baptist prepared the way for the salvation offered by the Savior. Many people seem to have made an easy profession of faith in Christ but without any repentance, without any intention to turn from wickedness. Again, "Let everyone who names the name of Christ depart from iniquity." (2 Timothy 2:19) So many seem to have no problem whatsoever being a supportive part—with their votes and their membership in the Democrat Party of baby mutilation—in the most vile, despicable, heinous, disgusting, putrid evil ever known to man. They "name the name of Christ" but they refuse to "depart from iniquity." Why?

"You search the Scriptures, for in them **you think you have eternal life**; and these are they which testify of Me." (John 5:39) Jesus warns those of us who *think* we have eternal life but do not. The Jews of Jesus' day were covered in their own righteousness and had no interest in the righteousness of Christ or in the salvation he offered. They "thought they had eternal life." They were quite full of themselves as "children of Abraham" (when in truth they were children of their father, Satan- see John 8: 31-59 and in particularly verse 44) and how superior they thought they were. Today how many Americans who call themselves *Christians*, even go to church on Sunday and practice good charitable works, like the leaders of the Jews 2000 years ago, are also covered in their own righteousness and have no real interest in repenting of that foolishness and humbling themselves and embracing the righteousness of Christ and the salvation he offers? They are those who are wearing the Christian religion on the outside. "They think they have eternal life" yet they indicate otherwise by refusing to depart from iniquity. God is a God of justice and mercy. He will have mercy upon those who repent and he will execute justice on those who do not.

"He who overcomes shall inherit all things, and I will be his God and he shall be my son. But the cowardly, and unbelieving, the abominable, and <u>murderers</u>, and whoremongers, and sorcerers, and idolaters, and all liars shall have their part in the lake which burns with fire and brimstone, which is the second death." (Revelation 21:7-8)

There will be no unrepentant murderers, or baby mutilators, or their voters in heaven. They will be agonizing in hell. So I would suggest that if anyone out there within earshot falls into that category, <u>Repent</u>! Change! Actually embrace Christ as your Savior and follow Him! And turn from this disgusting evil before you wake up in eternal horror.

We are all sinners, some saved by grace, some still lost in their sins and in their own *righteousness*. We're not "standing in judgment," only issuing a warning. But if we have indeed spoken the truth here about so-called Christians and their demonic voting patterns, and someone is offended by this[862] then their offense is

---

862 (We would also respectfully submit to anyone who actually *is* offended by this to stop and think for just a moment how those little babies who are getting mutilated to death while they are still alive every single day in this country might just be a tad more offended at those who could care less about their terrible little plight as to go out and support this evil with their votes, religiously, every two years. It's all a matter of perspective, and frankly we'd much rather offend a baby-mutilator voter, hopefully into repentance, than to offend one of these little ones. Because, again... "It would be better for him if a millstone were hung around his

really not with us, but with the truth Himself. For Jesus and the truth are one and the same. "Jesus said unto him, '**I am** the way, **the truth**, and the life: no man comes unto the Father, but through me." (John 14:6)

There is absolutely nothing we can do to *earn* eternal life. There's nothing we can do to *deserve* eternal life. It's a gift. And no one can come to Jesus for salvation, "except the Father …draws him." (John 6:44) And when we think that the vast majority of people who have ever lived on this earth and the vast majority of people who are alive today have not been blessed like us to receive the gift of eternal life, and therefore they are to suffer horribly for all of eternity, there's a whole lot that we should be very thankful for, as true Christians. But here's the thing. We can't earn or deserve our salvation, but what we can do after we get saved is do our best to thank Him by serving him, and following him and by learning the truth about him and his commandments. We can seek to become rooted and grounded in the truth, to study to show ourselves approved, to do our best not to sin, and we can also actually depart from iniquity rather than blissfully going along with and supporting it! Think of what a slap in the face it is to Jesus Christ, to the Lord of all creation who came down here on this earth and died on a cross and suffered humiliation so we can spend eternity in Paradise here on this earth with Him, and be saved from the flames of eternal fire and horrible suffering; think of what a slap in the face it is to Him that millions of people in this country who name His name can't even find it in their minds and hearts to depart from the disgusting evil of voting for and supporting baby mutilators?! I find that almost incomprehensible. Nobody's asking us to do what Paul did; to leave our wives and our children and our family and our homeland and wander around the world and be beaten and stoned and whipped and imprisoned for the cause of Christ. Nobody's asking us to go to Africa or to Saudi Arabia or to South America in a boat to be a missionary for Christ. All most of us are asked to do, in return for being given the greatest gift of all time, is to struggle with sin as best we can and to depart from iniquity, and we can't even do that. We can't even tear our minds away from the arrogance and the ignorance and the deceit of Satan just long enough to depart from the iniquity of being complicit in supporting the mutilation of millions of babies to death. Wake up, *Christian* America! Your behavior is nauseating!

The good news in Paul's communication with the Corinthians is that he could report in his second letter his delight in Christ that they had repented and confronted the offender and that he had in turn repented. Our hope here is twofold. First that Christian America can wake up and find the courage to do what they are commanded to do; to confront in love all of those *Christians* "who name the name of Christ" but who also refuse to depart from this horrible iniquity. And secondly that those *Christians* who are guilty of this iniquity might then repent and come to the *saving* knowledge of our Lord and Savior Jesus Christ, and not the kind of peripheral knowledge and deluded self-righteousness that would allow them to be complicit in the most vile, disgusting, demonic, and Nazi-like horror of our time. WWJD: What Would Jesus Do? Well you can be as sure as the sun's going to rise tomorrow that he would never vote for baby mutilators.

I would encourage everyone, and especially any *fence sitters*, to Google "Dura Realidad" (hard truth) and watch Mexican actor/producer Eduardo Verastegui's

---

neck and he were cast into the sea, than he should **offend one of these little ones**." Luke 17:2)

video. It shows the hideous results of the baby mutilation *procedure*. I know of no one who has been able to watch the entire clip. It is the unvarnished, brutal truth of what you are accomplice to if you go out and vote for the baby mutilators.

Here are the words of Eduardo Verastegui @ www.durarealidad.com, speaking at the end of his video:

> "I know watching this video has been difficult and painful. It's not easy seeing babies that have been killed like this. There are people who after watching this video wonder, how is this legal? I think that abortion is legal because there are not enough men and women to speak out against it. We need to put an end to abortion, and political candidates play a very important role in this matter. If a president is not willing to defend the most innocent of his own country, which are the babies in their mother's wombs, then my question is, who is he willing to defend? It's time to wake up. Now you know the truth. The abortion holocaust represents an evil so terrible that words alone cannot describe it."[863]

Planned Parenthood's Stealth Genocide of Blacks

(Margaret Sanger's extermination of African-Americans is still proceeding as she planned, even decades after her death. And sadly it is her very victims who go out and support her quest, in higher percentages than any other group!?!)

There is another aspect to this "abortion" issue which we would be not only remiss, but in fact racist if we didn't report it; although in doing so the lying, fascist Democrat-Left will no doubt accuse us of being just that- racist. But that is what they do best, lie.

Let's start at the beginning. Margaret Sanger, Democrat Hillary Clinton's self-proclaimed hero, and the founder, inspiration, guiding light and beloved idol of the Planned Parenthood (Baby-Mutilation-hood) organization, was nothing less than an American female version of Adolph Hitler. She was…

> …"one of the worst bigots in American history. …She was nothing more than a sexual deviant who considered herself and the white race as the only race on earth that deserved to partake in the riches of this world."[864]

(Not surprisingly, Hillary's "hero" was also a supporter of the Ku Klux Klan.) And whereas Hitler hated the Jews and wanted to exterminate them in order to speed up the human race's *advancement* by eliminating a *weaker race* (Darwinism in practice), Sanger wanted to do the same here in America, by exterminating "the Negro." They both wanted to carry out Darwin's "survival of the fittest" on human beings. And whereas Hitler deemed the Jews as "unfit," Sanger felt the same about "the Negro." They were two peas in a pod, but whereas we look upon the racist, genocidal Hitler with complete scorn, disgust and contempt, the Democrat-Left looks upon the racist, genocidal Sanger with love, respect and admiration. And African-Americans flock to vote almost exclusively for her and her Democrat-Planned-Genocide-of-the-Blacks Party. Which is the same as if Jews today were to flock to vote almost exclusively for a resurrected Nazi Party, which was still in the business of exterminating them.

"In the *Birth Control Review*, Sanger wrote frankly about the eugenic agenda behind the planned parenthood movement:

---

863   (Eduardo Verastegui's talk at the end of his video @ durarealidad.com)
864   (From "Margaret Sanger, Baby Killer," @ deathbyabortion.org)

"Birth Control is thus the entering wedge for the Eugenic educator... the unbalance between the birth rate of the "unfit" and the "fit" is admittedly the greatest present menace to civilization... The most urgent problem today is how to limit and discourage the overfertility of the mentally and physically defective."

"For Sanger, among the most unfit of all were minority populations, particularly African Americans, whom she believed exhibited an unfortunate tendency to 'breed' excessively.

"From the very beginning, Planned Parenthood crusaders deliberately "infiltrated" Christian churches in order to inculcate their ideas. They found useful idiots—such as Planned Parenthood supporter Robert Drinan, a Jesuit priest... who could spread the anti-child gospel among the churches and synagogues in ways that they, avowed atheists, could not.

..."Slowly, step by step, one by one, over about 20 or 30 years, the various Christian denominations and liberal Jewish organizations caved in. With the advent of oral contraceptives in the early 1960s, the sexual nirvana prophesied by Margaret Sanger—which she defined as "unlimited sexual gratification without the burden of unwanted children"—had arrived.

"Within a few short years, Ellen Peck would publish *The Baby Trap* in 1971 and proclaim that having a baby is "the biggest mistake of your life" and Paul Ehrlich, author of the 1968 book *The Population Bomb*, would convince government and media elites that the greatest problem the world faced was too many children.

"Even though virtually every single one of Ehrlich's dire predictions would be proven dead wrong—and the developed countries of Europe now face a disastrous "birth dearth," not a population bomb—his conclusions about "over-population" are still unquestioned dogma among the older, more insensate members of the media, government, the law courts, academia, and the clergy."[865]

God says, "Be fruitful and multiply." (Genesis 1:28) Satan and many of his liberal followers say that babies are a problem, a hindrance and a "choice" to be mutilated to death at the whim of the mother. Kind of reminds you of the Carthaginians, allowing strangers to slice the throats of their own children for a few trinkets. But Margaret Sanger's primary focus for her Planned Parenthood organization was on eliminating the Black race...

"We should hire three or four colored ministers, preferably with social-service backgrounds, and with engaging personalities. The most successful educational approach to the Negro is through a religious appeal. We don't want the word to go out that **we want to exterminate the Negro population**, and the minister is the man who can straighten out that idea if it ever occurs to any of their more rebellious members."[866] (Emphasis mine.)

Keep in mind that in her own words she definitely wanted "to exterminate the Negro population," she just didn't want the word about it to get out. And it is no accident that today her baby-mutilation clinics are primarily located in poor Black and Hispanic neighborhoods.

---

865 (Pgs. 84-86, *The Politically Incorrect Guide to the Bible*, by Robert J. Hutchinson)

866 (*"Margaret Sanger's December 19, 1939 letter to Dr. Clarence Gamble, 255 Adams Street, Milton, Massachusetts. Original source: Sophia Smith Collection, Smith College, North Hampton, Massachusetts. Also described in Linda Gordon's Woman's Body, Woman's Right: A Social History of Birth Control in America. New York: Grossman Publishers, 1976."* ...From *EadsHome Ministries* online)

"To deal with the problem of resistance among the black population, Sanger recruited black doctors, nurses, ministers and social workers 'in order to gain black patients' trust' in order 'to limit or even erase the black presence in America'"[867]

It is also interesting that the primary influence that compels the vast majority of Blacks to vote Democrat every two years comes down sadly from their own pulpits, just as Sanger had planned: "The most successful educational approach to the Negro is through a religious appeal." Listen to Brenda Battle-Jordan, director of Black Americans for Life:

"I love working with Black Americans for Life to save the lives of our little black babies who are aborted **three times more often than babies of all other races combined**. [Emphasis mine] I read a book entitled: "The Pivot of Civilization." (1922). It is a biography of Margaret Sanger, the founder of Planned Parenthood. The book revealed that Sanger was into eugenics, and she claimed that some races were genetically superior and more fit to survive than others. Sanger called for segregation of feebleminded, mental defects, imbeciles and morons stating that they should be prevented from reproducing their kind. We can still hear her statements today as her mission is still being carried out by Planned Parenthood in their facilities across America. All abortion clinics in our county are located in poor neighborhoods targeting poor Black and Hispanic women, and that was her goal."[868]

Three times as many Black babies are mutilated to death every year than "all other races combined!" 37% of all babies mutilated to death every year are African-American babies, yet Blacks are only 12% of the overall population. Sanger, if she wasn't suffering in hell, would be rolling around in her grave, dying for a second time, this time from laughter. But here is the real killer: 90 percent of African-Americans go out and vote for this every two years, religiously! Incredibly, they have been bamboozled into going out and voting for their own genocide! (And 70% of Hispanics do the same, even though they too were, and are, a target of Margaret's extermination policy.) And this is not to accuse them of being "stupid," or "ignorant," or "not as bright" as other races. (For one, I don't remember seeing any blacks joining the Moonies or the Hare Krishna's.) But it is to accurately report the truth about this situation. And yet if this truth *had* been accurately reported over the years instead of being swept under the rug by a lying, left-wing, racist Democrat media, then Blacks would certainly not be in this situation. But then they also wouldn't be going out in record percentages and voting for the Democrats, religiously, every two years, and that is a potential situation that is anathema to the white liberals in control of the dissemination of information in our culture. So the truth is suppressed, and Blacks as a voting bloc are still kept in the dark.

It is more important for the Democrat-Left to see the genocidal dreams of their racist-bigot-hero-feminist friend, Margaret Sanger, being fulfilled, than it is to report the truth about this ongoing genocide to the African-American community so they might be able to make a more informed decision as to who they vote for. This is "all the news" that the left-wing racists at the New York Times do <u>not</u> think is "fit to print." Chris Matthews (the disgusting racist) is not only thrilled; he has thrills going up his leg over the election of someone who will ignorantly continue

---

867 (Pgs. 197-198, *Medical Apartheid: The Dark History of Medical Experimentation on Black Americans from Colonial Times to the Present*, by H.A. Washington)
868 (Brenda Battle-Jordan, director, Black Americans for Life, @flintrtl.com)

Margaret Sanger's, and Planned Parenthoods, genocide of the Blacks. He and his racist kind in the media will do everything in their power to keep Blacks in the dark; and to lie about, demean, disparage and mock anyone who would attempt to enlighten them. Democrat broadcasters, entertainers, professors, comics and politicians, who put themselves forth as being the great liberal defenders against racism, are in fact the very racist bastards who purposely keep Black Americans in the dark. And sadly, the last guilty party are the many Democrat, false Christian preachers who have lied to their own Black congregations, not only keeping them in the dark, but actively campaigning for them to go out and vote Democrat, religiously, every two years, no matter what. (Hello again, Dr. Anderson.) It is sickening, unless you're a lying racist of the Democrat-Left.

Listen to black presidential candidate and former U.N Ambassador Alan Keyes, speaking on the occasion of the opening of one of Planned Racist-hood's new super-clinics...

> "The number one taker of black life is abortion, and it's time people woke up to that fact. The location of this latest Planned Parenthood facility in yet another minority neighborhood is a part of the continuing fulfillment of the racist, eugenicist dream of their founder, Margaret Sanger, but it is a travesty of the American dream."[869]

Is this really the legacy of the civil rights movement? Is this the end result where left-wing black leaders, false-Christian black ministers, and ignorant uneducated left-wing entertainers and sports figures like Oprah Winfrey and Charles Barkley would lead their own people down the road of slow genocide, fulfilling the dreams of the white, feminist hero and disgusting bigot, Sanger?[870] Wow! According to the idiots on the Left, Barack Obama is a genius. Yet what level of "genius" does it actually take to head up a political party dedicated to the genocide of your own race? I mean, how bamboozled and uneducated do you have to be to perpetrate your own races genocide? Shouldn't that "genius" actually read "ignoramus?" (Or is it the half-white part of Obama that hates blacks and is a racist him-half-self??)

It is also important to note that the leaders of the civil rights movement in the 60's were all firmly against the legalization of murdering babies, because they feared, correctly, that it would be used to eugeni-cise Blacks. Yet modern day black leaders like Jesse Jackson and the "Reverend" Al Sharpton, who would never ever even entertain the thought of straying from the Democrat plantation, have been well-compensated for the selling out of their own race. The racists on the Democrat-Left like to call any black who is not a Democrat-voter like them, an "Uncle Tom." But who are the real "Uncle Tom's?" The leaders of the modern-day civil rights, vote-Democrat movement? The Black Democrat Congressional Caucus? Black, left-wing sell-outs in our lame-stream media and the nation's newspapers? Black, left-wing, Democrat-voting professors like our good friend Dr. Marc Lamont Hill? And finally all of those Black Democrat turncoat preachers and ministers? Will the real Uncle Toms please stand up? I report, you decide.

Ann Coulter, in her book *Godless: The Church of Liberalism* compares Hitler and his program to exterminate the Jews, with the program of a soul mate of his over here:

---

869 (Alan Keyes)
870 (See *Killer Angel: A Short Biography of Planned Parenthood's Founder, Margaret Sanger*, by George Grant)

"In America, Margaret Sanger, founder of Planned Parenthood and early proponent of "positive eugenics," also cited Darwinism to promote her "religion of birth control." She believed the theory of evolution provided grounds for eliminating the "unfit." In her 1922 book *Pivot of Civilization*, she advocated the elimination of "weeds ... overriding the human garden"; the segregation of "morons, misfits, and the maladjusted"; and the sterilization of "genetically inferior races." She was not oblique in identifying the "weeds" of humanity. In a 1939 manifesto titled "Birth Control and the Negro," she noted that "the poorer areas, particularly in the South ... are producing alarmingly more than their share of future generations." Sanger recommended birth control to lessen the financial burden of caring for such weeds, "destined to become a burden to themselves, to their family, and ultimately to the nation." Undoubtedly, she would be delighted to know that today (1) Planned Parenthood is the leading provider of abortion in the United States, and (2) about 36 percent of our aborted babies are black, almost three times their percentage in the American population. Mission accomplished, Margaret!"[871]

Yea, boy! Vote Democrat! Vote Planned Parenthood! Vote Margaret Sanger! Vote for the slow but relentless genocide of the black race! Incredible how the wool has been pulled over an entire nation's eyes, isn't it?

"If we say that we have fellowship with him, and walk in darkness, we lie, and do not walk in the Truth." (1 John 1:6)

Today blacks are enslaved on the Democrat plantation. They blindly march out and do the bidding of their white, liberal masters, and in doing so they also support Margaret Sanger's and Planned Parenthood's slow genocide of their own race. Many blacks are already aware of this, and a number of blacks *and* whites have come out and said so, yet many more are afraid to speak up for obvious reasons. But the only way African-Americans will ever be freed from this modern-day enslavement is to be properly informed as to the truth, which they will never get as long as the Democrat-Left continues to control our media, entertainment industry, educational system, late-night and Comedy Central bozos, and far too many black pulpits. Again, the lies of Satan shackle the minds of the people of this earth, and they believe what they are taught, and sadly in this instance they vote for who they are taught to.

Margaret Sanger was a disgusting racist and evolutionist who had as one of her goals in setting up Planned Parenthood to "exterminate" (her words) what she called the "unfit and mentally and physically defective[872] Negro race." And she is accomplishing that goal today as we speak, from the grave, with the help of her Democrat Party and all of those deceived by the ruler of this world into going out and voting for them. That's the truth. Do with it what you will, America.

The toll of abortion on its female victims

"Your iniquities have separated between you and your God, and your sins have hid his face from you, so that he will not hear. For your hands are defiled with blood, and your fingers with iniquity; your lips have spoken lies; your tongue has muttered perversity." (Isaiah 59:2-3)

One of the lies that the progressive Left used in order to rationalize the legalization of baby mutilation was the life of the mother. (Like baby mutilators really give a damn about life!) The story was told over and over back in the 60's

---

871   (Pg. 271, *Godless: The Church of Liberalism*, Ann Coulter)
872   (Obviously she never watched a basketball game.)

and the early 70's of how 5,000 to 10,000 women per year were killed by "back alley abortions." It turns out this was a total lie, the figure being closer to 200 to 300. Now any death is a tragedy, but for the Left to intentionally inflate the actual deaths of two to three hundred into five to ten thousand (a 3 thousand percent exaggeration), in order to legalize the mutilation of 1.2 million babies per year, is an atrocity. Again, do the math. Also, the phrase "back alley abortion" was touted over and over to engender sympathy, but the fact of the matter is that the vast majority of illegal abortions were still performed by doctors, and no doctor to my knowledge has ever set up their practice in a "back alley." They pretty much unanimously prefer to practice "indoors." But don't confuse a Democrat and his nifty slogan with something as inconsequential as the truth. The funny thing is, although it's not that funny, today more women are killed per year due to legalized abortion than ever died in the past, before Roe V. Wade.[873] My, "you've come a long way baby."

"Most people today …think of abortionists during the decades of abortion illegality in a different way. A Lexis-Nexis check of *The New York Times* shows "back-alley" linked to "abortion" 155 times as in, "the threat of the back-alley abortionist with a coat hanger . . . the back-alley abortion era . . . specter of a return to back-alley abortions . . . drive women—as in the pre-Roe days—to risk their lives to end pregnancies with illegal back-alley abortions." Abortion through the 1930s was terribly dangerous for women and fatal for their children—but not because of back-alley practice. The maternal danger came because of infections that could occur at the hands of careful practitioners and, before the creation of penicillin, the absence of ways to fight them... But it was during the antibiotics era from the 1940s onward that the possibility of maternal death, although decreasing, became a powerful propaganda tool for those in academia and the media who greased abortion attitudes in a way that made *Roe v. Wade* possible."

…"From 1962 through 1972 a full-court press sold abortion on both the front pages and the editorial pages of leading newspapers, at a time when those gazettes still had clout. [Isn't it nice to know that even back then the Left was lying to the American voter through its control over our media? And now they've had so much more time to hone their craft.] The editorials were somber, while the news pages made abortion seem easy. By the end of the decade, with abortion legal in several states, the *Omaha World-Herald* was quoting "Betty" describing her abortion: "I had to stay quiet for 15 minutes. When I got up, I felt like a brand-new woman. I felt so happy." The *Long Island Press* quoted "Susan" telling the abortionist when the operation was over, "Oh, thank you, thank you." The reporter added, "Within the next half hour she will have some cookies and a soft drink in the recovery lounge, fill out a few forms, pay a fee of $200 and be on her way back home"—probably skipping, the article seemed to suggest.

[Skipping along, no doubt, while whistling the more anatomically challenging version of Zip-a-Dee-Doo-Dah made famous by a Clark Griswold (Chevy Chase) line in National Lampoon's *Vacation*.]

"Stories on abortion typically portrayed pro-abortionists as merciful and anti-abortionists as closed-minded. The *Memphis Commercial Appeal*, under a headline "Hand of Mercy Extends in Abortions," indicated that pro-abortionists counseling pregnant women were "answering these women's needs." The *Houston Post* quoted this line: "People say an aborted child might have grown up to be President. There's a better chance

---

873   (From article, "Myths of the abortion lobby," by Marvin Olasky, World Magazine, January 2009)

he would have grown up to be the one who shot the President." [How nice! Talk about flat out racism. Can we get a comment from Obama on that? Or is that "a little above his pay grade" as well?] The article attacked anti-abortion laws "passed before women could vote, based on ideologies conceived by men.""

[And how about that Nazi propaganda! "If you shout a lie loud enough and long enough, enough people will believe you." Even in the 60s Satan was using his media to accomplish his goals, in this case bringing back child sacrifice, just like the way things were thousands of years ago in those "primitive" cultures around the Mediterranean.]

"Even when bombarded with such propaganda, Americans who had the chance to vote on abortion resisted."[874]

Which is why the secular-progressive propagandists needed to circumvent the will of the voter by using their allies on the court to pervert and distort the Constitution of the United States. Of course nothing has changed today. Why is it that the Left has to lie in order to "sell" the American people on any of their policies? Whether it's baby-mutilation, open borders for illegal immigration, Cap and Tax, Spendulous, destroying health care, downplaying the War on Islamic psychos or socializing our economy: why do they always have to lie? Why is the truth such a sworn enemy to them and their ideology and their policies? Could it be as we have said for the hundredth time that their god is Satan, the *father* of lies?

They had to lie to make it legal, and they have to lie to keep it legal...

"As women continue to be victimized by abortion, mainstream medicine ignores the long-term negative mental health risk it poses. ...Published, peer-reviewed studies from New Zealand, Australia, and the United States show a link between abortion and alcoholism, substance abuse, depression and anxiety, among other less common negative effects. Personally, I can attest to the despair, depression, what-ifs, and regrets woman have after induced abortion. ...The politics of abortion advocacy has tainted medical research and medical researchers into claiming safety and no long-term adverse mental effects for a procedure that takes a pre-born human life. Excellent studies contradict their conclusions, conclusions that fundamentally lack the ring of truth."[875]

"*Safe* and legal"?! Certainly not for the baby, and not for the mother as well. Maybe feminists should change their slogan to "Dangerous and harmful, but what the f**k, it's legal!"

"Any abortion is psychologically and spiritually hard for all but the most hardened."[876]

And then there are our friends at Planned Mutilate-Your-Baby-hood who really have the best interests of the distraught pregnant woman in mind, have no vested or financial interest in coercing her into killing her baby for their profit, and are just overall some of the most kind and decent people you could ever imagine. Right up there with the concentration camp guards. Just doing their job. Just "protecting women's rights." Just another paycheck. Another day, another few thousand babies slaughtered.

---

874 (From article, "Selling Abortion: Abortion Past: Through academia, the press, and TV, pro-abortion forces peddled their cause," by Marvin Olasky, World Magazine, January 2009)
875 (From article, "Hidden Pain: Politics suppresses the problems of many post-abortive women," by Matt Anderson, a practicing OB/GYN in Minnesota, in World Magazine Sept. 2009)
876 (From article, "Myths of the abortion lobby," by Marvin Olasky, World Magazine, January 2009)

"The segment of the American public most devoted to unlimited abortion rights is young women, right? *Wrong.* Surveys reveal that *young single men aged 18 to 34* favor unlimited abortion more than any other segment."[877]

Hmm. It seems the end result of the feminists "fight" for the reproductive rights of women has been to give young single men the pleasure of taking advantage of them without any consequence. The mother of the aborted baby, however, must live the rest of her life with the physical and emotional consequences while the young penis…er…man can saunter off to freely take advantage of another poor young sucker. Yea, "You've come a long way, baby." Young single men should at least have the decency to get together en masse and send off a few million thank you notes to the National Organization of Women.

And then there's this…

"The Elliot Institute has released a report that exposes America's forced abortion epidemic.

Elliot Institute spokesperson Amy Solby tells OneNewsNow that one study found 64 percent of women who had abortions reported they felt pressured to abort by others. "Something like 80 percent of them said that they didn't get the counseling they needed to make a good decision, that often they were not given counseling at all, or that the counseling they had was inadequate," she explains.

"Solby also mentions forced abortions, which are not widely discussed in the U.S. An article released from the Institute cites two cases in 2006 in which teenage girls were violently persuaded to have abortions. In Maine, a couple abducted their 19-year-old daughter, bound and gagged her, and drove her to New York for an abortion. However, she escaped from her parents in the parking lot of a store and called police from her cell phone."[878]

But at least women have an advocate in the National Cancer Institute. Yea, right…

"A national cancer information organization wants a congressional probe of the National Cancer Institute (NCI) for avoiding to report a major cancer link.

"The Coalition on Abortion/Breast Cancer, which has been fighting a battle for years for recognition of the link between abortion and breast cancer, is sending a letter to President Obama and congressional leadership.

""There's a difference between what the National Cancer Institute says about the link between abortion and breast cancer, and the breast cancer risk of using oral contraceptives or the pill," explains Karen Malec, Coalition president.

"She adds that extensive research, now recognized by one of the National Cancer Institute's leading researchers, shows a definite link between the two. The NCI, however, continues to deny it, so Malec and her organization are demanding a congressional investigation by sending the letter signed by doctors and pro-family organizations to Capitol Hill.

""If they want to sit back and watch women die because they don't have the political courage to do the right thing, to clean house at this corrupt agency, well then, both parties are going to have to answer to angry women for their failure to intercede here on women's behalf," the Coalition president warns.

---

877   (From Human Events circular, Tom Winter, Editor-in-Chief)

878   (From article, "Forced abortions – America's secret epidemic," by Charlie Butts @ OneNewsNow.com 1/11/2009)

"She further mentions that her organization has clear evidence of misconduct on the part of the NCI, and she feels it is necessary to get information out on the cancer connection so women can make informed decisions."[879]

Isn't it truly remarkable in America today that we are burdened with a Democrat media and political party that pretends to care so much about woman and their rights that they would allow information about the health risks to woman to be repressed because it might cast a bad light on their one great love- mutilating babies to death? How else do you explain it? A new word needs to be invented because hypocrites doesn't quite do them justice. How about skankocrites?

If you possess the *morality* that allows you to mutilate 50 million babies to death, then lying is as nothing. Baby mutilation itself comes from the father of lies, and those who side with him have been given over by God to a reprobate mind, controlled by the lies that they so love. The Democrat-Left tells themselves how *righteous* they are for "protecting women's rights." But that is all a lie. There is nothing righteous about what they do. There is nothing righteous about killing babies for the convenience of the mother. The truly righteous thing is protecting a babies right to live, like those "eeevil," "mean-spirited" Christians, conservatives and Republicans want to do; so they can actually live their life (whether with their birth mother or an adoptive family) like the liars on the Democrat-Left have been allowed to do. And now, in addition to all of their other numerous falsehoods on this subject, we find that these same "champions" of women and their "rights" have been repressing vital information about the very real health risks associated with a very demonic ritual. (I guess, unlike the right to choose, the right to be informed is just one that women are not entitled to.)

Forgiveness

If you are a woman who has suffered through having an abortion and has lived to regret it and repent of it, then know that God of course loves you. Enough to die on a cross for you and your sins so if you respond to His grace those sins will be washed away, and put under the blood of Christ. Paid for. Forgiven. And know that your baby is in heaven praying for your repentance and conversion and salvation so the two of you can be reunited one day. He or she is waiting with a big hug. God is very quick to forgive the little people who make big mistakes. Even grievous ones. He knows the deceiving nature of the evil one and his lies. And He knows the stress and the strain and the pressures and the struggles of daily existence in a fallen world. Indeed he is more aware of it than you or I could ever be. We have very limited awareness of just how much He loves and cares for us, and how much His very Presence is right here with us, all of the time.

But what is to be said to the poor girl or young woman today who finds herself pregnant but who doesn't want the baby? Well, for starters, let's think of the baby, shall we? Let's ask the baby if he or she would rather be born, or be mutilated to death? To be adopted by a nice family, or to be brutally slaughtered before she breathes her first breath? And even if being born would entail living in poverty with a struggling single mom. I would also like to be able to get a message out to the young women of this country and this world that, if you get pregnant, you've got yourself a baby. Accidents will happen, but murder is not an option. Not a sane

---

879   (From article, *Important cancer findings remain overlooked*, by Charlie Butts @ OneNewsNow.com 1/29/2010)

option at least. And not to be thought crass but my suggestion therefore is, if you don't want a baby then don't stick a penis in your vagina. Call me crazy but I think that works every time. I mean, despite what the Democrat-Left tells us, we are not powerless buffoons. We are not chimpanzees without any conscious control over the desires of our loins. We have power over our bodies if we are raised that way and if we choose to act in a moral and an intelligent way, with some discretion, and in the way that all women in the past acted in order to protect themselves from being used for the sexual pleasure of a young man who had no intention of "buying the cow when he could get the milk for free." Which of course was a very wise and common sense saying from the past, repeated for millennia, that young woman were taught by older intelligent and experienced women that if he doesn't want to marry you and support you and your baby then do NOT be so stupid as to let him f**k you for free, and then go on his merry way. "OH, but I love you so much!" "If you "really" loved me, baby, you would let me do it!" If you ever hear that, young lady, make sure you also hear this next word as well, that Satan and his demons are snickering behind your back. A word that they and the Democrat-Left do NOT want you to hear... "SUCKER!!"[880] But that's not how we are raising teenage girls in this society today. They are being raised by the sexually "liberated" liars in our Democrat-Left media, educational system and entertainment industry, who tell them to follow their sexual urges without restraint and if "oops" they get pregnant, they always have the "easy" baby-disposal option in their little purses.

Also, for the record, we personally think that a woman who becomes pregnant due to rape or incest should have the option of killing the offspring of the rapist. Also, if a woman's <u>life</u> is <u>truly</u> in jeopardy then that decision to kill the baby in her womb in order to save her own life, is between her and her Maker. But these very rare situations are not really a part of the debate. It's like telling the Nazi's it is wrong for them to slaughter six million Jews, and for them to reply, "Well what if the Jew is a murderer? Can't we kill him then?" Again, it's another red herring and has nothing to do with the *debate*.

And finally, we want to end this chapter with the powerful lyrics of the song *Happy Birthday* by the rapper, Piper. You can Google this and watch it on You Tube. It's a beautiful message of repentance, and of hope for forgiveness and reconciliation...

> Happy Birthday...so make a wish
> Please accept my apologies; I wonder what would have been.
> Would you have been a little angel or an angel of sin?
> Tom-boy running around, hanging with all the guys.
> Or a little tough boy with beautiful brown eyes?
> I paid for the murder before they determined the sex,
> choosing our life over your life meant your death.
> And you never got a chance to even open your eyes,
> sometimes I wonder as a fetus if you fought for your life.

---

880 (And let me encourage, nay plead with, every young lady to protect themselves against being a sucker of the dark side and of our Democrat-Left indoctrination system by reading and following the advice of a book previously recommended, *Seduced by Sex: Saved by Love: A Journey out of False Intimacy*, by Jan Kern. And may God richly bless you.)

Would you have been a little genius? In love with math?
Would you have played in your school clothes and made me mad?
Would you have been a little rapper like your poppa The Piper?
Would you have made me quit smoking by finding one of my lighters?
I wonder about your skin tone and shape of your nose,
and the way you would have laughed and talked fast or slow?
I think about it every year, so I picked up a pen
Happy birthday, I love you whoever you would've been.
Happy Birthday

What I thought was a dream (Make a wish)
Was as real as it seemed (Happy Birthday)
What I thought was a dream (Make a wish)
Was as real as it seemed
I made a mistake

I got a million excuses, as to why you died.
And other people got their own reasons for homicide.
Who's to say it would've worked, and who's to say it wouldn't have
I was young and struggling, but old enough to be your dad.
The fear of being my father has never disappeared,
I ponder it frequently while I'm sippin' on my beer.
My vision of a family was artificial and fake
so when it came time to create, I made a mistake.
But now you got a little brother, maybe it's really you.
Maybe you really forgave us knowing we were confused.
Maybe, every time that he smiles it's you proudly
knowing that your father's doing the right thing now.
I'll never tell a woman what to do with her body,
but if she don't love children, then we can't party.
I think about it every year, so I picked up a pen.
Happy birthday, I love you whoever you would've been.
Happy birthday.

What I thought was a dream (Make a wish)
Was as real as it seemed (Happy Birthday)
What I thought was a dream (Make a wish)
Was as real as it seemed
I made a mistake

From the Heavens to the womb to the Heavens again.
From the ending to the ending, never got to begin.
Maybe one day we could meet face to face,
in a place without time and space.
Happy birthday.
From the Heavens to the womb to the Heavens again.
From the ending to the ending, never got to begin.
Maybe one day we could meet face to face,
in a place without time and space.
Happy birthday.

> What I thought was a dream (Make a wish)
> Was as real as it seemed (Happy birthday)
> What I thought was a dream (Make a wish)
> Was as real as it seemed
> I made a mistake[881]

"From the Heavens to the womb to the Heavens again. From the ending to the ending, never got to begin. Maybe one day we could meet face to face, in a place without time and space." Wow. Amen, brother. I pray that you do. I pray that you do.

In the book of 2 Kings the history is recounted of when Elisha the prophet was surrounded by the king's army of horses and chariots, and his young servant, seeing this great host that outnumbered them by thousands to just two, was exceedingly afraid. Until Elisha prayed that the young man's eyes would be opened so he could see that the number of those who were with them was greater than those who were against them.

"And when the servant of the man of God awoke early, and went outside, behold, an army compassed the city both with horses and chariots. And his servant said unto him, 'Alas, my master! What shall we do?' And he answered, 'Fear not, for those who are with us are more than those who are with them. And Elisha prayed, and said, 'Lord, I pray, open his eyes that he may see.' Then the Lord opened the eyes of the young man, and he saw. And, behold, the mountain was full of horses and chariots of fire round about Elisha." (2 Kings 6:15-17)

We can only hope and pray that the Lord will open the eyes of this nation so they too, like the servant of Elisha, will see. Not only the armies of angels with horses and chariots of fire, poised to judge this nation and this earth for allowing unspeakable evil to run rampant over it. A judgment that will include more earthquakes and floods and tornadoes and hurricanes and tsunamis and fires and wars and plagues and financial turmoil and economic collapse; that will continue to increase in frequency and severity in the coming years as our Mother Earth prepares to cleanse herself of this wickedness that weighs so heavily upon her; and as we get closer and closer to the end times and that final, even more terrible, seven-year tribulation period. But also to open our eyes so we can see just how ugly and disgusting and putrid and inhuman and monstrous the mutilation to death of millions upon millions of His innocent babies actually is, so we might wake up as a nation and put an end to it. Before it's too late.

See:
The LIE: Evolution, by Ken Ham
Sanctity of Life, a short film by Beamer Films
The Politically Incorrect Guide to the Bible, by Robert J. Hutchinson
"A Guide to the Political Left," @ *discoverthenetworks.org*
United in Hate: The Left's Romance with Tyranny and Terror, by Jamie Glazov
EadsHome Ministries online
Black Americans for Life, @flintrtl.com
Killer Angel: A Short Biography of Planned Parenthood's Founder, Margaret Sanger, by George Grant

---

881   (Flipsyde – Happy Birthday, featuring The Piper and t.A.T.u.)

www.durarealidad.com
www.prolife.com
www.hispanicsforlife.com

Unplanned: The Dramatic True Story of a Former Planned Parenthood Leader's Eye-Opening Journey across the Life Line, by Abby Johnson and Cindy Lambert

CHAPTER 12

HOMOSEXUALITY

"A new commandment I give unto you, That you love one another; as I have loved you, that you also love one another." (John 13:34) (Love one another, yes. But that doesn't mean *sleep* with one another…)

Before we get started in this chapter we want to state that, similar to what we said at the beginning of the last chapter, our intention here is to uncover the truth about homosexuality and lesbianism, and not to condemn, put down, demean or vilify any man or woman who is trapped in that behavior, or who is struggling to be freed from it. We are all sinners, some of us saved by grace, and repentance and forgiveness (again, in that order) are open to all. Christ died for sinners, and we are all of us sinners. Some of us struggle with drugs, some with alcohol; some with fear, and some with worry. Many struggle through life under the weight of the lies that they were taught, and some struggle with coming to the knowledge of the truth when presented with it. Some struggle with lust and adultery, and some with homosexuality. We are all of us strugglers in one way or another, so we certainly don't want to come across in the wrong way in this most sensitive, and politically incorrect, of subjects. And because exposing the truth about homosexuality and lesbianism *is* so politically incorrect and therefore protected from any discussion, rational thought or opposing viewpoints by the Democrat-Left that again controls the dissemination of information in our culture, we know that no matter what we say here, and no matter how absolutely true it is, and no matter how kind and loving and sensitive we are in its presentation, they will still attack us like a pack of rabid dogs. For the very reason that it is the truth. (If it were lies they would love it.) "If the world hates you, know that it hated me first." (John 25:18) So we will just seek the truth here, and let the chips fall where they may. But again, this is a book about the truth, so we will assume that the reader—whatever their background may be, and especially after delving this far—is interested in seeking the truth more than they are seeking "not to be offended." Also, we will let the reader decide for him-or-herself whether we have presented the truth the whole truth and nothing but the truth here or not. (So help me, God.)

Also, by using certain analogies in this chapter we are not trying to equate homosexuality or lesbianism with other far more heinous sins. Like bestiality, for example, or the totally disgusting and putrid iniquity of child molestation. (Or supporting baby-mutilation with your votes for that matter.) We are using these analogies only to make a point, and again not to equate the two parts of the analogy. They are *analogies*. They are used as a teaching tool, and that is all.

What is really *normal* and what is really not

"You shall not lie with mankind, as with womankind: it is abomination." (Leviticus 18:22) (And don't confuse that with *Obama-nation* which is a totally different subject, albeit related… "You shall not lie with the left, as with the right: it is *Obama-nation*.")

Is homosexuality *normal*? Good question. But before we answer it, let's first ask another one… Is sheep f***ing *normal*? Answer… Of course not. Everyone agrees with that. (Well, almost everyone. Everyone except the sheep f***ers.) And whereas we can all thank God that we still live in a society where people actually do have that kind of sense, there is no guarantee that they will still have that kind of sense tomorrow, or next month, or next year. (Give our liberal television producers a few more years and all bets are off… "The critics are just raving about the upcoming new comedy for the 2015 fall season, *Two Men and a Sheep!*" Baaa!) Because, like morality, if what is considered *normal* comes from the consensus of the majority then anything goes. And what is disgusting one day—like the aforementioned *Two Men and a Sheep* or the equally distasteful *Will and Grace and Fido* and *Desperate Housewives and their German Shepherds*, or the altogether repugnant *NAMBLA Goes to Romper Room*—can be completely acceptable the next.

And that is the point of the modern day *enlightened* claim of *normalcy* for homosexuality and lesbianism. There was a time when these sexual practices were considered abnormal and unacceptable. But now with decades of help from our liberal media, entertainment industry, educational system and judiciary, and the radical homosexual movement itself, the tide is turning. But those same people who consider themselves progressive and tolerant, who are being turned by this tide, might want to take a sneak peek at what the next step is going to be. We have seen earlier in chapter 2 how the ACLU—as evidenced by their pro bono defense of the repulsive North-American-Man-Boy-Rape Association (NAMBRA)[882]— thinks <u>raping little boys</u> should now be thought of as *normal*, and those who encourage the same should be vigorously defended for free. If what was *abnormal* yesterday is *normal* today, then what is *abnormal* and unacceptable today might very well be quite normal tomorrow. Seeing as how we are already headed in that direction. (PETA is already there.)

No, the fact of the matter is that just as what is right and what is wrong comes from above and does not ooze up from below; so too the concept of what is normal and acceptable sexually, and what is not, comes from the decree on high as well. (And in addition to His decree, it stems from his very Creation. He <u>created</u> them male and female from the beginning, for the purpose of union and procreation and child rearing.) Again, people believe what they are taught, and the Democrat-Left in this country has been shouting the lie loud enough and long enough in our media, our schools, our entertainment industry and through our courts, that homosexuality is normal and wonderful and should be thought of in the same light as heterosexuality; that enough weak-minded folks have bought into it. These people are easily influenced. They just want to fit in; to conform to the crowd and not make any waves. They don't possess any great desire to actually think for themselves especially if that would entail being *different*, and being the target of the Left's hatred, and our late-night imbecile's merciless ridicule. They are people who end up letting the great deceiver do their thinking for them.

The Left makes up their own morals (although in truth they are the morals of the wicked one) so why wouldn't they make up what is *normal* sexually for

---

882   (You're right, they call themselves the North American Man Boy *Love* Association, but that is such a stinking lie let's not repeat it anymore. They are full of nothing but anger and hatred for the boys that they hope to shower their perversion on; and to mentally, emotionally and spiritually torment for the rest of their lives in the process. Their father, Satan does however *love* what those disgusting pigs are doing.)

themselves as well, and then try to shove that *normalcy* down the throats of everyone else in the process? (In the name of *tolerance*, of course.) Welcome to the new normal, which has been around since the days of Sodom and Gomorrah, and ancient Rome's very own sweetheart, Caligula, but since the left needs to see themselves as progressive, as on the cutting edge and as morally superior to everyone else, let's try not to reduce them to the same mental level of those barbarians of old. Conversely, they have yet to start shouting the whopper that bestiality or sleeping with children is normal, so we are safe from vast numbers of liberal "no-think-ums" embracing that, at least for now.

This goes back to what we discussed in the chapter on Moral Absolutes. Either there is a universal moral code that comes from the very moral character of the Creator, or anything goes. What is *right* then is whatever we can get a consensus on. Mutilating babies to death? No problem, if you're a Democrat. Blowing up Jewish woman and children in the streets of Israel. To be applauded, if you're a Mohammedan. Slicing the clitoris off of hundreds upon hundreds of millions of little girls over the centuries? Praise Allah! Slaughtering black babies at three times the rate of their percentage of the overall population to satisfy the evolutionary, African-American-extermination mission of America's premier racist, Margaret Sanger? No problem, and let's all run out and vote Democrat to keep making her dreams come true. Well, the same holds true for what is normal and what is not when it comes to sexual activity. If normal is whatever we can get a consensus on, then not only homosexuality and lesbianism, but bestiality and child perversion and even rape and forced sodomy can all be pronounced normal and right as long as enough of our progressive, conformist citizens can be convinced to "open up their minds" and embrace it. And if you go back and re-visit the section in chapter 2 on "The Art World;" and if you also revisit our discussion a few pages prior to that on Postmodernism and how it has declared war on truth itself and decency itself; calling them the "great oppressors;" and that therefore they must subvert all truth and must tear down all boundaries, moral and otherwise, in order to show that boundaries are arbitrary; and if you also realize that these Postmodernists are the progressive, ACLU, Democrat-Left that controls our entire culture and even much of our government; you will see that this is not so farfetched after all. (You might want to keep Lassie on a tighter leash.)

What does the Creator think?

"And the angels who kept not their proper domain, but left their own habitation, He has reserved in everlasting chains under darkness for the judgment of the great day. Even as Sodom and Gomorrah, and the cities around them in like manner, having given themselves over to sexual immorality and going after strange flesh [homosexual and lesbian activities], are set forth as an example, suffering the vengeance of eternal fire." (Jude 1:6-7) (But at the same time as we share that difficult verse keep in mind that Jesus suffered and died on a cross for the sins of all of us, including our homosexual and lesbian brothers and sisters. And anyone can repent and turn from their sins and embrace him as Lord and Savior.)

It should be self-evident to anyone not indoctrinated with the abnormal lies of the Left that homosexuality and lesbianism are perversions of Gods intended purpose for human sexuality. And again, not perversions in the sense of child perverts or bestiality, but perversions nonetheless. God created us male and female, and created sex itself for the purpose of procreation, child-rearing and sexual intimacy between a married man and a woman. That was his intended purpose, and not for "hooking up" or "shacking up" or running around on your wife or husband or molesting children or assaulting barnyard animals or for men to have sexual congress with other men or for women to get it on with other ladies. Those are all things that God did not intend for human sexuality to be used for. That was not His purpose. All that is the purpose of the Prince of Darkness. And again, just a short journey through His Word, which we know now is absolutely true, confirms all of that.

And in addition, our souls confirm it as well. Morality comes down from above, from the Creator of this universe. And it is written into the fabric of his universe, and into our consciences. That is why the world-wide aversion to homosexual sex is natural and universal and doesn't have to be taught. It is also why the Democrat-far-left and their militant homosexual activists (like GLSEN- see chapter 2) are trying to force their propaganda and lies down the throats of our children in our public schools, and at a very young age. Under the guise of "acceptance" and "tolerance" they seek to distort the thinking of our youth so they will accept sexual perversion as "normal." And so they will believe that anyone who does not agree with this twisted vision is "abnormal." And these child molesters have a name that they and their media use to label anyone who disagrees with them. It is called *homophobe*.

But don't worry. If you do get labeled one you will find yourself in the most excellent company. For according to these child abusers, God himself is a homophobe. Just read what he clearly says in His Word. But in reality, the only "phobes" we have here are truth-o-phobes. And they are the very people who falsely accuse the rest of us—those who refuse to go along with their perverted notions about human sexuality—of being phobic. They deny the truth and the wisdom of their Creator and instead try to substitute their own "wisdom," the wisdom that comes up from the pit, which is nothing but foolishness. "For the wisdom of this world is foolishness with God." (1 Corinthians 3:19) (We should have that one memorized by now.) They are truth-a-phobes because they have an unnatural aversion towards, and hatred of, the truth. They are bigoted against the Truth. (And remember that God (Jesus) and the truth are one and the same. John 14:6) So don't be intimidated by them, or their media. Rather rejoice! That your name is written down in heaven...

"Blessed are you, when men shall revile you, and persecute you, and shall say all manner of evil against you falsely, [and when truth-a-phobes shall falsely accuse you of being a homophobe] for my sake [for the Truths sake]. Rejoice, and be exceeding glad: for great is your reward in heaven: for so persecuted they the prophets who were before you." (Matthew 5:11-12)

Besides, there is nothing more cruel and more hateful and more unloving than to lie to someone who is lost in sin. To encourage a woman to go and mutilate her baby to death as the truth-a-phobes in our schools, our media, our entertainment industry and at Planned Mutilate-Your-Baby-Hood do thousands of times every day. To cower from telling our lost Islamic brothers and sisters the truth about their clitoral-mutilating, satanic religion because they might cut your head off or because you are one of their useful idiots, as the truth-a-phobes in our left-wing, Democrat, Islamo-embracing schools and media do every day. And to allow our brothers and sisters who are lost in the sin of homosexuality to remain there because we are too afraid of being labeled *homophobes* by the liars on the Left, who lie to everyone every day. Nay, rather we should have the courage and the love and the compassion to speak the truth, and of course to speak it in love, to our homosexual and lesbian brothers and sisters. We are all cut from the same cloth. We are all made in the image and likeness of our Creator. (Although David Letterman does bear a striking resemblance to a Tyrannosaurus rex.) (There are exceptions to every rule.) And "we have <u>all</u> gone astray. Every one of us has turned to our own way. And the Lord has laid upon Jesus the iniquity of us <u>all</u>." (Isaiah 53:6)

There is a difference between homosexuals (the person), and homosexuality (the act). Homosexuals (the person) are lost, but they can be found. Homosexuality (the act) is a perversion of Gods intended purpose for human sexuality. The sinner can be brought to salvation. The sin is to be rejected. To have an aversion to the thought of homosexual sex, is natural. To have an aversion to those lost in the homosexual lifestyle, is un-Christian.

Homosexual *marriage*?

"Have you not read that He who made them at the beginning made them male and female, and said, For this reason shall a man leave his father and mother, and shall be joined to his wife, and the two shall become one flesh? Wherefore they are no longer two but one flesh. What therefore God has joined together, let not man put asunder." (Matthew 19:4-6)

The above verse in particular and God's word in general makes plain that the sexual union between a man and a woman is something that our Creator takes very seriously. It is sacred to Him and not something to be trifled with. And that there is much more going on there than just some physical sex. There is something going on there between the *souls* of the married man and woman. And that we should probably take the whole sex thing a little more seriously ourselves than to teach our children that satisfying their wanton sexual urges is one of their highest callings in life and should not be hindered, stifled or controlled in any way (except to throw a condom into the mix), as the Democrat-Left has been doing for decades now through their control of our entertainment industry and our schools. And that we should probably think twice before we spit in the face of that same Creator, as we have already profusely done with baby mutilation and the theory of evolution, for a third time by defiling His sacred institution of marriage. Why? Three strikes and you're out.

Which brings us to the whole notion of homosexual *marriage*. What can be said about this satanic ploy? Oh. I just said it. Homosexual marriage is Lucifer's political and judicial ploy to further attack the sacred, God-ordained institution of marriage, the bulwark of any society. I say further attack because his assault was started years ago with the feminist movement and the sexual revolution of the 60's. His first salvo was to denigrate women from their high moral status that they had enjoyed for six thousand years, by *liberating* far too many of them of one of their greatest gifts; their sexual purity and virtue. As we mentioned in our segment on "Feminism" in chapter 2, the Left Elites have succeeded in all but destroying that for far too many of our young women. They have been in charge of the raising of our children for decades now thanks to their takeover of our educational system and our entertainment industry, and have used that power to turn many of our young woman today into what previous generations, undeterred by a political correctness that protects the cesspool values of the Left, would have called sluts and unpaid whores. Look at some of the popular TV shows of today like *Desperate Whores* (aka *Desperate Housewives*) and *Hosebags on the Hunt* (aka *Sex and the City*) and imagine the outcry if they been aired alongside *Ozzie and Harriet* a few decades ago. And those are just two of countless examples of the triumph of Satan's Democrat-Left in duping this culture into allowing the once high moral standard of women in our society to be mocked, disparaged, ridiculed, denigrated and then brought down to the level of raw sewage where the totalitarian-left is most comfortable residing.

His second barrage was against men, the head of the traditional household, and his goal was to reduce them from someone to be respected, admired and obeyed, and whose authority in the home was to be taken very seriously, into clueless buffoons whose authority is now held in complete contempt. Compare *Father Knows Best*, a successful show from the 60's, to anything being vomited out of the mouth of our entertainment industry today: *Married with Children*, *The Simpson's*, *Family Guy*, these shows along with every other commercial portray men as idiots. (And since most men today don't have a clue or don't seem to care about what's being done to them, maybe they are.) But curiously they never portray women in the same way. (And if one gave an IQ test to Joy Behar or Tina Fey, just for starters, they would find that it's not for lack of material.) (With all due respect of course.) And every other movie being spewed out of Hollywood has at least one scene where the actor or actress reminisces about their father, and of course he was either an abusive, violent drunk or a child molester. Score another one for Lucifer; and his left.

But now Satan is going for the final blow. He has successfully trashed the two necessary components of a marriage. The man, whose respect and authority in the home is of foundational importance to both the marriage and to the proper development of the children. And the woman, whose virtue and moral character is also of foundational importance, again not only for the success of the marriage but for the proper emotional and moral instruction of the children as well. And emboldened by these two successes, he now seeks to destroy the very institution itself. If it becomes permissible in this society to allow two men or two women to be substituted for a man and a woman as an acceptable combination to constitute what will then be recognized as a *marriage*, then marriage no longer exists. It is then a joke. Worse yet, it will have become just another pile of excrement, defecated onto the American culture by the godless, Democrat-Left. God created marriage and He intended it to be a sacred institution. Satan and his duped human accomplices want to turn it into a farce and a stinking pile of manure. (Like our federal government has been turned into under the deft leadership of Obama, Pelosi, Weiner, Schumer, Durbin and Reid et al.)

And it is no accident that a great many of the radical homosexual activists who are leading the charge to turn marriage into a pile of excrement enjoy wallowing in the very thing themselves. But then our left-wing, lying, progressive, Democrat media has never reported on those homosexual excrement parties that occur regularly in San Francisco and elsewhere. They are called "poo poo parties" and here is one description of what occurs (get the bathroom spray back out)…

"An undisclosed number of homosexual men form an ass-to-face circle and release fecal matter onto the awaiting face of the men behind them. The fecal matter is then used to condition the skin and hair. Any remaining feces is used for **party games** like pin the tail on the donkey and bobbing for apples (except feces is used instead of apples)."[883]

(Takes the idea of having a couple of friends over for dinner to a whole new level.) I call them *Postmodernist parties*- "subvert all truth, tear down all moral boundaries for they are nothing but *arbitrary*." (Like the *arbitrary* boundary that most decent, un-debased humans keep between their ass and their face, or their mouth and their toilet for that matter.) And uninformed, uneducated, kept-blissfully-in-the-dark America sits back while radical homosexuals shove their agenda and their lifestyle on their elementary school children, and then saunter ignorantly out and vote for it to boot.

In addition, as we have mentioned previously, if you are going to allow "marriage" to occur between a man and a man, or a woman and a woman, then you will have absolutely no choice but to allow, and no legal framework from which to block, further "marriages" between a man and his sheep, and a woman and her German Shepherd. Just ask PETA's bastard child, PMRA (People for the Marital Rights of Animals). Again, God created marriage and He intended it to be a sacred institution. One that was holy, separate from the profane. Satan, with the witless assistance of his duped, Democrat-Left, liberal accomplices, seeks to turn it into a farce and a stinking pile of dogs**t.

Also, this political movement has nothing whatsoever to do with "Constitutional rights." That is just a lie that the ACLU and their Democrat-Left shout in order to give the cover that will allow their Constitution-perverting friends on the courts to legislate in their favor. To accomplish through judicial fiat what they could never do in the legislature. And keep again in mind that it is the prerogative of the legislatures alone to pass laws, not the courts. It is the courts job to rule on whether a law is <u>unconstitutional</u>, not on whether or not it coincides with one's ideological and wholly <u>un</u>-Constitutional agenda. To do this is to be a traitor to the very Constitution which they swore an oath to uphold and defend. Yet they purposely pervert the clear meaning of our Constitution and the obvious intent of the founders who wrote it. And they do this every day, with impunity. Because their left-wing, lying friends in the media cover it all up. (Are we starting to see a pattern here?)

The other lie being shoved down our throats, and indoctrinated into the minds of our young people through the Democrat-Left's control of our media, entertainment industry and educational system is that to refuse to allow homosexuals or lesbians to marry is to "discriminate" against them. Like refusing to allow homosexuals to serve as scoutmasters is to "discriminate" against them as well. Sure. Homosexuals don't <u>qualify</u> to serve as the scoutmasters of young boys, just as Bill Clinton doesn't <u>qualify</u> to serve as tent prefect on the next Girl Scout camping trip. And two men or two women also do not <u>qualify</u> to be married; only one man and one woman do. Two men or two women do not meet the <u>requirements</u> of a God–ordained marriage; only one man and one woman do. (And that *is* where marriage came from, God. If left to his own devices man would have come up with a naked version of Twister.) But the forces of left-wing darkness in our culture are busy at work poisoning the minds of the next generation of Americans into believing that if we don't allow homosexuals and lesbians to destroy the institution of marriage then we are "discriminating" against them. And the ruler of this world is riding

---

883   (From Urban Dictionary @ urbandictionary.com) (And of course we report this under-reported fact with all due apologies to those homosexuals who are not that debased.)

this lie all the way to the bank: to the hammering in of the final nail in the coffin of God-ordained marriage.

Like moral absolutes, real marriage comes down from the heavens above. Counterfeit marriage oozes up from Satan's sewer below. With all due respect of course to those homosexuals and lesbians who find themselves in a deep and sexually intimate relationship with someone whom they may love very much; (an emotional bond we will address shortly).

Nature versus nurture

"A woman shall not wear anything that pertains unto a man, neither shall a man put on a woman's garment, for all who do so are an abomination unto the Lord your God." (Deuteronomy 22:5) But the Democrat-Left celebrates cross-dressers and transvestites, even in our public schools, rather than offering these poor confused individuals some desperately needed help. Which is like responding to the cries of a young person who is threatening suicide by handing them a pack of razor blades, a couple bottles of sleeping pills and a loaded gun. And we have allowed these liberal jackasses to take over our culture.

God does not condemn anyone for being born a woman, nor does he condemn anyone for being born a man. And nor would he condemn the sin of homosexuality, if people were "born that way." But they are not. That is just another lie of the evil one, perpetrated by radical homosexuals and spread by their willing accomplices in our left wing, lying, Democrat media, entertainment industry and educational system. This untruth is also shouted loud enough and long enough, not only so enough weak-minded people will believe it, but also so those ex-homosexuals who would try to help others escape out of that sad and destructive trap, can be effectively ridiculed and legally blocked. Satan doesn't let go easily. He is far more tenacious in keeping people trapped than Hussein Obama is in shoving his socialized, health-care slop down America's throat. That old serpent, the devil, covers this world with his distortions like excrement covers the faces of participants at a poo-poo party, and he will do everything in his power, and in the social and political power wielded by his human dupes, to keep the world from being set free.

In the culture war for decades the Left had the American Psychological Association as one of their willing accomplices in pushing the *normalcy* of homosexuality...

"**Homosexuality, Mental Illness, and the American Psychiatric Association**: "In 1973, the APA removed homosexuality from its diagnostic category of mental illnesses. This action came not as a result of new research and findings, but was ultimately brought about by a militant protest staged by activists at the APA annual convention. In other words, intimidation was a key motivation. In fact, only 16% of the entire APA membership actually voted in favor of the radical change. — Dr. James Mallory, Head of Psychiatric unit – Rapha Center, Atlanta, Ga., *Homosexuality*; Ronald Bayer, *Homosexuality and American Psychiatry: The Politics of Diagnosis* (NY: Basic Books, 1981), 101- 54; William Dannemeyer, *Shadow in the Land* (San Francisco: Ignatius Press, 1989), pp. 24-39."[884]

Fortunately, though, that is one ally the Left just lost...

---

884   (From "Fast Facts: Homosexuality," from *Center for Reclaiming America*, an outreach of Coral Ridge Ministries)

"The attempt to prove that homosexuality is determined biologically has been dealt a knockout punch. An American Psychological Association publication includes an admission that there's no homosexual "gene" -- meaning it's not likely that homosexuals are born that way.

For decades, the APA has not considered homosexuality a psychological disorder, while other professionals in the field consider it to be a "gender-identity" problem. But the new statement, which appears in a brochure called "Answers to Your Questions for a Better Understanding of Sexual Orientation & Homosexuality," states the following:

""There is no consensus among scientists about the exact reasons that an individual develops a heterosexual, bisexual, gay or lesbian orientation. Although much research has examined the possible genetic, hormonal, developmental, social, and cultural influences on sexual orientation, no findings have emerged that permit scientists to conclude that sexual orientation is determined by any particular factor or factors. Many think that nature and nurture both play complex roles...."

"That contrasts with the APA's statement in 1998: "There is considerable recent evidence to suggest that biology, including genetic or inborn hormonal factors, play a significant role in a person's sexuality."

"Peter LaBarbera, who heads <u>Americans for Truth About Homosexuality</u>, believes the more recent statement is an important admission because it undermines a popular theory.

""People need to understand that the "gay gene" theory has been one of the biggest propaganda boons of the homosexual movement over the last 10 [or] 15 years," he points out. "Studies show that if people think that people are born homosexual they're much less likely to resist the gay agenda."

"Matt Barber with <u>Liberty Counsel</u> feels the pronouncement may have something to do with saving face. "Well, I think here the American Psychological Association is finally trying to restore some credibility that they've lost over the years by having become a clearly political organization as opposed to an objective, scientific organization," he states. ...With the new information from the APA, Barber wonders if the organization will admit that homosexuals who want to change *can* change."[885]

"Barber wonders if the organization will **admit that homosexuals who want to change *can* change**." They already **know** it, mind you, he just wonders when they're going to **admit** it.

There are also well-documented studies of identical twins where one grows up to be a normal heterosexual man or woman, and yet the other develops into a practicing homosexual or lesbian. If "homosexuals are born that way" then clearly both of the identical twins would be homosexual, but studies show that if one of the twins becomes homosexual only about 38% of the time does his identical brother become one too. "For lesbianism the concordance" is about 3 out of 10. So 62 percent of the time with identical twin men and 70 percent with identical twin women, when one becomes a homosexual or lesbian the other does <u>not</u> follow suit. So much for homosexuals and lesbians being "born that way."

"Identical twins have identical genes. If homosexuality was a biological condition produced inescapably by the genes (e.g. eye color), then if one identical twin was homosexual, in 100% of the cases his brother would be too. But we know that only about 38% of the time is the identical twin brother homosexual. Genes are responsible for an indirect influence, but on average, they do not force people into homosexuality. This

---

885   (From article, "APA revises "gay gene" theory," by Charlie Butts @ onenewsnow.com, 5/14/2009)

conclusion has been well known in the scientific community for a few decades (e.g. 6) but has not reached the general public. Indeed, the public increasingly believes the opposite.

"Identical twins had essentially the same upbringing. Suppose homosexuality resulted from some interaction with parents that infallibly made children homosexual. Then if one twin was homosexual, the other would also always be homosexual. But as we saw above, if one is homosexual, the other is usually not. Family factors may be an influence, but on average do not compel people to be homosexual.

"Twin studies suggest that as a class, events unique to each twin—neither genetic nor family influences—are more frequent than genetic influences or family influences. But many individual family factors (such as the distant father) are commoner than the individual unique factors. Unique events would include seduction, sexual abuse, chance sexual encounters, or particular reactions to sensitive events, when young. Everyone has their own unique path which only partly follows that of the theoreticians!

"A fascinating sidelight on all this comes from the work of Bailey. His team asked non-concordant identical twins (one was homosexual, one not) about their early family environment, and found that the same family environment was experienced or perceived by the twins in quite different ways. These differences led later to homosexuality in one twin, but not in the other."[886]

Satan wants to take choice out of every bad decision, out of every sin decision, so people can blame their sins on everything else instead of on themselves, and therefore escape having to take responsibility for their own actions. And the Left is Satan's eager accomplice in this deception. Criminals… it's not their fault, it's society's fault. (See "Restorative justice" section in chapter 13.) Homosexuals… they were "born that way." But the fact is that homosexuality *is* a choice. And even though it is true that, like all bad choices, most are *compelled* into making them (for a myriad of reasons and influences and experiences, including ones myriad of reactions *to* those experiences and influences), nevertheless each person who makes that bad choice (or *sin* choice) is still *responsible* for making it, and will be held accountable in the last day if they have refused to repent and turn to the only real source of love and intimacy, as well as salvation and eternal life, Jesus Christ.

"Acting upon one's homosexuality ("acting out") is a choice. However, same-sex attraction itself is not always a choice."[887]

Islamic jihadists are compelled, many from the earliest age, to think and believe what they do. That compulsion was not their choice. However, to remain trapped there, to continue to embrace that disgusting, clitoral-mutilating insanity throughout their life, and to refuse the cry of their own consciences and the truth of their own Creator in the process, is their choice. And the same is true of remaining trapped in voting Democrat, or in adultery, or homosexuality or lesbianism, etc. The original compulsion or attraction or ideological lies that landed one there is not an excuse for remaining there. Especially when one comes in contact with the truth.

---

886   (From article, "The Importance of Twin Studies," by N. E. Whitehead, Ph.D. @ Narth. com - National Association for Research and Therapy of Homosexuality)
887   (Pg. 258, *God's Grace and the Homosexual Next Door: Reaching the Heart of the Gay Men and Women in Your World*, by Alan Chambers and the Leadership Team at Exodus International)

"John had never known a loving, stable family. When he was just five, his biological parents divorced, beginning a revolving door of different father figures, none of whom lasted. What he learned from these men was that they were something he didn't want to be.

"By the time he was eighteen, John's mother and father were each on their third spouses, and he had learned that men "were unstable, they hurt you, and they put your mother down." He spent his childhood as his mother's caretaker, and as a result, he bonded and attached himself to her.

"John grew up ashamed of himself and his male identity. As a young man, dressing himself up as "Candy," he would masquerade as a female impersonator. This disguise was a cover-up for the inner hatred he had for himself and the misery he felt as a result. He was crying out for love, the love that only Christ could completely provide, but he had never heard. He said, "When I would come home from a bar with a partner I didn't know, I would break down in tears ... feel like a piece of meat. ... I was just a hollow shell. ... I was twenty-four but felt like eighty. I tried to take my life. ... I was tired and worn out. ... I didn't want to die; I wanted to escape. ... I wanted someone to tell me that 'I love you. There is something of value in you.'"

"This tortured young man did not know at the time that God had a different plan for his life, a plan that would slowly be revealed through the quiet, consistent witness of a Christian couple."[888]

And through their consistent witness and Christ-like love this couple slowly led John to the Lord! And now his life—once a waste of "looking for love in all the wrong places," and all the wrong sexual perversions, which will never, ever provide it—has been turned completely around. But keep in mind that if the Democrat-Left and their militant homosexuals have their way, and continue to have success with their unconstitutional "hate crimes" legislation, then it will be against the law to free John and many others like him trapped in a sad, emotionally-devastating sexual perversion that instead of being exposed to the light of truth is celebrated in darkness by the same Democrat-Left and their uneducated voters and supporters.

And then there is the testimony of Melissa, a woman who was living in a lesbian relationship but nevertheless became involved in a local church:

"There was a couple in their seventies, L. J. and Doris Crain, who took me under their wings and into their hearts. ...And although they knew about me and the life I was living, they never said a word about my homosexual lifestyle. Instead, they met me where I was, accepted me with grace, loved me unconditionally, and prayed for me fervently. It was through relationship with them that I was led to make the most important decision of my life. One afternoon, sitting alone on the edge of my bed in the stillness of my bedroom, I said quietly in my heart, "Jesus, Jesus, please, please come and be the Savior of my life and the Lord of my life." He did; and a new seed was planted.

"Mt partner at the time actually gave me my first Bible. ...I read again the Scripture in Leviticus... I also found verses in Deuteronomy, Romans, and 1 Corinthians. And this is when the wrestling ensued.

"For months I went around and around with the Lord. ...I argued with him; I pleaded with him; I yelled at him; and I opposed him. ...The Bible says the Word of God is living and active, sharper than a double-edged sword; it penetrates even to dividing soul and spirit, joints and marrow. It judges the thoughts and attitudes of the heart. Isaiah 55:11

---

888   (Pg. 1, *The Homosexual Agenda: Exposing the Principle Threat to Religious Freedom Today*, by Alan Sears and Craig Osten)

says that God's Word will not return empty, but it will accomplish what God desires and achieve the purposes for which He sends it. For the next four, eight, and twelve months the Holy Spirit of God continued to work those truths into my heart until I came to the point of convicting revelation where I knew that I knew that I knew what I was doing was sin. ... During those months, even in my confusion, even in my anger, even in my rebellion, God showered me with His love. In 1992, by His grace, I repented of my years of sexual sin and the new seed sprouted.

"The separation, though, from my partner at the time was not immediate. Although at the time of repentance all sexual immorality stopped, we still shared a bed together, a home together, and a life together. That was one of the first tangible experiences I had of God's patient and tender mercy. He knew how incredibly enmeshed we were. It was as if we were two infected wounds that had been crudely bandaged together with dirty and soiled gauze. But like a sensitive and compassionate doctor, the Lord began to slowly and gently unwrap that dressing. Over the next months, He continued to move in our lives and we eventually separated completely."[889]

Both of those testimonies came from the pages of Alan Sears and Craig Osten's book, *The Homosexual Agenda: Exposing the Principle Threat to Religious Freedom Today*. The authors comment...

"Can you imagine if Melissa had never had a chance to hear and respond to the Gospel? If the law had forbidden it? If it were considered hate speech to tell her? She would still be trapped in a behavior that brought her so much shame. She would have no idea that Christ loves her and can forgive her for her past."[890]

In spreading their politically-correct lie throughout our culture that homosexuals are "born that way" and therefore "cannot change," the Left has another crime against humanity to add to their portfolio: which includes mutilating 50 million babies to death and going for another 50 million more (yes they can!); shoving the lie of evolution down our school children's throats; failing to report on their Islamic friends propensity to hack up the genitals of a hundred and thirty million of their little girls; etc. In this case their crime is mainly against homosexuals by seeking to keep them trapped in a miserable lifestyle; one not intended for them by their Creator, but one designed by the father of lies.

"Unfortunately, if many homosexual activists have their way, Christ's message of redemptive love will be silenced and those who share it through the preaching of the uncensored words of Scripture will be punished. Thus, those who need to respond to the Gospel will never have the opportunity to hear it. [Which of course is Beelzebub's whole plan.] The effort of homosexual activists to convince Americans to tolerate (i.e. "affirm") homosexual behavior tramples religious freedom and leaves a trail of broken bodies in the dust. Broken bodies, broken souls who without the chance to hear and respond to the gospel will never know that there is a way out of a lifestyle, and its accompanying behaviors, that falls far short of the joy their Creator intended and leads to despair, disease, and early death. Yet despite these dangers, it is a behavior that is being promoted as nothing more than an alternative lifestyle, and any dissent is ridiculed, vilified, and censored."[891]

---

889   (Pgs. 5-7, *The Homosexual Agenda*, by Alan Sears and Craig Osten)
890   (Pg. 7, *The Homosexual Agenda*, by Alan Sears and Craig Osten)
891   (Pgs. 2-3, *The Homosexual Agenda: Exposing the Principle Threat to Religious Freedom Today*, by Alan Sears and Craig Osten)

Another lie perpetrated by the Left is the percentage of the population that is "homosexually oriented." The answer is only about 2 to 3%. The ten percent figure we have all heard stated over and over again is an intentional fabrication. It was based on the exaggerations of the biased *Kinsey Report* (1948 and 1953) which has since been completely discredited, but like all of the non-existent evidence for evolution, it will continue to be reported as if it were fact.[892]

But what about those who feel such a strong emotional and physical attraction to members of the same sex? Doesn't that in and of itself *prove* that they can't control it and therefore cannot be held accountable? In fact, shouldn't they be celebrated for "following their heart?" Well, not according to their Creator:

> "[The biblical] condemnation of homosexuality—like the biblical condemnation of adultery or incest—rejects the argument that the parties involved feel a sincere love for one another. The act is simply forbidden."[893]

And the reason that our Creator never bought into the "strong feelings" argument for either adultery or homosexuality is that he knew all along that "strong feelings" and "sincere love" are after all about as fickle as the wind. They blow hot today and cold tomorrow. They are all-consuming today, and impossible to find the next. Not to be a little jaded but give me a nickel for every angry break-up or divorce that followed close on the heels of "sincere love"' ("I love you sooo much") and I'll pay off the national debt. (The national debt after The Three Stooges—Obama, Pelosi, Reid—get through with it is another story entirely.) No, God knows what is best for us, and that is why He has forbidden illicit sex, adultery, shacking up, hooking up, and homosexuality and lesbianism. It is not because he is a big meanee and doesn't want us to find fulfillment and love and sexual satisfaction and emotional gratification. It's just that he knows we will never find it by being sucked into one of the many sexual perversions, distortions and sins that come from the enemy of fulfillment and love and long lasting emotional gratification, that old dragon, El Diablo.

If people are not born homosexual or lesbian, then how do they end up with strong emotional and physical attractions to the same sex? That's a good question and one that also has some good answers, but to elaborate fully on them is a little beyond the scope of this chapter. (See recommended reading list at end.) But I think there are a few simple insights worth sharing...

Imprinting is a psychological term which refers to the social bonding that occurs very early in animal or childhood development. If a boy, for whatever reason, develops a very strong emotional attachment to his mother (or a girl to her father)—a very strong love for her and need for her, perhaps because of the absence of a strong father figure, or because of a rejection by the same—and this attachment progresses to the point of imitation and emulation; then in the child's subconscious mind they begin to desire to be like the mother or father, which would of course entail acting and talking like the opposite sex and desiring what the mother desires; or in the case of the girl, being attracted to what the father is attracted to. Hence the subconscious, and perhaps even conscious, desire to switch the natural desire with an unnatural one, but one which nevertheless fits with their

---

892   (Pg. 258, *God's Grace and the Homosexual Next Door: Reaching the Heart of the Gay Men and Women in Your World*, by Alan Chambers and the Leadership Team at Exodus International)

893   (Pg. 134, *The Politically Incorrect Guide to the Bible*, by Robert J. Hutchinson)

strong desire to emulate, imitate and therefore outwardly express their strong love for the parent of the opposite sex.

"Girls will be boys and boys will be girls, it's a mixed-up muddled-up shook-up world, except for Lola; L-O-L-A Lola."[894]

Love is the most powerful force in the universe, which might have something to do with the fact that "God is love." (1 John 4:16) And the need to be loved, and to give love, is the most powerful and the most basic of all needs. (Which is why hell is so horrible. It is the complete absence of love.) (And water.)

And of course the exact opposite of that can be true. A boy could be driven by his hatred for his father (whether that be because the father left and rejected him or any number of other reasons) to not want to grow up to be like him, as Johns above case illustrated. And the same can be true of women...

"Most lesbians failed to connect in some important way with their mothers...and in fact, in surveys, they confirm that they did *not* have a desire to be like their mothers. The incidence of sexual abuse among lesbians is much higher than among women in general."[895]

And then there is this insight...

"A developing homosexual has trouble relating to members of his or her own sex... often identifying instead with the opposite sex. That is, many pre-homosexual boys enjoy playing with girls more than boys. And pre-lesbian girls often find it hard to relate to other girls as being the same as they are. Instead, they identify with the boys they know and would prefer to be with them. This process continues into adulthood. And when a homosexual becomes a Christian and begins to accept his or her gender, there is a time of learning to be "one of the guys" or "one of the girls." Overcoming homosexuality, then, is in part learning to relate properly to one's own gender."[896]

At the root of the matter a homosexual is a man who has come to <u>think</u> of himself, for whatever reason, as a woman, on the inside. Who has come to <u>feel</u> that he is a woman, on the inside. Or who <u>desires</u> to be a woman on the inside and on the outside. And at the heart of the matter a lesbian is a woman who has come to <u>think</u> of herself, for whatever reason, as a man, on the inside. Or who has come to <u>feel</u>, perhaps even from an early age, that she is a man, on the inside. Or who <u>desires</u> to be a man on the inside and on the outside. But this is not inherited. Thoughts, feelings and desires are not inherited. And neither is homosexuality or lesbianism. They are all of them acquired. They are learned. And they can be unlearned. If the desire is there to do so. And that desire is available through a relationship with Jesus Christ.

"The Lord bless you, and keep you. The Lord make his face shine upon you, and be gracious unto you. The Lord lift up his countenance upon you, and give you peace." (Numbers 6:24-26)

---

894   (From song, Lola, by the Kinks)
895   (Pg. 265, *God's Grace and the Homosexual Next Door: Reaching the Heart of the Gay Men and Women in Your World*, by Alan Chambers and the Leadership Team at Exodus International)
896   (Pg. 261, *God's Grace and the Homosexual Next Door: Reaching the Heart of the Gay Men and Women in Your World*, by Alan Chambers and the Leadership Team at Exodus International)

The Radical Homosexual Agenda

"Therefore God also gave them up to uncleanness through the lusts of their own hearts, to dishonor their own bodies between themselves, who exchanged the truth of God for the lie, and worshiped and served the creature rather than the Creator, who is blessed forever. Amen. For this reason God gave them up unto vile affections. For even their women did change the natural use into that which is against nature. And likewise also the men, leaving the natural use of the woman, burned in their lust one toward another. Men with men working that which is unseemly, and receiving in themselves the just penalty for their perversions." (Romans 1:24-27)

In the first chapter of the book of Romans Paul talks about some of the consequences of refusing to glorify God and give thanks to Him, and of treating His truth as if it were a lie. The above verses describe homosexuality and lesbianism as one of the "rewards" man receives for that refusal. The Left lacks the sense to see it for what it is, and instead they celebrate it as something that it is not.

"For the wrath of God is revealed from heaven against all ungodliness and unrighteousness of men, who suppress the truth in unrighteousness; Because that which may be known of God is manifest in them, for God has shown it unto them." (Romans 1:18-19) Therefore there is no excuse.

> "**Homosexual Deception**: "In the early stages of any campaign to reach straight America, the masses should not be shocked and repelled by premature exposure to homosexual behavior itself. Instead, the imagery of sex should be downplayed and gay-rights should be reduced to an abstract social question as much as possible. First, let the camel get his nose inside the tent—and only later his unsightly derriere!" — Homosexual Activists Kirk Marshall and Pill Erastes, *The Overhauling of Straight America*"[897]

"But the men of Sodom were wicked and sinful exceedingly before the Lord." (Genesis 13:13) They were also very clever…

Radicalized, militant homosexuals have targeted our children for indoctrination. They came to the realization of the difficulty of indoctrinating them with the lie of homosexuality's *normalcy* once those children passed the age of puberty and experienced the natural revulsion inside of them toward homosexual sex, and after their parents had taught them the truth. So they formulated a plan to get to them as early as possible, in our public schools; while at the same time making an end run around their parents by keeping them in the dark as to what was actually going on. The goal was to get the young children to believe that their own parents were lying to them, and that the homosexual activists and their teacher accomplices were telling them the truth. (Kind of reminds you of Satan's original lie in the Garden, that God was lying and the prince of devils was the one telling the truth.) So they now use our elementary schools to shove the lie that homosexuality and lesbianism are *normal*—spitting in the face of the Creator of the universe in the process—and that those who do not agree, including their parents, are just ignorant, biased "homophobes" who are not as "smart" and "enlightened" as the indoctrinator and their gullible young charges. I could be wrong but I believe you'd have to go back to Mao's Cultural Revolution or Hitler's Youth Movement to find an historical

---

897   (From "Fast Facts: Homosexuality," from *Center for Reclaiming America*, an outreach of Coral Ridge Ministries)

comparison where Nazi's[898] taught children that they were smarter than their own parents. And this indoctrination is being spread in what was once the *safety* and *security* of our elementary and high schools...

"One day during his biology class, Kyle's teacher stated that homosexual behavior was genetic. Kyle immediately raised his hand to disagree. The teacher, a self-professed bisexual who had testified in support of civil unions in Vermont, immediately ridiculed and humiliated him in front of the entire high school class.

""What's the matter, Kyle?" she said mockingly. "Are you unsure of your sexuality? Did you know that the people who scream the loudest turn out to be gay themselves?"

[Can you imagine if a Christian teacher had mocked a Muslim student in the same manner: "What's the matter, Ahmed? Are you unsure of your beliefs? Don't you know that Mohammed was a disgusting child molester and your false religion is responsible for hacking the clitoris off of 130 million little girls in this generation alone?" He would be fired in an instant. Or if a properly educated high school science teacher had said, "Are you kidding me, Brian? Are you actually so stupid that you believe that life just happened accidentally by random chance? HA! HA! What's the matter? Are you afraid of your own Creator?" He would be fired as well. And rightly so for, even though both would be telling the truth, there is no place in our public school system for a teacher who mocks or degrades anyone. Kyle's teacher, however, was not only mocking and degrading, she was lying as well. But teachers like her are well-protected by the NEA (National Education Association) and their Democrat media, and by all of those uninformed people who still go out and vote for both.]

"The other students in the class, who had been subjected to homosexual indoctrination for years, laughed at him.

"A few weeks later, in the middle of a discussion on genealogy, the teacher again digressed into the subject of homosexual behavior. Kyle asked again what homosexuality had to do with the subject.

"The teacher again questioned his sexuality and implied that he might be covering up the fact that he was a homosexual. Kyle stood up and denied the accusation, stating that he had a girlfriend.

"The class snickered at him. One classmate went as far to suggest the girlfriend was a cover-up and that Kyle was a closet homosexual.

"Devastated and humiliated, Kyle's grades dropped from a 3.70 grade point average to 2.10 in the months following the incident.

[But of what concern are grades to a teacher whose primary goal is left-wing indoctrination? (Remember those examples across this nation of elementary school kids being made to sing praises to Obama after the 2008 election? If a teacher had done that for Bush they would have been hanged.) Our children have fallen desperately behind all other industrialized countries in math, science and speaking in complete sentences, but they are well-educated when it comes to both homosexuality and voting Democrat.]

"That story is just one of numerous examples of how the homosexual agenda is being pushed in public schools and how those who don't toe the line are being humiliated and punished.

"Every fall, millions of parents drop their children off at taxpayer-funded public

---

898   (I know, the Chinese were Communists, not Nazi's, but it doesn't matter: Nazi's, Communists, Baby Mutilators, Clitoris-Hackers... they're all cut from the same cloth.)

schools, assuming that their children's education will provide what they need to be successful in life: strong academics, civility, and responsibility.

"Unfortunately, many of these same parents have little or no idea what is happening to their children once they pass through the classroom door. Instead of learning the three Rs or how to be good citizens like many of us were taught, they are learning how to reject the common values that many of their parents have tried hard to instill in them, the values that built America. Sadly, many of these parents refuse to believe that this is happening, even when you produce evidence of how the radical homosexual activists are targeting children in public schools to accept, affirm, and be recruited into homosexual behavior.

"On a daily basis, all across America (but more prevalent in some areas of the country than others), children as young as kindergarten are being told that their parents are "stupid" or "bigots" or "intolerant" if they do not accept and embrace homosexual behavior as normal, or even something to be celebrated. In some classes, children are even recruited to promote gay pride marches and events.

"For young children, the open promotion of homosexual behavior in the media and the classroom has led to preteen children announcing to their parents that they are homosexual. Why? **Because they are taught that if you are a girl who doesn't like boys, you are a lesbian, and if you are a boy who doesn't like girls, you are a homosexual**. [Emphasis mine.] Yet, for most eight-year-olds, members of the opposite sex have "cooties" or similar perceived afflictions. This is the time when children bond and form their identity as members of their own sex, and it is just a natural part of childhood. To introduce homosexual behavior in this stage of development is only confusing for children. [In addition to being a crime against humanity.]

"Finally, once they reach their teenage years, sexually confused teenagers are turned over to homosexual counselors who assist them in determining how they can come out to their families and friends.

"Homosexual activists know that the best time to reach children is during the earliest, most impressionable ages. Christian researcher George Barna has documented that the chance for an individual to make a personal decision to believe in Christ greatly diminishes after the age of fourteen. Just as children are more receptive to the gospel and religious instruction at an early age, they are also more susceptible to homosexual indoctrination. Is it any wonder why homosexual activists and their allies have tried so hard to keep the gospel out of and get homosexual indoctrination into the public schools?

[That is the ACLU's gift to America: God is now out and Satan is in. (Their way of saying thanks for all of the misguided support they've gotten over the decades, by an America that has no idea of what they've really been up to.) (And see chapter 13.)]

"In their private meetings, homosexual activists boldly proclaim their goal to get children to reject their parents' beliefs. At a 1999 Gay, Lesbian, Straight, Educational Network (GLSEN) conference in Atlanta, the following comments were made: "The fear of the religious right is that the schools of today are the governments of tomorrow. And you know what, they're right" and "If we do our jobs right, we're going to raise a generations of kids who don't believe the claims of the religious right."

"Former President Bill Clinton [the lecher] has chimed in on the need for children to be reeducated to reject their parents' beliefs. In a 1997 hate crimes conference at George Washington University, Clinton said, "Children have to be taught hate. We want to teach them a different way. Don't you think you almost have to have an organized effort to do it? There would almost have to be some sort of club or organization at school ... because if you think about it, your parents are pretty well separated.""[899]

---

[899]   (Pgs. 45-47, *The Homosexual Agenda: Exposing the Principle Threat to Religious Freedom Today*, by Alan Sears and Craig Osten)

(Maybe Clinton could also drum up an "organized effort" for "some club or organization" in Arkansas to put an end to disgusting rapists biting the lip of unsuspecting Democrat supporters while she is getting his unwanted bent member crammed into her most private area.) (Just a thought.)

The distorting of a child's mind at an early age to get them to except as normal a deviant, sinful and sexually perverted lifestyle is nothing less than satanic. But "hey! as long as we do it under the guise of promoting "tolerance," we can fool enough dimwits into letting us get away with it!"

Again, if a Christian teacher had mocked a student for believing in the scientifically-impossible lie of evolution, or because he was born into the hideous religion of Islam, it would have made national headlines and he or she would have been fired. But homosexual advocates in our schools get away with mocking Christians every day, because they are protected by the Democrat-Left's control of both our educational system and our media.

> "There is a gay and secular fascism in this country that wants to impose its will on the rest of us; and is prepared to use violence, and harassment. It is prepared to use the government if it can get control of it. It is a very dangerous threat to anybody who believes in traditional religion. And I think if you believe in historic Christianity... or Judaism, you have to confront the reality that the secular extremists are determined to impose on you, acceptance of a series of values that are antithetical, they are the opposite of [everything you believe in]."[900]

There is nothing inherently objectionable in the teaching of "*tolerance* for other lifestyles," even if one particular "other lifestyle" involves the immersion into a sinful perversion of God's intended use and purpose for human sexuality. Homosexuals and lesbians, in other words, should not be hated, or vilified, demeaned, disparaged or marginalized.[901] They should be loved, and prayed for. We are all sinners and all should be treated with love and respect. (Except of course in the case of criminal perversions like child molestation or rape. Those should be treated with arrest and removal from society. Unless of course you are a Democrat ex-president. Then that rapist should be celebrated with lucrative speaking engagements and the continued adulation of the Democrat-Left and its media.) But homosexual militants use "tolerance" as a ploy to gain access to elementary school children for the purposes of indoctrinating them into not only accepting homosexuality itself as *normal*, but into embracing their perverted lifestyle for themselves as well. *Normalcy*, and recruitment into their lifestyle, is the goal. The concept of *tolerance* merely comes in handy as the clever camouflage for reaching that goal.

Indeed if true tolerance were really their goal then homosexual militants might be something other than some of the most hate-filled and intolerant bigots on this planet. Just ask Anita Bryant of Florida orange juice fame; or Dr. Laura Schlessinger of "The Left Wing" fame; or revisit the vile, hate-crime antics of Perez Hilton which we reported on in chapter 2. They could not tolerate Anita Bryant doing Florida Orange Juice commercials because she refused to agree with

---

900   (Newt Gingrich on *The O'Reilly Factor*, Fox News, 11/14/2008)
901   (Like the "tolerant" folks on the Democrat-Left have done to their Christian and conservative opponents for decades, and with impunity)

the "normality" of their lifestyle. Nor could they tolerate Dr. Laura having her own TV show as the only viewpoint homosexual militants will tolerate being heard is their own distorted one. They could care less about tolerance. That is just a word that they use. They are only interested in the dominance of their viewpoint and the acceptance of their lifestyle as *normal* by everyone. And anyone who disagrees can look forward to being bludgeoned to death with their club of tolerance; to being "homophobed" to death by truth-a-phobes.

> "It's becoming more and more obvious that the homosexual agenda isn't really about tolerance after all. It's about imposing a radical agenda on our society and using nasty rhetoric and the force of law to suppress opinions that oppose their agenda. Anyone who dares to stand for Biblical notions of sexuality will face the intolerance of homosexual activism."[902]

There's a psychological condition known as *projection*. It is where one projects their own ugly thoughts, feelings, shortcomings, faults or imperfections on the people around them, thus alleviating their own guilt and removing responsibility for their own sins. We can see this in how the shameless liars on the Democrat-Left falsely accuse their opponents of *lying*. "Bush lied and soldiers died." They project their own rampant dishonesty onto their opponents for both political gain and to enhance their delusional feelings of superiority. And in how they can falsely accuse the Bush administration of "torture" because water was poured on the faces of three Muslim terrorists, when the Democrat-Left actually does torture millions of precious babies to death every year. Another example is how a number of communist-indoctrinated Hitler youth in Canada the other day, 3/23/2010, at the University of Ottawa loudly and incessantly chanted "No hate-speech on our campus" in order to block Ann Coulter from speaking. (The irony of hate-filled students shouting a "no hate-speech" slogan in order to silence free speech was lost on them. To them speech should only be "free" if it agrees with their backward ideology, no matter how hate-filled their type of speech may actually be. Critical thought is the primary casualty at a left-wing university. Goebbels, however, would be damn proud.) And this projection can also be seen in how the wildly intolerant Democrat-Left and their militant homosexuals falsely accuse those who disagree with them of intolerance. ("If you shout a lie loud enough and long enough, you can get enough uninformed liberals to swallow it.")

> "Radical and intolerant homosexual activists are working overtime to equate distinctions and differentiations based on sex/gender, sexual orientation, and various forms of sexual behavior with the public and thus legal public policy equivalent of racial discrimination, despite the dramatic differences."[903]

Another prong of the Democrat-Lefts attack on that part of America which has resisted embracing the *normalcy* of homosexuality is the old "gays in the military" ploy. What they can't yet shove down the throats of the adult population

---

902   (Focus on the Family)
903   (Pg. 214, *The Homosexual Agenda: Exposing the Principle Threat to Religious Freedom Today*, by Alan Sears and Craig Osten)

in general, they feel they can get away with in our military. For our brave men and women there are in no position to refuse. Clinton instituted the "Don't Ask, Don't Tell" policy which was basically a compromise. It kept homosexuals and lesbians from openly flaunting their sexuality, but it also protected them from the fear of being *outed* and discharged. But now (Spring 2010) the Obama administration is preparing to remove this policy. Emboldened by the ease in which they have been able to gain access to our first and second graders, and the absence of any measurable objection from the purposely uninformed American public, they are now expanding their playing field to include our military. This is reprehensible, for in everyday life a person is free to avoid an employment situation, which are actually hard to find (unless you live in San Francisco), where homosexual men or women flaunt their sexuality openly. And if they do find themselves in one then they are free to leave and find another job. So they are not forced to be around the celebration of something that they may find immoral and offensive, or that at the very least might make them feel very uncomfortable. But not in the military. They are not free to leave. They must go where they are assigned, and stay where they are told to. And if the militant homosexuals and their Democrat-Left supporters get their way, many of our brave, decent fighting men and woman will be stuck in situations where the "normalcy" of a sexual perversion will be shoved down their throats whether they like it or not. "Civil rights," Democrat style. Jesus' name you cannot even mention—the military has forbidden Christian chaplains from even praying in Jesus' name (<u>it might make someone feel **uncomfortable**</u>)—but homosexuality you can shove down a soldiers throat, no matter how uncomfortable it might make him or her feel; and no matter how offensive it might be to their deeply-held moral or religious convictions.

"The homosexual agenda has as its primary aim to "trump" the rights of all other groups, especially those of people of faith."[904]

Which includes the rights of people of faith in our military.

Homosexuals need our love and our prayers, in order to help them free themselves from Satan's sexually perverted and hate-filled clutch. What they do not need is to be allowed to flaunt that sin, which is an affront to God, inside of our military. The same ignorant, irrational political correctness that allowed an avowed Islamic-jihadist-traitor to infiltrate our military, to waltz around in it with impunity and with a disdain for our country, and to slaughter 12 soldiers and wound 31 more at Fort Hood Military Base in Texas, is the same ignorant, irrational political correctness that would allow homosexuals and lesbians to be as open in our military as normal heterosexuals. And keep in mind that the Islamic-jihadist-traitor at Fort Hood had even previously announced to his fellow officers in one of his military lectures that suicide bombings were justified, and he was in email contact with an Al Qaeda terrorist leader overseas. (Would our WWII generation have been so catatonic as to sit back and watch and say nothing as an American Army officer and German immigrant proclaimed that he supported both Hitler's military aggression and his slaughter of the Jews? No, they would have taken the traitor out and shot him in the head.) (Like our Navy Seals just did to bin Laden!) But the military didn't see any red flags because he was a Muslim and if anyone dared to report his behavior they would have been labeled a racist or a Muslim-phobe; and would have had their careers destroyed. So instead 43

904   (From back cover, *The Homosexual Agenda*, by Alan Sears and Craig Osten)

of our decent young soldiers lay slaughtered or wounded and an Islamic traitor sits in a military hospital, smiling proudly and spitting in the face of them and their families. Allah Akbar! And this is the same sick, ignorant, irrational political correctness that is behind allowing homosexuals to serve openly in our military.

"If a man lies with a man, as he lies with a woman, both of them have committed an abomination." (Leviticus 20:13) Welcome to the abomination of Obama-nation.

Most homosexuals are not radical or militant, but the small minority who are have the blessing, endorsement, encouragement and the cover of our national media. So their agenda rolls on unabated. Also, the majority of homosexuals who are not militant have nevertheless been conspicuously silent about either the indoctrination of our young school children or the forcing of open homosexuality upon our United States military personnel. And they have also been conspicuously silent on a related issue, the decision of our Democrat-Left media to allow militant homosexuals to spread politically-correct lies rather than saving many of their own brothers' lives…

"Thousands of gay men died and their blood is on the hands of the so-called AIDS activists who thought it was more important to push their political and social agendas than it was to educate gay men about the dangers of public, anonymous, promiscuous, multiple-partner unprotected sex. Or as it's known in West Hollywood, "Friday night.""[905]

Rather than reporting the truth about the deadly danger of promiscuous homosexual activity, they thought instead that it would be cute to further advance their agenda to achieve "normalcy" in the minds of enough clueless liberals by pretending that AIDS was just as equally a heterosexual threat. By spreading the fear of AIDS to the heterosexual population in general they were able to deflect the heat away from their own dangerously promiscuous behavior that they were unwilling to change, and they did it at the expense of the lives of many homosexual men. All now very much deceased, in the name of tolerance of course.

Help for the homosexual

"Although homosexual activity is condemned in both the Old and New Testaments, a very important positive reference is made in 1 Corinthians 6:9-11 to *former* homosexuals who had been washed, sanctified, and justified by their faith in Christ."[906]

"Know you not that the unrighteous shall not inherit the kingdom of God? Be not deceived: neither fornicators, nor idolaters, nor adulterers, nor homosexuals, nor sodomites, nor thieves, nor covetous, nor drunkards, nor revilers, nor extortioners shall inherit the kingdom of God. **And such were some of you**. But you are washed, but you are sanctified, but you are justified in the name of the Lord Jesus, and by the Spirit of our God." (1 Corinthians 6:9-11)

"**There is Hope for Change**! How do we know? Because thousands have done so!

"For those homosexuals who are unhappy with their life and find effective therapy, it is curable." — Robert Kronemeyer, *Overcoming Homosexuality*, 1980.

". . . all the existing evidence suggests strongly that homosexuality is quite

---

905 (Pg. 114, *If Democrats Had Any Brains, They'd Be Republicans*, by Ann Coulter)
906 (Pg. 260, *God's Grace and the Homosexual Next Door: Reaching the Heart of the Gay Men and Women in Your World*, by Alan Chambers and the Leadership Team at Exodus International)

changeable." — Dr. Jeffrey Satinover, former Fellow of Psychiatry & Child Psychiatry at Yale University,

Homosexuality and the Politics of Truth, 1996.

"If patients were motivated, whatever procedure is adopted, a large percentage will give up their homosexuality. In connection with this, public information is of the greatest importance. The misinformation spread by certain circles that homosexuality is untreatable by psychotherapy does [enormous] harm to thousands of men and women." — Dr. Ruben Fine, Director of the New York Center for Psychoanalytic Training, *Male & Female Homosexuality: Psychological Approaches*, 1987."[907]

"I am waiting. I am waiting. Oh yeah. Oh yeah. Waiting for some One to come out of somewhere." That's a slight play on the Stones song *I Am Waiting*. Mick and Co. were waiting, as most of us are, for some*one*, not some One. They were *seeking*, as most of us are, for some*one*, not some One. And whereas there is nothing wrong with searching and seeking and waiting for that right someone to "come out of somewhere," there is a great deal to be said for seeking "first the kingdom of heaven, and all else shall be added unto you." (Matthew 6:33) For what we tend to forget, if we were ever aware of it in the first place, is that as absolutely fantastic as the gifts of sex, and love, and physical, mental and emotional attraction and the accompanying intimacy are in this physical realm, they only exist because of the boundless love, infinite imagination and creativity of the One behind those gifts. So our mistake lies in our passionate striving for the gift itself while neglecting to seek the Giver of the gift, or in even bothering to thank him for giving us this wonderful gift in the first place. The Creator of this gift of physical love and intimacy is obviously far greater than the gift itself; as awesome as it may be. He is also the fulfillment of a love that is also worlds apart from and far greater than the love that any two people in this life can experience.

Let's say hypothetically that a builder in this life made a small rowboat and you wanted that rowboat very bad, but he said you couldn't have it, that you should wait, and that he had something even better in mind for you; but your desire got the better of you and while he was away you stole the rowboat and set out across the lake. Of course you wrecked in a storm (they always come along in relationships, whether hetero or homo) and barely made it to the shore alive. As you lay there miserable, word came from across the lake that the builder had returned and he had with him a 70 foot, twenty million dollar mega-yacht for you, but you had already made your choice and now you are stuck on the other shore with a leaky rowboat. Now, if I had to guess, as far as analogies go, that one is probably a minus 5 on the scale of one to ten, but it's the best I could come up with. But the point is that we get very obsessed with fulfilling our needs in this life and as quickly as possible with the mere gifts of the Infinite Giver, without seeking Him who is the Source of all that is lovely and attractive and beautiful and fulfilling and deeply satisfying; and without desiring to do His will and to trust that what, and who, he wants for us is so much better than what we could end up with by trying to fulfill our wants in our own way while ignoring his direction and wise council (as if we knew better than Him), and thus we end up in a big mess. (Yes, that is the world's longest sentence.) And getting sucked into and remaining

---

907 (From "Fast Facts: Homosexuality," from *Center for Reclaiming America*, an outreach of Coral Ridge Ministries)

trapped in homosexuality or lesbianism is about as big a mess relationally and spiritually as you could end up with. That is why we use the term *homosexual* instead of the cheerful term *gay*. Because we don't want to put a happy face on a very unhappy decision, and situation, for those trapped in it. We would no more call homosexuals *gay* than we would call Jon Stewart *educated*.

> "*Gay* is primarily a political term used by radical homosexual activists to take attention away from their sexual behavior. In a conference in their early 1970s, a decision was made by a group of activists to purposely label homosexuals and their behaviors as gay in order to reposition them politically. One of their goals was to get the general public to use the word *gay* instead of *homosexual* since they believed *gay* would take the onus off homosexual behavior. That stated, homosexuality is 'intrinsically disordered' and contrary to Scripture and natural law."[908]

Our lost homosexual brothers and sisters need to hear the truth, so they can have the opportunity to leave the hatred, distortions and confusion of the great deceiver behind, and embrace Jesus as their Lord and Savior; and thus come to know His infinite and unconditional love which can overcome the desire for involvement in sexual perversions. And all of the "enlightened" yet confused progressives among us who have been duped into swallowing the lie that homosexuality is *just as normal* as heterosexuality, need to wake up and embrace the truth as well, and stop helping the father of lies in doing irreparable harm to the very homosexuals and lesbians they supposedly "care" so much about.

The decision that a person who is trapped in the desire for homosexual or lesbian sex has to make, is the same decision that we all have to make. Whether we will serve our Creator, or ourselves. "As for me and my house, we will serve the Lord." (Joshua 24:15) And in serving ourselves we invariably end up serving, unawares, the adversary of God. If you are a homosexual and you choose to serve God then the path is clear. And He will be there to assist you no matter how difficult or even impossible it may now appear. (See the earlier testimonies of John and Mary.) One intimacy can always be replaced with another. Breaking up is hard to do, but we have all done it at one time or another now haven't we. And besides, there is no real substitute for the Love of God, for "God **IS** love, and he who abides in love abides in God, and God in him." (1 John 4:16) And it is only in seeking Him that "all else can be added unto you." (Luke 12:31)

"I dreamed I searched heaven for you. In vain I searched heaven for you. Friend, won't you prepare to meet me up there, lest I should search heaven for you?"[909]

My personal message to anyone who has been swept into the lifestyle of homosexuality or lesbianism is, I'm praying for you. (And that doesn't imply that I am any better than you, as there are those who are praying for me as well, for we are all sinners and we all struggle with one sin or another.) I'm rooting for you, to be swept out! Because I don't want to search heaven in vain for you. I want to meet you in paradise here on this earth in the near future so we can share together our stories of how we were drawn from darkness into the light of Christ. How through the grace of God we were "delivered from the power of darkness, and translated

---

908   (Pgs. 2-3, *The Homosexual Agenda: Exposing the Principle Threat to Religious Freedom Today*, by Alan Sears and Craig Osten)
909   (Bluegrass gospel hymn, lyrics by Kitty Wells)

into the kingdom of his dear Son." (Colossians 1:13) Here's to meeting on the other side, my sisters and brothers from some other mothers.

See:
- *The Homosexual Agenda: Exposing the Principal Threat to Religious Freedom Today*, by Alan Sears and Craig Osten
- "Fast Facts: Homosexuality," from *Center for Reclaiming America*, an outreach of Coral Ridge Ministries
- *101 Frequently Asked Questions about Homosexuality*, by Mike Haley
- *The Politically Incorrect Guide to the Bible*, by Robert J. Hutchinson (see chapter 8, "Sodom and Gomorrah")
- *Leaving Homosexuality: A Practical Guide for Men and Women Looking for a Way Out*, by Alan Chambers
- *God's Grace and the Homosexual Next Door: Reaching the Heart of the Gay Men and Women in Your World*, by Alan Chambers

BOOK IV

EARTH

CHAPTER 13

SATAN, THE GREAT DECEIVER

"You are of your father the devil, and the desires of your father you want to do. He was a murderer from the beginning, and abode not in the truth, because there is no truth in him. When he speaks a lie, he speaks from his own resources, for he is a liar and the father of it." (John 8:44)

In the chapter on Bible Prophecy we outlined God's plan for this planet and for man, the crown of his creation, in the coming years. In this chapter we will outline Satan's plan for this earth, which includes the destruction of the nation of Israel, and the destruction of the United States of America, Israel's greatest ally. We know of course that Satan doesn't stand a chance to thwart the will of God, but he doesn't know this, so he plunges forth hell-bent in a vain attempt to escape his fate, eternal judgment in the Lake of Fire. And he instills in his dupes—the Islamics, the Democrat-Left, and most of the nations of this world—a deluded and self-righteous fervor to carry out his will. The Muslim-Arabs obsession with their hatred of Israel—and the many lies they have spread that blame the Jews for the *Middle East Problem* instead of the obvious obstacle to peace, themselves—stem from the very heart and mind of the father of lies himself. He is pure hatred, pure evil, a despiser of the truth, and the ultimate source of the Muslims hatred of the Jews and their tiny nation of Israel.

It is impossible to understand what is happening on this planet without the knowledge of the underlying spiritual struggle going on behind the scenes, which is the driving force behind the outward conflict enfolding all around us. "For we wrestle not against flesh and blood, but against principalities, against powers, against the rulers of the darkness of this world, against spiritual wickedness in high places." (Ephesians 6:12) We have already seen in earlier chapters (1, 3, 5 and 10) that to study the physical world without a fundamental understanding of the metaphysical is to be of all men most ignorant. But as a Christian, to be unaware of the spiritual warfare we are immersed in, even so much as to be duped into supporting the other side, is truly sad indeed. In chapter 11 we have shown how foolish it is, especially for a "Christian," to vote for and support the morally-perverse, baby-mutilating Democrat-Left. And in Part II of this chapter we will show how decidedly foolish it is, especially for Christians, to support those same folks who now under Obama have begun the first stages of withdrawing what was once this nations strong unwavering defense of the beleaguered and slandered nation of Israel. So Satan and his Muslims, Palestinians and the rest of the Arab nations in the Middle East can try to finish what Hitler started, in a *second* Jewish holocaust.

But first, let's study a little more about the instigator himself…

Know your enemy

"Please allow me to introduce myself, I'm a man of wealth and taste. I've been around for a long, long year, stole many a man's soul and faith. …Pleased to meet you, hope you guess my name, but what's puzzling you is the nature of my game. ...If you meet me have some courtesy, have some sympathy, have some taste, *hold onto my lies and when you die*, I will lay your soul to waste. "[910]

Satan is the "ruler of the darkness of this world" (Ephesians 6:12) and "the father of lies" (John 8:44). He is "that great dragon, that old serpent, called the Devil and Satan who goes forth to deceive the whole world" (Revelation 12:9). He is Beelzebub, Belial, and "the evil one" (Matthew 12:24, 2 Corinthians 6:15, John 17:15). He was "a murderer from the beginning" and is "our adversary, the power of darkness and the prince of devils" (John 8:44, 1 Peter 5:8, Colossians 1:13, Matthew 12:24). He is Abaddon, "the angel of the bottomless pit" (Revelation 9:11).[911] He is "the god of this world who has blinded the minds of those who believe not, lest the light of the glorious gospel of Christ, who is the image of God, should shine unto them" (2 Corinthians 4:4). And how many of our teachers and professors, comics, celebrities and entertainers, media and intellectual elites believe that their own father and the inspiration behind their ideological beliefs and political agenda is nothing more than a figment of other, more "primitive" folk's imagination. Wow! Satan's "greatest deception" really does lie "in convincing the world that he doesn't even exist."

> "Satan is the archenemy of God. Satan is an incurable liar and a deceiver and the Bible says he is the father of lies. Satan hates the truth of God. And he dominates the world that he rules with falsehood. In fact, Romans 1 says that civilization in general has exchanged the truth of God for the lie. In other words, the world lives under pervasive deception and falsehood. And Satan's lies literally pervade all human thought, governing all intellectual work, all science, all philosophy, all sociology, all psychology and everything else."[912]

> "Satan will hide wherever he can—in religion, in music, in art [as we saw in chapter 2], in psychology and psychiatry, or in philosophies and ideologies. Satan is not down in hell heaping coals into the furnace. No, he's right in step with the philosophies of this world, infiltrating them wherever he can. Remember Satan's answer to God when he was asked, "From where do you come?" Satan replied, "From going to and fro on the earth, and from walking back and forth on it" (Job 2:2)."[913]

The lies of Satan cover this earth like a blanket... *The truth is relative. There are no moral absolutes. The physical world is all there is. Man is just a physical animal, not much different from the apes. There is no God. The Bible is to be mocked. Jesus was something other than who he clearly said he was. False religions. Everything continues as it always has and there is no coming judgment of the wicked. Everything was created by a big explosion and random accidental chance. Baby mutilation is a woman's "right" to choose. Homosexuality is "normal."* And all that just scratches the surface, because for each major lie that he has perpetrated, there are hundreds of smaller, related lies that are there to prop up the main ones. The lies of the devil truly cover this earth like a fog and they are

---

910 (From "Sympathy for the Devil," by the Rolling Stones, slight return)

911 (Of course, Patrick Heron argues that Abaddon is actually the antichrist, another fallen angel, and not Satan himself. See our discussion in chapter 9, "Bible Prophecy," and chapters 6 and 7 in *Return of the AntiChrist and the New World Order*, by Patrick Heron)

912 (From "The Battle for the Beginning," by John MacArthur)

913 (Pg. 22, *Spiritual Warfare*, by Dr. David Jeremiah)

specifically designed to keep the Light from ever reaching the minds, and hearts, of men. He surely is the god of this world. Satan's desire from the very beginning was to be worshipped as God (it's one of the things that caused him to fall in the first place), and unsuspecting billions are already obliging him.

> "Everything Satan does is a copy of something God has done. For instance, God has a city that is a bride, Jerusalem (Revelation 21:2), and Satan has a city that is a harlot, Babylon (Revelation 17:5, 18). Jesus is "the light of the world" (John 9:5) and Satan masquerades as an angel of light (2 Corinthians 11:14). Jesus is the Kings of Kings (1 Timothy 6:15) and Satan is "king over all the children of pride" (Job 41:34). Jesus is the Prince of Peace (Isaiah 9:6) and Satan is the "ruler of this world" (John 14:30) and the "prince of the power of the air" (Ephesians 2:2). Jesus is called "the Lord my God" (Zechariah 14:5) and Satan is called "the god of this age" (2 Corinthians 4:4). Jesus is called "the Lion of the tribe of Judah" (Revelation 5:5) and Satan is a "roaring lion, seeking whom he may devour" (1 Peter 5:8).
> "Satan deceives today basically the same way he began deceiving in the Garden of Eden: by speaking lies. Only today he speaks through false prophets and false teachers."[914]

Yet the devil was originally the highest of all of God's angels, before his fall:
"The word of the LORD came to me: This is what the Sovereign Lord says, You [the devil] were the seal of perfection, full of wisdom and perfect in beauty. You were in Eden, the garden of God; every precious stone was your covering: the ruby, topaz, and diamond, the beryl, onyx, and jasper, the sapphire, turquoise, the emerald, and gold. Your settings and mountings were made of gold; on the day you were created they were prepared. You were anointed as a guardian cherub, for so I ordained you. You were upon the holy mountain of God; you have walked back and forth in the midst of the stones of fire. You were perfect in your ways from the day you were created, till iniquity was found in you. ...Therefore I have cast you as a profane thing out of the mountain of God; and I destroyed you, O covering cherub, from the midst of the stones of fire. Your heart was lifted up because of your beauty; you have corrupted your wisdom for the sake of your splendor; I will cast you to the ground, I will lay you before kings, that they may behold you. All who knew you among the peoples are astonished at you; you have become a horror, and never shall you be any more." (Ezekiel 28:11-19)

And in the book of Isaiah we learn more about the descent into evil of this once-brightest of all of God's angelic creations, and the beginning of his role as the prince of darkness:

"How you are fallen from heaven, O Lucifer, star of the morning, son of the dawn! You have been cut down to the earth, you who weakened the nations! For you have said in your heart, 'I will ascend into heaven, I will exalt my throne above the stars of God; I will also sit upon the mount of the congregation, in the farthest reaches of the north. I will ascend above the heights of the clouds; **I will make myself like the Most High**.'[915] Yet you shall be brought down to hell, to the lowest depths of the Pit." (Isaiah 14:12-15)

---

914   (Pg. 25, *Spiritual Warfare*, by Dr. David Jeremiah)
915   (And isn't it interesting that Lucifer's second lie to Eve in the Garden was that they also would be "like the Most High" if they ate of the forbidden fruit... "and you shall be as gods" Genesis 3:5)

And this is also confirmed in the Book of Revelation:

"So the great dragon was cast out, that old serpent, called the Devil, and Satan, who deceives the whole world; he was cast to the earth, and his angels were cast out with him." (Revelation 12:9)

Satan took a third of all of the angelic hosts with him in his fall from grace:

"And another sign appeared in heaven: and behold, a great red dragon having seven heads and ten horns, and on his heads were seven crowns. And his tail drew **the third part of the stars of heaven**, and did cast them to the earth." (Revelation 12:3-4)

These fallen *stars* or fallen angels became demons. Some of these spirits roam the earth, and some are already trapped in hell, but all will end up there soon. The ones still in this realm work with the devil in his quest to prevail over the will of God. (Just as most of mankind refuses to seek or to follow God's will, and instead follows the will of the god of this world.) And all of them will suffer the same eternal fate as Lucifer: "And the angels who kept not their first estate, but left their own habitation, he has reserved in everlasting chains under darkness unto the judgment of the great day." (Jude 1:6)

Those verses from the books of Isaiah, Ezekiel, Jude and Revelation tell the story of Satan's fall from the highest heights (from being "the anointed cherub") to the lowest of lows (to becoming "the evil one"). He was the brightest of all of God's angelic host before he was overcome with pride and narcissism. And through that self-love and arrogance he turned from the worship of God to the worship of himself. He entertained the thought and then followed through with the desire to "make himself like the Most High." And how many of us, through greed and pride and blindness and self-love and un-thankfulness for all that we do have, fall into the same trap. Not necessarily trying to be like the Most High (unless one is a follower of Oprah and the *New* Age religion, which is nothing more than the same *old* lie from the Garden: "You shall be like God" Genesis 3:5), but overlooking the really important things in life like praising, thanking and serving our Creator instead of ignoring him or taking him for granted; like focusing primarily on being a Christ-like servant to our wives and family and children and friends instead of what material things and selfish fulfillments we can get out of life. There is a lesson here somewhere to be learned from the fall of Satan. He had it all, and he "gave it away for the sake of a dream in a penny arcade. If you know what I mean.[916]"[917] He was overcome by pride in his own beauty and intelligence, and he foolishly gave himself, rather than God, the glory for it; as we too can be seduced by our own pride in our abilities, looks and intelligence, rather than giving God the glory. And because of that sinful nature, that inbred arrogance, the great deceiver is also able to keep us trapped in his eternally destructive lies. We are too proud of and too in love with our own thoughts, opinions and beliefs, and for far too many of us that ends up delivering us to the same horrible eternal fate as the one who beguiles us.

"Satan is that old deceiver, and remember that his battle with mankind is fought with deception. He uses the weapon of deception; he deceives the minds of men. He is a liar from the beginning; he deceived Eve with a lie. And he deceives people with lies; the lie

---

916   (No, not really.)
917   (From song, *If You Know What I Mean*, by Neil Diamond)

that they will find their greatest joy apart from God is perhaps his most successful lie of all time. That it is away from God and in the things of the world that you will find your joy and your fulfillment and your satisfaction. That it is by the exalting of self and the seeking of your own that you will find the meaning and purpose and significance of life. This lie has been so incredibly successful; how many hundreds of millions of people have been deceived by that lie."[918]

Just as God has a wonderful plan of salvation for all of mankind, for those who turn from darkness to the Light of Christ, so too does God's adversary, the great deceiver have a counterproductive, diabolical plan himself. And that of course is to deceive as many souls as possible to follow him down to hell, through the lies of false religion, and through distorting and mocking the truth so it will have no effect on the hearts and minds of lost mankind. And he has been very successful in this endeavor over the centuries. Remember the warning that "many are called, but few are chosen." (Matthew 22:14) This is a prophecy that the "way that seems right to most men (and women) is actually the way of eternal death." (Proverbs 16:25)

"Satan's hatred of God for having been cast out of heaven is behind his purposes. He is intent on disrupting God's plan and dethroning God, to include consigning all of God's human creation into a place of torment. Satan has subordinates and an army of demons who work to get a foothold into your life through the avenue of the flesh. The fallen human nature is Satan's primary "hook" to gain entrance into our lives and attempt to ruin us, thereby foiling God's attempt to grant us forgiveness and eternal life. We should not underestimate the ability of Satan to accomplish his purposes. ...He beguiles, seduces, opposes, resists, deceives, sows tares, hinders, buffets, temps, persecutes, and blasphemes."[919]

And we should not underestimate him because "next to God, Satan is the most powerful being in the universe."[920] Which is why we need to follow the command to "put on the whole armor of God, in order to withstand all of the fiery darts of the wicked one." (See Ephesians 6:10-18, and our discussion of it in the "Walking" section of chapter 8.) And why we should "be sober; be vigilant; because your adversary the devil walks about like a roaring lion, seeking whom he may devour." (1 Peter 5:8)

"Whether we like it or not, we are caught in the midst of the spiritual battle. It has been that way since Adam. Our position in this battle depends upon our choice as individuals. This choice determines who we will follow as our ultimate leader: Christ or Satan. It doesn't matter how good our personal character may be. If we reject God's plan of escape, we are automatically on the side of Satan. There is no middle ground. The only exceptions to the rule are children under the age of accountability, those persons who are mentally impaired, and those of very remote areas where the gospel has not penetrated, that is to say, those individuals of remote areas who have the law of God in their hearts (Rom. 2:7-16)."[921]

Note that sentence, "It doesn't matter how good our personal character may

---

918   (Pgs.121-122, *The Real Meaning of The Zodiac*, by D. James Kennedy, Ph.D.)
919   (Pg. 24, *Spiritual Warfare*, by Dr. David Jeremiah)
920   (Pg. 24, *Spiritual Warfare*, by Dr. David Jeremiah)
921   (Pgs. 243-244, *The Unveiling: A Journey Through the Book of Revelation*, by Keith Harris)

be." This goes back to what we have talked about a number of times before, but it is such a sly deception that it is worth visiting one more time. We can't get to heaven covered in our own righteousness, for indeed, "all of our righteousness is as filthy rags" (Isaiah 64:6) We must be covered in the righteousness of Christ. We must be washed in the "soul-cleansing blood of the Lamb." ..."Are your garments spotless? Are they white as snow? Are you washed in the blood of the Lamb?"[922] You must have on a wedding garment. (See parable of the Marriage Feast in Matthew 22:1-14) A lost person's personal righteousness or character many times can fool even some true Christians into making the false assumption that they are saved; that they must be headed for heaven, but that is also one of Satan's more deceptive traps. Which is why Jesus warned of ending up at the gates of heaven ill prepared, "Many will say to me in that day, Lord, Lord, have we not prophesied in your name? and in your name have cast out devils? and in your name done many wonderful works? And then will I profess unto them, I never knew you: depart from me, you who work iniquity." (Matthew 7:22-23) Don't be deceived into not witnessing to someone because of their apparent goodness, or kindness, or personal righteousness or character. Even to those who say they are Christians but who may be lost in the error of Roman Catholicism, or a part of liberal Churchianity, or deceived into supporting the Democrat party; and therefore show themselves, despite their apparent goodness, character and righteousness, to be clothed in their own righteousness and not the righteousness of Christ. You will only be doing them a great disservice in not witnessing to them, for they may be very ripe for becoming a true disciple of Christ and one who will then hear these words, "Well done, you good and faithful servant. You have been faithful over a few things, I will make you ruler over many things: enter you into the joy of your lord." (Matthew 25:21)

Satan's primary mission is in trapping the lost. But he also has another very important task at hand, one that is meant to thwart his own eternal fate (see chapter 9, Bible Prophecy), and that is the destruction of the nation of Israel. Indeed, over the last two millennia and well before Israel's miraculous rebirth in 1948, he used all of his power and influence to direct his followers to hate, persecute, torment and oppress the Jews in every nation and on every continent where they were dispersed. His purpose was to wipe them out as a distinct, separate group of people; to harass and persecute them to the point where it would become too painful and too dangerous of a prospect to continue to pass on and preserve from one generation to the next their ancient heritage, culture, traditions and religion. No other group could have sustained such a relentless, satanically-inspired effort to destroy them. In the past two thousand years the Jews as a distinct people should have long ago disappeared, and been absorbed into the countries and societies of their dispersion. And if Satan had gotten his way he would have accomplished just that, and the nation of Israel therefore could have never been reborn; for there would have been no distinct and separate Jewish people left to come back and re-form it. But miraculously and against all odds (and because of the hand of an all-powerful Creator), he failed.

"Israel coming back as a nation to me is the most astonishing event in history. And there's a lot of astonishing events in history. But it's never happened before. It's never happened in the history of civilization, that a nation could be completely dispersed and

---

922   (From gospel song, "Are you washed in the blood," by Elisha A. Hoffman)

1800 years later come back together as a nation with its original national identity. This has never happened in the history of the world."[923]

The Muslims were his most loyal allies, his most fervent religious devotees, over the centuries. Their persecution of the Jews is legendary and continues of course unabated to this day, while the Catholic Church and the Russians have come in a close second. But the most monstrous single occurrence in Satan's goal to eradicate the Jews was brought about by the Nazi's during World War II. In the Holocaust the Germans exterminated six million Jews. (As Margaret Sanger, her Planned Parenthood and her Democrats have exterminated millions upon millions of blacks, before they even had a shot at life.) And again as we mentioned in chapter 10, Hitler was inspired not only by the great deceiver but by his genesis account as well. The theory of evolution, accompanied by its sacred tenant, *The Survival of The Fittest* gave Hitler the "scientific" authority (the Jews being "unfit" according to Hitler) to carry out his gruesome mission, and add substantially to Darwin's "Deadly Legacy." But God, who is of course all-powerful and who knows the evil designs of man before they are even conceived, took Hitler's maniacal quest and let it become the actual driving force behind the nation of Israel's miraculous rebirth in 1948, just three years after the end of that holocaust. What Satan had designed as the end of the Jewish people, God used for His own purposes of re-forming their ancient nation in Palestine.

> "Not the least of Satan's goals is to destroy Israel and the Jewish people. …the Jewish people and their beloved homeland face unprecedented horrors at the hands of Islamic terrorists and our own [Democrat-Left] politicians. The world seems to bow and scrape before Islam and religious fanaticism and to accept all beliefs as legitimate- except for the only two based on God's Word: Judaism and Christianity."[924]

Lucifer knows now that his time is running out and he is desperate to annihilate the nation of Israel before the Lord returns to rule over the earth from Jerusalem, and to send him to torment in the depths of hell. And for the rest of this chapter we are going to explore the various aspects of what Satan is doing to try to realize this goal. In "Part I- Destroy Israel" we will expose the many lies and subterfuge that he and his Middle-Eastern Muslims have concocted to fool the rest of the world into allowing them to "push the Jews into the sea." And then in "Part II- Destroy America, Israel's Protector" we will see what he has been doing to destroy Israel's primary defender, America. Indeed with the election of one Indonesian non-Muslim, Barack Hussein Mohammed Obama, his dreams for the fate of this country are almost within reach.

Again, lies are Satan's primary weapon, and because of this he has amassed quite an impressive number over the years. And besides his main ones of false religion (Roman Catholicism, Jehovah's False Witnesses, Mormonism, Islam, Hinduism, Buddhism, etc.), false baby-mutilating ideologies (liberalism and secular-progressivism), and false-scientific theories (evolution and the big bang), there are layer upon layers of other ones that are used to prop up the main ones. And as we go through the rest of this chapter we thought it would be a good idea to demonstrate their bounty and how "mainstream" they have become by notating each one with the elevated word "**Lie**"[(LIE)] in parentheses. As we shall see, the god

---

923 (Jay McCarl, Middle Eastern Customs and Prophecy, on The Final Prophecies DVD, an Ingenuity Films Production)
924 (From article "The Master Deceiver's Handiwork," by Thomas C. Simcox in *Israel My Glory Magazine* July/August 2007. A Ministry of The Friends of Israel Gospel Ministry Inc.)

of this world is a prolific prevaricator, and with the help of both his Islamic and left-wing adherents, he has become quite successful in drowning this country, and this earth, in his falsehoods...

Part I- Destroy Israel

"Therefore when you see the abomination of desolation, spoken of by Daniel the prophet, standing in the holy place (whosoever reads, let him understand) then let those who are in Judea flee to the mountains." (Matthew 24:15-16)

> "Some say Israel's existence threatens world stability. Eliminate Israel and the world will achieve peace. So say the misinformed. Yet this worldview is thriving. In fact, a day is coming when the entire world will want to eliminate Israel. But the world will find itself fighting against God himself."[925] ("Many are called. Few are chosen.")

As we mentioned previously, and covered in chapter 9, "Bible Prophecy," for Satan to thwart God's ultimate plan for the redemption of man, and save his own hide, he must destroy the tiny nation of Israel before the Lord returns. He began his assault militarily as soon as the nation was formed. In 1948 Israel legitimately declared its independence in accordance with the Balfour Declaration and the Resolution of the United Nations General Assembly. However, the next day six Arab-Muslim armies from Egypt, Lebanon, Syria, Jordan, Iraq and Saudi Arabia launched an attack with overwhelming odds. Allah Akbar! But miraculously for Israel, and pathetically for them, they lost! Glory and honor to the true God of Israel! The chance of the infant state of Israel's survival against the combined, devastating force of the armies from these countries was close to nil but, because God is all-powerful, she prevailed. This was Satan's first attempt at their annihilation. Since then there have been other unprovoked attacks against Israel, but the great deceiver and his Arab-Muslims have been stymied militarily for now.

"Let God arise, let his enemies be scattered. Let them also who hate him flee before him. As smoke is driven away, so drive them away; as wax melts before the fire, so let the wicked perish at the presence of God." (Psalms 68:1-2)

Land for Peace[(LIE)]

So Satan has gone to plan B. His design now is to weaken the nation of Israel to the point where she is absolutely indefensible. The Muslim-Arabs have been embarrassed by their losses to the tiny Jewish state in every previous war, but it is important to point out that this should ultimately be attributed to the hand of the Almighty than to their own ineptitude, or to the superior military acumen of Israel, (although both of those certainly have played their part). Indeed in the Yom Kippur War of 1973 the Arab armies were on the verge of complete victory, and there is no other explanation for the sudden and miraculous turn of events that saved Israel from complete annihilation than the intervention of the God of Israel.

---

925   (From article "Victory at Armageddon," by Steve Herzig, *Israel My Glory* Magazine May/ June 2009)

But these defeats have not been lost on the Arabs, so for the decades since their last humiliation it has been theirs and their father's plan to reduce Israel's size to the place where she cannot hope to defend herself, to where even the Girl Scouts of America could prevail. And this is where the absurd idea of Israeli "land for peace" has come into play. The Arabs dwarf the Israelis in the size of *their* land. The next time you go shopping drop a penny next to your car. The penny is Israel. The rest of the parking lot is the Muslims land.

> "In size, the tiny nation of Israel stands surrounded by Arab countries that make up about 650 times its land area. [Emphasis mine.] Those Arab nations together are the size of the USA, Mexico, and Central America combined. Israel, however, is about the size of New Jersey."[926]

But for some strange reason it is the *Israeli's* who must give up a part of their measly parcel of land to the Arabs in order to achieve peace. What's wrong with that picture? It's painted by Beelzebub, that's what, and the idiot nations of the world have bought into it. (And sadly so have many liberal American, and even liberal Israeli, Jews.) The Arab–Muslims have somehow convinced a gullible world, and even unfortunately far too many Israeli voters, that if they just give up more and more of their tiny country (to the point where it is absolutely indefensible) then the great, generous, illusive Arab genii of "peace" will suddenly pop out of his bottle.[LIE] (Right after Obama passes Economics 101.) Land for Peace is such an obvious lie because the stated goal of the Palestinians and other Muslim nations surrounding Israel is to "push the Jews into the sea." They will only have their "peace" when they have removed this nation, and every single Jew, from the Middle East. In fact, if you look at many Arab/Muslim maps you will find that the country of Israel does not even appear. Now isn't that revealing.

> "One would expect that the concept of "land for peace" would work both ways. After all, should not the Arabs also make some territorial sacrifices for peace? Unfortunately, that is not the case. Every inch of land held by the Arabs is considered "holy Arab soil" and its possession by the "infidels" (Christians or Jews) is inadmissible, intolerable, a blasphemy and a case for "jihad" (holy war). No compromise, no concession is ever possible."[927] (Of course, just about everything else on the face of this earth is a "case for jihad" for these morons, including farting in a mosque.)

The Arab-Muslims do not really want any "land for peace," because they do not want peace. They want all of the land that Israel has and they want to exterminate the Jews from off of that land. Yet our Democrat media has repeated the Muslims "Land for Peace" lie so often that enough gullible people have now come to believe it.

In 1948 then Arab League Secretary-General Azzam Pasha said this:

> "This will be a war of extermination and a momentous massacre which will be spoken of like the Mongolian massacres and the Crusades."[928] (You know, nothing gets those Muslims worked up like a good old-fashioned Jew-extermination. Sieg Heil!)

Thirty two years later, in 1980, PLO Chairman Yasser Arafat (now dearly-departed) said this:

---

926 (From circular, by Dr. Mitch Glaser, President, *Chosen People Ministries*)
927 (From ad, "Land for Peace? Can it solve the problems of the Middle East?" published by FLAME: Facts and Logic About the Middle East. Go to factsandlogic.org)
928 (As quoted in a circular from the *Zionist Organization of America*, by Morton A. Klein and Dr. Daniel Mandel)

"Peace for us means the destruction of Israel. We are preparing for an all-out war, a war which will last for generations."[929]

And nothing has changed to this day. In 2005, the Egyptian Muslim Brotherhood leader Muhammad Mehdi Akef said...

"I declare that we will not recognize Israel which is an alien entity in the region. [Akef is not only a Jew-hater, but a historical ignoramus as well, for the Jews were in Palestine long before the Muslims.] And we expect the demise of this cancer soon."[930]

If he had been referring to the disease that is his own hate-filled, clitoral-mutilating religion, then he would have been prophetic.[931]

One recent example of what a canard *Land for Peace* actually is came right after Israel was again snookered into parting with more of her precious land. This time it was the 2005 pull-out from Gaza.

"In 2005, Israel unilaterally pulled out of the Gaza Strip—painfully extricated thousands of men, women and children from the beautiful cities and communities they had planted and nurtured, to give the Arabs a completely *judenrein* (free of Jews) Gaza—a sacrifice to achieve peace. They left homes and neighborhoods and lovely synagogues they had built, as well as thriving agricultural businesses—businesses that employed thousands of Arab Palestinians.

"Through a special Israeli grant, the greenhouses were prepared to continue operating so that the Gazans would not lose this source of employment, revenue and agriculture.

"They were burned to the ground.

"The beautiful cities the now homeless Jews left behind were ransacked, the synagogues burned, and hateful anti-Semitic graffiti painted all over the rubble.

"And surprisingly, on the day Israel gave the Palestinians the entirety of the land of Gaza—*for the sake of peace*—they began sending rockets into Israel's homes and schools.

"There was no world outcry."[932]

Let's review here for a minute...

Number one, why couldn't Israel have just given the Palestinians control over Gaza without the Jews being forced to leave? In 1948 the Muslim-Arab (Musrab) population that remained in Israel after the Musrab countries surrounding it tried to annihilate it, were never forced to move out!? They're still there, six decades later. Why aren't the Jews still in Gaza? Answer: because the violent, racist Palestinian Muslims cannot stand to live with Jews. They are ignorant racists, by definition, but whereas ignorant racists in America and elsewhere in the world are heartily condemned (except for all the liberal racists in our media—see the case of Travon Martin), the Muslims get a pass. The Nazi's collaborators in our Democrat-Left media just ignore it.

Number two, what kind of hate-infested mob of Muslim cretins would destroy homes and businesses that they could have moved into themselves and appropriated for their own livelihood and nourishment? (No answer necessary.)

---

929 (As quoted in a circular from the *Zionist Organization of America*, by Morton A. Klein and Dr. Daniel Mandel)
930 (As quoted in a circular from the *Zionist Organization of America*, by Morton A. Klein and Dr. Daniel Mandel)
931 (And yes, that would be the same Muslim Brotherhood that our genius, non-Muslim president is sending billions of our tax dollars to support.) (Can't wait to reelect him)
932 (From article "The Real War Israel is Fighting," by Sarah Weiner in *Jewish Voice Today*, March/April 2009)

I mean, it's not like they themselves were enterprising enough to build decent homes and businesses on their own over the years. At least their despicable Nazi friends in Germany 70 years ago had the sense not to destroy the Jewish homes and businesses but just to take them over. But in the spirit of "suffering fools gently," the fact is that the lies of Satan curse this earth like an Islamic jihad, and people believe what they are taught. (Which is not an excuse for their ignorance and blind hatred, but a source of understanding.) And the Palestinians believe the pure satanic, Islamic, Jew-hating drivel that they have been fed. Their anthropologic behavior when taking over Gaza is just one of the results.

But the Musrabs are not the only idiots in this fiasco. The gullible Israelis who gave up Gaza—and even more incredibly tore innocent Jewish families from their homes and businesses and livelihoods and communities and worship centers just to satisfy the Nazi-Islamic Palestinians racist demand to remove all Jews from their land—did it all for nothing. They did not get their Land for Peace. They got "Rockets for Peace" instead:

> "The Rockets began falling on the tiny border city of Sderot, Israel, and life changed forever for the residents. Death, terror, listening for the daily air-raid siren, running for the bomb shelter, children with post-traumatic stress syndrome... this was the result of Israel's supreme sacrifice of land. ...The extreme demand to provide *judenrein* [Jewish-free] land had been granted in Gaza, (which in itself is an unconscionable request) and yet, rather than granting peace, they began attacking Israeli citizens on a daily basis from their position of strength. ...This is what Nazi Germany was all about—Hitler's dream—creating a *judenrein* Europe, and then a *judenrein* world. Hitler exterminated 6 million Jews in his quest before he was stopped. [The Arab-Muslims, conversely, have yet to even be discouraged. Funny how history repeats itself.]
>
> "The Jews who survived thought they finally had a tiny place on the earth where they would be safe. And yet they are asked—*and the U.S. is involved in helping*—to create *judenrein* territories for Arabs who call themselves "peace partners"—territories *where Jews are not allowed* or safe to enter...
>
> "Conversely, Arabs enjoy full citizenship in Israel—with full voting rights. In fact, they have more rights and freedoms as Israeli citizens than in almost any Arab country."[933]

In addition to the land for peace scam, every one of the so-called peace initiatives over the years have done nothing to advance that cause, because the Musrabs do not want peace. They want to get rid of the Jews and their nation of Israel from the Middle East. And they have used these "peace initiatives" to sucker liberal Jews in Israel (with the help of enormous pressure from the Muslims useful idiots who control most of the nations and the medias of the Western world) to give away one concession after another, while they the Musrabs give up absolutely nothing in return, except of course for their empty promise of peace. Just listen to what PLO Chairman Yasser Arafat said in 1993, on the very day that the world heralded an historic "milestone" in the Arab-Israeli conflict, the Oslo "Peace" Accords...

> "Since we cannot defeat Israel in war we do this in stages. We take any and every territory that we can of Palestine, and establish sovereignty there and we use it as a springboard to take more. When the time comes, we can get the Arab nations to join us for

---

933   (From article "The Real War Israel is Fighting," by Sarah Weiner in *Jewish Voice Today*, March/April 2009)

the final blow."[934] - PLO Chairman Yasser Arafat.

My only question is why the Nazi collaborators…er…journalists in the liberal Western medias didn't scream this from the front pages, instead of ignoring it and going on and on about "peace" this and "peace" that. Can you imagine how these same media outlets would have reacted if the Israeli Prime Minister had said, on the very day of the signing of this "momentous" peace accord, that "since we cannot defeat the Arabs in war we do it in stages. We take any and every territory that we can from the Muslims, and establish sovereignty there and we use it as a springboard to take more. When the time comes, we can get the United States to join us for the final blow." There would have been a world-wide conniption fit to rival the union riots in Wisconsin (and that's not any small feat) as the Left screamed for the dissolution of the State of Israel, to protect the poor Arabs from the Jews stated goals to exterminate them. And they would have screamed how the Oslo Peace Accords were an obvious sham because the two-faced Israeli's were talking out of both sides of their mouth. But not a word about the real two-faced Musrabs. Wow!

> "Ever since the failed Oslo agreement in 1993, supporters of Israel have placed a false sense of hope with every new peace initiative that is part of a process that would weaken, dismantle, and ultimately destroy the Jewish state."[935]

And Satan is quite happy with the cooperation he has gotten from liberal Israelis, in addition to many liberal Jews in this country. They hide their heads in the sand and ignore reality as they aid and abet their fiercest enemy.

> "The Arab nations should sacrifice up to 10 million of their 50 million people, if necessary, to wipe out Israel. Israel is to the Arab world a cancer to the human body."[936] - Saudi's King Saud.

There are some Arab countries that are supposedly more moderate in their hatred of the Jews, countries like Egypt (of course now that Mubarak has been forced out, and the Muslim Brotherhood has gotten in, all bets are off) and Saudi Arabia for example. They will speak one way out of one side of their mouth when the Western World is listening, but on their own state run radio and TV, in their Newspapers and magazines, and in their Mosques, the truth concerning their feelings and intentions toward Israel is broadcast loud and clear for all of *their* people to hear. Underneath the benign surface that gullible Western journalists allow themselves to see and report on, is a cauldron of anti-Semitic racism, a river of hatred, waiting for its chance to erupt. (See "The battle of Gog and Magog" in Chapter 9, "Bible Prophecy.") Other countries like Iran, Syria, Iraq prior to the 2nd Gulf War and the Palestinians are much more honest about their stated intentions to annihilate the nation of Israel and slaughter all of its Jewish inhabitants.

> "If the Arabs laid down their arms there would be no more war. If the Israelis laid down their arms there would be no more Israel."[937]

---

934 (PLO Chairman Yasser Arafat "in a taped interview on Jordan TV." Source: "Know Them by Their Words," article @ shalomjerusalem.com)

935 (From ad, "It doesn't take a genius to figure it out. There is peace only through sanity and strength." With Columbo's—Peter Falk's—picture/endorsement at the top)

936 (Saudi King Saud in 1954, as quoted in a circular from the *Zionist Organization of America*, by Morton A. Klein and Dr. Daniel Mandel)

937 (Dick Morris on Fox News *The O'Reilly Factor*, 6/2/2010)

There are other outrageous lies that Satan, his Muslims and the Western media have spread that undermine, and are intended to ultimately destroy, Israel. Here are some of the more popular ones…

Palestinian "refugees"

The Democrat media tell us that Israel is responsible for the Palestinian refugee problem.(LIE) But in reality, it is not. The Musrabs are. The Jews absorbed all of the Muslims who remained in Israel after the 1948 war, and gave them full rights of citizenship. Conversely, the Musrabs mercilessly persecuted, killed and drove nearly all of the Jews from their homes and businesses in most of their countries, while at the same time refusing to absorb those Musrab refugees who had been told by the invading Arabs to flee Israel. They instead kept them separated and virtually imprisoned, in poverty and misery over the decades, because they were more valuable to them as a tool to wear down the nation of Israel and to make certain that no viable peace plan and therefore no real or lasting peace could ever be reached. The Muslims have gotten away with falsely blaming their prisoners… er…refugees existence on the Jews(LIE) due to the cover from the complicit medias of the Western world, including of course our own. And they have inhumanely kept millions of these refugees in poverty, misery and squalor for decades in the hope of one day using them to fulfill their demonic desire for a second Jewish holocaust.

Palestinian refugees have a "right of return"(LIE)

The way the Musrabs hope to use their refugees to accomplish this holocaust is by eventually duping enough idiots, or Nazi's, in this world to force Israel into accepting the implementation of the Muslims absurd demand that all of these refugees and millions of their descendants somehow have the "right of return"(LIE) to the land of Israel:

> "The **Palestinian right of return** …is a political position or principle asserting that Palestinian refugees, both first-generation refugees and their descendants, have a right to return to the property they left …in the former British Mandate of Palestine (currently Israel and Palestinian territories), as part of the 1948 Palestinian exodus, a result of the 1948 Palestine War and due to the 1967 Six-Day war."[938]

And if they could foist this abomination on Israel then the Jews would suddenly become the minority population and the Musrab majority could then do to all of them what they did to all of the Jews in their countries for millennia, and including right after the 1948 war: persecute, kill, rob, rape and then drive them out. Those darn peaceful Muslims! Aren't they clever? Or is this world so absolutely stupid? The right of return claim is of course an asinine and laughable one. The original "refugees" were not driven out of Israel by the Jews. They left voluntarily because they were told to leave for their own safety by the invading Arab/Muslim armies, and that after the Jews were exterminated they could return not only for their own property but they could take that of the dead Jews as well! Only the problem with that scheme was that the Mohammedans got their arses kicked. And the "refugees" found themselves stuck out in the desert, where the Arabs have kept them imprisoned for 63 years. Allah Akbar!

---

938   (From Wikipedia, the free encyclopedia)

"When Israel declared its statehood in 1948, pogroms[939] broke out across the entire Arab-Muslim world. Thousands died in this violence. Their homes and businesses were destroyed, their women violated. The vast majority of those Jews fled from where they had lived for centuries. They had to leave everything behind. Most of those who were able to escape found their way to the just-created state of Israel."[940]

Does anyone ever remember hearing the Arabs mention the Jews right of reimbursement for what was stolen from them? Not a chance. The *Jews* "right of return," not that they would ever want to, is also never discussed. Nearly one million of these Jews, forcibly ejected from Arab-Muslim lands after the 1948 and 1967 wars, were absorbed and integrated into Israeli society. Conversely, here is what happened to those Muslim-Arabs who, again, voluntarily left Israel in 1948, the origin of the so-called "Palestinian Problem." About 650,000 Arabs...

...."fled from Israel during its War of Independence. Most left following the strident invocations of their leaders, who urged them to leave, so as to make room for the invading Arab armies. After victory was to be achieved, they could return to reclaim their property and that of the Jews, all of whom would have been killed or would have fled."[941]

Again, no one forced them to leave Israel. Their homes and businesses were not destroyed or stolen from them as was the case with those Jews living in Arab-Muslim lands. And how did their Muslim brothers treat these refugees? Were they absorbed and integrated into the Arab countries surrounding Israel by the Musrabs who possess territory that is hundreds of times larger than the tiny nation of Israel? Of course not.

"They confined them into so-called refugee camps ...essentially extended slum cities, where their descendants - now the fourth generation - have been living ever since. **The reason for the Arabs' refusal to accept them was and still is the desire to keep them as a festering sore and to make solution of the Arab/Israel conflict impossible.** [emphasis mine] These "refugees," whose number has by now miraculously increased from their original 650,000 to 5 million, are seething with hatred toward Israel and provide the cadres of terrorists and suicide bombers."[942]

But should we expect anything less despicable from a *religion* that hacks the clitoris off and mutilates the womanhood of hundreds of millions of its own little girls? Their minds have been poisoned with "the word of Lucifer" ...to the point of insanity.

"Eternal Refugees... The Arabs, rather than settling those who fled in their own lands, put their Arab brethren into camps. Their sympathy value on the world stage was worth more than the quality of their lives. ...Jews are not welcome in most Arab states—especially in the Palestinian territories...but Arabs have full citizenship and full rights in

---

939   ("A planned campaign of persecution or extermination sanctioned by a government and directed against an ethnic group, especially against the Jewish people in tsarist Russia." Encarta)

940   (From ad, "The Forgotten Refugees: Why does nobody care about the Jewish refugees from Arab lands?" published by FLAME: Facts and Logic About the Middle East. Go to www.factsandlogic.org)

941   (From ad, "The Forgotten Refugees: Why does nobody care about the Jewish refugees from Arab lands?" published by FLAME: Facts and Logic About the Middle East. Go to www.factsandlogic.org)

942   (From ad, "The Forgotten Refugees: Why does nobody care about the Jewish refugees from Arab lands?" published by FLAME: Facts and Logic About the Middle East. Go to www.factsandlogic.org)

Israel. The Arabs refuse to resettle their own refugees, *and the world blames Israel?* Why *are* there still "Palestinian refugees" after 60 years?"[943]

Good question. And maybe some western journalist might get around to actually asking the Arabs that, sometime before hell actually does freeze over.

"The Palestinian refugees occupy a unique place in the annals of warfare and migration of people. Such migrations are common and unavoidable in the long history of human conflict. Following WW II, millions of Germans were expelled from Czechoslovakia and the eastern regions of Germany that were turned over to Poland. In the upheavals that followed the withdrawal of the British from the Indian subcontinent, millions fled or were driven from their homes. Following France's loss of its Algerian possessions, hundreds of thousands fled that North African country. These millions of refugees and others were absorbed into their new countries and were no longer the concern of the world. It is only the Palestinians who, for almost sixty years, have been considered "refugees." In fact, a special branch to the United Nations (UNWRA) exists only for the maintenance of these "refugees," at a cost of billions of dollars.

"The problem of the "Palestinian refugees" is a red herring, fostered for almost sixty years in order to maintain the fiction that these "refugees" have the right to return to their "homeland.""[944]

Palestinian statehood[(LIE)]

Palestinian statehood is another red herring. The Arab-Muslims don't give a damn about the Palestinians or their statehood. In fact, this concept didn't even arise until 1967, after the Six Day War...

"The concept of Palestinian nationhood is a new one and had not been heard of until after the Six-Day War (1967), when Israel, by its victory, came into the administration of the territories of Judea and Samaria (the "West Bank") and the Gaza Strip. The so-called "Palestinians" are no more different from the Arabs living in the neighboring countries of Lebanon, Syria and Jordan, than Wisconsinites are from Iowans."[945]

Yet our liberal media constantly feeds us the lie that "The *Palestinians* are a nation[LIE] and therefore deserve a homeland."[(LIE)946] The goal of the Musrabs is the complete eradication of the Jewish state. Period. And they will use any means of violence, lying and subterfuge to accomplish that goal. If they had to slaughter every single man, woman, and child in Palestine tomorrow in order to wipe out the nation of Israel, those terrorist groups in alliance with the nations surrounding Israel would do it in a heartbeat. Besides, the Arabs had their chance at a Palestinian state sixty years ago, but they rejected it because that is not what they were interested in:

"There is not an Arab state or Arab-controlled piece of territory in the Middle East that will allow one Jew to live in it. That is why in 1948 the Arab states rejected the two-state

---

943　(From article, "The Real War Israel is Fighting," by Sarah Weiner in *Jewish Voice Today*, March/April 2009)

944　(From ad, "The Big Lie (II): What about those Arab "refugees"?" published by FLAME: Facts and Logic About the Middle East. Go to www.factsandlogic.org)

945　(From ad, "Myths About Israel and the Middle East (I): Do the media feed us fiction, instead of fact?" published by FLAME: Facts and Logic About the Middle East. Go to www.factsandlogic.org)

946　(From ad, "Myths About Israel and the Middle East (I): Do the media feed us fiction, instead of fact?" published by FLAME: Facts and Logic About the Middle East. Go to www.factsandlogic.org)

solution that would have created a Palestinian state in the West Bank and Gaza alongside the State of Israel. They wanted to destroy the Jewish state more than they wanted to create a Palestinian one."[947]

They rejected the two-state idea back then because they were completely confidant that they would march in and destroy the infant nation of Israel. They were shocked, shamed and embarrassed when, despite their overwhelming odds, they failed miserably to do so. But now it is expedient for them to advance their Jew-killing obsession by demanding a Palestinian state. Whether they get it or not, they will never stop salivating over the destruction of Israel and the extermination of all Jews.

"The battle is not about Palestinians gaining a sovereign state. It is about terrorists groups in evil alliance with surrounding nations determined to eradicate the Jewish State. It is about the demonic entity Allah daring to contend with the Creator of the Universe. But globally-networked and enraged hordes of hell are no match for our Champion and Savior, the Holy and Righteous Lord of Hosts."[948]

Who is also of course the ultimate Protector of Israel.

"Israel can call itself whatever it wants, but the Palestinians will never recognize it as a Jewish state."[949]

"Allah, take the Jews and their allies, take the Americans and their allies... kill them to the last one."[950]

Oh those peace-loving Muslims, finding another practical application for their sacred religion of peace: mass murder.

"Why is Israel under attack? Because Hezbollah and other Middle Eastern terrorists hate Jews and want to kill every Jew possible! **They believe that in killing Jews and Christians they can usher in the second coming of the 12th Imam and the entire world will bend its knee to Islam.** Hezbollah has more than 7000 missiles that remain hidden. It is believed they also have weapons of mass destruction from Iran that can be loaded into the missiles."[951] (As most likely does Syria, from Iraq.)

Israeli "occupation"[(LIE)]

"The wicked plot against the just, and gnash upon him with their teeth." (Psalms 37:12)

Our Democrat media continues to uncritically repeat the Muslim propaganda that Israel must end its "occupation" of Arab territory in order for there to be peace.[(LIE)] But that is just another lie. First of all, the leaders of the Palestinians have no interest in peace. Peace is just a word they use. The word "peace" spoken

---

947   (Pg. 4, "Why Israel is the Victim and The Arabs are the Indefensible Aggressors in the Middle East," by David Horowitz)

948   (From article "In Times of Peace... In Times of War... How to Pray for Israel in Difficult Times," in *Jewish Voice Today*, Sept/Oct 2006)

949   (Palestinian Authority Prime Minister Salam Fayad in interview with *Al-Arabiya* TV as quoted in mailer, "Deceived by our own false hopes," by Morton A. Klein, President, *Zionist Organization of America*)

950   (Dr. Ahmad Bahar, *Speaker of the Palestinian Legislative Council* (PA TV, April 20, 2007) As quoted in a Zionist Organization of America circular.)

951   (From "Save Jerusalem" newsletter, by Mike Evans)

by the Arab-Muslim-Palestinians translates to "exterminate the Jews and destroy Israel." Only then will they have their peace.

"Even if agreements were signed [regarding] Gaza and the West Bank, we will not forget Haifa, Acre, Jaffa, the Galilee Triangle, and the Negev. It is only a question of time..."[952] - From Palestinian TV.

And that is just one of thousands of incriminating statements from Musrab leaders and their media. Obviously they don't just want the "occupied territories" of Gaza and the West Bank. That's just a ruse. They consider all of Israel "occupied territory" and they won't rest until they acquire all of it. Then they will be at "peace." So the big stink about occupied territories is just another red herring. But while we're on the subject. let's explore whether these territories are even "occupied." Let's take a look at the Muslims invalid claims to the West Bank and Gaza:

"The very concept of "occupation," as applied to Gaza and Judea/Samaria (the "West Bank"), is a myth. When the Egyptians were in possession of Gaza for almost twenty years nobody accused them of being occupiers. The Jordanians, who were in possession of Judea/Samaria (the "West Bank") and much of Jerusalem for the same length of time, were not considered occupiers by the nations of the world and by the inhabitants of the area. The Israelis were considered "occupiers" when, after a war of extermination against them, they prevailed and came into possession of these territories. The oldest rule of warfare is that to the victor belong the spoils."[953]

Unless of course you are a Jewish nation trapped in a satanically-influenced Jew-hating world. Then "to the *loser* go the spoils," so that the Muslim losers will be free to attack you over and over and over again, and when that doesn't work, to constantly beat you over the head with the "occupation" lie.

"The concept that to the loser, rather than to the victor, belong the spoils is a radically new one, never before thought of in world history. Israel has emerged victorious in the five wars imposed on it by the Arabs. In order to make peace, it has returned over 90% of the territory occupied by it, specifically the vast Sinai Peninsula, to Egypt. That territory contained some of the most advanced military installations in the world, prosperous cities and settlements, and oil fields developed entirely by Israel that made it independent of petroleum imports."[954]

That questionable sacrifice was followed more recently by another ill-advised one: the withdrawal from Gaza that we mentioned previously.

"The abandonment of Gaza was ...a trial balloon... Instead of being grateful to Israel for having left them alone and for having lifted the "occupation," the "thanks" was that in the course of just a few years they launched close to 10,000 rockets on Israel, mostly on the close-by city of Sderot, whose inhabitants have to spend much of their time in bomb shelters."[955]

---

952   (Palestinian TV, September 8, 2000, from aish.com)

953   (From ad, "Myths About the Israel-Arab Conflict (I): Is 'occupation' the cause of this conflict?" published by FLAME: Facts and Logic About the Middle East. Go to www.factsandlogic.org)

954   (From ad, "Myths About Israel and the Middle East (I): Do the media feed us fiction, instead of fact?" published by FLAME: Facts and Logic About the Middle East. Go to www.factsandlogic.org)

955   (From newsletter, March 2009, published by FLAME: Facts and Logic About the Middle East)

"Palestinian Authority president and Fatah chairman Mahmoud Abbas stated unequivocally Monday that he does not accept the Jewish state. "I say this clearly: I do not accept the Jewish State, call it what you will," he said at a preliminary conference of the Palestinian Youth Parliament in Ramallah. At the end of the conference, Abbas was presented with a large framed map of "Palestine," covering the entire area of Israel. The photo of the map being held aloft by a smiling Abbas was featured in a prominent front-page position in both PA daily newspapers. ...the word "Palestine" appears on the map in English.

"Abbas... [is] depicted by the West as a moderate to whom Israel is supposed to offer statehood. But the TV channel directly controlled by Abbas's office regularly runs educational programs to teach Palestinian adults and children alike that there is no state called Israel, and that all Israel's land is actually "occupied Palestine." Palestinian children are taught that Israeli cities throughout the entire country - from Haifa in the North, to Jaffa (part of Tel Aviv), to Eilat in the south, are all actually Palestinian cities. Videos feature songs about a "Palestine" that erases Israel and a future when the Israeli cities Jaffa and Haifa will be "liberated." "[956]

With all of the astronomical evidence at hand showing the true, Nazi-like, genocidal intentions of the Arab-Muslims towards the nation of Israel, the level of willful ignorance of their liberal, useful-idiot supporters in our country is stunning. In June of 2010 Israel defended itself by stopping and boarding a flotilla of boats from Turkey, one of which was filled with terrorists who were trying to sway worldwide opinion against the naval blockade of Gaza; so they could then be free to smuggle in more weapons and rockets from Iran, by sea. It was particularly disgusting listening to the world's Jew-hating response to this, full of lies and complete mischaracterizations of the situation, and even more nauseating watching those angry, uneducated genius' shouting and marching through the streets in our own country. (And, yes, those genius' vote Democrat, religiously, whenever they do bother to vote.)

"The Palestinians, along with most other Arabs and Moslems, are single-mindedly focused on the destruction of Israel, to "wipe it off the map." If Israel were pressured into ending the "occupation" of the territories, it would not bring peace. On the contrary, it would bring about bloody warfare, just as in Gaza and Lebanon, and could well be the end of Israel. Those who advocate the ending of the "occupation" do so either out of ignorance or because they have a death wish for Israel."[957]

Or both.

"Israel acquired the territories (the "West Bank" and Gaza) in defense of an aggressive war waged against it. No country in history has ever been asked to return such territories. Do the Poles return the huge chunk of Germany that they acquired in the wake of World War II? Do the Czechs return the Sudetenland, do the French return Alsace-Lorraine? [Or should we return the thirteen original states to England? Texas to Mexico? Louisiana to the French? The entirety of North America to the Indians?] Of course not! Only Israel is being asked to return such territories. The last sovereign of the "West Bank" and of Gaza were the Ottomans. The "West Bank" and Gaza are unallocated territories. To speak of Israel as "occupier" is preposterous; to speak of it, as Kofi Annan, the Secretary

---

956   (From article by Itamar Marcus and Barbara Crook, Palestinian Media Watch, @ pmw. org)

957   (From ad, "Myths About the Israel-Arab Conflict (I): Is "occupation" the cause of this conflict?" published by FLAME: Facts and Logic About the Middle East. Go to www.factsandlogic.org)

General of the UN does, as "illegal occupiers" is poisonous slander. He knows better. But unfortunately, the Big Lie of Israel's "occupation" has been repeated so long and so often that even people of good faith have come to believe it and to accept it."[958]

Are they really "people of *good* faith?" Or are they people who are far too comfortable with the lies they have been fed; who have no interest or intention of waking up to the reality of the situation? What were the people in Germany who sat around and watched as the Jews were being rounded up, their homes confiscated, their businesses stolen, their wives raped? Were they "people of *good* faith" too? I report, you decide...

> "The time will come, by Allah's will, when their property will be destroyed and their children will be exterminated, and no Jew or Zionist will be left on the face of this earth."[959]
> - Hamas (Al-Aqsa) TV, April 3, 2009.

Peace partner, anyone? Israelis would be better off climbing in bed with poisonous snakes and rabid dogs.

Jewish settlements, stumbling blocks to peace[(LIE)]

Again, the Arab-Muslim-Palestinians have no interest in peace. Peace is just a word they use. Peace to them means "exterminate the Jews and destroy Israel." Then and only then will they have their peace. "Settlements" have absolutely nothing to do with it one way or the other. But again, since we're on the subject, let's explore this lie as well...

> "It should be clear that the "settlements" [in the "West Bank"] – about 300,000 Jews in a sea of over 3 million Arabs – cannot be an obstacle to peace, since the over 1 million Arabs living in Israel are not considered a problem."[960]

But that is because the Jews are not Arab-hating racists, whereas, as we have seen, the Musrabs *are* Jew-hating racists. (A documented fact conveniently ignored by the Western press.)

> "Why can't Jews have settlements in the West Bank? The answer is because the Palestinians Arabs are filled with a racist and theocratic hate towards the Jews. They can't tolerate a non-Muslim, non-Arab people -- however small a minority -- living in their midst. (The 7000 Jews of Gaza – out of a population of 1.2 million – were law-abiding and peaceful and created a horticultural industry that produced ten percent of Gaza's gross national product. But they were Jews. And that was intolerable to Palestine's Nazis. So they had to be removed.)"[961]

I wonder what the world would say if Israel demanded that all of the Arab "settlements" in Israel be removed, because they are an "obstacle to peace;" that all of the Arabs who have "settled" in Israel just get out? And leave all of their homes and businesses. The world would have an epileptic fit. And clueless media morons like Chris Matthews, Wolf Blitzer, Anderson Cooper, Christiane Amanpour, Jon

---

958 (From ad, "The Big Lie (I): Are the "occupied territories" really occupied territories?" published by FLAME: Facts and Logic About the Middle East. Go to www.factsandlogic.org)

959 (From article by Itamar Marcus and Barbara Crook, Palestinian Media Watch, @ pmw.org)

960 (From ad, "Mr. Netanyahu's Offer (II): Are the objections of the Palestinians justified?" published by FLAME: Facts and Logic About the Middle East. Go to www.factsandlogic.org)

961 (From article, "Jimmy Carter: Jew-Hater, Genocide-Enabler, Liar," by David Horowitz, @ FrontPageMagazine.com, posted 12/14/06)

Stewart and David Letterman would call for the immediate dissolution of the state. But nary a word from those clowns in the other direction; not a peep about the disgusting, Jew-hating racism of "Palestine's Nazis."

"There are more than a million Arabs living safely in Israel where they enjoy more citizen rights than the Arabs living in any Arab country – or for that matter the Muslims living in any Muslim country. If the Arab states and the Palestinians were willing to treat Jews the way Jews treat them, there would be no Middle East problem, and in particular no problem of Jewish "settlements" in the West Bank and Gaza. The Jews in those settlements would be accepted as citizens by the host population in the same way the Arabs in Israel are accepted as citizens by the Jews. ...Jew-hatred is the only reason the Jews in Gaza are an issue at all, and is the real cause of the Middle East conflict."[962]

End of story. This world is an insane asylum, and some of its worse inmates, the clitoral-mutilating Muslims and their useful-idiots on the baby-mutilating Democrat-Left are in charge of how it disseminates, and distorts, its information. You too can be another pathetic dupe of Satan and his lies. Just believe what "they say" about "The Middle East Problem."

"Oh Allah, take this oppressive, Jewish, Zionist band of people and kill them, down to the very last one."[963]

That's from the Muslim Brotherhood, who Obama and his Democrats don't seemed to be concerned about their being a part of the new "democracy" in Egypt, now that the decades-long dictatorship of one Hosni Mubarak has been removed. Interesting.

Zionism is racism[(LIE)]

"Lying lips are an abomination to the Lord; but they who deal truthfully are his delight." (Proverbs 12:22)

"Today *Zionism* has become a dirty word. Israel's enemies are convincing the world Zionism is racism. Contrary to all logic they claim it is racist for the Jewish people to possess a single, miniscule country the size of the state of New Jersey but not racist for Muslims to possess the 22 surrounding nations totaling 640 times Israel's land mass. Former U.S. President Jimmy Carter joined the chorus when he claimed Israel practices apartheid even though its Arab citizens have the same rights as Jewish citizens, but he said nothing about Muslim countries that overtly discriminate against Jews. [Which might have something to do with the fact that he is an incredibly dishonest Jew-hater in addition to being a world-class, uneducated imbecile. Or, maybe there's no connection.] Saudi Arabia, in fact, bans Jewish people from entering its borders even as tourists."[964]

The United Nations, that corrupt, anti-American and anti-Israel organization, recently "declared that Zionism [the age-old desire of the Jewish people for a persecution-free nation in their ancient homeland of Israel] is racism." Yet in this meeting, "which was dominated by representatives of Islamic and Arab states

---

962   (Pg. 4, "Why Israel is the Victim and The Arabs are the Indefensible Aggressors in the Middle East," by David Horowitz)
963   (Muslim Brotherhood Leader Sheikh Yusuf Al-Qaradawi, January 2009, as quoted in *Wall Street Journal* ad, June 11, 2009, run by the *American Jewish Committee*, Richard J. Sideman, President)
964   (From article, "Why Christians Should Be Zionists," by Thomas C. Simcox, *Israel My Glory* magazine, July/August 2010)

and other anti-Israel forces,"[965] no other nation was accused of racism. Certainly not one of the Arab-Muslim countries, the world's premier racists as we have documented earlier, were so accused. Yet...

"In truth, Israel is perhaps the most racially and ethnically diverse and tolerant country in the world. More than half of Israel's Jewish population consists of people of color—blacks from Ethiopia and Yemen, as well as brown-skinned people from Morocco, Iran, Syria, Egypt and Israel itself. In addition, Israel's population includes more than one million Arabs, who enjoy the same civil rights as Jewish Israelis. In Israel hate speech is banned, and it is against the law to discriminate based on race or religion."[966]

Conversely, as we stated earlier but it bears repeating here, "there is not an Arab state or Arab-controlled piece of territory in the Middle East that will allow one Jew to live in it."[967] Yes, will the real *Arab* racists please stand up? (Of course this begs the question of why the brain-dead liberal politicians in our government who refuse to cut-off funds to the UN are elected in the first place, and then reelected, and reelected again.) (Couldn't have anything to do with our Democrat media, could it?) (Of course not.)

"In contrast, anti-Semitism—a poisonous form of racism directed specifically against the Jewish people—is rampant in most all Islamic societies. ...it is propagated shamelessly by their leaders, in state-sponsored media, and by Muslim clergy.

..."In response to a terrorist attack in Saudi Arabia in May 2004, Crown Prince Abdullah declared that "Zionism is behind [these] terrorists actions in the kingdom." [Sure, and Howdy Doody was behind 9/11, you Crown Putz.] (Zionism is the code word often used by Islamic anti-Semites for Jews.)"

..."**Anti-Semitism is expressed so freely and ubiquitously** in most Islamic societies that no citizen can escape it. During Ramadan in 2002, Egypt's state-controlled TV aired "Horsemen Without a Horse," a program based on the notorious forgery, *The Protocols of the Elders of Zion*, in which Jews allegedly use the blood of non-Jews to make Passover matzot. In Iran, a TV series, "Zahra's Blue Eyes," portrays "Zionists" kidnapping Palestinian children and harvesting their organs."[968]

"Perhaps nowhere is the hatred of Jews more virulent than among the Palestinians. Most perniciously, Palestinian children are taught in school that Jews are descended from apes and pigs [whereas American children are taught in school that everyone is descended from apes and fish. Go figure] and that the most noble thing they can do is to kill Jews. Muslim clerics like Imam Ibrahim Madiras, an employee of the Palestinian Authority, declared in a television sermon, "Jews are a cancer" and later that, "Muslims will kill the Jews ... [and] rejoice in Allah's victory.""[969]

Unfortunately for them, that's a "victory" they're going to be celebrating in hell. Unless they repent, and turn from their disgusting, Jew-hating, satanic lies, and embrace the truth of Jesus Christ as Lord and Savior.

---

965   (From ad, "Racism in the Islamic World: How can peace prevail in the Middle East in the face of Islamic bigotry and hate? When will moderate Muslims speak out?" published by FLAME: Facts and Logic About the Middle East. Go to www.factsandlogic.org)

966   (From ad, "Racism in the Islamic World," published by FLAME)

967   (Pg. 4, "Why Israel is the Victim and The Arabs are the Indefensible Aggressors in the Middle East," by David Horowitz)

968   (From ad, "Racism in the Islamic World," published by FLAME)

969   (From ad, "Racism in the Islamic World," published by FLAME)

Jimmy Carter in his book *Palestine: Peace Not Apartheid* falsely accuses the Israelis of being the cause of the problem in the Middle East, and even more incredibly, of *apartheid* against the Arabs.[LIE] Here is one "follower of Christ" who is wholly devoted to his father Satan's program of spreading Jew-hating lies and falsehoods in order to isolate and weaken the tiny nation of Israel. Jimmy has also been well-compensated for his treachery, receiving tens of millions of dollars in donations from his Saudi-Arabian friends. But yet he's *born-again*?!#*? (Can you be born twice into the kingdom of Satan? ..."Nicodemus said unto Jesus, How can a man be born when he is old? can he enter the second time into his mother's womb, and be born?" (John 3:4) Nicodemus needs to ask Jimmy. It would appear he's found a way.)

Addressing the incessant demands this world has been placing on the Israelis to make even further "gestures for peace" towards the Musrabs, the editors of FLAME had this suggestion...

> "Most of the 22 Arab countries consider themselves in a state of war with Israel and don't even recognize its "existence." That has been going on for almost sixty years. Isn't it about time that the **Arabs** made some kind of a "gesture?"[970] (Emphasis mine.)

But yet, with all that said, I do think it's about time, probably long overdue, that the lying, Jew-hating, racist Musrabs who constantly whine and cry that their Israel-destroying and Jew-exterminating demands be met, actually do receive some kind of a "gesture." May I suggest the middle finger?

Disproportionate force[LIE]

Another false charge, this time the one of *disproportionate force*, is leveled against Israel by the Arab-Muslims and their complicit medias of the Western world every time she is forced to defend herself, and of course to retaliate, against another violent unprovoked attack. It is another in what has become a long line of tired old lies, but this one is especially clever in that it is used to condemn "Israel to being the perpetual victim of violence, while denying it recourse to defend itself."[971] One can only imagine the response if, say, Russia, or China, or even the United States was constantly having rockets raining down on its citizenry in its border towns. God help the poor dumb bastards who would ever be so foolish as to perpetrate that folly. Entire cities would be leveled before any of these or countless other sovereign nations would tolerate the kind of torture that the Palestinians, Hamas and Hezbollah have perpetrated on the Jewish people and their homeland for decades. But poor, pathetic, liberal Israel—so concerned with a world opinion that in reality they don't stand a chance of swaying no matter what they were to do or not to do (unless they declared a national holiday and they all went to the beach and swam way, way out into the Mediterranean sea)—goes out of its way in its attempt to limit civilian casualties in its response. But all to no avail. Their efforts are made especially difficult by the Muslims cowardly use of "human shields"—civilians, including women and children, who are deliberately placed next to or in close proximity to the terrorists and their rocket launching sites (or vice versa). And Israel's world approval rating continues to plummet along with its ability to defend itself.

---

970   (From ad, "Myths About Israel and the Middle East (2): Should we re-examine endlessly repeated clichés?" published by FLAME: Facts and Logic About the Middle East. Go to www.factsandlogic.org)
971   (Richard S. Gordon, President, American Jewish Congress)

"There is no legal equivalence between the deliberate killing of civilians, which is what Hamas is doing by lobbing its rockets into Israeli cities without strategic significance, and the targeted killing of Hamas militants. …In its air and ground operations against Gaza Israel went to unprecedented lengths to avoid killing civilians. In an area such as Gaza, one of the most densely populated areas in the world – and in view of Hamas's custom of locating its rocket launchers and other military installations in the middle of residential areas and even in mosques, using civilians as shields – that becomes particularly difficult. [Whereas carpet-bombing the entire place, like we did to Dresden, would be a little less difficult.] In what is certainly unique in the history of warfare, Israel, in its respect for human rights, dropped tens of thousands of leaflets over Gaza and placed telephone calls to warn residents of non-military installations to get out of the way of military action. The accusation that Israel is using "disproportionate force" is absurd."[972]

It is also a bald-faced lie, but that won't stop our media, and the rest of the Western press, every time Israel is forced to respond militarily to another attack, from continuing to spread this falsehood, in order to continue to cloud the brains of their naïve and gullible listeners. The lies of Satan scurry over this earth like a cockroach infestation, and too many of its inhabitants swallow the horses**t they are fed.

Finally, here are the words of Israeli Prime Minister Benjamin Netanyahu, addressing a joint session of congress in May of 2011:

"We're proud in Israel that over one million Arab citizens of Israel have been enjoying these [Democratic] rights for decades. Of the 300 million Arabs in the Middle East and North Africa only Israel's Arab citizens enjoy real Democratic rights. Now I want you to stop for a second and think about that, of those 300 million Arabs, less than one half of one percent are truly free, and they're all citizens of Israel. This startling fact reveals a basic truth: Israel is not what is wrong about the Middle East. Israel is what is right about the Middle East."[973] (Emphasis mine)

Amen, brother. And all God's people said, Amen.

* * *

The preceding is only a partial list of Satan's propaganda, lies and slander against Israel. In fact he and his Muslim followers can spew out any unsubstantiated absurdity whatsoever and it will still be bought hook, line and sinker and endlessly repeated by his useful left-wing idiots in the medias all across the Western world. And the goal behind these lies is to continue to make Israel out as the bad guy, and the "occupier," and the one guilty of "racism" and "apartheid" and using "disproportionate force," in order to continue to sway the opinion of this world so they will pressure Israel into foolishly giving up even more of its tiny parcel of land to the point where there is left no possibility whatsoever of defending it.[974]

---

972   (From ad, "Israel's Defensive Response to Gaza: Was Israel using "disproportionate force?" " published by FLAME: Facts and Logic About the Middle East. Go to www.factsandlogic.org)

973   (Israeli Prime Minister Benjamin Netanyahu addressing a joint session of Congress, May 24, 2011)

974   (And now, in May of 2011, isn't it fascinating that our non-Muslim president has just turned his back on our only true ally in the Middle East by announcing that he thinks Israel should now meet the preposterous demands of the Palestinians and their terrorist-leaders

But not to be discouraged. The enemies of Israel will soon be dealt a serious blow in the coming battle of Gog and Magog when Russia and the Muslim nations attack Israel and are miraculously defeated. And then shortly thereafter in the battle of Armageddon at the end of the final seven-year tribulation period that will culminate this sin-filled age, all of Israel's enemies will be crushed in the valley of Megiddo in Northern Israel. The Lord himself will accomplish this. In the meantime we should not only keep the Jews in our prayers, but also their enemies in the Arab-Muslim world and in the totalitarian-left-wing medias, educational systems and political parties of the Western world. They are not the enemy. Satan is. They are just held under his powerful spell because they have embraced his bald-faced lies. And there but for the grace of God go you or I. Pray for their deliverance. For the Jews from her enemies which will be accomplished shortly. And for their enemies from the imprisoning lies of their father Satan before they perish eternally.

- Footnote: A nuclear Iran? It is May of 2010 and since Neville Hussein Chamberlain and his administration have proven to be completely impotent (or uninterested) in stopping Iran from getting nukes,[975] the question remains: When will Israel attack Iran's nuclear facilities in order to keep their fanatical leaders from getting nukes and using them to annihilate them? It should be only a matter of time because they really have no choice. Iran's mentally unstable rulers—Ahmadinejad and the Mullahs and Ayatollahs—would sacrifice every man woman and child in their country if the payoff was "pushing the Jews into the sea," or nuking them into history. So what other choice does Israel have? With that said the only other question that remains is whether that attack will set off the prophesied battle of Gog and Magog. We're leaning in that direction but only time will tell. (And now it is April of 2011 and we do seem to be accelerating toward this conflict, as riots and regime change have flooded the Arab nations of the Middle East, with the Jew-exterminating regime of Iran and the Israel-annihilating organization of the Muslim Brotherhood being the primary beneficiaries of these developments. Stay tuned for the latest updates…)

Also, as we will cover later in this chapter in section 13 "Control the Media," we do not use the term *Nazi* lightly or without serious consideration; unlike our friends on the Democrat-Left who have used it liberally over the years to falsely accuse conservatives, Christians and Republicans. But we need to keep the facts of this section in mind later on in this chapter when we credibly accuse our media (and the medias of the rest of the Western world for that matter) of being not just liberal, not just biased, not just Democrat-Left and not just corrupt, but truly being a Nazi media. We will document our case fully later but for now understand that what the Nazi's in Germany did to the Jews, the Musrabs in the Middle East are attempting to repeat. So it is not without historic precedent to accuse these

Hamas by giving back all of the land that they gained 44 years ago in the 1967 War, and thus return themselves to the indefensible borders that existed at that time, and which *invited* that unprovoked attack. How clever of him)

975   (Yet in all fairness, the previous two administrations of Clinton and Bush didn't do much better. But now the Iranian problem has reached critical mass and demands to be addressed aggressively; by someone with a little more spine than our Colonel Klink.)

Muslims of being *Nazis*, not only with regard to their genocidal intentions toward the nation of Israel and all Jews in general, but also with the number of lies (Nazi *propaganda*) that they broadcast daily to assist them in their goal. And our own left-wing Democrat media, in repeating their lies and concealing the Muslims intentions are supporting, aiding and abetting them in their genocidal quest. That would make them, at the very least, Nazi *collaborators*. In fact the Muslims are being supported, aided and abetted, and their true genocidal intentions concealed, not only by our media, but by the rest of the medias of not only the West but pretty much the entire world as well. Now if that's not Nazism I'd like to know what is. If that's not Nazism I'd like to know what it is. Journalism? Journalism that aids and abets with its propaganda and lies the imminent slaughter of nearly six million Jews in Israel by the deranged followers of Mohammad and his god Satan is not journalism. It's Nazism.

Seventy years ago in Nazi Germany the Big Lie was that the "Jews were the cause of all of Germany's problems." The Big Lie now and that will be shouted even more so in the near future, is that "the Jews and their nation of Israel are the cause of all of the *world's* problems." "The more things change, the more they stay the same." But do not let your hearts be troubled, for like Belshazzar the king of the Chaldeans who saw the handwriting on the wall ("MENE, MENE, TEKEL, UPHARSIN." Daniel 5:25) and knew his time was up; the Jew-haters days are also numbered—"He who sits in the heavens shall laugh. The Lord shall hold them in derision." (Psalms 2:4)—for attempting to exterminate his Chosen People from off of their Promised Land which He himself gave to them, in perpetuity.

Addendum

Replacement Theology

"Put away from you a deceitful mouth, and perverse lips put far from you." (Proverbs 4:24)

We mentioned this previously in chapter 7 as a question to ask a prospective pastor, but we will elaborate further on it here. Replacement Theology is a view that we would ascribe more to <u>Churchianity</u>, to those Christians-in-name-only, than to actual Christianity. And it believes, erroneously…

> …"that the Church is the **New** Israel, a continuation of the concept of Israel from the Old Testament. This view teaches that the Church is the replacement for Israel and that the many promises made to Israel in the Bible are fulfilled in the Christian Church, not in biblical, literal Israel. So, the prophecies in Scripture concerning the blessing and restoration of Israel to the Land of Promise are "spiritualized" into promises of God's blessing for the Church. The prophecies of condemnation and judgment, however, still remain for national Israel and the Jewish people. [How convenient.] This view has been called Replacement Theology because the Church replaces Israel in the program of God. Major problems exist with this view, such as the continuing existence of the Jewish people throughout the centuries and especially with the revival of the modern state of Israel. If Israel has been condemned by God, there being no future for the Jewish nation, how do we account for the supernatural survival of the Jewish people, Israel's rebirth among the gentile nations, victories in major wars with the Arabs and a flourishing modern democratic Jewish state?"[976]

---

976   (From article "Replacement Theology" by Alan Torres @ biblicist.org)

How do Replacement Theologists account for that? They can't. The "most widely held view of the apostolic church," of the early church, however, was that "The Church is Totally Different and Distinct from Israel."

"Although being suppressed throughout the history of the Church, the view that Israel and the Church are different is clearly taught in the New Testament. In this view, the Church is completely different and distinct from Israel and the two are never to be confused or used interchangeably. We are taught from Scripture that the Church is an entirely new creation, that came into being on the Day of Pentecost and will continue until it is translated to Heaven at the Rapture (Eph. 1:9-11). The Church has no relationship to the curses and blessings for Israel, the covenants, promises and warnings are valid only for Israel. Israel has been set aside in God's program during these past 2,000 years of dispersion. The Lord has preserved the Jewish people through great persecutions, though they are largely in unbelief."[977]

It is also revealing that many of the Christians today who fall into the Replacement Theology camp have a callous disregard for the plight of the Jews in Israel, while at the same time sympathizing with the Palestinians, and regurgitating most if not all of the Arab-Muslim lies that we have exposed to the light of truth in this section. A light of truth we might add that these professed Christians purport to love and worship and follow; Jesus himself being that Light of Truth. So something seems terribly wrong with their picture. But you can decide for yourself the true nature of these usurpers of Gods promises to Israel after reading the following excerpt from an article in *Israel My Glory* Magazine by James A. Showers (and, if needed, after reading the rest of that article and the other recommended articles in the footnote below):

"If you watch the news headlines, you may have read these: "Episcopal Church Is the Next to Shun Israel," "Presbyterian Church to Justify Israel Divestment," and "Methodist Church Renews Drive for Divestment from Israel."

"*Divestment* involves withdrawing investments from companies doing business with a particular nation in order to put economic pressure on the government. It was a technique used against South Africa to break apartheid. Over the years, some Protestant churches have asked people to stop investing in companies doing business with Israel based on their claim that Israel is a racist nation. But nothing could be further from the truth.

"Israel is a democracy that freely gives the vote to both Jews and Arabs. Apartheid occurs when a minority race uses its power to take economic advantage of the majority of another race. Such is not the case in Israel.

"In 2007 a Methodist women's group wrote a report, sponsored by and paid for by the Methodist Church that referred to the founding of the state of Israel as the "original sin," thus equating modern Israel with Adam and Eve's rebellion against God. The report implied humanity sinned against God in creating the modern nation. It was a shocking statement.

"The same report claims, "The Holocaust, and the impact of the Holocaust on Israel's Society, has caused hysteria and paranoia amongst Israelis." The report not only treated the murder of 6 million people casually, as if it were a minor event, but also claimed Jewish people have blown it out of proportion. If an event in history had put 6 million Methodists to death, these women probably would have a different outlook on the Holocaust.

"Then, of course, there is Jimmy Carter. The former U.S. president's 2006 book,

---

977   (From article "Replacement Theology" by Alan Torres @ www.biblicist.org)

Palestine: Peace Not Apartheid, distorted facts, demonized Israel, and contain hundreds of falsehoods and fictional statements.

In light of all these events, perhaps you've asked yourself, *Why are all of these Christians picking on Israel?* [Uh, maybe because they're *not* Christians!?] *What's behind these attacks?* [Uh, Satan!?]

"The answer is Replacement Theology, the belief the church has replaced physical Israel (the Jewish people) in the plan of God. Those who hold this view believe the church has become "spiritual Israel" and the inheritor of all the covenant promises God made to Israel. But they say the Jewish people retain all the curses."[978]

See:
- Article, "Replacement Theology: The Black Sheep of Christendom (Part 1)" by James A. Showers, *Israel My Glory* Magazine, March/April 2007
- Article, "The Roots of Replacement Theology" by William L. Krewson in *Israel My Glory* Magazine, May/June 2007
- Article, "Upholding the Truth" article by Richard D. Emmons in *Israel My Glory* Magazine, March/April 2007
- Article "Replacement Theology" by Alan Torres @ www.biblicist.org

* * *

"O Jerusalem, Jerusalem, who kills the prophets and stones those who are sent unto her! How often would I have gathered your children together, as a hen gathers her brood under her wings, and you would not! Behold, your house is left unto you desolate: and verily I say unto you, You shall not see me here again, not until you learn to cry, Blessed is he who comes in the name of the Lord." (Luke 13:34-35)

See:
- "Why Israel is the Victim and The Arabs are the Indefensible Aggressors in the Middle East," by David Horowitz
- FLAME: Facts and Logic About the Middle East. Go to www.factsandlogic. org and click on "Our Ads and Positions" to see a complete list of their educational ads ("educational" in that they overcome the many lies of the Muslims and their Left, with the documented truth) over the years and to learn the truth about the Arab-Israeli conflict.
- *The Terrorism Awareness Project* @ terrorismawareness.org
- "Big Lies: Demolishing the Myths of the Propaganda War Against Israel," by David Meir - Levi
- "Jihad in America! Militant Islam: The Religious War Against America," by Martin Mawyer
- "What Americans Need To Know About Jihad," by Robert Spencer
- *Unholy Alliance: Radical Islam and the American Left*, by David Horowitz
- *The Grand Jihad: How Islam and the Left Sabotage America*, by Andrew C McCarthy
- *In Defense of Israel* by John Hagee

---

978 (From article "Replacement Theology: The Black Sheep of Christendom (Part 1)" by James A. Showers, *Israel My Glory* Magazine, March/April 2007.)

- *Unshaken: A Story of Faith and Hope*, by ACLJ Films
- *The Politically Incorrect Guide to the Middle East*, by Martin Sieff.

Part II- Destroy America, Israel's Protector

"This matter is by the decree of the watchers, and the demand by the word of the holy ones: to the intent that the living may know that the Most High rules in the kingdom of men, and gives it to whomsoever he will, and sets up over it the basest [most ignorant] of men." (Daniel 4:17) (In case anyone is wondering how Obama got elected.)

In the remainder of this chapter we are going to focus on Satan's attack on America, which in many respects includes his attack on the entire Western world and on the world in general, both East and West. He's already pretty much taken over Europe, and he has owned the Arab and Muslim worlds for centuries, ever since his religion of Mohammad subdued it by the sword, but he is still stymied here in America, although he has made tremendous inroads in two of the previous elections, '06 and '08. Satan wants to destroy Israel by destroying America, and with the help of his Democrat party under the leadership of Barack Hussein Obama, he is well on his way to doing both.

Satan's hatred for America is nearly as strong as that of Israel because without our support, Israel would probably have already been "pushed into the sea." America is a uniquely Christian nation, founded on Judeo-Christian principles and the Christian religion. (Somebody might want to explain that to Wilbur.) It is because of this fact that America has become the greatest nation in history, the most prosperous, the most generous, the most sacrificial in terms of the price our young men have paid in blood for the liberation and freedom of so many of the peoples and nations of this earth. No other nation is owed such a debt of gratitude, yet we are now hated and despised by an ever increasing and alarming percentage of these same people, otherwise known as uneducated, clueless ingrates whose "thinking" is fubar (f**ked-up-beyond-all-recognition) by the lies they have been fed. (Having a jackass jetting around the world apologizing for America hasn't helped the ingrates perception either. It just reinforces, from the "highest" of America's own internal source, their belief that they are justified in resenting us. Thanks again, Mr. Ed.)[979] But nevertheless the historical fact remains that America is a great nation, and a Christian nation, and because of this we have also been Israel's greatest friend and protector. And Satan is none too thrilled about it. He wants our nation weakened to the point where we will no longer be able to stop him and his Muslims from finishing off Israel. And he has been tirelessly seeking to accomplish this goal for decades, by tearing us down from within.

Again, Satan is the father of lies and they are his primary weapon in this battle. He is also "the prince of the power of the air" and whereas the exact meaning or perhaps multiples of meanings of that phrase can be debated, it should be pretty clear at least for these modern, electronic times that we find ourselves in that the prince of darkness has a powerful hold over the radio and television air *waves* of this globe. One would expect no less control from the god of this world.

"Wherein in time past you walked according to the course of this world,

---

979   (Wilbur's talking horse)

according to the prince of the power of the air, the spirit who now works in the children of disobedience." (Ephesians 2:2)

The great deceiver has also had a near stranglehold on the dissemination of information in our country for decades with his control of our media, our educational system, and our entertainment industry. Each carry water for him, they spew his lies and help him to advance closer towards his goal of sabotaging the economic and military power of America from within. Most are ignorant of course of whom, and of whose plans, they serve, having bought into the far-left, baby mutilating ideology that they were taught, thinking that they are "smarter" and "wiser" and more "superior" and far more "righteous" (albeit self-righteous) than the traditional Americans who built this great country and whose values they hold in virtual contempt. For traditional Americans are the ones who stand in the way of their, and their father Satan's, goals. Our nation is at great risk. Our freedom is at great risk. Our economy, our health care system, our very livelihoods are all at great risk. And Israel is at even greater risk.

In Part II we will focus on Satan's assault on the success and prosperity of our nation involving 13 different areas. From destroying Christianity, the family and the historical foundations of this country, to weakening America's economy, her respect abroad and controlling our media, we will seek to expose what he has been up to so we might no longer "be ignorant of his devices." Today far too many Americans, and far too many American voters—including millions of people who name the name of Christ but have failed to depart from iniquity (2 Timothy 2:19) (and see chapter 11)—seem completely ignorant of his devices, even to the point of lending him their full support. Hopefully we can all wake up before it's too late; before Lucifer has drug us from our once lofty and admirable post as the world's greatest nation, down into the dregs of history, alongside the likes of Cuba, Western Europe, California or Greece.

1- Destroy Christianity

"But if our gospel be hid, it is hid to those who are lost. In whom the god of this world has blinded the minds of those who believe not, lest the light of the glorious gospel of Christ, who is the image of God, should shine unto them." (2 Corinthians 4:3-4)

If you can destroy Christianity in America then you will destroy America herself. For the two are inexorably linked. Tear down Christianity here and America will be no different than any other nation. American Exceptionalism will be dead—traded for Obama-style European socialism. We will no longer have any inclination to risk the condemnation of the UN and the international community by defending Israel, for our government will have fallen permanently into the hands of those who have bought into all of Satan's anti-Israel, Muslim lies (like Obama and his administration have today). And that is exactly how Satan wants it. So his onslaught against Christ and Christianity marches on…

"Satan is a liar, deceiver, and destroyer. In his fruitless battle to "be like the Most High," he uses all the evil talents and tricks at his disposal in a multi-pronged attack that has discouraged and deceived the vast majority of the human race (Isaiah 14:14).

"Today, God is denied, Jesus is rejected and mocked, and Christians are viewed as intellectual pygmies. Such is the Evil One's master plan: to denigrate God and make Him totally irrelevant, to have Him openly denied in our schools and rejected by the vast majority of our populace, and to convince people- who constitute the crowning achievement of God's creation- that they merely evolved.

"Satan also wants to disgrace God's Son, Jesus. Hollywood portrays Him as a weakling or a mere human who struggled with lust and sexual problems. Christmas has been turned into the "winter holiday." Manger scenes and even secular symbols, such as Christmas trees, have become weapons in Satan's battle to expunge Christmas from the human psyche.

"We who name the name of Christ are among Satan's most hated targets. Christians are often parodied; laughed at; and portrayed as uninformed, unenlightened, and almost utterly clueless human beings."[980]

We have already covered earlier, in chapter 2 "Moral Absolutes," how the constitution-perverting ACLU-Left has taken prayer, the Ten Commandments, The Bible, the Creator's scientifically-accurate Genesis and any mention of God from our public schools, and replaced them with the complete package of their father Satan's lies: "the truth is relative,"[(LIE)] "morals are relative,"[(LIE)] "the theory of evolution is a *fact*,"[(LIE)] etc. But one of his most vocal lies, the one that has been used to justify his judicial assault on Christianity in the first place; one that he has shouted the loudest and the longest through his media and educational-indoctrination system, and therefore one which the vast majority of mis-educated Americans have bought into, is that of "The Separation of Church and State."

The Separation of Church and State[(LIE)]

"It is impossible to rightly govern the world without God and the Bible." President George Washington, Sept. 17th, 1776. ...(So much for the "Separation of Church and State.")

This whopper has been repeated so often and so authoritatively by the Democrat-Left who control our media, educational system and entertainment industry and therefore has become so rooted in our nations popular consciousness that the majority of Americans haven't the slightest clue as to the extent that they have been misled. Far left, anti-American, anti-Christian organizations like the ACLU and Americans United for the Separation of *People from their Right Mind* (oops, that should read: *of Church and State*) have used this phrase, with the help of their constitution-perverting, left-wing plants on our courts and a complicit media, to justify their all-out assault on Christianity in America. The weapon of choice for these domestic terrorists is the lawsuit, which they use to bludgeon, coerce and threaten schools, towns and municipalities into succumbing to their demands to remove everything from the Ten Commandments to Christmas nativity scenes. Most of their victims cannot afford to defend themselves against the costly litigation. In the past these folks, brandishing their "separation" battle cry, have succeeded in removing the Bible and prayer from our public schools. But curiously enough in recent times they have had no problem with the Koran or Muslim teachings being brought into these same schools. And that is because "The ACLU doesn't hate religion, they just hate Christianity." Just like their father Satan. They have forced their will down the throats of cities and towns by stopping them from continuing their centuries-old tradition of having nativity scenes displayed on public property. They have oppressed those who want to pray

---

980   (From article "The Master Deceiver's," by Thomas C. Simcox, *Israel My Glory* Magazine, July/August 2007)

at civic gatherings, as Americans have done for hundreds of years; and have made it illegal to teach the truth of creation, instead coercing our schools to indoctrinate our students with the scientifically-impossible theory of evolution, leaving this bald-faced lie to fester in their minds, unchallenged by the truth. And now even our military chaplains are being told they cannot pray in Jesus' name.

"These "elites" use the fallacious cry of "separation of church and state"—which cannot be found anywhere in our Constitution, but is actually part of the former Russian Constitution (1936)—to silence all critics "[981]

But their attacks are all built on a lie. The phrase "separation of church and state" is not found in the Constitution of the United States. It is not found in our countries Declaration of Independence, nor in our Bill of Rights. Neither can it be found anywhere in any of this nation's founding documents. So from whence did this all-important, pivotal, foundational lie...er... "doctrine" of the Democrat-Left arise? It is to be found in a <u>letter</u> written by Thomas Jefferson to the Baptists Association of Danbury, Connecticut. Wow! At least it was more than a passing remark during a conversation at a local bar. The enemies of Christianity who seek to rewrite the foundations of this country have based a half century of anti-Christian attacks on what? A comment in a letter! But wait, it gets better! Thomas Jefferson, who was the author of that letter with that comment, immediately after sending it, <u>attended a Christian church service in our **government's** Capitol building itself</u>?! (Obviously, old Tom didn't quite comprehend exactly what he himself had written. He needed to wait two hundred years for the ACLU and their judges to "figure it out" for him.)

"Two or three days after he (Thomas Jefferson) sent that famous letter to the Danbury Baptists, he began to attend public church services in the Capitol building of the United States. Jefferson also permitted church services to be held in the executive office buildings, the treasury office, the war office, there were services in the Supreme Court. On a Sunday in Washington in, say, 1810, you would have had services going on in all branches of government. You could actually say that the state became the church on a Sunday in Washington. Some of these arguments that we're having about religion in the public square are certainly not grounded in the historical record."[982]

No, they're not. But they are grounded in something else- Satan, his ACLU and his Democrat-Left's effort to remove Christianity from America.

"Is the wall of separation between church and state mandated by the Constitution? No, it is not! It is an anti-Christian lie, nothing more- and it is being promoted by federal judges in an unprecedented wave of judicial tyranny! Church-state separation is a lie, and it is deadly. It was this lie in 1962 that led the Warren Court to ban prayer in public schools. ...It is as though reverence for God, so evident throughout so much of our history as a nation, was quietly and tragically withdrawn on that day. Yet subsequent sessions of the Court followed with more disturbing rulings- such as banning Bible readings from public education and banishing the Ten Commandments from public school displays."[983]

This lie of the separation of church and state is now rooted in the public consciousness to the point where most people erroneously believe that it is an integral part of our Constitution and Founding Documents, and that any attempt to cross this holy, yea "sacred" line of separation is blatantly "<u>unconstitutional</u>."[(LIE)]

---

981   (Tom De Rosa, *Creations Studies Institute*, in a March 2009 Newsletter)
982   (James Hutson, Library of Congress Historian)
983   (From a *Coral Ridge Ministries* newsletter)

But that is totally false. What the opponents of Christianity have been doing for decades is what is actually unconstitutional- thwarting the will of the people in this Christian nation from exercising their right of freedom of worship and religious expression.

> "Everyone appointed to public office must say, "I do profess faith in God the Father, and in the Lord Jesus Christ His only Son, and in the Holy Ghost. In God who is blessed forevermore I do acknowledge the Holy Scriptures and the Old and New testaments which are given by divine inspiration."[984]

That's from the Delaware State Constitution of 1776. So much for the lie of the "separation of church and state." Christianity was intricately married to the state from the beginning of this republic. Indeed, this country was actually <u>founded</u> as a Christian nation, placing its trust and allegiance in the God of the Bible.

There is absolutely nothing whatsoever in our Constitution that says it is illegal for the Ten Commandments to be displayed, or for the Bible to be taught, or for Christian prayers to be heartily invoked either in our public schools, or in our court houses, or in our state legislators, or in our government buildings, etc. And the fact that a number of unelected left-wing lawyers, domestic enemies all, who have perverted and distorted the very Constitution that they swore an oath to uphold and defend (treason?), have declared these actions unconstitutional does not change the fact that they indeed are not. When the ACLU-Left declares that something is "unconstitutional" that is just their code word for "un-LEFT-itutional;" for they could care less about our Constitution. In fact, the very suggestion that our founding fathers wrote into the Constitution the idea that anything whatsoever to do with the Christian religion cannot be discussed or displayed by any government agency, in any government building or in any government school is idiotic. It also betrays a complete lack of historical knowledge (aka ignorance) by these "separation of church and state" zealots, about the beliefs, practices, daily activities, and Constitutional intent of those same founding fathers. (And if not ignorance then the deliberate intent to deceive.)

Our Founding Fathers wrote into the Constitution, as the First Amendment in the Bill of Rights, these words, "Congress shall make no law respecting an establishment of religion, or prohibiting the free exercise thereof." These men were aware that our nation was originally colonized by people who had come to these shores to escape the blatant religious oppression of European countries where there **was an established Christian religion**, whether it was Catholicism, the Church of England or another state-mandated form of Christianity. And if you dared to worship differently or to follow any Christian religion (such as Anabaptist, Lutheran, Huguenot, etc.) other than the one established by that state then you could be arrested, imprisoned and even executed. <u>That</u> is what they wanted to protect against. And <u>not</u> people praying in school, or reading the Bible in class, or printing the word Christmas in school bulletins, or setting up Nativity scenes, or posting the Ten Commandments, or teaching God's Genesis, anywhere in America, both in public and private places. Protecting against this kind of Christian religious expression is the intention, <u>not</u> of our founding fathers who actually wrote the Constitution, but of godless, left-wing, Christian-hating, lying, ACLU, baby-mutilating (coincidentally) lawyers and judges who are following

---

984   (Excerpt from the Delaware State Constitution of 1776)

the desires and the design of their father, the great deceiver.

Nearly all of the founders of this country were devout, practicing Christians. The idea that they would have written into the Constitution the prohibition of the practice of Christianity in federal, state, city, or local government is patently absurd. It is also a bald-faced lie. The ACLU has had as its stated goal from its inception to destroy Christianity in America. (See section 7, "Taking over the Judiciary.") But they also knew that they could never accomplish this openly. So instead they hide like cowards, masking their true intentions, while shoving a lie down Americas throat.

Another historical fact that exposes the lie of "the separation of church and state" is the fact that Thomas Jefferson himself (whom the ACLU claims as their great mentor and the one supposedly behind this mythical separation) while chairman of the committee on education for the <u>public</u> schools in Washington D.C. actually ordered that two particular books were to be <u>required</u> reading for the children. What were they? *Mein Kampf* and *Heather has Two Mommies*? (No, the ACLU wasn't around back then.) The two books were the <u>Bible</u> and a Christian Hymnal! Imagine that. In a public school system no less. Separation of church and state, my asbestos. I don't think so. Try separation of the American people from the truth. ("If you shout a lie loud enough and...") Thomas Jefferson also said that "if a nation expects to be ignorant and free it expects what never was and what will never be." But unfortunately that is what we have now become, a nation of mis-educated-by-the-Democrat-Left ignoramuses. Which has reached its zenith in the abomination of Obama-nation. (Hopefully, like the Confederacy at Gettysburg, that was the Democrats "high water mark." Because if they get any higher we're all screwed.)

George Washington himself said that without the Bible it would be impossible to govern America. Impossible to govern America! Without the Bible! "Separation of church and state?" Wow. It would almost be funny if it wasn't so pervasive and so subversive. So many Americans have been duped into embracing this lie that it has become "common knowledge," instead of the common deceit that it is. James Madison, the architect of the federal constitution and 4th president of the United States, said in 1788...

> "We have staked the whole future of American civilization, not upon the power of government, far from it. We have staked the future upon the capacity of each and all of us to govern ourselves and to sustain ourselves according to the Ten Commandments of God."[985]

Which are now illegal to teach to our children in our public schools thanks to Satan and his ACLU-Democrat-Left. Obviously our founders had no problem with the Ten Commandments being taught to our public school pupils, in fact they demanded it.

> "In Germany, as here in the United States, one of the most clever tools in the enemy's arsenal, used to silence and intimidate Christians, to drive them out of the public square, was the lie of the separation of church and state."[986]

Germany had the Nazis. America has the Lefties. It is also revealing to note that it is only Christianity that seems to catch the ire of our friends on the left; it is only Christianity that needs "separating" from the state. The Muslim religion

---

985　(James Madison)
986　(Dr. Lawrence White, Senior Pastor of Our Savior Lutheran Church in Houston, Texas, from *Focus on the Family* broadcast, November 3, 2008)

has escaped their notice, for there is a curious silence from the ACLU-Left when it comes to the separation of *that* religion from the state. We mentioned this in the chapter on false religion but it is worth revisiting here…

"In California, "Muslim Understanding Week" is an approved activity in school districts across the state. Students memorize verses from the Koran (the Muslim holy book), dress in traditional Muslim garb, bring prayer rugs to class, pray Muslim prayers, and learn about Mohammed's journey to Mecca. Can you imagine children spending a week memorizing the Lord's Prayer and learning about Jesus' Sermon on the Mount?"[987]

No "separation of church and state" outcry there, thank you. Because it is all a lie. They're only looking to destroy Christianity. And with the success of their lie being swallowed by America, and with the numbers of foolish people who continue to go out and vote for them, they are well on the way to accomplishing their, and their father's, goal.

As we said in the chapter on "Moral Absolutes," if you reject the morals that come down from above you will free yourself from the constraints of those protective morals, but at the same you will end up embracing, and falling under the vile unrestraint, of the evil that oozes up from below. Using their separation lie and aided by the courts, the Democrat-Left have outlawed our Father God's Ten Commandments from our public schools and in their place they have put their father Satan's "Ten Prohibitions…"

"1. *Thou shalt not pray in public schools.*

2. *Thou shalt not allow Nativity scenes on public property.*

3. *Thou shalt not display the Ten Commandments.*

4. *Thou shalt not pray at graduation ceremonies.*

5. *Thou shalt not permit teachers to have even their personal Bibles in class.*

6. *Thou shalt not allow Boy Scouts on public property because they promote religious and moral values.*

7. *Thou shalt not pray at school sporting events.*

8. *Thou shalt not allow crosses in Korean War memorials.*

9. *Thou shalt not say the words "under God" in the Pledge of Allegiance.*

10. *Thou shalt not distribute Bibles to students.*"[988]

(And keep in mind that none of these ten things that the Democrat-Left have prohibited are "unconstitutional." They are un-liberal, that is all. Again, they could care less about our actual Constitution and what it clearly says.)

Some of the champions of the separation of church and state myth even want to take "In God We Trust" off of our money. But in this case I think a compromise could be in order. Let's allow them to remove that as long as we can put the following in its place, "The wicked shall be turned into hell, and all the nations that forget God." (Psalms 9:17) May He have mercy on their souls.

The theory of evolution[(LIE)]
"As a dog returns to his vomit, so a fool returns to his folly." (Proverbs 26:11)

"Evolution is an impossibility. If you remember that you'll understand that we're left with no alternative but creation. Evolution cannot occur.

987 (From The David Horowitz Freedom Center)
988 (by Rev. Don Swarthout, of *Christians Reviving America's Values*)

..."The more science looks at life, the more complex it becomes. The body, for example, is made up of trillions of cells. In just one of those cells ...the amount of genetic information ...has been estimated to fill at least one thousand books of 500 pages. That's to run [just] one cell out of trillions in one human body. And most scientists think that is an underestimation of the complexity. Where did all this information come from? Better, from *whom* did all this information come?

"To make evolution the answer is ridiculous ...so ridiculous as to qualify someone for a trip to the mental institution. Why then do scientists continue to advocate this ridiculous theory of evolution motivated by chance? Why do they do that?

"Well, the bottom line is they do it to avoid God. They do that to push God out of their lives, to avoid His law, to avoid His standards, to avoid His will, to avoid His Word and to avoid His judgment on their lives. Evolution is nothing more than what Henry Morris so aptly called it, "The long war against God."

"Evolution is the contemporary expression of the long war against God. The Old Testament says the fool has said in his heart there is no God. ...It is not rational to reject a creator. It is not rational to empower chance. It is not rational to assume that one kind of living organism can become another. It is not wise to reject God's law and God's Word and God's gospel.

"If it is neither rational nor wise, then why do men do it? And the answer is that men do it because they love sin and they love darkness because their deeds are evil. They love themselves and they love their sin and they refuse to worship God or submit to His Word or His law. They will not recognize Scripture. And by the way, Scripture shows us that what is in God's *world* is in God's Word. All we know about creation from nothing is what the creator has told us and the only place He's told us is in the Scripture. Evolution is a war on God. It is the ...contemporary modern attack in the long, long war that Satan has carried on against God. ...[It] is empty philosophy. It is vain deceit. It is designed to attack the creator and His glory. It denies His glorious revelation in Scripture. It denies His authority over the universe of man. It denies the dignity of man. It denies the image of God in man. It is a cunningly devised fable. It is religious harlotry. It is the latest abomination of the earth spawned by the father of lies, Satan."[989]

We have already covered this fully in chapter 10, but to reiterate, the theory of evolution is Satan's genesis. It is an unscientific, bald-faced deception but nevertheless one that poisons the minds of the Democrat-Left elites who control our educational system, media and entertainment industry. And it stands in direct contradiction to God's Genesis account, the one that *is* scientifically accurate. The Left has closed their ears to hearing the words of their own Creator and instead follow the asinine lies of their father Satan, the god of this world. That same father of lies who is the power and influence behind the ascension of his own laughable version of our beginnings. His goal was and is to destroy Christianity by destroying its foundations, the Bible. And the best way to destroy the Bible is by destroying *its* foundations, the Book of Genesis.

The ACLU are champions of Free Speech(LIE)

Actually the exact opposite is true. The ACLU and their Democrat-Left friends are only interested in protecting left-wing free speech. The free speech rights of their ideological opponents they not only could care less about defending, but they will actively seek to stifle at any opportunity. We have documented this previously...

---

989 (From a Grace to You radio broadcast "God: Creator and Redeemer" by John MacArthur)

- In chapter 2 "Moral Absolutes" we saw:

- How they have cut off the rights of the Christian majority in this nation to read and teach the Bible in our public schools (but teaching Satan's un-holy book, the Koran, is just fine), to post the Ten Commandments or even to pray.

- How they attempt to squelch the free speech of their political opponents by lying about and distorting their message, and by denying their rightful airtime; as evidenced in just one recent example by their distorted and downright vulgar coverage, or total lack thereof, of the Tea Party protests.

- How they have now called their political opponents—like anti-abortion activists, fundamental Christians, border security protestors, people who attend Bible Prophecy Conferences, conservatives in general, and most incredibly of all, even returning **veterans** from Iraq—"terrorists"[(LIE)] in order to intimidate and marginalize them and to demean their right to speak out in a free society; but are too incredibly stupid or ideologically enslaved to call the crazed Islamic killers who actually want to terrorize us, "terrorists."

- How they lie about Rush Limbaugh, and every other popular conservative, Christian and Republican, in order to marginalize them and their ability to communicate their ideas in a free society; mainly because they, the Democrat-Left, find it impossible to debate their ideological opponents on the substance and merits of the issues.

- How they are chomping at the bit to enact their Nazi...er...Fairness Doctrine in order to shut off conservative and Christian talk radio.

- How they have criminalized the right of Christian chaplains in our military to pray in Jesus' name.

- And in chapter 10 "The Theory of Evolution is a Joke" we saw how they squelch the free speech rights of any scientist or academic who would attempt to expose the world's greatest hoax.

And throughout this chapter, and in particularly in section 13, "Control the Media," we will further document the Democrat-Left's extensive attempts at denying the free speech rights of their political and ideological opponents. But just from what we have already pointed out it is clear that the Left has used its stranglehold on our media, educational system and entertainment industry to successfully advance the lie that they care so much about free speech that they can barely stand it,[(LIE)] when in reality the exact opposite is true.

Six decades ago, in his quest to destroy Christianity in America, Satan was successful in slipping in an unconstitutional amendment inside a Congressional bill that has stayed on the books unchallenged ever since. It was meant to stifle the free speech of Christian Preachers and Pastors. What it does is prohibit churches from speaking out against whatever they may disagree with politically, or risk losing their tax exempt status, and [shocker] it has almost exclusively been used to silence those Christian churches and pastors who are not in the hip pockets of Margaret Sanger and the Democrats.

"Free speech is one of the fundamental rights America was built on. It's enshrined in the First Amendment. We as Americans were given the right to have an opinion and to express it freely. We were given the right to call others to action. This is one of our nation's most distinguishing features. But in 1954, Senator Lyndon B. Johnson slipped a little-noticed amendment into a tax bill which barred all churches and other tax-exempt groups from participating in political activity- or else face the loss of their tax-exempt status. This law is a disaster. Why? - It directly violates the First Amendment rights of people of faith. - The IRS is often selective in its enforcement (ignoring liberals, targeting conservatives). - Your pastor should not have to put your church's tax-exempt status at risk every time he speaks out on abortion or same-sex marriage- or any issue whatsoever. - Americans United and other like-minded groups intimidate pastors and Christian leaders, particularly in election years, causing them to tiptoe around key issues which every church should be free to discuss."[990]

Satan wants to destroy Christianity and what better way than to silence the free speech rights of true Christian pastors? And he has been getting away with this for almost 70 years, while Christian pastors and organizations (with the exception of Jay Sekulow and the ACLJ quoted above, and a few others) have sat back silently, failed to challenge this unconstitutional ploy and allowed him to do it. We're still trying to figure out why.

Spanking is child-abuse[(LIE)]

Actually, the exact opposite is true. Let's see what He who actually created husbands and wives and children and the raising of the latter has to say...

"He who spares his rod hates his son: but he who loves him chastens him promptly." (Proverbs 13:24)

Satan and his followers say that spanking is child abuse, but mutilating babies to death in the womb is not.[(LIE)] It is self-evident from this comparison alone that they are, of course, insane. God has given them over to a reprobate mind. One that is not only capable of sinking to this level of absolute illogic, but is also still able to be absolutely full of itself in the process. (What time does The Daily Show come on?)

"Withhold not correction from the child: for if you beat him with the rod, he shall not die. You shall beat him with the rod, and shall deliver his soul from hell." (Proverbs 23:13-14)

But Satan and his followers are not interested in delivering anyone from hell. He is interested in sucking as many down there with him as possible, and also in destroying Christian America so our country will collapse, and so the extermination of Israel and the Jews can then easily be accomplished.

"My son, despise not the chastening of the Lord; neither be weary of his correction: For whom the Lord loves he corrects; even as a father the son in whom he delights." (Proverbs 3:11-12)

> "Is spanking an effective means of discipline for kids, or does it merely teach them to be violent? ...what does the law say? Is it illegal to spank your kids? The answer is no-but parents who spank must be very careful to avoid running afoul of the law. ...[Because] of the child abuse law[s] [of certain states], parental discipline through spanking may not be justifiable if the child is bruised or otherwise injured. Thus, spanking is not illegal [for now], but injuring a child is. [Unless of course that child is still in the womb, then all bets are off.]
>
> "Apart from the legalities, is spanking a good idea? Does it work? According to the American Academy of Pediatrics, about 90 percent of U.S. parents spank, and about 59 percent of pediatricians in a 1992 survey said they support the practice. According

---

990   (Jay Sekulow, ACLJ American Center for Law and Justice)

to the academy, effective discipline has three key components: first, a loving, supportive relationship between parent and child; second, use of positive reinforcement when children behave well; and third, use of punishment when children misbehave. Many parents these days are fearful of using spanking as punishment, either because of the law or because they fear it teaches violence to their kids.

"Some professional organizations of physicians and psychologists have suggested that spanking is detrimental and leads to family violence and child abuse.[LIE] They have suggested that spanking teaches physically aggressive behavior which the child will imitate.[LIE] But does the research support these assertions? According to the National Institute for Healthcare Research, more than 80 percent of the professional publications attacking spanking were reviews and commentaries, rather than quantitative research. When analyzing the small portion of quantitative studies that included spanking, more than 90 percent of these studies lumped together mild forms of spanking with severe forms of physical abuse without discussing why they did so. Thus, **the professional organizations which advocated outlawing spanking evidently made their decisions without the benefit of the facts**. [Emphasis mine.] Mild spanking and severe child abuse are not the same thing."[991]

Yet an internet search of "jail time for spanking" results in endless examples across this nation where Christians are being arrested and tried for properly raising, and not *abusing*, their children. But the same totalitarians of the Democrat-Left can't seem to find the energy to stop Muslims from slicing the clitoris off of their little girls. Can we conduct a survey of Muslim women to see what percentage of them would rather have been spanked that day instead of having their sexual organs hacked off?

Outlawing spanking in this country—as most godless, baby-mutilating, progressive, and morally bankrupt European socialist countries have already done—is Satan's next move in banning the free exercise of Christianity in America; so he can destroy America herself. Banning the free exercise of Satanism—like mutilating babies to death, teaching evolution or having sexual relations with barnyard animals—is not on their agenda anytime soon.

Hate crime laws are fair and necessary[(LIE)]

The truth is that hate crimes legislation is meant to further stifle the free exercise of the Christian religion by making it a crime to preach the truth about homosexuality and to share the gospel with them, which has already been done in a number of European and Scandinavian countries. It is also going to be used in the future to criminalize those who speak the truth about the vile religion of Islam.

"The media touts "hate crime" legislation as fair and proper — but it's an affront to the very Constitution of the United States, and a danger to the proclamation of the Gospel.

"What is a "hate crime?" Any "criminal offense ... motivated, in whole or in part, by the offender's bias against race, religion, disability, sexual orientation, or ethnicity/national origin." Any crime should be punished appropriately — but "hate crime" legislation *increases* the penalty artificially, to punish the offender *even more*, if the victim turns out to be a homosexual, and the offender is believed to have any kind of "bias against" homosexual activity.

---

991   (From article "To Spank or Not to Spank? (Fact sheet from the Rocky Mountain Family Council)" @ tldm.org)

"**The myth** is that "hate crime" laws are fair.[LIE] **The truth** is that they single out homosexuals for special protection on the basis of their behavior, and single out Bible-believing Christians for their biblically based views on sexual morality.

"**The myth** is that "hate crime" laws are needed because of a floodtide of violent crimes against homosexuals.[LIE] **The truth** is that existing laws already protect *everyone* against crime; discrimination is already outlawed. Furthermore, the purported flood of crime isn't actually happening. The FBI reports so-called "hate" crimes are on the wane!

"**The myth** is that "hate crime" legislation will keep people safe.[LIE] **The truth** is that they will chill or silence free speech, including proclamation of the Gospel. Sharing your faith with a homosexual will make you criminally liable.

"**The myth** is that how people behave sexually is dictated by their DNA, and/or that it doesn't ultimately matter.[LIE] **The truth** is that people *decide* how to behave sexually, and the Bible warns against sex outside of marriage because it's unhealthy on a number of levels.

"**The myth** is that Christians who take a stand against homosexual behavior or special rights for homosexuals are motivated by hatred.[LIE] **The truth** is that Christians are called to love everyone, and calling anyone to sexual morality is an act of compassion. [As opposed to, say, the Democrat-Left's call for everyone to join them in an eternity in The Lake of Fire, which is an act of demonic lunacy.]

"**The myth** is that the young homosexual Matthew Shepard was brutally murdered because of his sexual orientation.[LIE] **The truth**, as proven by an ABC-TV *20/20* investigation, is that he was only targeted for drug money. [But why should Satan and his followers let the truth get in the way of a good excuse to legislate further against Christianity?]

"**The myth** is that "hate crime" legislation represents the true American spirit of fair play.[LIE] **The truth** is that it criminalizes viewpoints — ideas — thoughts — in the single most shocking departure from the doctrine of "free speech" since the First Amendment was ratified by the states!"[992]

"Hate crimes" legislation is another one of Satan's very clever maneuvers to destroy Christianity in America. It is meant to portray *his* followers as the concerned and caring protectors of any "persecuted" minority,[LIE] when in reality they are the suppressors of free speech, and the ones who seek to persecute and criminalize Christians, in order to eradicate the one true religion, Christianity.

And Satan's beat goes on...

"O full of all deceit and all mischief, you child of the devil, you enemy of all righteousness, will you not cease to pervert the right ways of the Lord?" (Acts 13:10)

Satan and his Democrat followers have also attempted to force Christian churches to hire openly homosexual individuals. But curiously they have <u>not</u> attempted to force Islamic Mosques to hire the same, being as how cleaning up the mess after a homosexual or lesbian job applicant gets their head hacked off is not high on their priority list; not like bashing Christians, or disparaging Jesus.

"What becomes of a culture that denigrates the Christian faith? The "God is dead" philosopher Frederick Nietzsche hated Jesus with a passion. Adolph Hitler was Nietzsche's disciple. Hitler likened the Christian faith to "slave morals" and called the God of the Bible "crazed," "stupid," a "despot." [Of course Hitler considered himself an unparalleled

---

992   (From a 2009 *Coral Ridge Ministries* circular)

genius.] Hitler killed 15 million people. [An unparalleled genius who is now on fire in hell.] Mao was an atheist. His system killed 72 million."[993]

"Christians are branded as preachers of hate, fundamentalist, and enemies of democracy and human rights.[LIE] It will not be long before the Bible is called "discriminating and dangerous" literature and forbidden."[994]

The forces of the satanic Democrat-Left are spread across this country and they are assaulting Christianity and Christians at every turn:
> "Here in the culture of the U.S.A., Christians are being subjected to a ceaseless drumbeat of negative bias in the media, and the courts, in the legislatures, and in many classrooms and offices. It is often vicious, malicious, and very deliberate.
> …"The side-effects are extremely serious. Not long ago, a pastor was shot dead in the pulpit on a Sunday morning in Illinois. …some 14 other deadly shootings have occurred in churches in the past decade (most unreported by the national media)."[995]

That would be the national, left wing, lying, state-run, Democrat media. Imagine how this same media would have responded if 14 Islamic Imams had been killed in the same time frame? They would have gone apoplectic, jumped at the chance to portray Islam as a persecuted religion in this country, and of course blamed Christians for their plight.

Here are just a few other manifestations of Satan's anti-Christian movement in our country today:
> "- Elementary school officials in Tennessee ordered the phrases "God Bless the USA" and "In God We Trust" covered up on children's handmade posters.

[But the phrase "God d**n America" on a child's handmade poster of Jeremiah Wright, the imbecile who further poisoned the mind of our president, was displayed on every wall and in every classroom, alongside Stalin's and Chairman Mao's. (You're right, I made that up, but it's really not all that far-fetched.)]
> "- A policeman who refused to arrest Christians for sharing their faith - insisting to his bosses that this was indeed not "disorderly conduct" - was suspended from his job without pay and threatened with termination.
> "- A Los Angeles college student decided to deliver a required speech on the subject of faith. Midway through it, his professor interrupted, calling him a "fascist bastard" for quoting two Bible verses."

[And yes, the above actual "fascist bastard" is still being paid handsomely to indoctrinate our kids. He has tenure, you see.]
> "- In Washington State, college students planning a pro-life event were threatened with expulsion because they weren't including a pro-abortion viewpoint."

[But that's alright, the next week the "fascist bastard" administrators (thank you professor), to be fair and balanced, threatened Muslims with expulsion because they refused to include the Christian viewpoint in their newsletter. (No, I made that up too. You know that would never happen; because the Democrat-Left and their father Satan love clitoral-mutilating Islam. It's just Christianity that they hate.)]

---

993   (From a *Coral Ridge Ministries* mailer)
994   ("Israel, A Stumbling Block for Churchianity," by Reinhold Fedorolf, *News From Israel Magazine*, April 2009)
995   (From a Coral Ridge Ministries communiqué)

"- A Texas judge warned that any student mentioning Jesus at graduation would be jailed for six months."

[Welcome to Nazi America! When is the next election? Get ready to proudly (read: ignorantly) display your "Obama Biden" bumper stickers, America!]

"- A major computer company fired a Christian for displaying Bible verses in his cubicle."

[But another employee was given a raise for displaying an array of oversized, ethnically-diverse dildos in hers.]

"- A fourth-grade student in Missouri was placed on detention for a week for saying grace silently before lunch."[996]

Isn't that the kind of thing you'd expect to see in Nazi Germany, or North Korea? And not in Nazi…er…North America? If all that doesn't make you sick to your stomach then I don't know what to say. And keep in mind that "all that" is just the tip of the fascist bastard's iceberg, indeed most of their attacks go unreported. Satan truly is the god of this world, and the god of the Democrat-Left in this country. His lies delude the Left like a mind-altering drug, and they have ingested into their heart the poison that they were taught. And he is determined to destroy Christianity from across this land, so he can destroy America herself and then deal with Israel unimpeded. But not to worry, for "greater is He [Jesus] who is in us, than he [Satan] who is in the world." (1 John 4:4)

"If the world hates you [true followers of Christ], you know that it hated me [Jesus Christ] before it hated you. If you were of the world, the world would love its own, but because you are not of the world, but I have chosen you out of the world, therefore the world hates you." (John 15:18-19)

"…but be of good cheer; I [Jesus] have overcome the world." (John 16:33)

Amen. "Even so, come quickly, Lord Jesus."

See:

- *One Nation Under God: Ten Things Every Christian Should Know About the Founding of America,* by Dr. David C. Gibbs, Jr., President *Christian Law Association,* with Jerry Newcombe

- *Persecution: How Liberals are Waging War Against Christians,* by David Limbaugh

- Article, "The True Wall of Separation," by M. Stanton Evans, *The American Spectator* Magazine, April 2007

- *Indefensible: 10 Ways the ACLU is Destroying America,* by Sam Kastensmidt

- *The ACLU VS America: Exposing the Agenda to Redefine Moral Values,* by Alan Sears and Craig Osten

2- Destroy the family

"Put on the whole armor of God that you may be able to stand against the wiles of the devil. For we do not wrestle against flesh and blood, but against principalities, against powers, against the rulers of the darkness of this world, against spiritual wickedness in high places." (Ephesians 6:11-12)

"Once upon a time, men wore the pants, and wore them well. Women rarely had to open doors and little old ladies never crossed the street alone. Men took charge because

---

996   (Those last 7 quotes were from a *Coral Ridge Ministries* newsletter, April 28, 2009.)

that's what they did. But somewhere along the way, the world decided it no longer needed men. Disco by disco, latte after foamy non-fat latte, men were stripped of their khakis and left stranded on the road between boyhood and androgyny. But today, there are questions our genderless society has no answers for. The world sits idly by as cities crumble, children misbehave and those little old ladies remain on one side of the street. For the first time since bad guys, we need heroes. We need grownups. We need men to put down the plastic forks and step away from the salad bar and untie the world from the tracks of complacency. It's time to get your hands dirty. It's time to answer the call of manhood. It's time to wear the pants."[997]

Another area of Satan's attack on America is the family. Destroy the family and you can destroy America. Indeed if you tear down the basic family structure then you can destroy any nation or any society. Satan is well aware of this.

Marriage, doggie style

In listing his "5 very bad ideas that came out of the 60's sexual revolution and gained acceptance at that time, and that are still very much with us today," Dr. Dobson from Focus on The Family lists very bad idea #5: "Divorce is an easy way out for the frustrated, disappointed or adventuresome."[(LIE)] Thus began the age of no-fault divorce buoyed by another Herculean epiphany, "if it feels good, do it." And in the case of marriage, if it no longer "felt good" then it was your right and your duty not to "do it" anymore for the sake of your own happiness[(LIE)] and that of your children.[(LIE)] Satan used his *experts* to convince us that "it is better for the children to go through a divorce than to be forced to live in an unhappy situation."[(LIE)] But it was all a lie. It turns out the exact opposite is true. All of the most recent studies have shown that children are emotionally devastated by divorce, and are much better off if their parents tough it out and find a way to make their marriage work. Also it turns out that according to recently concluded long-term studies, the couples in those marriages that were on the verge of divorce but opted to stay together ended up much happier than couples in the same situation who opted to move on to *greener* pastures. But Satan's family-destroying lies are still broadcast, and the truth is still hidden by his eager little elves in our feminazi[998] media and entertainment industry.

Homosexual "marriage"[(LIE)]

In addition to making divorce more socially acceptable and more readily available, Satan has sought to destroy the family by tearing down the very institution of marriage itself, without which there is no family.

"Sociologist David Popenoe and Barbara Dafoe Whitehead at Rutgers University conducted a longitudinal study of the family between 1960 and 1999, and concluded that the institution of marriage appeared to be dying. There is accumulating evidence now that they were right. A handful of power-obsessed judges are determined to impose the homosexual agenda on the nation and thereby change forever the legal definition of marriage."[999]

---

997   (From a Dockers ad, believe it or not! Maybe there's still some hope left for our popular culture after all...)

998   (Term originally coined by one Rush Limbaugh I believe)

999   (From article, "Marriage on the Ropes," by Dr. James Dobson, Focus on the Family)

We covered this substantially in chapter 12, "Homosexuality," so we will not repeat it here, but to say that the militant left's push for homosexual "marriage"[(LIE)] is a large part of the prince of darkness' assault on the American family.

"The gay agenda made significant progress in America, with homosexuality depicted as "normal" in TV sitcoms, movies, educational institutions, and promoted as an accepted lifestyle–even in some churches. Five states legalized "same-sex marriage," a number that could grow as pro-gay propaganda continues to saturate our culture, despite the fact that as a constitutional republic founded on biblical principles, the United States has always defined marriage as between a man and a woman.

"A surge of other immoral forces further eroded the walls of the sacred institution of marriage. As the Pew Research Center reported in a July 1, 2007, study entitled, "As Marriage and Parenthood Drift Apart, Public Is Concerned about Social Impact," many young adults have lost the perspective that traditional marriage is the most important social institution for the well-being of a society. The study revealed a wide generation gap in behaviors and values, stating:

"Younger adults attach far less moral stigma than do their elders to out-of-wedlock births and cohabitation without marriage. They engage in these behaviors at rates unprecedented in U.S. history. Nearly four-in-ten (36.8%) births in this country are to an unmarried woman. Nearly half (47%) of adults in their 30s and 40s have spent a portion of their lives in a cohabitating relationship."

"It also showed that the link between parenthood and marriage suffered significant weakening: "In perhaps the single most striking finding from the survey, just 41% of Americans now say that children are "very important" to a successful marriage, down sharply from the 65% who said this in a 1990 survey." The younger generation was said to be more self-indulgent, throwing away the moral values of their elders about "sex, marriage, and parenthood." The study concluded by stating, "In the United States today, marriage exerts less influence over how adults organize their lives and how children are born and raised than at any time in the nation's history.""[1000]

And all of this of course is devastating to the basic family structure.

Parental notification

He has also weakened the family by using his Democrat-Left and their control of our courts to usurp the rights of parents. One of the most obvious of these is parental lack-of-notification laws that allow minor girls to mutilate their babies to death without notifying their parents, stripping them of all rights in this life-altering decision. Conversely, if they want to take an aspirin at school their parents do have to be contacted. The level of stupidity and ignorance of a nation of parents that would allow this evil and dangerous outrage to be foisted upon them by a small minority of ACLU-Planned Parenthood, Now gang, baby-killing fanatics who control our courts is truly stunning. And even more so is their ignorance in continuing to go out and vote for the outrage. But then, who can blame them, Oprah, and the gals on *The View*, have yet to inform them.

"What Democrats mean by "civil rights" is the civil right of a woman not to inform her husband she's aborting his baby; the civil rights of a minor to have an abortion without notifying her parents; the civil right of a woman to plunge a fork in the head of a child as it struggles through the birth canal because it has a cleft lip. That's "civil rights.""[1001]

1000   (From a Creation Studies Institute newsletter, Tom Rosa, Executive Director and Founder)
1001   (Ann Coulter)

Spanking, the sequel

"Foolishness is bound in the heart of a child; but the rod of correction shall drive it far from him." (Proverbs 22:15)

As we mentioned in the previous section, Satan's goal is to use his Democrat-Left-leaning courts to further usurp parental rights and to further weaken and destroy the family by making spanking illegal. He has already succeeded in this endeavor in many European and Scandinavian countries. The whopper is that if you spank your children it is "child abuse,"[(LIE)] but, conversely, if you mutilate them to death in the womb, it is not.[(LIE)] Their idiocy would actually be hilarious if it were not so dangerous.

"Parents who don't use spanking, invariably get extremely frustrated with their children for not listening to them, but they have absolutely no recourse to alter the behavior of their children except to yell louder and louder and get more and more angry. And this in of itself is a far greater "child-abuse" than smacking them on the fat of their behind. There has not been one scientific study that has documented any harm from corporal punishment when it is administered by loving parents properly educated on how to use spanking to alter the defiant behavior of a young child.

"Doctors at the University of Nebraska Medical Center reviewed 38 studies, and found that in children under the age of seven non-abusive spanking produced no harmful effects. While at the same time it substantially reduced misbehavior. I am absolutely convinced that child-abuse is most likely to occur when the parents have no answer for rebellious, irritating behavior that's motivated by willful defiance. And as a result, all they can do is talk and warn and threaten and scream and blow up. By contrast, those who handle the misbehavior early, they know how to deal with it, avoid that volatile level where anything can happen. You see child-abuse is not usually premeditated. It's a result of out-of-control parents who become so angry that they want to hurt the child who is frustrating them out of their minds. Corporal punishment when it's used rightly therefore actually reduces the incidence of harm to children.

[As a child goes through life at an early age he finds out by experience and through pain what to avoid—a hot stove for example, it burns...]

..."So for three or four years [the child] accumulates bumps and bruises, scratches and burns and each one teaches him something about life's boundaries. Now do these experiences make him a violent person? No. That's what people say about spanking: "it makes them violent." No it doesn't make them violent, unless it is done abusively. The minor pain associated with these events teaches a child to avoid making the same mistakes again. God created this mechanism as a valuable vehicle for instruction. Now when a parent administers a reasonable spanking in response to willful defiance, a similar nonverbal message is being given to the child. He must understand that there are not only dangers in the physical world to be avoided, he should also be wary of dangers in the social world such as defiance and sassiness and selfishness and temper tantrums and behavior that puts his life in danger. The minor pain that's associated with this deliberate misbehavior, when it's willful, tends to inhibit it, just as the discomfort works to shape behavior in the physical world. Neither conveys hatred. Neither results in rejection. Neither makes the child more violent. In fact, children who have experienced corporal punishment from loving parents do not have trouble understanding its meaning. And they get the picture even if some of their critics sometimes don't."[1002]

1002   (From "Parenting Tips for a New Generation," Dr. James Dobson, Focus on the Family)

The rod and discipline give wisdom; but a child left to himself brings shame to his mother." (Proverbs 29:15) …In addition to driving her out of her happy mind.

Satan's other goal in criminalizing the spanking, and therefore proper discipline, of children is to criminalize all true Christian parents who raise their children in the love of Christ, but who also know that love without discipline is fear and cowardice. God says that as a truly loving parent you must spank your child occasionally when they directly disobey, in order to <u>discipline</u> them, so they will learn to associate <u>pain</u> with <u>disobedience</u>. So that when they are older and when their friends tempt them to disobey the laws of God by doing drugs or by stealing, or cheating, or having promiscuous sex, or to think that it is cool to drop out of school, that they will be able to see all that slop for the <u>painful</u> choices that they actually are, and therefore won't be deceived into thinking it is an attractive choice that they "just can't resist." *If* they have been properly disciplined as a child to associate disobedience with pain. The hope is also that when disciplined children grow up they will also see the deceitful call of Satan to reject God's free gift of eternal life for the slop that it is as well, because they will also have learned to <u>associate pain</u> (and in this case extreme eternal pain and anguish) with the foolish choice of disobeying God's call to an eternity of peace and love. Undisciplined children lack this advantage.

Divorce, Satan style

"For this cause shall a man leave father and mother, and shall be joined to his wife: and they two shall be one flesh." (Matthew 19:5) That is the ultimate intent behind God's design of sex. "And the glory which you gave me I have given them; that they may be one, even as we are one." (John 17:22) (Yes, Jesus was referring to his followers there, but the same oneness was intended for marriage.)

Satan has also successfully attacked marriage and the family by being the driving force behind the far-left feminist ideology that has taken control over our courts and their divorce laws. Together they have made divorce extremely attractive to the wife. Their laws actually encourage women to divorce their husbands. The incentives are almost impossible to resist. For the slightest whim—if she finds her husband temporarily physically or emotionally unattractive to her, or she finds someone at work who is more stimulating for the moment, or for whatever reason whatsoever and at the drop of a hat—she can get rid of him, take his money, take his belongings, take his kids, take their house, and destroy his life. In the meantime, she can go out and remarry, get a job if she doesn't already have one, and with his child support enjoy a triple income and a ticket to the high life. Depending upon how much heart, or lack thereof, that a particular woman possesses, that's a deal that's almost too good to pass up. And this country is filled with millions of poor unfortunate men who have had this evil foisted upon them, in many cases through no fault of their own, and now they are sentenced to spend the rest of their lives watching their children being raised by another man, in another home, while they themselves are strapped financially and can barely afford to remarry and start a life of their own. And yet millions of these same men go out and vote for this every two years religiously, ignorant of the Democrat-Left behind this evil. And Satan of course is laughing at them raucously behind their backs.

Conversely, the women's movement would go berserk if the same thing were reversed. If men were given the same legal right—to take half of everything, the house, and all of the children, go off and jump in bed with another woman while enjoying the financial fruits of child support from the poor victimized ex-wife. The Democrat-feminist-Left movement is not about "equality of the sexes." It is about the advancement of their man-hating, marriage-hating ideology which fits hand-in-glove with Satan's desire to destroy America by destroying the family.

The Great Lie

Then there is the devastation caused by Linden B. Johnson's, and the Democrat Party's, "Great Society"[(LIE)] welfare programs of the 60's. They were meant to eliminate poverty but instead they have greatly exacerbated it while trapping millions in its clutches for entire lifetimes. And as a result today only a mere 34% of African American children are raised in a two parent household. In 1966 that number was 85%. The corresponding illegitimacy rate is 70%! And yet blacks continue to vote Democrat in record percentages—in the last (2008) presidential election they went over 90% for the baby-mutilating Democrats. Would I be flip in asking my brothers and sisters what exactly they are trying to do, get their illegitimacy rate up to 100%? Satan, of course, is very pleased with their, as well as others, loyal assistance to his ongoing effort to destroy America by destroying the family.

"Right up until the 1950s, blacks, who were struggling against undeniable prejudice and discrimination and were economically disadvantaged as a result, held together as families, often set a model of good behavior, and were hardly more disposed to crime than their white contemporaries. Yet today, 40 years on from the Civil Rights Act, the picture is very different—horrifyingly so—with 70 percent of black children born out of wedlock, escalating crime rates, and a steep relative decline in school performance. ...A vast investment has been made in (the) official explanation, which defines a comfortable whitewash for the liberal conscience, as well as a lucrative source of income for the rent seekers of the social services. Those who suggest that welfare benefits might be the cause of the problem, rather than the solution to it, risk having their heads bitten off, and will certainly find themselves marginalized in any academic community."[1003]

"For even when we were with you, this we commanded you, that if any would not work, neither should he eat." (2 Thessalonians 3:10)

For a society to have some sort of safety net in place is one thing. Destroying the family while advancing your socialist, nanny-state political ideology in order to get as many voters as possible dependent upon the government's massive teat (more Democrat votes), and taking the illegitimacy rate of one particularly targeted minority to astronomical levels, is quite another. Here are some conclusions from African-American author and economist Walter E. Williams:

"There's a huge segment of the black population for whom upward mobility is illusive. And I believe it's illusive because of the welfare state. It's because of government."[1004]

"They may never learn to pull themselves out of poverty one step at a time. Like some giant drug pusher, their government has lured them into dependency on a system that will maintain them in permanent poverty. In every respect, welfare has backfired."[1005]

1003   (From article "Truth and Self-Censorship," by Roger Scruton, the American Spectator, Dec 2007/Jan 2008)
1004   (African American author Walter E. Williams, Economist @ George Mason University, on John Stossel's "The War on Poverty," Fox News Reporting, 6/5/2011)
1005   (Walter E. Williams, from the 1985 documentary "Good Intentions", from Free To

"The Welfare State has done to Black Americans what slavery could not have done, the harshest Jim Crow Laws and racism could not have done, namely, break up the black family. That is, today just slightly over 30% of black kids live in two-parent families. Historically, from 1870s on up to about 1940s, depending on the city, 75 to 90% of black kids lived in two-parent families. [The] illegitimacy rate is 70% among blacks, where that is unprecedented in our history."[1006]

That's the gift that the Democrat Party has given to the blacks in America, in return for their blind and overwhelmingly loyal support, that's what they gave them, they destroyed their family. Wake up!! My poor, deluded black brothers and sisters, and free yourselves from the shackles of the black-exterminating Democrat Party and their wicked, lying media, for both really do not give a damn about you or your race.

- The minimum wage Associated with the liberal Democrats Great Society and its fantasy of ending poverty is the minimum wage. They constantly fight to keep increasing it in a supposed attempt to assist those less fortunate by giving them a higher, "fairer" income so they can be better equipped to "make ends meet." Certainly at face value it would seem a noble intention, but like just about all of the liberals programs of the last half century it has had the exact opposite effect...

"The minimum wage, rather than helping the poor, has hurt them, because now far fewer kids can get jobs—entry-level positions—where they could work their way up the ladder. Instead, now 70 % of blacks in poor neighborhoods are chronically unemployed. That is the result of liberal good intentions."[1007]

"Back in those days [before the minimum wage] just about any kid who looked for a job could find one. Today, in Ghettos like I grew up in, 70% of black children who look for jobs cannot find them."[1008]

"90 % of economists say the minimum wage laws increase unemployment."[1009]

Yet despite the abject failure of their minimum wage program the Democrat-Left continues to foment to make it even higher, and even more destructive; while at the same time accusing anyone with common sense and a real heart for the poor who opposes them of being "cruel" and "mean-spirited," of "caring only for the rich," and of "not wanting to pay their fair share." And backed up by a Democrat media, they continue to get away with it.

Choose Media, as reported on John Stossel's "The War on Poverty," Fox News Reporting, 6/5/2011)

1006   (Walter E. Williams, on John Stossel's "The War on Poverty," Fox News Reporting, 6/5/2011)

1007   (From John Stossel's "The War on Poverty," Fox News Reporting, 6/5/2011)

1008   (Walter E. Williams, from the 1985 documentary "Good Intentions," from Free To Choose Media)

1009   (From John Stossel's "The War on Poverty" on Fox News Reporting, 6/5/2011)

See:
- *The State Against Blacks*, by Walter E. Williams, Economist @ George Mason University. His book is an exposé of the horrors visited upon the black community by the good intentions of the minimum wage and the welfare state.
- *Race and Economics*, by Walter E. Williams

The Homosexual agenda

Satan also goes about to destroy and weaken the traditional family by shoving the radical homosexual agenda down the throats of our children in our public schools, starting as young as kindergarten and, naturally, without parental consent. As we discussed in the previous chapter, "Homosexuality," this is done under the guise (the big lie) of "tolerance,"[LIE] which of course means tolerating every aberrant and perverted behavior including violent crime as long as you're careful not to tolerate any of the values or morals of Christianity or traditional America. What better way to destroy the family than attempting to sign up children at a young age for the homosexual lifestyle? They should be arrested and put on trial for child abuse, but instead we allow them the authority to set the curriculum for our children in our government-run schools.

All of these things—marriage with no commitment, outlawing spanking, homosexual *marriage*, laws encouraging women to divorce, usurping parental rights, Democrat welfare programs, the radical homosexual agenda—all the pet projects of the "progressive" Democrat-Left—are ways that Satan has used and is using to destroy the family in America. Wake up America! Or you're going to end up looking more like a drug-infested alley in Holland, and not what was once the greatest nation in the history of the world.

3- Destroy the morals of the nation

"He who sins is of the devil, for the devil sinned from the beginning. For this purpose the Son of God was manifested, that he might destroy the works of the devil." (1 John 3:8)

We have covered this subject extensively in the second chapter, and throughout the book, so we will only touch on it here. Satan and his Democrat-Left have succeeded in legislating from the bench the vile immorality of murdering babies and this evil now permeates the land, staining it with innocent blood. They have also succeeded in outlawing discipline in our elementary and high schools, turning them into hellholes of foul language, disrespect and immorality in many places, especially the poorest. The art world, entertainment and music industry have descended into a stinking cesspool that glorifies evil in all of its forms, while disparaging and mocking the values and faith of decent Americans. Satan has used his feminists and radical homosexuals to further tear down the moral fabric of this nation. He has taken over the courts where judges ignore the Constitution that they have sworn an oath to uphold and defend, instead they illegally legislate the Left's agenda and every form of immorality from the bench. The bald-faced, scientifically-impossible and asinine lies that are the theories of evolution and the big bang are shoved down the throats of our unsuspecting and naïve school children in our colleges, universities, elementary and high schools. People like

Oprah Winfrey are put forth as the great magnanimous champion of abused and underprivileged children in Africa, while here in her own country she religiously votes for and supports the political Party of the worst child-abuse ever known to man—baby-mutilation—as well as the Party that is carrying out the mandate of Planned Parenthood's founder, Margaret Sanger, and slowly exterminating Ms. Winfrey's own race. Good grief!

"Billy Graham's daughter was interviewed on the *Early Show* and Jane Clayson asked her, regarding Katrina, "How could God let something like this happen?" Anne Graham gave an extremely profound and insightful response. She said, "I believe God is deeply saddened by this, just as we are, but for years we've been telling God to get out of our schools, to get out of our government and to get out of our lives. And being the gentleman He is, I believe He has calmly backed out. How can we expect God to give us His blessing and His protection if we demand He leave us alone?" In light of recent events...terrorists attacks, school shootings, etc. I think it started when Madeleine Murray O'Hare (she was murdered, her body found recently) complained she didn't want prayer in our schools, and we said OK. Then someone said you better not read the Bible in school. The Bible says thou shalt not kill, thou shalt not steal, and love your neighbor as yourself. And we said OK. Then Dr. Benjamin Spock said we shouldn't spank our children when they misbehave because their little personalities would be warped and we might damage their self-esteem (Dr. Spock's son committed suicide). We said an expert should know what he's talking about and we said OK. Now we're asking ourselves why our children have no conscience, why they don't know right from wrong, and why it doesn't bother them to kill strangers, their classmates, and themselves. Probably, if we think about it long and hard enough, we can figure it out. I think it has a great deal to do with "WE REAP WHAT WE SOW." Funny how simple it is for people to trash God and then wonder why the world's going to hell. Funny how we believe what the newspapers say, but question what the Bible says."[1010]

"Know you not that the unrighteous shall not inherit the kingdom of God? Be not deceived; neither fornicators, nor idolaters, nor adulterers, nor effeminate, nor abusers of themselves with mankind, nor thieves, nor covetous, nor drunkards, nor revilers, nor extortioners, shall inherit the kingdom of God. And such were some of you: but you are washed, but you are sanctified, but you are justified in the name of the Lord Jesus, and by the Spirit of our God." (1 Corinthians 6:9-11)

Years ago we had Ozzie and Harriet. Now we have Ozzie Osborne on a show where children cuss out their parents on national TV. And this foul mouthed group was applauded for its "openness" and "freedom of expression" by the malignant narcissists on the Democrat-Left. And even more revealing, in the 1996 Presidential election the utterly depraved nature of Bill Clinton's moral character had become common knowledge due to his numerous sexual indiscretions, assaults, rapes and prevarications which had come to light during his first term. Even Democrat Senator Bob Kerry accurately stated, "Clinton's an unusually good liar". So how did those clever Democrats, striving to still somehow get the skank reelected to a second term, overcome this obstacle? Did they try to defend him? Did they try to make excuses for him? (He had a tough childhood so it wasn't his fault.) No, they just trotted out another one of their terribly witty and this time rather revealing

1010   (From email circulating on the net)

slogans, "Character doesn't count." And to a Democrat voter it obviously didn't because they trotted out obediently and reelected him. And that alone tells you all you need to know about the morals, or lack thereof, of the Democrat-Left. With their eager assistance, we have become a culture and a nation that calls evil good and good evil (and says that rape "doesn't count"). And Satan celebrates our descent into his immorality from hell, another rung on the ladder of his destruction of America.

4- Destroy the historical foundations of America

"Therefore rejoice, you heavens, and you who dwell in them. Woe to the inhabitants of the earth and of the sea! for the devil is come down unto you, having great wrath, because he knows that he has but a short time." (Revelation 12:12)

> "We hold these truths to be self-evident, that all men are created equal, that they are endowed by their Creator with certain unalienable Rights, that among these are Life, Liberty and the pursuit of Happiness.—that to secure these rights, Governments are instituted among Men, deriving their just powers from the consent of the governed."[1011]

Most Americans today take those words for granted, or are wholly ignorant of their very existence. We are all so far removed from that time 230 some years ago when they were first proclaimed and splashed onto the world's consciousness as The Declaration of Independence of the then thirteen United States of America. We have no idea how extraordinary they were...

> "When the United States of America was founded, it was unlike anything anyone had ever seen before. Never throughout world history had there been a country of the people, by the people, and for the people—with religious liberty, freedom of speech, and confidence in the common man. This was because out of the fifty-six signatories to the Declaration of Independence, fifty-four of them were believers—men who understood the powerful influence the Savior could have on each and every person.
>
> "They sought God's wisdom in creating the framework that would become the great nation we know and love today. They sowed faith, prayer, trust in His Word, and obedience to His commands—and the Lord blessed them because of their faithfulness to Him. America's Founding Fathers were not perfect—they were people just like you and me—but what God did through them was absolutely astounding."[1012]

American Exceptionalism

> "America is an exceptional nation, but not because of what it has achieved or accomplished. America is exceptional because, unlike any other nation, it is dedicated to the principles of human liberty, grounded on the truths that all men are created equal and endowed with equal rights. These permanent truths are "applicable to all men and all times," as Abraham Lincoln once said.
>
> "America's principles have created a prosperous and just nation unlike any other nation in history. They explain why Americans strongly defend their country, look fondly to their nation's origins, vigilantly assert their political rights and civic responsibilities, and remain convinced of the special meaning of their country and its role in the world. It is

---

1011   (From "The Declaration of Independence" of The United States of America)
1012   (Excerpt from "From the Pastor's Heart" by Charles F. Stanley, July 2010, In Touch Ministries)

because of its principles, not despite them, that America has achieved greatness.

"To this day, so many years after the American Revolution, these principles—proclaimed in the Declaration of Independence and promulgated by the United States Constitution—still define America as a nation and a people. Which is why friends of freedom the world over look to the United States not only as an ally against tyrants and despots but also as a powerful beacon to all those who strive to be free."[1013]

America stands out as unique among the nations of this world. God truly has "shed his grace on thee" ...in the past (today, all bets are off). And that past, the history of the founding of our great nation is also unique and a potential source of great pride and love and patriotism for our citizenry. And that is exactly why Satan and his deceptive Left must destroy it; why they must rewrite history to fit their America- and Christian- hating vision. Satan and his Democrat-Left do not want our children to know how great and how unique our country is. Nor do they want them to know how great and unique and devoutly <u>Christian</u> most of our founders were. So they lie to them. They tell them that our founders were racist[(LIE)] homophobes[(LIE)] who denied women their rights.[(LIE)] And they tell them that the America that resulted from these Christian homophobe racists is bad[(LIE)] and "imperialist"[(LIE)] and one that is in need of being fundamentally transformed. [(LIE)] And who better to "radically transform" America than themselves, these "morally-upright," "righteous," baby-killing, lying, history-distorting reprobates of the far Left. And that is why so many left-wing-indoctrinated college students bounded out in November of 2008 in near orgasmic delight to vote for Obama. Satan and his academic Left had done their job all too well.

Governor Mike Huckabee, on his show on Fox News, talked about becoming...

..."embroiled in a little controversy this week from liberal blogs and TV shows because of my involvement in the new project called "Learn Our History." ...I was very fortunate to have great teachers at Brookwood elementary in Hope, Arkansas. They encouraged me to read biographies and books about great Americans, and American history. I grew up believing that America is the greatest country in the world, blessed by God, and it's a place where even a kid like me from humble beginnings in a small little town would have no limits. I was shocked to learn the results of a recent citizenship test of high school students that only 25% of them could even name our first president. Or that less than 3% of those students could correctly answer six out of 10 questions on the US citizenship test. And six out of 10 of those questions are required to become a nationalized citizen. Now for all the talk about illegal immigration, I wonder, maybe we ought to be talking about illiterate citizenship.

[And could somebody please explain to me what is the value of these idiots going out and voting? "Rock the Vote!?#%*?" It's pathetic. We have liberals controlling our educational system and graduating uneducated youths who know next to nothing about our history or our country, but who have been well-brainwashed into obediently voting Democrat, no matter what.]

Now, I believe history is to a culture what memory is to the individual. Because if we don't know where we've come from, we have no idea where were going. Having no memory would leave a person lost and disoriented, confused and uncertain. When a nation loses its sense of history, and it doesn't know where it came from, it's lost and confused,

---

1013   (From "Why is America Exceptional?" by Matthew Spalding, Ph.D., Director of the B. Kenneth Simon Center for American Studies at The Heritage Foundation, @ heritage.org)

and its leaders are uncertain. [Think Obama] …Folks, this is tragic! Look, I don't mind if the left-wing scoffs at the "Learn Our History" project. After all, it is a series of animated videos for kids, age 7 and above, and it might be beyond the intellectual capacity of some of the critics. But more importantly, because it's based on the notion that America is a great place with a great history that kids need to know about, it might be beyond the capacity of the cynics. …Look, I do love this country. It has been really, really good to me. But I want future generations to love it too. And they will if they learn the truth about it."[1014] (Which is why it is so important for the Left to keep it from them.)

Larry Schweikart, in *48 Liberal Lies about American History (That You Probably Learned in School)*, …
…"takes on authors of popular textbooks who, in an effort to be politically correct, tarnish America's image in the eyes of our own students.
[Which of course is Satan's design. But…]
"The problem isn't that liberal authors present their opinions or interpretations of history from an obvious left-wing bias. The problem is authors who actually distort facts and manipulate data in an effort to appear objective and unbiased.
"As a result of this deceptive behavior, students in both high school and college learn that, since its creation, the United States has been driven by hate, greed, violence, and intolerance.[LIE] They learn that the Founding Fathers were elitists who drafted the Constitution to protect their own economic interests.[LIE] They learn that racist groups such as the KKK represented our society in the early twentieth century.[LIE] They learn that America's countless victories in battle were merely the designs of war-mongering presidents trying to take over the world.[LIE]
[If we allowed Al Qaeda to teach our children they couldn't come up with any more of a rank distortion than our Democrat-Left friends have done. And America is kept in the dark about this academic subversion by the *selective* reporting of a complicit media.]
"Armed with this false knowledge, our students graduate with misunderstandings about economics, foreign policy, war, religion, social issues, and many other subjects. And then [like so many of their parents] they carry their ignorance into the voting booth."[1015]
Here are a few of the "insidious lies" that these "politically-correct" (albeit factually incorrect) textbook authors and their accomplices at the front of many of our classrooms indoctrinate our youth with…
"- Lincoln issued the Emancipation Proclamation only because he needed black soldiers.[LIE]
- The failures of capitalism caused the Great Depression.[LIE]
- Joseph McCarthy concocted the Red Scare, and there was nothing to fear from Communist subversives.[LIE]
- Mikhail Gorbachev, not Reagan, ended the Cold War."[LIE][1016]
And in addition, they teach these falsehoods:
"FDR Knew in Advance About the Japanese Attack on Pearl Harbor[LIE] …Harry Truman Ordered the Atomic Bombing of Japan to Intimidate the Soviets with "Atomic

---

1014   (Governor Mike Huckabee, on Fox News' Huckabee, 5/14/2011. And to order Mike's "Learn Our History" video series go to learnourhistory.com)
1015   (From inside cover, *48 Liberal Lies about American History (That You Probably Learned in School)*, by Larry Schweikart)
1016   (From inside cover, *48 Liberal Lies about American History*, by Larry Schweikart)

Diplomacy"[LIE] ...The "Peace Movement" Activists Were Not Dupes of the KGB[LIE] ... September 11 Was Not the Work of Terrorists: It Was a Government Conspiracy[LIE] ...No Terrorists or Weapons of Mass Destruction Were Hiding in Iraq[LIE] ...Lee Harvey Oswald Shot JFK Because He Was a Deranged Marine [fits their deranged, anti-military narrative], Not Because He Was a Communist[LIE] ...Columbus Was Responsible for Killing Millions of Indians[LIE] ..The Early Colonies Were Intolerant and Racist[LIE] ...Early America Was Home to Few Guns and Gun Owners[LIE] ...Neither Ronald Reagan's Election nor the "Contract with America" Proved the Triumph of Conservative Ideas[LIE] ...George W. Bush Was Selected, Not Elected, in 2000, and Votes Were Stolen on His Behalf[LIE] [conversely, ACORN has never committed voter fraud and has the highest possible rating with the Better Business Bureau] ...Global Warming Is a Fact, and It's a Man-made, American-Driven Problem[LIE] ...Northern Capitalist Greed—not Slavery—Drove the Civil War[LIE] ...LBJ's Great Society Had a Positive Impact on the Poor[LIE] ...The Reagan Tax Cuts Caused Massive Deficits and the National Debt."[LIE][1017]

(Note that the above list of historical distortions is only a summary. For the facts and evidence that prove these distortions for what they are, as well as an in-depth exposé, see Mr. Schweikarts books, *48 Liberal Lies about American History (That You Probably Learned in School).*)

The subtle message of course behind all of this rewriting of history is that the America that we all know and love is "bad" and can't be trusted by the student. Whereas the left-wing indoctrinator, their new best friend, at the front of the classroom is a veritable Einstein. And all of our Founders, and all of the Christians, conservatives, traditionalists and Republicans who are intelligent enough to admire those founders, are also "bad" and not to be trusted as well. "So listen to us, young students, because we, your lying, historical-revisionist, left-wing, baby-killing indoctrinators, are the *really good guys*, and those *evil, racist, imperialist* Founding Fathers, Republican presidents and all those who respect them, are the bad folks." And of course those kids, just like everybody else, believe what they are taught. And we wonder why the Hitler Youth Movement is alive and well and thriving on American college campuses everywhere. "O-BA-<u>MA</u>! O-BA-<u>MA</u>!"

> "While other nations have built their governments upon the shaky foundations of communism, socialism, and countless other anti-God philosophies, only to see those foundations crumble, America stands without equal as a beacon of hope and freedom in a hurting world. Our Founding Fathers delivered to us a system of government that has enjoyed unprecedented success: we are now the world's longest ongoing constitutional republic. Well over two hundred years under one form of government is an accomplishment unknown among contemporary nations."[1018]

You don't have to go far to see the pernicious results of the Left's rewriting of our history, just ask any elementary, high school or college student in America what he or she thinks about George Washington, or Benjamin Franklin, or any of our Founding Fathers. If you can find one who even recognizes their names you will be lucky, and what you will most likely hear from those few who do is

---

1017   (Pgs. ix-xiv, *48 Liberal Lies about American History (That You Probably Learned in School)*, by Larry Schweikart)

1018   (Pg. 7, "In God We Still Trust," by Dr. Richard G. Lee)

something like this: "Oh, you mean those slave holders?" "They were racists, weren't they?" "Didn't they oppress women?" "How many witches did they burn?" Yet the real history of the character of those who founded this nation is quite a different tale from what these students are being fed:

"On July 4, 1776, the forefathers of our nation set upon a course and forged a direction that would reverberate throughout the world and lead this nation to offer unforeseen levels of hope, prosperity, and freedom to an amazing tapestry of people. The American Revolution was truly revolutionary.

""Government of the people, by the people, for the people" was a very radical concept. No one could have dreamed the impact it would have. In our founding document, the Declaration of Independence, one of the most profound ideals set forth was that "all men are created equal." Today, it is hard to truly understand how radical the introduction of that concept was. It helps to go back to the eighteenth century and gain a greater understanding of what the world was like. Kings and queens were the rulers and conquerors of the day. Justice and wealth was held in their hands. Our forefathers sought to take some of that tightly bound power and distribute it so that many who could never dream of hope and opportunity would find peace and prosperity through a freedom that was built upon the principles of God."[1019]

Our great nation has an awe-inspiring story to tell. It is being fubared (f**ked-up-beyond-all-recognition) by the liars on the Democrat-Left...

"What happens to a tree when its root system is destroyed? The tree can no longer access the nutrients it needs, and it begins to weaken. The leaves wilt, the branches droop, the fruit ceases to grow. Eventually the tree dies. America was formed and developed from biblical roots, but today our leaders chop away at our foundation. They deny the core of our purpose. They allow invasive routes to choke out the Truth. Yet what do Christians do? We cling tightly to the rotting fruit on the drooping branches, while underneath us the truth of our country's past and the hope for our present and future are being destroyed. Without the root system, there can be no fruit.

"The secular thinkers of this country would like us to think our country was founded on humanistic values and ideals. They view our country's birth as a philosophical and political event. However, the original words from the colony charters, the Declaration of Independence, and early political speeches all point directly to our country's foundation upon the Christian faith and biblical Truth. Our Founding Fathers believed that this is a nation with its roots in the Bible.

"The Mayflower Compact specifies that the colonies were established "for the glory of God, and advancement of the Christian faith." The Virginia Charter instructs the colonists to help "in the propagation of the Christian religion to such people as yet live in ignorance of the true knowledge and worship of God." The Delaware Charter defines one of the purposes for its settlements as the "further propagation of the Holy Gospel." The charter of Rhode Island commits to "the true Christian faith and worship of God." The Charter of Maryland explained "a pious Zeal for extending the Christian Religion."[1020] (Someone fire off a tweet to Barbara Mikulski.)

America is "the greatest country that God gave man," yet it has been turned into dross for a generation of our hapless youth. A number of them could be

---

1019   (Pg. 7, *Under God*, by Toby Mac and Michael Tait)
1020   (From feature, "Returning to Our Roots," in "God Save America: My Journal July 2010" by Dr. Michael Youssef and Leading The Way International)

seen this past weekend, April 17, 2011, disrupting Tea Party rallies by shouting obscenities, holding signs and screaming slogans that displayed both their historical and political ignorance, as well as chanting the "God D*** America" of Obama's pastor and mentor. But the most telling display came by way of one graduate who held an American flag while shouting proudly that he used it "to wipe his ass every morning." (And then his face.) (Hopefully he had the sense to wash it before wearing it to the rally…) (Or maybe he flew in from San Francisco.) Another example of what happens to a mind when it's been poisoned by the lies of the Left. Satan's inventions cover our school system like a Che Guevara t-shirt (that our friend above mistook for his flag), and too many of our kids swallow, uncritically, the progressive horses**t that has been shoved down their throats.

One of the Left's favorite ways to tear down this nation is to accuse it of being founded by "racists" because slavery was tolerated in the new nation. Of course they fail to point out the fact that slavery was commonplace throughout the world at that time, as it had been for thousands of years prior. Here are some facts that are purposely kept hidden from our school children: 29 million slaves were taken out of Africa. 17 million went to Muslim countries. 12 million went to Brazil. Yet only a comparatively few <u>645 thousand</u> came to the United States (which is <u>less than</u> 1 million; for the mathematically challenged). That's a ratio of 26 to 1 less than the Muslim countries, and 18 to 1 less than Brazil. But you never hear the Left trashing Brazil or all of those Muslim countries, only the United States. In addition, the United States *inherited* the problem, because slavery first came to America in 1619, 157 years <u>before</u> our country was founded. So that would be another egregious lie of the Left,[(LIE)] this time another one of omission.

But the fact that the founders of this country did not end, and indeed could not have ended, slavery at the inception of this nation is not a condemnation of them and their character, nor of the extraordinary experiment in democracy that they *were* able to forge. But that this fact is brought up constantly by the Left in order to demean them and what they did accomplish, is in reality a condemnation of the Left, and an exposé of themselves for the history-distorters that they are. By way of example, the only way Ronald Reagan in the 80's was able to push through a Democrat-controlled Congress the kind of tax cuts that brought us out of the malaise of the Carter recession, was to compromise by giving those Democrats the massive spending increases that they demanded in return for their critical votes. And the tax-cuts that resulted from that tough compromise gave us one of the greatest periods of economic growth and prosperity that this country has ever known. (As opposed to this period of economic depression exacerbated by Obama and the Democrat's unrestrained socialist spending and business-strangling government-over-regulations.) In the same way, it would have been impossible for our founders to even form The United States of America if they had tried to do it while abolishing slavery at the same time. There would have then been a war on two fronts, one against the English for independence, and another internal war among the colonies. And the chances of success in either endeavor—fighting a Revolutionary War and a Civil War at the same time—would have been nil. But what they were able to accomplish was not only establish the free-est, most prosperous and most generous country that ever was, but in the process they also allowed for the very democratic, moral, Christian and political conditions that would end slavery as well, just 87 years later.

"The Founders were, first and foremost, politicians willing and able to compromise in order to accomplish **what was doable under the circumstances**."[1021] (Emphasis mine.)

Those are the facts that the Democrat-Left for some strange reason has refused to tell our unsuspecting and gullible school children in our colleges, universities, elementary and high schools. And that is because not too long ago these teachers on the Left *were* the unsuspecting and gullible school children whose minds were poisoned with Satan's left-wing, America- and Christian-hating lies. And the curse goes on... "The LORD is longsuffering, and of great mercy, forgiving iniquity and transgression, but by no means clearing the guilty, **visiting the iniquity of the fathers upon the children unto the third and fourth generation**." (Numbers 14:18) Isn't it about time that we as a nation find the guts to end this particular curse?

Of course, the not-so-subtle insinuation is that these academics, if they had been alive back then, would have done a *so much better job*. Certainly none of them would have been slave-owners, being as how *they* are so morally superior to that.[(LIE)] (They might have mutilated a few hundred million babies to death but, unlike slavery, killing babies is their *acceptable* form of moral repugnancy.) They also, I am sure, would have formed a much better nation.[(LIE)] Assuming, of course, that they would have had the courage to risk their lives, their families and their fortunes, as all of our Founders did, to form any nation at all. No, we'd still be an English colony. And the "much better nation" that *they* would have formed (from the safety of their ivy-covered halls) would have looked a lot like Cuba, or Venezuela, or the United *Socialist* States of Amerika, than the greatest nation in the history of the world. But divorced from this reality, they will continue to spew their self-righteous, anti-American rants from the lecterns and podiums of our high school and college classrooms. And America's parents will continue to remain clueless as to what is actually going on. Congratulations are in order to both Lucifer and his media.

In the heat of the 2008 presidential election, John McCain made an appearance on *The View*, that feminine gabfest presided over by the usual suspects in our predictably Democrat media (and one token conservative, Elisabeth Hasslebeck). Like their friends on the "hard news" side of the media, *The View* has never interviewed a liberal that they didn't fawn all over. Conversely, any conservative, Christian or Republican is routinely subjected to their thinly-veiled scorn. In that vein McCain was asked a question about what type of Justices, if elected President, he would nominate to the Supreme Court. When he responded with Justices who would respect the original intent of the founders, Whoopee Goldberg replied, "Oh, so you want to bring back slavery?"[(LIE)] She was implying of course that following the actual Constitution, instead of ignoring it to suit your liberal ideology, would somehow "bring back slavery." (No, but it might free a whole lot of little innocent babies from getting mutilated to death, Ms. Goldberg, as well as ending the slow extermination of your own race.) McCain could have replied, "No, I want to bring back America from the hands of history-distorting Democrats like yourself, who also as judges ignore the Constitution while legislating their anti-American agenda from the bench, which, the last time I looked, has absolutely nothing to do with "bringing back slavery."'" He could have then asked a follow-

---

1021   (From "Faith of our fathers," by Rev. Michael P. Orsi, The American Spectator, Dec 2008/Jan 2009)

up question as to why Whoopi marches out, religiously, every two years and votes for the genocide of her own race. (Now that would have stumped the band big time.) Of course as we have already stated it would have been impossible for our Founders to have ended slavery, and if they had attempted to the result would not have been the end of slavery, it would have been the end of the United States of America before it even began. But unfortunately McCain didn't seem to possess the knowledge to point that out to his history-distorting, antagonistic foil. Instead he mumbled something pathetic and slumped down deeper in his chair. He also could have said that without that Constitution that her and her Democrat friends hold in contempt, slavery might still exist today not only in this country but in many other countries across the earth as well. And that the force for ultimate good and freedom in this world for the last two hundred and thirty four years, the United States of America, which that same Constitution gave birth to, would also probably not exist. And therefore Jews would still be getting exterminated in Germany and Europe and Poland, and women would still be getting stoned in Afghanistan by the Taliban, and apartheid would still be going on in South Africa, and Saddam's sons, Uday and Qusay, would still be raping eight and nine-year-old little school girls in Baghdad. (Of course, if Whoopee and her anti-Iraq-War friends had gotten *their* way, they still would.) He also could have pointed out that our Constitution is the same document that *led* to the abolition of slavery. And then he could have shown Ms. Goldberg's oh-so-clever gotcha question to be the unschooled propaganda of the Left that it was.

Progressives goal is to "change history so that [they] can justify what [they] want to do in the present..."

"They reshape the past in order to make policies for the present. That's the definition of historicism ... [which] was brought over from Germany by the progressives. "Let's change history so that we can justify what we want to do in the present.""[1022]

And what those progressive Democrats want to "justify doing" in the present is to "radically transform" America into something far different than what she has always been. And that is what Satan wants as well, because that "radical transformation" will result in a country that is far less powerful economically and militarily and will then no longer be in any position of strength to continue its defense of the nation of Israel. And that is also why they are, and have been for quite some time, deliberately lying to our school children in our elementary and high schools, colleges and universities about the history of the founding and the progression of our great nation. Wake up, America. You are being torn down from within your own *educational* system, and you're waltzing out every two years, religiously, and voting for it to boot.

See:
- *Under God*, by Toby Mac and Michael Tait
- *The 10 Big Lies About America: Combating Destructive Distortions About Our Nation*, by Michael Medved
- *48 Liberal Lies about American History (That You Probably Learned in School)*, by Larry Schweikart
- "In God We Still Trust," by Dr. Richard G. Lee

1022 (Larry Schweikart on the *Glenn Beck Show*, Fox News Channel, 7/9/2010)

- "The Declaration of Independence"
- "The Constitution of The United States of America"
- "The Bill of Rights"
- "The Gettysburg Address"
- "The Ten Commandments"
- "God Save America: My Journal July 2010" by Dr. Michael Youssef and Leading The Way International
- *Seven Events That Made America America: And Proved That the Founding Fathers Were Right All Along*, by Larry Schweikart
- *A Patriot's History of the United States: From Columbus's Great Discovery to the War on Terror*, by Larry Schweikart
- *New Deal or Raw Deal? How FDR's Economic Legacy Has Damaged America*, by Robert P. Murphy
- *The Politically Incorrect Guide to the Great Depression and the New Deal*, by Robert Murphy
- *The Walls Came Tumbling Down: A Christian Perspective on the Fall of Communism in our Time*, by Peter J. Leithart & George Grant

5- Take over the educational system

"But unto the wicked God says, What right do you have to declare my statutes, or that you should take my covenant in your mouth? Seeing you hate instruction, and cast my words behind you." (Psalms 50:16-17)

In the Miss Teen USA pageant in 2007, Miss South Carolina was asked "Why can't one out of five Americans find the US on a map?" This was her answer:

"I personally believe that U.S. Americans are unable to do so because, uh, some people out there in our nation don't have maps, and, uh, I believe that our education like such as in South Africa and, uh, the Iraq everywhere like, such as and I believe that they should, our education over here in the U.S. should help the U.S., er, should help South Africa and should help the Iraq and the Asian countries, so we will be able to build up our future for our children."

Of course, not only is she remarkably stupid as in <u>uneducated</u> stupid, but the correct answer to the question is "because one out of five Americans (or perhaps a much higher percentage) are also just like her- remarkably stupid as in <u>uneducated</u> stupid." But the point here is that the reason they are all so uneducated stupid is: it is by design. It is not by accident. It is the <u>intent</u> of Satan and his Democrat-Left who have taken control of our public education system. The uneducated stupid, you see, can be reliably counted on to go out and vote Democrat. For the less informed an electorate is, the easier it is to manipulate them, to fool them, to pull the wool over their eyes, and to impress them with absolutely meaningless slogans like "Yes We Can!" And Satan can reliably count on the Democrats to do his will by proceeding along with their legislation of the destruction of this nation. (The "accomplishments" of the Obama, Pelosi and Reed administration, just in their first year and a half together, should be enough to permanently cripple this nation economically, and that's <u>before</u> they destroyed our health care system.)

"Ilya Somin, a George Mason University law professor and author of the eye-opening study *When Ignorance Isn't Bliss: How Political Ignorance Threatens Democracy*, said, "If you greatly increase voter turnout, especially among the young, who have particularly low levels of political knowledge, you will have an electorate that is on average even more ignorant that the one we have now." Ignorance, it just so happens, skews in a partisan direction, with new young voters overwhelming voting for Democrats."[1023]

Yea, makes you want to rush right out and "Rock the Vote," now doesn't it? We'd be far better off if someone came up with a program that paid them all to just stay home.

"Public schools are the Left's madrassas. ...At least the crazy Muslims get funding from Saudi Arabia for their madrassas. Liberals force normal Americans to pay for their religious schools."[1024]

"Not long ago, a young man named Dan Gelernter wrote a piece for the *Weekly Standard* about what it felt like to be a conservative in his upscale, overwhelmingly liberal public high school in Connecticut. "The teachers are predictable liberals," he wrote, but "the students are more worrying. ...Most lunch table liberals say they do not love America, and would not defend it." Try telling one of them that you get goose bumps when you hear "God Bless America," and he'll think you're hopelessly uncool (at best), maybe even a borderline fascist."

"Why? Because in Dan Gelernter's school, as in many others around the country, teachers trash America and its values as a matter of course. "As computer geeks used to say," Gelernter writes, "garbage in, garbage out.""[1025]

"Hopelessness is one of the dishes that the public schools serve our children. The secular schools serve up some serious atheistic-stew from the kitchen of Darwin. This is one of the reasons we have so many school shootings. Kids killing kids is a very complicated issue. Yet if students are told that they are no more than a "grown up germ," ultimately birthed by the slime of a primordial soup, they will act like slime. The schools that teach our kids that they are just "apes with Reeboks" are indirectly encouraging them to behave like the Gorilla's in the *Planet of the Apes*."[1026]

Our public schools, from the elementary level on up, have become reliably-left-wing indoctrination centers. And the end game of Satan and his Democrat-Left who are in control of the indoctrination is to graduate a generation of uneducated youth who will mindlessly support them and their America-destroying agenda. They brainwash them into seeing their political party as a positive force in the country,[(LIE)] and any political party or individual who opposes them as a highly negative, destructive force.[(LIE)]

"Public schools and universities — with curricula controlled by the social agenda of the teacher's union, the Democrats' most powerful activist bloc — are hotbeds of social engineering, anti-capitalist indoctrination, "diversity" training (where primacy is put on looking different but thinking alike), speech codes, and political correctness. Honest inquiry, debate, and dissent are rarely tolerated, a liberal failure of conscience that defrauds

---

1023   (From article "Suffer the (Political) Children," by Shawn Macomber, the American Spectator, October 2008)
1024   (Pg. 12, *Godless: The Church of Liberalism*, by Ann Coulter)
1025   (Pg. 293, *110 People Who Are Screwing Up America*, by Bernard Goldberg)
1026   (Pg. 57, *There Are Moral Absolutes* by Michael A. Robinson)

millions of young minds ... while pocketing their parents' tuition dollars.

"Portraying the country as a repository of injustice[LIE] American teachers pressure American students to despise America. The incomparable Founders (if they are mentioned at all) are reduced to and dismissed as racist, sexist "dead white guys" who inflicted pollution, consumerism, inequality, imperialism, and hegemony on the rest of the world. [LIE] Since the 70s, liberals have run the schools as anti-American propaganda factories, cheating generations of U.S. kids of their own profound, complex, and often heroic history — an unforgiveable moral failure."[1027]

In the beginning, just about every university in America was a Christian university. They were founded by Christian leaders and Christian philanthropists, who donated vast amounts of their resources and their money and their time to start these universities that would be honoring to God and that would study His world and His Word and pass on truth to their students. (Just as most hospitals were originally started by Christians.) But slowly over the years the Left has come to power in these universities and has turned them more into temples of Satan worship, where his many lies are also exalted.

"If you think of America's college campuses as a "marketplace of ideas" ...a kind of philosophical and cultural crossroads were young people like your children and grandchildren can listen, observe, and freely share their own beliefs with their peers and professors ...it's been a long time since you went to college. Many of today's university campuses are effectively closed to the Gospel, and aggressively intolerant of anyone who testifies to faith in Jesus Christ or speaks up—however respectfully—for the truth He lived and taught. Persecution of Christians is alive and well and growing in America's universities..."[1028]

The following is from an article by Abraham H. Miller, an emeritus professor of political science. It was written for *The Conservative Voice* on the subject of Evan Maloney's documentary "Indoctrination U," a scathing indictment of our "Orwellian reeducation camps" that pass for modern universities:

"Why send your child to a prestigious school that will cost a small fortune? For less than a quarter of the cost, your child can receive the same meaningless education at a local college or university. They have all the politically correct courses that the prestigious schools have. And in the shadow university, they have diversity and sensitivity training as mindless as any you can imagine. They boast a speech code (often euphemistically called a decency code) probably designed by some human rights authority that will rival that of any institution committed to stifling debate. ...Campus administrators may be venal, but their stupidity is grossly overestimated ...[Their] Political correctness and indoctrination are not for everyone. ...[for] the real work of the university continues in science, mathematics, engineering and even in music and art. [But in other departments it is a different story...] There is no difference between a department of political science, for example, and a department of ethnic studies. The dissemination of social myths based on political ideology is no different than the dissemination of myths based on race or gender. Administrators don't really care what goes on in these departments as long as the students don't riot. Students learn little of value in such departments and that fact is widely known. When a colleague once bemoaned to several prominent businessmen the egregious violations

---

1027   (From article "Liberal Failures," the Limbaugh Letter, Rush Limbaugh)
1028   (From an Alliance Defense Fund circular)

of academic standards in one such department, his despondency was greeted with hearty laughter. One of his area's leading employers simply looked at him and said, "we don't take degrees from that department seriously." ...The English department fights over what politically correct textbook is going to be adopted, and women's studies fight over whether they should recruit a visiting scholar who is an expert on the greater or lesser lesbian poets, and black studies insist that Jesus was black... After four years of reciting leftist nostrums, which could easily be gleaned from inanities offered free by Rosie O'Donnell, not only aren't liberal arts students prepared for the real world, but also potential employers know they are unprepared. What capitalist wants to hire someone who has spent four years learning that capitalism is the root cause of all the worlds' problems? OK, I grant that students do learn how to cheat, drink, and have casual sex, but are these really skills that are in demand in the economic marketplace? Wouldn't it be cheaper to give your child a bottle of Jack Daniels, a credit card and directions to a motel?

"Imagine the scholarships that could be underwritten if every politically correct division of a college or university administration were removed and imagine what could be done if all the ethnic and feminist "studies" courses were appropriately consigned to the dung heap where they belong. Indeed, it would be a step toward a real liberal arts curriculum where students learn to grapple with ideas and actually question what professors have to say. But that is not going to happen, because the interest groups using academic departments for identity politics, masquerading as legitimate fields of inquiry, won't let it happen. In the meantime, parents will find that they are viewed as the "great unwashed," as administrators refer to them. Their children are seen as the progeny of a racist, sexist, homophobic and imperialist society.[LIE] The political and religious values their children bring to campus are viewed as rooted in bigotry[LIE] (especially if they are Christian) and decidedly in need of change. And for some obscene amount of money, a group of ultra-leftist Stalinist wannabes at a prestigious university are going to indoctrinate them in the virtues of a dysfunctional political and economic system that no one except Fidel Castro and Jimmy Carter take seriously."[1029]

"I have more understanding than all my teachers; for your testimonies are my meditation." (Psalms 119:99)

And this is from an *Intercollegiate Studies Institute* "Problem Statement," January, 2010. It talks about our friends, the 60s radicals, and their subsequent takeover of academia:

"By the 1980s, these radicals had grown old without ever having grown up. They entered teaching and administrative positions within colleges across the country, and put their schemes into action at great expense both to the people who fund higher education and to the students who are supposed to benefit from it. Throughout these years of turmoil ... colleges across the country spent huge fortunes to jettison traditional courses of study and to replace them with radical propaganda. In one representative example, student journalists at Duke University... wrote about this trend at their schools English Department. ... [They] published a piece warning of "an onslaught of newly-moneyed nihilists whose tendentious epistemology is the academic equivalent of Nazi book-burning."

..."And, in addition to hijacking traditional fields of study such as English, campuses across this country created new disciplines from scratch, and staffed them with their radical comrades. The shotgun-toting Black Panthers of the '60s became today's "African American Studies" professors, while the feminists of yesteryear now populate today's

---

1029   (From article by Abraham H. Miller, for *The Conservative Voice*)

"Women's Studies" departments. Of course these departments have little use for an open-minded and disinterested pursuit of truth and knowledge. Instead they serve thinly veiled political agendas that allow no dissent from the party line. The result has been a bloated and costly university system that is fundamentally at odds with the ideals of American freedom."

The "Problem Statement" goes on to quote survey data from pollster Frank Luntz which documents…

…"the full extent of the political and intellectual bias that has gripped our universities. In his survey of Ivy League professors, Luntz found that members of left-leaning parties (Democrats, Greens, etc.) outnumbered members of right-leaning parties by more than 19 to 1 – even though the two sides were separated by just a few points among the general population in the most recent Presidential election. [And keep in mind that the title of this section is Satan's takeover of our educational system. 19 to 1! That's a takeover if ever there was one.] Liberal and somewhat liberal Ivy League professors outnumbered conservative and somewhat conservative professors 64% to 6%, or nearly 11 to 1. With such deeply rooted political bias infecting our academy, it has become nearly impossible for students to receive a fair and balanced education in civics or American history.

""Today's colleges and universities are not, to use the current buzzword, "diverse" places," reported the American Enterprise Institute. "Quite the opposite: they are virtual one-party states, ideological monopolies, badly unbalanced ecosystems. They are utterly flightless birds with only one wing to flap. They do not, when it comes to political and cultural ideas, look like America."

"In addition to creating a flourishing environment for careerists and politicized radicals, another result of [the Lefts takeover of] academia has been a sustained assault on the institutions and courses of study that traditionally have cultivated American leaders.

"In years past, a core curriculum of required courses served as a medium for teaching students about the underpinnings of American democracy. Students emerged from this course of study with a basic level of competency in subjects that every citizen should know about, including American history and civics.

"However, as America's investment in higher education has increased, radicals who dismissed such standards as "snobbery" actively worked to undermine this course of study– and they have largely been successful. A landmark study by the National Association of Scholars (NAS) called *The Dissolution of General Education*: 1914-1993 tracks the decline over the course of the 20th century."

…"The study concluded: The general education programs of most of our best institutions have ceased to demand that students become familiar with the basic facts of their country's history, political and economic systems, philosophic traditions, and literary and artistic legacies that were once conveyed through mandated and preferred survey courses."

"The consequences of this trend–both intended and unintended–have been devastating. In the absence of a core curriculum, it has been possible–indeed, quite common–for students to emerge from top universities with little or no knowledge of their own country."[1030]

Is it any wonder then that record numbers of brainwashed young people ran out in droves to vote for an unqualified jackass like Obama? Anyone with a rudimentary grasp of the issues could see right through him and his disingenuous rhetoric. But that was the natural end result of their Leftist programming. They did exactly what they were told to do. And of course the reason Obama is the leftist

---

1030   (From *Problem Statement*, Intercollegiate Studies Institute. January, 2010)

that he is, is also a result of the same America-hating, Leftist programming that he received. And the beat goes on...

"Socialists re-write Preamble to the Constitution... Ah, public schools - now many of them across the country are teaching our children that the Preamble to the Constitution mandates *"The People's basic needs must be met in a country. Needs for housing, education, transportation, and health care overseen by our government system."* This comes from an 'educational' program called Building Fluency Through Practice and Performance."[1031]

"Our children are being indoctrinated into the world of Marxism."[1032]

The following is from the Intercollegiate Studies Institute:

"In America's college classrooms, hostility towards traditional Western principles is threatening to rip our culture apart. With the brilliance of the Founders ridiculed and the traditions of Western Civilization rarely taught, the tipping point is closer than many realize. The following used to be unthinkable: Professor Robert Jensen of the University of Texas said "The United States has lost the war in Iraq, and that's a good thing. I welcome the U.S. defeat... it's essential that the American empire be defeated and dismantled." North Carolina State University professor Kamau Kambon told a gathering at Howard University, "We are going to exterminate white people because that in my estimation is the only conclusion I have come to. We have to exterminate white people off the face of the planet."

[Traitors and genocidal racists, yet they hold tenured professorships in our colleges and universities, and are paid handsomely for filling our children's heads with their psychotic rants. And there are many more where they came from.]

"These views are "normal" for far too many professors." We could go on about Yale University banning ROTC on campus, yet recruiting "a high-ranking ex-member of the Taliban" as a student; and "Edgewood College professor Kevin Barrett teaching what he calls the 9/11 Big Lie—that the 9/11 attacks were not orchestrated by Osama Bin Laden or murderous jihadists, but that the World Trade Center buildings were likely blown up by the U.S. government using controlled demolitions to start a war with radical Islam."[LIE] And there's professor Ward Churchill of the University of Colorado with his infamous "little Eichmann's" comment directed at the innocent victims of the 9/11 attacks. And Professor Nicholas DeGenova of Columbia University who "told the audience at a faculty led, anti-war "teach in" that he "wished for a million Mogadishu's" to visit our soldiers and Marines in Iraq."[1033]

And our Democrat media will never report this to the American people. If these professors were Republicans, conservatives, or Christians, and were spewing comparable lies, then it would be headline news, 24/7.

"Every day, radical leftists posing as professors at American universities and colleges, large and small, fill their classrooms with hate-filled, anti-American rants. Leftists in the media are becoming more and more strident in promoting a far left agenda under the guise of "objective news reporting." The goal of these leftists is undeniable: They are blatantly attempting to influence the outcome of elections by helping elect liberals. They are

---

1031   (From a Glenn Beck email)
1032   (Glenn Beck, on Fox News' the Glenn Beck Show, 05/04/2011)
1033   (From *The Intercollegiate Studies Institute*)

working feverishly to influence government policies by promoting one left wing program after another. They are committed to destroying conservatives while bolstering those on the far left who share their radical views."[1034]

These leftists, whatever they think their goals may be, are working hand-in-hand with their father Satan in his goal of "radically transforming" (aka destroying) America. And using their control of our educational system to confuse and deceive the minds of our youth is one of the most insidious and effective tools in his shed.

"Former football-star-turned-Army Ranger Pat Tillman died in combat in Afghanistan April, 2004. A University of Massachusetts newspaper published an editorial saying, "Pat Tillman is not a hero. He got what he deserved." It gets worse. The article continued: "He did die in vain, because in the years to come, we will realize the irrationality of the War on Terror and the American reaction to September 11."

What can I say. A mind is a terrible thing to waste; or to poison to the point where it actually thinks that evil is good, and good evil.

"A University of Northern Colorado class was told to write essays on "Why President Bush is a war criminal."[LIE] The student who refused and instead wrote how Saddam Hussein was a war criminal got an 'F.'"[1035]

The following is a conversation between Sean Hannity and David Horowitz about his book, *One-Party Classroom: How Radical Professors at America's Top Colleges Indoctrinate Students and Undermine Our Democracy*, by David Horowitz and Jacob Laskin:

David: "What's happened to our universities is …I estimate there are between 10 and 30,000 indoctrination courses, that is, courses designed that when you come out of them you have a radical, Marxist, anti-American, anti-capitalist, anti-free market point of view. You only get books on one side.

…"An Italian Marxist named Antonio Gramsci …saw in the 1930s that the proletariat, the working class, was not going to make the revolution. And so what he said was, you have to change the culture, the ideas. And that's why radicals went into the universities and became professors and then redesigned the courses. I described at the University of California, Bettina Abchecker, 20 years on the Central committee of the Communist Party, took over the Introduction to Woman's Studies course, made it political, the whole department of Feminist Studies. [She] was told by her other communist professors it's your revolutionary duty to do this.

…"[William Ayers] is an unrepentant terrorist [and a] compulsive liar. To this day, he says he was protesting the Vietnam War. [Yet] he was [still] setting bombs when Jimmy Carter was president, three years after all American troops were withdrawn.

Hannity: "Barack Obama said he's a well-respected professor."

David: "He was a moron. I knew him when he was a radical. He was an idiot, very shallow person, very dedicated to the destruction of this country."

…"They [leftist professors and administrators] recruit each other. You want a job? There's no more demonstrations? Come to academia. You can work six hours a week. You can make $100,000 a year. You have a lifetime job. You get to have four months paid vacation. Why not? And you have all these captive students that you can indoctrinate."

Hannity: "And this is the position the students are in, they've got to regurgitate the lines that they feed you, otherwise they risk getting poor grades, they risk getting into law school, medical school."

1034　(From *The Media Research Center*)
1035　(From *The Leadership Institute*)

David: "In every Women's Studies program in the country you have to believe in order to get a decent grade that gender, that is the differences between men and women that are not anatomical, is socially constructed, that is determined by society,[LIE] not by biology. Well, if you go to the biology department or the neuroscience department you'll find out that men and women are different, they're hardwired different [which] any five year old can tell you, but not in our universities."

Hannity: "But it gets serious as it relates to the issue of terrorism. You have a chapter *Uptown Madrassa Columbia University* where you go over some of the people that have been teaching there for many years."

David: "The Middle Eastern Studies departments are supporters of Hamas and Hezbollah during the Gaza war. Hamas is a genocidal, Jew-hating, America-hating, terrorist organization. Professors on campuses across the country, in conjunction with the Muslim Student Association, which is part of the Muslim Brotherhood and the Jihad, organized demonstrations to attack Israel. This is almost universal in Middle Eastern Studies departments and it's part of the political recruitment structure in the University."

Hannity: "And the net result of this is now we condition a generation of people in this country to not read the Constitution, the words of our Framers and Founders, but to actually believe that it is the responsibility of society to provide everything for them."

David: "And if the government provides everything, it can take everything away. It's a totalitarian plan. And if you wonder where all those Obama activists, all the ACORN activists, came from; they came through American universities, and were trained in courses like we describe."[1036]

"Why boast yourself in evil, O mighty man? The goodness of God endures continually. Your tongue devises mischief; like a sharp razor, working deceitfully. You love evil more than good; and lying rather than to speak righteousness. You love all devouring words, you deceitful tongue. God shall likewise destroy you for ever, he shall take you away, and pluck you out of your dwelling place, and root you out of the land of the living." (Psalms 52:1-5)

The Honorable Rick Santorum writing in an Intercollegiate Studies Institute memo reports:

"Following Sharia law, Harvard has a gym that now bans male students during set regular hours so that Muslim coeds can be segregated from males. Can you imagine the ruckus if exclusive gym time was set aside for Christians! ...Liberal revisionists are rapidly disconnecting young Americans from examining and learning the Judeo-Christian values of the West while simultaneously teaching incomplete, biased, and white-washed views of Islam- and much of this propaganda is paid for directly by Saudis who want their propaganda validated by major universities, such as Georgetown and Harvard to whom they have given 40 million dollars. ...While they teach a "tidy" view of radical Islam and Sharia law, they vilify America's traditional culture and heritage through wildly popular books such as Howard Zinn's Marxist inspired *A People's History of the United States*."[1037]

Israel Apartheid Week[LIE]

The following excerpt is taken from a speech delivered by David Horowitz. It

---

1036   (David Horowitz on Fox News *Hannity*, 03/09/2009)
1037   (From an *Intercollegiate Studies Institute* memo, by Rick Santorum)

is entitled The War Against the Jews at UC San Diego and appeared in a newsletter from Front Page Magazine.com.

"The last time I visited this campus was 40 years ago when I came to have lunch with Professor Herbert Marcuse, a malicious Marxist who made a reputation for himself and became a hero to the left by advancing the proposition that if you can claim to speak on behalf of the oppressed you have the right and the obligation to silence, and, if necessary, to obliterate people who disagree with you. This, of course, is the creed of every run-of-the-mill dictator on earth, and it is the creed of the progressive left today.

"It is appropriate — and not coincidental of course — that I arrive here during "Israel Apartheid Week" or, as I prefer to call it, "Hitler Youth Week." I use that reference because — as should be evident — what is going on here is part of a globally organized movement centered in the Arab Muslim Middle East to finish the job that Hitler started. Unlike the Nazis, who hid their "final solution" from ordinary Germans and the world at large, the Muslim radicals shout it from the rooftops, put it in their organizational charters and get applauded for doing so by millions of Muslims who want to see it happen. And the left does its job, as during the Cold War years, of conducting auxiliary campaigns to help the genocide along.

"Along with Israel Apartheid Week, these leftists will commemorate what the genocide-minded call the "Nakba," which is Arabic for "catastrophe" and is the name they have given to the creation of the only Jewish state in existence. The protests are genocidal intentions in themselves, seeking the obliteration of a people's homeland; and they are supported by this university which prides itself on its diversity and respect for others – apparently excepting Jews. The "Israel Apartheid Wall," which is the centerpiece of this genocidal week at UC San Diego, was created with $8,500 in student funds. Another $32,000 in student funds went to the sponsoring organization, the Muslim Students Association, for the hate speakers it brought in and other activities this spring. The hate week itself is supported by several university departments, including Ethnic Studies, Visual Arts and Literature, and by one of the colleges of the university, as it happens Thurgood Marshall College, named for a civil rights activist.

"It is also fitting that one of the speakers should be Angela Davis, an academic icon and a lifelong Communist who devoted herself to furthering the agendas of the most oppressive, blood-soaked regime in recorded history; and received for her services a Lenin Prize from the East German police state, the most wretched and ruthless of all the Soviet satellite regimes; who bought an arsenal of weapons for a young Black Panther, who attempted to take a judge and his courtroom hostage, blew off the head of the judge and got himself killed in the process.

"Today Angela Davis comes as the leader of a 20-year campaign to end what she calls the "Prison-Industrial Complex." This is a racist campaign to free every incarcerated criminal in the United States who happens to be of a darker skin tone. Her movement is actually doubly racist. First, because it singles out black and Hispanic criminals for "liberation," and second, because ninety-five percent of their crimes are committed against law-abiding black and Hispanic people. Angela Davis is a "University Professor," which means that she makes a six-figure income, is provided with a personal staff and is honored in a way that only a select few of University of California faculty are. A revered figure among the academic left, she makes scores of appearances every academic year and receiving $10,000 honoraria an appearance to spread her hatred of white people, of Jews, and of America to college audiences. This evening we have been thrust into a very sad situation.

"Ecclesiastes famously said, "There is nothing new under the sun," and the older I get, the more impressed I am by this observation. My own parents were card-carrying members of the American Communist Party, which meant that they were pawns, witting and unwitting, of the Soviet empire. They called themselves progressives and believed they had joined a movement to bring social justice to all mankind. In 1956, the leader of world Communism gave a secret speech about the crimes that Stalin had committed in the name of social justice, which included the murders of millions of people, including millions of Communists. He gave the speech because the post-Stalin leaders of the Soviet Union were afraid a new Stalin would emerge and slaughter them too. Like today's leftists, my parents hated the anti-Communists on the political right and called them liars when they confronted progressives with the crimes their hero had committed. The anti-Communists claimed that Stalin had killed 10 million people. In fact, when the truth was revealed, the figure was 40 million. That's interesting in itself: The right understates the truth, and the left overstates the lies, in the name of social justice.

"When the speech appeared on the front page of *The New York Times*, its impact was devastating to the progressive cause to which my parents had devoted themselves. Progressives were demoralized and unable to regroup. So their children — my generation — launched a new left. We fatuously believed that there could be a *new* left – one untainted by the crimes of the old and unlikely to repeat them despite the fact that our understanding of the world and our agenda for the future — social justice, meaning equality enforced by government — were fundamentally the same. The American founders understood and asserted that politically enforced equality would mean the end of liberty for all, a wisdom we ignored as the new left quickly revealed itself to be no different from the old, embracing Communists in Vietnam and Central America and eventually Islamic totalitarians in Gaza and the Middle East.

"About the time I was having lunch with Herbert Marcuse, I was also raising a considerable sum of money for the Black Panther Party in order to buy a church in East Oakland to house a Panther school. I was then the editor of *Ramparts*, the largest magazine of the left, and had recruited our bookkeeper, Betty Van Patter, to maintain the school's financial records because I foolishly believed our own propaganda that a "racist" federal government would shut down the school if the Panthers didn't keep their books properly. In December 1974, Betty Van Patter disappeared, and by the time the police fished her body out of San Francisco Bay, I knew the Panthers had killed her, and I also knew that every progressive friend I had was a potential threat to me and my four children, because if I said what I thought, members of that community would denounce me as a CIA agent and a racist, and I couldn't know what consequences might ensue.

"So for me Israel Apartheid Week where secular progressives join hands with theocratic fascists and disregard the realities in front of their eyes is a case of *déjà vu* all over again. On this campus we are witnessing a "protest" in support of a movement worse than the Nazis, which unlike Hitler broadcasts from the rooftops its plan to obliterate the Jews, to obliterate Israel and to obliterate America. Have you noticed a single Muslim country distancing itself from this plan or denouncing Ahmadinejad and Hizbollah and Hamas for proclaiming this solution to the problem posed by infidel Jews and Americans? Even *one* Muslim state out of the 57 that exist? When people say Islam is a religion of peace, I'm sure that's true for some individual Muslims, and have said so many times. But when you look at the organized Muslim community, the element that is willing to stand up for peace, that will defend the right of the Jewish state to exist, that is willing to pass a UN resolution condemning the genocidal terrorists of Hizbollah and Hamas is so inconsequential that it might as well not exist."[1038]

1038 (Excerpts from a speech delivered by David Horowitz, "The War Against the Jews at UC San Diego" printed in a newsletter from Front Page Magazine.com)

To say that what is happening, and has been going on for quite some time now on our college campuses, and in the name of *education*, is disgusting and alarming is a vast understatement. And it continues because of the cover afforded by a Democrat-Left media; a media full of people who were indoctrinated the same way and swallowed it all uncritically; and because of the millions of uninformed folks who go out and support the entire web of deceit with their votes.

Mr. Horowitz goes on in his speech to expose the fabrications of the Muslims and their Democrat-Left that permeate our universities. We have covered these Jew-exterminating lies earlier in this chapter, but his is such a good summary of the entire Arab-Israeli situation, and it exposes the ultimate purpose of Satan's control over our educational system—turning the next generation of American youth into Jew-hating useful-idiots of his who will grow up, take over the reins of power in this country and then side with whatever their Muslim deceivers tell them is the "right" thing to do about the Middle East's "Jewish problem." (Kind of smacks of Nazi Germany, does it not?) Because of this, and despite its length, we have decided to include it here in its entirety... (If only it were included as a part of the required curriculum in our universities, the tide might be able to be turned.)

"Now to the conflict in the Middle East which Israel Apartheid Week is supposed to illuminate. Since there are young people in this room I'm going to focus this lesson on a recent event — the 2005 Israeli withdrawal from Gaza. Gaza is a little strip of land bounded on one side by the Mediterranean and on the other by Israel. It is about six miles wide and 25 miles long. Across this stretch of land, Israel was attacked by Egyptian tanks in three aggressive wars supported by all the Arabs in the region – in 1948, 1967 and 1973. Since Gaza is a long corridor the support of its population for the anti-Israel aggressor gave Egyptian forces the opportunity to come within striking range of important Israeli cities. This is why in the aftermath of those aggressions Israel decided it was necessary for security reasons to station Israeli forces in Gaza.

"Not only did Egypt invade Israel three times across the Gaza strip, but after the first invasion it annexed Gaza, and did so with no protest from Palestinians or the Arab world, which now pretend to be so concerned about Palestinian land and a Palestinian state. Following the 1948 war, when Gaza was under Egyptian control, the 25 mile strip of land became the base for the first terrorist attacks against Israeli civilians. These were conducted by the *Fedayeen* who were created by Egypt's dictator, Gamel Abdel Nasser, and were led by the head of his secret police.

"These and later terrorist attacks by the PLO and Hamas strengthened Israel's resolve to take the step that any country would take to protect its citizens by stationing its own military in the Gaza strip until such a time as the Arab states and the Palestinians were ready to accept Israel's existence and live in peace. It has been more than sixty years since the creation of Israel and the Arabs have never been willing to accept Israel's existence and consequently there is no peace. No Palestinian entity recognizes the state of Israel. None. Israel has been continuously under attack by Arab regimes and Arab terrorists since its creation. The Islamic Republic of Iran, armed with Chinese rockets and soon to possess nuclear weapons has declared its intention to wipe Israel from the face of the earth. The Palestinian Authority, Hizbollah, and Hamas have made the identical intention clear so that no one can mistake it. These calls for genocide have not been repudiated by a single Muslim state, and no Muslim state has offered Israel's threatened Jews their support.

"Nonetheless, in 2005, Israel, in the interests of peace, unilaterally withdrew its military forces from Gaza, taking a risk unprecedented in the annals of nations. The events that followed the Israeli withdrawal tell you everything you need to know about the Palestinians and their cause, and about the nature of the conflict in the Middle East.

"Prior to Israel's withdrawal, there were 9,000 Jews living in Gaza along the Mediterranean coast. Nine thousand Jews out of a population of 1.2 million Arabs; nine thousand *law-abiding* Jews. These 9,000 law-abiding Jews created a horticulture industry, which employed 12,000 Palestinians and whose product accounted for 10% of the entire gross national product of Gaza. Consider that 9,000 Jews are only .02% of the Gaza population. Two hundredths of 1%. As it happens the entire state of Israel also exists on only .02% of the entire Arab land mass. But that is too much for the Arabs. And nine thousand Jews out of a population of 1.2 million Arabs was too much for them as well.

"Think about it. If the Muslims of Gaza – or if the leaders they have elected — were rational, decent, normal human beings their reaction to the law-abiding Jews living among them could have been: Let's import more Jews! Let's encourage more Jews to live here. Let's make Jews welcome in our country; they are creating so much wealth for our citizens.

"But that did not happen. Instead the Israeli government out of concern for the safety of the 9,000 productive, law-abiding Jews who lived there, evacuated them from Gaza. Some of these Jews were so attached to their homes in Gaza that they had to be removed at gun point. Why did the Israeli government do that? Because if they left them behind they would have been slaughtered. Everybody, including those individuals in this audience who disagree with everything else I say, should understand that. Palestinians elect regimes – both the PLO and Hamas – which organize terrorist attacks against men, women and children; which *target* men, women and children. No civilized people in the world does that. Everybody understands that if the Palestinians were disarmed today, there would be peace in the Middle East. Everybody understands that if the Jews in Israel disarm, they will be massacred. These are simple facts that every rational person can see provided their vision isn't impaired by religious hatred.

"It was even worse than these bare facts would suggest. A philanthropic Jew in America, Mortimer Zuckerman, the publisher of *U.S. News and World Report*, fully aware that the Jews of Gaza were going to be removed forcibly to save their lives because the Palestinians are so bloodthirsty and genocidal that Israel couldn't guarantee their safety, collected $14 million from American Jews to buy the 3,000 greenhouses and all the plumbing that went into this horticulture industry. They then *gave* it to the Palestinians as a gesture of peace. And on the day that Hamas took control of Gaza, the Palestinians destroyed all the greenhouses, all the plumbing, and started firing thousands of rockets into unarmed towns and schoolyards in Israel.

"Every Palestinian rocket is an anti-civilian rocket because they are rockets that can't be aimed. The Palestinians fire their rockets at random in the hopes that they will kill Jews. Israel's rockets, by contrast, are precision weapons that are aimed solely at military targets in an effort to save civilian lives.

"The lesson of the Gaza Strip is that the Palestinian jihadists are so full of hate that they are willing to destroy 10% of their own gross national product because it was created and donated by Jews. Israel, on the other hand, was forced to evacuate the people who created this wealth to save them from being massacred because they were Jews. That tells you everything you need to know about the conflict in the Middle East.

"Here's another instructive fact about Gaza. After the 1948 war, the entire Gaza Strip was annexed by Egypt. It disappeared from the maps. The same thing happened to the West Bank, which was annexed by Jordan. Yet there was no protest from the Arabs or the "Palestinians." So much for the desire for Palestinian self-determination.

"The so-called Palestine Liberation Organization[LIE] was created in 1964, fifteen years *after* the creation of Israel. This was the first time since the Arabs invaded the region in the 7ᵗʰ Century that there was any hint there was a people who identified themselves as "Palestinians" and desired a Palestinian state.

"The charter of the newly created PLO which proclaimed the goals of the "liberation" movement failed to mention either the annexation of the West Bank or the Gaza strip. There was not a word in it about the liberation of Gaza from Egypt. There was not a word about the liberation of the West Bank from Jordan. Instead the PLO charter called for the elimination of the "Zionist entity." From its beginning the Palestinian liberation movement has been a movement driven by Jew hatred. It is not designed to liberate the Palestinians by winning them a state; it is designed to "liberate" *Israel* and the Arab Middle East from the Jews, to expunge an infidel presence from the Muslim *umma*. Hamas and Hizbollah and the PLO do not want a Palestinian state. They want an Islamic empire, an infidel-free zone stretching "from the Jordan to the sea."

"Now allow me to deal with some of the lies perpetrated by the Muslim Students Association and their "Israel Apartheid Week" on this campus. On the "apartheid" wall of hate funded by this university there is a map that pictures Israel, Gaza and the West Bank. It is labeled "occupied Palestine."[LIE] It is a lie.

"There is no occupation of a country called "Palestine." There were no Palestinian lands originally to be stolen. Israel was created in the same way that Jordan, Syria, Lebanon, and Iraq were created – out of the ruins of the Turkish Empire. The Turks are not Arabs. They had ruled the entire region for 400 years since the 16ᵗʰ Century, until they joined the powers that were defeated in World War I.

"At the end of the war, the victors – Britain and France — divided up the spoils, in this case the defeated Turkish empire. The "Palestine Mandate," which was part of these Turkish spoils did not refer to a people but to a geographical region. The people in the region for the previous thousand years called themselves "Arabs" not "Palestinians." The word "Palestine" is not even an Arab word. It is Roman in origin. When the Romans drove the Jews out of their homeland, Judea, in the first century CE, they renamed it after the Jews' enemies, the Philistines, who were Europeans, not Arabs. Hence the name "Palestine."

"The claim that Israel is "occupied Palestine"[LIE] is a lie of Hitlerian proportions and has the same genocidal intention. And it is funded by the University of California on this campus.

"A second version of this lie is focused on Jewish settlements in the West Bank, as though these are an extension of the so-called "occupation." The hate week displays put up by the Muslim Students Association, show the West Bank dotted with the settlements of Jewish interlopers.[LIE] Let me put to you this question: If you are not a xenophobe or a racist what is wrong with Jewish settlements – with Jews settling on the West Bank or in Gaza? Why is a Jewish settlement a problem? There are more than a million Arab Muslims settled permanently in Israel as Israeli citizens. They have more rights than the Muslims of Gaza, more rights than the Muslims in the West Bank, more rights than the Muslims in Iran, more rights than the Muslims in *any* Muslim country. Jews welcome Muslims as settlers in their country. But Arab Muslims do not welcome Jews. They are xenophobes and Judeo-phobes, and that is the only reason why Jewish settlements are a problem for them.

"Consider that the Jews of Israel have been attacked in a single generation in three aggressive wars designed to push them into the sea. By every precept of international law they had – and have — the right to annex those territories from which they were attacked and expel the entire population which joined the aggressions.

"That's exactly what happened to the Germans after World War II. East Prussia was the industrial heartland of Germany. Ethnic Germans had inhabited it for 1,000 years. But when Germany invaded Poland twice in a generation, the allied powers said, enough. As part of the peace that followed the war, they expelled 12 million Germans from East Prussia and gave the land they had inhabited to Poland. There was no international outcry over that.

"The Israelis had every right to take the same measures to secure peace in the Middle East in 1948 and 1967. In 1948 they could have expelled all the Arabs who had not left Israel, the way the Arabs expelled the Jews living in their countries in the Middle East. In 1967 they could have expelled all the Arabs living in Gaza and all the Arabs on the West Bank. But they didn't. They preferred to live in peace with their Arab neighbors. The Israelis extended their hands to them and even when their peace offers were rejected built universities on the West Bank and made it a booming economy – an economy the PLO has since destroyed. And now the Israelis are being threatened with extinction by these same people as a reward for their generosity.

"A third lie featured in the Muslim-sponsored hate fest this week is that there are five million Palestinian refugees who should be returned to their homes in Israel. The Arab records show quite clearly that when the Arab armies invaded Israel, the Arab governments called on those Arabs living in the area to flee and promised them they would be returned to their homes by the victorious Arab armies. In war, people sometimes flee because they are scared; some were undoubtedly driven out. Eventually there were an estimated five or six hundred thousand Arabs who fled. But there were also six hundred thousand Jews who were driven out of Iraq, Morocco, Tunisia and other Arab lands where they had lived for centuries. The largest ethnic community in Baghdad until the 30's, when Iraq joined the Nazis, was the Jews.

"The Arab propagandists and their University of California supporters now claim that there are more than five million Palestinian refugees,[LIE] even though there were only a tenth that many originally and most of them have died. But why are there any Palestinian refugees? It is sixty years later. There are no longer any Jewish refugees, because they were all re-settled by Israel. The Palestinian refugees were not re-settled by the Arabs, not even in Jordan whose territory consists of 80% of the original Palestine Mandate and whose population is 70% made up of Palestinian Arabs. Jordan by the way is ruled by a Hashemite minority, yet there are no calls for the self-determination of Palestinians in Jordan. Palestinian Arabs are not *allowed* to settle in other Arab states. They are barred from entering.

"In other words, the reason there are Palestinian refugees is because the PLO, Hamas and the Arab states want them to be refugees — want them to be confined to miserable camps, where they can use them for propaganda and as cannon fodder for the war to cleanse the Jews from the Muslim Middle East. Indeed the refugee camps are bases for the terrorist armies of the Islamic jihad. The United States and Israel, along with other countries, have poured billions of dollars into the refugee cause and it ended up in the Swiss bank accounts of the real oppressors of Palestinians – the leaders of the PLO and Hamas. Yasser Arafat died with a Swiss bank account worth $35 billion. That was money donated by Jews and Americans and taken out of the pockets of Palestinians.

"The centerpiece of the Muslim sponsored hate week on this campus is the so-called Apartheid Wall[LIE, LIE] which is itself two lies. The wall being protested – the "wall" in Israel — is not a wall but a security fence, and it is not an "apartheid" fence because it is not designed to enforce a separation of races or ethnicities. There are, as noted, a million Arabs living in Israel with more rights than the Arabs of any Arab country.

"The fence was not erected to keep Arabs out. It was erected to keep terrorists out.

The West Bank and Gaza are terrorist camps whose governments honor, support, and orchestrate a terrorist war against the ordinary citizens of Israel. What if the Jews took down the wall? We know from the past exactly what would happen. Israel's security fence did not appear out of the blue. The wall was a desperate last measure the Israelis devised after years of watching their children slaughtered by Palestinians in discos and at Sbarro pizza parlors who were then honored as heroes and martyrs in the wake of their crimes. The fence has stopped virtually all of these attacks. So what the hate week on this campus is really calling for when it calls for the removal of the fence is a renewal of terrorist attacks against the children of Israel.

"There are whole departments of this university that are sponsoring this hate week and thus the war against the Jews it encourages, including the Visual Arts Department, the Literature Department and the Ethnic Studies Department. The Thurgood Marshall College is another official entity sponsoring these incitements and lies. If you look at the codes this university claims to live by, you will see that chief among them is respect for diversity – for the ethnicities of students who attend this school. There is no respect for Jewish students at this campus when a week of hate like this is thrust in their faces courtesy of university faculties and administrators.

"There are thirty campuses across the nation hosting Israel Apartheid Weeks this spring, including the University of California — Irvine, UC Berkeley, Boston University, Brandeis. Brown, University of Wisconsin, University of Houston, Brooklyn College, University of Chicago, UC Santa Barbara, UC Santa Cruz, UCLA, DePaul, Columbia, University of Illinois, University of Minnesota, University of Washington and others.

"Behind each and every one of these hate weeks against the Jews is the Muslim Students Association. Many people on this and other campuses mistake the Muslim Students Association for a cultural organization that represents all Muslims. It is no such thing. The Muslim Students Association is a sister organization of the terrorist organization Hamas, and like Hamas, is part of the Muslim Brotherhood network.

"Hasan al-Banna, the founder of the Muslim Brotherhood and the architect of terrorist jihad was an admirer of Adolf Hitler, whose organization translated *Mein Kampf* into Arabic. The father of Palestinian nationalism, Haj Amin al-Husseini, was one of Al-Banna's heroes and is revered to this day by the Palestinian Authority and Hamas as the father of Palestinian nationalism. Haj Amin al-Husseini was a Nazi. In the twenties and the thirties he preached the extermination of the Jews and inspired two celebrated massacres of Jewish settlers. During the Second World War he went to Berlin to work with Hitler to recruit Arabs to Nazism. He devised his own plan to create an Auschwitz in the Middle East and was thwarted in setting up his death camps only because Rommel was defeated at El-Alamein. After the war, he and al-Banna led the Arab crusade against the creation of the Jewish state.

"Why is the Muslim Students Association that violates the diversity principles and ethical codes of every one of these universities allowed to sponsor hate weeks against Israel and the Jews on these campuses? Where is the outrage over the lies the Muslim Students Association spreads along with its incitements against the Jewish state? Shame on the University of California for its role in this event. Shame on Thurgood Marshall College and the faculties that sponsored it. And shame on the Muslim students who use the shield of their religion to advance the Islamic war against the Jews."[1039]

1039   (From "The War Against the Jews at UC San Diego," David Horowitz, frontpagemagazine. com)

Again, we include this part of Mr. Horowitz' speech in this section on Satan's takeover of our educational system because it clearly shows his primary goal of destroying Israel, and his secondary goal to destroy America in order to eliminate that protection. It also clearly documents the level of stupidity and naiveté that exists in much of our universities faculty, staff, administrations, and in many of their students. Satan and his Muslims and his Democrat-Left have taken over our universities and with their bald-faced, racist, Jew-hating lies are indoctrinating the next generation of Americans to side with them in their quest to eliminate Israel and exterminate all Jews. "Heil Hitler!" Thank God we have the promise of the Creator of the universe, Jesus Christ, that he will never allow them to succeed.

See:

- *One-Party Classroom: How Radical Professors at America's Top Colleges Indoctrinate Students and Undermine Our Democracy*, by David Horowitz;
- *There are Moral Absolutes*, Michael A. Robinson
- *Indoctrination U.: The Left's War Against Academic Freedom*, by David Horowitz
- *The Professors: The 101 Most Dangerous Academics in America*, by David Horowitz
- *The New Thought police: Inside the Left's Assault on Free Speech and Free Minds*, by Tammy Bruce
- *The Church of Liberalism: Godless*, by Ann Coulter
- *110 People Who are Screwing up America*, by Bernard Goldberg
- *Class Warfare: Inside the Fight to Fix America's Schools*, by Steven Brill
- "The Political Assault On America's Universities," by David Horowitz
- The *Intercollegiate Studies Institute*
- The *Media Research Center*
- "Stupid in America," A Fox News Special with John Stossel. It exposes the government education bureaucracy and the teachers unions (and not a lack of $funds$) as the very cause of the problems in our public schools and the failure of our students there to learn.
- *IndoctriNation: Public Schools and the Decline of Christianity in America*, A Film by Joaquin Fernandez and the Gunn Brothers. Wow! This film is overflowing with fascinating yet little-known historical information about our public school system and how it relates to today that will surely blow you away.

6- Take over the entertainment industry

"God spared not the angels who sinned, but cast them down to hell, and delivered them into chains of darkness, to be reserved unto judgment; And spared not the old world, but saved Noah the eighth person, a preacher of righteousness, bringing in the flood upon the world of the ungodly; And turning the cities of Sodom and Gomorrah into ashes condemned them with an overthrow, making them an example unto those who afterward should live ungodly." (2 Peter 2:4-6)

"Would you like to swing on a star? Carry moonbeams home in a jar? And be better off than you are? Or would you rather be a pig? A pig is an animal with dirt on his face.

His shoes are a terrible disgrace. He has no manners when he eats his food. He's fat and lazy and extremely rude. But if you don't care a feather or a fig. You may grow up to be a pig"[1040]

Those are some lyrics from a Bing Crosby song three quarters of a century ago, but they are just as timely today. I would add, however, one more verse for today's youth, "A pig's attitude is ignorant and crass, and his pants hang down below his ass." Wouldn't it make a great rap song? Or is there something wrong with acting like an actual civilized human being? (In between dustin' cops and bustin' ho's of course.) And no, sorry, can't accuse me of racism, for an equal percentage of white youth have fallen into the same beltless pit.

We have covered the results of Satan and his Democrat-Left's takeover of the entertainment industry extensively in three sections of chapter 2—the "Art" world, the Entertainment Industry and the Music Industry—so we would refer the reader back to that information, and the recommended books at the end of that chapter.

Satan's takeover of this industry is tied in closely with his takeover of the educational system. His goal in both is to indoctrinate the youth of this nation with his left-wing lies and his immorality so they will grow up to be in no position to carry on the traditions and the Christian culture which have made this country exceptional and unique. Movie and TV stars, comedians and celebrities are a great attraction to, and therefore have a great influence on, many in our culture, especially on our naïve and impressionable young people. And the primary deception is that it gets them to think of what they hear and see constantly in our entertainment culture—the immorality, casual sex, filthy language, ignorant attitudes, disdain of decency, disrespect for parents (especially the father), contempt for authority, mockery of God, his Christ and his followers, and outright hatred for traditional Americans—as normal and acceptable. This is the great deceiver's powerful subliminal message to our children that sinks deeply into their subconscious, thanks to the Democrat-Lefts control of our entertainment industry. Again, we've come a long way from the days of Ozzie and Harriet, Andy of Mayberry, and Leave it to Beaver. Now we have Ozzie and His Foul-mouthed, Undisciplined Family, Will of Gayberry, and Love it and Leave the Beaver. Along with Saturday Night Live and all the other totalitarian-left, late-night comedy clowns whose underlying "hilarious" message is that you are a retard if you don't vote baby-mutilating Democrat, religiously, just like them. "And just look at all of those stupid, ignorant Christians, Republicans and conservatives.[(LIE)] Hardy-har-har."

What a stark comparison we also have between our old movie stars...

...and the anti-American movie stars of today. ...In contrast to the ideals, opinions and feelings of today's Hollywonk, the real actors of yesteryear loved the United States. They had both class and integrity. With the advent of World War II, many of our actors went to fight rather than stand and rant against this country we all love. They gave up their wealth, position and fame to become servicemen and women, many as simple enlisted men. 18 of these men came home with over 70 medals in honor of their valor, spanning from Bronze Stars, Silver Stars, Distinguished Service Crosses, Purple Hearts, and one Congressional Medal of Honor. So how do you feel the real heroes of the Silver Screen

---

1040 (From song "Swinging On a Star," lyrics by Johnny Burke & Jimmy Van Heusen, performed by Bing Crosby, 1944)

acted when compared to the Hollywonks today who spew out anti-American drivel as they bite the hand that feeds them? Can you imagine these stars of yesteryear saying they hate our flag, making anti-war speeches, marching in anti-American parades, and saying they hate our president [Bush]?"[1041]

Of course you can't. And neither can I.

"Hollywood no longer reflects—or even respects—the values of most American families. On many of the important issues in contemporary life, popular entertainment seems to go out of its way to challenge conventional notions of decency. For example:

"-Our fellow citizens cherish the institution of marriage and consider religion an important priority in life; but the entertainment industry promotes every form of sexual adventurism and regularly ridicules religious believers as crooks or crazies.

"- In our private lives, most of us deplore violence and feel little sympathy for the criminals who perpetrate it; but movies, TV, and popular music all revel in graphic brutality, glorifying vicious and sadistic characters who treat killing as a joke.

"- Americans are passionately patriotic, and consider themselves enormously lucky to live here; but Hollywood conveys a view of the nation's history, future, and major institutions that is dark, cynical, and often nightmarish.

"- Nearly all parents want to convey to their children the importance of self-discipline, hard work, and decent manners; but the entertainment media celebrate vulgar behavior, contempt for all authority, and obscene language—which is inserted even in "family fare" where it is least expected."[1042]

"And you, Hollywood, which art exalted unto heaven, shall be brought down to hell." (Matthew 11:23) (Did that say *Hollywood*? Oops, *Capernaum*, it was supposed to say Capernaum. Typo.)

Remember from our discussion in chapter 1, "Truth," that *Postmodernism declares that truth itself is the "great oppressor." And therefore they must subvert all truth. It also demands the tearing down, the "transgression," of all boundaries, moral and otherwise, in order to show that boundaries are arbitrary. "Absolute liberation from suppression" is their battle cry.* And Satan is the inspiration behind this academic gobbledygook, graduating "artists," comedians and entertainment moguls not only incapable of comprehending morality, but having an outright, self-righteous aversion to it. And it is all a part of his master plan to destroy America.

See:
- *The Death of Right and Wrong: Exposing the Left's Assault on Our Culture and Values*, by Tammy Bruce
- *Hollywood VS. America*, by Michael Medved
- *Slouching Towards Gomorrah: Modern Liberalism and American Decline*, by Robert H. Bork
- *Culture Warrior*, by Bill O'Reilly
- *The Church of Liberalism: Godless*, by Ann Coulter
- *The ACLU VS America: Exposing the Agenda to Redefine Moral Values*, by Alan Sears and Craig Osten

---

1041   (Source: internet email)
1042   (Pg. 10, *Hollywood vs. America*, by Michael Medved)

- *110 People Who are Screwing up America,* by Bernard Goldberg
- *Primetime Propaganda: The True Hollywood Story of How the Left Took Over Your TV,* by Ben Shapiro

7- Take over the criminal justice system

"Now is the judgment of this world come; now the ruler of this world will be cast out." (John 12:31)

But before he is cast out, Satan will continue to have his way with the help of so many of his unwitting followers. In our own country his far-left inductees have assisted his rule by infiltrating our legal system and twisting it to serve both of their designs. Starting in Law school, potential judges are taught the precepts of the far-left. And because of that programming many turn out to be baby-mutilation supporters (the vast percentage of lawyers and judges vote for the Democrat Party) having bought into the lie that killing a baby is a woman's *right.*[(LIE)] And God therefore gives them over to a reprobate mind, one committed to the advancement of Satan's Democrat-Left agenda, and one lacking the common sense or the rationality to dispense justice or, in many cases, even sanity from the bench.

"The biggest myth about judges is that they are somehow imbued with greater insight, wisdom and vision than the rest of us; that for some reason God Almighty has endowed them with superior judgment about justice and fairness. But the truth is that judges are men and women with human imperfections and frailties. Some have been brilliant, principled and moral. Others have been mentally impaired, venal, and even racist."[1043]

Attorney General Eric Holder comes immediately to mind. Although he is not a judge but a left-wing lawyer—which is a Constitution-perverting judge in the larvae stage. That he has remained our Atty. General after his irrational escapade in office[1044] is a testament to the ignorance and arrogance of both Obama and his media.

Trial lawyers are known as the parasite class, and too many actually are. Preying upon innocent victims and denying them justice by perverting the system to improve their win/loss percentages. (See "The Courts" section in chapter 2) But what is even worse is that many of these same ethically challenged lawyers become our judges! And then they are free to pervert the system from a position of far greater power.

"Federal judges are seizing power in this country – writing laws from the bench, instead of interpreting laws as the Constitution mandates.

"This is judicial tyranny – nothing less. Federal judges are appointed for life – and because they are not elected to office, they often become unaccountable to the electorate and out of control in their decisions."[1045]

1043   (Pg. 1, *Men in Black: How the Supreme Court is Destroying America*, by Mark R. Levin)

1044   (Which now, winter of 2012, includes lying to congress about his lack of knowledge of the "Fast and Furious" Mexican gun-running affair, which as it turns out was nothing more than another left-wing ploy to build a case for more gun control; and that an ATF agent is dead because of it will not deter "I-could-care-less" Holder or his Obama in the least, being as how their satanic agenda trumps all other concerns, including those of a grieving family)

1045   (From "Truth and Tyranny Report," a pamphlet from Coral Ridge Ministries)

Judicial activism

"He who walks in uprightness fears the Lord; but he who is perverse in his ways despises him." (Proverbs 14:2)

Judges take an oath to uphold and defend the Constitution of the United States. There are those who take that oath seriously. There are far too many others, on the Democrat-Left-progressive side, who make a mockery of it. The word "unconstitutional" when used by left-wing judges in order to strike down a whole host of legitimate and completely constitutional laws passed either by state or congressional legislatures, or voted into law by state ballot propositions, has nothing whatsoever to do with the actual Constitution. That is just a ruse. It has to do with their left-wing, progressive, Democrat ideology straight up and nothing more. It would be far more accurate and honest if it were reported that the constitutional laws and ballot propositions that were struck down by them were not really unconstitutional according to the clear meaning of that document; but that they were struck down because they were "un-liberal" and "un-progressive" and "un-Democrat-Left." But that would take some accurate reporting from something other than a state-run, Democrat media. "Constitution" is just a word that they throw around to make them sound like something they are not. They want people to think of them as Americans who respect the Constitution of the United States and care about upholding and defending it, rather than who they actually are - traitors to that same Constitution who want to fundamentally transform this country into something that the founders never intended it to be, and who will trample that same Constitution whenever it gets in the way of their goals.

There is nothing in our Constitution that gives homosexuals or lesbians the "right" to be married. Nor women the "right" to mutilate babies to death in their womb. But these are declared "constitutional"(LIE) by ACLU-Democrat-Left judges. Nor is there anything in our Constitution prohibiting our children from praying, or reading the Bible, in school. Nor is it in there that the Ten Commandments cannot be posted in Judge Roy Moore's courthouse in Alabama. Nor can it be found in our Constitution that terrorists have the same legal rights as US citizens. (But it can be found in the ACLU's playbook.) Nor that it is "unconstitutional" for a state, Arizona, to defend itself and its citizens by securing its border and by arresting those individuals who are not there legally. All of these things are clearly constitutional. They are all, however, clearly un-liberal, and un-far-left, and un-*progressive*, and un-Democrat-Left.

To take an oath to uphold and defend the Constitution of the United States against all enemies, foreign and domestic, and to then deliberately break that oath in order to further your father's Democrat-Left agenda, is treason; plain and simple. These judges are not just "activists"—which they clearly are—they are domestic enemies, attacking and shredding and perverting the Constitution to fit their *superior* ideological mindset.

""The practice of judicial activism – legislating from the bench – is now a standard for many federal judges," says Congressman Ron Paul of Texas.

""[They treat] the Constitution as fluid and malleable, to create the desired outcome in any given case ... with the federal judiciary focused more on promoting a social agenda than upholding the rule of law, Americans find themselves increasingly governed by men they did not elect and cannot remove from office."[1046]

1046   (From "Truth and Tyranny Report," a pamphlet from Coral Ridge Ministries)

These judges also regularly make laws that go against the will of the American people and that therefore would have little or no chance of being passed by our legislature, the ones who are endowed by the Constitution with the power to pass laws. 82% of the people are OK with prayer at public school graduation ceremonies. 76% are OK with the Ten Commandments being displayed on government property.[1047] But both have been deemed "unconstitutional" by Democrat-Left judges.

"Is judge-made law legal?" No, it is not. Not according to the Constitution. Nevertheless, we are ruled by judge-made laws in this country.

> "There is a faction within the country that believes the Constitution is a living document. And because it's a living document, some believe you can torture it 'till it says whatever you want it to."[1048]

And since that faction is unwilling to go to the people to change our constitution with amendments, because they would never win, they prefer instead to appoint judges who will merely ignore it in order to suit their anti-American and anti-Constitutional agenda. And of course half of this country continues to go out like they are told, and vote for it.

> "The Supreme Court endorses terrorists' rights, flag burning, and importing foreign law. Is that in the Constitution? But these days the Constitution is no restraint on our out-of-control Supreme Court. The Court imperiously strikes down laws and imposes new ones purely on its own arbitrary whims. There's a word for this: tyranny."[1049]

There's another word for it as well: treason.

"The Constitution is whatever the judges *say* it is,"[(LIE)] said former justice of the Supreme Court, Charles Evans Hughes. Senator Charles Schumer, a Democrat, said: "The Supreme Court makes law."[(LIE)] (Unfortunately, Chuck, that *has* been the case all too often, but they're not *supposed* to do that, not according to our Constitution. Call your local community college. They might be offering a refresher course in civics.) And Democrat Al Gore, former Vice President under Bill Clinton, added this when he was unsuccessfully (Hallelujah!!) running for the presidency: "I would look for justices of the Supreme Court who understand that our Constitution is a living and breathing document,[(LIE)] that it was intended by our founders to be interpreted in the light of the constantly evolving experience of the American People."[(LIE)] No, Alvin, you pompous, condescending ass, it wasn't. It was intended to be interpreted <u>exactly</u> as it was written. Otherwise, why bother writing it? Constitutional <u>Amendments</u> are intended to be used if you want to <u>re-write</u> it "in the light of the constantly evolving experience of the American People."

> "Liberalism has failed to keep faith with the U.S. Constitution, preferring "emanations" and "penumbras" (that liberals pretend support abortion) to the clear words in the document — the right to bear arms, for instance —designed to preserve the liberty of the citizens and limit the power of the government."[1050]

Thomas Jefferson himself warned that: "The Constitution will become a mere thing of wax in the hands of the judiciary which they may twist and shape into

---

1047   (From a Fox News/Opinion Dynamics Poll, 11/30/2005, as reported in "Truth and Tyranny Report," a pamphlet from Coral Ridge Ministries)
1048   (Jack Kinsella, The Omega Letter, from *Shadow Government DVD*)
1049   (From back cover, *Men in Black: How the Supreme Court is Destroying America*, by Mark R. Levin)
1050   (From article "Liberal Failures," The Limbaugh Letter, Rush Limbaugh)

any form they please." That is, of course, if unscrupulous, Constitution-ignoring judges with an agenda diametrically opposed to that Constitution were to be appointed.

"Most Americans believe the Constitution is to be interpreted literally, not as a *living document* open to the whims and vagaries of a panel of judges who have never worked an honest day in their lives."[1051]

"Since the judicial heyday of Earl Warren, Americans have grown accustomed to federal court decrees that seemingly come out of nowhere to impose radical social changes on a baffled nation. ...Typically, the outcomes mandated by the jurists couldn't have been gained through legislative methods, which is why the cases wound up in the courtroom. ...In many instances, they dictate a single top-down standard for the country, overriding state and local customs. ...And they have less evident linkage to the Constitution allegedly being construed then to axe-grinding academic theories as to how the nation should be governed. ...In a nutshell, our ingenious jurists and their civil-libertarian allies have stood the First Amendment on its head. Using provisos intended to protect the states from federal interference as a pretext for such interference, they have established as our national religious creed a cult of secularist irreligion."[1052]

A liberal judge told e-harmony, an online dating service started by an evangelical Christian, that it has to also set up a service for homosexual dating. (Called "e-bominy.") To not do so would be, you guessed it, *un*-constitutional.[(LIE)]
And now in March, 2010, Barack Obama is in the process of nominating a federal judge, one of his more "moderate" selections, that ruled it was "unconstitutional" to have an 18-hour cooling off period between the time that someone seeks an abortion to when it can actually be performed so the prospective victim has some time to look at all of the facts and consider all of the ramifications. But this "moderate" judge said that was "unconstitutional."[(LIE)] Of course it is not. It's un-liberal. It's un-Democrat-baby-mutilating-Left. After all, women might then have time to change their mind and actually give birth to the baby, and that is an "evil" that the baby-aborters cannot allow to happen. Yet Obama would have us believe that he wants "fewer abortions." He must think we're stupid. (Either that or he thinks his voters are.)

Judicial Tyranny

"By the time Ronald Reagan came to Washington, the Supreme Court was perhaps the most powerful agent of social change in America. It redrew state electoral maps, bussed students away from home to meet racial goals, discovered new rights for criminals, banned organized prayer in public schools, and in Roe vs. Wade declared that the Constitution contained a right to an abortion."[1053]

Notice that statement, "the Supreme Court was perhaps the most powerful agent of social change in America." The only problem is, IT'S NOT SUPPOSED

1051   (Source unknown)
1052   (From article, "The True Wall of Separation," by M. Stanton Evans, the American Spectator, April 2007)
1053   (From, "Fox News Reporting: The Right, All Along, The Rise, Fall and Future of Conservatism, Part 4" with Brit Hume)

TO BE! The founders never intended for a tiny little group of nine unelected lawyers to have the power to dictate to the entire nation what they were and were not going to be able to do or say. That is not "government of the people, by the people and for the people." That is the same government by the elitist few that the original colonists risked life and limb, and family and friends, and all of their worldly possessions, and fled the old country in droves to get away from. And that government by the elitists few imposed upon this nation by Constitutional traitors and their Democrat-Left supporters has been going on for far too long. And it is getting worse every year with each and every Democrat appointment of another judicial tyrannist like Darth Bader Ginsburg, Sonia Sotomayor and Elena Kagan who have no business being appointed as judge of the local traffic court much less the Supreme Court, where they can continue to reap incalculable harm to our system of government by imposing their will, the will of the tiny leftist minority, on an entire nation of "we the people." And of course we can all thank the Democrat-Left media for never reporting to the American people just what was being stolen from them over the last half century—their power that was given to them by our Founders. And that is because our "main-stream," totalitarian-left media could care less about the will of the people or their elected legislators. They care only about the radical transformation of America into something it has never been, and was never intended to be; a communist satellite.

Impositionalist judges are those who want to <u>impose</u> their will on the American people from the bench. They are those who base their legal opinions on what they <u>want</u> the constitution to say, rather than on what it actually says. Our courts are full of them. And now increasingly our judiciary is being infested with what are called *trans-nationalists*. These folks believe that we should make judicial decisions based on *international* legal opinions—the legal decisions of other countries. Both of these judges, loved and appointed by the Democrat-Left, are traitors to the Constitution of the United States which they must sware an oath to uphold and defend against all enemies foreign and <u>domestic</u>.[1054] So they actually sware an oath to defend the Constitution against *themselves*. Boy if there ever was a case of the fox guarding the hen house...

Precedence

"Precedent that must be applied or followed is known as *binding precedent* (alternately *mandatory precedent, mandatory* or *binding authority,* etc.). Under the doctrine of *<u>stare decisis</u>*, a <u>lower court</u> must honor findings of law made by a higher court that is within the appeals path of cases the court hears."[1055]

Precedence is another technique used by activist judges to further pervert, water down and destroy the Constitution of the United States. If some activist left-wing judges in the past made some idiotic ruling that clearly went against the meaning of the Constitution and the clear intent of the founders—like Roe V Wade, prayer in schools, or posting the Ten Commandments in courthouses for example—then, SHAZAM! that ruling now carries more weight than the original Constitution itself, and it is that bizarre ruling and not the original Constitution

---

1054   (But then again, their fingers could have been crossed)
1055   (From Wikipedia, The Free Encyclopedia)

which must be adhered to for all future related court decisions. What a beautiful scam to incrementally destroy the Constitution. These guys are brilliant. They are to our Constitution what termites are to a house. And their media will never expose them.

Restorative justice

"He who justifies the wicked, and he who condemns the just, even they both are abomination to the Lord." (Proverbs 17:15)

"Restorative justice" is the latest unintelligible theory to come out of the reprobate-mind of the Left. It empowers judges to go out of their way to *restore* criminals—even heinous ones, including child molesters who can never be rehabilitated, and violent criminals, rapists and murderers—to society as soon as possible with no thought, care or concern for justice for their victims and the victims' families, or with any regard for all of their future victims; just so they can toy with another one of their stupid, liberal social experiments. Every year we see mothers crying their hearts out in anguish on national TV over the disappearance of their little girl, and yet this country still continues to go out and vote for Democrats who continue to appoint these criminally-insane judges who continue to let these perverts go free to continue to kidnap, rape and kill our children. (Maybe we could put that into a simple flow chart so even Democrat voters could understand it.) And our Democrat media doesn't say a word. That wouldn't be "news." To report the truth might hurt their chances of getting their friends elected, and that in turn would hamper Satan's agenda to destroy America aided by his ownership of our criminal justice system.

> "The liberal approach to crime eschews punishment, preferring to believe the fantasy that bad behavior results from poverty or other deprivation arising from the "root cause" of American inequality. Liberals despise traditional American morality-based justice, with its confident judgments on right and wrong and its unapologetic drive to punish wrong. With liberals' preference for "rehabilitation" and coddling of criminality gaining ground from 1960 to the early 1990s, America experienced a more than 500 percent increase in violent crime."[1056]

Yet the poor, the group most victimized by this explosion in crime, continue to be duped into going out and religiously voting for the liberal Democrats and their policies that are at the root cause of their misery.

> "Last month, researchers from the JFA institute ["Justice For All" ...Yea, sure] — a George Soros funded group of libs — announced that keeping people in jail comes at great cost to society. JFA President James Austin asserts: "There is no evidence that keeping people in prison longer makes us any safer." (Despite declining crime statistics.) [Sure, and there's no evidence that keeping Muslim terrorists in Guantanamo makes us any safer either, you genius. Let's let them roam our streets as well.] JFA recommends shorter sentences, alternative punishments, more parole, relaxed drug laws, and more taxpayer resources for criminals released from the pokey."[1057]

The next thing they're going to recommend is that child-rapist-murderers be furloughed each day (remember Dukakis?) to babysit the children of single moms

---

1056   (From article "Liberal Failures," The Limbaugh Letter, Rush Limbaugh)
1057   (From article "Liberal Failures," The Limbaugh Letter, Rush Limbaugh)

while they're away at work. HEY!! THAT'S A GREAT IDEA FOR ANOTHER LIBERAL PROGRAM!

Hate Crime legislation

As we mentioned in a previous section, "hate crime" legislation is a sly tactic enacted by Satan for use in the near future to criminalize Christianity in general, and the preaching of the gospel specifically. He has already accomplished this in many European countries, where preaching about homosexuality, and exposing the false religion of Islam, can land you in jail. And this is what is coming to America. Interestingly though, the vicious murder of Matthew Shepard, the case that was used to spur the enactment of hate crimes legislation in the first place, as it turns out was a dispute over unpaid drug money, and had nothing to do with any hatred or bias towards homosexuals. But then again, that's the truth, which has nothing to do with the Democrat-Left, their media, or their agenda.

"The media touts "hate crime" legislation as fair and proper – but it's an affront to the very Constitution of the United States, and a danger to the proclamation of the Gospel. ...The truth is that it criminalizes viewpoints–ideas–thoughts–in the single most shocking departure from the doctrine of "free speech" since the First Amendment was ratified by the states!"[1058]

The Insanity Defense

The insanity defense is insane. It was conjured up by some lawyers and judges who couldn't comprehend evil, so they had to call it something else. It is just another ruse. It allows parasitic lawyers to game the system by letting more heinous criminals escape justice, and just punishment. Again, they spit in the face of the victims, who actually deserve to see justice rendered. Next left-wing lawyers will come up with the "stupidity defense," a get-out-of-jail-free card which they'll be able to use for themselves to escape justice whenever they get in trouble with the law. "Sorry, I can't be held accountable, 'cause I'm f**king stupid." Brilliant.

Capital Punishment

"Because sentence against an evil work is not executed speedily, therefore the heart of the sons of men is fully set in them to do evil." (Ecclesiastes 8:11)

It is actually possible to treat criminals with the love of Christ and still punish them. In fact, to do otherwise is to hate their past, present and future victims. The kind of "love" and "compassion" that the malignant narcissists of the Left have for criminals is actually a disguised hatred for law-abiding society. They hate traditional Americans, and blame them ("society") for the heinous crimes that are committed. They therefore get delight in sending violent offenders back into that same society as soon as they can get away with it. It was societies fault after all,[(LIE)] and not the poor, misunderstood burglars, home-invaders, or child-kidnapping, and child-killing, rapists.

"The American public has lost patience with the failures of liberalism's anti-death-penalty strain, its softness on crime, its obsession with gun control, its lack of judicial

---

1058   (From a Coral Ridge Ministries publication)

support for police. But liberalism, which cannot win at the ballot box, still injects its poison into the American judicial system through the courts — the machinations of activist liberal judges, the ACLU, and the radical pro-criminal agenda of the defense bar. The result has been disastrous for the law-abiding in every segment of society."[1059]

The self-actualized individuals who march outside of a prisons walls when a criminal who committed a heinous act is about to get what he actually deserves, pride themselves in how "caring" and "enlightened" and "compassionate" they are.[(LIE)] But they in truth are mentally disturbed. They spit in the face of the innocent victims and their grieving families. Their father Satan also loves nothing more than for the guilty to escape punishment in order to encourage others to follow in their path. He also hates for their innocent victims to at least find some comfort in seeing that justice gets served. These protestors think that they are so "forgiving,"[(LIE)] but they are self-deluded. They are protesting to remove the consequences for evil action. And in so doing they cultivate more evil. They should learn to forgive like God forgives. He forgives those who repent and ask for his forgiveness. But in forgiving them he does not remove the consequences for their evil actions in this life. He forgave one of the thieves on the cross and granted him an eternity in paradise, but the man was still executed. And to those who refuse to repent, he does not forgive. He allows them to spend an eternity in an excruciatingly pain-filled place of absolute horror called The Lake of Fire. He is very long-suffering and patient and is "not willing that any should perish, but that all should come to repentance," (2 Peter 3:9) but if that repentance doesn't come, then the Lake of Fire does.

Life is Gods to give or to take, because he alone created it. But God gave authority to the state to take life. Not innocent life. But to take the life of the guilty. If they have murdered the innocent, or committed a heinous crime worthy of death. God gave this authority to the state because he is not a left-wing indoctrinated idiot, and because he wanted the innocent protected by the state, and the guilty to reap what they had sowed. Amen.

The ACLU

"These are wells without water, clouds that are carried with a tempest; to whom the blackness of darkness is reserved for ever." (2 Peter 2:17)

The ACLU is an anti-American, anti-Christian, NAZI organization. They are domestic terrorists, terrorizing America with their anti-constitutional agenda, and their costly, countless and frivolous lawsuits. Under the phrase "all enemies foreign and domestic" they fall under the domestic enemy heading. But most folks are clueless as to what they are all about because they have been purposely kept in the dark by our Democrat-Left media, educational system and entertainment industry. And, of course, they own the courts as too many judges are ACLU members.

"Despite proclaiming itself the nation's 'guardian of liberty'—a claim swallowed hook, line, and sinker by the mainstream media—the American Civil Liberties Union

---

1059   (From article "Liberal Failures," The Limbaugh Letter, Rush Limbaugh)

(ACLU) is actively striving to eliminate the freedoms of millions of Americans."[1060]

The ACLU, again, is a domestic terrorist organization. And their means of terrorizing is not with the strap-on bomb, it is with the strap-on lawsuit (which hangs out of their collective zipper as they parade around the country seeking decent, traditional, Constitution-respecting and Christian Americans to skewer). In the Liberty Counsel's fact sheet about the ACLU titled, "Some Amazing Facts about America's "Enemy Within,"" it reports that:

> "The ACLU was founded in 1920 by communism enthusiast Roger Baldwin. "I am for socialism, disarmament, and, ultimately, for abolishing the state itself... I seek the social ownership of property, the abolition of the propertied class, and the sole control of those who produce wealth. Communism is the goal.""[1061]

America has been fooled into thinking (by a lying, left-wing media and educational system of course) that the ACLU stands for their "civil rights"[(LIE)] but that is just a ruse. Their real goal is to radically transform America into something that our founders and authors of our Constitution did not have in mind. In fact, what the ACLU has in mind would be about the furthest thing from what our founders had in theirs. As such, the Constitution of the United States is their worst enemy, no matter how often they hide behind the ruse of their "great concern" for the "constitutionality" of the laws they oppose. The Constitution actually stands in the way of what they want to accomplish.

> "Here is what Mr. Baldwin wrote in 1917 about future strategies: "We want also to look like patriots in everything we do. We want to get a good lot of flags, talk a good deal about the Constitution and what our forefathers wanted to make of this country, and to show that we are really the folks that stand for the spirit of our institutions.""[1062]

What a bunch of subversives! Communist subversives! And the Democrats best friends. The truth is that they stand for the destruction of everything that this country was founded on, and that most Americans hold dear. And this "enemy within" has taken over our judiciary.

> "The ACLU is determined that your children and grandchildren MUST have the right to internet pornography. They defend the right of NAMBLA (North American Man/Boy Love Association) to rape little boys. And they are trying their best to get "under God" removed from the Pledge of Allegiance."[1063]

But conversely they defended the taxpayer funding of religious "art" that put the crucifix of Christ in a jar of urine, and that smeared elephant dung on a picture of the Virgin Mary.

Liberty Council lists... "Ten Campaigns of the ACLU:

> "1. Erasing our Christian Heritage. 2. Attacking Religious Liberties. 3. Silencing the Church. 4. Advancing Sexual Anarchy. 5. Sexualizing Our Children. 6. Redefining Marriage and Family. 7. Promoting Obscenity. 8. Promoting a Culture of Death. 9. Impeding the War on Terror. 10. Looting the American Taxpayer."

That last one, "looting the American taxpayer," is a result of a law that was slipped in during the 1960s that allows them to collect monetary rewards every time one of their comrades on the courts rules in their favor. So they in essence

---

1060   (From "The ACLU vs. America," by Alan E. Sears, President of the Alliance Defense Fund, "the nation's largest religious liberty legal alliance.")

1061   (From "Some Amazing Facts about America's "Enemy Within,"" published by Liberty Counsel)

1062   (Allen E. Sears, President of the Alliance Defense Fund)

1063   (From Liberty Counsel)

are able to get paid by the taxpayer to terrorize this country with their lawsuits. What a sweet deal. (And what a bunch of stupid suckers we are for going out and voting for it.) They raked in $549,000 in Alabama in the Justice Roy Moore Ten Commandments lawsuit. And another $74,462 in Georgia, on a Ten Commandments lawsuit against the Habersham County Courthouse. And they got $6 million from American taxpayers when the Supreme Court declared the Nebraska partial birth abortion ban unconstitutional.[1064] And this is just a partial list. The followers of the America- and Christian- hating communist enthusiast, Roger Baldwin, are getting rich suing America while systematically stripping the suckers of their Constitutional rights; and sadly, America continues to go out and support this scam every time they vote Democrat. For as Planned Parenthood and the Democrat Party are one and the same, so are the Democrats and the ACLU.

And consider this…

"The Center for Constitutional Rights [the Left is always so clever to hide their contempt for our Constitution under the cloak of their supposed "concern" for it], which is a Marxist organization, has been Al Qaeda's lawyer basically since …9/11 and possibly before. There's a reason for that. There's a reason the ACLU and CAIR, the Council on American Islamic Relations, work together. Both the Islamists and the Left see as their main obstacles to the aspirations that they have, for the society they'd like to see, the destruction of individual liberty. And they see as a major obstacle American constitutional democracy. They don't agree on all the particulars but they know that what we cherish as Americans is something they need to take out."[1065]

Again, as we stated before, the female-genital-mutilating Muslims and the baby-mutilating, ACLU-Democrat-Left are one and the same. And now Obama has filled the justice department with these same traditional-America-hating ACLU fanatics from Eric Holder on down. One minute they're defending our enemies for free, in the middle of a war no doubt, advancing irrational arguments in favor of their release so they can be free to terrorize us openly and kill our soldiers and innocent civilians once again; and the next minute they are infesting our own justice department at every level. What type of sucker does it take to march out and vote for the enemies of his or her own country, in the middle of a war no less? One who has been hornswoggled by the lies of a Democrat, state-run media, their educational system, and their entertainment industry with its late-night bozos.

See:
- *The ACLU VS America: Exposing the Agenda to Redefine Moral Values*, by Alan Sears and Craig Osten
- *Indefensible: 10 Ways the ACLU is Destroying America*, by Sam Kastensmidt
- *Men in Black: How the Supreme Court is Destroying America*, by Mark R. Levin
- *Judicial Tyranny: the new kings of America?* By Mark I. Sutherland

8- Take over the government
"When the righteous are in authority, the people rejoice: but when the wicked man rules, the people groan." (Proverbs 29:2)

1064   (From Sen. Malcolm Wallop (Ret.), Founder and Chairman, Frontiers of Freedom.)
1065   (Andrew McCarthy, author of *The Grand Jihad: How Islam and the Left Sabotage America*, on Fox News' Hannity, 5/24/2010)

"There is such a thing as demonic activity in the governments of the world. The secular world sees only the visible human agent, the king, the ruthless dictator, the tyrants of history, but there are evil workings behind the scenes."[1066]

The policies, decisions and the direction of the governments of this world, and of course of our own government, are heavily swayed by "spiritual wickedness in high places." For Satan is the god of this world and he uses his influence over its governments and over the unsaved people who are in charge of them to help him accomplish his goals and bring to fruition his plans. Without this knowledge in the forefront of our thinking many of the decisions of the governments of this world, including many of our own, will do nothing but frustrate, disappoint, anger and even flabbergast. But if we are fully aware of what is going on behind the scenes—and if we have a basic knowledge of Bible Prophecy (see chapter 9)— then it all makes perfect sense. It is also important to keep in mind that God is still ultimately in control, He is still sovereign, and despite Satan's temporary power and influence, it is God's perfect will that ultimately, and shortly, will be carried out.

"Then I saw an angel coming down from heaven, having the key to the bottomless pit and a great chain in his hand. And he laid hold on the dragon, that old serpent, who is the Devil, or Satan, and bound him for a thousand years, and he cast him into the bottomless pit, and locked him up, and sealed it over him, so that he should <u>deceive the nations no more</u>, till the thousand years should be fulfilled." (Revelation 20:1-3)

Bye, bye, Miss American Pie

...(cast my ballot for an idiot, now our economies goin' dry...)

"Once we had Reagan, Johnny Cash and Bob Hope. Now we have Obama, no cash and no hope."[1067]

But that's what happens when a left-wing, lying, Democrat media and entertainment industry selectively covers up the news in order to elect a clueless, inexperienced, socialist "genius" ...you know, one just like them. Benjamin Franklin said that "man will ultimately be governed by God or by tyrants," and America, cowering under the tyranny of the ACLU and its totalitarian-left partners-in-crime has rejected God from the public square, has mandated the teaching of the bald-faced, scientifically-impossible lie of evolution that spits in the face of the God of Creation, and now we have been given tyrants. Tyrants in the form of left-wing, lying, true-believing socialists who could care less about our Constitution. Those who hold the traditional, exceptional America that we all know and love in complete contempt, feeling the intense need as they do to "radically transform" it. Into Cuba.

""We the People..." So starts off the Constitution, a document written by a group of white male elitists, but elitists who understood that a government that would endure and serve all of the people of the new United States had to be responsive to the people it served. The framers may have been elite, but now we have elitists who, by definition, believe they are smarter and better suited to make decisions for the rest of us than we are, and are quite

---

1066   (Pgs.197-198, *The Unveiling: A Journey Through the Book of Revelation*, by Keith Harris)

1067   (From internet, source unknown)

happy to impose their values, for the "common good," on society. The president and his elitist henchmen have largely abandoned every other part of the Constitution, so why not the first three words? "Elitist"—usually preceded by "liberal"—is a word much used these days about our dear leaders in Washington, as well it might be. They are very much in charge, and they display about the same concern for the opinions of their subjects as the kings and dictators of times gone by."[1068]

Satan and his Left had already infiltrated our government long before Barack Obama was elected. Only now his destruction (i.e. bankrupting) of our country by way of government policies has advanced exponentially. America's modern day Three Stooges—Obama, Pelosi and Reid—have shown a complete disregard for the will of the people, including those who foolishly elected them, and for what is best for this nation economically. They are slaves to their leftist, socialist ideology and they will tax and spend and redistribute wealth at every opportunity, even if the end result—like Greece—is the financial collapse of our nation. Indeed, they have put us so far in debt now that bankruptcy is nearly inevitable.

"Reader, suppose you were an idiot. And suppose you were a member of Congress. But I repeat myself."[1069]

"They were and are arrogant. And they were and are clueless; which generally goes hand-in-hand with being arrogant. Because when you are arrogant, you haven't a clue about what matters to your subjects. Only that you keep subjecting them to what matters only to you. That is arrogant. So you challenge in court an illegal immigration crackdown that voters overwhelmingly support in poll after poll. Now that's arrogant. You push trillions of dollars of stimulus on tax payers who consistently told you not to, but you did. Now that's arrogant. You take over scores of industries amid overwhelming protest, and you're subverting the Constitution in the process. Now that's arrogant."[1070]

From the start Obama showed his complete disregard for the will of the people and for the economic needs of the nation by issuing an executive order that sent our tax dollars to kill babies in foreign countries. Now I'm sure that was something that every taxpayer had as their top priority in voting to "turn this country around." He did it of course just to pay off the far-left feminists and his friends at Planned Parenthood. (That he is also helping them to slowly exterminate his own race is another issue that appears slightly "above his pay-grade.") Next he pushed through an obscene, trillion-dollar spending bill that had nothing whatsoever to do with stimulating our economy. It was designed to satisfy every liberal pork barrel spending project—no matter how bizarre, wasteful or ridiculous—that had been waiting for years for this "golden opportunity" (a "crisis" that, in the words of Rahm Emmanuel, "they couldn't let go to waste"); in addition to paying off the many state and municipal public service unions that had supported him, but now in this terrible economy were threatened with layoffs. So he took more money out of an already stifled economy, disregarded all of the millions of taxpayers who had lost jobs in the private sector, and gave much of the spendulus money to prop up the jobs of union employees. That's the change we can all believe in.

1068 (Alfred S. Regnery, Publisher, *The American Spectator* 7/16/2010)
1069 (Mark Twain, quoted by Albert Bigelow Paine in *Mark Twain: A Biography*, 1912)
1070 (Neil Cavuto, on Fox News' "Your World with Neil Cavuto," July 13, 2010)

"OBAMA LIED AND FREEDOM DIED."[1071]

And all that was just the beginning...

He has refused to secure our borders, but instead finds it a more attractive alternative to attack and sue Arizona for attempting to secure theirs. Arizonans lives are being threatened, their ranchers and sheriffs are being killed and their public services are being overrun by an illegal immigration influx that threatens to bankrupt their state. But why help them when our president and the complete imbecile that passes for an Attorney General can go out and make fools of themselves by criticizing Arizona's bill without ever having read it. Then, being ignorant of what was actually in it, they lied about it, saying it would lead to "racial profiling"[(LIE)] even though the actual bill expressly forbade that; and even though the actual bill was identical to the federal bill and exponentially less stringent than Mexico's immigration laws which actually <u>do</u>, not only <u>allow</u> for express racial profiling, but actually <u>require</u> it of *their* law enforcement personnel. We must live in an insane asylum, and the idiots must have taken over.

> "I love Arizona. I've spent a lot of time there. Beautiful trails I used to hike down the Huachuca Mountains on the Southern Arizona border with Mexico, and you can't even hike them in the daytime anymore. They're not safe. ...Those trails are now used by drug smugglers, by immigrant traffickers. They're strewn with garbage. They're filthy. And they're full of gunmen and killers. It's a sad case where one State of the Union, Arizona, feels it is being abandoned by the federal government and the other 49. It's all well that you yak about "Oh, human rights for illegal immigrants." [But] what about the human rights of our citizens, of our ranchers, of the victims of crime in Phoenix and elsewhere? When is the federal government going to protect American citizens?"[1072]

(Uh, maybe when we fumigate the Democrat-Left infestation there?)

The Democrats also say they want "Comprehensive Immigration Reform," but that is just code speak for open borders and amnesty, a ticket to a whole lot more ignorant, uninformed voters which assures a whole lot more Democrat votes. They also broadcast the lie, with the eager aid of their "Obama-mania" media that these people are "only coming here to do the jobs that Americans refuse to do."[(LIE)] Sure, tell that to all of the millions of black youths who can't find a job, and to all of the other unemployed people in this terrible Obam-economy.

"THE ONLY THING WE HAVE TO FEAR IS OBAMA HIMSELF."[1073]

Obama had no problem calling the Massachusetts police "stupid"[(LIE)] just for doing their job, but yet he doesn't want us to call the <u>Muslim</u> Army major in Texas who killed 13 soldiers, a terrorist, which of course he is. Now that, Wilbur, is what is truly stupid.

> "This is the story of two young men. One [Barry Obama] fell in with left-wing radicals. The other [Les Phillips] immigrated to America. While one played with terrorists and allowed an America-hating bastard [I think that might have been "pastor"] to baptize his children, the other joined the Navy to defend his country. I love America, but president Obama is ashamed of it. I'm going to Congress to help stop him from destroying our nation, and they're not going to call me a racist. [Because Les is black.] I'm Les Phillips and I took an oath to defend this country against enemies foreign and domestic, and I approve this message."[1074]

1071   (Sign at tea Party Rally, Washington D.C. June, 2010)
1072   (Colonel Ralph Peters on Fox News *Hannity*, 04/28/2010)
1073   (Sign at tea Party Rally, Washington D.C. June, 2010)
1074   (Ad from African American Republican candidate Les Phillips, running for Alabama's

Amen, brother. And why couldn't we have elected someone like Les as the first black president instead of this ignoramus? Answer: because a left-wing, lying, Democrat media and equally ignorant late night clowns would never have stood for it. They would have thrown the usual tantrum that they do whenever a decent, non-baby-mutilating candidate has a chance of getting elected.

Obama has trashed the Constitution by appointing dozens of "Czars" who are unaccountable to Congress, only to him. But he's not a socialist even though so many of his followers, including many of those White House czars, are. Some are even avowed communists, but that must just be an amazing coincidence that somehow escaped his notice, or again was above his pay-grade.

"The First Amendment- they're attacking talk radio, they're coming for it. The Second Amendment, the Right to Bear Arms- they're coming after that. The Fifth Amendment, the Property Rights. The Tenth Amendment, State Authority. Almost every aspect of the Constitution is under attack, unless you're a terrorist, then they want to give you due process."[1075]

He is also shredding the Constitution by letting the government take over control of many of America's businesses. The banks, the insurance companies, the car companies, the health care system.

"This administration has problems with the rule of law, and it doesn't regard the law as that which is written to protect people. It regards it as an obstacle to the accomplishment of its political goals. ... Its behavior in the Gulf has been one of authoritarian, "I don't care what the law is, I'm gonna demonize BP;" when they shook down $20 billion from BP with no legal authority to do so whatsoever; when they said to BP, "Before you sell any assets or any stock clear it with us first," with no legal authority to do so whatsoever. They don't care about the law. They care about the political response to their behavior."[1076]

He wants to pass a bill that's called "Cap and Trade" which will drastically raise energy costs for both residential and business customers, but it will appease the global-warming wackos who want to send us back to the dark ages to satisfy their poorly disguised hatred of both America and capitalism.

He shoved the convoluted, incomprehensibly stupid, catastrophic nightmare monstrosity of Obama care, otherwise known as the destruction of the greatest health care system the world has ever known, down the throats of the American voter against their will, and we are only now finding out what a business-crippling and job-destroying pile of dung it is. For a fraction of the money the congressional Democrats allocated for Obama's idiotic spendulus bill they could have insured every single needy American family—not <u>illegals</u> who don't warrant free health care and not millions of young people who choose not to buy it, but every truly needy family—and left the greatest health-care system on the face of the earth intact. But that would not have satisfied their ideological obsession to redistribute wealth, nor their socialist obsession for the government to take over our private industries; and nor would it have satisfied their father Satan's goal of destroying America.

"America has just witnessed an unconscionable abuse of power. President Obama has betrayed his oath to the nation — rather than bringing us together, ushering in a new kind

5th Congressional District)

1075   (Mark Levin on Fox News *Hannity*, 3/23/2009, on his new book, *Liberty and Tyranny: A Conservative Manifesto*)

1076   (Judge Andrew Napolitano on Your World with Neil Cavuto, July 13, 2010)

of politics, and rising above raw partisanship, he has succumbed to the lowest denominator of incumbent power: justifying the means by extolling the ends. He promised better; we deserved better. He calls his accomplishment "historic"[LIE] — in this he is correct, although not for the reason he intends. Rather, it is an historic usurpation of the legislative process — he unleashed the nuclear option, enlisted not a single Republican vote in either chamber, bribed reluctant members of his own party, paid-off his union backers, scapegoated insurers, and justified his act with patently fraudulent accounting. What Barack Obama has ushered into the American political landscape is not good for our country; in the words of an ancient maxim, "what starts twisted, ends twisted." His health-care bill is unhealthy for America. It raises taxes, slashes the more private side of Medicare, installs price controls, and puts a new federal bureaucracy in charge of health care. It will create a new entitlement even as the ones we already have are bankrupt. For these reasons and more, the act should be repealed. That campaign begins today."[1077]

(Of course we wouldn't have needed this campaign if we hadn't elected the idiot in the first place.)

"Medicare: broke. Social Security: broke. Medicaid: broke. Amtrak: broke. The Post Office: broke. Who in their right mind with that track record would give health care to the same people that have broken everything they have tried to manage? [Uh, Democrat voters?] ...Where in the Constitution is it authorized for the federal government to regulate health care?"[1078] - Judge Andrew Napolitano. (The same place where woman are authorized to mutilate their babies to death, Judge.)

According to Obama, Afghanistan was the "good war," but when his field commander asked for 40,000 more troops in order to achieve victory there, Obama dithered around for months until finally approving only 30,000. And Iraq was "George Bush's war,"[LIE] the "bad war,"[LIE] the one that he and all of his Democrat friends tried to undermine; and then they vehemently opposed the surge that ultimately proved a brilliant and successful maneuver. But now that things are going well his administration hasn't hesitated to take credit for it. How's that for shameless hypocrisy? And his state-run media will let him get away with it, unchallenged and unexposed, so the no-think-ums will still go out and vote for him.

"The two enemies of the people are criminals and government. Let us tie the second down with the chains of the Constitution, so the second will not become the legalized version of the first." -Thomas Jefferson. (Which is why the second enemy of we the people, the government, and in this case government by the Democrats, trashes our Constitution. Its "chains" tie them, and their agenda, down.)

Obama took the funds that had been going to adult stem cell research, which has shown results, and instead he diverted them into embryonic stem cell research, which has shown no results. Why? Why would he do that? Because he is a rabid baby mutilator, and he could not let the chance to destroy baby embryos slip away. That's the only explanation. They are Nazis. Ask the Jews. The Nazis would scour every nook and cranny of every occupied country in order to uncover, and kill, one more Jewish man, woman or little child. They were obsessed with the hatred of their father Satan. And so are Obama and the Democrat-Left.

"It has been said that politics is the second oldest profession. I have learned that it bears a striking resemblance to the first." -Ronald Reagan.

1077   (From article "A Campaign Begins Today," by Mitt Romney on National Review Online, 3/22/10)

1078   (From a speech in 2010 by Judge Andrew Napolitano)

Obama, in contrast to all of the good things he constantly finds to say about his clitoral-mutilating Muslim friends, commented after 9/11 that "America hasn't lived up to her ideals."(LIE) Which is so false and defamatory that it is stomach-turning; coming from our own genius-president no less. The truth is that America has actually done a better job of living up to its ideals than most of the other countries on this earth combined (including all of Mr. Ed's Islamic countries). Harold Estes, a 95 year-old "venerable and much honored WW II vet …well known in Hawaii for his seventy-plus years of service to patriotic organizations and causes all over the country," took exception to Obama's bilge…

> "Which [ideals] did you mean? Was it the notion of personal liberty that 11,000 farmers and shopkeepers died for to win independence from the British? Or maybe the ideal that no man should be a slave to another man, that 500,000 men died for in the Civil War? I hope you didn't mean the ideal 470,000 fathers, brothers, husbands, and a lot of fellas I knew personally died for in WWII, because we felt real strongly about not letting any nation push us around, because we stand for freedom. I don't think you mean the ideal that says equality is better than discrimination. You know the one that a whole lot of white people understood when they helped to get you elected."[1079]

No, Mr. Estes, I think Obama meant the "ideals" of him and his America-hating, baby-mutilating, left-wing-indoctrinated, clueless buffoons who infest our media and our entertainment industry and who don't have the slightest idea of what America, and her ideals, are all about. Obama spent too many years learning the ideals of the Muslims he was surrounded by as a boy in Indonesia, those of the mentally disturbed, Black Liberation Theology preacher, Jeremiah Wright, who delighted in saying right after 9/11 that "America's chickens have come home to roost," in addition to this profound homiletic: "God d**n America."

> "DO YOU MISS ME NOW?" – query on a billboard picturing the previous president, George W. Bush.

> "Obama is a vampire who doesn't live on blood - he lives on applause." - Lt. Col. Ralph Peters

Ex-leftist radical David Horowitz uses his similar upbringing and indoctrination to shed some light on the true nature of our president:

> "When Barack Obama won the 2008 election, I knew America was in deep trouble. Here's why. I was a leftist as early as I can remember. Raised in a Communist family and surrounded by radicals my entire childhood, I could hardly be anything else.

> "As an adult in the 1960's I was a leader of the New Left and editor of *Ramparts*, a fanatically anti-war, anti-capitalist magazine. My friends were some of the best-known radicals in America. But something happened that changed my life.

> "A friend of mine named Betty Van Patter was murdered by the Black Panthers in 1974. I had referred her for a job as bookkeeper for the Panthers in Oakland, California. When she discovered things they didn't want outsiders to know, she ended up floating in a river, dead.

> "Like so many of today's leftists, I had thought of people like the Black Panthers as especially virtuous because of their "victim" status and their so-called fight for social justice. It was hard to believe they were just murderers and thugs. But when Betty was killed, the evidence against them was too strong for me to resist.

> "Most of my friends refused to believe it. When I told Betty's own adult daughter who had killed her mother, she replied, "The Panthers are good people." As a leftist, she could not accept the truth.

1079   (From a letter Mr. Estes dictated to a friend, circulating on the net)

"I was forced to question my most basic beliefs, and that began my long and difficult journey to sanity. I lost most of my friends and became the target of their vicious hatred, which lasts to this day. But I gradually gained an appreciation for America's culture and history, for our economic system and the worship of God, and especially for our freedom...

How leftists think

"[The Left's] goal is to seize power, no matter what they claim. All of their agenda items from healthcare to environmental protection to saving the economy are just their means of accumulating power.

"They see the world in terms of "haves" and "have-nots" and they believe wealth must be redistributed to make everyone equal. The Obama machine has already spent trillions of taxpayer dollars to finance a takeover of the American workplace and to stifle the independence of the American people. Towards that end, and in the service of that belief, they are willing to commit hideous atrocities, or to avoid seeing the atrocities committed by others.

"I've seen how easy it is for leftists to believe terrible lies, and to follow people who are moral monsters while denying the evidence of their own eyes and ears.

"Why didn't my leftist comrades believe the Panthers had murdered my friend Betty? Dozens, maybe hundreds, of local activists were aware of what had happened, but no one raised any question about it – none of these people who are usually so quick to protest injustice. No one seemed to care about Betty's murder. The Panthers' victim status and usefulness to the left outweighed the horrible crime they had committed.

"I remember the feeling. Like all radicals, I was intoxicated by my own virtue. It was my noble intentions that counted, not flesh and blood people. I'm sure you'll recognize today's leftists in that description.

Saul Alinsky, community organizer, and his disciple Barack Obama

..."*Rules for Radicals* [is] a 1971 book written by the guru of community organizers Saul Alinsky. Barack Obama learned and taught Alinsky's methods as a community organizer in Chicago. ...of all the outrageous things in Alinsky's book, perhaps the most horrifying is his dedication. It's "to the first radical known to man who rebelled against the establishment ... Lucifer." Yes that's right: Alinsky dedicated his book to Satan.

Who are the real "haters"?

"I wonder how much that dedication would bother today's radicals. Perhaps not much. I thought the left was brutal when I belonged to it, but I've been stunned by how vicious they have become – and not only the admitted left, but also the media who still claim to be neutral.

"The attempted destruction of Rush Limbaugh by creating completely false racist quotes which the media repeat ad nauseam is something I couldn't have imagined even a few years ago. Look how they successfully demonized George W. Bush; look what they're doing to Glenn Beck, Fox News and Sarah Palin. Look at Nancy Pelosi's description of ordinary protesters as Nazi's. They have long understood Saul Alinsky's rule: *Pick the target, freeze it, personalize it, and polarize it*.

"They do this claiming to be fighting "hate" – defined as everything they disagree with. So when they claim to be fighting against hate speech, they really mean shutting down criticism. Present Obama's "Diversity Czar" at the Federal Communications Commission praised the Venezuelan dictator Hugo Chavez for his crackdown on the media, and favors regulations on radio stations here that would wipe out national conservative talk radio.

"Accustomed to the malice of the left, when I left my revolutionary comrades and moved to the right, I was stunned at how <u>nice</u> most conservatives are. They believe in decency and civilization, in laws and customs, and very few want to breach the rules of civility."[1080]

1080 (From The David Horowitz Freedom Center newsletter, David Horowitz Founder and

Question: Does Obama love or hate this country? Well, he did say that he wanted to "radically transform" America, didn't he? And...

> ..."a person doesn't aspire to fundamentally change something he loves. So that alone is all the proof we need that Obama does not love America."[1081]

(Go tell your wife you want to radically transform her and watch where you'll be sleeping tonight.) If he hates America (and we're not saying that he does, we report, you decide) then what does he love? Himself? Is he a narcissist? ...

> "In psychiatry, [narcissism is] a personality disorder characterized by the patient's overestimation of his or her own appearance and abilities and an excessive need for admiration."[1082]

Well, that definition does have a certain "Obama-ring" to it, does it not?

> "Obama's narcissism manifests itself in ... [his] grandiose delusions about his own importance, about his own historic role in the world. He, as you recall, said, "We are the ones we've been waiting for," meaning, "I am the one you've been waiting for." He suggested that he [would] cause the oceans to subside and this would mark the moment when the planet would begin to heal. ...I think he believes it and he is completely conceited. He has been sheltered all his life. He's been pampered by all his handlers and been told that he can do all these things, and he has never really had to do anything except street agitation and give speeches behind a teleprompter. When you couple his extreme leftist ideology with his narcissism and his messianic complex, you find a guy who believes he's on an urgent mission to implement his leftist agenda [translate: to destroy America] and will not be deterred from that mission no matter what kind of public opposition he encounters."[1083]

May God help us if he is not sent out to pasture in November of 2012.

"As it is written... Their throat is an open sepulcher; with their tongues they have practiced deceit; the poison of asps is under their lips: Whose mouth is full of cursing and bitterness. Their feet are swift to shed the blood [of innocent babies]. Destruction and misery are in their ways: And the way of peace have they not known: There is no fear of God before their eyes." (Romans 3:13-18)

...And they are called the true-believing, totalitarian-liberals of the Democrat-Left.

But is he a Muslim?

"For the wrath of God is revealed from heaven against all ungodliness and unrighteousness of men, who hold the truth in unrighteousness." (Romans 1:18)

A recent poll showed that 1 out of 4 Americans believe that Obama is a Muslim, which, incredibly, is an *increase* from before the election. Why? Are the American people stupid for thinking that,(LIE) as our Democrat media says? Or is the Democrat-Left media stupid for thinking they're stupid?(THE TRUTH) Well let's see. Here's a conversation between Sean Hannity and ex-Muslim Brigitte Gabriel on Fox News Hannity, 8/19/2010:

> Brigitte: "His actions since he became president have been pandering to the Islamic world, apologizing on behalf of America to the Islamic world, praising the Islamic world while putting America down. His first television exclusive was to El Arabia television where he bragged about how his family members were Muslims. The video address that

---

CEO)
1081   (David Limbaugh, from interview in The Limbaugh Letter, October, 2010)
1082   (From Microsoft Encarta Dictionary)
1083   (David Limbaugh, from interview in The Limbaugh Letter, October, 2010)

he sent to the Iranians congratulating them on their spring holiday. The speech in Cairo. So the American public is seeing a pattern."

Hannity: "He said we're not a Christian nation[LIE] ...that America is arrogant.[LIE] He went on the apology tour. His first major speech was on El Arabia TV. ...His next 2 major speeches were in Turkey and Cairo. ...The NASA chief said the top priority was Muslim outreach. The money, 900 million, to Hamas. His [ignorant] treatment of Benjamin Netanyahu [when he visited the White House]. His lack of outspokenness on the Iranian democracy movement. The refusal to acknowledge the Fort Hood shooter was a terrorist. ...He went to a Muslim school. [Where he studied the "holy" (his words) Koran.] He said that the call to prayer at sunset was one of the most beautiful things on the face of the earth. [Tell me Wilbur, what does slicing off little girl's clitorises come in, a close second?]

Brigitte: "Finally people are paying attention to things after the fog has been lifted off of their eyes as to who did we really elect as president. The signs and the information were all out there. [That a lying, Democrat media never reported.] President Obama was born into the Muslim faith, raised as a Muslim as a child, to a father who was a Muslim. He attended Islamic schools."[1084]

It is also interesting to note that the Muslim world considers him a Muslim because he was born to a Muslim father. As far as they are concerned that is the end of the story. Period.

In his Cairo speech where he displayed either his utter ignorance or complete dishonesty by apologizing to them for America, Obama also, incredibly, said this, "I know, too, that Islam has always been a part of America's story."(LIE) Oh really? It has? Were there any Muslims when the Pilgrims landed? No. At the first Thanksgiving? No. Any Muslim signatures on the US Constitution? The Declaration of Independence? The Bill of Rights. No, no and no again. How about Muslims who fought for our independence against England? None? How many Muslims fought to preserve the union and free the slaves during the Civil War? Zip. Nada. Bupkis. Talk about rewriting history, this guy is the clown prince! Islam's only "part of America's story" was in our war against the Barbary pirates, who were Muslim mariners that operated out of North America and attacked American merchant vessels at the turn of the 18th century. (The ideological ancestors of the Somali pirates. "The more things change, the more they stay the same.")

And speaking of freeing the slaves during our Civil War, which no Muslim had a part in; they themselves seem to still have a problem in that area:

"In fact, Muslims to this day are still the largest traffickers in human slavery. [Obama's] own half-brother, a devout Muslim, still advocates slavery himself, even though Muslims of Arabic descent refer to black Muslims as "pug nosed slaves." Says a lot of what the Muslim world really thinks of your family's "rich Islamic heritage," doesn't it, Mr. Obama?

[And were there any Muslims fighting for Woman's suffrage, or participating in the Civil Rights battles of the 1960s? No?]

"In fact, devout Muslims demand that women are subservient to men in the Islamic culture. So much so, that often they are beaten for not wearing the 'hijab' or for talking to a man who is not a direct family member or their husband. [Or for looking between their legs and trying to find their genitals.] Yep, the Muslims are all for women's rights, aren't they? Where were Muslims during World War II? They were aligned with Adolf Hitler.

1084 (Brigitte Gabriel, Lebanese American author and outspoken critic of Islam, on Fox News Hannity, 8/19/2010)

The Muslim grand mufti himself met with Adolf Hitler, reviewed the troops and accepted support from the Nazi's in killing Jews. [Hitler is dead, but it is the Muslims who are alive and well and still seeking the extermination of the Jews and the annihilation of their nation of Israel.]

"Finally, Mr. Obama, where were Muslims on Sept. 11th, 2001? If they weren't flying planes into the World Trade Center , the Pentagon or a field in Pennsylvania killing nearly 3,000 people on our own soil, they were rejoicing in the Middle East . No one can dispute the pictures shown from all parts of the Muslim world celebrating on CNN, Fox News, MSNBC and other cable news networks that day. Strangely, the very "moderate" Muslims who's asses you bent over backwards to kiss in Cairo, Egypt on June 4th were stone cold silent post 9-11. To many Americans, their silence has meant approval for the acts of that day. [Amen.]

"And THAT, Mr. Obama, is the "rich heritage" Muslims have here in America.

"Oh, I'm sorry, I forgot to mention the Barbary Pirates. They were Muslim. And now we can add November 5, 2009 - the slaughter of American soldiers at Fort Hood by a Muslim major who was a doctor and a psychiatrist who was supposed to be counseling soldiers returning from battle in Iraq and Afghanistan.

"That, Mr. Obama is the "Muslim heritage" in America."[1085]

And that's also pretty much the "Muslim heritage" in every other part of the world that has had the distinct displeasure of being overrun by this repulsive, malevolent and feeble-minded religion. In addition, consider this from some Arab Americans who explained to Dr. Jim Murk, a Middle Eastern Scholar and expert on Islam, why Obama never takes his wife on his Middle East trips:

"An orthodox Muslim man would never take his wife on a politically oriented trip to any nation which practices Sharia law, particularly Saudi Arabia where the Wahhabi sect is dominant."

Dr. Murk: "This is true and it is why Obama left Michelle in Europe. She will stay home when he visits Arab countries. He knows Muslim protocol; this includes, bowing to the Saudi King. Obama is regarded as a Muslim in the Arab world, because he was born to a Muslim father; he acknowledged his Muslim faith with George Stephanopoulos. Note that he downplays his involvement with Christianity, by not publicly joining a Christian church in D.C. ...He also played down the fact that America is a Christian country and said, unbelievably, that it was one of the largest Muslim nations in the world, which is nonsense. He has publicly taken the side of the Palestinians in the conflict with Israel and he ignored the National Day of Prayer, something no other President has ever done. He is bad news! He conceals his true faith to the detriment of the American people."[1086]

I also find it highly interesting that when it came to the protests in the streets of Iran in 2009, that very probably could have overthrown that putrid regime, if the United States had supported the opposition, our president instead said that "we shouldn't get involved in Iran's internal affairs." Wow! And now, in March of 2011, his Secretary of State, Hillary Clinton, rather than supporting those risking their lives and getting shot dead protesting for freedom in the streets of Syria, instead said that the brutal Syrian dictator Assad was a "reformer." Which is such a bald-faced lie that the only larger one I can think of was the one told by her

---

1085   (Taken from "An American Citizen's Response" to Obama's absurd Cairo claim, from an email circulating on the net)
1086   (Jim Murk, Doctor of Philosophy in Middle Eastern Culture & Religion.)

husband, "I didn't have sex with that woman, Monika Lewinski." But the point here is that these two brutal regimes, Iran and Syria, are ours and of course Israel's worst enemies in the Middle East and are also, coincidentally, the two that Mr. Muslim man...er...I mean Mr. Obama has no interest it would appear in seeing overthrown. For those countries are already controlled by Islamic jihadees sworn to the destruction of both Israel and America so, if you happen to be an Islamic sympathizer cleverly disguised as the president of the United States who the idiots went out and elected, why would you want them overthrown? And conversely, in the two countries that are not controlled by Islamic jihadees, Egypt and Libya, he *does* want *their* dictators overthrown, even though the Muslim Brotherhood in Egypt and Al Qaeda in Libya are the two Islamic terrorist organizations which are most likely to gain in power or take over completely. Hosni Mubarak of Egypt (who now at this writing has been removed), and Muammar Gaddafi of Libya, both were and are[1087] brutal dictators but certainly no less brutal than Assad of Syria and Iran's *I'm-a-nut-job* (Ahmadinejad). And even more incredibly, in Libya we may actually be supporting the very Islamic jihadists who killed our young fighting men and woman in Afghanistan and Iraq. The rebels there are tied to Al Qaeda and the Muslim Brotherhood, and in the wars in Iraq and Afghanistan 20 percent of the foreign fighters captured were Libyans! Are we stupid or what? Or are we governed by a colossal fool, or a cleverly-disguised, Islamic-jihadist-sympathizing, Muslim plant? Or, heck, maybe it's all just a strange coincidence, like all of those communists and socialists he accidentally surrounded himself with in his administration. We report, you decide. What say you, O'Reilly?

But in truth, we are not saying that he actually does sympathize with the Jihad. That would be almost too stunning to comprehend. [Of course now, in May of 2011, when he has just stabbed Israel in the back by stating that he thinks they should return to the indefensible borders that existed before they were attacked once again by the Musrabs in 1967, it's becoming a little less incomprehensible...] But the above pattern concerning his glaringly different treatment of Iran and Syria, and Egypt and Libya is at the very least eyebrow-raising, and his ignorant treatment of Israel, revealing. Also, I don't know, or care, whether he actually is a Muslim or not. But I do know that with all of the above in mind, one out of four Americans are obviously not "stupid" for thinking he just might be. And it is our lame-stream, Democrat media, however, that *is* stupid for just blowing it off. And it is the thoughtlessness of a non-Muslim nation, founded on the principles and the faith of Christianity that would nevertheless obey that same stupid Democrat media and rush out orgasmically, like the Germans who hung on Hitler's every idiotic word, and elect someone like Osama Bin Obama less than 9 years after 9/11... that is what is beyond stupid.

1087 (This is being penned in March of 2011 and Gaddafi still remains in power in Libya although Barack Hussein Obama has said that he must go.)

"IF YOU VOTED FOR OUR CURRENT PRESIDENT IN 2008
TO PROVE YOU WEREN'T A RACIST
PLEASE VOTE FOR SOMEONE ELSE IN 2012
TO PROVE YOU'RE NOT AN IDIOT."[1088]

Voting Democrat, and other mental diseases

"There is a generation that is pure in their own eyes, and yet is not washed from their filthiness." (Proverbs 30:12)

Again, as we covered in chapter eleven, voting for baby mutilators is a choice that comes with terrible eternal consequences, including for those "Christians" who do so and yet are deceived into thinking that they are "serving the Lord," because you cannot know Christ as Savior and support those who murder the most innocent. The two, Jesus and iniquity, just do not equate. ("Let everyone who names the name of Christ depart from iniquity." 2 Timothy 2:19) But in addition to it being spiritually stupid, it is also not too bright politically. For the evil consequences of voting for the Democrat-Left go far beyond just slaughtering innocent babies. (As vile as that is in and of itself.) And that is because as a politician you don't kill babies on the one hand, and love truth, justice and the American way on the other. You kill babies on the one hand and are a liar, a cheat, a propagandist, an opportunist and a false accuser of your opponents on the other. If you are a baby mutilator then you have no moral character (no matter how adept you are at deceiving yourself to the contrary). You have sunk to the depths of human depravity, and lying and deceit are as nothing. And being a reprobate minded participant in Satan's "radical transformation"—aka destruction—of America is also of no concern to your conscience.

But an even further disconnect lies with the American people and the values that they hold dear, and the Democrat party that holds those same values in complete contempt, who so many of them nevertheless are deceived into going out and voting for. The following is a partial list:

Most people oppose partial-birth abortion and live-birth abortion, yet that's what they vote for every time they go out and vote Democrat.

Most people do not want the ancient and sacred institution of marriage denigrated to the point where a man can "marry" a man, a woman can "marry" a woman, or a woman can marry her German Shepherd, yet that's what they vote for every time they go out and vote Democrat.

Most people do not think innocent children should be allowed to be adopted by homosexuals, they think they deserve a normal family, yet that's what they vote for every time they go out and vote Democrat.

Most people do not agree with hate crimes legislation, which is merely the first step in allowing left-wing, politically-correct thought police to jail people—like they already do in Europe and the Middle East—who believe and preach what the Bible says about homosexuality and false religions like Islam, yet that's what they vote for every time they go out and vote Democrat.

Most people don't think that the radical homosexual agenda should be shoved down the throats of kindergartners, first and second graders, or any age schoolchild for that matter, yet that's what they vote for every time they go out and vote Democrat.

1088   (Billboard by *Concerned Citizens of America*)

Most people don't think that an insignificantly tiny, constitution-trashing, ACLU-minority should be allowed to eradicate Christianity from public life in this country, yet that's what they vote for every time they go out and vote Democrat.

Most people do not want the words "under God" removed from the Pledge of Allegiance or "in God we trust" removed from our national motto, yet that's what they vote for every time they go out and vote Democrat.

Most people think discipline in our schools is a good thing, and the lack of it is the cause of the insanity we have going on there right now, yet that ACLU-litigated insanity is what they go out and vote for every time they vote Democrat.

Most people think that America's religious heritage and true history should be taught in our public schools, and not a false, America-hating, left-wing, rewritten version, yet the false version is what they go out and vote for every time they vote Democrat.

Most people have no problem with voluntary prayers in our public schools or in any other public setting, but they vote for the exact opposite every time they vote Democrat.

Most people think school vouchers are a good idea, yet they vote for the exact opposite of that every time they go out and vote Democrat.

Most people do not want a very few unelected lawyers in black dresses overturning sensible, Constitutionally-valid laws just because they would hamper their far-left ideological agenda, yet that's what they vote for every time they vote Democrat.

Most people are repulsed at the idea of shoving scissors in the back of a living babies head and sucking its brains out with a vacuum, yet that's what they vote for every time they go out and vote Democrat.

Most people think that parents should be notified before their minor child has an abortion, yet they vote for the exact opposite of that every time they vote Democrat.

Most people do not think that illegal immigrants should be allowed to stream across our southern border unencumbered and then given sanctuary in certain cities, driver's licenses in certain states like Maryland, and eventual amnesty just so they can vote Democrat, yet that's what they vote for every time they go out and vote Democrat.

Most people do not want heinous child perverts who kidnap, rape and murder their children to be set free by demented liberal lawyers in black dresses disguised as actual judges, yet that's what they vote for every time they go out and vote Democrat.

Most people do not think it's fair for state, local and federal government employees to have jobs for life with nearly double the salary and benefits of their counterparts in the private sector who pay for those jobs, yet that's what they vote for every time they go out and vote Democrat.

Most people, if they were ever informed about it, would not want their tax dollars taken away from *adult* stem cell research which has produced many cures, and given to embryonic stem cell research which has produced none, yet that's what they vote for every time they go out and vote Democrat.

Most people, if they were ever informed about it, would also not want Constitution-distorting and -perverting judges to be given lifetime appointments so they can legislate a liberal, far-left agenda from the bench which would not and could not be enacted by Congress, yet that's what they vote for every time they go out and vote Democrat.

And we could go on and on. Our disgraceful, national, state-run, Democrat media has done an excellent job of keeping half of the voters in this country from becoming cognizant of the total disconnect between their actual values, their stand on the issues, and what they religiously go out and vote for every two years. Not since the Bernie Madoff con—where people thought they were making really smart investments (think *votes*) yet lost all of their life savings (think *country*)—has a scam been so far reaching and devastating in its effect. Take a minute to congratulate our friend, Comedy Central's arrogant imbecile, Jon Stewart and all of his totalitarian-left friends in the entertainment industry and our media the next time you see them.

Far too many life-long Democrats think they are still voting for the party of JFK, but that party doesn't exist anymore. Now they are going out and voting for the party of Satan. And speaking of Lucifer, here's what one of the primary financial backers of the Democratic Party, billionaire George Soros, thinks of himself...

"It is sort of a disease when you consider yourself some kind of god, the creator of everything, but I feel comfortable about it now since I began to live it out." -George Soros

What a surprisingly candid statement, that he has a mental disease, from one who is helping to create, both in this country and abroad, in accordance with the will of his father the wicked one, the conditions necessary for the rise of the anti-Christ and his one world government...

"George Soros said, "America is the biggest roadblock to an Open Society" (borderless, one world government)."[1089]

And keep in mind that this megalomaniac is a good portion of the purse strings behind both the Democratic Party and many organizations like move on.org which are their propaganda arm. Democrat voters need to wake up to just what they are marching out and voting for. They've been duped into investing with Barack "Bernie Madoff" Obama. And they're darn close to losing their country.

John Kerry said it perfectly about five weeks before the 2010 midterm elections in his consternation over the coming Republican landslide:

"We have an electorate that doesn't always pay that much attention to what's going on, so people are influenced by a simple slogan rather than by the facts and the truth."[1090]

Only Mr. Nincompoop didn't realize that he was actually referring to the <u>2008</u> elections and his <u>own</u> voters. For an uneducated electorate, "paying little attention to what's actually going on" and influenced by "simple slogans" like "YES WE CAN!" is the Democrats best friend. Thank you, Mr. Kerry, we couldn't have said it better ourselves. But, seriously though, and with all due respect, who goes out and votes for someone for president of the United States, the most powerful position of leadership (or lack thereof) in the world, who had as one of his best friends and as one of his close political mentors an unrepentant domestic terrorist, Bill Ayers? After 9/11!?!#* Obama started his political career in the terrorist's kitchen for goodness sake. And one who also deeply admired and sat in the pew for 20 years of another mentor, a racist, black-theology preacher who spewed anti-American bilge? What kind of an idiot, Mr. Kerry, goes out and votes for someone like that? (One deceived by a lying, Democrat media, that's who.) That's dumber than the Palestinians going out and electing Hamas, and they had guns to their

---

1089   (Glenn Beck Program, March 31, 2011)
1090   (John Kerry, 9/25/2010)

heads. Conversely, the vast majority of the Tea Party folks who helped bring about the Republican landslide and the repudiation of Obama's asinine, socialist, tax-and-spend policies in the November, 2010 elections were highly informed and quite fluent on most of the issues.

"The heart of the wise inclines to the right, but the heart of the fool to the left." (Ecclesiastes 10:2)

> "Norman Mattoon Thomas (1884 – 1986) was a leading American socialist, pacifist, and six-time presidential candidate for the Socialist party of America. [He] said this in a 1944 speech: 'The American people will never knowingly adopt socialism. But, under the name of "liberalism," they will adopt every fragment of the socialist program, until one day America will be a socialist nation, without knowing how it happened.' He went on to say: 'I no longer need to run as a Presidential Candidate for the Socialist Party. The Democrat Party has adopted our platform.'"[1091]

In November of 2010, the midterm elections came, the voters spoke, and the Messiah's policies that are precipitating the radical transformation of America into a third world country were totally rejected, thank God! In a historical landslide, not seen in half a century, the electorate said "enough is enough" to the far-left Obama administrations agenda. The Republicans picked up 7 seats in the Senate and an unprecedented 63 seats to give them enough of a majority in the House of Representatives that they can halt any further damage, and hopefully slash the Democrats spending and reverse some of Obama's idiotic legislation, including Obamacare. But the incredible thing is that even in the midst of this tsunami there were still a number of pockets, mostly in Senate elections, where the Democrats barely hung on. There was Richard Blumenthal who won the Senate seat in Disconnecticut despite lying about serving in Vietnam. Barney Frank was reelected in Taxachusetts despite his being responsible for helping to orchestrate the housing market collapse. The "bearded Marxist" Chris Coons won the Senate in Delaware even though Harry Reid, the far-left Democrat Senate majority leader, affectionately called him his "pet." Another tax and spender, Martin O'Malley, was reelected to the governorship in Md. in a time when tax and spenders are bankrupting our country. In West Virginia they elected Joe Manchin who, to his credit, in a campaign ad shot a hole in Obama's asinine "Cap and Tax" bill, but he's still a Democrat, unfortunately. (Maybe he'll switch.) Harry Reid, the eager orchestrator of Obama's job-killing and economy-destroying policies as the Senates Majority Leader was rewarded with another 6 years by his obedient Democrats and Hispanics in Nevada. And California—bankrupt to the point of 130 billion dollars in debt which they will never be able to pay back and which was caused by the profligate spending of the Democrats who control the state house—in their eternal wisdom or their smoke-filled brains ("Hey dude, I forgot, who are we supposed to vote for, the Democrats?") elected Jerry Brown for another stint as governor. (The only positive thing to say there is that they deserve each other.) And, oblivious to what is actually happening in the United States of America (California being a separate country), elected ultra-liberal Barbara Boxer for another out-of-control-spending term in the Senate. Likewise in Oregon, Patty Murray was reelected to the Senate.

1091 (Posted on wordpress.com on May 4, 2009 by 84rules)

And my only question is, in this election, with all that the congressional Democrats have done to screw America, wipe out job creation and cripple our economy by rubber stamping Obama's feeble-minded, socialist agenda: why would anyone in their right mind go out and vote for them? Answer: They're not in their right mind politically or spiritually. They are in the mind of the great deceiver and they are doing his political bidding, controlled as they are by the evil one's media, entertainment industry and late-night clowns. Congratulations again, Mr. Stewart.

"The democracy will cease to exist when you take away from those who are willing to work and give to those who would not."

Why do blacks vote Democrat?

The same reason why Muslims worship Mohammed; because they are deceived into doing so. We already saw in chapter 11 how blacks have been misinformed into voting for their own slow genocide; how Margaret Sanger said she wanted to "exterminate the blacks" and how she is accomplishing that, even from the grave; how most abortion clinics are in poor black neighborhoods, and how 33 % of all babies exterminated each year are black babies even though blacks are only 12% of the overall population. The Democrat Party enjoys the black vote, but they do not deserve it. Like a pimp who enjoys the loyalty of the women in his stable, the left enjoys keeping their African-American voting block obediently on the Democrat plantation. But they do not deserve it. (See *Bamboozled: How Americans are being Exploited by the Lies of the Liberal Agenda*, by black author Angela McGlowan.) Satan truly is the great deceiver, who goes forth to deceive the whole world. And Black America, like all the rest of us, is not immune to his deceptions.

"A recent report by the Guttmacher Institute shows that black women now have abortions at five times the rate of white women. Among the 1.2 million babies aborted in the US each year—at least it's down from 1.6 million in 1990—37 percent are black, 34 percent are non-Hispanic white, and 22 percent are Hispanic.

"As an African-American, I am saddened by evidence that black women continue to be targeted by the abortion industry," said Bishop Martin Holley. "The loss of any child from abortion is a tragedy, but we must ask: Why are minority children being aborted at such disproportional rates?"

"An answer of sorts was given years ago by the Black Panthers, a militant group (since defunct), who were vehemently opposed to abortion. When New York State legalized abortion in 1970, the Panthers warned that the "oppressive ruling class" would use the new law to unleash a black "holocaust."

"It's hard to recall that many black leaders were once pro-life, among them the Rev. Jesse Jackson...

"Bishop Holley says that abortion is the leading cause of death for blacks, with 13 million abortions so far. That is one third of the present African-American population. With most Planned Parenthood clinics concentrated in inner cities, it is reasonable for Bishop Holley to believe that blacks are being targeted.

"In 1950, about 15 percent of black children were born to unmarried mothers... About 70% of black children are born to single mothers today, and before birth they are a constant target for abortion.

"Meanwhile, someone I know wondered out loud: How would Jesse Jackson like it if Klan members showed up at [Obama's] inauguration with signs reading: "Keep Abortion Legal"?"[1092]

When black America sold its soul to the Democrat Party they sold their soul to the devil.

"I predict future happiness for Americans if they can prevent the government from wasting the labors of the people under the pretense of taking care of them."[1093]

And in addition to the extermination-through-abortion issue, and all of the other reasons for not voting Democrat listed above, for blacks there also is this:

President Johnson passed "The Great Society" legislation with the help of a Democrat Congress back in the 60's, and I'm sure with the greatest of intentions. This "War on Poverty" was supposed to take us to a new utopia where "the poor" would be a thing of the past and liberal white rich people could fully enjoy all of their luxuries without those nagging feelings of guilt over "others less fortunate." But, as mentioned previously, the results of these liberal "good intentions" have been disastrous because Satan had something else entirely in mind, and as the god of this world he was able to further his plan to destroy America:

"To date, this has meant a $6.6 trillion transfer of wealth. The result; destruction of the black family, with illegitimacy rates at 75% (85% in some inner cities); an increase in the number of poor people (nearly 15 million on welfare and a tripling in the percentage of children on welfare); higher crime rates among the poor; families trapped in dependency."[1094]

And no ethnic group has been more directly, and adversely, affected than the blacks.

The lies of Lucifer blanket this earth, and people believe what they are taught. And one reason so many blacks vote Democrat, despite confessing Christianity in overwhelming numbers, is they have not been taught the true gospel of Jesus Christ but instead have been taught one form or another of the *social gospel*, one that merely uses Christianity as a cover, and one that is spread by those religious organizations that do seek to help the poor and that do try to right the perceived wrongs in our society. Unfortunately these social gospelites have been deceived into thinking that they and the Democrat Party are the great compassionate protectors of the poor and the downtrodden,[(LIE)] rather than the great destroyers of these folks and their lives that they in truth are. For they—liberals, Democrats, leftists, social gospelites—steal other people's money in the form of taxes, only a tiny portion of which actually ends up in the pockets of the poor and the needy, and then these handouts serve more to keep those poor people trapped in poverty and dependent on the government dole. The bulk of the money goes to bloated government bureaucracies that serve to further take this country down the disastrous road of Marxism-socialism, expanded government control over our lives, less personal freedom, less business freedom and therefore more of a drain on our economy, and more government centralized power and planning which is the communist and not

---

1092   (From article, "A Culture Mired in Callousness," by Tom Bethel, The American Spectator, February 2009)
1093   (Thomas Jefferson, source - brainyquote.com)
1094   (Rush Limbaugh, from "The Limbaugh letter," May, 2007)

the American dream. And of course the poor are not helped by this stolen largesse like they would have been if they had received aid from local Christian charities that could have also counseled and assisted them in overcoming the obstacles (drugs, alcohol, lack of education, lack of sexual abstinence as a teenage girl) that put them in and keep them trapped in their situation. Instead the poor under the liberals programs are trapped and imprisoned in poverty by making them and then keeping them wards of the Democrats ever-expanding socialist state. That is the end result of their perverting the true gospel of Jesus Christ and instead following a social gospel, a gospel of man's own making, one that is designed to soothe the conscience while inflating ones feelings of self-righteousness and superiority. A gospel that also keeps one trapped in the lies of the great deceiver, covered in one's own righteousness, supportive of the iniquity of murdering little babies, voting for the slow extermination of the black race, and separated from the salvation offered by Jesus Christ.

Again, the Democrat Party **enjoys** the African-American vote, but they do not **deserve** that vote. And they have enjoyed it for free for quite some time, having given nothing but misery and false hope in return. Wake up! My African American brothers and sisters!! Let's all make our Great Escape from the Democrat plantation.

And what of the Jewish vote?

"He has rescued us from the power of darkness, and transferred us into the kingdom of his dear Son." (Colossians 1:13-14)

Ben Stein had an interesting observation in his column in the American Spectator:

"Why don't all Jews vote GOP all the time? What's wrong with us? Bush is the most pro-Israel president there has ever been. He is so solid with Eretz Israel it's just magnificent. Stands up for Israel all the time. The Democrats? ...the party is ambivalent about Israel. Yet (Jews) voted something like 85% Democrat in the 2006 elections. Why?"[1095]

Why? Because they too are deceived. Like the black and the Hispanic vote, the Democrats enjoy the Jewish vote, but they do not deserve it. Like myself in 1980 when I was the stupid (spiritually), uneducated (politically) dupe who ran out and voted for Carter—"We gotta all go out and vote for Jimmy or that monster Reagan will become president! Oh no! What are we gonna do?"—the Jews and the blacks and the Hispanics run out mindlessly and vote Democrat because that is what they have been taught to do, and that is what liberals are supposed to do. And unfortunately, like myself years ago who also did not know any better, they obey. Let us pray for them; and for our country that is in serious danger because of the uninformed voting habits of these groups.

Government regulations

"Blessed is the nation whose God is the Lord." (Psalm 33:12)

Conversely, cursed is the nation whose god is their government.

"Government's multiple regulatory agencies (EEOC, FCC, FTC, FAA, FEC, FHA, TSA — which issues instructions how to pack your suitcase — FERC, etc.) attract liberals, who live to direct the (in their view) lesser lives of their fellow citizens. Though liberals

---

1095   (From article by Ben Stein in the American Spectator)

in general have little expertise in the industries they regulate, they flood the marketplace with millions of rules, attempting to "improve" society, "level the playing field," and all manner of social engineering. They regulate virtually every aspect of the production and distribution of food and goods, buying, selling, lending, trading, earning, hiring, profit making, and movement of capital. Today's mind-boggling tangle of regulations — nectar to liberals — shackles the nation and fails America's great spirit."[1096]

And Obama, the ultimate liberal, has taken the shackling of America's free, entrepreneurial spirit to a whole new level, disguised of course as some idiots long-awaited radical transformation. Oh happy day!

"My reading of history convinces me that most bad government results from too much government."[1097]

The Nazi Doctrine

"Our massive strategy was to use the Fairness Doctrine to challenge and harass right-wing broadcasters and hope the challenges would be so costly to them that they would be inhibited and decide it was too expensive to continue." -- President John F. Kennedy's commerce secretary, Bill Ruder.[1098]

As we have seen throughout this book, and will continue to document for the remainder of this chapter, the truth is not the Left's best friend. In fact, the truth exposes all of the many lies and distortions not only of the Democrat-Left, but of their father Satan as well. So the Left's problem is, "How do we shut down the voices of truth that are exposing us for the charlatans, deceivers and liars that we are?" (I know, they don't state it that way, that would be telling the truth, and again, they have a real problem with it.) One way the Obama administration has been attempting this is by reinstituting an old law known as the "Fairness Doctrine" which was created in 1949...

..."a time when the media had to compete for limited access to the broadcast spectrum. But rather than ensuring fairness it discouraged the media from airing controversial programming at all. [And of course "controversial" and "fair" is always in the eye of the beholder, which in this case was the eye of the progressive-left-controlled state.] Stations had no way of knowing what the government would consider "reasonable" or "vitally important controversial issues," or which viewpoints needed to be presented, so they stayed away from anything that might attract government attention.

"In 1989 the FCC under Ronald Reagan eliminated the Fairness Doctrine. The following year Rush Limbaugh's radio show went national. And soon AM radio, which had been given up for dead, was revitalized as hundreds of talk shows filled the air, most of them conservative-leaning.

"The left did not like this development. They were used to dominating the media. It wasn't long before calls began to reinstate the Fairness Doctrine."[1099]

"Did you know Obama has an FCC Diversity czar? He's Mark Lloyd and his real job is to silence talk radio. Lloyd, an ardent admirer of Hugo Chavez, believes that the

---

1096   (From article "Nailing the Left: Liberal Failures," by Rush Limbaugh, "The Limbaugh letter")
1097   (Thomas Jefferson, source - brainyquote.com)
1098   (From article, "Liberals Spew Propaganda and Push 'Fairness Doctrine' to Stop Conservative Talk Radio," by Brent Bozell, August 2007 issue of "The Watchdog," a monthly publication of the Media Research Center)
1099   (From a "Heritage Foundation Special Report")

fact Americans vastly prefer conservative talk radio is a "problem" that must be fixed by government. His solution - forced diversity through fees, fines and regulations that will drive private media out and bring public (meaning government) broadcasting in. Adolf Hitler had a man whose job description was much the same. His name: Joseph Goebbels, Minister of Propaganda for the Nazi Party."[1100]

Mitch McConnell, Republican Senate Minority Leader, had this to say in response to the Democrats most recent attempts at reinstatement:

"I will not support efforts to restrict free speech, silence political voices, and limit the free flow of information through legislative "fixes" like the so-called "Fairness Doctrine." Unlike the totalitarian state depicted in Orwell's 1984, ours is not a state controlled by a Ministry of Truth with a speech police telling our citizens what we can and cannot say and hear."[1101]

At least not yet. (But then again, Mitch must not have made any recent visits to a college campus.)

And Jay Sekulow of the ACLJ added...

"The reintroduction of the so-called "Fairness Doctrine" is a slap in the face of our constitutional right of guaranteed free speech. ...It's an unconstitutional restriction of conservatives' free speech, and an unconstitutional compulsion of opposing viewpoints ... government-mandated speech. It's an outrage! ...[And] a dangerous sign of the times."[1102]

But our friend Nancy Pelosi—"stop-an-electric-fan-with-her-tongue stupid" according to pundit Dennis Miller—has a different take. She provides a glimpse into the Democrat leadership's fascist strain by way of this comment from one of her senior advisors:

"Conservative radio is a huge threat and political advantage for Republicans and we have to find a way to limit it."[1103]

Limit free speech? Really, Nancy? So the Left, the great champions of free speech,[(LIE)] who have had complete control of the main-stream, lame-stream media, our entertainment industry and our educational system for decades, need to stifle conservative and Christian speech in order to keep themselves and their lies from being exposed. Wow! Move over Goebbels, there's a new sheriff in town.

The government-sector union scam

"America is staring into an economic crisis, and government unions stand in the way of a solution. As the nation plunges deeper and deeper in debt, governments at all levels find themselves unable to meet their obligations, yet public employees continue to press their demands for unreasonable compensation and benefits. All across the country, rapacious government unions are eating taxpayers alive, while working feverishly to shape the political battlefield to their own advantage—under cover of a crusade for "social justice."[1104]

1100   (Rush Limbaugh, from circular Sept 2010)
1101   (Senator Mitch McConnell)
1102   (Jay Sekulow, American Center for Law and Justice)
1103   (Senior advisor to Speaker of the House, Nancy Pelosi)
1104   (Pg. 1, Government Unions: How They Rob Taxpayers, Terrorize Workers, and

"In the private sector, only 7% of the workforce is unionized. In the public sector, the fastest-growing sector of unionization, you've got 40% and growing with the NEA, with the SEIU, with AFSCME and other government unions. That's the real 800 pound gorilla in most state capitals. Think of the conflict of interest here. You've got governors negotiating with state unions that contribute to their campaign, so if the governor wants to do a collective bargaining agreement with the state employee union, that very same party to that transaction is buying off the governor who is doing the negotiating. So there's an inherent problem here, and frankly the country ought to take a real good look as to whether or not public sector union membership is a good idea. …State and local workers make about 45% more than private-sector workers when you include health benefits and pensions. That's not fair."[1105]

"Collective bargaining [by public sector unions] has established in the state of New Jersey as well as many other states across the country over the last 50 years this whole new 4th branch of government. The unions in N. J. are the most potent powerful political force in the state. They raise millions of dollars from the payroll of their membership. They then use that money to influence elections. They then in essence elect the very people that they're negotiating with across the bargaining table. …Most of the people that are elected to office in New Jersey are beholden to the unions and that's really corrupted the whole Democratic system."[1106]

In February of 2011 all across the country newly sworn-in Republican governors and their new Republican-controlled state houses attempted to do what they were elected to do, save their states from bankruptcy. A financial crisis brought on in large measure by the obscene spending of the previous Democrat administrations, which included the outrageous and unsustainable pay and benefit packages given to their public service unions in return for lucrative campaign contributions and political support-ads for, of course, those same Democrats. This happened in New Jersey, Ohio, Illinois and Wisconsin just to name a few. But it is in Wisconsin where the budget battle reached the level of hysteria. There the 14 Democratic state senators fled the state in order to delay a vote on the new budget which merely required the public service unions to pay just a little more for their health care and pension benefits, yet still at levels far below that of their private sector counterparts. You'd have thought they were told to sacrifice their children. (Oh, that's right; Democrats have no problem with that.) The state capitol was flooded with throngs of crybaby protestors livid over being asked to give up some of their outrageous benefits in order to do a small part in helping reverse the states 3½ billion dollar budget shortfall; and in doing so to also keep thousands of their fellow workers from being laid off.

Their larger complaint, though, was that the new budget bill also ended their collective bargaining "rights," which allowed them to secure their outrageous benefit and pay packages in the first place, and even though many other states in the country do not allow their public service unions to collectively bargain, and despite the fact that those other states are attracting businesses and jobs. Without their unfair collective bargaining advantage, public service unions could no longer

---

Threaten Our Democracy, by Matthew Vadum)
1105   (Jason Lewis, Radio Talk Show Host on America Live with Megyn Kelly, 7/8/2010)
1106   (Steve Lonegan, from Americans for Prosperity, on Your World with Neil Cavuto, 2/25/2011)

force government employees to join, and then pay dues against their will. Dues that those unions in turn give almost exclusively to their Democrat friends, and against the will and political affiliation of a great many (roughly half) of their members. That's fair, if you're a Democrat politician at the receiving end of the scam, or a union member who has no problem being grossly overpaid at the expense of taxpaying citizens in the private sector who enjoy none of the special pay and benefit treatment that you have been getting.
" (page 1 of 3)

> As Wisconsin public employees do their intimidating, self-pitying best to assure continued advantages at the potentially catastrophic expense of everyone else, Franklin D. Roosevelt's is a voice to heed. Very nearly worshipped by liberals for his New Deal interventionism, this president was conservative enough to understand how the over-empowerment of government workers could distort democracy and cheat the common good. Government, he said, is supposed to be answerable to "the whole people" instead of some segment that comes along with the binding tool of collective bargaining. His gist was that government employees should understand they have a distinct obligation to serve the public, not force the public to serve them. The last thing permissible, Roosevelt added, is "militant tactics." With the shrillest possible militancy – the in-your-face, wild-eyed, rant-and-rave kind – the unionists protesting in Madison have screeched instead that collective bargaining is their right and that all they want is to negotiate."[1107]

Neil Cavuto was also unimpressed with their self-serving and disingenuous whimpering:

> "You really want to know the quickest way to stop change? Say it'll make things worse. *Much* worse. Put the fear of God into people. Exaggerate. If you must, lie if you have to. Take these budget cutting measures in so many states. Now they're about destroying unions,[LIE] not about saving money. Now they're about forcing teachers out,[LIE] not bringing accountability back in. Now they're trying to take their pensions away from them,[LIE] not simply ask that they pay a little bit more for them, not remotely close to what we pay, for example, just *closer* to what we pay for example. But no, now we're just decimating our schools,[LIE] not actually helping our kids. Now we're on a vendetta against the working man,[LIE] not trying to provide a break for the cash-strapped working *taxpayer*. Now we're just callous,[LIE] not just going after benefits that in this day and age seem clueless. That is what is unfair, my friends, equating cutting benefits from public workers with all but killing public workers. No one is dying here or remotely that many being let go here. Just making the same modest adjustments that scores of private workers have been making for years here. It is called being fair, and it is about getting real, and it's real unfair saying it's about anything else. Because when teachers use kids as pawns to protect *their* benefits, game over, reality over, our country over. Because you want to know the real truth here? It's about covering their costs. It's not about covering their asses. There is a difference."[1108]

And the following comes from an article by Michael Barone titled "Public-sector unions bleed taxpayers:"

> "Public-sector unionism is a very different animal from private-sector unionism. It is not adversarial but collusive. Public-sector unions strive to elect their management, which

---

1107   (From article, "FDR knew the danger of public unions" by Jay Ambrose, 2/28/2011 @ newschief.com)
1108   (Neil Cavuto's "Common Sense" commentary, from Your World with Neil Cavuto, 02/28/2011)

in turn can extract money from taxpayers to increase wages and benefits -- and can promise pensions that future taxpayers will have to fund.

"The results are plain to see. States such as New York, New Jersey and California, where public-sector unions are strong, now face enormous budget deficits and pension liabilities. In such states, the public sector has become a parasite sucking the life out of the private-sector economy. Not surprisingly, Americans have been steadily migrating out of such states and into states like Texas, where public-sector unions are weak and taxes are much lower.

"Barack Obama is probably the most union-friendly president since Lyndon Johnson. He has obviously been unable to stop the decline of private-sector unionism. But he is doing his best to increase the power -- and dues income -- of public-sector unions.

"One-third of last year's $787 billion stimulus package was aid to state and local governments -- an obvious attempt to bolster public-sector unions. And a successful one: While the private sector has lost 7 million jobs, the number of public-sector jobs has risen. The number of federal government jobs has been increasing by 10,000 a month, and the percentage of federal employees earning over $100,000 has jumped to 19 percent during the recession.

"Obama and his party are acting in collusion with unions that contributed something like $400,000,000 [$400 MILLION] to Democrats in the 2008 campaign cycle. Public-sector unionism tends to be a self-perpetuating machine that extracts money from taxpayers and then puts it on a conveyor belt to the Democratic party."[1109]

As with their phony concern about the poor, the Democrats pretend to care so much about "the rights of workers" that they can barely stand that as well. But it's all a ruse. It's really about the money. Listen to Wisconsin Governor Scott Walker...

"They [the public union bosses] were willing to throw their members under the bus as the process went along because what they wanted more than anything was an automatic guarantee that every public employee in the state be forced to be a part of a union, and that they get those union dues. [Emphasis mine] We give those employees the right to choose whether or not they want to be a part of a union or not. And ultimately we saw the real fact is they're fighting not for worker rights, they're fighting for union dues. ...That's the politics in play. It's not about protecting workers rights. It's ultimately about grabbing those dollars for their power plays."[1110]

Before the 1950s there were no state and local government unions. As mentioned, Franklin Delano Roosevelt, the Left's hero, was against them. And another Democrat, Jimmy Carter, actually banned collective bargaining by federal union employees during his thankfully-single term. So today there are 2 million federal workers who do not have collective bargaining rights like some state unions are clamoring to keep in place. And interestingly, even though president Obama has done nothing to change that reality in the last two years when he controlled both houses of Congress and could have easily done so, this disingenuous rascal has come out and injected himself, again, into something he should not have and declared that the Wisconsin budget bill is "an assault" on the unions.[(LIE)] He of

---

1109   (From article, "Public-sector unions bleed taxpayers" by Michael Barone on 2/07/2010@ the washingtonexaminer.com)

1110   (Wisconsin Governor Scott Walker on Your World with Neil Cavuto, 5/27/2011)

course is a liar. The bill is a move in the right direction to get the state solvent again; so it can pay its bills; and its salaries; and the bill merely does to Wisconsin's state unions what Jimmy Carter had already done to federal workers decades ago. (And of course what Obama had also done by doing nothing to reverse Carter's ban.) And that's the truth that our Democrat-Left media forgot to tell us. (They're still looking for it, out on the golf course with OJ.)

> ..."Texas does not have collective bargaining and their state's doing terrifically well. I don't think there's any starving workers in Texas, and Texas is leading the country in economic growth."[1111]

One union employee in Wisconsin, a bus driver, made $157,000 in one year! And bus drivers there make on average over $100,000 per year![1112] And that's because with collective bargaining in place, the public service unions are negotiating with the same people who are beholden to them and their campaign contributions in order to get re-elected! That's bribery! It's extortion! It's a quid pro quo. You scratch our back, and we'll scratch yours. You meet our demands for excessive pay and benefits, and we'll in turn support your re-election. And screw the taxpaying suckers in the private sector who make 42% <u>less</u> than us, but are forced to subsidize our largesse! And if our state goes bankrupt, tough beans.

In the private sector employees negotiate their salary and benefits package according to their worth. If the company fails to pay its employees what they are worth then they will go elsewhere. But the increasingly more and more outrageous pay and benefits that these government sector unions have negotiated over the years has nothing to do with the worth of their employees. It has everything to do with extorting outrageous levels of salary and benefits, far above what the employees are worth, because their union bosses have the Democrat politicians in their hip pockets. Again, it's a quid pro quo. And that is what the union protestors in Wisconsin are rioting over. It's enough to make you sick, especially if you are someone who lost your job in the private sector in this terrible Obam-economy, or a taxpayer forced to pay for their largesse, or someone who will be adversely affected when their state goes bankrupt. And their union leaders even went so far as to call upon the memory of Martin Luther King and his assassination in Memphis to somehow support their own crybaby protest and selfish demands. But Peter Kirsanow, a member of the US Commission on Civil rights, saw through that one:

> "It is delusion, bordering on abomination, to try to equate what Martin Luther King was doing in Memphis to public workers getting Cadillac benefits for which they contribute very little, or nothing, at taxpayers' expense."[1113]

And the Reverend Jesse Lee Peterson, a non-race-baiting black preacher who our lame-stream media of course ignores, had this to add not only about the Union rioters attempt to misappropriate the memory of Dr. King, but also about how the Democrats did the same thing during the health care debate when Nancy Pelosi was seen walking arm-in-arm with black congressional leaders in order to evoke memories of the civil rights movement:

> "Some 50 years ago, the government came in under Lyndon B. Johnson, along

---

1111   (Steve Lonegan, from Americans for Prosperity, on Your World with Neil Cavuto, 2/25/2011)

1112   (Source, Ann Coulter on Hannity, 03/10/2011)

1113   (Special Report with Brett Baer, Fox News Channel, 03/18/2011)

with these so-called civil rights leaders, and they said to black Americans, 'You've been discriminated against. The government is going to help you, but you can't have a father in the home.' So they destroyed the family by taking the fathers away and now many if not most black Americans are relying on the government. I now realize that the reason that they did that, because if they could cause black Americans to rely on the government, rely on the Democratic Party, then it would give power to the Democrats, to the liberals, and that's what you saw with this socialized health care bill. It was about using black Americans in order to gain power and wealth. The unfortunate thing is that my grandparents and my great grandparents, they suffered because they were black, they fought for freedom, so did Dr. King. They just wanted to have the right to work hard, to live in a country as free men and women. They never intended for the family to be destroyed, and these so-called civil rights leaders, along with the liberal Democrat Party to use black Americans such as they're doing now. Whenever they want to destroy America and the values of this great country they use the civil rights movement in order to do that. We've seen them do it with this idea that America is a racist society. We've seen them do it by saying that black folks are not capable of working for themselves, so they need *programs*. It's just a *misuse* of a people who needed the government to protect them. This is a sad occasion, and for them to do this on the day that Dr. King was assassinated, it shows that they have no love for black Americans, nor do they have love for our great country."[1114]

It is also revealing that it took the current, record financial crisis to wake the American people up to what has been going on. Which begs the question why our lame-stream, Democrat media never brought any of this to their attention before? And why they are still lying about these union abuses, covering them up, and even outright supporting the corrupt, financially-devastating gravy train to this day?! Answer? We just said it! It's a <u>Democrat</u> media. That's why! (Shazaam.)

"Our nations unions are becoming increasingly difficult to distinguish from the Brown Shirts or the Communist thugs of the last century."[1115]

The union goons in Wisconsin have also given our Democrat Hitler youth a new mantra to spew, "Fox News lies."[(LIE)] (In the muddled brains of these youths if you actually report the news, you're "lying!?") The Democrat's fascist youth shout their new slogan into that network's cameras at every opportunity in order to try to silence real journalism. (In the name of free speech of course.) The children appear incensed that anyone would dare report the news. For them, "fair" reporting comes only from the state-run networks, those that do the bidding of Obama, the Democrats and the unions by skewing the reportage to favor their obnoxious and unsupportable demands. Yea, *Fox News Lies*! Now the Democrats have another nifty slogan that they can put up there on the propaganda mantle right next to "It's the economy, stupid," "Tax cuts for the rich,"[(LIE)] "Bush lied and soldiers died"[(LIE)] and their classic bilge: "YES WE CAN!"[(LIE)]

See:
- "Government Unions: How They Rob Taxpayers, Terrorize Workers, and Threaten Our Democracy," by Matthew Vadum

---

1114 (The Reverend Jesse Lee Peterson on Your World with Neil Cavuto 04/04/2011)
1115 (Glenn Beck, on Fox News' The Glenn Beck Show, 05/04/2011)

The Battle of the Budget

"To destroy, or not to destroy our economy, that is the question. Whether 'tis nobler to suffer the slings and arrows of outrageous lies from the Democrats and their media. Or to take arms against the lying skanks, and by exposing, end them. So they can tax, spend and regulate us into oblivion, no more…"[1116]

In February of 2011, only 3 months since they lost the House, the Democrats started screaming and lying about the Republicans attempts to cut just *a tiny fraction* of their outrageous spending of the last two years. They're hoping for a repeat of 1994 when the Republicans, after winning both houses of Congress in the midterm elections and then focusing on cutting some of the liberal Democrats worthless, costly and unnecessary programs, were assaulted by the Democrats and their media with one lie and one distortion after another. They were accused of wanting to "cancel Big Bird" by attempting to remove taxpayer funding from the decidedly liberal PBS (the Public Broadcasting System); of wanting grandma to eat dog food, of starving children and herding the homeless into slave-labor camps. They were also successfully and falsely saddled with the blame for the temporary government shutdown that occurred. So, emboldened by that previous success, the Democrat-Left is hoping for a repeat performance. They've called on all of their congressional and media dogs to demagogue the issue, force another government shutdown, lie to the predictably stupid American voter, and hopefully get them back into power. And screw both the country and our economy in the process. Harry Reid, Anthony Weiner and Nancy Pelosi have made some of the more absurd pronouncements, with Pelosi saying that a 5% cut in a meals program for the homebound elderly would starve millions of seniors. And where exactly it is written in our Constitution that feeding seniors is the role of the federal government, she couldn't say. That has traditionally been the job of local churches and charities that would be more than happy to feed the needy before Republicans got around to "starving them to death." But facts won't deter them and their media from falsely disparaging any attempt by the opposition to get our federal budget in order. Every sensible and necessary budget cut will be met with one hysterical claim after another. Because the Democrats and their Left don't want to cut spending (if they did, then why would they have so irresponsibly run up such an enormous debt and deficits in the last two years in the first place?!); and they don't care if the country goes bankrupt; or turns into a third-world country; as long as they can get the house of representatives back.

"Here's all you need to know how tough it will be for Republicans to cut spending. The media won't cut them a break if they so much as try, even if what they're trying really isn't that much at all. Take today's New York Times. (Please!) … Big editorial entitled "The High Price of Rigidity." And guess who's rigid? Tada! *Clueless, selfish* Republicans. It goes on to say that the Grand Old Party will butcher, their words, the current year's budget with $61 billion of what the newspaper calls, "radical" cuts. 61 billion is radical in a deficit of more than 1 & 1/2 trillion. Silly me, the 1 1/2 trillion number is the number that really strikes me as radical, but then again, I'm not the Times. Just like I'm not on the same page about what's the real urgency here; not whether the government shuts down for a while, but our country and very system shuts down, period. Because if the New York Times thinks 61 billion is being rigid, please tell me what is being soft? …$61 billion is pocket change. …In a more than 3 1/2 trillion dollar budget it's like a rounding error. …

1116 (From *Hamlet's* opening soliloquy, by William Shakespeare)

So back to the headline, "The High Price of Rigidity," at first I thought the good folks at the good Times were just talking about themselves."[1117]

The Times *were* talking about themselves, Neil, they're just too stupid to understand that, and of course are counting on their readers being the same. But here's an even more egregious example of what the Democrat-Left has in store for the Republicans, and the nation:

"Perhaps the most surreal moment in the budget crisis came when Senate Majority Leader Harry Reid said, concerning Republican efforts to pass another temporary measure, "they'll say it's short-term but what that really means is it's a short-cut around doing our jobs.[LIE] Instead of solving problems, they're stalling. [LIE] They're procrastinating. [LIE] That's not just bad policy, it's a fantasy." "Reid, of course, belongs to the party that couldn't produce a budget despite years of total control in both houses of Congress. Is he out of his mind to suggest the *Republicans* are somehow "procrastinating" because they haven't corrected his party's dereliction of duty yet?"[1118]

No, he is in complete control of his reprobate mind. Reid and the Democrats are playing politics with the solvency of our economy. And their lying, pious, pompous-ass-in-chief came out and issued a "dire warning" concerning the possible shutdown (as if he really cared):

"The Obama administration warned Wednesday that a federal shutdown would undermine the economic recovery, delay pay to U.S. troops fighting in three wars, slow the processing of tax returns and limit small business loans and government-backed mortgages during peak home buying season."[1119]

But the question his Democrat media never posed to him was, "Then why are you shutting it down, you idiot?" For it is Obama and his Democrats who are responsible for there not being a budget at this eleventh hour. This budget was supposed to be passed by him last October (2010) when he controlled <u>both</u> houses of Congress. But they were derelict in their duty because they were afraid it would be used as a campaign issue against them. So they did nothing. And now they are still doing nothing, except falsely smearing, disparaging, vilifying and lying through their teeth about whatever budget the Republicans who control the House of Representatives now actually <u>do</u> put forward.

Besides, if Obama were honest, or possessed a rudimentary level of common sense, he might be a tad more concerned about his outrageous spending that got us here in the first place. As Neil Cavuto opined, he should "save the feigned outrage" for those who are fooled by it:

"I'm not worried about our government shutting down, it might happen. I'm more worried about our fiscal sanity shutting down because it already *has* happened. I'm more worried about our comprehension of *basic math* shutting down. I'm worried more about the disasters government creates when it is *open*, not when it is closed. ...I'm not impressed with those warning of financial Armageddon if we shut down, but didn't say boo when we were piling all this debt up. So quit whining about a shutdown and for God's sake just man up; about your spending that has us spent; and your new found concern for fiscal responsibility that has us at a loss. You say there is hell to pay if we close the government?"[1120] ...There was far more "hell to pay" when Obama had it open for the last two years.

1117   (From Neil Cavuto's Common Sense, Your World with Neil Cavuto, 03/30/2011)
1118   (John Hayward, from a "Daily Events" email, 4/07/11)
1119   (From online article "White House says shutdown will delay pay to troops," by Richard Lardner And Jim Kuhnhenn, Associated Press, 04/07/2011, @ yahoo.com)
1120   (From Neil Cavuto's "Common Sense" on Your World with Neil Cavuto, 04/07/2011)

Yet the Republicans response to the Democrats demagogue-ing of the deficit issue has been less than stellar:

> "It is increasingly clear that the Republicans are a bunch of wimps. In the past week, Democrats have said the Republicans want women to die and old people to starve. They've accused the GOP of hurting kids. Barack Obama himself has entered the fray. He says the GOP would turn the nation into a third world state. He says the Republicans want to hurt the elderly with their Medicare reforms.
>
> ..."All the while, the Republicans keep talking about wanting to reach across the aisle. They are perfectly willing to go along with the Democrats in raising the debt ceiling.
>
> ..."The Republicans have gotten it in their head that if they just play nice and look leaderly, the nation will embrace them. Instead, the GOP is getting defined by the Democrats. They are getting pinned down by a media willing to push the Democrat talking points and their polling is correspondingly going downhill.
>
> "John Boehner and Mitch McConnell don't seem to understand that we need fighters not lovers right now. The Republican Party has turned into the Party of Wimps. If they don't start throwing punches soon, they'll be undone and so will the nation."[1121]

Mark Levin calls it "a gutless, Rino" (Republicans in name only) party with a "mush-headed mentality."[1122] Or like we said earlier in chapter 2 ...*what is certain is that just as the 9/11 report said that Islamic psychos were at war with us but we weren't at war with them, the Democrat-Left and their media declared all-out war on Republicans, conservatives and Christians a long time ago. But still a number of naïve Republican leaders seem to think they're in a pie-eating contest at their churches Sunday social. And they can't figure out why, as they're bent over slurping away on their lemon meringues' like good little boys and girls, the opposition across the aisle are bashing them over their heads with the picnic benches.*

And Casino Mogul Steve Wynn, whom many consider to be the father of modern Las Vegas, has had about all he can stomach from our deceiver-in-chief:

> "While the president and his minions vilify any attempt to straighten out America and act responsibly, I end up shouting at the television set. [How] outrageous the lies and the misrepresentations; the deliberate lies are incredible. And you ask yourself, when will the American public have enough of this jazz that's being dealt to them? Haven't we gotten it yet?"[1123]

Apparently not, Steve, because despite the "wild, reckless, irresponsible spending of this government" our main stream moron-media nevertheless has continued to make certain that there are still enough uninformed voters out there to reelect the jackass. They have failed to inform their captive and obedient audience that if the massive deficit that his spending has created is not immediately reversed then the shitake is really going to hit the fan. Our credit rating has already been lowered and our astronomical debt is now poised to knock the dollar from its lofty perch as the world's reserve currency, which in turn will likely lead to hyper-inflation which will destroy our economy and our very way of life. But at least that will allow the Democrats to take advantage of another golden opportunity of never letting a crisis go to waste. They'll just blame it all on Bush, declare Marshall Law, and ride the destruction they caused to even more political power.

1121   (By Erick Erickson @ Daily Events online, 04/15/2011)

1122   (Mark Levin on Your World with Neil Cavuto, 5/18/2011)

1123   (Steve Wynn, on Your World with Neil Cavuto, 04/20/2011)

"A reserve currency, or anchor currency, is a currency that is held in significant quantities by many governments and institutions as part of their foreign exchange reserves. It also tends to be the international pricing currency for products traded on a global market, and commodities such as oil, gold, etc.

"This permits the issuing country to purchase the commodities at a marginally lower rate than other nations, which must exchange their currencies with each purchase and pay a transaction cost. For major currencies, this transaction cost is negligible with respect to the price of the commodity. It also permits the government issuing the currency to borrow money at a better rate, as there will always be a larger market for that currency than others."[1124]

But that advantage that the United States has enjoyed since shortly after World War II is at risk of disappearing as some of the other major countries in the world have already stated their desire to see a different reserve currency than the dollar. A currency that is not tied to a country where half of its populace still supports a president and a federal government that has just spent them into oblivion.

"Standard & Poor's on Monday warned the United States that it could lose its AAA bond rating if the wild spending continues. If that AAA rating goes away, your dollars will be worth less and many people will not buy U.S. bonds, a disaster in the making.

"So let me ask you something. How much more evidence do we need for Americans to understand that the country is in dire trouble? I mean, if you are a liberal American, you've got to wake up. Your dollars are in jeopardy, too."

[That was from Bill O'Reilly's "Talking Points Memo" in April of 2011, and that downgrade did happen just 4 months later in August of that year. O'Reilly goes on to quote the president on his early campaign trail:]

"PRESIDENT OBAMA: 'I want to live in a society that's fair ... why wouldn't I want to have a society where I knew that the American dream was available for everybody?'

[The American dream is already available for everybody, Wilbur, only it is only **attainable** by those who will go out and work their asses off for it. It is NOT attainable, however, by those who expect you, Mr. Ed, and your government to try to GIVE it to them by stealing from others. O'Reilly replies to Barack's sophomore-in-college level of naïveté...]

"Respectfully, Mr. President, it is impossible to have the American dream available for everybody. It can't happen. **Stop chasing leprechauns**. [Emphasis mine.] Some people will not work for a living, will not go to school, will not obey the law. That is reality. Responsible safety nets are needed, but the nanny state mentality is going to bring this country down unless we get rid of it."[1125]

Which would entail first getting rid of a certain non-Muslim president, his "nanny state mentality" and his entourage that infests much of Capitol Hill, would it not?

H.L Mencken prophetically warned us more than 90 years ago...

"As democracy is perfected, the office of President represents, more and more closely, the inner soul of the people. On some great and glorious day the plain folks of the land will reach their heart's desire at last, and the White House will be occupied by a downright moron."[1126]

That happy day has finally arrived.

1124   (From Wikipedia, the free encyclopedia)

1125   (From The O'Reilly Factor's "Talking Points Memo" 04/19/2011)

1126   (H.L. Mencken, The Baltimore Evening Sun, July 26, 1920)

Obama came out recently in one of his campaign rallies (he's interested in getting reelected, period, and not in curtailing his spending) and lied about how the Republicans were cutting spending by "breaking the backs of the poor."[(LIE)] Bill O'Reilly exposes the disingenuous rascal...

"Talking Points does not believe that cutting federal spending is breaking the backs of the poor. Quite the contrary, if the president continues racking up debt *all* of our backs will be broken because the American economy will collapse. But there *is* merit in Mr. Obama's theme, that a just society should help the downtrodden. ...But what's happened in America is that many social justice programs simply don't work. And we have lost trillions of dollars trying to force-feed, pardon the pun, folks who are not doing well. Americans *are* the most generous people in the world and studies show conservative Americans give more to charity than liberals. So it is unfair of the president to label Republicans and conservatives as uncaring. That's simply not true. And the president should *get off it*, with all due respect. What needs to be done now is massive budget cutting while at the same time instituting *smart* programs. One vivid example: education. The feds pour billions of dollars into classrooms every year and much of the money is wasted on touchy-feely programs that do little.

"Here's what we need: Every public school kid in a uniform, strict code of behavior on school grounds, mandated test scores for every grade and special-help programs for kids who fall behind or have learning disabilities. Those ideas don't cost a ton of money. Discipline is free. Taking political correctness out of the system is free. So there you go. I just saved the country a trillion dollars. Smart policy, not lavish spending, is the key to America's future success."[1127]

But before you stand a chance at instituting any reforms in our schools that actually make sense, Bill, you'd have to ship the ACLU to Cuba. Getting rid of them, and the policies of the Democrat-Left, "is the key to America's future success." Florida congressman Allen West explains to Greta Van Susteren his comment that president Obama was showing "third world dictator-like arrogance" in his latest campaign speech:

"The truth needs to be said. ...I am sick and tired of this class warfare, this Marxist, demagogic rhetoric that is coming from the president of the United States of America. It is not helpful for this country, and it's not gonna move the ball forward as far as rectifying the economic situation in our country. I'm not going to back away from telling what the truth is.

..."I don't think it's very presidential when Barack Hussein Obama refers to my colleague, Paul Ryan, as a simple little accountant either. So I think that when you look at what a community organizer is turning out to be, it does seem to be like a low-level socialist agitator. When you look at the economic situation that we have, to have a gentleman in the White House who really has never run any type of business or organization... When I talk about the chickens coming home to roost when we continue to play this election-cycle American-idol in the United States of America, this is what we end up with, with someone that really is not in tune with the American people. ... [Someone who] never even ran a lemonade stand."[1128]

And Mark Levin had this response to another round of their predictable race-baiting:

1127   (From The O'Reilly Factor's "Talking Points Memo" 04/21/2011)
1128   (Allen West, "On the Record with Greta Van Susteren," 04/22/2011)

"I'm sick and tired of liberals in this country, most of whom are white, telling conservatives in this country that they're racist because we don't adopt their big government, authoritarian, top-down policies that seek to divide this nation on every front, including race."[1129]

The Democrats fought hard for all of their outrageous spending, and they will not give it up without a fight. (Screw the country, its businesses, jobs and the economy.) Here is what Erskine Bowles, himself a Democrat, had to say about our looming economic disaster:

"I think we face the most predictable economic crisis in history. A lot of us sitting in this room didn't see this last crisis as it came upon us. [Referring to the housing bubble collapse of 2008 and the banking industry meltdown that helped get Obama elected.] But this one is really easy to see. This debt and these deficits that we are incurring on an annual basis are like a cancer. And they are truly going to destroy this country from within, unless we have the common sense to do something about it."[1130]

And that's exactly Satan's design, "to destroy this country from within." Back in 1994 the electorate listened to a lying, left-wing media and erroneously blamed the Republicans for the government shutdown, even though that was an early attempt by them to divert the government, and the country, away from what we are now facing today. The voters then went out and reelected a rapist for another four years in the White House. And the Democrat-Left figured if they could get away with that, they could get away with anything. And now, a decade and a half later, its déjà vu all over again. Obama and the Democrat-Left are again going for pure political gain and playing rank partisan politics despite the fact that this time they are doing it in the face of "the most predictable economic crisis in history" that is "truly going to destroy this country from within." They could care less. It is a crisis that anyone with a brain, and some common sense, can see coming. But one that idiots cannot. And so our president, our media, the idiots in our entertainment industry, the late-night bozos and the Democrat weasels in congress like Weiner and Schumer, Pelosi and Reid are banking on their voters listening to them again. And Satan and his minions are cheering them all on. God help America.

"And the devil that deceived them was cast into the lake of fire and brimstone, where the beast and the false prophet are, and they shall be tormented day and night for ever and ever." (Revelation 20:10)

Conclusion

Satan is relentless in his quest to destroy Americas economy by directing the path and the policies of our federal government. In election after election, and in law after business-stifling law, he continues his steps forward toward that goal. And even though he suffers some little setbacks along the way, like the 2010 congressional elections that saw him lose the house, he is not deterred in the least. His very eternal fate hangs in the balance. In the last 50 years for every two steps he has taken forward, he has only been pushed back one. Those true-believing

---

1129 (Mark Levin on Your World with Neil Cavuto, 5/18/2011)
1130 (Erskine Bowles, co-chairman of President Obama's own Deficit Commission, speaking before a congressional committee on March 8, 2011)

liberals who control our media, educational system, entertainment industry and presently 2/3's of our government, think they are fighting the good fight, to redistribute wealth to help the poor and the downtrodden, to institute a *better* economic system in America; but in reality they are working hand-in-hand with the evil one to destroy America economically (and the poor will be the ones hurt the most). This of course will weaken us militarily so we will no longer be able to come to the aid of the deceivers primary target, Israel. And the beat goes on...

See:
- *If Democrats had any Brains, They'd be Republicans*, by Ann Coulter
- *Uncle Sam's Plantation: How Big Government Enslaves America's Poor and What You Can Do About It*, by Star Parker
- *Bamboozled: How Americans are being Exploited by the Lies of the Liberal Agenda*, by Angela McGlowan
- *Culture of Corruption: Obama and His Team of Tax Cheats, Crooks, and Cronies*, by Michelle Malkin
- *Radical-in-Chief: Barack Obama and the Untold Story of American Socialism*, by Stanley Kurtz
- *Liberal Fascism: The Secret History of the American Left from Mussolini to the Politics of Meaning*, by Jonah Goldberg
- *Liberty and Tyranny: A Conservative Manifesto*, by Mark R. Levin
- *Real Change: From the World That Fails to the World That Works*, by Newt Gingrich
- *Showdown: Confronting Bias, Lies, and the Special Interests that Divide America*, by Larry Elder
- *White Guilt: How Blacks and Whites Together Destroyed the Promise of the Civil Rights Era*, by Shelby Steele
- *Hating Whitey and other Progressive Causes*, by David Horowitz
- *Stupid Black Men: How to Play the Race Card—and Lose*, by Larry Elder
- *Crimes Against Liberty: An Indictment of President Barack Obama*, by David Limbaugh
- *Fed Up!: Our Fight to Save America from Washington*, by Rick Perry and Newt Gingrich
- *The Obama Diaries*, by Laura Ingraham
- *Demonic: How the Liberal Mob Is Endangering America*, by Ann Coulter
- *Gangster Government: Barack Obama and the New Washington Thugocracy*, by David Freddoso
- "From Shadow Party to Shadow Government: George Soros and the Effort to Radically Change America," by David Horowitz and John Perazzo
- *The Great Destroyer: Barack Obama's War on the Republic*, by David Limbaugh
- "Barack Obama's Rules for Revolution: The Alinsky Model," by David Horowitz

9- Weaken America from without
"Even him, whose coming will be in accordance with the work of Satan, with all kinds of counterfeit miracles, signs and lying wonders, and in every sort of evil

that deceives those who are perishing; because they did not receive the love of the truth, that they might be saved." (2 Thessalonians 2:9-10)

Radical Islam's war on America

"You believe that there is one God. You do well. The devils also believe, <u>and tremble</u>." (James 2:19) (A special verse for those religiously devout folks trapped in the world's most heinous false religion. Pray for them.)

Satan has sought to destroy America from without primarily by using the more fanatical of his Islamic followers to wage jihad against the United States. Like the confederates at Gettysburg, one can only hope that 9/11 was their high watermark. For there is still the possibility of their succeeding in significantly weakening America if they are able to get their hands on a nuclear weapon and can sneak it into the United States to blow up Los Angeles, New York, or Washington DC. It may not be possible for our nation to ever recover from the havoc that would wreak, especially since it is very probable that God has already removed his hand of blessing and stopped "shedding his grace on thee" in large part due to our ignorance in continuing to vote for those who mutilate his innocent little babies to death and also for spitting in his face by teaching Satan's idiotic genesis account over His scientifically-accurate one.

> "Since 9/11, Islamic jihadists have perpetrated not 500 not 2,500, but more than 5,000 terror attacks. Liberals hush-up this number, claiming the truth would provoke "reprisals" by "violent" Americans against "peaceful" Muslims."[1131]

("Liberals hush-up this number" because, again, they and their clitoral-mutilating friends are one and the same.)

This global conflict has been misnamed the "War on Terror" because the more politically correct among us in the Bush administration didn't have the sense or the courage to call it what it actually is, not a war on "terror"—we're not fighting an emotion after all—but a war against Muslim psychotics, whose minds have been obliterated by the lies of Mohammed. But then in calling it what it actually is, you have to use the word "Muslim," and that truth is unacceptable both to the adherents of the "religion of peace" and the politically-correct among us.

> "Islamofascism is the successor to Nazism and communism as a totalitarian challenge to the Western Democratic way of life."[1132] -Norman Podhoretz

After 9/11 Ann Coulter had this observation:

> "We know who the homicidal maniacs are. They are the ones cheering and dancing right now. We should invade their countries, kill their leaders, and convert them to Christianity. We weren't punctilious about locating and punishing only Hitler and his top officers. We carpet-bombed German cities; we killed civilians. That's war. And this is war."[1133]

They were at war with us for quite some time, but we didn't get it. And even now the Democrat-Left still doesn't get it. Their ACLU is waging its own war to hinder us from prosecuting this one. They have us fighting on two fronts. Why they haven't been locked up as traitors is beyond me, but it has a lot to do

---

1131   (From a Human Events circular)
1132   (Norman Podhoretz, author of *World War IV: The Long Struggle Against Islamofascism*)
1133   (From "This Is War" 9-12-01, pg. 7, *If Democrats Had Any Brains, They'd Be Republicans*, by Ann Coulter,)

with the Democrat-Left itself which controls our media, entertainment industry, educational system, courts and now our federal government, who are on the same side as those traitors. Our enemy within is just as dangerous, perhaps more so, than the Islamic enemy we fight all across the globe. The Democrat-Left is more interested in pretending to be more "compassionate" and "loving" then the more *barbaric* among us who are more interested in killing them before they kill us. They are also interested in having "a dialogue" with our enemies despite the fact that our Muslim enemies have no interest whatsoever in having a dialogue with them. (Unless that conversation were to occur during one of their head-removal ceremonies.)

But as far as real compassion and real love is concerned, it is important to treat even our fiercest enemies, and even in a life and death global war like this, with the love of Christ. And Americans of course do for the most part. There are countless stories of our brave men and woman overseas risking their own lives to save the lives of Islam's jihadists who, moments before, were trying to kill them but who now lay wounded and in need of aid. Aid that comes in the form of American soldiers and medics. God bless them.

Rick Mathes, a well-known leader in the Christian prison ministry, had this to say:

"Last month I attended my annual training session that's required for maintaining my state prison security clearance. During the training session there was a presentation by three speakers representing the Roman Catholic, Protestant and Muslim faiths, who explained each of their beliefs. I was particularly interested in what the Islamic Imam had to say. The Imam gave a great presentation of the basics of Islam, complete with a video. After the presentations, time was provided for questions and answers. When it was my turn, I directed my question to the Imam and asked: "Please, correct me if I'm wrong, but I understand that most Imams and clerics of Islam have declared a holy jihad [holy war] against the infidels of the world and, that by killing an infidel, (which is a command to all Muslims) they are assured of a place in heaven. If that's the case, can you give me the definition of an infidel?"

"There was no disagreement with my statements and, without hesitation, he replied, "Non-believers!"

"I responded, "So, let me make sure I have this straight. All followers of Allah have been commanded to kill everyone who is not of your faith so they can have a place in heaven. Is that correct?"

"The expression on his face changed from one of authority and command to that of "a little boy who had just been caught with his hand in the cookie jar."

"He sheepishly replied, "Yes."

"I then stated, "Well sir, I have a real problem trying to imagine Pope Benedict commanding all Catholics to kill those of your faith or Dr. Charles Stanley ordering all Protestants to do the same in order to guarantee them a place in heaven!"

"The Imam was speechless! and I continued, "I also have a problem with being your "friend" when you and your brother clerics are telling your followers to kill me! Let me ask you a question. Would you rather have your Allah, who tells you to kill me in order for you to go to heaven, or my Jesus who tells me to love you because I am going to heaven and He wants you to be there with me?" You could have heard a pin drop as the Imam hung his head in shame.

"Needless to say, the organizers and/or promoters of the "diversification" training seminar were not happy with Rick's way of dealing with the Islamic Imam and exposing the truth about the Muslims' beliefs. In twenty years there will be enough Muslim voters in the U.S. to elect the President! [Hmmm... In 2008, didn't we already do that for them?] I think everyone in the U.S. should be required to read this, but with our liberal justice system, liberal media and the ACLU, there is no way this will be widely publicized. The most puzzling thing to me is... once Islam takes over, there will no longer be a liberal media or an ACLU, or a liberal justice system. So in effect, they are slitting their own throats by not publicizing this."[1134]

Ultimately we are fighting a war, not only against brainwashed Islamic jihadists, but against their father Satan who has stripped them, like the Imam above, of any logic, of any common sense, of any ability to smell the obvious stench of the false-religious lies that inhabit their brain and direct their paths. Like the theory of evolution that is so scientifically impossible, so stupid, idiotic and absurd that it literally borders on the insane; so too the religion of Islam is so disgusting, so vile, so idiotic, mutilating the genitals of a hundred and thirty million of their own little girls, worshipping a known child molester and his demonic god, Allah, so bereft of any common sense or logic or truth that it also is literally insane. People believe what they are taught and there but for the grace of God and the geography of our birth, go you or I. Pray for them.

Earlier in this book we said that the baby-mutilating Democrat-Left in this country and the clitoral-mutilating Muslims across this globe are one and the same. They have the same father, Satan, the father of lies, the great deceiver, the one who goes forth to deceive the whole world. And with this in mind it is easy to understand why the Left in this country is doing what it can, not in all cases but certainly in many, to hinder the war against their Muslim terrorist brethren. Like their take on the responsibility (or lack thereof) of criminals, their crimes being not their own but societies fault,[(LIE)] the Left sees America as the one guilty for the Muslims crime of jihad.[(LIE)] They are upset over our "preferential"[(LIE)] treatment of Israel, and the "unfair"[(LIE)] treatment of the Palestinians. They're poor. They live in a desert and got sand in their eyes. Why do you think Obama went around right off the bat apologizing for America to his Islamic friends? Because he too sees America as the problem[(LIE)] in the war on terror rather than the solution; a "problem" that needs to be "radically transformed" by him and the rest of the idiots on the left.

See:
- "Jihad in America! Militant Islam: The Religious War Against America," by Martin Mawyer
- "What Americans Need To Know About Jihad," by Robert Spencer
- *The Politically Incorrect Guide to Islam (and The Crusades)*, also by Robert Spencer
- *Unholy Alliance: Radical Islam and the American Left*, by David Horowitz
- *The Grand Jihad: How Islam and the Left Sabotage America*, by Andrew C McCarthy
- *They Must Be Stopped: Why We Must Defeat Radical Islam and How We Can Do It*, by Brigitte Gabriel
- *Militant Islam Reaches America*, by Daniel Pipes

1134   (From "Allah or The Lord Jesus Christ?" by Rick Mathes)

- *Unveiling Islam*, by Ergun and Emir Caner
- *The Truth About Islam*, by Anees Zaka & Diane Coleman
- *World War IV: The Long Struggle Against Islamofascism*, by Norman Podhoretz
- *Endless War: Middle-Eastern Islam vs. Western Civilization*, by Lt. Col. Ralph Peters

10- Weaken America from within

"I will deliver you from the people, and from the Gentiles, unto whom I now send you, to open their eyes, and to turn them from darkness to light, and from the power of Satan unto God, that they may receive forgiveness of sins, and an inheritance among those who are sanctified by faith that is in me." (Acts 26:17-19)

There are a number of ways that the prince of darkness is weakening America from within. Indeed all that we have covered in the first eight sections, from destroying Christianity to taking over the government, falls under this heading. But in this section we will focus on how he is using his Democrat-Left to hinder the war effort against the Islamic jihadees. We have already mentioned a number of times that we are fighting a war on two fronts: one against the Muslims on a global front, and one here in our own country against their sympathizers in the ACLU, moveon.org, our colleges, universities and public schools, the courts, our entertainment industry, the Democrat media and of course the Democrat Party; who aid, abet and give comfort to our enemy by erecting non-existent legal obstacles, denying that a war even exists and hindering our efforts to defeat them at most every turn.

"You serpents, you generation of vipers, how can you escape the damnation of hell?" (Matthew 23:33)

The enemy among us

First, let's explore briefly how those wiley Democrats got back into the majority in congress in 2006, after a 12 year hiatus. Their comeback started in the aftermath of the war in Iraq, in 2003. President Bush had become very popular after 9/11, the country had rallied behind their leader, but when weapons of mass destruction—one of the reasons for deposing Saddam—were not found in Iraq they saw a political opening to tear down that popularity. So they began falsely accusing Bush of "lying"[(LIE)] about these "undiscovered" WMDs. This is right out of the Nazi playbook. "If you can get away with it, falsely accuse your opponents of lying and it will start a backfire, so when they accurately accuse you of doing the same their accusations will burn out in the smoke of confusion before they can do any damage." In the world of Nazi propaganda, it is known as a preemptive strike. The reason their accusations were false is because Bush merely stated what virtually every member of the Democrat Party, the Clinton administration before him, and virtually every intelligence agency in the Western world including Israel's had also said and believed. And, as it turns out, they weren't wrong...

"President Bush didn't lie about Saddam Hussein's weapons of mass destruction. Not only did Iraq stockpile chemical and biological weapons, Baghdad used them [on the Kurds for one] and Bush's political opponents had no doubts about the existence of those arms before Bush came to power. And now Hussein's top military adviser, Georges Sada, has written a persuasive and well-documented book detailing what happened to those weapons. Based on information obtained through his close relationships with other Iraqi pilots, Sada reveals in "Saddam's Secrets" that Iraqi military personnel stripped out the seats in commercial airliners, loaded them with the weapons and flew them to Damascus just prior to the U.S. invasion. It's a blockbuster revelation, but you didn't read about it in the New York Times, Washington Post or in any reports by the Associated Press. Only talk radio, cable television and the alternative news media gave the book and its disclosures any attention. Of course, that means Bush's opportunistic political opponents will continue to hammer away with the "Bush-lied-kids-died" line right up through the congressional midterm elections later this year [2006]. In other words, they will continue to accuse Bush of doing exactly what they are doing – lying.

...."President Bush made a tactical error when he announced to the world that U.S. intelligence on Iraqi weapons of mass destruction was wrong – and that Saddam Hussein had no such weapons."[1135]

Why Bush lacked the common sense or the advice from his advisors to stand up like a man and expose his detractors for the bald-faced liars that <u>they</u> actually were is beyond me. But *if* he had he would not have had to search very far to find examples of Democrats who said the exact same thing that he did before the Iraq War concerning Saddam's WMDs. Here is just a small sampling of quotes from Democrats in the years leading up to the decision to prosecute the War in Iraq. Many of these same people turned around and falsely accused Bush of lying. So, will the real liars please stand up:

"[We] urge you ...to take necessary actions (including, if appropriate, air and missile strikes on suspected Iraqi sites) to respond effectively to the threat posed by Iraq's refusal to end its weapons of mass destruction programs." - Democratic Senators Carl Levin, Tom Daschle, John Kerry, and others, in a letter to then President Clinton, Oct. 9, 1998.

"Saddam Hussein has been engaged in the development of weapons of mass destruction technology which is a threat to countries in the region." - Democrat Rep. Nancy Pelosi, Dec. 16, 1998.

"Hussein has ... chosen to spend his money on building weapons of mass destruction and palaces for his cronies." - Secretary of State Madeline Albright under Democrat president Clinton, Nov. 10, 1999.

"There is no doubt that ... Saddam Hussein has reinvigorated his weapons programs. Reports indicate that biological, chemical and nuclear programs continue." - Letter to President Bush, signed by Democrat Senator Bob Graham and others, Dec 5, 2001.

"We begin with the common belief that Saddam Hussein is a tyrant and a threat to the peace and stability of the region. He has ignored the mandate of the United Nations and is building weapons of mass destruction and the means of delivering them." - Dem. Sen. Carl Levin, Sept. 19, 2002.

"We know that he has stored secret supplies of biological and chemical weapons throughout his country. ...Iraq's search for weapons of mass destruction has proven impossible to deter and we should assume that it will continue for as long as Saddam is in power." - Democrat presidential candidate and Crockumentary producer, Al Gore, Sept. 23, 2002.

1135   (From article "Syria's weapons of mass destruction," by Joseph Farah @ WorldNetDaily. com, posted February 01, 2006)

"We have known for many years that Saddam Hussein is seeking and developing weapons of mass destruction." - Dem. Sen. Ted Kennedy, Sept. 27, 2002.

"We are confident that Saddam Hussein retains some stockpiles of chemical and biological weapons, and that he has since embarked on a crash course to build up his chemical and biological warfare capabilities. Intelligence reports indicate that he is seeking nuclear weapons." - Dem. Sen. Robert Byrd, Oct. 3, 2002.

"I will be voting to give the President of the United States [George W. Bush] the authority to use force -- if necessary -- to disarm Saddam Hussein because I believe that a deadly arsenal of weapons of mass destruction in his hands is a real and grave threat to our security." - Democrat Sen. John F. Kerry, Oct. 9, 2002.

"There is unmistakable evidence that Saddam Hussein is working aggressively to develop nuclear weapons and will likely have nuclear weapons within the next five years ... We also should remember we have always underestimated the progress Saddam has made in development of weapons of mass destruction." - Dem. Sen. Jay Rockefeller, Oct 10, 2002.

"He [Saddam] has systematically violated, over the course of the past 11 years, every significant UN resolution that has demanded that he disarm and destroy his chemical and biological weapons, and any nuclear capacity. This he has refused to do." - Democrat Rep. Henry Waxman, Oct. 10, 2002.

"In the four years since the inspectors left, intelligence reports show that Saddam Hussein has worked to rebuild his chemical and biological weapons stock, his missile delivery capability, and his nuclear program. He has also given aid, comfort, and sanctuary to terrorists, including al Qaeda members ... It is clear, however, that if left unchecked, Saddam Hussein will continue to increase his capacity to wage biological and chemical warfare, and will keep trying to develop nuclear weapons." - Dem. Sen. Hillary Clinton, Oct 10, 2002.

"We are in possession of what I think to be compelling evidence that Saddam Hussein has, and has had for a number of years, a developing capacity for the production and storage of weapons of mass destruction." - Dem. Sen. Bob Graham, Dec. 8, 2002.

"Without question, we need to disarm Saddam Hussein. He is a brutal, murderous dictator, leading an oppressive regime ... He presents a particularly grievous threat because ...[of] his continued deceit and his consistent grasp for weapons of mass destruction ... So the threat of Saddam Hussein with weapons of mass destruction is real." - Democrat Sen. John F. Kerry, Jan. 23. 2003.[1136]

A total of 67 Democrats in all stood up on the floor of congress and said that Saddam had WMDs. Yea, "Bush lied, and soldiers died."[(LIE)] What disgusting liars. Many of these Democrats were the most vocal in falsely accusing Bush of lying even though they clearly told the same "lie." Talk about shamelessness! And they got away with it! Their media never exposed them. It just repeated their lie, over and over again. And equipped with this dis-information an army of savvy voters went out and gave them congress in 2006, as reward for their treachery.

"We do not need low-life scum-bags like this aiding and abetting the enemy! This is a fight of will, and when you deliberately go out and try to undermine the will of the American people by putting out lies and falsehoods, then you are aiding and abetting the enemy."[1137]

1136   (Source: davidstuff.com/political/wmdquotes.htm)
1137   (Lt. Gen Thomas McInerney on Fox News)

So the truth is that it was the Democrats who were the liars when they falsely accused their Commander in Chief, in a time of war, of lying to the nation in order to bring us into that war. And that intentionally orchestrated propaganda campaign, for the sole purpose of benefiting themselves politically, had the added benefit of encouraging, emboldening and therefore aiding our enemies in Iraq. And that resulted in no less than the killing of even more of our soldiers in that theatre. So all of those Democrats quoted above, and others, who then went out years later when it was politically expedient and falsely accused George Bush of lying, are not only liars, and traitors to this country, but murderers as well. ...And now you know just how these hapless, socialist clowns, who have just destroyed our economy, got there in the first place; by stomping over the graves of our soldiers. Congratulations again, America.

The David Horowitz Freedom Center compares two quotes, one ancient and one modern. The first is from the *Art of War*, by Sun Tzu:

"To fight and conquer in all your battles is not supreme excellence; supreme excellence consists in breaking the enemy's resistance without fighting."

The second is from Democrat Harry Reid:

""This war is lost, and the surge is not accomplishing anything."[LIE] -Senate majority leader Harry Reid, April 19, 2007, less than 60 days after the surge was announced and more than 100 days before it was fully implemented."

""Bush Lied. People Died."[LIE] This tired saw has been repeated so many times by elements of the radical left and embraced by the Democratic Party that many Americans, sadly, believe it. The truth is, however, the President hasn't lied. Instead, he's sought to protect this nation *and* protect international law. But under the guise of "patriotic dissent," rabid hate-America leftists have labored long and hard to discredit our President, our troops and our nation."[1138]

In the book *Party of Defeat: How Democrats and Radicals Undermined America's War on Terror Before and After 9-11*, authors David Horowitz and Ben Johnson document...

..."the unprecedented attacks by leaders of the Democratic Party on a war they supported and then turned against. In a democracy like ours, criticism of war policy is legitimate and necessary. But deliberate undermining of a war policy ...is a different matter."[1139]

Their book should be required reading, for every clueless Democrat voter. Rush Limbaugh added this:

"Throughout the Iraq War it was Barack Obama and the Democrat Party which actively sought the defeat of the U.S. military. They convened hearings and accused General Petraeus of lying.[LIE] [In addition to George Soros' moveon.org running a full page ad in the New York Times slandering General Petraeus, calling him "General *Betray*us."[(LIE)] How clever.] They said the surge would not work. Harry Reid stands up, waves the white flag: "This War is lost!"[LIE] Jack Murtha is out saying our marines at Haditha are guilty of rape.[LIE] John Kerry is accusing our marines of committing terrorism acts by going into the homes of Iraqis at midnight in the dark terrorizing, looking for Al Qaeda or whoever

---

1138   (From the David Horowitz Freedom Center)
1139   (From cover, *Party of Defeat: How Democrats and Radicals Undermined America's War on Terror Before and After 9-11*, by David Horowitz and Ben Johnson)

was there.[LIE] Yea... I do question their commitment to National Security. I question their commitment to the U.S. military. They'll put their political survival and their political power being gained over anything else. They'll use anybody, and throw any away in order to achieve it."[1140]

Chris Matthews, one of the resident Democrat apologists at MSNBC, dragging them deeper into the ratings toilet, called our President a "criminal."[LIE] It is a shame that scumbag wasn't around during World War II, the Japanese wouldn't have needed Tokyo Rose. They could have just sent him a few yen, along with some pictures of the Bataan Death March to proudly hang on his wall.

The Democrat-Left saw how their treasonous strategy was working, and advancing their poll numbers, so they ramped it up even more and went on to tear down the war effort at every opportunity; knowing they had the eager protection of their media. Here's a small sampling:

"The Bush administration approves of "torture"!"[LIE] (Sure, but what the Democrats approve of doing to millions of little babies every year is not torture.)

"The Iraq war is an "illegal" war!"[LIE] It's Bush's war, and it's a mistake! [LIE] (I'm sure the soldiers over there prosecuting that war, and the families of those who died were just thrilled to hear that.) "We're losing! It's not a part of the war on terror."[LIE] (No, it's a part of the school lunch program, you jackasses.)

"No blood for oil!"[LIE] (How about a more accurate slogan... "No congressional terms for traitors!")

Hillary Clinton said she couldn't "believe anything this President says."[LIE] (But then again, this is the same woman who believed everything her husband said—"I never had sex with that woman, Monica Lewinsky, and I love you so much, Hillary"—so you make the call.) It turns out unfortunately for her that like her husband, "Bent Willy," she is also "quite an accomplished liar."

"There is no terrorist threat against America."[LIE] -Michael Moore. (Right! and there is no Iranian threat against Israel either, you colossal ass.[1141] It must all be in our heads.)

"The War in Iraq is a "fraud" made up in Texas."[LIE] – The Late Democrat Senator Ted Kennedy. (Who should know about that because, coincidentally, he was himself a fraud made up in Massachusetts.)

"The President's decision to go into Iraq and stop Saddam Hussein from funding more terrorist attacks was an "appalling mistake."[LIE] -Democrat Rep. Nancy Pelosi.[1142]

Keith Olbermann (the one shoving scissors into the backs of the heads of nine-month old little babies in chapter 11), who also just got fired from MSNBC (which is quite an accomplishment in and of itself), went for the hat trick. He called George Bush a liar[LIE] a traitor[LIE] and demanded that he be tried for war crimes. (His mistake was in not also wishing that Bush and his grandchildren get AIDS, like Nina Totenberg over at PBS did to Sen. Jesse Helms, or he'd still have a job.)

And then there was the incessant phony outcry to close Guantanamo Bay. Phony because after 2 and 1/2 years into his catastrophic administration Obama

---

1140 (Rush Limbaugh on *Fox News Sunday* with Chris Wallace, 11/01/09)
1141 (And no, we do not mean that to be taken literally)
1142 (The last 3 quotes taken from a Citizens United circular)

still has it open, and the outcry has disappeared. Obama campaigned to close it but now that he's been elected it turns out that even he has enough sense to know that it is a great place to house captured Muslim terrorists. (He has finally capitulated that it is also the best place to try them. It took him a while, because, like Holder, he suffers from ACLU-disease.) The "Guantanamo lie" was meant to further hinder and disparage the effort to prosecute not only the war on "terror," but the actual captured terrorists themselves. Ann Coulter, however, saw a silver lining…

> "On the bright side, at least liberals have finally found a group of people in Cuba whom they think deserve to be rescued."[1143]

And then there is the heinous lie that "the Bush administration used torture."[(LIE)] And it is heinous because, as we mentioned in the 2nd chapter, a whopping three (count 'em, three) Muslim lunatics had water poured on their face, with no lasting deleterious effects (except for some soiled undies); yet scores of our military personnel have been water boarded as a part of their training; and the policies of the Democrat-Left have tortured 50 million babies, to death, while they were still alive, some having scissors jammed into the backs of their neck at the fully developed age of 9 months, but that's not torture. Yes, they are some "accomplished liars." But they get away with it because they own the media. And their lies did the trick. In 2006 the enhanced interrogation program of the Bush administration was effectively…

> …"stopped by a combination of the Supreme Court and the efforts of liberals in Congress. … I sat down in a room with the people, the real Jack Bauer's, who actually interrogated these people, and they told me, they walked me through, how they stopped terrorist attacks. How they interrogated Khalid Sheik Mohammed. How they got him to tell us his plans for follow-on attacks after 9/11.
>
> "[We] stopped them from committing attacks, to fly planes into the library tower (in LA), to blow up our consulate in Karachi, to blow up our Marine camp in Djibouti, …to do 9/11 in London – fly planes into the London financial district and Heathrow airport. And so all of these things were stopped.
>
> "Now fast forward to 2009. [There's] a new terror network … Al Qaeda in the Arabian Peninsula. By the Obama administration's admission, we know nothing about them. We didn't even know they were planning attacks in the United States. Why? Because we are not capturing terrorists and interrogating them anymore. So we are blind because we've taken away the most important tool of intelligence we have. … We're courting disaster. We are in danger because we do not have this capability.
>
> "These [CIA interrogators] are heroes. They're not torturers. They're heroes who stopped the next 9/11, and they're vilified."[1144]

Newt Gingrich on Fox News' Hannity (May, 2011) commented on the possibility of these un-American investigations, spearheaded by the ACLU's Justice Dpt. plant, Eric Holder, ending up with prosecutions of Bush administration CIA interrogators. Interrogations we should add that exacted information that was instrumental in finding and killing Bin Laden:

> "Eric Holder should never have been approved by the US Senate. He *volunteered* to write papers for terrorists. He volunteered to try to help *terrorists* get out of jail. His record out of office …was such that it should have disqualified him from serving as Attorney

---

1143   (Ann Coulter)
1144   (Marc Thiessen, author of *Courting Disaster: How the CIA Kept America Safe and How Barack Obama Is Inviting the Next Attack*, on Fox News' Hannity, 01/22/2010)

General and in fact the US House should be investigating what his beliefs are because he is consistently doing the wrong things as it relates to Americans, and every American should be outraged that CIA officers who are risking their lives to try to save this country from attack are being attacked by their own Atty. General."[1145]

I have to disagree with one thing, Newt. Holder's *not* our "own Atty. General," he's Al Qaeda's.

Then there's the Obama administration's decision to try Khalid Sheikh Mohammed, the disturbed master-mind behind 9/11, and some of his friends, in civilian courts in New York City. On *The O'Reilly Factor* (Nov. 2009) Karl Rove relates just how ignorant that decision actually was:

Karl Rove: "This was an utter, unmitigated disaster for the security... and for the interests of the United States in the dangerous world we live in. The sanctimonious comments by Attorney General Holder today are... self-serving and self-centered. What we ought to be concerned about here is what is in the interests of the United States of America. ...What good is there to be gained by trying... these mass murderers, these war criminals in civilian courts that have rules designed for conventional crimes committed by conventional criminals? ...They're going to challenge the way they've been treated. They didn't get a Miranda warning. ...Did we have all the legal constitutional niceties in a war? They're going to claim outrageous treatment and lie about it because they've got the world stage, and this is going to serve to recruit additional jihadists because they are going to attack America throughout this entire... sorry episode. And they're going to try to get support throughout the Muslim world for their cause by excoriating the United States of America, our values, and our views. And why we have given [them] a stage... is beyond me. And the President of the United States and his sanctimonious Attorney General... they did this on a Friday afternoon, because it was a cowardly act. It was not in the interests of the United States of America."

O'Reilly: "Some people feel that the reason they did this, particularly Attorney General Holder, is because he wants to put you and the rest of the Bush administration on trial... [That] this is a political play as well because they'll be able to embarrass the Bush administration, embarrass the Central Intelligence Agency... and discredit you guys."

Rove: "It is a sorry, sordid, little petty political maneuver if that's what they want to do.

[Of course they do. Read chapter 2, they have no values. Hell, they've killed our soldiers just for their own political gain, and they rejoiced over jamming scissors into little babies skulls. Compared to that, what is trying terrorists in an improper and dangerous venue in order to score a few more partisan political points?]

...The United States of America's cause will suffer in the Muslim world, because the truth runs at about 20 miles an hour, lies go at about 500 miles an hour and these lies are going to course through the Muslim world and inflame resentment.

[Rove then goes on to answer O'Reilly's question as to why the Bush administration failed to have already tried these terrorists in military tribunals...]

"Even military tribunals are subject to outside pressures. And a lot of outside pressures were put by lawyers who have been recruited by Holder to serve inside this Justice Department. This is a continuation of a long-standing plot by a bunch of left-wing lawyers who do not love America, who want to undermine our cause in the global war on terror."[1146]

1145   (Newt Gingrich on Fox News' Hannity, 5/11/2011)

1146   (Karl Rove on *The O'Reilly Factor*, Fox News Channel, 11/13/2009)

Like we said, he's Al Qaeda's Atty. General. Michelle Malkin adds this…

"The members of team Obama are masters at deflecting blame and then demonizing their enemies, and that certainly is the strategy here. They demonize everyone except for the demons who are in our midst, and that really is the problem with how they are prosecuting this war on terror. They see it as a law-enforcement approach rather than being on a war footing, and everything flows from that. If you do not name and identify your enemy how are you going to beat them and defeat them? Instead what we have is the demonizing of peaceful American citizens. We saw it in the Arizona debate, we've seen it with the tea party over the last year. And I think one of the ultimate signs that this administration has not learned from its mistakes in how it's waging this war on terror is the fact that we still have Janet Clownitano and Corruptocrat Eric Holder still in place!"[1147]

"**Appeasing Dictators.** The President has changed America's course in the world by making nice with Hugo Chavez and bowing down to King Fahd, while cold-shouldering our Israeli allies and ignoring the brave struggle of Iranian dissidents. America's promotion of Democracy and human rights have been replaced by the eagerness to hold "talks" with tyrants."[1148]

"Obama has abandoned our allies and engaged our enemies." Why, America? Is that because he is the unparalleled genius that the morons in our media and entertainment industry say he is? Or is it because he is either a traitor or the world's largest jackass? Who else "abandons their allies and engages their enemies?" (An America-hating Muslim plant?) We ask the question, you decide.

"Obama's National Defense Strategy – Who needs national defense when the "Messiah" is going to *charm* our enemies into giving up violence and war? Rejoice! No more 9/11 terror attacks. No more Mumbai massacres. The simple truth is "Obamacrats" see defense as a fuzzy social issue, as in – "Oh, you attacked us – what can we do to make you forgive us?" Liberals always blame America first. Unfortunately, Iran's soon-to-be-nuclear-armed fanatic, Mahmoud Ahmadinejad, and others of his ilk, understand well the liberal instinct to appease and fully intend to use it against us."[1149]

It's Jimmy Carter all over again, with bigger ears and a lot more charisma.

"The left has been wrong about just about everything internationally since the 1960s. Its trust in negotiation, diplomacy, appeasement rather than American strength has led it to coddle our enemies (including communists during the Cold War), eschew national protection (anti-SDI), undermine the military, and invest in defeat in the War on Terror. Liberals' greatest failure is their betrayal of the Constitution's prime directive for government: to provide for the common defense."[1150]

As opposed to, say, destroying our economy? But again, what is the Constitution to the Democrat-Left, Rush, but a document to be twisted, perverted, distorted, lied about and ignored when it doesn't suit their ideological agenda?

"The worldview of the American Left is fundamentally out of touch with the reality of the real world."[1151]

1147 (Michelle Malkin, on Fox News *Hannity* 5/5/2010)
1148 (From a *David Horowitz Freedom Center* Circular)
1149 (Rush Limbaugh, from The Limbaugh Letter)
1150 (From article: "Nailing the Left: Liberal Failures," Rush Limbaugh, the Limbaugh Letter)
1151 (Newt Gingrich on Fox News' Hannity, 5/11/2011)

Then there was Obama's speech in Oslo where he finally found a little spine and called our enemies evil. (Terrorists? No, not yet. But at least he's heading in the right direction. Give him some time. He's growing. He's evolving.) But whereas his Democrat-Left thought he was brilliant, Charles Krauthammer had a different take…

> "I was less impressed with the dawn of this new realism. What the president had stood up and said was, "there is evil in the world and Gandhi would not have done well against Hitler." Well, for most of us, you dispose of those issues in the first week in the freshman dorm in college, after a couple of late night discussions. And to elevate it as a great philosophical achievement, for him to say that, is quite astonishing. It's emperor's new clothes. It's obvious. The fact that we were all impressed is to tell you how unrealistic, idealistic and naïve were all the previous speeches; starting with the speech he gave in 08 in Berlin, [where] he said that the wall had come down in Berlin because the world stood as one. Well that's not why the wall had come down. It came down because America stood fast for 50 years on the ramparts of freedom and didn't flinch. And in the end, the other guy conceded and collapsed. So this kind of globalism, universalism, naïveté, runs through all of his policies."[1152]

On Iran's quest to nuke Israel

Liz Cheney, hosting Fox News' *Hannity*, 08/06/2010, had this to say about the Iran situation:

> "For years Iran has armed our enemies in Iraq and Afghanistan. It continues to support terrorism around the world, and develop a nuclear weapons program. Nonetheless the president [Obama] remains bent on negotiating with the regime."

And despite the fact that Iran has no interest whatsoever in negotiating with him. They do however have a keen interest in laughing at him. Liz continues:

> "We have a real disconnect here. Iran supplies our enemies with the technology and the materials they need to make these IED's that are killing American troops, and yet the president seems to think that we are on the same side with Iran. Do you think he's possibly that misinformed?"

Elliot Abrams, from the Council on Foreign Relations, replies:

> "I think he's certainly very mistaken. …Negotiating with the Iranians …is something the Iranians don't seem to want. The problem is that for years and years they've been getting away with this. …They've been killing Americans for years in terrorist attacks, going back decades, and they've never actually been punished by the United States. So they continue to think they can get away with it."

And they of course think they can get away with it because it's <u>true</u>! They <u>can</u>! And <u>have</u>! Because we are a bunch of spineless knuckleheads. We've lacked the guts as a nation to retaliate; like dropping a daisy cutter on their nut job's (Ahmadinejad's) head during one of his military review ceremonies for example; or how about hanging him from a tree in front of the UN (in honor of his visit) with a sign "DEATH TO AMERICA" nailed to his chest, rather than letting him speak at one of our idiot-infested, Marxist universities. America had already become a paper tiger when it came to Iran, but Obama has taken us to new heights; to that of a toilet paper tiger. John Bolton, former ambassador to the UN, adds this:

> "We're past the point on the nuclear weapons issue when diplomacy or sanctions are going to have any effect, so to hear the president still saying that he holds out hope he can

---

1152   (Charles Krauthammer on Fox News' Special Report with Brett Baier, 12/25/09)

negotiate with Iran on nuclear weapons is past the point of naïve. It's gotten to the point of being dangerous."

Mrs. Cheney then asks him, tongue in cheek, if the White House has some historical model where they can say, "Well here's a moment when appeasement worked?"

Bolton's answer: "I think the president believes very firmly if he could just find some Iranian to sit down with that he could convince them to give up their nuclear weapons program. I think that has gotten to the point of mindlessness."[1153]

Either that or he could care less if Iran gets nuclear weapons, and whether they then use them on Israel, as is their stated intention. (But don't forget, stupid, he's not a Muslim.) (And keep looking in the car.) The enemy within is in our Oval Office. But then again, he did kill Bin Laden. And don't you ever forget it. (Not that he would ever let you.) And the sad thing is that the majority of Jews were bamboozled into going out and electing him. (But at least they didn't vote for Arafat. He wasn't on the ballot.)

Charles Krauthammer agrees with Bolton's assessment...

..."to stretch out a hand [to Iran's Ahmadinejad] that's been spat on for 20 months is simply unbelievable. It betrays a misunderstanding of the nature of the international community that is not even a law professors, it's an adolescents."[1154]

On her show, "America Live," Megan Kelly plays devil's advocate concerning Obama's groveling approach to Iran, and asks Ambassador Bolton whether that approach is better than his predecessor's (Bush), who more aggressively engaged in two wars in Afghanistan and then Iraq which caused most of the world to remove him from their Christmas card lists. John Bolton replies...

"The president of the United States is not supposed to follow a popularity contest in the rest of the world. The decisions to topple Saddam Hussein and to depose the Taliban in Afghanistan were based on critical American national interests; interests that are still valid to this day. There's no doubt that having tried for nearly 10 years through diplomacy and sanctions to stop Iran from getting nuclear weapons that the policies have failed. The most likely outcome now is that Iran gets nuclear weapons. That's extraordinarily unattractive. The only thing that stands between Iran and nuclear weapons is the possibility that some outside power, Israel, the United States, would strike those nuclear facilities. That's a very unattractive option, but it's even more unattractive to contemplate Iran with nuclear weapons."[1155]

And on two of our former allies pictured "holding hands triumphantly" with Iran's Ahmadinejad, Charles Krauthammer laments:

"This is a disaster for our stopping Iran on nukes, but it's worse than that. This is a collapse of the Obama foreign policy of the last year and a half. There's a picture that we saw today of the leaders of Brazil, our biggest ally in Latin America, and Turkey, the Muslim anchor of NATO- our most important anchor in the Mediterranean area. Here they're holding hands triumphantly with the most anti-American leader in the world. Now that is a picture that is absolutely astonishing. What it tells you is that American allies have watched this administration in action over a year and a half and decided that there is no profit in being an ally of this administration. After all the apologies, the excuses, the appeasement. They watch Russia exert its influence over Eastern Europe, over Ukraine,

1153   (Elliot Abrams, John Bolton and guest host Liz Cheney on Fox News *Hannity*, 08/06/2010)

1154   (Charles Krauthammer on Fox News' *Special report with Brett Baer*, 9/23/2010)

1155   (Former UN Ambassador John Bolton on Fox News' "America Live with Megan Kelly 09/23/2010)

over Georgia. They watch the administration appease Iran fruitlessly, being insulted time and again, and returning as if nothing had happened. They watch our appeasement of Syria, sending an Ambassador to Syria, as it's re-exerting its influence over Lebanon in supplying Hezbollah with SCUDs. And Brazil is watching in Latin America as Hugo Chavez organizes an anti-American coalition. And in Honduras where a Chavez ally reaches to become a dictator and the US supports him against the rest of the country. All of that and they look around and say there is no profit in being an ally of the United States. We are going to realign ourselves, and there's no danger in raising hands with the Iranian leader and being on the other side. This is not just America in decline, this is America in retreat. The world is seeing it and acting rationally, accordingly."[1156]

(Would that our voters could find it in them to act a little rationally, accordingly...)

But, if possible, it gets even worse...

"Can you imagine the U.S. military shooting down Israeli jets or bombers? Brace yourself. It could happen. One of Obama's closest advisers, Zbigniew Brzezinski, has already said the U.S. should blast Israeli planes out of the sky if they attempt to take out Iran's nuclear installations. Can you imagine an American president actually doing that? I can, if the president's name is Barack Obama who learned his anti-Semitism from the master, Rev. Jeremiah Wright."[1157]

Remember, Satan loves Iran and hates Israel. He must destroy Israel soon before the Lord returns at the battle of Armageddon at the end of the seven-year tribulation period. His time is running out. Iran may be his last best hope. Whose side do you think his lying, baby-mutilating, reprobate-minded, America-hating Democrat-Left are on? Especially now that we see them being commanded by someone who said not too long ago that he thought the Muslim call to prayer at sunset was one of the most beautiful things on the face of the earth (the screams of little girls when they get their genitals sliced off was right next to it), which I guess was a while before he converted to "Christianity;" Reverend Wright's "GOD D**N AMERICA" kind. We're in an insane asylum, pure and simple.

Peace protestors

"Know you not, that to whom you yield yourselves servants to obey, his servants you are to whom you obey; whether of sin unto death, or of obedience unto righteousness?" (Romans 6:16)

Another way that the left undermines America's will to defend herself is by ringing Pavlov's bell and salivating their "peace protestors" anytime there is a war and a Republican in the White House. This was certainly true of George Bush and the Iraq War. And their Democrat media gave them front-page coverage, 24/7, even when there were only 7 of them. But both the peace protestors and their media curiously disappear anytime a Democrat is in the Oval Office, and even when he is waging the same war that his predecessor did. Hypocrisy, liberal peace-advocate style.

"If you claim you're for peace, it's the same thing as being for breathing. It's hard to find someone willing to debate it. And that's what has made pacifists so lazy in articulating anything beyond their initial proclamations. No one shows up to say, "Hey, peace sucks." But pacifists aren't just lazy, they are also stupid: stupid to the ways of history and to the ways of people.

1156 (Charles Krauthammer on the Fox News' Fox Report with Bret Baer, 5/17/10)
1157 (Rush Limbaugh, from circular Sept 2010)

...."male pacifists, in my mind, are as bad as traitors. No matter how evil an adversary is—Hitler, Tojo, or Saddam—peaceniks will vehemently come out against doing anything to stop them. Because stopping them means war. In the end, not stopping them only means that sooner or later we all die at the hands of those who rule by ruthless force. For peace to work, both sides need to have it. And when the other guy wants domination, and you want peace, there is only one option: War.

"America has done a good job defeating people who want to rule the world, and if we had listened to the pacifists the world would be a different place.

"First of all, all of the pacifists would be dead."[1158]

What reader hasn't had the experience occasionally of driving behind another naïve, left-wing dupe sporting the embarrassingly stupid slogan on their bumper, "War is Not the Answer!"[(LIE)] Yet in reality it is their own moral equivalency born out of the typically-liberal feeling of self-righteous superiority that is actually "Not the Answer!" Pull up next to them at a light some time and try explaining that.

The Muslim massacre @ Fort Hood

On November 5, 2009 the Democrat-Left's political correctness gone wild in our own military caused the death of 12 of our soldiers and wounded 31 more at Ft. Hood in Texas. And that is because the Muslim infiltrator, murderer and traitor, Army Major Nidal Hasan, was known to be a disturbed, radical Islamic jihadist well before that infamous day when he perpetrated his Moham<u>mad</u>-inspired slaughter. Years before in lectures to fellow Army officers he had said that non-Muslims should be beheaded, forced to drink boiling oil and set on fire. He then spent time on the website of Islamic jihadist Imam Anwar al-Awlaki (now expired after the car he was driving ran into one of our drones), asking him for *spiritual* guidance on killing infidels.[1159] Fellow officers had questioned his loyalty in the past, but it all fell on deaf ears. And now, to add politically correct insult to injury, we have the cover up...

Obama and his idiot Atty. General refused to call the massacre what it was, Islamic terrorism, even though the sick traitor shouted "Allah Akbar" (which translates roughly as "Allah Sucks Donkey Dicks!")[1160] (with all due apologies to the donkeys) as he sprayed his bullets. Instead, Obama said it was a "tragedy."[(LIE)] But that of course is a lie. A "tragedy" occurs, for example, when someone dies suddenly and unexpectedly from an unavoidable and unforeseeable accidental cause, like a child who drowns, or when someone dies in a car crash—not when a deranged Islamic killer is allowed by Obama and his politically correct, progressive, terrorist-appeasing, treasonous Democrat-Left to infiltrate our military, to flourish there with impunity, and to use that as a secure base to slaughter our brave men and women. Charles Krauthammer got it right, "This wasn't a tragedy. It was a

---

1158   (From "Peace Signs," an article by Greg Gutfeld in The American Spectator, November 2007)

1159   (By the way, if Allah is the all-powerful god that they pretend he is, then why doesn't he just kill all the infidels himself and be done with it? If he's god. Of course if Allah is just Satan, which he is, then he would only have the power to dement the most violent of his Muslim followers into trying to kill "all of the infidels." Pull up next to one of *them* at a light and try asking that.)

1160   (Keep that in mind the next time you hear them shouting that on your TV, and they'll be a little easier to stomach)

travesty."[1161] But Obama needs to lie about it in order to diffuse blame away from himself and his Democrat-Left, the ones ultimately responsible for allowing this "tragedy" to happen.

And unfortunately that same military, poisoned with political correctness at the highest levels, has learned nothing from this attack. The same denial that killed and wounded all of those soldiers in November of 2009, is still on display today:

> "On 10 August 2010, the Department of Defense (DoD) released <u>23 pages</u> of recommendations from an independent review of the Fort Hood massacre. It is not as pure a whitewash as the first Fort Hood report, called "<u>Protecting the Force</u>." It actually has some relevant recommendations. However, it so assiduously avoids the central issue that it becomes a masterpiece of political correctness."
>
> …"**There is not a single mention of Islam, Islamic radicalization, jihad, jihadists, or religious extremism in the entire document**. [Emphasis mine.] Nor is there any other word or phrase that would indicate to the reader that Hasan was anything other than a mentally disturbed murderer."[1162]

> "The most prominent Islamic activist leader in America at the time [Abdulrahman Alamoudi] …had infiltrated the highest levels of political power. … [He was asked] by the Defense Department to establish the military's Muslim chaplain corps, and appointed by the State Department to serve as a civilian ambassador, taking six taxpayer-funded trips to the Middle East. … Just days after the 9/11 attacks, he appeared with President Bush and other Muslim leaders at a press conference at the Islamic Center of Washington, D.C. despite his public comments a year earlier at a rally just steps from the White House identifying himself as a supporter of the Hamas and Hezbollah terrorist organizations. In July 2005 the Treasury Department revealed that Alamoudi had been one of al-Qaeda's top fundraisers ….
>
> …"We've heard a lot lately about al-Qaeda's new star, Anwar al-Awlaki, who has been behind many of the recent terrorist attacks on America. But did you know… Despite being subject to an FBI investigation initiated in 1999, and having been interviewed by the FBI at least four times after 9/11 for his contacts with two of the hijackers, Al-Awlaki was leading prayers for congressional Muslim staffers inside the U.S. Capitol. … Al-Awlaki was also feted at a luncheon inside the still-smoldering Pentagon following the 9/11 attacks."[1163]

We can only be thankful that our WWII generation wasn't burdened with today's kind of fools that we have for leaders, or after the Japanese attack on Pearl Harbor they would have had to watch as Hirohito was wined and dined at the Waikiki Hilton, with the USS Arizona still smoking in the background. How Al-Awlaki was able to keep from rolling around on the floor laughing at the horses asses who invited him still remains a mystery, but it might have something to do with the Muslims taqiyya.

> "Obama's National Defense Policy" – It's a dream come true for America's enemies. By now, after Obama's lame responses to Fort Hood and the underwear bomber, every radical Islamist (think Ahmadinejad) and tin-point dictator (think Chavez) on earth knows

---

1161   (Charles Krauthammer on the *Fox Report with Brett Baier*, 11/11/09)

1162   (From article "Excusing Islam for the Fort Hood Massacre," by Lance Fairchok in the American Thinker online, August 25, 2010)

1163   (From article, "Political Correctness Kills: Study Shows How Terrorists Infiltrate U.S. Government" by Barry Rubin @ Act For America)

that Obama would rather be attending to health care or cap-and-trade than defending U.S. citizens. What better incentive for attacking America and killing her citizens than the knowledge that, in "retaliation," the U.S. Commander-in-Chief will dither for weeks, form a commission… and finally issue more empty threats? My friend, we are in mortal danger."[1164]

The enemy within. Earlier we accused the Democrat-Left of being traitors by aiding, abetting and encouraging the Islamic terrorists in Iraq with their deliberate lies, including their favorite, "Bush lied,[(LIE)] and soldiers died." Fort Hood, and its aftermath, is just one more example of how they will willingly sacrifice the lives of our soldiers on the altar of their twisted ideology. In this case, their asinine political correctness.

Oh Joy! A Victory Mosque at Ground Zero
"The eyes of the Lord are in every place, beholding the evil and the good." (Proverbs 15:3)
Over the centuries Muslims have been known for establishing *victory* mosques in many of the cities and countries that they conquered. Many times they would even destroy an existing church or synagogue, and build their temples to Beelzebub in its place. The Dome of the Rock now stands on the Temple Mount in Jerusalem (why the Jews haven't bulldozed it off yet is beyond me, but I think it has something to do with the expression "why don't you grow a pair") where for centuries stood Solomon's, and then the second, Temple of God. In Damascus they erected their "Grand Mosque" after they had destroyed the Church of John the Baptist. And there are countless other examples in Armenia; in Turkey where they built a mosque on the site where they had destroyed a Basilica; in Egypt Muslim-destroyed churches and synagogues now are the site of mosques; and in the unholy city of Mecca itself Mohammed slaughtered his own Jewish relatives before appropriating their synagogue for his own satanic rituals.

And that is what they are trying to pull off at Ground Zero in New York City, the site of the worst attack on our own soil in the history of our country, and hallowed ground for not only countless thousands of the victims' families and close friends but for most of the rest of non-Muslim, non-Democrat-Left America as well. But now an Imam with questionable radical-Islamic ties wants to spit in our faces and pour salt in the wounds of the victim's families by building a mosque there. It is akin to allowing a violent convicted rapist to erect a big-dildo shrine, glorifying his "conquest," in the yard across the street from the deceased woman's home, despite the daily pain it would cause her grieving husband and children.

And we are told by the radical-Islamist's useful-idiots on the left like Mayor Bloomberg who haven't a clue what is actually going on (or even more troubling, don't care to know) that this mosque should not be opposed because we need to be "tolerant" of the Muslim religion.[(LIE)] Really? That's what you think, Bloomberg? Well I think we've been more than tolerant in this country of that foreign, clitoral-mutilating, woman-degrading false religion, and I think that rather it is high time for it to be tolerant of everyone else. Like how about for starters let them show a little tolerance for the Jewish people and the nation of Israel. The former they wish

---

1164   (Rush Limbaugh, from circular Sept 2010)

to exterminate and the latter they want to destroy. How about welcoming Jews into Arab-Muslim lands where now they are virtually non-existent, having been driven out years ago by this pork-rind of a religion that now calls on us to be "tolerant?" What a hoot! How about tolerating the Jewish religion, the Christian religion and the real holy book, the Bible, in Musrab lands? Furthermore did you know that no Christian or Jew is even allowed to step foot in their unholy cities of Mecca and Medina, much less build even one single church or synagogue there? I think we should take Mr. Bloomberg, his idiot friends on the Democrat-Left and their Ground-Zero-mosque Imam up on their calls for tolerance. Let's start by declaring that not one more mosque will be allowed to be built anywhere in the United States of America until both a church and a synagogue are built in both Mecca and Medina, and until the Jews are welcomed into all Musrab lands like the one million Arab-Muslims who are not only welcomed today in Israel but live there will full equal rights—more rights than they would enjoy in any Islamic nation. How about that for tolerance, Mr. Ground-Zero-mosque Muslim man? (Merry Flippin' Christmas.) But the Democrat-Left wouldn't like that, for it would rub the little doggie noses of their legendarily intolerant Muslim friends in their own deceitful doo-doo. Nor would they ever get around to actually reporting the fact that Muslims are possessed of the most IN-tolerant religion ever to stink up the face of the earth.

> "We are told endlessly that Islam is a "Religion of Peace" in the face of overwhelming evidence to the contrary. After a decade of the violent reality of Islam, the claim is so absurd that it becomes a parody of delusional political correctness. It borders on collective insanity. The proposed mosque and "Islamic Center" near Ground Zero is a case in point. Parroting the White House backdoor talking points, leftist publications from Slate to the New York Times tell us that opposing the Ground Zero mosque is racist and bigoted.[LIE] It is a version of the same tired rhetoric used to promote racial Exceptionalism and "social justice." Americans are mightily sick of it.
>
> "What evidence makes us wary of Islamic promises of dialogue? Perhaps 15,000 terror attacks since 9-11 against Muslims and non-Muslims alike -- vicious, indiscriminate, brutal attacks meant to terrify and horrify. No other religious faith is guilty of so many atrocities against so many innocents in so many disparate lands."[1165]

Yet we are supposed to be "tolerant" of a victory mosque being shoved in our faces at Ground Zero. And the Imam behind this travesty would like us to believe that he's building it as a beacon for religious "tolerance." Yea, *Muslim tolerance*. That's an oxymoron to beat all oxymoron's. But the clowns who inhabit the Democrat-Left buy it without question, and without so much as a single, solitary, critical thought.

Peter King's Muslim hearings

In March of 2011 the Democrats control over our federal government was reduced by a third, the Republicans having won the majority in the House of Representatives four months prior. And so representative Peter King, Republican from New York and new Chairman of the Homeland Security Committee decided to have some long overdue public hearings in order to shine some well-needed light on the problem of Muslim recruitment and radicalization within our own

---

1165   (From article "Excusing Islam for the Fort Hood Massacre," by Lance Fairchok in the American Thinker online, August 25, 2010)

borders. The Muslim jihadees among us have been poisoning the minds and hearts of some of our children from within their American mosques, and from within our own high schools, colleges and prisons, and have turned them into hate-filled, suicidal monsters who will do the bidding of the prophet-of-hell, Mohammed, and his demonic entity, Allah. And one of the benefits of having these hearings and exposing this enemy-among-us is that we might save some other poor unfortunate American children from having their minds mutilated by the same brainwashing, and their lives ruined; and also so we might better defend ourselves, our country and our homeland. Wow! You'd of thought the left had been told they must sacrifice their children to the god Moloch. (Oh, dog gone it! I forgot again. They do that already.) They went, and continue to go, berserk…

> "Today a Congressional hearing detailing how Islamic extremists are radicalizing American Muslims is taking place. Simply put, terrorist-related organizations are turning American citizens into home-grown terrorists. There is no greater threat to our safety and the safety of our homeland than a radicalized American citizen intent on doing us harm. For having the boldness to bring this emerging threat to light and trying to conduct a civil and rational discussion about this significant threat, the Homeland Security Committee Chairman, Representative Peter King, has been accused of racism, McCarthyism, and more. The ACLU and several other liberal groups have been trying to stop this hearing in the name of political correctness. This is not about offending a group of American citizens - this is about understanding the clear and present danger that radicalized Muslim American citizens pose to our nation."[1166]

And the stealth-jihad Muslims in this country (see "Stealth jihad" section of chapter 8) also went berserk. This included CAIR (the Council on American-Islamic Relations) which has a strong vested interest in the truth not coming out in these hearings…

> "Why Do Groups Like CAIR Fear Hearings on Islamic Radicalization? …Because they are the radicalization problem."[1167] (Emphasis mine.)

That's CAIR, and then we have the Muslim Brotherhood, Obama's favorite Egyptian political organization. The Muslim Brotherhood is a known terrorist organization sworn to both jihad and the annihilation of Israel, but nevertheless they were still allowed to bequeath a particularly subversive gift to American "higher" education:

> "The first organization the international Muslim Brotherhood founded in America was the Muslim Students Association (MSA), in order to recruit young people into its fold."[1168]

This Muslim Student Association, a "virtual terror factory" according to terrorism expert Erick Stakelbeck,[1169] has injected itself onto hundreds of American college campuses. Their pledge is "Jihad is my spirit!" and yet we have allowed them to operate on these campuses in the middle of a war against that Islamic jihad no less. Are we stupid, or what? CAIR and their enablers on the Democrat-Left are doing everything in their power to keep this radical organization and others from being exposed in these hearings. We'd have to ask the Nazi collaborators in our Democrat-Left media why they haven't already been exposed, but obviously that's not their job. Lying to America, by commission and omission, however is.

1166   (From ACLJ email, 03/10/2011)
1167   (From Act for America email, 03/10/2011)
1168   (From an Act for America email, 3/31/2011)
1169   (Author of the book *The Terrorist Next Door: How the Government is Deceiving You About the Islamist Threat*)

America needs to rise up and support men like Congressman King, who has the non-politically-correct, non-cowardly-Democrat-Left guts to hold those hearings despite all of the rabid attacks from the far-left and their organizations, so the enemy among us can be smoked out and imprisoned. That is what you actually do in a war, isn't it? And so we can also protect more of our gullible, naïve, youthful targets who have already been set-up for treason by the American-hating lies of the left, from being sucked into turning on their own country by the world's most disgusting excuse for a religion. And let those disingenuous Muslim-*American* whiners who would try to block this vital information from coming out, be exposed for the infiltrators that they are; and let their useful morons on the Democrat-Left also be exposed for their dangerous attempt to cover-up reality; and then let those devious little bastards over at CAIR, practicing their time-honored Muslim tradition of taqiyya (deception), put a sock in it, right before boarding a flight to Mecca. One way. Good riddance. And God bless you, Peter King.

Burning Beelzebub's bible

In April of 2011 mobs of Muslim cretins across the globe rioted over the burning of a Koran by a Florida preacher. In Afghanistan they slaughtered dozens, throwing in a few ritual beheadings just for sport. But the full story, the one behind the story, hasn't been told by our media. Naturally. So here it is:

"...I don't find the burning any more offensive in principle than I do its opposite extreme: the bizarre hyper-reverence with which the Koran is handled by the Defense Department.

"Down at Gitmo, the Defense Department gives the Koran to each of the terrorists even though DoD knows they interpret it (not without reason) to command them to kill the people who gave it to them. To underscore our precious sensitivity to Muslims, standard procedure calls for the book to be handled only by Muslim military personnel. Sometimes, though, that is not possible for various reasons. If, as a last resort, one of our non-Muslim troops must handle or transport the book, he must wear white gloves, and he is further instructed primarily to use the right hand (indulging Muslim culture's taboo about the sinister left hand). The book is to be conveyed to the prisoners in a "reverent manner" inside a "clean dry towel." This is a nod to Islamic teaching that infidels are so low a form of life that they should not be touched (as Ayatollah Ali Sistani teaches, non-Muslims are "considered in the same category as urine, feces, semen, dead bodies, blood, dogs, pigs, *women's genitals* [yes, we added that], alcoholic liquors," and "the sweat of an animal who persistently eats [unclean things]."

"This is every bit as indecent as torching the Koran, implicitly endorsing as it does the very dehumanization of non-Muslims that leads to terrorism. Furthermore, there is hypocrisy to consider: the Defense Department now piously condemning Koran burnings is the same Defense Department that *itself* did not give a second thought to confiscating and burning bibles in Afghanistan.

"Quite consciously, U.S. commanders ordered this purge in deference to sharia proscriptions against the proselytism of faiths other than Islam. And as General Petraeus well knows, his chain of command is not the only one destroying bibles. Non-Muslim religious artifacts, including bibles, are torched or otherwise destroyed in Islamic countries every single day as a matter of standard operating procedure."[1170]

1170 (From article, "More On Koran Burning" by Andrew C. McCarthy, 04/05/2011 @ actforamerica.org)

Like we said previously, Islam is the world's most vile, disgusting religion. And it is nauseating to listen to our media, including Fox News, tell only part of the story. They have fallen all over themselves to condemn the Florida preacher, each trying to outdo the other in how many ignorant names they can call him, but do you remember them even reporting any of the thousands of Bible-burnings by the Muslims, and by our own clueless, politically-correct military? No, I didn't either.

Personally, though, being as green-conscious and fastidious about recycling as we are, we do not approve of the burning of the Koran. That could be "harmful to the environment." We suggest using it for toilet paper instead.

* * *

In the next 3 sub-sections we will deal with some issues that are not directly a part of the undermining of our prosecution of the War on <u>Islamic</u> terrorists (as opposed to <u>domestic</u> terrorists like the ACLU), but ones that still highlight Satan's ongoing attempts to weaken America from within. These subjects are: "Arizona's immigration law," "Gun control" and "The Gulf Oil Spill"…

Arizona's immigration law

"What is desired in a man is his kindness; and a poor man is better than a liar." (Proverbs 19:22)

What better way for our Democrat-controlled government to destroy America than to flat out throw off all pretenses and attack her openly from within? How clever. What else, besides utter stupidity, could be behind Obama's unprecedented attack on Arizona for actually trying to defend her own borders against a massive armed invasion—the nerve of Arizona—which our Democrat-controlled government has no real interest in doing, even though <u>that</u> is what our Constitution mandates for them to do (as opposed to destroying our health-care system, which it does not). Illegal aliens are streaming across Arizona's southern border, despoiling the countryside, and some of the more violent are murdering, kidnapping and assaulting <u>legal</u> Americans, including legal <u>Mexican-Hispanic</u>-Americans, but the Democrat-Left is far more interested in encouraging as many illegal immigrants as possible to sneak in, by keeping our borders operating like a sieve and by relentlessly posturing politically for the granting of amnesty, citizenship and voting rights for those already here (under the guise of "comprehensive immigration reform").

"Rewarding Illegal Immigrants (AKA Democrat Votes) Obama's immigration reform translates as citizenship for illegal's, a weakening of border enforcement, and a huge toll on taxpayers in the form of new "social programs.""[1171]

What started this unprecedented internal assault by our own government was Arizona's passing of a border enforcement law that merely mirrored the federal law, as well as other states immigration laws, and that was also far less severe than the immigration laws of Mexico right on the other side of the Rio Grande. But none of those facts stopped the laws critics—mostly those who want open borders and Democrat votes—from immediately attacking it on the grounds of "racial profiling,"[LIE] even though the bill went to great lengths to specifically legislate

---

1171　(From a *David Horowitz Freedom Center* Circular)

against that. Obama and Holder (see movie, *Dumb and Dumber*) did this without having even read the bill. (Of course, the Democrats passed a 2,500 page health-care atrocity without ever having read that either, so what should one expect?) (They're brain-dead.)

Next they sued Arizona, using tax dollars that we no longer have, in order to try to get their Constitution-ignoring friends on the court to strike the law down by declaring it un-constitutional,(LIE) or more accurately, un-liberal. But revealingly, they didn't sue on the grounds of racial profiling (Obama and Holder knew that was just a lie, meant only to get their base all worked up), instead they said in court that Arizona was "usurping" the federal government's role in border security;(LIE) which was also a crock because the states are supposed to assist the federal government in all of their legislative and constitutional mandates. But again the Constitution is no more of a concern to Obama, Holder and the Democrats than is the safety and security of the citizens of Arizona.

"This administrations idea of a rogue state is not Iran, or North Korea, or Cuba, or Venezuela, but Arizona."[1172]

Then, almost even more incredibly, the president of Mexico, Felipe Calderon, was allowed to come here in May of 2010 and lambast Arizona for its new immigration law right from the podium of our own Congress, and even though his own countries immigration laws are vastly more severe than either our federal law or Arizona's. What a hypocrite. And Obama and his Democrats sat there and not only let him get away with his blatant hypocrisy, but stood up and applauded him. If it had been a movie instead of reality nobody would have gone to see it because it would have strained credulity. Who would believe that anyone could be *that* stupid? (Unless they really do hate America and want to radically transform—destroy—it, just like their father.)

Here is Mexico's immigration law:

"Immigrants must have the means to sustain themselves economically- they can't be an economic burden. They must be healthy and have no criminal record. Immigrants must show a birth certificate and provide for their own health care. Also, the Mexican government can ban foreigners due to race. (Hmm, racial profiling?) Illegally entering Mexico is a felony and will land you in jail. Document fraud will get you a fine and/or jail, and deportation can occur without due process. Also, their police must enforce their immigration laws."[1173]

And that's the kicker because in contrast Arizona's law merely allowed their police to ask someone about their immigration status; and only if that individual was already in their custody for a violation not related to immigration. So Calderon lied to America from the floor of our own Congress and the Democrats applauded him. He should have stood up and condemned his countries harsh immigration laws while applauding Arizona for their relative restraint. But that would have had something to do with the actual truth, which Mr. Calderon and his Democrat cheer leaders have a problem locating. Arizona should have adopted Mexico's much more severe immigration laws, then what would the jackasses have said?

Obama called the Arizona immigration law "misguided."(LIE) But that is a lie. Obama is the one who is misguided.

1172   (From unidentified contributor on Fox News)

1173   (Taken from "The Truths about Mexico's immigration law" @ Michelle Malkin.com May 2010, and from "Mexico's Immigration Law: Let's Try It Here at Home," by J. Michael Waller, 05/08/2006 @ humanevents.com)

Sean Hannity had this to say:

"Benjamin Netanyahu; our ally; the only democracy in the region; he comes to the United States and look at how poorly he is treated [by Obama]. No state dinner. No fanfare. No photography. Nothing. And he's left to his own devices in the White House for a long period of time. ...Now we've gotten to the point here in America where leaders of other nations come here, stand next to our president, they dump on America and Americans, and our [lame-brain] president joins in."

Stephen Hayes from The Weekly Standard replies: "I think this is truly disgraceful. ... When Felipe Calderon comes and lectures the United States, in the United States, on the floor of Congress, chastises us for laws that were passed in the state of Arizona that a governor signed; people who were duly elected; Obama not only said nothing to criticize Calderon but senior members of his administration stand and applaud this kind of criticism. It's a rather shocking turn. [Enough to turn ones stomach.]

Hannity: "Before the president of Mexico comes here and lectures us on the South lawn of the White House, you think he ought to maybe talk about changing his own laws first? ... Ronald Reagan never would have allowed this to happen, if he were president of the United States. Ronald Reagan would have stood up for the people of Arizona. He would have stood up for the rule of law. He would have stood up for the respect and sovereignty of the United States of America, and this president again and again and again has shown an unwillingness to do so. The question I have for you is, why?"[1174]

(The answer I have for you is... he's a jackass.)

"The cost of providing social, medical, law enforcement and educational services to illegal aliens from every level of government is estimated at an astounding **$113 BILLION** per year!"[1175]

And now our president has taken his assault on Arizona to the UN. Obama sent his Secretary of State, Hillary Clinton, to the UN Human Rights Council with a report that, shockingly, includes Arizona's immigration law as an example of a human rights "violation"[(LIE)] here in America. ("Mea Culpa! We're sorry, world, for being America! Please forgive us!" ...That guttural sound you hear under your feet is The United States of America, puking her guts out.) Of course, The Human Rights Council at the UN is a joke. It includes such human rights luminaries as Saudi Arabia, Libya, Uganda, and Pakistan. We have never been a part of it. Its sole purpose is to condemn Israel while letting all of their heavily oppressive friends off the hook. And therefore we have always ignored it and its infantile shenanigans; that is, until Obama. (Infantile shenanigans. It seems Obama finally has found something that is *not* above his pay grade.)

"This institution [the UN] is the most anti-American, anti-Israeli, anti-freedom institution in the history of the world."[1176]

"President Obama has launched a new offensive against Arizona's immigration law - this time at the United Nations (U.N.). He has filed an unprecedented report with the U.N. Human Rights Council, citing the Arizona immigration law as a "human rights problem" in this country -

---

1174   (Stephen Hayes from *The Weekly Standard* on Fox News' *Hannity*, May 2010)
1175   (From a Judicial Watch circular)
1176   (Monica Crowley, on the O'Reilly Factor, 5/10/2011)

Effectively handing over review of a domestic issue to some of the *WORST HUMAN RIGHTS OFFENDERS IN THE WORLD.*"[1177]

Now wrap your mind around this, America: Arizona's immigration law is identical to the federal law. Rhode Island, Missouri and Virginia have virtually identical laws, and many other states have followed Arizona's lead and passed similar laws. And Mexico's immigration laws are drastically more severe than Arizona's, and they <u>do</u> allow for racial profiling. But Arizona's law (and <u>not</u> the feds, or Rhode Islands, or Missouri's, or Virginias, or Mexico's) is a "human rights violation!!"[(LIE)] I'm sorry, and with all due respect to the office of the presidency, either these folks are bald-faced liars, or they are the stupidest people to ever walk the face of the earth. (Or they really do want to radically destroy America, from within.) God help us. And God help America if the uninformed, or rather intentionally <u>mis</u>informed, voters continue to go out religiously and obediently, every two years, and vote for them. Just like the "genius," Jon Stewart.

Gun control

"The strongest reason for the people to retain the right to keep and bear arms is, as a last resort, to protect themselves against tyranny in government."[1178] - Thomas Jefferson

The Democrat-Left is a big fan of Gun control, but what their media fails to report is that Hitler was a big fan as well. The Nazi's "Weapons Act of 1938" legislated that…

"All citizens who wished to purchase firearms had to register with the Nazi officials and have a background check. [It] presumed German citizens were hostile and thereby exempted Nazis from the gun control law. [It] gave Nazis unrestricted power to decide what kinds of firearms could, or could not be owned by private persons. The types of ammunition that were legal were subject to control by bureaucrats. Juveniles under 18 years could not buy firearms and ammunition."[1179]

And Hitler himself said this:

"The most foolish mistake we could possibly make would be to allow the subject races to possess arms. History shows that all conquerors who have allowed their subject races to carry arms have prepared their own downfall by so doing. Indeed, I would go so far as to say that the supply of arms to the underdogs is a *sine qua non* for the overthrow of any sovereignty. So let's not have any native militia or native police."[1180]

In our own country, fortunately, we have a Constitution that protects our individual right to bear arms; as clearly enumerated in the Second Amendment. Unfortunately, much of our country has fallen under the control of the Democrat-Left who could care less about our Constitution. And their efforts at controlling firearms are another glaring example of that disregard. In addition, the reasons they give for it are all just a ruse. They have no real interest in "protecting the citizenry from gun violence." If that were really the case then they would not be

---

1177   (From Jay Sekulow @ ACLJ.org 9/23/2010)

1178   (Thomas Jefferson, source - brainyquote.com)

1179   (From article, "Hitler was a leftist: Nazi Gun Control," posted on tripod.com)

1180   (Adolf Hitler, dinner talk on April 11, 1942, quoted in Hitler's Table Talk 1941-44: His Private Conversations, Second Edition (1973), Pg. 425-426. Translated by Norman Cameron and R. H. Stevens. Introduced and with a new preface by H. R. Trevor-Roper. The original German papers were known as *Bormann-Vermerke*. From tripod.com)

trying to disarm them and leave them defenseless against the hordes of armed and violent criminals that their left-wing criminal injustice system continues to dump back onto the streets—criminals by the way who don't obey the law to begin with and certainly wouldn't be inclined to obey the liberals stupid gun-control ones. They might also show their great "care" and "concern" by being a whole lot tougher on crime and violent criminals to begin with.

"No free man shall ever be debarred the use of arms."[1181]

Here is Ted Nugent's take, as only he could put it:

"I believe that a person's moral compass can be determined by how he references freemen the right to defend themselves. The Second Amendment is so obvious to me it's insane that there's an argument. ...Let's pretend there is no document. Let's pretend brave families didn't leave the tyrants and the slave drivers of Europe so that they could practice the religion of their choice; so they could speak up without being murdered; that they could produce wool without the kings men coming and taking it from them every season of harvest. Let's pretend none of that happened. Let's just pretend this guy named Ted Nugent parachuted onto Earth and woke up one morning and saw these wonderful resources and had dreams of excellence and being the best that I can be. I don't need a document [referring to our Constitution], and I don't need another man to explain to me that I have the right to defend my gift of life. And that there is an argument in America, from Hillary Clinton, from Barbara Boxer, Dianne Feinstein, from a whole gaggle of numb-nuts who would try to tell me they will dictate where, how and *if* I can defend myself. I find that preposterous. I find it unacceptable, and I will not accept it. I am a free man. Don't tread on me. A good law-abiding citizen not convicted of a felon—the second amendment of our bill of rights is my concealed weapons permit. Period!

..."Instead of arresting people for molesting children 24 times, I would rather the dad walked into the room, found the person molesting that child and blew his brains out. I would rather that the lady in Massachusetts last month who was taking her daughter to soccer, who was carjacked by a recidivistic maggot who had been in the prison system all his life but was let out again, because [sarcastically] "we feel sorry for him. Maybe he had a bad childhood." Instead of her being hijacked and murdered I'd rather she just shot the bastard dead. But in Massachusetts, somebody decided she can't do that. So she's dead. I would rather she was alive and the carjacker was dead. I'm weird. [Supportive laughter is heard in the background.] I would rather that the guy who beat this lady to within an inch of her life in Waco, on parole was he—phenomenal—and beat her to within an inch of her life in front of her grandchildren with a whiskey bottle. I would rather she fell to the ground, pulled out a 38 and shot him six times in the chest and killed him. Am I weird? Because the guy is gonna get out again. I don't like repeat offenders. I like dead offenders."[1182]

And all of God's people said... AMEN! Thank you, Ted.

The Gulf Oil Spill

In April of 2010 British Petroleum's Deepwater Horizon oil rig exploded and gushed 206 million gallons of crude oil into the Gulf of Mexico over the next three months before it was finally capped. But what we need to know for our purposes here is that the primary cause of this catastrophe can be firmly laid at the

---

1181   (Thomas Jefferson, source - brainyquote.com)
1182   (Ted Nugent being interviewed on the Second Amendment, The Right to Bear Arms, by Evan Smith, Editor Texas Monthly, on klru radio, Austin Texas)

feet of the Democrats and the influence that environmental wackos have had on them and their policy decisions over the years. You see, BP wanted to drill in 500 feet of water, closer to shore, where divers could much more easily have handled any disaster because of the relatively low pressures at that depth, and probably would have capped that well in a few days. The state of Louisiana gave them permission to do so but the Feds stepped in and said "No Way!" you're gonna drill @ 5 thousand feet instead. Why? There is no logical reason. Unless they wanted to cause a disaster so they could use it as an excuse to further hamper our ability to tap into our own extensive oil wealth, rather than making the Saudis and their terrorists even richer by continuing to tap into theirs. And, interestingly this is exactly what the Obama administration did. He employed the strategy of his Chief of Staff, Rahm Emanuel, "Never let a serious crisis go to waste," and they stopped all drilling in the Gulf. The Saudi's and the terrorists could be heard laughing all the way over here. Now obviously the Democrat-Left didn't plan this oil spill disaster in advance (unlike Bush and 9/11) but they got their convenient catastrophe nevertheless and they used it to continue to strip America of her ability to produce her own energy, and thereby accelerate her destruction from within. And now you know the rest of the story that a lying Democrat media will never tell you.

Judge Andrew Napolitano summed it all up perfectly:

"Last night in an Oval Office speech to the nation President Obama took the gloves off. He lashed out at BP over the monumental oil spill in the Gulf of Mexico. He vowed that he would make BP pay for all the long and short term damage it caused, and he argued that the battle against this nearly 2-month old raging gusher of oil is tantamount to a war. And he asked the country to allow him and the Congress to regulate all businesses and private homes in the name of going Green. For most Americans, it was the first time they saw him angry, although a controlled anger, from the "no-drama-Obama," as his campaign staff labeled [him] two years ago. But will the president's feigned anger get the oil well plugged? Will it get cash into the hands of those truly harmed? Will it prevent future disasters? No, no, and no. Here are the facts. After the Exxon Valdez disaster off Alaska in 1989 had been cleaned up and nearly paid for by Exxon, the oil companies lobbied the Congress for liability limits. That would be maximum amounts that they could be held to pay in the event of a disaster. A Republican Congress and President Bill Clinton together made it into law that oil companies will be limited to pay $75 million for the cleanups. And the taxpayers, that would be you and me, would pay for the rest. In return, the feds will be able to tell the oil companies where to drill. In this case of BP it asked the state of Louisiana if it could drill in 500 feet of water and Louisiana said it could. But the federal government vetoed that and told BP it could only drill in 5000 feet of water. Never mind that no oil company had ever cleaned up a broken well at that depth. And never mind that the feds had never monitored a broken well at that depth. And never mind that BP only needed to set aside 75 million in case something went wrong. The feds trumped BP engineers, and the feds trumped the wishes of the folks who live along the coast, and the feds decided where this oil well would be drilled. Disaster struck. The feds did nothing. Oil gushed out in an amount so great as to be immeasurable. Political pressure grew. The president eventually panicked, because he believes that his federal government can right every wrong, regulate every activity, and protect us from every catastrophe. He's wrong. Louisiana Governor Bobby Jindal was ready to build barriers to protect his state's coastline and the Feds incredibly said no. The president even invoked powers that allowed him to supervise the cleanup using BP personnel and equipment, and the oil still gushes. Last week the president stopped all drilling in the Gulf, putting thousands of Americans out of work. Last night he demanded billions from BP, so his team could decide who gets it. And today, a terrified BP gave him all the cash he asked for. So the government, the government that foolishly limited BP's maximum liability, the government that claimed it knew where best to drill, the government that actually stopped locals from protecting their own shoreline, that would be the same government that has bankrupted Social Security, Medicare, Medicaid, the post office, Amtrak, and virtually everything it has ever managed; the government now wants to decide who gets BP's cash. The last time this government had this much private cash to give away during the GM and Chrysler bankruptcies, it disregarded well-settled law, and it gave it to its friends in the labor unions. To whom will it give this cash? The innocent injured, or its political buddies? The government cannot protect us from every catastrophe, especially ones its rules have facilitated. How about this, "that government is best which governs least." The people have a right to a government that obeys the laws of economics and the laws of physics and the Constitution."[1183]

The people may have a right to that, Andrew, but they're never going to get it when they continue to march out like obedient little lemmings and vote for the kind of "progressive," Constitution-despising politicians you accurately described above. As far as the part about Obama stopping "all drilling in the Gulf" which in turn put tens of thousands more "Americans out of work," that was a perfect example of the Left taking political advantage of a catastrophe that they caused. In this case using it to limit drilling for oil here so we can do what? …send more money to terrorists overseas. Newt Gingrich added this:

"The president's drilling moratorium, which …the *Wall Street Journal* proved conclusively last week [that] every expert that they claim supported it has said they did not support it. They think it actually makes drilling more dangerous. They think it means that the best people and the best rigs are going to leave the Gulf. Governor Jindal [of Louisiana] estimates it's going to kill 84,000 jobs. We have an unnecessary publicity stunt that kills 80,000 jobs and actually weakens safety in the Gulf. …This administration has failed to be competent at the most basic aspects of responding to a crisis."[1184]

And that's because, Newt, their primary concern is not so much "responding to a crisis" as it is what kind of long term political and environmental hay they can make out of it. As Charles Krauthammer observed:

"[Barack Obama] made a small opening before the [Gulf oil] spill to have a slight relaxation of the moratorium on drilling in the Atlantic and Pacific and in Alaska. And of course all of that is retracted as a result of the Gulf spill. He took advantage of that and he's governing exactly the opposite way. Our problem is that …it's a national security issue and it's also an economic issue. We are shipping a third of a trillion dollars overseas on imported oil which we don't have to do because we have oil here. The Democrats have insisted for two decades and more on not drilling—and that's Anwar in Alaska, it's the national petroleum reserve in Alaska, drilling on the coasts, natural gas and nuclear energy—and this administration, Democrats, are wedded to a policy of solar and wind which are sweet and nice and theoretical but they are extremely uneconomical."[1185]

And speaking of Anwar—that barren, oil-rich patch of Martian landscape in Alaska—in 1996 the Republicans passed a law to allow drilling there but President Clinton vetoed it, saying that "besides it would be another ten years before we

---

1183   (Judge Andrew Napolitano, filling in on the Glenn Beck show on Fox News, 6/16/2010)
1184   (Newt Gingrich on Fox News *Hannity*, 6/21/2010)
1185   (Charles Krauthammer, on Fox News' Special Report with Brett Baer, 10/5/2010)

would get that oil." Well, it's now 2011 and we could have been getting 3 million barrels of oil a day for the last 5 years if we hadn't elected *that* particular rapist in the first place. And you can multiply Anwar by the thousands with all of the other places the Democrat-Left and their environmental wackos have blocked our drilling for oil, like in the Gulf of Mexico and offshore in the Atlantic and the Pacific, and all of the natural gas that we're not allowed to tap into.[1186] And on that subject, Senator David Vitter had this to say:

> "We are *the* single most energy rich country on the globe, bar none. The only one who comes close is Russia, who's second. But we're the only energy rich country that takes 95% of those resources [and] puts them off limits. We can't afford to do that. It's absolutely ridiculous. We need to open up that access for all sorts of energy, not just in the Gulf [but in] other offshore areas, offshore Alaska, Anwar, onshore Alaska, and Western Shale."[1187]

And the Democrats are still standing up in Congress to this day and saying "why drill for oil, because it will be 5 to 7 years before we get that oil." They've been using this argument for decades, and they'll still be using it decades from now. (If the Lord hasn't returned.) In addition to robbing Americans of hundreds of thousands of high-paying jobs in the oil and natural gas industry; wasting billions of our taxpayer dollars on their pie-in-the-sky "green energy" policies; sending trillions of our dollars to enrich the terrorist-supporting Saudis; forcing us to spend <u>eight</u> <u>times</u> more for their oil than if we produced our own; trying to be the first customer in line for Brazil's oil, again, rather than producing our own; etc. etc. It is a crime against America what the Democrat-Left has done to our energy production with the help of their smiley-faced liars in our media and the uninformed folk who march out obediently every two years and vote for them. Congratulate yourself, America, the next time you go broke pulling up to a gas station.

America is being destroyed from within; and Americans are going out and voting for it; incredibly, for their own destruction. Satan is hell bent on tearing down America, in order to make sure it is too weak to stop him from destroying Israel, and he has enlisted the help of his eager beavers on the Democrat-Left. They think they are real smart, and real progressive, and are radically transforming our country; but they are nothing more than real duped. The lies of Satan poison their minds like a cloud of industrial smog, and they believe the liberal garbage that they have been taught. Pray for them. And us.

See:
- *Courting Disaster: How the CIA Kept America Safe and How Barack Obama Is Inviting the Next Attack*, by Marc A. Thiessen
- "Who's Responsible for America's Security Crisis?" by David Horowitz
- *Party of Defeat*, by David Horowitz & Ben Johnson
- *America Alone: The End of the World as We Know It*, by Mark Steyn
- *Power to the People*, by Laura Ingraham

1186 (From Governor Haley Barbour, Republican, Mississippi, on Fox News' Your World with Neil Cavuto, 03/30/2011)
1187 (Senator David Vitter, Republican, Louisiana, on Your World with Neil Cavuto, 04/01/2011)

- *The Roots of Obama's Rage*, by Dinesh D'Souza
- *The Terrorist Next Door: How the Government is Deceiving You About the Islamist Threat*, by Erick Stakelbeck
- *Slander: Liberal Lies about the American Right*, by Ann Coulter
- *Fleeced*, by Dick Morris and Eileen McGann
- *Catastrophe*, by Dick Morris and Eileen McGann)
- *Party of Defeat*, by David Horowitz and Ben Johnson
- *Treason*, by Ann Coulter
- *Useful Idiots*, by Mona Charen
- "The Great Betrayal: Obama's Wars and the War in Iraq," by Daniel Greenfield

11- Weaken America's Economic Strength

"Here is wisdom. Let he who has understanding calculate the number of the beast, for it is the number of a man; and his number is six hundred threescore and six." (Revelation 13:18)

President Obama on the campaign trail, fall 2010:

"Republicans spent the last decade driving this economy into a ditch.[LIE] [Actually, stupid, they spent the last decade trying to fight a war against Islamic extremists abroad, and their Democrat-Left sympathizers like you here in America.] And so, for the last 20 months, Democrats have gotten down into the ditch, put on our boots.[LIE] [For the last 20 months Democrats were down in the sewer, where they've been for decades, and shoveled so much s**t the rest of us had to put on boots.] We're down there pushing, pushing, pushing on the car.[LIE] [You're up here lying, lying, lying to the American people.] Every once in a while we'd look up and see the Republicans standing there, sipping on a Slurpee. [LIE] [Republicans were standing there telling you to listen to the American people who were screaming at you to STOP!] And we'd say, 'Come on down and help.'[LIE] [Obama actually rejected every single idea the Republicans and the American people had and at every turn in favor of his radical and idiotic socialist policies.] They say, 'You're not pushing the right way.'[LIE] [No, Wilbur, they said you idiots were pushing our country into the river.] And we just kept on pushing and pushing. [You sure did.] And finally we got this car up on level ground.[LIE] [Finally, you destroyed our health care system too.] Now, this car is a little beat up now because they drove it into the ditch.[LIE] It's got some dents, needs a tune-up. But it's pointing in the right direction.[LIE]" [If you consider Cuba the "right direction."] — President Obama, standard riff in over 20 campaign speeches in fall 2010."

Does this guy even know that there is such a thing as *the truth*? No, that's right, to him and his friends the truth doesn't even exist. It's "relative." (See chapter 1.)

Rush Limbaugh had this response to Obama's delusional car routine:

"The truth is, Democrats began steering the American economy toward the precipice once they won control of Congress in 2006. And ever since Barack Obama climbed into the driver's seat in January 2009, it's been "Thelma and Louise." Obama and his Party put pedal to the metal and drove the U.S. economy off the cliff — and now it's crash-landed, battered and broken, not in a mere ditch, but at the bottom of a desolate canyon called "Obamaville.""[1188]

1188   (From "Welcome to Obamaville," an article by Rush Limbaugh in The Limbaugh Letter,

Whatever they may say, and however they may try to disguise their many sweet-sounding programs by wrapping them around deceptive slogans like "helping the poor"[(LIE)] and "assisting the middle class"[(LIE)] and "leveling the playing field"[(LIE)] etc. etc., it is the "progressive" Democrat-Left's ultimate goal to replace capitalism and our free-enterprise system with a quasi-European- or outright- socialist one. And they could care less how many of the "poor" and the "middle class" get their jobs, the quality of their lives and their finances trampled on that "level playing field" in the process. That is the translation into English of Obama's famous "radical transformation" slogan. His "change," that there are now very few intelligent people left on earth to still "believe in," would destroy the greatness and Exceptionalism of America along with her economy. And that is also of course exactly what Satan wants, so we will no longer be in a position of military strength in which to defend Israel. (Of course, with Obama in the White House, one could argue that Satan doesn't need to destroy our economy in order to keep America from defending Israel. (But we're not saying he's a Muslim. Maybe he's a Rastafarian...)

Free-enterprise, coupled with the Judea-Christian value system and biblical morality that our society has traditionally been based on, is the source of the strength and the unparalleled success of America's economic engine. And both are also the foundation of the American Exceptionalism that has led to our phenomenal growth and unparalleled prosperity as a nation; which is why they both have been heavily under attack for years by the great deceiver and his followers in our educational-left-wing-indoctrination system. And now (early 2010) Obama and his Democrats, products of that system, have us on a Baton death march to full blown European socialism or Marxism-Lite, and that will be the death of the prosperity and the Exceptionalism of this nation. Satan works behind the scenes in our classrooms to raise a generation of Americans who are brainwashed in his lies, and wholly ignorant of just what has made this country great. (See section 4, "Destroy the Historical Foundations of America," and section 5, "Take over our educational system.") And he is already well on his way to accomplishing his goal of annihilating the moral fiber of this once-great, Christian nation. (See chapter 2, "Moral Absolutes," and sections 1, 2 and 3 of this chapter.) The end game of all of this is the raising up of a generation of good little left-wing-indoctrinated Barackers who, ignorant of the truth, will mindlessly give his Democrat-Left the political power necessary to destroy the American economy. Just look at what Obama, Pelosi and Reid have "accomplished" with their total control of our federal government for the last 2 years (Jan 2009-Oct 2010): an economic Armageddon and a country tottering on the verge of insolvency as a result of their profligate spending.

> "**Bankrupting the Economy.** The deficit is nearly $2 trillion, 19% of our economy, and growing. Trillions have already been spent to grow government, wither individual initiative, and diminish the size of the American Dream."[1189]

The Democrat Housing Bubble

To understand fully what is going on today economically (October of 2010) it is important to go back to the beginning of this recession and uncover its root

---

causes. Obama has repeatedly accused George Bush of being responsible for it,[LIE] but that of course is just another of his many lies, like "America is not a Christian nation,"[LIE] "I am not a socialist"[LIE] (nor a Muslim) and "I did NOT rape that woman, Juanita Broderick, in a hotel room in Arkansas!"[LIE] (Woops! Sorry, that was Clinton, not Obama.) So what was the real cause of the 2008 financial meltdown that provided an opportune crisis for the Democrats to exploit in order to gain further control of Congress and for their media to help them place an unqalified novice in the presidency? The answer of course is the "housing bubble," and its subsequent collapse. But *what was the* cause *of the housing bubble*?

> "Two narratives seem to be forming to describe the underlying causes of the financial crisis. One, as outlined in the *New York Times* front-page story on Sunday, December 21, is that President Bush excessively promoted growth in home ownership without sufficiently regulating the banks and other mortgage lenders that made the bad loans. The result was a banking system suffused with junk mortgages, the continuing losses on which are dragging down the banks and the economy. The other narrative is that government policy over many years—particularly the use of the Community Reinvestment Act and Fannie Mae and Freddie Mac to distort the housing credit system—underlies the current crisis. The stakes in the competing narratives are high. The diagnosis determines the prescription. If the *Times* diagnosis prevails, the prescription is more regulation of the financial system; if instead government policy is to blame [which it is], the prescription is to terminate those government policies that distort mortgage lending.
>
> ..."There are two key examples of this misguided government policy. One is the Community Reinvestment Act (CRA). The other is the affordable housing "mission" that the government-sponsored enterprises (GSE's) Fannie Mae and Freddie Mac were charged with fulfilling."[1190]

And because of this "mission" banks were essentially forced into making bad loans. Many of the safeguards in lending that followed the stock market collapse of 1929 (which led to the Great Depression) were thrown out the window in favor of the Democrat-Left's obsession with transferring wealth (in this case home ownership) to those "less fortunate." The end result of course was that the "less fortunate" couldn't afford the mortgage, lost their homes and became even less fortunate. Again, the Democrat-Left's sweet-sounding yet deceptive slogans that they use to legislate their "help the poor"[LIE]- "assist the middle class"[LIE]- "level the playing field"[LIE]- programs are nothing more than a disguise for their father's America-hating, prosperity-despising and free-enterprise-killing socialism. Besides, if they really gave even the slightest damn for the "less fortunate" then they might be the least bit concerned for the millions of "less fortunate" babies that they mutilate to death every year. But that logic of course, is a little above their notorious pay grade.

> "There was fraud. There was an attempt here to use the Community Redevelopment act as a way to extort banks into giving ...home loans to people who couldn't afford them. And this was a strong Democratic initiative that carried through ACORN. ...Barack Obama was a lawyer for ACORN, helping ACORN extort banks in the subprime area. ... They should be ...investigated. Again, these private deals that Senator Dodd and others had to their benefit, while the economy tanked, and there's millions of people now facing

---

1190   (From article "The True Origins of This Financial Crisis: As opposed to a desperate liberal legend." By Peter J. Wallison, in The American Spectator, February 2009)

losing their homes. Senator Dodd and the other Democrats should be held accountable for this."[1191] (And that'll happen about the same time hell freezes over.)

This trend from the Democrat-Left to intimidate, cajole and ultimately force banks through legislation into making these risky home loans to unqualified buyers began in the Carter administration, and then steamrolled over the next few decades through the power of high placed Democrat advocates of this financial insanity in Congress like Chris Dodd and Barney Frank. On Fox News' *The O'Reilly Factor*, 10/10/08, Bill O'Reilly and Newt Gingrich elaborate further:

O'Reilly: "It's basically about people like Barney Frank [Democrat congressman and head of the House banking committee]. Frank wanted all of the government lending agencies [Fannie Mae and Freddie Mac] and indeed the private banks to lend money to poor people, because Frank's philosophy is that poor people get hosed in America. And we have to be proactive in getting them entitlements, which includes mortgages. So Frank's whole philosophy and whole career has been built on that. And he's not alone. But he's the head of the House finance committee. So that's where the pressure was, social engineering, ACORN [Association of Community Organizers for Reform Now], and Obama is tied into that whole social engineering philosophy if you will. Now it's not a bad philosophy. [Emphasis mine.] It's humane. It's compassionate, but it's irresponsible in a capitalist system because the system can't afford it."

Gingrich replies: "I disagree with you. It is a bad philosophy. It's bad to tell people they can have something for nothing. It's bad to get people in a mortgage they can't pay for. It's bad to put them in a house they're going to lose. It's bad to set the entire housing system up to fail. It is bad to have ACORN go out as they did in Chicago, where Obama was one of their trainers, and have them go out to bankers homes and humiliate bankers in their neighborhoods and pressure them into making bad loans. [Emphasis mine.] It is bad to take your tax money and my tax money as a Democrat Congress did this summer and give it to ACORN- 500 million dollars a year of your tax money and my tax money- is going to go to left-wing organizing groups. These things are bad. They're not just value neutral. And I think they've been bad for America."[1192]

ACORN, now exposed as a corrupt organization focused on defrauding the election process to favor their Democrats, was at the forefront of this campaign, and our friend Obama was one of their early new recruit instructors.

The following is from Peter J. Wallison in an article for The American Spectator:

"For banks, simply proving that they were looking for qualified buyers wasn't enough. Banks now had to show that they had actually made a requisite number of loans to low- and moderate-income (LMI) borrowers. ...In other words, it called for the relaxation of lending standards."

[The article then quotes Stan Liebowitz of the University of Texas at Dallas:] "...In fact, it was the [federal government] regulators who relaxed these standards—at the behest of community groups and "progressive" political forces...For years, rising house prices hid the default problem since quick refinances were possible. But now that house prices have stopped rising, we can clearly see the damage done by relaxed loan standards."

..."If we are really serious about preventing a recurrence of this crisis, rather than increasing the power of the government over the economy, our first order of business

---

1191 (Jerome R. Corsi, PhD. author of *America for Sale: Fighting the New World Order, Surviving a Global Depression, and Preserving USA Sovereignty*, on Fox News' Hannity, 10/13/09)

1192 (Newt Gingrich on Fox News' "The O'Reilly Factor," 10/10/08)

should be to correct the destructive housing policies of the U.S. government."[1193]

Satan, however, <u>is</u> very serious about doing nothing to prevent its recurrence and his Democrat-Left's first order of business is to use this crisis as a convenient means to amass more and more power for the government over our economy, while at the same time, with the eager assistance of their media, covering up their responsibility for the entire disaster.

And speaking of culpability, there's the lovable Barney Frank, the powerful Democrat Congressman from Massachusetts and sex-educator for young male interns.

> "What Barney Frank did was in the 1980s; he and the Democrats went to Fannie Mae and said you are not buying up enough mortgages of poor people. We want you to buy up 42% of your mortgages to be of poor people, and then they raised it to 50. And Fannie Mae said they can't put any money down. And Barney Frank and Chris Dodd said its okay, don't let that matter, don't require any money down. So they gave mortgages they shouldn't have given, where people didn't have any investment. And then in the 2000's, when Bush proposed measures to rein in Fannie Mae, Barney Frank killed it."[1194]

Thank you so much, Massachusetts, for another Democrat "gift" to the nation. Perhaps you could do all of us a favor and just succeed from the union. (And take Maryland and California with you.)

Thomas Fitton, President of Judicial Watch, follows the scam earlier through the Clinton administration:

> "Beginning in the late 1990's, the Clinton administration pressured lenders to lessen home loan requirements in order to increase home ownership for those unable to afford homes and to curry political favor with left-wing activists, like the ACORN crowd!

> "As far back as September 30, 1991, the New York Times reported that: "Fannie Mae, the nation's biggest underwriter of home mortgages, has been under increasing pressure from the Clinton administration to expand mortgage loans among low and moderate income people and felt pressure from its stock holders to maintain its phenomenal growth in profits." The article went on to predict this strategy would run Fannie Mae into serious financial difficulty during an economic downturn. <u>We were forewarned</u>!

> "Nevertheless, Clintonite hacks at Fannie and Freddie like Franklin Raines, Jamie Gorelick [no relation to Global Warming suck-ups in our lame-stream media] and Jim Johnson (later a key adviser to Barack Obama) took Clinton's risky gamble and then doubled down. ...Encouraged by Fannie and Freddie, banks began making tens of thousands of "subprime" loans to individuals who under prudent lending rules never would have qualified for them. This ruinous game was played for ten years under both the Clinton and George W. Bush administrations.

> "As the Washington Post noted in a December 9, 2008, story: "These new products included home loans made to people with blemished credit histories, called subprime loans, and mortgages made without verification of income, assets or employment ... The loans required borrowers to state their incomes and assets, but not prove them." The decision to violate sound lending principles to advance the Clinton administration's political agenda was bad enough. But that's not all.

> "Executives at Fannie and Freddie received huge bonuses if "loan targets," including targets for subprime loans, were met. <u>Franklin Raines earned $100 million and Jamie</u>

1193   (From article "The True Origins of This Financial Crisis: As opposed to a desperate liberal legend." By Peter J. Wallison, in The American Spectator, February 2009)
1194   (Dick Morris on the *O'Reilly Factor*, April 8, 2009)

<u>Gorelick earned $75 million in bonuses from Fannie Mae</u>! [But take heart, Obama has told them to pay the money back.] [Yea, right.]

"In other words, these "government-sponsored entities," not only trashed well-established lending standards -- they also encouraged their executives to hustle as many bad loans as possible. In fact, we now know that at Fannie Mae in 1998, it's then-head (and later Obama advisor) Jim Johnson cooked the books by deferring $200 million in expenses in order to fraudulently inflate profits so he and other senior Fannie executives could receive massive bonuses that year. What a racket!

"Despite their accounting problems and warnings of crisis, Fannie and Freddie, with the full support of liberals in Congress and their special-interest community organizing friends like ACORN <u>pushed for more subprime lending</u>. ...Overall, Fannie and Freddie spent more than $180 million over the last ten years to lobby Congress, and make sure that no one rocked their highly profitable boats.

"...Over the same ten year period, Fannie and Freddie contributed lavishly to the politicians who chaired or had senior positions on the committees that had oversight responsibility for their institutions. ...The top four recipients of money from Fannie Mae in order of cash magnitude were: **Sen. Christopher Dodd (D-CT), Sen. Barack Obama (D-IL), Sen. Chuck Schumer (D-NY), Rep. Barney Frank (D-MA)**"[1195]

And isn't it amazing that Fannie and Freddie, at the heart of this housing-bubble-caused-recession, are exempt from the Democrat's Dodd-Frank financial "reform" legislation. What a joke. A bad one, but a joke nonetheless.[1196]

"Mortgage risks could be passed on to the secondary mortgage market by Fannie Mae and Freddy Mac in Washington, D.C., backed by an implicit federal guarantee. They would buy up these mortgages, call them "assets," and sell them to investment bankers who believed that they were worth a lot more than they turned out to be. Like a Ponzi scheme, it seemed to work for a while, but when house prices began to decline everything fell apart.

"In recent years, the great political promoter of the Ponzi scheme has been the chairman of the House Financial Services Committee, Barney Frank. Fannie and Freddie showered Democrats, and to a lesser extent Republicans, with campaign contributions. They have invested in politicians. Since 1989, Connecticut's Sen. Chris Dodd has been the top recipient; more recently, one Barack Obama."[1197]

And for Barack's part in the scheme, he was elected president. (Who said the Palestinians were stupid for electing Hamas, a terrorist organization?)

And now you know the truth about the "Housing Bubble" and just who was responsible for putting our economy in the tank. Yet Obama will continue to blame Bush, and his Democrat media will continue to allow him to get away with it, and far too many people will continue to blindly and ignorantly support him, no matter what.

1195   (From Judicial Watch circular, "Stop The Cover Up of the Fannie Mae and Freddie Mac Scandals," by Thomas Fitton, Pres.)

1196   (We don't have the space here to cover this "other" Obama nightmare monstrosity legislation—in addition to his health-care slop—but suffice it to say that "Dodd-Frank" is to business and our economy what "Jim Crow" was to blacks. And whereas Jim Crow laws segregated blacks from the rest of society, the Dodd-Frank regulations will segregate businesses from any hope of profits or the hiring of additional employees)

1197   (From article "Crediting the Uncreditworthy," by Tom Bethel, The American Spectator, November 2008)

Stimulate this

"Ah, sinful nation, a people laden with iniquity, a brood of evildoers, children given over to corruption! They have forsaken the Lord. They have provoked the Holy One of Israel unto anger, and turned their backs to him." (Isaiah 1:4) (In case anyone is wondering why our economy is in a free fall.)

President Obama's stimulus[LIE] package was a lie. It was a spending package that created more unnecessary government jobs that steal money from taxpayers; bailed out the state public employee unions (Democrat votes and Democrat campaign contributions secured with your tax dollars) and threw billions into every idiotic and inane pork-barrel project that the fevered minds of the Democrats in Congress could come up with on very short notice. (I'm surprised they didn't just cut through the BS and pass a bill that would study the environmental effects of burning hundreds of billions of one dollar bills out on the Washington Mall.) And it sucked money OUT of the economy, which is the exact opposite of what is needed to stimulate it. But then Obama hasn't given us any reason to think he has even the slightest idea of how our economy works, much less how to "stimulate" it. If he did he would begin by getting himself, his taxes, his bloated government and its regulations the hell out of its way, and stop choking it to death with more brainless, socialized, expensive and redundant liberal programs that don't work and are bankrupting the nation in the process.

Since the stimulus was rammed through Congress in February of 2009, 2.05 million jobs have been lost despite the spending of $800 billion dollars which we do not have. (Makes a lot of sense doesn't it? ...Spending money you don't have.) The unemployment rate has risen from around 8% to around 10% nationwide despite the fact that we were told (lied to) that if this record trillion-dollar, spendulus bill wasn't immediately passed without any argument, dissent, debate, opposition or so much as a single critical thought then the unemployment rate would rise to 10%.[LIE] Which it did anyway. Go figure. 831,000 construction workers also have lost their jobs despite all of those mythical "shovel-ready" projects that were coming right down the pike. (Obama now admits, in October of 2010, that "there is actually no such thing as a "shovel-ready" project," so I guess the only thing he was shoveling was more of his bulls**t.)

Here is what retiring Senator Evan Bayh had to say in February of 2010 about his own parties "stimulus" package:

> "If I could create one job in the private sector by helping to grow a business, that would be one more than Congress has created in the last 6 months."[1198] And he's a Democrat!

(And at least he's honest.) The same is true of Obama's "jobs" bill a year later. It has nothing to do with jobs, and everything to do with government expansion. He wants to spend another 20 to 30 billion or so, again money that we don't have, in order to "keep teachers and police officers and firefighters from losing their jobs." It sounds good, but he doesn't give a damn about those folks. What he does care about is the public service unions they belong to, and if he can take billions of dollars and give it to those unions he knows that they will in turn give back hundreds of millions of dollars to his Democrats in the form of campaign ads and campaign contributions. (And this is despite the fact that a great many of their members are not Democrat supporters and do not want their forced union dues used in this manner.) They are thieves. They steal money from taxpayers

---

1198   (Senator Evan Bayh 2/16/2010)

and give it to unions in order to get it back for themselves—via union campaign contributions and advocacy ads—in order to get elected. And we voted for it by electing the thieves in the first place.

Let's be honest. Obama is a far-left indoctrinated ideologue. He's a true believer. If he had to choose between: A) wildly successful economic growth and prosperity but doing it with tax cuts and less government and less regulation and less spending so that Reagan Conservatism would get all the credit; or B) another great depression with 25% unemployment with massive poverty and hunger and suffering like never before but so his far-left ideology did not only not suffer but could continue with even more massive spending and tax hikes; he would choose B. It's left-wing, income-redistributing, socialist ideology uber alles. The real battle here is between the American way and the Marxist way. It's between capitalism and socialism.

> "Liberalism's love of control and consequent attraction to a Soviet-style planned/ command economy and revulsion toward capitalism leads it to demonize wealth and achievement. Hence, liberals war on the most successful American companies: Wal-Mart, McDonald's; its hatred of lucrative industry: pharmaceuticals, Big Oil, business in general. The left's inability to respect, appreciate, encourage, or support the entrepreneurial creativity, innovation, and productive genius of the private sector is one of liberalism's biggest failures."[1199]

Here is an excerpt from Sean Hannity's interview with Mark Levin concerning his new book, *Liberty and Tyranny: A Conservative Manifesto*:

> Hannity: "You write about the free market. I'm a huge believer in capitalism; it brings the best out of people. While we're pursuing a standard of living for ourselves, we're advancing the human condition and creating wealth that's been unprecedented in the history of mankind, and it is constantly under attack [from the Democrat-Left Elite]. You call the free market the most transformative of economic systems. It fosters creativity and inventiveness that we have never seen before."

> Mark Levin: "Most of the problems in the free market are created by government; whether it's the government compelling banks to give out bad loans, as they did in the 70s, 80s and 90s that led to the housing bust; first the boom and then the bust. Even the automobile industry …isn't an example of a free market. The government controls the labor force. It controls the product that they make. It controls management. I mean, I don't know how these poor guys manage these companies to begin with. On the other hand, failure in the free market is not necessarily a good thing, but it's a productive thing."[1200]

And when the government picks winners and losers by determining that a company is "too big to fail" it destroys the free market. (And Bush was guilty of this as well with his tarp bailout of the large banks and insurance companies in the fall of 2008. Although there is an argument by respected economists that if he hadn't bailed them out we would have plunged into another Great Depression… but of course now we'll never know.) It also is extremely unfair to all of the smaller competitors of those bailed-out companies who did not get any help, who were not "too big to fail." Who were "little enough to get screwed." It is not the government's job to pick economic winners and losers, and it's unconstitutional as well.

> "I think that the unearned transfer of wealth to The United Auto Workers in the GM

1199   (From article: "Nailing the Left: Liberal Failures," the Limbaugh Letter, Rush Limbaugh)
1200   (Mark Levin on Fox News' *Hannity*, 3/23/2009)

and Chrysler case is a scandal. (…look at how much money the United Auto Workers gave the Democrats over the last four elections) It is an absurdity; it is in my judgment a corrupt act; …the idea that you are going to turn over to the Union that helped bankrupt the company that level of money; for the Union to in effect be rewarded for helping to bankrupt the company is a fundamental violation of the American way of doing business and the American way of life."[1201]

But it fits right in with the Democrat way of doing business and the Communist way of life.

"Cut taxes and give small business and entrepreneurs a chance to create real jobs. Don't keep pouring money into failed bureaucracies that for a year have thrown away our money and our children and grandchildren's money and accomplished nothing, except to further weaken the economy.

…"People are increasingly coming to realize that with Obama what you get is great rhetoric but no reality. You get somebody who can read the teleprompter pretty well, but he doesn't actually have a real plan …[or] a real approach. …Bob Bennett …the Senator from Utah, had a great line … "spend more, tax more, borrow more, that's the six word Obama plan and it doesn't work.""[1202]

Obama said that GM paid back their bailout money.(LIE) But that was a lie. GM got a 4.7 billion dollar loan which was used to "screw over the bond holders and give the unions a sweetheart deal." They then got 52 billion in tarp money, most of which "went for stock to give to the government so they own 61% of GM now."[1203] GM paid back the 4.7 billion but not the 52 billion, but Obama made it sound like they paid all the money back. What a kidder. And his Democrat media let him get away with it. "Obama lied and our economy died!"

Campaigning in October, 2010 in San Francisco, Nancy Pelosi said, "It is the biggest bang for the buck when you do food stamps and unemployment insurance —the biggest bang for the buck." (Does that s**t-4-brains actually think the American people are that stupid?) (No, she thinks her voters are that stupid.) I'm not sure why San Franciscans don't just throw off all pretenses and go out and elect a pig to Congress. (Or a plate of dogs**t.) (Neither could do as much harm as Pelosi.) It would be a first. And it certainly would get a lot of support from the folks at PETA.

A local rock radio station (which leans far-left, naturally—rock and roll having not caught up with the real world yet) has been running an ad which says: "We're the only radio station in town that gives you commercial-free Mondays!" And our response is: Yea! That's because you got no advertisers, you stupid marble flippers! That's because all you ignoramus' voted for O-BA-MA! O-BA-MA! And pretty soon if you keep supporting O-BA-MA! O-BA-MA! you're gonna have commercial free Mondays, Tuesdays, Wednesdays, Thursdays, Fridays, Saturdays and Sundays! (But then maybe you can qualify for a bailout.)

1201   (Newt Gingrich on Fox News' Hannity 6/03/09)
1202   (Newt Gingrich, on Fox News *Hannity*, 2/04/2010)
1203   (Sorry, we lost the source of those two quotes. I think it was from Judge Andrew Napolitano on Fox News…)

In George W. Bush's 8 years in office Congress racked up 4 trillion dollars of debt. Then Senator and presidential candidate, Barack Hussein Obama, said that was unpatriotic and un-American. Now he's equaled Bush's "unpatriotic" level of debt, but he's done it in only 2 and ½ years. Obviously Dumbo will say anything, and then do the exact opposite; and his Democrat media will let him get away with it.

"The US deficit for fiscal year 2010 is expected to be $1.3 trillion, compared to a 2007 deficit of $160.7 billion."[1204] (But yet Bush was the one who "drove our car into the ditch!?" ...Actually you *can* fool all of the people all of the time. The Democrats have certainly gotten away with it with their voters.)

That's an explosion of over 8 times that Obama has foisted on America under the guise of a "stimulus" plan and "change all kinds of dimwits can believe in."

"From the moment George Washington took the oath of office until the moment Barack Obama took the oath of office [219 years later] we borrowed 9 trillion dollars. In the last 19 months [1.6 years] we've borrowed 3.5 trillion more. What else do you need to know?"[1205]

The thing is, the first 43 presidents weren't trying to radically "transform" (aka "destroy") America, where Barry Obama and his father Satan are. He's spending us into bankruptcy. Now isn't that stimulating?

Americas Adventures in Obameconomyland

On Election Day, November 4, 2008, while going for a walk in the woods, America stepped into a voting booth and fell into a hole that they have yet to climb out of. (Would that *we* were only hallucinating.)

"Hey, Obama has just nationalized nothing more and nothing less than General Motors! Comrade Obama! Fidel, careful or we are just going to end up to his right."[1206]

That was Hugo Chavez gloating to his Marxist buddy, Castro. And doesn't that about say it all, America?

"The Troubled Economy - Liberals love it. Rep. Barney Frank hails the arrival of "resurgent government activism" similar to FDR's New Deal. Hide your wallet! Bad economic times gladden the heart of every power-hungry government planner. They can't wait to increase state power, enact harsh regulations, fatten up the bureaucracy, and pile on the tax hikes. Don't believe their phony sorrow, folks. These people don't "feel your pain." They delight in it!"[1207]

"They believe in a fantasy world of giant government, giant taxes, Washington controlling everything. And if you disagree with them it's a sign you're dumb."[1208]

"Obama wants to return the nation's wealth to its rightful owners. He comes from a belief that those in America who have succeeded have done so on the backs of the poor and the disadvantaged. They've had their wealth stolen from them."[1209]

1204   (From Fox News' *The O'Reilly Factor*, 9/01/2010)
1205   (Dick Morris on Fox News' *The O'Reilly Factor*, 9/01/2010)
1206   (Hugo Chavez on 6/02/09)
1207   (Rush Limbaugh, The Limbaugh Letter)
1208   (Newt Gingrich)
1209   (Rush Limbaugh on Fox News' Hannity, 6/04/09)

"If Al Qaeda wants to demolish the America we know and love, they better hurry, because Obama's beating them to it."[1210]

I can go you one better; if we had elected Osama Bin Laden in 2008 instead of Osama Bin Obama, he could not have done a better job of destroying America. If Bin Laden had been elected and had to stay within the confines of "extra"-constitutional legislation like Osama Bin Obama has done, he could not have come up with, nor passed, anything more destructive than Obama's catastrophic, economy- and job-destroying, socialist policies. He wouldn't have launched all those drones, but he couldn't have done anything worse economically than Obama and his other Two Stooges have done.

"I believe that the economy is under siege. It's being destroyed. Anybody with any economic literacy would not do one thing this administration's done to try to revitalize the private sector. They're destroying it."[1211]

(Uh, that would assume that he even *has* the goal of "revitalizing the private sector?")

Barack Obama "had a summit at the White House and he brought in business leaders [and asked] "Why aren't you hiring?" ...They said, it's [the] increasing taxes. It's the responsibility that we may have to pay for all this increased health care. It's the issue of energy costs if they do these mandates with the Environmental Protection Agency with carbon regulation. It's the union voting so that ... they'll lose an opportunity to have a secret ballot. ...Those are the things that are in the president's agenda."[1212]

Why aren't business leaders hiring, Mr. President? Because of your job-killing agenda, that's why.

"Business' are tired of all the uncertainty created by this administration: higher health care costs, higher social security taxes, payroll taxes. They want the Bush tax cuts extended permanently. ...They're tired of being charmed to death. It's not working because of all the real things that are being thrown at them as soon as they walk out the room. ... All the town hall meetings. All the phone calls. All the press conferences. All of these things where the president is trying to show that he cares belie the fact that all his actions show that he is attacking small business too frequently. [Small business owners do tend to vote Republican, after all.] Too much uncertainty and uncertainty is the enemy of job creation. Do something that business can hang their hat on, cut their tax rates. Give them certainty."[1213]

Obama, give the "tax cuts for the rich" mantra a break. It's called "tax cuts for the job creators" so they will have the seed money to take risks and actually *create* some new jobs, you uneducated putz. (With all due respect, of course.)

"For a perfect example of what actually does work just look at Germany... They reduced their unemployment benefits, they cut spending, they cut their deficit and they loosened up on crippling business regulations. In the second quarter (2010) they had 9% growth, and their unemployment is down to 7.6%. Compare that to the Obama administration which did the exact opposite of all that: we have 9.5% unemployment, 1.6% growth and their "summer of recovery" is a joke."[1214]

1210   (Source unknown)

1211   (Rush Limbaugh on Fox News Sunday with Chris Wallace, 11/01/09)

1212   (John Barrasso, Republican Senator from Wyoming, on Fox News Hannity, 08/06/2010)

1213   (Jim LaCamp from Macro Portfolio Wealth Management, on Your World with Neil Cavuto, 11/23/2010)

1214   (Fred Barnes of the Weekly Standard on Fox News' Special Report with Brett Baer,

"One of the riskiest things you can do in America is to hire somebody. And because of that reason, because of all the liabilities from government, from lawsuits that you have put on employers, most small business' main concern is how <u>not</u> to hire people. How can I grow my business and hire as few people as possible? That is not something that happens in the market. That's something that happens as a consequence of government."[1215]

"It's very risky [to hire people right now in America]. Unfortunately, for politics, there are a lot more people who are employees than employers. So politicians want to get votes so they promise all sorts of special privileges for employees, but when they do that they put obligations and risks on employers. It is so easy now to sue your boss, if he doesn't create the job in the exact manner that the government specifies they have to do it. And because of all these costs, not just the taxes, but the potential litigation, a lot of businesses do what they can not to hire. ...The government makes it too expensive and too risky to hire a human being." ...So they use machines instead whenever they can."[1216]

When you raise taxes on the rich you take money away from them that they would use to invest in their business or take a risk in other startup businesses, which in turn hire people. Instead you suck that money out of the economy and you send it to Washington to be squandered. And then there's this:

"An economics professor at a local college made a statement that he had never failed a single student before, but had once failed an entire class. That class had insisted that Obama's socialism worked and that no one would be poor and no one would be rich, a great equalizer. The professor then said, "OK, we will have an experiment in this class on Obama's plan." All grades would be averaged and everyone would receive the same grade so no one would fail and no one would receive an A... After the first test, the grades were averaged and everyone got a B. The students who studied hard were upset and the students who studied little were happy. As the second test rolled around, the students who studied little had studied even less and the ones who studied hard decided they wanted a free ride too so they studied little. The second test average was a D! No one was happy. When the 3rd test rolled around, the average was an F. The scores never increased as bickering, blame and name-calling all resulted in hard feelings and no one would study for the benefit of anyone else. All failed, to their great surprise, and the professor told them that socialism would also ultimately fail because when the reward is great, the effort to succeed is great but when government takes all the reward away, no one will try or want to succeed."[1217]

Whether this actually happened or not I cannot determine. For one, what college would allow such an intelligent, non-left-wing-indoctrinated-dupe-of-a-professor on its staff? He would have been fired before he ever approached tenure. But the point is valid either way.

And here's an interesting idea: let's hang a 50 by 100 ft. banner across the street from the entrance to the White House that says:

8/30/10)

1215   (Peter Schiff, CEO of Euro Pacific Cap tol, testifying at a congressional subcommittee hearing on government regulations and creating jobs, 09/13/2011)

1216   (Peter D. Schiff, CEO of Euro Pacific Capitol and author of the book *How an Economy Grows and Why it Crashes*)

1217   (Internet post. Source unknown.)

SOCIALISM DOESN'T WORK, STUPID!

Not that he would get it. Above his pay grade.

Look at the case of Maryland Governor, Democrat Martin O'Malley. He created a special tax for millionaires that was supposed to bring in an additional $106 million dollars in revenues to pay for the usual Democrat reckless over-spending. But instead he lost $257 million. Why? Because the millionaires changed their residency to other states which didn't have O'Malley's punitive taxation. They can do that because the wealthy own homes in other states, or they can afford to move. It happens every time but they never learn.

But the liberals and their Democrat media will continue to preach their class warfare lies: "Everyone needs to pay their fair share."[LIE] (Actually the top 1% of wage earners in this country pay 37% of our tax bill. The top 5% pay 59%. And the top 50% pay 98%.[1218] If that's not "paying their fair share" I'd like to know just what the hell it is. God, those Democrats are such lying bastards. Of course, as we pointed out previously, if you can hack up babies then lying is as nothing for you. And their media will run cover for them all day long.) "It's a zero sum game."[LIE] (Which means that if you have more then I have less, or vice versa.) "Tax cuts for the rich."[LIE] (As opposed to what, "tax cuts for the poor" who pay no taxes to begin with!?#&@)) Progressives are clueless when it comes to prosperity, how to create it, and where it comes from. They are real good, however, at mutilating babies to death.

""The Obama Economic Recovery"[LIE] - Day after day, Obama declares "our economy is growing again."[LIE] The "swing to expansion,"[LIE] he claims, is "the largest in nearly three decades."[LIE] You can choke on the sheer audacity of the lies. Meanwhile, job-killing, growth-stifling taxes are heaped on businesses... and laid-off workers lose hope. There is no expansion – zip, zero, nada - except in the size of government and the deficit. The White House mantra of "jobs created or saved"[LIE] is nothing but a *tour de force* of lies crafted by the maestros of the "art.""[1219]

"Obama is making Americans less free. Austrian philosopher Friedrich Von Hayek issued the same warning half a century ago. Having witnessed the horrors of Nazi Germany and Fascist Italy, Hayek argued that centralizing the economy is a first step towards fascism. "In order to achieve their ends," Hayek wrote, "the planners must create power, power over men wielded by other men of a magnitude never before known. Democracy is an obstacle to the suppression of freedom which the centralized direction of economic activity requires.""[1220]

Friedrich Von Hayek issued that warning in a book titled, *The Road to Serfdom*. Daniel Hannan, British member of EU Parliament, is reissuing the warning in his book, The New Road to Serfdom: A Letter of Warning to America. Here is what he had to say concerning it on Fox News' Hannity:

1218   (Source: National Taxpayers Union)

1219   (Rush Limbaugh, from circular Sept 2010)

1220   (Friedrich Von Hayek, author of *The Road to Serfdom*, as quoted by Sean Hannity on Fox News' Hannity, 9/27/2010)

"I'm writing as a British politician who loves his own country very much but who values, as people do, something special about this country. The US isn't just a nation like any other. It's the embodiment of an ideal. We're all involved in the success of America and if we see the US becoming poorer, less democratic, less free than that's everybody's problem. ...Changing America into a different society will mean forsaking the most successful constitutional model in the world. ... [The American Constitution] was the highest encapsulation of our philosophy of freedom. And it's worked! ...It has served to make this country successful and strong and free and prosperous. And it did those things because it dispersed power, it democratized power, it recognized localism and states' rights. And when I see the US going down this road towards federal czars, more ...spending, centralization of power in the White House, it becomes less American and therefore less prosperous and less free."

"I've been 11 years in the European Parliament. So I've seen it firsthand the kind of model towards which your current administration is taking you. European healthcare. European day care. European social security. European welfare. European unemployment rates. And believe me you're not going to like it."

"The bits of Europe which are most obviously failing, the welfare system, the high spending, they're the bits that you're copying."[1221]

"Changing America into a different society will mean forsaking the most successful constitutional model in the world." And that is exactly where Obama and his Democrat-Left—the enemy within—are taking us. The uninformed were led to believe that he was going to "radically transform" us into some kind of mythical, utopian, "free-lunch" nation. But we already were the greatest nation in the history of the world, and rather than actually trying to improve upon and tweak that success, which he and his lefties have absolutely no intelligent concept of how to do, he has drug us in the exact opposite direction.

"The way of getting rid of the deficit ultimately is through growth and stimulating the economy. The best way of doing that is to give people tax cuts so that they can spend their own money, rather than taking money out of the economy, which is what the stimulus package does, and putting it in the hands of the bureaucracy. If that were a better way of running the economy, we'd have lost the Cold War and you and I would be having this interview in Russia."

"I think the Tea Party is a remarkable phenomenon ... looking at the main message, the thrust of their argument, which is that you can't carry on raising more money, spending more money, and borrowing more money, and that is a very reasonable proposition. I'm really struck by the way that we keep on reading in the media that this is a mob of racists, a gaggle of stump-toothed, Appalachian mountain men, and yet the American voters regard the Tea Party as more moderate and more reasonable than the Democratic Party."[1222] Hmmm. Could that be because, unlike the media, those American voters actually have a brain? Time for an MRI...

And I'm sure that most of us thought debtor's prison, that shameful practice of throwing people in jail who couldn't pay their bills, had been done away with a century ago. We thought wrong:

1221   (Daniel Hannan, member of British Parliament, on Fox News' Hannity, 9/27/2010)

1222   (Daniel Hannan, British member of EU Parliament 9/28/2010 on Fox News' Your World with Neil Cavuto)

"Prior to the mid-19th century debtors' prisons were a common way to deal with unpaid debt. Currently, the practice of giving prison sentences for unpaid debts has been mostly eliminated, with a few exceptions such as inability to pay child support and certain taxes, and some specific countries, such as the United Arab Emirates."[1223]

…But now it's making a comeback in Alice's Obameconomyland:

"Eclipsed by media focus on the foreclosure mess, personal bankruptcy filings have totaled over 1.1 million since the beginning of 2010, according to Foxbusiness.com, an increase of 11 percent over last year. And as reported by UPI.com, debt imprisonment has risen dramatically as poor people unable to pay legal fees are jailed. Cash-strapped states are imposing financial penalties on poor defenders to raise revenue, and incarcerating them — on top of their criminal sentences — when they are unable to pay it. Yes, debtor's prison — horrors from what we all thought was a bygone area."[1224]

"YES WE CAN! Yes you can what, stupid? Yes we can throw those poor folk in jail rather than taking a dime from the public service unions, even though we care so much about the poor[LIE] we can hardly stand it. (Or, at least, that's what we want our no-think-um voters to keep on a-thinkin')."

And then there's the unemployment fiasco:

"Close to 10 million Americans receive unemployment compensation, almost quadruple the 2007 number, thanks to benefits being extended eight times by Congress. More than a million unemployed are "99ers," who have gone through a full 99 weeks of unemployment benefits, and yet are demanding another extension. Since Obama has been in office, unemployment benefits have jumped from $43 billion to $160 billion, as reported by *USA Today*.

"In less than two years, Obama and the Democrats have taken a reasonable program that temporarily aided laid-off employees as they transitioned to a new job — and turned it into just another form of welfare. At the same time, the continual extensions have not only rewarded idleness, but have encouraged widespread fraud, such as double-dipping and working off the books. Like so much the libs do, it's all under the guise of "compassion" — which actually hurts those it purports to help, as it undermines society."[1225]

And don't be fooled when Obama tries to disguise more of his out-of-control spending by using the word "investment" as he did numerous times in the annual State of the Union Address (Jan. 25, 2011). As in, "We need to *invest* in our children's education." Or, "We need to *invest* in our infrastructure" or "*invest* in our future." Sounds great, doesn't it?! Investing! That's what rich people do with their money! Why shouldn't our government be just as smart as all of those rich people? Because they're not "investing," no matter how many times they repeat that word. They're only spending; and now they're trying to continue to do it, like drunken monkeys, with money they do not have. As Mike Huckabee pointed out:

"Some time ago I took some of my own money and I bought some stock in Apple computers. It was my money and if Apple continues to do well and I get a profit …I get the increase. But if Apple mismanages what people like me and others give them …then I lose. But that's the dynamics of an investment. But what if I took your money against your will

1223   (From Wikipedia, the free encyclopedia)
1224   (From article, "Welcome to Obamaville," by Rush Limbaugh, The Limbaugh Letter, November 2010)
1225   (From article, "Welcome to Obamaville," by Rush Limbaugh, The Limbaugh Letter, November 2010)

and then I spent it? That's not an investment. I'm tempted to say that by most definitions it would be theft. But government does have the legal authority to exact money from you …and it's called a tax. But when they spend it it's not an "investment" unless you get the return. Now when government's expenditure succeeds, they get the benefit. If it fails, you eat it. Follow me so far? So when the president tells us that his proposed spending is an investment […don't buy it]."[1226]

Every time the American people hear them talk about "investing," they need to understand that it's just more of their code speak for "taxing and spending," "taxing and spending;" which at this point is nothing more than "spending and bankrupting," "spending and bankrupting."

"Obama's 'Vision' for America" It's Greece! Thanks to the president, the ranks of government employees are exploding. Recession? There is none, if you have a federal, state, county or municipal job. Public-sector union employees now outnumber private sector union workers. In this recession, over 8 million private sector jobs have been lost – while over 100,000 government jobs have been added. And federal workers make in pay and benefits $30,000 more than their private sector counterparts. Happy days are here again – if you've got a government job in Obama's European style socialist America.

"America's most successful criminal? For my money, it's Algore. He – aided by corrupt leftist scientists, the state-controlled media and actor/wackos like Ed Begley, Jr., Ted Danson and others – have caused *trillions* of taxpayer dollars to be confiscated in the name of the global-warming fraud. Result: Algore is now the world's first "carbon billionaire." I think he's the Al Capone of our era and should be investigated – in between his hotel massages, of course.

"What's the difference between Obama-nomics and Bernie Madoff's Ponzi scheme? Nothing. Take a look: Your taxes are about to skyrocket because of health care, the VAT, energy tax increases, and the sunsetting of the Bush tax cuts. That money is being redistributed to people who are not paying any taxes at all, but who remain the darlings of the Regime. My friend, that's a million times worse than anything Madoff ever did – and Bernie is serving a 150-year prison sentence!

"Were a long way down that dreary, one-way road to socialism. Fact is, more of us are getting from the government than are paying for government. The number of Americans who aren't paying taxes is frighteningly close to 50 percent. Those are people who won't care how high Obama (or any president) raises taxes. What's more, 52 percent of the citizenry now get their income directly or indirectly from a government source. That's called "incremental socialism," my friend."[1227]

It's also called "the great deceiver is well on his way to completely destroying America's economy," my friend.

Eric Bolling, of Fox News "The Five," opines on the presidents attempt, in Sept of 2011 to burn another half trillion dollars in another one of his brain-dead stimulus' (disguised as a "jobs" plan)…

"The president said his plan will create 1.9 million jobs. Only in Obamaland, just a short monorail ride from Fantasyland, would spending a half a trillion dollars create jobs because, in the real world, we've already spent a trillion and a half dollars on stimulus and unemployment skyrocketed. …Apparently I failed Obamamath."[1228]

1226 (Mike Huckabee, on his show on Fox News channel, Feb. 19, 2011)
1227 (Rush Limbaugh, from a Circular, Sept. 2010)
1228 (Eric Bolling on Fox News' "The Five," 09/16/2011)

Apparently.

"If Only Obama Would Abdicate," by Quin Hillyer:

"This boringly ineffective would-be demagogue in the Oval Office keeps trotting out the same tired, petty, counterfactual lines in every one of his pompous, detached-from-reality speeches -- and he did so again in his Saturday radio address. So shopworn is his refrain, and so counterproductive are his policies (both enacted and proposed), that the markets tank just about every time he opens his mouth, and the economy suffers with each minute he continues to occupy the White House. His resignation from office, in abject embarrassment at his failures, would be a great first step toward economic recovery, not to mention a balm to the souls of tens of millions of Americans sick of his condescension, his prevarications, and his incompetence."[1229]

That is a miracle we can only hope for in November of 2012...

And finally, there was a pathetic spectacle on the news recently (Oct. 2010) that again demonstrates what a miserable failure socialism is: rioters in France were throwing a tantrum because they were asked to give up a tiny portion of their socialist goodies, by way of a "whopping" two year increase "in their countries retirement age. France is attempting this "draconian" measure because, like the United States, it has a no-longer-financially-sustainable teat-sucking class and is headed towards insolvency and bankruptcy. Daniel Hannan of the British Parliament had this observation:

"If you make somebody dependent on the state; if you create a welfare system that drives out the private sphere, that takes over for the family—education, health care, social care—you infantilize people; you turn people into children. And we all know what sort children are like. They don't have to see the responsibility as well as the rights. They want their parents to give them things. It's only when we get older that we appreciate that you have to pay for things. A country can become that way if you take away people's sense of responsibility and independence."[1230]

Today the infantilized, rioting socialists in France are demanding that their financially unsustainable largesse remain untouched, no matter what the consequences to everyone else and to their entire country. Tomorrow, if we are able to bring our country back from Obama's brink and start turning this country around, many of the Democrat's voters will be doing the same here in the United States. "O-BA-<u>MA</u>! O-BA-<u>MA</u>!" "Gimmee! Gimmee! Gimmee!" "Waah! Waah! Waah!"

...In fact it's already happening: First in Wisconsin (February 2011) where rioters from the teachers union have called in sick and even taken some of their students to the state capitol along with them to protest a small change in their benefits package. They are asked to shoulder this tiny inconvenience in order for their nearly bankrupt state to remain solvent, and in lieu of the fact that they pay a small fraction for their benefits compared to what folks in the private sector have to pay. "Waah! Waah! Waah!" What a bunch of spoiled brats. ...Democrat voters all.

And now our country is being subjected to the "Occupy Wall Street" movement, the Democrat-Left's incoherent response to the Tea Party. A small mob of young

---

1229   (From American Spectator online, 08/25/2011)
1230   (Daniel Hannan, on Your World with Neil Cavuto, 10/20/2010)

unemployed urban campers, whose minds have been all but destroyed by the lies of the left, many of whom are either rapists, drug addicts, anarchists or flat-out America-despisers, want the rest of us to pay their way in life. Not surprisingly, this group of embarrassing losers have been fawned all over by Nancy Pelosi, the Lord of the Idiots and the Democrat media despite the fact that they have no intelligible message, other than "Gimmee! Gimmee! Gimmee!" and "If you don't we're gonna poop on your police cars!"

> "The Occupy Movement starts with the premise that we all owe *them* everything. They take over a public park they didn't pay for, to go nearby to use bathrooms they didn't pay for, to beg for food from places they don't want to pay for, to instruct those who are going to work to pay the taxes to sustain the bathrooms and to sustain the park so they can self-righteously explain that they are the paragons of virtue, to which we owe everything. Now that is a pretty good symptom of how much the Left has collapsed as a moral system in this country, and why you need to reassert something as simple as saying to them, 'Go get a job! Right after you take a bath.' "[1231]

Yes, Newt, but think how damaging that would be to their self-esteem.

See:

- *How an Economy Grows and Why It Crashes* by Peter D. Schiff and Andrew J. Schiff
- *The New Road to Serfdom: A Letter of Warning to America* by Daniel Hannan

Global Warming

"For, behold, the day comes, that shall burn as an oven; and all the proud, yes, and all who do wickedly, shall be stubble; and the day that comes shall burn them up, says the Lord of hosts." (Malachi 4:1)

> "Environmentalism, the new home of socialism, considers human beings the enemy of the earth,[LIE] and the internal combustion engine the greatest threat to the planet.[LIE] For years, the left has targeted SUVs in a failed attempt to control American vehicle purchases and steer us toward glorified lawnmowers.
>
> "Thanks to the EPA and the Endangered Species Act, normal development is constantly impeded. Wetlands, spotted owls, a clump of native grasses, or the existence of a handful of sand flies have permanently shut down building projects across the fruited plain, rendering land and investments worthless.
>
> "In what may be a first in the history of human civilization, our nation has been cowed by activist liberals from using our own natural resources. To thwart forward-looking — one could even call them "progressive" — Americans from drilling for oil or building new refineries in order to ease our dependence on foreign sources of energy, liberals sue, enact anti-development regulations, or produce shameful congressional testimony of a weeping teenager who's been made to believe her culture is being destroyed. For decades, liberals have grossly failed the nation by checkmating us in every attempt to meet our own energy needs."[1232]
>
> "Around the mid-1980s the environmental movement was basically high jacked by the political Left. And, at the same time the Berlin Wall came down, communism

---

1231   (Newt Gingrich, reported on the O'Reilly Factor, 11/21/2011)
1232   (From article: "Nailing the Left: Liberal Failures," the Limbaugh Letter, Rush Limbaugh)

ended, and a lot of peaceniks who were …anti-American and leftist in their orientation, moved into the environmental movement, bringing their neo-Marxism with them. And they learned to use *green language* in a clever way to cloak agendas that …have more to do with anti-capitalism and globalization than anything to do with science or ecology."[1233]

That quote was from Patrick Moore, one of the founders of Greenpeace, who left the movement when he saw what its socialist hijackers were turning it into. He goes on to say how these folks have… "taken the science out of it, and instead there is this mixture of sensationalism, disinformation and fear." Like the theory of evolution, science has again taken a back seat to the agenda, and the lies, of the great deceiver…

Satan is attempting to destroy America's financial strength, and he has found a reliable and zealous ally in the modern-day environmental movement. But in the beginning the conservation movement in America was led by some of our countries greatest visionary heroes—like John Muir and Teddy Roosevelt—and we are in their debt for what they accomplished in establishing many National Parks and Monuments and preserving many of our countries greatest natural treasures and wonders for posterity. That was then, this is now. Today's environmental movement has been taken over by America- and capitalist- hating leftists who have goals far removed from simply protecting the environment. The "radical transformation" of our society from freedom and capitalism to one of servitude and communism is the pot of gold they seek at the end of the rainbow. And the hysterical global warming scam is their latest, and most promising, scheme.

"Liberals have told us for years that if we don't adopt their policies and give them more control over our lives, environmental Armageddon will be just around the bend.[LIE] They say science is on their side and there's nothing left to debate.[LIE] …Scientists, politicians and big business have turned global warming hysteria into a multi-billion industry. …[and its] hysteria is spreading across the country. People live in fear that the planet will perish unless they drastically alter how they go about their daily lives."[LIE][1234]

Algore and his friends, however, are exempt from these draconian measures. They can continue to fly around the globe in their private jets, enjoy their hundred-foot-long houseboats and live in their ninety-room mansions, spewing more carbon in a day than one of the common folk do in a year. But they don't have to change. Only their suckers…er…I mean their voters do.

"Hollywood hypocrites …drive most of the way in from Malibu in Lincoln Navigators … but arrive on camera at the Academy Awards in hybrid, eco-friendly Priuses."[1235]

But should we expect anything less from Hollow-wood?

"Future generations will wonder in bemused amazement that the early twenty-first century's developed world went into hysterical panic over a globally averaged temperature increase of a few tenths of a degree, and, on the basis of gross exaggerations of highly uncertain computer projections combined into implausible chains of inference, preceded to contemplate a rollback of the industrial age…"[1236]

1233   (Patrick Moore as quoted on "The Green $windle," a special addition of Fox News Hannity, 8/27/10)

1234   (From "The Green $windle," a special addition of Fox News Hannity, 8/27/10)

1235   (From Human Events circular, Tom Winter, Editor-in-Chief)

1236   (Dr. Richard S. Lindzen, Alfred P. Sloan Professor of Atmospheric Sciences, Massachusetts Institute of Technology (MIT); member of the National Academy of Sciences; and former lead author, UN Intergovernmental Panel on Climate Change.) (And Algore and

And especially, the industrial age in America.

If one wants to get at the real "inconvenient truth" behind the global warming hysteria that is being shoved down the throats of our naïve and unsuspecting school children in our colleges, universities, elementary and high schools, and down the throats of naïve and misinformed adults by the Democrat-Left's control over our media and entertainment industry, you can't rely on that same left-wing educational system and lame-stream media, or an Algore Crockumentary. We would recommend one of Christopher Horner's books, *The Politically Incorrect Guide to Global Warming*, in which he documents "the junk science—and hidden political motives—behind the phoniest environmental scare in decades." Horner explains that...

"For decades, , environmentalism has been the Left's best excuse for increasing government control over our actions. ...It's for Mother Earth! It's for the children! It's for the whales! But until now, the doomsday-scenario environmental scares they've trumped up haven't been large enough to give the sinister prize they want most of all: total control of American politics, economic activity, and even individual behavior. With global warming, however, greenhouse gasbags can argue that auto emissions in Ohio threaten people in Paris, and that only "global governance" (Jacques Chirac's words) can tackle such problems.

"[Horner] tears the cover off the Left's manipulation of environmental issues for political purposes—and lays out incontrovertible evidence for the fact that catastrophic man-made global warming is just more Chicken-Little hysteria, not actual science. ... Although Al Gore and his cronies among the media elites and UN globalists endlessly bleat that "global warming" is an unprecedented global crisis, they really think of it as a dream come true. It's the ideal scare campaign for those who hate capitalism and love big government. ...[Horner] reveals the full anti-American, anti-capitalist, and anti-human agenda of today's environmentalists, dubbing them "green on the outside, red to the core."[1237]

The author goes on to point out:

- "Proof that greenhouse gas concentrations do not determine temperatures-notwithstanding media hype and deceptive Al Gore slide shows ...[and that] only a tiny portion of greenhouse gases are man-made." - "The hole in the ozone layer—the 1980s man made environmental crisis—was caused by the Antarctic atmosphere being too *cold*." - "Most of Antarctica is getting colder." - "The Earth has often been hotter than it is now." - "The media only recently abandoned the "global cooling" scare." - "Hurricanes are not getting worse—our tendency to build houses in their path is getting greater." - "Many big businesses lobby for global warming policies that will increase their profits—and our costs." - "The Medieval Warm Period was significantly warmer than temperatures today—and was a golden age for agriculture, innovation, and lifespan." - "Environmentalists throughout modern history have instilled fear about one looming "crisis" or another with the aim of increasing government control." - "The environmental alarmists do whatever they can to avoid actual debate."[1238]

One particularly nefarious thing they do in order to "avoid actual debate"

---

his friends would have us believe that "the debate is over" and "anyone who disagrees with our hysterical, bogus take on global warming is no different than a holocaust denier.")

1237   (From inside cover, *The Politically Incorrect Guide to Global Warming*, by Christopher C. Horner)

1238   (From cover, *The Politically Incorrect Guide to Global Warming*, by Christopher C. Horner)

is by labeling anyone who refuses to go along with them as being the same as a "Holocaust denier." The difference of course is that the Holocaust actually happened; whereas the Democrat-Left's unscientific fantasy of catastrophic <u>man-made</u> global warming has not. Man has at best only an extremely miniscule effect on the climate compared to the forces of the sun and nature, but that inconvenient truth won't fill the coffers of Algore and his friends, nor will it assist Satan in his desire to dismantle capitalism and cripple America economically.

The most famous way that Algore has padded his own pocket is through his aforementioned, critically-acclaimed (by the Left) yet outright dishonest Crockumentary, *An Inconvenient Truth*:

> "Liberals are now ...attempting to panic the population with the latest political jeremiad, Algore's *Inconvenient Truth* — a junk-science propaganda vehicle for the global warming hoax, which has been adopted by the left with all the fervor and dogma of religious faith. Combining a paranoid fear of carbon, a hatred of progress (especially due to eeevil American prosperity) with a lust for control of others' behavior that is so characteristic, liberals flush with virtue and self-righteous Puritanism are attempting to inflict guilt and asceticism society-wide — absurdly blaming the American lifestyle for planetary temperature. [As opposed to say, just off the top of my head, the <u>sun</u>.] Environmentalists constantly harangue Americans about what they drive and purchase, how goods are packaged and disposed of, attempting to make life as inconvenient as possible. Americans stubbornly choose progress, development, innovation, and refuse the left's constant attempts to get them to live in huts as simple peasants."[1239]

"The planet has a fever." – Algore. Algore has a fever; in his head.

One of the more bald-faced lies that the global warming zealots along with their Democrat-Left media have broadcast is...

> ..."the so-called "Hockey Stick Diagram," which states that temperatures stayed the same until the 1800s, when the industrial revolution took place, and then [rose] sharply like the bottom of a hockey stick. ...[However, this] is bogus. When you include the Medieval Warming Period, which was far warmer than today and was followed by the Little Ice Age in the 14th century, the temperature diagram dramatically changes and looks like climate variations occur in a cyclical manner. There is growing contention about GW [Global Warming] within the scientific community that is not being reported. [Shocker!] ... nearly 50% of them do not agree that it is a man-made problem. [That means that half of all of our scientists are damn "Holocaust Deniers!" Call the thought police!] In fact, rather than being a pollutant, CO2 is an extremely beneficial gas that could increase food production by 20 to 40 percent and dramatically improve the health of earth's ecosystems if its concentration doubled."[1240]

We also know now that the celebrated author of the Hockey Stick Diagram, Dr. Michael Mann of the Department of Geosciences, University of Massachusetts, fudged the entire thing. As just stated, for decades climatologists had documented both a Medieval Warming Period (warmer than today) from 1100 to 1500 AD, which was followed by the Little Ice Age that continued thru the 1900's. All of that changed in 1999 when...

> ..."a new paper published in 'Geophysical Research Letters' altered the whole landscape of how past climate history was to be interpreted by the greenhouse sciences. ...

1239   (From article: "Nailing the Left: Liberal Failures," the Limbaugh Letter, Rush Limbaugh)
1240   (From Fox News Special "Global Warming, or a lot of Hot Air?" 1/01/10)

Dr. Michael Mann …was the primary author of the GRL paper, and in one scientific coup overturned the whole of climate history. …Using tree rings as a basis for assessing past temperature changes back to the year 1,000 AD, supplemented by other proxies from more recent centuries, Mann completely redrew the history, turning the Medieval Warm Period and Little Ice Age into non-events, consigned to a kind of Orwellian 'memory hole'."[1241]

This "memory hole" was required of course because that information didn't support the financially lucrative—think government $grants$—global warming scare these environmentalists and their scientists were concocting. (And nor did it fit in with the designs of the capitalist-hating communist ideology of those true believers who had taken over the environmental movement.) In order to accomplish this they had to show that global temperatures for the last millennia were fairly constant[(LIE)] (even though in reality they were not) and that temperatures have shot up only recently[(LIE)] due to eeevil man and his industrialization.

- Climategate But then, like a pie in the face of the GW crowd but a breath of fresh air to true science, came Climategate, the name given to the scandal surrounding a series of hacked emails in November, 2009 from the Climate Research Unit (CRU) of a British University that is closely allied financially, in the form of research $grant$, with the "all-important" United Nations Intergovernmental Panel on Climate Change. This UN panel has itself been vigorously spewing out man-made global warming propaganda for years. Finally there was conclusive proof that what the global-warming skeptics had been saying all along was true after all. For the emails clearly exposed in their own words that these British scientists were deliberately manipulating the data in order to advance their bogus theory.[1242] They "revealed a plot among the world's top climate scientists to hide the real inconvenient truth that the evidence supporting man-made global warming is far from conclusive."[1243] The emails showed:

"Conspiracy, collusion in exaggerating warming data, possibly illegal destruction of embarrassing information, organized resistance to disclosure, manipulation of data, private admissions of flaws in their public claims and much more. …But perhaps the most damaging revelations …are those concerning the way Warmist scientists may variously have manipulated or suppressed evidence in order to support their cause."[1244]

"Bottom line, [man-made Global Warming is] a corrupt ideology based on government expansion, and they [the GW propagandists] got found out."[1245]

In one of the more famous emails, a scientist admitted having to "hide the decline" of worldwide temperatures because the real data did not fit their computer models. The inconvenient facts threatened their lucrative "save the planet" funding. Sean Hannity, in recommending the book *Climategate* by Brian Sussman, said this:

1241   (From article, "The 'Hockey Stick': A New Low in Climate Science," by John L. Daly @ www.john-daly.com)

1242   (From article, "UN panel promises to investigate leaked 'climategate' e-mails," by Philippe Naughton and Ben Webster, December 4, 2009, timesonline.co.uk)

1243   (Sean Hannity on "The Green $windle," a special addition of Fox News Hannity, 8/27/10)

1244   (From article, "Climategate: the final nail in the coffin of 'Anthropogenic Global Warming'?" by James Delingpole, November 20th, 2009 @telegraph.co.uk)

1245   (Greg Gutfeld on Fox News Hannity, 2/10/10, speaking of record cold, record snowfalls, and the climategate email disclosures.)

"In the last decade with the help of poster boy Al Gore the global warming movement has become an international concern. But in the wake of the climate gate scandal many people are questioning the motives of the people pushing the global warming crisis, and asking whether the world has been duped by a complex cover-up of something that does not exist. In his new book *Climategate* Brian Sussman asserts that the global warming hype is really just a form of communism rooted in the principles of Karl Marx. Now Sussman argues that the communist tenants of moral relativism, the belief in science over God and the idea of de-developing successful nations to fill the needs of underdeveloped countries all describe the global warming movement. Climate gate, he argues, highlighted these principles by showing that the scientists were purposely hiding information that would prove their theories wrong. And knowingly deceiving people by manipulating the science, the "Green Movement" has made billions for international corporations, governments, and elitist individuals—prime example, the jet-setting, green-hero Al Gore. Between his Oscar-winning film, bestselling books, comfy board positions on Google and Apple, and his investments in "Green company startups," some estimates are out there that Gore's made over $500 million off the global warming hype, and stands to make much more. And let's not forget the current administration, President Obama, has pushed hard for Cap and Tax, and has the most aggressive green agenda to date."[1246]

Of course the email scandal did not put so much as a dent in our president's environmental agenda. Obama traipsed off to the UN Climate Change conference in Copenhagen in December of 2009, just one month after the fraud was exposed, as if the emails never existed, and continued to spout the man-made-global-warming-crisis[(LIE)] party line. True believers never let the facts get in their way.

"The threat from climate change is serious, it is urgent, and it has grown."[(LIE)] - Obama. (Let's get real, the threat from <u>Obama</u>'s kind of change "is serious, it is urgent, and it has grown.")

Speaking with Laura Ingraham about Obama's religious pilgrimage to Copenhagen, Christopher Horner added:

"Nothing proposed [at Copenhagen] would under any scenario, detectively impact the climate, so it's not reasonable to conclude that this is about the climate. ...At some point, however, as you ration energy, and the noose has to continually tighten, you're talking about a massive, in the double-digit trillions, cost to the economy; massive human suffering of course from the dislocation. That's inevitable. We don't have pixie dust. There is no bridge off of the energy sources that work. This isn't about climate. This is about getting people off of energy, because it leads to all of those other things that our moral superiors don't like, and they generally involve our freedoms."

Ingraham: "It's about money, is it not? In the end it's about reducing America's standard of living so we can then be at parity with other nations who don't have the standard of living. People just don't like the fact that Americans have lived pretty well; [we've] worked hard ... we can buy these big cars, you can have big families, and that's what they don't like. They want that to end."

Horner: "I was in the room in the Hague in November, 2000 in one of these meetings when Jacques Chirac said *Kyoto was the first component of an authentic global governance.* That caused quite a stir, but only among the Americans in the audience. Margo Wahlstrom was Europe's Environment Commissioner, and she said, this is a quote, "Kyoto was not

---

1246   (Sean Hannity, on Fox News *Hannity*, 04/22/2010)

about whether scientists agree. It's about leveling the playing field for big business." They can't liberate their economy. They have to bring ours down."[1247]

And that is the ultimate goal of Satan's man-made global warming lie; strangling the American economy, so he will be free to destroy Israel. And that is also Satan's hidden intent behind the Obama administration and his Democrat-Left's pursuit of an insane policy known as Cap and Trade, or more accurately called "Cap and Tax."

"Cap and Trade is a mechanism for punishing the use of fossil fuels, so other sources of energy which are inherently more expensive will become more attractive."[1248]

What the free market cannot afford to do because of the prohibitive expense, the Democrats and their global-warming alarmists would shove down our throats. And it will cost the American consumer 2 trillion dollars in increased gas and energy costs as well as its increased costs to businesses. Which will further stifle our already faltering socialist economy.

"The whole point of cap and trade is to hike the price of electricity and gas so that Americans will use less. These higher prices will show up not just in electricity bills or at the gas station but in every manufactured good, from food to cars. Consumers will cut back on spending, which in turn will cut back on production, which results in fewer jobs created or higher unemployment. Some companies will instead move their operations overseas, with the same result."[1249]

Yet the mindless hysteria surrounding the global warming scam is not at all new. It is just the most recent in a long list of false alarms that have come out of the wacko wing of the environmental movement. Here are some excerpts from a Time Magazine article from June 24, 1974. The headline read, "Another Ice Age?" ...

"The weather aberrations they are studying may be the harbinger of another ice age." "Telltale signs are everywhere—from the unexpected persistence and thickness of pack ice in the waters around Iceland..." "Whatever the cause of the cooling trend, its effects could be extremely serious if not catastrophic."[250]

Oooh! "<u>Extremely serious</u>!" Nay, "<u>catastrophic</u>." And less than a few decades later the Global Freezing scare has reversed course 180 degrees and has now become the Global Warming crisis! What will their next crisis be, moderate temperatures?!#@

Walter E. Williams, in an article in The Baltimore Examiner, 5/8/2008, titled, "Environmentalists' wild predictions," lists a number of their more outrageous prognostications over the decades. Here are a few:

"At the first Earth Day celebration, in 1969, environmentalist Nigel Calder warned, "The threat of a new ice age must now stand alongside nuclear war as a likely source of wholesale death and misery for mankind." C.C. Wallen of the World Meteorological Organization said, "The cooling since 1940 has been large enough and consistent enough

1247 (Christopher Horner, author of the book *Red Hot Lies: How Global Warming Alarmists Use Threats, Fraud and Deception to Keep You Misinformed*, speaking with Laura Ingraham, guest host on the O'Reilly Factor, 12/21/2009)
1248 (Roy Spencer, PhD. And former NASA scientist, on "The Green $windle," a special addition of Fox News Hannity, 8/27/10)
1249 (From article, "The Cap and Tax Fiction," June 26, 2009, the Wall Street Journal online)
1250 (From Time Magazine article, "Another Ice Age?" June 24, 1974)

that it will not soon be reversed."

"In 1968, professor Paul Ehrlich, Vice President Al Gore's hero and mentor, predicted there would be a major food shortage in the U.S. and "in the 1970s ... hundreds of millions of people are going to starve to death."

..."Harvard University biologist George Wald in 1970 warned, "Civilization will end within 15 or 30 years unless immediate action is taken against problems facing mankind." [Sound familiar?] That was the same year that Sen. Gaylord Nelson warned, in Look Magazine, that by 1995 "somewhere between 75 and 85 percent of all the species of living animals will be extinct."

[Mr. Williams concludes...]

"Here are a few facts: More than 95 percent of the greenhouse effect is the result of water vapor in Earth's atmosphere. [Forget moderate temps, Global Humidity is next.] Without the greenhouse effect, Earth's average temperature would be zero degrees Fahrenheit. Most climate change is a result of the orbital eccentricities of Earth and variations in the sun's output. On top of that, natural wetlands produce more greenhouse gas contributions annually than all human sources combined."[1251]

Brace yourself for the wackos next outcry, "Save the Earth! Drain the Wetlands!"

In addition there are the following facts:

"Every year, an eevil polluter spills 620,000 barrels of oil off North America's coasts. Who? Mother Nature, through naturally occurring ocean floor seepage. On the other hand, thanks to new safety technologies, oil companies have spilled only 852 barrels off California's coast in the last 40 years. Does this mean liberals will cease blocking America's access to the estimated 10 billion barrels of California offshore oil? Not a chance."[1252]

And speaking of hysteria, we all remember Rachel Carson's famous book, *Silent Spring*, which back in the early 60s catapulted the environmental movement into the national conscience. What we don't all remember is how it has since been exposed for the false-alarm that it was. (And the reason we don't all remember that is because our Democrat media felt it was information that, unlike them, we wouldn't be sophisticated enough to ignore, so it was best that they just didn't tell us in the first place. Isn't it wonderful how they're always looking to protect us? ...from the truth.)

"Rachel Carson's 1962 book, *Silent Spring*, claimed dichlorodiphenyl-trichloroethane, DDT, the tremendously effective pesticide, could kill off the world's birds by thinning their eggs, thus upsetting our fragile ecosystem and poisoning the food chain. Policymakers swallowed her every claim, labeling DDT "the world's deadliest poison" and a carcinogen. It was banned here and abroad. Result: malaria, which was nearly eradicated before *Silent Spring*, spread like wildfire throughout the world's poorest countries. Turns out Carson's anti-DDT jeremiad was poppycock. As *City Journal* summarized, "test after test has shown that DDT isn't dangerous to humans." Malaria, a preventable disease, now kills about 2.5 million people a year, mostly children, 90 percent of them in Africa. That's easily over 30 million deaths because 1962 liberals were panicked by rumors of thinning bird eggs."[1253]

1251   (From article, "Environmentalists' wild predictions," by Walter E. Williams, in The Baltimore Examiner, 5/8/2008)
1252   (Rush Limbaugh, The Limbaugh Letter)
1253   (From article: "Nailing the Left: Liberal Failures," the Limbaugh Letter, Rush Limbaugh)

And now the Democrat-Left's environmental hysteria is beginning to turn violent where, like the Jews in Germany 75 years ago, those who oppose them will no longer be safe. In October of this year (2010) on his Fox News program Glenn Beck quoted from his internet site, the blaze.com, which is committed to, among other things, documenting some of the more outrageous rants that have issued from the fevered minds of the environmentalist's wacko-fringe:

"They [blaze.com] have begun gathering a list of all those ridiculous, implausible ideas that have been said by environmentalists recently. Remember the ends justify the means, they're saving the planet. In June of last year liberal blog site TPM posted an article and then quickly had to take it down when people started to protest. It read, "At what point do we jail or execute global warming deniers?" In 2006 the eco-magazine Grist called for Nuremberg-style trials for skeptics. In 2008 the Guardian reported that NASA's James Hansen called for the trials of climate skeptics for high crimes against humanity. In 2008 Canadian environmentalist David Suzuki called for government leaders skeptical of global warming to be thrown into jail. In 2007 Robert F. Kennedy Jr. said of skeptics, "This is treason. We need to start treating them as traitors." There's more, including the environmentalists that said, "Soon another generation will come and strangle you in your beds."[1254]

If you listen carefully you can hear chants from the not-so-distant past… Sieg Heil! Sieg Heil! It was the Jews then; who will be next? Global Warming *deniers*? Evolution *deniers*? Christians? (O-ba-ma! O-ba-ma!) Yet we should know that whatever the misguided intentions of the hysterical Global Warmists, Satan's obvious intention is to use their indoctrination to further destroy America economically so she will no longer be in any position of strength to protect Israel from annihilation. But even if Satan is successful with his scheme; even if the American voter continues to elect the Democrat-Left; and even if they then are able to shove their Cap and Tax legislation down our unwilling throats; and even if America is crippled economically and militarily to the point of helplessness in aiding Israel; know that the Lord Himself will protect the Jewish State and will not allow Satan and his Islamic Mongol-hordes to destroy her. So it is really only America, her economy, and the prosperity of her people that hangs in the balance.

And, when you think about it, maybe this is the time for God's judgment on America, and maybe we are beginning to reap now what we have been sowing. For He has blessed our country far above all other nations, and yet in repayment we have mutilated 50 million of his babies to death and then went out in 2008 and voted for the killing of 50 million more (Yes We Can!); and we spit in his face by indoctrinating our school children with the idiotic lie of Satan's genesis. Think about it. How would you repay America if you were Him? Read the Old Testament. How did He repay Israel for her blatant disobedience, worshiping of idols, involving themselves in the revolting child-sacrifice of neighboring nations, and ignoring the One who gave them all of their incredible blessings in their Promised Land? God is longsuffering and slow to anger, America, but He is not an idiot, a sucker or a fool. And when He has finally had just about enough, and the shit really starts to hit the fan, then you will look back as a nation on the Great Depression of the 30s as the days of wine and roses. Wake-up! My idiot voter friends!

1254 (From the Glenn Beck show, 10/06/2010)

For further information on the man-made global warming scam, see: *The Politically Incorrect Guide to Global Warming*, by Christopher C. Horner; *Climategate: A Veteran Meteorologist Exposes the Global Warming Scam*, by Brian Sussman; and *Red Hot Lies: How Global Warming Alarmists Use Threats, Fraud, and Deception to Keep You Misinformed*, also by Christopher C. Horner

Destroying Health Care

...and America along with it.

"HILTHCARE IS MAAH RAAT!"[LIE] (...Health care is my right.) "HILTHCARE SHOODBEE FUREE!"[LIE] (...Health care should be free.) As the embarrassingly stupid (thank our Democrat-Left educational system) woman on TV the other day (October 2009) shouted indignantly into the cameras to the delight of lame-stream, Obama-mania journalists all across the land. And whereas suffering fools gently is a worthy virtue, we nevertheless do not think they should be framing the debate. Lemme 'splain, Lucy (as Ricky Ricardo was fond of saying), "Hilthcare" actually is <u>not</u> "furee," nor can it be. It doesn't fall from the trees like leaves, and you cain't go for a walk through the woods and gather it in your basket like acorns, for "furee." Someone actually has to <u>pay</u> (a novel concept, we realize, for the "shoodbee furee" citizens of this country) the Doctor. Someone actually has to <u>pay</u> his nurses. Someone actually has to <u>pay</u> his secretary. Someone actually has to <u>pay</u> the landlord rent for the Doctor's office. Someone actually has to <u>pay</u> for the drugs and the medical equipment and the bandages and the sutures and the MRI's, etc. etc. Get it? Also, madam, "raats" come from God, and not from the state. And the last time I looked God was not giving anyone the "raat" to steal a Doctors services for "furee." Nor did he give any government the "raat" to steal other people's money in order to give someone else "hilthcare" for "furee." That's what Christian charity is for. As previously stated, we know you are a result of the Democrat-Left's educational system, or rather lack of education thereof, so you are to be excused and, like all other fools, to be suffered gently. But health care is not your right, madam, nor can it ever be free. Someone has to pay for it.

But isn't Canadian- and European-style socialized medicine better for everybody? Hardly. Here are some excerpts from an article "Socialized Medicine's Waiting Room" by Rush Limbaugh which appeared in a recent edition of The Limbaugh Letter:

> "In his propaganda movie, "Sicko," Michael Moore raved about universal health care as found in Canada and Britain. You know, "free" health care. Those are the models for American liberal politicians' health care schemes. But before leaping to copy these medically socialized countries, let's take a look at how they work.
>
> "The truth is, they don't. In the real world, nationalized medicine fails every time it's tried; it breaks the bank, limits choice, and punishes entrepreneurs and pioneers. Government control over services always, by definition, limits supply. When that immutable economic principle is applied to health care, the inevitable result is deadly. When medical services are limited, lives are lost."[1255]

Rush goes on to document the case of a British woman who, according to *The UK Daily Mail*, "was monitored every day because doctors were worried about

---

1255 (From article, "Socialized Medicine's Waiting Room," by Rush Limbaugh, the Limbaugh Letter)

the health of her baby." Sounds great, right? Except on "the day of the birth, [the woman] was twice turned away from the hospital because it was full — forcing her partner to deliver the baby himself at their home." Then there's the case of a woman who, after finding a lump in her breast, was told she would have to wait 17 weeks for the mammogram. Thank God she was able to immediately get a private test which showed she had cancer. If she hadn't, the "cancer would have had 17 weeks to spread before the government even gave her a diagnosis." The article continues…

"The UK is so shoddy at screening for breast cancer — where early detection is the key — that the government actually tells women to schedule mammograms once every three years, instead of annually, because they can't handle the demand any faster. [But at least they all wear their pink ribbons.]

"The stories go on. Waiting *72 weeks* to be fitted with a hearing aid in Edinburg. More than 1.5 million Ontarians (or 12 percent of that province's population) unable to find family physicians. Brits pulling their own teeth. Canadians hopping over the border for life-saving surgery (including Canadian Liberal party Member of Parliament and Bill Clinton squeeze Belinda Stronach, who according to Canadian television CTV "travelled to California last June for an operation that was recommended as part of her [later-stage] treatment" for breast cancer.)

""But Rush, but Rush," you say. "Isn't this just cherry-picking worst-case anecdotes to help your argument?" Hardly. The anecdotes are backed by cold, hard statistics. As Dr. David Gratzer, Canadian doctor and author of *The Cure: How Capitalism Can Save American Health Care*, reports: "A survey in 2000 involving 1,500 people suggested that a full eight out of ten Canadians consider their health care system to be 'in crisis.'" And that was in 2000; waiting times, by the governments own admission, have gotten worse since. (Ontario actually has a "wait-times czar.")

"The wait times for emergency surgery …are mind-boggling to any American used to our stellar system: 5 to 6 weeks for heart surgery; 10.7 weeks in Canada for neurosurgery. Can you imagine people in excruciating hip pain waiting 10 to 24 weeks for a doctor? These wait times represent innumerable lost opportunities to prevent strokes and heart attacks and the spread of cancers. They represent lives lost — *that would not have been lost in America*. [Of course, keep in mind that this article was written prior to the passing of the travesty of Obamacrap…]

…"We don't have a government-run system that puts a cap on supplies — medicine, hospitals, doctors — to save costs, and which thus can't help but form waiting lists.

"Back in April of 2004, Mrs. Bill Clinton wrote a near-6,000 word manifesto in *The New York Times Magazine* entitled, "Now Can We Talk About Health Care?" She shamelessly said: "Instead of putting consumers in the driver's seat, [our private system] actually leaves consumers at the mercy of a broken market."

"What painful ignorance and arrogance, Mrs. Clinton. Sadly, there's no pill for it."[1256]

1256   (From article, "Socialized Medicine's Waiting Room," by Rush Limbaugh, The Limbaugh Letter) (We have used a number of quotes from Rush Limbaugh in this section, and in a number of other places throughout this chapter. And I think it needs to be addressed here the unfortunate reality that a whole lot of mis-educated and easily duped people in this country have been instructed by the liars on the Left to vilify him and to turn a deaf ear to anything he has to say, if they ever accidentally come in ear shot at all. "Hear no evil, See no evil… Hear no truth, See no truth." The reality is that Rush is hated (like George W. Bush, Bill O'Reilly, Dick Cheney, Sarah Palin, Sean Hannity, Carl Rove, etc.) by the liars on the Democrat-Left exactly because he tells the truth every day and exposes them for the hate-filled, humorless,

- Obamacrap And now the nation has been cursed with the Democrat's socialist health care plan, "Obamacare." But it's really Obamacrap, because it doesn't take "care" of anyone. Instead, it craps on our economy, our freedom of choice, the ability of our business' to show a profit, as well as on our health care system itself. And we're stuck with it unless it can be repealed, hopefully after January of 2013, or overturned by the Supreme Court if it has the Constitutional allegiance to do it. (And with justices like Sonia Sotomayor, Elena Kagan and Darth Bader Ginsburg sitting there we shouldn't be holding our breath.) Obamacrap is not a legislative bill. It is an evil spell cast upon America's health care system, and upon her economy. The puppet-master, Satan, is using his dupes to absolute perfection. He has them pursuing their socialist dreams, but the reality they will end up foisting on all of us will be a nightmare.

> **"Disastrous, UNCONSTITUTIONAL Health Care Plan.** Obamacare was forced down our throats! Using unconstitutional procedures, Obama and leftists Nancy Pelosi and Harry Reid conspired and passed a plan that dictates to Americans how, when and who will care for them ... and if we don't like it, we face fines, jail or both! The plan costs $1.3 trillion when payoffs to the American Medical Association and unions are added in. The Congressional Budget Office says the cost will soar over the next 10 years. This bill will throttle the medical innovation that makes America the best place to be sick, and throw the best health care system in the world into chaos. [Or into what is otherwise known as "Obama's toilet" where every other industry he has touched has ended up.] Bureaucrats sitting on a federal Health Board will be in our examining rooms – hanging over us and our doctors [Bill Clinton has requested the chairmanship of the Gynecological Board], making life and death choices for us."[1257]

Obamacrap is an absolute nightmare monstrosity. It is the most complicated and convoluted piece of idiotic legislation to ever come down the pike (and there have been some doozies). "It will ruin our health care system and bankrupt our country."[1258] It is the communists Central Planning come to America; centralizing power and control over the people and the economy into the hands of the few "elites" in government. And they didn't even read the thing. Nancy Pelosi, the Madam of the Whore House... er ...Speaker of the House said, "We have to pass the bill so that you can find out what is in it." (Miller wasn't exaggerating when he said she was "stick-her-tongue-in-an-electric-fan stupid.") It was never even meant to be passed as a final bill. But the problem for them occurred when Republican Scott Brown was elected in January of 2010 to fill the late Ted Kennedy's Senate seat, in a shocking upset that occurred in the reliably Democrat state of Taxachusetts (it's just to the right of North Korea). With Brown's election the Obamacrats in Congress lost their 60 seat majority which had allowed them to pass anything they wished because the Republicans did not have enough votes to stop them. So they had to come up with a deceitful legislative maneuver to ram it through, as it was, without any further debate that would have then, with only 59 seats to their name, have killed it. It was not democracy in action. It was "a crime against democracy." It was what you might expect from an arrogant, fascist, third-

---

clueless liars that they are. So it is with great pride that we were able to use some of his fact-based assessments here.)
1257   (From a *David Horowitz Freedom Center* Circular)
1258   (John Boehner)

world dictator high on coke and his legislative, arse-licking lackeys hoping just to survive till another day, and not from an American president and his Congress.

"Let me get this straight . . . We're going to be "gifted" with a health care plan we are <u>forced to purchase</u> **and** <u>fined</u> if we don't. Which purportedly covers at least <u>ten million more people</u>, without adding <u>a single new doctor</u>, but provides for <u>16,000 new IRS agents</u>, written by a committee whose chairman says he <u>doesn't understand it, passed</u> **by a Congress** <u>that didn't read</u> **it but** <u>exempted</u> **themselves from it, and signed by a Dumbo President** who <u>smokes</u>, with <u>funding</u> administered by a treasury chief who <u>didn't pay his taxes</u>, for which we'll be <u>taxed for four years before any benefits take effect</u>, by a government which has <u>already bankrupted Social Security and Medicare</u>, all to be overseen by a surgeon general who is <u>obese</u>, and <u>financed</u> by a country that's <u>broke!!!!!</u> 'What could *possibly* go wrong?'" - Donald Trump

Yes, they are insane.

Obamacrap has nothing whatsoever to do with helping the poor, or giving health care to all. The poor will be made even poorer with its further sinking of our already listing economy. That is one guarantee that is sure to follow if it is ever implemented. Obamacrap has everything to do with taking over a major part of our economy; of socializing it, which is what socialists do. It is about vastly increasing the power of the federal government, unconstitutionally, over the states and the individual, and it will destroy the greatest health care system in the world. For ours is the system that all of those folks in Canada, Britain and Europe—trapped in terrible, socialized ones—have been coming over here to take advantage of; the ones who can afford it, that is. And that says it all. If the Democrats actually wanted to help those who can't afford health care they could simply take a few tens of billions out of their near-trillion dollar Spendulous package and *buy* those people health insurance. Not illegal aliens, and not young people who can afford insurance but choose not to spend their money on it; but they could buy insurance for all of those who are truly in need, struggling single moms for example; and leave the greatest health care system in the world intact. But they don't do that because that is not what they want. They also could pass a bill that would allow health insurance companies to sell in all states, like auto insurance, instead of each state limiting coverage to just a few companies who meet their own unique, schizophrenic requirements. This would greatly increase competition and lower premiums, which is one of the Democrats stated goals, but which in reality they have no real interest in doing. Again, lowering costs is just a canard, what they really want is to take over the health care industry, like Cuba. Also, if they really wanted to lower costs, they would pass tort reform which would limit the windfall profits of their trial lawyer friends, and greatly reduce the now astronomical costs of liability insurance that doctors must bear, and then pass on to their patients.

"It has come to my attention that several health insurer carriers are sending letters to their enrollees falsely[LIE] blaming premium increases for 2011 on the patient protections in the Affordable Care Act ... there will be zero tolerance for this type of misinformation[LIE] and unjustified[LIE] rate increases..." — Kathleen Sebelius, Obama Health and Human Services Secretary, warning letter to America's Health Insurance Plans [AHIP], The National Association of Health Insurers, for telling the truth, 9/9/10.

[Ms. Sebelius must have bumped her head. She's not sure where she's at, the United States of America, or Nazi Germany.]

"Obamacare is already a disaster. Although most of it isn't even due to kick in until 2014, the first elements became law on Sept. 23. The results: double-digit insurance premium hikes, slashed benefits, and people being dumped from insurance plans all across the country."[1259]

You were lied to America. You were lied to. And again it's not that Obama and the Democrats actually think the American people are that stupid. They know they're not. They saw how they rose up in record numbers from the grass roots against their economically-disastrous, stupid monstrosity (it's inaccurate to even call it a "plan"). No, the Barackers think that their voters, at least the hard core ones (the Jon Stewart wing), *are* that stupid, and will keep on voting for them no matter what;[1260] no matter how many silly, outrageous, bald-faced lies they keep throwing at them; and no matter how much disastrous, fascist legislation they jam through. And that's what they are counting on in 2012. They are playing to their base, and hoping that the morons in their media can pull it off, once again.

"This health care law is bad for patients. It's bad for providers—the nurses and the doctors. It's bad for payers—people paying their health bills, and the tax payers. We need to replace it, after we repeal it."[1261]

But it's just fine for the Prince of devils and for his plan to decimate the economy of the once-great United States of America.

"And I beheld another beast coming up out of the earth, and he had two horns like a lamb, and he spoke like a dragon. And he exercises all the authority of the first beast in his presence, and causes the earth and those who dwell in it to worship the first beast, whose deadly wound was healed." (Revelation 13:11-12)

See:
- *Economic Facts and Fallacies*, by Thomas Sowell
- *Basic Economics 4th Ed: A Common Sense Guide to the Economy*, by Thomas Sowell
- *Power to the People*, by Laura Ingraham
- *Catastrophe*, by Dick Morris and Eileen McGann)
- *Liberty and Tyranny: A Conservative Manifesto*, by Mark R. Levin
- *Real Change: From the World That Fails to the World That Works*, by Newt Gingrich
- *How an Economy Grows and Why It Crashes* by Peter D. Schiff and Andrew J. Schiff
- *The Road to Serfdom*, by Friedrich Von Hayek
- *The New Road to Serfdom: A Letter of Warning to America*, by Daniel Hannan, member of British Parliament

---

1259   (From article "Obamacare: Already Unraveling," by Rush Limbaugh, The Limbaugh Letter, November 2010)
1260   (The "Democrat" Party? The truth would be better served if they were renamed Slaves-R-Us. And no, that doesn't make us "racist" because blacks are only a small percentage of the folks who are mentally imprisoned on Satan's white, liberal, progressive, Christian-bashing, European-endarkenment, baby-mutilating, America-hating, black-exterminating, Postmodern plantation. Pray for them)
1261   (John Barrasso, M.D. and Republican Senator from Wyoming, on Fox News Hannity, 08/06/2010)

- America for Sale: Fighting the New World Order, Surviving a Global Depression, and Preserving USA Sovereignty, by Jerome R. Corsi, PhD.

- The Politically Incorrect Guide to Global Warming, by Christopher C. Horner

- Climategate: A Veteran Meteorologist Exposes the Global Warming Scam, by Brian Sussman

- Red Hot Lies: How Global Warming Alarmists Use Threats, Fraud, and Deception to Keep You Misinformed, by Christopher C. Horner

- New Deal or Raw Deal? How FDR's Economic Legacy Has Damaged America, by Burton W. Folsom, Jr.

12- Weaken America's respect abroad

"Now there was a day when the sons of God came to present themselves before the Lord, and Satan also came among them. And the Lord said unto Satan, 'From where do you come?' So Satan answered the Lord and said, 'From going to and fro on the earth, and from walking back and forth on it.'" (Job 1:6-7) ...And from spreading his lies about Israel and America everywhere he goes.

What better way to tear down a country than to turn the entire world against it? And that is the recipe the great deceiver has been preparing for some time. A little deception here; some rewriting of history there; throw in a lot of left-wing lies spread by our own academic establishment; add a generous helping of unfounded, illiterate hatred from the Muslim lands; and then top it off with a lying, left-wing, America-hating media that poisons the entire Western world and he has his ingredients for a stinking dish of unfounded animosity and uneducated resentment stewing on the world stage.

"And they [the nations of the world] had a king over them, the angel of the bottomless pit, whose name in the Hebrew tongue is Abaddon, but in Greek he has the name Apollyon." (Revelation 9:11)

> "How all business phones should be answered: 'Good morning. Welcome to the United States of America. Press one for English. Press two to disconnect until you learn to speak English. And remember, only two defining forces have ever offered to die for you: Jesus Christ, and the American soldier. One died for your soul. The other for your freedom.'"[1262]

Amen. And now we have the problem exacerbated by our own president travelling around the world apologizing for us, for what I haven't the slightest idea, and honestly in his distorted thinking I don't think he has either. He just does it because that's what he was taught to do. The end result of left-wing ideological indoctrination. (Thank a teacher, or a left-wing professor, if you too can disparage America for no reason.)

> "Maybe if president Obama were to visit a few American military cemeteries, and realized that after we liberated Europe all we asked for was enough ground to bury our young men who had died to free Europe. We didn't seek to conquer. We didn't seek to dominate. We don't dominate today. And I think it is not just insulting, but is historically dangerous to have an American president who so fundamentally misunderstands the world.

1262   (From email travelling around the internet. Source unknown.)

It is not America which is dangerous, it is North Korea. It is not America which is arrogant, it is Iran."[1263]

I don't know the source of this next piece as it came through the internet email rounds, but it highlights the ignorance of America's bashers abroad:

"JFK'S Secretary of State, Dean Rusk, was in France in the early 60's when DeGaulle decided to pull out of NATO. DeGaulle said he wanted all US military out of France as soon as possible. Rusk responded "does that include those who are buried here?" DeGaulle did not respond. You could have heard a pin drop.

"When in England, at a fairly large conference, Colin Powell was asked by the Archbishop of Canterbury if our plans for Iraq were just an example of empire building by George Bush. He answered by saying, "Over the years, the United States has sent many of its fine young men and women into great peril to fight for freedom beyond our borders. The only amount of land we have ever asked for in return is enough to bury those that did not return." You could have heard a pin drop.

"There was a conference in France where a number of international engineers were taking part, including French and American. During a break, one of the French engineers came back into the room saying "Have you heard the latest dumb stunt Bush has done? He has sent an aircraft carrier to Indonesia to help the tsunami victims. What does he intend to do, bomb them?" A Boeing engineer stood up and replied quietly: "Our carriers have three hospitals on board that can treat several hundred people; they are nuclear powered and can supply emergency electrical power to shore facilities; they have three cafeterias with the capacity to feed 3,000 people three meals a day, they can produce several thousand gallons of fresh water from sea water each day, and they carry half a dozen helicopters for use in transporting victims and injured to and from their flight deck. We have eleven such ships; how many does France have?" You could have heard a pin drop.

"A U.S. Navy Admiral was attending a naval conference that included Admirals from the U.S, English, Canadian, Australian and French Navies. At a cocktail reception, he found himself standing with a large group of Officers that included personnel from most of those countries. Everyone was chatting away in English as they sipped their drinks but a French admiral suddenly complained that, whereas Europeans learn many languages, Americans learn only English. He then asked, "Why is it that we always have to speak English in these conferences rather than speaking French?" Without hesitating, the American Admiral replied, "Maybe it's because the Brit's, Canadians, Aussie's and Americans arranged it so you wouldn't have to speak German."

You could have heard a pin drop.

"Robert Whiting, an elderly gentleman of 83, arrived in Paris by plane. At French Customs, he took a few minutes to locate his passport in his carry on. "You have been to France before, monsieur?" the customs officer asked sarcastically. Mr. Whiting admitted that he had been to France previously. "Then you should know enough to have your passport ready." The American said, "The last time I was here, I didn't have to show it." Impossible. Americans always have to show your passports on arrival in France!" The American senior gave the Frenchman a long hard look. Then he quietly explained, "Well, when I came ashore at Omaha Beach on D-Day in 1944 to help liberate this country, I couldn't find a single Frenchmen to show a passport to." You could have heard a pin drop."[1264]

1263   (Newt Gingrich on the presidents international apology tour, on Fox News' Hannity, April 9, 2009)
1264   (From internet circulating email. Source unknown.)

732

If ignorance actually was bliss then many of the "peoples, multitudes, nations and languages"[1265] on this earth would have already made it to paradise. Yet now, with the groveling talk from our current president, disrespect for the U. S. is reaching an all-time low, down to that of complete contempt. Iran, for example, unimpressed with this administrations overtures of appeasement and cowardice towards them, literally flips Obama the bird while flaunting its relentless pursuit of nuclear weapons. And North Korea, a country despoiled by six decades of communist rule by a Mickey Mouse pair of father and son imbeciles, outright laughs at the United States:

> "I think this is eerily reminiscent of Jimmy Carter going to Notre Dame for the commencement in 1977, and announcing that we have had an inordinate fear of communism, that the world is really safe. And if you look at Carter's policies which were all a fantasy—they set the stage for the 444 day Iranian hostage crisis; they set the stage for the Soviet invasion of Afghanistan; they set the stage for the Nicaraguan communists and then the battle in El Salvador and Grenada—we have a similar pattern beginning to build of a fantasy foreign policy that has no connection with reality. You couldn't have had a better moment than the North Koreans deliberately firing a missile, just before president [Obama] gave a speech on nuclear disarmament, as though to show their total contempt for his fantasy world view."[1266]

The United Nations

This body of Jew-haters and America-bashers has done more to destroy our respect around the world than any other single source. And yet we are the purse strings behind it and its anti-American activities! Governor Mike Huckabee nailed it...

> "[The United Nations] seems to be unraveling from its mission of being a forum for peace and understanding. Among the many of its transgressions are giving a microphone, a podium and a platform to a murdering, racist tyrant like Iranian president Mahmoud Ach-ma-nut-job, so he can go around spewing his vile and insulting hate: blaming America for 9/11, denying the Holocaust and defiantly calling for the death of Jews and the destruction of Israel. ...That the UN would give the Iranian president a forum as if he's some kind of legitimate leader makes me think it's time to rethink the US participation in this worthless, toothless white elephant. ...There's nothing *united* about the UN, [it is an] impotent and corrupt organization..."[1267]

Foreign Aid

"I will punish the world for their evil, and the wicked for their iniquity; and I will cause the arrogance of the proud to cease, and will humble the haughtiness of the ruthless." (Isaiah 13:11)

Excuse me, where do I go to sign the petition to give billions of our hard-earned tax dollars, in the middle of the second great depression, to foreign countries, some of whom are openly hostile to us, and many of whom treat us with complete contempt? Israel and a handful of other friendly nations who actually are our allies should be the only recipients. For the rest, like arguably we should

---

1265　(Revelation 17:15)
1266　(Newt Gingrich on Fox News' Hannity, April 9, 2009)
1267　(Governor Mike Huckabee, host of Fox News' Huckabee, 09/24/2011)

do with social security, *means-test* it.[1268] If any country is interested in getting a piece of (what once was) our largesse then have every member of its federal government and 70 percent of its citizenry swear their love and allegiance to the United States of America, to freedom and democracy, and to our quest to eradicate jihad worldwide.[1269] Otherwise, don't give 'em a nickel.

And then there's this… In July of 2010 in Hawaii a bunch of left-wing-indoctrinated, uneducated (but I repeat myself) professors organized a conference. It was paid for by US taxpayers through $140,000 in funding from the NEH (National Endowment for the Humanities), another one of our many, redundant, expensive and basically worthless government agencies that are costing us our own financial and economic ruin but can't be gotten rid of because of the deafening decibel level of hysteria that blasts from the Democrats and their media every time it is attempted. The conference was supposed to be about WWII, but instead it spewed outrageous lies about our military and our veterans, and rewrote history to fit the far-left, America-hating fantasy world of its organizers and speakers. Fortunately, the unseemly nature of the conference was exposed by one attendee, a Professor Penelope Blake. In a letter to her Congressman, requesting that funding be cut for any future such events, she stated the following:

"In my thirty years as a professor in upper education, I have never witnessed nor participated in a more extremist, agenda-driven, revisionist conference, nearly devoid of rhetorical balance and historical context for the arguments presented. In both the required preparatory readings for the conference, as well as the scholarly presentations, I found the overriding messages to include the following:

"1. The U.S. military and its veterans constitute an imperialistic, oppressive force which has created and perpetuated its own mythology of liberation and heroism, insisting on a "pristine collective memory" of the war[LIE]…

"2. The Japanese attack on Pearl Harbor should be seen from the perspective of Japan being a victim of western oppression[LIE] (one speaker likened the attack to 9-11, saying that the U.S. could be seen as "both victim and aggressor" in both attacks)[LIE]…

"3. War memorials, such as the Punchbowl National Memorial Cemetery (where many WWII dead are buried, including those executed by the Japanese on Wake Island and the beloved American journalist Ernie Pyle), are symbols of military aggression and brutality "that pacify death, sanitize war and enable future wars to be fought."[LIE] (Ferguson and Turnbull, 1). One author stated that the memorials represent American propaganda,[LIE] "the right to alter a story" (Camacho 201).

"4. The U.S. military has repeatedly committed rapes and other violent crimes throughout its past through the present day. Cited here was the handful of cases of attacks by Marines in Okinawa (Fujitani, et al, 13ff). (What was not cited were the mass-murders, rapes, mutilations of hundreds of thousands of Chinese at the hands of the Japanese throughout the 1930s and 40s. This issue is a perfect example of the numerous instances of assertions made without balance or historical context.) Another author stated that the

---

1268   (Although with social security those who can afford to forgo it were still forced to pay into it for all of those years, so there is a fairness issue at play as well as the solvency of the whole broken system. With foreign aid, however, there is no such problem)

1269   (Maybe we can require our university professors to do the same, *before* they're given tenure. Otherwise, no cushy, secure job which they can use to poison the minds of their young students.)

segregation in place within our military and our "occupation" of Germany after the war was comparable to Nazism ("we were as capable of as much evil as the Germans")[LIE] even though the author admits, with some incredulity, that he "saw no genuine torture, despite all the [American] arrogance, xenophobia and insensitivity." He attributes American kindness towards conquered Germans to our "wealth and power"[LIE] which allowed us to "forego the extreme kinds of barbarism" (Davis 586)...

"5. Those misguided members of the WWII generation on islands like Guam and Saipan who feel gratitude to the Americans for saving them from the Japanese are blinded by propaganda supporting "the image of a compassionate America" or by their own advanced age.[LIE]

"6. It was "the practice" of the U.S. military in WWII to desecrate and disrespect the bodies of dead Japanese[LIE] (Camacho 186). (Knowing this to be absolutely false, I challenged the speaker/author, who then admitted that this was not the "practice" of our military. Still, the word remains in his publication. As he obviously knew this to be false, I can only assume that his objective was not scholarship but anti-military propaganda.)

"7. Conservatives and veterans in the U.S. have had an undue and corrupt influence on how WWII is remembered..."[LIE][1270]

Professor Blake's letter went on to document even more disturbing lies, propaganda and lunacy from this taxpayer-funded conference. In reading the above, it is clear that we are dealing with people who are not in their right minds. They are the equivalent of those false-religious-indoctrinated loons who tried to get laid by flying planes into tall buildings. It's an insane asylum, and as previously stated, the inmates have taken over our colleges and universities. These academics should not be indoctrinating our youth. They should be committed. ... It's little wonder that the rest of the misinformed masses around this world have little respect for America when our own jackass university professors can put on a conference like that.

In addition to our educational system helping to tear down the world's opinion of America, we can also blame our entertainment industry for being a contributing factor as well. We shared these thoughts from Greg Gutfeld of Fox News show "The Five" in the section on "The Entertainment Industry" in chapter 2, but they are equally as relevant here...

"...here's some folks that you never see in the movies: An American soldier who's not a psycho. A Christian not portrayed as a wild-eyed nut. A corporate head who isn't corrupt. An Italian who's not a mobster. A community activist who's really just a protestor living off the government. A journalist who's a lefty propagandist. An academic who's the same. All of these represent reality far more than movies because movies are now defined by the fake edginess of their attitudes, and how cleverly you can dis America. This matters. Hollywood is how America talks to the world. Why put this bunch of coddled geezers in charge of that? Thanks to the relentless drone since the 1960s it's no wonder the world hates us. If America really reflected what's in our movies, wouldn't you? But these films don't reflect us at all. They reflect an America existing in the Viagra-addled minds of Starbucks socialists, who hate our country, [its] values and themselves."[1271]

Amen, brother Greg. Satan is working overtime to weaken America's respect abroad, and he has many allies within our own government, media, entertainment industry and educational system to assist him in his goal.

1270   (As reported by libertynewsonline.)
1271   (Greg Gutfeld, on Fox News The Five, 02/24/2012)

"And out of his mouth goes a sharp sword, that with it he should smite the nations. And he shall rule them with a rod of iron. And he treads the winepress of the fierceness and wrath of Almighty God." (Revelation 19:15)

13- Control the media

"And you He made alive, who were dead in trespasses and sins, in which you once walked according to the course of this world, <u>according to the prince of the power of the air, the spirit who now works in the sons of disobedience</u>." (Ephesians 2:1-2)

The prince of devils, who is the prince of the power of the air, is also the prince of the power of the air *waves*, as he controls most of the TV and much of the radio, certainly in this country but around the world as well.

We live in an insane asylum, and the inmates have also taken over our national media. In the world they live in, the truth is a lie, and lies are true. Again, the distortions of Satan cover the minds of most of the people on this earth, because they have believed what they were taught. And our Democrat-Left national media has swallowed the deception of Satan hook, line and sinker and they disseminate it at every opportunity, whether their venue is commentary or "hard news" reporting. They control most of our TV networks, cable news channels and newspapers. Serving their baby-mutilating, evolutionary, satanic, progressive, liberal, godless, Democrat-Left ideology and belief system with their "slanted, spun, mangled and perverted disinformation"[1272] takes precedence over any journalistic ethics or even the faintest attempt at fairness and honesty. It also takes precedence over ratings and profits.

> "By *liberal media bias* ...I mean the mainstream national news outlets consistently convey impressions or opinions contrary to the Judeo-Christian view on such ethical, religious, and social issues as the sanctity of human life; the definition of marriage; the right of people to be conservative, biblical Christians and to communicate those beliefs without discrimination; and the right to believe judges should adjudicate those basic questions under the Constitutions in a way that secures our religious liberties and basic civil rights.
>
> "Furthermore, I define *media bias* as conveying impressions and opinions of support for those who oppose Judeo-Christian views."[1273]

The ascendancy of Fox News is one of the greatest evidences of not only the blatant liberal bias of the Democrat's media, but of their willingness to put that bias over ratings and therefore money. Money is everything, but not to true believers. Their religion-ideology is far more important to them. Fox News has become so wildly successful and profitable in such a short period of time that you would think that someone with an actual brain at the other cable news channels or even the network news would copy their success and steal some of their profits. Like, for example, the first person to think up the mini-storage center idea has since been copied by countless competitors with a brain and an eye for financial return. Not so with the Democrat-Left media. They are more than willing to drive

---

1272   (Rush Limbaugh)
1273   (From article, "Facing the Goliath of the Liberal Media," by Craig L. Parshall, Israel My Glory Magazine, March/April 2009)

their networks into the ratings commode than admit that they have been blatantly biased, get rid of some of the worst offenders, hire some Christians, constitutional conservatives and Republicans to actually *report* the news, apologize to America, and reap the ratings and financial rewards. Not so for religious proselytes. They will fly their planes into the twin towers of irrelevancy and lack of credibility, just for the self-satisfaction of knowing that they refused to sway from the ideological lies they have believed. They find that much more palatable than telling the truth. For they, like their father Satan, are allergic to it.

"The purpose of a media in a democracy is to inform the electorate. Their job is to be a watchdog on the government. But we have a media that has prostituted itself totally to an agenda. And not only in terms of what it promotes but in terms of what it hides. The real truth was hidden from the American people, and still is. It's an incredibly well-organized propaganda machine."[1274]

Nothing can happen; nothing can occur; nothing can succeed in Satan's multi-pronged attack to tear down America without near-total control of the media; without near-total control of the means of information dissemination in our culture. The Nazis understood this in Germany some 80 years ago. And since the media in a free society is supposed to be the great exposer of lies, liars and their corruption, it is that which you must first control before you can be free to disseminate *your own* lies through that media, and subsequently through all of the other institutions: education, entertainment, the courts, the government, etc. You must first be in the powerful position of running cover for those other institutions. Otherwise the lies and liars would be immediately exposed, and the gig would be up. And Satan has already accomplished this. His Democrat-Left has controlled the media for a number of decades. And they are getting increasingly more and more brazen in their unapologetic support for Obama and the Democrats; in spewing the propaganda and lies necessary to try to lend credibility to that support; and in their ongoing campaign to demonize, vilify, marginalize, demean and disparage their Christian, conservative and Republican opponents. Their brazenness is also seen in the fact that the ever-widening list of their ideological opponents includes the majority of the American people. And this is also why Satan's Democrat-Left, both in and out of the media, hate Fox News and conservative talk radio with such a demonic and irrational passion. They know, and their father knows, that their total control of the media has been challenged by that ascendency, and that they are now being exposed for the corrupt, Democrat-Left aligned *manufacturers* and *suppressers* of the news (as opposed to *reporters* of the news) that they are. And this cannot be tolerated. So they continue in their attempt to marginalize Fox news at every turn, and even though they are increasingly less successful in this with a great many folks, they don't care. It is their *voters* that they want to keep on the plantation, so they just ramp up the propaganda that they spew at them even more in the desperate hope that they will never pick up the remote. Remember, they have been told over and over again for years and in a million overt and subtle ways that "Fox News lies," and that only "stupid" people watch it, and they submissively obey, to the detriment of their own ability to make informed decisions on the crucially important issues of our time. And that is <u>exactly</u> how Satan wants those voters, in his hip pocket.

1274 (Chuck Missler, founder Koinonia Institute, from *Shadow Government DVD*)

Our left-wing media deliberately hides information unfavorable to their Democrat candidates and to their liberal yet America-destroying agenda. They refuse to even cover many stories that are of national importance, are already of great interest to the majority of the nation, and would increase their failing ratings and profitability, because those stories expose their ideology for the corruption that it is. And they have become increasingly more irrational, more unhinged, and more vicious in their attacks of any media outlet or any individual in the media that exposes their game. Our Founders knew how important a free press would be for a free society, especially if that society were to stay free, and as such they gave the press unequaled rights. And to willingly betray that trust by becoming an arm of the state, (the Democrat state), without even being coerced, and to the extent that they have, is treason of a high order. "All enemies foreign and domestic." Our lame-stream, lying, Democrat media has become one of the greatest of this countries domestic enemies; one of our most dangerous enemy within. For without them, all of our other enemies in our courts, entertainment industry and educational system would be exposed, discredited and their influence greatly diminished.

The Nazi Media

"The wicked shall be cut off from the earth, and the transgressors shall be rooted out of it." (Proverbs 2:22)

Over the last nearly decade and a half it has been interesting to watch how one very prominent cable news commentator has slowly evolved in his assessment of the fairness or lack thereof of the national media. In the early days of his show, circa 1996, he did not buy into the whole "media bias" thing, being himself a product of the "main-stream" media. Then, after a number of years of seeing their incredibly vicious and unfounded attacks against his own network, merely because it actually let the other side have a voice, his eyes began to be opened. And by the time of the publication in 2001 of Bernard Goldberg's exposé, *Bias: A CBS Insider Exposes How the Media Distort the News*, he concluded that there was indeed a liberal slant out there. Soon thereafter, as his show and his network soared in the ratings and the onslaught of the attacks against them became even more rabid, numerous, often and dishonest, he agreed that there was some serious bias in the media. More recently that bias has turned into outright corruption in his eyes, and one of his fellow commentators on that network has even declared that journalism in America is dead. (Having given way to outright, unabashed Democrat propaganda.) But there were millions of us who long before the fair and balanced network of that commentator even existed could plainly see that our national media was blatantly liberal and one-sided; outright disdainful and consistently repressive of their ideological opponents and their views; corrupt to rival the state-run news operations of communist dictatorships; and actively involved in a campaign not to report the news but to *manufacture* it, and to vilify, demean, disparage, tear down and marginalize any Christian, conservative or Republican who came to any position of prominence in the eyes of the nation. And we say all this not so much to criticize Mr. O'Reilly but to congratulate him on finally getting up to speed.

However, as we stated earlier at the end of Part I of this chapter, we go even a step further in coming to the sad and disturbing conclusion that what we have here in this country is not just a liberal, or a biased, or a Democrat, or a lame-stream, or a corrupt, or a state-run, a dead or an "Obama-mania" media, but truly a <u>Nazi</u> media. (Two words, Keith Olberman.) And we don't use that word lightly (the Nazi word, not the Olberman one), nor do we use it like those liars on the left do to falsely disparage and unfairly label their ideological opponents.[1275] We use it for three very good reasons:

The first is that the propaganda and lies of our national "mainstream" media (they are so removed from the mainstream of American society as to be openly hostile to it) has so far exceeded that of their mentors from Germany 75 years ago that it cannot be compared. The Nazi's in Germany had one major lie: that the Jews were the cause of all of their problems. The Nazi media in this country has too many lies to cover. (See the [LIE] symbol throughout this chapter.) It is also interesting to note that as the German Nazi's had the one major lie that the Jews were the cause of all of their problems, the Nazi's in our media propagate the Democrat-Left's biggest lie that the Christians-conservatives-Republicans are the cause of all of our problems, rather than themselves. And this lie is shouted loud enough and long enough so their misinformed, bamboozled voters will continue to swallow it. ("Blame it all on Bush." – Obama's theme song.)

The second reason is that they use their media power to advance and sustain the agenda of the baby mutilators—Planned Parenthood, the ACLU, the Democrat Party, the Hollywood crowd, our educational establishment. The Nazi's in Germany *only* killed 6 million Jews. (And we use the word *only* with just a hint of sarcasm and certainly not to diminish in any way what the Jews suffered at the barbaric hands of the German Nazis.) Whereas the proud Nazi's in our media and their Democrat Party have already mutilated **50** million babies to death and they are still going strong! ("YES WE CAN!") The word *Nazi* no longer refers to just a political party in Germany's past. It refers to heinous, genocidal, demonic Jew-killers in Germany's past (who are actually alive and well and living inside the heinous, genocidal, demonic Jew-hating Muslims of today). And along the same lines it is accurate to use it to refer to the disturbed, genocidal, left-wing-indoctrinated baby-killers who control our media today.

And thirdly, what we stated earlier at the end of Part I we will repeat here…

What the Nazi's in Germany did to the Jews, the Arab-Muslims in the Middle East are attempting to repeat. So it is not without historic precedent to accuse these Muslims of being *Nazis*, not only with regard to their genocidal intentions toward the nation of Israel and all Jews in general, but also with the number of lies (Nazi *propaganda*) that they broadcast daily to assist them in their goal. And our own left-wing Democrat media, by repeating their lies and concealing the Muslims intentions are supporting, aiding and abetting them in their genocidal quest. That would make them, at the very least, Nazi *collaborators*. In fact the Muslims are being supported, aided and abetted, and their true genocidal intentions concealed, not only by our media, but the rest of the medias of not only the West but pretty much the entire world as well. Now if that's not Nazism I'd like to know what is. If that's not Nazism I'd like to know what <u>it</u> is. Journalism? Journalism

---

1275   (Remember what we said in section 10 concerning the Nazi playbook of "falsely accusing your opponents of lying so it will start a backfire, then when you are accurately accused of lying, it will burn out in the smoke of confusion before it can stick." Well the same is true of why, over the years, they have falsely accused their opponents of being Nazis (a Nazi in their mind being one who refuses to mutilate babies to death) so they could deflect themselves from being accused of what they actually were, and are)

that aids and abets with its propaganda and lies the imminent slaughter of nearly six million Jews in Israel by the deranged followers of Mohammad and his god Satan is not journalism. It's Nazism.

Seventy years ago in Nazi Germany the Big Lie was that the "Jews were the cause of all of Germany's problems." The Big Lie now and that will be shouted even more so in the near future, is that "the Jews and their nation of Israel are the cause of all of the *world's* problems." "The more things change, the more they stay the same."
And with those three talking points we will rest our case.

"In an age of reckless slander, no charge is so vicious as the one that has become so common on the American Left: that their political and ideological opponents are Nazis."[1276]
Which is nothing more than the actual Nazi's falsely accusing their opponents of being what they actually are. (Quite a technique.) Again, what do we refer to Jew-killers as? Nazis. And what should we refer to baby-killers as? Exactly. Nazis. But for our purposes here we will leave the Nazi word aside for the most part because of its painful history in particularly to our Jewish brethren and will instead continue to use (now that we've made our point) the term Democrat-Left in describing our truly dangerous, <u>Nazi</u> media. (Except for an occasional slip when we get the urge.)

And along the lines of this accurate comparison you'll find a book by Jonah Goldberg, *Liberal Fascism: The Secret History of the American Left*, very instructive…

""Fascists!" "Brownshirts!" "Jackbooted storm troopers!" Such are the insults typically hurled at conservatives by their liberal opponents. But who are the real fascists in our midst? In *Liberal Fascism, National Review* columnist Jonah Goldberg shows that the original fascists were really on the left- and that liberals, from Woodrow Wilson to FDR to Hillary Clinton, have advocated policies and principles remarkably similar to those of Hitler and Mussolini. Jonah Goldberg Reveals: How the Nazis declared war on smoking; supported abortion, euthanasia, and gun control; and maintained a strict racial quota system in their universities—where campus speech codes were all the rage. How the Nazis were ardent socialists (hence the term "National Socialism") who loathed the free market, believed in free healthcare, opposed inherited wealth, spent vast sums on public education, purged Christianity from public policy, and inserted the authority of the state into every nook and cranny of daily life."[1277]
Not to mention the ethical similarities and accurate moral equivalency between the slaughter of 6 million Jews and the mutilation of 50 million babies.

"The press has become the greatest power within the Western countries, exceeding that of the legislature, the executive, and the judiciary. Yet one would like to ask: According to what law has it been elected and to whom is it responsible? …Who has voted Western journalists into their position of power, for how long a time, and with what prerogatives? …Unrestrained freedom exists for the press, but not for the readership, because newspapers mostly transmit in a forceful and emphatic way those opinions which do not too openly contradict their own. …Without any censorship in the West, fashionable trends of thoughts

---

1276 (Pg. 256, *110 People Who are Screwing Up America*, by Bernard Goldberg)
1277 (From, *Liberal Fascism: The Secret History of the American Left, From Mussolini to the Politics of Meaning*, by Jonah Goldberg)

and ideas are fastidiously separated from those that are not fashionable, and the latter, without ever being forbidden, have little chance of finding their way into periodicals or books or being heard in colleges."[1278]

Those prophetic words were spoken in 1978 by Aleksandr Solzhenitsyn, someone who had experienced Pravda first-hand. He is not alive today, and a number of years before he passed he was able to return to his beloved Russia so he was spared the pain of experiencing Pravda right here in our own country. If we could only be so lucky.

""The State-Controlled Media" – That's my name for the mainstream media – ABC, CBS, NBC, all but a few of the major newspapers plus the two "zombie" cable channels (they're dead but don't know it yet), MSNBC and CNN. I call them "state-controlled" because they are no different than the old Soviet media. The lapdog devotion bestowed upon Obama by network anchors resembles the slobbery admiration given to the "Dear leader" Kim Jong Il by North Korea's bogus "journalists." The liberal agenda - abortion, global warming, socialized medicine, affirmative action - is sacred to the robot journalists of "Obama News, Inc.""[1279]

"Surveys over the past 30 years have consistently found that journalists — especially those at the highest ranks of their profession — are much more liberal than the rest of America. They are more likely to vote liberal, more likely to describe themselves as liberal, and more likely to agree with the liberal position on policy matters than members of the general public."[1280]

"[Since Watergate] we've gone from a media that challenged authority while seeking the truth, to a media that champions ideology."[1281]

…While eagerly spreading lies, including their main one: "Democrats good! Republicans bad!" "Baby mutilators good! Christians and conservatives bad!" ("See spot run!" "Me Tarzan, you Jane!")

"In 1981, S. Robert Lichter, then with George Washington University, and Stanley Rothman of Smith College, released a groundbreaking survey of 240 journalists at top media outlets — including the *New York Times, Washington Post, Wall Street Journal, Time, Newsweek, U.S. News & World Report,* ABC, CBS, NBC, and PBS — on their political attitudes and voting patterns. The data showed journalists hold liberal positions on a wide range of social and political issues. Lichter and Rothman's book, *The Media Elite,* became the most widely quoted media study of the 1980s. Key findings: More than four-fifths of the journalists interviewed voted for the Democratic presidential candidate in every election between 1964 and 1976. "Fifty-four percent placed themselves to the left of center, compared to only 19 percent who chose the right side of the spectrum," Lichter and Rothman's survey of journalists discovered. "Fifty-six percent said the people they worked with were mostly on the left, and only 8 percent on the right — a margin of seven-to-one." Nearly half of the journalists surveyed agreed that "the very structure of our

---

1278   (Aleksandr Solzhenitsyn)
1279   (Rush Limbaugh, from circular Sept 2010)
1280   (From Media Research Center article, "Media Bias 101" @ mrc.org)
1281   (Bill O'Reilly, *O'Reilly Factor,* Monday 07/20/09)

society causes people to feel alienated,"[LIE] while the authors found "five out of six believe our legal system mainly favors the wealthy. 30 percent disagreed that "private enterprise is fair to workers;" 28 percent agreed that "all political systems are repressive." 54 percent did not regard adultery as wrong, compared to only 15 percent who regarded it as wrong. "Ninety percent agree that a woman has the right to decide for herself whether to have an abortion; 79 percent agree strongly with this pro-choice position." Majorities of journalists agreed with the statements: "U.S. exploits Third World, causes poverty"[LIE] (56 percent); and "U.S. use of resources immoral"[LIE] (57 percent). Three-fourths disagreed that the "West had helped Third World.""[1282]

And since 1981 it has gotten even worse. What the false beliefs and erroneous assumptions of our media-elite cited above also show is that a lack of proper education, and an abundance of far-left ideological indoctrination, is a prerequisite to their disturbing bias. They were not born that way. They had to be taught.

A History of Bias

"The Lord preserves all those who love him; but all the wicked will he destroy." (Psalms 145:20)

In April of 2011 Sean Hannity hosted an hour-long special on Fox News focusing on our Democrat-Left news media titled, "Behind the Bias: The History of Liberal Media." In it he exposed "the Obama-mania-media's liberal bias," stating that "it's common knowledge that the mainstream media from the major television networks to the country's most influential newspapers are biased against the GOP." (*Common knowledge,* unfortunately, to everyone except that same Democrat media and their followers.)

Sean traces the history of "the media's hostility towards the GOP" and explains how that conflict "first emerged [due to the liberal media's] bizarre tendency to apologize for America's enemies." There was Walter Duranty of the New York Times who, in 1933, "told his readers ...during the Ukrainian famine that killed millions" that "Any report of a famine in Russia is today an exaggeration or malignant propaganda."[LIE] (As opposed to the malignant propaganda that he had just spewed.) On CBS's "60 Minutes" in Feb, 1990, Mike Wallace said: "Many Soviets, viewing the kind of chaos and nationalist unrest under Gorbachev, look back almost longingly to the era of brutal order under Stalin."[LIE] (Sure, and I guess unemployed blacks in today's Obam-economy look back longingly to the era of slavery, Mike.) And on CBS News, May 9, 1990, Bert Quint opined: "Communism is being swept away, but so too is the social safety net it provided." (Yea, and Mussolini got the trains to run on time.) Not to be outdone in the Praise for Communism category, one of ABC News Correspondents, Jerry King, on "World News Tonight" April 6, 1992 threw a bone to their feminist viewers: "Under the communists, women in the workplace were glorified." How nice. And Barbara Wawa, on ABC's 20/20 in October of 2002 gushed the following embarrassment: "For Castro freedom starts with education, and if literacy alone were the yardstick Cuba would rank as one of the freest nations on earth." (If

---

1282   (From Media Research Center article, "The Media Elite" @ mrc.org)

"literacy alone were the yardstick," you putz, Hitler would rank right up there with Mother Theresa.)

"To have a fundamentally conflicted attitude about not just the United States but whether the United States is a constructive force in this world, *that* is something that I think really does separate a lot of liberals and conservatives." - Gerard Alexander, University of Virginia

"There was an element on the left, particularly in the news media, that saw the Soviet Union as a wonderful experiment. They were just simply blind to the tens of millions who were massacred." [1283] - Brent Bozell

Of course, conservatives, fundamental Christians and non-Rino Republicans were offended by the Communist-fawning, anti-American bilge coming out of the liberals who even then controlled the media, and thus became a target for their ridicule and scorn. And that is because the conservative dissenters from the Democrat party line had, according to Juan Williams, "the ability to puncture a lot of liberal myths." And that ability incensed many of those myth-filled liberals in the media. Their credibility and the integrity of their liberal positions had to be defended, which they could not do by debating the issues on their substance and merits (which they still fail to do to this day), so they instead began their decades-long crusade to falsely vilify, marginalize, demean and disparage their ideological opponents. They figured that if they destroyed their opponent's credibility in the eyes and minds of their misinformed listeners, they would "win" the debate without ever having to. (Kind of like what the Darwinists and Man-made-global-warmists are also doing.) They used their media power and freedom, granted by the same Constitution that they also hold in contempt, to protect their sacred liberal views from well-deserved exposure, ridicule and attack. (This is also why they hated Rush Limbaugh with such an irrational passion when he came along in the 1980s and became a national talk-radio hit with millions of listeners. They, their policies and their ideology were now being exposed, and deservedly ridiculed, on a national level.)

Starting in the middle of the 20th century...

"Liberals dominated most of American intellectual and public life ... [and] that dominance shaped the media's portrayal of the GOP. ...The GOP's most influential figures [i.e. Nixon and Reagan were portrayed] as either evil or stupid.

"The Reagan years oppressed me because of the callousness and the greed and the hardhearted attitude toward people who have very little in this society."[LIE] – Howell Raines, former NY Times exec editor, on PBS's "Charlie Rose, Nov 17, 2003 [Translation: Reagan's not a stupid liberal who's compassionate with other people's money, like me.]

"They would not allow that conservatives might be wrong but are driven by noble intentions. They wouldn't even allow that. Conservatives have to be described as being wrong and being evil at the same time." - Brent Bozell

"There is a self-righteous streak that runs through a great deal of liberal and progressive policy making from the New Deal through the 1960s and 1970s in which they were convinced that they really had the solutions to problems, in which case if you disagreed with them, you probably didn't want to solve that problem. And for some reason, you were uncaring." - Gerard Alexander

---

1283   (Gerard Alexander, University of Virginia, and Brent Bozell on "Behind the Bias: The History of Liberal Media" a Sean Hannity/Fox News Special, 04/22/2011)

Hannity: "And if conservatives aren't evil and heartless, well then they're just stupid."

Tom Friedman, NY Times columnist on CBS's "Face the Nation," March 14, 1999 said this: "When I listen to the Republicans in Congress on foreign policy, there's such a 'I'm stupid and proud-of-it attitude.'"[LIE] [Tom, there's a psychological malady known as "projection." Ever heard of it?]

And Howard Fineman of Newsweek on MSNBC's "Countdown," Aug 29, 2008 said: "Sarah Palin makes Barack Obama look like John Adams."[LIE] [No, Howard, Sarah Palin makes Barack Obama look like the inexperienced, unqualified, brainwashed dupe of the Left that he is. But then again, John Adams has been dead and his skull completely empty for some time now, Howard, so you might be on to something.]

Brent Bozell: "They see themselves as enlightened.[BUT THEY'RE NOT] They see themselves as intelligent.[THEY'RE NOT] And they see themselves as good.[GOOD LITTLE BABY-KILLERS?] And therefore conservatives are the opposite."[(LIE)]

Juan Williams: "The story, time and again, is Republicans don't know what they're talking about. They're ill-informed, i.e. they're dummies.[LIE] They're a bunch of stupid people.[LIE] And they're a bunch of ignoramuses[LIE] who react on the basis of emotion,[LIE] say the most outrageous and provocative things,[LIE] but really don't know what they're talking about.[LIE]

And, oh gee wiz, Democrats …they may not exactly get it right [the world's greatest understatement] but they're really the *smart* ones."[LIE][1284] (And Juan should know what he's talking about because he spent his formative years surrounded by actual ignoramuses over at PBS.) (He's since taken a number of showers.)

Like we've said before… "Democrats Good! Republicans Bad! Baby mutilators real smart! Conservatives and Christians real stupid!" The simple message that our Democrat media's "Me Tarzan, You Jane" intellect can wrap itself around.

Sean goes on to expose how "the main-stream media holds conservatives to different standards than their liberal colleagues." And then asks… "How does this hypocrisy effect American politics?"

"There is so much left-wing bias in the main-stream media that if you give them a sword *they will use it*, and they will *relish* using it. [However] if a Democrat gives them a sword, more often than not, they will lay it down." - Hugh Hewitt, radio talk-show host

Hannity: "As a result of their bias the liberal media imposes outrageous double-standards on the GOP. From racial issues to matters of intellect, Republicans are routinely held to a much higher standard than their Democrat counterparts."

Brent Bozell: "They are always willing to forgive Democrats and liberals, or ignore the sins of Democrats and liberals, but will never forgive conservatives for what they say."

Juan Williams, himself an African-American, takes exception to how the Naz…er…liberal media will play the race card, but only in one direction…

"Anybody with any smarts who is paying attention says, 'You know what? This process lacks integrity. It's not equal. It's not balanced. You're not really concerned about race and about intimidation and offense, what you're concerned with is scoring political points.'"

Hannity: "The roots of this hypocrisy go all the way back to the Civil Rights movement and the Democrat Parties shameful past."

Hugh Hewitt: "The Democrat Party has a lot to apologize for when it comes to race.

1284 (From "Behind the Bias: The History of Liberal Media" a Sean Hannity/Fox News Special, 04/22/2011)

Robert Byrd [KKK enthusiast] ...Al Gore Sr. who opposed the Civil Rights Act of 1964. Sam Ervin, the hero of Watergate. These were all deep-dyed segregationists. And so, if you dig very deep into Democratic dirt you'll be turning over just a treasure chest of old racist comments and attitudes."

Hannity: "And the liberal media has allowed the Democrats to bury this history."

Hewitt: "Along about the time that the media went left they began to assist the Democrat Party in the airbrushing of their racist past."

They had to keep the narrative consistent with their desire to keep as many voters as possible uninformed, unaware and firmly rooted on their Democrat plantation, willingly handing liberals their votes, and getting nothing but deceit in return.

Hannity: "But perhaps no media double standard is as outrageous as the one surrounding intelligence and stupidity."

Brent Bozell: "The default position for the liberal media ...is that conservatives are dumb. So that when a Dan Quayle makes a faux pas they jump on it because "AH HA! See, I told you he was dumb." When a liberal makes the same kind of faux pas, the medias reaction is completely the opposite. They see these people as uber-enlightened and therefore it was *just* a faux pas and therefore it doesn't merit coverage, and so they ignore it."

Hannity: "And ignore it they have, from Al Gore's infamous inquiry at Monticello [where he couldn't recognize the busts of George Washington and Benjamin Franklin and asked "Who are these people?"] to the presidents [Obama] and vice presidents [Biden] greatest hits... [which included Obama's comment that he had campaigned "in 57 states" and had "one left to go."]

... "And the double standards imposed by the media shape the news. As Americans turn on the television day after day, they also shape the nation's political narrative."

Juan Williams: "The assumption is, Republicans don't know what they're talking about,[LIE] don't understand because they haven't taken the time to study the issue.[LIE] Democrats know so much about this issue, [LIE] are so well-schooled,[LIE] are so erudite,[LIE] and have the experience to deal with this issue,[LIE] but most people just can't keep up with them."[LIE]

And since the media's assumptions are all lies, it renders them incapable of delivering political news in any semblance of a fair and balanced manner.

Hugh Hewitt: "Once they've defined somebody as either outside the mainstream, or lunatic, or stupid, the circuit breaker throws the minute you see them on television and their ability to persuade is over, and *that's* really the most damaging thing to a democratic republic is the inability to hear people who your programmed to disregard."[1285]

Which of course today is their entire agenda, keeping their people safely on the plantation, with a high enough fence so they will never see the freedom that lies on the other side, and never have the opportunity to entertain any "contrarian" thoughts.

"Behind the Bias" also reported on the now-famous story by Dan Rather that aired on CBS news' "60 Minutes" right before the 2004 presidential election, in which he falsely disparaged President George Bush's exemplary military service.

Hannity: "[This] explosive segment would later become known as one of the most obscene examples of media bias in American history."

1285   (From "Behind the Bias: The History of Liberal Media" a Sean Hannity/Fox News Special, 04/22/2011)

(And that's saying a lot because it had to win out over thousands of other very impressive entries.) It resulted in the firing of both Dan Rather and his Republican-hating producer Mary Mapes.

"I don't think there is any serious doubt that what was going on here was an effort by at least some people at CBS news to influence the 2004 presidential election."

Rather produced a few defamatory documents upon which the entire story was based. Yet he never bothered to check them out. They turned out to be obvious forgeries. (Teenagers faking driver's licenses could have done a better job.) The day after the story aired, bloggers were already reporting that the documents, dated 1977, used a type font that was not available until many years later. But it wasn't just the typing that exposed the documents as being bogus...

"The truth is these documents ...completely misrepresented what was in fact a very honorable career by Lieutenant Bush in the Texas Air National Guard. He got glowing evaluations from his superiors, and they're publicly available." – John Hinderaker, Power Line Blog

Hannity: "And it was later revealed that the producer ...Mary Mapes, knew about those evaluations all along."

But she just accidentally failed to include them in Dan Rather's on-air smear. It was just a slip of the mind. It could've happened to any other loyal Democrat disguised as a legitimate producer of national "news." Unbelievable. But even more incredible than their lying documents and their ignoring of the real evaluations, was the fact that John Kerry's campaign manager, Joe Lockhart, was linked to this Bush-bashing hoax.

"The Kerry campaign started an advertising campaign on television the morning after the "60 Minutes" show ran [completely coincidentally, of course], on this same theme of president Bush's service in the Texas Air National Guard."[1286] – John Hinderaker, Power Line Blog

"It's pretty breathtaking to see that kind of hand-in-glove operation between a presidential campaign and a media outlet that's supposed to be covering both campaigns but was clearly in collusion with one of the campaigns." – Ed Gillespie, Former RNC Chairman

And pathetically, even though it is a historical fact that George W. Bush "was smeared by 60 Minutes, by Mary Mapes and Dan Rather, and even though we now know that those allegations were proven to be false"[1287] there are still many "true believers" on the Democrat-Left who refuse to let facts get in the way of a good story, and who still believe the smear rather than accepting the truth. O-BA-<u>MA</u>! O-BA-<u>MA</u>!

Hannity: "And it is *that* tragic reality that helps put the sheer power of the left's control over main-stream media into perspective."

The media again displayed its liberal arrogance in the Joseph Wilson-Valerie Plame affair. In January of 2003 then-president George W. Bush made this statement, "The British Government has learned that Saddam Hussein recently sought significant quantities of uranium from Africa," which turned out to be accurate.

1286   (From "Behind the Bias: The History of Liberal Media" a Sean Hannity/Fox News Special, 04/22/2011)
1287   (John Hinderaker, Power Line Blog, on "Behind the Bias: The History of Liberal Media")

Hannity: "Those 16 words sparked a political firestorm and, thanks to the mainstream media, they also sparked what became known as the Plame scandal, in which the media sacrificed principle at the altar of politics and exposed their blind hatred for the Bush administration."

After Bush's State of the Union address, "left-wing diplomat Joe Wilson, who had been sent to Niger on an intelligence gathering mission in 2002," wrote an op-ed for the New York Times in which he claimed that...

..."the president's statement was false,[LIE] and charged that the Bush administration had manipulated intelligence in order to launch the Iraq War.[LIE]" The main-stream media then "energetically peddled Wilson's story about the administrations alleged lies. ... [Indeed they were] so inclined to believe Wilson that they ignored a bipartisan Senate Intelligence Report that exposed his lies." - Hannity

Byron York: "I think a lot of reporters just didn't want to consider the idea that Wilson had actually gone to Niger and essentially found that there *had been* an earlier attempt by Iraq to purchase yellow cake uranium."[1288]

Of course Wilson *had* found just that and had put it in his report, but the distorters ...er...reporters from our main-stream media just ignored those facts. Then after that report Wilson reversed himself when the opportunity arose for him to lie in order to damage the Bush administrations credibility. And the IZAN media curiously never picked up on this massive discrepancy which destroyed Wilson's (and not Bush's) credibility. They just missed it. Somehow. It slipped by them unnoticed. They're not biased, just selectively unobservant.

And then there's the New York Times...

"Astute readers over the years understand that [the NY Times] *operational* motto is 'All the news that's fit to distort ...to suppress and ...to ignore." - Michelle Malkin

Countless examples could be given that expose this left-wing rag for the Democrat's propaganda machine that it has become, but one of the more glaring is when they immediately jumped to a false conclusion just because it fit their liberal, race-baiting narrative, and smeared three *white* Duke lacrosse players who had been, it turned out, underlined falsely accused by a *black* stripper. A year after her accusations of rape she admitted she made up the entire story. But "their reputations had already been tarnished" by liberal papers like the Times, as well as by 88 Duke University professors (2/3 of whom were either black or female or both) who also rushed to condemn the 3 players without any evidence. And whereas the Duke Lacrosse coach was forced to resign, these 88 false accusers were not. Words like "hypocrite" and "double standard" haven't yet made it onto Liberal America's vocabulary list.

Another way that the Times Democrat bias shone forth was with their slanted coverage of the Arizona Immigration Law...

..."When that law passed you really would have thought it was something akin to Hitler's Nuremberg Laws,"

...and in the immediate aftermath of the Gabrielle Giffords shooting...

..."When Arizona Congresswoman Gabrielle Giffords was shot by a deranged man earlier this year the NY Times rushed to judgment once again."

Michelle Malkin: "When you have the likes of Paul Krugman, Frank Rich, every hired and paid NY Times columnist trying to link those acts of violence that were committed by

---

1288 (From "Behind the Bias: The History of Liberal Media")

a single, lone, deranged individual, and using it to try to undermine the free speech of conservatives; their explicit, overt agenda was to try and chill conservative thought and criminalize conservative speech."

Brent Bozell: "The NY Times lost its journalistic mind. [See "Reprobate Mind" discussion in chapter 11.] It went from being an unquestioned left-of-center newspaper to a radical left-wing newspaper. And it might as well be produced by the Obama administration."[1289]

(That's funny, I thought it already was. Or hasn't he taken over *that* industry yet?)

Newt Gingrich, speaking on the bigotry of the liberal networks that was clearly on display when they took turns hosting the Republican primary debates in 2011, said the following:

"They [the liberal media] have this automatic mindset that they want to define us [Republican candidates for the presidency] in ways that will make us unacceptable [to the American voter] and they just can't help themselves.

..."The liberals can't defend their policies. ...So, obviously, rather than defend their failures, they attack us. ...We have to have the courage to make fun of them, to go straight at them, to tell the truth, to stick with the facts. And I think that frankly we're winning the argument in the country at large, and people in the country overall increasingly realize how shrill and how fundamentally dishonest the left-wing media is."[1290]

Would that their voters might start realizing this as well, because if America doesn't find a way to go around the media and reach their no-think-ums, we're screwed.

Me Tarzan, You Jane

Vote Democrat! That is the primary message of our Democrat-Left media. Democrats good, Republicans bad. Democrats smart, Republicans, Christians and conservatives stupid. Of course the exact opposite is true, but they're not interested in the truth. To the "enlightened ones," it doesn't even exist, so therefore it can't get in the way of all the decidedly untrue things that they believe. Like their father, Beelzebub, they despise the truth and those who embrace it. And with this as their foundation they will distort, mangle, ignore, suppress and lie about the news in order to continue broadcasting their primary message; so the evil one can continue to radically transform, aka destroy, this nation from within its own voting booths.

The Media Research Center reported that after...

..."President Bush's re-election in 2004, ABC morning host Diane Sawyer relayed what she said were the concerns of *Kerry voters* [including, obviously, herself]: "There's a definite sense this morning on the part of the Kerry voters that perhaps 'moral values' is code for something else.[LIE] It's code for taking a different position about gays in America, an exclusionary position, a code about abortion, a code about imposing Christianity over other faiths.[LIE]"

[Isn't it interesting how Diane Sawyer sees any position other than hers about

---

1289   (From "Behind the Bias: The History of Liberal Media")
1290   (Newt Gingrich, on Fox News' Hannity, 01/20/2012)

"gays in America" as "exclusionary." And of course she's lying when she implies that Christians want to "impose" their religion on the nation. What *is* being imposed on the nation is the Democrat-Left's godless religion which includes baby-killing and evolution.]

And a few days later, NBC's *Meet the Press* host Tim Russert announced: "One Democrat said to me, "Are we on the verge of a theocracy, where if you don't agree with the president and evangelical Christians on abortion or on gays, there really is no room for you to practice what you believe in the United States?"[LIE]"[1291]

Tim Russert has since passed away, and he was one of the more honest members of the Democrat-Left media, but it is still interesting to note how misinformed he was because, unfortunately, the facts clearly show that increasingly "there really is no room for you to practice *Christianity* in the United States" anymore. Not in our "public" schools, not in our court houses, not in our universities, not even outside during Christmas.

And then there is the case of all of those *homosexual* Catholic priests that used their position of power and trust to prey on a number of young boys over a number of decades in a number of dioceses across the land. The media of course reported it as if it were a problem of *celibate* Catholic priests,[(LIE)] but that was just another lie. The problem was not celibacy; the problem was having homosexuals who were attracted to young boys being in a position of authority and close unsupervised contact over them...

"Bill Donohue, president of the Catholic League for Religious and Civil Rights, spoke out in an article on the Washington Post's website on Thursday, defending his assertion that the widespread perception of a "pedophilia crisis in the Catholic Church" is not supported by data and research. The more significant problem, Donohue argued, is the incidence of homosexuality among priests. [Bingo.]

"Citing a number of medical journals in the field of human sexuality research, Donohue explained in his submission to the Washington Post's "On Faith" section that "homosexuals are disproportionately represented among child molesters." Statistically, he said, the evidence for a "link between homosexuality and the sexual abuse of minors" in the general population is "overwhelming."

"This link is borne out in the majority of sex offenses committed by priests, according to Donohue. "As I have said many times, most gay priests are not molesters, but most of the molesters have been gay."[1292]

But the Democrat-Left media is not interested in the truth, not if it shines a bad light on one of their protected, minority, "victim" groups; in this case, homosexuals. Then they find it more palatable to cover up the truth and continue to spew propaganda and distortions that fit their ideology. (Just like you would expect the N___S to.)

Speaking on the occasion of Don Imus' firing in April, 2007, for his insensitive comments about the Rutgers women's basketball team, Brent Bozell had these thoughts:

"Liberal reporters and liberal leaders in America have been making hateful, bigoted, and racist comments against conservatives for decades, and the liberal media establishment

1291 (The Media Research Center)
1292 (From article "Abuse scandal rooted in homosexuality, not pedophilia, says Catholic League president," @ catholicnewsagency.com)

has let them slide. Yet when a conservative makes an off-color remark -- in passing, by mistake, or even if distorted by enemies -- the liberal media attacks. This double standard is outrageous and sickening when one considers how hateful the left really is. Consider some of the following quotes and consider that the liberal media did **nothing** to condemn them and, in many cases, applauded them.

- "If there is retributive justice," Sen. Jesse Helms "will get AIDS from a transfusion, or one of his grandchildren will get it." -- Nina Totenberg, National Public Radio's legal affairs correspondent...

- "For hypocrisy, for sheer gall, [Newt] Gingrich should be hanged." -- *Washington Post* syndicated columnist Richard Cohen. [Translation: "He's not a baby mutilator like me, dag nab it."]

- Evangelical Christians are "poor, uneducated, and easy to command."[LIE] -- Reporter Michael Weisskopf, *Washington Post*. [Unlike his readers, of course.]

- Vladimir Putin "is perceived to be an effective dictator. What we have in this country is a dictator [speaking of George W. Bush] who is ineffective." -- *Newsweek* contributing editor Eleanor Clift. [Would that Obama were as "ineffective," we might still have an economy.]

- After interviewing the MRC's Robert Knight, who criticized homosexual scoutmasters in the Boy Scouts, host Bryant Gumbel exclaimed, "What a f***ing idiot!"[LIE] -- CBS *Early Show*. [Gumbel was reprimanded, and agreed that in the future, and for the sake of honesty in journalism, he'll make certain he's looking in a mirror every time he makes that statement.] [Until it sinks in.]

- About Supreme Court Justice Clarence Thomas: "I hope his wife feeds him lots of eggs and butter and he dies early like many black men do, of heart disease." -- Syndicated columnist Julianne Malveaux. [But it's OK, she was wearing her pink breast-cancer ribbon.]

- Justice Clarence Thomas is "a handkerchief-head, chicken-and-biscuit-eating Uncle Tom." -- Movie director Spike Lee. [Speaking of biscuits, how's the food on that Democrat plantation, Spike?]

- "We would stone Henry Hyde to death! ... I'm not finished. We would stone Henry Hyde to death and we would go to their homes and we'd kill their wives and children! We would kill their families!" -- Actor Alec Baldwin. [Hmmm. Another convert to Islam?]

- "I hate Republicans and everything they stand for." -- Democratic National Committee Chairman Howard Dean. [Which is not mutilating babies to death.]

- "What it looks like is going to happen is that [Lewis] Libby and Karl Rove are going to be executed. ...I don't know how I feel about it because I'm basically against the death penalty, but they are going to be executed." -- Former Air America radio host Al Franken, now a Democratic senatorial candidate."[1293] (And now an actual Senator! Good grief! Howdy Doody wasn't running, however, so what choice did the poor Minnesotans have?)

"The tongue of the wise uses knowledge correctly, but the mouth of fools pours out foolishness." (Proverbs 15:2)

And then there's the media's vile treatment of the death of Christian leader, Jerry Falwell. (He didn't mutilate any babies to death either. What an awful man.)

"I think it's a pity there isn't a hell for him to go to ... The empty life of this ugly little

---

1293   (From article, "Liberal Media Hypocrisy: Slam Imus But Give Pass to Olbermann, Totenberg, Huffington, and Other Liberal Heroes," The Watchdog, a Media Research Center publication, May 2007)

charlatan ... Such a little toad[LIE] ... This horrible little person[LIE] ... I'm glad to see he skipped the rapture, just found on the floor of his office." -- *Vanity Fair* Contributing Editor Christopher Hitchens on CNN's *Anderson Cooper 360*, May 15 [2007]

[If you're having a hard time realizing just how disgusting that is—because we are so used to them vilifying Christians and conservatives—imagine for a moment that Barack Obama dropped dead and a reporter said all that about him! That he was "such a little toad," and an "ugly little charlatan." There would be unending hysterics and he would not only be fired, but executed. But no condemnation of Hitchens. And he's still invited back on. And you thought I was kidding when I said we have a Nazi media.]

"The reaction from the reporters? Grins and chuckles mostly. One grizzled veteran journalist said, 'I hope they (CNN) remember all the horrible things he said.' [Like "killing babies is wrong"?] Another reporter simply stated, 'It is a good day.'" -- A Congressional press gallery reporter e-mails National Review Online about gallery reporters' reactions to Falwell's death, May 15 [2007]

"My very first thought upon hearing of the Rev. Falwell's passing was: Good. And I didn't mean 'good' in a oh-good-he's-gone-home-to-be-with-the-Lord kind of way. I meant 'good' as in 'Ding-dong, the witch is dead.'" -- *Chicago Sun-Times* columnist Cathleen Falsani, May 18 [2007]"[1294] (As opposed to, say, "Ding-dong, your brain is dead.")

These people are more than just biased, irrational and hate-filled; they are seriously disturbed. But that's exactly how their father Satan has grown them.

"Educate a child in the ways of the evil one, and when he is old, he will not depart from it." (Proverbs 22:6 ...slight return)

At the Winter Olympics in 2006, HBO's *Real Sports* host Bryant Gumbel said this, "Try not to laugh when someone says these are the world's greatest athletes, despite a paucity of blacks that makes the Winter Games look like a GOP convention."[1295] Yea, Gumbel, it's always about race; your race. And I guess the paucity of whites in the NBA makes it look like an NAACP convention. My only question is, How did a man this ignorant ever get to be a reporter on a national level? (Oh, I forgot, that's one of the *requirements* for being a reporter on the national level.) And don't forget to pick up a copy of Bryant's latest book, *It's OK to be an Ignorant Racist: as long as You're a Liberal, and Black Like Me*. Oh, by the way, Gumbel, what "paucity" of education and intelligence does it take for you to go out and vote religiously every two years for the slow extermination of your own race? If you're having trouble coming up with an answer, ask Margaret Sanger.

Indeed, in the two-faced world of what Rush calls the "Democrats Ministry of Misinformation" it actually *is* OK to be a bigot, as long as you're black or liberal and you're attacking a Christian or a conservative. Condoleezza Rice was called "Aunt Jemima" and a "black trophy of the Bush administration" by white radio host John Sylvester, yet whereas they made sure Imus got canned, he just gained a lot of street cred' from the rest of his pals in the liberal media. She was portrayed as President bush's thick-lipped parrot by cartoonist Pat Oliphant; referred to as President Bush's "House Nigga" by nationally-syndicated commentator Ted Rall;

1294   (From exposé, "How the Liberal Media Said Goodbye to Jerry Falwell," The Watchdog, a Media Research Center publication, May 2007)
1295   (From The Watchdog, a Media Research Center publication)

and Gary Trudeau showed President Bush calling her "Brown Sugar" in his comic strip, "Doonesbury."[1296]   In none of these examples did these left-wing racists suffer any consequences for their actions. Can you imagine the spastic uproar, and the career ending consequences, if a conservative called Barack Obama a "House Nigga," or Michelle Obama "Aunt Jemima"?

Tammy Bruce said that a "Gestapo has emerged in America." (Could she be suggesting that we have, I shudder to say, a not see Media?!) "You have a media Gestapo in *Media Matters*, and then you have the political Gestapo in MoveOn. org."[1297] And of course the Gestapo in Germany was just an arm of the larger <u>Nazi</u> Party.

> "Imagine that you're watching a news program on broadcast or cable TV and in a story about a politician or political organization, you hear descriptions like "mean-spirited" …"extreme" …"hate speech" …"politics of personal destruction" …Do you think that news story is likely to be about: (a) A Democratic politician/liberal organization? Or (b) A Republican politician/conservative organization? <u>You know the answer without a doubt, don't you</u>? They're talking about a Republican politician or conservative organization."[1298]

We are all familiar with the Reverend Wright scandal and the Democrat medias cover-up of it, but it is worth revisiting here… In the May 2008 edition of the Media Research Center's monthly report, "The Watchdog," their lead article, "Liberal Media Hide Facts and Spin the Truth To Protect Barack Obama and Political Liberals," documented how that liberal media glossed over the damning Reverend Wright revelations. Their coverage "reveals their gross political biases and how far they will go to spin the facts and to keep Americans in the dark." And especially, to keep their voters, the no-think-ums, in the dark.

> "Wright is a flaming black nationalist whose writing, sermons, and rhetoric frequently fall into anti-white and anti-American rants. Typical was a sermon he gave after 9/11 in which he blamed America for the terrorist attacks[LIE] and in another talk railed about the "U.S. of K-K-K-A." There are countless examples of his lunatic speech. He's claimed that the U.S. government invented AIDS as a "genocide" against blacks.[LIE] Five times in one speech he roared 'God d--- America!"
>
> "Despite Wright's maniacal ramblings and that he is Obama's spiritual adviser and close friend, the liberal media has tried to spin the acidic truth away. …Can you imagine the news stories and endless drip, drip, drip of reports if Sen. John McCain (R-Ariz.) had attended an anti-American "white liberation" church headed by a wacky, racist pastor? It would be the top story every day in every major paper and news show until McCain explained himself, left the church and totally rejected his pastor- and then the liberal media would still hound him."[1299]

And that's because the media can't have their voters beginning to think for themselves, and making their own informed decisions based on <u>all</u> the facts and evidence. If that happened they might stray off the plantation. So they must do their "thinking" for them.

1296   (From The Clare Boothe Luce Policy Institute)
1297   (Tammy Bruce ON Fox News host Bill O'Reilly's radio show, Sept, 2007)
1298   (The Media Research Center)
1299   (From "The Watchdog," a monthly report from Brent Bozell's The Media Research Center, May 2008)

"Chris Matthews, Katie Couric, The New York Times, CNN, Newsweek, NPR, and the liberal media elites are ...working overtime to convince the nation that they should support impeaching the President [Bush], surrendering in Iraq, raising taxes, appeasing our Islamofascist enemies, expanding government, and abandoning our traditional family values."[1300]

And sadly it seems, at least judging from the 2006 and 2008 elections, they are having their way. And even after the 2010 elections where America rejected the far-left policies of Obama-Pelosi-Reid and their Democrat media, it is still sad and disturbing to see how many Democrat Senators, and the Democrat governor of Maryland, O'Malley, still somehow were reelected. Again, "if you shout a lie loud enough, and long enough, enough uninformed people will believe you." The Nazi media understands this and that is why they continue their ways despite losing ratings and profits. Keeping as many people as they can in the dark is of far greater importance to them. They are not about to surrender to the truth, no matter how much it costs. Indeed...

"Our media and the anchors and journalists who make up that media are enemies of the truth, they are at war with the truth."[1301]

When Ann Coulter was asked (June 25, 2007) by ABC's Chris Cuomo...

..."about criticism she had encountered because of previous comments she had made about John Edwards ... Miss Coulter replied:

""...Bill Maher was not joking [when] saying he wished Dick Cheney had been killed in a terrorist attack. So, I've learned my lesson. If I'm gonna' say anything about John Edwards in the future, I'll just wish he had been killed in a terrorist assassination plot."

"With that, the liberal media were off to the races. [They only reported the last sentence in Ann's response to make it look like she was actually wishing for Kerry's death,[LIE] rather than making a point about the medias nauseating double standard.] ... This is precisely how the liberal media operate: distort, twist, spin, cut and paste, and lie about their enemies regardless of the facts."[1302]

"A June 21 [2007] report by Bill Dedman of MSNBC.com documented that out of 144 journalists surveyed, 125 of them (87 percent) gave financial contributions to Democrats and liberal causes; only 17 gave to Republicans."[1303]

And keep in mind that Talking Points is not alleging a slight under-reporting of negative things about the Democrats, and a slight over-doing it for possible or even falsified transgressions concerning Christian-conservative-Republicans. We're saying that the difference is night and day between their nearly complete cover-up and white-washing on the one hand of any story that puts their Democrats or their leftist ideology in a bad light; and going 24/7 ballistic on anything that hurts their ideological opponents on the other. It's not a liberal media. It's a N*** media, and it is covered wall-to-wall, 24/7 with their Nazi propaganda.

1300 (The Media Research Center)
1301 (Source unknown)
1302 (From article, "Liberal Media Deliberately Distort and Coulter's Comment," in The Watchdog, a monthly report from The Media Research Center, August 2007)
1303 (From "The Watchdog," a monthly report from The Media Research Center, August 2007)

"Ignoring 9/11, the attack on the USS Cole and the 1993 bombing at the World Trade Center ... some in the liberal media now declare that the U.S. presence in Iraq is the cause of terrorism!⁽ᴸᴵᴱ⁾ Go figure.

"But even worse than the liberal "news" media are Hollywood and "entertainment" television. They are pushing an even more radical anti-American message to millions of viewers every day."

[Here are a few examples:]

"This President invaded a sovereign nation in defiance of the UN.[LIE] He is basically a war criminal.[LIE] Honestly. He should be tried at The Hague." -- Rosie O'Donnell, Fox's At Large with Geraldo Rivera, Apr. 30, 2005

"Cheney is a terrorist.[LIE] He terrorizes our enemies abroad and innocent citizens here at home[LIE] indiscriminately." -- Alec Baldwin, HuffingtonPost.com, Feb. 17, 2006 [Uh, Alec, you moron, aren't we supposed to "terrorize our enemies abroad?" I mean, they're our enemies, right? Maybe you could go over and give them all massages.]

"No wonder Bush is intimidated. ... His hands are covered in the blood of Cindy Sheehan's son.[LIE] [Actually, the memory of Cindy Sheehan's son is being soiled, sadly, by the irrational antics of his mom. May he rest in peace.] They are dripping with the blood of all who have died there."[LIE] -- Actress Christine Lahti, HuffingtonPost.com, Aug. 11, 2005. [Don't look now, Ms. Lahti, but your hands are dripping with the blood of 50 million innocent babies. And counting...]

"The president is a moron![LIE] I'm saying it. I don't care. He's an idiot.[LIE] Cheney is evil.[LIE] I'm sick of it, impeach them, get them out! I hate them! I hate them. Get them out." -- Kathy Griffin, Comedy Central's Weekends at the DL, Sept. 10, 2005 [A mind is a terrible thing to waste.]

"All their reasons for waging war on Iraq have been proven to be manipulation of facts,[LIE] untruths,[LIE] and lies, lies, and more lies."[LIE][LIE][LIE] -- Jessica Lange, C-SPAN, Sept. 24, 2005 [Actually, Ms. Lange, it is your statement that "has been proven to be lies, lies, and more lies."]

[And now for a welcomed change...]

"President <u>OBAMA</u> is a rube.[TRUE] He is a dolt.[TRUE] He is a yokel on the world stage.[VERY TRUE] He is a Gilligan[TRUE] who cannot find his ass with two hands.[NO, HE CAN FIND HIS ASS] He is a vain half-wit[TRUE] who interrupts one incoherent sentence [teleprompter malfunction] with another incoherent sentence." -- Bill Maher, NBC's *Tonight Show*, Feb. 20, 2007"[1304]

(I know, he was talking about Bush. But those false accusations of GW have nonetheless turned out to be an accurate prophecy of the current occupant of the White House! Thank you, Mr. Maher. It just goes to show that even a godless imbecile like you can stumble upon an ounce of truth every once in a while.)

Newt Gingrich goes *On the Record* with Greta Van Susteren on October 22, 2008, concerning that year's election coverage:

Newt, "I think the elite media's attack on Governor Palin again and again has been factually wrong, intellectually dishonest, totally biased, worthy of the Polish State news media attacking Lech Walenza back in the 1980s. I mean, this is the kind of deliberate, vicious, dishonest, total distortion of who Governor Palin is, including by the way the Saturday Night Live skits, some of which I think were slander, and were worthy of a

---

1304   (From article, "Hollywood, Hand in Hand with Liberal Media, Slams War on Terrorism, Bush and Conservatism," by Brent Bozell, The Watchdog, July 2008)

lawsuit. And I think that the American people should realize that the elite media on the left is so desperate to elect Barack Obama that the view they are giving you of governor Palin is fundamentally a falsehood, and the one you saw from CNN is so outrageous that they owe her an absolute apology. ...Again and again you've seen the lead reporters do things that were false in order to try and make Governor Palin look bad."

[Sarah Palin isn't an ignorant, baby-mutilating reprobate, so why shouldn't she be crucified? Fair's fair, after all.]

Greta asks, "When the elite media gets these quotes wrong is it deliberate, or just incompetence." [Or are they a bunch of Nazi propagandists?]

Gingrich, "No, it's deliberate. Look, you and I have a fundamentally different view, Greta. When the New York Times goes out of its way for the second time in three weeks to run a page one attack piece on Cindy McCain, to have a totally biased coverage; if you were to look at the coverage that Michelle Obama has gotten and the coverage that Cindy McCain has gotten, you have to believe that the fix is in. When you look at the magazine covers, you look at the pictures that are taken, again and again, over and over, for the last year, we have been brainwashed, propagandized, insultingly lectured by the news media. If you look at what happened to Joe the plumber: Joe Biden, in the middle of that vice presidential debate was totally wrong about the restaurant he claimed to go to, to learn about the middle class, because it went out of business in 1986. Now, I haven't seen anybody in the elite media doing a live broadcast from the non-existent Katie's restaurant to point out that Joe Biden is just out of touch with reality. He's either dishonest, or has a total memory lapse about his hometown. Yet nobody notices it. I think there were 14 factual mistakes by Biden in the vice presidential debate. Nobody said that disqualified him. This has been a totally one-sided campaign in which the news media [the ____ news media] has been the best ally that Barack Obama has gotten. There is no pretense, not a shred, of neutrality on the part of the major networks or the New York Times."

Greta, "Today on Greta wire I put a blog entry which says 'The Arrogance,' and then I posed the question, 'Who's the Dummy?' [Palin, or the not c's?] And I guess for me it's really hard to understand how arrogant the media can be. When you think about it, Governor Palin had to be elected by the city to become mayor, elected by the state to become governor. And to become a journalist, all you have to do is say, 'I'm a journalist.' You don't even have to take a test. And if the journalists don't agree with her, they tag her 'stupid' when she has done nothing but do her job very effectively." [But hey! Gotta keep them no-think-ums on the plantation.]

Gingrich, "I've been thinking about having one of my researchers look at every single elite media interview of Governor Palin to see if any of them asked her how she gave $1300 a person back to Alaskans as a tax break. How she negotiated and got so much money out of big oil. How she worked out the deal for the natural gas pipeline, or anything of substance about how she shaped an 11 billion dollar budget, involving 29,000 employees of the state government. I don't believe to the best of my knowledge that there has been a single question by an elite television journalist about her actual career in Alaska. And I think it is the most insulting. And, I can't say this too strongly, this is like watching Pravda. This is a one-sided, vicious, unending and dishonest campaign. And then we look at the poll numbers and all they tell you is that in the short run, being brutalized by the national media has an impact. And, by the way, I want to go back to Joe the Plummer. Here's an innocent guy whose only mistake was he took on Barack Obama in a way that was very painful, and maybe the reason you have two polls this evening that show this race too close to call, because Joe the plumber pointed out he didn't want some politician spreading his wealth to that politicians cronies and allies "

Greta, "[Barack Obama complained that] he would be about three points higher in the polls if it were not for Fox news. A viewer e-mailed me, 'what would his numbers be if it weren't for MSNBC and CNN, where would his poll numbers be?' That question was not posed."

Gingrich, "And where would he be without the New York Times, which has been the most unendingly dishonest newspaper in America this year, which is really a disgrace. This was once a great newspaper, and it is now a totally one-sided and totally distorted publication that does remarkably bad coverage of the race. That hasn't looked into even something as simple as what happened to Senator Obama when he was at Columbia University, which is in New York City, where the New York Times is published. So it just tells you how one-sided their coverage is. …At what point do you just shrug your shoulders and say look, the fix is in, the liberal elite's in Hollywood and New York are desperate for you to elect Obama. The labor union bosses are desperate for you to elect Obama. The New York Times editorial page can't wait for you to elect Obama. … They want to do everything they can to discredit Governor Palin and to help elect Senator Obama. Let's just be honest about what's going on here. The fix is in."[1305]

"An unjust man is an abomination to the just; and he who is upright in the way is abomination to the wicked." (Proverbs 29:27) …In case anyone was wondering why the Democrat-Left despised Bush and Palin; and why they adore Obama. Baby-mutilators do love their own.

And then there are the comics…

…"When you use comedy for malicious reasons, you become a propagandist. Tina Fey is a propagandist."[1306]

With all due respect, Tina Fey is an idiot. She disguises as *comedy* her pathological obsession to demean, disparage, vilify and marginalize Sarah Palin—a decent, honest, patriotic woman—merely because Palin is not an ignorant, uneducated, left-wing-indoctrinated, baby-mutilating psychotic like her; and conversely she applauds all things Democrat-Left including their propensity to torture little human beings to death. That is beyond the mere stupid that she falsely accuses Sarah Palin of being. It is pathological. She needs professional help. (I would recommend Dr. Jesus, Ms. Fey. There's nothing like the truth to fumigate the deceit of the evil one from one's mind, and the blood of millions of innocent little babies from ones soul. Call Him. He's still available. Before you die.)

And then we have the following, from the category of Most Repulsive…

Bill Maher called Michelle Obama a cunt, and can anyone figure out why he's still on the air? OH!! That's right! He didn't call Michelle Obama a cunt, or he *would* be off the air. He called Sarah Palin one. And David Letterman said that Michelle Obama's little girl got "knocked-up" by Alex Rodriguez during the 7th inning stretch of a Yankees baseball game, and can anyone figure out why *he's* still on the air? OH!! That's right! He didn't say that about Michelle Obama's daughter, or he *would* be off the air. He said that about Sarah Palin. And Michelle

---

1305   (Newt Gingrich, *On the Record* with Greta Van Susteren, October 22, 2008)

1306   (Phillip Perea, Dallas, Texas, O'Reilly Factor email, 11/17/2010)

Obama herself, such a "great defender of civil discourse" (along with her two-faced husband), actually went on Letterman's show after that and yukked it up with the pig. If he *had* said it about her daughter she wouldn't have gone on his show in a million years. The Democrat-Left's hypocrisy is quite remarkable, from Obama and our First Lady, right down the line. How does that expression go, that the fish always rots starting at the head?

"Only with your eyes shall you behold and see the reward of the wicked." (Psalms 91:8)

And then there are the following quotes: (If Sarah Palin had been responsible for any of them our late-night and Comedy Central clowns would still be laughing. But they can't laugh at president Obama because then they would be laughing at themselves.)

"Let me be absolutely clear. Israel is a strong friend of Israel's." – Barack Obama

"I've now been in 57 states. I think one left to go." – Barack Obama [There *are* 57 states in Islam.]

"On this Memorial Day, as our nation honors its unbroken line of fallen heroes, and I see many of them in the audience here today." - Barack Obama

"What they'll say is, 'Well it costs too much money,' but you know what? It would cost, about. It, it, it would cost about the same as what we would spend. It. Over the course of 10 years it would cost what it would costs us. (Nervous laugh) All right. Okay. We're going to. It. It would cost us about the same as it would cost for about, hold on one second. I can't hear myself. But I'm glad you're fired up, though. I'm glad." - Barack Obama [It's hard keeping all those stories straight, as every prevaricator knows.]

"The reforms we seek would bring greater competition, choice, savings and inefficiencies to our health care system." Barack Obama – ["Out of the mouths of babes and children…"]

"I bowled a 129. It's like - it was like the Special Olympics, or something." - Barack Obama [If Bush had said something so callous and insensitive, they would have impeached him.]

"Everybody knows that it makes no sense that you send a kid to the emergency room for a treatable illness like asthma, they end up taking up a hospital bed, it costs, when, if you, they just gave, you gave them treatment early, and they got some treatment, and a, a breathalyzer, or inhalator, not a breathalyzer. I haven't had much sleep in the last 48 hours." - Barack Obama [Teleprompter malfunction.]

…"And now you know why he brings his teleprompter with him everywhere he goes…even when talking to a 6th grade class! How many of these quotes did you read about or hear from the mainstream media? Do you think the mainstream media might have given more coverage to these statements had they been made by Sarah Palin? The leader of the free world was elected because he reads well. The man is an idiot!"[1307] (An idiot that nevertheless is a genius to the Democrat-Left.)

But all of that doesn't matter, you see. Not to the left. All that matters is that he's a liberal, just like them. It doesn't matter how stupid, unqualified and inexperienced he actually is, and how idiotic and destructive his policies are. He's one of them and therefore he must be protected and defended at all costs.

In October of the 2008 presidential election The O'Reilly Factor reported on a study by the Project for Excellence in Journalism concerning media coverage. It

---

1307   (Source: internet email)

came up with these figures: overall media coverage for McCain was 57% negative, but for Obama was only 29% negative. Almost 2 to 1. Newspaper coverage was 69% negative for McCain, while only 28% negative for Obama. NBC news came in at 54% negative for McCain and 21% negative for Obama. MSNBC, that private joke featuring an anchor that has a problem with thrills running up his leg, was an absurd 73% negative for McCain, and only 14% negative for Mr. Ed. And Fox News itself, falsely disparaged by the Democrat-Left for being "in the tank" for the Republicans, came in at 40% negative for both candidates. (But that's not fair and balanced. That's "in the tank" for the Republicans.) (If you're an idiot.) (Or a main-stream journalist.) And to top it off, six days before America went to the polls, the LA Times refused to release a tape that showed one Barack Hussein Obama hugging and yucking it up with a known PLO terrorist official.[1308] (But he's not a Muslim. He was born a Muslim; to a Muslim father, and raised as a Muslim, and has never, ever renounced his, and I quote, "Muslim faith." But he's not a Muslim. Don't be silly! He can't be! Because then America would be the dumbest nation to *ever* walk the face of the earth.)

And then there's this from the comic strip Mallard Fillmore:
"Mallards 2009 New Year's prediction, the Secret Service will have a new challenge… keeping the media from smooching President Obama's rear. … "NBC's Lee Conan is in mid-pucker. Let's move!""

Jay Rockefeller, liberal Democrat Senator from West Virginia, had this to say:
"I hunger for quality news.[LIE] I'm tired of the right and the left.[LIE] [Bulls**t, he's only tired of the right.] There's a little bug inside of me which wants to get the FCC to say to Fox and to MSNBC "Out! Off! End! Goodbye!""[1309]

Translation…
"I hunger for Nazi propaganda. I'm tired of my liberal philosophy being openly challenged. There's a big fascist bastard inside of me that wants to get the government to censure Fox News and the squirrel network [Rockefeller just threw in MSNBC to make him look fair. Did he fool you?] And to say to free speech "Out! Off! End! Goodbye!""
…It's so important to hear what they're really saying.
Rockefeller "is the typical rich American supporting liberal causes with money he never earned."[1310]
And not to jump on the old stump-toothed, mountain-man-bashing bandwagon but repeatedly electing and reelecting an ass like Rockefeller to the United States Senate is a much more insulting joke on West Virginia than inbreeding could ever be.

Wow, Fox News! Now there's a novel idea! … "Hey, I've got it, let's just not lie to them anymore! We can call it... NEWS!" Get it? The lame-stream media sure doesn't.
"Look, the great sin of Fox News is that it broke the monopoly of the liberal media. That's the reason why it's so wildly successful. It was once said years ago that the genius

---

1308   (The O'Reilly Factor October 30, 2008)
1309   (Senator Jay Rockefeller, Democrat, West Virginia)
1310   (From blog post on politico.com 11/17/2010)

of Roger Ailes and Rupert Murdoch is to have discovered a niche American broadcasting audience in news, namely half of the American people. And the other consequence is that it angers the Obama administration which is used to, particularly after last year, wall-to-wall adulation. I mean, this is almost comical if you look at the lineup. On the one hand, in the *tank* are ABC, NBC, CBS, NPR, PBS, CNN and MSNBC. Some of these like MSNBC are so *in the tank* they need scuba gear. Some of them occasionally emerge for a breath of air, but only occasionally. And Fox stands up and refuses to bend a knee, and that's what they can't stand. Look, CNN is patted on the head by the Obama administration as *objective*. CNN is an organization that a few weeks ago did fact checking [on] a Saturday Night Live skit that was mildly critical of Obama, but did *no* fact checking on wildly, grotesquely-libelous racist statements *allegedly* made by Rush Limbaugh which were *not* made by Rush Limbaugh. It gives you an idea of the difference in how they treat things, and that's not a matter of sloppiness, that's a matter of ideology."[1311]

"The way of the wicked is an abomination unto the Lord; but he loves those who follow after righteousness." (Proverbs 15:9)

O'Reilly: "The Conservative Media Research Center watched the three nightly network newscasts from September 1 to October 25 [2010] to see how they were handling the political situation in America. Well here's the headline: only conservatives and tea party people were labeled extremists by the networks. No liberal person was given that description. Not one. ...It was so blatantly obvious. ...That's a pretty long period of time, September 1 to October 25, and George Soros and everybody else—nobody is an extremist on the left—but all these people are extremists on the right.

Bernie Goldberg: "Yeah, but that's pretty easy to explain, because to many liberal journalists, they don't think there's any such thing as an extremist on the left. Barney Frank isn't an extremist on the left.[LIE] Dennis Kucinich isn't extreme.[LIE] Alan Grayson, the most embarrassing member of Congress, who thinks that Republicans want you to die quickly, isn't extreme.[LIE] But a conservative politician, especially if he or she is a member of the tea party, is extreme.[LIE] And that's because, as far as liberals in the media are concerned, every politician to the right of center is conservative. And that's true. But any politician to the left—no matter how far to the left—is middle-of-the-road; is mainstream. [LIE] Because they're observing this campaign sitting in the left-field bleachers."[1312]

This is what Keith Olbermann thought (and I use that term lightly) of Scott Brown when he shocked the Northeast by winning Ted Kennedy's vacated Senate Seat in that reliably Democrat stronghold, The People's Republic of Massachusetts:

"In short in Scott Brown we have an irresponsible,[LIE] homophobic,[LIE] racist,[LIE] reactionary,[LIE] ex-nude-model, "tea bagging"[LIE] [a not-so-subtle reference to what Keith and Chris do to each other when no one is looking] supporter of violence against woman[LIE] and against politicians[LIE] with whom he disagrees."[1313]

1311   (Charles Krauthammer on Special Report with Brett Baier, 10/19/2009, speaking on the Hussein "ACORN" Obama administrations Hugo Chavez-like attacks on Fox News starting in October, 2009. ...Chris Wallace nailed it a few weeks earlier when he said that he has never seen such a bunch of crybabies in the White House.)

1312   (Bernie Goldberger on Fox News' The O'Reilly Factor, 11/01/2010, the night before the midterm elections)

1313   (Keith Olberman, 01/18/2010, on The Cartoon Network, cleverly disguised as a cable

(...Translation, "He's not on the Democrat plantation, like me and my tea-bagging buddy, Chris.")

And here's another member in good standing of what Fox News' Tony Blankley calls "The Mainstream Moron Media." ... A female reporter in a radio interview with a General Cosgrove was giving him a hard time because he actually planned to sponsor a Boy Scout Troop visit to his military headquarters in which they would be taught "climbing, canoeing, archery, and [God forbid] shooting." But when she foolishly accused him of "equipping them to become violent killers,"(LIE) he responded: "Well, Ma'am, you're equipped to be a prostitute, but you're not one, are you?" "The radio went momentarily silent and the interview ended."[1314] There is some satisfaction to be had from a left-wing-indoctrinated journalist getting her little nose succinctly rubbed in her own do-do. Thank you so much, General Cosgrove. We are eternally in your debt, for both your service and your wit.

On the Arizona Immigration Issue the Media Research Center "did a review of the morning and evening newscasts of ABC, NBC and CBS" and found that out of 120 stories, 77 were anti-Arizona law, 35 were neutral and only a paltry 8 were pro-Arizona law...

> "You've got two problems going on here. Number one, it's the quantity. In a four month period, 128 broadcast news stories at nighttime. That's an avalanche. That's more than propaganda. [Could it be, I dare say it, *Nazi* propaganda?] You're talking about brainwashing, when by a 10 to 1 margin they attack Arizona. ...Here's the reality of the situation: the American people by 10 to 1 want border security. So the media are doing the exact opposite of what the American people want."[1315]

I can't believe it. Who hired these people, the Russians?

The media also essentially blacked-out the climategate scandal; those devastating emails that exposed the global warming fraud for what it is (but that's another part of all the news that is <u>not</u> fit to print because it might confuse those marching orders that have been planted in the heads of Democrat voters):

> "In the last 12 months, ever since this massive story came out the media have made a deliberate decision not to cover it. They've covered it 12 times all year; that's once a month for the big three networks; that's no coverage—virtually no coverage—whatsoever. When they do cover it, it's to dismiss the scandal that exists ...a scandal that blew the entire environmental movement to shreds because it proved that climate [change] was a political operation to control world economies."[1316]

Let's see here; on the one hand we have "128 broadcast news stories at nighttime in a four month period" concerning the Arizona Immigration Law, almost all negative. And on the other we have only 12 news stories in an entire year concerning Climategate, the Watergate-style exposition of the wholly

---

news outlet, MSNBC)

1314   (As reported by forums.military.com)
1315   (Brent Bozell, president, Media Research Center, on Fox News *Hannity*, 7/30/2010)
1316   (Brent Bozell on Fox News' Hannity, 11/19/2010)

corrupt global warming scam. ...And they still call it "news" ...and themselves "journalists."

And my, how the Democrat media loves to protect the racists they call their own:

> "The late Senator Robert Byrd: here's a guy who was in the KKK; he was recruiting people to join; 150 people he convinced to join the KKK. Now people like Bill Clinton and Barack Obama are coming out of the woodwork to defend this guy because he was "just trying to get elected." Wait a minute, it's okay to stand with people who are lynching people; convincing people to go burn a cross in some dudes yard because of the color of their skin, because you're just trying to get elected? And yet conservatives are now being considered racist because they support Arizona's immigration policy. This guy gets a pass for the KKK, because he's trying to get elected. Byrd also filibustered against the Civil Rights Act when he was 47 years old—this isn't a youthful indiscretion—somehow getting a pass, but conservatives can't even say something nice about Strom Thurmond or they get disciplined, just like Trent Lott did. He got thrown out."[1317]

Without double standards, the mainstream media would have no standards at all!"[1318]

> "Look at how the New York Times portrays Thurmond and Byrd. ... Here it is, the headline, New York Times [June 27, 2003], "Strom Thurmond, foe of integration, dies at 100." Same paper, New York Times, on the death of KKK member and filibuster of the civil rights movement, Robert Byrd: "Robert C. Byrd, pillar of the Senate, dies at 92." [June 28, 2010] ...Both obituaries were written by the same person, Adam Clymer."[1319]

And Clymer calls himself a journalist! Not a liar, mind you, nor a propagandist, nor a Democrat shill, but a *journalist*. Wow! That's like calling a rapist a sex-education instructor.

The Democrat-Left's dishonest smear of Clarence Thomas, led by the now-departed Theodore Kennedy, the "conscience of the Senate" (and equally acclaimed lifeguard), is truly one of the most ignorant, disgusting and shameful parts of the history of this country. And that the vast majority of blacks were somehow suckered into quietly and obediently going along with it, is equally as disturbing. Virginia Thomas, Clarence's wife, was right today (October 19, 2010) in demanding an apology from Anita Hill, the disingenuous woman whom the Democrat's marched out to falsely slander Mr. Thomas with baseless charges of sexual harassment in a not-so-subtle attempt to derail his Supreme Court confirmation in the fall of 1991. But the ones who *should* be apologizing are the Democrat-Left and their media, and not their convenient pawn, Ms. Hill.

The media coverage of that fiasco, as one-sided as it was though, is still overshadowed by the spectacle of the Senate's "conscience" (who was quite the impressive female-harasser in his day) presiding over it. If any Republican or Christian with such a womanizing past (including that covered-up drowning fatality) had tried to preside over a hearing where one of the Democrats far-left, constitution-perverting appointees like Ruth Bader Ginsburg or Sonia Sotomayor were accused of some past sexual indiscretions, the media would have had a

---

1317 (Glenn Beck, Glenn Beck Show on the Fox News Channel, 7/9/2010)
1318 (Brent Bozell, founder and president of the Media Research Center)
1319 (Glenn Beck, Glenn Beck Show on the Fox News Channel, 7/9/2010)

collective hysterical fit that would have rivaled the Muslims worldwide rioting over the Danish cartoons of Mohammed. The left would have screamed their demands 24/7 and from every nook and cranny of their print, TV and radio media (and with the aid of the selective derision from their late-night clowns) for that conservative "conscience" to be removed. And they would have continued their onslaught every day until he was, while at the same time explaining over and over again that the nominees past was "irrelevant" today, and how "character doesn't count." But there wasn't a peep over the sheer hypocrisy of having Ted Kennedy preside over that "high-tech lynching."

It is also a sad and dangerous thing to have in our country a lying, Democrat media that is so persuasive and influential that it can—with the help of its comrades in the entertainment industry, throughout our educational system and in the pulpits of black churches—turn a decent, honest, self-made, intelligent, moral man who should be a highly admired hero to every single black American in this country, and especially to every single black youth, instead into a pariah who is falsely demeaned, disparaged marginalized and derided (like every other decent, moral, non-baby-mutilating Christian and conservative) all because he had the balls to throw down his spade and walk off the white liberal Democrat plantation. (Actually, in Thomas' case, thanks to his father's strict, excellent parenting, he never picked up the spade.) One of the byproducts of the civil rights struggle was supposed to be that we would never have to publicly hear another black man called a n***er. And thankfully this has all but held true, except for the case of Clarence Thomas (and to a lesser extent Condoleezza Rice) where he has been repeatedly called nothing less than that, even in those countless cases where his white, liberal, racist detractors in the media and the Democratic Party didn't have the guts to actually use the term. And the sad reality is that we have had to sit back and watch the African-American community go along with it, and for all these years, just because Clarence Thomas refused to support the politics of race-baiters and baby-mutilators, like 90% of them still unfortunately do. WAKE UP! my sadly politically misinformed brothers and sisters!

"It really is striking how the mainstream media have become guard dogs for the [Obama] administration, rather than the guard dogs for the people."[1320]
I would say it really is typical, and to be expected.

In April of 2010 president Obama, our "cry-baby president" as Sean Hannity has accurately labeled him, whined about the "venom" directed at him from his political opponents and their supposed responsibility for his diving poll numbers. (The real reason of course is the nation can't take anymore of him diving our nation into the toilet with all of his job-killing, economic-stifling, socialist policies.) But Karl Rove points out how he conveniently ignores the real venom from the Left, including from his own Democrat National Committee, directed at the Right. These are just a sampling:
"Vile 2-bit wing nuts." "Narrow-minded nut jobs." "Slimy thugs." "Gas bags." "Hate spewing venom." "They should slink back into whatever century they crawled out from."[1321]

1320 (Ann Coulter on Fox News' *Hannity*, 9/02/2009)
1321 (Karl Rove on Fox News *The O'Reilly Factor*, 4/2/2010)

But he probably never heard any of those spewing's from the left, because his Democrat media didn't consider them newsworthy. He also had no problem with his media, and his party's henchmen, slandering all of those decent Americans who made up the Tea Party who didn't think it was such a good idea for left-wing-indoctrinated socialist baboons to be over-spending our country into bankruptcy, and over-regulating our economy into oblivion...

> "It is its own special kind of hate to smear hundreds of thousands of Americans with charges of racism and hate and violence [when] the Left and the Democrats in the media are mired in it far more than Tea Partiers will ever be mired in actual racism. ... The point is to make the Right look bad: it's not about reporting. ...Democrats should have some shame and shut up every now and then about how every conservative is evil and hateful and violent."[1322]

Yes, they should, and that'll happen right after Wilbur has enough shame to stop mutilating little babies to death, and the intelligence and education to stop presiding over the party of Margaret Sanger and her Planned Parenthood which is perpetrating the genocide of his own race. Vegas has slightly better odds on hell freezing over.

* * *

This section documents only a tiny fraction of the volume of examples of lame-stream media bias in favor of their Democrats. Indeed a 50,000 page exposé (only a few pages longer than their healthcare bill) would still fall way short. But to any fair observer the evidence is thorough, documented, conclusive and overwhelming: we have a national, lame-stream, liberal, biased, corrupt, dead, Democrat-Left, moronic, state-run, "Obama-mania," Nazi-propaganda media whose minds are poisoned by the lies of the great deceiver, and who are driven to advance his leftist, anti-American agenda by manufacturing, as opposed to actually reporting, the news. Those underserved both within that media and throughout the rest of the Democrat-Left, who refuse to see this are no different than the holocaust deniers that they falsely accuse their man-made-global-warming critics of being. And unfortunately, around half of the voters in this country are still not aware of this. And because of this that same media will persevere along its current path without so much as a hiccup, or even a momentary flash of introspection, self-awareness or honesty. Because it is far more important to keep as many Americans as possible in the dark about what they, and their Democrat party, are up to. And Satan, the diabolical genius behind their mission, knows that with their all-important cover he can continue virtually unchallenged disseminating his disinformation and lies through our educational system and entertainment industry. And that in turn will allow him to further orchestrate the destruction (aka "radical transformation") of our once-great country. Again, as those few honest commentators, politicians and journalists keep asking over and over again concerning some of the more obvious, downright stupid and laughable lies of the Democrats and their media, "Do they really think the American people are that stupid?!" No! They think their *voters* are that stupid (politically and spiritually), and they will do everything to keep them that way, including willingly sacrificing ratings and profit. They, after all, are on

---

1322   (Commentator Mary Catherine Ham on The O'Reilly Factor, 3/29/2010)

a mission from "god," to "save" the world. That is their religion. And they will remain loyal to it with the same unswerving devotion as an Islamic lunatic, or a Moonie, all the way to the end. (If only they could be kidnapped, taken to a hotel room, and deprogrammed.)

See:
- *Bias: A CBS Insider Exposes How the Media Distort the News*, by Bernard Goldberg
- *Arrogance: Rescuing America from the Media Elite*, by Bernard Goldberg
- *Coloring the News: How Political Correctness Has Corrupted American Journalism*, by William McGowan
- *A Slobbering Love Affair: The True (And Pathetic) Story of the Torrid Romance Between Barack Obama and the Mainstream Media*, by Bernard Goldberg
- *Media Malpractice: How Obama got Elected and Palin was Targeted*, a film by John Ziegler, go to YouTube or howobamagotlected.com
- *Slander: Liberal Lies About the American Right*, by Ann Coulter
- Brent Bozell's The Media Research Center
- David Horowitz's Freedom Center
- *Gray Lady Down: What the Decline and Fall of the New York Times Means for America*, by William McGowan
- "Behind the Bias: The History of Liberal Media," a Sean Hannity Fox News Special, 04/22/2011

Addendum

The Vessels of Wrath

"What if God, in order to show his wrath, and to make his power known, endured with much longsuffering the vessels of wrath prepared for destruction; And that he might make known the riches of his glory on the vessels of mercy, which he had before prepared for glory, even us, whom he has called." (Romans 9:22-24)

Coming to the knowledge of the truth, accepting Christ as Lord and Savior, and having your mind freed of all of the lies of the deceptive one, is the greatest gift that you can receive in this life. But it does have a downside. For we live in a fallen and very wicked world, and whereas a certain amount of ignorance *can* be bliss, conversely it can also be very frustrating at times knowing the truth and being surrounded by the lies of the evil one, and having those lies constantly spewed at you every time you pick up a newspaper or turn on your TV. It's hard not to shout at the set, "IT'S A LIE!" (As in Gene Wilder's "IT'S ALIVE!" from the movie *Young Frankenstein*.) And that's where the knowledge in the above verse from Romans can be of value. For what God is saying is that there are only two kinds of people (vessels) on this earth, the lost and the saved. The "vessels of wrath prepared for destruction" and the "vessels of mercy ...prepared for glory." And when that truth takes hold of you then those feelings of anger and frustration give way to a great thankfulness on the one hand, for the incredible, undeserved gift that we have been given. And on the other hand to feelings of pity and compassion for those who are still lost in Satan's deceptions, because of the horrible fate that they are about to suffer. As well as a strong desire to reach as many of those lost as possible, to turn as many as possible from the terrible eternal fate of being "cast into hell, into the fire that shall never be quenched." (Mark 9:45)

"They are wells without water, clouds that are carried with a tempest; to whom the mist of darkness is reserved for ever." (2 Peter 2:17)

But what should we make of God saying in those verses from Romans that he indeed <u>prepared</u> these people for that eternal horror, and conversely that He chose to "make known the riches of his glory on the vessels of mercy, even us, who he had before <u>prepared</u> for glory." And if that is the case, then where is the free will? What choice do they have? Isn't their blindness of no fault of their own? And what choice did we have to benefit from His grace? The answer lies in the verses of Romans 9:13-21, that precede 22-24, where, incredibly, God begins the discussion by saying that...

"As it is written, Jacob have I loved, but Esau have I hated." (Romans 9:13)

What shot did Esau have then? If God hated him? The answer is forthcoming in the very next verse...

"What shall we say then? Is there unrighteousness with God? God forbid." (Romans 9:14)

And if there is no unrighteousness with God, and there isn't, then there is also no *injustice* with God either. And for there to be justice in the fate of the vessels of wrath then they must also be free to choose or to reject Jesus's call to repentance. And they of course are, but that choice also exists simultaneously with the absolute sovereignty and will and foreknowledge and yes even the choice of God which is coupled with each vessels choice (as hard as that might be for our limited minds to comprehend).

"For he says to Moses, I will have mercy on who I will have mercy, and I will have compassion on who I will have compassion. So then it is not dependent on the will of man, or on his effort, but on God who shows mercy. For the scripture says unto Pharaoh, Even for this same purpose have I raised you up, that I might show my power in you, and that my name might be proclaimed throughout all the earth." (Romans 9:15-17)

As God hardened Pharaohs heart in order that he might show his glory to the children of Israel and throughout all of Egypt, so he now hardens the hearts and the minds of those vessels of wrath fitted for destruction. This is not to say that these vessels do not still exercise their free will in choosing the lies of Satan, but God hardens their hearts and their minds, as he did to Pharaoh, so they will be *free* to exercise that choice until they die in their sins. His hardening therefore is only an accommodation of their own choice. Paul goes on...

"Therefore he has mercy on whom he wants to have mercy, and whom he wants to he hardens. Will you say then unto me, Why does he still find fault with us? For who has resisted his will? No, O man, but who are you to talk back to God? Shall the thing formed say to him who formed it, Why have you made me thus? Has not the potter power over the clay, of the same lump to make one vessel unto honor, and another unto dishonor?" (Romans 9:18-21)

Fully reconciling the sovereignty of God with the free will of man ultimately is something that requires an intellect a number of pay grades above that of any of us. Suffice it to say, like God himself declares, that He is perfectly just and that

Danny Erhardt

his absolute sovereignty also allows for the completely free will of man, so we are both fully culpable in our decisions and God is totally sovereign in His will at the same time. The clearest way to understand the apparent conflict is to realize that we are all sinners, that we are all born vessels of wrath, that we are all bound for an eternity in that Lake of Fire just like every single fallen angel is. But that in our case, the case of humanity, God *chose* to redeem us—some of us ("For many are called, but few are chosen" Matthew 22:14)—to turn us from that path of destruction, to shed his grace upon us so that "there but for the grace of God go you and I." "For by grace are you saved through faith; and that not of yourselves: it is the gift of God." (Ephesians 2:8) So the vessels of wrath are there because of their choosing, and because of God's choosing. And we are where we are because of the exercise of our free will, and because of God's grace.

> "The truth is that God does not violate our wills by choosing us and redeeming us. Rather, He changes our hearts so that our wills choose Him. "We love Him because He first loved us" (1John 4:19), and "You did not choose me, but I chose you" (John 15:16)."[1323]

And in the same way "God does not violate" the will of others by *not* choosing them and by not redeeming them, even though at the same time he is "not willing that any should perish, but that all should come to repentance." (2 Peter 3:9) Again, from our feeble viewpoint an apparent contradiction. But from God's perspective, perfect justice and His sovereign will exist without in any way hindering the free will of man.

Also keep again in mind that in the end we will all realize this, including any vessel of wrath that would now in this life attempt to use any apparent contradiction in God's word because it is too difficult for us to fully comprehend, as a golden opportunity to mock both Him and it. For in the end "every knee shall bow to [Jesus], and every tongue shall confess to God." (Romans 14:11) "At the name of Jesus every knee should bow, of things in heaven, and things in earth, and things under the earth; and every tongue should confess that Jesus Christ is Lord, to the glory of God the Father." (Philippians 2:10-11) So, in the end, every vessel of wrath and every mocker will know and cry out that God is both sovereign and completely just, even though they will also know that they are about to be thrown into the eternal fires and horror of hell. For then their realization will be too late. "Behold, now is the accepted time [for repentance]; behold, now is the day of salvation." (2 Corinthians 6:2) Which is why it is such a good idea for everyone, no matter their background or indoctrination, to seek the truth now while there is still a little time.

"They are spots in your love feasts, when they feast with you, feeding themselves without fear: clouds they are without water, carried about of winds; trees whose fruit withers, without fruit, twice dead, plucked up by the roots; raging waves of the sea, foaming out their own shame; wandering stars, to whom is reserved the blackness of darkness for ever." (Jude 1:12-13)

And with all of this in mind, those feelings of anger, frustration, consternation and/or hatred for the vessels of wrath, for their reprobate minds constant, confident yet illogical spewing's of untruths, and their rote, irrational defending of the indefensible, and for their blind allegiance to the dark side can be overcome by pity, compassion and prayer for them, knowing where they are heading, as well as

---

1323 (From "How does God's sovereignty work together with free will?" @ gotquestions.org)

766

prompting any "vessel of mercy" to do whatever he or she can to bring them from the devil's sleight of mind to the knowledge of the truth, and to their humbling under the righteousness of Christ, and acceptance of their Creator as their Lord and Savior.

"I have become all things to all men, that by all means I might save some." (1 Corinthians 9:22)

And speaking of Paul, he killed Christians. And yet God turned even him from darkness to the light and to a follower of Jesus Christ. (Of course it doesn't hurt in one's conversion process to be knocked off your horse by a voice as loud as thunder and blinded by a light brighter than the noonday sun. For the rest of us, it might take a little more faith. "Jesus said unto him, Thomas, because you have seen me, you have believed: blessed are those who have not seen, and yet have believed." John 20:29) Could there be any greater candidate for being a vessel of wrath than someone who persecuted and even killed followers of God and His Christ? Of course not. Yet Paul turned out to be the very vessel of mercy who wrote (by the inspiration and guidance of the Holy Spirit) the Book of Romans that we are discussing here.

"The wrath of God is revealed from heaven against all ungodliness and unrighteousness of men, who hold the truth in unrighteousness. Because that which may be known of God is manifest in them; for God has shown it to them. For the invisible things of Him from the creation of the world are clearly seen, being understood by the things that are made, even his eternal power and Godhead; so that they are without excuse." (Romans 1:18-20)

So, the next time you're watching TV and are subjected to the preposterous claims of, and to the facts getting twisted by, Democrat journalists, apologists and their politicians in order to prop up their false beliefs and support their senseless liberal positions; folks like James Carvel, Paul Begala, Tina Fey or Nancy Pelosi, Jon Stewart, Rosie O'Donnell, Barack Obama, Keith Olberman and Bill Maher, Joy Behar, Bill Moyers, David Letterman, Jimmy Carter or George Soros, Janeane Garofalo, Al Franken, Eric Holder, Anthony Weiner, Harry Reid and Chuck Schumer, Ed Asner, Debbie Wasserman Schultz, the occupiers of Wall Street, Algore, Barbara Boxer, Barney Frank, Kathleen Sebelius, Ward Churchill, Wisconsin union protesters, Jeremiah Wright, Bill and Hillary Clinton, Patricia Ireland, Chris Matthews, etc.[1324] ...remember to suffer these fools gently, and remember where they are heading, and that the only reason we are headed in a far better direction is because of the grace of God. Pray for them, knowing the final destination that the path they now choose is taking them to. And warn them by speaking the truth, in love, whenever you get the chance.

"He has delivered us from the power of darkness, and has translated us into the kingdom of his dear Son, in whom we have redemption through his blood, even the forgiveness of sins." (Colossians 1:13-14)

We have everlasting life to look forward to, in paradise, in the restored Garden of Eden here on this planet. The vessels of wrath have nothing to look forward to except for an eternity in hell, and they don't even know it. Pray for their translation.

1324   (If we overlooked anyone, please accept our apologies)

See:
- *Demonic: How the Liberal Mob Is Endangering America*, by Ann Coulter

* * *

"The Lord reigns; let the earth be glad; let the multitude of islands rejoice." (Psalms 97:1)

What Satan is up to on this planet—his obsession with drawing as many individuals down to an eternity in hell; his goal to destroy Israel; and his quest to overthrow America—is a well-kept secret from the great majority of people living today. So is the fact that his lies—false religions, false ideologies and false scientific theories—cover this earth like a shroud and, because people believe what they are taught and the truth is not high on the list of things imparted, most folks are trapped in the web of his deceit. He is the great deceiver and the father of lies who goes forth to deceive the whole world. His plans therefore are in great need of being exposed to the folks so they can "know their enemy" first and foremost in order to "not be ignorant of his many devices." And in order to be made aware of their pressing need of coming from his darkness to the Creator's Light; from his lies to God's perfect truth; and for the true righteousness of the Messiah to cover their sins. Jesus Christ is also their only hope of freeing their minds, and hearts, from the power of Satan's deception. ("Behold, I stand at the door, and knock: if any man hear my voice, and open the door, I will come in to him, and will dine with him, and he with me." (Revelation 3:20) And when Christ comes in, the devil and his lies go out.) For the devil is far too powerful and cunning (he is the 2nd most powerful being in the universe after all), and his lies have far too strong a grip, for one to set themselves free on their own. Satan is the father of lies but Jesus Christ, who is the most powerful being in the universe and who is not only the Creator of this universe but the Creator of Lucifer as well, is all truth. Against Him, Satan doesn't stand a chance.

"I am the way, the truth, and the life.
No man comes to the Father, but by me."
(John 14:6)

"I am the door.
By me if any man enter in, he shall be saved,
and shall go in and out, and find pasture."
(John 10:9)

See:
- *Spiritual Warfare*, by Dr. David Jeremiah
- "Why Israel is the Victim and The Arabs are the Indefensible Aggressors in the Middle East," by David Horowitz
- "Big Lies: Demolishing the Myths of the Propaganda War Against Israel," by David Meir - Levi
- "Jihad in America! Militant Islam: The Religious War Against America," by Martin Mawyer
- "What Americans Need To Know About Jihad," by Robert Spencer

- *Unholy Alliance: Radical Islam and the American Left*, by David Horowitz
- *The Grand Jihad: How Islam and the Left Sabotage America*, by Andrew C McCarthy
- *Unshaken: A Story of Faith and Hope*, by ACLJ Films
- *The Politically Incorrect Guide to the Middle East*, by Martin Sieff.
- *One Nation Under God: Ten Things Every Christian Should Know About the Founding of America*, by Dr. David C. Gibbs, Jr., President *Christian Law Association*, with Jerry Newcombe
- *Persecution: How Liberals are Waging War Against Christians*, by David Limbaugh
- *Indefensible: 10 Ways the ACLU is Destroying America*, by Sam Kastensmidt
- *The ACLU VS America: Exposing the Agenda to Redefine Moral Values*, by Alan Sears and Craig Osten
- *Under God*, by Toby Mac and Michael Tait
- *The 10 Big Lies About America: Combating Destructive Distortions About Our Nation*, by Michael Medved
- *48 Liberal Lies about American History (That You Probably Learned in School)*, by Larry Schweikart
- *Seven Events That Made America America: And Proved That the Founding Fathers Were Right All Along*, by Larry Schweikart
- *A Patriot's History of the United States: From Columbus's Great Discovery to the War on Terror*, by Larry Schweikart
- *New Deal or Raw Deal? How FDR's Economic Legacy Has Damaged America*, by Robert P. Murphy
- *The Politically Incorrect Guide to the Great Depression and the New Deal*, by Robert Murphy
- *The Walls Came Tumbling Down: A Christian Perspective on the Fall of Communism in our Time*, by Peter J. Leithart & George Grant
- *There are Moral Absolutes*, Michael A. Robinson
- *Indoctrination U.: The Left's War Against Academic Freedom*, by David Horowitz
- *The Professors: The 101 Most Dangerous Academics in America*, by David Horowitz
- *The New Thought police: Inside the Left's Assault on Free Speech and Free Minds*, by Tammy Bruce
- *The Church of Liberalism: Godless*, by Ann Coulter
- *110 People Who are Screwing up America*, by Bernard Goldberg
- "The Political Assault On America's Universities," by David Horowitz
- *The Death of Right and Wrong: Exposing the Left's Assault on Our Culture and Values*, by Tammy Bruce
- *Hollywood VS. America*, by Michael Medved
- *Slouching Towards Gomorrah: Modern Liberalism and American Decline*, by Robert H. Bork
- *Culture Warrior*, by Bill O'Reilly
- *Men in Black: How the Supreme Court is Destroying America*, by Mark R. Levin
- *Judicial Tyranny: the new kings of America?* By Mark I. Sutherland
- *If Democrats had any Brains, They'd be Republicans*, by Ann Coulter
- *Uncle Sam's Plantation: How Big Government Enslaves America's Poor and*

What You Can Do About It, by Star Parker
- *Bamboozled: How Americans are being Exploited by the Lies of the Liberal Agenda*, by Angela McGlowan
- *Culture of Corruption: Obama and His Team of Tax Cheats, Crooks, and Cronies*, by Michelle Malkin
- *Radical-in-Chief: Barack Obama and the Untold Story of American Socialism*, by Stanley Kurtz
- *Liberal Fascism: The Secret History of the American Left from Mussolini to the Politics of Meaning*, by Jonah Goldberg
- *Liberty and Tyranny: A Conservative Manifesto*, by Mark R. Levin
- *Real Change: From the World That Fails to the World That Works*, by Newt Gingrich
- *Showdown: Confronting Bias, Lies, and the Special Interests that Divide America*, by Larry Elder
- *White Guilt: How Blacks and Whites Together Destroyed the Promise of the Civil Rights Era*, by Shelby Steele
- *Hating Whitey and other Progressive Causes*, by David Horowitz
- *Stupid Black Men: How to Play the Race Card—and Lose*, by Larry Elder
- *Crimes Against Liberty: An Indictment of President Barack Obama*, by David Limbaugh
- *Fed Up!: Our Fight to Save America from Washington*, by Rick Perry and Newt Gingrich
- *The Politically Incorrect Guide to Islam (and The Crusades)*, also by Robert Spencer
- *They Must Be Stopped: Why We Must Defeat Radical Islam and How We Can Do It*, by Brigitte Gabriel
- *Militant Islam Reaches America*, by Daniel Pipes
- *Unveiling Islam*, by Ergun and Emir Caner
- *The Truth About Islam*, by Anees Zaka & Diane Coleman
- *Courting Disaster: How the CIA Kept America Safe and How Barack Obama Is Inviting the Next Attack*, by Marc A. Thiessen
- "Who's Responsible for America's Security Crisis?" by David Horowitz
- *Party of Defeat: How Democrats and Radicals Undermined America's War on Terror Before and After 9-11*, by David Horowitz & Ben Johnson
- *America Alone: The End of the World as We Know It*, by Mark Steyn
- *Power to the People*, by Laura Ingraham
- *Crazies to the Left of Me, Wimps to the Right: How One Side Lost Its Mind and the Other Lost Its Nerve*, by Bernard Goldberg
- *The Roots of Obama's Rage*, by Dinesh D'Souza
- *Slander: Liberal Lies about the American Right*, by Ann Coulter
- *Fleeced*, by Dick Morris and Eileen McGann
- *Catastrophe*, by Dick Morris and Eileen McGann)
- *Treason: Liberal Treachery from the Cold War to the War on Terrorism*, by Ann Coulter
- *Useful Idiots: How Liberals Got it Wrong in the Cold War and Still Blame America First*, by Mona Charen
- *America for Sale: Fighting the New World Order, Surviving a Global Depression, and Preserving USA Sovereignty*, by Jerome R. Corsi, PhD
- *The Road to Serfdom*, by Friedrich Von Hayek

- *The New Road to Serfdom: A Letter of Warning to America*, by Daniel Hannan
- *The Politically Incorrect Guide to Global Warming*, by Christopher C. Horner
- *Climategate: A Veteran Meteorologist Exposes the Global Warming Scam*, by Brian
Sussman
- *Red Hot Lies: How Global Warming Alarmists Use Threats, Fraud, and Deception to Keep You Misinformed*, by Christopher C. Horner
- *Economic Facts and Fallacies*, by Thomas Sowell
- *The Great Destroyer: Barack Obama's War on the Republic*, by David Limbaugh
- *Bias: A CBS Insider Exposes How the Media Distort the News*, by Bernard Goldberg
- *Arrogance: Rescuing America from the Media Elite*, by Bernard Goldberg
- *A Slobbering Love Affair*, by Bernard Goldberg
- *Media Malpractice*, a film by John Ziegler, go to YouTube or howobamagotlected.com
- *The New Leviathan*, by David Horowitz and Jacob Laskin

CHAPTER 14

FASTING AND PRAYER

"And he said unto them, This kind [of faith] [...the kind that could move mountains] can come forth by nothing, but by prayer and fasting." (Mark 9:29)

AND EATING AT THE TABLE

OF THE LIVING GOD

"And God said, Behold, I have given you every herb bearing seed, which is upon the face of all the earth, and every tree, in the which is the fruit of a tree yielding seed; to you it shall be for food. And to every beast of the earth, and to every fowl of the air, and to every thing that creeps upon the earth, wherein there is life, I have given every green herb for food: and it was so." (Genesis 1:29-30)

Before the fall of man, 6000 years ago in the Greeks Garden of the Hesperides and in the Bibles Garden of Eden, that was what we ate and it was all we desired. Before we were cast out. It was what our ancient father and mother ate. It was the perfect diet, the diet God created for us, for our nourishment and for our health, when he created us. (And one would think that he would have known what he was doing; and what was optimum for us to eat.) And it remains so to this day. It is a diet of food that is made only by the fire of life, sunlight, and not by the fire of death, the fire that burns outside of our bodies and destroys the flesh of the living, both animal and vegetable. It is the food that is *cooked*, prepared, by God and his holy angels of sunlight and of air and of water and of earth. As opposed to those foods that are prepared, cooked, by man.

"For I tell you truly, the God of the living is richer than all the rich of the earth, and his abundant table is richer than the richest table of feasting of all the rich upon the earth."[1325]

Now I know that this is a radical concept to most people and we're certainly not trying to tell anyone what they should or should not eat. (You yourself, and God himself, can do that.) But this is a book about the truth and we would be remiss if we did not include this subject, because our health, after all, is so fundamental to our lives. The old saying goes that we take our health for granted until we lose it, and only then do we realize just what we had, and just how important it is. So we thought, if you seek the truth which we would assume you do if you've gotten this far, then you might be interested in being exposed to this information. What you do with it after that is entirely up to you. For you can't get to heaven by eating your way there, or by fasting your way there, but you certainly can go a long way towards avoiding the hospital, as well as putting off for as long as possible that inevitable trip to the undertakers.

1325   (Page 41, *The Essene Gospel of Peace: Book One*, The Third Century Aramaic Manuscript and Old Slavonic Texts, Compared, Edited and Translated by Edmond Bordeaux Szekely)

The secret to vibrant health

Does that secret even exist? And if so where does it lie? Well it does exist and it lies in fasting, and in the eating of only (or at the very least, primarily) raw foods, unheated to the point of killing the enzymes or the life inside of it. The secret lies in the original diet of man. See Genesis 1:29 above. Fasting cleans your body and its hundreds of trillions of cells of disease-causing material and toxins that are a drag on each cells metabolic processes, as well as a clog and a hindrance to the blood stream, lymph system and organs of the body. Eating at the table of the living God (eating foods prepared only by Him) gives your body the necessary ingredients for vibrant health including, and most importantly, the enzymes which are the life of those foods. That diet also goes a long way toward doing what fasting does but on a daily and less intense scale, keeping your bodies tissues and its blood and lymph systems free of unwanted chemicals and material by not ingesting them in the first place, and by keeping each cells metabolism—which is so essential for the life and activity and performance of that cell and for the health of your overall body which is merely the composite of all of your cells—in optimum shape, free of constantly trying to clean itself of the harmful and clogging by-products of a less than optimum diet.

"The foods which you eat from the abundant table of God give strength and youth to your body, and you will never see disease."[1326]

There is a great deal of confusion and misinformation out there concerning the matter of diet. The truth remains a secret; unknown to most folks. And for the most part, like the truth about false religions, it is an unwanted secret. Because we are addicted to what we eat. Indeed our addiction to food so far surpasses the addiction to any other substance or drug that it is in a class all by itself. Do you think it's tough for a crack or heroin addict to get off those drugs, no matter how bad they desire to be free? Or for a drunk to go cold turkey? Well try not eating any cooked food from this moment forward and for the rest of your life. Eating only raw fruits, nuts and vegetables, without so much as cheating on even a tiny morsel of steak or pizza. Ever. Try to do that and you will laugh at crack. You will hold heroin in complete derision. For you'd sell your soul for a slice of moldy bread. In less than a week.

Or try fasting for 40 days and 40 nights without letting anything whatsoever pass your lips (or even enter into your blood stream intravenously) except for pure water. Compared to that, a drunk going cold turkey while working full time as a bartender would be a walk in the park. But let's not get lost in hyperbole. The point is that we are very heavily addicted to food, as anyone who has dealt with a weight problem is certainly aware. But unless you have tried to walk away from food entirely (which is what most of us would consider the equal to walking away from any and all cooked foods entirely) then you don't have any idea just how powerful our addiction to food actually is.

But nevertheless, and despite the fact that many people have a very powerful built-in resistance to hearing the truth about fasting and diet, we have decided to include it here because, for one, it is a secret of the universe that most folks are unaware of. And secondly, it is a secret that is very worthwhile knowing,

---

1326 (Page 41, *The Essene Gospel of Peace: Book One*, The Third Century Aramaic Manuscript and Old Slavonic Texts, Compared, Edited and Translated by Edmond Bordeaux Szekely)

especially for anyone who is suffering from a terrible illness, or who has been sick and in poor health for most of their life, or who has a child or loved one suffering from leukemia, asthma or any number of other debilitating diseases; and who has lost hope and is crying out to God, "Why? Oh why?" Hopefully in the remainder of this chapter they can get not only a glimpse of the why, but also some hope as to what they might be capable of doing about it if they so choose.

Cleaning house

"Know you not that you are the temple of God, and that the Spirit of God dwells in you? If anyone defiles the temple of God, him shall God destroy; for the temple of God is holy, which temple you are." (1 Corinthians 3:16-17)

(Most of us could not stand to live in a filthy, cluttered house with trash and rotting garbage as a tripping hazard in every room and the stench of cat urine and dog poop constantly assaulting our nostrils, but we have no problem dwelling in a filthy, cluttered body.)

We all know the importance of supplying our bodies with all the necessary ingredients—vitamins, minerals, protein, carbohydrates, etc.—that are required for maintaining our health and vitality. And these necessary "building blocks" are very important. But they are only half of the picture. The other part lies in keeping our bodies clean. Free of toxins and poisons, and sludge in the form of pathological mucus—the byproducts of a less-than-clean diet—that clog our cells and tissues, lymph and blood stream and digestive track. The body in many ways is a massive pipe system and the more sludge you have in your pipes the less healthy you will be. (Think of what comes out of you when you have a cold or sinus infection.) Also your body is built of a hundred trillion cells and these powerful little cities work best when they are clean, free of poisons and foreign material, and thus when they don't have to deal constantly with expending energy to rid themselves of that burden. And all of this has everything to do with what you eat and don't eat, and with whether or not you periodically give your body a rest by not eating at all so it can clean itself out, again by ridding itself of poisons, harmful chemicals and unwanted material.

The body has a number of miraculous built-in cleaning systems. The lungs not only provide oxygen, and to a lesser extent nitrogen, to the cells, but they expel the by-products of our cells metabolism: CO_2, along with other chemical and gaseous wastes. Our sinuses, our urinary apparatus, our skin itself, the digestive track, the blood system that carries waste products as well as unwanted material from an improper diet away from the cells in order to be discarded, are all mechanisms, in addition to their other functions, that strive to keep our bodies as clean as possible depending upon the make-up of the foods we eat and what we therefore give them to deal with on a regular basis. And all of these internal cleansing systems are ratcheted up to high gear when one stops eating, and drinks only water, or fresh fruit juices. Also when you lighten your diet and eat only fruits and/or raw vegetables the same cleansing regimen will kick in but to a lesser degree than when you abstain from food altogether.

When you first start down the road of fasting and a cleaner diet regimen, years of accumulated wastes and poisons throughout the body will begin to be expelled primarily through the blood stream. During this initial cleansing period, and depending upon ones age and health level, it is natural to feel a little sick, because the actual cause of sickness is being removed from the body. But better to get rid of it that way then to have it accumulate and build up until your body can't take it anymore and you get really sick, with the likes of pneumonia, tuberculosis, asthma, influenza, migraines, serious bacterial and viral infections, cancer, heart disease, diabetes, and death.

The Germ Theory of Disease

Most of us are also aware that viruses and bacteria play a significant role in many of the diseases that plague us; but just as many of us, including the same percentage of the medical community, are also not aware of the <u>actual</u> roll that they play. Bacteria were created by God to dispose of the dead remains of both animals and vegetables. They turn it all back into soil so the whole cyclical process can be continuously repeated. If it were not for their presence, the entire planet would be covered with the dead remains of animal bodies and vegetable matter to a depth of hundreds of feet. But because of the existence of these ever-present microbes, it is not.

The germ theory of disease therefore, which conversely puts forth that bacteria are the bad guys and the cause of disease, has missed the point. Bacteria are a *symptom* but not the *cause* of disease. The cause is the dead matter inside of you that bacteria and germs naturally attack as a part of their pre-programming, placed in their DNA by their Creator. When you fill your stomach with dead food you are not only clogging your system and setting yourself up for many varieties of dis-ease, but you are also inviting these microbes to come inside and flourish on what you have been filling your body with. So, as such, bacteria are not the <u>cause</u> of disease; they are a <u>symptom</u> of it. And certainly they do need to be dealt with, but if you hadn't filled your body—your tissues, cells, digestive tract and blood stream—with dead material you would not have been attacked by them in the first place. If your body is clean then you will be able to live in the world surrounded by germs and bacteria and viruses without them attacking you. If your body is full of rotting garbage and dead material in your intestines, tissues and blood stream then you are a number of "diseases" waiting to happen. These micro-organisms at that point are a <u>symptom</u> of your dis-ease. They are a symptom of your lack of health. They are a symptom of your lack of internal cleanliness. (*"Internal* cleanliness is next to godliness.") They are definitely a major part of the problem and need to be addressed by whatever method you and/or your physician deem necessary. But they are a symptom and not a cause. In the same way that rats and seagulls (flying rats) are drawn to garbage dumps and landfills and if you want to get rid of them you can keep shooting them day after day after month after year but they will always come back, <u>unless you remove the garbage</u>. Removing the garbage inside of you (by fasting) and keeping it out (by eating at the table of the living God—eating raw foods) is the secret to health, and to the absence of disease.

This is also the reason why we've been looking for the cause of cancer and heart disease and diabetes and leukemia etc. etc. for all these years, and yet have been unable to find it. We've been looking in all the wrong places. We've been looking at all of the potential microbial "causes" outside of the body rather than the real cause inside of the body. A cause that got there in the first place through our open mouths. If we address the real cause and take care of that, the symptoms will take care of themselves.

WWJS (What Would Jesus Say)

But what if anything does Jesus, the Word made flesh, the Author of the Word of God, the Creator of the universe and all of life, and the Author and Finisher of our faith have to say on this subject? Well, in addition to the bible verses quoted at the beginning of this chapter, and other Bible verses we will also look at, there is a very interesting and controversial extra-biblical source, one that is not found in the canon of Scripture, that addresses this subject and also claims to be authored by Jesus himself. It reportedly comes by way of an ancient Aramaic Manuscript, dated to the Third Century and found in the archives of the Vatican by its translator, Edmond Bordeaux Szekely. He titled the work, *The Essene Gospel of Peace*. The accuracy of the text as far as what it teaches about both diet and fasting cannot be argued. It is confirmed by all that we now know about nutrition and diet and of the real cause and prevention of disease. But the controversy is over its authorship, and we are in no position to determine conclusively one way or the other on that subject. We'll let each reader decide for themselves, and whereas we will address this controversy to a certain extent a little later in the chapter, for now we will only say that "My sheep hear my voice, and I know them, and they follow me." (John 10:27) If you have accepted Christ as your Savior, are a true Christian who has departed from iniquity and is clothed in His righteousness, then you are one of His sheep. Whether the following is His voice or not, we'll let you decide...

Here then are some selected passages from Mr. Szekely's (or Jesus'?) *Essene Gospel of Peace*...

"And then many sick and maimed came to Jesus, asking him: "If you know all things, tell us, why do we suffer with these grievous plagues? Why are we not whole like other men? Master, heal us, that we too may be made strong, and need abide no longer in our misery. We know that you have it in your power to heal all manner of disease. Free us from Satan and from all his great afflictions. Master, have compassion on us."

And Jesus answered: "Happy are you, that you hunger for the truth, for I will satisfy you with the bread of wisdom. Happy are you, that you knock, for I will open to you the door of life. Happy are you, that you would cast off the power of Satan, for I will lead you into the kingdom of our Mother's angel's, where the power of Satan cannot enter."

And they asked him in amazement: "Who is our Mother and which her angels? And where is her kingdom?"

"Your Mother is in you, and you in her. She bore you: she gives you life. It was she who gave to you your body, and to her shall you one day give it back again. Happy are you when you come to know her and her kingdom; if you receive your Mother's angels and if you do her laws. I tell you truly, he who does these things shall never see disease. For the power of our Mother is above all. And it destroys Satan and his kingdom; and has rule over all your bodies and all living things.

"The blood which runs in us is born of the blood of our Earthly Mother. Her blood falls from the clouds; leaps up from the womb of the earth; babbles in the brooks of the mountains; flows wide in the rivers of the plains; sleeps in the lakes; rages mightily in tempestuous seas.

"The air which we breathe is born of the breath of our Earthly Mother. Her breath is azure in the heights of the heavens; soughs in the tops of the mountains; whispers in the leaves of the forest; billows over the cornfields; slumbers in the deep valleys; burns hot in the desert.

"The hardness of our bones is born of the bones of our Earthly Mother, of the rocks and of the stones. They stand naked to the heavens on the tops of mountains; are as giants that lie sleeping on the sides of the mountains, as idols set in the desert, and are hidden in the deepness of the earth.

"The tenderness of our flesh is born of the flesh of our Earthly Mother; whose flesh waxes yellow and red in the fruits of the trees, and nurtures us in the furrows of the fields.

"Our bowels are born of the bowels of our Earthly Mother, and are hid from our eyes, like the invisible depths of the earth.

"The light of our eyes, the hearing of our ears, both are born of the colors and the sounds of our Earthly Mother; which enclose us about, as the waves of the sea a fish, as the eddying air a bird.

"I tell you in very truth, Man is the Son of the Earthly Mother, and from her did the Son of Man receive his whole body, even as the body of the newborn babe is born of the womb of his mother. I tell you truly, you are one with the Earthly Mother; she is in you, and you in her, of her were you born, in her do you live, and to her shall you return again. Keep, therefore, her laws, for none can live long, neither be happy, but he who honors his Earthly Mother and does her laws. For your breath is her breath: your blood her blood; your bone her bone; your flesh her flesh; your bowels her bowels; your eyes and your ears are her eyes and her ears.

"I tell you truly, should you fail to keep but one only of all these laws, should you harm but one only of all your body's members, you shall be utterly lost in your grievous sickness, and there shall be weeping and gnashing of teeth. I tell you, unless you follow the laws of your Mother, you can in no wise escape death. And he who clings to the laws of his Mother, to him shall his Mother cling also. She shall heal all his plagues, and he shall never become sick. She gives him long life, and protects him from all afflictions; from fire, from water, from the bite of venomous serpents. For your Mother bore you, keeps life within you. She has given you her body, and none but she heals you. Happy is he who loves his Mother and lies quietly in her bosom. For your Mother loves you, even when you turn away from her. And how much more shall she love you, if you turn to her again? I tell you truly, very great is her love, greater than the greatest of mountains, deeper than the deepest seas. And those who love their Mother, she never deserts them. As the hen protects her chickens, as the lioness her cubs, as the mother her newborn babe, so does the Earthly Mother protect the Son of Man from all danger and from all evils.

"For I tell you truly, evils and dangers innumerable lie in wait for the Sons of Men. Beelzebub, the prince of all devils, the source of every evil, lies in wait in the body of all the Sons of Men. He is death, the lord of every plague, and taking upon him a pleasing raiment, he tempts and entices the Sons of Men. Riches does he promise, and power, and splendid palaces, and garments of gold and silver, and a multitude of servants, all these; he promises renown and glory, fornication and lustfulness, gluttony and wine-bibbing, riotous living, and slothfulness and idle days. And he entices every one by that to which their heart is most inclined. And in the day that the Sons of Men have already become the slaves of all these vanities and abominations, then in payment thereof he snatches from the Sons of Men all those things which the Earthly Mother gave them so abundantly. He takes from them their breath, their blood, their bone, their flesh, their bowels, their eyes and their ears. And the breath of the Son of Man becomes short and stifled, full of pain and evil-smelling, like the breath of unclean beasts. And his blood becomes thick and evil-smelling, like the water of the swamps; it clots and blackens, like the night of death. And his bone becomes hard and knotted; it melts away within and breaks asunder, as a stone falling down upon a rock. And his flesh waxes fat and watery; it rots and putrefies, with scabs and boils that are an abomination. And his bowels become full with abominable filthiness, with oozing streams of decay; and multitudes of abominable worms have their habitation there. And his eyes grow dim, till dark night enshrouds them, and his ears become stopped, like the silence of the grave. And last of all shall the erring Son of Man lose life. For he kept not the laws of his Mother, and added sin to sin. Therefore, are taken from him all the gifts of the Earthly Mother: breath, blood, bone, flesh, bowels, eyes and ears, and after all else, life, with which the Earthly Mother crowned his body.

"But if the erring Son of Man be sorry for his sins and undo them, and return again to his Earthly Mother; and if he do his Earthly Mother's laws and free himself from Satan's clutches, resisting his temptations, then does the Earthly Mother receive again her erring Son with love and sends him her angels that they may serve him. I tell you truly, when the Son of Man resists the Satan that dwells in him and does not his will, in the same hour are found the Mother's angels there, that they may serve him with all their power and free utterly the Son of Man from the power of Satan.

"For no man can serve two masters. For either he serves Beelzebub and his devils or else he serves our Earthly Mother and her angels. Either he serves death or he serves life. I tell you truly, happy are those that do the laws of life and wander not upon the paths of death. For in them the forces of life wax strong and they escape the plagues of death."

And all those round about him listened to his words with amazement, for his word was with power, and he taught quite otherwise than the priests and scribes.

..."I tell you truly, God and his laws are not in that which you do. They are not in gluttony and in wine-bibbing, neither in riotous living, nor in lustfulness, nor in seeking after riches, nor yet in hatred of your enemies. For all these things are far from the true God and from his angels. But all these things come from the kingdom of darkness and the lord of all evils. And all these things do you carry in yourselves; and so the word and the power of God enter not into you, because all manner of evil and all manner of abominations have their dwelling in your body and your spirit. If you will that the living God's word and his power may enter you, defile not your body and your spirit; for the body is the temple of the spirit, and the spirit is the temple of God. Purify, therefore, the temple, that the Lord of the temple may dwell therein and occupy a place that is worthy of him.

"And from all temptations of your body and your spirit, coming from Satan, withdraw beneath the shadow of God's heaven.

"Renew yourselves and fast. For I tell you truly, that Satan and his plagues may only be cast out by fasting and by prayer. Go by yourself and fast alone, and show your fasting to no man. The living God shall see it and great shall be your reward. And fast till Beelzebub and all his evils depart from you, and all the angels of our Earthly Mother come and serve you. For I tell you truly, except you fast, you shall never be freed from the power of Satan and from all diseases that come from Satan. Fast and pray fervently, seeking the power of the living God for your healing. While you fast, eschew the Sons of Men and seek our Earthly Mother's angels, for he that seeks shall find.

"'Seek the fresh air of the forest and of the fields, and there in the midst of them shall you find the angel of air. Put off your shoes and your clothing and suffer the angel of air to embrace all your body. Then breathe long and deeply, that the angel of air may be brought within you. I tell you truly, the angel of air shall cast out of your body all uncleannesses which defiled it without and within. And thus shall all evil-smelling and unclean things rise out of you, as the smoke of fire curls upwards and is lost in the sea of the air. For I tell you truly, holy is the angel of air, who cleanses all that is unclean and makes all evil-smelling things of a sweet odor. No man may come before the face of God, whom the angel of air lets not pass. Truly, all must be born again by air and by truth, for your body breathes the air of the Earthly Mother, and your spirit breathes the truth of the Heavenly Father.

"After the angel of air, seek the angel of water. Put off your shoes and your clothing and suffer the angel of water to embrace all your body. Cast yourselves wholly into his enfolding arms, and as often as you move the air with your breath, move with your body the water also. I tell you truly, the angel of water shall cast out of your body all uncleannesses which defiled it without and within. And all unclean and evil-smelling things shall flow out of you, even as the uncleannesses of garments washed in water flow away and are lost in the stream of the river. I tell you truly, holy is the angel of water who cleanses all that is unclean and makes all evil-smelling things of a sweet odor. No man may come before the face of God whom the angel of water lets not pass. In very truth, all must be born again of water and of truth, for your body bathes in the river of earthly life, and your spirit bathes in the river of life ever-lasting. For you receive your blood from our Earthly Mother and the truth from our Heavenly Father."

... "And if afterward there remain within you aught of your past sins and uncleannesses, seek the angel of sunlight. Put off your shoes and your clothing and suffer the angel of sunlight to embrace all your body. Then breathe long and deeply, that the angel of sunlight may be brought within you. And the angel of sunlight shall cast out of your body all evil-smelling and unclean things which defiled it without and within. And all unclean and evil-smelling things shall rise from you, even as the darkness of night fades before the brightness of the rising sun. For I tell you truly, holy is the angel of sunlight who cleans out all uncleannesses and makes all evil-smelling things of a sweet odor. None may come before the face of God, whom the angel of sunlight lets not pass. Truly, all must be born again of sun and of truth, for your body basks in the sunlight of the Earthly Mother, and your spirit basks in the sunlight of the truth of the Heavenly Father.

"The angels of air and of water and of sunlight are brethren, They were given to the Son of Man that they might serve him, and that he might go always from one to the other.

"Holy, likewise, is their embrace. They are indivisible children of the Earthly Mother, so do not you put asunder those whom earth and heaven have made one. Let these three brother angels enfold you every day and let them abide with you through all your fasting.

"For I tell you truly, the power of devils, all sins and uncleannesses shall depart in haste from that body which is embraced by these three angels. As thieves flee from a deserted house at the coming of the lord of the house, one by the door, one by the window, and the third by the roof, each where he is found, and whither he is able, even so shall flee from your bodies all devils of evil, all past sins, and all uncleannesses and diseases which defiled the temple of your bodies. When the Earthly Mother's angels enter into your bodies, in such wise that the lords of the temple repossess it again, then shall all evil smells depart in haste by your breath and by your skin, corrupt waters by your mouth and by your skin, by your hinder and your privy parts. And all these things you shall see with your eyes and smell with your nose and touch with your hands. And when all sins and uncleannesses are gone from your body, your blood shall become as pure as our Earthly Mother's blood and as the river's foam sporting in the sunlight. And your breath shall become as pure as the breath of odorous flowers; your flesh as pure as the flesh of fruits reddening upon the leaves of trees; the light of your eye as clear and bright as the brightness of the sun shining upon the blue sky. And now shall all the angels of the Earthly Mother serve you. And your breath, your blood, your flesh shall be one with the breath, the blood and the flesh of the Earthly Mother, that your spirit also may become one with the spirit of your Heavenly Father. For truly, no one can reach the Heavenly Father unless through the Earthly Mother. Even as no newborn babe can understand the teaching of his father till his mother has suckled him, bathed him, nursed him, put him to sleep and nurtured him. While the child is yet small, his place is with his mother and he must obey his mother. When the child is grown up, his father takes him to work at his side in the field, and the child comes back to his mother only when the hour of dinner and supper is come. And now his father teaches him, that he may become skilled in the works of his father. And when the father sees that his son understands his teaching and does his work well, he gives him all his possessions, that they may belong to his beloved son, and that his son may continue his father's work. I tell you truly, happy is that son who accepts the counsel of his mother and walks therein. And a hundred times more happy is that son who accepts and walks also in the counsel of his father, for it was said to you: 'Honor thy father and thy mother that thy days may be long upon the earth.' But I say to you, Sons of Man: Honor your Earthly Mother and keep all her laws, that your days may be long on this earth, and honor your Heavenly Father that Eternal Life may be yours in the heavens. For the Heavenly Father is a hundred times greater than all fathers by seed and by blood, and greater is the Earthly Mother than all mothers by the body. And dearer is the Son of Man in the eyes of his Heavenly Father and of his Earthly Mother than are children in the eyes of their fathers by seed and by blood and of their mothers by the body. And more wise are the words and laws of your Heavenly Father and of your Earthly Mother than the words and the will of all fathers by seed and by blood, and of all mothers by the body. And of more worth also is the inheritance of your Heavenly Father and of your Earthly Mother, the everlasting kingdom of earthly and heavenly life, than all the inheritances of your fathers by seed and by blood, and of your mothers by the body."

FOR YOUR HEAVENLY FATHER IS LOVE.
FOR YOUR EARTHLY MOTHER IS LOVE.
FOR THE SON OF MAN IS LOVE.

...."Happy are those that persevere to the end, for the devils of Satan write all your evil deeds in a book, in the book of your body and your spirit. I tell you truly, there is not one sinful deed, but it is written, even from the beginning of the world, before our Heavenly Father. For you may escape the laws made by kings, but the laws of your God, these may none of the Sons of Man escape. ...But if you repent of your sins, and by fasting and prayer you seek the angels of God, then each day that you continue to fast and to pray, God's angels blot out one year of your evil deeds from the book of your body and your spirit. And when the last page is also blotted out and cleansed from all your sins, you stand before the face of God, and God rejoices in his heart and forgets all your sins. He frees you from the clutches of Satan and from suffering; he takes you within his house and commands that all his servants, all his angels serve you. Long life does he give you, and you shall never see disease."

..."Give thanks to God that his angels have made you free, and sin no more, lest Satan return to you again. Let your body be henceforth a temple dedicated to your God.

..."And feed not Satan, for the wages of sin is death."

...And it was by the bed of a stream, many sick fasted and prayed with God's angels for seven days and seven nights. And great was their reward, because they followed Jesus' words. And with the passing of the seventh day, all their pains left them. And when the sun rose over the earth's rim they saw Jesus coming towards them from the mountain, with the brightness of the rising sun about his head.

"Peace be with you."

And they said no word at all, but only cast themselves down before him, and touched the hem of his garment in token of their healing.

"Give thanks not to me, but to your Earthly Mother, who sent you her healing angels. Go, and sin no more, that you may never again see disease. And let the healing angels become your guardians."

But they answered him: "Whither should we go, Master, for with you are the words of eternal life? Tell us, what are the sins which we must shun, that we may nevermore see disease?"

Jesus answered: "Be it so according to your faith," and he sat down among them, saying:

"It was said to them of old time, 'Honor thy Heavenly Father and thy Earthly Mother, and do their commandments, that thy days may be long upon the earth.' And next afterward was given this commandment, 'Thou shall not kill,' for life is given to all by God, and that which God has given, let not man take away. For I tell you truly, from one Mother proceeds all that lives upon the earth. Therefore, he who kills, kills his brother. And from him will the Earthly Mother turn away, and will pluck from him her quickening breasts. And he will be shunned by her angels, and Satan will have his dwelling in his body. And the flesh of slain beasts in his body will become his own tomb. For I tell you truly, he who kills, kills himself, and whoso eats the flesh of slain beasts, eats of the body of death. For in his blood every drop of their blood turns to poison; in his breath their breath to stink; in his flesh their flesh to boils; in his bones their bones to chalk; in his bowels their bowels to decay; in his eyes their eyes to scales; in his ears their ears to waxy tissue. And their death will become his death. For only in the service of your Heavenly Father are your debts of seven years forgiven in seven days. But Satan forgives you nothing and you must pay him for all. 'Eye for eye, tooth for tooth, hand for hand, foot for foot; burning for burning, wound for wound; life for life, death for death.' For the wages of sin is death. Kill not, neither eat the flesh of your innocent prey, lest you become the slaves of Satan. For that is the path of sufferings, and it leads unto death. But do the will of God, that his angels may serve you on the way of life. Obey, therefore, the words of God: 'Behold, I have given you every herb bearing seed, which is upon the face of all the earth, and every tree, in the which is the fruit of a tree yielding seed; to you it shall be for food. And to every beast of the earth, and to every fowl of the air, and to everything that creeps upon the earth, wherein there is breath of life, I give every green herb for food. Also the milk of every thing that moves and lives upon earth shall be food for you; even as the green herb have I given unto them, so I give their milk unto you. But flesh, and the blood which quickens it, shall ye not eat. And, surely, your spurting blood will I require, your blood wherein is your soul; I will require all slain beasts, and the souls of all slain men. For I the Lord thy God am a God strong and jealous, visiting the iniquity of the fathers upon the children unto the third and fourth generation of them that hate me; and showing mercy unto thousands of them that love me, and keep my commandments. Love the Lord thy God with all thy heart, and with all thy soul, and with all thy strength: this is the first and greatest commandment.' And the second is like unto it: 'Love thy neighbor as thyself.' There is none other commandment greater than these."

...And Jesus continued: "God commanded your forefathers: 'Thou shall not kill.' But their heart was hardened and they killed. Then Moses desired that at least they should not kill men, and he suffered them to kill beasts. And then the heart of your forefathers was hardened yet more, and they killed men and beasts likewise. But I do say to you: **Kill neither men, nor beasts, <u>nor yet the food which goes into your mouth</u>**. [Emphasis mine.] For if you eat living food, the same will quicken you, but if you kill your food, the dead food will kill you also. For life comes only from life, and from death comes always death. For everything which kills your foods, kills your bodies also. And everything which kills your bodies kills your souls also. And your bodies become what your foods are, even as your spirits, likewise, become what your thoughts are. Therefore, eat not anything which fire, or frost, or water has destroyed. For burned, frozen and rotted foods will burn, freeze and rot your body also. Be not like the foolish husbandman who sowed in his ground cooked, and frozen, and rotten seeds. And the autumn came, and his fields bore nothing. And great was his distress. But be like that husbandman who sowed in his field living seed, and whose field bore living ears of wheat, paying a hundredfold for the seeds which he planted. For I tell you truly, live only by the fire of life, and prepare not your foods with the fire of death, which kills your foods, your bodies and your souls also."

"Master, where is the fire of life?" asked some of them.

"In you, in your blood, and in your bodies."

"And the fire of death?" asked others.

"It is the fire which blazes outside your body, which is hotter than your blood. With that fire of death you cook your foods in your homes and in your fields. I tell you truly, it is the same fire which destroys your foods and your bodies, even as the fire of malice, which ravages your thoughts, ravages your spirits. For your body is that which you eat, and your spirit is that which you think. Eat nothing, therefore, which a stronger fire than the fire of life has killed. Wherefore, prepare and eat all fruits of trees, and all grasses of the fields, and all milk of beasts good for eating. For all these are fed and ripened by the fire of life; all are the gift of the angels of our Earthly Mother. But eat nothing to which only the fire of death gives savor, for such is of Satan."

…"And the living angels of the living God serve only living men. For God is the God of the living, and not the God of the dead."

"So eat always from the table of God: the fruits of the trees, the grain and grasses of the field, the milk of beasts, and the honey of bees. For everything beyond these is of Satan, and leads by the way of sins and of diseases unto death. But the foods which you eat from the abundant table of God give strength and youth to your body, and you will never see disease."

…"For I tell you truly, the God of the living is richer than all the rich of the earth, and his abundant table is richer than the richest table of feasting of all the rich upon the earth. Eat, therefore, all your life at the table of our Earthly Mother, and you will never see want. And when you eat at her table, eat all things even as they are found on the table of the Earthly Mother. Cook not, neither mix all things one with another, lest your bowels become as steaming bogs."

…"For I tell you truly, the power of God enters into you, if you eat after this manner at his table. But Satan turns into a steaming bog the body of him upon whom the angels of air and water do not descend at his repasts. And the Lord suffers him no longer at his table. For the table of the Lord is an altar, and he who eats at the table of God is in a temple. For I tell you truly, the body of the Son of Man is turned into a temple, and his inwards into an altar, if he does the commandments of God. Wherefore, put naught upon the altar of the Lord when your spirit is vexed, neither think upon any one with anger in the temple of God. And enter only into the Lord's sanctuary when you feel in yourselves the call of his angels, for all that you eat in sorrow, or in anger, or without desire, becomes a poison in your body. For the breath of Satan defiles all these. Place with joy your offerings upon the altar of your body, and let all evil thoughts depart from you when you receive into your body the power of God from his table. And never sit at the table of God before he call you by the angel of appetite.

"Rejoice, therefore, always with God's angels at their royal table, for this is pleasing to the heart of the Lord. And your life will be long upon the earth, for the most precious of God's servants will serve you all your days: the angel of joy.

"And forget not that every seventh day is holy and consecrated to God. On six days feed your body with the gifts of the Earthly Mother, but on the seventh day [Saturday] sanctify your body for your Heavenly Father. On the seventh day eat not any earthly food, but live only on the words of God, and be all the day with the angels of the Lord in the kingdom of the Heavenly Father. And on the seventh day let the angels of God build the kingdom of the heavens in your body; as you labor for six days in the kingdom of the Earthly Mother. And let not food trouble the work of the angels in your body throughout the seventh day."

...."Come to me, all that are weary and that suffer in strife and affliction! For my peace will strengthen you and comfort you. For me peace is exceeding full of joy. ...Go, and sin no more. And give to every one your peace, even as I have given my peace unto you. For my peace is of God. Peace be with you."

And he left them.[1327]

Very interesting at the very least, don't you think? Certainly those are some powerful words; perhaps the most powerful ever found outside of the Bible. But for those who are in love with their current diet, those words can also be found offensive (and I have experienced some folks like that over the years). But nevertheless, the real question that remains, for those who seek the truth is, Are those the words of Jesus? And *my* answer is, I don't know. On the one hand, how could they not be? Who else could have written that? For they are definitely true, as far as the dietary and fasting information goes. There is no argument there. Cooking food destroys the enzymes, the "life" of the food. And by eating food devoid of enzymes you are robbing your body of their healing, metabolic, life-giving and digestive properties. You may still be getting a certain amount of nutrition, but you are also slowly killing your body by slowly killing the cells within your body by eating lifeless food. (As far as the nutritional and spiritual debate over the deleterious effects of "eating the flesh of slain beasts," that is a subject a little too involved for our time here. The reader can search that out for himself.) But as far as the validity of the raw food diet described in that ancient manuscript, nutritionally and scientifically, it is beyond reproach, argument or debate. That doesn't mean you have to follow it. It just means it can't be blown off by saying it's untrue.

Understanding therefore its truthfulness, the question that remains is, why would someone else write it and yet deceive as far as its author is concerned? What would be the motivation? Wouldn't you gain far more fame and wealth (the translator has always sold the book at cost) if you put your own name on the work? It is poetry and prose to rival Shakespeare, is it not? The words are arguably eternal, and definitely inspirational and transportive in their wisdom, insight and truth. Why write it yourself and then lie about who wrote it? It doesn't make any sense. Yet on the other hand, like we mentioned earlier, there is some controversy

---

1327   (Pgs. 9-12, 14-15, 16-19, 28-29, 34, 35, 36-37, 39-41, 44, 48, *The Essene Gospel of Peace: Book One, The Third Century Aramaic Manuscript and Old Slavonic Texts*, Compared, Edited and Translated by Edmond Bordeaux Szekely)

surrounding Mr. Szekely and his "discovery" of this manuscript. We will try to deal with both sides of the issue here, briefly, and hopefully in a fair and balanced way. We'll first look at the controversy surrounding its discovery:

"The Essene Gospel of Peace was brought to the 20th century world by Edmond Szekeley [sic]. Szekeley says that a Catholic priest led him into secret vaults at The Vatican, telling him there he would find the 'Source.' Szekeley then found the ancient manuscripts of the teachings of the Essenes and translated the writings.

"The Catholic Church says The Essene Gospel of Peace is a fraud, that the manuscripts/secret vaults never existed. However, those who believe EGOP to be authentic point to its similarities to other writings that predate The Bible. Most shockingly for many Christians, it portrays Jesus as a vegetarian. EGOP also includes a beautiful passage on the nature of love - in EGOP attributed to Jesus, but in the Bible attributed to Paul.

"Obviously, someone here is lying.

"Did Szekely lie? If so, we have to wonder what his motivation for doing so would have been. Career ambitions? Money? He was adamant that EGOP should not be sold for more than a token price.

"Did the Catholic Church lie? If so, we have to wonder what their motivation for doing so might have been. I will point out that, if anyone senior in the Catholic Church were to say that EGOP was authentic, this would obviously create a huge problem for the Catholic Church and in fact millions of Christians, as, if the Church decided that the words in EGOP attributed to Jesus were actually His words, all Christians would have to be directed to stop eating meat. Tricky."[1328]

Tricky yes, but not evidence in and of itself that the Catholic Church is "spinning" the truth and not Szekely.

Another point is that, if they are indeed the words of Jesus, why are they not included in the canon of scripture, those writings that were near universally accepted as the inspired writings of God by the early church and therefore included in the Bible. And that is for a very good reason, and one that only God knows. But a very good reason nonetheless for we can be certain that, no matter what the good or bad intentions of man might have been, if God had wanted it in the Bible, it would have been there. No human being, or council of men, could have opposed His will. So it is very safe to assume that even if these are the words of Jesus, they are not included in the canon of Scripture for a very good, divine reason.

Also, and along those lines, there are literally hundreds of extra-biblical writings from the first few centuries that are not included in the Bible and many for very good reason, but that this Essene Gospel is not there does not mean in and of itself that these are not the words of Jesus, because the Bible clearly says at the end of the Gospel of John that... "there are also many other things which Jesus did [and by extension, I'm sure, said], the which, if they should be written every one, I suppose that even the world itself could not contain the books that should be written. Amen." (John 21:25)

And then, in addition to Genesis 1:29-30 quoted earlier, there's this...

"And he showed me a pure river of water of life, clear as crystal, proceeding out of the throne of God and of the Lamb. In the midst of the street of it, and on either side of the river, was there the tree of life, which bare twelve kinds of fruits, and yielded her fruit every month." (Revelation 22:1-2)

1328   (From article, The Essene Gospel of Peace: Pt. 1 (Background), @debbietookrawforlife. blogspot.com)

Which would seem to indicate what the diet in the new heaven and the new earth will be, the same as the original diet in the Garden. And this of course would also make perfect sense. If the original diet of man, Created by God himself was perfect, as it obviously was, then why would the final diet of man be anything less than that?

There is also the point that Jesus himself made when he was being accused by the scribes of casting out demons by the power of the devil, that "a house divided against itself" cannot stand:

"And the scribes who came down from Jerusalem said, He has Beelzebub, and by the ruler of the devils he casts out demons. So he called them to himself, and said unto them in parables, How can Satan cast out Satan? If a kingdom be divided against itself, that kingdom cannot stand. And if a house be divided against itself, that house cannot stand." (Mark 3:22-25)

In the Essene Gospel Jesus is not casting out demons but he is evidently exposing the works, the dietary works, of Satan. If Jesus is not the author, why then would Satan, by inspiring some liar (someone pretending to be Jesus when he was not) to pen it, have his works exposed? Again, it's "a house divided against itself," and it doesn't add up.

And finally there are these two verses from the New Testament:

"It is good neither to eat flesh, nor to drink wine, nor any thing whereby your brother stumbles, or is offended, or is made weak." (Romans 14:21)

"Wherefore, if meat makes my brother to offend, I will eat no flesh while the world stands, lest I make my brother to stumble." (1 Corinthians 8:13)

And whereas those two verses are not a definitive argument for abstaining from meat, the eating of just living foods or the authorship of the EGOP, they at the very least show us that this subject was not foreign to the author of those letters.

Conversely though, the same author, Paul, said this:

"Now the Spirit says expressly that in the latter times some shall depart from the faith, giving heed to seducing spirits, and doctrines of demons; Speaking lies in hypocrisy; having their conscience seared with a hot iron; Forbidding to marry, <u>and commanding to abstain from certain foods, which God has created to be received with thanksgiving by those who believe and know the truth. For every creature of God is good, and nothing is to be refused if it be received with thanksgiving: For it is sanctified by the word of God and prayer.</u>" (1 Timothy 4:1-5)

The "certain foods" that some were commanding the early Christians to abstain from were those unclean animals forbidden by the dietary restrictions given to the Jews by God through Moses. What God was saying is that Christians were not under these Hebrew dietary restrictions, and they were not to listen to those Jews who were trying to command them to follow those laws. But what God was also saying was that the eating of any meat, of any animal or bird or "moving thing" was also not forbidden to Christians, as God had previously allowed it when Noah and his family left the ark:

"Every moving thing that lives shall be meat [food] for you; even as I gave you the green herbs, I **now** have given you all things." (Genesis 9:3-4)

And throughout the Old Testament as well there are countless examples of the eating of meat including its consumption by the Jewish temple priests following the sacrifice of animals on the altar. Therefore I think the only conclusion you can come to after all of this is that whoever the author of the Essene Gospel ultimately is, eating either cooked food or meat is obviously not a sin. How could it be? God himself allowed it. Also in this fallen world there are a million examples of where one would starve to death if they tried to eat only raw food. For one, in the middle of the winter in any century but the most recent ones you would not have had much access to raw foods in many colder climes.

"And you shall know the truth, and the truth shall make you free." (John 8:32)

So we are free as Christians to eat whatever we choose to eat, even though because of our lack of knowledge, and because of our powerful habits and addictions, we do not always choose so freely. And that is where we get back to the priceless information and instructions contained in the Essene Gospel. There is a lot to be said for the wisdom in coming to the knowledge of the truth about diet and fasting, and in then seeking to overcome deeply ingrained habits and addictions, so one will be truly free to choose when and what they eat, and to reap the physical and spiritual benefits that come from choosing wisely.

"If the Son therefore shall make you free, you shall be free indeed." (John 8:36)

Fasting

"The thief [Satan] comes not, but for to steal, and to kill, and to destroy. I am come that they might have life, and that they might have it more abundantly." (John 10:10)

That might seem a strange verse to put at the beginning of a discussion on fasting because most of us would not relate it to the abundant life. We would more equate it to mourning and sackcloth and ashes, or starving to death. But that is because most of us are unfamiliar with the many benefits to be derived from it. The first is cleansing, as mentioned before, or detoxifying the body. "Detoxification is a normal body process of eliminating or neutralizing toxins through the colon, liver, kidneys, lungs, lymph glands, and skin."[1329] The removal of these toxins, poisons, sludge and dead material is also the removal of the cause of many of our diseases. Fasting also frees up the energy of the body to more aggressively heal itself; energy that because of the fast is no longer devoted to the digestive system. Also, there is a quickening of the speed of metabolism, and therefore an increase in the energy and vitality, for each individual cell in your body that comes from a cleaner, healthier cell; one that is unburdened by the junk that otherwise slows its functions. There are many documented cases where extraordinary healings have occurred after a serious fast, cases where all other methods had failed and all medical hope was lost. (For one, see the book *Cancer Winner*, by Jaquie Davison, recommended at the end of this chapter.) There is also a greater feeling of health and well-being that follows a fast, and fasting has been shown to increase one's life expectancy. (And I found a ten dollar bill in the pocket of an old sports coat after a three day fast!)

And then there are the spiritual benefits:

> "*Fasting and prayer often lead to victory over sin.* The world has many Christians who have trusted Christ, who sincerely love Him, who are going to Heaven—yet Christians who have no daily victory over sin. Everywhere I go I find Christians who say they cannot quit cigarettes, they cannot control their tempers, they have trouble in surrendering even enough to give God regularly the tithe. Christians find it hard to forgive one another and

---

1329   (From article "Detox Diets: Cleansing the Body," by Jeanie Lerche Davis @ WebMD)

are constantly falling under the temptation of Satan. Is there victory for such Christians? Yes, there is. But sometimes it is found only in the time of fasting and prayer, waiting on God and laying aside every weight, every duty, every pleasure that might interfere with our wholehearted prayers."[1330]

Another spiritual benefit is revealed in the verse quoted at this chapters beginning that the kind of faith that can move mountains can only come forth "by prayer and fasting." (Mark 9:29) And since "without faith it is impossible to please God" (Hebrews 11:6), that benefit should not be one that is taken lightly. For pleasing God should be high on the priority list of any Christian. Right up there with glorifying him and praising his name. (And not voting for baby-mutilators.) (I know, I just had to throw that in.)

And finally, as a precautionary note, do not attempt any fasts of any length without first knowing exactly what to expect and especially how to properly break a fast and start eating again. There is the story of an ancient ruler who after fasting for a number of days ate a large bowl of dates and promptly expired. So we would recommend reading a good book on the subject. (See recommended books at the end of this chapter.) One of the best is Paul Bragg's, *The Miracle of Fasting*.[1331]

Further thoughts

"Listen, listen to me, and eat what is good, and your soul will delight in the richest of fare." (Isaiah 55:2-3)

If you choose to start down the path of improving your diet then it is important to understand the difference between living to eat, and eating only to live. If you live to eat, as far too many of us are guilty of, looking forward to one meal after another, habitually answering the thrice or four-time daily call to eat, then you are a slave to food. And when that is the case you have a huge psychological obstacle standing between you and your own health. But if instead, eating is merely something that is necessary to do to keep your body running, like refueling your car, and there is no psychological and physical addictions involved, then eating becomes more like breathing, or like drinking a glass of water. You enjoy the water. It's refreshing. But you don't lust after it. You don't think you must have it in order to be satisfied, fulfilled or at peace. You merely need it because you happen to be thirsty. And you don't overdrink.

"We were meant to consume food. But food was never meant to consume us."[1332]

Also, just as the lies of Beelzebub envelop this earth and people believe what they are taught, and will defend those lies vehemently no matter how absurd they may be; in the same way people come to enjoy whatever foods their particular culture has fed them, even though people from other cultures very often find those "delicacies" undesirable and in some cases even revolting. A partial list includes squirrel brains, spam, rats, cats, dogs, horsemeat, blood, camel's feet, partridges and other birds hung till they are rotten and then eaten by the British, and even placenta, a delicacy of, who else, radical feminists.[1333] The point is that people

---

1330   (Pg. 218, *Prayer: Asking and Receiving*, by John R. Rice)
1331   (By the way, that mythical Fountain of Youth was just a water fountain. That simple.)
1332   (Lysa TerKeurst, author of *Made to Crave: Satisfying Your Deepest Desire with God, Not Food*)
1333   (From weird-food.com)

can learn to relish virtually any kind of food, and this should give one pause to consider that what they now desire is actually all in their head, and as powerful as the desires may now be, just as one *learned* to desire these foods in the first place they can also *re-learn* not to desire them if they so will, and to replace them with other foods more profitable to their health. As addicted and as "in love" as one may now be with their current diet, know that the lust for the cooked foods from the "table of Satan" can be replaced with a real love and strong desire for the delicacies prepared at the table of the living God. Either way, it is of some value to go through the process, however difficult it may be or seem at first, to free yourself from being addicted to and enslaved by what you have learned to desire to eat.

If tomorrow you could feed your body with the same mental freedom that you, say, fuel your car, then you would find it a thing of absolute ease to eat right and be in perfect health. But imagine what it would be like if when your car was getting close to empty it demanded that you put dirty fuel into its tank. Even though it knows that it runs much better on clean gasoline but it just really desires the taste of that good ole dirty gas with chunks of unrefined crude oil and mud floating in it. Or when you try to put some Quaker State in its crank case, it refuses and insists on some sludge from the bottom of the oil tanker instead. Your car would have some real trouble running smoothly if it was afflicted by such idiotic desires as these. But your car has no such afflictions, or addictions or desires, so you put only the best gasoline and oil into its mouth. Our cars have no such afflictions, but we do.

"You shall have no other gods before me." (Exodus 20:3) And that includes food...

"Is it possible we love and rely on food more than we love and rely on God? ...I had to get honest enough to admit it: I relied on food more than I relied on God. Food was my comfort. Food was my reward. Food was my joy. Food was what I turned to in times of stress, sadness, and even in times of happiness."[1334]

The author of those words, Lysa TerKeurst, turned to prayer for the power to overcome her daily addiction to unhealthy foods...

"I determined to make God, rather than food, my focus. Each time I craved something I knew wasn't part of my plan I used that craving as a prompt to pray. I craved a lot. So I found myself praying a lot.

"...[That] was my way of tearing down the tower of impossibility before me and building something new. My tower of impossibility was food. Brick by brick, I imagined myself dismantling the food tower and using those same bricks to build a walkway of prayer, paving the way to victory.

"Did this simple visualization make it easier? Sometimes it did. And other times my cravings for unhealthy food made me cry. ...Sometimes I wound up on the floor in my closet, praying with tears running down my face."[1335]

But through that type of determination, she did find victory. And we can too. If we seek him with all of our heart. "And you shall seek me, and you shall find me, when you shall search for me with all your heart." If we desire him like the deer dying of thirst searches for the mountain stream... "As the deer pants after

---

1334   (Pgs. 28-29, *Made to Crave: Satisfying Your Deepest Desire with God, Not Food*, by Lysa TerKeurst)
1335   (Pgs. 29-30, *Made to Crave: Satisfying Your Deepest Desire with God, Not Food*, by Lysa TerKeurst)

the water brooks, so pants my soul after thee, O God." (Psalms 42:1)   God is waiting to see if we want him more than we want the "other gods," ("You shall have no other gods beside me") and food has definitely become another god for far too many of us. How badly do we want him? More than eating at the table of Satan? As much as we want our next breath?

When you clean up your diet you are not giving up food, you are giving up the addiction to food. And to be addicted is to be enslaved. So you are merely "giving up" your own slavery. Also, it is Satan himself who is the deceiver and the liar behind every enslavement and addiction of man. And his lies will scream through the head of most anyone who begins traveling down this path. Like the dog that whines and yelps at the back door until he is let in, the dog inside your head will whine and yelp and torment you each day until he is fed; or starved to death. And the lies in your head will tell you how much you have to eat to be happy, and how much you have to eat right now tonight in order to be comforted, be rewarded and be at peace. (Whereas "resisting the devil so he will flee from you" is the only way to find real, lasting peace. For giving in to the screams of Satan means that the slave master will just be back screaming inside your head just as powerfully tomorrow.) It is no different than the lies that scream through the mind of an alcoholic or crack addict when they try to quit. Satan does not give up easily, but "greater is he who is in me than he who is in the world." And "I can do all things through Christ who strengthens me." (Philippians 4:13) Also, someone once said that you have to go through hell before you get to heaven.

If you travel down this path and if you finally do free yourself from the addiction to eating food in general and cooked food in particular then you will be free from the constant cravings and longings for what you once had. It's like, a person used to live in Siberia. They were born and raised there. They loved it. It was all they ever knew. But now they have lived in Hawaii for a number of years and, frankly, they wouldn't go back to Siberia if you paid them.

And finally, there is only one diet that has ever and will ever work. And that is the diet where you eat whatever you want. The conflict arises of course when one part of you wants to eat only those foods which are for your health, your youth, your vitality, your energy, your peace and your well-being. And another part of you wants to continue eating those foods which you have learned to desire that you now know are for your addiction, your disease, your suffering, your agitation and eventually for your untimely death. Those who have struggled with being overweight and trying to change their diet in order to lose that weight, know of this conflict. But very few of us have known the ultimate conflict of struggling to go from death to life, from the table of Satan to the table of the living God. Of struggling to get to the point where you eat whatever you want, but all you want is what your body, and not your addictions, desires. Anyone up to the challenge? Then good "luck" and may God richly bless…

"Prove all things; hold fast that which is good. Abstain from all appearance of evil." (1 Thessalonians 5:21-22)

See:

- *The Essene Gospel of Peace*, by Edmond Bordeaux Szekely
- *The Miracle of Fasting*, by Paul C. Bragg
- *The Miracle of Fasting: Proven Throughout History for Physical, Mental, & Spiritual Rejuvenation*, by Patricia Bragg
- *The Miracle of Living Foods*, by Dr. Kristine Nolfi, M.D.
- *Eating in the Raw: A Beginner's Guide to Getting Slimmer, Feeling Healthier, and Looking Younger the Raw-Food Way*, by Carol Alt
- *The Power of Prayer and Fasting: God's Gateway to Spiritual Breakthroughs*, by Ronnie W. Floyd
- *Made to Crave: Satisfying Your Deepest Desire with God, Not Food*, by Lysa TerKeurst
- *Cancer winner: How I purged myself of melanoma*, by Jaquie Davison
- *Mucusless Diet Healing System*, by Arnold Ehret
- *The Miracle Results of Fasting: Discover the Amazing Benefits in Your Spirit, Soul and Body (Christian Living Series)*, by Dave Williams
- "Diabetes – No More! Discover and Heal Its True Causes," by Andreas Moritz

THE BOOK OF ETERNAL LIFE

"And I heard a great voice out of heaven saying, 'Behold, the tabernacle of God is with men, and he will dwell with them, and they shall be his people, and God himself shall be with them, and be their God. And God shall wipe away every tear from their eyes; and there shall be no more death, neither sorrow, nor crying, neither shall there be any more pain: for the former things have passed away.' And He who sat upon the throne said, 'Behold, I make all things new.' And he said unto me, 'Write, for these words are true and faithful.' And he said unto me, 'It is done! I am Alpha and Omega, the Beginning and the End. I will give unto him who thirsts of the fountain of the water of life freely. He who overcomes shall inherit all things; **and I will be his God, and he shall be my son.** But the fearful, and unbelieving, and the abominable, and murderers [and all of those who *vote* for the murderers], and whoremongers, and sorcerers, and idolaters, and all liars shall have their part in the lake which burns with fire and boiling sulfur, which is the second death.'" (Revelation 21:3-8)

In the book of Acts we read of the arrest and imprisonment of the Apostle Paul and his subsequent hearing before King Agrippa. Paul took some time to relate his miraculous and life-changing encounter with the risen Christ on the road to Damascus and his conversion from persecutor and murderer of Christians to a follower himself. "Then Agrippa said to Paul, 'You <u>almost</u> persuade me to become a Christian.'" (Acts 26:28) The operative word there is *almost*. And because of that *almost*, think for a moment of where that King has been for the last 2000 years, and the excruciating pain and suffering he has been heir to. I pray that no one who reads this book will be like King Agrippa, *almost* convinced of the truth.

Now you know the truth. "And you shall know the truth, and the truth shall make you free." (John 8:32) If you have ears to hear. "And He said unto them, he who has ears to hear, let him hear." (Mark 4:9) But for those who refuse to listen to the Truth, and instead turn their backs on Him, and continue to embrace the lies of the great deceiver, for them there is no hope. "And the devil who deceived them was cast into the lake of fire and boiling sulfur, where the beast and the false prophet are, and they shall be tormented day and night for ever and ever." (Revelation 20:10)

The lies of Satan do cover this earth and people believe what they are taught, and this world is full of people who desperately need at the very least to be given the opportunity to make a decision between what they have been taught, and the Truth. This has been the purpose and the work of countless numbers of brave true-Christian missionaries who have spread the gospel all around the globe for the last two thousand years. And it is for this ultimate purpose, to give people the opportunity to come from Satan's lies to the knowledge of the Truth, that this book was written. Peace be with you, and may God richly bless.

"I have no greater joy than to hear that my children walk in truth." (3 John 1:4)

* * *

"Then I saw a great white throne and Him who sat on it, from whose face the earth and the heaven fled away. And there was found no place for them. And I saw the dead, small and great, standing before God, and the books were opened. And another book was opened, which is the Book of Life. And the dead were judged according to their works, by the things which were written in the books. And the sea gave up the dead who were in it, and death and hell delivered up the dead who were in them. And they were judged, each one according to his works. Then death and hell were cast into the lake of fire. This is the second death. **And whosoever was not found written in the Book of Life was cast into the lake of fire**." (Revelation 20:11-15)

"I Jesus have sent my angel to testify unto you these things in the churches.
I am the Root and the Offspring of David,
and the Bright and Morning Star."
(Revelation 22:16)